Dried Fruits

Phytochemicals and Health Effects

WILEY-BLACKWELL
A John Wiley & Sons, Inc., Publication

Functional Food Science and Technology series

Functional foods resemble traditional foods but are designed to confer physiological benefits beyond their nutritional function. Sources, ingredients, product development, processing and international regulatory issues are among the topics addressed in Wiley-Blackwell's new Functional Food Science and Technology book series. Coverage extends to the improvement of traditional foods by cultivation, biotechnological and other means, including novel physical fortification techniques and delivery systems such as nanotechnology. Extraction, isolation, identification and application of bioactives from food and food processing by-products are among other subjects considered for inclusion in the series.

Series Editor: **Professor Fereidoon Shahidi**, PhD, Department of Biochemistry, Memorial University of Newfoundland, St John's, Newfoundland, Canada.

Titles in the series

Nutrigenomics and Proteomics in Health and Disease: Food Factors and Gene Interactions
Editors: Yoshinori Mine, Kazuo Miyashita and Fereidoon Shahidi
ISBN 978-0-8138-0033-2

Functional Food Product Development
Editors: Jim Smith and Edward Charter
ISBN 978-1-4051-7876-1

Cereals and Pulses: Nutraceutical Properties and Health Benefits
Editors: Liangli (Lucy) Yu, Rong Tsao and Fereidoon Shahidi
ISBN 978-0-8138-1839-9

Dried Fruits
Phytochemicals and Health Effects

Edited by

Cesarettin Alasalvar
Food Institute, TÜBİTAK Marmara Research Center, Turkey

Fereidoon Shahidi
Department of Biochemistry,
Memorial University of Newfoundland, Canada

WILEY-BLACKWELL
A John Wiley & Sons, Inc., Publication

Wiley-Blackwell is an imprint of John Wiley & Sons, formed by the merger of Wiley's global Scientific, Technical and Medical business with Blackwell Publishing.

Editorial offices: 2121 State Avenue, Ames, Iowa 50014-8300, USA
The Atrium, Southern Gate, Chichester, West Sussex, PO19 8SQ, UK
9600 Garsington Road, Oxford, OX4 2DQ, UK

For details of our global editorial offices, for customer services and for information about how to apply for permission to reuse the copyright material in this book please see our website at www.wiley.com/wiley-blackwell.

Library of Congress Cataloging-in-Publication Data

Dried fruits : phytochemicals and health effects / edited by Cesarettin Alasalvar and Fereidoon Shahidi.
p. cm. – (Functional food science & technology)
Includes bibliographical references and index.
ISBN 978-0-8138-1173-4 (hardback : alk. paper) 1. Dried fruit–Health aspects.
2. Phytochemicals–Health aspects. I. Alasalvar, Cesarettin. II. Title: Phytochemicals and health effects.
TX397.D75 2013
615.3′21–dc23 2012023960

A catalogue record for this book is available from the British Library.

Wiley also publishes its books in a variety of electronic formats. Some content that appears in print may not be available in electronic books.

Cover image: Dried fruit © Yeko Photo Studio/Shutterstock.com; Human body © Sebastian Kaulitzki/Shutterstock.com
Cover design by His and Hers Design

Set in 10/12pt Times by Aptara® Inc., New Delhi, India
Printed and bound in Malaysia by Vivar Printing Sdn Bhd

1 2013

Contents

List of Contributors

Cesarettin Alasalvar
Food Institute
TÜBİTAK Marmara Research Centre
Gebze-Kocaeli
Turkey

Sadeq Hasan Al-Sheraji
Department of Nutrition and Dietetics
University Putra Malaysia
Selangor
Malaysia

Department of Food Science
Ibb University
Ibb
Yemen

Emilio Alvarez-Parrilla
Departamento de Ciencias Químico Biológicas
Universidad Autónoma de Ciudad Juárez
Ciudad Juárez-Chihuahua
Mexico

Jesús F. Ayala-Zavala
Centro de Investigación en Alimentación y Desarrollo, A. C.
Dirección de Tecnología de Alimentos de Origen Vegetal
Hermosillo-Sonora
Mexico

Azrina Azlan
Department of Nutrition and Dietetics
University Putra Malaysia
Selangor
Malaysia

Halal Products Research Institute
University Putra Malaysia
Selangor
Malaysia

Debasis Bagchi
Department of Pharmacological and Pharmaceutical Sciences
University of Houston College of Pharmacy
Houston, TX
USA

Manashi Bagchi
NutriToday LLC
Boston, MA
USA

Alessandra Del Caro
Dipartimento di Agraria
Università degli Studi
Sassari
Italy

Arianna Carughi
Sun-Maid Growers of California
Kingsburg, CA
USA

Carter D. Clary
Department of Horticulture
Washington State University
Pullman, WA
USA

Manuel A. Coimbra
QOPNA
Department of Chemistry
University of Aveiro
Aveiro
Portugal

Neal M. Davies
Faculty of Pharmacy
University of Manitoba
Winnipeg, MB
Canada

Huertas María Díaz-Mula
Department of Applied Biology, EPSO
University Miguel Hernández
Orihuela-Alicante
Spain

Vural Gökmen
Department of Food Engineering
Hacettepe University
Beytepe-Ankara
Turkey

Neslihan Göncüoğlu
Department of Food Engineering
Hacettepe University
Beytepe-Ankara
Turkey

Gustavo A. González-Aguilar
Centro de Investigación en Alimentación y Desarrollo, A. C.
Dirección de Tecnología de Alimentos de Origen Vegetal
Hermosillo-Sonora
Mexico

Fouad Abdulrahman Hassan
Department of Nutrition and Dietetics
University Putra Malaysia
Selangor
Malaysia

Department of Food Science
Ibb University
Ibb
Yemen

Chi-Tang Ho
Department of Food Science
Rutgers University
New Brunswick, NJ
USA

Tzou-Chi Huang
Department of Food Science
National Pingtung University of Science and Technology
Pingtung
Taiwan

Amin Ismail
Department of Nutrition and Dietetics
University Putra Malaysia
Selangor
Malaysia

Halal Products Research Institutes
University Putra Malaysia
Selangor
Malaysia

Ajit P.K. Joshi
Department of Environmental Sciences
Dalhousie University
Truro, NS
Canada

William L. Kerr
Department of Food Science and Technology
University of Georgia
Athens, GA
USA

Joydeb Kumar Kundu
College of Pharmacy
Keimyung University
Daegu
South Korea

Changbao Li
Institute of Agro-Food Science & Technology
Guangxi Academy of Agricultural Sciences
Nanning
China

Hongyan Li
State Key Laboratory of Food Science and Technology
Nanchang University
Nanchang-Jiangxi
China
Guelph Food Research Centre
Agriculture and Agri-Food Canada
Guelph, ON
Canada

Li Li
Institute of Agro-Food Science & Technology
Guangxi Academy of Agricultural Sciences
Nanning
China

Zhichun Li
Institute of Agro-Food Science & Technology
Guangxi Academy of Agricultural Sciences
Nanning
China

Fen Liao
Institute of Agro-Food Science & Technology
Guangxi Academy of Agricultural Sciences
Nanning
China

Letitia McCune
BotanyDoc Inc.
Tucson, AZ
USA

Esteban I. Mejia-Meza
PepsiCo Global R&D
Valhalla, NY
USA

Burçe Ataç Mogol
Department of Food Engineering
Hacettepe University
Beytepe-Ankara
Turkey

Marian Naczk
Department of Human Nutrition
St. Francis Xavier University
Antigonish, NS
Canada

Beraat Ozcelik
Department of Food Engineering
Istanbul Technical University
Maslak-Istanbul
Turkey

Mine Gultekin Ozguven
Department of Food Engineering
Istanbul Technical University
Maslak-Istanbul
Turkey

Antonio Piga
Dipartimento di Agraria
Università degli Studi di Sassari
Sassari
Italy

K. Nagendra Prasad
Chemical Engineering Discipline
School of Engineering
Monash University
Selangor
Malaysia

Laura A. de la Rosa
Departamento de Ciencias Químico Biológicas
Universidad Autónoma de Ciudad Juárez
Ciudad Juárez-Chihuahua
Mexico

H.P. Vasantha Rupasinghe
Department of Environmental Sciences
Dalhousie University
Truro, NS
Canada

K.M. Schaich
Department of Food Science
Rutgers University
New Brunswick, NJ
USA

Alexander G. Schauss
Natural and Medicinal Products Research
AIBMR Life Sciences
Puyallup, WA
USA

María Serrano
Department of Applied Biology, EPSO
University Miguel Hernández
Orihuela-Alicante
Spain

Fereidoon Shahidi
Department of Biochemistry
Memorial University of Newfoundland
St. John's, NL
Canada

Haiming Shi
Institute of Food and Nutraceutical Science
School of Agriculture & Biology
Shanghai Jiao Tong University
Shanghai
China

Lisete Silva
QOPNA
Department of Chemistry
University of Aveiro
Aveiro
Portugal

Jian Sun
Institute of Agro-Food Science & Technology
Guangxi Academy of Agricultural Sciences
Nanning
China

Guangxi Crop Genetic Improvement
Laboratory Nanning
China

Young-Joon Surh
Tumor Microenvironment Global Core Research Center
College of Pharmacy
Seoul National University
Seoul
South Korea

Zhuliang Tan
Department of Biochemistry
Memorial University of Newfoundland
St. John's, NL
Canada

Rong Tsao
Guelph Food Research Centre
Agriculture and Agri-Food Canada
Guelph, ON
Canada

Daniel Valero
Department of Food Technology, EPSO
University Miguel Hernández
Orihuela-Alicante
Spain

Jaime A. Yáñez
Drug Metabolism and Pharmacokinetics
Alcon Research Ltd., a Novartis Company
Fort Worth, TX
USA

Xiangrong You
Institute of Agro-Food Science & Technology
Guangxi Academy of Agricultural Sciences
Nanning
China

Liangli (Lucy) Yu
Department of Nutrition and Food Science
University of Maryland
College Park, MD
USA

Barakatun Nisak Mohd Yusof
Department of Nutrition and Dietetics
University Putra Malaysia
Selangor
Malaysia

Shirley Zafra-Stone
Products Solution Research Inc.
Davis, CA
USA

Ying Zhong
Department of Biochemistry
Memorial University of Newfoundland
St. John's, NL
Canada

Preface

Dried fruits serve as important healthful snacks worldwide. They provide a concentrated form of fresh fruits, prepared by different drying techniques. Dried fruits, with their unique combination of taste/aroma, essential nutrients, fiber, and phytochemicals or bioactive compounds, are convenient for healthy eating and bridge the gap between recommended intake of fruits and actual consumption. Dried fruits are nutritionally equivalent to fresh fruits in smaller serving sizes, ranging from 30 to 43 g depending on the fruit, in current dietary recommendations in different countries. Numerous scientific evidences suggest that individuals who regularly consume generous amounts of dried fruits have lower rates of cardiovascular diseases, obesity, various types of cancer, type 2 diabetes, and other chronic diseases. Therefore, daily consumption of dried fruits is recommended in order to get full benefit of nutrients, health-promoting phytochemicals, and antioxidants that they contain, together with their desirable taste and aroma. Dried fruits also have the advantage of being easy to store and distribute, available around the year, readily incorporated into other foods and recipes, and present a healthy alternative to salty or sugary snacks.

This book examines most popular dried berries (blackberries, blackcurrants, blueberries, cranberries, goji berries, mulberries, raspberries, and strawberries), nontropical dried fruits (apples, apricots, cherries, citrus fruits, figs, nectarines, peaches, pears, prunes, and raisins), and tropical dried fruits (açai fruits, bananas, dates, guavas, papayas, mangoes, passion fruits, and pineapples). It is divided into three sections preceded by an introductory chapter (Chapter 1) providing an overview of dried fruits: composition, phytochemicals and health effects as well as cancer chemopreventive effects of selected dried fruits (amla fruits or Indian gooseberries, avocados, berries, mangoes, mangosteens, persimmons, prunes, raisins, kiwi fruits, and other dried fruits) (Chapter 2). The first section (Chapters 3–10) covers dried berries; the second section (Chapters 11–20) discusses nontropical dried fruits; and the final section (Chapters 21–26) includes tropical dried fruits.

Contributors to this volume are internationally renowned researchers who have provided a comprehensive account of the global perspectives of the issues of concern to phytochemicals and health effects of dried fruits. The book will serve as a resource for those interested in the potential application of new developments in dried fruits' nutraceuticals and functional foods. Biochemists, chemists, food scientists/technologists, nutritionists, and health professionals, from academia, government laboratories, and industry will benefit from this publication. Although this book is intended primarily as a reference book, it also summarizes the current state of knowledge in key research areas and contains ideas for future work. In addition, it provides easy-to-read text suitable for teaching senior undergraduate and post-graduate students.

We are indebted to the participating authors for their state-of-the-art contributions and dedication in providing authoritative views resulting from their latest investigations on nutritional significance, phytochemical composition, and potential health benefits of dried fruit consumption.

Cesarettin Alasalvar and Fereidoon Shahidi

1 Composition, phytochemicals, and beneficial health effects of dried fruits: an overview

Cesarettin Alasalvar and Fereidoon Shahidi

1.1 Introduction

Dried fruits serve as a concentrated form of fresh fruits prepared by different drying techniques. In other words, dried fruits possess much lower moisture content as a large proportion of their original water has been removed, either naturally through sun drying or through the use of specialized dryers or dehydrators. Considering the 2011 global production of commercially important dried fruits (Table 1.1), dates rank first on a global basis with a production of 6,598,000 metric tonnes (MT), followed by raisins (1,170,999 MT), prunes (236,500 MT), apricots (198,917 MT), and figs (105,453 MT) [1]. To the best of our knowledge, little information is available about the production of other dried fruits (açai berries, apples, bananas, black currants, blackberries, cherries, citrus fruits, cranberries, gingers, goji berries, guavas, kiwis, mangoes, mulberries, nectarines, papayas, passion fruits, peaches, pears, pineapples, raspberries, star apples, and strawberries, among others).

Dates, figs, prunes, raisins, apricots, peaches, apples, and pears are referred to as "conventional" or "traditional" dried fruits. On the other hand, some fruits such as blueberries, cranberries, cherries, strawberries, and mangoes are infused with sugar solutions (e.g., sucrose syrup) or fruit juice concentrates prior to drying. Some products sold as dried fruit, such as papayas and pineapples, are actually candied fruit [2].

Epidemiologic studies have found an association between dried fruit consumption and diet quality. Raisins may be among the most researched of all dried fruits showing a health benefit [3], followed by dates, prunes, figs, apricots, peaches, apples, pears, and other fruits, which together constitute nearly half of all dried fruits produced in the world each year [2].

This overview chapter summarizes the nutritional significance, phytochemical composition, and potential health benefits of dried fruit consumption and discusses their great potential as medicinal or healthy foods for a number of diseases inflicting human beings.

Dried Fruits: Phytochemicals and Health Effects, First Edition. Edited by Cesarettin Alasalvar and Fereidoon Shahidi.

Table 1.1 World dried fruits production (metric tonnes)

Production	2007	2008	2009	2010	2011
Apricots	137,100	146,950	164,350	159,100	198,917
Dates	6,400,000	6,323,000	6,599,000	6,708,000	6,598,000
Figs	100,600	92,000	102,000	107,562	105,453
Prunes	199,204	229,942	253,851	245,630	236,500
Raisins	1,053,500	1,042,450	1,061,600	1,083,547	1,170,999

Source: Adapted from INC [1].

1.2 Compositional and nutritional characteristics of dried fruits

Dried fruits come in almost as many varieties as fresh fruits. Although raisins, figs, dates, prunes, and apricots are the most common dried fruits in the marketplace, health food stores and local markets offer many more choices such as dried apples, pineapples, berries, mangoes, papayas, and even the exotic dragon fruit. They are rich sources of essential nutrients and health-promoting bioactive compounds. Table 1.2 summarizes the nutritional composition of some dried fruits (apples, apricots, dates, figs, peaches, pears, prunes, and raisins) [4]. Dried fruits are rich in carbohydrates (61.33–79.18 g/100 g) and devoid of fat (0.32–0.93 g/100 g). The most calorie-rich of these fruits are raisins (299 kcal/100 g), followed by dates (282 kcal/100 g). Dried fruits are excellent sources of sugar ranging from 38.13 g/100 g in prunes to 63.35 g/100 g in dates. Fructose and glucose are the main sugars found in all dried fruits, followed by sucrose. Trace amounts of maltose and galactose are found in some dried fruits. Levels of sugar may differ according to drying methods and regional and varietal factors.

It is important to note that the high content of dietary fiber (3.7–9.8 g/100 g) found in dried fruits is an important source that helps meet our dietary recommendations (14 g of fiber for every 1000 calories of food consumed each day).This becomes 25–38 g of fiber per day depending on age and gender [5]. On a per serving basis (40 g), dried fruits deliver more than 9% of the daily value of fiber, depending on the fruit [4]. It has been reported that dried fruits (40 g/serving) compare favorably in their fiber content with common fresh fruit (one cup or one fruit serving) options [4, 6].

With respect to nutritional aspects, percentage of recommended dietary allowances (RDA) or adequate intake (AI) of minerals for adult males and females (aged 15–50 years) are also given in Table 1.3. Dried fruits, in general, serve as a reasonable source of copper, iron, magnesium, manganese, phosphorus, and potassium. Among the eight dried fruits listed in Table 1.3, peaches possess the highest mineral content, whereas apples contain the lowest. Consuming 40 g (on a per serving basis) of dried fruits (Table 1.3) supplies 0.6–6.5% of calcium, 8.4–16.4% of copper, 2.1–20.3% of iron, 1.6–8.6% of magnesium, 1.6–11.3% of manganese, 2.2–6.8% of phosphorus, and 3.8–9.9% of potassium for RDA or AI for adults [4, 7–9]. Based on RDA and AI values, dried figs are high in calcium, magnesium, and manganese, whereas dried peaches are good sources of iron and phosphorus. Moreover, apricots are an important source of potassium among the eight dried fruits listed in Table 1.3. On a per serving basis (40 g or about one-fourth cup), dried fruits rank among the top potassium sources in diets around the world [6]. Moreover, on a per serving basis, different dried fruits such as apricots, currants, dates, figs, peaches, prunes, and raisins (40 g serving)

Table 1.2 Compositional and nutritional characteristics of some dried fruits (values in per 100 g edible portion)

Nutrient	Units	Apples	Apricots	Dates[a]	Figs	Peaches	Pears	Prunes	Raisins[b]
Proximate composition									
Water	g	31.76	30.89	20.53	30.05	31.80	26.69	30.92	15.43
Energy	kcal	243	241	282	249	239	262	240	299
Protein	g	0.93	3.39	2.45	3.30	3.61	1.87	2.18	3.07
Lipid	g	0.32	0.51	0.39	0.93	0.76	0.63	0.38	0.46
Ash	g	1.10	2.57	1.60	1.86	2.50	1.11	2.64	1.85
Carbohydrate	g	65.89	62.64	75.03	63.87	61.33	69.70	63.88	79.18
Dietary fiber	g	8.7	7.3	8.0	9.8	8.2	7.5	7.1	3.7
Sugars	g	57.19	53.54	63.35	47.92	41.74	62.20	38.13	59.19
Minerals									
Calcium	mg	14	55	39	162	28	34	43	50
Copper	mg	0.19	0.34	0.21	0.29	0.36	0.37	0.28	0.32
Fluoride	μg	nd	nd	nd	nd	nd	nd	4.0	234
Iron	mg	1.40	2.66	1.02	2.03	4.06	2.10	0.93	1.88
Magnesium	mg	16	32	43	68	42	33	41	32
Manganese	mg	0.09	0.24	0.26	0.51	0.31	0.33	0.30	0.30
Phosphorus	mg	38	71	62	67	119	59	69	101
Potassium	mg	450	1162	656	680	996	533	732	749
Selenium	μg	1.3	2.2	3.0	0.6	0.5	0.2	0.3	0.6
Sodium	mg	87	10	2	10	7	6	2	11
Zinc	mg	0.20	0.39	0.29	0.55	0.57	0.39	0.44	0.22
Vitamins									
Betaine	mg	nd	0.30	0.4	0.7	nd	nd	0.4	0.3
Choline	mg	17.6	13.9	6.3	15.8	12.7	23.0	10.1	11.1
Folate	μg	nd	10.0	19.0	9.0	nd	nd	4.0	5.0
Niacin	mg	0.93	2.59	1.27	0.62	4.38	1.37	1.88	0.77
Pantothenic acid	mg	0.25	0.52	0.59	0.43	0.56	0.15	0.42	0.10
Pyridoxine	mg	0.13	0.14	0.17	0.11	0.07	0.07	0.21	0.17
Riboflavin	mg	0.16	0.07	0.07	0.08	0.21	0.15	0.19	0.13

(continued)

Table 1.2 *(Continued)*

Nutrient	Units	Apples	Apricots	Dates[a]	Figs	Peaches	Pears	Prunes	Raisins[b]
***Vitamins* (*cont.*)**									
Thiamin	mg	nd	0.02	0.05	0.09	tr	0.01	0.05	0.11
Vitamin A (RAE)	µg	nd	180	tr	tr	108	nd	39	nd
Vitamin C	mg	3.9	1.0	0.4	1.2	4.8	7.0	0.6	2.3
Vitamin E (ATE)	mg	0.53	4.33	0.05	0.35	0.19	0.06	0.43	0.12
Vitamin K	µg	3.00	3.10	2.70	15.6	15.7	20.4	59.5	3.5
Amino acids									
Alanine	g	0.033	0.110	0.083	0.134	0.215	0.062	0.066	0.105
Arginine	g	0.029	0.066	0.136	0.077	0.092	0.032	0.037	0.413
Aspartic acid	g	0.162	0.937	0.213	0.645	0.602	0.368	0.801	0.110
Cystine	g	0.012	0.019	0.067	0.036	0.029	0.018	0.011	0.019
Glutamic acid	g	0.097	0.188	0.359	0.295	0.548	0.135	0.114	0.164
Glycine	g	0.037	0.070	0.101	0.108	0.126	0.054	0.047	0.080
Histidine[c]	g	0.015	0.047	0.032	0.037	0.067	0.020	0.027	0.072
Isoleucine[c]	g	0.037	0.063	0.049	0.089	0.104	0.054	0.041	0.057
Leucine[c]	g	0.057	0.105	0.084	0.128	0.204	0.094	0.066	0.096
Lysine[c]	g	0.058	0.083	0.066	0.088	0.116	0.066	0.050	0.084
Methionine[c]	g	0.009	0.015	0.022	0.034	0.087	0.022	0.016	0.021
Phenylalanine[c]	g	0.026	0.062	0.050	0.076	0.114	0.049	0.052	0.065
Proline	g	0.032	0.821	0.130	0.610	0.152	0.051	0.130	0.254
Serine	g	0.038	0.087	0.057	0.128	0.167	0.067	0.059	0.070
Threonine[c]	g	0.033	0.073	0.043	0.085	0.141	0.049	0.049	0.077
Tryptophan[c]	g	0.009	0.016	0.012	0.020	0.010	nd	0.025	0.050
Tyrosine	g	0.017	0.039	0.015	0.041	0.094	0.016	0.021	0.012
Valine[c]	g	0.043	0.078	0.071	0.122	0.197	0.066	0.056	0.083

Source: Adapted from USDA [4].
Note: Some numbers are rounded to the second digit after decimal point.
RAE, retinol activity equivalents; ATE, α-tocopherol equivalents; nd, not detected; tr, trace.
[a]Deglet noor.
[b]Seedless.
[c]Indispensable amino acids.

Table 1.3 Percentage of RDA values for adults (aged 19–50) in 40 g of dried fruits (per serving basis)

Mineral	RDA or AI*	Unit	Apples	Apricots	Dates[a]	Figs	Peaches	Pears	Prunes	Raisins[b]	Reference
Males											
Calcium	1000 mg/day*	mg	0.6	2.2	1.6	6.5	1.1	1.4	1.7	2.0	[4, 7]
Copper	0.9 mg/day	mg	8.4	15.1	9.3	12.9	16.0	16.4	12.4	14.2	[4, 8]
Fluoride	4000 μg/day*	μg	nd	nd	nd	nd	nd	nd	tr	2.3	[4, 7]
Iron	8 mg/day	mg	7.0	13.3	5.1	10.2	20.3	10.5	4.7	9.4	[4, 8]
Magnesium	400–420 mg/day	mg	1.6	3.1	4.2	6.6	4.1	3.2	4.0	3.1	[4, 7]
Manganese	2.3 mg/day*	mg	1.6	4.2	4.5	8.9	5.4	5.7	5.2	5.2	[4, 8]
Phosphorus	700 mg/day	mg	2.2	4.1	3.5	3.8	6.8	3.4	3.9	5.8	[4, 7]
Potassium	4700 mg/day	mg	3.8	9.9	5.6	5.8	8.5	4.5	6.2	6.4	[4, 9]
Selenium	55 μg/day	μg	0.9	1.6	2.2	0.4	0.4	0.1	0.2	0.4	[4, 10]
Sodium	1500 mg/day	mg	2.3	0.3	0.1	0.3	0.2	0.2	0.1	0.3	[4, 9]
Zinc	11 mg/day	mg	0.7	1.4	1.1	2.0	2.1	1.4	1.6	0.8	[4, 8]
Females											
Calcium	1000 mg/day*	mg	0.6	2.2	1.6	6.5	1.1	1.4	1.7	2.0	[4, 7]
Copper	0.9 mg/day	mg	8.4	15.1	9.3	12.9	16.0	16.4	12.4	14.2	[4, 8]
Fluoride	3000 μg/day*	μg	nd	nd	nd	nd	nd	nd	tr	3.1	[4, 7]
Iron	18 mg/day	mg	3.1	5.9	2.3	4.5	9.0	4.7	2.1	4.2	[4, 8]
Magnesium	310–320 mg/day	mg	2.0	4.1	5.5	8.6	5.3	4.2	5.2	4.1	[4, 7]
Manganese	1.8 mg/day*	mg	2.0	5.3	5.8	11.3	6.9	7.3	6.7	6.7	[4, 8]
Phosphorus	700 mg/day	mg	2.2	4.1	3.5	3.8	6.8	3.4	3.9	5.8	[4, 7]
Potassium	4700 mg/day	mg	3.8	9.9	5.6	5.8	8.5	4.5	6.2	6.4	[4, 9]
Selenium	55 μg/day	μg	0.9	1.6	2.2	0.4	0.4	0.1	0.2	0.4	[4, 10]
Sodium	1500 mg/day	mg	2.3	0.3	0.1	0.3	0.2	0.2	0.1	0.3	[4, 9]
Zinc	8 mg/day	mg	1.0	2.0	1.5	2.8	2.9	2.0	2.2	1.1	[4, 8]

RDA, recommended dietary allowances; AI*, adequate intake; nd, not detected; tr, trace.
[a]Deglet noor.
[b]Seedless.

compare positively in their potassium content with the 10 most common fresh fruit options such as apples, bananas, grapes, mangos, oranges, peaches, pears, pineapples, strawberries, and watermelons (one cup or one fruit serving) [4, 6].

Dried fruits contain both water-soluble (betaine, choline, folate, niacin, pantothenic acid, pyridoxine, riboflavin, thiamine, and vitamin C) and fat-soluble vitamins (A, E, and K) (Table 1.2). Among the eight dried fruits listed, prunes are the richest source of vitamin K (59.5 μg/100 g), whereas apricots are the richest source of vitamin A (180 μg/100 g) and vitamin E (4.33 mg/100 g) [4]. Dried fruits, in general, contain a small amount of vitamin C. With regard to RDA of vitamins, 40 g dried fruits provide up to 1.6–12.5% of niacin, 0.8–4.7% of pantothenic acid, 2.2–6.5% of pyridoxine, 2.2–7.6% of riboflavin, and 0.9–26.4% of vitamin K for RDA or AI for adults [4, 8, 10, 11]. Prunes are particularly high in vitamin K. Among these eight dried fruits, prunes, apricot, and peaches contain higher amounts of vitamins than other dried fruits (Tables 1.2 and 1.4).

Despite the fact that dried fruits contain all indispensable amino acids (except tryptophan in pears), in general, they are not good sources of amino acids due to their low protein content (Table 1.2).

In summary, the following are some nutritional facts about dried fruits [12]:

- Dried fruits are low in their fat and sodium content and, as expected, devoid of trans-fats and cholesterol [4].
- Dried fruits are good sources of dietary fiber and potassium. Among all fruits, they are among the top five contributors of fiber and potassium [5].
- Dried fruits provide essential nutrients that are otherwise low in today's diet, such as vitamin A (apricots and peaches), calcium (figs), vitamin K (prunes), boron (raisins and prunes), iron, and copper [4, 13].
- Traditional dried fruits have no added sugars. Most traditional dried fruits contain low amounts of sucrose; their sugar content is in the form of fructose and glucose [4].

1.3 Phytochemicals in dried fruits

Phytochemicals are defined as nonnutritive, naturally occurring, biologically active, and chemically derived compounds found in the plant kingdom. More than several thousands of individual phytochemicals have been identified in plant-derived foods and their by-products, but a large percentage of phytochemicals still remain unknown and need to be identified before we can fully understand the health benefits of phytochemicals in whole foods. Dried fruits are highly nutritious and provide a range of phytochemicals such as phenolic acids, flavonoids (anthocyanidins, flavan-3-ols, flavones, flavonols, and isoflavones), phytoestrogens, and carotenoids, among others [3, 4, 14–28]. They, in general, contain traces or undetectable amounts of proanthocyanidins [29]. Proanthocyanidins detected in plums and grapes are absent in prunes and raisins, which suggests that these compounds are degraded during the drying process [30].

Dried fruits are excellent sources of phenolic compounds in the diet [31–35]. These make up the largest group of plant phytochemicals in the diet and they appear to be, at least in part, responsible for the health benefits associated with diets abundant in fruits and vegetables. Phenolic compounds contribute most to the antioxidant activity of fruits and vegetables [36] and have a multitude of functional capacities, which may have a beneficial effect on health [6].

Table 1.4 Percentage of RDA values for adults (aged 19–50 years) in 40 g of dried fruits (per serving basis)

Vitamin	RDA or AI*	Unit	Apples	Apricots	Dates[a]	Figs	Peaches	Pears	Prunes	Raisins[b]	Reference
Males											
Choline	550 mg/day*	mg	1.3	1.0	0.5	1.1	0.9	1.7	0.7	0.8	[4, 11]
Folate	400 μg/day	μg	nd	1.0	1.9	0.9	nd	nd	0.4	0.5	[4, 11]
Niacin	16 mg/day	mg	2.3	6.5	3.2	1.6	11.0	3.4	4.7	1.9	[4, 11]
Pantothenic acid	5 mg/day*	mg	2.0	4.2	4.7	3.4	4.5	1.2	3.4	0.8	[4, 11]
Pyridoxine	1.3 mg/day	mg	4.0	4.3	5.2	3.4	2.2	2.2	6.5	5.2	[4, 11]
Riboflavin	1.3 mg/day	mg	4.9	2.2	2.2	2.5	6.5	4.6	5.8	4.0	[4, 11]
Thiamin	1.2 mg/day	mg	nd	0.7	1.7	3.0	tr	0.3	1.7	3.7	[4, 11]
Vitamin A (RAE)	900 μg/day	μg	nd	8.0	tr	tr	4.8	nd	1.7	nd	[4, 8]
Vitamin C	90 mg/day	mg	1.7	0.4	0.2	0.5	2.1	3.1	0.3	1.0	[4, 10]
Vitamin E (ATE)	15 mg/day	mg	1.4	11.5	0.1	0.9	0.5	0.2	1.1	0.3	[4, 10]
Vitamin K	120 μg/day*	μg	1.0	1.0	0.9	5.2	5.2	6.8	19.8	1.2	[4, 8]
Females											
Choline	425 mg/day*	mg	1.7	1.3	0.6	1.5	1.2	2.2	1.0	1.0	[4, 11]
Folate	400 μg/day	μg	nd	1.0	1.9	0.9	nd	nd	0.4	0.5	[4, 11]
Niacin	14 mg/day	mg	2.7	7.4	3.6	1.8	12.5	3.9	5.4	2.2	[4, 11]
Pantothenic acid	5 mg/day*	mg	2.0	4.2	4.7	3.4	4.5	1.2	3.4	0.8	[4, 11]
Pyridoxine	1.3 mg/day	mg	4.0	4.3	5.2	3.4	2.2	2.2	6.5	5.2	[4, 11]
Riboflavin	1.1 mg/day	mg	5.8	2.5	2.5	2.9	7.6	5.5	6.9	4.7	[4, 11]
Thiamin	1.1 mg/day	mg	nd	0.7	1.8	3.3	tr	0.4	1.8	4.0	[4, 11]
Vitamin A (RAE)	700 μg/day	μg	nd	10.3	tr	tr	6.2	nd	2.2	nd	[4, 8]
Vitamin C	75 mg/day	mg	2.1	0.5	0.2	0.6	2.6	3.7	0.3	1.2	[4, 10]
Vitamin E (ATE)	15 mg/day	mg	1.4	11.5	0.1	0.9	0.5	0.2	1.1	0.3	[4, 10]
Vitamin K	90 μg/day*	μg	1.3	1.4	1.2	6.9	7.0	9.1	26.4	1.6	[4, 8]

RDA, recommended dietary allowances; AI*, adequate intake; RAE, retinol activity equivalents; ATE, α-tocopherol equivalents; nd, not detected; tr, trace.
[a]Deglet noor.
[b]Seedless.

Table 1.5 Comparison of total phenolics and ORAC values of some dried fruits (values in per 100 g edible portion)

Dried fruits	Total phenolics (mg of GAE/100 g)	Total ORAC (μmol of TE/100 g)	Reference
Apples[a]	324	6,681	[14, 17]
Apricots[a]	248	3,234	[14, 17]
Dates (Deglet noor)	661	3,895	[14, 21]
Dates (Medjool)	572	2,387	[14, 21]
Figs	960	3,383	[14, 21]
Peaches[a]	283	4,222	[14, 17]
Pears[a]	679	9,496	[14, 17]
Prunes	1195	8,578	[14, 21]
Raisins (golden seedless)	–	10,450	[14, 19]
Raisin (seedless)	1065	3,037	[14, 21]
Raisins (white)[a]	372	4,188	[14, 17]

GAE, gallic acid equivalents; TE, trolox equivalents.
[a]Dried to 40% moisture (purchased in Italy).

Values of the total phenolic content and oxygen radical absorbance capacity (ORAC) for a selection of dried fruit are given in Table 1.5. Prunes have the highest total phenolic content (1195 mg of gallic acid equivalents (GAE)/100 g), whereas raisins (golden seedless) have the highest ORAC value (10,450 μmol trolox equivalents (TE)/100 g). Significant differences in the total phenolic content and the ORAC value exist among raisin varieties, being lowest in white raisins and highest in golden seedless raisins [14, 17, 19, 21]. Values are much higher for dried fruits than the corresponding values for their fresh counterparts because antioxidants become concentrated after the drying or dehydration process. While there is a loss or modification of some specific phytochemicals during drying, antioxidant activity and the total phenolic content remain relatively unchanged during the process, implying that many of the phenolic compounds are yet unidentified [37]. This could include oligomeric or polymeric products that are difficult to characterize. Pellergrini *et al.* [38] measured the total antioxidant capacity (using three different *in vitro* assays) of food including four dried fruits (apricots, figs, prunes, and raisins). Among these fruits, prunes exhibited the highest value followed by apricots. Little information is available on the phenolic profiles and antioxidant components of the other dried fruits.

The 1,1-diphenyl-2-picrylhydrazyl (DPPH) radical-scavenging activity and polyphenol content of 22 dried fruits (apples, apricots, bananas, blueberries, cherries, cranberries, figs, raisins, hawthorns, jujube, kiwis, kumquats, mangoes, melons, muscats, papayas, peaches, pears, pineapples, prunes, rakankas, and strawberries) have been evaluated and compared with fresh fruits by Ishiwata *et al.* [39]. Among the dried fruits examined, hawthorns, apricots, and blueberries exhibited the highest DPPH radical-scavenging activity. The polyphenol content of dried fruits and DPPH radical-scavenging activity were highly correlated. On a fresh weight basis, dried fruits, in general, contain higher radical-scavenging activity than fresh fruits. In contrast, the radical-scavenging activity of dried fruits is lower than that of the corresponding fresh fruits on a fresh weight basis [39]. A similar pattern was also reported by Vinson *et al.* [40].

Vinson *et al.* [40] reported the total phenolic content of fresh and dried fruits (apricots, cranberries, dates, figs, grapes/raisins, and plums/prunes) (Figure 1.1a). Dates had the highest

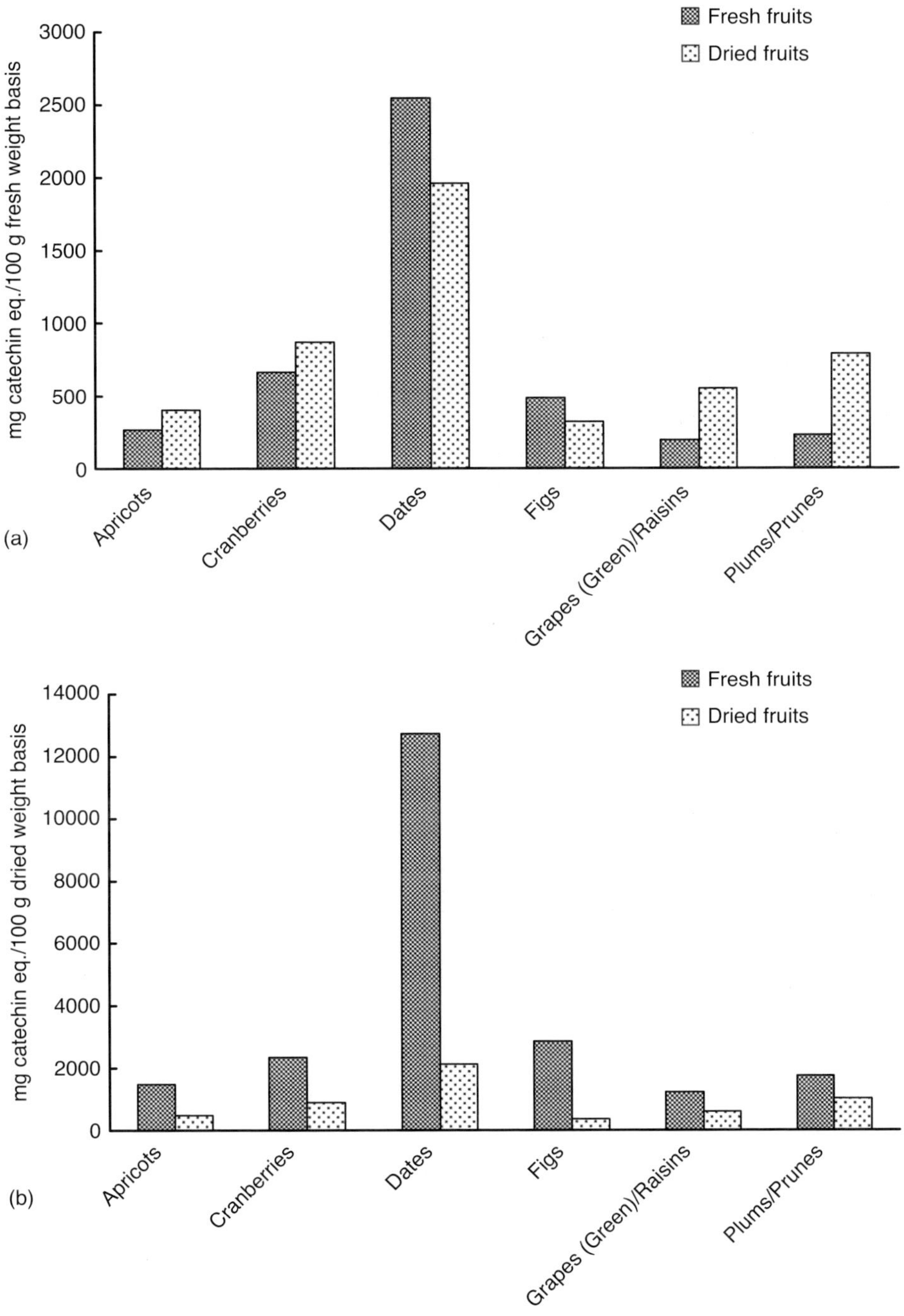

Figure 1.1 Comparison of quantity of total phenol in fresh and the corresponding dried fruit on a fresh weight basis (a) and dry weight basis (b). (Adapted with permission from Vinson *et al.* [40]).

concentration of total phenolics in both fresh and dried versions (2546 and 1959 mg catechin equivalents (CE)/100 g on a fresh weight basis, respectively). The total phenolic content averaged 731 mg of CE/100 g for fresh fruits and 815 mg of CE/100 g for dried fruits [40]. A comparison of the quantity of total phenolics in fresh and dried fruit pairs on a dry weight basis is illustrated in Figure 1.1b. The average is 3730 mg of CE/100 g for fresh fruits and only 910 mg of CE/100 g for dried varieties [40]. The process of producing dried fruit significantly decreases the total phenol content of the fruits on a dry weight basis. Dried fruits are significantly higher ($P < 0.005$) in total phenols than 20 fresh fruits, 815 versus 173 mg/100 g, respectively [41].

Flavonoids are another group of phenolic compounds that can be classified into seven groups: flavanones, flavones, isoflavones, anthocyanidins, flavonols, flavononols, and flavanols or flavan-3-ols [42]. Flavonoids, which are the most common and widely distributed group of plant phenolics, are increasingly appreciated as an important component of the human diet. Humans consume approximately 1 g of flavonoids per day [43]. Different classes (anthocyanidins, flavan-3-ols, flavones, and flavonols) and amounts of flavonoids have been reported for different dried fruits [15, 18, 23, 24]. Despite the fact that raisins contain the above-mentioned classes of flavonoids, the total content of flavonoids among the four dried fruits listed in Table 1.6 varies between 0.85 mg/100 g in raisins and 7.66 mg/100 g in cranberries. Dried fruits contain traces or undetectable amounts of anthocyanins, which are likely degraded to phenolic acids. Dates contain one anthocyanidin (such as cyanidin) and one flavonol (such as quercetin). Flavan-3-ols are only present in raisins.

The available data show that dried fruits have a unique spectrum of phenols, polyphenols, and tannins. For example, in raisins, the most abundant phenolic compounds are the

Table 1.6 Comparison of flavonoids (mg/100 g edible portion) of some dried fruits

Flavonoid	Cranberries [15, 23]	Dates[a] [15, 24]	Prunes [15, 23, 24]	Raisins[b] [15, 18, 23, 24]
Anthocyanidins				
Cyanidin	0.60	1.70	0.71	0.03
Delphinidin	0.10	–	0.04	0.01
Pelargonidin	0.02	–	–	0.01
Flavan-3-ols				
(–)-Epicatechin	–	–	–	0.10
(+)-Catechin	–	–	–	0.42
Flavones				
Apigenin	0.01	–	–	–
Luteolin	0.02	–	0.01	0.01
Flavonols				
Kaempferol	0.01	–	0.01	0.01
Myricetin	2.40	–	0.01	0.01
Quercetin	4.50	0.93	1.80	0.25
Total	7.66	2.63	2.58	0.85

[a] Deglet noor.
[b] Seedless.

flavonoids quercetin (Table 1.6) and phenolic acids caftaric and coutaric acids [3]. The predominant phenolic compounds in Greek currants (raisins) are vanillic, caffeic, gallic, syringic, *p*-coumaric, and protocatechuic acids and the flavonoid quercetin [44]. Hydroxycinnamic acids, especially chlorogenic acid isomers, are the major phenolics in prunes, representing more than 94% of the total [45]. Rutin is the predominant flavonol in prunes and prune juice [26]. Prunes also contain quinic acid that is metabolized to hippuric acid, which, as some research suggests, helps prevent urinary tract infections [35, 46]. Four free phenolic acids (protocatechuic, vanillic, syringic, and ferulic) and nine bound phenolic acids (gallic, protocatechuic, *p*-hydroxybenzoic, vanillic, caffeic, syringic, *p*-coumaric, ferulic, and *o*-coumaric) have been reported in fresh and sun-dried Omani dates of three native varieties [47]. Nutritional and functional properties as well as phytochemical characteristics (carotenoids, phytosterols, polyphenols, phenolic acids, flavonoids, anthocyanins, and phytoestrogens) of dates have been extensively reviewed [28, 48].

Phytoestrogens comprise three major classes: isoflavones, lignans, and coumestan. Some dried fruits (such as apricots, currants, dates, prunes, and raisins) have been reported to contain phytoestrogens [such as isoflavones (formononetin, daidzein, genistein, and glycitein), lignans (matairesinol, lariciresinol, pinoresinol, and secoisolariciresinol), and coumestan (coumestrol)]. Apricots contain the highest concentration of total phytoestrogens (444.5 μg/100 g) among the five dried fruits listed in Table 1.7, followed by dates (329.5 μg/100 g), prunes (183.5 μg/100 g), currants (34.1 μg/100 g), and raisins (30.2 μg/100 g) [25]. Coumestan, measured as coumestrol, is generally present in low concentrations within dried fruit groups. Dried fruits have higher concentration of lignans (ranging from 20.9 to 400.5 μg/100 g) than isoflavones (ranging from 4.2 to 39.8 μg/100 g) [25].

Five carotenoids (namely, α-carotene, β-carotene, β-cryptoxanthin, lutein, and zeaxanthin) are present in some dried fruits. Of these, β-carotene, which acts as provitamin A, is the

Table 1.7 Comparison of phytoestrogen content of some dried fruits (μg/100 g edible portion)

Phytoestrogen	Apricots[a]	Currants	Dates[b]	Prunes[c]	Raisins[d]
Formononetin	12.5	0.6	0.4	0.5	0.4
Daidzein	6.4	2.2	1.2	2.6	1.5
Genistein	19.8	10.0	3.4	0.2	5.2
Glycitein	1.1	0.2	0.2	0.9	1.0
Total isoflavones	39.8	13.1	5.1	4.2	8.1
Matairesinol	0.6	1.1	0.3	0.2	0.4
Lariciresinol	62.1	5.8	116.9	2.1	9.2
Pinoresinol	190.1	3.0	100.2	71.5	0.8
Secoisolariciresinol	147.6	10.9	106.2	103.8	11.5
Total lignans	400.5	20.9	323.6	177.5	22.0
Coumestrol	4.2	0.1	0.8	1.8	0.2
Total coumestan	4.2	0.1	0.8	1.8	0.2
Total phytoestrogens	444.5	34.1	329.5	183.5	30.2

Source: Adapted with permission from Thompson *et al.* [25].
[a]Turkish.
[b]Whole pitted.
[c]Whole pitted.
[d]California seedless.

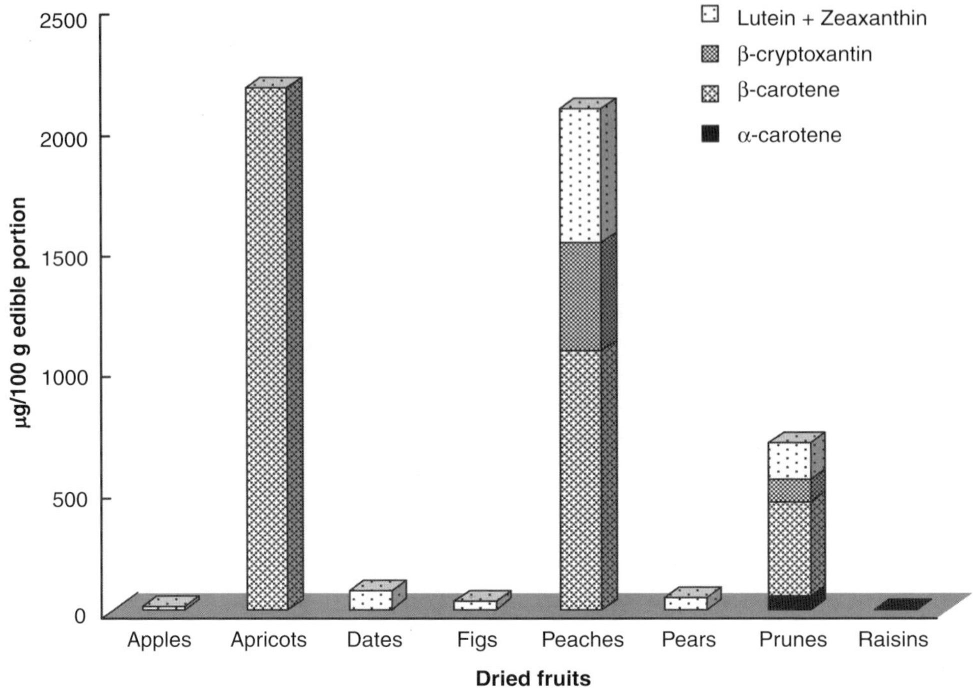

Figure 1.2 Carotenoid content of selected dried fruits on a fresh weight basis. (Adapted from USDA [4]).

most abundant in dried apricots (2163 μg/100 g), peaches (1074 μg/100 g), and prunes (394 μg/100 g), followed by lutein + zeaxanthin in peaches (559 μg/100 g), and β-cryptoxanthin in peaches (444 μg/100 g) [4]. No carotenoids are present in raisins and small and/or trace amounts of carotenoids are found in apples, dates, figs, and pears (Figure 1.2).

In summary, the following are some of the health protective components in dried fruits [12]:

- Dried fruits are good sources of phytochemicals [3, 4, 14–28].
- By virtue of their high phytochemical content, dried fruits are an important source of antioxidants in the diet [21, 28, 40]. Dried apricots and peaches are good sources of carotenoids [4].
- Dried fruits, such as prunes, provide pectin, a soluble fiber that may lower blood cholesterol [49].
- Dried fruits, such as raisins, are a source of prebiotic compounds in the diet. They contain fructooligosaccharides such as inulin that contributes to colon health [50, 51].
- Dried fruits contain organic acids such as tartaric acid (raisins) and sugar alcohol such as sorbitol (prunes). These compounds and fiber appear to work synergistically to maintain a healthy digestive system. They may also help increase the bioavailability of minerals in the diet, such as calcium and iron [52].

1.4 Beneficial health effects of dried fruits

As shown by multiple epidemiological studies, fruit and vegetable consumption reduces the risk of many chronic diseases such as cancer [53–55], heart disease [56, 57], stroke [58], obesity [59], and type 2 diabetes [59, 60], among others [28]. Additionally, there is an inverse relationship between fruit and vegetable intake and blood pressure [61]. The US National Cancer Institute (NCI) and National Research Council (NRC) recommend at least five servings of fruits and vegetables daily. Similarly, the World Health Organization (WHO) recommends 400 g of fruits and vegetables per day or the equivalent of five servings of 80 g each [62].

Numerous health benefits of dried fruits have been reported [6, 28, 40, 51, 55, 63, 64] and reviewed individually throughout this book. The health benefits of dried fruits mainly originate from their essential nutrients and phytochemicals (such as anthocyanidins, carotenoids, phytoestrogen, flavan-3-ols, flavones, flavonols, and phenolic acids, among others) as well as their antioxidant activities. The intake of flavonoids (major part of phytochemicals) has been associated with a lower incidence of various diseases such as cancer, stroke, cardiovascular disease (CVD), and other chronic disorders [65–67]. The positive health effects of flavonoids are probably related to their strong antioxidant properties, among other mechanisms and effects [68, 69]. There is considerable research supporting the role of dried fruits, particularly, in promoting digestive health [51, 70, 71]. Dried fruits, particularly prunes, play a role in supporting bone health [63, 64, 72]. Finally, dried fruits, such as raisins, may promote healthy teeth and gums [73–75].

Dried fruits are important sources of potassium and dietary fiber. Increasing dietary potassium intake can lower blood pressure [76]. Higher fiber diets are recommended to reduce the risk of developing various conditions including constipation, type 2 diabetes, obesity, diverticulitis, colorectal cancer, and CVD [6]. Dried fruits may contribute to healthy body weights. According to the National Health and Nutrition Examination Survey (1999–2004), data showed that intake of dried fruit was associated with a lower body mass index (BMI), reduced waist circumference, abdominal obesity, and improved diet quality [77]. Emerging data suggest that dried fruit promotes satiety by affecting the levels of hormones such as leptin and regulates appetite [78].

Because of the sweetness of dried fruits, it is expected to exert a high glycemic index (70 and above) and insulin response. Recent studies have shown that dried fruits have a low (55 and under) to moderate (56–69) glycemic and insulin index (Table 1.8) and glycemic and

Table 1.8 Glycemic index (GI) of some dried fruits

Dried fruits	GI[a]	Reference
Apples	29	[79]
Apricots	30	[79]
Dates	39	[80]
Figs	61	[79]
Peaches	35	[79]
Prunes	29	[79]
Raisins	52	[81]

[a]High GI (70 and above), moderate GI (56–69), and low GI (55 and under).

insulin response comparable to those in fresh fruits [80–83]. This could be due to the presence of fiber, polyphenols, and tannins that can modify the response [84–87]. Foods with a low glycemic index may help decrease the risk of diabetes and are useful in the management of the established condition [6].

As part of a diet study involving 13,292 participants, dried fruit consumers were defined as those who consume at least one-eighth cup-equivalent of fruit per day [88]. Dried fruit consumption is associated with a lower body weight, improved adiposity measures, higher overall diet quality, and higher nutrient intake of vitamins A, E, and K, phosphorus, magnesium, and potassium. These benefits are attributed to a higher fiber content, reduced intake of solid fats, alcohol, and added sugars. However, only 7% of the subjects in the study consumed significant amounts of dried fruits in their diet, which questions the effectiveness of ongoing public health campaigns around the world that encourage an increase in fruit consumption.

Given many of the benefits of dried fruits in terms of health maintenance, what might dried fruits contribute to an increase in fresh and dried fruit consumption while also offering meaningful health benefits beyond their nutritive value? The answer may lie in focusing on foods that offer significant antioxidant and anti-inflammatory benefits.

1.5 Commercial products and industrial applications of dried fruits

Dried fruits are widely used as ingredients in packaged snacks, confectionary products, baked goods, cereals, energy and nutritional bars, ready-to-eat salads, and sweet industries, among many other specialty foods [89].

Fruits can be dried whole (e.g., grapes, apricots, and plums), in halves, as slices, or diced (e.g., mangoes, papayas, and kiwis). Alternatively, they can be chopped after drying (e.g., dates), made into pastes or concentrated juices. Fruits can also be dried in puree form, as leathers, or as a powder, by spray drying. Some fruits can be freeze-dried (e.g., strawberries, raspberries, cherries, apples, and mangoes, among others). The freeze-dried fruits become very light and crispy and retain much of their original flavor (taste and aroma) and phytochemicals [12].

1.6 Conclusions

Dried fruits, with their unique combination of taste and aroma, essential nutrients, fiber, and phytochemicals or bioactive compounds, are a convenient step toward healthier eating and a means to bridge the gap between recommended intake of fruits and actual consumption. They should be included together with fresh fruit recommendations around the world since they help meet dietary guidelines for daily fruit serving (recommended five servings per day) and address barriers to fruit intake. Dried fruits in smaller serving sizes, ranging from 30 to 43 g depending on the fruit, are considered nutritionally equivalent to fresh fruits in current dietary recommendations in different countries [6, 90, 91]. Numerous scientific evidences suggest that individuals who regularly consume generous amounts of dried fruits have a lower rate of CVD, obesity, various types of cancer, type 2 diabetes, and other chronic diseases. Therefore, dried fruits should be consumed daily in order to get full benefit of nutrients, health-promoting phytochemicals, and antioxidants they contain, together with their unique and desirable taste and aroma.

References

1. INC (2012) *World Dried Fruits Production*. International Nut and Dried Fruit Council Foundation, Reus, Spain.
2. Barta, J. (2006) Fruit drying principles. In: *Handbook of Fruits and Fruit Processing* (ed. Y.H. Hui). Blackwell Publishing, Oxford, UK, pp. 81–94.
3. Willimason, G. & Carughi, A. (2010) Polyphenol content and health benefits of raisins. *Nutrition Research*, **30**, 511–519.
4. USDA (2010) *National Nutrient Database for Standard Reference, Release 23*. Published online at: http://www.nal.usda.gov/fnic/foodcomp/search/, last accessed March 7, 2011.
5. USDA (2005) *Carbohydrates. Dietary Guidelines for Americans*. US Department of Agriculture and US Department of Health and Human Services. Government Printing Office, Washington, DC.
6. INC (2011) Traditional dried fruits: valuable tools to meet dietary recommendations for fruit intake. Paper given at the *XXX World Nut and Dried Fruit Congress*, Budapest, Hungary, May 20–22, 2011 (Dried Fruit Round Table).
7. DRIs (1997) *Dietary Reference Intakes for Calcium, Phosphorus, Magnesium, Vitamin D, and Fluoride*. The National Academies Press, Washington, DC.
8. DRIs (2001) *Dietary Reference Intakes for Vitamin A, Vitamin K, Arsenic, Boron, Chromium, Copper, Iodine, Iron, Manganese, Molybdenum, Nickel, Silicon, Vanadium, and Zinc*. The National Academies Press, Washington, DC.
9. DRIs (2004) *Dietary Reference Intakes for Water, Potassium, Sodium, Chloride, and Sulfate*. The National Academies Press, Washington, DC.
10. DRIs (2000) *Dietary Reference Intakes for Vitamin C, Vitamin E, Selenium, and Carotenoids*. The National Academies Press, Washington, DC.
11. DRIs (1998) *Dietary Reference Intakes for Thiamin, Riboflavin, Niacin, Vitamin B_6, Folate, Vitamin B_{12}, Pantothenic Acid, Biotin, and Choline*. The National Academies Press, Washington, DC.
12. Dried Fruit (2011) Published online at: http://en.wikipedia.org/wiki/Dried_fruit, last accessed July 18, 2011.
13. Rainey, C.J., Nyquist, L.A., Christensen, R.E., Strong, P.L., Culver, B.D. & Coughlin, R. (1999) Daily boron intake from the American diet. *Journal of the American Dietetic Associations*, **99**, 335–340.
14. USDA (2010) *USDA Database for the Oxygen Radical Absorbance Capacity (ORAC) of Selected Foods, Release 2.0*. Published online at: http://www.ars.usda.gov/nutrientdata, last accessed March 9, 2011.
15. USDA (2011) *USDA Database for the Flavonoid Content of Selected Foods, Release 3*. Published online at: http://www.ars.usda.gov/nutrientdata, last accessed November 2, 2011.
16. USDA (2008) *USDA Database for the Isoflavone Content of Selected Foods, Release 2.0*. Published online at: http://www.ars.usda.gov/nutrientdata, last accessed March 9, 2011.
17. Bacchiocca, M., Biagiotti, E. & Ninfali, P. (2006) Nutritional and technological reasons for evaluating the antioxidant capacity of vegetable products. *Italian Journal of Food Science*, **18**, 209–217.
18. Arts, I.C.W., van de Putte, B. & Hollman, P.C.H. (2000) Catechin contents of foods commonly consumed in the Netherlands. 1. Fruits, vegetables, staple foods, and processed foods. *Journal of Agricultural and Food Chemistry*, **48**, 1746–1751.
19. Parker, T.L., Wang, X.-H., Pazmiño, J. & Engeseth, N.J. (2007) Antioxidant capacity and phenolic content of grapes, sun-dried raisins, and golden raisins and their effect on *ex vivo* serum antioxidant capacity. *Journal of Agricultural and Food Chemistry*, **55**, 8472–8477.
20. Horn-Ross, P.L., Barnes, S., Lee, M., Coward, L., Mandel, J.E., Koo, J., John, E.M., Smith, M. (2000) Assessing phytoestrogen exposure in epidemiologic studies: development of a database (United States). *Cancer Causes and Control*, **11**, 289–298.
21. Wu, X., Beecher, G.R., Holden, J.M., Haytowitz, D.B., Gebhardt, S.E. & Prior, R.L. (2004) Lipophilic and hydrophilic antioxidant capacities of common foods in the United States. *Journal of Agricultural and Food Chemistry*, **52**, 4026–4037.

22. Liggins, J., Bluck, L.J.C., Runswick, S., Atkinson, C., Coward, W.A. & Bingham, S.A. (2000) Daidzein and genistein content of fruits and nuts. *Journal of Nutritional Biochemistry*, **11**, 326–331.
23. Franke, A.A., Custer, L.J., Arakaki, C. & Murphy, S.P. (2004) Vitamin C and flavonoid levels of fruits and vegetables consumed in Hawaii. *Journal of Food Composition and Analysis*, **17**, 1–35.
24. Harnly, J.M., Doherty, R.F., Beecher, G.R., Holden, J.M., Haytowitz, D.B., Bhagwat, S., Gebhardt, S. (2006) Flavonoid content of U.S. fruits, vegetables, and nuts. *Journal of Agricultural and Food Chemistry*, **54**, 9966–9977.
25. Thompson, L.U., Boucher, B.A., Liu, Z., Cotterchio, M. & Kreiger, N. (2006) Phytoesterogen content of foods consumed in Canada, including isoflavones, lignans, and coumestan. *Nutrition and Cancer*, **54**, 184–201.
26. Donovan, J.L., Meyer, A.S. & Waterhouse, A.L. (1998) Phenolic composition and antioxidant activity of prunes and prune juice (*Prunus domestica*). *Journal of Agricultural and Food Chemistry*, **46**, 1247–1252.
27. Slatnar, A., Klancar, U., Stampar, F. & Veberic, R. (2011) Effect of drying of figs (*Ficus carica* L.) on the contents of sugars, organic acids, and phenolic compounds. *Journal of Agricultural and Food Chemistry*, **59**, 11696–11702.
28. Vayalil, P.K. (2012) Date fruits (*Phoenix dactylifera* Linn): an emerging medicinal food. *Critical Reviews in Food Science and Nutrition*, **52**, 249–271.
29. USDA (2004) *USDA Database for the Proanthocyanidin Content of Selected Foods*. Published online at: http://www.ars.usda.gov/nutrientdata, last accessed March 9, 2011.
30. Gu, L., Kelm, M.A., Hammerstone, J.F., Beecher, G., Holden, J., Haytowitz, D. & Prior, R.L. (2003) Screening of foods containing proanthocyanidins and their structural characterization using LC-MS/MS and thiolytic degradation. *Journal of Agricultural and Food Chemistry*, **51**, 7513–7521.
31. Wu, X., Beecher, G.R., Holden, J.M., Haytowitz, D.B., Gebhardt, S.E. & Prior, R.L. (2004) Lipophilic and hydrophilic antioxidant capacities of common foods in the United States. *Journal of Agricultural and Food Chemistry*, **52**, 4026–4037.
32. Vinson, J.A., Zubik, L., Bose, P., Samman, N. & Proch, J. (2005) Dried fruits: Excellent *in vitro* and *in vivo* antioxidants. *Journal of the American College of Nutrition*, **24**, 44–50.
33. Vinson, J.A., Su, X., Zubik, L. & Bose, P. (2001) Phenol antioxidant quality and quantity in foods: fruits. *Journal of Agricultural and Food Chemistry*, **49**, 5315–5321.
34. Kaliora, A.C., Kountouri, A.M. & Karathanos, V.T. (2009) Antioxidant properties of raisins (*Vitis vinifera* L.). *Journal of Medicinal Food*, **12**, 1302–1309.
35. Kayano, S.-I., Kikuzaki, H., Fukutsuka, N., Mitani, T. & Nakatani, N. (2002) Antioxidant activity of prune (*Prunus domestica* L.) constituents and a new synergist. *Journal of Agricultural and Food Chemistry*, **50**, 3708–3712.
36. Chun, O.K., Kim, D.-O., Smith, N., Schroeder, D., Han, J.T. & Lee, C.Y. (2005) Daily consumption of phenolics and total antioxidant capacity from fruits and vegetables in the American diet. *Journal of the Science of Food and Agriculture*, **85**, 1715–1724.
37. Madrau, M.A., Sanguinetti, A.M., Del Caro, A., Fadda, C. & Piga, A. (2010) Contribution of melanoidins to the antioxidant activity of prunes. *Journal of Food Quality*, **33**, 155–170.
38. Pellegrini, N., Serafini, M., Salvatore, S., Del Rio, D., Bianchi, M. & Brighenti, F. (2006) Total antioxidant capacity of spices, dried fruits, nuts, pulses, cereals and sweets consumed in Italy assessed by three different *in vitro* assays. *Molecular Nutrition & Food Research*, **50**, 1030–1038.
39. Ishiwata, K., Yamaguchi, T., Takamura, H. & Matoba, T. (2004) DPPH radical-scavenging activity and polyphenol content of dried fruits. *Food Science and Technology Research*, **10**, 152–156.
40. Vinson, J.A., Zubik, L., Bose, P., Samman, N. & Proch, J. (2005) Dried fruits: Excellent *in vitro* and *in vivo* antioxidants. *Journal of the American College of Nutrition*, **24**, 44–50.
41. Vinson, J.A., Su, X., Zubik, L. & Boase, P. (2001) Phenol antioxidant quantity and quality in foods: Fruits. *Journal of Agricultural and Food Chemistry*, **49**, 5315–5321.
42. Shahidi, F. & Ho, C.-T. (2005) Phenolics in food and natural health products: an overview. In: *Phenolic Compounds in Foods and Natural Products* (eds F. Shahidi & C.-T. Ho). ACS Symposium Series 909, American Chemical Society, Washington, DC, pp. 1–8.
43. Kuhnau, J. (1976) The flavonoids: a class of semi-essential food components: their role in human nutrition. *World Review of Nutrition and Dietetics*, **24**, 117–191.
44. Chiou, A., Karathanos, V.T., Mylona, A., Salta, F.N., Preventi, F. & Andrikopoulos, N.K. (2007) Currants (*Vitis vinifera* L.) content of simple phenolics and antioxidant activity. *Food Chemistry*, **102**, 516–522.

45. Nakatani, N., Kayano, S.-I., Kikuzaki, H., Sumino, K., Katagiri, K. & Mitani, T. (2000) Identification, quantitative determination, and antioxidative activities of chlorogenic acid isomers in prune (*Prunus domestica* L.). *Journal of Agricultural and Food Chemistry*, **48**, 5512–5516.
46. Fang, N., Shanggong, Y. & Prior, R.L. (2002) LC/MS/MS characterization of phenolic constituents in dried plums. *Journal of Agricultural and Food Chemistry*, **50**, 3579–3585.
47. Al-Farsi, M., Alasalvar, C., Morris, A., Barron, M. & Shahidi, F. (2005) Comparison of antioxidant activity, anthocyanins, carotenoids, and phenolics of three native fresh and sun-dried date (*Phoenix dactylifera* L.) varieties grown in Oman. *Journal of Agricultural and Food Chemistry*, **53**, 7592–7599.
48. Al-Farsi, M. & Lee, C.Y. (2008) Nutritional and functional properties of dates: a review. *Critical Reviews in Food Science and Nutrition*, **48**, 877–887.
49. Tinker, L.F., Davis, P.A., Schneeman, B.O., Gallaher, D.D. & Waggoner, C.R. (1991) Consumption of prunes as a source of dietary fiber in men with mild hypercholesterolemia. *American Journal of Clinical Nutrition*, **53**, 1259–1265.
50. Camire, M.E. & Dougherty, M.P. (2003) Raisin dietary fiber composition and *in vitro* bile acid binding. *Journal of Agricultural and Food Chemistry*, **51**, 834–837.
51. Carughi, A. (2009) Raisins as a source of prebiotic compounds in the diet. *FASEB Journal*, **23**, 716.9.
52. Spiller, G.A., Story, J.A., Furumoto, E.J., Chezem, J.C. & Spiller, M. (2003) Effect of tartaric acids and dietary fiber from sun-dried raisins on colonic function and on bile acid and volatile fatty acid excretion in healthy adults. *British Journal of Nutrition*, **90**, 803–807.
53. Block, G., Patterson, B. & Subar, A. (1992) Fruit, vegetables, and cancer prevention: a review of the epidemiological evidence. *Nutrition and Cancer*, **18**, 1–29.
54. Steinmetz, K.A. & Potter, J.D. (1996) Vegetables, fruits and cancer prevention: a review. *Journal of the American Dietetic Association*, **10**, 1027–1039.
55. Tantamango, Y.M., Knutsen, S.F., Beeson, W.L., Fraser, G. & Sabate, J. (2011) Food and food groups associated with the incidence of colorectal polyps: the Adventist health study. *Nutrition and Cancer*, **63**, 565–572.
56. Rimm, E.B., Ascherio, A., Giovanucci, E., Speigelman, D., Stampfer, M.J. & Willett, W.C. (1996) Vegetable, fruit and cereal fiber intake and risk of coronary heart disease among men. *Journal of the American Medical Association*, **275**, 447–451.
57. Hu, B. (2003) Plant-based foods and prevention of cardiovascular disease: an overview. *American Journal of Clinical Nutrition*, **78** (Suppl.), 544S–551S.
58. Joshipura, K.J., Ascherio, A., Manson, J.E., Stampfer, M.J., Rimm, E.B., Speizer, F.E., Hennekens, C.H., Spiegelman, D., Willett, W.C. (1999) Fruit and vegetable intake in relation to risk of ischemic stroke. *Journal of the American Medical Association*, **282**, 1233–1239.
59. Schröder, H. (2007) Protective mechanisms of the Mediterranean diet in obesity and type 2 diabetes. *Journal of Nutritional Biochemistry*, **18**, 149–160.
60. Ford, E.S. & Mokdad, A.H. (2001) Fruit and vegetable consumption and diabetes mellitus incidence among U.S. adults. *Preventive Medicine*, **32**, 33–39.
61. Ascherio, A., Stamper, M.J., Colditz, G.A., Willett, W.C. & McKinlay, J. (1991) Nutrient intakes and blood pressure in normotensive males. *International Journal of Epidemiology*, **20**, 886–891.
62. WHO/FAO (2003) *Diet, Nutrition and the Prevention of Chronic Disease. Report of a Joint FAO/WHO Expert Consultation*. WHO, Geneva, Switzerland.
63. Halloran, B.P., Wronski, T.J., VonHerzen, D.C., Chu, V., Xia, X., Pingel, J.E., Williams, A.A. & Smith, B. (2010) Dietary dried plum increases bone mass in adult and aged male mice. *Journal of Nutrition*, **140**, 1781–1787.
64. Hooshmand, S. & Arjmandi, B.H. (2009) Viewpoint: dried plum, an emerging functional food that may effectively improve bone health. *Aging Research Reviews*, **8**, 122–127.
65. Hertog, M.G.L., Feskens, E.J.M., Hollman, P.C.H., Katan, M.B. & Kromhout, D. (1993) Dietary antioxidant flavonoids and risk of coronary heart disease: the Zutphen Elderly Study. *Lancet*, **342**, 1007–1011.
66. Keli, S.O., Hertog, M.G., Feskens, E.J. & Kromhout, D. (1996) Dietary flavonoids, antioxidant vitamins, and incidence of stroke: the Zutphen study. *Archives of Internal Medicine*, **156**, 637–642.
67. Neuhouser, M.L. (2004) Flavonoids and cancer prevention: what is the evidence in humans? *Pharmaceutical Biology*, **42**, 36–45.
68. Wang, H., Cao, G., & Prior, R.L. (1996) Total antioxidant capacity of fruits. *Journal of Agricultural and Food Chemistry*, **44**, 701–705.
69. Pietta, P.-G. (2000) Flavonoids as antioxidants. *Journal of Natural Products*, **63**, 1035–1042.

70. Spiller, G.A., Story, J.A., Lodics, T.A., Pollack, M., Monyan, S., Butterfield, G., Spiller, M. (2003) Effect of sun-dried raisins on bile acid excretion, intestinal transit time, and fecal weight: a dose response study. *Journal of Medicinal Food*, **6**, 87–91.
71. Attaluri, A., Donahoe, R., Valestin, J., Brown, K. & Rao, S.S.C. (2011) Randomised clinical trial: dried plums (prunes) vs. psyllium for constipation. *Alimentary Pharmacology & Therapeutics*, **33**, 822–828.
72. Nielsen, F.H. (2008) Is boron nutritionally relevant? *Nutrition Reviews*, **66**, 183–191. http://www.cnpp.usda.gov/dietaryguidelines.htm
73. Kashket, S., Van Houte, J., Lopez, L.R. & Stocks, S. (1991) Lack of correlation between food retention on the human dentition and consumer perception of food stickiness. *Journal of Dental Research*, **70**, 1314–1319.
74. Rivero-Cruz, J.F., Zhu, M., Kinghorn, A.D. & Wu, C.D (2008) Antimicrobial constituents of Thomson seedless raisins (*Vitis vinifera*) against selected oral pathogens. *Phytochemistry Letters*, **1**, 151–154.
75. Utreja, A., Lingström, P., Evans, C.A., Salzmann, L.B. & Wu, C.D. (2009) The effect of raisins-containing cereals on the PH of dental plaque in young children. *Pediatic Dentistry*, **31**, 498–503.
76. Appel, L.J. (2003) Lifestyle modifications as a means to prevent and treat high blood pressure. *Journal of the American Society of Nephrology*, **14**, S99–S102.
77. Keast, D.R. & Jones, J.M. (2009) Dried fruit consumption associated with reduced improved overweight or obesity in adults: NHANES, 1999–2004. *FASEB Journal*, **23**, LB511.
78. Puglisi, M.J., Mutungi, G., Brun, P.J., McGrane, M.M., Labonte, C., Volek, J.S. & Fernandez, M.L. (2009) Raisins and walking alter appetite hormones and plasma lipids by modifications in lipoprotein metabolism and up-regulation of the low-density lipoprotein receptor. *Metabolism—Clinical and Experimental*, **58**, 120–128.
79. Glycemic Index Database (2011) *Glycemic Index Databses: Sydney University's Glycemic Index Research Service (SUGIRS)*. Published online at: http://www.glycemicindex.com/, last accessed May 28, 2011.
80. Miller, C.J., Dunn, E.V. & Hashim, I.B. (2002) Glycemic Index of 3 varieties of dates. *Saudi Medical Journal*, **23**, 536–538.
81. Kim, Y., Hertzler, S.R., Byrne, H.K. & Mattern, C.O. (2008) Raisins are a low to moderate glycemic index food with a correspondingly low insulin index. *Nutrition Research*, **28**, 304–308.
82. Anderson, J.A., Huth, H.A., Larson, M.M., Colby, A.J., Krieg, E.J., Golbach, L.P., Simon, K.A., Wasmundt, S.L., Malone, C.J. & Wilson, T. (2011) Glycemic and insulin response to raisins, grapes and bananas in college aged students. *FASEB Journal*, **25**, 587.4
83. Anderson, J.A., Andersen, K.F., Heimerman, R.A., Larson, M.M., Baker, S.E., Freeman, M.R., Carughi, A. & Wilson, T. (2011) Glycemic response of type 2 diabetics to raisins. *FASEB Journal*, **25**, 587.5
84. Zunino, S.J. (2009) Type 2 diabetes and glycemic response to grapes or grape products. *Journal of Nutrition*, **139** (Suppl.), 1794S–8000S.
85. Johnston, K.L., Clifford, M.N. & Morgan, L.M. (2003) Coffee acutely modifies gastrointestinal hormone secretion and glucose tolerance in humans: glycemic effects of chlorogenic acid and caffeine. *American Journal of Clinical Nutrition*, **78**, 728–733.
86. Björck, I. & Elmståhl, H.L. (2003) The glycemic index: importance of dietary fibre and other food properties. *Proceedings of the Nutrition Society*, **62**, 201–206.
87. Widanagamage, R.D., Ekanayake, S. & Welihinda, J. (2009) Carbohydrate-rich foods: glycemic indices and the effect of constituent macronutrients. *International Journal of Food Science and Nutrition*, **60**, 215–223.
88. Keast, D.R., O'Neil, C.E. & Jones, J.M. (2011) Dried fruit consumption is associated with improved diet quality and reduced obesity in US adults: National Health and Nutrition Examination Survey, 1999–2004. *Nutrition Research*, **31**, 460–467.
89. Meduri Farm Inc. (2011) *Specialty Dried Fruits*. Published online at: http://www.medurifarms.com, last accessed July 18, 2011.
90. USDA (2010) *Dietary Guidelines for Americans*. Published online at: http://www.cnpp.usda.gov/dietaryguidelines.htm, last accessed June 1, 2011.
91. FAO (2011) *Food Based Dietary Guidelines*. Published online at: http://www.fao.org/ag/humannutrition/nutritioneducation/fbdg/en, last accessed June 1, 2011.

2 Cancer chemopreventive effects of selected dried fruits

Joydeb Kumar Kundu and Young-Joon Surh

2.1 Chemoprevention: an overview

Cancer still imposes a huge health and economic burden throughout the world. The number of new cancer cases, which was recorded as about 3 million in the year 2000, is expected to increase to 7.1 million by the year 2020 [1, 2]. Global cancer mortality is also projected to be doubled in the next 50 years [3]. Such alarming statistics have revitalized our ever-lost war against cancer. Even though the research done over the last few decades failed to generate a magical cure for this dreaded disease, we have learned a lesson that cancer is largely a preventable disease. Since the majority (90–95%) of all cancers are linked to infections, exposure to environmental pollutants, and different lifestyle factors, such as smoking, diet, alcohol consumption, physical inactivity, obesity, and solar exposure, there are ample opportunities for preventing cancer [4].

The cancer prevention research conceptualized by Lee Wattenberg in the 1960s [5] is currently at the forefront in the fight against cancer. "Chemoprevention," the term coined by Michael B. Sporn in 1976, refers to the use of nontoxic compounds from natural or synthetic sources to inhibit, retard, or reverse carcinogenesis [6]. This definition of chemoprevention has recently been revised to assimilate the clinical status of cancer patients. Thus, chemoprevention now encompasses primary, secondary, or tertiary prevention of cancer. The strategy for primary chemoprevention is to prevent carcinogenesis in healthy individuals, who belong to a low-risk group, while secondary chemoprevention is aimed at intervening in the progression of premalignant lesions into malignancy. Tertiary prevention of cancer refers to the blockade of the recurrence of primary tumors [3, 7, 8].

2.2 The promise of dried fruits in cancer prevention

Convincing results from a wide spectrum of preclinical and epidemiological studies suggest chemoprevention as one of the most practical approaches for reducing the global burden of cancer [9–11]. A report from the World Cancer Research Fund (WCRF) [12] indicates that

Dried Fruits: Phytochemicals and Health Effects, First Edition. Edited by Cesarettin Alasalvar and Fereidoon Shahidi.

about 30–40% of cancers are preventable by proper intake of food and appropriate nutrition and physical activity and by avoiding obesity. An inverse association exists between the regular consumption of fruits and the risk of various organ-specific cancers [13, 14]. For instance, a meta-analysis based on 14 prospective studies and 5838 cases found a significant reduction in the incidence of distal colon cancer that was attributable to frequent consumption of fruits [15]. Pooled analysis of eight cohort studies revealed that the intake of fruits protects against lung cancer as well [16]. The European Prospective Investigation into Cancer and Nutrition (EPIC) study has also shown that adequate consumption of fruits is associated with the reduced risk of colon cancer [17] and lung cancer [18]. The chemopreventive potential of fruits and fruit ingredients has also been documented in a wide range of preclinical studies [19–23].

Common fruits with cancer chemopreventive activity include, but are not limited to, apples, avocados, berries, citrus fruits, kiwi fruits, litchis, mangoes, mangosteens, and persimmons, among others. Many of these fruits are produced on a seasonal basis and are not available in fresh conditions throughout the year. Fresh fruits are, therefore, processed by various techniques to prolong their shelf life. A conventional way of fruit preservation adopted since ancient times is the drying of fruits to reduce moisture content. Fruits can be dried by natural sun drying or by using mechanical devices, such as dryers and a microwave. Freeze drying is another technique to preserve fruits for long-term use. Osmo-convective dewatering of fresh fruits before drying may prevent substantial loss of nutritional ingredients in dried fruits [24]. Appropriate drying methods and temperature are selected depending on the type of fruit. Details of the drying process and its impact on the chemical composition of dried fruits are beyond the scope of this chapter.

Although drying alters the chemical composition to some extent, most of the dried fruits retain the biological properties of their fresh counterparts [25]. For example, vacuum-microwave-dried cranberries retain the antioxidant property of the fresh fruits [26]. Several other studies have reported the antioxidant capacity of dried fruits [27–29]. Among the selected dried fruits, prunes exhibited the highest antioxidant activity followed by dried apricots [27, 29, 30]. Conversely, Halvorsen *et al.* [31] reported the highest antioxidant activity of dried apricots followed by prunes. Such discrepancies are attributable to seasonal variation, and differences in the harvesting stage and the drying process. Raisins exhibited higher oxygen radical scavenging capacity (ORAC) than did fresh grapes [32]. Extracts of different varieties of freeze-dried berries showed antioxidant and cancer chemopreventive activities [23, 33]. Administration of super *CitriMax*, a calcium/potassium salt of (-)-hydroxycitric acid derived from the dried fruit rind of Malabar tamarind (*Garcinia cambogia*), reduced oxidative stress and inflammation in male obese Zucker rats [34]. Heat-treated citrus fruit peel extracts attenuated lipopolysaccharide (LPS)-induced expression of cyclooxygenase-2 (COX-2) and inducible nitric oxide synthase (iNOS), two representative pro-inflammatory enzymes, in murine macrophages [35]. Since oxidative stress and chronic inflammation play key roles in carcinogenesis [36], dried fruits with antioxidative and anti-inflammatory properties hold the promise of cancer chemoprevention. The antioxidant, anti-inflammatory, and chemopreventive activities of dried fruits are largely attributed to their polyphenolic compounds [30, 37]. Vinson *et al.* [28] reported higher polyphenol contents in dried apricots, cranberries, figs, raisins, dates, and prunes as compared with their fresh counterparts. Because of their ready availability and preservation, dried fruits are widely consumed in different societies [38] and can be an excellent natural resource for preventing cancer. The purpose of this chapter is to shed light on the cancer chemopreventive potential of selected dried fruits and their active ingredients.

2.3 Dried fruits as a potential source of chemopreventive phytochemicals

Fruits are rich in antioxidants and anti-inflammatory phytochemicals that help to reduce the incidence of various chronic degenerative diseases including cancer. Vitamins and polyphenols are major chemopreventive phytochemicals present in dried fruits. Drying of fruits by sun exposure or at high temperature in dryers often causes loss of bioactive phytochemicals. For example, fresh grapes contain carotenoids, which are lost in raisins. Certain vitamins are thermolabile and can be degraded during high-temperature drying of fruits. Many volatile compounds present in fresh fruits may be decomposed, while new volatile compounds can be generated during the drying process [39]. Freeze drying is often used to preserve thermolabile components, such as vitamins, carotenoids, and certain phenolics, that are abundant in fresh fruits. Freeze drying extends the shelf life of foods by preventing microbial growth and retarding lipid oxidation [40, 41]. Comparison between *in natura* and freeze-dried pulps of pineapples, cherries, guavas, papayas, and mangoes revealed that freeze drying provides products with higher nutritive value [42]. The effects of freeze drying on the total phenolics and the antioxidant activity of five tropical fruits, namely, starfruits, mangoes, papayas, muskmelons, and watermelons, were investigated [43]. According to this study, freeze drying did not cause any remarkable change in the β-carotene concentration, except for mangoes and watermelons. As compared with fresh fruits, freeze-dried mangoes and starfruits showed a significant decrease in total phenolics and antioxidant activity [43]. Freezing may cause decompartmentalization of polyphenol oxidase, thereby enhancing oxidation of freeze-dried fruit polyphenols. Despite such process-induced loss of active ingredients, dried fruits still contain adequate amount of bioactive compounds, including flavonols, anthocyanins, xanthones, coumarins, phenolic acids, acetogenins, and terpenes, among others (Figure 2.1).

2.4 Biochemical basis of chemoprevention with dried fruits

Because of the presence of antioxidant and anti-inflammatory substances, dried fruits can prevent carcinogenesis. Since oxidative stress and chronic inflammation are the key pathologic events for neoplastic transformation of cells [36, 44], the biochemical basis of cancer chemoprevention with dried fruits might be the maintenance of cellular redox balance and attenuation of inflammatory tissue damage. Though reactive oxygen species (ROS) are physiologic by-products of cellular metabolism and are often essential for many biochemical reactions and cell signaling pathways, excessive generation of ROS can cause oxidative damage to cellular macromolecules, such as proteins, lipids, and nucleic acids [44, 45]. ROS-induced damage of biomolecules perturbs normal cell functions and predisposes cells to acquire a premalignant phenotype. While oxidative stress can cause tissue inflammation,

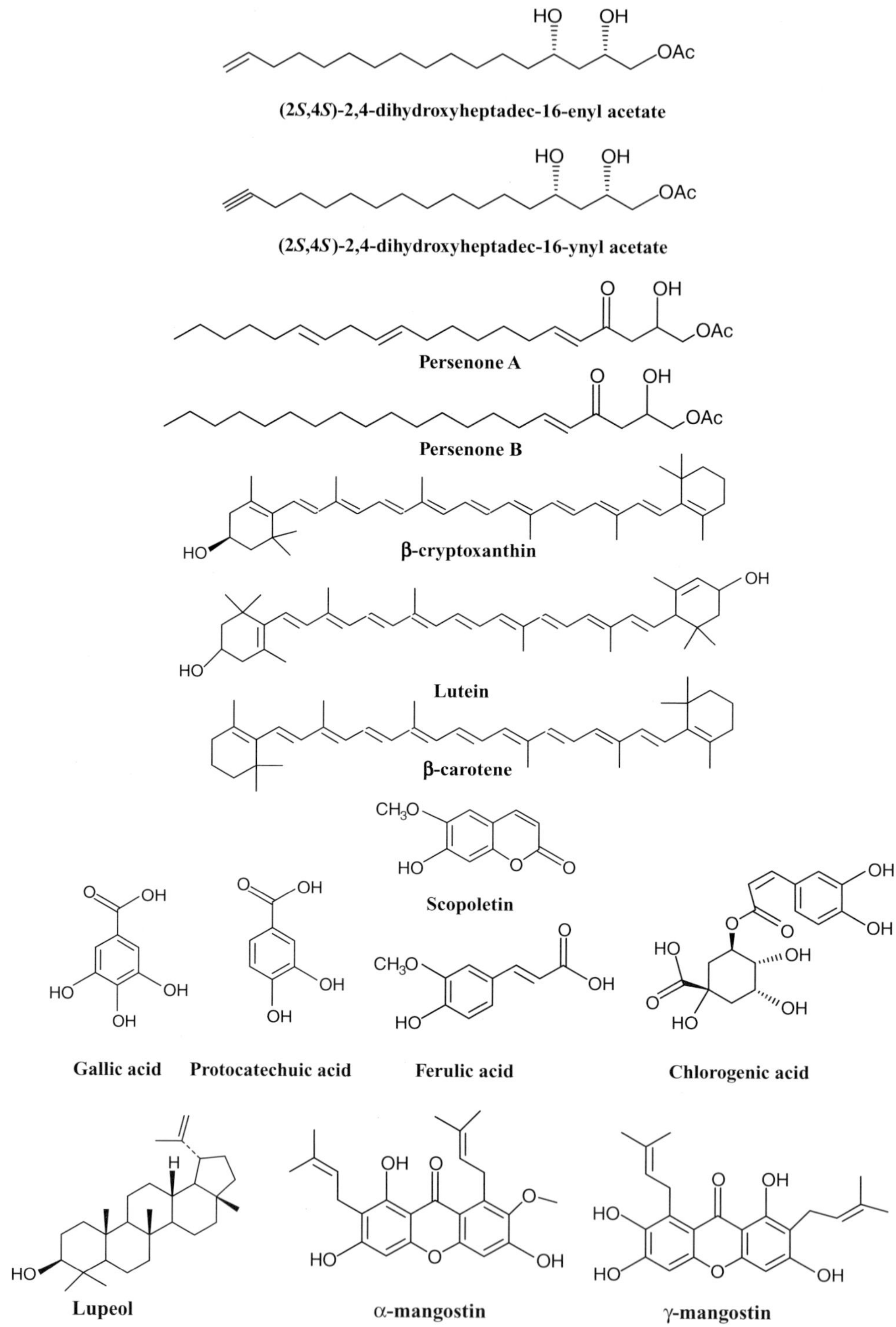

Figure 2.1 Common chemopreventive phytochemicals present in dry-processed fruits.

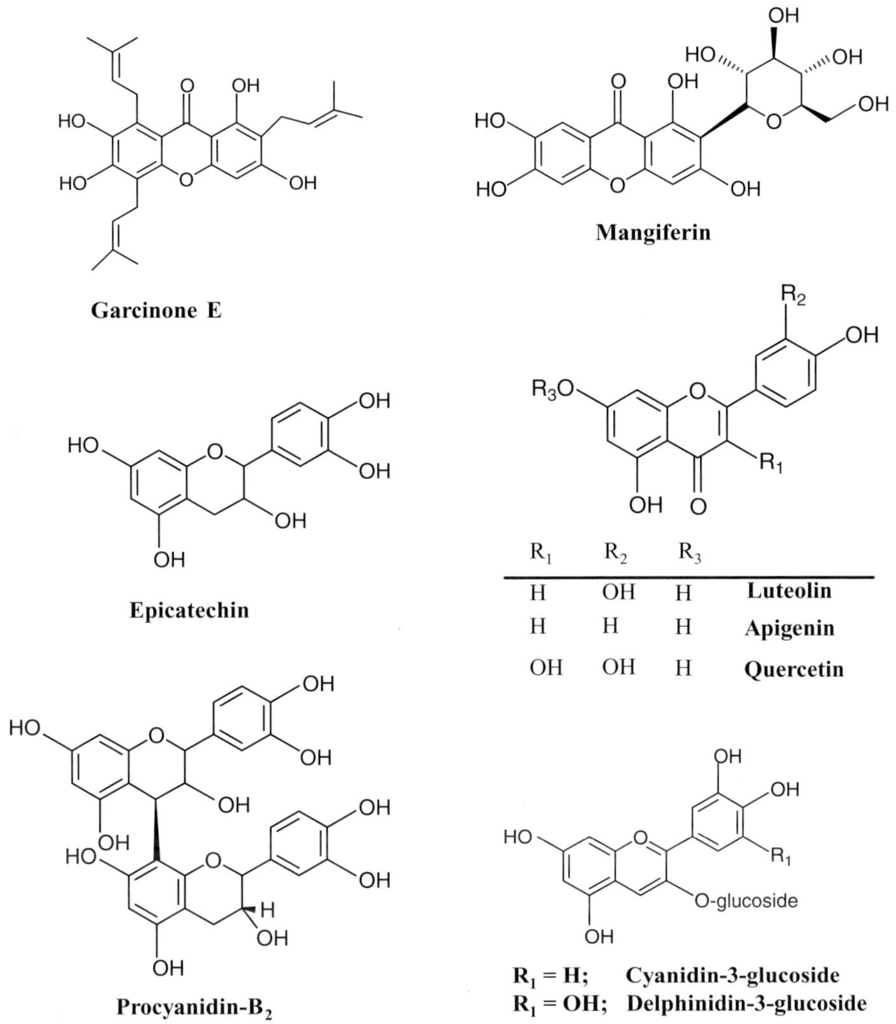

Figure 2.1 (*Continued*)

persistent inflammation can trigger the generation of ROS [44, 45]. Thus, fruit phytochemicals having both antioxidative and anti-inflammatory properties represent a good choice for cancer chemoprevention.

Carcinogenesis is a heterogenous disease process that consists of several distinct, but closely linked stages of initiation, promotion, and progression. Initiation is a rapid phase of cell transformation often caused by DNA damage. Promotion, the clonal expansion of transformed cells to form a benign tumor, spans several years and involves altered biochemical processes, such as uncontrolled cell proliferation, autonomic growth, lack of differentiation potential, and escape from programmed cell death. Progression is the rapid march of premalignant cells toward complete malignancy featuring invasion and metastasis of cancer cells [11, 46]. Although this is oversimplification of the complex cancer biology, such stepwise

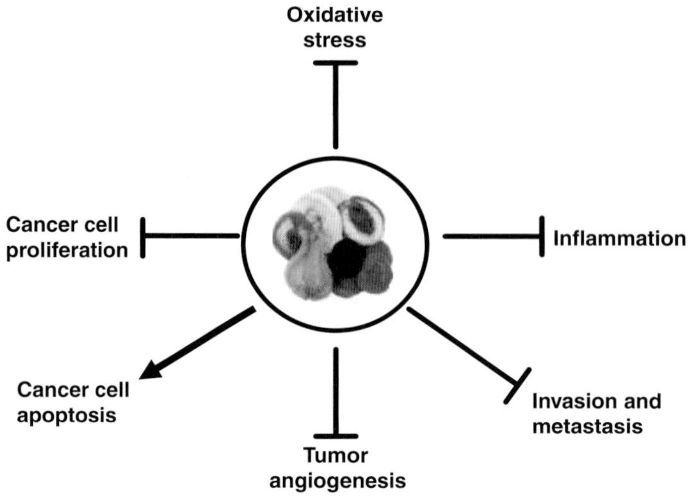

Figure 2.2 Biochemical mechanisms of chemoprevention with fruit phytochemicals. For color detail, see color plate section.

division of carcinogenesis may help to understand the molecular basis of cancer chemoprevention with natural compounds, especially fruit phenolics. Using this simplified model of carcinogenesis, Wattenberg [46] classified chemopreventive agents into two main categories as *blocking* agents and *suppressing* agents. Blocking agents prevent metabolic activation of carcinogens and stimulate cellular detoxification pathways to help eliminate carcinogens, thereby protecting cellular macromolecules, such as DNA, RNA, and proteins, from undesired oxidative or covalent modifications. Suppressing agents, on the other hand, inhibit cell proliferation, induce apoptosis, and block angiogenesis, thereby halting the neoplastic conversion of premalignant cells.

The biochemical basis of cancer chemoprevention with diverse classes of fruit phytochemicals has been extensively investigated. Figure 2.2 represents a general mechanistic aspect of chemoprevention with fruit phytochemicals, which enhance cellular antioxidant capacity and prevent oxidative damage of cellular macromolecules, block the activation of carcinogens, and stimulate carcinogen detoxification, reduce inflammatory responses, inhibit proliferation of cancer cells and induce apoptosis in cancer cells, and block angiogenesis and metastasis.

2.5 Chemopreventive properties of bioactive substances derived from selected dried fruits

Fruits processed by drying constitute an important part of our regular diet. Anticancer activity of fresh fruits and various fruit ingredients has been widely investigated. Since numerous health-beneficial fruit phytochemicals are conserved after processing of fruits, regular intake of dried fruits can help prevent cancer. Table 2.1 summarizes the chemopreventive effects of selected dried fruits and their active constituents.

Table 2.1 Chemopreventive effects of dry-processed fruits and their active ingredients

Treatment	Experimental model	Major findings	Reference
Amla fruits			
Ethanol extract	DMBA-induced genotoxicity	Decreased bone marrow nuclei formation Increased activities of GPx, GR, and GST Reduced the activities of hepatic cytochrome P450 and cytochrome b5 in DMBA-challenged mice	[50]
Methanol extract	DEN-initiated and 2-AAF-promoted rat hepatocarcinogenesis	Inhibited liver carcinogenesis	[54]
Aqueous extract	DMBA-initiated and croton oil-promoted mouse skin carcinogenesis	Reduced skin tumor incidence and multiplicity	[52]
	Six human cancer cells: A549, HeLa, MDA-MB-231, SK-OV3, SW620, chemically induced mouse skin carcinogenesis	Induced DNA fragmentation and apoptosis Activated caspase 3/7 and caspase 8 Increased the expression of Fas Suppressed breast cancer cell invasion Inhibited the average number of skin tumors	[53]
	Mouse bearing Dalton's lymphoma ascites	Reduced the number of ascites and solid tumors Increased life span of tumor-bearing mice Decreased Cdc25 phosphatase activity	[51]
Polyphenol fraction	Dalton's lymphoma ascites NDEA-induced liver carcinogenesis	Induced apoptosis in Dalton's lymphoma ascites Reduced hepatic tumor burden Decreased the activities of γ-glutamyltranspeptidase, serum alkaline phosphatase, and glutamate pyruvate transaminase Inhibited lipid peroxidation	[57]
Avocados			
Chloroform extract	Premalignant and malignant human oral squamous carcinoma cells	Induced ROS production Activated caspase-8 Induced apoptosis	[59]
Acetone extract (carotenoid-rich fraction)	Human prostate cancer LNCaP and PC3 cells	Attenuated cell proliferation; Induced G2/M cell cycle arrest and p27 expression	[60]
Methanol extract, persenone A or persenone B	LPS plus IFNγ-stimulated murine macrophages	Inhibited generation of superoxide and NO; Diminished TPA-induced H_2O_2 generation in HL60 cells	[61]

(continued)

Table 2.1 *(Continued)*

Treatment	Experimental model	Major findings	Reference
Persenone A	LPS plus IFNγ-stimulated raw 264.7 cells, and TPA-treated mouse skin	Inhibited the expression of iNOS and COX-2	[62]
Acetogenins	Human oral cancer cells	Inhibited cell proliferation Reduced phosphorylation of EGFR, c-RAF, and ERK	[21]
Scopoletin	DMBA-initiated and croton oil-promoted mouse skin papillomagenesis	Decreased the incidence and multiplicity of mouse skin papillomas	[63]
	Prostate cancer (PC3) cells	Induced apoptosis	[64]
	Multidrug resistant human leukemia cells	Reduced cell proliferation Induced apoptosis	[65]
	Human melanoma A375 cells	Inhibited cell proliferation Decreased expression of survivin, cyclin D1, PCNA, and phosphorylation of STAT3 Induced apoptosis through activation of p53 and caspase-3	[66]
Berries			
Ethanol extract of freeze-dried strawberry and blueberry	Human cervical and breast cancer cells	Inhibited cell proliferation	[71]
Blueberry powder in diet	Breast cancer cells xenograft in nude mice	Reduced the size of xenograft tumor Decreased expression of β-catenin Reduced phosphorylation of GSK-3β and APC Inhibited activation of Akt and NF-κB Induced caspase-3 cleavage Decreased MMP-9 activity Inhibited secretion of uPA Attenuated metastasis of xenografted MDA-MB-231 cells	[72, 73]
Blueberry extract	Hemangioendothelioma formation in nude mice inoculated with transformed murine endothelial cells	Decreased the incidence and size of hemangioendotheliomas Increased survival Reduced activation of JNK and NF-κB Decreased MCP-1 expression Inhibited angiogenesis	[75]
Diet rich with 10% extract of freeze-dried bilberry, lingonberry, and cloudberry	APC^{min+} mice	Reduced the development of intestinal adenomas Decreased the expression of β-catenin and cyclin D1	[76]

Blueberry polyphenol-rich diet	MMTv-Wnt1-transgenic mice fed with blueberry	Sera from MMTv-Wnt1 transgenic mice blocked mamosphere formation in MDA-MB-231 cells in culture	[74]
Dietary freeze-dried white currant	APC^{min+} mice	Reduced the development of intestinal adenomas Decreased the expression of β-catenin and NF-κB	[77]
Aqueous extract of black currant	HepG2 cells	Inhibited cell proliferation	[78]
Anthocyanin-rich black currant in diet	DEN-induced rat hepatocarcinogenesis	Decreased the formation of neoplastic foci Reduced hepatic NF-κB activation and COX-2 expression	[79]
Freeze-dried black raspberry extract in diet	C57BL/6 mice challenged with DSS	Reduced colitis Decreased production of TNFα, IL-1β, and expression of COX-2 in mouse colon	[86]
Freeze-dried black raspberry extract in diet	NMBA-induced rat oesophageal cancer	Decreased esophageal tumors Reduced expression of COX-2, iNOS, and VEGF Decreased microvessel density	[87]
Mangoes			
Polyphenols of mango fruit	Human colon cancer cells	Induced apoptosis Increased expression of p21, Bax, and caspase-8	[106]
Supplementation with mango pulp extract or lupeol	Androgen-treated mice Human prostate cancer (LNCaP) cells	Inhibited enlargement of the prostate Induced apoptosis Activated caspase-3 Inhibited Bcl-2 and Bcl-xl	[107–109]
Lupeol	Chemically induced mouse skin carcinogenesis	Decreased the incidence and multiplicity of skin papillomas Decreased ODC activity Reduced the expression of COX-2 and iNOS Attenuated activation of Akt and NF-κB	[110]
Mangiferin in the diet	B[*a*]P-induced mouse lung tumorigenesis	Decreased lung tumor formation Increased activities of detoxification enzymes	[101]
	AOM-treated F344 rats	Decreased the incidence and multiplicity of colon tumors	[102]
	LPS plus IFNγ-stimulated rat macrophages	Inhibited xanthine oxidase activity Reduced superoxide generation Inhibited expression of iNOS	[103]
Mangosteens			
Ethanol extract of fruit rind	*In vitro* study	Free radical scavenging effect	[119]
α-Mangostin	*In vitro* study, DMBA-treated mouse mammary organ culture	Antioxidant activity Decreased formation of preneoplastic lesions	[120]

(continued)

Table 2.1 *(Continued)*

Treatment	Experimental model	Major findings	Reference
	Prostate cancer cells	G1 cell cycle arrest Decreased the CDK4 activity Inhibited the growth of prostate cancer cells xenograft in nude mice	[124]
	Mammary cancer cells	G1 and S phase arrest Inhibited Akt phosphorylation Activated caspases Attenuated metastasis of breast cancer cells in tumor-bearing mice	[123]
γ-Mangostin	C6 rat glioma cells	Attenuated constitutive or LPS-induced expression of COX-2; Inhibited IKK activity, and NF-κB activation	[126]
Garcinone E	Hepatocellular, gastric, and lung cancer cells	Induced apoptosis	[121]
Magostenone (C–E)	Epidermoid carcinoma, breast cancer, lung cancer cells	Induced cytotoxicity	[122]
Persimmons			
Aqueous extract and flavonoids	Human leukemia Molt 4B cells	Inhibited the ODC activity Induced apoptosis	[19]
Ethylacetate extract	Human monocyte (THP-1)cells	Decreased LPS-induced TNFα production	[143]
Prunes			
Ethanol extract	Human colon cancer, Caco-2 and KATO III cells	Decreased the cell viability	[185]
Kiwi fruits			
Hydroalcoholic extract	Human squamous carcinoma (HSC-2) cells	Induced cytotoxicity	[196]
Methanol extract of freeze-dried fruit and quercetin	Rat liver epithelial cells exposed to H_2O_2	Restored connexin-43 expression Inhibited ERK phosphorylation Inhibited gap-junction intercellular communication	[199]

2-AFF, 2-acetylaminofluorine; AOM, azoxymethane; APC, adenomatous polyposis coli; B[*a*]P, benzo[*a*]pyrene; CDK4, cyclin-dependent kinase-4; COX-2, cyclooxygenase-2; c-RAF, the protein product of *raf* oncogene; DEN, diethylnitrosoamine; DMBA, 7,12-dimethylbenz[*a*]anthracene; EGFR, epidermal growth factor receptor; ERK, extracellular signal-regulated kinases; GPx, glutathione peroxidase; GR, glutathione reductase; GSK-3β, glycogen synthase kinase-3β; GST, glutathione-*S*-transferase; IFNγ, interferon-γ; IKK, IκB kinase; IL-1β, interleukin-1β; iNOS, inducible nitric oxide synthase; JNK, c-Jun N-terminal kinase; LPS, lipoloysachaaride; MCP-1, monocyte chemoattractant protein-1; MMP-9, matrix metalloproteinase-9; NDEA, *N*-nitroso diethylamine; NF-κB, nuclear factor-kappaB; NO, nitric oxide; ODC, ornithine decarboxylase; PCNA, proliferating cell nuclear antigen; ROS, reactive oxygen species; STAT3, signal transducer and activator of transcription-3; TNFα, tumor necrosis factor-α; TPA, 12-*O*-tetradecanoylphorbol-13-acetate; VEGF, vascular endothelial growth factor.

2.5.1 Amla (Indian gooseberries) fruits

Fruits of *Phyllanthus emblica* Linn. or *Emblica officinalis* Gaertn. (Euphobiaceae) are commonly known as Indian gooseberry or amla or amlaki, which are consumed in either fresh or in dried form. Sun-dried amla fruits are important components of Ayurvedic medicine. Amla fruits are rich sources of vitamin C, different amino acids, and minerals. Besides these micronutrients, amla fruits contain a wide variety of secondary metabolites, such as alkaloids, flavonoids, terpenoids, and tannins [47–49]. The methanol extract of dried rind of amla fruits has a strong nitric oxide (NO) scavenging activity [48]. Major constituents with the NO scavenging property were identified as gallic acid, methyl gallate, corilagin, furosin, and geranin [48]. Poltanov *et al.* [49] reported that ellagic acid, gallic acid, and corilagin were common antioxidants present in four commercial varieties of dried amla fruits. The ethanol extract of dried amla fruits protected against 7,12-dimethylbenz[*a*]anthracene (DMBA)-induced genotoxicity in Swiss albino mice [50]. Aqueous extract of dried fruits of *E. officinalis* induced cytotoxicity in L929 cells and reduced the tumor volume in mice implanted with Dalton's lymphoma ascites (DLA). The antitumor effects of amla fruits were associated with the decreased activity of the cell cycle regulating enzyme cell division cycle (Cdc)-25 phosphatase. Moreover, amla fruit extracts increased the life span of DLA tumor-bearing mice [51].

Administration of the aqueous extract of dried amla fruits by gavage significantly inhibited the incidence and the multiplicity of DMBA-initiated and croton oil [52]- or 12-*O*-tetradecanoylphorbol-13-acetate (TPA)-promoted [53] mouse skin tumorigenesis. Pretreatment with defatted methanolic extract of dried amla fruits attenuated diethylnitrosoamine (DEN)-initiated and 2-acetylaminofluorine (2-AAF)-promoted hepatic carcinogenesis in Wister rats [54]. Incubation with aqueous extract of dried amla fruits induced apoptosis in human cervical cancer (HeLa) cells. According to this study, the amla extracts increased DNA fragmentation, induced the activity of caspase-3, -7, and -8, and elevated the expression of Fas protein [53]. Moreover, the same extracts attenuated the invasiveness of human breast cancer (MDA-MB-231) cells [53]. Arsenic is a ubiquitous environmental carcinogen that causes cancer of the skin, bladder, liver, lung, and kidney [55]. Treatment with the aqueous extract of dried amla fruits inhibited arsenic-induced lipid peroxidation and the activity of serum transaminase in livers of Swiss albino mice [56]. In addition, dried amla fruit extracts restored arsenic-induced depletion of hepatic superoxide dismutase (SOD), catalase, and serum alkaline phosphatase activity in mouse liver [56]. A polyphenolic fraction prepared from dried fruits of *E. officinalis* induced apoptosis in DLA cells and reduced *N*-nitrosodiethylamine (NDEA)-induced liver tumor formation in rats [57]. Amla fruits contain substantial amounts of tannins and flavonoids. Treatment of human lung cancer (H441 and H520) cells with pyrogallol, a major catechin compound present in amla fruits, induced G2/M phase arrest and apoptosis, which was associated with increased expression of Bax and reduced expression of Bcl-2, cyclin B1, and Cdc25. In addition, dried amla fruit extracts caused the regression of lung cancer cells xenograft in nude mice [58].

2.5.2 Avocados

Avocados (*Persea americana* Mill., Lauraceae), also known as alligator pear, are widely consumed throughout the world. They are high in nutrients and low in calories, sodium, and fat. More than 25 chemopreventive phytochemicals have been isolated from avocados [59]. These include alkanols or aliphatic acetogenins (e.g., persin and persenone A and B), flavonoids (e.g., catechin, epicatechin [EC], luteolin, apigenin, and quercetin), carotenoids (e.g., zeaxanthin, lutein, and β-carotene), terpenoid glycosides, and coumarins [59, 60]. Phytochemicals isolated from avocados induced cell cycle arrest and apoptosis in various cancer cells. Some of the compounds present in avocados, for example, catechin, quercetin, apigenin, and luteolin are well-known cancer chemopreventive agents. Acetone extract of avocados, rich in carotenoids and tocopherols, inhibited the growth of both androgen-dependent (LNCaP) and androgen-independent (PC3) prostate cancer cells. This study also demonstrated that treatment with avocado extracts induced the expression of p27 and arrested the growth of PC3 cells at the G2/M phase of the cell cycle [60]. Kim *et al.* [61] isolated aliphatic acetogenins, persenone A and B, from the methanol extract of fresh avocados. Both persenone A and B showed inhibitory effects on the generation of superoxide and NO in murine macrophage RAW 264.7 cells stimulated with a combination of LPS and interferon-γ (IFNγ), and attenuated TPA-induced superoxide generation in human promyelocytic leukemia HL60 cells [61]. Incubation with persenone A caused significant inhibition of iNOS and COX-2 expression in LPS plus IFNγ-stimulated RAW 264.7 murine macrophages, and topical application of persenone A attenuated TPA-induced generation of hydrogen peroxide in mouse skin [62]. Chloroform extract of avocados inhibited the growth of premalignant and malignant human oral cancer cells, partly by inducing apoptosis through generation of ROS and FAS (a cysteine-rich transmembrane protein belonging to tumor necrosis factor family of cytokines)-mediated caspase-8 activation [59].

Avocado extracts also inhibited the phosphorylation of epidermal growth factor receptor (EGFR), the protein product of *raf* oncogene (c-RAF), and extracellular signal-regulated kinases (ERK). Bioactivity-guided fractionation of chloroform extract yielded two active constituents, (2*S*,4*S*)-2,4-dihydroxyheptadec-16-enyl acetate and (2*S*,4*S*)-2,4-dihydroxyheptadec-16-ynyl acetate [21]. Scopoletin (7-hydroxy, 6-methoxy-coumarin) is a bioactive compound present in avocados. Mice fed scopoletin at a daily dose of 50 or 100 mg/kg body weight for 24 weeks reduced the size of skin papillomas induced by DMBA plus croton oil [63]. Moreover, scopoletin induced apoptosis in human prostate cancer (PC3) cells [64] and multidrug-resistant human leukemia (CEM/ADR500) cells [65]. Incubation of human melanoma A375 cells with polylactide-co-glycolide-encapsulated scopoletin inhibited cell proliferation and induced apoptosis, which was associated with downregulation/inactivation of survivin, cyclin D1, proliferating cell nuclear antigen (PCNA), and signal transducer and activator of transcription-3 (STAT3), and also with induction of p53 and caspase-3 activity [66].

Chemical and biological investigations on dried avocados are still limited. Whether the aforementioned chemopreventive phytochemicals reported from fresh avocados are present in dried ones is yet to be confirmed. However, a recent study demonstrated the presence of carotenoids and procyanidins in freeze-dried peels and pulps of avocados. Furthermore, the antioxidant activity of dried avocado parts was correlated with the procyanidin content, but not with the total carotenoids [67]. Considering the presence of anticancer properties in fresh avocados, the chemopreventive activity of dried avocados merits further investigation.

2.5.3 Berries

A wide variety of berry fruits are known to possess diverse health-beneficial effects [68, 69]. Dietary intake of different berry fruits, either fresh or in the processed form, can prevent multistep carcinogenesis. Commonly consumed berry fruits include blackberries (Rubus spp.), black raspberries (*Rubus occidentalis*), blueberries (*Vaccinium corymbosum*), cranberries (*Vaccinium macrocarpon*, *Vaccinium oxycoccus*), red raspberries (*Rubus idaeus*), and strawberries (*Fragaria x ananassa*) [69]. Berries contain a diverse range of phytochemicals with health benefits, such as antioxidant, neuroprotective, anti-inflammatory, and anticancer activities. The major classes of berry phenolics are anthocyanins, flavonols, flavanols, ellagitannins, gallotannins, proanthocyanidins, and phenolic acids [70]. Methanol extract of six popularly consumed berries, including blackberries, black raspberries, blueberries, cranberries, red raspberries, and strawberries, inhibited the growth of human oral, breast, colon, and prostate cancer cells at concentrations ranging from 25 to 200 μg/mL. Ethanol extracts of freeze-dried strawberries and blueberries inhibited the growth of human cervical (Caski and SiHa) and breast (MCF-7 and T47D) cancer cells in culture [71]. Administration of whole blueberry powders in diet significantly decreased the tumor volume in female nude mice implanted with human breast cancer MDA-MB231 cells and reduced the liver metastasis of these cells. Tumors from blueberry-fed mice showed decreased expression of β-catenin, reduced phosphorylation of glycogen synthase kinase-3β (GSK-3β), and elevated expression of adenomatous polyposis coli (APC) [72]. Moreover, tumors from blueberry-fed mice showed decreased activation of Akt and nuclear factor-kappaB (NF-κB), and induced caspase-3 cleavage [73]. The antimetastatic effect of blueberries was associated with the decreased activity of matrix metalloproteinase (MMP)-9, reduced secretion of urokinase plasminogen activator (uPA), and increased secretion of tissue inhibitor of metalloproteinase-1 and plasminogen activator inhibitor-1 in MDA-MB231 cells [73]. A recent study demonstrated that blueberry polyphenol-containing diet repressed mammosphere formation of human breast cancer cells, suggesting that blueberry extracts may target cancer stem-like cells in preventing breast carcinogenesis [74]. Treatment with blueberry extracts significantly reduced the incidence and the size of hemangioendotheliomas in nude mice subcutaneously inoculated with transformed murine endothelial (EOMA) cells and prolonged the survival of tumor-bearing mice. Blueberry extracts blocked the signaling mediated by c-Jun N-terminal kinase (JNK) and NF-κB, thereby decreasing the expression of monocyte chemoattractant protein-1 (MCP-1) that is required for hemangioendothelioma development. The antiangiogenic effect of blueberry extracts was evident from decreased sprouting on matrigel and transwell migration of EOMA cells [75].

Diets containing 10% freeze-dried bilberries, lingonberries, or cloudberries significantly inhibited the formation of intestinal adenomas in APC^{min+} mice with decreased expression of β-catenin and cyclin D1. Affimetrix microarray analysis of adenoma tissues from berry-fed mice revealed the decreased expression of adenosine deaminase, ecto-5′-nucleotidase, and prostaglandin E_2 receptor-4 (EP4), which are involved in intestinal tumorigenesis [76]. Likewise, dietary administration of freeze-dried white currants attenuated the formation of intestinal adenomas in association with decreased β-catenin expression and NF-κB activation in APC^{min+} mice [77]. An aqueous extract of black currants' skin, rich in cyanidin-3-*O*-rutinoside, inhibited proliferation of hepatocellular carcinoma HepG2 cells, and its effect was more pronounced than that of delphinidin and cyanidin, two anthocyanin aglycones present in black currants [78]. Moreover, the same extract inhibited the formation of DEN-induced hepatic gammaglutamyl transpeptidase-positive preneoplastic foci and decreased the expression of COX-2 and activation of NF-κB in rat liver [79].

Seeram *et al.* [70] examined the effects of blackberries, black raspberries, blueberries, cranberries, red raspberries, and strawberries on the growth of various cancer cells and found that extracts of black raspberries and strawberries were most active in arresting growth and inducing apoptosis in human colon cancer (HT-29) cells. The induction of apoptosis in HT-29 cells with different berry extracts was associated with increased expression of p21 and Bax [80]. The multiplicity and the volume of 17β-estradiol-induced mammary tumors were significantly reduced in female ACI rats given diet supplemented with either blueberry or black raspberry powders or their active constituent ellagic acid [81]. 17β-Estradiol causes oxidative DNA damage via redox cycling of its metabolite 4-hydroxyestradiol. Dietary administration of blueberry extracts or ellagic acid diminished 4-hydroxyestradiol-induced formation of 8-oxo-7,8-dihydroguanine (8-oxodG), a marker of oxidative DNA damage, in rats [81]. Diet containing 5 or 10% freeze-dried strawberries inhibited the multiplicity of *N*-nitrosomethylbenzylamine (NMBA)-induced esophageal cancer in F344 rats, presumably by blocking the metabolic activation of NMBA [82]. A significant decrease in O^6-methylguanine levels was observed in the esophageal DNA of animals fed strawberries, suggesting that one or more components in strawberries influence(s) the metabolism of NMBA to generate DNA-damaging species. However, according to another study, lyophilized strawberries in the diet failed to inhibit 4-(methylnitrosamino)-1-(3-pyridyl)-1-butanone (NNK)- and benzo[*a*]pyrene (B[*a*]P)-induced mouse lung tumorigenesis [83].

Administration of lyophilized black raspberries inhibited the formation of the promutagenic adduct O^6-methylguanine and significantly reduced the multiplicity of NMBA-induced esophageal tumors [84] and inhibited azoxymethane (AOM)-induced aberrant crypt foci (ACF) formation in colon [85] of F344 rats. Treatment of C57BL/6 mice with freeze-dried black raspberry extracts attenuated dextran sulfate sodium (DSS)-induced colitis, which often progress to colon cancer. This study also demonstrated that black raspberry extracts decreased DSS-induced production of tumor necrosis factor-α (TNFα) and interleukin (IL)-1β and the expression of COX-2 in mouse colon [86]. Black raspberry extracts reduced the mRNA and protein expression of COX-2, iNOS, and c-Jun as well as the level of prostaglandin E_2 (PGE_2) in preneoplastic lesions of the NMBA-treated rat esophagus [87]. In addition, black raspberry extracts inhibited NMBA-induced expression of vascular endothelial growth factor (VEGF)-C and reduced the microvessel density in NMBA-treated rat esophagus [87], indicative of the antiangiogenic potential of black raspberry. Ethanol extract of freeze-dried black raspberries inhibited the growth of premalignant and malignant, but not normal, human oral epithelial cells. This ethanol fraction reduced the expression of cyclin A and cell division cycle gene 2 (Cdc2) in premalignant cells and that of cyclin B1, cyclin D1, and Cdc2 in the malignant cell lines. Ferulic acid, isolated from ethanol fraction of freeze-dried black raspberries, arrested the growth of premalignant and malignant oral epithelial cells at the G2/M phase of the cell cycle and increased levels of cyclin B1 and Cdc2 in both cell lines [88]. In a subsequent study, dietary administration of an anthocyanin-rich fraction of freeze-dried black raspberries inhibited cell proliferation, inflammation, and angiogenesis and induced apoptosis in rat preneoplastic and papillomatous esophageal tissues [89]. These preclinical chemopreventive effects were supported by significant reduction in the urinary excretion of two oxidative stress markers, 8-epi-prostaglandin F_{2alpha} (8-Iso-PGF_2) and 8-hydroxy-2′-deoxyguanosine (8-OHdG), in patients with Barrett's esophagitis who received lyophilized black raspberry extracts [90]. Incubation of mouse epidermal JB6 cells with different fractions of methanol extract of freeze-dried black raspberries attenuated B[*a*]P-diol-epoxide (BPDE)-induced activation of NF-κB and activator protein-1 (AP-1) through inhibition of phosphorylation of mitogen-activated protein (MAP) kinases and IκB kinase (IKK) [91]. Moreover, treatment with different solvent fractions of freeze-dried black raspberries decreased the expression of VEGF and iNOS in BPDE-stimulated JB6 cells by blocking phosphatidylinositol-3-kinase/Akt signaling [22]. Major constituents of the most active fractions were identified as cyanidin-3-*O*-glucoside, cyanidin 3-*O*-xylosylrutinoside, and cyanidin 3-*O*-rutinoside, which also attenuated BPDE-induced NF-κB activation in JB6 cells [92]. Topical application of the standardized black raspberry extracts significantly inhibited ultraviolet-B (UVB) radiation-induced inflammation and tumor promotion in mouse skin [93].

2.5.4 Mangoes

Mangoes (*Mangifera indica* L., Anacardiaceae) are grown in tropical regions. While green and ripe mangoes are consumed as seasonal fruits, green mangoes are often sun-dried and preserved for off-season use. Mango fruit leathers are made by sun drying a very thin layer of fruit puree to obtain a chewable product [94]. Dried mango peels contain pectin and polyphenols [95]. Several gallotannins have been isolated from both dried peels and pulps of mango [96]. Mangoes also contain high amounts of flavonol-*O*- and xanthone-*C*-glycosides [97, 98]. Moreover, carotenoids constitute a major part of mangoes [99, 100]. Chen *et al.* [100] investigated the effect of different drying conditions on the carotenoid content of Taiwanese mangoes. According to this study, hot-air drying or freeze drying of sliced mangoes reduced the carotenoid content. However mango slices soaked in either $NaHSO_3$ or ascorbic acid for 30 minutes prior to freeze drying or hot-air drying retained high level of carotenoids. All-*trans*-β-carotene and its *cis* isomers are the major carotenoids isolated from mangoes. Other carotenoids in mango include neochrome, violaxanthin, zeaxanthin, luteoxanthin, neoxanthin, and *cis*-lutein. Since all-*trans*-β-carotene possesses high antioxidant activity, treatment of mango slices with ascorbic acid or $NaHSO_3$ before drying is important to protect all-*trans*-β-carotene from thermal decomposition.

A major chemopreventive phytochemical present in mangoes is mangiferin (1,3,6,7-tetrahydroxyxanthone-C2-beta-D-glucoside), which is a C-glucosylxanthone. Dietary administration of mangiferin inhibited B[*a*]P-induced lung carcinogenesis in Swiss albino mice. Mangiferin increased the activities of detoxification enzymes and restored antioxidant enzyme activities in lung and liver of tumor-bearing mice [101]. Diet containing 0.1% mangiferin reduced the incidence and the multiplicity of colon tumors in AOM-treated F344 rats [102]. Mangiferin attenuated the mRNA expression of iNOS, TNFα, and transforming growth factor-β, and diminished the generation of superoxide by blocking xanthine oxidase activity in LPS plus IFNγ-stimulated rat macrophages [103]. Norathyriol (1,3,6,7-tetrahydroxy-9H-xanthen-9-one), an aglycone of mangiferin, has been reported to possess antioxidant and anti-inflammatory properties [104]. Treatment with norathyriol inhibited calcium ionophore-induced COX expression and lipoxygenase activity in rat neutrophils [104]. Topical application of norathyriol suppressed UVB-induced mouse skin papillomagenesis by blocking the activation of NF-κB and AP-1 [105].

Polyphenolic fractions obtained from several varieties of mangoes induced mRNA expression of p21, Bax, and caspase-8, and arrested the growth of human colon cancer (SW480) cells, but did not affect the proliferation of normal colon epithelial cells [106]. Supplementation with mango pulp extracts or their triterpene constituent lupeol inhibited prostate enlargement in androgen-treated male Swiss albino mice [107]. Moreover, incubation with mango pulp extracts or lupeol induced apoptosis in LNCaP cells in culture through activation of caspase-3 and downregulation of B-cell lymphoma-2 (Bcl-2) and B-cell lymphoma-extra large (Bcl-xl) [108, 109]. Topical application of lupeol inhibited TPA-induced skin inflammation and tumor promotion, which was associated with the decreased expression of ornithine decarboxylase (ODC), COX-2, and iNOS, and the inhibition of NF-κB and Akt [110]. Other polyphenols present in mangoes include ellagitannins and ellagic acid [106, 111], which also exhibited chemopreventive properties [112–114].

2.5.5 Mangosteens

Mangosteens (*Garcinia mangostana* L., Guttiferae) also known as the "queen of fruits," are widely grown in tropical countries. The pericarp of mangosteens has long been used in Ayurvedic medicine for the treatment of infection, pain, inflammation, and gastrointestinal disorders [115]. The fruits contain dark purple or reddish pericarp (peels and rinds) with white, soft, and juicy edible pulps (arils). Mangosteens contain xanthone derivatives [116]. More than 50 xanthone compounds have been isolated from pericarps and pulps of mangosteen. Examples are α-, β-, or γ-mangostin, gartanin, garcimangosone (A–D), tovophyllin (A and B), garcinone (A–E), mangostenone (A and B), caloxanthone-A, macluraxanthone, euxanthone, cudraxanthone, 8-desoxygarnatin, calabaxanthone, demtheylcalabaxanthone, 9-hydroxycalabaxanthone, and 1- or 3-isomangostin [115]. The xanthone content of fruit rinds was significantly reduced during drying at elevated temperatures. However, drying by low-pressure superheated steam has been proposed as an appropriate condition of drying mangosteens for the retention of their xanthone contents and antioxidant property [117]. Phytochemical analysis of freeze-dried mangosteen peels, rinds, and aril parts revealed a high content of phenolic acids, such as hydroxybenzoic acid derivatives (e.g., *m*-hydroxybenzoic acid, *p*-hydroxybenzoic acid, protocatechuic acid, vanillic acid, and veratric acid), hydroxycinnamic acid derivatives (e.g., *p*-coumaric acid, caffeic acid, and ferulic acid), and other phenolic acids (e.g., benzoic acid, cinnamic acid, and *p*-hydroxyphenylacetic acid) [118].

The ethanol extract of mangosteen rinds exhibited free radical scavenging property [119]. Xanthones including 8-hydroxycudraxanthone-G, garcinone D, garcinone E, gartanin, 8-deoxygartanin, garcimangosone B, 1-isomangostin, smeathxanthone A, tovophyllin A, γ-mangostin, and α-mangostin were isolated from dried pericarps of mangosteen. These xanthones showed antioxidant activity, with α-mangostin being the most active [120]. In a mouse mammary organ culture assay, α-mangostin was found to inhibit DMBA-induced formation of preneoplastic lesions with an inhibitory concentration 50% (IC_{50}) value of 2.44 μM [120]. Garcinone E elicited cytotoxic effects against human hepatocellular carcinoma, gastric cancer, and lung cancer cells [121]. Suksamrarn *et al.* [122] isolated prenylated xanthones, such as mangostenone (C–E), along with several known xanthones including α-mangostin. Mangostenone-C elicited a potent cytotoxic effect against human epidermoid carcinoma of mouth (KB), breast cancer (BC-1), and lung cancer (NCI-187) cells. The cytotoxic effects of α-mangostin and gartanin were observed against BC-1 and NCI-187 cells with IC_{50} values of 0.92 and 2.08 μg/mL, respectively [122]. Treatment with α-mangostin inhibited lymph node metastasis in mice inoculated with metastatic murine mammary adenocarcinoma cells stably transfected with Luc2 gene (BJMC3879luc2) and increased the survival of tumor-bearing mice. This study also demonstrated that α-mangostin induced G1 and S phase cell cycle arrest and induced apoptosis in cultured mammary carcinoma cells, which was associated with increased activation of caspase-9 and -3, and the inhibition of Akt phosphorylation [123]. Likewise, α-mangostin attenuated the proliferation of human prostate cancer cells by inducing G1 cell cycle arrest and inhibiting cyclin-dependent kinase-4 (CDK4) activity.

Moreover, administration of α-mangostin by gavage significantly reduced the growth of 22Rv1 human prostate cancer cells inoculated to athymic nude mice [124]. Incubation of human melanoma SK-MEL-28 cells with α-mangostin, γ-mangostin, or 8-deoxygartanin induced apoptosis. The apoptotic effect of α-mangostin was mediated through the induction of caspase-3 activity and decreased mitochondrial membrane potential [125]. Nakatani *et al.* [126] reported that γ-mangostin attenuated both constitutive and LPS-induced expression and the activity of COX-2 in C6 rat glioma cells by blocking the IKK activity and NF-κB-mediated transcription.

Besides xanthones, phenolic acids and anthocyanins present in mangosteens possess cancer chemopreventive properties. For example, oral administration of either caffeic acid or ferulic acid decreased B[*a*]P-induced forestomach tumor formation in ICR/Ha mice [127]. Caffeic acid induced apoptosis in human fibrosarcoma HT1080 cells by generating ROS [128]. Ferulic acid, given by gavage, significantly inhibited DMBA-induced formation of rat mammary tumor [129], hamster buccal pouch carcinomas [130], and mouse skin papillomas [131]. Dietary administration of ferulic acid attenuated 4-nitroquinoline-1-oxide (4NQO)-induced rat tongue carcinogenesis [132] and AOM-induced ACF formation in rat colon [133]. Ferulic acid also inhibited proliferation and induced apoptosis in cancer cells in culture [134, 135]. Protocatechuic acid is another chemopreventive phenolic acid present in mangosteens. Lin *et al.* [136] reported that protocatechuic acid inhibited the migration of gastric cancer cells and attenuated the liver metastasis of B16/F10 melanoma xenograft in nude mice. According to this study, protocatechuic acid suppressed the expression and the activity of MMP-2 and downregulated the NF-κB activity by blocking Ras/Akt-mediated signaling [136]. Incubation of human cancer cells with protocatechuic acid induced apoptosis, which was associated with the activation of caspase-3 and -8 [137]. Feeding with protocatechuic acid containing diet significantly decreased the occurrence of advanced tongue squamous cell carcinomas in rats [138]. Dietary administration of protocatechuic acid also suppressed DMBA-induced hamster cheek pouch carcinogenesis [139] and *N*-nitroso-*bis*(2-oxopropyl)amine-induced hamster pancreatic carcinogenesis [140].

2.5.6 Persimmons

Persimmons (*Diospyros kaki* Thumb., Ebenaceae), widely grown in the northeastern part of Asia, are consumed in either a fresh or a dried form. Several varieties of persimmons are commonly known as "kham" in Korea or "kaki" in Japan. Fresh persimmons are available only during autumn and winter. Sun-dried persimmons, known as "kot-kham," are consumed as a popular snack in Korea. The alcoholic extracts of dried persimmon retained almost equal quantities of dietary fibers, trace elements, and polyphenolic substances as that of fresh fruits [141, 142]. The methanol extracts of dried persimmon exhibited almost an equal antioxidant and free radical scavenging property as that of fresh fruits [141]. Kim *et al.* [143] demonstrated that the ethylacetate fraction of dried persimmons conserved the highest amount of phenolic compounds and elicited the most potent antioxidant activity. The polyphenol-rich ethylacetate extracts of dried persimmon attenuated LPS-induced TNFα production in human monocyte THP-1 cells. Catechins are the major polyphenolic compounds present in persimmons. Treatment of human leukemia Molt 4B cells with persimmon extracts or its flavonoid constituents, such as epicatechin (EC), epicatechin gallate (ECG), epigallocatechin (EGC), or epigallocatechin gallate (EGCG), inhibited ODC expression and induced apoptosis [19]. Among these catechin polyphenols, EGCG has been extensively investigated for its cancer chemopreventive activity [144]. Topical application of EGCG inhibited UVB-induced DNA damage and mouse skin papillomagenesis [145] and induced apoptosis in mouse skin tumors [146]. Administration of EGCG by gavage attenuated TPA-induced COX-2 expression in mouse skin by blocking the activation of MAP kinases and the DNA binding of NF-κB and cyclic adenosine monophosphate (AMP) response element binding (CREB) protein [147, 148]. Likewise, EGCG diminished TPA-induced NF-κB activation in mouse epidermal JB6 cells [149]. Treatment with EGCG induced apoptosis of various human cancer cells including chondrosarcoma [150], fibrosarcoma [151], human ovarian [152], colon [153], cervical [154], and prostate [155] cancer cells.

Other major chemopreventive phytochemicals present in persimmons are triterpenoids, such as oleanolic acid and ursolic acid [156]. Multiple lines of evidence suggest that these compounds possess cancer chemopreventive potential. Both oleanolic acid and ursolic acid inhibited the proliferation of either parent or multidrug-resistant human colon cancer cells (SW480), and leukemia (HL60), and breast cancer (MCF-7) cells and induced apoptosis by downregulating the expression of Bcl-xl, Bcl-2, and survivin [157]. Treatment of human hepatocellular carcinoma (HuH7) cells with oleanolic acid or ursolic acid induced sub-G1 arrest. According to this study, oleanolic acid- or ursolic acid-induced cell death was associated with induction of caspase-9 and -3 activity, the cleavage of polyadiporibosyl polymerase (PARP), the downregulation of the NF-κB activity, and the expression of X-linked inhibitor of apoptosis (XIAP) [158]. Oleanolic acid and ursolic acid elevated the activity of caspase-3 and -8, and diminished the expression of MMP and VEGF in human liver cancer (HepG2, Hep3B, HuH7, and HA22T) cells [159]. Incubation of human breast cancer (MDA-MB231) cells with ursolic acid resulted in mitochondria-dependent as well as death receptor-mediated apoptosis [160]. Treatment with ursolic acid attenuated the growth of human prostate cancer cell xenograft in nude mice and inhibited the proliferation of prostate cancer (DU-145 and LNCaP) cells by blocking the activation of NF-κB and STAT3 signaling pathways [161]. In a transgenic adenocarcinoma of mouse prostate (TRAMP) model, ursolic acid was found to suppress metastatic progression of prostate cancer, partly by blocking the expression of C-X-C chemokine receptor-4 (CXCR4) and the activation of NF-κB [162].

2.5.7 Prunes

Prunes are dried plums of *Prunus domestica* L., Rosaceae). They are produced by dehydrating fresh plums in a hot-air (85–90°C) tunnel dehydrator to reduce the moisture content from 75 to 21% and can be stored at ambient temperature for at least 1 year. Because of their hardness, prunes are subjected to rehydrate before consumption. The total antioxidant capacity of prunes extracts was highest among dried fruits [27, 29]. Despite partial loss of carotenoids present in fresh plums during prune processing, considerable amounts of lutein, β-carotene, and α-carotene were detected in dried prunes [39]. The antioxidative [163] and chemopreventive activity [164–166] of carotenoids, such as β-carotene and lutein, have been well documented. Prunes also contain a trace amount of β-carboline alkaloids [167]. β-Carboline exerts anticancer effects by acting as an inhibitor of the cell cycle regulatory protein CDK4 [168], inducing cancer cell death [169], and blocking DNA synthesis selectively in cancer cells [170].

Prunes also contain large amounts of phenolic compounds, such as chlorogenic acid, neochlorogenic acid, and hydroxycinnamic acids [37]. Kayano *et al.* [171] identified chlorogenic acid, neochlorogenic acid, and cryptochlorogenic acids as the major antioxidants in prunes. Topical application of chlorogenic acid inhibited TPA-induced ODC activity and chemically induced papillomagenesis in mouse skin [172]. A diet containing chlorogenic acid protected against *N*-methyl-*N*-nitrosourea (MNU)-induced glandular stomach carcinogenesis in rats [173]. Treatment with chlorogenic acid inhibited TPA-induced transformation of mouse epidermal JB6 P+ cells by blocking phosphorylation of MAP kinases and activation of AP-1 and NF-κB [174]. Moreover, chlorogenic acid activated nuclear factor erythroid related factor-2 (Nrf2) and induced the expression and activity of antioxidant enzyme glutathione-*S*-transferase (GST)-A in JB6 cells [174]. Incubation with chlorogenic acid induced apoptosis in chronic myelogenous leukemia cells by suppressing Bcr-Abl kinase and inducing p38 MAP kinase [175]. Thus prunes, because of their chlorogenic acid content, hold the potential to be a good chemopreventive agent.

Kimura and colleagues isolated oligomeric proanthocyanidins from the methanol extract of prunes. These proanthocyanidins showed a better *in vitro* antioxidative effect than chlorogenic acid [176]. The oligomeric proanthocyanidins comprising of EC and catechin monomers might contribute to the anticancer activity of prunes. Increasing intake of proanthocyanidins reduced the risk of pancreatic [177], gastric [178], and colorectal [179] cancers. Proanthocyanidins inhibited the growth of non-small-cell-lung cancer cells by suppressing PGE_2 production and blocking PGE_2 receptor, subtype-2 (EP2) receptor signaling [180]. The pro-apoptotic and antiangiogenic effects of proanthocyanidins are associated with their ability to activate caspases and to inhibit the activation or expression of NF-κB, MAP kinases, PI3K/Akt, and cell cycle regulatory proteins, and the release of cytokines and growth factors [181].

Prunes contain high dietary fiber, which is known to reduce the risk of colorectal cancer [182]. One of the risk factors for colorectal cancer is the generation of secondary bile acids from primary bile acids by large intestinal microflora [183]. In a pilot human study comprising 41 men, intake of prunes reduced the fecal content of secondary bile acid and lithocholic acid [184], indicating the possible beneficial effects of prunes against colorectal carcinogenesis. The ethanol fraction of concentrated prune juices reduced the viability of human colon cancer cells (Caco-2 and KATO III), but not that of normal colon epithelial (CCD-18Co) cells [185]. In contrast, dietary administration of prune powders did not affect AOM-induced ACF formation in rats. However, this study demonstrated that prune diets reduced total fecal and secondary bile acid concentrations [186].

2.5.8 Raisins

Raisins are dried grapes of four different varieties, including Thompson seedless, Muscat, Sultana, and Black Corinth. About 90% of the total raisin supply in the United States is in the form of dark raisins made from Thompson seedless grapes. Other popular varieties of raisins include golden raisins prepared from Muscat grapes, Zante currants made from Black Corinth grapes, and European Sultanas raisins produced from seedless yellow grapes. Due to the loss of moisture during the drying process, the total protein, sugars, and fiber content of fresh grapes become concentrated in raisins [187] and a high sugar, low moisture condition prevents spoilage of raisins during storage. Besides macronutrients, the concentration of minerals is usually 3- to 7-fold higher in raisins than in fresh grapes. However, the drying process leads to the loss of certain vitamins, such as vitamin A, C, and K [188].

Other nonnutritive phytochemicals, especially polyphenols, in raisins are considered to be similar to those present in fresh grapes. However, a remarkable difference in the phenolic contents in raisins and fresh grapes has been noted. Raisins contain higher levels of phenolics than do their fresh counterparts [29, 32]. Major phenolic compounds present in raisins and fresh grapes are phenolic acids and flavonols. As compared with fresh and frozen grapes, sun-dried raisins and golden raisins exhibited better ORAC and a higher content of *trans*-caftaric acid, *trans*-coutaric acid, rutin, quercetin glycosides, and kaempferol glycosides [32]. The increased concentration of total phenolics in raisins may result from the lack of polyphenol oxidase in raisins, and the loss of moisture and modification of certain phenolic compounds during drying processes [29]. In contrast, Karadeniz *et al.* [189] reported that 90% of phenolic acids (e.g., *trans*-caftaric acid and *trans*-coutaric acid) in fresh grapes are lost during processing of grapes into sun-dried raisins. These authors have also demonstrated that procyanidins and flavonols are completely decomposed during the processing of raisins. The reason for such discrepancies may be due to the different grape cultivars, drying temperature, the drying process, and the method of solvent extraction of raisins to obtain phenolic compounds. While stilbenoids, in particular resveratrol, are present in grapes and wine, resveratrol was not detected in raisins [32, 189]. Zhao and Hall [190] reported the presence of catechins and phenolic acids, such as gallic acid, ferulic acid, and chlorogenic acid, in raisins. Thus, more rigorous analysis of the chemical composition of raisins may help to better understand the chemopreventive potential of raisins.

Despite inconsistent data on the phenolic compounds present in raisins as compared to fresh grapes [191], the presence of trace amount of isoflavones, such as daidzein and genistein [192], and chlorogenic acid [190] in raisins suggests the cancer chemopreventive potential of these dried fruits. Moreover, because of the high dietary fiber content, raisins may protect against colon carcinogenesis. Daily consumption of raisins have been shown to decrease urinary excretion of 8-epi-$PGF_{2\alpha}$ [193] and 8-hydroxy-2-deoxyguanosine [194], markers of inflammation and oxidative DNA damage, respectively. Thus, the assessment of anticancer effects of raisins merits further investigation.

2.5.9 Kiwi fruits

Kiwi (*Actinidia chinensis* L. syn. *Actinidia deliciosa*) fruits are traditionally used for the treatment of many different types of cancers, for example, stomach, lung, and liver cancer in folk medicine. Kiwi fruit extracts inhibit the growth of cancer cells. Kiwi fruits protected against oxidative DNA damage, thereby inhibiting neoplastic cell transformation [195], and the growth of sarcoma 180 in mice [196]. Chemical analysis of kiwi fruit peels and pulps revealed that the fruits are rich in vitamin E, tocopherols, sterols, ursolic acid, chlorogenic acid, and several flavonoids [197]. Freeze-dried pulp extracts of kiwi fruit attenuated H_2O_2-induced lymphocyte DNA damage [198]. Hydrophilic fractions of 70% methanol extracts of dried kiwi fruit elicited free radical scavenging activity and inhibited growth of human squamous carcinoma (HSC-2) cells [196]. Moreover, methanol extracts of freeze-dried kiwi fruit powder and their bioactive flavonoid quercetin showed antioxidant activity and inhibited H_2O_2-induced disruption of gap-junction intercellular communication in rat liver epithelial (WB-F344) cells by restoring connexin-43 expression and blocking ERK phosphorylation [199].

2.5.10 Other dried fruits

Several other dried fruits are commercially available and widely consumed throughout the world. These include apricots, apples, figs, snake fruits, star fruits, guavas, bananas, etc. Apricots are important for their minerals and other nutrients. Sulfurized (sulfur-dried) apricots retain selenium, though to a lesser extent than the fresh fruits [200], which possesses anticancer properties [201]. MK615, a formulation developed from Japanese apricots, decreased the viability of MDA-MB231 and MCF-7 cells [202]. Intake of dried Japanese apricots protected against *Helicobacter pylori*-induced chronic atrophic gastritis, a pathological condition often leading to gastric cancer [203]. Moreover, diet containing 15 or 30% sun-dried apricot powders diminished ethanol-induced liver damage in rats by suppressing malondialdehyde production and activating cytoprotective enzymes such as quinone reductase, SOD, GST, and glutathione peroxidase, indicating its potential to prevent liver carcinogenesis [204].

The latex of fresh figs contains polyphenolic compounds. Incubation of human glioblastoma and hepatocellular carcinoma cells with fig's latex reduced proliferation and colony formation, and induced apoptosis in these cells [205]. Significantly higher concentrations of organic acids and phenolic compounds, such as chlorogenic acid, catechin, EC, kaempferol-3-*O*-glucoside, and luteolin-8-C-glucosides, were detected in sun-dried or oven-dried figs as compared with fresh figs [206]. Thus, dried figs may be utilized for the prevention of cancer. The cancer chemopreventive activity of these dried fruits merits further investigation.

2.6 Conclusions

Fruits are nature's gift to mankind both for providing nutrition and for preventing diseases. Although research done until now indicates that consumption of dried fruits can reduce the risk of cancer, more extensive studies are required to examine the chemopreventive potential of diverse classes of dried fruits and to unravel their molecular mechanisms. One of the major safety concerns with regard to dried fruits is the microbial spoilage during storage and the potential health hazard resulting from microbial toxins [207, 208]. Dried fruits can be contaminated with aflatoxins, ochratoxin-A, kojic acid, and, occasionally patulin or zearalenone [209]. Thus, dried fruits must go through routine quality assessment before consumption. Moreover, drying of fresh fruits at high temperature may generate Maillard reaction products, which are potentially genotoxic. Metabolites of 5-(hydroxymethyl)-2-furfural, a Maillard reaction product, have been detected in the urine of human volunteers receiving dried prune juices [210]. The aqueous extract of several commercially available dried fruits induced clastogenic activity as revealed by increased chromosome breaks or exchanges per metaphase plate in cultured Chinese hamster ovary (CHO) cells. Incubation with a liver microsomal S9 mixture reduced this clastogenic effect of dried fruits [211]. To avoid non-enzymatic browning reactions, processing of dried fruits should be carried out at relatively low temperatures. Besides heat-catalyzed Maillard reactions, an enzymatic browning reaction occurs in dried fruits. Appropriate processing can prevent browning of dry-processed fruits. For instance, pretreatment of sliced apples by pineapple juices reduced the rate and extent of enzymatic browning reaction. Other preprocessing techniques include bisulfate and ascorbic acid treatment of fresh fruits prior to hot-air or oven drying. With the advancement in food processing technology, a wide variety of dried fruits and their combinations are now available in the shelves of supermarkets, indicating the increased popularity of dried fruits. Someday, dried fruits may be exploited to develop functional foods for the prevention of cancer.

Acknowledgments

This work was supported by the grant for the Global Core Research Center (GCRC) grant from the National Research Foundation, Ministry of Education, Science and Technology, Republic of Korea.

References

1. Mackay, J., Jemal, A., Lee, N. & Parkin, D. (2006) *The Cancer Atlas*. American Cancer Society, Atlanta, GA.
2. Park, S., Bae, J., Nam, B.H. & Yoo, K.Y. (2008) Aetiology of cancer in Asia. *Asian Pacific Journal of Cancer Prevention*, **9**, 371–380.
3. Mann, J.R., Backlund, M.G. & DuBois, R.N. (2005) Mechanisms of disease: inflammatory mediators and cancer prevention. *Nature Clinical Practice Oncology*, **2**, 202–210.
4. Anand, P., Kunnumakara, A.B., Sundaram, C., Harikumar, K.B., Tharakan, S.T., Lai, O.S., Sung, B.Y. & Aggarwal, B.B. (2008) Cancer is a preventable disease that requires major lifestyle changes. *Pharmaceutical Research*, **25**, 2097–2116.
5. Wattenberg, L.W. (1966) Chemoprophylaxis of carcinogenesis: a review. *Cancer Research*, **26**, 1520–1526.
6. Sporn, M.B. (1976) Approaches to prevention of epithelial cancer during the preneoplastic period. *Cancer Research*, **36**, 2699–2702.
7. Tsuda, H., Ohshima, Y., Nomoto, H., Fujita, K., Matsuda, E., Ligo, M., Takasuka, N. & Moore, M.A. (2004) Cancer prevention by natural compounds. *Drug Metabolism and Pharmacokinetics*, **19**, 245–263.
8. De Flora, S. & Ferguson, L.R. (2005) Overview of mechanisms of cancer chemopreventive agents. *Mutatation Research*, **591**, 8–15.
9. Aggarwal, B.B. & Shishodia, S. (2006) Molecular targets of dietary agents for prevention and therapy of cancer. *Biochemical Pharmacology*, **71**, 1397–1421.
10. Baer-Dubowska, W. (2006) Cancer chemopreventive agents-drugs for the 21st century? *Acta Poloniae Pharmaceutica*, **63**, 369–373.
11. Surh, Y.-J. (2003) Cancer chemoprevention with dietary phytochemicals. *Nature Reviews Cancer*, **3**, 768–780.
12. WCRF. (2007) World Cancer Research Fund/American Institute for Cancer Research: Food, Nutrition, Physical Activity and the Prevention of Cancer: A Global Perspective.
13. Temple, N.J. & Gladwin, K.K. (2003) Fruit, vegetables, and the prevention of cancer: research challenges. *Nutrition*, **19**, 467–470.
14. Terry, P., Terry, J.B. & Wolk, A. (2001) Fruit and vegetable consumption in the prevention of cancer: an update. *Journal of Internal Medicine*, **250**, 280–290.
15. Koushik, A., Hunter, D.J., Spiegelman, D., Beeson, W.L., van den Brandt, P.A., Buring, J.E., Calle, E.E., Cho, E., Fraser, G.E., Freudenheim, J.L., Fuchs, C.S., Giovannucci, E.L., Goldbohm, R.A., Harnack, L., Jacobs, D.R.Jr., Kato, I., Krogh, V., Larsson, S.C., Leitzmann, M.F., Marshall, J.R., McCullough, M.L., Miller, A.B., Pietinen, P., Rohan, T.E., Schatzkin, A., Sieri, S., Virtanen, M.J., Wolk, A., Zeleniuch-Jacquotte, A., Zhang, S.M. & Smith-Warner, S.A. (2007) Fruits, vegetables, and colon cancer risk in a pooled analysis of 14 cohort studies. *Journal of National Cancer Institute*, **99**, 1471–1483.
16. Smith-Warner, S.A., Spiegelman, D., Yaun, S.S., Albanes, D., Beeson, W.L., van den Brandt, P.A., Feskanich, D., Folsom, A.R., Fraser, G.E., Freudenheim, J.L., Giovannucci, E., Goldbohm, R.A., Graham, S., Kushi, L.H., Miller, A.B., Pietinen, P., Rohan, T.E., Speizer, F.E., Willett, W.C. & Hunter,

D.J. (2003) Fruits, vegetables and lung cancer: a pooled analysis of cohort studies. *International Journal of Cancer*, **107**, 1001–1011.

17. van Duijnhoven, F.J., Bueno-De-Mesquita, H.B., Ferrari, P., Jenab, M., Boshuizen, H.C., Ros, M.M., Casagrande, C., Tjonneland, A., Olsen, A., Overvad, K., Thorlacius-Ussing, O., Clavel-Chapelon, F., Boutron-Ruault, M.C., Morois, S., Kaaks, R., Linseisen, J., Boeing, H., Nothlings, U., Trichopoulou, A., Trichopoulos, D., Misirli, G., Palli, D., Sieri, S., Panico, S., Tumino, R., Vineis, P., Peeters, P.H., van Gils, C.H., Ocke, M.C., Lund, E., Engeset, D., Skeie, G., Suarez, L.R., Gonzalez, C.A., Sanchez, M.J., Dorronsoro, M., Navarro, C., Barricarte, A., Berglund, G., Manjer, J., Hallmans, G., Palmqvist, R., Bingham, S.A., Khaw, K.T., Key, T.J., Allen, N.E., Boffetta, P., Slimani, N., Rinaldi, S., Gallo, V., Norat, T. & Riboli, E. (2009) Fruit, vegetables, and colorectal cancer risk: the European Prospective Investigation into Cancer and Nutrition. *Americal Journal of Clinical Nutrition*, **89**, 1441–1452.
18. Miller, A.B., Altenburg, H.P., Bueno-de-Mesquita, B., Boshuizen, H.C., Agudo, A., Berrino, F., Gram, I.T., Janson, L., Linseisen, J., Overvad, K., Rasmuson, T., Vineis, P., Lukanova, A., Allen, N., Amiano, P., Barricarte, A., Berglund, G., Boeing, H., Clavel-Chapelon, F., Day, N.E., Hallmans, G., Lund, E., Martinez, C., Navarro, C., Palli, D., Panico, S., Peeters, P.H., Quiros, J.R., Tjonneland, A., Tumino, R., Trichopoulou, A., Trichopoulos, D., Slimani, N. & Riboli, E. (2004) Fruits and vegetables and lung cancer: findings from the European Prospective Investigation into Cancer and Nutrition. *International Journal of Cancer*, **108**, 269–276.
19. Achiwa, Y., Hibasami, H., Katsuzaki, H., Imai, K. & Komiya, T. (1997) Inhibitory effects of persimmon (*Diospyros kaki*) extract and related polyphenol compounds on growth of human lymphoid leukemia cells. *Bioscience, Biotechnology and Biochemistry*, **61**, 1099–1101.
20. Baliga, M.S. & Dsouza, J.J. (2011) Amla (*Emblica officinalis* Gaertn), a wonder berry in the treatment and prevention of cancer. *European Journal of Cancer Prevention*, **20**, 225–239.
21. D'Ambrosio, S.M., Han, C., Pan, L., Kinghorn, A.D. & Ding, H. (2011) Aliphatic acetogenin constituents of avocado fruits inhibit human oral cancer cell proliferation by targeting the EGFR/RAS/RAF/MEK/ERK1/2 pathway. *Biochemical and Biophysical Research Communications*, **409**, 465–469.
22. Huang, C., Li, J., Song, L., Zhang, D., Tong, Q., Ding, M., Bowman, L., Aziz, R. & Stoner, G.D. (2006) Black raspberry extracts inhibit benzo(a)pyrene diol-epoxide-induced activator protein 1 activation and VEGF transcription by targeting the phosphotidylinositol 3-kinase/Akt pathway. *Cancer Research*, **66**, 581–587.
23. Stoner, G.D., Wang, L.S., Zikri, N., Chen, T., Hecht, S.S., Huang, C., Sardo, C. & Lechner, J.F. (2007) Cancer prevention with freeze-dried berries and berry components. *Seminars in Cancer Biology*, **17**, 403–410.
24. Konopacka, D., Jesionkowska, K., Klewicki, R. & Bonazzi, C. (2009) The effect of different osmotic agents on the sensory perception of osmo-treated dried fruit. *Journal of Horticultural Science and Biotechnology*, **ISAFRUIT Special Issue**, 80–84.
25. Jesionkowska, K., Sijtsema, S.J., Konopacka, D. & Symoneaux, R. (2009) Dried fruit and its functional properties from a consumer's point of view. *Journal of Horticultural Science and Biotechnology*, **ISAFRUIT Special Issue**, 85–88.
26. Leusink, G.J., Kitts, D.D., Yaghmaee, P. & Durance, T. (2010) Retention of antioxidant capacity of vacuum microwave dried cranberry. *Journal of Food Science*, **75**, C311–C316.
27. Pellegrini, N., Serafini, M., Salvatore, S., Del Rio, D., Bianchi, M. & Brighenti, F. (2006) Total antioxidant capacity of spices, dried fruits, nuts, pulses, cereals and sweets consumed in Italy assessed by three different *in vitro* assays. *Molecular Nutrition & Food Research*, **50**, 1030–1038.
28. Vinson, J.A., Zubik, L., Bose, P., Samman, N. & Proch, J. (2005) Dried fruits: excellent *in vitro* and *in vivo* antioxidants. *Journal of American College of Nutrition*, **24**, 44–50.
29. Wu, X., Beecher, G.R., Holden, J.M., Haytowitz, D.B., Gebhardt, S.E. & Prior, R.L. (2004) Lipophilic and hydrophilic antioxidant capacities of common foods in the United States. *Journal of Agricultural and Food Chemistry*, **52**, 4026–4037.
30. Karakaya, S., El, S.N. & Tas, A.A. (2001) Antioxidant activity of some foods containing phenolic compounds. *International Journal of Food Science and Nutrition*, **52**, 501–508.
31. Halvorsen, B.L., Holte, K., Myhrstad, M.C., Barikmo, I., Hvattum, E., Remberg, S.F., Wold, A.B., Haffner, K., Baugerod, H., Andersen, L.F., Moskaug, O., Jacobs, D.R. Jr. & Blomhoff, R. (2002) A systematic screening of total antioxidants in dietary plants. *Journal of Nutrition*, **132**, 461–471.

32. Parker, T.L., Wang, X.H., Pazmino, J. & Engeseth, N.J. (2007) Antioxidant capacity and phenolic content of grapes, sun-dried raisins, and golden raisins and their effect on ex vivo serum antioxidant capacity. *Journal of Agricultural and Food Chemistry*, **55**, 8472–8477.
33. Schauss, A.G., Wu, X., Prior, R.L., Ou, B., Huang, D., Owens, J., Agarwal, A., Jensen, G.S., Hart, A.N. & Shanbrom, E. (2006) Antioxidant capacity and other bioactivities of the freeze-dried Amazonian palm berry, *Euterpe oleraceae* mart. (acai). *Journal of Agricultural and Food Chemistry*, **54**, 8604–8610.
34. Asghar, M., Monjok, E., Kouamou, G., Ohia, S.E., Bagchi, D. & Lokhandwala, M.F. (2007) Super CitriMax (HCA-SX) attenuates increases in oxidative stress, inflammation, insulin resistance, and body weight in developing obese Zucker rats. *Molecular and Cellular Biochemistry*, **304**, 93–99.
35. Ho, S.C. & Lin, C.C. (2008) Investigation of heat treating conditions for enhancing the anti-inflammatory activity of citrus fruit (*Citrus reticulata*) peels. *Journal of Agricultural and Food Chemistry*, **56**, 7976–7982.
36. Kundu, J.K. & Surh, Y.J. (2008) Inflammation: gearing the journey to cancer. *Mutation Research Reviews*, **659**, 15–30.
37. Donovan, J.L., Meyer, A.S. & Waterhouse, A.L. (1998) Phenolic composition and antioxidant activity of prunes and prune juice (*Prunus domestica*). *Journal of Agricultural and Food Chemistry*, **46**, 1247–1252.
38. Jesionkowska, K., Sijtsema, S., Simoneaux, R., Konopacka, D. & Plocharski, W. (2008) Preferences and consumption of dried fruit and dried fruit products among Dutch, French and Polish consumers. *Journal of Fruit and Ornamental Plant Research*, **16**, 261–274.
39. Stacewicz-Sapuntzakis, M., Bowen, P.E., Hussain, E.A., Damayanti-Wood, B.I. & Farnsworth, N.R. (2001) Chemical composition and potential health effects of prunes: a functional food? *Critical Reviews in Food Science and Nutrition*, **41**, 251–286.
40. Ratti, C. (2001) Hot-air and freeze-drying of high-value foods: a review. *Journal of Food Engineering*, **49**, 311–319.
41. Ciurzyńska, A. & Lenart, A. (2011) Freeze-drying—application in food processing and biotechnology—a review. *Polish Journal of Food and Nutrition Science*, **61**, 165–171.
42. Marques, L.G., Silveira, A.M. & Freire, J.T. (2006) Freeze-drying characteristics of tropical fruits. *Drying Technology*, **24**, 457–463.
43. Shofian, N.M., Hamid, A.A., Osman, A., Saari, N., Anwar, F., Dek, M.S. & Hairuddin, M.R. (2011) Effect of freeze-drying on the antioxidant compounds and antioxidant activity of selected tropical fruits. *International Journal of Molecular Sciences*, **12**, 4678–4692.
44. Reuter, S., Gupta, S.C., Chaturvedi, M.M. & Aggarwal, B.B. (2010) Oxidative stress, inflammation, and cancer: how are they linked? *Free Radical Biology and Medicine*, **49**, 1603–1616.
45. Federico, A., Morgillo, F., Tuccillo, C., Ciardiello, F. & Loguercio, C. (2007) Chronic inflammation and oxidative stress in human carcinogenesis. *International Journal of Cancer*, **121**, 2381–2386.
46. Wattenberg, L.W. (1985) Chemoprevention of cancer. *Cancer Research*, **45**, 1–8.
47. Krishnaveni, M. & Mirunalini, S. (2010) Therapeutic potential of *Phyllanthus emblica* (amla): the ayurvedic wonder. *Journal of Basic and Clinical Physiology and Pharmacology*, **21**, 93–105.
48. Kumaran, A. & Karunakaran, R.J. (2006) Nitric oxide radical scavenging active components from *Phyllanthus emblica* L. *Plant Foods for Human Nutrition*, **61**, 1–5.
49. Poltanov, E.A., Shikov, A.N., Dorman, H.J., Pozharitskaya, O.N., Makarov, V.G., Tikhonov, V.P. & Hiltunen, R. (2009) Chemical and antioxidant evaluation of Indian gooseberry (*Emblica officinalis* Gaertn., syn. *Phyllanthus emblica* L.) supplements. *Phytotherapy Research*, **23**, 1309–1315.
50. Banu, S.M., Selvendiran, K., Singh, J.P. & Sakthisekaran, D. (2004) Protective effect of *Emblica officinalis* ethanolic extract against 7,12-dimethylbenz(*a*) anthracene (DMBA) induced genotoxicity in Swiss albino mice. *Human Experimental Toxicology*, **23**, 527–531.
51. Jose, J.K., Kuttan, G. & Kuttan, R. (2001) Antitumour activity of *Emblica officinalis*. *Journal of Ethnopharmacology*, **75**, 65–69.
52. Sancheti, G., Jindal, A., Kumari, R. & Goyal, P.K. (2005) Chemopreventive action of *Emblica officinalis* on skin carcinogenesis in mice. *Asian Pacific Journal of Cancer Prevention*, **6**, 197–201.
53. Ngamkitidechakul, C., Jaijoy, K., Hansakul, P., Soonthornchareonnon, N. & Sireeratawong, S. (2010) Antitumour effects of *Phyllanthus emblica* L.: induction of cancer cell apoptosis and inhibition of *in vivo* tumour promotion and *in vitro* invasion of human cancer cells. *Phytotherapy Research*, **24**, 1405–1413.

54. Sultana, S., Ahmed, S. & Jahangir, T. (2008) *Emblica officinalis* and hepatocarcinogenesis: a chemopreventive study in Wistar rats. *Journal of Ethnopharmacology*, **118**, 1–6.
55. Tokar, E.J., Benbrahim-Tallaa, L., Ward, J.M., Lunn, R., Sams, R.L., 2nd. & Waalkes, M.P. (2010) Cancer in experimental animals exposed to arsenic and arsenic compounds. *Critical Reviews in Toxicology*, **40**, 912–927.
56. Sharma, A., Sharma, M.K. & Kumar, M. (2009) Modulatory role of *Emblica officinalis* fruit extract against arsenic induced oxidative stress in Swiss albino mice. *Chemico-Biological Interactions*, **180**, 20–30.
57. Rajeshkumar, N.V., Pillai, M.R. & Kuttan, R. (2003) Induction of apoptosis in mouse and human carcinoma cell lines by *Emblica officinalis* polyphenols and its effect on chemical carcinogenesis. *Journal of Experimental and Clinical Cancer Research*, **22**, 201–212.
58. Yang, C.J., Wang, C.S., Hung, J.Y., Huang, H.W., Chia, Y.C., Wang, P.H., Weng, C.F. & Huang, M.S. (2009) Pyrogallol induces G2-M arrest in human lung cancer cells and inhibits tumor growth in an animal model. *Lung Cancer*, **66**, 162–168.
59. Ding, H., Chin, Y.W., Kinghorn, A.D. & D'Ambrosio, S.M. (2007) Chemopreventive characteristics of avocado fruit. *Seminars in Cancer Biology*, **17**, 386–394.
60. Lu, Q.Y., Arteaga, J.R., Zhang, Q., Huerta, S., Go, V.L. & Heber, D. (2005) Inhibition of prostate cancer cell growth by an avocado extract: role of lipid-soluble bioactive substances. *Journal of Nutritional Biochemistry*, **16**, 23–30.
61. Kim, O.K., Murakami, A., Nakamura, Y., Takeda, N., Yoshizumi, H. & Ohigashi, H. (2000) Novel nitric oxide and superoxide generation inhibitors, persenone A and B, from avocado fruit. *Journal of Agricultural and Food Chemistry*, **48**, 1557–1563.
62. Kim, O.K., Murakami, A., Takahashi, D., Nakamura, Y., Torikai, K., Kim, H.W. & Ohigashi, H. (2000) An avocado constituent, persenone A, suppresses expression of inducible forms of nitric oxide synthase and cyclooxygenase in macrophages, and hydrogen peroxide generation in mouse skin. *Bioscience, Biotechnology, and Biochemistry*, **64**, 2504–2507.
63. Bhattacharyya, S.S., Paul, S., Dutta, S., Boujedaini, N. & Khuda-Bukhsh, A.R. (2010) Anti-oncogenic potentials of a plant coumarin (7-hydroxy-6-methoxy coumarin) against 7,12-dimethylbenz [*a*] anthracene-induced skin papilloma in mice: the possible role of several key signal proteins. *Zhong Xi Yi Jie He Xue Bao (Journal of Chinese Integrative Medicine)*, **8**, 645–654.
64. Liu, X.L., Zhang, L., Fu, X.L., Chen, K. & Qian, B.C. (2001) Effect of scopoletin on PC3 cell proliferation and apoptosis. *Acta Pharmacologica Sinica*, **22**, 929–933.
65. Adams, M., Efferth, T. & Bauer, R. (2006) Activity-guided isolation of scopoletin and isoscopoletin, the inhibitory active principles towards CCRF-CEM leukaemia cells and multi-drug resistant CEM/ADR5000 cells, from *Artemisia argyi*. *Planta Medica*, **72**, 862–864.
66. Khuda-Bukhsh, A.R., Bhattacharyya, S.S., Paul, S. & Boujedaini, N. (2010) Polymeric nanoparticle encapsulation of a naturally occurring plant scopoletin and its effects on human melanoma cell A375. *Zhong Xi Yi Jie He Xue Bao (Journal of Chinese Integrative Medicine)*, **8**, 853–862.
67. Wang, W., Bostic, T.R. & Gu, L.W. (2010) Antioxidant capacities, procyanidins and pigments in avocados of different strains and cultivars. *Food Chemistry*, **122**, 1193–1198.
68. Del Rio, D., Borges, G. & Crozier, A. (2010) Berry flavonoids and phenolics: bioavailability and evidence of protective effects. *British Journal of Nutrition*, **104**, S67–S90.
69. Seeram, N.P. (2008) Berry fruits: compositional elements, biochemical activities, and the impact of their intake on human health, performance, and disease. *Journal of Agricultural and Food Chemistry*, **56**, 627–629.
70. Seeram, N.P., Adams, L.S., Zhang, Y., Lee, R., Sand, D., Scheuller, H.S. & Heber, D. (2006) Blackberry, black raspberry, blueberry, cranberry, red raspberry, and strawberry extracts inhibit growth and stimulate apoptosis of human cancer cells *in vitro*. *Journal of Agricultural and Food Chemistry*, **54**, 9329–9339.
71. Wedge, D.E., Meepagala, K.M., Magee, J.B., Smith, S.H., Huang, G. & Larcom, L.L. (2001) Anticarcinogenic activity of strawberry, blueberry, and raspberry extracts to breast and cervical cancer cells. *Journal of Medicinal Food*, **4**, 49–51.
72. Adams, L.S., Kanaya, N., Phung, S., Liu, Z. & Chen, S. (2011) Whole blueberry powder modulates the growth and metastasis of MDA-MB-231 triple negative breast tumors in nude mice. *Journal of Nutrition*, **141**, 1805–1812.

73. Adams, L.S., Phung, S., Yee, N., Seeram, N.P., Li, L. & Chen, S. (2010) Blueberry phytochemicals inhibit growth and metastatic potential of MDA-MB-231 breast cancer cells through modulation of the phosphatidylinositol 3-kinase pathway. *Cancer Research*, **70**, 3594–3605.
74. Montales, M.T., Rahal, O.M., Kang, J., Rogers, T.J., Prior, R.L., Wu, X. & Simmen, R.C. (2012) Repression of mammosphere formation of human breast cancer cells by soy isoflavone genistein and blueberry polyphenolic acids suggests diet-mediated targeting of cancer stem-like/progenitor cells. *Carcinogenesis*, **33**, 652–660.
75. Gordillo, G., Fang, H., Khanna, S., Harper, J., Phillips, G. & Sen, C.K. (2009) Oral administration of blueberry inhibits angiogenic tumor growth and enhances survival of mice with endothelial cell neoplasm. *Antioxidants & Redox Signaling*, **11**, 47–58.
76. Misikangas, M., Pajari, A.M., Paivarinta, E., Oikarinen, S.I., Rajakangas, J., Marttinen, M., Tanayama, H., Torronen, R. & Mutanen, M. (2007) Three Nordic berries inhibit intestinal tumorigenesis in multiple intestinal neoplasia/+mice by modulating beta-catenin signaling in the tumor and transcription in the mucosa. *Journal of Nutrition*, **137**, 2285–2290.
77. Rajakangas, J., Misikangas, M., Paivarinta, E. & Mutanen, M. (2008) Chemoprevention by white currant is mediated by the reduction of nuclear beta-catenin and NF-κB levels in Min mice adenomas. *European Journal of Nutrition*, **47**, 115–122.
78. Bishayee, A., Haznagy-Radnai, E., Mbimba, T., Sipos, P., Morazzoni, P., Darvesh, A.S., Bhatia, D. & Hohmann, J. (2010) Anthocyanin-rich black currant extract suppresses the growth of human hepatocellular carcinoma cells. *Natural Product Communications*, **5**, 1613–1618.
79. Bishayee, A., Mbimba, T., Thoppil, R.J., Haznagy-Radnai, E., Sipos, P., Darvesh, A.S., Folkesson, H.G. & Hohmann, J. (2011) Anthocyanin-rich black currant (*Ribes nigrum* L.) extract affords chemoprevention against diethylnitrosamine-induced hepatocellular carcinogenesis in rats. *Journal of Nutritional Biochemistry*, **22**, 1035–1046.
80. Wu, Q.K., Koponen, J.M., Mykkanen, H.M. & Torronen, A.R. (2007) Berry phenolic extracts modulate the expression of p21(WAF1) and Bax but not Bcl-2 in HT-29 colon cancer cells. *Journal of Agricultural and Food Chemistry*, **55**, 1156–1163.
81. Aiyer, H.S., Srinivasan, C. & Gupta, R.C. (2008) Dietary berries and ellagic acid diminish estrogen-mediated mammary tumorigenesis in ACI rats. *Nutrition & Cancer*, **60**, 227–234.
82. Stoner, G.D., Kresty, L.A., Carlton, P.S., Siglin, J.C. & Morse, M.A. (1999) Isothiocyanates and freeze-dried strawberries as inhibitors of esophageal cancer. *Toxicological Sciences*, **52**, 95–100.
83. Carlton, P.S., Kresty, L.A. & Stoner, G.D. (2000) Failure of dietary lyophilized strawberries to inhibit 4-(methylnitrosamino)-1-(3-pyridyl)-1-butanone-and benzo[a]pyrene-induced lung tumorigenesis in strain A/J mice. *Cancer Letters*, **159**, 113–117.
84. Kresty, L.A., Morse, M.A., Morgan, C., Carlton, P.S., Lu, J., Gupta, A., Blackwood, M. & Stoner, G.D. (2001) Chemoprevention of esophageal tumorigenesis by dietary administration of lyophilized black raspberries. *Cancer Research*, **61**, 6112–6119.
85. Harris, G.K., Gupta, A., Nines, R.G., Kresty, L.A., Habib, S.G., Frankel, W.L., LaPerle, K., Gallaher, D.D., Schwartz, S.J. & Stoner, G.D. (2001) Effects of lyophilized black raspberries on azoxymethane-induced colon cancer and 8-hydroxy-2′-deoxyguanosine levels in the Fischer 344 rat. *Nutrition & Cancer*, **40**, 125–133.
86. Montrose, D.C., Horelik, N.A., Madigan, J.P., Stoner, G.D., Wang, L.S., Bruno, R.S., Park, H J., Giardina, C. & Rosenberg, D.W. (2011) Anti-inflammatory effects of freeze-dried black raspberry powder in ulcerative colitis. *Carcinogenesis*, **32**, 343–350.
87. Chen, T., Hwang, H., Rose, M.E., Nines, R.G. & Stoner, G.D. (2006) Chemopreventive properties of black raspberries in *N*-nitrosomethylbenzylamine-induced rat esophageal tumorigenesis: down-regulation of cyclooxygenase-2, inducible nitric oxide synthase, and c-Jun. *Cancer Research*, **66**, 2853–2859.
88. Han, C., Ding, H., Casto, B., Stoner, G.D. & D'Ambrosio, S.M. (2005) Inhibition of the growth of premalignant and malignant human oral cell lines by extracts and components of black raspberries. *Nutrition & Cancer*, **51**, 207–217.
89. Wang, L.S., Hecht, S.S., Carmella, S.G., Yu, N., Larue, B., Henry, C., McIntyre, C., Rocha, C., Lechner, J.F. & Stoner, G.D. (2009) Anthocyanins in black raspberries prevent esophageal tumors in rats. *Cancer Prevention Research*, **2**, 84–93.
90. Kresty, L.A., Frankel, W.L., Hammond, C.D., Baird, M.E., Mele, J.M., Stoner, G.D. & Fromkes, J.J. (2006) Transitioning from preclinical to clinical chemopreventive assessments of lyophilized black

raspberries: interim results show berries modulate markers of oxidative stress in Barrett's esophagus patients. *Nutrition & Cancer*, **54**, 148–156.

91. Huang, C., Huang, Y., Li, J., Hu, W., Aziz, R., Tang, M.S., Sun, N., Cassady, J. & Stoner, G.D. (2002) Inhibition of benzo(*a*)pyrene diol-epoxide-induced transactivation of activated protein 1 and nuclear factor kappaB by black raspberry extracts. *Cancer Research*, **62**, 6857–6863.
92. Hecht, S.S., Huang, C., Stoner, G.D., Li, J., Kenney, P.M., Sturla, S.J. & Carmella, S.G. (2006) Identification of cyanidin glycosides as constituents of freeze-dried black raspberries which inhibit anti-benzo[*a*]pyrene-7,8-diol-9,10-epoxide induced NF-κB and AP-1 activity. *Carcinogenesis*, **27**, 1617–1626.
93. Duncan, F.J., Martin, J.R., Wulff, B.C., Stoner, G.D., Tober, K.L., Oberyszyn, T.M., Kusewitt, D.F. & Van Buskirk, A.M. (2009) Topical treatment with black raspberry extract reduces cutaneous UVB-induced carcinogenesis and inflammation. *Cancer Prevention Research*, **2**, 665–672.
94. Azeredo, H.M.C., Brito, E.S., Moreira, G.E.G., Farias, V.L. & Bruno, L.M. (2006) Effect of drying and storage time on the physico-chemical properties of mango leathers. *International Journal of Food Science & Technology*, **41**, 635–638.
95. Berardini, N., Knodler, M., Schieber, A. & Carle, R. (2005) Utilization of mango peels as a source of pectin and polyphenolics. *Innovative Food Science & Emerging Technologies*, **6**, 442–452.
96. Berardini, N., Carle, R. & Schieber, A. (2004) Characterization of gallotannins and benzophenone derivatives from mango (*Mangifera indica* L. cv. 'Tommy Atkins') peels, pulp and kernels by high-performance liquid chromatography electrospray ionization mass spectrometry. *Rapid Communication in Mass Spectrometry*, **18**, 2208–2216.
97. Berardini, N., Fezer, R., Conrad, J., Beifuss, U., Carle, R. & Schieber, A. (2005) Screening of mango (*Mangifera indica* L.) cultivars for their contents of flavonol O- and xanthone C-glycosides, anthocyanins, and pectin. *Journal of Agricultural and Food Chemistry*, **53**, 1563–1570.
98. Berardini, N., Schieber, A., Klaiber, I., Beifuss, U., Carle, R. & Conrad, J. (2005) 7-*O*-Methylcyanidin 3-*O*-beta-D-galactopyranoside, a novel anthocyanin from mango (*Mangifera indica* L. cv. 'Tommy Atkins') peels. *Zeitschrift für Naturforschun-B*, **60**, 801–804.
99. Chen, J.P., Tai, C.Y. & Chen, B.H. (2004) Improved liquid chromatographic method for determination of carotenoids in Taiwanese mango (*Mangifera indica* L.). *Journal of Chromatography A*, **1054**, 261–268.
100. Chen, J.P., Tai, C.Y. & Chen, B.H. (2007) Effects of different drying treatments on the stability of carotenoids in Taiwanese mango (*Mangifera indica* L.). *Food Chemistry*, **100**, 1005–1010.
101. Rajendran, P., Ekambaram, G. & Sakthisekaran, D. (2008) Effect of mangiferin on benzo(*a*)pyrene induced lung carcinogenesis in experimental Swiss albino mice. *Natural Product Researh*, **22**, 672–680.
102. Yoshimi, N., Matsunaga, K., Katayama, M., Yamada, Y., Kuno, T., Qiao, Z., Hara, A., Yamahara, J. & Mori, H. (2001) The inhibitory effects of mangiferin, a naturally occurring glucosylxanthone, in bowel carcinogenesis of male F344 rats. *Cancer Letters*, **163**, 163–170.
103. Leiro, J.M., Alvarez, E., Arranz, J.A., Siso, I.G. & Orallo, F. (2003) *In vitro* effects of mangiferin on superoxide concentrations and expression of the inducible nitric oxide synthase, tumour necrosis factor-alpha and transforming growth factor-beta genes. *Biochemical Pharmacology*, **65**, 1361–1371.
104. Hsu, M.F., Lin, C.N., Lu, M.C. & Wang, J.P. (2004) Inhibition of the arachidonic acid cascade by norathyriol via blockade of cyclooxygenase and lipoxygenase activity in neutrophils. *Naunyn Schmiedebergs Archives of Pharmacology*, **369**, 507–515.
105. Li, J., Malakhova, M., Mottamal, M., Reddy, K., Kurinov, I., Carper, A., Langfald, A., Oi, N., Kim, M.O., Zhu, F., Sosa, C. P., Zhou, K., Bode, A.M. & Dong, Z. (2012) Norathyriol suppresses skin cancers induced by solar ultraviolet radiation by targeting ERK kinases. *Cancer Research*, **72**, 260–270.
106. Noratto, G.D., Bertoldi, M.C., Krenek, K., Talcott, S.T., Stringheta, P.C. & Mertens-Talcott, S.U. (2010) Anticarcinogenic effects of polyphenolics from mango (*Mangifera indica*) varieties. *Journal of the Agricultural and Food Chemistry*, **58**, 4104–4112.
107. Prasad, S., Kalra, N., Singh, M. & Shukla, Y. (2008) Protective effects of lupeol and mango extract against androgen induced oxidative stress in Swiss albino mice. *Asian Journal of Andrology*, **10**, 313–318.
108. Prasad, S., Kalra, N. & Shukla, Y. (2008) Induction of apoptosis by lupeol and mango extract in mouse prostate and LNCaP cells. *Nutrition & Cancer*, **60**, 120–130.

109. Prasad, S., Nigam, N., Kalra, N. & Shukla, Y. (2008) Regulation of signaling pathways involved in lupeol induced inhibition of proliferation and induction of apoptosis in human prostate cancer cells. *Molecular Carcinogenesis*, **47**, 916–924.
110. Saleem, M., Afaq, F., Adhami, V.M. & Mukhtar, H. (2004) Lupeol modulates NF-kappaB and PI3K/Akt pathways and inhibits skin cancer in CD-1 mice. *Oncogene*, **23**, 5203–5214.
111. Masibo, M. & He, Q. (2008) Major mango polyphenols and their potential significance to human health. *Comprehensive Reviews in Food Science and Food Safety*, **7**, 309–319.
112. Bell, C. & Hawthorne, S. (2008) Ellagic acid, pomegranate and prostate cancer: a mini review. *Journal of Pharmacy and Pharmacology*, **60**, 139–144.
113. Edderkaoui, M., Odinokova, I., Ohno, I., Gukovsky, I., Go, V.L., Pandol, S.J. & Gukovskaya, A.S. (2008) Ellagic acid induces apoptosis through inhibition of nuclear factor kappa B in pancreatic cancer cells. *World Journal of Gastroenterology*, **14**, 3672–3680.
114. Mukhtar, H., Das, M. & Bickers, D.R. (1986) Inhibition of 3-methylcholanthrene-induced skin tumorigenicity in BALB/c mice by chronic oral feeding of trace amounts of ellagic acid in drinking water. *Cancer Research*, **46**, 2262–2265.
115. Pedraza-Chaverri, J., Cardenas-Rodriguez, N., Orozco-Ibarra, M. & Perez-Rojas, J.M. (2008) Medicinal properties of mangosteen (*Garcinia mangostana*). *Food and Chemical Toxicology*, **46**, 3227–3239.
116. Mahabusarakam, W., Kuaha, K., Wilairat, P. & Taylor, W.C. (2006) Prenylated xanthones as potential antiplasmodial substances. *Planta Medica*, **72**, 912–916.
117. Suvarnakuta, P., Chaweerungrat, C. & Devahastin, S. (2011) Effects of drying methods on assay and antioxidant activity of xanthones in mangosteen rind. *Food Chemistry*, **125**, 240–247.
118. Zadernowski, R., Czaplicki, S. & Naczk, M. (2009) Phenolic acid profiles of mangosteen fruits (*Garcinia mangostana*). *Food Chemistry*, **112**, 685–689.
119. Pothitirat, W., Chomnawang, M.T., Supabphol, R. & Gritsanapan, W. (2010) Free radical scavenging and anti-acne activities of mangosteen fruit rind extracts prepared by different extraction methods. *Pharmaceutical Biology*, **48**, 182–186.
120. Jung, H.A., Su, B.N., Keller, W.J., Mehta, R.G. & Kinghorn, A.D. (2006) Antioxidant xanthones from the pericarp of *Garcinia mangostana* (mangosteen). *Journal of Agricultural and Food Chemistry*, **54**, 2077–2082.
121. Ho, C.K., Huang, Y.L. & Chen, C.C. (2002) Garcinone E, a xanthone derivative, has potent cytotoxic effect against hepatocellular carcinoma cell lines. *Planta Medica*, **68**, 975–979.
122. Suksamrarn, S., Komutiban, O., Ratananukul, P., Chimnoi, N., Lartpornmatulee, N. & Suksamrarn, A. (2006) Cytotoxic prenylated xanthones from the young fruit of *Garcinia mangostana*. *Chemical and Pharmaceutical Bulletin*, **54**, 301–305.
123. Shibata, M.A., Iinuma, M., Morimoto, J., Kurose, H., Akamatsu, K., Okuno, Y., Akao, Y. & Otsuki, Y. (2011) Alpha-mangostin extracted from the pericarp of the mangosteen (*Garcinia mangostana* Linn) reduces tumor growth and lymph node metastasis in an immunocompetent xenograft model of metastatic mammary cancer carrying a p53 mutation. *BMC Medicine*, **9**, 69.
124. Johnson, J.J., Petiwala, S.M., Syed, D.N., Rasmussen, J.T., Adhami, V.M., Siddiqui, I.A., Kohl, A.M. & Mukhtar, H. (2012) Alpha-mangostin, a xanthone from mangosteen fruit, promotes cell cycle arrest in prostate cancer and decreases xenograft tumor growth. *Carcinogenesis*, **33**, 413–419.
125. Wang, J.J., Sanderson, B.J. & Zhang, W. (2011) Cytotoxic effect of xanthones from pericarp of the tropical fruit mangosteen (*Garcinia mangostana* Linn.) on human melanoma cells. *Food and Chemical Toxicology*, **49**, 2385–2391.
126. Nakatani, K., Yamakuni, T., Kondo, N., Arakawa, T., Oosawa, K., Shimura, S., Inoue, H. & Ohizumi, Y. (2004) Gamma-mangostin inhibits inhibitor-kappaB kinase activity and decreases lipopolysaccharide-induced cyclooxygenase-2 gene expression in C6 rat glioma cells. *Molecular Pharmacology*, **66**, 667–674.
127. Wattenberg, L.W., Coccia, J.B. & Lam, L.K. (1980) Inhibitory effects of phenolic compounds on benzo(*a*)pyrene-induced neoplasia. *Cancer Research*, **40**, 2820–2823.
128. Rajendra-Prasad, N., Karthikeyan, A., Karthikeyan, S. & Reddy, B.V. (2011) Inhibitory effect of caffeic acid on cancer cell proliferation by oxidative mechanism in human HT-1080 fibrosarcoma cell line. *Molecular and Cellular Biochemistry*, **349**, 11–19.
129. Baskaran, N., Manoharan, S., Balakrishnan, S. & Pugalendhi, P. (2010) Chemopreventive potential of ferulic acid in 7,12-dimethylbenz[*a*]anthracene-induced mammary carcinogenesis in Sprague-Dawley rats. *European Journal of Pharmacology*, **637**, 22–29.

130. Balakrishnan, S., Menon, V.P. & Manoharan, S. (2008) Ferulic acid inhibits 7,12-dimethylbenz[*a*]anthracene-induced hamster buccal pouch carcinogenesis. *Journal of Medicinal Food*, **11**, 693–700.
131. Alias, L.M., Manoharan, S., Vellaichamy, L., Balakrishnan, S. & Ramachandran, C.R. (2009) Protective effect of ferulic acid on 7,12-dimethylbenz[*a*]anthracene-induced skin carcinogenesis in Swiss albino mice. *Experimental and Toxicologic Pathology*, **61**, 205–214.
132. Tanaka, T., Kojima, T., Kawamori, T., Wang, A., Suzui, M., Okamoto, K. & Mori, H. (1993) Inhibition of 4-nitroquinoline-1-oxide-induced rat tongue carcinogenesis by the naturally occurring plant phenolics caffeic, ellagic, chlorogenic and ferulic acids. *Carcinogenesis*, **14**, 1321–1325.
133. Kawabata, K., Yamamoto, T., Hara, A., Shimizu, M., Yamada, Y., Matsunaga, K., Tanaka, T. & Mori, H. (2000) Modifying effects of ferulic acid on azoxymethane-induced colon carcinogenesis in F344 rats. *Cancer Letters*, **157**, 15–21.
134. Janicke, B., Hegardt, C., Krogh, M., Onning, G., Akesson, B., Cirenajwis, H.M. & Oredsson, S.M. (2011) The antiproliferative effect of dietary fiber phenolic compounds ferulic acid and p-coumaric acid on the cell cycle of Caco-2 cells. *Nutrition & Cancer*, **63**, 611–622.
135. Serafim, T.L., Carvalho, F.S., Marques, M.P., Calheiros, R., Silva, T., Garrido, J., Milhazes, N., Borges, F., Roleira, F., Silva, E.T., Holy, J. & Oliveira, P.J. (2011) Lipophilic caffeic and ferulic acid derivatives presenting cytotoxicity against human breast cancer cells. *Chemical Research in Toxicology*, **24**, 763–774.
136. Lin, H.H., Chen, J.H., Chou, F.P. & Wang, C.J. (2011) Protocatechuic acid inhibits cancer cell metastasis involving the down-regulation of Ras/Akt/NF-kappaB pathway and MMP-2 production by targeting RhoB activation. *British Journal of Pharmacology*, **162**, 237–254.
137. Yin, M.C., Lin, C.C., Wu, H.C., Tsao, S.M. & Hsu, C.K. (2009) Apoptotic effects of protocatechuic acid in human breast, lung, liver, cervix, and prostate cancer cells: potential mechanisms of action. *Journal of Agricultural and Food Chemistry*, **57**, 6468–6473.
138. Suzuki, R., Kohno, H., Sugie, S. & Tanaka, T. (2003) Dietary protocatechuic acid during the progression phase exerts chemopreventive effects on chemically induced rat tongue carcinogenesis. *Asian Pacific Journal of Cancer Prevention*, **4**, 319–326.
139. Ohnishi, M., Yoshimi, N., Kawamori, T., Ino, N., Hirose, Y., Tanaka, T., Yamahara, J., Miyata, H., & Mori, H. (1997) Inhibitory effects of dietary protocatechuic acid and costunolide on 7,12-dimethylbenz[*a*]anthracene-induced hamster cheek pouch carcinogenesis. *Japan Journal of Cancer Research*, **88**, 111–119.
140. Nakamura, H., Nishikawa, A., Furukawa, F., Kasahara, K., Miyauchi, M., Son, H.Y. & Hirose, M. (2000) Inhibitory effects of protocatechuic acid on the post-initiation phase of hamster pancreatic carcinogenesis induced by *N*-nitrosobis(2-oxopropyl)amine. *Anticancer Research*, **20**, 3423–3427.
141. Jung, S.T., Park, Y.S., Zachwieja, Z., Folta, M., Barton, H., Piotrowicz, J., Katrich, E., Trakhtenberg, S. & Gorinstein, S. (2005) Some essential phytochemicals and the antioxidant potential in fresh and dried persimmon. *International Journal of Food Sciences and Nutrition*, **56**, 105–113.
142. Park, Y.S., Jung, S.T., Kang, S.G., Delgado-Licon, E., Ayala, A.L.M., Tapia, M.S., Martin-Belloso, O., Trakhtenbergh, S. & Gorinstein, S. (2006) Drying of persimmons (*Diospyros kaki* L.) and the following changes in the studied bioactive compounds and the total radical scavenging activities. *Lwt-Food Science and Technology*, **39**, 748–755.
143. Kim, E.O., Lee, H., Cho, C.H., Kim, Y.J. & Kim, D.O. (2011) Antioxidant capacity and anti-inflammatory effect of the ethyl acetate fraction of dried persimmon (*Diospyros kaki* Thumb.) on THP-1 human acute monocytic leukemia cell line. *Journal of Korean Society of Applied Biological Chemistry*, **54**, 606–611.
144. Park, O.J. & Surh, Y.-J. (2004) Chemopreventive potential of epigallocatechin gallate and genistein: evidence from epidemiological and laboratory studies. *Toxicology Letters*, **150**, 43–56.
145. Meeran, S.M., Mantena, S.K., Elmets, C.A. & Katiyar, S.K. (2006) (-)-Epigallocatechin-3-gallate prevents photocarcinogenesis in mice through interleukin-12-dependent DNA repair. *Cancer Research*, **66**, 5512–5520.
146. Lu, Y.P., Lou, Y.R., Xie, J.G., Peng, Q.Y., Liao, J., Yang, C.S., Huang, M.T. & Conney, A.H. (2002) Topical applications of caffeine or (-)-epigallocatechin gallate (EGCG) inhibit carcinogenesis and selectively increase apoptosis in UVB-induced skin tumors in mice. *Proceedings of the National Academy of Sciences USA*, **99**, 12455–12460.

147. Kundu, J.K., Na, H.K., Chun, K.S., Kim, Y.K., Lee, S.J., Lee, S.S., Lee, O.S., Sim, Y.C. & Surh, Y.-J. (2003) Inhibition of phorbol ester-induced COX-2 expression by epigallocatechin gallate in mouse skin and cultured human mammary epithelial cells. *Journal of Nutrition*, **133**, 3805S–3810S.
148. Kundu, J.K. & Surh, Y.-J. (2007) Epigallocatechin gallate inhibits phorbol ester-induced activation of NF-κB and CREB in mouse skin: role of p38 MAPK. *Annals of the New York Academy of Sciences*, **1095**, 504–512.
149. Nomura, M., Ma, W., Chen, N., Bode, A.M. & Dong, Z. (2000) Inhibition of 12-*O*-tetradecanoylphorbol-13-acetate-induced NF-κB activation by tea polyphenols, (-)-epigallocatechin gallate and theaflavins. *Carcinogenesis*, **21**, 1885–1890.
150. Yang, W.H., Fong, Y.C., Lee, C.Y., Jin, T.R., Tzen, J.T., Li, T.M. & Tang, C.-H. (2011) Epigallocatechin-3-gallate induces cell apoptosis of human chondrosarcoma cells through apoptosis signal-regulating kinase 1 pathway. *Journal of Cellular Biochemistry*, **112**, 1601–1611.
151. Lee, M.H., Han, D.W., Hyon, S.H. & Park, J.C. (2011) Apoptosis of human fibrosarcoma HT-1080 cells by epigallocatechin-3-*O*-gallate via induction of p53 and caspases as well as suppression of Bcl-2 and phosphorylated nuclear factor-kappaB. *Apoptosis*, **16**, 75–85.
152. Rao, S.D. & Pagidas, K. (2010) Epigallocatechin-3-gallate, a natural polyphenol, inhibits cell proliferation and induces apoptosis in human ovarian cancer cells. *Anticancer Research*, **30**, 2519–2523.
153. Hwang, J.T., Ha, J., Park, I.J., Lee, S.K., Baik, H.W., Kim, Y.M. & Park, O.J. (2007) Apoptotic effect of EGCG in HT-29 colon cancer cells via AMPK signal pathway. *Cancer Letters*, **247**, 115–121.
154. Ahn, W.S., Huh, S.W. & Bae, S.M. (2003) A major constituent of green tea, EGCG, inhibits the growth of a human cervical cancer cell line, CaSki cells, through apoptosis, G1 arrest, and regulation of gene expression. *DNA Cell Biology*, **22**, 217–224.
155. Brusselmans, K., De Schrijver, E., Heyns, W., Verhoeven, G. & Swinnen, J.V. (2003) Epigallocatechin-3-gallate is a potent natural inhibitor of fatty acid synthase in intact cells and selectively induces apoptosis in prostate cancer cells. *International Journal of Cancer*, **106**, 856–862.
156. Zhou, C., Sheng, Y., Zhao, D., Wang, Z. & Tao, J. (2010) Variation of oleanolic and ursolic acid in the flesh of persimmon fruit among different cultivars. *Molecules*, **15**, 6580–6587.
157. Shan, J.Z., Xuan, Y.Y., Ruan, S.Q. & Sun, M. (2011) Proliferation-inhibiting and apoptosis-inducing effects of ursolic acid and oleanolic acid on multi-drug resistance cancer cells *in vitro*. *Chinese Journal of Integrative Medicine*, **17**, 607–611.
158. Shyu, M.H., Kao, T.C. & Yen, G.C. (2010) Oleanolic acid and ursolic acid induce apoptosis in HuH7 human hepatocellular carcinoma cells through a mitochondrial-dependent pathway and downregulation of XIAP. *Journal of Agricultural and Food Chemistry*, **58**, 6110–6118.
159. Yan, S.L., Huang, C.Y., Wu, S.T. & Yin, M.C. (2010) Oleanolic acid and ursolic acid induce apoptosis in four human liver cancer cell lines. *Toxicology In Vitro*, **24**, 842–848.
160. Kim, K.H., Seo, H.S., Choi, H.S., Choi, I., Shin, Y.C. & Ko, S.G. (2011) Induction of apoptotic cell death by ursolic acid through mitochondrial death pathway and extrinsic death receptor pathway in MDA-MB-231 cells. *Archives of Pharmcal Research*, **34**, 1363–1372.
161. Shanmugam, M.K., Rajendran, P., Li, F., Nema, T., Vali, S., Abbasi, T., Kapoor, S., Sharma, A., Kumar, A.P., Ho, P.C., Hui, K.M. & Sethi, G. (2011) Ursolic acid inhibits multiple cell survival pathways leading to suppression of growth of prostate cancer xenograft in nude mice. *Journal of Molecular Medicine*, **89**, 713–727.
162. Shanmugam, M.K., Manu, K.A., Ong, T.H., Ramachandran, L., Surana, R., Bist, P., Lim, L.H., Kumar, A.P., Hui, K.M. & Sethi, G. (2011) Inhibition of CXCR4/CXCL12 signaling axis by ursolic acid leads to suppression of metastasis in transgenic adenocarcinoma of mouse prostate model. *International Journal of Cancer*, **129**, 1552–1563.
163. Sindhu, E.R., Preethi, K.C. & Kuttan, R. (2010) Antioxidant activity of carotenoid lutein *in vitro* and *in vivo*. *Indian Journal of Experimental Biology*, **48**, 843–848.
164. Gunasekera, R.S., Sewgobind, K., Desai, S., Dunn, L., Black, H.S., McKeehan, W.L. & Patil, B. (2007) Lycopene and lutein inhibit proliferation in rat prostate carcinoma cells. *Nutrition & Cancer*, **58**, 171–177.
165. Huang, J.P., Zhang, M., Holman, C.D. & Xie, X. (2007) Dietary carotenoids and risk of breast cancer in Chinese women. *Asia Pacific Journal of Clinical Nutrition*, **16**, 437–442.
166. Larsson, S.C., Bergkvist, L., Naslund, I., Rutegard, J. & Wolk, A. (2007) Vitamin A, retinol, and carotenoids and the risk of gastric cancer: a prospective cohort study. *American Journal of Clinical Nutrition*, **85**, 497–503.

167. Tsuchiya, H., Yamada, K., Kato, H., Hayashi, H., Miyazaki, T. & Hayashi, T. (1995) High-performance liquid-chromatographic analysis of tetrahydro-β-carbolines in food plants. *Phytochemical Analysis*, **6**, 297–301.
168. Jenkins, P.R., Wilson, J., Emmerson, D., Garcia, M.D., Smith, M.R., Gray, S.J., Britton, R.G., Mahale, S. & Chaudhuri, B. (2008) Design, synthesis and biological evaluation of new tryptamine and tetrahydro-β-carboline-based selective inhibitors of CDK4. *Bioorganic Medicinal Chemistry*, **16**, 7728–7739.
169. Bemis, D.L., Capodice, J.L., Gorroochurn, P., Katz, A.E. & Buttyan, R. (2006) Anti-prostate cancer activity of a beta-carboline alkaloid enriched extract from *Rauwolfia vomitoria*. *International Journal of Oncology*, **29**, 1065–1073.
170. Beljanski, M. & Beljanski, M.S. (1982) Selective inhibition of *in vitro* synthesis of cancer DNA by alkaloids of β-carboline class. *Experimental Cell Biology*, **50**, 79–87.
171. Kayano, S., Kikuzaki, H., Fukutsuka, N., Mitani, T. & Nakatani, N. (2002) Antioxidant activity of prune (*Prunus domestica* L.) constituents and a new synergist. *Journal of Agricultural and Food Chemistry*, **50**, 3708–3712.
172. Huang, M.T., Smart, R.C., Wong, C.Q. & Conney, A.H. (1988) Inhibitory effect of curcumin, chlorogenic acid, caffeic acid, and ferulic acid on tumor promotion in mouse skin by 12-*O*-tetradecanoylphorbol-13-acetate. *Cancer Research*, **48**, 5941–5946.
173. Shimizu, M., Yoshimi, N., Yamada, Y., Matsunaga, K., Kawabata, K., Hara, A., Moriwaki, H. & Mori, H. (1999) Suppressive effects of chlorogenic acid on *N*-methyl-*N*-nitrosourea-induced glandular stomach carcinogenesis in male F344 rats. *Journal of Toxicological Sciences*, **24**, 433–439.
174. Feng, R., Lu, Y., Bowman, L.L., Qian, Y., Castranova, V. & Ding, M. (2005) Inhibition of activator protein-1, NF-κB, and MAPKs and induction of phase 2 detoxifying enzyme activity by chlorogenic acid. *Journal of Biological Chemistry*, **280**, 27888–27895.
175. Bandyopadhyay, G., Biswas, T., Roy, K.C., Mandal, S., Mandal, C., Pal, B.C., Bhattacharya, S., Rakshit, S., Bhattacharya, D.K., Chaudhuri, U., Konar, A. & Bandyopadhyay, S. (2004) Chlorogenic acid inhibits Bcr-Abl tyrosine kinase and triggers p38 mitogen-activated protein kinase-dependent apoptosis in chronic myelogenous leukemic cells. *Blood*, **104**, 2514–2522.
176. Kimura, Y., Ito, H., Kawaji, M., Ikami, T. & Hatano, T. (2008) Characterization and antioxidative properties of oligomeric proanthocyanidin from prunes, dried fruit of *Prunus domestica* L. *Bioscience, Biotechnology and Biochemistry*, **72**, 1615–1618.
177. Rossi, M., Lugo, A., Lagiou, P., Zucchetto, A., Polesel, J., Serraino, D., Negri, E., Trichopoulos, D. & La Vecchia, C. (2012) Proanthocyanidins and other flavonoids in relation to pancreatic cancer: a case-control study in Italy. *Annals of Oncology*, **23**, 1488–1493.
178. Rossi, M., Rosato, V., Bosetti, C., Lagiou, P., Parpinel, M., Bertuccio, P., Negri, E. & La Vecchia, C. (2010) Flavonoids, proanthocyanidins, and the risk of stomach cancer. *Cancer Causes and Control*, **21**, 1597–1604.
179. Rossi, M., Negri, E., Parpinel, M., Lagiou, P., Bosetti, C., Talamini, R., Montella, M., Giacosa, A., Franceschi, S. & La Vecchia, C. (2010) Proanthocyanidins and the risk of colorectal cancer in Italy. *Cancer Causes and Control*, **21**, 243–250.
180. Sharma, S.D., Meeran, S.M. & Katiyar, S.K. (2010) Proanthocyanidins inhibit *in vitro* and *in vivo* growth of human non-small cell lung cancer cells by inhibiting the prostaglandin E_2 and prostaglandin E_2 receptors. *Molecular Cancer Therapeutics*, **9**, 569–580.
181. Nandakumar, V., Singh, T. & Katiyar, S.K. (2008) Multi-targeted prevention and therapy of cancer by proanthocyanidins. *Cancer Letters*, **269**, 378–387.
182. Howe, G.R., Benito, E., Castelleto, R., Cornee, J., Esteve, J., Gallagher, R.P., Iscovich, J.M., Deng-ao, J., Kaaks, R. & Kune, G.A. (1992) Dietary intake of fiber and decreased risk of cancers of the colon and rectum: evidence from the combined analysis of 13 case-control studies. *Journal of the National Cancer Institute*, **84**, 1887–1896.
183. Reddy, B.S., Engle, A., Simi, B., O'Brien, L.T., Barnard, R.J., Pritikin, N. & Wynder, E.L. (1988) Effect of low-fat, high-carbohydrate, high-fiber diet on fecal bile acids and neutral sterols. *Preventive Medicine*, **17**, 432–439.
184. Tinker, L.F., Schneeman, B.O., Davis, P.A., Gallaher, D.D. & Waggoner, C.R. (1991) Consumption of prunes as a source of dietary fiber in men with mild hypercholesterolemia. *American Journal of Clinical Nutrition*, **53**, 1259–1265.

185. Fujii, T., Ikami, T., Xu, J.W. & Ikeda, K. (2006) Prune extract (*Prunus domestica* L.) suppresses the proliferation and induces the apoptosis of human colon carcinoma Caco-2. *Journal of Nutritional Science and Vitaminology*, **52**, 389–391.
186. Yang, Y. & Gallaher, D.D. (2005) Effect of dried plums on colon cancer risk factors in rats. *Nutrition & Cancer*, **53**, 117–125.
187. Camire, M.E. & Dougherty, M.P. (2003) Raisin dietary fiber composition and *in vitro* bile acid binding. *Journal of Agricultural and Food Chemistry*, **51**, 834–837.
188. Morgan, A.F., Kimmel, L., Field, A. & Nichols, P.F. (1935) The vitamin content of sultanina (Thompson seedless) grapes and raisins. *Journal of Nutrition*, **9**, 369–382.
189. Karadeniz, F., Durst, R.W. & Wrolstad, R.E. (2000) Polyphenolic composition of raisins. *Journal of Agricultural and Food Chemistry*, **48**, 5343–5350.
190. Zhao, B. & Hall, C.A. (2008) Composition and antioxidant activity of raisin extracts obtained from various solvents. *Food Chemistry*, **108**, 511–518.
191. Pezzuto, J.M. (2008) Grapes and human health: a perspective. *Journal of Agricultural and Food Chemistry*, **56**, 6777–6784.
192. Reinli, K. & Block, G. (1996) Phytoestrogen content of foods: a compendium of literature values. *Nutrition & Cancer*, **26**, 123–148.
193. Rankin, J.W., Andreae, M.C., Chen, C.Y.O. & O'Keefe, S.F. (2008) Effect of raisin consumption on oxidative stress and inflammation in obesity. *Diabetes, Obesity and Metabolism*, **10**, 1086–1096.
194. Spiller, G.A., Story, J.A., Furumoto, E.J., Chezem, J.C. & Spiller, M. (2003) Effect of tartaric acid and dietary fibre from sun-dried raisins on colonic function and on bile acid and volatile fatty acid excretion in healthy adults. *British Journal of Nutrition*, **90**, 803–807.
195. Collins, A.R., Harrington, V., Drew, J. & Melvin, R. (2003) Nutritional modulation of DNA repair in a human intervention study. *Carcinogenesis*, **24**, 511–515.
196. Motohashi, N., Shirataki, Y., Kawase, M., Tani, S., Sakagami, H., Satoh, K., Kurihara, T., Nakashima, H., Mucsi, I., Varga, A. & Molnar, J. (2002) Cancer prevention and therapy with kiwifruit in Chinese folklore medicine: a study of kiwifruit extracts. *Journal of Ethnopharmacology*, **81**, 357–364.
197. Fiorentino, A., D'Abrosca, B., Pacifico, S., Mastellone, C., Scognamiglio, M. & Monaco, P. (2009) Identification and assessment of antioxidant capacity of phytochemicals from kiwi fruits. *Journal of Agricultural and Food Chemistry*, **57**, 4148–4155.
198. Collins, B.H., Horska, A., Hotten, P.M., Riddoch, C. & Collins, A.R. (2001) Kiwifruit protects against oxidative DNA damage in human cells and *in vitro*. *Nutrition & Cancer*, **39**, 148–153.
199. Lee, D.E., Shin, B.J., Hur, H.J., Kim, J.H., Kim, J., Kang, N.J., Kim, D.O., Lee, C.Y., Lee, K.W. & Lee, H.J. (2010) Quercetin, the active phenolic component in kiwifruit, prevents hydrogen peroxide-induced inhibition of gap-junction intercellular communication. *British Journal of Nutrition*, **104**, 164–170.
200. Munzuroglu, O., Karatas, F. & Geckil, H. (2003) The vitamin and selenium contents of apricot fruit of different varieties cultivated in different geographical regions. *Food Chemistry*, **83**, 205–212.
201. Jackson, M.I. & Combs, G.F. Jr. (2008) Selenium and anticarcinogenesis: underlying mechanisms. *Current Opinion in Clinical Nutrition & Metabolic Care*, **11**, 718–726.
202. Nakagawa, A., Sawada, T., Okada, T., Ohsawa, T., Adachi, M. & Kubota, K. (2007) New antineoplastic agent, MK615, from UME (a variety of) Japanese apricot inhibits growth of breast cancer cells *in vitro*. *Breast Journal*, **13**, 44–49.
203. Enomoto, S., Yanaoka, K., Utsunomiya, H., Niwa, T., Inada, K., Deguchi, H., Ueda, K., Mukoubayashi, C., Inoue, I., Maekita, T., Nakazawa, K., Iguchi, M., Arii, K., Tamai, H., Yoshimura, N., Fujishiro, M., Oka, M. & Ichinose, M. (2010) Inhibitory effects of Japanese apricot (*Prunus mume* Siebold et Zucc.; Ume) on *Helicobacter pylori*-related chronic gastritis. *European Journal of Clinical Nutrition*, **64**, 714–719.
204. Yurt, B. & Celik, I. (2011) Hepatoprotective effect and antioxidant role of sun, sulphited-dried apricot (*Prunus armeniaca* L.) and its kernel against ethanol-induced oxidative stress in rats. *Food and Chemical Toxicology*, **49**, 508–513.
205. Wang, J., Wang, X., Jiang, S., Lin, P., Zhang, J., Lu, Y., Wang, Q., Xiong, Z., Wu, Y., Ren, J. & Yang, H. (2008) Cytotoxicity of fig fruit latex against human cancer cells. *Food and Chemical Toxicology*, **46**, 1025–1033.

206. Slatnar, A., Klancar, U., Stampar, F. & Veberic, R. (2011) Effect of drying of figs (*Ficus carica* L.) on the contents of sugars, organic acids, and phenolic compounds. *Journal of Agricultural and Food Chemistry*, **59**, 11696–11702.
207. Iamanaka, B.T., Taniwaki, M.H., Menezes, H.C., Vicente, E. & Fungaro, M.H. (2005) Incidence of toxigenic fungi and ochratoxin A in dried fruits sold in Brazil. *Food Additives and Contaminants*, **22**, 1258–1263.
208. Zinedine, A., Soriano, J.M., Juan, C., Mojemmi, B., Molto, J.C., Bouklouze, A., Cherrah, Y., Idrissi, L., El-Aouad, R. & Manes, J. (2007) Incidence of ochratoxin A in rice and dried fruits from Rabat and Sale area, Morocco. *Food Additives and Contaminants*, **24**, 285–291.
209. Trucksess, M.W. & Scott, P.M. (2008) Mycotoxins in botanicals and dried fruits: a review. *Food Additives and Contaminants. Part A, Chemistry, Analysis, Control, Exposure and Risk Assessments*, **25**, 181–192.
210. Prior, R.L., Wu, X. & Gu, L. (2006) Identification and urinary excretion of metabolites of 5-(hydroxymethyl)-2-furfural in human subjects following consumption of dried plums or dried plum juice. *Journal of Agricultural and Food Chemistry*, **54**, 3744–3749.
211. Stich, H.F., Rosin, M.P., Wu, C.H. & Powrie, W.D. (1981) Clastogenic activity of dried fruits. *Cancer Letters*, **12**, 1–8.

Part 1
Dried Berries

3 Phytochemicals and health benefits of blackberries and black currants

Haiming Shi and Liangli (Lucy) Yu

3.1 Introduction

Berry fruits are widely consumed not only in fresh and frozen forms but also as processed or derived products including juices, jams, wines, jellies, and canned fruits. Recently, there has been a growing trend toward utilization of berry extracts in functional foods and dietary supplements for health promotion and disease risk reduction [1–3].

Blackberries (genus *Rubus*, Rosaceae) are widely distributed worldwide and their overall consumption has increased rapidly in the past 10 years [4]. Besides vitamins and dietary fiber, blackberries are rich in phenolic compounds including anthocyanins, flavonols, tannins, and phenolic acids [5]. Numerous studies have shown that these phenolic compounds may possess various biological benefits associated with their antioxidant, anticancer, antineurodegenerative, and anti-inflammatory properties [1, 3, 5]. On the other hand, black currants (*Ribes nigrum* L.) are a species of *Ribes* berry native to central and northern Europe and northern Asia. Black currants are one of the richest sources of antioxidants, notably anthocyanins, hydroxycinnamic acids, and vitamin C [6]. Their extracts have been shown to inhibit the development of certain types of cancers, cardiovascular disease (CVD), and chronic inflammation-related diseases [7, 8].

This chapter highlights the compositional and nutritional characteristics, phytochemicals, health effects, and food applications of fresh blackberries and black currants with particular references to their dried counterparts when data are available.

3.2 Compositional and nutritional characteristics of blackberries and black currants

The nutrient value in fresh blackberries and black currants has been reported by United States Department of Agriculture (USDA) [9] (Table 3.1). The values may vary in fruits growing at different locations and in different years.

Dried Fruits: Phytochemicals and Health Effects, First Edition. Edited by Cesarettin Alasalvar and Fereidoon Shahidi.

Table 3.1 Compositional and nutritional characteristics of blackberries and black currants (values in per 100 grams edible portion)

Nutrient	Units	Fresh blackberries	Fresh black currants
Proximate composition			
Water	g	88.15	81.96
Energy	kcal	43	63
Protein	g	1.39	1.40
Lipid	g	0.49	0.41
Ash	g	na	na
Carbohydrate	g	9.61	15.38
Dietary fiber	g	5.3	na
Sugars	g	4.88	na
Minerals			
Calcium	mg	29.0	55
Copper	mg	na	na
Iron	mg	0.62	1.54
Magnesium	mg	20.0	24.0
Manganese	mg	na	na
Phosphorus	mg	22.0	59.0
Potassium	mg	162	322
Selenium	μg	na	na
Sodium	mg	1.0	2.0
Zinc	mg	0.53	0.27
Vitamins			
Betaine	mg	na	na
Choline	mg	na	na
Folate (DFE)	μg	25.0	na
Niacin	mg	0.65	0.30
Pantothenic acid	μg	nd	nd
Pyridoxine	mg	0.03	0.07
Riboflavin	mg	0.03	0.05
Thiamin	mg	0.02	0.05
Vitamin A (RAE)	μg	11.0	12.0
Vitamin C	mg	21.0	181
Vitamin E (ATE)	mg	1.17	1.0
Vitamin K	μg	19.8	na

Source: Adapted from USDA database [9].
Note: Some numbers are rounded to the second digit after decimal point.
DFE, dietary folate equivalents; RAE, retinol activity equivalents; ATE, alpha-tocopherol equivalents; na, not available; nd, not detected.

3.2.1 Dietary fiber

Blackberries and black currants contain high levels of dietary fiber. The total fiber content of frozen blackberries is 7.1 g/100 g fresh weight, which is higher than other berry species [10]. Forty-two food products, including vegetables and fruits from Belgium, were examined for dietary fiber and the second highest fiber content in fruits was detected in blackberries (4.87 g/100 g) [11]. Chokeberries, bilberries, and black currants collected at maturity were analyzed for acid-detergent fiber [12]. The highest fiber concentration was found in black currants.

3.2.2 Vitamin C

Vitamin C is widely distributed in many fruits and vegetables and is well known for its antioxidant activity, as it acts as a reducing agent to reverse oxidation in food and biological systems. Kafkas *et al.* [13] reported the vitamin C content of five blackberry genotypes from Turkey by high-performance liquid chromatographic (HPLC) analysis. Vitamin C was detected in "C. Thornless" (2.5 mg/g extract), "Bursa 2" (4.6 mg/g extract), and "Loch Ness" (14.9 mg/g extract), but not in "Navaho" or "Jumbo" cultivars. In addition, 11 types of blackberries, harvested in different seasons and regions in Mexico and in the United States, were collected and their vitamin C content was determined [14]. Levels of vitamin C ranged from 0.82 mg/g in "Brazos" (from Zitácuaro harvested in spring) to 0.08 mg/g in "Comanche" (from autumn harvest in Zirahuen). Several studies reported the vitamin C content in black currants. Seventeen UK-grown black currant cultivars were analyzed for their vitamin C content [6]. The concentration of vitamin C ranged from 1.92 to 5.41 mg/g fresh weight, which is much higher compared with that in other commonly consumed berry fruits in which the vitamin C content is commonly >1 mg/g fresh weight, confirming that black currants are a good source of vitamin C. In summary, both blackberries and black currants are excellent dietary sources of vitamin C and different cultivars may differ significantly in their vitamin C concentrations.

3.2.3 Vitamin E

Vitamin E is found in many fruits including blackberries and black currants. Chun *et al.* [15] determined tocopherol and tocotrienol levels in fruits and vegetables from the United States. The α-, β-, γ-, and δ-tocopherol levels of blackberries were 1.43, 0.04, 1.42, and 0.85 mg/100 g edible weight, respectively, while tocotrienols were not detected. However, the information on vitamin E composition of black currants is lacking. Overall, blackberries contained much higher levels of vitamin E than some other commonly consumed berries.

3.2.4 Sugars

Kafkas *et al.* [13] analyzed the major soluble sugar contents of five blackberry genotypes from Turkey by HPLC. They found that fructose was the most abundant sugar and its level was greatest in the "Navaho." Sugar was also examined for 52 blackberry samples [16]. Sucrose, glucose, and fructose were found in blackberries. Sucrose levels varied greatly, ranging from 0 to 12.9% of total sugars, with a mean of 4.6%. Glucose and fructose were the dominant sugars present in nearly equal quantities, with the overall glucose to fructose ratio ranging from 0.81 to 1.17. In addition, Bordonaba and Terry [6] reported the sugar composition of 17 UK-grown black currant cultivars. Sucrose, glucose, and fructose were identified in all cultivars, and concentrations ranged between 3.38 and 36.89, 36.86 and 82.78, and 43.43 and 85.73 mg/g fresh weight, respectively. The proportion of each sugar also varied with genotype. Glucose and fructose were the dominant sugars found in all black currant cultivars, making up 40 and 49% of total sugars, respectively. Sucrose, on the other hand, made up 10% of the total sugar concentration but was more variable among cultivars. Impact of latitude and weather conditions on the sugar composition of black currants in Finland was reported [17]. Thus, the level of sugars in blackberries and black currants is influenced by genetic and environmental factors, including cultivar, maturity, and weather conditions.

3.2.5 Minerals

Wu *et al.* [18] determined the content of minerals, including Ca, Fe, K, Mg, Na, Se, and Zn, in five blackberry cultivars ("Navaho," "Young," "Brazos," "Boysen," and "Triple Crown") in China. The levels of Ca, K, and Mg were higher than other determined minerals, and K was present in the highest amount and above 1.0 mg/g fresh weight in all tested blackberry samples, except "Boysen." There was no significant difference ($P > 0.05$) in the mineral contents among the five blackberry cultivars. Furthermore, Ilkay *et al.* [19] reported little change in Ca, K, Mg, and Zn concentrations during ripening of blackberries. In addition, Huo *et al.* [20] reported that black currants contained Ca (0.38 mg/kg), Cu (1.1 mg/kg), Fe (8.7 mg/kg), K (3.04 mg/kg), Mg (0.07 mg/kg), Mn (0.7 mg/kg), Na (0.047 mg/kg), and Zn (23.6 mg/kg).

3.2.6 Amino acids

Wu *et al.* [18] reported the amino acid content of five blackberry cultivars collected in China. All samples contained eight essential amino acids and had different total amino acid contents, ranging from 7.91 to 15.21 mg/g fresh weight. According to the review by Huo *et al.* [20], black currants also contained eight essential amino acids, and histidine and lysine concentrations were 108.2 and 64.6 mg/kg fresh weight, respectively.

3.3 Phytochemicals in blackberries and black currants

3.3.1 Flavonoids

Flavonoids are polyphenolic compounds and represent a large group of secondary plant metabolites. Structurally, flavonoids are usually characterized by a C_6-C_3-C_6 carbon skeleton, which may occur as aglycones or glycosides. An inverse association between the intake of flavonoids and the risk of chronic diseases has been shown in epidemiological studies [21]. Flavonoids are found in a wide variety of fruits and vegetables. Blackberries and black currants are good dietary sources of flavonoids.

3.3.1.1 *Anthocyanins*

Anthocyanins are glycosides of polyhydroxy and polymethoxy derivatives of 2-phenylbenzopyrylium or flavylium salts. They have been extensively studied in fruits and vegetables due to their health benefits and potential as natural food colorants. Blackberries and black currants are rich sources of anthocyanins. There is growing evidence that various biological activities of blackberries and black currants are correlated with their anthocyanins contents.

Cyanidin-3-*O*-glucoside (**1**) and cyanidin-3-*O*-rutinoside (**2**) have long been recognized as the respective major and minor anthocyanins in blackberries [22] (Figure 3.1). Sapers *et al.* [23] reported the presence of cyanidin-3-*O*-xyloside (**3**) along with two cyanidin derivatives

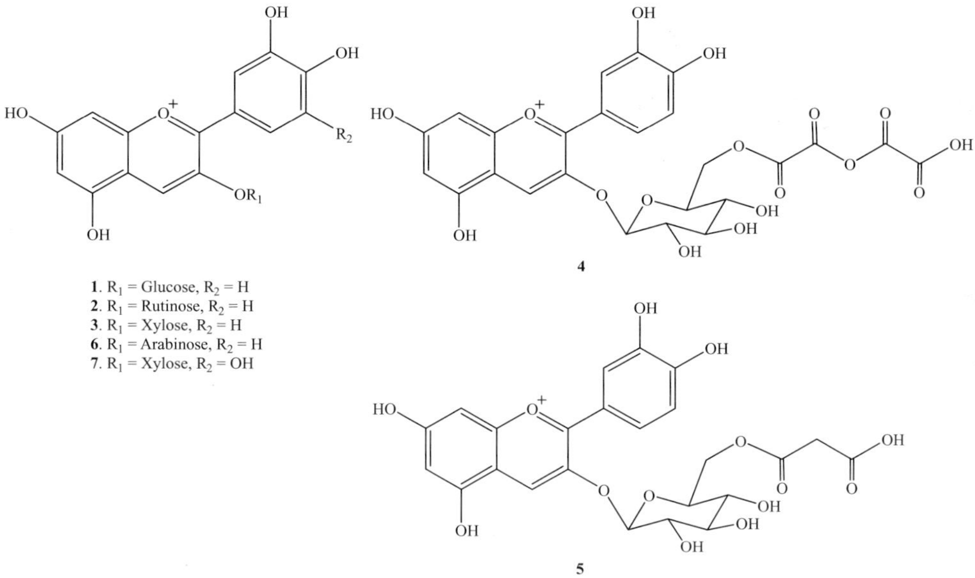

Figure 3.1 Structures of anthocyanidins isolated from blackberries.

that were vaguely characterized as being substituted with dicarboxylic acids. Later, Fan-Chiang [24] identified one of these anthocyanins as cyanidin 3-*O*-glucoside acylated with malonic acid (**4**) using HPLC–mass spectrometry (HPLC-MS) and chemical reactions. A novel zwitterionic anthocyanin was also isolated from blackberries and structurally characterized as cyanidin 3-*O*-(6″-*O*-dioxalylglucoside) (**5**) [25]. Recently, two minor anthocyanins were also reported in blackberry extracts: one was identified as cyanidin-3-*O*-arabinoside (**6**) and the other one was tentatively deduced as delphinidin-3-*O*-xyloside (**7**) by HPLC-MS analysis [26].

Anthocyanins were also reported in black currants. 3-*O*-glucosides of pelargonidin (**8**), cyanidin (**9**), peonidin (**10**), delphinidin (**11**), petunidin (**12**), and malvidin (**13**); the 3-*O*-rutinosides of pelargonidin (**14**), cyanidin (**15**), peonidin (**16**), delphinidin (**17**), petunidin (**18**), and malvidin (**19**); cyanidin 3-*O*-arabinoside (**6**); and the 3-*O*-(6″-*p*-coumaroylglucoside)s of cyanidin (**20**) and delphinidin (**21**) were found in black currants (Figure 3.2). The four major anthocyanins, the 3-*O*-glucosides and the 3-*O*-rutinosides of delphinidin and cyanidin, made up >97% of the total anthocyanin content [27]. In addition, Borges *et al.* [28] found 11 anthocyanins in a commercial black currant extract including three minor anthocyanins delphinidin-3-*O*-galactoside (**22**), peonidin-3-*O*-galactoside (**23**), and malvidin-3-*O*-galactoside (**24**), which were tentatively identified on the basis of their MS fragmentation profiles. Recently, two anthocyanins petunidin-3-*O*-(6″-coumaroyl) glucoside (**25**) and peonidin-3-*O*-(6″-coumaroyl) glucoside (**26**) were detected in black currants for the first time [29].

The anthocyanin contents in blackberries and black currants might vary with genotype, harvest season, growing location, maturity stage, and other preharvest factors. Bowen-Forbes *et al.* [30] reported that a wild Jamaican-grown blackberry species, *Rubus jamaicensis*,

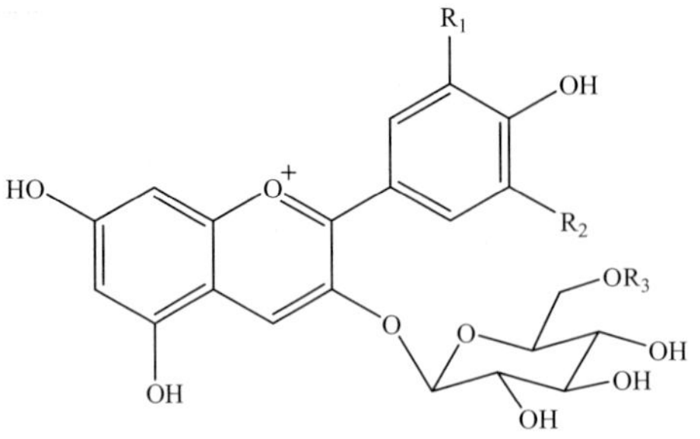

8. $R_1 = H, R_2 = H, R_3 = H$
9. $R_1 = OH, R_2 = H, R_3 = H$
10. $R_1 = OCH_3, R_2 = H, R_3 = H$
11. $R_1 = OH, R_2 = OH, R_3 = H$
12. $R_1 = OCH_3, R_2 = OH, R_3 = H$
13. $R_1 = OCH_3, R_2 = OCH_3, R_3 = H$
14. $R_1 = H, R_2 = H, R_3 =$ Rhamnose
15. $R_1 = OH, R_2 = H, R_3 =$ Rhamnose
16. $R_1 = OCH_3, R_2 = H, R_3 =$ Rhamnose
17. $R_1 = OH, R_2 = OH, R_3 =$ Rhamnose
18. $R_1 = OCH_3, R_2 = OH, R_3 =$ Rhamnose
19. $R_1 = OCH_3, R_2 = OCH_3, R_3 =$ Rhamnose

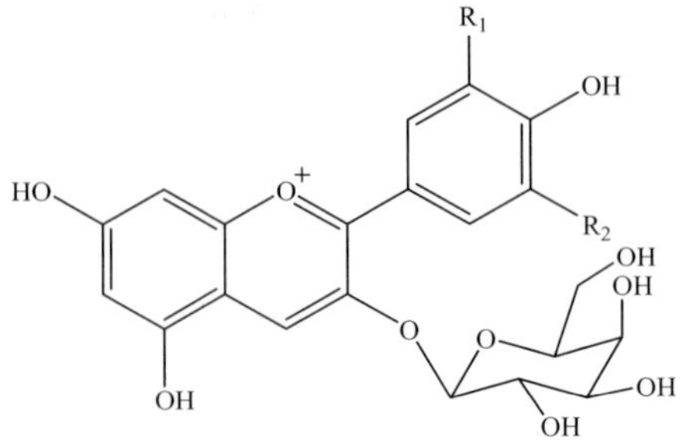

22. $R_1 = OH, R_2 = OH$
23. $R_1 = OCH_3, R_2 = H$
24. $R_1 = OCH_3, R_2 = OCH_3$

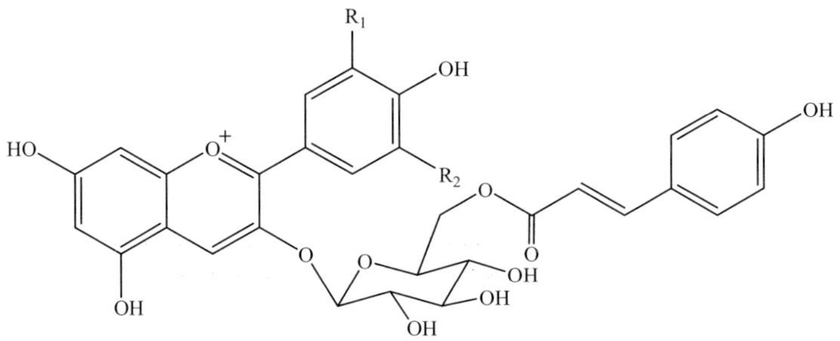

20. $R_1 = OH, R_2 = H$
21. $R_1 = OH, R_2 = OH$
25. $R_1 = OCH_3, R_2 = OH$
26. $R_1 = OCH_3, R_2 = H$

Figure 3.2 Structures of anthocyanidins isolated from black currants.

possessed 1673 mg anthocyanins/100 g fresh weight. Fan-Chiang and Wrolstad [31] reported the anthocyanin contents of 51 blackberry samples at 70.3–201 mg of cyanidin-3-*O*-glucoside equivalents (C3GE)/100 g with a mean value of 137 mg of C3GE/100 g, representing 18 different blackberry varieties and 20 selections from five different geographic locations and three harvest seasons. In addition, Siriwoharn *et al.* [32] observed that the anthocyanin contents in Marion and Evergreen blackberries increased considerably during the course of ripening, which is consistent with previous investigations [23].

The anthocyanin contents of 17 UK-grown black currant cultivars were analyzed [6]. Relative concentrations of cyaniding-3-*O*-glucoside, cyaniding-3-*O*-rutinoside, delphinidin-3-*O*-glucoside, and delphinidin-3-*O*-rutinoside were 3.1–7.9, 35.4–47.0, 7.6–12.5, and 36.9–50.9%, respectively. In an earlier study, Anttonen and Karjalainen [33] compared the anthocyanin profiles of organically and conventionally grown black currants from commercial farms within a climatically similar area. Significant differences ($P < 0.05$) were found between farms for all other three compounds but not for cyanidin-3-*O*-rutinoside. The data from these research findings also showed that cultivation technique was not a major factor in determining the anthocyanin contents of black currants.

3.3.1.2 *Flavonols*

Flavonols are a class of flavonoids that have the 3-hydroxyflavone backbone. Besides anthocyanins, flavonols from berries are considered as another group of compounds that play an important role against oxidative damage in biological systems due to their potential ability to scavenge free radicals. Cho *et al.* [34] identified eight flavonols of quercetin and quercetin–sugar conjugates in Kiowa blackberries, namely, rutinoside (**27**), galactoside (**28**), methoxyhexoside (**29**), glucoside (**30**), pentoside (**31**), [6″-(3-hydroxy-3-methylglutaroyl)]-β-galactoside (**32**), glucosylpentoside (**33**), and oxalylpentoside (**34**) (Figure 3.3). They also quantified the contents of individual flavonols in six blackberry genotypes. Quercetin 3-*O*-galactoside was the predominant flavonol in the "Apache," "Arapaho," "Kiowa," and "Navaho," while quercetin 3-*O*-glucoside was the primary flavonol in the "Prime-Jim" and "Chickasaw." Eight flavonols were identified as myricetin-3-*O*-rutinoside (**35**), myricetin-3-*O*-glucuronide (**36**), myricetin-3-*O*-(6″-malonyl)glucoside (**37**), quercetin-3-*O*-rutinoside (**27**), quercetin-3-*O*-glucoside (**30**), quercetin-3-*O*-(6″-malonyl)glucoside (**38**), kaempferol-3-*O*-rutinoside (**39**), and kaempferol-3-*O*-galactoside (**40**) in black currant extracts by HPLC-MS [28] (Figure 3.3). The HPLC-based quantitative procedure, with improved extraction and hydrolysis, was used to analyze the content of quercetin, myricetin, and kaempferol in 10 black currant cultivars from organic farms and in 5 cultivars from conventional farms [35]. Myricetin was the most abundant flavonol, and its amount varied significantly among cultivars, from 8.9 to 24.5 mg/100 g fresh weight. The quercetin level in black currants

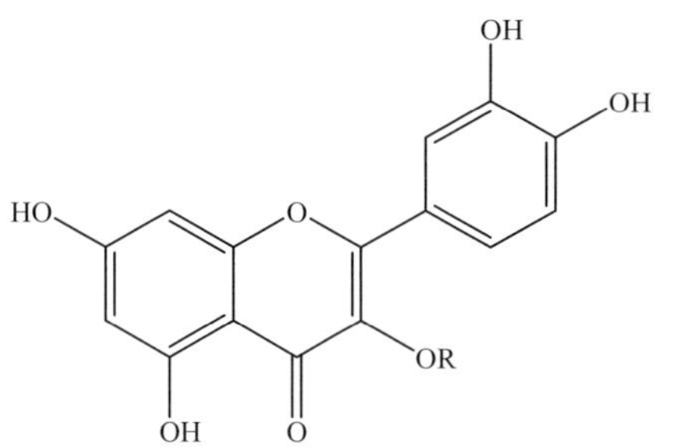

27. R = Rutinose
28. R = Galactose
29. R = Methoxyhexose
30. R = Glucose
31. R = Pentose
32. R = [6"-(3-hydroxy-3-methylglutaroyl)]-galactose
33. R = Glucosylpentose
34. R = Oxalylpentose
38. R = (6"-malonyl)-glucose

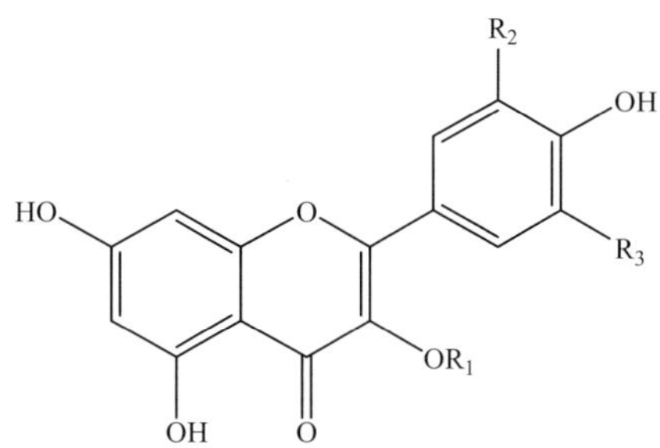

35. R_1 = Rutinose, R_2= OH, R_3= OH
36. R_1 = Glucuronic acid, R_2= OH, R_3= OH
37. R_1 = (6"-malonyl)-glucose, R_2= OH, R_3= OH
39. R_1 = Rutinose, R_2= H, R_3= H
40. R_1 = Galactose, R_2= H, R_3= H
41. R_1 = Glucose, R_2= OH, R_3= OH
42. R_1 = Rutinose, R_2= OCH_3, R_3= H
43. R_1 = Glucose, R_2= H, R_3= H
44. R_1 = Arabinose, R_2= OH, R_3= OH
45. R_1 = Glucose, R_2= OCH_3, R_3= H

Figure 3.3 Structures of flavonols isolated from blackberries and black currants.

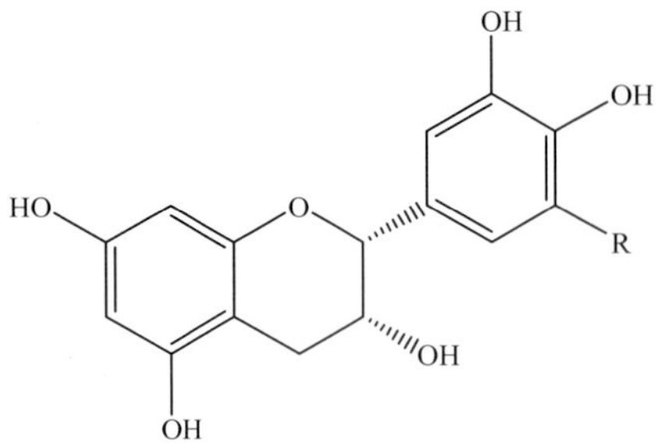

46. R = H
48. R = OH

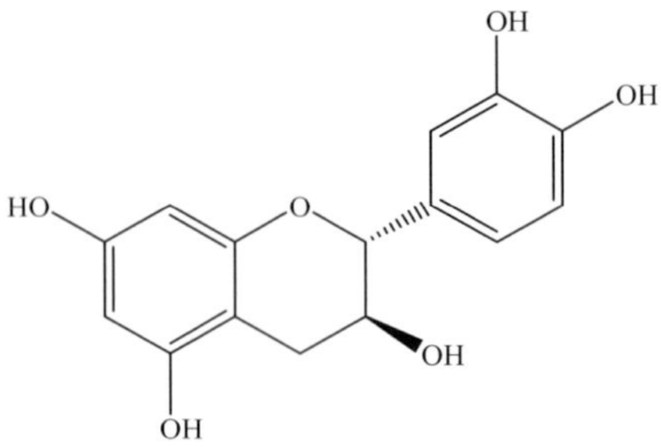

47

Figure 3.4 Structures of flavanols isolated from blackberries and black currants.

also varied widely among the cultivars, from 5.2 to 12.2 mg/100 g fresh weight. The levels of kaempferol in black currant cultivars were low, ranging from 0.9 to 2.3 mg/100 g. Koponen *et al.* [36] reported the existence of three flavonols myricetin-3-*O*-glucoside (**41**), isorhamnetin-3-*O*-rutinoside (**42**), and kaempferol-3-*O*- glucoside (**43**), together with two unreported flavonols, myricetin-3-*O*-arabinoside (**44**) and isorhamnetin-3-*O*-glucoside (**45**), in black currants (Figure 3.3).

3.3.1.3 *Flavanols*

Unlike the other classes of flavonoids, flavanols are usually found in free rather than glycosylated or esterified form in fruits. Mertz *et al.* [37] identified and quantified (-)-epicatechin (**46**) in two blackberry species (*Rubus glaucus* and *Rubus adenotrichus*) (Figure 3.4). Catechin was detected in two blackberry cultivars: "Choctaw" (313 mg/100 g fresh weight) and "Kiowa" (266 mg/100 g fresh weight) [38]. Black currants contain low levels of the flavanols, (+)-catechin (**47**), (-)-epicatechin (**46**), and (-)-epigallocatechin (**48**) [39], all identified by HPLC-MS analysis [40] (Figure 3.4).

3.3.2 Phenolic acids

Phenolic acids constitute about one-third of the dietary phenols, and may occur in plants in the free and bound forms. Bound phenolic acids may be linked to various plant components through ester, ether, and/or acetal bonds [41]. Phenolic acids can be subdivided into two major groups: hydroxybenzoic and hydroxycinnamic acid derivatives. They have various biological activities such as antioxidant and antimutagenic activities [42, 43].

Five phenolic acids such as *p*-hydroxybenzoic (**49**), gallic (**50**), *p*-coumaric (**51**), caffeic (**52**), and ferulic acids (**53**) were quantified in two blackberry cultivars [38] (Figure 3.5). The concentrations were 4.12–6.42, 0.4–2.08, 1.38–3.64, and 2.99–3.51 mg/100 g fresh weight for gallic, *p*-coumaric, caffeic, and ferulic acids, respectively, whereas *p*-hydroxybenzoic acid was not detectable. Later, gallic acid together with gallic acid esters, caffeic acid ester, *p*-coumaric acid esters, and ferulic acid esters were also reported in two blackberry species [37]. In both blackberries, *p*-coumaric acid esters were the most abundant and caffeic acid ester was the second primary one.

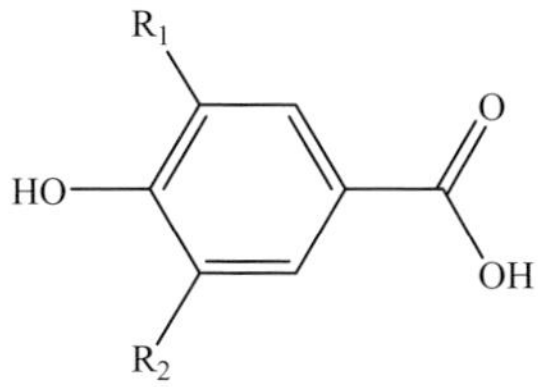

49. $R_1 = H$, $R_2 = H$
50. $R_1 = OH$, $R_2 = OH$

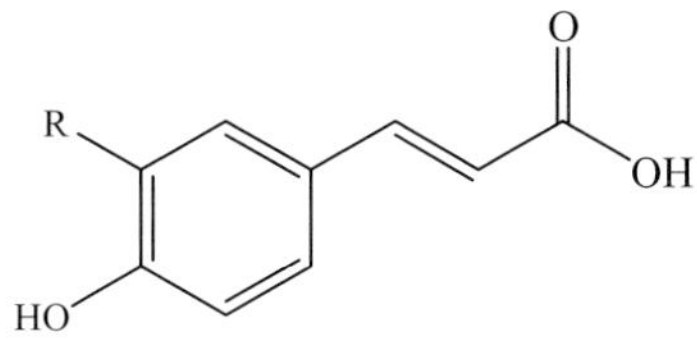

51. $R = H$
52. $R = OH$
53. $R = OCH_3$

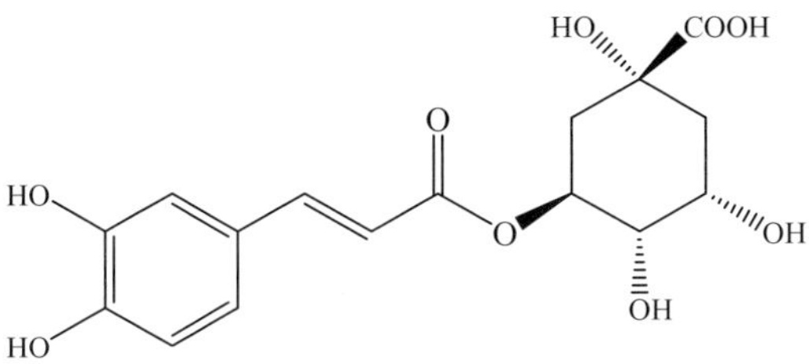

54

Figure 3.5 Structures of phenolic acids isolated from blackberries and black currants.

Phenolic acids are also present in black currants. Gavrilova *et al.* [29] reported that a caffeic acid derivative was the major component in black currants; chlorogenic acid (**54**) (Figure 3.5) and derivatives of *p*-coumaric and ferulic acids were also detected, which is in accordance with previous reports [40, 44]. Meanwhile, two minor peaks with *m/z* 421 and 451 in negative ion mode may indicate the presence of *p*-coumaric acid hexose derivative and ferulic acid hexose derivative, respectively. In addition, Zadernowski *et al.* [42] reported the detailed distribution of free, ester-bonded, and glycoside-bonded phenolic acids in blackberries and black currants.

3.3.3 Tannins

Tannins are oligomeric and polymeric forms of phenolics and may provide astringency in fruits and vegetables. They are traditionally divided into two classes of condensed and hydrolysable tannins. Hydrolysable tannins contain one or more hexahydroxydiphenic acid moieties, esterified to a polyol, most often to β-D-glucose. Like other berries, blackberries are abundant in hydrolysable tannins. Mertz *et al.* [37] tentatively identified six ellagic acid derivatives and two ellagitannins, sanguin H-6 (**55**) and lambertianin C (**56**), by HPLC-MS in two blackberry species (Figure 3.6). Two isomers of lambertianin A/sanguiin H-6, two isomers of lambertianin C, two isomers of pedunculagin, two isomers of castalagin/vescalagin, a isomer of galloyl-hexahydroxydiphenic (galloyl-HHDP) glucose, an isomer of lambertianin D, an isomer of galloyl-*bis*-HHDP glucose, ellagic acid (**57**), and two unknown ellagitannins were detected in blackberry fruits using a combination of HPLC–electrospray ionization–MS (HPLC-ESI-MS) and matrix-assisted laser desorption ionization time-of-flight mass spectrometry (MALDI-TOF-MS) [45]. However, only detailed nuclear magnetic resonance

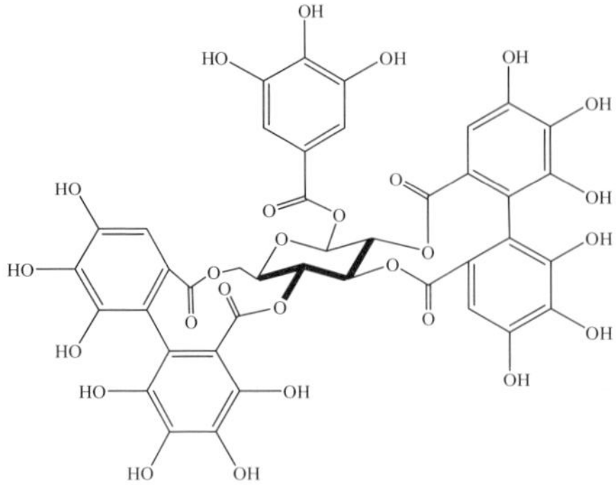

55

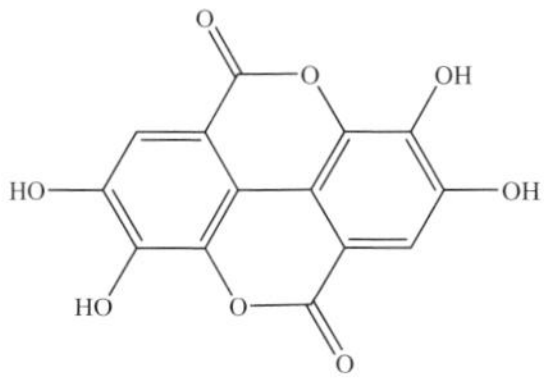

57

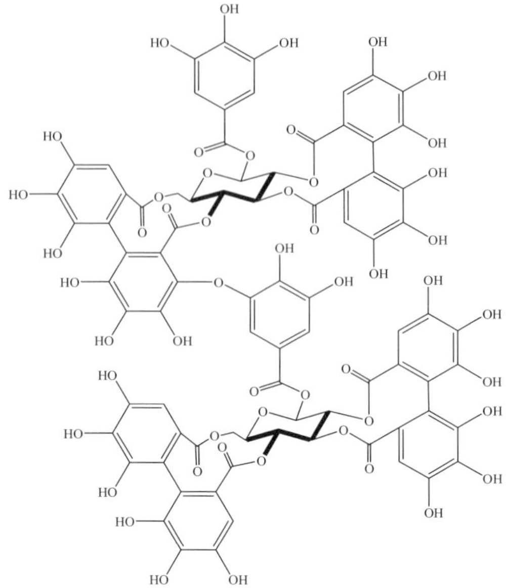

56

Figure 3.6 Structures of representative tannins isolated from blackberries and black currants.

(NMR) analysis allows unambiguous identification due to the diverse and complex nature of tannins. The structural complexity of tannins also makes it difficult to quantify them effectively. Quantitative data on tannins in blackberries have been reported in the last decade by determining acid-hydrolyzed tannins [46, 47].

3.3.4 Carotenoids

The common carotenoids in blackberries and black currants include lutein, zeaxanthin, β-cryptoxanthin, β-carotene, and α-carotene [48, 49]. Although carotenoids are a group of effective natural singlet-oxygen quenchers, data on the type and levels of carotenoids in blackberries and black currants are limited, probably due to the difficulties in separating individual carotenoids. Marinova and Ribarova [48] determined the content of lutein, zeaxanthin, β-cryptoxanthin, β-carotene, and α-carotene in Bulgarian blackberries and black currants. Blackberries had high lutein, zeaxanthin, β-cryptoxanthin, β-carotene, and α-carotene contents at 270, 29, 30, 9.2, and 100 μg/100 g, respectively, while only lutein, zeaxanthin, and α-carotene were detected in black currants at 210, 5.7, and 62 μg/100 g, respectively. In addition, Heinonen *et al.* [49] reported lutein (440 μg/100 g), β-carotene (99 μg/100 g), and α-carotene (trace) in Finnish black currants, whereas Mertz *et al.* [50] detected trace levels of carotenoids in French-grown blackberries. Although berries cannot be considered as good dietary sources of carotenoids, blackberries had a higher level of carotenoids than commonly consumed berries, whereas black currants contained low levels of carotenoids.

3.3.5 Other phytochemicals

Besides the above-mentioned compounds, three phytosterols were isolated from freeze-dried blackberries and identified as stigmasta-5,22-dien-3-ol (**58**), β-sitosterol (**59**), and β-sitosterol-3-β-D-glucose (**60**) [51] (Figure 3.7). Blackberries also contained secoisolariciresinol (**61**), which is a lignin resulting from the interactions between two cinnamic acids [52].

McDougall *et al.* [53] reported that black currants might contain a family of flavanol–anthocyanin condensation products (Figure 3.7). Two major condensation products, (E) (+)-gallocatechin-cyanidin-3-*O*-rutinoside (**62**) and (E) (+)-gallocatechin-delphinidin-3-*O*-rutinoside (**63**), could be identified in the fresh black currant extracts. Anttonen and Karjalainen [33] found an aureusidin glucoside in black currants. The fruits of black currant were found to contain cassis polysaccharide (CAPS), consisting of rhamnose, mannose, arabinose, galactose, xylose, and glucose [54]. CAPS might consist of a soluble component and a precipitable component with mean molecular weights of 80 and 600 kDa, respectively, in

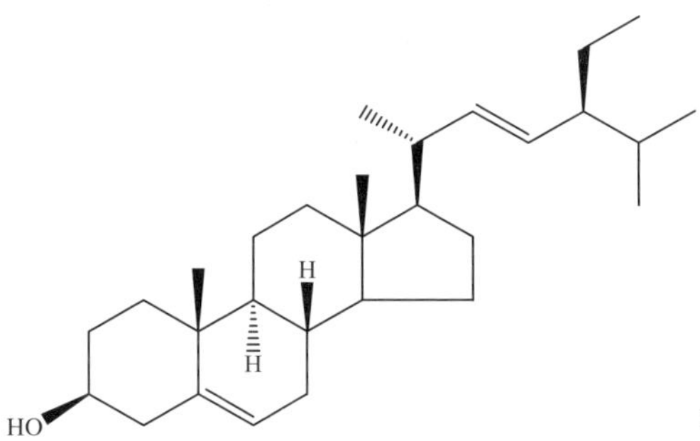

58

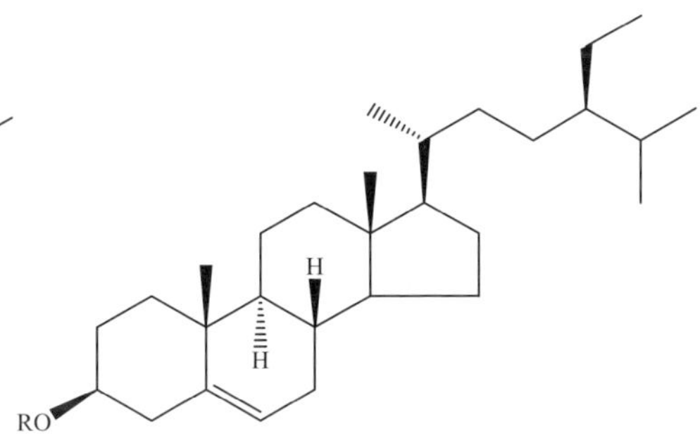

59. R = H
60. R = Glucose

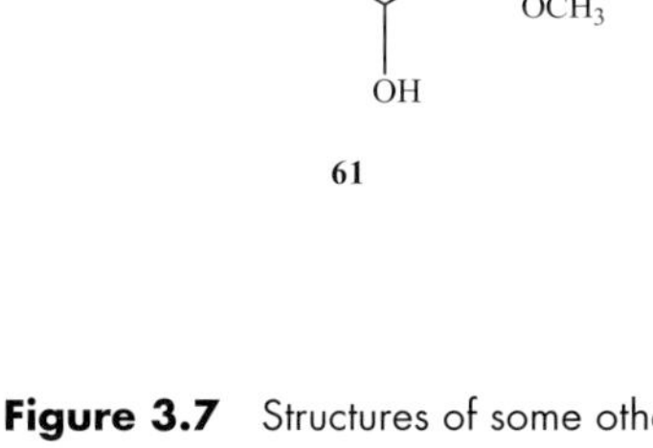

61

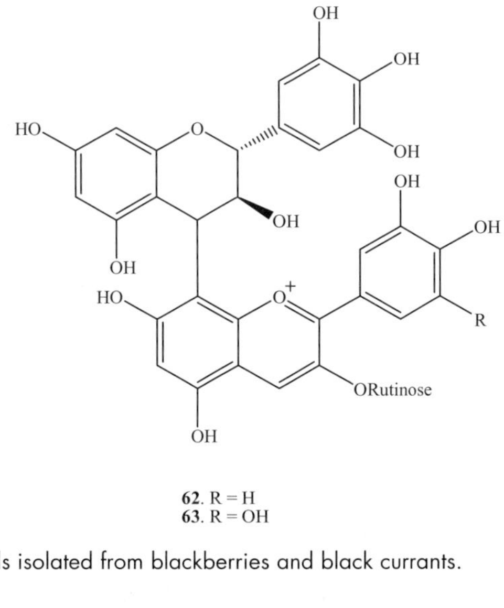

62. R = H
63. R = OH

Figure 3.7 Structures of some other compounds isolated from blackberries and black currants.

45% (v/v) ethanol solution. In addition, Hilz *et al.* [55] reported the presence of xyloglucans in black currants, which had a rather simple XXXG-type structure with galactose- and fucose-containing side chains.

3.4 Health benefits of blackberries and black currants

3.4.1 Antioxidant activities

Blackberries, black currants, and other small fruits are excellent sources of natural antioxidants, which might contribute to their increased popularity in the human diet [5, 9, 56]. Hassimotto *et al.* [57] reported the *in vivo* antioxidant activity of an aqueous extract of blackberries and its two derived fractions, the anthocyanin-enriched fraction (AF) and the ellagitannin-enriched fraction (EF). After 35 days of administration, the AF and EF extracts significantly reduced thiobarbituric acid reactive substance (TBARS) levels and increased glutathione levels in the liver, kidney, and brain of male Wistar rats. The hexane, methanol, and ethanol extracts of blackberries (*R. jamaicensis*) inhibited lipid peroxidation by 74, 53, and 48%, respectively [30]. Antioxidant activity was measured for the low- and high-molecular-weight phenolic fractions (LMWPF and HMWPF) from blackberries by the classical ferric reducing antioxidant potential (FRAP) assay. FRAP values ranged from 2.03 to 3.92 mmol of Fe^{2+} equivalents/100 mg of respective fractions [58]. Antioxidant activities of both the crude and anthocyanin-enriched blackberry extracts were determined with oxygen radical absorbance capacity (ORAC) values of 674.2 and 4885 μmol trolox equivalents (TE)/g, respectively [59]. Meanwhile, a cellular antioxidant assay was also determined by the 2,2′-azobis(2-amidinopropane) dihydrochloride (AAPH) generated free radicals in INT-407 cells. Similar results indicated that the anthocyanins might contribute a major part of the antioxidant ability to suppress both peroxyl radical-induced chemical and intracellular oxidation. The TE antioxidant capacity (TEAC) value of freeze-dried blackberries on 2,2′-azinobis-(3-ethylbenzothiazoline-6-sulfonic acid) (ABTS) assay was 406 mmol of TE/g [60]. The antioxidant activities of blackberries ranked fourth and second in the TBARS and ABTS assays in 28 different fruits including cherries, apples, peaches, pears, and grapes, respectively. Results from these studies suggested that anthocyanins were the major contributors to antioxidant capacity in blackberries. Tabart *et al.* [61] reported the antioxidant capacities of different black currant cultivars based on 2,2-diphenyl-1-picrylhydrazyl (DPPH) radical scavenging capacity and observed only small differences among the eight tested cultivars. The antioxidant capacity of black currant extracts was also determined using the FRAP assay [28], and it was found that anthocyanins were the major contributors to it. Thus, blackberries and black currants contain antioxidants at a level comparable or greater than that in the commonly consumed fruits such as apples, peaches, pears, and grapes.

3.4.2 Anticancer activities

A number of *in vitro* and animal studies have shown that berries might have anticancer properties. The lyophilized and powdered fruits of blackberries (*R. jamaicensis*) were extracted successively with hexane, ethyl acetate and methanol, and the corresponding extracts were evaluated for their tumor cell proliferation inhibitory activities [30]. The hexane extract had the greatest overall capacity to inhibit the progression of tumor cell growth, inhibiting the proliferation of colon, breast, lung, and gastric human tumor cells by 50, 24, 54, and 37%, respectively. In another study, the blackberry extracts were evaluated for their ability to inhibit the growth of human oral (KB, CAL-27), breast (MCF-7), colon (HT-29, HCT116), and prostate (LNCaP) tumor cell lines with IC_{50} values ranging from 49.61 to 122.00 μg/mL. The blackberry extracts were also evaluated for their ability to stimulate apoptosis of the cyclooxygenase-2 (COX-2) expressing HT-29 colon cancer cells and it was found that they induced apoptosis 1.8-fold over untreated controls [5]. In addition, Dai *et al.* [26] reported that blackberry extracts inhibited HT-29 colon cancer cell growth in a concentration-dependent manner with 49.2 μg of total anthocyanins/mL inhibiting HT-29 cell growth up to 66% in 72 hours.

Polysaccharides from black currants showed certain cytotoxicity directly against ehrlich ascite cells, and the IC_{50} value was estimated to be about 760 μg/mL. Oral administration of black currant juices and polysaccharide to ehrlich carcinoma-bearing mice retarded the growth of the solid tumor by 45 and 51%, respectively [54]. These research findings suggested that additive and synergistic effects of phytochemicals in blackberries and black currants may also be responsible for their potent anticancer activities.

3.4.3 Anti-inflammatory activities

Chronic inflammation has been associated with the increased risk of several human diseases including cancer and CVD. Consumption of dietary anti-inflammatory components may reduce the risk of these human chronic health problems. The anti-inflammatory activities of LMWPF and HMWPF, isolated from three blackberry cultivars (e.g. Navaho, Kiowa, and Ouachita), were investigated by an *in vivo* mouse ear edema model [58]. All fractions significantly ($P < 0.05$) reduced 12-*O*-tetradecanoylphorbol-13-acetate (TPA)-induced irritation injury. Furthermore, mouse ear myeloperoxidase (MPO) activity, an indicator of polymorphonuclear leukocyte infiltration, was assessed and found to be significantly reduced after topical application of all blackberry preparations ($P < 0.05$). In another study, the hexane extract of blackberries (*R. jamaicensis*) selectively inhibited COX-2 at 100 μg/mL [30]. Dai *et al.* [26] reported that an anthocyanin-rich extract from Hull blackberries grown in Kentucky inhibited interleukin-12 (IL-12) release from dendritic cells in both background and stimulated levels, which demonstrated the anti-inflammatory properties of the anthocyanin-rich extract of blackberries. Recently, total polyphenolic and anthocyanin- and proanthocyanidin-rich fractions from four wild blackberry genotypes, a domesticated noncommercial breeding line, and a commercial cultivar were reported to inhibit pro-inflammatory responses including nitric oxide (NO) production, inducible nitric oxide synthase (iNOS) expression, COX-2 expression, and prostaglandin E2 (PGE2) in RAW 264.7 macrophages stimulated by lipopolysaccharide (LPS) [62].

Polyphenolic compounds of black currants had the potential to reduce eosinophil recruitment and alleviate eosinophilic-driven airway inflammation [63]. Another study by Balstad *et al.* [64] suggested that high intakes of lyophilized black currants modulated *in vivo* nuclear factor-kappa B (NF-κB) signaling in the liver following LPS-induced stress. In addition, the study by Lyall *et al.* [65] showed that consumption of short-term black currant extracts alleviates exercise-induced oxidative stress and LPS-stimulated inflammatory responses. These research findings suggested the potential of blackberries and black currants in reducing the risk of chronic inflammation and its related health problems.

3.4.4 Other bioactivities

A study by Shukitt-Hale *et al.* [66] demonstrated that the polyphenolics in a 2% blackberry-supplemented diet, when fed to rats from 19 to 21 months of age, retarded and even reversed age-related decrements in motor and cognitive performance. The blackcurrant extracts inhibited herpes simplex virus type 1 attachment on the cell membrane completely at a 100-fold dilution, as well as the plaque formation of herpes simplex virus types 1 and 2, and varicella zoster virus by 50% at a 400-fold dilution or lower concentrations [67]. The blackcurrant extracts also showed obvious inhibition activities on influenza virus types A and B [68]. In addition, Takata *et al.* [54] reported the immunostimulatory effects of blackcurrant extracts on the release of IL-2, IL-10, interferon-γ, and IL-4 from splenocytes using a phosphate-buffered saline (PBS) control in tumor-bearing mice. Interestingly, black currant extracts could activate endothelial nitric oxide synthase (eNOS) via the Akt/PI3 kinase pathway *in vitro* in human umbilical vein endothelial cells (HUVEC) and the effect was not dependent on vitamin C [69]. Black currant extracts could alleviate physical fatigue with increasing the time of loading swimming, the hepatic glycogen content, and decreasing blood urea nitrogen in mice [70].

3.5 Commercial products and industrial applications of blackberries and black currants

Blackberries and black currants, like other berries, are widely distributed as processed products (such as jellies, juices, jams, and wines) because fresh fruits are extremely perishable with a very short shelf life. Many researches have conducted investigations to determine the nutraceutical and health properties of blackberries and black currants and their derived commercial products.

3.5.1 Juices

Juices obtained by squeezing fruit without any pasteurization treatments represent an alternative way of consuming fresh fruits. Hager *et al.* [71] reported an approximate 67% loss of total monomeric anthocyanins, 55% loss of ORAC value, as well as a small increase in polymeric color occurred in juice processing of blackberries. Over the storage of six months at room temperature, ORAC value of blackberry juices showed no significant changes ($P > 0.05$), while total monomeric anthocyanins were lost by 69–75% [71]. Later, Hager *et al.* [72] reported that total ellagitannin concentration of juices changed markedly during juice processing steps. The low levels of ellagitannins in juice products were attributed to their loss to the presscake. The anthocyanin pigment profiles of commercial blackberry and black currant juices were characterized by Hong and Wrolstad [73]. They found out that cyanidin 3-*O*-rutinoside, 3-*O*-sophoroside, 3-*O*-glucosylrutinoside, and 3-*O*-glucoside existed in the blackberry samples, while cyanidin and delphinidin 3-*O*-rutinosides and delphinidin 3-*O*-glucoside were present in the black currant samples. Other properties including tinctoral strength, total anthocyanin concentration, browning, titratable acidity, and Hunter tristimulus color values were also determined. More recently, Serraino *et al.* [74] showed that blackberry juices containing cyanidin-3-*O*-glucoside scavenged peroxynitrite and exerted a protective effect against endothelial dysfunction and vascular failure induced by peroxynitrite.

3.5.2 Jams

Berry jams are an important dietary form of berry fruits [75]. HPLC spectral patterns of blackberry jams showed that the predominant anthocyanin was cyanidin-3-*O*-glucoside, together with traces of cyanidin-3-*O*-rutinoside [76]. The commercial blackcurrant jams contained cyanidin and delphinidin 3-*O*-rutinosides as the major anthocyanins [76]. Blackberry jams had the lowest total phenolic content and total anthocyanins among all processed products including frozen, freeze-dried, hot-air dried, canned-in-water, canned-in-syrup blackberries, and blackberry jams retained only 33% of total phenolic content and 20% of total anthocyanins compared with their frozen controls [77]. This may be explained by the acceleration in condensation and polymerization reactions of phenolic compounds after cellular disruption and the increase in fruit temperature combined with oxygen exposure [78].

3.5.3 Wines

Blackberry wines possess stronger superoxide and hydroxyl radical scavenging properties and are more effective in inhibiting calmodulin-promoted phosphodiesterase than red grape wines [79]. Increased antioxidant activity and high total phenolic content were detected in blackberry wines [60%, 1232 mg of gallic acid equivalents (GAE)/L] [80]. Total antioxidant activity, total phenolic content, and mineral contents of black currant wines have been reported [81].

3.5.4 Canned fruits

Wu *et al.* [77] reported that canning significantly reduced total monomeric anthocyanins in blackberries. Furthermore, compared with canning in water, canning in sucrose syrup doubled the loss of total monomeric anthocyanins and decreased radical scavenging activities. In addition, 22 and 27% reduction in ORAC and 10.5 and 17.8% reduction in total monomeric anthocyanins accompanied with darkening of polymeric colour were observed in 40° Brix Sweetose syrup and water canned shiny black 'Apache' blackberries harvested in Arkansas in 2005, respectively [71]. Additionally, a 6-month postprocessing storage at 25°C led to 60.6 and 65.8% loss in total monomeric anthocyanins, respectively, but no significant changes ($P >$ 0.05) in ORAC values were observed. Taken together, these previous studies suggested the potential effects of fruit processing on the components and health properties of blackberries and black currants. This information could be important for better use of blackberries and black currants, as well as other fruits.

3.6 Drying effects on antioxidant capacities and phenolics of blackberries and black currants

Drying is a common method to preserve fresh berries by reducing their water activity. Different drying technologies showed different impacts on the bioactive compounds and antioxidant capacities of blackberries (Table 3.2). Freeze drying was recognized as the best technology to make high-quality dried products, resulting in slight changes in bioactive constituents and antioxidant activities. However, obvious decreases in total phenolics, anthocyanins, and antioxidant capacities of blackberries were observed after hot-air drying. Significant decrease

Table 3.2 Percent change of bioactive compounds and antioxidant activities of two blackberry varieties compared with frozen controls after drying processing

	Total phenolics (mg of GAE/g)	Anthocyanins (mg of C3GE/g)	RSA (mg of AAE/g)	ORAC (μmol of TE/g)	FRAP (μmol of TE/g)
Marion					
Freeze drying	+27%	−25%	nc	nc	nc
Hot–air drying	na	−56%	na	−37%	−27%
Evergreen					
Freeze drying	+21%	+5.5%	+14%	nc	nc
Hot–air drying	−37%	−84%	−13%	na	na

Source: Adapted from Wu *et al.* [77].
RSA, radical scavenging activity; ORAC, oxygen radical absorbance capacity; FRAP, ferric reducing antioxidant potential; GAE, gallic acid equivalents; C3GE, cyaniding-3-glucoside equivalents; AAE, ascorbic acid equivalents; TE, trolox equivalents; na, not available; nc, not changed.

($P < 0.05$) in vitamin C content was observed in freeze-dried and air-dried blackberry samples as compared with their frozen counterparts [82]. The total phenolic levels in air-dried blackberries were 15.6–21.1% lower than those found in frozen fruits [82]. There is limited literature about drying effects on phenolics and antioxidant activities of black currants.

3.7 Conclusions

Blackberries and black currants have considerable amounts of nutrients and nutraceutical components as well as potential health benefits. Meanwhile, dried blackberries and black currants retain significant levels of the nutritional value of fresh ones, and are also recommended in daily dietary intake to promote human health. Scientific data suggest that their intake may reduce the risk of several aging-associated human chronic diseases. Additional research is needed to better understand the mechanism behind the health properties of dried blackberries and black currants. Further research is also required to identify and quantify the individual beneficial components in dried blackberries and black currants, as well as the effects of different dehydration procedures and conditions on health components in berry fruits. In addition, the absorption, distribution, metabolism, and excretion of dried blackberries and black currants or their bioactive constituents should be investigated, along with a better understanding of their bioavailability in human.

References

1. Zhao, Y.Y. (2007) *Berry Fruit: Value-Added Products for Health Promotion.* CRC Press, Taylor & Francis Group, Boca Raton, FL.
2. Seeram, N.P. (2008) Berry fruits for cancer prevention: current status and future prospects. *Journal of Agricultural and Food Chemistry*, **56**, 630–635.
3. Szajdek, A. & Borowska, E. (2008) Bioactive compounds and health-promoting properties of berry fruits: a review. *Plant Foods for Human Nutrition (Formerly Qualitas Plantarum)*, **63**, 147–156.

4. Strik, B.C., Clark, J.R., Finn, C.E. & Banados, M.P. (2007) Worldwide blackberry production. *HortTechnology*, **17**, 205–213.
5. Seeram, N.P., Adams, L.S., Zhang, Y.J., Lee, R., Sand, D., Scheuller, H.S. & Heber, D. (2006) Blackberry, black raspberry, blueberry, cranberry, red raspberry, and strawberry extracts inhibit growth and stimulate apoptosis of human cancer cells *in vitro*. *Journal of Agricultural and Food Chemistry*, **54**, 9329–9339.
6. Bordonaba, J.G. & Terry, L.A. (2008) Biochemical profiling and chemometric analysis of seventeen UK-grown black currant cultivars. *Journal of Agricultural and Food Chemistry*, **56**, 7422–7430.
7. Karjalainen, R., Anttonen, M., Saviranta, N., Stewart, D., McDougall, G.J., Hilz, H., Mattila, P. & Törrönen, R. (2009) A review on bioactive compounds in black currants (*Ribes nigrum* L.) and their potential health-promoting properties. *Acta Horticulturae*, **839**, 301–307.
8. Lister, C., Wilson, P., Sutton, K. & Morrison, S. (2002) Understanding the health benefits of blackcurrants. *Acta Horticulturae*, **585**, 443–450.
9. USDA (2011). *USDA National Nutrient Database for Standard Reference, Release 24*. Published on-line at: http://www.ars.usda.gov/ba/bhnrc/ndl, last accessed 6 January 2012.
10. Vollendorf, J.A.M.N.W. (1994) Dietary fiber content and composition of different forms of fruits. *Food Chemistry*, **51**, 39–44.
11. Gyurova, D. & Ribarova, F. (2005) Dietary fibres in Bulgarian vegetables and fruits. *Dokladi na Bulgarskata Akademiya na Naukite*, **58**, 107–110.
12. Borowska, E.J. & Szajdek, A. (2005) Dietary components and bioactive substances in chokeberry, bilberry, and black currant fruits. *Bromatologia i Chemia Toksykologiczna*, **38**, 181–184.
13. Kafkas, E., Koşar, M., Türemiş, N. & Başer, K.H.C. (2006) Analysis of sugars, organic acids and vitamin C contents of blackberry genotypes from Turkey. *Food Chemistry*, **97**, 732–736.
14. Reyes-Carmona, J., Yousef, G.G., Martínez-Peniche, R.A. & Lila, M.A. (2005) Antioxidant capacity of fruit extracts of blackberry (*Rubus* sp.) produced in different climatic regions. *Journal of Food Science*, **70**, 198–202.
15. Chun, J., Lee, J., Ye, L., Exler, J. & Eitenmiller, R. (2006) Tocopherol and tocotrienol contents of raw and processed fruits and vegetables in the United States diet. *Journal of Food Composition and Analysis*, **19**, 196–204.
16. Fang-Chiang, H.J. & Wrolstad, R.E. (2010) Sugar and nonvolatile acid composition of blackberries. *Journal of AOAC International*, **93**, 956–965.
17. Zheng, J., Yang, B.R., Tuomasjukka, S., Ou, S.Y. & Kallio, H. (2009) Effects of latitude and weather conditions on contents of sugars, fruit acids, and ascorbic acid in black currant (*Ribes nigrum* L.) juice. *Journal of Agricultural and Food Chemistry*, **57**, 2977–2987.
18. Wu, W.L., Li, W.L., Lu, L.F. & Wang, X.M. (2007) Comparison of nutrient constituents in fresh fruit of different cultivars of blackberry. *Journal of Plant Resources and Environment*, **16**, 58–61.
19. Ilkay, N., Ustun, S. & Tekguler, B. (2008) Physical and chemical changes during ripening of blackberry fruits. *Scientia Agricola*, **65**, 87–90.
20. Huo, J.W., Li, Z.H. & Qin, D. (2011) Review of nutritional ingredients and health protectal function of black currant fruit and its prospect in industrial development. *Journal of Northeast Agricultural University*, **42**, 139–144.
21. Knekt, P., Kumpulainen, J., Järvinen, R., Rissanen, H., Heliövaara, M., Reunanen, A., Hakulinen, T. & Aromaa, A. (2002) Flavonoid intake and risk of chronic diseases. *American Journal of Clinical Nutrition*, **76**, 560–568.
22. Mazza, G. & Miniati, E. (1993) *Anthocyanins in Fruits, Vegetables, and Grains*. CRC Press, Taylor & Francis Group, Boca Raton, FL.
23. Sapers, G.M., Hicks, K.B., Burgher, A.M., Hargrave, D.L., Sondey, S.M. & Bilyk, A. (1986) Anthocyanin patterns in ripening thornless blackberries. *Journal of the American Society for Horticultural Science*, **111**, 945–950.
24. Fang-Chiang, H.J. (2000) *Anthocyanin Pigments, Nonvolatile Acid and Sugar Composition of Blackberries*. MSc Thesis, Oregon State University, Portland, OR.
25. Stintzing, F.C., Stintzing, A.S., Carle, R. & Wrolstad, R.E. (2002) A novel zwitterionic anthocyanin from evergreen blackberry (*Rubus laciniatus* Willd.). *Journal of Agricultural and Food Chemistry*, **50**, 396–399.
26. Dai, J., Patel, J.D. & Mumper, R.J. (2007) Characterization of blackberry extract and its antiproliferative and anti-inflammatory properties. *Journal of Medicinal Food*, **10**, 258–265.

27. Slimestad, R. & Solheim, H. (2002) Anthocyanins from black currants (*Ribes nigrum* L.). *Journal of Agricultural and Food Chemistry*, **50**, 3228–3231.
28. Borges, G., Degeneve, A., Mullen, W. & Crozier, A. (2010) Identification of flavonoid and phenolic antioxidants in black currants, blueberries, raspberries, red currants, and cranberries. *Journal of Agricultural and Food Chemistry*, **58**, 3901–3909.
29. Gavrilova, V., Kajdžanoska, M., Gjamovski, V. & Stefova, M. (2011) Separation, characterization and quantification of phenolic compounds in blueberries and red and black currants by HPLC-DAD-ESI-MS^n. *Journal of Agricultural and Food Chemistry*, **59**, 4009–4018.
30. Bowen-Forbes, C.S., Zhang, Y. & Nair, M.G. (2010) Anthocyanin content, antioxidant, anti-inflammatory and anticancer properties of blackberry and raspberry fruits. *Journal of Food Composition and Analysis*, **23**, 554–560.
31. Fang-Chiang, H.J. & Wrolstad, R.E. (2005) Anthocyanin pigment composition of blackberries. *Journal of Food Science*, **70**, 198–202.
32. Siriwoharn, T., Wrolstad, R.E., Finn, C.E. & Pereira, C.B. (2004) Influence of cultivar, maturity, and sampling on blackberry (*Rubus* L. Hybrids) anthocyanins, polyphenolics, and antioxidant properties. *Journal of Agricultural and Food Chemistry*, **52**, 8021–8030.
33. Anttonen, M.J. & Karjalainen, R.O. (2006) High-performance liquid chromatography analysis of black currant (*Ribes nigrum* L.) fruit phenolics grown either conventionally or organically. *Journal of Agricultural and Food Chemistry*, **54**, 7530–7538.
34. Cho, M.J., Howard, L.R., Prior, R.L. & Clark, J.R. (2005) Flavonol glycosides and antioxidant capacity of various blackberry and blueberry genotypes determined by high-performance liquid chromatography/mass spectrometry. *Journal of the Science of Food and Agriculture*, **85**, 2149–2158.
35. Mikkonen, T.P., Määttä, K.R., Hukkanen, A.T., Kokko, H.I., Törrönen, A.R., Kärenlampi, S.O. & Karjalainen, R.O. (2001) Flavonol content varies among black currant cultivars. *Journal of Agricultural and Food Chemistry*, **49**, 3274–3277.
36. Koponen, J.M., Happonen, A.M., Auriola, S., Kontkanen, H., Buchert, J., Poutanen, K.S. & Törrönen, A.R. (2008) Characterization and fate of black currant and bilberry flavonols in enzyme-aided processing. *Journal of Agricultural and Food Chemistry*, **56**, 3136–3144.
37. Mertz, C., Cheynier, V., Günata, Z. & Brat, P. (2007) Analysis of phenolic compounds in two blackberry species (*Rubus glaucus* and *Rubus adenotrichus*) by high-performance liquid chromatography with diode array detection and electrospray ion trap mass spectrometry. *Journal of Agricultural and Food Chemistry*, **55**, 8616–8624.
38. Sellappan, S., Akoh, C.C. & Krewer, G. (2002) Phenolic compounds and antioxidant capacity of Georgia-grown blueberries and blackberries. *Journal of Agricultural and Food Chemistry*, **50**, 2432–2438.
39. Macheix, J.J., Fleuriet, A. & Billot, J. (1990) *Fruit Phenolics*. CRC Press, Taylor & Francis Group, Boca Raton, FL.
40. Määttä, K.R., Kamal-Eldin, A. & Törrönen, A.R. (2003) High-performance liquid chromatography (HPLC) analysis of phenolic compounds in berries with diode array and electrospray ionization mass spectrometric (MS) detection: *Ribes* species. *Journal of Agricultural and Food Chemistry*, **51**, 6736–6744.
41. Robbins, R.J. (2003) Phenolic acids in foods: an overview of analytical methodology. *Journal of Agricultural and Food Chemistry*, **51**, 2866–2887.
42. Zadernowski, R., Naczk, M. & Nesterowicz, J. (2005) Phenolic acid profiles in some small berries. *Journal of Agricultural and Food Chemistry*, **53**, 2118–2124.
43. Kikuzaki, H., Hisamoto, M., Hirose, K., Akiyama, K. & Taniguchi, H. (2002) Antioxidant properties of ferulic acid and its related compounds. *Journal of Agricultural and Food Chemistry*, **50**, 2161–2168.
44. Hakkinen, S. & Auriola, S. (1998) High-performance liquid chromatography with electrospray ionization mass spectrometry and diode array ultraviolet detection in the identification of flavonol aglycones and glycosides in berries. *Journal of Chromatography A*, **829**, 91–100.
45. Hager, T.J., Howard, L.R., Liyanage, R., Lay, J.O. & Prior, R.L. (2008) Ellagitannin composition of blackberry as determined by HPLC-ESI-MS and MALDI-TOF-MS. *Journal of Agricultural and Food Chemistry*, **56**, 661–669.
46. Vrhovsek, U., Palchetti, A., Reniero, F., Guillou, C., Masuero, D. & Mattivi, F. (2006) Concentration and mean degree of polymerization of Rubus ellagitannins evaluated by optimized acid methanolysis. *Journal of Agricultural and Food Chemistry*, **54**, 4469–4475.

47. Vrhovsek, U., Giongo, L., Mattivi, F. & Viola, R. (2007) A survey of ellagitannin content in raspberry and blackberry cultivars grown in Trentino (Italy). *European Food Research and Technology*, **226**, 817–824.
48. Marinova, D. & Ribarova, F. (2007) HPLC determination of carotenoids in Bulgarian berries. *Journal of Food Composition and Analysis*, **20**, 370–374.
49. Heinonen, M.I., Ollilainen, V., Linkola, E.K., Varo, P.T. & Koivistoinen, P.E. (1989) Carotenoids in Finnish foods: vegetables, fruits, and berries. *Journal of Agricultural and Food Chemistry*, **37**, 655–659.
50. Mertz, C., Gancel, A.L., Gunata, Z., Alter, P., Dhuique-Mayer, C., Vaillant, F., Perez, A. M., Ruales, J. & Brat, P. (2009) Phenolic compounds, carotenoids and antioxidant activity of three tropical fruits. *Journal of Food Composition and Analysis*, **22**, 381–387.
51. Ruan, J.L., Zhao, X.Z., Cassady, J.M. & Stoner, G.D. (2001) Study on the constituents from freeze-dried power of blackberries (*Rubus ursinus*). *Journal of Chinese Medicinal Materials*, **24**, 645–647.
52. Thompson, L.U. (1999) Role of lignans in carcinogenesis. In: *Phytochemicals in Human Health Protection, Nutrition and Plant Defense* (ed. J.T. Romeo). Kluwer Academic/Plenum Publishers, New York, NY, pp. 51–61.
53. McDougall, G.J., Gordon, S., Brennan, R. & Stewart, D. (2005) Anthocyanin-flavanol condensation products from black currant (*Ribes nigrum* L.). *Journal of Agricultural and Food Chemistry*, **53**, 7878–7885.
54. Takata, R., Yamamoto, R., Yanai, T., Konno, T. & Okubo, T. (2005) Immunostimulatory effects of a polysaccharide-rich substance with antitumor activity isolated from black currant (*Ribes nigrum* L.). *Bioscience, Biotechnology, and Biochemistry*, **69**, 2042–2050.
55. Hilz, H., Dejong, L., Kabel, M., Schols, H. & Voragen, A. (2006) A comparison of liquid chromatography, capillary electrophoresis, and mass spectrometry methods to determine xyloglucan structures in black currants. *Journal of Chromatography A*, **1133**, 275–286.
56. Moyer, R.A., Hummer, K.E., Finn, C.E., Frei, B. & Wrolstad, R.E. (2002) Anthocyanins, phenolics, and antioxidant capacity in diverse small fruits: *Vaccinium*, *Rubus*, and *Ribes*. *Journal of Agricultural and Food Chemistry*, **50**, 519–525.
57. Hassimotto, N.M.A. & Lajolo, F.M. (2011) Antioxidant status in rats after long-term intake of anthocyanins and ellagitannins from blackberries. *Journal of the Science of Food and Agriculture*, **91**, 523–531.
58. Srivastava, A., Greenspan, P., Hartle, D.K., Hargrove, J.L., Amarowicz, R. & Pegg, R.B. (2010) Antioxidant and anti-inflammatory activities of polyphenolics from southeastern U.S. range blackberry cultivars. *Journal of Agricultural and Food Chemistry*, **58**, 6102–6109.
59. Elisia, I., Hu, C., Popovich, D.G. & Kitts, D.D. (2007) Antioxidant assessment of an anthocyanin-enriched blackberry extract. *Food Chemistry*, **101**, 1052–1058.
60. Garcia-Alonso, M., de Pascual-Teresa, S., Santos-Buelga, C. & Rivas-Gonzalo, J.C. (2004) Evaluation of the antioxidant properties of fruits. *Food Chemistry*, **84**, 13–18.
61. Tabart, J., Kevers, C., Pincemail, J., Defraigne, J.O. & Dommes, J. (2006) Antioxidant capacity of black currant varies with organ, season, and cultivar. *Journal of Agricultural and Food Chemistry*, **54**, 6271–6276.
62. Cuevas-Rodríguez, E.O., Dia, V.P., Yousef, G.G., García-Saucedo, P.A., López-Medina, J., Paredes-López, O., Gonzalez de Mejia, E. & Lila, M.A. (2010) Inhibition of pro-inflammatory responses and antioxidant capacity of Mexican blackberry (*Rubus* spp.) extracts. *Journal of Agricultural and Food Chemistry*, **58**, 9542–9548.
63. Hurst, S.M., McGhie, T.K., Cooney, J.M., Jensen, D.J., Gould, E.M., Lyall, K.A. & Hurst, R.D. (2010) Blackcurrant proanthocyanidins augment IFN-g -induced suppression of IL-4 stimulated CCL26 secretion in alveolar epithelial cells. *Molecular Nutrition & Food Research*, **54**, S159–S170.
64. Balstad, T.R., Paur, I., Poulsen, M., Markowski, J., Kolodziejczyk, K., Dragsted, L.O., Myhrstad, M.C.W. & Blomhoff, R. (2010) Apple, cherry, and blackcurrant increases nuclear factor kappa B activation in liver of transgenic mice. *Nutrition and Cancer*, **62**, 841–848.
65. Lyall, K.A., Hurst, S.M., Cooney, J., Jensen, D., Lo, K., Hurst, R.D. & Stevenson, L.M. (2009) Short-term blackcurrant extract consumption modulates exercise-induced oxidative stress and lipopolysaccharide-stimulated inflammatory responses. *American Journal of Physiology—Regulatory, Integrative and Comparative Physiology*, **297**, R70–R81.
66. Shukitt-Hale, B., Cheng, V. & Joseph, J.A. (2009) Effects of blackberries on motor and cognitive function in aged rats. *Nutritional Neuroscience*, **12**, 135–140.

67. Suzutani, T., Ogasawara, M., Yoshida, I., Azuma, M. & Knox, Y.M. (2003) Anti-herpesvirus activity of an extract of *Ribes nigrum* L. *Phytotherapy Research*, **17**, 609–613.
68. Knox, Y.M., Suzutani, T., Yosida, I. & Azuma, M. (2003) Anti-influenza virus activity of crude extract of *Ribes nigrum* L. *Phytotherapy Research*, **17**, 120–122.
69. Edirisinghe, I., Banaszewski, K., Cappozzo, J., McCarthy, D. & Burton-Freeman, B.M. (2011) Effect of black currant anthocyanins on the activation of endothelial nitric oxide synthase (eNOS) *in vitro* in human endothelial cells. *Journal of Agricultural and Food Chemistry*, **59**, 8616–8624.
70. Li, L., Zhao, X.G., Ma, L., Gu, Y.J. & Ma, X.D. (2008) The anti-fatigue effect of black currant extracts in mice. *Acta Nutrimenta Sinica*, **30**, 499–501.
71. Hager, T.J., Howard, L.R. & Prior, R.L. (2008) Processing and storage effects on monomeric anthocyanins, percent polymeric color, and antioxidant capacity of processed blackberry products. *Journal of Agricultural and Food Chemistry*, **56**, 689–695.
72. Hager, T.J., Howard, L.R. & Prior, R.L. (2010) Processing and storage effects on the ellagitannin composition of processed blackberry products. *Journal of Agricultural and Food Chemistry*, **58**, 11749–11754.
73. Hong, V. & Wrolstad, R.E. (1990) Characterization of anthocyanin-containing colorants and fruit juices by HPLC/photodiode array detection. *Journal of Agricultural and Food Chemistry*, **38**, 698–708.
74. Serraino, I., Dugo, L., Dugo, P., Mondello, L., Mazzon, E., Dugo, G., Caputi, A.P. & Cuzzocrea, S. (2003) Protective effects of cyanidin-3-*O*-glucoside from blackberry extract against peroxynitrite-induced endothelial dysfunction and vascular failure. *Life Sciences*, **73**, 1097–1114.
75. Figuerola, F.E. (2007). Berry jams and jellies. In: *Berry Fruit: Value-Added Products for Health Promotion* (ed. Y.Y. Zhao). CRC Press, Taylor & Francis Group, Boca Raton, FL, pp. 367–386.
76. García-Viguera, C., Zafrilla, P. & Tomás-Barberán, F.A. (1997). Determination of authenticity of fruit jams by HPLC analysis of anthocyanins. *Journal of the Science of Food and Agriculture*, **73**, 207–213.
77. Wu, R.Y., Frei, B., Kennedy, J.A. & Zhao, Y.Y. (2010) Effects of refrigerated storage and processing technologies on the bioactive compounds and antioxidant capacities of "Marion" and "Evergreen" blackberries. *LWT—Food Science and Technology*, **43**, 1253–1264.
78. Amakura, Y., Umino, Y., Tsuji, S. & Tonogai, Y. (2000). Influence of jam processing on the radical scavenging activity and phenolic content in berries. *Journal of Agricultural and Food Chemistry*, **48**, 6292–6297.
79. Pinhero, R.G. & Paliyath, G. (2001). Antioxidant and calmodulin-inhibitory activities of phenolic components in fruit wines and its biotechnological implications. *Food Biotechnology*, **15**, 179–192.
80. Yildirim, H.K. (2006). Evaluation of colour parameters and antioxidant activities of fruit wines. *International Journal of Food Sciences and Nutrition*, **57**, 47–63.
81. Rupasinghe, H. & Clegg, S. (2007). Total antioxidant capacity, total phenolic content, mineral elements, and histamine concentrations in wines of different fruit sources. *Journal of Food Composition and Analysis*, **20**, 133–137.
82. Asami, D.K., Hong, Y.J., Barrett, D.M. & Mitchell, A.E. (2003) Comparison of the total phenolic and ascorbic acid content of freeze-dried and air-dried marionberry, strawberry, and corn grown using conventional, organic, and sustainable agricultural practices. *Journal of Agricultural and Food Chemistry*, **51**, 1237–1241.

4 Dried blueberries: the effects of processing on health-promoting compounds

William L. Kerr

4.1 Introduction

Blueberries are a sweet, succulent fruit grown in many parts of the world. In 2010, 186,000 metric tonnes (MT) of blueberries were produced in the United States [1]. Maine is the largest grower of wild blueberries, producing some 40,000 MT in 2008. Several other states are major producers of cultivated blueberries including Michigan (49,000 MT), Georgia (28,000 MT), Oregon (24,000 MT), Washington (19,000 MT), North Carolina (18,000 MT), California (13,000 MT), and Florida (8000 MT). Canada is another major producer of blueberries, yielding 42,000 MT in 2010 [2]. It is Canada's largest fruit export, and over half of the fruit-growing area is devoted to blueberries. British Columbia is the main supplier of highbush blueberries, while Nova Scotia and Quebec produce a large number of wild blueberry plants.

South America has more recently developed a major blueberry industry, with some 16,000 hectares devoted to blueberry cultivation. Chile is the largest producer and is a major exporter to North America and Europe, shipping 21,000 MT of fresh fruit and 1000 MT of frozen fruit. Other South American growers include Argentina and Uruguay. Several European countries also produce highbush blueberries, originally in the acidic sandy soils of Germany and the Netherlands, but since spreading to Poland, France, Austria, and Italy [3]. In 2008, an estimated 7300 hectares were planted with blueberries in Europe. Since the 1970s, Australia, New Zealand, and South Africa have also become major blueberry producers.

Recent interest in health-promoting compounds has spurred the production of blueberries. In particular, blueberries have been recognized as a major source of antioxidants and other phytochemicals including anthocyanins, proanthocyanidins, and flavonols that may have potential antidisease effects [4]. In addition, the desire to eat blueberries during off-seasons has encouraged imports from countries such as Chile and Argentina. Worldwide blueberry acreage has nearly doubled from 34,000 to 66,000 hectares from the period 2003 to 2008 [5]. In the United States, production of Maine wild blueberries increased 58%, while cultivated blueberries increased 24,000 hectares in 2008 to 28,000 hectares in 2010 [1].

Blueberries do have a rather short growing season, and this necessitates the development of processes to preserve them. In the United States, approximately 111,000 MT of blueberries

Dried Fruits: Phytochemicals and Health Effects, First Edition. Edited by Cesarettin Alasalvar and Fereidoon Shahidi.

enter the fresh market, while 75,000 MT are processed further [1]. Of processed fruit, about half are individually quick frozen (IQF). Others may be produced into juices, purees, and dried fruit, with the latter often infused with sugars. A wide variety of products may be produced from processed berries including blueberry pies, jams and jellies, baked goods, and snack foods.

This chapter reviews some of the blueberry varieties that are available, the important nutrients and phytochemicals they contain, current research on beneficial health effects, and some of the effects of processing on these components, with particular attention to drying technologies.

4.2 Varieties and composition

Blueberries are flowering plants from the family Ericaceae, which contain a variety of shrubs and trees that thrive in acidic soils [6]. They belong to the genus of *Vaccinium*, which consists of over 450 species of deciduous shrubs that thrive in cooler areas. Other *Vaccinium* berries include cranberries, bilberries, cowberries, crowberries, and farkleberries. Blueberries are "false" or epigenous berries and develop from an inferior ovary, and include tissue that develops from parts of the flower. They are perennial plants that live for several years, grow, and develop fruit in spring and summer and die off in the autumn.

Many varieties of blueberries exist, some of which are listed in Table 4.1. Of these, only a few are of major commercial importance in North America. The northern highbush blueberry (*Vaccinium corymbosum* L.) is the most abundant and grows in northeastern United States,

Table 4.1 Species of blueberries

Species	Common names	Distribution
Vaccinium angustifolium	Wild lowbush	Eastern and Central Canada, and Northeast United States
Vaccinium ashei	Southern rabbiteye	Southeastern United States
Vaccinium boreale	Northern	New England, Newfoundland, and Quebec
Vaccinium caesariense	New Jersey	New Jersey and parts of Eastern United States
Vaccinium corymbosum	Northern highbush	Nova Scotia, Ontario, Southeastern United States, Wisconsin, and Texarkana
Vaccinium cyanococcus	American	North America
Vaccinium darrowii	Southern highbush	Southeastern United States
Vaccinium elliotti	Elliott	Southeastern United States and Texarkana
Vaccinium formosum	Southern	Southeastern United States and coastal plains
Vaccinium fuscatum	Black highbush	New England, Southeastern United States, and Texarkana
Vaccinium hirsutum	Hairy fruited	Georgia, North Carolina, and Tennessee
Vaccinium myrtilloides	Canadian, velvetleaf	Canada, Northeastern, and Northwestern United States
Vaccinium operium	Cyan-fruited	Nova Scotia, Montana to Virginia
Vaccinium pallidum	Dryland, blue ridge	Eastern United States and Ontario
Vaccinium simulatum	Upland highbush	Southeastern United States, Ohio, and Kentucky
Vaccinium tenellum	Small black	Southeastern United States
Vaccinium virgatum	Small flower	Southeastern United States and Texarkana

Quebec, Ontario, parts of the Midwest, southeastern United States, Washington, and British Columbia, and accounts for over half of the world production of blueberries [7]. They can be 1.8 to 3.6 meters in height with two to five stems rising from the base. At least 50 cultivars exist including Bluejay, Bluecrop, Bluetta, Bluerat, Duke, Jersey, and Patriot. These are variations of the basic species selected from the wild, or bred to attain desirable properties, and are maintained by propagation through cuttings.

Another important variety is the Southern Rabbiteye (*Vaccinium ashei*) and includes many cultivars such as Bluebelle, Blue Gem, Brightwell, Climax, Delight, Powderblue, Southland, Tifblue, and Woodard. The Southern Rabbiteye grows primarily in the southern United States in a region stretching from North Carolina down to Florida, and westward to Texas, as it is more adapted to hot, humid summers, and mild winters. Rabbiteyes are also grown in Chile, Brazil, Australia, and South Africa and account for 15% of world production [8]. Plants are typically 1.8–2.4 meters tall, spread 0.9–3.0 meters, and have 5–10 stems developing from the base of the plant. The plants are productive and the berries ship well, but the fruit does tend to ripen a bit late for the fresh market. A little over half of the fruit is destined for further processing.

The lowbush blueberry (*Vaccinium angustfolium*) is a primarily wild-type bush, although a few improved varieties have been developed. It is a low spreading shrub forming dense colonies, and typically less than 0.9 meters in height. It is grown primarily in Maine, Massachusetts, and Canada. While accounting for 40% of the North American production, and 31% of world production, most of the fruit is used for further processing [8].

4.3 Compositional and nutritional characteristics of blueberries

4.3.1 Macro- and micronutrients

As with other berries, blueberries contain a variety of nutrients and related compounds beneficial to health. They are a particularly good source of vitamin C, with a one cup serving providing about one-fourth the recommended dietary allowances (RDA) value needed for an adult (Table 4.2) [9]. Blueberries and similar berries have been touted as a good source for at-risk populations with low vitamin C levels in their diet [10]. While fruits and nuts are not typically rich sources of vitamin K, blueberries (and a few others such as blackberries, grapes, and figs) do provide a significant amount of phylloquinone (vitamin K_1) [11]. Blueberries can also be a reasonable source of folic acid. Deficiencies of vitamin B have been related to neural tube defect in newborns, so there is concern that expectant mothers have sufficient levels [12]. Some studies have also suggested that folic acid can reduce the risk of heart disease and some types of cancer [13].

Amongst the mineral nutrients, manganese is the most prevalent in blueberries in terms of its contribution to RDA values. A one cup serving provides 0.5 mg or 25% of RDA. Manganese plays a role in forming connective tissue and bones, blood clotting, production of sex hormones, calcium absorption, and blood sugar regulation. It is a component or cofactor of several enzymes, including the antioxidant enzyme superoxide dismutase [14]. There is generally not much concern about adequate amounts of manganese in the diet, although some studies have suggested as many as 37% of Americans may be deficient in manganese, due to an overreliance on cereal grains in the diet in lieu of fruits and vegetables.

Table 4.2 Compositional and nutritional characteristics of raw blueberries (values in per cup = 148 grams edible portion) and percentage of RDA

	Units	Blueberries	Percent of RDA
Proximate composition			
Water	%	84.2	–
Energy	kcal	84	–
Protein	g	1.1	2
Lipid	g	0.49	1
Starch	g	0.0	–
Carbohydrate	g	21.45	16
Dietary fibre	g	3.6	14
Sugars	g	14.74	–
Minerals			
Calcium	mg	9.0	1
Copper	mg	0.08	4
Iron	mg	0.41	5
Magnesium	mg	9.0	2
Manganese	mg	0.50	22
Phosphorus	mg	18.0	2
Potassium	mg	114	3
Selenium	µg	0.1	0.2
Sodium	mg	1.0	0
Zinc	mg	0.24	2
Vitamins			
Folate	µg	6.0	2
Niacin	mg	0.42	3
Pantothenic acid	mg	0.12	2
Pyridoxine	mg	0.05	4
Riboflavin	mg	0.04	3
Thiamin	mg	0.04	3
Vitamin A (RAE)	µg	3.0	0.3
Vitamin C	mg	9.7	0.5
Vitamin E (ATE)	mg	0.57	4
Vitamin K	µg	19.3	16

Source: Adapted from USDA [9].
Note: Some numbers are rounded to the second digit after decimal point.
RDA, recommended dietary allowances; RAE, retinol activity equivalents; ATE, alpha-tocopherol equivalents.

Blueberries also provide small amounts of copper (4% of RDA), potassium (3% of RDA) magnesium (2% of RDA), and phosphorous (2% of RDA).

4.3.2 Dietary fiber

Blueberries are also a good source of dietary fiber. Although an exact definition is lacking, dietary fiber usually refers to plant components that are indigestible in the small intestine. These consist of remnants of the plant cell, lignins, and cell wall polysaccharides that are not digested. In fruits, the parenchymous flesh contains substantial pectin as part of the primary cell wall. Vascular tissue contains more of the hemicellulose xyloglucan. Epidermal tissue also contains cellulose and lignin along with cutin and other waxy materials [15]. There are two primary types of fiber, which is soluble and insoluble. According to USDA data, one cup of berries supplies 3.6 g of total dietary fiber (TDF), or 14% of RDA. While the database does not specify variety, there seem to be some variations in fiber depending on the cultivar, and indeed on the analytical procedure. Lowbush blueberries from Maine were found to have 4.06–4.39 g/100 g TDF depending on storage time, puree method, and concentration of citric acid used to extract fiber [16]. Pectin content ranged from 0.27 to 0.54 g/100 g. A Canadian study showed that in 100 g fresh blueberries with 2.6 g TDF, there was 0.69 g of soluble fiber and 1.9 g insoluble fiber [17]. In a comparison study of two Rabbiteye varieties (Tifblue and Premier) and two southern highbush varieties (Pearl River and Magnolia), rabbiteyes were found to have significantly higher neutral detergent fiber (NDF) and acid detergent fiber (ADF) than southern highbush varieties [18]. In 2002, the Rabbiteye varieties contained 17.5–23.7% NDF and 8.7–13.6% ADF. In comparison, southern highbush varieties contained 6.9–8.3% NDF and 4.3–4.6% ADF. The NDF fractions were determined to contain cellulose, hemicellulose, and lignin originating from the cell walls. It was also noted that fiber decreased as the fruit changed from a purple to a fully ripe stage.

4.3.3 Sugars and organic acids

As noted, 148 g of ripe blueberries have on the order of 45 g carbohydrate, with 14.74 g as sugar. In ripe berries, the predominant sugars are fructose [32.9 g/100 g dry weight (DW)], glucose (32.9 g/100 g DW), and sucrose (1.81 g/100 g DW) [19]. During ripening, the total amount of sugars increases from 35 to 63% of dry weight basis, and the ratio of glucose to fructose increases from 0.85 to 1.02. In addition to being sweet, blueberries are tart in nature, but the degree and type of acid vary with the variety. Lowbush varieties have approximately 36 μeq. citric acid, 31 μeq. malic acid, and 20 μeq. quinic acid per gram of dry weight, and also contain 10 μeq./g DW of the phenolic chlorogenic acid [20]. Highbush blueberries have approximately 75% citric acid and 17% succinic acid, while Rabbiteyes have 50% succinic, 37% malic, and 10% citric acids [21].

4.4 Phytochemicals

While blueberries are a good source of several nutrients, recent research has focused on the many phytochemicals in blueberries that may positively affect health. Blueberries are a good source of many polyphenolic compounds, including anthocyanins, flavonols, hydroxycinnamic acids, flavon-3-ols, and procyanidins. Of these, the class of pigments known as anthocyanins are most well known as they provide the characteristic color to *Vaccinium* spp. and related fruits. Many red, violet, and blue colors in most fruits, vegetables, and cereals arise from anthocyanins [22, 23]. They can occur in various parts of the plant, but are particularly concentrated in fruits and flowers. Anthocyanins are produced by the phenylpropanoid pathway and stored in cell vacuoles, and can occur in fruits, stems, tubers, and leaves. In fruits such as blueberries, the anthocyanins reside mostly in the epidermal layer of the skin. They seem to have several beneficial functions (or consequences) for plants although their exact roles are still debated [23]. In flowers, the color can attract pollinating insects for better pollination, while in the fruit skins the colors attract animals that eat the fruit and consequently disperse the seed. In leaf tissue, anthocyanins may provide protection against damaging UV radiation. Some research has suggested anthocyanins may modulate reactive oxygen signaling that mediates plant growth and development [24]. While the color of anthocyanidins and anthocyanins are pH dependent, in the low pH environment of fruit, a variety of characteristic colors are possible. Delphinidin typically contributes to blue and blue-red colors; cyanidin, peonidin, and pelargonidin to red-orange and purple colors; petunidin to dark red and purple colors; and malvidin to blue and red colors. In the intact fruit, the colors are modified or enhanced by several copigmentation systems [25].

The separation and identification of anthocyanins from plants can be difficult, particularly due to limited standards and variations caused by different solvent extractions [31]. Thus, many researchers report only on total anthocyanins or total phenolics. In addition, the type, distribution, and glycosylation of anthocyanidins depend upon cultivar, degree of maturity, soil, and weather conditions. This, in turn, determines bioactivity, as, for example, cyanidin-3-glucoside and -rhamnoglucoside have relatively high oxygen radical absorbance capacity (ORAC) values, while pelargonidin-3,5,-diglucoside and malvidin-3-glucoside have relatively low ORAC values [26]. Only five or six anthocyanidins are common in blueberry fruits [27]. Using paper chromatography, researchers isolated 15 anthocyanins from *V. angustifolium* [28] and showed the presence of delphinidin, petunidin, malvidin, peonidin, and cyanidin along with the sugars glucose, galactose, and arabinose (Figure 4.1). More recent studies have been based on high-performance liquid chromatography (HPLC) analysis including identification with standards or mass spectrometry (MS) [29–31]. Table 4.3 shows typical values for a few select lowbush, highbush, and Rabbiteye varieties. Most varieties contain the above aglycones, although varieties of lowbush blueberries do not contain pelargonidin [31, 32]. Most tend to have relatively high levels of delphinidin and malvidin, and lesser amounts of petunidin, cyanidin, and peonidin [32, 33]. Highbush varieties have higher levels of malvidin-3-galactoside, delphinidin-3-galactoside, delphinidin-3-arabinoside, petunidin-3-galactoside, petunidin-3-arabinoside, and malvidin-3-arabinoside than other varieties [34]. Most varieties also contain glucose, galactose, and arabinose, although the disaccharide 6-*O*-L-rhamnosyl-D-glucose (rutinoside) has been reported, particularly in the form of delphinidin-3-*O*-rutinoside [35]. Others have shown that many of the glycosyl residues may be acetylated. For example, "Chignecto" cultivars of *V. angustifolium* had 35% of the total anthocyanins in the acetylated form, while other cultivars had lesser amounts [30]. More recent studies have identified 49 different anthocyanins in wild blueberry [31]. Of those, only 19 had been previously identified. The other 30 included compounds such as delphinidin-3-hexose, delphinidin-3-oxalyl-hexose, cyanidin-malyl-pentose, and cyanidin-3-propionyl-galactoside.

Many researchers have preferred to report on the total anthocyanins or total phenolic compounds. Total anthocyanins are often measured by a pH differential method, while total phenolics are determined by reaction with the Folin–Ciocalteu reagent [36]. Table 4.3 shows total anthocyanin and phenolic content for select varieties of blueberries. Some varieties have as low as 33 mg cyanidin-3-glucoside equivalents (C3GE)/100 g ("Bluecrop") and some as high as 822.7 mg C3GE/100 g. It should be noted that much variation could arise due to the method of extraction. For example, higher levels of anthocyanins are attained with acidified solutions of methanol [37].

As phenolic compounds are concentrated in the skins of most blueberries, it has been hypothesized that smaller berries that have greater surface-to-volume ratios would contain more of these compounds when calculated on a per weight basis. Some research suggests that

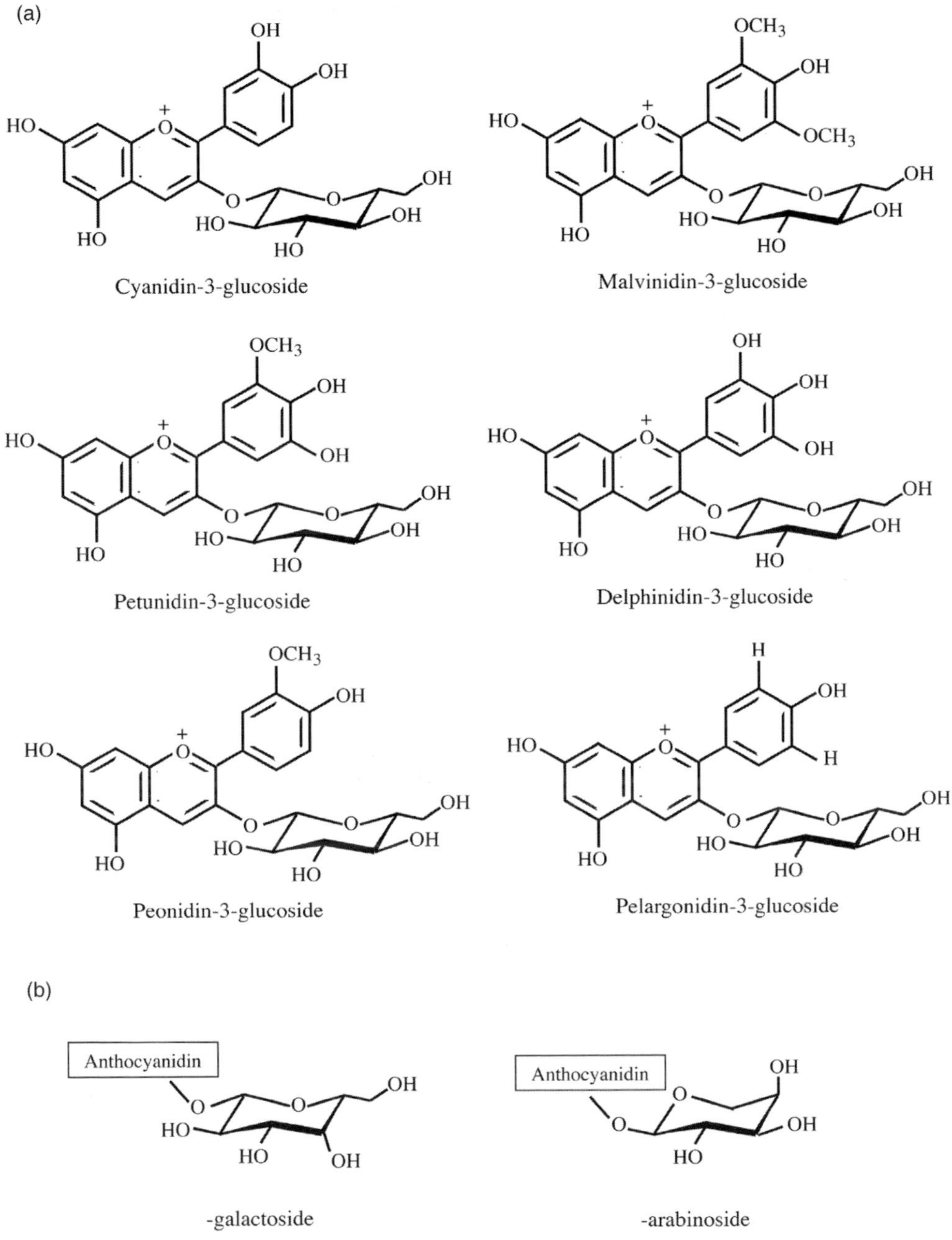

Figure 4.1 Major anthocyanins in blueberries (a) in glucoside forms. Anthocyanidins may also be attached to (b) galactoside or araboniside groups.

Table 4.3 Anthocyanins in various blueberry cultivars

	Wild [35]	Fundy lowbush [30]	Crunchie highbush [29]	Bluecrop highbush [35]	Highbush [33]	Bluecrop highbush [32]	Ozarkblue highbush [32]
Delphinidin-3-glucoside	84.4	16.2	7.7	7.9	6.3	12.6	0.8
Delphinidin-3-galactoside	25.8	15.9	17.5	14.1	11.0	18.8	38.9
Delphinidin-3-arabinoside	nd	7.1	10.4	14.2	11.9	16.3	18.6
Delphinidin-3-glucoside	17.9	nd	nd	nd	nd	nd	nd
Petunidin-3-glucoside	30.3	11.5	6.7	5.1	4.3	11.1	0.8
Petunidin-3-galactoside	25.5	8.0	9.5	7.7	7.4	11.3	22.8
Petunidin-3-arabinoside	nd	3.3	5.8	5.0	4.9	8.2	11.3
Malvidin-3-glucoside	139.6	19.9	nd	11.2	15.2	15.3	1.8
Malvidin-3-galactoside	101.2	17.4	nd	13.7	25.4	15.9	19.6
Malvidin-3-arabinoside	27.4	nd	8.5	nd	23.5	14.4	105.2
Peonidin-3-glucoside	17.4	4.9	17.6	0.5	0.9	nd	nd
Peonidin-3-galactoside	16.4	2.5	0.7	0.6	0.8	1.4	1.8
Peonidin-3-arabinoside	1.1	nd	11.4	nd	nd	0.6	0.2
Cyanidin-3-glucoside	27.5	9.1	0.9	0.7	1.0	2.0	0.4
Cyanidin-3-galactoside	nd	9.4	2.1	2.7	1.0	3.6	11.3
Cyanidin-3-arabinoside	nd	5.8	1.1	2.0	1.0	2.6	4.7
Pelargonidin-3-glucoside	29.6	nd	nd	nd	nd	nd	nd
Others	14.3	nd	0.1	1.2		9.1 (acylated)	0.7 (acylated)
Total	558.3	131.1	79	101	114.6	143.5	144.3

Note: Some numbers are rounded to the first digit after decimal point.
nd, not detected.
Values are expressed as mg of cyanidin-3-glucoside per 100 g sample.

lowbush blueberries contain higher levels of anthocyanins, total phenolics, and antioxidant capacity than highbush varieties, but there is no correlation between fruit size and anthocyanin content [37]. Other studies have shown that anthocyanin content is correlated with size in 15 cultivars of *V. corymbosum* ($r^2 = 0.84$), but not in other *Vaccinium* species [38].

It has been shown that total phenolics are highest for unripe green berries and decrease during color break and ripening, particularly at the unripe purple stage [39]. For example, unripe "Bluecrop" cultivars decreased from 60.7 to 33 mg gallic acid equivalents (GAE)/g DW at first harvest. Some varieties stabilized, some increased somewhat.

Often, antioxidant activity, total anthocyanins, and total phenolics content are well correlated. In a study on eight blueberry varieties, the antioxidant activity was correlated with total phenolics ($r^2 = 0.99$) and anthocyanins ($r^2 = 0.91$). Total anthocyanins were also correlated with total phenolics ($r^2 = 0.87$) [40]. However, others have shown modest correlation between ORAC values and total phenolics ($r^2 = 0.78$), but not with total anthocyanins.

4.5 Health effects related to blueberries

4.5.1 Micronutrients and health

Much of the health-promoting effects of blueberries can be traced to their micronutrient constituents. Many of these are not specific to blueberries and have been reviewed elsewhere. It has been suggested that blueberries can be a good source of vitamin C for at-risk populations [10]. Vitamin C is important for collagen synthesis and plays a role in hormone synthesis, immune system, iron absorption, platelet aggregation, and thrombus formation. It may also play a role in disease prevention by mediating oxidation in the body [41].

Vitamin K is needed as a cofactor for conversion of glutamic acid to γ-carboxyglutamic acid [42] and is important for blood coagulation through prothrombin and related factors, bone metabolism through osteocalcin, MGP, and periostin, and stimulation of vascular cells through growth arrest-specific protein 6. Some evidence suggests vitamin K helps with bone growth and density. Deficiencies of vitamin K are not common, especially as colonic bacteria helps produce vitamin K, so this vitamin does not receive much attention. Some have suggested that newborns are at most risk for vitamin K deficiency, and in the United States it is recommended that 0.5–1.0 mg vitamin K_1 be administered to newborns [43], although this practice is not without its critics. Foods containing vitamin K often get a lot of attention due to interactions with "blood-thinning" (anticoagulant) drugs such as warfarin and coumarin, which are used in the treatment of blood clots and embolisms. These drugs are actually vitamin K antagonists, as they decrease blood coagulation by inhibiting vitamin K epoxide reductase [44]. Relatively large amounts of vitamin K can interfere with these drugs, as they promote a secondary pathway for clotting [45]. Patients on anticoagulant medicines are usually monitored to ensure stable clotting times as measured by the international normalized ratio (INR), and there is some concern that eating foods high in vitamin K, particularly at irregular intervals, can interfere with stabilized clotting times.

4.5.2 Blueberry phytochemicals and health

In addition to the roles that various vitamins and minerals play in human nutrition, blueberries contain many antioxidants, phytosterols, enzymes, and other phytochemicals that may benefit human health [46]. These may work independent of, or in concert with, existing nutrients. In blueberries, the anthocyanins have garnered particular attention, but other compounds including pterostilbene, proanthocyanidins, flavonols, tannins, and reseveratrol may contribute to bioactivity. In a broad sense, many of these compounds have been related to reduced risk for coronary heart disease (CHD) and cancer, as well as reducing inflammatory processes. There have been many studies linking phytochemicals in blueberries with reduction of risk for different diseases. It should be noted in context, however, that few have reached the state of clinical trials and scientific consensus needed to make a health claim.

4.5.2.1 *Role of oxidative stress*

Some research suggests that free radicals can damage cells. Aging leads to reduced ability to combat oxidative stress, a state that contributes to cell aging, cancer, cardiovascular disease (CVD), cognitive impairment, cataracts, and macular degeneration [47–49]. Some studies have suggested that blueberries are a leading source of antioxidants, as measured by ORAC values, with values ranging from 13.9 to 44.6 mmol trolox equivalents (TE)/g fresh berries.

4.5.2.2 *Role in brain health and cognition*

Several research studies have indicated that blueberry phytochemicals can benefit the brain and reduce cognitive impairment from aging and memory loss. One theory is that antioxidants can limit oxidative stresses and inflammation that contribute to cell aging [50]. Polyphenolic antioxidants may also improve communication amongst neurons, mediate neuroprotective shock proteins, aid neuronal plasticity, and inhibit stress producing pathways [51]. Injections of rats with *Vaccinium* extracts showed that it facilitated transport of triiodothyronine (a T3 thyroid hormone that regulates metabolism) into the brain, and into specific areas associated with memory, vision, and sensory input [52]. When aged rats had a diet supplemented with 2% dried blueberry extract, they were able to navigate a maze that tests working memory as well as younger rats, and rats fed blueberry supplement performed better than those supplemented with strawberry or spinach [53]. Other biochemical markers related to neuronal function were also improved, including evoked dopamine uptake by striatal slices, stimulated guanosine triphosphatase (GTPase) activity, and calcium buffering. Related studies tested diet supplementation and its effect on β-adrenergic receptor function, which has been correlated with decline in the ability to learn new motor skills [54, 55]. Researchers reported that after subjecting rats to hyperoxia, a form of oxidative stress, antioxidants in blueberry and spinach improved the ability to navigate mazes and reduced declines in β-adrenergic receptor function in rat Purkinje neurons. They also showed that the diet supplements were able to decrease other age-related deterioration of the brain such as decreases in nerve growth factor [56].

Subsequent studies were conducted on rats predisposed to Alzheimer's disease. Older rats fed a diet chow containing 2% dried blueberry extract were able to learn to navigate a *Y* maze as well as young rats and better than those without supplementation [57]. In addition, studies on excised brain tissue showed that supplementation enhanced memory-associated neuronal signaling and altered sphingomyelin phospholipase C activity. Related studies showed that rats fed blueberry extracts showed improvement in balance, coordination, working memory, and reference memory [58]. The ability to regenerate hippocampal neurons is associated with memory decline. Studies on aged rats fed 20 g dried blueberry/kg showed improved biochemical markers for neuronal plasticity including hippocampal neurogenesis and extracellular kinase activity [59]. In addition, rats had fewer memory errors related to spatial tasks.

In other studies, young rats and those supplemented with blueberry extract performed better on tests of object recognition than aged control diet rats [60]. In addition, young rats and blueberry-supplemented aged rats had lower levels of the oxidative stress-responsive protein NF-κB (nuclear factor kappa-light-chain-enhancer of activated B cells) in four brain regions, indicating that these areas were under less oxidative stress. Another study investigated the brain's ability to generate heat shock protein 70 (HSP70), which acts to protect cells from thermal or oxidative stress [51]. Young and old rats were fed diets supplemented with blueberry, then hippocampal tissue was subject to an *in vitro* inflammatory challenge. Old rats with blueberry-supplemented diets had restored HSP70 levels as compared with those without supplementation, again suggesting protection against neurodegenerative processes.

One question of course is whether anthocyanins and related compounds can act directly in brain tissue. In a study on F344 rats fed with 2% blueberry supplementation, liquid chromatography–mass spectrometry (LC-MS) was used to identify anthocyanins in different regions of the brain [61]. Several anthocyanins were found in the cerebellum, cortex, hippocampus, or striatum of supplemented rats, including cyanidin-3-*O*-β-galactoside, cyanidin-3-*O*-β-glucoside, cyanidin-3-*O*-β-arabinose, malvidin-3-*O*-β-galactoside, malvidin-3-*O*-β-glucoside, malvidin-3-*O*-β-arabinose, peonidin-3-*O*-β-arabinose, and delphinidin-3-*O*-β-galactoside. These compounds were not found in the control groups. In addition, the levels of these compounds in the tissue were correlated with increased spatial learning and memory. This indicates that these anthocyanins can cross the blood–brain barrier and into regions associated with learning and memory. In a related study, in which European greenfinches were fed blackberries or blueberries (one per day for 2 weeks), cyanidin-3-glucoside was measured in homogenized brain tissue [62].

Few trials have been done on human subjects in which cognitive function has been tested after administration of blueberry products. In one medium-term trial, nine older adults with early memory changes drank 2–2.5 cups of *V. angustifolium* blueberry juice for 12 weeks, and were evaluated using the Verbal Paired Associate Learning Test, the California Verbal Learning Test, and the Geriatric Depression Scale [63]. Improvement was noted in paired associate learning and word list recall, and there was a trend toward lower symptoms of depression. Similar positive findings have been found with daily supplementation of Concord grape juice [64].

4.5.2.3 *Urinary tract infections*

Extensive work has been done on the treatment of urinary tract infections (UTIs) with diet or extracts from the *Vaccinium* family, particularly cranberries, bilberries, and blueberries. By and large, most studies including clinical trials have been carried out with cranberry and suggest that cranberry is beneficial for preventing UTI but less helpful for treating existing UTI [65]. For example, one clinical trial showed that daily consumption of cranberry juice reduced UTIs by nearly half in elderly women [66]. One theory has been that cranberries help reduce the pH of urine from near neutral to 5.5 or lower, which helps prevent bacteria from growing. While urinary pH can be lowered by daily regimens of cranberry juice [67], the role of lower pH in limiting UTI has been questioned [68]. Current hypotheses suggest that particular "A Type" proanthocyanidins in cranberry, blueberry, and peanut skins inhibit bacterial adhesins that allow bacteria, particularly *Escherichia coli* with fimbrae, from attaching to the mucosal surfaces of the urinary tract, oral cavities, or stomach [69]. One group reported on a compound in blueberry and cranberry juices that inhibits the mannose-resistant (MR) adhesin of *E. coli* [70]. Others studied extracts of wild blueberry that were separated into two fractions: one with a 3.25 and one with a 5.65 average degree of polymerization [71]. The former was able to suppress uropathogenic *E. coli* strains from agglutinating human red blood cells. Evidence also suggests that proanthocyanidins in berries can suppress gene expression that allows *E. coli* to produce fimbrae for attachment [72]. While implications for blueberries in preventing UTI are promising, it should be emphasized that more studies are needed to support their efficacy as a treatment. Indeed, a systematic review of the use of cranberries and blueberries in UTI prevention showed that while several studies indicated that these products significantly reduced the incidence of UTIs, "no trials were identified which evaluated the effectiveness of blueberry products for the prevention or treatment of UTIs" [73].

4.5.2.4 *Cancer*

A substantial number of studies have investigated whether polyphenolic compounds in fruits and vegetables can reduce cancer risks. In general, epidemiological studies have shown that people with high fruit intake are at reduced risk for cancer mortality [74], and that fruits and vegetables provide significant protective effects against lung, colon, and many other forms of cancer [75]. These effects seem to be related to plant flavonoids. A study on 9959 Finnish men and women reported that persons with higher levels of flavonoids in their diet had a decreased risk of lung cancer [76]. Subsequent work suggested that men with higher intake of quercetin had lower incidence of lung cancer and those with higher intake of myricetin had lower incidence of prostate cancer [77]. Although many studies suggest polyphenolics have chemoprotective activity, some indicate there is no effect or even cancer-inducing effects [78]. The mode of action has been attributed to effects on cellular differentiation, proliferation, and apoptosis, as well as enzymes involved with the cellular processes.

Anthocyanin-rich fractions from several Rabbiteye and highbush blueberries were studied for their ability to regulate apoptosis and phase II enzymes in HT-29 (human colona adenocarcinoma) cell lines [79]. Fractions from Tifblue and Powderblue cultivars at 50–150 μg/mL caused increased DNA fragmentation characteristic of cell apotosis, suggesting that the cells can then prevent unregulated growth.

Pterostilbene found in blueberries was found to suppress aberrant crypt foci formation in a rat model for colon cancer [80]. These foci are clusters that would eventually form into polyps or colon cancer. Blueberry and other *Vaccinium* species were shown to bind bile acids *in vitro* [81]. The authors suggested this might be beneficial in the prevention of colon cancer by removing toxic substances.

Proanthocyanidin-rich fractions from wild and cultivated blueberries were investigated for their ability to inhibit growth of two prostate cancer cell lines [82]. Two fractions from both types of berries were effective against one cell line but not the other. Nuclear magnetic resonance (NMR) analysis suggested that the active fractions contained a mixture of oligomeric proanthocyanidin molecules. The authors speculated that the proanthocyanidins are most effective against androgen-dependent growth of prostate cancer cells. In another study, fractions prepared from lowbush blueberries were able to inhibit the matrix metalloprotein in

DU145 human prostate cancer cells [83]. This protein is an enzyme that may allow cancer cells to metastasize.

In a comprehensive study on anticancer activity of several berries, extracts were prepared from blackberries, black raspberries, blueberries, cranberries, red raspberries, and strawberries [84]. These extracts were tested for their ability to limit proliferation of human oral, breast, colon, and prostate cancer cell lines. All extracts were able to limit proliferation in the different cell lines and the effect was shown to be dose dependent. Tests were also conducted to see if extracts could stimulate apoptosis in colon caner (HT-29) cells. Black raspberry and strawberry extracts were the most pro-apoptotic.

4.5.2.5 *Cardiovascular disease*

Several epidemiological studies have linked diets rich in fruit, vegetable, and whole-grain cereals with reduced risk for CVD [85]. These benefits have been attributed to the dietary fiber and numerous phytochemicals, and of the latter flavonoids have been given particular attention. As berry crops have substantial numbers of polyphenolic compounds, they have been investigated for their health-promoting potential [86]. Several mechanisms have been suggested for the cardioprotective attributes of flavonoids. As they are known antioxidants, they may act *in vivo* to limit the formation of antherogenic oxidized low-density lipoprotein (LDL) [87]. Plaque formation is also related to arterial inflammation and flavonoids may function by reducing this vascular inflammation [88]. Some research also suggests that flavonoids can limit the platelet aggregation and adhesion that accompany arterial plaque formation [89].

Using pigs as a model for human cardiovascular attributes, researchers fed the animals a plant-based diet (70% soy, oats, and barley) supplemented with 1–4% freeze-dried blueberry powder [90]. LDL and high-density lipoprotein (HDL) cholesterol levels were reduced by up to 5% in the diet supplemented with 2% blueberry. The effects of supplementation were reduced when the basal diet contained only 20% plant matter, suggesting a synergistic effect between the blueberry and cereal-based ingredients.

Studies on humans have also been promising. In a feeding trial, 48 obese participants with metabolic syndrome had diets supplemented with 50 g freeze-dried blueberry powder for 8 weeks [91]. Supplemented participants had 6 and 4% lower systolic and diastolic blood pressures than the control group, 28% less oxidized LDL, and 17% lower levels of plasma malondialdehyde.

4.5.2.6 *Vision health*

A few studies have suggested that extracts from *Vaccinium* spp. can improve eyesight. Most work has been done on bilberry (*Vaccinium myrtillus*), particularly as it relates to improving eyestrain and night vision. For example, rats supplemented with bilberry extract were better able to regenerate the "visual purple" pigment (rhodopsin) in their photoreceptor cells [92]. Human subjects given bilberry extract were also better able to adapt to dark surroundings, with the benefits lasting for 24 hours [93]. Others have suggested that antioxidants including anthocyanins can alleviate underlying causes of macular degeneration [94]. A 2004 review of 30 human trials concluded that while some animal studies and weaker trials showed promising results, more rigorous clinical studies do not support the hypothesis that bilberry anthocyanins improve night vision [95].

4.6 Effects of processing on blueberry components

4.6.1 Importance of blueberry processing

Blueberries generally have a short harvest season of several weeks. They typically can be stored refrigerated for only 4–6 weeks and those fruit not destined for the fresh market are frozen, processed into juices, and canned products, or dried into intermediate-moisture fruits for hand eating, or into powders for ingredients or supplements.

4.6.2 General effects of heat, oxygen, and enzymes

Much food processing relies on the application of heat to destroy microorganisms, inactivate enzymes, or promote the removal of water. Heating is known to be detrimental to anthocyanins, and the temperature and duration of heating determine the degree of degradation. When blueberry juice was heated at 40, 50, 60, 70, and 80°C, the resulting half-life for anthocyanins was 180.5, 42.3, 25.3, 8.6, and 5.1 hours, respectively [96]. Thermal degradation followed a first-order kinetic model. Q10 values were found in the range of 1.6–4.3, which is typical of many chemical reactions. The thermal breakdown of anthocyanins is commodity-specific and depends on solid content, pH and copigment factors. Some anthocyanins may be more susceptible to heat, such as cyanidin-3-glucoside and pelargonidin-3-glucoside in blackberries and strawberries [97]. At pH 3.5, breakdown products include chalcone glycosides with an opening of the flavylium backbone [98]. Phenolic acids and phloroglucinaldehyde were also found as remainders of the B- and A-rings. Acylation and methoxylation improved the stability of anthocyanins to heating [99].

Heating is required, however, to inactivate polyphenol oxidases (PPOs) and catechol oxidases. In plants, the exact function of these enzymes is not fully understood, but they may be related to protection against pests and other stresses [100]. In fruit subject to handling and processing, these enzymes can catalyze the hydroxylation of monophenols to diphenols, with the subsequent formation of *o*-quinones which polymerize to produce brown pigments. PPOs are processed in chloroplasts and cutting, grinding, or bruising actions disrupt compartmentalization, allowing the enzymes to act on phenolic substrates. In addition, catechols and other diphenols are converted to *o*-quinones in the presence of oxygen and cell disruption encourages this exposure. It has been noted that when blueberries and other berries are not blanched, anthocyanin degradation occurs, even when the samples are stored frozen [101].

4.6.3 Juices, purees, and canning

Blueberries are processed into several high-moisture products including juices, purees, pie fillings, and concentrates. This requires several operations that may introduce high shear and cell disruption (grinding), heating (blanching and pasteurization), and storage at various temperatures. When frozen highbush blueberries with 99.9 mg C3GE/100 g (total anthocyanins) were processed into juice and concentrate, 33 mg C3GE/100 g were retained in the initial juice, 178 mg C3GE/100 g in the concentrate, and 184 mg C3GE/100 g in the pressed cake, accounting for 18% of the anthocyanins in the original fruit [101]. Interestingly, for juice pasteurized at 90°C for 1 min, anthocyanin levels were higher (38 mg C3GE/100 g). The authors felt that losses were incurred from enzymatic degradation during storage and that these were inactivated during pasteurization. This theory was reinforced as the addition of blanched blueberry puree to juice resulted in no further loss of anthocyanins, while addition of unblanched puree resulted in a 50% loss. The proportion of malvidin glycosides increased to 80% in the pressed juice and 63% in the pasteurized juice. Delphidin glycosides were least stable, followed by cyanidin and petunidin glycosides. The authors suggested that malvidin glycosides are stabilized by the two methyl groups on the B-ring, while delphinidin has three ortho phenolic groups. Cinnamic acid is more water soluble, thus more of it is recovered in the juice (13.2 mg/100 g as compared to 27.4 mg/100 g in the fruit), while little was found in the pressed cake (3.2 mg/100 g). Flavonol glycosides totaled 40.1 mg/100 g in the fruit, 24.9 mg/100 g in the juice, 16.7 mg/100 g in the pasteurized juice, and 26.8 mg/100 g in the press cake. The authors noted that these compounds are less susceptible to enzymatic degradation. Procyanidins totaled 9.9 mg/100g in the fruit, 4.5 mg/100g in the initial juice, and 5.0 mg/100g in the pasteurized juice. Little was found in the press cake.

Subsequent studies reported on the effects of pureeing, canning, and juicing on polyphenolics in highbush blueberries [102]. Chlorogenic acid was retained in pasteurized canned berries and nonclarified juice, but not as well in purees and clarified juice. In purees, chlorogenic acid was likely destroyed by PPO prior to blanching. High levels of flavonols were retained in berries canned in water and in purees (>97%). Only 57% remained in clarified juice, which the authors attributed to physical removal or loss of coprotective compounds. After 6 months at 25°C, 75–82% were retained in berries canned in syrup and in nonclarified juice. Anthocyanins were not retained as well during processing (41–72%), with greatest losses in clarified juice followed by purees. All incurred significant losses during six months storage. Related work on procyanidins showed a range of retention levels for nonclarified juice (19%), clarified juice (23%), purees (41%), berries canned in syrup (65%), and berries canned in water (78%) [103]. Larger oligomers were less well extracted during juicing than mono- and dimers. After 6 months storage, only 8–32% of the original levels were present and again mono- and dimers were better retained during storage than oligomers.

Other studies examined the effects of processing on antioxidant and antiproliferation activity [104]. For wild blueberries, IQF had the highest total phenolics followed by fresh, freeze-dried, canned, and juice concentrates. Differences were smaller for cultivated varieties and products that were heat-treated had slightly lower levels. Interestingly, antioxidant assays [ferric reducing antioxidant power full name required (FRAP) and full name required 2,2-diphenyl-1-picrylhydrazyl radical scavenging activity (DPPH)] did not show distinctions between heated and nonheated products. However, heat-treated products had much lower antiproliferation activity against liver cancer cell cultures.

4.6.4 Freezing

Freezing is perhaps one of the best ways to preserve chemical constituents of berries [105]. Studies on *V. corymbosum* L. compared the effects of freezing versus storage of fresh samples at 5°C and drying in a cabinet dryer. For the dried products, anthocyanin content was 41–49% lower than in fresh berries. In contrast, anthocyanin levels in frozen blueberries did not change even after three months storage. Similar studies on *V. corymbosum* L. were conducted in which berries were stored at –18 and –35°C for six months. Derivatives of malvidin and delphinidin accounted for most of the anthocyanins [106]. No significant changes ($P > 0.05$) were found in total anthocyanins, individual anthocyanin components, or chlorogenic acid at either temperature. In addition, antioxidant activity remained unchanged compared to values measured just after freezing. In comparison, studies on raspberries and blackberries showed some small changes after 12 months storage at –24°C [107]. For example, for wild blackberries, vitamin C decreased from 25.6 to 16.6 mg/100 g, total phenolics from 978 to 756 mg GAE/100 g, and anthocyanins from 306 to 248 mg C3GE/100 g.

Another study examined the effects of storage temperature on blanched extracts from Tifblue and Powderblue cultivars [108]. At –20°C, there were no changes on total phenols, anthocyanins, or antioxidant activity after 30 days storage. At 6°C, some changes were noted after 15 days. Greater losses were noted at 23 and 35°C, especially after 15 days storage and anthocyanins were not detected after 60 days storage at 35°C.

4.6.5 Drying

Drying is one of the primary means of preserving fruits and vegetables and functions by lowering the water activity below a point at which microorganisms can proliferate, and hopefully below a point at which chemical degradation occurs. There are several types of drying, however, each utilizing a particular temperature and means of applying heat and removing water. In turn, these result in a variety of products ranging from chewy intermediate-moisture fruits to dried powders, with differing recovery of phytochemicals (Table 4.4).

Table 4.4 Affects of select drying technologies on blueberry phytochemicals

	Material	Results	Reference
Fresh	Whole highbush (87% moisture)	670 mg GAE/g total phenolics (glycosides)	[113]
Freeze dried	Whole highbush (5% moisture)	342 mg GAE/g total phenolics (glycosides)	[113]
Hot air (76°C)	Whole highbush (5% moisture)	79 mg GAE/g total phenolics (glycosides)	[113]
Microwave/vacuum	Whole highbush (5% moisture)	234 mg GAE/g total phenoics	[113]
Fresh	Highbush	7.2 mg C3GE/g total anthocyanins	[105]
Hot air (90°C)	Highbush	4.3 mg C3GE/g total anthocyanins	[105]
Hot air (90°C), 60% sugar	Highbush	3.7 mg C3GE/g total anthocyanins	[105]
Extract	Rabbiteye (35% EtOH extract)	9.2 mg C3GE/g total anthocyanins 30 mg GAE/g total phenolics	[120]
Spray dried	Rabbiteye (35% EtOH extract)	3.7 mg C3GE/g total anthocyanins 15 mg GAE/g total phenolics	[120]
Dough	Blueberry concentrate/corn	0.40 mg C3GE/g total anthocyanins	[125]
Extrudate	Blueberry concentrate/corn	0.04 mg C3GE/g total anthocyanins	[125]
Radiant zone	Highbush mix extract Highbush mix dried extract	11.7 mg GAE/g total phenolics 11.2 mg GAE/g total phenolics	[127]
Vacuum belt	Whole rabbiteye (fresh)	16.7 mg C3GE/g total anthocyanins 41.8 mg GAE/g total phenolics	[128]
	Whole rabbiteye (dried)	16.0 mg C3GE/g total anthocyanins 41.7 mg GAE/g total phenolics	[128]

GAE, gallic acid equivalents; C3GE, cyanidin-3-glucoside equivalents; EtOH, ethanol.

4.6.5.1 *Freeze drying*

Freeze drying occurs at low vacuum and relatively low temperature, and water is removed as vapor from the product directly from the solid state. It is believed to be one of the least detrimental to chemical constituents in foods and pharmaceuticals, although loss of volatiles can occur. Indeed, it is often used as part of the procedure for preparing berries and other food materials for chemical analysis [109]. It has been noted, however, that spray-dried and freeze-dried blueberries can be exceptionally hygroscopic. This is somewhat ameliorated by depectinization, drying with a carrier and the addition of free-flow agents.

When compared to forced air, vacuum oven and microconvection, freeze-dried lowbush blueberries were darker and redder, had higher soluble solids, and higher vitamin C retention [109]. In addition, freeze-dried samples had the lowest bulk density and highest rehydration ratio. In studies on bioactive properties of wild blueberries, it was found that antioxidant activity, galvinoxyl free radical quenching, ability to limit lipid oxidation, and activity against Hepa 1c1c7 cell proliferation were maintained in freeze-dried samples [110].

4.6.5.2 *Hot-air drying*

Drying berries by subjecting them to a steady stream of hot air is one of the most common technologies in practice in the industry. To improve both heat and mass transfer, various means are used to insure good airflow around individual pieces. In fluidized bed driers, the airflow is sufficient to suspend individual berries. In rotary driers, a drum is used to constantly turn pieces and bring buried pieces to the surface. One impediment to drying is the waxy epidermal layer, which inhibits transfer of moisture in and out of the fruit. Several methods have been investigated for removing this prior to drying. In a research setting, berries have been tumbled with an abrasive sidewall to scuff the surface. In a high-tech variant, researchers have investigated the use of CO_2 lasers to perforate the skin prior to infusion [111]. In commercial practice, berries are usually passed through a scarifier, which consists of a series of blades which score the surface and somewhat flatten the product. In addition, berries destined to be eaten as dried fruits are often pretreated in solutions containing high sugar levels. This osmoconcentration step provides some initial drying and infuses the product, making it more flexible and less dense. The combination of reduced flavor loss, added sugar, and lower tartness often makes for more acceptable dried fruit products.

Hot-air drying was compared with microwave-assisted drying, freeze drying, and vacuum drying of cranberries (*Vaccinium macrocarpon*) infused with 76°C Brix sugar solution [112]. While freeze-dried products had the best rehydration properties and microwave-assisted processes dried the fastest, there were no differences in appearance or taste, save that hot-air-dried berries had the best appearance. In comparative studies on blueberries, the degree of phytochemical and bioactivity retained during freeze, microwave-vacuum (MIVAC), hot-air (76°C), and combined hot-air MIVAC (HAMIVAC) were determined [113]. Hot-air-dried blueberries had highest retention of ellagic acid, quercetin, and kaempferol. Freeze-dried and HAMIVAC drying had best retention of total phenolics and anthocyanins. For example, fresh berries had 66.8 mg GAE/100 g total polyphenols while freeze dried berries had 34.2 g GAE/100 g, HAMIVAC berries had 23.4 mg GAE/100 g, and hot-air-dried berries had 7.99 g GAE/100 g. Corresponding values for total anthocyanins were 113, 291,147, and 67.9 mg pelargonidin-3-glucoside equivalents (P3GE)/100 g, respectively. Freeze-dried blueberries had the highest antioxidant capacity. Berries dried only with hot air had the lowest retention of individual polyphenols. A similar study found that a combination of pretreatment in ethyl oleate/NaOH (to make the skin more permeable), osmotic dehydration in sucrose, and microwave-assisted convection drying produced blueberries that were comparable to freeze-dried berries in a relatively short time [114]. The effects of combining microwave with (spouted) fluidized drying (MWSB) of prefrozen blueberries have also been examined [115]. The assisted drying time was reduced up to 1/24th of that for tray drying. MWSB berries had low bulk density, better rehydration properties, and slightly redder color than those produced from noncombined spouted bed or tray drying.

The effects of cabinet drying (90°C for 90 min) on highbush blueberries, with and without an osmotic pretreatment (60% sucrose and 1% NaCl), were examined [105]. The total anthocyanins was less in dried samples compared to fresh blueberries (7.2 mg C3R/g dry matter). Untreated dried berries had slightly higher values (4.3 mg C3R/g) than osmotically treated berries (3.7 mg C3R/g). It has been noted that dewaxing of the berry cuticle can lead to leaching of anthocyanins when blueberries are placed in osmotic solutions [116].

Several commercially available blueberry products that had been hot-air-dried, spray-dried, or freeze-dried were analyzed for antioxidant and antiproliferation activity [71]. Products that were heat processed generally had slightly lower levels of phenolic compounds and antioxidant activity. However, they had much lower antiproliferation activity against Hepa-1c1c7 cells. Another survey of commercially processed blueberry products found that products that had received less processing had higher antioxidant activity [117]. Compared to fresh fruit with ORAC values of 52.9 mmol TE/100 g dry weight basis, IQF frozen berries had 31.2, intermediate moisture fruit 25.5, low-moisture fruit 15.1, sugar-infused fruit 11.3, and blended powder 7.44 mmol TE/100 g dry weight basis.

The effects of hot air and freeze drying were compared for conventional and organic blueberries [118]. No differences ($P > 0.05$) in phytochemicals were measured between conventional and organic berries. Blanching prior to drying helped to improve the drying rate and the retention of phytochemicals, but air drying resulted in substantial losses in anthocyanins, total phenolic compounds, and antioxidant activity. Freeze drying had lowest losses in phytochemicals.

In an attempt to improve drying properties, Rabbiteye blueberries were subjected to high-frequency ultrasound during osmoconcentration in 55°C Brix sucrose, prior to drying in a cabinet dryer at 70°C for 10 hours [119]. Osmoconcentration decreased titratable acidity and resulted in losses as high as 60% of anthocyanins and phenolics. They attributed this to leakage into the sugar media, facilitated by prior disruption of the skin, and cuticle by freezing. Subsequent air drying decreased levels up to 69% further. The samples treated with ultrasound had even greater losses than those with no treatment. The combination of high temperature, high sugar levels, and oxygen were reported to be most detrimental to color and antioxidant properties.

4.6.5.3 *Spray drying*

Spray drying is a convenient way to produce dry powders as it allows high production rates and relatively low production costs, particularly as compared to freeze drying. However, spray drying of fruit extracts or purees can be complicated by the presence of high sugar levels. These result in relatively high glass transition temperatures and sticky points. Often, high molecular weight carriers such as maltodextrins are added, which can also help reduce oxidation by coating the fruit particles [120]. Despite an important processing method, little research has been conducted on spray drying of blueberry products.

Studies showed that for spray drying of several fruit juice concentrates, best results were attained with maltodextrin DE6 at a 65:35 ratio for blackcurrants and 55:45 ratio for raspberries, coupled with relatively low air temperatures (160°C inlet/90°C exit) [121]. Others have noted that spray drying of fruit juices is quite complex, due to the interplay of initial solids, type of carrier, feed flow, and air temperatures [122]. For orange juice, the inlet air temperature and feed flow rate had the greatest effect on dryer yield and wall deposit. In studies on spray-dried acerola pomace extract, the best processing conditions were obtained using an inlet temperature of 194°C, a 4:1 ratio of drying aid (maltodextrin or cashew tree gum) to extract, and replacement of the maltodextrin with up to 80% cashew tree gum [123]. The resulting powders were least hygroscopic and with good flow ability. Follow-up studies showed that there is a trade-off concerning the retention of nutrients and bioactive compounds [124]. Maximum retention of the acerola extract ascorbic acid (93%) and total anthocyanins (90%) were attained at an inlet temperature of 170°C and drying aid-to-extract ratio of 5:1.

As previously noted, spray-dried blueberry products, as well as other products subject to heating, had relatively low antiproliferation activity as compared to fresh, IQF frozen, or freeze-dried products [103]. Total phenolic compounds retained were also lower in spray-dried blueberries as compared to non-heat-treated products, and canned blueberries or juice concentrates.

Response surface methodology was used to optimize a method for producing polyphenolic nutraceuticals from Muscadine grapes and Rabbiteye blueberries [120]. Aqueous 35% ethanol extracts were best obtained at 60°C after depectinization, then fermented with *Saccharomyces cerevisiae* at 25°C. The extract was filtered, then evaporated under vacuum. Concentrated extracts were spray-dried at 150°C (inlet) and 90°C (outlet). Although only 22–52% of the total phenolics and anthocyanins were recovered from the raw material, the authors felt the process was promising for commercial production.

4.6.5.4 *Extrusion*

Another interesting potential for blueberries, grapes, and other fruit products is as an ingredient for extruded products. Extrusion can technically be considered as a form of dehydration, although the amount of water removal is usually not large. For cooked and puffed snacks and cereals, dough of approximately 15–20% moisture is forced through the extruder barrel by a series of screws that mix, shear, and compress the product. The product exits the die, often with some expansion, at moisture content between 6 and 12%. In one study, white corn meal was extruded with corn syrup, blueberry concentrate, or grape juice concentrate to form a potential breakfast cereal [125]. Extrusion reduced the level of anthocyanins from 40 mg malvidin-3-glucoside equivalents (M3GE)/100 g DW to 4 mg M3GE/100 g DW in the extruded product, but they were stable in subsequent storage for up to three months. Part of the loss was attributed to pigment polymerization and subsequent browning.

In subsequent studies, fruit powders, including blueberries, cranberries, concord grapes, and raspberries, were used to form extruded breakfast cereals [126]. The fruit cereals were slightly denser and redder, and contained higher phenolics and anthocyanins, and blueberry cereals had the highest level of anthocyanins. During 3–6 weeks storage, fruit cereal also had lower levels of hexanal and similar headspace volatiles, which indicated that less lipid oxidation had occurred.

4.6.5.5 *Radiant zone and vacuum belt drying*

Blueberry powders were prepared from extract, juice, and puree using a novel radiant zone dryer [127]. In this system, liquid product was deposited on a moving polyester belt, and moved through five temperature zones (45–90°C) produced by radiant heaters, allowing for maximum water removal during early drying. No significant changes ($P > 0.05$) in total anthocyanins were noted between liquid and powders. Highest levels were found in the extract and powder made therefrom (11.7 and 11.2 mg C3GE/g dry weight basis). Total phenolics also did not vary between wet and dry materials, with most found in the extract and extract powders (97.1 and 113.1 mg GAE/g dry weight basis). Total antioxidant activity, as determined by a proprietary assay, also did not vary from the wet and dried products.

Continuous vacuum belt drying has been studied in our labs as a means of rapid and low-temperature drying [128]. In this approach, fruit or purees are introduced through an air lock onto a belt held under vacuum. The material is conveyed over several conduction heaters, providing independent temperature during different drying periods. With this approach, Rabbiteye blueberries could be dried to very low moisture levels within 60–90 min. In addition, total monomeric anthocyanins, total phenolics, and ORAC values were not impacted with conduction temperatures 110°C or below.

4.7 Conclusions

Several varieties of blueberries exist that contain a combination of traditional nutrients and phytochemicals that are linked to good health promotion. Many studies have examined the phenolic compounds in blueberries, their bioavailability, and their potential for improving brain function, limiting infections, and reducing risk of cancer and CVD. As a significant amount of blueberries are processed, the role of processing in degrading or reducing bioavailability is of considerable interest. Processes that include heating or initiate enzymatic activity are most detrimental. Thus, freezing or freeze drying is least harmful to anthocyanins and related compounds, while juicing, canning, and hot-air drying is most harmful. Amongst drying technologies, freeze drying, radiant zone, and vacuum belt drying cause least changes in blueberry phytochemicals.

References

1. USDA (2011) *Noncitrus Fruits and Nuts: 2010 Summary*. USDA, National Agricultural Statistics Service, Washington, DC.
2. The National Highbush Blueberry Promotion and Research Agency (2011) *Facts Regarding Blueberry Production in Canada*. Published on-line at: http://www.bcblueberry.com/media/documents/national_council/faqs_blueberry_web.pdf, last accessed September 5, 2011.
3. Naumann, W.D. (1993) Overview of the *Vaccinium* industry in Western Europe. In: *Proceedings of the Fifth International Symposium on Vaccinium Culture* (ed. KA Clayton-Greene). International Society for Horticultural Science, Wageningen, The Netherlands, pp. 53–58.
4. Seeram, N.P., Adams, L.S., Zhang, Y., Lee, R., Sand, D., Scheuller, H.S. & Heber, D. (2006) Blackberry, black raspberry, blueberry, cranberry, red raspberry, and strawberry extracts inhibit growth and stimulate apoptosis of human cancer cells *in vitro*. *Journal of Agricultural and Food Chemistry*, **54**, 9329–9339.
5. US Highbush Blueberry Council (2009) *Blueberry Trends and Market Challenges*. Published on-line at: http://sfp.ucdavis.edu/events/09blueberries/villata.pdf, last accessed September 6, 2011.

6. Kron, K., Powell, E.A. & Luteyn, J.L. (2002) Phylogenetic relationships within the blueberry tribe (*Vaccinieae*, *Ericaceae*) based on sequence data from MATK and nuclear ribosomal ITS regions, with comments on the placement of Satyria. *American Journal of Botany*, **89**, 327–336.
7. USDA (2010) *Plant Fact Sheet: Highbush blueberry (Vaccinium corymbosum L.)*. Published on-line at: http://plants.usda.gov/ factsheet/pdf/fs_vaco.pdf, last accessed September 5, 2011.
8. Ballington, J. (2005) Blueberry varieties around the world. Paper given at the *Ciclo de Seminarios Frutícolas de Actualización Técnico Comercial. Berries: Arándanos - Frambuesas*, Santiago, Chile, June 21–22, 2005
9. USDA (2011) *National Nutrient Database for Standard Reference, Release 24*. Published on-line at: http://www.nal.usda.gov/fnic/foodcomp/search/, last accessed September 7, 2011.
10. Beattie, J., Crozier, A. & Duthie, G.G. (2005) Potential health benefits of berries. *Current Nutrition and Food Science*, **1**, 71–86.
11. Dismore, M.L., Haytowitz, D.B., Gebhardt, S.E., Peterson, J.W. & Booth, S.L. (2003) Vitamin K content of nuts and fruits in the US diet. *Journal of the American Dietetic Association*, **103**, 1650–1652.
12. Honein, M.A., Paulozzi, L.J., Mathews, T.J., Erickson, J.D. & Wong, L.Y.C. (2001) Impact of folic acid fortification of the US food supply on the occurrence of neural tube defects. *Journal of the American Medical Association*, **285**, 2981–2986.
13. Verhaar, M.C., Stroes, E. & Rabelink, T.J. (2002) Folates and cardiovascular disease. *Arteriosclerosis Thrombosis and Vascular Biology*, **22**, 6–13.
14. Crowley, J.D., Traynor, D.A. & Weatherburn, D.C. (2000) Enzymes and proteins containing manganese: an overview. *Metal Ions in Biological Systems*, **37**, 209–278.
15. Dreher, M.L. (2001) Dietary fiber overview. In: *Handbook of Dietary Fiber* (eds S. Cho & M.L. Dreher). CRC Press, Boca Raton, FL, pp. 1–17.
16. Chen, H.C. & Camire, M.E. (1997) Recovery of anthocyanins, pectin, and dietary fiber from cull lowbush blueberries. *Journal of Food Quality*, **20**, 199–209.
17. Mongeau, R., Brassard, R. & Verdier, P. (1989) Measurement of dietary fiber in a total diet study. *Journal of Food Composition and Analysis*, **2**, 317–326.
18. Marshall, D., Spiers, J.M., Silva, J. & Curry, K.J. (2006) Fiber content of two rabbiteye and two southern highbush blueberry cultivars. In: *Proceedings of the 10th North American Blueberry Research and Extension Workers' Conference* (ed. D.S. NeSmith). Tifton, GA, pp. 176–179.
19. Ayaz, F.A., Kadioglu, A., Bertoft, E., Acar, C. & Turna, I. (2001) Effect of fruit maturation on sugar and organic acid composition in two blueberries (*Vaccinium arctostaphylos* and *V. myrtillus)* native to Turkey. *New Zealand Journal of Crop and Horticultural Science*, **29**, 137–141.
20. Kalt, W. & McDonald, J.E. (1996) Chemical composition of lowbush blueberry cultivars. *Journal of the American Society of Horticultural Science*, **121**, 142–146.
21. Ehlenfeldt, M.K., Meredith, F.I. & Ballington, J.R. (1994) Unique organic acid profile of rabbiteye vs highbush blueberries. *HortScience*, **29**, 321–323.
22. Naczk, M. & Shahidi, F. (2006) Phenolics in cereals, fruits and vegetables: occurrence, extraction and analysis. *Journal of Pharmaceutical and Biomedical Analysis*, **41**, 1523–1542.
23. Hatier, J.H.B. & Gould, K.S. (2009) Anthocyanin function in vegetative organs. In: *Anthocyanins: Biosynthesis, Functions, an Applications* (eds K. Gould, K. Davies & C. Winefield). Spring Science + Business Media, LLC, New York, NY, pp. 1–19.
24. Foyer, C.H. & Noctor, G. (2005) Oxidant and antioxidant signaling in plants: a re-evaluation of the concept of oxidative stress in a physiological context. *Plant, Cell and Environment*, **28**, 1056–1071.
25. Rein, M.J. & Heinonen, M. (2004) Stability and enhancement of berry juice color. *Journal of Agricultural and Food Chemistry*, **52**, 3106–3114.
26. Wang, H., Cao, G. & Prior, R.L. (1997) Oxygen radical absorbing capacity of anthocyanins. *Journal of Agricultural and Food Chemistry*, **45**, 304–309.
27. Camire, M.E. (2000) Bilberries and blueberries as functional foods and nutraceuticals. In: *Herbs, Botanicals and Teas* (eds G. Mazza & B.D. Oomah). Technomic Publishing Co., Lacaster, PA, pp. 289–314.
28. Francis, F.J., Harborne, J.B. & Barker, W.G. (1966) Anthocyanins in lowbush blueberry *Vaccinium angustifolium*. *Journal Food Science*, **31**, 583–587.
29. Lohachoompol, V., Mulholland, M., Srzednicki, G. & Craske, J. (2008) Determination of anthocyanins in various cultivars of highbush and rabbiteye blueberries. *Food Chemistry*, **111**, 249–254.

30. Gao, L. & Mazza, G. (1994) Quantitation and distribution of simple and acylated anthocyanins and other phenolics in blueberries. *Journal Food Science*, **59**, 1057–1059.
31. Nicoue, E.E., Savard, S. & Belkacemi, K. (2007) Anthocyanins in wild blueberries of Quebec: extraction and identification. *Journal of Agricultural and Food Chemistry*, **55**, 5626–5635.
32. Cho, M.J., Howard, L.R., Prior, R.L. & Clark, J.R. (2004) Flavonoid glycosides and antioxidant capacity of various blackberry, blueberry and red grape genotypes determined by high-performance liquid chromatography/mass spectrometry. *Journal of the Science of Food and Agriculture*, **84**, 1771–1782.
33. Scibisz, I. & Mitek, M. (2007) The changes of antioxidant properties in highbush blueberries (*Vaccinium corymbosun* L.) during freezing and long-term frozen storage. *ACTA Scientiarum Polonorum Technologia Alimentaria*, **6**, 75–82.
34. Wang, M., Li, J., Shao, Y., Huang, T.-C., Huang, M.-T, Chin, C.-K, Rosen, R.T. & Ho, C.-T. (2000) Antioxidative and cytotoxic components of highbush blueberry (*Vaccinium corymbosum* L.). In: *Phytochemicals and Phytopharmaceuticals* (eds F. Shahidi & C.-T. Ho). AOCS Press, Champaign, IL, pp. 271–277.
35. Hosseinian, F.S. & Beta, T. (2007) Saskatoon and wild blueberries have higher anthocyanin contents than other Manitoba berries. *Journal of Agricultural and Food Chemistry*, **55**, 10832–10838.
36. Singleton, V.L., Orthofer, R. & Lamuela-Raventós, R.M. (1999) Analysis of total phenols and other oxidation substrates and antioxidants by means of Folin–Ciocalteu reagent. *Methods in Enzymology*, **299**, 152–178.
37. Kalt, W., Ryan, D.A.J., Duy, J.C., Prior, R.L., Ehlenfeldt, M.K. & Vander Kloet, S.P. (2001) Interspecific variation in anthocyaninins, phenolics and antioxidant capacity among genotypes of highbush and lowbush blueberries (*Vaccinium cyanococcus* spp.). *Journal of Agricultural and Food Chemistry*, **49**, 4761–4767.
38. Moyer, R.A., Hummer, K.E., Finn, C.E., Frei, B. & Wrolstad, R.E. (2001) Anthocyanins, phenolics and antioxidant capacity in diverse small fruits: *Vaccinium*, *Rubus* and *Ribes*. *Journal of Agricultural and Food Chemistry*, **50**, 519–525.
39. Castrejohn, A.D.R., Elchholz, I., Rohn, S., Kroh, L.W. & Huyskens-Keil, S. (2008) Phenolic profile and antioxidant activity of highbush blueberry (*Vaccinium corymbosum* L.) during fruit maturation and ripening. *Food Chemistry*, **109**, 564–572.
40. Connor, A.M., Luby, J.J., Hancock, J.F., Berkheimer, S. & Hanson, E.J. (2002) Changes in fruit antioxidant activity among blueberry cultivars during cold-temperature storage. *Journal of Agricultural and Food Chemistry*, **50**, 893–898.
41. Padayatty, S.J., Katz, A., Wang, Y., Eck, P., Kwon, O., Lee, J.-H., Chen, S., Corpe, C., Dutta, A., Dutta, A.D. & Levine, M. (2003) Vitamin C as an antioxidant: evaluation of its role in disease prevention. *Journal of the American College of Nutrition*, **22**, 18–35.
42. Furie, B., Bouchard, B.A. & Furie, B.C. (1999) Vitamin K-dependent biosynthesis of gamma-carboxyglutamic acid. *Blood*, **93**, 1798–1808.
43. American Academy of Pediatrics Committee on Fetus and Newborn (2003) Controversies concerning vitamin K and the newborn. *Pediatrics*, **112**, 191–192.
44. Ansell, J., Hirsh, J., Poller, L., Bussey, H., Jacobson, A. & Hylek, E. (2004) The pharmacology and management of the vitamin K antagonists: the seventh ACCP conference on antithrombotic and thrombolytic therapy. *Chest*, **126**, 204S–233S.
45. Booth, S.L., Charnley, J.M., Sadowski, J.A., Saltzman, E., Bovill, E.G. & Cushman, M. (1997) Dietary vitamin K1 and stability of oral anticoagulation: proposal of a diet with constant vitamin K1 content. *Thrombosis and Haemostasis*, **77**, 504–509.
46. Zafra-Stones, S., Yasmin, T., Bagchi, M., Chatterjee, A., Vinson, J.A. & Bagchi, D. (2007) Berry anthocyanins as novel antioxidants in human health and disease prevention. *Molecular Nutrition & Food Research*, **51**, 675–683.
47. Knight, J.A. (2000) The biochemistry of aging. *Advances in Clinical Chemistry*, **35**, 1–62.
48. Valko, M., Izakovic, M., Mazur, M., Rhodes, C.J. & Telser, J. (2004) Role of oxygen radicals in DNA damage and cancer incidence. *Molecular Cell*, **266**, 37–56.
49. Meyer, C.H. & Sekundo, W. (2005) Nutritional supplementation to prevent cataract formation. *Developments in Ophthalmology*, **38**, 103–111.
50. Lau, F.C., Bielinski, D.F. & Joseph, J.A. (2007) Inhibitory effects of blueberry extract on the production of inflammatory mediators in lipopolysaccharide-activated BV2 microglia. *Journal Neuroscience Research*, **85**, 1010–1017.

51. Galli, R.L., Bielinski, D.F., Szprengiel, A., Shukitt-Hale, B. & Joseph, J.A. (2006) Blueberry supplemented diet reverses age-related decline in hippocampal HSP70 neuroprotection. *Neurobiology of Aging*, **27**, 344–350.
52. Saija, A., Prici, P., D'Amico, N., De Pasquale, R. & Costa, G. (1990) Effect of *Vaccinium myrtillus* anthocyanins on triiodothyronine transport into brain in the rat. *Pharmacological Research*, **22** (Suppl. 3), 59–60.
53. Joseph, J.A., Shukitt-Hale, B., Denisova, N.A., Bielinski, D., Martin, A., McEwen, J.J. & Bickford, P.C. (1999) Reversals of age-related declines in neuronal signal transduction, cognitive, and motor behavior deficits with blueberry, spinach, or strawberry dietary supplementation. *Journal of Neuroscience*, **19**, 8114–8121.
54. Bickford, P.C., Shukitt-Hale, B. & Joseph, J. (1999) Effects of aging on cerebellar noradrenergic function and motor learning: nutritional interventions. *Mechanisms of Aging and Development*, **11**, 141–154.
55. Bickford, P.C., Gould, T., Briederick, L., Chadman, K., Pollock, A., Young, D., Shukitt-Hale, B. & Joseph, J. (2000) Antioxidant-rich diets improve cerebellar physiology and motor learning in aged rats. *Brain Research*, **866**, 211–217.
56. Joseph, J.A., Denisova, D., Fisher, B., Shukitt-Hale, B., Bickford, P., Prior, R. & Cao, G. (1998) Membrane and receptor modifications of oxidative stress vulnerability in aging: nutritional considerations. *Annals of the New York Academy of Sciences*, **854**, 268–276.
57. Joseph, J.A., Arendash, G., Gordon, M., Diamond, D., Shukitt-Hale, B., Morgan, D. & Denisova, N.A. (2003) Blueberry supplementation enhances signaling and prevents behavioral deficits in an Alzheimer disease model. *Nutritional Neuroscience*, **6**, 153–162.
58. Youdim, K.A., Shukitt-Hale, B., Martin, A., Wang, H., Denisova, N., Bickford, P.C. & Joseph, J.A. (2000) Short-term dietary supplementation of blueberry phenolics: beneficial effects on aging brain performance and peripheral tissue function. *Nutritional Neuroscience*, **3**, 383–397.
59. Casadesus, G., Shukitt-Hale, B., Stellwagen, H.M., Stellwagon, H.M., Zhu, X., Lee, H-G., Smith, M. & James, J.A. (2004) Modulation of hippocampal plasticity and cognitive behavior by short-term blueberry supplementation in aged rats. *Nutritional Neuroscience*, **7**, 309–316.
60. Goyarzu, P., Malin, D.H., Lau, F.C., Taglialatela, G., Moon, W.D., Jennings, R., Moy, E., Moy, D., Lippold, S., Shukitt-Hale, B. & Joseph, J.A.(2004) Blueberry supplemented diet: effects on object recognition memory and nuclear factor-kappa B levels in aged rats. *Nutritional Neuroscience*, **7**, 75–83.
61. Andres-Lacueva, C., Shukitt-Hale, B., Galli, R.L., Juaregui, O., Lamuela-Raventos, R.M. & Joseph, J.A. (2005) Anthocyanins in aged blueberry-fed rats are found centrally and may enhance memory. *Nutritional Neuroscience*, **8**, 111–120.
62. Mullen, W., Larcombe, S., Arnold, K., Welchman, H. & Crozier, A. (2010) Use of accurate mass full scan mass spectrometry for the analysis of anthocyanins in berries and berry-fed tissues. *Journal of Agricultural and Food Chemistry*, **58**, 3910–3915.
63. Krikorian, R., Nash, T.A., Shidler, M.D., Shukitt-Hale, B. & Joseph, J.A. (2010) Blueberry supplementation improves memory in older adults. *Journal of Agricultural and Food Chemistry*, **58**, 3996–4000.
64. Krikorian, R., Nash, T.A., Shidler, M.D., Shukitt-Hale, B. & Joseph, J.A. (2010) Concord grape juice supplementation improves memory function in older adults with mild cognitive impairment. *British Journal of Nutrition*, **103**, 730–734.
65. Howell, A.B. (2007) Bioactive compounds in cranberries and their role in prevention of urinary tract infections. *Molecular Nutrition and Food Research*, **51**, 732–737.
66. Avorn, J., Monane, M., Gurwitz, J.H., Glynn, R.J., Choodnovsky, I. & Lipsitz, L. (1998) Reduction of bacteriura and pyuria after ingestion of cranberry juice. *Journal of the American Medical Association*, **271**, 751–754.
67. Jackson, B. & Hicks, L.E. (1991) Effect of cranberry juice on urinary pH in older adults. *Home Healthcare Nurse*, **15**, 199–202.
68. Howell, A.B. (2009) Update on health benefits of cranberry and blueberry. *Acta Horticulturae*, **810**, 779–785.
69. Foo, L.Y., Lu, Y., Howell, A.B. & Vorsa, N. (2000) A-type proanthocyanidin trimmers from cranberry that inhibit adherence of uropathogenic P-fimbriated *Escherichia coli*. *Journal of Natural Products*, **63**, 1225–1228.
70. Ofek, I., Goldhar, J., Zafriri, D., Lis, H., Adar, R. & Sharon, N. (1991) Anti-*Escherichia coli* adhesion activity of cranberry and blueberry juices. *New England Journal of Medicine*, **324**, 1599–1600.

71. Schmidt, B.M., Howell, A.B., McEniry, B., Knight, C.T., Siegler, D., Erdman, J.W. & Lila, M.A. (2004) Effective separation of potent antiproliferation and antiadhesion components from wild blueberry (*Vaccinium angustifolium* ait.) fruits. *Journal of Agricultural and Food Chemistry*, **52**, 6433–6442.
72. Lin, B., Johnson, B.J., Rubin, R.A., Malanoski, A.P. & Ligler, F.S. (2010) Iron chelation by cranberry juice and its impact on *Escherichia coli* growth. *BioFactors*, **37**, 121–130.
73. Jepson, R. & Craig, J.C. (2007) Review: A systematic review of the evidence for cranberries and blueberries in UTI prevention. *Molecular Nutrition & Food Research*, **51**, 738–745.
74. Hertog, M.G., Bueno-de-Mesquite, H.B. & Fehily, A.M. (1996) Fruit and vegetable consumption and cancer mortality in the Caerphilly study. *Cancer Epidemiology, Biomarkers & Prevention*, **5**, 673–677.
75. Block, G., Patterson, B. & Subar, A. (1992) Fruit, vegetables, and cancer prevention: a review of epidemiological evidence. *Nutrition and Cancer*, **18**, 1–29.
76. Knekt, P., Jarvinen, R., Seppanen, R., Heliovaara, M., Teppo, L, Pukkala, E. & Aromaa, A. (1997) Dietary flavonoids and the risk of lung cancer and other malignant neoplasms. *American Journal of Epidemiology*, **146**, 223–230.
77. Knekt, P., Kumpulainen, J., Järvinen, R., Rissanen, H., Heliovaara, M., Reunanen, A., Hakulinen, T. & Aromaa, A.(2002) Flavonoid intake and risk of chronic disease. *American Journal of Clinical Nutrition*, **76**, 560–568.
78. Nichenmetla, S.N., Taruscio, T.G., Barney, D.L. & Exon, J.H. (2006) A review of the effects and mechanisms of polyphenolics and cancer. *Critical Reviews in Food Science and Nutrition*, **46**, 161–183.
79. Srivastava, A., Akoh, C.C., Fischer, J. & Krewer, G. (2007) Effect of anthocyanin fractions from selected cultivars of Georgia-grown blueberries on apoptosis and phase II enzymes. *Journal of Agricultural and Food Chemistry*, **55**, 3180–3185.
80. Suh, N., Paul, S., Hao, X., Simi, B., Xiao, H., Rimando, A.M. & Reddy, B.S. (2007) Pterostilbene, an active constituent of blueberries, suppresses aberrant crypt foci formation in the azoxymethane-induced colon carcinogenesis model in rats. *Clinical Cancer Research*, **13**, 350–355.
81. Kahlon, T.S. & Smith, G.E. (2007) *In vitro* binding of bile acids by blueberries (*Vaccinium* spp.), plums (*Prunus* spp.), prunes (*Prunus* spp.), strawberries (*Fragaria x ananassa*), cherries (*Malpighia punicifolia*), cranberries (*Vaccinium macrocarpon*) and apples (*Malus sylvestris*). *Food Chemistry*, **100**, 1182–1187.
82. Schmidt, B.M., Erdman Jr, J.W. & Lila, M.A. (2006) Differential effects of blueberry proanthocyanidins on androgen sensitive and insensitive human prostate cancer cell lines. *Cancer Letters*, **231**, 240–246.
83. Matchett, M.D., MacKinnon, S.L., Sweeney, M.I., Gottschall-Pass, K.T. & Hurta, R.A.R. (2005) Blueberry flavonoids inhibit matrix metalloproteinase activity in DU145 human prostate cancer cells. *Biochemistry and Cell Biology*, **83**, 637–643.
84. Seeram, N.P., Adams, L.S., Zhang, Y., Lee, R., Sand, D., Schueller, H.S. & Heber, D. (2006) Blackberry, black raspberry, blueberry, cranberry, red raspberry, and strawberry extracts inhibit growth and stimulate apoptosis of human cancer cells *in vitro*. *Journal of Agricultural and Food Chemistry*, **54**, 9329–9339.
85. Joshipura, K.J., Hu, F.B., Manson, J.E., Stampfer, M.J., Rimm, E.B., Speizer, F.E., Colditz, G., Ascheria, A., Rosner, B., Spiegelman, D. & Willett, W.C. (2001) The effect of fruit and vegetable intake on risk of coronary heart disease. *Annals of Internal Medicine*, **134**, 1106–1114.
86. Manach, C., Mazur, A. & Scalbert, A. (2005) Polyphenols and prevention of cardiovascular diseases. *Current Opinion in Lipidology*, **16**, 77–84.
87. Castellani, W. (2004) Metabolic and nutritional aspects of the atherogenic atypical lipoproteins: lipoprotein(a), remnant lipoproteins, and oxidized low-density lipoprotein. *Nutrition Research*, **24**, 681–693.
88. Folts, J.D. (2002) Potential health benefits from the flavonoids in grape products on vascular disease. In: *Flavonoids in Cell Function* (eds B. Buslig & J. Manthey). Kluwer Academic/Plenum Publishing, New York, NY, pp. 95–111.
89. Keevil, J.G., Osman, H.E., Reed, J.D. & Folts, J.D. (2000) Grape juice, but not orange juice or grapefruit juice, inhibits human platelet aggregation. *Journal of Nutrition*, **130**, 53–56.
90. Kalt, W., Foote, K., Fillmore, S.A.E., Lyon, M., Van Lunen, T.A. & McRae, K.B. (2008) Effect of blueberry feeding on plasma lipids in pigs. *British Journal of Nutrition*, **100**, 70–78.
91. Basu, A., Du, M., Leyva, M.J., Sanchez, K., Betts, N.M., Wu, M., Aston, C.E. & Lyons, T.J. (2010) Blueberries decrease cardiovascular risk factors in obese men and women with metabolic syndrome. *Journal of Nutrition*, **140**, 1582–1587.

92. Bastide, P., Rouher, F. & Tronche, P. (1968) Rhodopsin et anthocyanosides. *Bulletin des Sociétés d'ophtalmologie de France*, **9**, 801–807.
93. Jayle, G.E. & Aubert, L. (1964) Action of anthocyanin glucosides on scotopic and mesopic vision in normal subjects. *Therapie*, **19**, 171–185.
94. Trevithick, J.R. & Mitton, K.P. (1999) Antioxidants and diseases of the eye. In: *Antioxidant Status, Diet, Nutrition and Health* (ed. A.M. Pappas). CRC Press, Boca Raton, FL, pp. 545–565.
95. Canter, P.H. & Ernst, E. (2004) Anthocyanosides of *Vaccinium myrtillus* (bilberry) for night vision—A systematic review of placebo-controlled trials. *Therapeutic Reviews*, **49**, 38–50.
96. Kechinski, C.P., Guimarães, P.V.R., Noreña, C.P.Z., Tessaro, I.C. & Marczak, L.D.F. (2010) Degradation kinetics of anthocyanin in blueberry juice during thermal treatment. *Journal of Food Science*, **75**, C173–C176.
97. Patras, A., Brunton, N.P., Gormely, T.R. & Butler, F. (2009) Impact of high pressure processing on antioxidant activity, ascorbic acid, anthocyanins and instrumental colour of blackberry and strawberry puree. *Innovative Food Science and Emerging Technologies*, **10**, 308–313.
98. Sadilova, E., Carle, R. & Florian, C. (2007) Thermal degradation of anthocyanins and its impact on color and *in vitro* antioxidant capacity. *Molecular Nutrition & Food Research*, **51**, 1461–1471.
99. Sadilova, E., Stintzing, F.C. & Carle, R. (2006) Thermal degradation of acylated and nonacylated anthocyanins. *Journal of Food Science*, **71**, C504–C512.
100. Mayer, A.M. (2006) Polyphenol oxidases in plants and fungi: going places? A review. *Phytochemistry*, **67**, 2318–2331.
101. Skrede, G., Wrolstad, R.E. & Durst, R.W. (2000) Changes in anthocyanins and polyphenolics during juice processing of highbush blueberries (*Vaccinium corymbosum* L.). *Journal of Food Science*, **65**, 357–364.
102. Brownmiller, C., Howard, L.R. & Prior, R.L. (2009) Processing and storage effects on blueberry (*Vaccinium corymbosum* L.) polyphenolics. *Acta Horticulturae*, **841**, 347–354.
103. Brownmiller, C., Howard, L.R. & Prior, R.L. (2009) Processing and storage effects on procyanidin composition and concentration of processed blueberry products. *Journal of Agricultural and Food Chemistry*, **57**, 1896–1902.
104. Schmidt, B.M., Erdman, J.W. & Lila, M.A. (2005) Effects of food processing on blueberry antiproliferation and antioxidant activity. *Journal of Food Science*, **70**, S389–S394.
105. Lohachoompol, V., Srednicki, G. & Craske, J. (2004) The change of total anthocyanins in blueberries and their antioxidant effect after drying and freezing. *Journal of Biomedicine and Biotechnology*, **5**, 248–252.
106. Scibisz, I. & Mitek, M. (2007) The changes of antioxidant properties in highbush blueberries (*Vaccinium corymbosum* L.) during freezing and long-term storage. *ACTA Scientiarum Polonorum Technologia Alimentaria*, **6**, 75–82.
107. Gonzalez, E.M., de Ancos, B. & Cano, M.P. (2003) Relation between bioactive compounds and free radical-scavenging capacity in berry fruits during frozen storage. *Journal of the Science of Food and Agriculture*, **83**, 722–726.
108. Srivastava, A., Akoh, C.C., Yi, W., Fischer, J. & Krewer, G. (2007) Effect of storage conditions on the biological activity of phenolic compounds of blueberry extract in glass bottles. *Journal of Agricultural and Food Chemistry*, **55**, 2705–2713.
109. Prior, R.L., Lazurus, S.A., Cao, G., Muccitelli, H. & Hammerstone, J.F. (2001) Identification of procyanidins and anthocyanins in blueberries and cranberries (*Vaccinium* spp.) using high-performance liquid chromatography/mass spectrometry. *Journal of Agricultural and Food Chemistry*, **49**, 1270–1276.
110. Smith, M.A.I., Marley, K.A., Siegler, D., Singletary, K.W. & Meline, B. (2000) Bioactive properties of wild blueberry fruits. *Journal of Food Science*, **65**, 352–356.
111. Fujimaru, T. (2010) *Experimental Investigation of Carbon Dioxide (CO_2) Laser Perforation as a Potential Skin Pretreatment for Sugar Infusion Process of Frozen Blueberries*. MSc Thesis, Oregon State University, Corvallis, OR.
112. Beaudry, C., Raghaven, G.S.V., Ratti, C. & Rennie, T.J. (2004) Effects of four drying methods on the quality of osmotically dehydrated cranberries. *Drying Technology*, **22**, 521–539.
113. Mejia-Meza, E.I., Yanez, J.A., Davies, N.M., Rasco, B., Younce, F., Remsberg, C.M. & Clary, C. (2008) Improving nutritional value of dried blueberries (*Vaccinium corymbosum* L.) combining microwave-vacuum, hot-air drying and freeze drying technologies. *International Journal of Food Engineering*, **4**, 1–6.

114. Venkatachalapathy, K. & Raghavan, G.S.V. (1998) Microwave drying of osmotically dehydrated blueberries. *Journal of Microwave Power and Electromagnetic Energy*, **33**, 95–102.
115. Feng, H. & Tang, J. (2000) Microwave and spouted bed drying of frozen blueberries: the effect of drying and pretreatment methods on physical properties and retention of flavor volatiles. *Journal of Food Processing and Preservation*, **23**, 463–479.
116. Sapers, G.M. & Phillips, J.G. (1985) Leakage of anthocyanins from skin of raw and cooked highbush blueberries (*Vaccinium corymbosum* L.). *Journal of Food Science*, **50**, 437–439, 443.
117. Kalt, W.E., McDonald, J.E. & Donner, H. (2000) Anthocyanins, phenolics, and antioxidant capacity of processed lowbush blueberry products. *Journal of Food Science*, **65**, 390–393.
118. Sablani, S.S., Andrews, P.K., Davies, N.M., Walters, T., Saez, H. & Bastarrachea, L. (2011) Effects of air and freeze drying on phytochemical content of conventional and organic berries. *Drying Technology*, **29**, 205–216.
119. Stojanovic, J. & Silva, J.L. (2007) Influence of osmotic concentration, continuous high frequency ultrasound and dehydration on antioxidants, colour and chemical properties of rabbiteye blueberries. *Food Chemistry*, **101**, 898–906.
120. Biswas, R. (2007) *Development of Technologies for the Production of Polyphenolic Nutraceuticals from Muscadine Grapes and Rabbiteye Blueberries [Electronic Resource]*. Available from University of Georgia Catalog, Ipswich, MA.
121. Bhandari, B.R., Senoussi, A., Dumoulin, E.D. & Lebert, A. (1993) Spray drying of concentrated fruit juices. *Drying Technology*, **11**, 1081–1092.
122. Chegini, G.R. & Ghobadian, B. (2007) Spray dryer parameters for fruit juice drying. *World Journal of Agricultural Sciences*, **3**, 230–236.
123. Moreira, G.E.G., Costa, M.G.M., de Souza, A.C.R., de Brito, E.S., de Medeiros, M.F.D. & de Azeredo, H.M.C. (2009) Physical properties of spray dried *Acerola pomace* extract as affected by temperature and drying aids. *LWT: Food Science and Technology*, **42**, 641–645.
124. Moreira, G.E.G., Azeredo, H.M.C., de Medeiros, M.F.D., de Brito, E.S. & de Souza, A.C.R. (2010) Ascorbic acid and anthocyanin retention during spray drying of acerola pomace extract. *Journal of Food Processing and Preservation*, **34**, 915–925.
125. Camire, M.E., Chaovanalikitt, A., Dougherty, M.P. & Briggs, J. (2002) Blueberry and grape anthocyanins as breakfast cereal colorants. *Journal of Food Science*, **67**, 438–441.
126. Camire, M.E., Dougherty, M.P. & Briggs, J. (2007) Functionality of fruit powders in extruded corn breakfast cereals. *Food Chemistry*, **101**, 765–770.
127. Chakraborty, M., Savarese, M., Harbertson, E., Harbertson, J. & Ringer, K.L. (2010) Effect of the novel radiant zone drying method on anthocyanins and phenolics of three blueberry liquids. *Journal of Agricultural and Food Chemistry*, **58**, 324–330.
128. Pallas, L.A. (2011) *Drying Methods and Their Effects on Bioactive Compounds and Quality of Georgian Blueberries*. MSc Thesis, University of Georgia, Athens, GA.

5 Functional characteristics of dried cranberries

K.M. Schaich

5.1 Introduction

Cranberries (*Vaccinium macrocarpon*) are one of the few fruits native to North America, and are still produced almost exclusively in the northern United States and Canada. At the time the New World was settled, cranberries were extremely important to both indigenous Americans and settling colonists for nutrition and medicine, especially providing a critical source of ascorbic acid (vitamin C) and a natural remedy for urinary tract infections (UTIs). After a very long time of being converted only to juice or reserved for jellied sauce to serve as a condiment with holiday roast meats, these tart little berries have recently returned to the forefront of nutraceuticals with their recognition as one of the "superfruits" that contain very high levels of natural nutrients and phytochemicals unusually protective for health. Today, even mainstream medicine accepts the value of cranberries in treating or preventing a wide range of disorders in addition to the traditional UTI.

Superfruit status [1] has now made dried sweetened cranberries a very popular "healthy" snack, and in addition has focused attention on the use of freeze-dried berries, dried cranberry powder, or other derivatives as supplement additives for a wide range of specialty foods and nutritional support products. All of these dried forms of cranberries offer the advantage of concentrating nutritional components. However, cranberry juice and raisin-like dried cranberry fruits have been criticized by some nutritionists for their high sugar content, while cranberry powder can be mixed with other ingredients using sweetening already present, or can be incorporated into capsules or powder blends for use with no sugar. In addition, cranberry powder is technologically easy to handle, providing a convenient medium for material transfer, portioning, mixing, and control, and is shelf-stable for extended periods [2]. It is not surprising, then, that even a cursory search on the Internet finds a huge number of websites advertising cranberry powders for incorporation in bases for on-the-go drink mixes, health supplements, urinary health products, antioxidant supplements, natural colorant/antioxidant in cosmetics, oral care products, and even pet food and nutrition products. For some years, there were great problems in comparing materials and research results, particularly with medical studies, because of the natural variability in cranberry products from different sources and growth history. To provide a common material of constant

Dried Fruits: Phytochemicals and Health Effects, First Edition. Edited by Cesarettin Alasalvar and Fereidoon Shahidi.

composition that would allow clear delineation of health effects, the US National Institutes of Health National Center for Complementary and Alternative Medicine contracted for development and production of a Standardized Dried Cranberry Powder for use in research on studies of cranberry bioavailability, pharmacokinetics, and physiological effects [3]. Specifications for these products are delineated in the US Code of Federal Regulations 21 CFR parts 110 & 111. Examples of cranberry powders developed under this mandate include the following:

From Ocean Spray Inc. (Lakeville-Middleboro, MA)

- Cranberry 90 MX Powder—spray-dried cranberry concentrate with magnesium hydroxide carrier and tricalcium phosphate anticaking agent, minimum 0.5% proanthocyanidins (PACs) and minimum 30% organic acids.
- Cranberry Extract Powder—freeze-dried liquid extract of cranberries (*Vaccinium macrocarpon*) containing ≥1.5% PACs and <10% organic acids.

From Decas Botanical Synergies (Wareham, MA)

- PACran®—freeze-dried whole cranberry powder, minimum 1.5% PACs.
- HI-PAC 4.0®—spray-dried cranberry juice water extract, minimum 4% PACs.
- HI-PAC POWDER®—spray-dried cranberry juice water extract, minimum 10% PACs.
- CystiCran®—spray-dried cranberry extract powder standardized to 30% PACs.
- CystiCran 40—spray-dried cranberry extract powder standardized to 40% PACs.

Cranberries and juice products have been reviewed [4, 5], but relatively little detailed consideration has been given to these new dried cranberry powders. To fill this gap, this chapter reviews composition and nutraceutical characteristics of cranberry powder as well as reported therapeutic effects and food applications.

5.2 Composition and nutritional characteristics of dried cranberry powder

5.2.1 Composition of fresh cranberries and modifications imposed by dehydration

Whole cranberries are well-known as a rich source of many physiological and phytochemical nutrients. Indeed, cranberries had the highest total phenol content of 20 common fruits tested by Vinson [6]. However, they are mostly consumed in some processed form rather than the exceptionally tart fresh raw fruit, so the nutrients that count are those remaining after juicing, gelling, and dehydrating. The number one nutrient loss during processing is ascorbic acid [2, 7] and some of the phytonutrients are damaged as well, by polyphenol oxidase action [8, 9], oxidation [10], hydrolysis [11], phenol condensation and polymerization into brown pigments [12], and thermal degradation [5, 13]. Anthocyanins (such as quercetin, myricetin, and kaempferol) are particularly sensitive, losing sometimes more than half their original content [14]. Nutrients also become diluted in cranberry juice. Overall, then, the nutrient composition of cranberry products actually consumed does not exactly match the composition of fresh whole berries, as was shown in a detailed tabulation of the phytonutrients (classes of compounds, individual components, and amounts) in whole fresh berries versus cranberry products in a recent review [5].

Dried cranberry powders present an interesting and complex paradox. On the one hand, a great concentration of nutrients occurs with dehydration of extracts, cranberry juice, or whole berries except for ascorbic acids (which is destroyed) [2]. Compounds that comprised only 10–20% of fresh berries now become 80–98% of the product, depending on the dehydration method and whether carriers and anticaking agents are added. In addition, dried whole cranberry powders retain the pomace, the skins, seeds, and solids left over after pressing fruit for juice. Cranberry pomace is normally discarded or recycled into sauce after juice extraction. However, it is a good source of dietary fiber and also contains appreciable amounts of polyphenols. White *et al.* [15] found 0.6 g polyphenols/100 g dry weight in lyophilized pomace; about 19% of this was distributed in 6 anthocyanins, 55% was in 13 flavonols, and 26% in 8 procyanidins (Table 5.1). Even if the PACs are bound to the pomace fiber (primarily pectins), they may be released by microbial metabolism in the intestines. Thus, overall, cranberry powders offer distinct advantages when the goal is to provide phytonutrients as an ingredient for food, personal care, and pharmaceutical products.

Table 5.1 Composition of polyphenols in freeze-dried cranberry pomace (material remaining after removal of juice)

	mg/100 g dry weight
Anthocyanins	
Cyanidin-3-galactoside	13.2
Cyanidin-3-glucoside	4.5
Cyanidin-3-arabinoside	49.6
Peonidin-3-galactoside	20.1
Peonidin-3-glucoside	7.4
Peonidin-3-arabinoside	26.6
Procyanidins	
DP1 cat/epi monomers	5.8
DP2A A-type dimer	82.6
DP2B B-type dimer	4.4
DP3A	30.8
DP3B	1.6
DP4B	22.9
DP5A	7.1
DP6A	12.1
Flavonols	
Myricetin 3-xyloside	1.5
Myricetin 3-arabinoside	1.8
Quercetin 3-galactoside	12.8
Quercetin 3-xyloside	5.5
Quercetin 3-arabinopyranoside	15.2
Quercetin 3-arabinofuranoside	16.7
Quercetin 3-rhamnoside	18.5
Methoxyquercetin 3-xyloside	11.4
Quercetin 3-coumaroyl galactoside	2.3
Quercetin 3-benzoyl galactoside	27.5
Myricetin	55.6
Quercetin	146.2
Unidentified	12.1

Source: Adapted with permission from White *et al.* [15].
DP, degrees of polymerization.

On the other hand, stresses during dehydration cause the same kinds of damage and loss of nutrients as processing of cranberry juice and sauce [16–19]. Consequently, how the cranberries are dried markedly affects the resulting phytochemical composition. High temperatures, air exposure, and berry disruption are three key factors that must be controlled to maintain nutrients in dried cranberry powder [20]. During heating steps, hydrolysis or thermal degradation of the matrix (cell wall polysaccharides) or polymeric phenols with concomitant release of phenolic acids, monomers, and small polymers are balanced against polymerization or oxidation of phenols. In general, the higher the temperature during drying and the longer the drying time, the greater the loss of phytonutrients [17]. Heat induces oxidation (including phenolic browning), isomerization (e.g., epimerization of catechins), and excision of the gallate group attached to the C-ring of flavonoids [16], and any moisture present during heating (e.g., early stages of dehydration) accelerates hydrolysis and dimerization [17, 21].

Any processing combinations that minimize heat, drying time, or air exposure protect nutrients in dried cranberries. For example, adding vacuum reduces loss of anthocyanins, ellagic acid, flavanols, and ascorbic acid compared to conventional air drying [16], and adding microwave treatments to the vacuum improves protection even further. When cranberries were pretreated by halving and steam blanching to inactivate enzymes, losses of ascorbic acid were 20–40% greater for convection drying than for vacuum microwave drying [19]. Under vacuum, the water boils at temperatures substantially below the atmospheric 100°C; at low microwave power (240 W), microwaves penetrate the center for the product to facilitate heating. The vacuum microwave combination facilitates water evaporation, reduces processing time, and retains acyl coenzyme A oxidase (AOX) activity [16]. At the same time, excessively high vacuums remove large proportions of volatile compounds involved in flavor and aroma. Thus "soft" conditions of vacuum microwave drying (e.g., 72–74 kPa vacuum and 360 W microwave power) have been recommended to retain highest concentrations of volatiles and the best sensory quality [18].

Given these observations, it is not surprising that lyophilization of whole berries under vacuum at 196°C gives dried products with the highest % cranberry solids and greatest retention of color, flavor, and nutrients. Lyophilization of whole or half berries minimizes tissue disruption that accelerates enzyme and oxidation reactions [2], and maintains the lowest temperature during drying, thus maximally immobilizing molecules, removing activation energies, and minimizing chemical and enzymatic reactions.

In spray drying of cranberry juices or extracts, the carriers also affect stability, at least in part due to moisture control. Anthocyanins in spray-dried açai degraded faster as water activity increased. In comparisons of 10 DE and 40 DE maltodextrins, gum Arabic, and tapioca starch as carriers, maltodextrin 10 DE provided the greatest anthocyanin protection.

5.2.2 Nutrient and phytonutrient composition of dried cranberry powder

Nutritionally, cranberries have long been recognized as rich sources of ascorbic acid, vitamin A, thiamine, riboflavin, niacin, dietary fiber, and the essential minerals such as calcium, magnesium, manganese, phosphorus, and potassium that should not be overshadowed by the emphasis on cranberry phytochemicals. Except for ascorbic acid, these are largely retained and concentrated in dried powders, as shown in the nutrient compositions of commercial cranberry powders from two major suppliers of standardized products (Table 5.2). Variations in

Table 5.2 Nutritional and functional characteristics of commercial dried cranberry powders (values in per 100 grams edible portion)

	Units	Freeze-dried	Spray-dried	Freeze-dried	Spray-dried cranberry juice concentrates			
		Extract powder [22, 23][a]	90 MX [24, 25][b]	PACran [26, 27][c]	NutriCran 90 [28, 29][d]	NutriCran 90S [30, 31][e]	Hi-PAC 10 [32, 33][f]	CystiCran 40 [34, 35][g]
Fruit solids	%	98	90	100	90	90	90	90
Water	g	5.5	2–5	4.6	3.85	3.85	3.85	1.7
Energy	kcal	367	360	432	356	356	356	394
Protein	g	1.15	<0.10	5.38	0.46	0.46	0.46	2.97
Fat	g	0.173	0.24	10.99	0.02	0.02	0.02	0.18
Calories from fat	g	1.56	2.16	99	0.0	0.0	0.0	2.0
Saturated fat	g	0.038	0.0	0.92	0.0	0.0	0.0	0.08
Trans fat	g	<0.007	0.0	0.0	0.0	0.0	0.0	0.0
Cholesterol	g	1.0	0.0	0.0	0.0	0.0	0.0	0.0
Ash	g	2.9	7.46	1.16	7.22	7.22	7.22	0.02
Carbohydrate	g	90.3	89.35	77.87	88.45	88.45	88.45	95.1
Dietary fibre	g	26.1	6.02	45.9	4.1	4.1	4.1	7.1
Sugars	g	6.4	36.9	13.21	37.91	37.91	37.91	0.0
Vitamins								
Niacin	mg	1.98	0.8	0.75	1.0	1.0	1.0	na
Riboflavin	mg	0.05	1.14	0.7	0.41	0.41	0.41	na
Thiamin	mg	0.18	0.22	0.24	0.18	0.18	0.18	na
Vitamin A	IU	na	na	892	<40	<40	<40	<40 IU
Vitamin C	mg	<1.0	4.9	19.4	2.8	2.8	2.8	0.0

(continued)

Table 5.2 *(Continued)*

		Freeze-dried	Spray-dried	Freeze-dried	Spray-dried cranberry juice concentrates			
	Units	Extract powder [22, 23][a]	90 MX [24, 25][b]	PACran [26, 27][c]	NutriCran 90 [28, 29][d]	NutriCran 90S [30, 31][e]	Hi-PAC 10 [32, 33][f]	CystiCran 40 [34, 35][g]
Minerals								
Calcium	mg	185	184	60.4	118.3	118.3	118.3	11.3
Copper	mg	0.44	0.5	0.4	0.3	0.3	0.3	
Iron	mg	8.63	4.28	6.61	10	10	10	na
Magnesium	mg	90.4	2877	49.5	1757	2757	2757	na
Phosphorus	mg	96.9	95	139.4	46.7	46.7	46.7	na
Potassium	mg	696	743	444.2	890	890	890	na
Sodium	mg	406	29	3.7	26.3	26.3	26.3	6.0
Zinc	mg	1.12	0.87	1.8	1.5	1.5	na	na
Organic acids	%	8–15	30	na	35	35	na	na
Total phytosterols	%	na	na	2.8	na	na	na	na
Total phenolics	%	na	na	2–5	2–5	2–5	2–5	na
Proanthocyanidins	%	>1.5	0.5	1.5	0.95	1.0	10	30
Anthocyanins	%	na	na	0.15–1.00	0.15	1.00	0.15	na
Quercetin	mg	na	na	na	3–4.4	3–4.4	3–4.4	na

IU, international unit; na, not available.
[a]Ocean Spray Extract Powder, freeze-dried whole berry powder, 2% tricalcium phosphate added to prevent caking.
[b]Ocean Spray 90 MX, spray-dried with magnesium hydroxide as carrier and tricalcium phosphate as anticaking agent.
[c]Decas PACran, freeze-dried whole berry powder.
[d]Decas NutriCran 90, spray-dried cranberry juice concentrate with magnesium hydroxide, tricalcium phosphate.
[e]Decas NutriCran 90S, spray-dried cranberry juice concentrate with magnesium hydroxide, tricalcium phosphate, 1% standardized proanthocyanidins.
[f]Decas Hi-PAC 10, spray-dried cranberry extract standardized to 10% proanthocyanidins, with magnesium hydroxide, tricalcium phosphate.
[g]Decas CystiCran 40, spray-dried cranberry extract standardized to 30% proanthocyanidins, with magnesium hydroxide, tricalcium phosphate.

Table 5.3 Classes of phytochemicals in dried cranberry power

Phytochemicals	Example	Function in cranberry products
Flavonoids		
Anthocyanins	Peonidin-3-*O*-galactoside	Primary red pigment and AOX
Flavonols	Quercetin-3-*O*-galactoside	Copigment and AOX stabilizer
Flavanols (catechins)	Epicatechin	Astringent flavors and AOX stabilizer
Proanthocyanidins	Proanthocyanin A2	Astringent flavors and stabilizer
Polymeric colour compounds	Cyanidin-pentoside-flavan-3-ol	Red-brown pigment
Non-flavonoids		
Non-flavonoid	Resveratrol	Astringent flavors and AOX stabilizer
Simple phenolics	Salicylic acid	Odours and AOX stabilizer
Non-phenolics		
Organic acids	Ascorbic acid	Sour flavors
Complex carbohydrates	Pectin	Gelation and edible films
Sugars	Fructose	Sweet flavors

Source: Adapted with permission from Pappas and Schaich [5].
AOX, antioxidant

composition, particularly in minerals, arise largely from additional ingredients incorporated as carriers and anticaking agents.

Phytochemical composition of dried cranberry powders is not as well documented as for whole berries and cranberry juice, particularly in terms of quantitation. Research is still largely at the stage of identifying total components so an extensive accounting of phytochemicals, as was done by Pappas [5], is not yet possible. The general classes of compounds are, of course, the same as in fresh whole cranberries (Table 5.3), although concentrations and, in some cases also product distribution (particularly PACs), can be quite different in dry powders. PACs are usually the dominant fraction in dry powders.

5.2.2.1 *Anthocyanins*

In one of the earliest studies using HPLC-ESI-MS-MS to determine structure and composition of cranberry anthocyanins, Wu and Prior [36] identified 13 anthocyanins in freeze-dried cranberries in methanol/water/acetic acid (85:15:0.5, v/v/v) extracts (Table 5.4). All were present as glycosides rather than aglycones, with galactose being the major sugar. Importantly, of 18 anthocyanin-rich fruits tested, only cranberries and Concord grapes contained all six anthocyanin backbones.

Instrumentation advances now make it possible to determine both what anthocyanins are present and in what quantities. Brown *et al.* [37] used ultraperformance liquid chromatography (UPLC) with time-of-flight mass spectrometry (MS) to identify and quantitate anthocyanins in five cultivars of freshly harvested and freeze-dried whole cranberries for metabolomic profiling. Peonidins were present in double the cyanidins concentrations, with peonidin-3-galactoside as the dominant anthocyanin in all cultivars tested (Table 5.5). Similar peonidin and peonidin-3-galactoside dominance was observed in anthocyanins of three commercial cranberry powders (Table 5.6) [38], although levels of all anthocyanins detected were considerably lower than in the fresh berry powders. Whether this difference is due to

Table 5.4 List of anthocyanin glycosides identified in freeze-dried cranberry powder by HPLC-ESI-MS-MS

Anthocyanins
Cyanidin 3-galactoside
Cyanidin 3-glucoside
Cyanidin 3-arabinoside
Delphinidin 3-arabinoside
Malvidin 3-arabinoside
Malvidin 3-galactoside
Pelargonidin 3-arabinoside
Pelargonidin 3-galactoside
Peonidin 3,5-digalactoside
Peonidin 3-arabinoside
Peonidin 3-galactoside
Peonidin 3-glucoside
Petunidin 3-galactoside

Source: Adapted from Wu and Prior [36].

actual concentrations, cultivars, or analytical methods cannot be ascertained since details were not provided in the Palikova *et al.* [38] study. In contrast, cranberries obtained from a local supermarket and freeze-dried showed 46% cyanidins (39.8 mg/100 g fresh weight) and 43% peonidins (36.8 mg /100 g fresh weight) [39].

These differences show clearly the need for standardization of materials used in medical studies and as food ingredients, as well as for high-level analytical instrumentation to obtain accurate quantitative and qualitative information. They also raise interesting questions about relative postharvest stability of the individual and total anthocyanins that must be considered in preparation of commercial standardized products.

Table 5.5 Anthocyanin contents (mg/100 g powder in dry weight) in five cultivars of fresh, freeze-dried cranberries ground to a powder

	Cyanidin 3-galactoside	Cyanidin 3-glucoside	Cyanidin 3-arabinoside	Pelargonidin 3-galactoside	Pelargonidin 3-arabinoside
Biological replicates[a]					
Ben Lear	149.4	14.6	135.2	367.2	131.9
Bergman	143.1	12.0	131.8	305.9	109.4
GH1	140.6	10.0	118.2	254.1	82.1
Pilgrim	70.0	8.3	61.7	138.5	49.6
Stevens	56.6	9.6	40.9	118.3	55.2
Analytical replicates[b]					
Ben Lear	111.2	10.0	85.4	301.3	121.5
Bergman	85.5	8.6	72.5	165.7	79.8
GH1	123.0	10.5	85.5	285.3	108.5
Pilgrim	68.5	8.4	58.2	124.1	58.7
Stevens	49.4	7.9	39.1	108.0	62.3

Source: Adapted with permission from Brown *et al.* [37].
[a]Biological replicates (five individual berries).
[b]Analytical replicates (pooled berries with replicate analyses).

Table 5.6 Anthocyanin composition of three commercial standardized cranberry powders (mg/100 g powder)

Anthocyanins	NUTRICRAN 90S	HI-PAC 4.0	PACran
Total anthocyanins	440	610	115
Cyanidin-3-arabinoside	24	70	12
Cyanidin-3-galactoside	54	81	21
Cyanidin-3-glucoside	31	10	3
Cyanidin-3-pentoside	13	na	1
Cyanidin-3,5-dihexoside	na	50	na
Delphinidin	8	na	3
Peonidin-3-arabinoside	25	109	16
Peonidin-3-galactoside	135	220	41
Peonidin-3,5-digalactoside	106	50	2
Peonidin-3-glucoside	45	30	6

Source: Adapted with permission from Palikova *et al.* [38].
na, not available.

5.2.2.2 *Flavonols and flavanols*

When product applications are health related, flavonols and flavanols in cranberry powders have attracted even more attention than anthocyanins, flavonols for their role in UTIs and flavanols because they provide building blocks for some PACs. Vvedenskaya *et al.* [40] extracted flavonols from Ocean Spray 90MX cranberry powder (freeze-dried) with acetone followed by ethyl acetate, separated fractions by high-performance liquid chromatography (HPLC), and determined structures of collected fractions by nuclear magnetic resonance (NMR). Twenty-two flavonoid (mostly quercetin) glycosides were identified (Table 5.7), six of them new (previously unidentified). In addition to standard flavonol glycosides,

Table 5.7 Flavonols of Ocean Spray 90 MX cranberry powder

Flavonols of Ocean Spray 90 MX[a]
3′-Methoxyquercetin-3-β-galactoside
Dimethoxymyricetin-hexoside
Methoxymyricetin-pentoside
Methoxyquercetin-pentoside
Myricetin-3-α-arabinofuranoside
Myricetin-3-α-xylopyranoside
Myricetin-3-β-galactoside
Quercetin-3-*O*-(6″-benzoyl)-β-galactoside
Quercetin-3-*O*-(6″-*p*-coumaroyl)-β-galactoside
Quercetin-3-rhamnopyranoside
Quercetin-3-α-arabinofuranoside
Quercetin-3-α-arabinopyranoside
Quercetin-3-α-xylopyranoside
Quercetin-3-β-galactoside
Quercetin-3-β-glucoside

Source: Adapted from Vvedenskaya *et al.* [40].
[a]It is a spray-dried cranberry concentrate with magnesium hydroxide as carrier and tricalcium phosphate as an anticaking agent.

Table 5.8 Comparison of flavonols reported for three commercial cranberry powders versus cranberries obtained from a local market and freeze-dried (mg/100 g dry weight)

	NUTRICran 90S [38]	HI-PAC 4.0 [38]	PACran [38]	Supermarket[a] [39]
Flavonols				
Quercetin	1352	2179	680	388
Myricetin				332
Flavan-3-ols				
Catechin				16
Catechins gallate				158
Epicatechin				90
Epicatechin gallate				0.0
Epigallocatechin				30
Gallocatechin				0.0
Gallocatechin gallate				8.0

[a]Fresh weight converted to dry wt assuming 95% moisture in whole cranberries.

three unique compounds were identified: two quercetin glycosides conjugated to benzoic or hydroxycinnamic acids and quercetin-3-arabinoside in both furanose and pyranose forms. Galactose and arabinose were the major conjugated sugars.

As with the anthocyanins, quantitation of flavonols by class in the three Decas commercial powders and supermarket cranberries showed large differences (Table 5.8); high levels of quercetin but no myricetin were reported for the Decas berry powders [38], while an order of magnitude lower quercetin and myricetin in about equal quantities were reported in the dried supermarket berries, even after adjusting values from fresh weight to dry weight basis [39]. Reasons for these huge differences were not clear from the papers, but once again argue for standardization of materials to be used in research or as ingredients. Less information is available about the flavanol content of cranberry powders. Flavan-3-ols contents determined in supermarket berries (unspecified variety) that were freeze-dried are also listed in Table 5.8.

5.2.2.3 *Proanthocyanidins*

Probably the most intense interest in composition of cranberry powders has been focused on the PAC fraction due to its reported effects on UTI. PACs (including degree of polymerization two and higher) are the dominant fraction in cranberry powders, accounting for as much as 63% of total phenols [41, 42] (Table 5.9).

The number of individual PAC species possible theoretically is astoundingly large due to the many kinds of linkages that can be formed, degrees of polymerization (DP), and monomer substrates that may be involved. Very few PAC structures have been elucidated fully because dealing with separation and structural analysis of such large molecules is technically quite challenging. However, with improvements in analytical instrumentation this limitation is changing. Until recently, it was assumed that the major (if not exclusive) monomers linked in cranberry PACs were catechin, epicatechin, epigallocatechin, and ellagic acid [43, 44], at least in part because these were the monomers identified in the few PACs whose structure

Table 5.9 Phenol distributions in Ocean Spray Cranberry OSA7 and freeze-dried whole cranberries

	OSA7 fraction[a] [41]						Freeze-dried whole cranberries[b] [42]
Fraction	**1 (mg/g solid)**	**2 (mg/g solid)**	**3 (mg/g solid)**	**4 (mg/g solid)**	**5 (mg/g solid)**	**6 (mg/g solid)**	**Total (mg/g dry weight)**
Phenolic acids	78						1.3
Flavonols	15	374	294	35	21	26	2.6
Anthocyanins	5	8	23	16	16	31	2.99
Proanthocyanidins	132	61	497	788	743	533	11.73
mDP[c]	1.2	1.6	5.8	6.1	12.7	11.3	2.4–35
% A-type bond	3.3	61.8	21.6	19.3	11.7	12.4	0.1–7.9

mDP, mean degree of polymerization.
[a]70% acetone fraction of extract from Stevens cranberries lyophilized and fractionated on a diol normal phase HPLC column. Six fractions collected and analyzed by reversed phase LC-MS.
[b]Cranberries purchased in a local supermarket lyophilized and fractions selectively extracted for analysis.
[c]Range over all fractions reported for whole berry powder.

had been determined. For example, Gu *et al.* [43] identified the following monobase PACs with characteristic A-type linkages in freshly harvested and freeze-dried cranberries:

(epi)cat→A→(epi)cat
(epi)cat→(epi)cat→AA→(epi)cat
(epi)cat→A→(epi)cat→(epi)cat
(epi)cat→(epi)cat→A→(epi)cat→A→(epi)cat
(epi)cat→(epi)cat→(epi)cat→A→(epi)cat
(epi)cat→(epi)cat→A→(epi)cat→(epi)cat
(epi)cat→A→(epi)cat→(epi)cat→(epi)cat
(epi)cat→(epi)cat→A→(epi)cat→(epi)cat→A→(epi)cat
(epi)cat→(epi)cat→(epi)cat→(epi)cat→A→(epi)cat
(epi)cat→(epi)cat→A→(epi)cat→(epi)cat→(epi)cat

It is extremely interesting, then, that recent application of thiolysis combined with advanced liquid chromatography–mass spectrometry (LC-MS) and UPLC-MS techniques identified mixed polymers with two types of subunits: anthocyanin–ethyl-flavanol (monomer to tetramer) and pyranoanthocyanin–flavanol (monomer to dimer) with DP up to 16 (Table 5.10). Sequential PAC extracts of OSA7 powder identified six fractions increasing in degree of polymerization from 1.2 to 11.3. The novel ethyl bridges were found only in later fractions 4–6 and so were clearly integral to PAC structures. Mixed PACs are not characteristic of fresh berries. Now that advanced methodology is available, it will be interesting to see whether these complex polymers are decomposition products generated during processing, dehydration, or oxidation, and also whether they retain the antioxidant activity associated with native monobase PACs.

5.2.2.4 *Simple phenols, phenolic acids, and nonaromatic organic acids*

Simple phenols and benzoates are quantitatively minor components of cranberries, but, nevertheless, contribute significantly to physiological effects, chemical reactivity, and innate stability of cranberries. Their effectiveness as antioxidants will be discussed in the next

Table 5.10 Mixed proanthocyanidins identified in extracts of OSA7 cranberry powder following thiolysis

Proanthocyanidins	Fraction	Proanthocyanidins (Continued)	Fraction
(epi)Cat-ethyl-cyanidin-arabinoside	5	Pyranocyanidin-hexoside-A2	6
Pyranopeonidin-arabinoside-(epi)cat	6	DP2B Pyranocyanidin-hexoside	6
(epi)Cat-ethyl-peonidin-arabinoside	4	A2-ethyl-cyanidin-hexoside	4–6
(epi)Cat-ethyl-cyanidin-hexoside	5, 6	DP2B-ethyl-cyanidin-hexoside	6
Pyranopeonidin-hexoside-(epi)cat	5	Pyranopeonidin-hexoside-A2	6
(epi)Cat-ethyl-peonidin-hexoside	4, 5	DP2B Pyranopeonidin-hexoside	6
Pyranocyanidin-arabinoside-A2	6	A2-ethyl-peonidin-hexoside	4–6
DP2B Pyranocyanidin-arabinoside	6	DP2B-ethyl-peonidin-hexoside	4, 5
A2-ethyl-cyanidin-arabinoside	4–6	DP3 1A-ethyl-peonidin-arabinoside	5, 6
DP2B-ethyl-cyanidin-arabinoside	6	DP3 1A-ethyl-cyanidin-hexoside	6
Pyranopeonidin-arabinoside-A2	6	DP3 1A-ethyl-peonidin-hexoside	5, 6
DP2B Pyranopeonidin-arabinoside	6	DP3B-ethyl-peonidin-hexoside	5
A2-ethyl-peonidin-arabinoside	4–6	DP4 1A-ethyl-cyanidin-hexoside	6
DP2B-ethyl-peonidin-arabinoside	5	DP4 1A-ethyl-peonidin-hexoside	5

Source: Adapted with permission from Tarascou *et al.* [41].

Table 5.11 Organic and phenolic acids in commercial cranberry powders (mg/100 g powder)

	NUTRICRAN 90S[a] [38, 49]	HI-PAC 4.0[a] [38, 49]	PACran[a] [38, 49]	CCP[b] [38, 49]
Organic acids				
Benzoic	372	311	160	
Citric	1980	1954	nd	
Hippuric				22.20
Malic	1892	1600	263	
Malonic	177	200	nd	
Phenylacetic				<0.2
Quinic	2570	2085	nd	
Phenolic acids				
3,4-Dihydroxyphenylacetic				<0.2
3-Hydroxybenzoic				1.90
3-Hydroxyphenylpropionic				1.53
4-Hydroxybenzoic				3.42
4-Hydroxyphenylacetic				3.21
Caffeic	10	14	6	8.42
Chlorogenic	25	35	nd	10.30
Dihydrocaffeic	121	129	53	
Ferulic				2.96
Gallic				14.50
Gentisic	8	23	nd	1.00
p-Coumaric	79	75	47	25.20
Protocatechuic	65	77	52	51.20
Vanillic	21	52	nd	

CCP, concentrated cranberry powders; nd, not detected.
[a]Products from Decas Botanicals, Inc., Carver, MA.
[b]Concentrated cranberry powder obtained from Decas Botanicals, Inc., Carver, MA.

section. In fresh berries, they are mostly in bound forms, esterified to sugars, cell wall polysaccharides, or other components; usually <10% of benzoates and simple phenols are present in free form (released without hydrolysis) [45]. The three primary nonvolatile organic acids (such as quinic, citric, and malic) of cranberry juice [46–48] are the major organic acids also in cranberry powders (Table 5.11). They contribute to tartness, maintenance of low pH for anthocyanin stabilization, and some metal complexation.

5.3 Natural antioxidants in dried cranberry powder

When asked about antioxidants in cranberries, most people immediately mention ascorbic acid first and anthocyanins second. Studies have shown that anthocyanins indeed are key antioxidants in cranberries [50–52], but in some cases flavonoids and PACs are stronger antioxidants than anthocyanins, and ascorbic acid contributes very little to antioxidant content and activity [6, 53]. Attempts to correlate antioxidant activity to total phenols or phenolic profiles in cranberries and other berries have been largely unsuccessful [54, 55]. Even more important is the observation that most of the phenol classes in cranberries have at least some antioxidant activity and they act synergistically with each other to enhance overall antioxidant effects [6, 56, 57], so a more appropriate question may be, what specific action does each of the molecular classes listed below contribute and how do these fractions interact (in parallel, in series, or in concert) to provide the overall effects of cranberry powders and cranberry extracts?

Fraction	Actions
Anthocyanins (dominant fraction)	Scavenge radicals, electron transfer (reducing agents), and recycle ascorbic acid
Flavonoids (flavonols and flavanols)	Inhibit lipid oxidation, chelate metals, scavenge radicals, and complex carbonyls
Tannins Condensed nonhydrolyzable (PACs) Hydrolysable: gallic/ellagic acid esters	Very high radical scavenging activity due to large numbers of phenol group
Phenolic acids	Scavenge radicals and decompose hydroperoxides
Resveratrol	Scavenge radicals

The relative importance of each class varies depending on concentrations and reaction environment. Chemically, this seems quite logical, especially when the phytochemical components exert different and complementary actions (e.g., radical scavenging, metal chelating, recycling other antioxidants, modulation of enzyme activity, and second messenger function) and have different solubilities in lipid versus aqueous phases. It is important to remember also that antioxidant activity is system specific. Many phenols can act predominantly as hydrogen atom transfer agents in one solvent but as an electron transfer agent in another, and which one dominates determines how the antioxidant interacts with target molecules. Reactivity in the common aqueous *in vitro* assays does not always accurately predict antioxidant activity in cell culture, foods, or *in vitro* where multiple phases, catalysts, and targets exist.

Overall, then, although the question of which fraction is the "most active" remains a controversial issue, it is fair to say that all of the following fractions work both separately and synergistically in contributing to antioxidant activity of cranberries, and that the relative effectiveness and active mechanism is determined by the reaction environment, target radicals active, and phase distribution of catalysts and antioxidants, as well as other factors [55].

How antioxidant activity is expressed makes a great difference in what molecules appear to be most active. Expressing activity on a weight basis (e.g., mmol quenching equivalents/mg dry matter) gives preference to molecules in highest concentration, while expressing activity on a molar basis (e.g., mmol quenching equivalents/mg or mol phenol) identifies which molecules or fractions are inherently more active in a given system. An example of this was shown dramatically in a study of hydrophilic and lipophilic radical scavenging by polar phenols (water-soluble, mostly phenolic acids), apolar phenols (flavonols, flavan-3-ols, and PACs), and anthocyanins [52]. When radical quenching was tested in an aqueous radical system (DPPD) and activity was expressed on a weight basis, anthocyanins were marginally the most active at acid pH but the hydrophilic fraction was twice as active at neutral pH. When expressed on a molar basis, the polar fraction was by far the most active under all conditions and anthocyanins were more active than apolar phenols. This behavior is not surprising considering that the radical test system was also aqueous with polar radicals. In a lipid oxidation test system, the apolar fraction surprisingly did not trade places with the polar fraction. The polar phenols were still the most active, but now only marginally more effective than anthocyanins at both acid and neutral pH. The pH dependence of the response suggests that protonated species are the most active forms. Cranberry anthocyanins are active only in acidic environments; they lose antioxidant activity when neutralized [58].

Structures of cranberry phenols within each class govern radical reactivities. Phenols are excellent radical scavengers because the electron from the phenoxyl radical generated when the hydrogen is abstracted becomes delocalized over the ring π system, thus lowering the electron density on the $-O^{\bullet}$ and hence the overall reactivity of the radical. As shown in Table 5.11, cranberry powders have appreciable contents of hydroxybenzoic and hydroxycinnamic acids that localize in polar extract fractions and phases of foods. Of these, (di)hydroxycinnamic acids (caffeic, chlorogenic, ferulic, sinapic, and *p*-coumaric) are more active radical scavengers than the corresponding (di)hydroxybenzoic acid derivatives such as *p*-hydroxybenzoic and vanillic acids [59]. The additional inductive effect of the double bond on the acid side chain in cinnamates pulls electrons away from the phenoxyl group, facilitating H release and stabilizing the phenoxyl radical.

In terms of flavonols, nine fractions isolated from dried cranberry extracts showed antioxidant activity in 1,1-diphenyl-2-picrylhydrazyl (DPPH) and oxidized low-density lipoprotein (LDL) assays [60]. Nearly all were glycosides: myricetin and its galactoside and arabinoside, quercetin galactoside, arabinoside, rhamnoside, and xyloside; and cyanidin-3 galactosides. In general, myricetin derivatives were less active than quercetin, and (in contrast to other studies), the glycosides were more active than their corresponding aglycones. Cyanidin-3-galactoside was dramatically strongest in its reaction with DPPH and inhibition of LDL oxidation.

Solubility and phase distribution affects how cranberry compounds respond in *in vitro* antioxidant tests and even more in actual applications, whether foods or tissues. Catechin monomers reacted more rapidly than procyanidins in quenching stable DPPH radicals, but procyanidins more actively inhibited lipid oxidation in emulsions [61]. This is not unexpected; PACs often show high radical scavenging activity because of the large number of phenolic

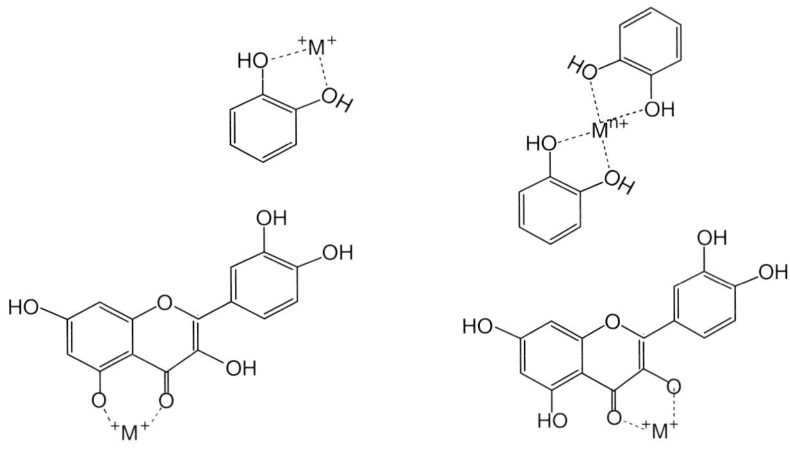

Figure 5.1 Examples of metal binding configurations for cranberry phenols.

groups distributed across the molecular surface. However, it is not clear whether these results reflect actual reactivity of the two materials, or just physical properties. PACs are more apolar, so it can be argued that these results stem primarily from their increased solubilization in the lipid. Similarly, the large size of PACs physically limits access to the sterically hindered the DPPH radical, so innate radical reactivity is not being measured. Lipid peroxyl radicals in methyl linoleate, on the other hand, are highly accessible. While this study has some interpretation questions, it does provide a good example of how compounds in cranberry powder may act synergistically by distributing differentially in aqueous and lipid phases.

No overview of cranberry antioxidant activity, even cursory ones such as this, can be complete without consideration of metal binding by cranberry phenols. Adjacent phenol groups (vicinal diphenols) and carbonyl-hydroxyl groups (Figure 5.1) are reasonably strong metal binders and mediate a second mechanism of antioxidant action by cranberry phenols. Of the anthocyanins, only delphinidin has the requisite structure (vicinal diphenol in the B-ring), but a number of flavonols and flavan-3-ols have both the B- and C-ring metal binding regions shown in Figure 5.1. Quercetin, the structure shown in Figure 5.1, is the major flavonol in cranberries. It uniquely contains all three metal binding positions [62] and is particularly well-recognized as a metal binder with the potential to inhibit metalloenzymes, inhibit metal-mediated lipid oxidation [63], remove critical metals from growth media, and in general interfere with any process involving or requiring metals [64]. Quercetin was identified as the most active compound in extracts of cranberry powder (Ocean Spray 90 MX) added to washed minced cod muscle to inhibit lipid oxidation; complexation of free iron and iron in heme was proposed as the active mechanism [63]. Metal binding may also contribute to the documented toxicity of quercetin [65].

This section will close with a note on the nonradical quenching antioxidant mechanism recently reported for catechins and procyanidins. Berry extracts were observed to inhibit formation of advanced glycation end-products (AGE) in serum models. Since procyanidins were found in all fractions of berry extracts, they were expected to be a major class of active compounds and a carbonyl complexation mechanism was proposed. Five different catechins–carbonyl adducts were identified in model systems of (+)-catechin and glyoxals

[66], verifying the proposed reaction. Catechins and other phenols can also work on the other side of AGE formation, binding to thiol and amine groups on proteins, thereby blocking carbonyl condensation sites [67, 68, 79]. The two reactions together provide a new antioxidant mechanism by which cranberry phenols may inhibit browning reactions in foods as well as prevent AGE-related chronic diseases *in vivo*.

5.4 Health effects of dried cranberry powders

The same antioxidant actions of radical quenching and metal complexation shown by cranberries in model reactions and foods can work in cells and tissues to inhibit progression of diseases that involve oxidative degradation. Cranberry phenols also bind to proteins, altering functionality and enzyme activity [5]. These three basic actions underlie numerous health benefits that have been associated with cranberry consumption and are now being targeted therapeutically with cranberry powders. These basic actions are also responsible for three general categories of cranberry actions in biological systems:

- Pathogen interactions.
- Antioxidant mechanisms or reduction of oxidative stress.
- Enzyme activity, signal transduction, and protein expression and activity.

General health effects of cranberries with specific focus on cranberry powders are covered briefly here. More detailed information is available in several excellent reviews [4, 5, 64, 69, 70].

5.4.1 Pathogen interactions

5.4.1.1 *Disruption of bacterial adhesion-prevention of UTIs, ulcers, and gum diseases*

Cranberries have been used in folk medicine for millennia to prevent UTI, and UTI remains the most-studied health effect of cranberries. Cranberries and cranberry juice have been the carriers in the past, with mixed results because concentrations could not be controlled. Dry cranberry powders offer distinct prophylactic advantages in being able to deliver high concentrations and standardized doses enriched in the most active components. In recent years, this approach has attracted increasing attention as development of antibiotic resistance has increased the rate of ineffective treatment and persistent reoccurrence. Dry cranberry powder has been particularly effective in reducing reoccurrence and severity of symptoms of UTIs.

To provide some examples, cranberry powder containing 200 mg of a concentrated cranberry extract standardized to 30% phenolics given in capsules twice daily for 12 weeks completely prevented UTIs in women who were subject to recurrent infections [71]. In a study of 65 women, 1200 mg cranberry powder/day was required to decrease oxidized proteins in serum (AGP). At that level, samples of subjects' urine inhibited adhesion of *Escherichia coli* strains *in vitro*. Hippuric acid, salicyluric acid and isomers, quercetin glucuronide, and dihydroxybenzoic acid isomers were the main metabolites, but the compounds responsible for antiadhesion activity could not be distinguished. Another study of 60 women aged 18 to 40 with a history of recurrent UTIs and the presence of *E. coli* and mild symptoms of UTI showed that 500 mg (low) or 1000 mg (high) whole cranberry powder per day reduced the number of *E. coli* detected in urine culture analysis between 25 and 45% after only 10 days of cranberry consumption [72]. This suppression was maintained over the 90 days of the study and, in addition, 40% of women receiving cranberry powder reported complete relief and remission from urological symptoms such as itching and burning sensation during regular and frequent urination. A daily dose of 500 mg or 1000 mg of standardized whole cranberry powder was thus recommended as an adjunct to antibiotic prophylactic therapy against recurrent UTIs. Similarly, in a prospective epidemiological study involving 120 recurrent cystitis patients of 43 urologists, treatment with concentrated cranberry powder reduced severity and prevalence of urinary problems from 98 to 28% after six months [73].

How do cranberries prevent development of UTIs? *E. coli* is the pathogen responsible for the most UTIs, accounting for 75–95% of infections [74, 75]. These bacteria are able to colonize in the urethra by expressing protein-based adhesins or lectins that bind them to glycoproteins and/or glycolipids on epithelial cell surfaces, allowing for subsequent colonization [76] in the urinary tract and finally in the bladder [75]. This binding is critical since free, unbound bacteria are washed out during urination. That cranberry products interfere with bacterial adhesion has been recognized for some time, and it is now generally accepted that the A-type cranberry PACs are the responsible compounds [77–81]. Repeated demonstration that PACs can inhibit adhesion of both resistant and sensitive strains of *E. coli* has led to development of cranberry powders standardized for high concentrations of PACs that are actively being evaluated in clinical tests. Results are still inconsistent due to differences in test systems, for example 50 μg/mL PACs was sufficient to inhibit adhesion of bacteria and hemagglutination of red blood cells *in vitro* [82], but in a study 70 mg PAC/day was required for maximum antiadhesion effects to be observed in the urine in humans [83]. Clearly, validating effects of cranberry powders to prevent UTIs, elucidating potential supporting roles for other cranberry phenols, and determining effective required doses will be a fruitful area for continued research.

Disruption of bacterial adhesion by cranberry compounds is not limited to the urinary tract, but is applicable to other body sites including the mouth and stomach which are constantly exposed to bacteria as well as to general resistance to viral and bacterial infections [see review (5)]. Addition of cranberry PACs to oral care interferes with bacterial adhesion to the tooth enamel, thereby preventing accumulation of the biofilms and plaque that lead to periodontal disease and tooth decay [84, 85]. Similarly, cranberry PACs interfere with the adhesion of *Helicobacter pylori*, a leading cause of stomach ulcers, which in turn are a major cause of stomach cancer. *H. pylori* has many adhesins that facilitate binding to gastric mucus and erythrocytes as well as epithelial cells [86, 87]. Cranberry juice [88] and cranberry juice powder (CJP) [89] have demonstrated generalized inhibition of bacterial cell adherence to surfaces, both glass and tissues, so it is not unreasonable to expect this mechanism to be active in the stomach lining. However, various combinations of blueberry, oregano, and grape seed extract powders with cranberry (dried juice) markedly increased anti-*H. pylori* activity in culture [90], showing apparent synergism between the various polyphenols. It was proposed that combinations of polyphenols can inactivate *H. pylori* by multiple mechanisms rather than a single mechanism of any one alone. Potential mechanisms include hyperacidification of the plasma membrane, disruption of H^+-ATPase, quenching electron flow in the electron transport cycle, disrupting energy metabolism, and destabilizing membrane structure and stability. Protein binding and metal chelation leading to enzyme inactivation should also be added to this list.

Cranberry compounds active against *H. pylori* have not been clearly identified, although, as with UTI, PACs or some high molecular weight polymers are considered most likely candidates. Much is still to be learned about the role of cranberry components in inhibiting oral diseases and ulcers. Delivering high concentrations of active compounds for prophylaxis or treatment through cranberry powders directly, as in a capsule, or indirectly by incorporation into oral products, foods, or beverages is currently an active area of research.

5.4.1.2 *Antimicrobial actions (metal binding, redox alterations, and protein binding)*

Several classes of cranberry compounds have shown antibacterial activity in broad contexts. Considering that phenol is commonly added to prevent mold growth in laboratory water baths and that polyphenols bind to proteins, this is not a strange observation. The question is, which classes are most active, and what are the mechanisms by which they inactivate microbial cells?

In comparisons of cranberry organic acid, phenolic, and anthocyanin effects on *E. coli* O157:H7, each class of compounds was shown to act by a different mechanism with different pH dependence [91]. Acids and anthocyanins were active only at low pH. Organic acid lowering of environmental and intracellular pH may create an electrical gradient that results in stacking organic acid ions against the cell membrane causing localized intense osmotic stress. Transmission electron microscopy detected completely collapsed cells when exposed to organic acids. In contrast, phenolics mediate redox reactions. Phenolic acids induce localized disruption of outer microbial membranes, and hydrophobic phenols bind to the outer membrane causing changes in membrane fluidity and shape. Once these disruptions occur, small phenolic compounds can enter the cell and disrupt metabolism. PACs bind to the lipopolysaccharide (LPS) on the outer membrane of Gram-negative bacteria by chelating metal ions. This destabilizes the outer membrane, releases LPS, and increases permeability. Metal binding by PACs and quercetin deplete the iron reserves necessary for growth. With these kinds of complementary interactions, it is easy to see why full extracts or combinations of polyphenols are more effective microbial inhibitors than single compounds.

5.4.2 Antioxidant mechanisms or reduction of oxidative stress and enzyme activity, signal transduction, and protein expression and activity

There is now substantial evidence that consumption of cranberries leads to increased plasma antioxidant capacity and decreased biomarkers of oxidative stress. What remains unclear is whether phytochemicals directly quench reactive oxygen species (ROS) and free radicals *in vivo* or whether they induce endogenous antioxidant responses through some alternate mechanism(s). It is often difficult to distinguish these two actions in complex systems because seldom is any process cleanly one or the other and production of radicals is difficult to separate from secondary effects of radicals, which include induction of enzymes and other cellular responses, directly or downstream. The brief overview below documents how cranberry phenols intervene in some pathological processes. In most cases, observations are still correlational, that is, feeding cranberry compounds is associated with some improvement in conditions, and the mechanisms have not yet been completely elucidated. Even so, the case for important positive contributions of cranberries to health continues to build.

5.4.2.1 *Promotion of cardiovascular health*

Since cardiovascular disease (CVD) is largely a condition driven by oxidative challenge and deterioration in blood vessel epithelia, in circulating lipids and proteins, and in circulating leukocytes, it seems logical to expect that the various antioxidant mechanisms of cranberry compounds should be able to reduce the stresses that lead to atherosclerosis and deterioration of heart muscle.

Investigating whether CJP could provide extra benefits when given to prevent UTIs, Valentova *et al.* [92] fed dried cranberry juice (NutriCran90) in capsules to 65 healthy women aged 19–28 at levels of 400 and 1200 mg/day for up to 56 days. Mixed effects were observed at both levels; 400 mg CJP lowered serum levels of LDL but also high-density lipoprotein (HDL) to the same degree ($\downarrow$ 0.2 mM), decreased aspartate aminotransferase (AST) significantly, and increased advanced protein oxidation products. With doses of 1200 mg cranberry powder/day, blood lipids and AST were not affected, but markers of enhanced antioxidant defenses were evident in increased serum uric acid and erythrocyte superoxide dismustase, glutathione peroxidase, and glutathione. Paradoxically, pro-oxidant action was evidenced at the same time in increased erythrocyte malondialdehyde, a lipid oxidation product, and advanced serum protein oxidation products plus decreased protein thiols. These results suggest that radical quenching and signal transduction/enzyme inhibition by cranberries are both active but not working synchronously. Since no polyphenols were detected in the serum but a number of endogenous antioxidants and enzymes were, it is clear that polyphenols in the intestines were triggering signal transduction cascades that activated endogenous antioxidants. The question is, which cranberry components activate antioxidant responses and which active pro-oxidant responses? Thus, while the idea of using cranberry powders for reducing systemic oxidative stress while also treating UTI is intriguing, caution must be advised until more is learned about cell controls that are activated by individual cranberry components.

A recent review of flavonoids also documented multiple pathways by which cranberry flavonoids influence development and progression of CVD, most of which did not involve radical scavenging [62]. While the review was generalized over flavonoids from all sources, the compounds forming the basis of the documentation are the same as those found in cranberries. Thus, it is reasonable to expect that cranberries also influence CVD by the following mechanisms:

(1) Abatement of oxidative stress by:
    Direct radical scavenging, for example, inhibition of LDL oxidation
    Metal complexation and chelation
    Inhibition of ROS producing enzymes such as xanthine oxidase ($O_2^{-\bullet}$), NADH oxidase ($O_2^{-\bullet}$), and lipoxygenase (LOX)
(2) Decreased expression of inflammatory signaling molecules:
    Inducible nitric oxide synthase (iNOS) and cyclooxygenase-2 (COX-2) expression
    Inhibition of leukocyte activation

(3) Miscellaneous:
Inhibition of platelet aggregation
Stimulation of vasodilation

5.4.2.2 *Modulation of inflammatory responses*

All of the mechanisms cited above for cranberry effects on CVD are equally applicable to inflammatory responses in general. At the present time, little is understood about the interplay between these mechanisms, and indeed whether individual classes of cranberry phenols act by alternate pathways.

Route of administration has critical effects on actions. For example, direct application to skin does not require absorption and transport to affected sites. Consequently, mouse ear edema induced by 12-*O*-tetradecanoylphorbol-13-acetate was reduced by 34.0 and 55.1% after topical application of 87.5 and 175 μg of quercetin-3-*O*-(6″-benzoyl)-β-galactoside isolated from processed cranberry powder [93]. This may have resulted from direct free radical quenching in the skin cells. Gavaging cranberry concentrate before phorbol challenge also reduced edema, but had to involve downstream signal transduction effects rather than direct radical quenching because no phenols were detected in serum before or during edema formation (unpublished data; Schaich, Singh, and Vorsa, from Rutgers University).

Phenol concentrations are also critical since nearly all antioxidants have multiple reactions that change in dominance with concentration. Of particular concern is the propensity for converting to pro-oxidant action as concentrations increase. For example, feeding rats freeze-dried high-lipid diets supplemented with 5 and 10% cranberry powder raised circulating levels of interleukin-10 (IL-10) and nitric oxide (NO) and decreased C-reactive protein (CRP) and interleukin-1β (IL-1β), but also increased tumor necrosis factor-α (TNF-α), IL-6, and IL-2, especially relative to lipid-injected controls [94]. In most cases, the 10% cranberry powder diet had less effect that the 5% diet, suggesting either saturation of effects or shifts from anti- to pro-oxidant mechanisms.

Clearly, the biochemical effects of cranberry components are complex and more detailed understanding about signal transduction effects, intervention in cytokine cascades, and the balance between various pathways is needed before using high levels of cranberry powder as treatments for any conditions involving inflammatory processes.

5.4.2.3 *Prevention and inhibition of cancer*

Cranberry phenols prevent initiation of cancer by at least two mechanisms: (1) quenching radicals that react with DNA or proteins to cause irreparable lesions that lead to tumor development [95] and (2) by inhibiting enzymes that metabolize toxicants into carcinogens [96]. Progression is similarly affected by multiple mechanisms, including inhibition of critical enzymes (e.g., ornithine decaroxylase), cell metabolism and proliferation, as well as enhancing unprogrammed cell death (apoptosis).

Doses of 5–30 mg/mL extracts of dried cranberries inhibited human breast cancer MCF-7 cell proliferation in culture by inducing apoptosis [97]. Fifty milligrams per milliliter led to 25% higher apoptosis, while doses of 10–50 mg/mL arrested cells in the G0/G1 phase. These results demonstrate that when cranberry phytochemical extracts come in contact with tumor cells, they have the capability of reducing cell proliferation, and this suppression is at least partly attributable to initiation of apoptosis and G1 phase arrest. Although the study was conducted with breast cancer cells, for this action to occur *in vivo* requires absorption and transport to breast tissue of sufficient quantities of intact active polyphenols. This is unlikely considering the very low absorption of these compounds. Nevertheless, these results do support a potential role for cranberries in preventing or slowing development of tumors in the gastrointestinal tract, for example, oral, esophageal, and colon cancers.

To determine potential use of cranberry powder in preventing development of esophageal cancer in patients with chronic acid reflux, SEG-1 and BIC-1 human esophageal cells were exposed as monolayers to dried PAC extract in concentrations ranging from 12.5 to 400 μg/mL [98], and cell viability and proliferation as well as acid-induced cell proliferation, cell cycle phase, and apoptosis were monitored for up to 96 hours. Fifty micrograms PACs per milliliter only moderately altered viability but significantly inhibited proliferation of both cell lines. Treatment with PACs before but not after acid challenge to model acid-reflux effects in the esophagus reduced cell proliferation, indicating that cranberry compounds must have been present before the acid challenge. Thus, maximum benefit can be gained by consumption of cranberry powder (in capsule or food) multiple times per day. In addition, PACs improved cell regulation, stopping cells at the G1 checkpoint and inhibiting progression to S phase, and increased apoptosis, unscheduled cell death. Both of these effects are critical in control of esophageal cancer.

Seeram *et al.* [99] fractionated Ocean Spray cranberry extract powder (90% fruit solids) to investigate which fractions contributed to antiproliferation effects of cranberries. Sugars, organic acids, total phenols, anthocyanins, and PACs were tested against human oral, colon, and prostate cancer cell lines in culture. Oral and colon cancer cells showed low responses to individual fractions but marked inhibition of cell proliferation by the total phenol fractions. All cranberry fractions (including phenolic acids) inhibited proliferation of prostate cancer cells by at least 60%, and total phenols almost completely stopped the process. Whether this augmentation was an additive or synergistic effect of multiple phenols could not be determined because doses were so different; total phenols were added at 200 μg/mL and the various fractions were added at their natural proportion of this. Antiproliferative effects toward all tumor cells were enhanced when sugars, organic acids, and phenolic acids were removed. These observations may provide some guidelines for developing cranberry products for cancer prevention, particularly for high-risk populations.

A very interesting series of studies on freeze-dried berries showed that blackberries and black raspberries inhibit both initiation and postinitiation events in rat esophageal and colon carcinogenesis as evidenced by decreases in tumor multiplicity, reduction in adduct formation, reduced proliferative indices, inhibition of preneoplastic lesion formation, and downregulation of COX-2 and iNOS [2, 96, 100–102]. Other genes in the esophagus associate with multiple cellular processes including apoptosis, angiogenesis, matrix formation, and cell cycle control. The dominant active fraction in these berries was anthocyanins. Cranberries have appreciable levels of anthocyanins (though not as high as in blackberries) [103], as well as flavonoids [39], and among the highest total phenols of all fruits [6]. Concentrating these phytonutrients in dried forms makes it feasible to administer the same levels as are present in other berries. Thus, there is every reason to expect that, as noted above, cranberry powders taken directly or dissolved in foods as carriers may offer significant potential for preventing, slowing, or lessening development of cancers all along the gastrointestinal tract.

5.4.2.4 *Protection of pancreatic cells and maintenance of insulin release*

Diet, genetics, and environment all play important roles in preserving pancreatic β-cell mass and function during aging. To investigate whether daily consumption of dried whole cranberry powder could prevent decline of pancreatic function and development of type 2 diabetes, rats were fed diets containing 2% whole cranberry from age 6 to 22 months, approximating full life span [104]. Cranberry diets delayed and diminished the decline in insulin production to the extent that after 22 months, the portal insulin concentrations were 7.6 times the control levels. Cranberry diets also improved β-cell glucose responsiveness to oral glucose challenge, dramatically increased β-cell mass and percentage of large islets, raised insulin and glucagon levels in β-cells, and enhanced production of pancreatic and duodenal homeobox-1 or insulin promoter factor-1 (PDX-1), a homeodomain transcription factor in β-cells. Insulin secretion was also increased without changing overall neogenesis patterns of pancreatic islets of leading to insulin resistance. These effects were attributed to flavonoids and proposed to result from a combination of mechanisms, including alteration of Ca^{2+} fluxes and cyclic nucleotide metabolism, decreased oxidative stress, and inhibition of α-amylase and α-glycosidase, which mediate release of glucose into the bloodstream. Extending this study to humans was recommended, given the positive results. Doses used in this study (~1000 mg/kg body weight) are equivalent to ~180 mg/kg body weight for humans, which can be achieved by consuming ~12 g cranberry powder daily. This quantity is too great to be administered by conventional capsules, but it might be feasible to deliver this in beverages.

5.4.3 Absorption of phenols from cranberry powder

Before closing this section, some note needs to be made regarding absorption and the physiological fate of cranberry phenols. Years of research were focused on free-radical quenching-based antioxidant assays of plant extracts, expecting to identify specific phenols that would be "magic bullets" for treating a broad range of pathologies involving oxidative deterioration. However, documentation of direct absorption of polyphenols, circulation in the blood, and distribution to tissues is skimpy and interpretations are often questionable when low nanomolar concentrations detected are converted to actual absorbed amounts. The issue remains highly controversial, but it is becoming increasingly clear that large polyphenols are very poorly absorbed as such so other mechanisms must be found to explain observed health effects. Two alternatives provide feasible explanations of observed physiological responses to cranberry consumption: (1) cranberry compounds, both monomers and PACs, are metabolized by intestinal flora to smaller compounds that can be absorbed and transported to tissues where chemistry is modified, and (2) cranberry compounds bind to proteins in the intestinal epithelium and thereby initiate a cascade of signal transduction that leads to observed responses.

Prior *et al.* [49] investigated microbial digestion and subsequent metabolite uptake of concentrated cranberry powder (Decas Botanicals) in the intestines of rats by following excretion of phenolic acids in urine. Concentrations of free phenolic acids in urine were rather low, in some cases being only a few percent of total material detected. In contrast, conjugated phenolic acids accounted for 50% (4-hydroxybenzoic acid) to as high as 96% (ferulic acid) of the detected phenolic acids. The major products were hippuric acid, 4-hydroxycinnamic acid, 4-hydroxyphenyl acetic acid, 3-hydroxyphenyl acetic acid, and 3-hydroxyphenylpropionic acid. 3,4-Dihydroxybenzoic acid was the major phenolic acid in cranberry powder; about 90% of 3,4-dihydroxybenzoic acid in the urine was conjugated. Thus, measuring only free forms of phenolic acids does not provide an accurate measure of phenolic acid excretion. PACs and anthocyanins are metabolized by microbial flora to phenolic acids or other lower weight compounds that are absorbed, conjugated by methylation or sulfonation, and excreted. 3,4-Dihydroxyphenyl acetic acid excretion was not affected by dietary levels of cranberry powder, and so was thought to be absorbed. This metabolite has been shown to have anti-inflammatory action.

5.5 Food applications of dried cranberry powders

The intense interest that has fired research in antioxidants for *in vivo* applications has also generated, in the process, a considerable body of information about how antioxidants work. This information, coupled with the current mandate for replacing synthetic antioxidants butylated hydroxyanisole (BHA) and butylated hydroxytoluene (BHT) with natural antioxidants that presumably have no toxicity, is now stimulating interest in using natural materials to stabilize foods, with mixed success. Natural antioxidant applications have been slow to develop due to lack of ability to predict behavior in foods and to strong colors and flavors that accompany antioxidants in extracts. The following brief survey of some recent results using cranberry powder show both the promise and the steep challenges and unpredictability in using natural antioxidants in complex multiphase foods.

5.5.1 Meats, poultry, and fish

Turkey and fish products are notoriously difficult to stabilize due to high unsaturation in the lipids and high levels of heme. Chloroform–methanol extracts of CJP mixed with comminuted turkey muscle inhibited lipid oxidation, gaining three weeks in frozen storage when added at 467 μmol/kg turkey. Quercetin and Q-3-*O*-(6′-benzoyl)-β-galactoside were both identified in the active fraction, though nearly all the protective action came from quercetin [105]. If the quercetin was glycosylated, inhibition of lipid oxidation was lost [106], indicating that the aglycone was the active form. Aglycones penetrated into model membranes but their glycosides did not, and ethylenediaminetetraacetic (EDTA) acid had no effect on lipid oxidation in the system. Thus, it was proposed that quercetin inhibits lipid oxidation in turkey meat by becoming embedded in muscle membranes and acting by free radical scavenging or other non-metal-chelating mechanism.

To elucidate actions of individual cranberry fractions in ground turkey, Ocean Spray 90MX powder (spray-dried concentrated juice) was separated into enriched fractions by column chromatography [107]. Fractions were (1) cinnamic acids, (2) anthocyanins, (3) flavonols, (4) flavonols, (5) PACs, and (6) PACs. This sequence parallels molecular size and increasing nonpolarity. Each fraction was mixed with turkey at a concentration of 200 μmol/kg meat. Polar cranberry phenols (e.g., simple phenols and phenolic acids) partitioned into the polar fractions at more than 50:50, but lipid oxidation was barely affected (Table 5.12). In contrast, as the percentage of phenol retained in the fat phase increased, protection correspondingly increased. Fraction 4 with flavonol aglycones was the most effective, while the PACS expected to be strong antioxidants behaved like the small polar compounds. A mechanism was proposed wherein inhibition of lipid oxidation required that the antioxidant

Table 5.12 Phase partition and inhibition of lipid oxidation in ground turkey by various fractions of cranberry powder prepared from spray-dried concentrated cranberry juice

Fraction	Composition	Percent in aqueous phase	Percent TBARS inhibition
1	Cinnamic acids	51.0	19.0
2	Anthocyanins	49.2	39.5
3	Flavonols (glycosides)	46.7	46.0
4	Flavonols (aglycones and quercetin)	31.7	74.2
5	Proanthocyanidins	44.6	36.2
6	Proanthocyanidins	35.9	21.8

Source: Adapted with permission from Lee *et al.* [107].
TBARS, thiobarbituric acid reactive substances.

be able to penetrate the lipid regions of membranes to align with acyl chains and be close to release points for heme iron and other catalysts. Small phenols are too polar to do this, while the PACs are too large to enter the membrane. This points out quite clearly how activity in *in vitro* antioxidant assays can be diverted in real systems where phase separations and cell/membrane compartmentalization do not always let the most reactive antioxidants get to their targets.

With green chemistry and the environmental movement, plant refuse, including cranberry press cake, is receiving intense scrutiny as a source of ingredient antioxidants. Press cake was extracted with water, acetone, and ethanol with and without microwave assistance, and these fractions were dissolved in ethanol and mixed with mechanically separated turkey breast [108]. Water extracts were close to inactive, while solvent extraction with acetone or solvent:microwave extraction with ethanol released compounds capable of inhibiting lipid oxidation. Lack of correlation with total phenols and poor correlation with quercetin suggested that other polyphenols in this material must have specific roles in addition to the action of quercetin in stabilizing ground turkey breast. Compounds in each extract were not identified.

Two approaches have been taken in stabilizing meats with natural antioxidants—treating the butchered meats (ground or sectioned) by adding antioxidant solutions as described above to stabilize products during marketing and storage or incorporating natural antioxidants into animal feeds to stabilize the meats endogenously, thus increasing the inherent resistance to oxidation during handling, storage, and cooking. One interesting study tested the second approach, feeding pigs a diet enriched with 15% Ocean Spray 90MX cranberry powder for six months [109]. Meat color was retained better but lipid oxidation [hydroperoxides and thiobarbituric acid reactive substances (TBARS)] was unexpectedly increased with cranberry diets. One explanation proposed was that cranberry compounds stabilize hydroperoxides, slowing their breakdown to reactive alkoxyl radicals. However, without measures of uptake and metabolism of cranberry polyphenols, it is not possible to know what compounds were absorbed in what levels and where they were localized. In addition, indirect effects of dietary cranberry compounds on antioxidant and metabolic enzymes have not been determined. Increasing dietary tocopherols has been shown to stabilize beef, but absorption and pharmacokinetics of these natural endogenous antioxidants are well-documented. While the idea of stabilizing meats by increasing levels of natural polyphenol antioxidants in animal diets is intriguing, phenols have complex actions in living systems, and much more information about microbial metabolism in the intestines; absorption, conjugation, and excretion versus distribution to tissues; binding to proteins; as well as signal transduction and enzyme induction processes is needed to determine whether this approach has promise for stabilizing meats.

Fish is an even greater challenge for stabilization than poultry and pork. Six fractions of OS 90MX cranberry powder (concentrated juice) were added to washed fish (cod) muscle and lipid oxidation was initiated by hemoglobin [63]. Phenolic acids (Fraction 1) and flavonols (Fractions 3 and 4) most effectively inhibited lipid oxidation, extending the lag period several days beyond the control, while the PACs and anthocyanins were relatively inactive, purportedly due to binding to insoluble muscle components. In contrast, reactivity with DPPH increased with fraction number and phenolic acids were the least reactive. Among individual compounds, propyl gallate (Fraction 1) was a stronger inhibitor of lipid oxidation than quercetin (Fraction 4). Propyl gallate is smaller and more lipid-soluble and so it more able to penetrate membranes than quercetin; it is a stronger metal binder so it can block electron transfer from the hemoglobin. Thus, as has been noted before in this chapter, effective antioxidant action in actual products depends most on getting the right antioxidant to the target.

Marinating fish fillets in solutions of cranberry powder appears to be an effective alternative for antioxidant treatment. Herring fillets marinated in a 50 g cranberry powder per liter solution for 24 hours before storage showed significant inhibition of lipid oxidation and production of ammonia as well as prolonged shelf life [110]. This seems to be a simple and cost-effective procedure that could be adapted for stabilizing both refrigerated and frozen commercial fish products.

Cranberry phenols have antimicrobial as well as antioxidant capabilities, as was discussed in Section 5.4. This offers potential for getting double benefits from using them in foods such as meats where microbial spoilage and lipid oxidation are both key problems. To test this possibility, ground beef in 1.5 g balls were autoclaved to sterilize and mimic cooking, soaked first in a 50:50 mixture of water-soluble phenolic extracts from oregano and cranberry extract powder (Decas Botanicals), 750 ppm each with 2% sodium lactate, and then dipped in a solution of *Listeria monocytogenes* and store at 4°C [111]. Log reductions after 20 days were 1.3 for phenols alone, 1.8 for lactate alone, and 2.1 for the total mixture. Proline-induced partial recovery of microbial outgrowth suggested that proline metabolism was being affected by the combination of phenols. Since muscle foods are good sources of proline, providing a ready supply for microbial recovery, meats may not be the optimum food form for application of the dual antioxidant–antimicrobial action.

5.5.2 Cereals

The pressure to "go natural," shifting away from synthetic additives, includes colors as well as antioxidants. Thus, interest in using natural pigments such as anthocyanins to gain a double benefit, antioxidant plus color, is growing. Anthocyanin addition to extruded breakfast cereals at levels ranging from 1 to 84.3% verified that cranberry anthocyanins can indeed serve this dual function [112]. However, there is not yet an experience base to guide formulations. More investigations are needed to determine levels of anthocyanins required to both create desired redness (not just pink) and prevent development of rancidity over desired shelf life.

5.6 Conclusions

Dried cranberry powders prepared from whole berries, cranberry juice, or various cranberry extracts deliver high concentrations of cranberry phenols in convenient form that is reasonably stable for long periods in storage. Because of demonstrated effectiveness of cranberries in preventing UTI, some cranberry powder products have been particularly formulated with high PAC concentrations, but fractions enriched in other cranberry classes are also available. Oxidation and heat during cranberry dehydration degrade phenolic compounds, particularly anthocyanins, so great care must be taken to protect sensitive compounds during processing. Lyophilization (freeze drying) is probably the method of choice for cranberry powders intended for nutraceutical applications since the very low temperatures and vacuum impose least stress.

Although studies with dried cranberry powders are still limited since standardized materials have been available commercially for only a few years, results to date show that the powders generally have beneficial health effects similar to those of fresh berries or juice, and in cases where high concentrations are needed, are even more useful. Based on research data, commercial products targeted at prevention of UTIs, prostate cancer, breast cancer, oral health have been developed and are available for use in capsules, beverages, gels, and other delivery modalities.

Although research was initially focused on direct effects of parent polyphenols in cranberry powder, there is increasing recognition that intestinal flora metabolize complex phenols to phenolic acids which may be even more important mediators of physiological effects. Most of these, such as polyphenols, are conjugated and eliminated in the urine, but at least one metabolite, 3-hydroxyphenyl acetic acid, is absorbed and reduces inflammation *in vivo*. Microbial metabolism and other mechanisms by which cranberry phenols can be activated and render their benefits promise to be hot areas of future research.

An interesting and potentially very profitable new direction for use of cranberry powders is inhibition of oxidative degradation in foods. Results have shown that cranberry phenols can limit troublesome lipid oxidation, particularly in muscle foods, but also interact with proteins and carbonyls to limit secondary co-oxidations and browning. For applications where the telltale red cranberry color is not an issue, cranberry powders may thus offer a productive new approach for stabilization of foods with natural compounds.

References

1. Gross, P.M. (2009) *Superfruits: Top 20 Fruits Packed with Nutrients and Phytochemicals, Best Ways to Eat Fruits for Maximum Nutrition, and 75 Simple and Delicious Recipes for Overall Wellness*. McGraw-Hill, New York, NY.
2. Stoner, G.D., Kresty, L.A., Carlton, P.S., Siglin, J.C. & Morse, M.A. (1990) Isothiocyanates and freeze-dried strawberries as inhibitors of esophageal cancer. *Toxicological Sciences*, **52**, S95–S100.
3. U.S. National Institutes of Health (2003) *National Center for Complementary and Alternative Medicine (NCCAM)*. National Institute of Health, Bethesda, MD.
4. Reed, J. (2002) Cranberry flavonoids, atherosclerosis, and cardiovascular health. *Critical Reviews in Food Science and Nutrition*, **42**, 301–316.
5. Pappas, E. & Schaich, K. (2009) Phytochemicals of cranberries and cranberry products: characterization, potential health effects, and processing stability. *Critical Reviews in Food Science and Nutrition*, **49**, 552–592.

6. Vinson, J.A., Su, X., Zubik, L. & Bose, P. (2001) Phenol antioxidant quantity and quality in foods: fruits. *Journal of Agricultural and Food Chemistry*, **49**, 5315–5321.
7. Licciardello, J.J., Esselen, J.R. & Fellers, C.R. (2006) Stability of ascorbic acid during the preparation of cranberry products. *Journal of Food Science*, **17**, 338–342.
8. Wrolstad, R.E., Durst, R.W. & Lee, J. (2005) Tracking color and pigment changes in anthocyanin products. *Trends in Food Science and Technology*, **16**, 423–428.
9. Rein, M.J. & Heinonen, M. (2004) Stability and enhancement of berry juice color. *Journal of Agricultural and Food Chemistry*, **51**, 3106–3114.
10. Starr, M.S. & Francis, F.J. (1968) Oxygen and ascorbic acid effect on the relative stability of four anthocyanin pigments in cranberry juice. *Food Technology*, **22**, 1293–1295.
11. Adams, J.B. (1973) Thermal degradation of anthocyanins with particular reference to the 3-glycosides of cyanidin. I. In acidified aqueous solution at 100°C. *Journal of the Science of Food and Agriculture*, **24**, 747–762.
12. Robards, K., Prenzler, P.D., Tucker, G., Swatsitang, P. & Glover, W. (1999) Phenolic compounds and their role in oxidative processes in fruits. *Food Chemistry*, **66**, 401–436.
13. Daravingas, G. & Cain, R.F. (1968) Thermal degradation of black raspberry anthocyanin pigments in model systems. *Journal of Food Science*, **33**, 138–142.
14. Hakkinen, S.H., Karenlampi, S.O., Mykkanen, H.M. & Torronen, A.R. (2000) Influence of domestic processing and storage on flavonol contents in berries. *Journal of Agricultural and Food Chemistry*, **48**, 2960–2965.
15. White, B.L., Howard, L.R. & Prior, R.L. (2010) Proximate and polyphenolic characterization of cranberry pomace. *Journal of Agricultural and Food Chemistry*, **58**, 4030–4036.
16. Wojdylo, A., Figiel, A. & Oszmianski, J. (2009) Effect of drying methods with the application of vacuum microwaves on the bioactive compounds, color, and antioxidant activity of strawberry fruits. *Journal of Agricultural and Food Chemistry*, **57**, 1337–1343.
17. Tonon, R.V., Brabet, C. & Hubinger, M.D. (2010) Anthocyanin stability and antioxidant activity of spray-dried acai (*Euterpe oleracea* mart.) juice produced with different carrier agents. *Food Research International*, **43**, 907–914.
18. Calin-Sanchez, A., Szumny, A.A., Figiel, A., Jałoszynski, K., Adamski, M. & Carbonell-Barrachina, A.A. (2011) Effects of vacuum level and microwave power on rosemary volatile composition during vacuum-microwave drying. *Journal of Food Engineering*, **103**, 219–227.
19. Dorofejeva, K.T., Rakcejevaa, T., Galoburda, R., Dukalska, L. & Kviesis, J. (2011) Vitamin C content in Latvian cranberries dried in convective and microwave vacuum driers. *Procedia Food Science*, **1**, 433–440.
20. Howard, L.R., Prior, R.L., Liyanage, R. & Lay, J.O. (2012) Processing and storage effect on berry polyphenols: challenges and implications for bioactive properties. *Journal of Agricultural and Food Chemistry*, **60**, 6678–6693.
21. Chamorro, S., Goni, I., Viveros, A., Hervert-Hernandez, D. & Brenes, A. (2012) Changes in polyphenolic content and antioxidant activity after thermal treatments of grape seed extract and grape pomace. *European Food Research and Technology*, **234**, 147–155.
22. Cranberry Extract Powder (2011) *Customer Specifications UPC #94815, Ocean Spray Ingredient Technology Group, Lakeville/Middleboro, MA*. Published online at: http://www.oceansprayitg.com/downloads/powder/specs/UPC94815, last accessed April 4, 2012.
23. Cranberry Extract Powder (2011) *Nutritional Analysis UPC #94815, Ocean Spray Ingredient Technology Group, Lakeville/Middleboro, MA*. Published online at: http://www.oceansprayitg.com/downloads/powder/nutritionals/UPC94815CEPNutritional.pdf, last accessed April 4, 2012.
24. 90 MX Cranberry Powder (2011) *Nutritional Analysis UPC #94513, Ocean Spray Ingredient Technology Group, Lakeville/Middleboro, MA*. Published online at: http://www.oceansprayitg.com/downloads/powder/nutritionals/UPC_94515-94513.pdf, last accessed April 4, 2012.
25. 90 MX Cranberry Powder (2011) *Customer Ingredient Specifications UPC #94513, Ocean Spray Ingredient Technology Group, Lakeville/Middleboro, MA*. Published online at: http://www.oceansprayitg.com/downloads/powder/specs/UPC_94513.pdf, last accessed April 4, 2012.
26. PACran (2010) *Ingredient Specifications, Product 06095, Decas Botanical Synergies, Carver, MA*. Published online at: https://www.decascranberry.com/sites/default/files/PACran%20Ingredient%20Specification%206095_0.pdf, last accessed April 4, 2012.

27. PACran (2010) *Nutritional Analysis, Product 06095, Decas Botanical Synergies, Carver, MA*. Published online at: https://www.decascranberry.com/sites/default/files/PACran%20Nutritional%20Analysis%206095_0.pdf, last accessed April 5, 2012.
28. NutriCran 90 (2010). *Nutritional Analysis, Product 06175, Decas Botanical Synergies, Carver, MA*. Published online at: http://www.decascranberry.com/sites/default/files/NutriCran%2090%20Nutritional%20Analysis.pdf, last accessed April 5, 2012.
29. NutriCran 90 (2010) *Ingredient Specifications, Product 06175, Decas Botanical Synergies, Carver, MA*. Published online at: http://www.decascranberry.com/sites/default/files/NutriCran%2090%20Ingredient%20Specification%206055_0.pdf, last accessed April 5, 2012.
30. NutriCran 90S (2010) *Ingredient Specifications, Product 06155, Decas Botanical Synergies, Carver, MA*. Published online at: http://www.decascranberry.com/sites/default/files/NutriCran%2090S%20Ingredient%20Specification%206155.pdf, last accessed April 5, 2012.
31. NutriCran 90S (2010) *Nutritional Analysis, Product 06155, Decas Botanical Synergies, Carver, MA*. Published online at: http://www.decascranberry.com/sites/default/files/NutriCran%2090S%20Nutritional%20Analysis.pdf, last accessed April 5, 2012.
32. CystiCran (2010) *Ingredient Specification, Product 06250, Decas Botanical Synergies, Carver, MA*. Published online at: http://www.decasbotanical.com/documents/Specifications%20010511/CystiCran%20Ingredient%20Specification%206250.pdf, last accessed April 6, 2012.
33. CystiCran (2010) *Nutritional Analysis, Product 06250, Decas Botanical Synergies, Carver, MA*. Published online at: http://www.decasbotanical.com/documents/Nutritional%20031308/CystiCran%20Nutritional%20Analysis%20121008.pdf, last accessed April 6, 2012.
34. Hi-PAC 10 (2010) *Nutritional Analysis, Product 06176, Decas Botanical Synergies, Carver, MA*. Published online at: https://www.decascranberry.com/sites/default/files/Hi-PAC%20Powder%20Nutritional%20Analysis%206176_0.pdf, last accessed April 6, 2012.
35. Hi-PAC 10 (2010) *Ingredient Specifications, Product 06176, Decas Botanical Synergies, Carver, MA*. Published online at: https://www.decascranberry.com/sites/default/files/Hi-PAC%20Powder%20Ingredient%20Specification%206176_0.pdf, last accessed April 6, 2012.
36. Wu, X. & Prior, R.L. (2005) Systematic identification and characterization of anthocyanins by HPLC-ESI-MS/MS in common foods in the United States: fruits and berries. *Journal of Agricultural and Food Chemistry*, **53**, 2589–2599.
37. Brown, P.N., Murch, S.J. & Shipley, P. (2012) Phytochemical diversity of cranberry (*Vaccinium macrocarpon* Aiton) cultivars by anthocyanin determination and metabolomic profiling with chemometric analysis. *Journal of Agricultural and Food Chemistry*, **60**, 261–271.
38. Palikova, I., Vostalova, J., Zdarilova, A., Svobodova, A., Kosina, P., Vecera, R., Stejskal, D., Proskova, J., Hrbac, J., Bednar, P., Maier, V., Cernochova, D., Simanek, V. & Ulrichova, J. (2010) Long-term effects of three commercial cranberry products on the antioxidative status in rats: a pilot study. *Journal of Agricultural and Food Chemistry*, **58**, 1672–1678.
39. Harnly, J.M., Doherty, R.F., Beecher, G.R., Holden, J.M., Haytowitz, D.B., Bhagwat, S. & Gebhardt, S. (2006) Flavonoid content of U.S. fruits, vegetables, and nuts. *Journal of Agricultural and Food Chemistry*, **54**, 9966–9977.
40. Vvedenskaya, I.O., Rosen, R.T., Guido, J.E., Russell, D.J., Mills, K.A. & Vorsa, N. (2004) Characterization of flavonoids in cranberry (*Vaccinium macrocarpon*) powder. *Journal of Agricultural and Food Chemistry*, **52**, 188–195.
41. Tarascou, I., Mazauric, J.-P., Meudec, E., Souquet, J.-M., Cunningham, D., Nojeim, S., Cheynier, V. & Fulcrand, H. (2011) Characterization of genuine and derived cranberry proanthocyanidins by LC-ESI-MS. *Food Chemistry*, **128**, 802–810.
42. Kylli, P., Nohynek, L., Puupponen-Pimia, R., Westerlund-Wikstrom, B., Leppanen, T., Welling, J., Moilanen, E. & Heinonen, M. (2011) Lingonberry (*Vaccinium vitis* Idaea) and European cranberry (*Vaccinium microcarpon*) proanthocyanidins: isolation, identification, and bioactivities. *Journal of Agricultural and Food Chemistry*, **59**, 3373–3384.
43. Gu, L., Kelm, M.A., Hammerstone, J.F., Beecher, G., Holden, J., Haytowitz, D. & Prior, R.L. (2003) Screening of foods containing proanthocyanidins and their structural characterization using LC-MS/MS and thiolytic degradation. *Journal of Agricultural and Food Chemistry*, **51**, 7513–7521.
44. Reed, J.D., Krueger, C.G. & Vestling, M.M. (2005) MALDI-TOF mass spectrometry of oligomeric food polyphenols. *Phytochemistry*, **66**, 2248–2263.

45. Zuo, Y., Wang, C. & Zhan, J. (2002) Separation, characterization, and quantitation of benzoic and phenolic antioxidants in American cranberry fruit by GC-MS. *Journal of Agricultural and Food Chemistry*, **50**, 3789–3794.
46. Coppola, E.D., Conrad, E.C. & Cotter, R. (1978) High pressure liquid chromatographic determination of major organic acids in cranberry juice. *Journal of the Association of Official Analytical Chemists*, **61**, 1490–1492.
47. Coppola, E.D. & Starr, M.S. (1986) Liquid chromatographic determination of major organic acids in apple juice and cranberry juice cocktail: collaborative study. *Journal of the Association of Official Analytical Chemists*, **69**, 594–599.
48. Hong, V. & Wrolstad, R.E. (1986) Cranberry juice composition. *Journal of the Association of Official Analytical Chemists*, **69**, 199–207.
49. Prior, R.L., Rogers, T.R., Khanal, R.C., Wilkes, S.E., Wu, X. & Howard, L.R. (2010) Urinary excretion of phenolic acids in rats fed cranberry. *Journal of Agricultural and Food Chemistry*, **58**, 3940–3949.
50. Wang, S.Y. & Stretch, A.W. (2001) Antioxidant capacity in cranberry is influenced by cultivar and storage temperature. *Journal of Agricultural and Food Chemistry*, **49**, 969–974.
51. Wu, X., Beecher, G.R., Holden, J.M., Haytowitz, D.B., Gebhardt, S.E. & Prior, R.L. (2004) Lipophilic and hydrophilic antioxidant capacities of common foods in the United States. *Journal of Agricultural and Food Chemistry*, **52**, 4026–4037.
52. Caillet, S., Cote, J., Doyon, G., Sylvain, J.F. & Lacroix, M. (2011) Antioxidant and antiradical properties of cranberry juice and extracts. *Food Research International*, **44**, 1408–1413.
53. Sun, J., Chu, Y., Wu, X. & Liu, R.H. (2002) Antioxidant and antiproliferative activities of common fruits. *Journal of Agricultural and Food Chemistry*, **50**, 7449–7454.
54. Heinonen, I.M., Lehtonen, P.J. & Hopia, A.I. (1998) Antioxidant activity of berry and fruit wines and liquors. *Journal of Agricultural and Food Chemistry*, **46**, 25–31.
55. Kahkonen, M.P., Hopia, A.I. & Heinonen, M. (2001) Berry phenolics and their antioxidant activity. *Journal of Agricultural and Food Chemistry*, **49**, 4076–4082.
56. Saucier, C.T. & Waterhouse, A.L. (1999) Synergistic activity of catechin and other antioxidants. *Journal of Agricultural and Food Chemistry*, **47**, 4491–4494.
57. Tabart, J., Kevers, C., Pincemail, J., Defraigne, J.O. & Dommes, J. (2009) Comparative antioxidant capacities of phenolic compounds measured by various tests. *Food Chemistry*, **113**, 1226–1233.
58. Cote, J., Caillet, S., Doyon, G., Dussault, D., Salmieri, S., Lorenzo, G., Sylvain, J.F. & Lacroix, M. (2011) Effects of juice processing on cranberry antioxidant properties. *Food Research International*, **44**, 2907–2914.
59. Marinova, E.M. & Yanishlieva, N.V. (2003) Antioxidant activity and mechanism of action of some phenolic acids at ambient and high temperatures. *Food Chemistry*, **81**, 189–197.
60. Yan, X., Murphy, B.T., Hammond, G.B. & Vinson, J.A. (2002) Antioxidant activities and antitumor screening of extracts from cranberry fruit (*Vacciniun macrocarpon*). *Journal of Agricultural and Food Chemistry*, **50**, 5844–5849.
61. Maatta-Riihinen, K.R., Kahkonen, M.P., Torronen, A.R. & Heinonen, I.M. (2005) Catechins and procyanidins in berries of Vaccinium species and their antioxidant activity. *Journal of Agricultural and Food Chemistry*, **53**, 8485–8491.
62. Mladenka, P., Zatloukalova, L., Filipsky, T. & Hrdina, R. (2010) Cardiovascular effects of flavonoids are not caused only by direct antioxidant activity. *Free Radical Biology and Medicine*, **49**, 963–975.
63. Lee, C.-H., Krueger, C.G., Reed, J.D. & Richards, M.P. (2006) Inhibition of hemoglobin-mediated lipid oxidation in washed fish muscle by cranberry components. *Food Chemistry*, **99**, 591–599.
64. Formica, J.V. & Regelson, W. (1995) Review of quercetin and related bioflavonoids. *Food Chemistry and Toxicology*, **33**, 1061–1080.
65. Boots, A.W., Haenen, G.R.M.M. & Bast, A. (2008) Health effects of quercetin: from antioxidant to nutraceutical. *European Journal of Pharmacology*, **585**, 325–337.
66. Wang, W., Yagiz, Y., Buran, T.J., Nascimento Nunes, C. & Gu, L. (2011) Phytochemicals from berries and grapes inhibited the formation of advanced glycation end-products by scavenging reactive carbonyls. *Food Research International*, **44**, 2666–2673.
67. Ezaki-Furuichi, E., Nonaka, G.I., Nishioka, I. & Hayashi, K. (1987) Affinity of procyanidins (condensed tannins) from the bark of *Rhaphiolepis umbellata* for proteins. *Agricultural and Biological Chemistry*, **51**, 115–120.

68. Arts, M.J.T.J., Haenen, G.R.M.M., Wilms, L.C., Beetstra, S.A.J.N., Heijnen, C.G.M., Voss, H.-P. & Bast, A. (2002) Interaction between flavonoids and proteins: effect on the total antioxidant capacity. *Journal of Agricultural and Food Chemistry*, **50**, 1184–1187.
69. Williams, R.J., Spencer, J.P.E. & Rice-Evans, C. (2004) Serial review: Flavonoids and isoflavones (phytoestrogens): absorption, metabolism, and bioactivity. *Free Radical Biology and Medicine*, **36**, 838–849.
70. Espin, J.C., Garcia-Conesa, M.T. & Tomas-Barberin, F.A. (2007) Nutraceuticals: facts and fiction. *Phytochemistry*, **68**, 2986–3008.
71. Bailey, D.T., Dalton, C., Joseph Daugherty, F. & Tempesta, M.S. (2007) Can a concentrated cranberry extract prevent recurrent urinary tract infections in women? A pilot study. *Phytomedicine*, **14**, 237–241.
72. Sengupta, K., Alluri, K., Golakoti, T., Gottumukkala, G., Raavi, J., Kotchrlakota, L., Sigalan, S., Dey, D., Ghosh, S. & Chatterjee, A. (2011) A randomized, double blind, controlled, dose dependent clinical trial to evaluate the efficacy of a proanthocyanidin standardized whole cranberry (*Vaccinium macrocarpon*) powder on infections of the urinary tract. *Current Bioactive Compounds*, **7**, 39–46.
73. Debre, B., Conquy, S., Amsallem-Ouazanal, D., Bruel, P., Rahhali, N. & Charles, J. (2010) Impact of preventive treatment with titrated dry cranberry extract in a patient suffering from recurrent cystitis. *Value in Health*, **13**, A186.
74. Gupta, K., Chou, M.Y., Howell, A., Wobbe, C., Grady, R. & Stapleton, A.E. (2007) Cranberry products inhibit adherence of *p*-fimbriated *Escherichia coli* to primary cultured bladder and vaginal epithelial cells. *Journal of Urology*, **177**, 2357–2360.
75. Jepson, R.G. & Craig, J.C. (2007) A systematic review of the evidence for cranberries and blueberries in UTI prevention. *Molecular Nutrition & Food Research*, **51**, 738–745.
76. Sharon, N. & Ofek, I. (2002) Fighting infectious diseases with inhibitors of microbial adhesion to host tissues. *Critical Reviews in Food Science and Nutrition*, **42**, 267–272.
77. Howell, A.B., Vorsa, N., Dermardersosian, A. & Foo, L.Y. (1998) Inhibition of the adherence of *p*-fimbriated *Escherichia coli* to uroepithelial-cell surfaces by proanthocyanidin extracts from cranberries. *New England Journal of Medicine*, **339**, 1085–1086.
78. Howell, A.B., Reed, J.D., Krueger, C.G., Winterbottom, R., Cunningham, D.G. & Leahy, M. (2005) A-type cranberry proanthocyanidins and uropathogenic bacterial anti-adhesion activity. *Phytochemistry*, **66**, 2281–2291.
79. Howell, A.B. (2007) Bioactive compounds in cranberries and their role in prevention of urinary tract infections. *Molecular Nutrition & Food Research*, **51**, 732–737.
80. Foo, L.Y., Lu, Y., Howell, A.B. & Vorsa, N. (2000) A-type proanthocyanidin trimers from cranberry that inhibit adherence of uropathogenic *p*-fimbriated *Escherischia coli*. *Journal of Natural Products*, **63**, 1225–1228.
81. Foo, L.Y., Lu, Y., Howell, A.B. & Vorsa, N. (2000) The structure of cranberry proanthocyanidins which inhibit adherence of uropathogenic *p*-fimbriated *Escherichia coli in vitro*. *Phytochemistry*, **54**, 173–181.
82. Gupta, A., Dwivedi, M., Ali Mahdi, A., Gowda, G.A.N., Khetrapal, C.L. & Bhandari, M. (2011) Inhibition of adherence of multi-drug resistant *E. coli* by proanthocyanidin. *Urology Research*, **40**, 143–150.
83. Howell, A.B., Botto, H., Combescure, C., Blanc-Potard, A.-B., Gausa, L., Matsumoto, T., Tenke, P., Sotto, A. & Lavigne, J.P. (2010) Dosage effect on uropathogenic *Escherichia coli* anti-adhesion activity in urine following consumption of cranberry powder standardized for proanthocyanidin content: a multicentric randomized double blind study. *BMC Infectious Diseases*, **10**, 94.
84. Duarte, S., Gregoire, S., Singh, A.P., Vorsa, N., Schaich, K., Bowen, W.H. & Koo, H. (2006) Inhibitory effects of cranberry polyphenols on formation and acidogenicity of *Streptococcus mutans* biofilms. *FEMS Microbiology Letters*, **257**, 50–56.
85. Petti, S. & Scully, C. (2009) Polyphenols, oral health and disease: a review. *Journal of Dentistry*, **37**, 413–423.
86. Burger, O., Ofek, I., Tabak, M., Weiss, E.I., Sharon, N. & Neeman, I. (2000) A high molecular mass constituent of cranberry juice inhibits *Helicobacter pylori* adhesion to human gastric mucus. *FEMS Immunology and Medical Microbiology*, **29**, 295–301.
87. Burger, O., Weiss, E., Sharon, N., Tabak, M., Neeman, I. & Ofek, I. (2002) Inhibition of *Helicobacter pylori* adhesion to human gastric mucus by a high-molecular-weight constituent of cranberry juice. *Critical Reviews in Food Science and Nutrition*, **42**, 279–284.

88. Johnson-White, B., Buquo, L., Zeinali, M. & Ligler, F.S. (2003) Prevention of nonspecific bacterial cell adhesion in immunoassays by use of cranberry juice. *Analytical Chemistry*, **78**, 853–857.
89. Cunningham, D.G., Vannozzi, S.A., Turk, R., Roderick, R., O'Shea, E. & Brilliant, K. (2003) Cranberry phytochemicals and their health benefits. In: *Nutraceutical Beverages: Chemistry, Nutrition, and Health Effects* (eds F. Shahidi & D.K. Weerasinghe). ACS Symposium Series 871, American Chemical Society, Washington, DC, pp. 35–51.
90. Vattem, D.A., Lin, Y.T., Ghaedian, R. & Shetty, K. (2005) Cranberry synergies for dietary management of *Helicobacter pylori* infections. *Process Biochemistry*, **40**, 1583–1592.
91. Lacombe, A., Wu, V.C.H., Tyler, S. & Edwards, K. (2010) Antimicrobial action of the American cranberry constituents; phenolics, anthocyanins, and organic acids, against *Escherichia coli* O157:H7. *International Journal of Food Microbiology*, **139**, 102–107.
92. Valentova, K., Stejskal, D., Bednãiå, P., Vostalova, J., Cihalik, C., Vecerova, R., Koukalova, D., Kolar, M., Reichenbach, R., Sknouril, L., Ulrichova, J. & Simanek, V. (2007) Biosafety, antioxidant status, and metabolites in urine after consumption of dried cranberry juice in healthy women: a pilot double-blind placebo-controlled trial. *Journal of Agricultural and Food Chemistry*, **55**, 3217–3224.
93. Vorsa, N., Vvedenskaya, I., Huang, M. & Rosen, R.T. (2007) *Anti-inflammatory Cranberry Flavonol Extract Preparations*. U.S. Patent #7270837.
94. Kim, M.J., Ohn, J., Kim, J.H. & Kwak, H.-K. (2011) Effects of freeze-dried cranberry powder on serum lipids and inflammatory markers in lipopolysaccharide treated rats fed an atherogenic diet. *Nutrition Research and Practice*, **5**, 404–411.
95. Vattem, D.A., Randhir, R. & Shetty, K. (2005) Cranberry phenolics-mediated antioxidant enzyme response in oxidatively stressed porcine muscle. *Process Biochemistry*, **40**, 2225–2238.
96. Stoner, G.D., Wang, L.-S., Zikri, N., Chen, T., Hecht, S.S., Huang, C., Sardo, C. & Lechner, J.F. (2007) Cancer prevention with freeze-dried berries and berry components. *Seminars in Cancer Biology*, **17**, 403–410.
97. Sun, J. & Liu, R.H (2006) Cranberry phytochemical extracts induce cell cycle arrest and apoptosis in human MCF-7 breast cancer cells. *Cancer Letters*, **241**, 124–134.
98. Kresty, L.A., Howell, A.B. & Baird, M. (2008) Cranberry proanthocyanidins induce apoptosis and inhibit acid-induced proliferation of human esophageal adenocarcinoma cells. *Journal of Agricultural and Food Chemistry*, **56**, 676–680.
99. Seeram, N.P., Adams, L.S., Hardy, M.L. & Heber, D. (2004) Total cranberry extract versus its phytochemical constituents: antiproliferative and synergistics effects against human tumor cell lines. *Journal of Agricultural and Food Chemistry*, **52**, 2512–2517.
100. Stoner, G.D., Chen, T., Kresty, L.A., Aziz, R.M., Reinemann, T. & Nines, R. (2006) Protection against esophageal cancer in rodents with lyophilized berries: potential mechanisms. *Nutrition & Cancer*, **54**, 33–46.
101. Casto, B., Kresty, L.A., Kraly, C.L., Pearl, D.K., Knobloch, T.J. & Schut, H.A. (2002) Chemoprevention of oral cancer by black raspberries. *Anticancer Research*, **6C**, 4005–4015.
102. Lu, H., Li, H., Zhang, D., Stoner, G.D. & Chuanshu, H. (2006) Molecular mechanisms involved in chemoprevention of black raspberry extracts: from transcription factors to their target genes. *Nutrition & Cancer*, **54**, 69–78.
103. Wu, X., Beecher, G.R., Holden, J.M., Haytowitz, D.B., Gebhardt, S.E. & Prior, R.L. (2006) Concentrations of anthocyanins in common foods in the United States and estimation of normal consumption. *Journal of Agricultural and Food Chemistry*, **54**, 4069–4075.
104. Zhu, M., Perez, E., Phillips, D., Kim, W., Ghaedian, R., Napora, J.K. & Zou, S. (2011) Effects of long-term cranberry supplementation on endocrine pancreas in aging rats. *Journal of Gerontology A: Biological Sciences & Medical Sciences*, **66A**, 1139–1151.
105. Kathirvel, P., Gong, Y. & Richards, M.P. (2009) Identification of the compound in a potent cranberry juice extract that inhibits lipid oxidation in comminuted fish muscle. *Food Chemistry*, **115**, 924–932.
106. Kathirvel, P. & Richards, M.P. (2009) Mechanisms by which flavonol aglycones inhibit lipid oxidation better than glycosylated flavonols in comminuted muscle tissue. *Food Chemistry*, **117**, 75–82.
107. Lee, C.-H., Reed, J.D. & Richards, M.P. (2006) Ability of various polyphenolic classes from cranberry to inhibit lipid oxidation in mechanically separated turkey and cooked ground pork. *Journal of Muscle Foods*, **17**, 248–266.

108. Raghavan, S. & Richards, M.P. (2006) Partitioning and inhibition of lipid oxidation in mechanically separated turkey by components of cranberry press cake. *Journal of Agricultural and Food Chemistry*, **54**, 6403–6408.
109. Larrain, R.E., Krueger, C., Richards, M.P. & Reed, J.D. (2008) Color changes and lipid oxidation in pork products made from pigs fed with cranberry juice powder. *Journal of Muscle Foods*, **19**, 17–33.
110. Sampels, S., Asli, M., Vogt, G. & Morkore, T. (2010) Berry marinades enhance oxidative stability of herring fillets. *Journal of Agricultural and Food Chemistry*, **58**, 12230–12237.
111. Apostolidis, E., Kwon, Y.I. & Shetty, K. (2008) Inhibition of *Listeria monocytogenes* by oregano, cranberry and sodium lactate combination in broth and cooked ground beef systems and likely mode of action through proline metabolism. *International Journal of Food Microbiology*, **128**, 317–324.
112. Camire, M.E., Dougherty, M.P. & Briggs, J.L. (2007) Functionality of fruit powders in extruded corn breakfast cereals. *Food Chemistry*, **101**, 765–770.

6 Phytochemicals and health benefits of goji berries

Ying Zhong, Fereidoon Shahidi, and Marian Naczk

6.1 Introduction

Goji berries, also known as Lycium fruits or wolfberries, are member of the Solanaceae family. The mature fruits are bright red, ellipsoid berries 1–2 cm long with high juice content. While goji berries are mainly used for decoration of gardens in most parts of the world, they are a well-known tonic food and alternative medicine in Asian countries, especially in China. Among the dozens of *Lycium* species existing around the world, about 90% of commercially available goji berries are *Lycium barbarum*, originating from the north central regions of China. *Lycium chinense* is a species closely related to *L. barbarum*. Both *L. barbarum* and *L. chinense* are medicinal plants native to China, but are also widely found in Korea, Japan, and other Asian countries.

The fruits of Lycium (goji berries), also referred to as *Lycii fructus*, are harvested in the summer and autumn and subjected to a drying process. The fruits are usually dried in the shade followed by exposure to sun for further drying until the skin turns hard, but the pulp remains soft [1]. Goji berries have a long history of application in health promotion in China as a core ingredient in many herbal remedies, especially those for nourishing the blood, enriching the yin, tonifying the kidney and liver, moistening the lungs, improving vision, and promoting longevity [2]. The emenagogue, diuretic, antipyretic, tonic, aphrodisiac, hypnotic, and hepataprotective effects of goji berries have been well recorded in ancient Chinese Pharmacopoeia 2300 years ago [3]. Goji berries are traditionally consumed as an ingredient in Chinese cuisine (prepared as broth of poultry) or as medicinal elixir (liquor or tea infusion) [1]. Goji berries in formulated Chinese medicine have been used in the treatment of eye diseases, skin rashes, psoriasis, allergies, insomnia, chronic liver disease, diabetes, tuberculosis, and kidney disorders [4]. The *Lycium* species are known for their sedative, diuretic, and digestive actions in Turkish folk medicine [5]. More recently, a trend to use goji berries as a dietary supplement and natural health product in the Western world including the United Kingdom and North America has emerged. This chapter provides a general overview of bioactive components of goji berries and summarizes the research findings on their identification and health effect evaluation, most of which were reported in Chinese or Japanese with a few publications in international journals.

Dried Fruits: Phytochemicals and Health Effects, First Edition. Edited by Cesarettin Alasalvar and Fereidoon Shahidi.

6.2 Functional components in goji berries

Goji berries having a significant nutritive value and serve as an excellent source of macronutrient, including carbohydrate (46%), protein (13%), fat (1.5%), and dietary fiber (16%). They also contain micronutrients such as vitamins and minerals. Every 100 g of goji berries contains up to 60 mg of calcium, 5.4 mg of iron, 434 mg of potassium, 1.48 mg of zinc, and 48 mg of vitamin C [6]. Goji berries also contain riboflavin, thiamine, nicotinic acid, and minerals (Cu, Mn, Mg, and Se), among others [7, 8]. In addition to the macro- and micronutrients present, the bioactive components responsible for the health benefits of goji berries have been isolated and evaluated. Functional components such as polysaccharides, carotenoids, and phenolics have been identified and associated with the health effects of goji berries [8, 9].

6.2.1 Polysaccharides

Polysaccharides are one of the most valuable functional constituents and active compounds responsible for various health effects of goji berries. High polysaccharide content (5–8% in dried fruits and 40% in many commercial extracts) is a unique compositional characteristic of goji berries. *L. barbarum* polysaccharides (LBP), in particular, have been the focus of much research on bioactives of goji berries. LBP content is considered to be an important indicator of the medicinal efficacy of *Lycium* products [10]. Wang *et al.* [9] reported that the content of crude polysaccharides in *L. barbarum* fruits was 57.2 μg/g, including 25.6 μg/g of neutral polysaccharides and 26.9 μg/g of acidic polysaccharides as well as a small amount of unidentified fractions. Polysaccharides in edible plants in Chinese herbs are generally composed of a minimum number of 100 monosaccharide units with their molecular weights ranging from 10 to 1449 kDa [11]. Several LBP were isolated and purified from their crude extracts by diethylaminoethyl (DEAE) ion-exchange cellulose and gel permeation chromatography, and the LBP fractions had molecular weights of 24–241 kDa [3, 12].

Polysaccharides are usually extracted from goji berries with boiling water and separated by various chromatographic techniques. In some cases, lipids are removed either prior to or after the hot water extraction of polysaccharides. Lipids may be removed by refluxing with chloroform/methanol or ethanol, and/or precipitation with ethanol or acetone [8, 13]. In a study by Wang *et al.* [11], crude polysaccharides were extracted from *L. barbarum* with boiling water, followed by precipitation with ethanol at −20 °C and protein hydrolysis with a proteinase from *Bacillus* L (type III). Subsequent separation of the crude extracts by high-performance size exclusion chromatography yielded five LBP factions with the two main fractions having a molecular weight of 24 and 79 kDa. Gas chromatographic (GC) analysis revealed the presence of rhamnose, arabinose, xylose, mannose, glucose, and galactose as the major monosaccharides [11]. In another study, the boiling water decoction of *L. barbarum* showed a carbohydrate content of 97.5%, mainly composed of rhamnose, xylose, arabinose, glucose, and fructose [14]. Luo *et al.* [8] separated the crude LBP using a DEAE cellulose column and obtained four pure polysaccharide fractions. The major fraction was analyzed by GC–mass spectrometry (MS) and found to contain six monosaccharides (namely, rhamnose, galactose, glucose, arabinose, mannose, and xylose in a mole ratio of 4.22:2.43:1.38:1.00:0.95:0.38). Similar to other botanical (plant and fungal-derived) polysaccharides, LBP mainly occur as water-soluble glycoconjugates, for example, conjugates of glycan with peptides or proteins [15, 16]. Peng and Tian [3] isolated five glycoconjugates which were composed of 2–6 monosaccharides and 17 amino acids.

Health effects of LBP have been reported by various *in vitro*, cellular, animal, and human clinical studies. LBP were found to possess numerous biological activities and health effects, including antioxidant, antiaging, antitumor, antidiabetic, cytoprotective, neuroprotective, and immunomodulatory effects, among others [16]. Consumption of LBP-standardized goji berry juice for 14 days significantly improved the neurologic/psychologic performance and gastrointestinal function in humans [17]. LBP protected cultured cortical neurons exposed to glutamate by inhibiting glutamate-induced phosphorylation of c-jun N-terminal kinase (JKK) [18]. Neuroprotective effect of LBP against β-amyloid peptide toxicity was also observed [19, 20]. LBP was effective in reducing blood glucose levels and serum total cholesterol (TC) and triacylglycerols (TAG) contents while increasing high-density lipoprotein (HDL) cholesterol levels in alloxan-induced diabetic or hyperlipidemic rabbits [8]. LBP treatment resulted in a significantly decreased level of fasting glucose, TC, and TAG in diabetic mice [21]. Immunomodulatory and antitumor activities of LBP have also been demonstrated. LBP were able to upregulate both innate and adaptive immune responses, increase ConA-triggered proliferation of splenocytes, and cytotoxicity of natural killer cells [22]. The polysaccharide–protein complex in goji berries, which is composed of 6 monosaccharides and 18 amino acid residues, was found to enhance immune function, induce lymphocyte proliferation and cytokine production, activate T lymphocytes, and reduce side effects of chemo- and radiotherapy [23, 24]. LBP have been shown to promote peripheral blood recovery in mice with irradiation or chemotherapy-induced myelosuppression through stimulating the production of granulocyte colony-stimulating factor (G-SF) by peripheral blood mononuclear cells (PBMC) [23]. *In vivo* evidence has shown that LBP can also act against cancer and suppress the growth of malignant tumor [25, 26]. The effects of LBP on cell proliferation and apoptosis in rat and human hepatocellular carcinoma (HCC) cells have been reported [27]. LBP inhibited the growth of transplantable sarcoma S180 and increased macrophage phagocytosis and spleen lymphocyte proliferation, suggesting their potential effect in controlling tumor size and improving immune system [28]. LBP treatment also exerted an inhibitory effect on the growth of human gastric cancer MGC-803 and SGC-7901 cells by inducing cell-cycle arrest at the G0/G1 and S phase, respectively [29]. The antiaging property of LBP has been investigated and found to involve multiple mechanisms [16]. LBP exhibited antidecrepit effect in brain and heart tissues in mice and prolonged the life span of Drosophila [30, 31].

Many of the biological activities of LBP may be directly or indirectly attributed to their antioxidant potential as many chronic diseases are oxidative stress-mediated [32, 33]. LBP can counteract oxidative stress by eliminating reactive oxygen species (ROS) such as free radicals and promoting endogenous antioxidant factors such as superoxide dismutase (SOD), glutathione peroxidase (GSH-Px), and catalase (CAT). *In vitro* assays have supported the antioxidant activities of LBP in scavenging 2,2-diphenyl-1-picrylhydrazyl (DPPH) and superoxide radicals, inhibiting β-carotene-linoleate oxidation, reducing and chelating metal ions [13]. Trolox equivalent antioxidant capacity (TEAC) and oxygen radical absorbance capacity (ORAC) values have also been reported for crude and purified LBP [8]. LBP effectively inhibited peroxyl radical-mediated erythrocyte hemolysis in mice [33]. Administration of LBP decreased malondialdehyde (MDA) levels and increased activities of antioxidant enzymes in mice [34]. LBP were able to restore increased lipid peroxidation and decrease activities of antioxidant enzymes (such as SOD, CAT, and GSH-Px) as well as impaired immune function in aged mice to normal levels [33]. Consumption of a LBP-standardized goji berry juice (1632 mg LBP/120 mL) in a human intervention study significantly enhanced the markers of serum antioxidant by 8.4% for SOD and 9.9% for GSH-Px, while decreasing the MDA levels by 8.7% [32].

6.2.2 Carotenoids

Carotenoids, including carotenes and xanthophylls, are 40-carbon isoprenoids that occur widely in plants, fruits, and vegetables as yellow, orange, and red lipid-soluble pigments. In addition to their use as natural nontoxic colorants in manufactured foods, drinks, and cosmetics, carotenoids possess various biological activities, among which their pro-vitamin A and antioxidant activities are well known. Goji berries are an excellent source of carotenoids both because of their high content and because of their unique profile of carotenoids. Peng *et al.* [35] reported that the total carotenoid content of different goji berries ranged from 0.03 to 0.5%. The carotenoid content varies with season, for example, total carotenoid content is higher in summer fruits than in fall fruits [36]. Carotenoids in goji berries may be in the free, partially esterified, or fully esterified forms. Zeaxanthin and its esters are major contributors to the carotenoids in *L. barbarum* [37, 38]. The content of zeaxanthin esters in ripening goji fruits can reach as high as >77.5% of total carotenoids [37, 39]. Zeaxanthin palmitate (phasalien), in particular, comprises 31–56% of the total carotenoids as detected by high-performance liquid chromatography–diode array detector (HPLC-DAD) [35]. Meanwhile, zeaxanthin dipalmitate, which is the characteristic carotenoid in goji berries, together with zeaxanthin, has been the focus of much research on the protective role of goji berries against atrophic age-related macular degeneration (AMD) [35]. Zeaxanthin as one of the main pigments of retina was found to be effective in preventing and treating AMD [40, 41]. In addition to its contribution to eye health, isolated zeaxanthin and its dipalmitate have been reported to show antihepatotoxic activity, comparable to that of silybin [42]. Zeaxanthin and other carotenoids are also potent antioxidants, which contribute to the health effects of goji berry against oxidative stress-mediated diseases.

Extraction of carotenoids from goji berries is usually carried out by first removing water-soluble components with water or 50% ethanol, followed by extraction with organic solvents, such as acetone, petroleum ether, hexane, toluene, or their mixture. In some cases, a saponification process is employed to obtain free carotenoids. Identification and quantification of goji berry carotenoids by chromatographic and mass spectrometric techniques have been reported. Li *et al.* [37] identified 10 carotenoids, including zeaxanthin, β-cryptoxanthin, violaxanthin, and their esters, using HPLC-DAD analysis. Stephen Inbaraj *et al.* [43] reported the presence of 11 free carotenoids and 7 carotenoid esters by HPLC–MS/GC–flame ionization detector (HPLC-MS/GC-FID). Zeaxanthin dipalmitate (1143.7 μg/g) was the predominant one among all carotenoids identified, followed by β-cryptoxanthin monopalmitate isomers (32.9–68.5 μg/g), zeaxanthin monopalmitate isomers (11.3–62.8 μg/g), all-*trans*-β-carotene (23.7 μg/g), and all-*trans*-zeaxanthin (1.4 μg/g). In another study, β-carotene, neoxanthin, cryptoxanthin, and zeaxanthin were identified in the saponified extract of goji berry carotenoids [9]. Major carotenoids found in goji berries are presented in Table 6.1and their structures are given in Figure 6.1.

The content and composition of carotenoids in goji berries may vary depending on the maturation stage of the fruits. For instance, the level of esterified carotenoids increased with maturation of the fruits, while free carotenoids such as zeaxanthin were less abundant in ripe fruits [44]. Goji berries are mostly consumed as dried fruits. The drying process is believed to affect the carotenoid profiles of goji berries. The conventional drying process and

Table 6.1 Carotenoids identified in goji berries

Carotenoid	Source	Reference
β-Carotene	Fruit of *Lycium barbarum*	[43]
Zeaxanthin	Fruit of *Lycium barbarum*	[43]
Neoxanthin	Fruit of *Lycium barbarum*	[9]
Zeaxanthin dipalmitate	Fruit of *Lycium barbarum*	[35, 43]
Zeaxanthin monopalmitate	Fruit of *Lycium barbarum*	[43]
β-Cryptoxanthin monopalmitate	Fruit of *Lycium barbarum*	[43]

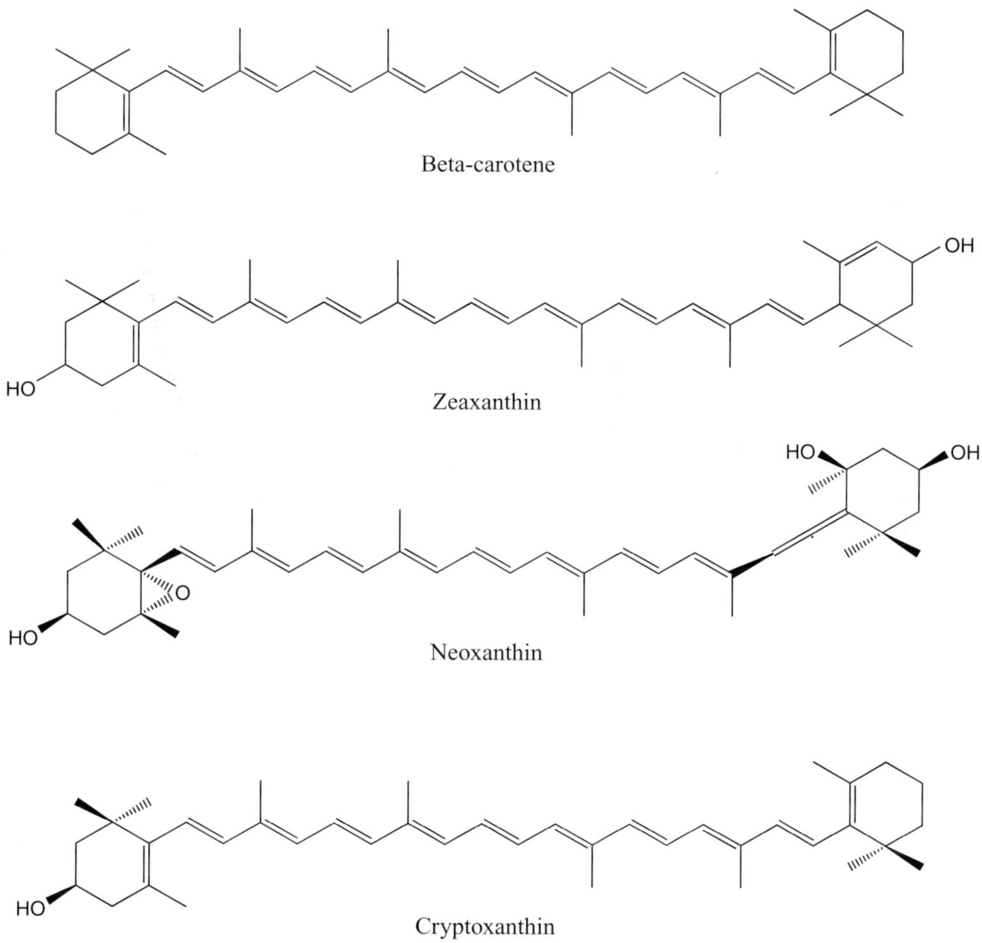

Figure 6.1 Structures of major carotenoids in goji berries.

storage usually lead to degradation or transformation of carotenoids. In general, esterified carotenoids undergo enzymatic transformation and oxidative degradation, which release the free form and oxidation products of carotenoids, respectively [36].

6.2.3 Phenolics

Phenolics are naturally occurring compounds widely distributed in the plant kingdom and beneficial components of human daily diet. They are important constituents of plants with multiple functions and as dietary phytochemicals for humans display a broad range of functional and biological activities. Phenolic compounds present in goji berries are mainly phenolic acids and flavonoids. Wang *et al.* [9] used 80% ethanol to extract phenolics from *L. barbarum*, followed by a separation with preparative chromatography. Four phenolic acids (chlorogenic, caffeoylquinic, caffeic, and *p*-coumaric) and three flavonoids (quercetin diglycoside, rutin, and kaempferol-3-*O*-rutinoside) have been identified as the major phenolic compounds. The flavonoid fraction obtained exhibited DPPH and 2,2′-azino-bis(3-ethylbenzothiazoline-6-sulfonic acid (ABTS) scavenging activity, metal ion chelation capacity, and reducing power. In another study, flavonoids were extracted with 95% ethanol from *L. barbarum* and evaluated for their antioxidant potential in terms of ABTS scavenging, metal chelation, and reducing power [1]. The extract contained 1.56 mg quercetin equivalents/g of total flavonoids, among which quercetin, myricetin, and kaempferol were predominant and comprised 43% of total flavonoids [1]. Stephen Inbaraj *et al.* [45] conducted a thorough investigation on the phenolic compounds in *L. barbarum*. The phenolics were extracted with 50% ethanol, isolated using a polymeric solid phase extraction cartridge and identified by HPLC-DAD–electrospray ionization-MS (HPLC-DAD-ESI-MS). The study revealed the presence of 52 phenolic acids and flavonoids. Quercetin-rhamno-di-hexoside (438.6 μg/g) was found to be the most abundant among all, followed by quercetin-3-*O*-rutinoside (281.3 μg/g), dicaffeoylquinic acid isomers (250.1 μg/g), and chlorogenic acid (237.0 μg/g). Caffeic, *p*-coumaric, and vanillic acids were also detected in the extract, although to a lesser extent. Other forms of phenolics, such as phenolic amides, have also been found in goji berries. The root bark of *L. barbarum* was reported to contain four phenolic amides, namely, dihydro-*N*-caffeoyltyramine, *trans*-*N*-feruloyloctopamine, *trans*-*N*-caffeoyltyramine, and *cis*-*N*-caffeoyltyramine [46]. These phenolic amides have been shown to exert antifungal effects at low concentrations (5–40 μg/mL) against dimorphic transition of the pathogen *Candida albicans* [46]. The main phenolic compounds identified in goji berries and their root barks are given in Table 6.2.

6.2.4 Other bioactive components

In addition to the major components discussed above, terpenes, alkaloids, cerebrosides, cyclic peptides, phytosterols, betaine, and a number of neutral volatile compounds have also been identified in goji berries [32, 47–50], which also make considerable contribution to the bioactivity and health effects of goji berries. Cerebrosides with hepatoprotective activity have been identified in *L. chinense*. Kim *et al.* [51] isolated a cerebroside from *L. chinense* that was effective in protecting primary cultured rat hepatocytes against galactosamine-induced hepatotoxicity. Kim *et al.* [42] isolated two cerebrosides from *L. chinense* fruits, which significantly suppressed the release of glutamic pyruvic transaminase (GPT) and sorbitol

Table 6.2 Phenolic compounds identified in goji berries and their root barks

Phenolic compound	Source	Reference
Quercetin	Fruit of *Lycium barbarum*	[1]
Myricetin	Fruit of *Lycium barbarum*	[1]
Kaempferol	Fruit of *Lycium barbarum*	[1]
Rutin	Fruit of *Lycium barbarum*	[9]
Caffeic acid	Fruit of *Lycium barbarum*	[9]
p-Coumaric acid	Fruit of *Lycium barbarum*	[9]
Caffeoylquinic acid	Fruit of *Lycium barbarum*	[9]
Kaempferol-3-*O*-rutinoside	Fruit of *Lycium barbarum*	[9]
Ferulic acid	Fruit of *Lycium barbarum*	[45]
Vanillic acid	Fruit of *Lycium barbarum*	[45]
Quercetin-rhamno-di-hexoside	Fruit of *Lycium barbarum*	[45]
Quercetin-3-*O*-rutinoside	Fruit of *Lycium barbarum*	[45]
Chlorogenic acid	Fruit of *Lycium barbarum*	[45]
dihydro-*N*-Caffeoyltyramine	Root bark of *Lycium barbarum*	[46]
trans-*N*-Feruloyloctopamine	Root bark of *Lycium barbarum*	[46]
trans-*N*-Caffeoyltyramine	Root bark of *Lycium barbarum*	[46]
cis-*N*-Caffeoyltyramine	Root bark of *Lycium barbarum*	[46]

dehydrogenase (SDH) by CCl_4-intoxicated hepatocytes. Other goji berry compounds with hepatoprotective effect have also been identified, such as pyrrole derivatives, betaine, and the arabinogalactan–protein complex [52–55]. Betaine is also a phytochemical that plays an important role in the prevention of cardiovascular disease (CVD) by decreasing homocysteine toxicity and hence reducing the risk of coronary heart disease (CHD) and stroke [56]. Its content was reported to be 1.5% in *L. chinense* leaves and 0.9–1.4% in *L. barbarum* whole plant on a dry weight basis [57, 58]. Lyciumins, a unique cyclic octapeptide in the roots of *L. chinense*, has shown angiotensin converting enzyme (ACE)-inhibitory activity, suggesting its antihypertensive potential [59]. The nonpolar constituents of goji berries have also been studied. The essential oil of goji berries have been reported to be mainly composed of ethyl hexadecanoate, 1-octadecanone, tetrapyrazine, 2-furan-carboxaldehyde, and ethyl linoleate from *L. chinense* and hexadecanoic acid, linoleic acid, β-elemene, myristic acid, and ethyl hexadecanoate from *L. barbarum* [60]. The oil exhibited excellent antioxidant activity in scavenging DPPH radical and inhibiting β-carotene bleaching, and its antioxidant activity was comparable to ascorbic acid and α-tocopherol [61]. In addition, a new analogue of ascorbic acid, 2-*O*-(β-D-glucopyranosyl) ascorbic acid, has been isolated from goji berries. This provitamin C compound accounts for approximately 0.5% of the dry weight of *L. barbarum*, and was thought to increase the level of blood ascorbic acid in rats and contribute to the antiaging properties of goji berries [62]. Minor bioactive components present in goji berries and their coproducts are summarized in Table 6.3.

6.3 Health benefits of goji berries

Goji berries are known for their broad array of biological activities and health effects. *L. barbarum* has been shown by modern studies to possess antioxidant, antiaging, antidiabetic,

Table 6.3 Other bioactive compounds identified in goji berries and their by-products

Compound	Source	Reference
2-*O*-(β-D-glucopyranosyl)-ascorbic acid	Fruit of *Lycium barbarum*	[62]
Hexadecanoic acid	Essential oil of *Lycium barbarum*	[75]
Linoleic acid	Essential oil of *Lycium barbarum*	[75]
β-Elemene	Essential oil of *Lycium barbarum*	[75]
Lyciumin	Root of *Lycium chinense*	[59]
Betaine	Leaf/stem/root of *Lycium barbarum*	[58]
Cerebrosides	Fruit of *Lycium chinense*	[74]
Glycerogalactolipids	Fruit of *Lycium chinense*	[4]
Pyrrole derivatives	Fruit of *Lycium chinense*	[53]

anticarcinogenic, neuroprotective, and immune enhancing properties [16, 17, 19, 28, 63]. They play a role in the prevention of diabetes, hepatitis, cancer, AMD, thrombosis, and male infertility, among other diseases and health conditions [7, 55]. Goji berry juice has been reported in a randomized double-blinded clinical study to increase subjective feelings of general well-being and improve neurologic/psychologic performance [17]. In addition to the goji berry fruits, the root bark and leaves of goji berries also contain pharmacologically active components with a variety of biological activities. The root bark of goji berries has been shown to possess hypotensive, hypoglycemic, antipyretic, and antiulceration effects in experimental animals and is used in oriental medicine as a tonic to relieve coughing, hypertension, and diabetes mellitus [64–68]. The leaves of *L. chinense* or *Lycii folium* are consumed as herbal tea in the oriental countries as a nourishing and tonic ingredient to reduce the risk of arteriosclerosis and essential arterial hypertension, diabetes, and night blindness [69]. The stamina-improving, tranquilizing, and thirst-quenching properties of goji berry leaf tea have also been well documented [42].

Although the beneficial effects of goji berries on human health have been known and practiced for centuries, only recent modern technology was able to unravel its mechanisms of action at the biochemical level. Many comprehensive studies have attributed the health effects of goji berries to their role as an antioxidant. *L. barbarum* is thought to elicit its protective effects in the body through counteracting free radical-induced oxidative stress [33]. The *L. barbarum* extract exhibited superoxide anion radical scavenging activity and inhibitory effect against oxidation products such as MDA in rat liver homogenates [70]. The aqueous extract of *L. chinense* fruit also inhibited hepatic MDA formation and depletion of reduced glutathione content and CAT activity in CCl_4-injected rats [55]. The ethanol (70%) extract of *L. chinense* protected liver cells against oxidative stress-induced cell damage by scavenging intracellular ROS, recovering SOD, CAT, and glutathione activity, decreasing lipid oxidation, DNA damage, and protein carbonyl values [71]. In addition to their antioxidant activity, goji berries may also render their hepatoprotective effect through other mechanisms, such as expressional regulation of cytochrome P450 and improving viability of liver cells via inhibition of apoptosis [55, 71]. Other bioactivities of goji berries not directly related to their antioxidant activity have also been reported. For example, the leaves of *L. chinense* have been shown to stimulate the growth of *Lactobacilus acidophilus* and other probiotic cells commonly used in the food industry [72, 73].

6.4 Conclusions

Goji berries, an ancient oriental medicinal food, have recently attracted much attention as a functional food and a source of natural health products. Modern research has revealed their multiple biological activities and active components responsible for their health benefits. More research is needed in order to take better advantage of goji berries for promoting human health.

References

1. Le, K., Chiu, F. & Ng, K. (2007) Identification and quantification of antioxidants in Fructus lycii. *Food Chemistry*, **105**, 353–363.
2. Pharmacopoeia of the People's Republic of China (2000) *Pharmacopoeia of the People's Republic of China*. Chemical Industry Press, Beijing, People's Republic of China.
3. Peng, X. & Tian, G. (2001) Structural characterization of the glycan part of glycoconjugate LbGp2 from *Lycium barbarum* L. *Carbohydrate Research*, **331**, 95–99.
4. Gao, Z., Ali, Z. & Khan, I.A. (2008) Glyceroglalctolipids from the fruit of *Lycium barbarum*. *Phytochemistry*, **69**, 2856–2861.
5. Kosar, M., Altintas, A., Kirimer, N. & Baser, K.H.C. (2003) Determination of the free radical scavenging activity of Lycium extracts. *Chemistry of Natural Compounds*, **39**, 531–535.
6. Centre for Food Safety (2011) *Nutrient*. Published online at: http://www.cfs.gov.hk/cgi-bin/cfsnew/nutrient/subgroups.cgi, last accessed October 28, 2011.
7. Li, Q.Y. (2001) *Healthy Functions and Medicinal Prescriptions of Lycium barbarum* (Gou Ji Zi). Jindun Press, Beijing, China.
8. Luo, W., Cai, Y., Yan, J., Sun, M. & Corke, H. (2004) Hypoglycemic and hypolipidemic effects and antioxidant activity of fruit extracts and polysaccharide fractions from *Lycium barbarum*. *Life Sciences*, **76**, 137–149.
9. Wang, C.C., Chang, S.C., Stephen Inbaraj, B. & Chen, B.H. (2010) Isolation of carotenoids, flavonoids and polysaccharides from *Lycium barbarum* L. and evaluation of antioxidant activity. *Food Chemistry*, **120**, 184–192.
10. Wong, C.K., Leung, K.N., Fung, K.P. & Choy, Y.M. (1994) Immunomodulatory and anti-tumour polysaccharides from medicinal plants. *Journal of International Medical Research*, **22**, 299–312.
11. Wang, C.C., Chang, S.C. & Chen, B.H. (2009) Chromatographic determination of polysaccharides in *Lycium barbarum* Linnaeus. *Food Chemistry*, **116**, 595–603.
12. Tian, M. & Wang, M. (2006) Studies on extraction, isolation and composition of *Lycium barbarum* polysaccharides. *China Journal of Chinese Materia Medica*, **31**, 1603–1607.
13. Li, X.L. & Zhou, A.G. (2007) Evaluation of the antioxidant effects of polysaccharides extracted from *Lycium barbarum*. *Medicinal Chemistry Research*, **15**, 471–482.
14. Li, X.M., Li, X.L. & Zhou, A.G. (2007) Evaluation of antioxidant activity of the polysaccharides extracted from *Lycium barbarum* fruits *in vitro*. *European Polymer Journal*, **43**, 488–497.
15. Wang, Q. (1991) Determination of polysaccharide contents in fructus Lycii. *Chinese Traditional Herbal Drugs*, **22**, 67–68.
16. Chang, R.C. & So, K.F. (2008) Use of anti-aging herbal medicine, *Lycium barbarum*, against aging-associated diseases. What do we know so far. *Cellular and Molecular Neurobiology*, **28**, 643–652.
17. Amagase, H. & Nance, D.M. (2008) A randomized, double-blind, placebo-controlled, clinical study of the general effects of a standardized *Lycium barbarum* (goji) juice, GoChi. *Journal of Alternative and Complementary Medicine*, **14**, 403–412.
18. Ho, Y.S., Yu, M.S., Yik, S.Y., So, K.F., Yuen, W.H. & Chang, R.C.C. (2009) Polysaccharides from wolfberry antagonizes glutamate excitotoxicity in rat cortical neurons. *Cellular and Molecular Neurobiology*, **29**, 1233–1244.
19. Yu, M.S., Leung, S.K.Y., Lai, S.W., Che, C.M., Zee, S.Y., So, K.F., Yuen, W.H. & Chang, R.C.C. (2005) Neuroprotective effects of anti-aging oriental medicine *Lycium barbarum* against beta-amyloid peptide neurotoxicity. *Experimental Gerontology*, **40**, 716–727.

20. Ho, Y.S., Yu, M.S., Lai, C.S., So, K.F., Yuen, W.H. & Chang, R.C.C. (2007) Characterizing the neuroprotective effects of alkaline extract of *Lycium barbarum* on beta-amyloid peptide neurotoxicity. *Brain Research*, **1158C**, 123–134.
21. Jing, L., Cui, G., Feng, Q. & Xiao, Y. (2009) Evaluation of hypoglycaemic activity of the polysaccharides extracted from *Lycium barbarum*. *African Journal of Traditional, Complementary and Alternative Medicines*, **6**, 579–584.
22. Wang, L., Zhao, C., Li, X. & Chien, Y. (1995) Isolation, purification, analysis and immunity effects of polysaccharide of *Lycium barbarum*. *Academic Journal of Kunming Medical College*, **16**, 29–32.
23. Gong, H., Shen, P., Jin, L., Xing, C. & Tang, F. (2005) Therapeutic effects of *Lycium barbarum* polysaccharide (LBP) on irradiation or chemotherapy-induced myelosuppressive mice. *Cancer Biotherapy and Radiopharmaceuticals*, **20**, 155–162.
24. Chen, Z., Tan, B.K.H. & Chan, S.H. (2008) Activation of T lymphocytes by polysaccharide–protein complex from *Lycium barbarum* L. *International Immunopharmacology*, **8**, 1663–1671.
25. Wang, B.K., Xing, S.T. & Zhou, J.H. (1988) Effect of *Lycium barbarum* polysaccharides on immune function of S180-bearing mice and their antitumor activity. *Chinese Journal of Pharmacology Toxicology*, **2**, 127–131.
26. Gu, S.N. & Jiao, K.Z. (1996) The analysis of *Lycium barbarum* polysaccharide on systemic treatment on 20 cases of primary liver cancer. *Chinese Medicine and Pharmacology Clinic*, **6**, 38–39.
27. Chao, J.C., Chiang, S.W., Wang, C.C., Tsai, Y.H. & Wu, M.S. (2006) Hot water-extracted *Lycium barbarum* and *Rehmannia glutinosa* inhibit proliferation and induce apoptosis of hepatocellular carcinoma cells. *World Journal of Gastroenterology*, **12**, 4478–4484.
28. Gan, L., Zhang, S.H., Yang, X.I. & Xu, H.B. (2004) Immunomodulation and antitumor activity by a polysaccharide–protein complex from *Lycium barbarum*. *International Immunopharmacology*, **4**, 563–569.
29. Miao, Y., Xiao, B., Jiang, Z., Guo, Y., Mao, F., Zhao, J., Huang, X. & Guo, J. (2010) Growth inhibition and cell-cycle arrest of human gastric cancer cells by *Lycium barbarum* polysaccharide. *Medical Oncology*, **27**, 785–790.
30. Xu, C.S. (2000) The effect of the *Lycium chinense* Mill. (LcM.) on the lifespan of *Drosophila melanogaster*. *Journal Taishan Medical College*, **3**, 184–186.
31. Xu, C. & Fang, Y. (2000) Experimental study on the anti-decrepit effect of *Lycium chinense* Mill. *Journal of JiLing Medical College*, **3**, 19–22.
32. Amagase, H., Sun, B.X. & Borek, C. (2009) *Lycium barbarum* (goji) juice improves *in vivo* antioxidant biomarkers in serum of healthy adults. *Nutrition Research*, **29**, 19–25.
33. Wang, J.M., Hu, Y.L., Wang, D.Y., Zhang, F., Zhao, X.N., Abula, S., Fan, Y.P. & Guo, L.W. (2010) *Lycium barbarum* polysaccharide inhibits the infectivity of Newcastle disease virus to chicken embryo fibroblast. *International Journal of Biological Macromolecules*, **46**, 212–216.
34. Wu, H.T., He, X.J., Hong, Y.K., Ma, T., Xu, Y.P. & Li, H.H. (2010) Chemical characterization of *Lycium barbarum* polysaccharides and its inhibition against liver oxidative injury of high-fat mice. *International Journal of Biological Macromolecules*, **46**, 540–543.
35. Peng, Y., Ma, C., Li, Y., Leung, K.S., Jiang, Z. & Zhao, Z. (2005) Quantification of zeaxanthin dipalmitate and total carotenoids in Lycium fruits (Fructus lycii). *Plant Foods for Human Nutrition*, **60**, 161–164.
36. Ma, W., Ni, Z., Li, H. & Chen, M. (2008) Changes of the main carotenoid pigment contents during the drying processes of the different harvest stage fruits of *Lycium barbarum* L. *Agricultural Sciences in China*, **7**, 363–369.
37. Li, Z., Peng, G. & Zhang, S. (1998) Separation and determination of carotenoids in *Fructus lycii* by isocratic non-aqueous reversed-phase liquid chromatography. *Chinese Journal of Chromatography*, **16**, 341–343.
38. Weller, P. & Breithaupt, E. (2003) Identification and quantification of zeaxanthin esters in plants using liquid chromatography–mass spectrometry. *Journal of Agricultural and Food Chemistry*, **51**, 7044–7049.
39. Peng, G.H., Li, Z. & Zhang, S.H. (1998) Separation and identification of carotenoids in *Fructus lycii* by thin-layer chromatography. *Acta Nutrimenta Sinica*, **20**, 76–78.
40. Snodderly, D.M. (1995) Evidence for protection against age-related macular degeneration by carotenoids and antioxidant vitamins. *American Journal of Clinical Nutrition*, **62**, 1448S–1461S.
41. Cooper, D.A., Eldridge, A.L. & Peters, J.C. (1999) Dietary carotenoids and certain cancers, heart disease, and age-related macular degeneration: a review of recent research. *Nutrition Review*, **57**, 201–214.

42. Kim, S.Y., Lee, K.H., Chang, K.S., Bock, J.Y. & Jung, M.Y. (1997) Taste and flavour compounds in box thron (*Lycium chinense* Miller) leaves. *Food Chemistry*, **58**, 297–303.
43. Stephen Inbaraj, B., Lu, H., Hung, C.F., Wu, W.B., Lin, C.L. & Chen, B.H. (2008) Determination of carotenoids and their esters in fruits of *Lycium barbarum* Linnaeus by HPLC-DAD-APCI-MS. *Journal of Pharmaceutical and Biomedical Analysis*, **47**, 812–818.
44. Li, H., Chen, M., Ma, W.P. & Wang, J.Y. (2006) Law of changes of carotenoids content in *Fructus lycii* of Chinese wolfberry (*Lycium barbarum* L.) at different mature periods. *Scientia Agricultura Sinica*, **39**, 599–605.
45. Stephen Inbaraj, B., Lu, H., Kao, T.H. & Chen, B.H. (2010) Simultaneous determination of phenolic acids and flavonoids in *Lycium barbarum* Linnaeus by HPLC-DAD-ESI-MS. *Journal of Pharmaceutical and Biomedical Analysis*, **51**, 549–556.
46. Lee, D.G., Park, Y., Kim, M.R., Jung, H.J., Seu, Y.B., Hahm, K.S. & Woo, E.R. (2004) Anti-fungal effects of phenolic amides isolated from the root bark of *Lycium chinense*. *Biotechnology Letters*, **26**, 1125–1130.
47. Itoh, T., Tamura, T. & Matsumoto, T. (1978) Four new and other 4α-methylsterols in the seeds of Solanaceae. *Phytochemistry*, **17**, 971–977.
48. Sannai, A., Fujimori, T. & Kato, K. (1982) Isolation of (-)-1,2-dehydro-α-crperone and solavetivone from *Lycium chinense*. *Phytochemistry*, **21**, 2986–2987.
49. Yahara, S., Shigeyama, C., Ura, T., Wakamastsu, K., Yasuhara, T. & Nohara, T. (1993) Cyclic peptides, acyclic diterpene glycoside and other compounds from *Lycium chinense* Mill. *Chemical and Pharmaceutical Bulletin*, **41**, 703–709.
50. Funayama, S., Zhang, G.R. & Nozoe, S. (1995) Kukoamin B, a spermine alkaloid from *Lycium chinense*. *Phytochemistry*, **38**, 1529–1531.
51. Kim, S.Y., Lee, J.E., Kim, P.H., Lee, H.S. & Kim, Y.C. (2000) LCC, a cerebroside from *Lycium chinense*, protects primary cultured rat hepatocytes exposed to galactoseamine. *Phytotherapy Research*, **14**, 448–451.
52. Kajimoto, Y. & Kurokawa, S. (1960) Pharmacological study on components of *Lycium chinense* Miller. *Folia Pharmacologica Japonica*, **56**, 151.
53. Chin, Y.W., Lim, S.W., Kim, S.H., Shin, D.Y., Suh, Y.G., Kim, Y.B., Kim, Y.C. & Kim, J. (2003) Hepataoprotective pyrrole derivatives of *Lycium chinense* fruits. *Bioorganic & Medicinal Chemistry Letters*, **13**, 79–81.
54. Zhang, M., Chen, H., Huang, J., Li, Z., Zhu, C. & Zhang, S. (2005) Effect of *Lycium barbarum* polysaccharide on human hepatoma QGY7703 cells: inhibition of proliferation and induction of apoptosis. *Life Sciences*, **76**, 2115–2124.
55. Ha, K.T., Yoon, S.J., Choi, D.Y., Kim, D.W., Kim, J.K. & Kim, C.H. (2005) Protective effect of *Lycium chinense* fruit on carbon tetrachloride-induced hepatotoxicity. *Journal of Ethnopharmacology*, **96**, 529–535.
56. Dalmeijer, G.W., Olthof, M.R., Verhoef, P., Bots, M.L. & van der Schouw, Y.T. (2008) Prospective study on dietary intakes of folate, betaine, and choline and cardiovascular disease risk in women. *European Journal of Clinical Nutrition*, **62**, 386–394.
57. Nishiyama, R. (1963) Betaine of *Lycium chinense*. *Nippon Shokuhin Kogyo Gakkai-Shi*, **10**, 517–519.
58. Chung, R.S., Chen, C.C. & Ng, L.T. (2010) Nitrogen fertilization affects the growth performance, betaine and polysaccharide concentration of *Lycium barbarum*. *Industrial Crops and Products*, **32**, 650–655.
59. Yahara, S., Shigeyama, C., Nohara, T., Okuda, H., Wakamatsu, K. & Yasuhara, T. (1989) Structures of anti-ace and -renin peptides from lycii radicis cortex. *Tetrahedron Letters*, **30**, 6041–6042.
60. Park, W.J., Bock, J.Y., Baik, S.O., Han, S.B. & Ju, H.K. (1997) Volatile components of *Lycium chinense* Miller. *Korean Society of Food Science and Nutrition*, **10**, 1–5.
61. Li, G., Shi, J., Suo, Y., Sun, Z., Xia, L., Zheng, J., You, J. & Liu, Y. (2011) Supercritical CO_2 cell breaking extraction of *Lycium barbarum* seed oil and determination of its chemical composition by HPLC/APCI/MS and antioxidant activity. *Food Science and Technology*, **44**, 1172–1178.
62. Toyoda-Ono, Y., Maeda, M., Nakao, M., Yoshimura, M., Sugiura-Tomimori, N. & Fukami, H. (2004) 2-*O*-(β-D-glucopyranosyl)ascorbic acid, a novel ascorbic acid analogue isolated from Lycium fruit. *Journal of Agricultural and Food Chemistry*, **52**, 2092–2096.
63. Qian, J.Y., Liu, D. & Huang, A.G. (2004) The efficiency of flavonoids in polar extracts of *Lycium chinense* Mill fruits as free radical scavenger. *Food Chemistry*, **87**, 283–288.

64. Funayama, S., Yoshida, K., Konno, C. & Hikino, H. (1980) Structure of kukoamine A, a hypotensive principle of *Lycium chinense* root barks. *Tetrahedron Letters*, **21**, 1355–1356.
65. Morota, T., Sasaki, H., Chin, M., Sato, T., Katayama, N., Fukuyama, K. & Mitsuhashi, H. (1987) Studies on the crude drug containing the angiotensin I converting enzyme inhibitors (I) on the active principles of *Lycium chinense* Miller. *Shoyakugaku Zasshi*, **41**, 169–173.
66. Zhu, Y.P. (1998) *Chinese Materia Medica—Chemistry, Pharmacology and Applications*. Hardwood Academic Publishers, Amsterdam, The Netherlands.
67. Huang, K.C. (1999) *The Pharmacognosy of Chinese Herbs*. CRC Press, Boca Raton, FL.
68. Chen, J.K. & Chen, T.T. (2004) *Chinese Medical Herbology and Pharmacology*. Art of Medicine Press, Inc., City of Industry, CA.
69. Mizobuchi, K.I., Taniguchi, H., Kita, Y., Inoue, K., Kimura, K. & Higashi, J. (1964) Studies on the chemical components of box thorn seasonal variation of vitamin C and rutin contents in the leaves of Japanese *Lycium chinense*. *Annual Report Faculty of Pharmacy, Tokushima University*, **18**, 27–30.
70. Wu, S.J., Ng, L.T. & Lin, C.C. (2004) Antioxidant activities of some common ingredients of traditional Chinese medicine, *Angelica sinensis*, *Lycium barbarum* and *Poria cocos*. *Phytotherapy Research*, **18**, 1008–1012.
71. Zhang, R., Kang, K.A., Piao, M.J., Kim, K.C., Kim, A.D., Chae, S., Park, J.S., Youn, Y.J. & Hyun, J.W. (2010) Cytoprotective effects of the fruits of *Lycium chinense* Miller against oxidative stress-induced hepatotoxicity. *Journal of Enthopharmacology*, **130**, 299–306.
72. Nishiyama, R. (1965) Studies on components of Kuko (*Lycium chinense* Miller). Part III. Growth effect for lactic acid bacteria by the components of Kuko (*Lycium chinense* Miller). *Journal of Agricultural Chemical Society of Japan*, **12**, 313–319.
73. Yeh, Y.C., Hahm, T.S., Sabliov, C.M. & Lo, Y.M. (2008) Effects of Chinese wolfberry (*Lycium chinense* P. Mill.) leaf hydrolysates on the growth of *Pediococcus acidilactici*. *Bioresource Technology*, **99**, 1383–1393.
74. Kim, S.Y., Choi, Y.H., Huh, K., Kim, J., Kim, Y.C. & Lee, H.S. (1997) New antihepatotoxic cerebroside from *Lycium chinense* fruits. *Journal of Natural Products*, **60**, 274–276.
75. Altintas, A., Kosar, M., Kirimer, N., Baser, K.H.C. & Demirci, B. (2006) Composition of the essential oils of *Lycium barbarum* and *L. ruthenicum* fruits. *Chemistry of Natural Compounds*, **42**, 24–25.

7 Dried mulberries: phytochemicals and health effects

Mine Gultekin Ozguven and Beraat Ozcelik

7.1 Introduction

Mulberry (*Morus* sp.) belongs to the family Moraceae which comprises about 40 genera and more than 1000 species [1]. *Morus* consists of 24 species and one subspecies, with more than 100 known varieties [2], with deciduous trees adapted to tropical, subtropical, and temperate zones of Asia, Europe, North and South America, and Africa [1, 3]. The optimum temperature for mulberry trees ranges from 20 to 24°C and the atmospheric humidity from 65 to 80% [4]. Mulberry trees, which are distributed widely, in particular in China and India [5], grow fast when young, but soon start growing slowly until reaching 10–15 meters in height [1]. Generally, mulberries are used for their foliage to feed the silkworm (*Bombyx mori* L.) in most mulberry-growing countries. On the other hand, mulberries are grown by European countries, including Turkey and Greece, for fruit production rather than foliage [2, 6, 7].

Mulberries are white to pale yellow with pink edges at the beginning. They then become red during ripening and dark purple to black when fully ripened [1]. There are three common mulberry species such as *Morus alba* (originated in Western Asia), *Morus rubra* (originated in North and South America), and *Morus nigra* (originated in Southern Russia) [3, 8, 9]. Fruits of white mulberries (*M. alba*) are perishable with a very sweet taste and low acidity and mostly consumed fresh. Purple mulberries (*M. rubra*) have a sweet taste and low acidity with high dry matter. Black mulberries (*M. nigra*) are juicy with extraordinary color and a refreshing acidic flavor [3, 8]. However, the color of mulberries cannot be used to identify the mulberry species since *M. alba* generates several colors of berries, including white, red, and purple [10].

Studies on mulberries have shown that they are rich sources of natural bioactive compounds or phytochemicals that exhibit biological effects on human health. This chapter highlights the compositional and nutritional characteristics, phytochemicals, health effects, and food applications of fresh mulberries with particular reference to their dried counterparts when data are available.

Dried Fruits: Phytochemicals and Health Effects, First Edition. Edited by Cesarettin Alasalvar and Fereidoon Shahidi.

7.2 Drying of mulberries

Drying is one of the most important preservation methods of agricultural products since it allows an extended storage period due to reduced moisture content [11]. Reducing water activity inhibits microbial growth, enzymatic reactions, and other deteriorative processes [12]. Since mulberries are highly sensitive to storage and their harvesting seasons are very short, they need to be preserved by drying [7, 9, 13].

Sun drying is traditionally used for mulberry drying. After harvesting mulberries by handpicking or by shaking tree branches, they are spread over a sheet on the ground to be dried [14, 15]. Although solar energy is abundant, cheap, renewable, and environmentally friendly, contamination with dust, soil, and insects may occur. Moreover, exposure to direct sun radiation results in undesirable color changes in the fruits [9, 11, 16, 17]. Due to longer drying time, drying rate should be increased by using pretreatment with ethyl oleate, potassium carbonate, and sodium hydroxide [9]. Therefore, faster, safer, and more controllable drying techniques such as mechanical (artificial) air drying or solar air drying should be used to obtain more uniform, hygienic, and attractive-colored fruits [14, 15, 18]. In a solar dryer, air is heated by the sun in a collector and then passed over the mulberries to be dried. Pangavhane *et al.* [18] reported that sensory qualities of solar dried grape fruits were superior to those of open sun dried ones. Furthermore, the extent of nonenzymatic browning was decreased when using solar drying method [18]. Meanwhile, solar drying needs negligible installation and energy cost. However, due to uncertain environmental conditions in late summer or early fall or climatic variations, inadequate dehydration and long processing periods may cause considerable product loss [19]. For this reason, mechanical air drying in air driers is a good alternative in obtaining a good quality product in a shorter time for the industry [16, 19]. In this method, air dryers are heated by burning a gas (usually propane) or fuel oil, or their combination [19]. The method allows minimizing the composition, texture, and color changes. Despite these advantages, initial investment and additional running costs are high [16, 19].

Freeze and microwave drying techniques for drying berries have also been reported in the literature [20]. Vacuum-assisted microwave drying is effective in preserving the heat- and oxygen-sensitive phenolic components and ascorbic acid, color, and antioxidant activity as well as textural attributes of berries. Moreover, it is a rapid and low-temperature drying process [20, 21]. Meanwhile, freeze-dried berries exhibit the greatest antioxidant effect [20]. Although freeze drying and microwave drying are highly expensive drying methods, they protect the shape of the product with minimal reduction of volume [20, 22].

Drying is completed by controlling the amount of moisture in the final product. Final moisture content must be at least 17% (w/w) for safe storage [9]. Moisture content, texture, uniform size and color, presence of mold, yeast, foreign materials, decay, and insect affect the quality of the final product [9].

7.3 Compositional and nutritional characteristics of mulberries

Literature on proximate composition of fresh mulberries indicates a moisture content of 85–88 g/100 g. After the drying process, the moisture content of dehydrated mulberries is

Table 7.1 Nutrient composition of fresh and freeze-dried mulberries (values in per 100 grams edible portion)

Nutrient	Unit	Fresh mulberries	Freeze-dried mulberries	Reference
Proximate composition				
Water	g	85–88	3.5–17	[24–26]
Energy	kcal	43	na	[24]
Protein	g	0.5–1.4	na	[24, 25]
Lipid	g	0.39–0.5	7.54	[2, 24–26]
Carbohydrate	g	7.8–9.8	na	[2, 24, 25]
Fiber	g	0.9–1.7	24.3	[24–26]
Sugars	g	1.8–16.2	72.7–80.2	[10, 25, 27]
Minerals				
Calcium	mg	39–443	na	[2, 25, 28, 29]
Copper	mg	0.06–0.5	1.18	[2, 24, 26, 29]
Iron	mg	1.85–190	48.1	[2, 24–26, 29]
Magnesium	mg	17–115	na	[2, 25, 28, 29]
Manganese	mg	3.8–4.2	7.45	[2, 26]
Phosphorus	mg	35–247	na	[2, 25, 28, 29]
Potassium	mg	194–1668	na	[2, 24, 25, 28, 29]
Selenium	µg	0.6	65	[24, 26]
Sodium	mg	10–61	na	[2, 25, 28, 29]
Zinc	mg	0.12–3.20	12.1	[2, 24, 26, 29]
Barium	mg	na	4.2	[26]
Molybdenum	mg	na	4.2	[26]
Chromium	mg	na	8.35	[26]

na, not available.
Values are expressed as (minimum–maximum).

decreased to 3.5–17 g/100 g [9, 23–25]. Therefore, nutrient content becomes concentrated, but essentially it is the same in dried mulberries as compared to the fresh ones. The nutrient composition, including proximate composition and minerals of fresh and freeze-dried mulberries, is given in Table 7.1. However, the composition of fruits depends on the species or varieties and the growing conditions, such as soil and geographical conditions [2].

Total sugar content of fresh and freeze-dried mulberries ranges from 1.8 to 16.2 g/100 g [10, 24, 27] and from 72.7 to 80.2 g/100 g [30], respectively (Table 7.1). Individual sugar content of dried mulberries changes from 34.2 to 38.3 g/100 g, from 36.9 to 40.8 g/100 g, and from 0.68 to 2.39 g/100 g for glucose, fructose and sucrose, respectively [30]. Sucrose is hydrolyzed to fructose and glucose during drying, which can participate in Maillard reaction [30]. Since, sugars are the initial precursors in the biosynthesis of anthocyanins, ripe mulberries containing high amounts of sugar include elevated anthocyanin content [10].

Table 7.1 shows the mineral contents in fresh and freeze-dried mulberries. In the literature, 10 minerals were found for fresh mulberries and 9 for freeze-dried ones [2, 24–26, 28, 29]. Mineral composition variation of fruits depends on the species or varieties and the growing conditions, such as soil and geographical conditions; however, it is possible to say that the drying process caused an increase in the mineral contents of mulberries [2, 24–26, 28, 29].

Table 7.2 Contents of total phenolics, flavonoids, and anthocyanins in fresh and freeze-dried mulberries (mg/g dry weight basis)

Nutrient	Unit	Fresh mulberries	Freeze-dried mulberries	Reference
Total phenolics	mg of GAE	0.96–3.48	23.0	[3, 11, 26, 31, 32]
Total flavonoids	mg of RE	0.18–3.90	3.90	[5, 26]
Total anthocyanins	mg of C3GE	0.01–96.1	0.87	[5, 11, 26, 31]
Resveratrol	mg	na	0.30	[26]
Rutin	mg	na	0.43	[26]
Morin	mg	na	0.15	[26]
Quercetin	mg	na	0.03	[26]
Myricetin	mg	na	0.01	[26]
Vitamin C	mg	11.0–36.4	1.20	[2, 10, 24–26]
Vitamin E	mg of ATE	0.87	0.32	[25, 26]

Note: Some numbers are rounded after decimal point.
GAE, gallic acid equivalents; RE, rutin equivalents; C3GE, cyanidin-3-glucoside equivalents; ATE, alpha-tocopherol equivalents; na; not available.
Values are expressed as (minimum–maximum).

Table 7.2 reports the composition of some vitamins for fresh and freeze-dried mulberries [2, 10, 24–26]. Although fresh mulberries contain the majority of vitamins (such as choline, thiamine, riboflavin, vitamin A, vitamin C, vitamin E, and vitamin K), carotenoids (α-carotene, β-carotene, and lutein + zeaxanthin), some heat-sensitive vitamins such as vitamin C and vitamin E may be lost during the drying process due to high temperature [33]. Significant loss in vitamin C and vitamin E occurs during the drying process (Table 7.2) [2, 10, 24–26]

7.4 Phytochemicals in mulberries and their by-products

Phytochemicals are essential bioactive substances present in plants. However, the human body is not able to produce them [34]. Therefore, plants rich in these bioactive substances should be consumed regularly. According to recent investigations, the fruits and leaves of mulberry plants contain healthful bioactive components such as flavonoids, anthocyanins, polyhydroxylated alkaloids, and hydroxystilbenes [5, 35].

7.4.1 Flavonoids

Mulberries are rich sources of flavonoids. The major flavonoids found in mulberries are quercetin-3-glucoside, rutin, isoquercetin, and morin (Figure 7.1) [36, 37], while the main flavonoids in mulberry leaves are quercetin-3-glucoside, kaempferol-3-glucoside (astragalin), and quercetin-3-(6-malonylglucoside) [1, 38]. Recently, Zhang *et al.* [39] isolated four new flavonoids, namely, 8-hydroxyethyl-7,2′,4′-trihydroxyflavone, (±)-7-methoxy-8-hydroxyethyl-2′,4′dihydroxyflavane, 2R*,4R*-8-hydroxyethyl-7,4′-dihydroxy-4,2′ epoxyflavane, and 2″E-3′(4″-hydroxyisopentenyl)-2,4,2′,4′-tetrahydroxychalcone from the leaves of *Morus mongolica*. Chang *et al.* [40] also identified five phenolics, namely,

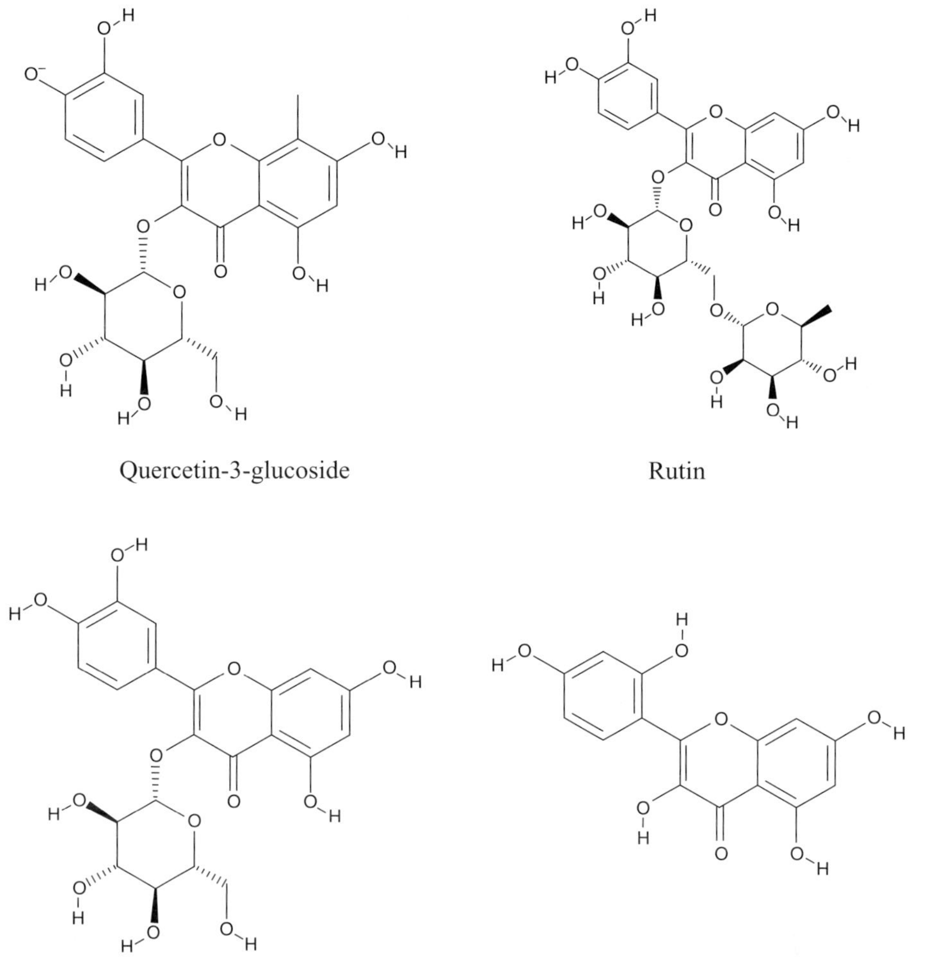

Quercetin-3-glucoside | Rutin

Isoquercetin | Morin

Figure 7.1 Structures of some major flavonoids found in mulberries.

maclurin (49 and 58.4%), rutin (9.3 and 12.5%), isoquercitrin (5.1 and 9.3%), resveratrol (15.5 and not detected), and morin (20.6 and 20.8%) in mulberry twigs and mulberry root bark, respectively. The contents of total phenolics, flavonoids, and anthocyanins in fresh and freeze-dried mulberries are summarized in Table 7.2 [3, 5, 10, 11, 26, 31, 32, 35].

Differences in total phenolic content between *M. nigra* and *M. rubra* are reported [3, 31]. Black mulberry genotypes have a higher phenolic content [1766–3488 μg gallic acid equivalents (GAE)/g] than their purple counterparts (1005–2388 μg GAE/g). However, differences in phenolic composition of the fruits depended on many factors such as growing conditions, genetic differences, and the degree of maturity at harvest [1, 2, 36]. In fact, Zhishen *et al.* [41] reported the total flavonoid content of mulberry leaves varied from 11.7 to 26.6 mg/g in spring and from 9.84 to 23.4 mg/g in autumn for the same mulberry variety;

thus flavonoid content changes depending on the season. The flavonoid content of the spring leaves was higher than that of the autumn leaves [41].

7.4.2 Anthocyanins

Anthocyanin profiles of red and black mulberries have been determined by different researchers. Hassimotto *et al.* [42] reported that 79 and 19% of anthocyanins in wild mulberries were cyanidin-3-glucoside and cyanidin-3-*O*-rutinoside, respectively. In contrast, Qin *et al.* [43] showed that most abundant anthocyanins in mulberries were cyanidin-3-*O*-rutinoside (60%) and cyanidin-3-*O*-glucoside (38%). The minor anthocyanins were pelargonidin 3-*O*-glucoside and pelargonidin 3-*O*-rutinoside (2%). These results are in agreement with those reported by others [1, 10, 37, 44]. Du *et al.* [45] also reported the anthocyanins of mulberries as cyanidin-3-*O*-(6″-*O*-α-rhamnopyranosyl-β-D-glucopyranoside), cyanidin-3-*O*-(6″-*O*-α-rhamnopyranosyl-β-D-galactopyranoside), cyanidin-3-*O*-β-D-galactopyranoside, and cyanidin-7-*O*-β-D-glucopyranoside. Meanwhile, Wu *et al.* [44] reported delphindin-3-(6″malonyl)-glucoside and peonodin-3-xylosylrhamnoside in mulberries. The total anthocyanin content of fresh and freeze-dried mulberries is summarized in Table 7.2 and the chemical structures of major anthocyanins found in mulberries are given in Figure 7.2.

Total anthocyanin contents of black and purple mulberry genotypes ranged from 693 to 787 μg and from 81 to 132 μg of cyanidin-3-glucoside equivalents (C3GE)/g, respectively [31]. In addition, Ozgen *et al.* [3] found that the anthocyanin contents in black and red mulberries ranged from 253 to 830 and from 3 to 200 μg C3GE/g, respectively. However, drying significantly destroyed the bioactive compounds such as anthocyanins, flavanols, and ascorbic acid [20]. Similarly, Lohachoompol *et al.* [46] found that total anthocyanins in blueberries were lost by 41–49% depending on different treatments.

7.4.3 Polyhydroxylated alkaloids

Polyhydroxylated alkaloids are widespread and isolated from plants, microbial filtrates, and insects. 1-Deoxynojirimycin was synthesized by removing the anomeric hydroxyl group of nojirimycin, but it was later isolated from bacterial cultures and plant sources [5, 47–49]. In fact, 1-deoxynojirimycin is more stable than nojirimycin; therefore, it has been used as a model glucosidase-inhibiting alkaloid [48]. 1-Deoxynojirimycin was isolated from root bark of the mulberry tree [50]. Nowadays, it can also be isolated from mulberry leaves [5, 48, 49]. The chemical structure of 1-deoxynojirimycin is given in Figure 7.3. Song *et al.* [5] found that the content of 1-deoxynojirimycin varied between 1.389 and 3.483 mg/g in leaves for 33 cultivars. The concentration of 1-deoxynojirimycin in three cultivars of mulberry leaves and their related products (tea leaf, tablet, and powder) was 100–480 mg/100 g in freeze-dried samples [51].

Asano *et al.* [49] described new polyhydroxylated alkaloids such as (2*R*, 3*R*, 4*R*)-2-hydroxymethyl-3,4-dihydroxypyrrolidine-*N*-propionamide isolated from the root bark of *M. alba* L., and 4-*O*-α-D-galactopyranosyl-calystegine B_2 and 3β,6β-dihydroxynortropane isolated from the fruits.

7.4.4 Hydroxystilbenes

Resveratrol (*trans*-3,40,5-trihydroxystilbene) and oxyresveratrol (*trans*-2,30,4,50-tetrahydroxystilbene) are hydroxystilbenes found in many plant species including grapes, peanuts,

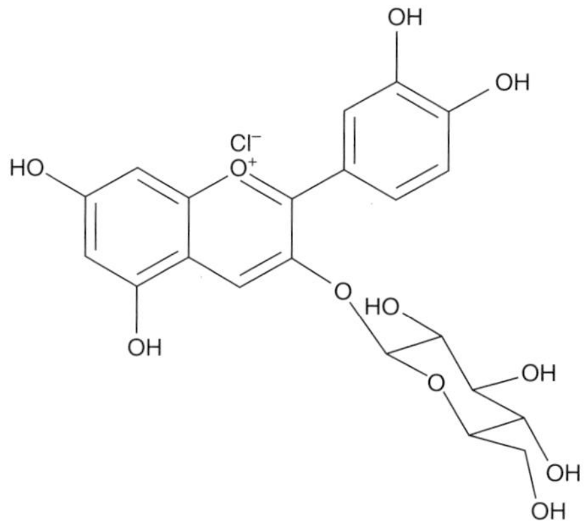

Cyanidin-3-glucoside

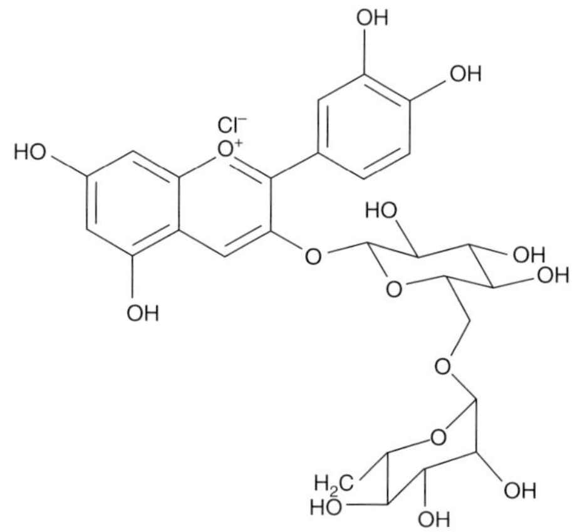

Cyanidin-3-rutinoside

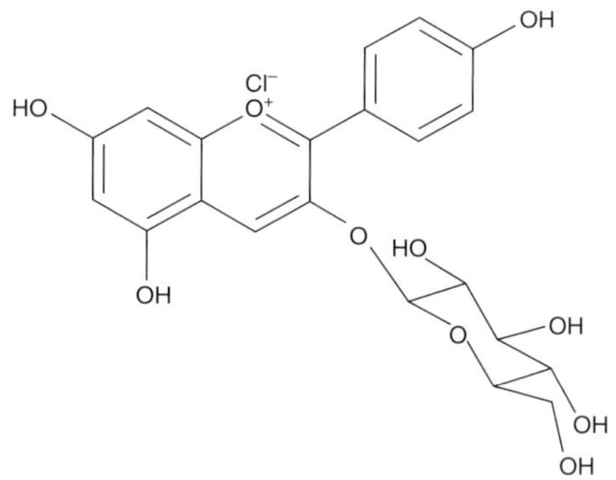

Pelargonidin-3-glucoside

Figure 7.2 Structures of major anthocyanins found in mulberries.

and mulberries (Figure 7.3). Song *et al.* [5] found oxyresveratrol and resveratrol in 38 cultivars of fresh mulberries, ranging from 2.4 to 29.5 μg/g and from 2.1 to 5.3 μg/g, respectively. In addition, oxyresveratrol varied between 5.3 and 179.9 μg/g in leaves of 33 cultivars [5].

7.5 Natural antioxidants in mulberries

Instead of synthetic antioxidants such as butylated hydroxytoluene (BHT) and butylated hydroxyanisole (BHA), naturally occurring antioxidants have gained importance in the food industry in recent years since they are considered safe [52, 53]. These natural antioxidants can be phenolic compounds, such as flavonoids and phenolic acids, as well as vitamins A, C, and E. Dark-colored mulberries exhibit strong antioxidant activity due primarily to

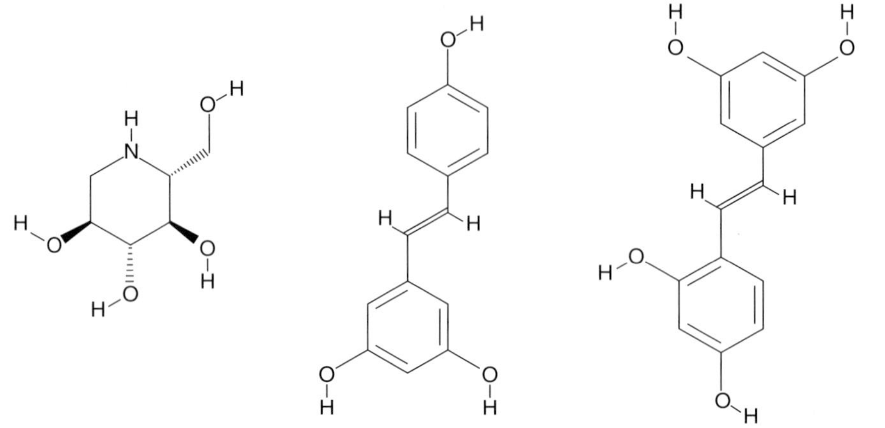

1-Deoxynojirimycin Resveratrol Oxyresveratrol

Figure 7.3 Structures of polyhydroxylated alkaloids found in mulberries.

flavonoids in particular anthocyanin pigments [32]. They contain high amounts of water-soluble anthocyanins, which are the most important natural antioxidative compounds in the plant kingdom [54]. Anthocyanins are excellent antioxidant agents due to their ability to inhibit lipid peroxidation via radical scavenging and metal chelating activities [42, 45]. In fact, it is known that *in vitro* antioxidant activities of anthocyanins are superior to vitamin E [32].

Various researchers have studied the antioxidant activities of mulberries. Du *et al.* [45] determined the antioxidant activities of five anthocyanin monomers [cyanidin 3-*O*-(600-*O*-α-rhamnopyranosyl-β-D-glucopyranoside), cyanidin-3-*O*-(600-*O*-arhamnopyranosyl-β-D-galactopyranoside), cyanidin-3-*O*-β-D-glucopyranoside, cyanidin-3-*O*-β-D-galactopyranoside, and cyanidin-7-*O*-β-D-glucopyranoside] present in *M. alba* and crude mulberry extracts by 1,1-diphenyl-2-picrylhydrazyl (DPPH) radical scavenging assay. Compared to vitamin C, crude mulberry extract (95.3%) exhibited the same DPPH radical scavenging activity as that of vitamin C (95.5%) or five anthocyanin monomers (ranging from 95.1 to 96.2%) at a concentration of 0.40 mg/mL, which indicates that mulberry extract is an excellent antioxidant.

Ercisli *et al.* [31] compared the antioxidant activities of black and purple mulberry species grown in Turkey. On the basis of ferric reducing antioxidant power (FRAP) assay, *M. nigra* L. genotypes (black color) had antioxidant activity ranging from 12.26 to 14.11 and *M. rubra* L. genotypes (purple color) ranging from 4.93 to 8.12 μmol of trolox equivalent (TE)/g. Similarly, Ozgen *et al.* [3] studied the antioxidant properties of *M. nigra* L. and *M. rubra* L. fruits harvested from Turkey. *M. nigra* L. fruits had the highest total antioxidant activity, within the range of 7.3–16.9 μmol TE/g, while that for *M. rubra* L. was 3.7–7.7 μmol TE/g by the FRAP method. These results indicate that black mulberries exhibited higher antioxidant activity than purple mulberries. Additionally, the antioxidant activity of fresh mulberries varies from 2.50 to 21.17 μmol TE/g on the basis of DPPH assay [31, 55], and from 0.30 and 1.73 μmol TE/g on the basis of oxygen radical absorbance capacity (ORAC) assay [35]. No data exist in the literature about antioxidant activity of dried mulberries.

Several studies have also reported the antioxidant activity of mulberry leaves. Antioxidant activity relies on six flavonol glycosides, namely, rutin, isoquercetin, quercetin-3-(6-acetylglucoside), astragalin, kaempferol-3-(6-acetylglucoside), quercetin-3-(6-malonylglucoside), and chlorogenic acid in mulberry leaves [56, 57]. Katsube *et al.* [56] determined the DPPH radical scavenging activity of each polyphenolic compound identified in mulberry leaves and their contributions to this scavenging activity in the extract of freeze-dried mulberry leaves. Chlorogenic acid exhibited 36.2% of the total DPPH radical scavenging activity which makes the largest contribution. Among quercetin glycosides, quercetin-3-(6-malonylglucoside) showed the highest contribution (21.4%), whereas kaempferol glycosides made a slight contribution only (0.2%). Quercetin, as one of the most abundant flavonoids in human diets, has been recovered in rat plasma as sulfate, glucuronide, and sulfoglucuronide conjugates after intragastric administration of quercetin aglycone and these quercetin conjugates protect low-density lipoprotein (LDL) from oxidation induced by copper ion [57]. However, phenolic content and antioxidant activity was not decreased significantly during air drying at 60°C and freeze drying, while both values were decreased during air drying at 70°C. Wanyo *et al.* [58] studied the antioxidant activity of fresh, combined far-infrared radiation with hot-air convection (FIR-HA) dried mulberry leaves and commercial mulberry tea dried by hot-air drying. The fresh leaves exhibited 71% of DPPH radical scavenging. Meanwhile, combined FIR-HA dried leaves provided higher DPPH radical scavenging (76%) than those of commercial tea products (42%). Based on these results, combined FIR-HA technique can be used instead of hot-air drying during commercial tea production.

Arabshahi-Delouee and Urooj [53] investigated the antioxidant properties of mulberry (*Morus indica* L.) leaves. EC_{50} value, 50% radical scavenging activity concentration of antioxidants, of methanolic extract of leaves was 79.53 μg/mL compared to values of 61.67 μg/mL for ascorbic acid and 41.07 μg/mL for BHT. Yen *et al.* [59] showed that *M. alba* leaf extracts exhibited strong antioxidant activity (78.2% inhibition on peroxidation of linoleic acid). The antioxidant activity of the extract was slightly less than that of BHA, but was stronger than that of α-tocopherol (72.1% inhibition on peroxidation of linoleic acid). Andallu *et al.* [60] also studied the antioxidant potential of *M. indica* L. leaves. The results indicated that mulberry leaf extract effectively scavenged nitric oxide, superoxide, and DPPH radicals and exhibited tremendous reducing power. Besides, mulberry leaf extract suppressed $FeSO_4$-induced lipid peroxidation and conjugated dienes/hydroperoxides in erythrocyte membrane very efficiently.

7.6 Health effects of mulberries

Mulberries are used effectively in folk medicines to treat fever, protect liver and kidney from damage, strengthen the joints, facilitate discharge of urine, improve eyesight, treat mouth lesions, lower blood pressure, treat sore throat, hypertension, and anemia, and prevent cardiovascular disease (CVD) [3, 26, 31, 32, 41, 53, 58]. Additionally, mulberry fruits can be used as a warming agent, as a remedy for dysentery, and as a tonic, sedative, laxative, odontalgic, anthelmintic, expectorant, and emetic, among others [1, 61]. Since mulberries are rich in naturally occurring flavonoids and anthocyanins, they exhibit anti-inflammatory, vasoprotective, antioxidantive, antiallergic, hepatoprotective, antithrombotic, antiviral, chemoprotective, and anticarcinogenic activities [5, 8, 32, 54, 62]. Therefore, besides anthocyanins are used as natural sources of food colorants, many consumers and food manufacturers have become interested in phenolic phytochemicals for their medicinal properties, especially their potential role in the prevention of cancers and CVD [32, 43].

In addition to fruits, almost all parts of the mulberry tree (leaves, bark of stems, and roots) are used for pharmacological purposes around the world, especially in Chinese folk medicine [63]. The leaves have been shown to possess diuretic, hypoglycemic, and hypotensive activities, and blood sugar- and blood pressure-reducing effects, whereas the root barks of mulberry trees have long been used for anti-inflammatory, antitussive, and antipyretic purposes [1, 52]. Meanwhile, mulberry leaf extracts were found to prevent amyloid fibril formation and showed neuroprotective effects [64].

7.6.1 Anticancer activities

Anthocyanins show greater inhibitory effect on suppressing the growth of tumor cells than other flavonoids [46]. However, biological effects of anthocyanins as well as their appearance depend on pH [65].

Major anthocyanins of colored mulberries, such as cyanidin-3-glucoside and cyanidin-3-rutinoside, are reported to have an inhibitory effect on migration and invasion of lung cancer cells [62]. These compounds help to suppress proliferation and angiogenesis, as well as the induction of cancer cell apoptosis [5, 26, 62]. In fact, effects of anthocyanins on antitumorigenesis in different mechanisms were reported by different researchers. Recently, Huang *et al.* [66] showed that mulberry extracts inhibited the growth of human gastric carcinoma cells. Moreover, they claimed that 0.2% mulberry extracts by both injection and oral gavages exhibited tumor inhibition effect. Therefore, it is suggested that mulberry anthocyanins can be used as potential therapeutic agents in preventing gastric carcinoma formation [66].

Huang *et al.* [8] also reported that mulberry anthocyanins had therapeutic potential on melanoma, a skin cancer type that originates in melanocytes, specialized pigment-producing cells found in both the basal layer of the epidermis and the eye. Mulberry anthocyanins control tumor metastasis based on the observation of its inhibitory effect on the motility of melanoma cancer cell line B16-F1. The researchers suggested that mulberry anthocyanins could suppress melanoma metastasis and may be used as potential candidates for cancer chemoprevention [8].

Resveratrol found in mulberries also exhibits anticancer activities because it acts as an antioxidant, inhibits lipid peroxidation, cyclooxygenases, and protein kinase C and prevents nuclear factor-kappa B (NF-κB) and activator protein-1 (AP-1) transcription factor induction and cytokine release [67].

7.6.2 Cardioprotection

Mulberry anthocyanins participate in CVD protection, which is strongly linked to reduction in oxidative stress. Anthocyanins protect from heart attacks since they have the ability of reducing inflammation, enhancing capillary strength and permeability, inhibiting platelet formation, and enhancing nitric oxide release [68]. Both mulberry anthocyanins and leaf extracts can scavenge reactive oxygen species (ROS), LDL cholesterol oxidation, prevent platelet aggregation, and decrease serum lipids, and therefore exert beneficial effects on blood lipid and atherosclerosis [5, 26, 58, 61, 68]. In particular, quercetin as a major representative of the flavonol group exhibits strong inhibitory effects on oxidative modification of human LDL *in vitro* [38, 56]. Resveratrol component of mulberries also helps cardioprotection [67].

7.6.3 Diabetes

The alkaloid component of mulberry leaves, 1-deoxynojirimycin, is known as one of the most potent α-glycosidase inhibitors that decreases blood sugar levels by establishing glycemic control in type 2 diabetes in humans [5, 49]. Tender mulberry leaves from the top of branches contain the highest amount of 1-deoxynojirimycin in the summer [4]. Hansawasdi and Kawabata [69] reported that hot-water extract of mulberry leaves exhibited inhibitory effect against α-glucosidases; therefore, mulberry tea can be consumed as an antidiabetic herbal tea. They concluded that infusion of 1 gram mulberry tea in 100 mL hot water (98°C) for 3–5 min exhibits an effective inhibitory activity against α-glucosidases, sucrase, and maltase enzymes. As an alternative, 1-deoxynojirimycin-enriched dietary supplement can be used for preventing diabetes mellitus [4].

7.6.4 Alzheimer's disease

Mulberry extracts help with the treatment of Alzheimer's disease by inhibiting amyloid β-peptide (1–42) fibril formation and showing neuroprotective effects [4, 61, 64], promoting age-dependent antioxidant protection, and reducing oxidative stress-induced damage [4, 64].

7.6.5 Hyperpigmentation disorders

The hydroxystilbene component of both mulberry fruits and leaves—oxyresveratrol—is an active compound in dermatology that has an inhibitory effect on tyrosinase to limit melanin biosynthesis [5, 40, 70]. Tyrosinase catalyzes both the hydroxylation reaction that converts tyrosine to 3-(3, 4-dihydroxyphenyl)-L-alanine (DOPA) and the oxidation reaction that converts DOPA into dopaquinone, which causes polymerizing of brown pigments [10]. Therefore, oxyresveratrol is used for whitening skin in cosmetics and treating hyperpigmentation disorders in medicine [5].

7.7 Food application of mulberries and their by-products

Food applications of mulberries include not only mulberry fruits themselves, but also their leaves and products.

7.7.1 Mulberry fruits and their products

The white, purple, and black mulberry fruits, which have a very pleasant taste, are consumed as fresh fruits. Additionally, black and purple mulberry fruits are consumed in the form of jam, marmalade, frozen desserts, pulp, juice, paste, ice cream, liquor, wine, and canned food, among others, since there is an increasing interest for mulberries in the food industry due to the presence of anthocyanins in purple and black cultivars [1, 3, 29, 38, 61, 62]. Moreover, mulberry fruits have been used for the production of some traditional Turkish confectionary products such as mulberry pekmez (a concentrated mulberry juice by boiling) and mulberry pestil [71]. Mulberry pekmez is generally consumed with breakfast [72]. It is produced from mulberry fruits by concentration of juices containing up to 70–80% soluble dry matter content [72, 73]. Due to the presence of high amounts of sugar, minerals, and organic acids, pekmez is preferred for babies, children, and sportspersons for supplying rapid energy [72]. Mulberry pestil is a dried fruit snack for consumption in the winter [73]. It is prepared by boiling mulberry juice, starch, and sugar before addition of pistachios and walnuts until a mild, crunchy, tasty, light, and chewable leathery product is obtained [74, 75]. Similar to pekmez, pestil is also a good source of energy containing high amounts of carbohydrates and minerals. Finally, colored mulberries are also used as natural dyes in the cosmetic industry [31, 76].

7.7.2 Mulberry leaves

Mulberry leaves are nutritious, palatable, and nontoxic [53]. The leaves have historically been used for feeding silkworms (*Bombyx mori* L.) in China and India [1, 5], the cocoon of which is used to make silk, and has big ecological and economical importance [5]. Fibroin protein solution, gel, and powder obtained from these cocoons have been processed into cosmetics, skin cream, face washing milk, shampoo, and bath agents in Japan [49]. Mulberry leaves are consumed in Thailand in different ways. Mulberry leaf tea has been very popular in Thailand for the last decade [57, 58]. In addition, the leaves are added in spicy soup preparation and the tips are used as a vegetable in light curry [58]. Furthermore, mulberry leaf carbonated beverage and mulberry leaf chrysanthemum beverage are known as functional beverages that are widely consumed in China [77]. The leaves are also used to feed dairy animals to improve milk yield [48]. Liu *et al.* [78] showed that it was possible to use mulberry branch barks to obtain pectin instead of discarding them as agricultural waste. Pectin is used as a gelling agent and stabilizer in the food industry in jams, jellies, and dairy products [78].

7.8 Conclusions

Mulberries are valuable horticultural products grown worldwide. They are rich in phytochemicals, in particular anthocyanins, which possess therapeutic and pharmacological benefits such as treating fever, protecting liver and kidney from damage, strengthening the joints and facilitating discharge of urine, among others. Besides fruits, mulberry leaves are used to feed silkworms and are edible.

Although growing conditions and cultural management techniques affect the nutritional value of the *Morus* species, more research is needed in the areas related to dried mulberry phytochemicals as well as their extraction, analyses, and bioavailability. Optimization of various drying techniques is needed to further study to minimize nutritional loss, undesirable texture, flavor, and color changes.

References

1. Pawlowska, A.M., Oleszek, W. & Braca, A. (2008) Quali-quantative analyses of flavonoids of *Morus nigra* L. and *Morus alba* L. (Moraceae) fruits. *Journal of Agricultural Food Chemistry*, **56**, 3377–3380.
2. Ercisli, S. & Orhan, E. (2007) Chemical composition of white (*Morus alba*), red (*Morus rubra*) and black (*Morus nigra*) mulberry fruits. *Food Chemistry*, **103**, 1380–1384.
3. Ozgen, M., Serce, S. & Kaya, C. (2009) Phytochemical and antioxidant properties of anthocyanin-rich *Morus nigra* and *Morus rubra* fruits. *Scientia Horticulturae*, **119**, 275–279.
4. Venkatesh Kumar, R. & Chauhan, S. (2008) Mulberry: life enhancer. *Journal of Medicinal Plants Research*, **2**, 271–278.
5. Song, W., Wang, H., Bucheli, P., Zhang, P., Wei, D. & Lu, Y. (2009) Phytochemical profiles of different mulberry (*Morus* sp.) species from China. *Journal of Agricultural Food Chemistry*, **57**, 9133–9140.
6. Sanchez, M.D. (2002) *World Distribution and Utilization of Mulberry and Its Potential for Animal Feeding*. FAO Animal Production and Health Paper, Rome, Italy.
7. Chen, Z., Zhu, C. & Han, Z. (2011) Effects of aqueous chlorine dioxide treatment on nutritional components and shelf-life of mulberry fruit (*Morus alba* L.). *Journal of Bioscience and Bioengineering*, **111**, 675–681.
8. Huang, H., Shih, Y., Chang, Y., Hung, C. & Wang, C. (2008) Chemoinhibitory effect of mulberry anthocyanins on melanoma metastasis involved in the Ras/PI3K pathway. *Journal of Agricultural Food Chemistry*, **56**, 9286–9293.
9. Doymaz, I. (2004) Drying kinetics of white mulberry. *Journal of Food Engineering*, **61**, 341–346.
10. Aramwit, P., Bang, N. & Srichana, T. (2010) The properties and stability of anthocyanins in mulberry fruits. *Food Research International*, **43**, 1093–1097.
11. Akbulut, M., Batu, A. & Coklar, H. (2007) Some physicochemical properties and production techniques of mulberry pekmez. *Teknolojik Araştırmalar*, **2**, 25–31 (in Turkish).
12. Mujumdar, A.S. & Law, C.L. (2010) Drying technology: trends and applications in postharvest processing. *Food Bioprocess Technology*, **3**, 843–852.
13. Doymaz, I. (2004) Pretreatment effect on sun drying of mulberry fruits (*Morus alba* L.). *Journal of Food Engineering*, **65**, 205–209.
14. Taser, O.F., Tarhan, S. & Ergunes, G. (2007) Effects of chemical pre-treatments on air-drying processing of black mulberry (*Morus nigra* L.). *Journal of Scientific & Industrial Research*, **66**, 472–487.
15. Ergunes, G., Gunes, M. & Cekic, C. (2003) The effect of various drying techniques on the quality of dried white mulberry. In: *Proceedings of National Kiwi and Grapes Symposium* (ed. T. Karadeniz). Ordu University, Ordu, Turkey, pp. 442–447.
16. Pangavhane, D.R. (2002) Comparative drying performance study of natural convection solar dryer with traditional grape drying methods. *International Energy Journal*, **1**, 13–23.
17. Akbulut, A. & Durmus, A. (2010) Energy and energy analyses of thin layer drying of mulberry in a forced solar dryer. *Energy*, **35**, 1754–1763.
18. Pangavhane, D.R. & Shawhney, R.L. (2002) Review of research and development work on solar dryers for grape drying. *Energy Conversion and Management*, **43**, 45–61.
19. Karathanos, V.T. & Belessiotis, V.G. (1997) Sun and artificial air drying kinetics of some agricultural products. *Journal of Food Engineering*, **31**, 35–46.
20. Wojdyło, A., Figiel, A. & Oszmianski, J. (2009) Effect of drying methods with the application of vacuum microwaves on the bioactive compounds, color, and antioxidant activity of strawberry fruits. *Journal of Agricultural and Food Chemistry*, **57**, 1337–1343.
21. Fazaeli, M., Yousefi, S. & Emam-Djomeh, Z. (2011) Investigation on the effects of microwave and conventional heating methods on the phytochemicals of pomegranate (*Punica granatum* L.) and black mulberry juices. *Food Research International*, doi:10.1016/j.foodres.2011.03.043.
22. Ratti, C. (2001) Hot air and freeze-drying of high value foods: a review. *Journal of Food Engineering*, **49**, 311–319.
23. TS 3570 ISO 7910 (2002) *Turkish Standard, Dried Mulberries—Specification*. Turkish Standard, Ankara, Turkey.
24. USDA. (2011) *National Nutrient Database for Standard Reference, Release 24*. Published online at: http://www.nal.usda.gov/fnic/foodcomp/search/, last accessed February 27, 2012.

25. Singhal, B.K., Khan, M.A., Dhar, A., Baqual, F.M. & Bindroo, B.B. (2010) Approaches to industrial exploitation of mulberry (*Mulberry* sp.) fruits. *Journal of Fruit and Ornamental Plant Research*, **1**, 83–99.
26. Yang, X., Yang, L. & Zheng, H. (2010) Hypolipidemic and antioxidant effects of mulberry (*Morus alba* L.) fruit in hyperlipidaemia rats. *Food and Chemical Toxicology*, **48**, 2374–2379.
27. Elmaci, Y. & Altug, T. (2002) Flavour evaluation of three black mulberry (*Morus nigra*) cultivars using GC/MS, chemical and sensory data. *Journal of the Science of Food and Agriculture*, **82**, 632–635.
28. Akbulut, M. & Ozcan, M.M. (2009) Comparison of mineral contents of mulberry (*Morus* spp.) fruits and their pekmez (boiled mulberry juice) samples. *International Journal of Food Sciences and Nutrition*, **3**, 231–239.
29. Eveleen Said, A. (2006) The biological benefits of black mulberry (*Morus nigra*) intake on diabetic and nondiabetic subjects. *Research Journal of Agriculture and Biological Sciences*, **6**, 349–357.
30. Bakkalbasi, E., Yemis, O. & Artik, N. (2004) Determination of physical and chemical properties with extraction conditions of dried mulberry. *Gıda*, **29**, 203–209 (in Turkish).
31. Ercisli, S., Tosun, M., Duralija, B., Voca, S., Sengul, M. & Turan, M. (2010) Phytochemical content of some black (*Morus nigra* L.) and purple mulberry genotypes. *Food Technology and Biotechnology*, **1**, 102–106.
32. Bae, S. & Suh, H. (2007) Antioxidant activities of five different mulberry cultivars in Korea. *LWT—Food Science and Technology*, **40**, 955–962.
33. Nindo, C.I., Sun, T., Wang, S.W., Tang, J. & Powers, J.R. (2003) Evaluation of drying technologies for retention of physical quality and antioxidants in asparagus (*Asparagus officinalis*, L.). *Lebensmittel-Wissenschaft und-Technologie*, **36**, 507–516.
34. Yahia, E.M. (2010) The contribution of fruit and vegetable consumption to human health. In: *Fruit and Vegetable Phytochemicals. Chemistry, Nutritional Value and Stability* (eds L. De la Rosa E. Alvarez-Parrilla & G.A. Gonzalez-Aguilar). Wiley-Blackwell, Oxford, UK, pp. 3–53.
35. Isabella, M., Lee, B.L., Ong, C.N., Liu, X. & Huang, D. (2008) Peroxyl radical scavenging capacity, polyphenolics, and lipophilic antioxidant profiles of mulberry fruits cultivated in southern China. *Journal of Agricultural and Food Chemistry*, **56**, 9410–9416.
36. Pérez-Gregorio, M.R., Regueiro, J., Alonso-González, E., Pastrana-Castro, L.M. & Simal-Gándara, J. (2011) Influence of alcoholic fermentation process on antioxidant activity and phenolic levels from mulberries (*Morus nigra* L.). *LWT—Food Science and Technology*, **44**, 1793–1801.
37. Liu, X., Xiao, G., Chen, W., Xu, Y. & Wu, J. (2004) Quantification and purification of mulberry anthocyanins with macroporous resins. *Journal of Biomedicine and Biotechnology*, **5**, 326–331.
38. Enkhmaa, B., Shiwaku, K., Katsube, T., Kitajima, K., Anuurad, E., Yamasaki, M. & Yamane, Y. (2005) Mulberry (*Morus alba* L.) leaves and their major flavonol quercetin 3-(6-Malonylglucoside) attenuate atherosclerotic lesion development in LDL receptor-deficient mice. *Journal of Nutrition*, **4**, 729–734.
39. Zhang, X., Jing, Y., Wanga, G., Wang, Y., Zhao, H. & Ye, W. (2010) Four new flavonoids from the leaves of *Morus mongolica*. *Fitoterapia*, **81**, 813–815.
40. Chang, L.-W., Juang, L.-J., Wanga, B., Wang, M.-Y., Tai, H.-M., Hung, W.-J., Chen, Y.-J. & Huang, M.-H. (2011) Antioxidant and antityrosinase activity of mulberry (*Morus alba* L.) twigs and root bark. *Food and Chemical Toxicology*, **49**, 785–790.
41. Zhishen, J., Mengcheng, T. & Jianming, W. (1999) The determination of flavonoid contents in mulberry and their scavenging effects on superoxide radicals. *Food Chemistry*, **64**, 555–559.
42. Hassimotto, N.M.A., Genovese, M.I. & Lajolo, F.M. (2008) Absorption and metabolism of cyanidin-3-glucoside and cyanidin-3-rutinoside extracted from wild mulberry (*Morus nigra* L.) in rats. *Nutrition Research*, **28**, 198–207.
43. Qin, C., Li, Y., Niu, W., Ding, Y., Zhang, R. & Shang, X. (2010) Analysis and characterisation of anthocyanins in mulberry fruit. *Czech Journal of Food Sciences*, **2**, 117–126.
44. Wu, X., Liang, L., Zou, Y., Zhao, T., Zhao, J., Li, F. & Yang, L. (2011) Aqueous two-phase extraction, identification and antioxidant activity of anthocyanins from mulberry (*Morus atropurpurea* Roxb.). *Food Chemistry*, **129**, 443–453.
45. Du, Q., Zheng, J. & Xu, Y. (2008) Composition of anthocyanins in mulberry and their antioxidant activity. *Journal of Food Composition and Analysis*, **21**, 390–395.
46. Lohachoompol, V., Srzednicki, G. & Craske, J. (2004) The change of total anthocyanins in blueberries and their antioxidant effect after drying and freezing. *Journal of Biomedicine and Biotechnology*, **5**, 248–252.

47. Watson, A.A., Fleet, W.J.G., Asano, N., Molyneux, R.J. & Nash, R.J. (2001) Polyhydroxylated alkaloids—natural occurrence and therapeutic applications. *Phytochemistry*, **56**, 265–295.
48. Nash, R.J., Watson, A.A. & Asano, N. (1996) Polyhydroxylated alkaloids that inhibit glycosidases. In: *Alkaloids: Chemical & Biological Perspectives*, Vol. 11 (ed. S.W. Pelletier). Elsevier, Oxford, UK, pp. 346–371.
49. Asano, N., Yamashita, T., Yasuda, K., Ikeda, K., Kizu, H., Kameda, Y., Kato, A., Nash, R.J., Lee, H.S. & Ryu, K.S. (2001) Polyhydroxylated alkaloids isolated from mulberry trees (*Morus alba* L.) and silkworms. *Journal of Agricultural and Food Chemistry*, **49**, 4208–4213.
50. Yagi, M., Kouno, T., Aoyagi, Y. & Murai, H. (1976) The structure of moraoline, a piperidine alkaloid from Morus species. *Nippon Nougei Kagaku Kaishi*, **50**, 571–572 (in Japanese).
51. Kimura, T., Nakagawa, K., Saito, Y., Yamagishi, K., Suzuki, M., Yamaki, K., Yamaki, K., Shinmoto, H. & Miyazawa, T. (2004) Determination of 1-deoxynojirimycin in mulberry leaves using hydrophilic interaction chromatography with evaporative light scattering detection. *Journal of Agricultural Food Chemistry*, **52**, 1415–1418.
52. Venkatesh Kumar, R., Chauhan, S. & Kumar, D. (2010) Nutritional composition in leaves of some mulberry varieties: a comparative study. In: *The 2010 IEEE International Conference on Bioinformatics and Biomedical Technology* (ed. V. Mahadevan). Chengdu, China, pp. 438–442.
53. Arabshahi-Delouee, S. & Urooj, A. (2007) Antioxidant properties of various solvent extracts of mulberry (*Morus indica* L.) leaves. *Food Chemistry*, **102**, 1233–1240.
54. Liu, L., Chou, F., Chen, Y., Chyau, C., Ho, H. & Wang, C. (2009) Effects of mulberry (*Morus alba* L.) extracts on lipid homeostasis *in vitro* and *in vivo*. *Journal of Agricultural Food Chemistry*, **57**, 7605–7611.
55. Oki, T., Kobayashi, M., Nakamura, T., Okuyama, A., Masuda, M., Shiratsuchi, H. & Suda, I. (2006) Changes in radical-scavenging activity and components of mulberry fruit during maturation. *Journal of Food Science*, **71**, C18–C22.
56. Katsube, T., Tsurunaga, Y., Sugiyama, M., Furuno, T. & Yamasaki, Y. (2009) Effect of air-drying temperature on antioxidant capacity and stability of polyphenolic compounds in mulberry (*Morus alba* L.) leaves. *Food Chemistry*, **113**, 964–969.
57. Katsube, T., Imawaka, N., Kawano, Y., Yamazaki, Y., Shiwaku, K. & Yamane, Y. (2006) Antioxidant flavonol glycosides in mulberry (*Morus alba* L.) leaves isolated based on LDL antioxidant activity. *Food Chemistry*, **97**, 25–31.
58. Wanyo, P., Siriamornpuna, S. & Meesob, N. (2011) Improvement of quality and antioxidant properties of dried mulberry leaves with combined far-infrared radiation and air convection in Thai tea process. *Food and Bioproducts Processing*, **89**, 22–30.
59. Yen, G., Wu, S. & Duh, P. (1996) Extraction and identification of antioxidant components from the leaves of mulberry (*Morus alba* L.). *Journal of Agricultural and Food Chemistry*, **44**, 1687–1690.
60. Andallu, B., Mahalakshmi, S. & Vinay Kumar, A.V. (2008) *In vitro* antioxidant activity of mulberry (*Morus indica* L.) leaf extract. *Journal of Pharmacy and Chemistry*, **2**, 89–93.
61. Kang, T.H., Hura, J.Y., Kim, H.B., Ryu, J.H. & Kim, S.Y. (2006) Neuroprotective effects of the cyanidin-3-O-β-D-glucopyranoside isolated from mulberry fruit against cerebral ischemia. *Neuroscience Letters*, **391**, 168–172.
62. Chen, P., Chu, S., Chiou, H., Kuo, W., Chiang, C. & Hsieh, Y. (2006) Mulberry anthocyanins, cyanidin-3-rutinoside and cyanidin-3-glucoside, exhibited an inhibitory effect on the migration and invasion of a human lung cancer cell line. *Cancer Letters*, **235**, 248–259.
63. Koyuncu, F. (2004) Organic acid composition of native black mulberry fruit. *Chemistry of Natural Compounds*, **4**, 367–369.
64. Shih, P., Chan, Y., Liao, J., Wang, M. & Yen, G. (2010) Antioxidant and cognitive promotion effects of anthocyanin-rich mulberry (*Morus atropurpurea* L.) on senescence-accelerated mice and prevention of Alzheimer's disease. *Journal of Nutritional Biochemistry*, **21**, 598–605.
65. Kutlu, T., Durmaz, G., Ates, B., Yılmaz, İ. & Cetin, M.S. (2011) Antioxidant properties of different extracts of black mulberry (*Morus nigra* L.). *Turkish Journal of Biology*, **35**, 103–110.
66. Huang, H., Chang, Y., Wud, C., Hung, C. & Wang, C. (2011) Anthocyanin-rich mulberry extract inhibit the gastric cancer cell growth *in vitro* and xenograft mice by inducing signals of p38/p53 and c-jun. *Food Chemistry*, **129**, 1703–1709.
67. Evers, D.L., Wang, X., Huong, S., Huang, D.Y. & Huang, E. (2004) 3,4′,5-Trihydroxy-*trans*-stilbene (resveratrol) inhibits human cytomegalovirus replication and virus-induced cellular signalling. *Antiviral Research*, **63**, 85–95.

68. Lila, M.A. (2004) Anthocyanins and human health: An *in vitro* investigative approach. *Journal of Biomedicine and Biotechnology*, **5**, 306–313.
69. Hansawasdi, C. & Kawabata, J. (2006) α-Glucosidase inhibitory effect of mulberry (*Morus alba*) leaves on Caco-2. *Fitoterapia*, **77**, 568–573.
70. Lorenz, P., Roychowdhury, S., Engelmann, M., Wolf, G.F.W. & Horn, T.F.W. (2003) Oxyresveratrol and resveratrol are potent antioxidants and free radical scavengers: Effect on nitrosative and oxidative stress derived from microglial cells. *Nitric Oxide*, **9**, 64–76.
71. Erdogan, U. & Pirlak, L. (2005) Production and evaluation of mulberries (*Morus* spp.) in our country. *Alatarim*, **2**, 38–43 (in Turkish).
72. Sengul, M., Ertugay, M.F. & Sengul, M. (2005) Rheological, physical and chemical characteristics of mulberry pekmez. *Food Control*, **16**, 73–76.
73. Yogurtcu, H. & Kamisli, F. (2006) Determination of rheological properties of some pekmez samples in Turkey. *Journal of Food Engineering*, **77**, 1064–1068.
74. Cagindi, O. & Otles, S. (2005) Comparison of some properties on the different types of pestil: A traditional product in Turkey. *International Journal of Food Science and Technology*, **40**, 897–901.
75. Zorba, O., Tas, C., Baytar, B. & Ciftci, G. (2010) The production of Gümüşhane mulberry pestil and köme. In: *Proceedings of the 1st International Symposium on Traditional Foods from Adriatic to Caucasus* (ed. M. Demirci). Tekirdağ, Turkey, pp. 851–852.
76. Gerasopoulos, D. & Stavroulakis, G. (1997) Quality characteristics of four mulberry (*Morus* sp.) cultivars in the area of Chania, Greece. *Journal of the Science of Food and Agriculture*, **73**, 261–264.
77. Harauma, A., Murayama, T., Ikeyama, K., Sano, H., Arai, H., Takano, R., Kita, T., Hara, S., Kamei, K. & Masayuki Yokode, M. (2007) Mulberry leaf powder prevents atherosclerosis in apolipoprotein E-deficient mice. *Biochemical and Biophysical Research Communications*, **358**, 751–756.
78. Liu, L., Cao, J., Huang, J., Cai, Y. & Yao, J. (2010) Extraction of pectins with different degrees of esterification from mulberry branch bark. *Bioresource Technology*, **101**, 3268–3273.

8 Dried raspberries: phytochemicals and health effects

Esteban I. Mejia-Meza, Jaime A. Yáñez, Neal M. Davies, and Carter D. Clary

8.1 Introduction

In recent years, consumer demand and awareness for health-promoting food has increased considerably in the United States and around the world. The interest has focused attention on a new group of nutritionally active compounds, principally antioxidants, present in fruits, and being promoted as important components in foods associated with a healthier lifestyle. Therefore, improvement in dietary habits, reducing the impact of sedentary jobs and long commutes, improving physical activity and exercise, and reducing stress are thought to have potential impacts on decreased risk of various degenerative diseases.

Raspberries are a good example of fruits rich in various nutrients and antioxidants [1], and are produced mainly in the Russian Federation [175,000 metric tonnes (MT)], Serbia and Montenegro (90,000 MT), the United States (62,000 MT), Poland (38,000 MT), and Ukraine (25,000 MT) [2]. This chapter provides information about phenolic compounds in fresh, dried, and processed raspberries and the impact of dehydration on their polyphenol constituents.

8.2 Dehydration of raspberries

Dehydration processes are commonly used to preserve fruits by near-complete removal of water. The displacement of water, mainly by evaporation, results in a reduction in moisture content and water activity, as well as inhibits deleterious biochemical reactions such as enzymatic browning, lipid oxidation, and growth of microorganisms [3]. However, traditional hot-air drying results in loss of phytochemicals due to oxidative reactions. Hot-air and drum drying adversely affect sensory and nutritional value of fruits may compromise their functional properties. Dehydration methods such as vacuum and freeze drying are potentially less damaging; however, these methods involve higher production costs [4].

Freeze and vacuum drying are commonly employed techniques that maintain highly desirable nutritional and functional properties of many foods, but the capital and operating costs are high. Microwave-vacuum drying is another technique that may result in the retention

Dried Fruits: Phytochemicals and Health Effects, First Edition. Edited by Cesarettin Alasalvar and Fereidoon Shahidi.

of flavor and color, less cell collapse and loss of tissue structure, and may potentially result in a reduced loss of bioactive components relative to hot-air drying, but it has received little commercial and academic attention until recently. Microwave-vacuum drying [5] takes significantly less drying time than freeze drying and hot-air drying. Nevertheless, the cost of using both techniques (e.g., a combination of freeze and microwave-vacuum drying) is often too high to be justified unless improvements in product quality warrant their use. Therefore, developing and evaluating dehydration technologies that produce high-quality health-promoting fruits at a reasonable cost is a desirable goal. Nevertheless, a combination of conventional drying techniques, such as hot-air drying with microwave-vacuum and/or freeze drying, may provide an opportunity to produce high-quality nutritive dried fruits with a relatively lower cost compared to freeze drying alone.

The moisture content of fruits is generally between 80 and 90% or higher [6]. Fruits are metabolically active; therefore, the shelf life of fresh fruits is very limited and makes handling and storage difficult. To extend the shelf life, to increase the availability of fruits over a longer period and to distant markets, and to preserve essential nutritive components, preservation techniques are needed. Dehydration is one technique that has been used since antiquity in the form of solar drying and sun drying (such as raisins), and more recently with industrial driers developed to preserve fruits and produce shelf-stable products with specific functional and sensory characteristics. The main purpose of dehydration is to reduce water activity, which inhibits the growth of microorganisms and retards certain deleterious biochemical reactions in fruits. Hot-air, freeze, and microwave-vacuum dryings are some of the dehydration techniques used to preserve fruits.

Hot-air drying, however, can alter the physical and chemical properties of fruits and may reduce the level of polyphenolic compounds because of their sensitivity to degradation at high temperatures for a relatively long drying time. Although freeze drying reduces the amount of heat damage and produces a dried product with generally better retention of cellular structure, functional, and quality properties, polyphenolics are exposed to oxidative conditions during freeze drying most likely due to the extended drying time. Microwave-vacuum drying is a relatively new dehydration technique that may reduce oxidation of important food components, such as antioxidants, since microwave-vacuum dehydration takes less time and is conducted at a lower temperature compared to hot-air drying. However, operating expenses of microwave-vacuum drying may be higher than for hot-air-dried foods. Therefore, a combination of conventional drying techniques such as hot-air drying with microwave-vacuum and/or freeze drying should reduce dehydration costs while preserving desirable physical and chemical properties of the fruit to a greater extent [5, 7, 8].

8.3 Phytochemicals in dried raspberries

Phenolic compounds are well distributed in fruits and their concentrations can vary tremendously between one fruit and another, even in cultivars of the same species [9, 10]. The content of total phenolics in fruits may be influenced by many factors, including genetics, environmental conditions, degree of ripeness, variety, storage, and analytical methods chosen for isolation, detection, and quantification of total phenolics.

Phytochemicals can be divided into different classes including simple phenols, phenolic acids, benzoquinones, phenylpropenes, cumarins, chromones, naphthoquinones, xanthones, stilbenes, flavonoids, tannins, and lignans, among others [11]. In fruits, polyphenols (Figure 8.1) are commonly found as flavonoids, phenolic acids, and tannins. Flavonoids

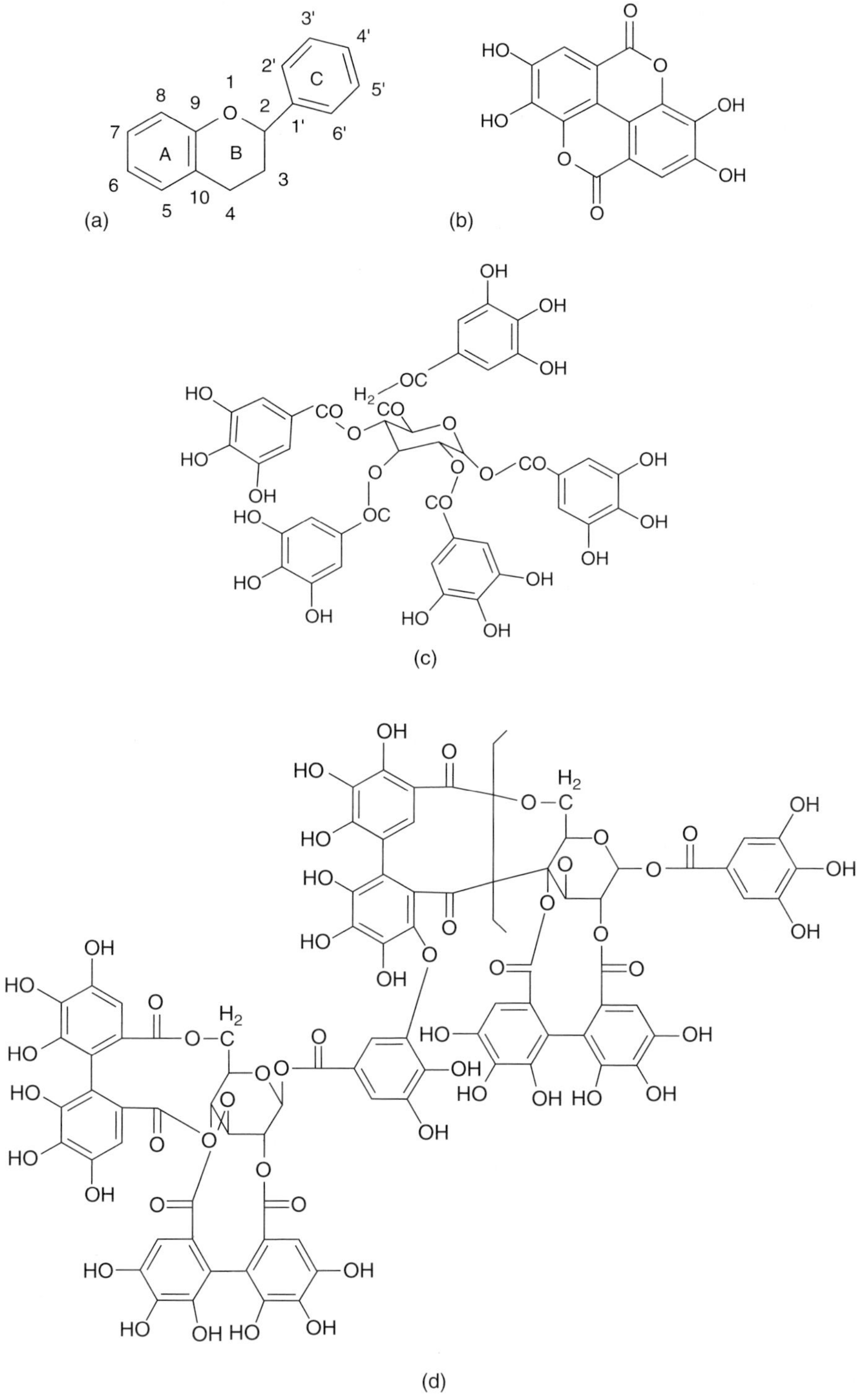

Figure 8.1 Common polyphenols found in raspberries: (a) generic structure of the flavonoid backbone, (b) ellagic acid, (c) tannic acid, and (d) ellagitannin sanguiin H6 [12].

are the most abundant phenolic compounds in fruits and vegetables with more than 5000 compounds identified to date [13, 14].

8.3.1 Flavonoids

Flavonoids, which are the largest group of phytochemicals, are classified into 13 subclasses: chalcones, dihydrochalcones, aurones, flavones, flavonols, dihydroflavonol, flavanones, flavanols, flavandiols, anthocyanidins, isoflavonoids, biflavonoids, and proanthocyanidins [11, 15]. They are associated with many physiological properties, including antioxidant, anti-inflammatory, antimicrobial, antihyperlipidemic, anticancer, antiviral, and antiallergenic, all of which are thought to play a role in reducing the risk of degenerative diseases such as cardiovascular disease (CVD).

The most abundant flavonoids in fruits are the flavonols (quercetin, kaempferol, and myricetin), flavones (apigenin and luteolin), flavanones (naringin, naringenin, hesperetin, and hesperidin), flavan-3-ols (catechin, catechin gallate, and proanthocyanidins), and anthocyanidins (cyanidin, dephinidin, pelargonidin, and glucosides anthocyanins) (Figure 8.2)

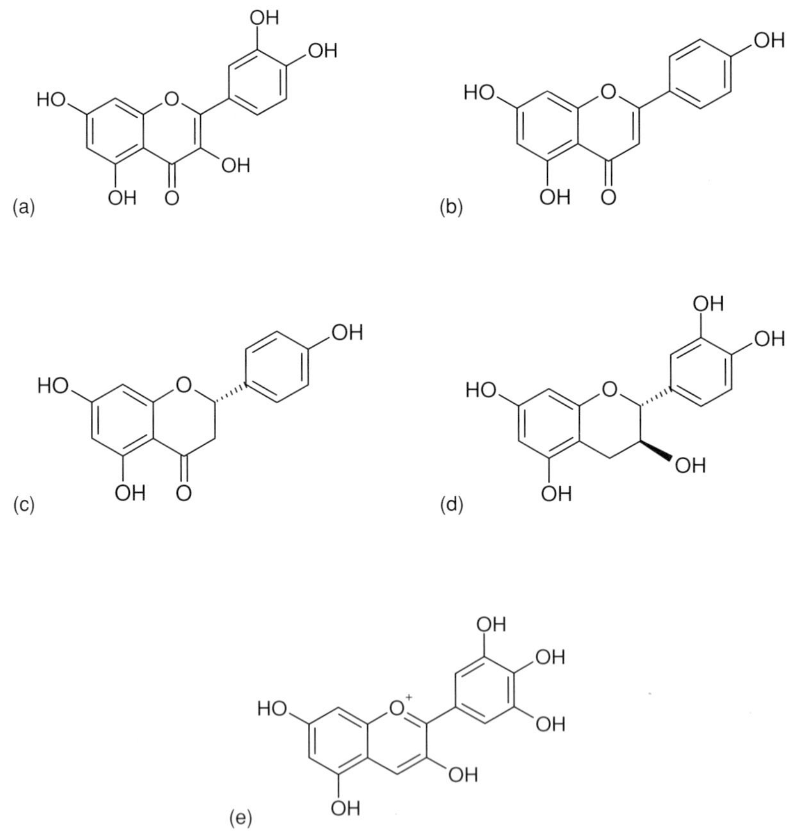

Figure 8.2 The most abundant flavonoids in raspberries: (a) quercetin, (b) apigenin, (c) naringenin, (d) (+)-catechin, and (e) cyanidin.

[11, 16, 17]. Flavonols are the most widely distributed flavonoids in fruits and exist mainly in the glycosidic form with a hydroxyl group conjugated most commonly at position 3 of the C-ring and with possible substitutions at 5, 7, 4′, 3′, and 5′ positions (Figure 8.2a) [13, 18]. Quercetin, kaempferol, and myricetin, the most common flavonols, are quantified in apples, berries, plums, tomatoes, peaches, and grapefruits [15, 17, 19]. Quercetin-3-glucoside and kaempferol-3-glucoside and their conjugates have been identified in raspberries [20], while quercetin and kaempferol aglycones and their glycosides have also been reported [21].

Flavones, another subclass of flavonoids, are common in fruits and exhibit a modified flavonol structure with an extensive range of carbon substitutions. The most common flavone, apigenin, exhibits possible substitutions at the A- and C-rings (Figure 8.2b) and is a constituent of fruits and vegetables including melons, watermelons, grapefruit, and bell peppers [15]. However, flavones have not been fully identified in raspberries [20].

Another subclass of flavonoids with high bioactivity is the flavanones. They occur as hydroxylated, glycosylated, and *o*-methylated derivatives. The most abundant members are naringin and hesperidin (flavanone glycoside), both common in citrus fruits, as naringenin (Figure 8.2c) and hesperetin (flavanone aglycones) [17, 22, 23]. Flavanones are prevalent in citrus fruits, particularly in grapefruits, oranges, and lemons [15, 19, 23]. Flavan-3-ols or flavanols, one of the most complex flavonoids, are characterized by two chiral centers at C2 and C3, in the heterocyclic C-ring for monomeric units such as (+)-catechin (Figure 8.2d) [18, 22, 24]. Flavanols such as (+)-catechin, epicatechin, epigallocatechin, and proanthocyanidins are abundant flavonoids in grapes, cacao, apples, peaches, nectarines, and berries [11, 15, 19, 23]. Flavanols are responsible for the bitterness and astringency of some fruits [17, 18] and raspberries have been reported to contain naringin enantiomers [21].

Anthocyanidins, another subclass of flavonoids, are widely distributed in food plants and are present in the sugar conjugated form, commonly known as anthocyanins [22]. The most common anthocyanidins are cyanidin, pelargonidin, delphinidin, petunidin, peonidin, and malvidin (Figure 8.3), generated from proanthocyanidins under acidic and high-temperature conditions [18]. Anthocyanidins are present in fruits such as red grapes, blueberries, raspberries, cranberries, and strawberries [17, 22].

Limited information is available about the phytochemical composition of dried raspberries. Recently, Mejia-Meza *et al.* [21] studied retention of polyphenol glycoside and aglycone compounds from dried raspberries by different drying techniques (Table 8.1).

Findings included the concentrations of individual polyphenols, namely, ellagic acid, quercetin, phlorizin, *R*-naringin, *S*-naringin, kaempferol, and their aglycones in fresh and dried raspberries (such as freeze drying, microwave-vacuum drying, hot-air drying, and a combination of hot-air drying and microwave-vacuum drying methods). The results demonstrated significantly ($P \leq 0.05$) higher concentrations of glycosides and aglycones compounds in fresh raspberries as compared to that of their dried counterparts. Ellagic acid and quercetin were present at the highest concentration in fresh and dried raspberries. Dehydration led to a reduction in polyphenolic content, but there was a tendency for microwave-vacuum and a combination of hot-air and microwave-vacuum drying to yield products with higher content of polyphenols (Table 8.1). Microwave-vacuum-dried raspberries retained significantly higher ($P \leq 0.05$) concentrations of kaempferol glycoside (0.660 mg/g basis) than other dehydrated samples. The concentration of glycosides and aglycone forms of ellagic acid, quercetin, phloretin, *R*-naringenin, *S*-naringenin, and kaempferol were also measured in dried raspberries. Both ellagic acid and kaempferol aglycones were retained to a greater extend in

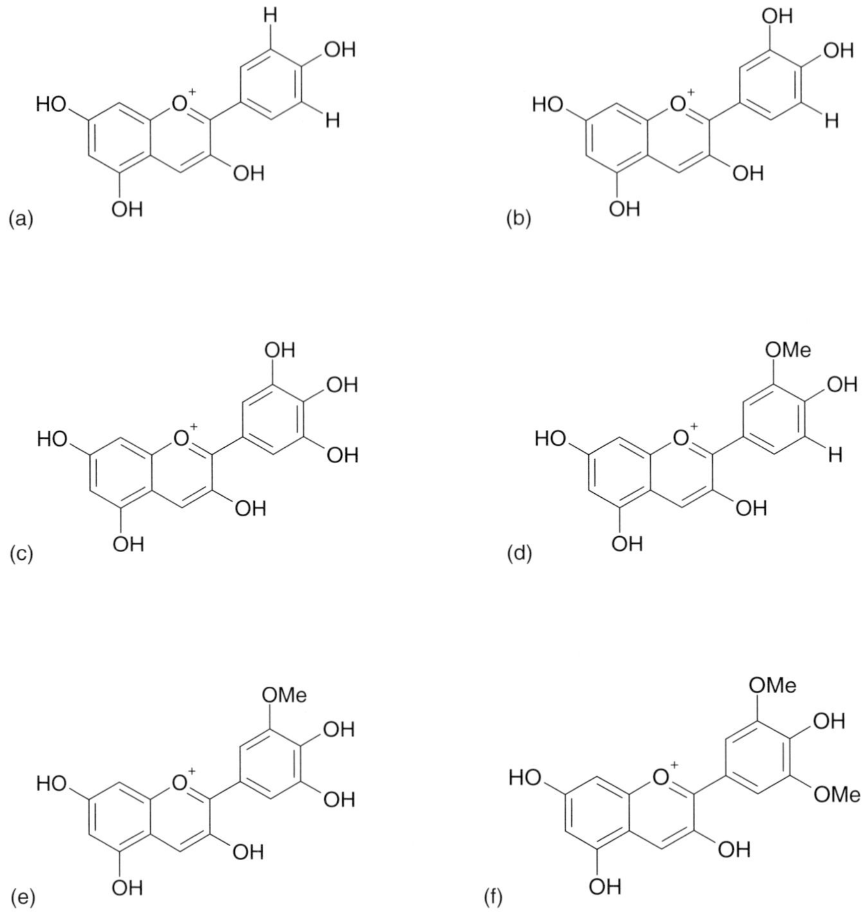

Figure 8.3 Anthocyanidins in raspberries: (a) pelargonidin, (b) cyanidin, (c) delphinidin, (d) peonidin, (e) petunidin, and (f) malvidin.

microwave-vacuum-dried raspberries (0.342 mg/g dry basis), and quercetin aglycone in hot-air-/microwave-vacuum-dried berries compared to other treatments. Raspberries are known to contain relatively high amounts of polyphenols, which are composed of 60% flavonoids, 30% phenolic acids, and 10% tannins [25]. These compounds have been extensively studied for their anticarcinogenic properties [26, 27].

Anthocyanins are responsible for the red and blue coloration of certain fruits, flowers, and leaves [22, 28] and are present in high concentrations in blueberries, raspberries, strawberries, and red grapes (Table 8.2) [19, 29–31]. Raspberries have been reported to contain cyanidin-3-glucoside, cyanidin-3-rutinoside, cyanidin-3-sophoroside, and pelargonidin-3-glucoside [20]. The popularity of anthocyanin-containing foods is increasing tremendously due to recent interest about their multiple health-promoting features including antioxidant, anti-inflammatory, and anticancer activities and more recently due to their chemoprotective, vasoprotective, and antineoplastic properties [29].

Table 8.1 Retention of polyphenol glycoside and aglycone compounds (mg/g dry basis) from dried raspberries by different drying technologies

Polyphenols	Fresh	Freeze drying	Microwave-vacuum drying	Hot-air drying	Hot-air drying/ microwave-vacuum drying
Glycosides					
Ellagic Acid	2.53 ± 0.29	0.66 ± 0.01[a]	2.14 ± 0.02	0.18 ± 0.03[a]	3.12 ± 0.36
Quercetin	2.18 ± 0.20	0.58 ± 0.03[a]	1.02 ± 0.06[a]	0.41 ± 0.06[a]	1.63 ± 0.22[a]
Phlorizin	0.02 ± 0.002	0.012 ± 0.007[a]	0.017 ± 0.001	0.03 ± 0.005[a]	0.09 ± 0.001[a]
R-Naringin	0.13 ± 0.01	0.019 ± 0.001[a]	0.017 ± 0.001[a]	0.018 ± 0.001[a]	0.42 ± 0.02[a]
S-Naringin	0.09 ± 0.01	0.048 ± 0.002[a]	0.051 ± 0.00[a]	0.138 ± 0.001[a]	0.032 ± 0.005[a]
Kaempferol	0.49 ± 0.04	0.260 ± 0.034[a]	0.660 ± 0.040[a]	0.287 ± 0.036[a]	0.048 ± 0.003[a]
Aglycones					
Ellagic acid	1.15 ± 0.10	0.142 ± 0.008[a]	0.342 ± 0.020[a]	0.152 ± 0.023[a]	0.068 ± 0.006[a]
Quercetin	0.188 ± 0.017	0.023 ± 0.001[a]	0.034 ± 0.001[a]	0.118 ± 0.023[a]	0.170 ± 0.018[a]
Phloretin	0.414 ± 0.041	0.002 ± 0.001[a]	0.002 ± 0.001[a]	0.029 ± 0.001[a]	0.014 ± 0.001[a]
R-Naringenin	0.05 ± 0.01	0.001 ± 0.000[a]	–	0.004 ± 0.001[a]	0.003 ± 0.001[a]
S-Naringenin	0.041 ± 0.005	0.020 ± 0.008[a]	–	0.051 ± 0.001[a]	0.066 ± 0.005[a]
Kaempferol	0.101 ± 0.011	0.05 ± 0.01[a]	0.068 ± 0.003[a]	0.008 ± 0.002[a]	0.022 ± 0.001[a]

Source: Adapted with permission from Mejía-Meza *et al.* [21].
Values are means ± SEM ($n = 9$).
[a]Signifcantly different from fresh raspberries ($P < 0.05$).

Table 8.2 Total anthocyanin content in selected fruits

Fruit	Method	Total anthocyanins	Unit	Reference
Mulberries	Spectrophotometric	137–2057	μg M3GE/g	[29]
Blackberries (raw)	HPLC	81–86	mg/100 g	[19]
Blueberries (raw)	HPLC	157–112	mg/100 g	[19]
Cherries (raw)	HPLC	35–120	mg/100 g	[19]
Cranberries (raw)	HPLC	86.1	mg/100 g	[19]
Raspberries (raw)	HPLC	26–38	mg/100 g	[19]
Strawberries (raw)	HPLC	8–40	mg/100 g	[19]
Plums (raw)	HPLC	12.5	mg/100 g	[19]
Nectarine (raw)	HPLC	3.5	mg/100 g	[19]
Peaches (raw)	HPLC	1.5	mg/100 g	[19]
Raspberries, (raw)	HPLC/colorimetric	1008–1350	mg C3GE/g	[45]
Aksu Kirizisi				
Water extract	Spectrophotometric	49.3	mg C3GE/100 g	[31]
Methanol extract	Spectrophotometric	24.8	mg C3GE/100 g	[31]
Newburgh				
Water extract	Spectrophotometric	69.5	mg C3GE/100 g	[31]
Methanol extract	Spectrophotometric	16.3	mg C3GE/100 g	[31]
Rubin				
Water extract	Spectrophotometric	60.3	mg C3GE/100 g	[31]
Methanol extract	Spectrophotometric	2424.1	mg C3GE/100 g	[31]
Heritage				
Water extract	Spectrophotometric	22.4	mg C3GE/100 g	[31]
Methanol extract	Spectrophotometric	12.4	mg C3GE/100 g	[31]
Hollanda Boduru				
Water extract	Spectrophotometric	45.6	mg C3GE/100 g	[31]
Methanol extract	Spectrophotometric	24.3	mg C3GE/100 g	[31]

HPLC, high performance liquid chromatography; M3GE, malvidin-3-glucoside equivalents; C3GE, cyanidin-3-glucoside equivalents.

Anthocyanins contribute about 25% to the total antioxidant capacity of fresh red raspberries [32], depending on the drying method employed. These phenolic compounds are sensitive to heat and oxidation, and dehydration or other heat preservation processes (e.g., pasteurization) frequently lead to considerable changes in their color and concomitant loss of polyphenols, depending on the drying method used [21, 33–35].

8.3.2 Phenolic acids

Another important class of polyphenols present in fruits is the phenolic acids, mainly derived from hydroxybenzoic and/or hydroxycynamic acids. Phenolic acids are commonly identified as gallic, vanillic, procatechuic, and syringic acids when derived from hydroxybenzoic acid, and as *p*-coumaric, caffeic, and ferulic acids when derived from hydroxycinnamic acid [18, 22]. Fruits, including blueberries, raspberries, cherries, apples, and cacao, contain large concentrations of phenolic acids [15]. Gallic acid is the most common phenolic acid forming the base unit of gallotannins and together with hexahydroxydiphenic acid forms subunits of the ellagitannins. Gallotannins and ellagitannins, also called hydrolysable tannins, release gallic acid and ellagic acid under dilute acid conditions while condensed tannins do not [22]. Caffeic acid is the most common hydroxycinnamic acid derivative. Dried raspberries have been reported to contain ellagic acid aglycone and glycoside [21], as well as 3′-caffeoylquinic acid, *p*-coumaric glucose, and cinnamoyl glucose [20].

8.3.3 Tannins

Tannins are one of the most widespread polyphenolics in fruits and are divided in two main classes of condensed tannins (proanthocyanidins) and hydrolysable tannins [36]. Condensed tannins are high-molecular-weight compounds formed from monomeric units of flavanols or flavan-3-ol, including (+)-catechin, (−)-epicatehin, (+)-gallocatechin, and (−)-epigallocatechin. Oxidative condensation occurs between the heterocyclic carbon C-4 of monomeric units of flavanols and the adjacent positions of carbons C-6 or C-8 of flavanols to create oligomeric and polymeric proanthocyanidins [5, 22, 36]. Condensed tannins are responsible for the astringency of many tannin-rich foods such as red wine and tea, resulting mainly from the precipitation of tannins with salivary proteins. No condensed tannins have been reported in raspberries. Rao *et al.* [12] state that raspberry polyphenols consist primarily of anthocyanins and hydrolyzable tannins. More specifically, raspberries are a particularly rich source of cyanidin glycosides and are unique among all berries for their high ellagitannin content, which when hydrolyzed yield ellagic acid.

8.4 Antioxidants in dried raspberries

Fruits are a good source of phenolics with antioxidant activity, specifically flavonoids, phenolic acids, and tannins. Individual analytical methods for the measurement of antioxidant activity can be difficult to interpret and often it is hard to determine what the interaction or synergistic effects are among different antioxidative components in a fruit and the overall antioxidant properties. As a result, numerous antioxidant methods, both *in vitro* and *in vivo*, have been developed to measure the antioxidant properties of food phenolics. There are two classes of *in vitro* assays involving either hydrogen atom transfer or single electron transfer reactions [37]. Single electron transfer assays are more commonly used to measure the antioxidant properties of polyphenolic compounds in fruit and involve redox reactions of an oxidant based indicator as the reaction end point. Types of these assays include the trolox equivalent antioxidant capacity (TEAC) assay, oxygen radical absorbance capacity (ORAC), the 2,2-di(4-tert-octylphenyl)-1-picrylhydrazyl (DPPH) assay, and the ferric ion reducing antioxidant parameter (FRAP) assay.

The antioxidant activity of fruits extracts are listed in Table 8.3 [9, 31, 30, 38–41]. Most of the fruits were analyzed in the fresh forms; however, data may be extrapolated to fruits such as raspberries in the dried form depending on the drying method employed. Although some fruits such as apples and berries are known to possess high antioxidant activity, it is difficult to compare the results for antioxidant activity between different methods because different assays can yield different results and may follow different mechanisms. Also, there is often an incomplete description of the fruit sample, how it had been stored and treated, particularly when results for processed food are presented. The wide variety of antioxidant activity methods, different units for reporting data, and other factors already known, such

Table 8.3 Antioxidant activity of polyphenol for select fruits

Fruit	Method	Antioxidant activity	Unit	Reference
Tomato	TOSC	10.4	μmol VCE/g	[38]
Blueberries				
Wild	FRAP	44	μM CE	[39]
Cultivated	FRAP	96	μM CE	[39]
Wild	FRAP	71	μM CE	[39]
Cultivated	FRAP	78	μM CE	[39]
Highbush blueberries	ORAC	50	μmol TE/g	[40]
Red huckleberries	ORAC	19 to 130	μmol TE/g	[40]
Raspberries	ESR	410	Number of Fremy's radical × 10^{16} reduced	[30]
Raspberries				
Aksu Kirizisi				
Water extract	TEAC	68.0	μmol TE/g	[31]
Methanol extract	TEAC	86.7	μmol TE/g	[31]
Newburgh				
Water extract	TEAC	86.7	μmol TE/g	[31]
Methanol extract	TEAC	98.6	μmol TE/g	[31]
Rubin				
Water extract	TEAC	64.4	μmol TE/g	[31]
Methanol extract	TEAC	72.9	μmol TE/g	[31]
Heritage				
Water extract	TEAC	65.4	μmol TE/g	[31]
Methanol extract	TEAC	74.3	μmol TE/g	[31]
Hollanda Boduru				
Water extract	TEAC	69.7	μmol TE/g	[31]
Methanol extract	TEAC	117	μmol TE/g	[31]
Raspberries				
Aksu Kirizisi				
Water extract	DPPH	76.6	μmol TE/g	[31]
Methanol extract	DPPH	122	μmol TE/g	[31]
Newburgh				
Water extract	DPPH	89.1	μmol TE/g	[31]
Methanol extract	DPPH	122	μmol TE/g	[31]
Rubin				
Water extract	DPPH	64.1	μmol TE/g	[31]
Methanol extract	DPPH	96.7	μmol TE/g	[31]
Heritage				
Water extract	DPPH	66.0	μmol TE/g	[31]
Methanol extract	DPPH	81.2	μmol TE/g	[31]
Hollanda Boduru				
Water extract	DPPH	77.6	μmol TE/g	[31]
Methanol extract	DPPH	142	μmol TE/g	[31]
Apples				
Red delicious (peel)	FRAP	17800	μM CE	[9]
Northern spy (peel)	FRAP	10040	μM CE	[9]
Ida red (peel)	FRAP	12100	μM CE	[9]
Red delicious (flesh)	FRAP	920	μM CE	[9]
Northern spy (flesh)	FRAP	6430	μM CE	[9]
Idared (flesh)	FRAP	2750	μM CE	[9]
Idared (peel)	TOSC	312	μmol VCE/g	[9]

Table 8.3 (*Continued*)

Fruit	Method	Antioxidant activity	Unit	Reference
Idared (flesh)	TOSC	47	μmol VCE/g	[9]
Idared (flesh + peel)	TOSC	72	μmol VCE/g	[9]
Rome beauty (peel)	TOSC	228	μmol VCE/g	[9]
Rome beauty (flesh)	TOSC	68	μmol VCE/g	[9]
Rome beauty (peel + flesh)	TOSC	132	μmol VCE/g	[9]
Golden delicious				
Integrated (peel)	DPPH	15.4	mM TE/kg	[41]
Organic (peel)	DPPH	13.2	mM TE/kg	[41]
Integrated (pulp)	DPPH	4.10	mM TE/kg	[41]
Organic (pulp)	DPPH	2.68	mM TE/kg	[41]

TOSC, total oxyradical scavenging capacity; ORAC, oxygen radical absorbance capacity; ESR, electron spin resonance; TEAC, trolox equivalent antioxidant capacity; FRAP, ferric reducing antioxidant power; DPPH, 2,2-diphenyl-1-picrylhydrazyl; VCE, vitamin C equivalents; TE, trolox equivalents; CE, catechin equivalents.

as the conditions of growth, harvest, handling, and storage, make it even more difficult to compare results between different studies.

Fresh raspberries had higher concentrations of individual, total polyphenols, and anthocyanins, and antioxidant activity than dried raspberries [21]. Microwave-vacuum drying and the combination of hot-air/microwave-vacuum drying of raspberries had higher retention of ellagic acid, quercetin, kaempferol, aglycones, and total polyphenols compared to other drying methods including hot-air drying and freeze drying alone. Total anthocyanins were less heat stable than polyphenols. Total aglycone polyphenol compounds may have higher resistance to drying temperatures than glycosides, and may be less affected by the combination of hot-air/microwave-vacuum drying method. Aglycone polyphenols caused a greater reduction in adipogenesis induction than glycoside compounds [21]. These results suggest that consumption of either fresh or dried raspberries can provide an important source of antioxidants and are also foods that could inhibit adipogenesis if these results are extrapolated to the *in vivo* situation after consumption.

8.5 Health benefits of dried raspberries

Polyphenolic compounds in fruits, including flavonoids (quercetin and kaempferol), phenolic acids (gallic and ellagic acids), and tannins (red wine polyphenols), exhibit some potential to inhibit the proliferation of certain cancer cell lines (colon and breast) *in vitro* [17]. They have strong antioxidant activities that are believed to inhibit the propagation of free radical reactions [11, 40]. *In vivo* antioxidants may be beneficial for inhibiting free radical-initiated reactions associated with aging, cancer, and CVD [29]. Recently, berries, including raspberries, have been studied to determine their effect in cancer proliferation and the results are summarized in Table 8.4 [42–44]. For instance, it has been reported that a raspberry extract at a concentration of 50 μg/mL gallic acid equivalents (GAE) reduces proliferation (quantified by cell invasion) of HT115 cells by 95% [44].

Table 8.4 Anticancer activity from berries *in vitro*

Fruit	Method	Proliferation reduction (%)	Concentration	Reference
Black-raspberries	HT-29/HCT116	78	200 μg/mL	[42]
Blueberries	HT-29/HCT116	78	200 μg/mL	[42]
Blackberries	HT-29/HCT116	78	200 μg/mL	[42]
Raspberry-extract	HT115	95 (cell invasion)	50 μg/mL GAE	[43]
Strawberry-extract	HT-29/MCF7	41–63/26–56	0–200 μg/mL	[44]

GAE, gallic acid equivalents.

8.6 Conclusions

Raspberries (fresh or dried) are known as a rich source of polyphenolic compounds with flavonoids, phenolic acids, and tannins being the most widely distributed subgroups. Study of these phenolic compounds is popular because of their extensive range of biological properties (antioxidant, anticancer, antimicrobial, antihyperlipidemic, and anti-inflammatory, among others) that may help decrease the risk of chronic health conditions such as obesity, type II diabetes, CVD, and cancer. Therefore, there is a movement among health professionals to encourage consumers to eat more antioxidant-rich foods as a phytopreventative strategy to reduce a variety of diseases. However, we still know very little about the composition and biological activity of polyphenolics in most fruits and how various types of processing may affect either the concentration or biological activity. As reported throughout this chapter, raspberries contain various polyphenols that contribute to their significant antioxidant capacity and observed *in vitro* biological activity. Currently, preservation techniques, including dehydration, as well as some processing techniques are applied to extend the shelf life of fruits but are known to have a negative impact on polyphenol retention depending on the drying method employed. In the case of dehydration, hot-air drying usually results in greater loss of phenolic compounds than other drying methods such as freeze drying or a combination of hot-air and microwave-vacuum drying. If responsible recommendations are to be made for promoting the consumption of fresh, dried, or processed fruits, it will be important to be able to provide a reasonable assessment of both the content and biological activity of these compounds. This conclusion extends to raspberries and any other fruit since the biological matrix and composition of fruits are grossly comparable.

References

1. The Worlds Healthiest Foods. (2012) *The World's Healthiest Foods*. Published online at: http://whfoods.org/genpage.php?tname=nutrientprofile&dbid=23, last accessed January 9, 2012.
2. FAO. (2012) *Agricultural Production: Countries by Commodity—Raspberries*. Published online at: http://faostat.fao.org/site/339/default.aspx, last accessed January 9, 2012.
3. Tang, J. & Yang, T. (2003) Dehydrated vegetables: principles and systems. In: *Handbook of Vegetable Preservation and Processing* (ed. Y.H. Huied). Marcel Dekker, New York, NY, pp. 335–372.
4. Mousa, M.F.A.N. (2002) Microwave vacuum drying of banana slices. *Drying Technology*, **20**, 2055–2066.

5. Clary, C.D., Wang, S. & Petrucci, V.F. (2005) Fixed and incremental levels of microwave power application on drying grapes under vacuum. *Journal of Food Science*, **70**, 344–349.
6. Reid, D.S. & Fennema, O.R. (2007) Water and ice In: *Food Chemistry*, 4th edn. (eds S. Damodaran, K. Parkin & O.R. Fennema). CRC Press Taylor & Francis Group, Boca Raton, FL, pp. 17–94.
7. Clary, C.D., Mejia-Meza, E., Wang, S. & Petrucci, V.E. (2007) Improving grape quality using microwave vacuum drying associated with temperature control. *Journal of Food Science*, **72**, E023–E028.
8. Giri, S.P.A.S. (2005) Drying kinetics and rehydration characteristics of microwave-vacuum and convective hot-air dried mushrooms. *Journal of Food Engineering*, **78**, 512–521.
9. Tsao, R., Yang, R., Xie, S., Sockovie, E. & Khanizadeh, S. (2005) Which polyphenolic compounds contribute to the total antioxidant activities of apple? *Journal of Agricultural and Food Chemistry*, **53**, 4989–4995.
10. Wolfe, K., Wu, X. & Liu, R.H. (2003) Antioxidant activity of apple peels. *Journal of Agricultural and Food Chemistry*, **51**, 609–614.
11. Bravo, L. (1998) Polyphenols: chemistry, dietary sources, metabolism, and nutritional significance. *Nutrition Reviews*, **56**, 317–333.
12. Rao, A.V. (2010) Raspberries and human health: a review. *Journal of Agricultural and Food Chemistry*, **58**, 3872–3883.
13. Kris-Etherton, P.M., Hecker, K.D., Bonanome, A., Coval, S.M., Binkoski, A.E., Hilpert, K.F., Griel, A.E. & Etherton, T.D. (2002) Bioactive compounds in foods: their role in the prevention of cardiovascular disease and cancer. *American Journal of Medicine*, **113**, 71S–88S.
14. Crespy, V., Morand, C., Besson, C., Cotelle, N., Vezin, H., Demigne, C. & Remesy, C. (2003) The splanchnic metabolism of flavonoids highly differed according to the nature of the compound. *American Journal of Physiology—Gastrointestinal and Liver Physiology*, **284**, G980–G988.
15. Sakakibara, H., Honda, Y., Nakagawa, S., Ashida, H. & Kanazawa, K. (2003) Simultaneous determination of all polyphenols in vegetables, fruits, and teas. *Journal of Agricultural and Food Chemistry*, **51**, 571–581.
16. Hertog, M.G., Feskens, E.J., Hollman, P.C., Katan, M.B. & Kromhout, D. (1993) Dietary antioxidant flavonoids and risk of coronary heart disease: the Zutphen Elderly Study. *Lancet*, **342**, 1007–1011.
17. Scalbert, A., Manach, C., Morand, C., Remesy, C. & Jimenez, L. (2005) Dietary polyphenols and the prevention of diseases. *Critical Reviews in Food Science and Nutrition*, **45**, 287–306.
18. Nichenametla, S.N., Taruscio, T.G., Barney, D.L. & Exon, J.H. (2006) A review of the effects and mechanisms of polyphenolics in cancer. *Critical Reviews in Food Science and Nutrition*, **46**,161–183.
19. Harnly, J.M., Doherty, R.F., Beecher, G.R., Holden, J.M., Haytowitz, D.B., Bhagwat, S. & Gebhardt, S. (2006) Flavonoid content of US fruits, vegetables, and nuts. *Journal of Agricultural and Food Chemistry*, **54**, 9966–9977.
20. Proteggente, A.R., Pannala, A.S., Paganga, G., Van Buren, L., Wagner, E., Wiseman, S., Van De Put, F., Dacombe, C. & Rice-Evans, C.A. (2002) The antioxidant activity of regularly consumed fruit and vegetables reflects their phenolic and vitamin C composition. *Free Radical Research*, **36**, 217–233.
21. Mejía-Meza, E.I., Yáñez, J.A., Remsberg, C.M., Takemoto, J.K., Davies, N.M., Rasco, B. & Clary, C. (2010) The effect of dehydration on raspberries: polyphenol and anthocyanin retention, anti-oxidant capacity and anti-adipongenic activity. *Journal of Food Science*, **75**, H5-H12.
22. Crozier, A., Yokota, T., Jaganath, I.B., Marks, S., Saltmarsh, M. & Clifford, M.N. (2006) Secondary metabolites in fruits, vegetables, beverages, and other plant-based dietary components. In: *Plant Secondary Metabolites: Occurrence, Structure and Role in the Human Diet* (eds A. Crozier, M.N. Clifford & H. Ashihara). Wiley-Blackwell Publishing, Oxford, UK, pp. 208–302.
23. Espin, J.C., Garcia-Conesa, M.T. & Tomas-Barberan, F.A. (2007) Nutraceuticals: facts and fiction. *Phytochemistry*, **68**, 2986–3008.
24. Iriti, M. & Faoro, F. (2006) Grape phytochemicals: A bouquet of old and new nutraceuticals for human health. *Medical Hypotheses*, **67**, 833–838.
25. Puupponen-Pimia, R., Nohynek, L., Alakomi, H.L. & Oksman-Caldentey, K.M. (2005) The action of berry phenolics against human intestinal pathogens. *Biofactors*, **23**, 243–251.
26. Juranic, Z. & Zizak, Z. (2005) Biological activities of berries: from antioxidant capacity to anti-cancer effects. *Biofactors*, **23**, 207–211.
27. Skupien, K., Oszmianski, J., Kostrzewa-Nowak, D. & Tarasiuk, J. (2006) *In vitro* antileukaemic activity of extracts from berry plant leaves against sensitive and multidrug resistant HL60 cells. *Cancer Letters*, **236**, 282–291.

28. McDougall, G.J., Dobson, P., Smith, P., Blake, A. & Stewart, D. (2005) Assessing potential bioavailability of raspberry anthocyanins using an *in vitro* digestion system. *Journal of Agricultural and Food Chemistry*, **53**, 5896–5904.
29. Bae Song-Hwan, S.H.-J. (2007) Antioxidant activities of five different mulberry cultivars in Korea. *LWT—Food Science and Technology*, **40**, 955–962.
30. Mullen, W., Marks, S.C. & Crozier, A. (2007) Evaluation of phenolic compounds in commercial fruit juices and fruit drinks. *Journal of Agricultural and Food Chemistry*, **55**, 3148–3157.
31. Sariburun, E., Şahin, S., Demir, C., Türkben, C. & Uylaşer, V. (2010) Phenolic content and antioxidant activity of raspberry and blackberry cultivars. *Journal of Food Science*, **75**, C328–C335.
32. Beekwilder, J., Hall, R.D. & de Vos, C.H. (2005) Identification and dietary relevance of antioxidants from raspberry. *Biofactors*, **23**, 197–205.
33. Lin, T.M., Durance, T.D. & Scaman, C.H. (1998) Characterization of vacuum microwave, air and freeze dried carrot slices. *Food Research International*, **31**, 111–117.
34. Yousif, A.N., Scaman, C.H., Durance, T.D. & Girard, B. (1999) Flavor volatiles and physical properties of vacuum-microwave- and air-dried sweet basil (*Ocimum basilicum* L.). *Journal of Agricultural and Food Chemistry*, **47**, 4777–4881.
35. Kwok, B.H.L., Hu, C., Durance, T. & Kitts, D.D. (2004) Dehydration techniques affect phytochemical contents and free radical scavenging activities of saskatoon berries (*Amelanchier alnifolia* Nutt.). *Journal of Food Science*, **69**, 122–126.
36. Singh, B., Bhat, T.K. & Singh, B. (2003) Potential therapeutic applications of some antinutritional plant secondary metabolites. *Journal of Agricultural and Food Chemistry*, **51**, 5579–5597.
37. Huang, D., Ou, B. & Prior, RL. (2005) The chemistry behind antioxidant capacity assays. *Journal of Agricultural and Food Chemistry*, **53**, 1841–1856.
38. Dewanto, V., Wu, X., Adom, K.K. & Liu, R.H. (2002) Thermal processing enhances the nutritional value of tomatoes by increasing total antioxidant activity. *Journal of Agricultural and Food Chemistry*, **50**, 3010–3014.
39. Schmidt, B.M., John, W., Erdman, J. & Lila, A.M.A. (2005) Effect of food processing on blueberry antiproliferation and antioxidant activity. *Journal of Food Science*, **70**, S389–S394.
40. Moyer, R.A., Hummer, K.E., Finn, C.E., Frei, B. & Wrolstad, R.E. (2002) Anthocyanins, phenolics, and antioxidant capacity in diverse small fruits: vaccinium, rubus, and ribes. *Journal of Agricultural and Food Chemistry*, **50**, 519–525.
41. Chinnici, F., Bendini, A., Gaiani, A. & Riponi, C. (2004) Radical scavenging activities of peels and pulps from cv. golden delicious apples as related to their phenolic composition. *Journal of Agricultural and Food Chemistry*, **52**, 4684–4689.
42. Seeram, N.P., Adams, L.S., Zhang, Y., Lee, R., Sand, D., Scheuller, H.S. & Heber, D. (2006) Blackberry, black raspberry, blueberry, cranberry, red raspberry, and strawberry extracts inhibit growth and stimulate apoptosis of human cancer cells *in vitro*. *Journal of Agricultural and Food Chemistry*, **54**, 9329–9339.
43. Coates, E.M., Popa, G., Gill, C.I., McCann, M.J., McDougall, G.J., Stewart, D. & Rowland, I. (2007) Colon-available raspberry polyphenols exhibit anti-cancer effects on *in vitro* models of colon cancer. *Journal of Carcinogenesis*, **6**, 4.
44. Olsson, M.E. (2006) Antioxidant levels and inhibition of cancer cell proliferation *in vitro* by extracts from organically and conventionally cultivated strawberries. *Journal of Agricultural and Food Chemistry*, **54**, 1248–1255.
45. Mullen, W., McGinn, J., Lean, M.E., MacLean, M.R., Gardner, P., Duthie, G.G., Yokota, T. & Crozier, A. (2002) Ellagitannins, flavonoids, and other phenolics in red raspberries and their contribution to antioxidant capacity and vasorelaxation properties. *Journal of Agricultural and Food Chemistry*, **50**, 5191–5196.

9 Phytochemical antioxidants and health benefits of dried strawberries

Rong Tsao and Hongyan Li

9.1 Introduction

Fruits and vegetables are good sources of antioxidant phytochemicals for reducing the incidence of diseases such as cardiovascular disease (CVD), diabetes, and certain types of cancer in humans [1–3]. Compared to other foods, fruits and vegetables are more beneficial to human health due to their safety, low cost, and oral bioavailability [4]. Strawberries contain a large amount of phytochemicals, in both fresh and processed forms [5]. They are widely known for their health-promoting properties such as antioxidant, anticancer, and antiatherosclerotic effects [6, 7]. Research has also shown that dried strawberries can lower the risks of oxidative damage, CVD, and esophageal cancer [8, 9].

Strawberries are available around the world. In the United States, 83% of the strawberries (6.35 million tonnes annually) are grown in California with the most common commercial varieties being *Camarosa*, *Diamonte*, *Chandler*, and *Selva*. They have a very short harvesting season (only last 3 to 4 weeks) and relatively short shelf life because of the natural ripening process and their sensitivity to fungal attack [10]. Postharvest storage conditions can also affect the texture, color, flavor, phytochemical content, and antioxidant activity of strawberries [11, 12]. For these reasons, a significant amount of strawberries is consumed in the forms of frozen and dried fruits, jams, yogurt, preserves, and juice concentrates, among others. The color, flavor (odor and taste), and storability of dried strawberry pieces are generally considered to be good. Several drying methods, including microwave drying and vacuum freeze drying, have been widely used to obtain high-quality dehydrated strawberries [13–15].

Strawberries are rich in fiber, potassium, vitamin C, folate, and phytochemicals and are relatively low in calories. The main phytochemicals in dried strawberries have been reported to be hydrolyzable tannins, anthocyanins, flavonols, flavanols, and coumaroyl glycosides [16]. These phytochemicals contribute significantly to the total antioxidant activity of fresh or dried strawberries. Several studies have reported that the high antioxidant activity of strawberries is linked to their phenolic contents [17]. However, the phytochemical contents can be affected by genetics, growing environment, processing and storage conditions, and

Dried Fruits: Phytochemicals and Health Effects, First Edition. Edited by Cesarettin Alasalvar and Fereidoon Shahidi.

by errors caused by different analytical methods [10, 18–20]. The phytochemical compositions of dried strawberries can, therefore, directly affect their antioxidant activity and other bioactivities related to human health.

This chapter reviews the latest research on strawberry phytochemicals and their roles in antioxidant and other human health potentials. Factors effecting the phytochemical composition and antioxidant activity during postharvest storage and drying/processing are also discussed. Readers who are interested in detailed information and in-depth discussions on the specific phytochemicals or functions of strawberries are referred to comprehensive reviews by others [9, 21, 22].

9.2 Phytochemicals

Strawberries are rich in micronutrients and phytochemicals, most of which are phenolic compounds. Extracts of freeze-dried whole strawberry fruit powder and fresh strawberry fruit extracts have been found to contain phenolic acids, hydrolyzable tannins (ellagitannins), coumaroyl glycosides, anthocyanins, flavonols, and flavanols [16]. Gallic, protocatechuic, *p*-hydroxybenzoic, caffeic, *p*-coumaric, *o*-coumaric, ferulic, *trans*-cinnamic, and ellagic acids constitute the phenolic acids in strawberries, with *trans*-cinnamic, *p*-coumaric, and gallic acids in the highest concentrations (566, 213, and 212 mg/kg, respectively) [23]. The major flavonols in dried strawberries are quercetin glycosides and anthocyanins. Pelargonidin and cyanidin glucosides or rutinosides are the main anthocyanins found in strawberries [16, 24, 25]. Major phenolic phytochemicals of strawberries are shown in Figure 9.1. The total phenolic and anthocyanin contents of freeze-dried strawberry powder were 33.74 g ferulic acid equivalents/kg and 3.51 g cyanidin-3-glucoside equivalents/kg, respectively [23]. Changes in major individual and total phytochemical contents before and after drying are listed in Table 9.1. These phytochemicals contributed to the antioxidant activity of strawberries. However, there are considerable differences in reported total and individual phytochemical contents due to the existing differences in genetics, growing environment, processing, handling after harvest, and the analytical methodology [9, 11].

9.2.1 Hydrolyzable tannins

Hydrolyzable tannins in strawberries are made up of ellagic acid, ellagic acid glycosides, ellagitannins, and gallotannins. They have been shown to have high antioxidant potential and are effective in preventing many oxidative stress-related chronic diseases [26, 27].

Ellagic acid has been reported to have antioxidant, antiproliferative, and antimalarial properties [28]. There are large differences in the content of ellagic acid in strawberry cultivars [29]. Sharma *et al.* [30] demonstrated that ellagic acid extracts obtained from strawberries were potential anticarcinogenic agents by using human 293T cell line and a luciferase reporter of canonical Wnt pathway-mediated transcriptional activation. Devipriya *et al.* [31] fed female albino Wistar rats (150–170 g) with different concentrations of ellagic acid and found that ellagic acid could inhibit alcohol-induced toxicity by improving body weight, restoring antioxidant status, modulating micronutrients, and attenuating the lipid levels in the circulation of rats.

Ellagitannins are hydrolyzable tannins formed through ester bond between ellagic acid or other phenolic acids and sugar molecules. Ellagitannins in dried strawberries are reported

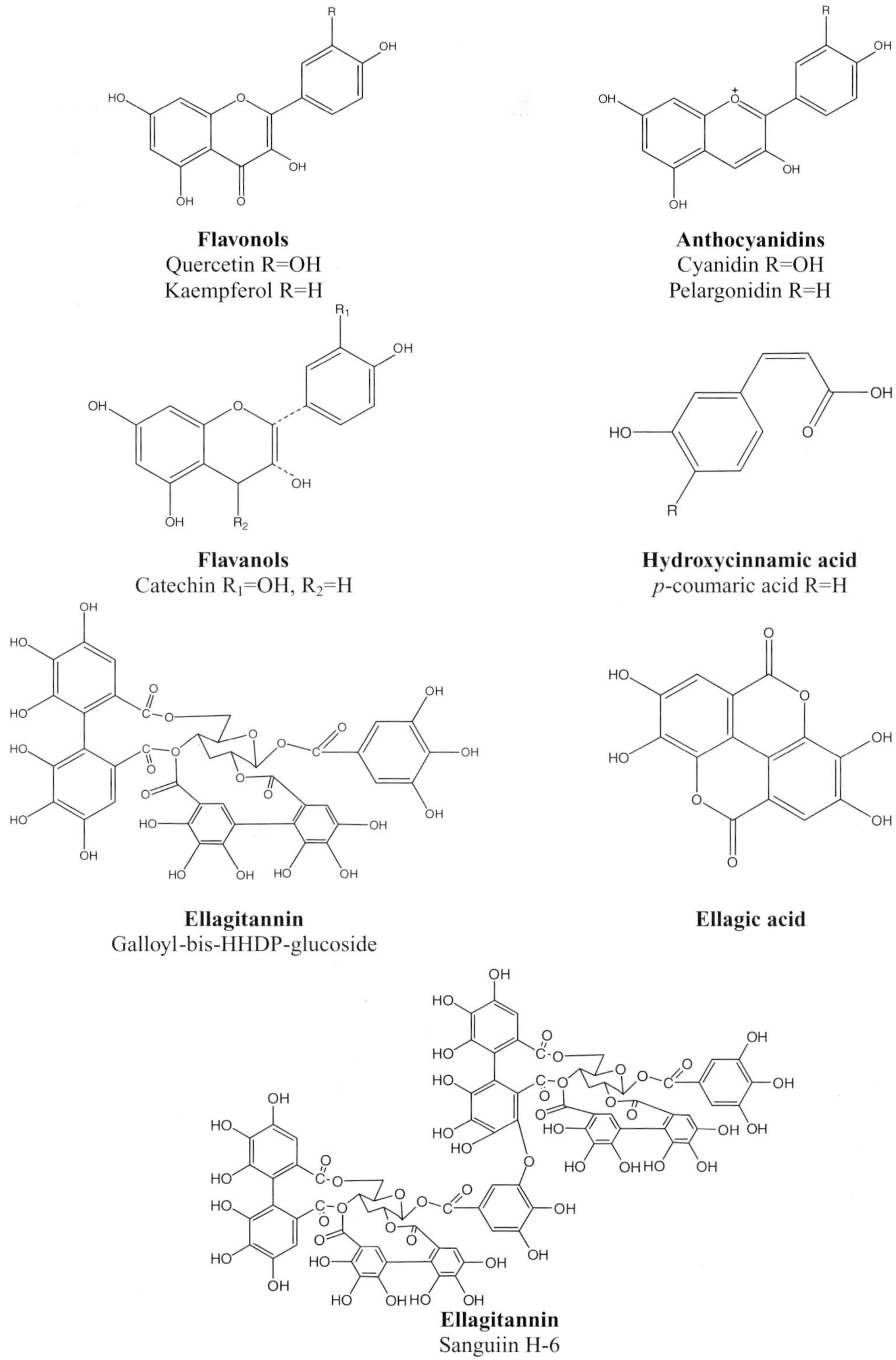

Figure 9.1 Major phytochemicals found in strawberries.

Table 9.1 Phytochemical changes before and after freeze drying of strawberries

Phytochemicals	Fresh strawberries (mg/100 g)	Dried strawberries (mg/100 g)	Changes	Reference
p-Coumaroyl glycoside	9.7	10.0	↑	[13]
Ellagic acid glycoside	13.6	9.0	↓	[13]
Quercetin	12.2	14.4	↑	[13]
Kaempferol	4.3	4.9	↑	[13]
Catechin	47.4	47.8	↑	[13]
Cyanidin-3-glucoside	8.3	9.9	↑	[13]
Pelargonidin-3-glucoside	215.4	218.9	↑	[13]
Pelargonidin-3-rutinoside	11.0	8.5	↓	[13]
Pelargonidin-3-malonyl-glucoside	59.7	64.8	↑	[13]
Total anthocyanins[a]	440	350	↓	[11, 23]
Total polyphenolic content	1906	1802	↓	[13]

All data were based on dried weight basis.
[a]mg cyanidin-3-glucoside equivalents/100 g.

to have antioxidant and cancer chemopreventive activities [26]. *In vitro* studies also indicate that ellagitannins (10–100 μM) have antiatherogenic, antithrombotic, anti-inflammatory, and antiangiogenic effects [26]. Pinto *et al.* [32] evaluated the inhibitory activities of purified ellagitannins from strawberries against α-amylase, α-glucosidase, and angiotensin I-converting enzyme (ACE) and found them to be effective in inhibiting cell proliferation and to have good potential for the management of hyperglycemia and hypertension linked to type 2 diabetes.

However, the bioavailability of the antioxidant ellagic acid and ellagitannins are low due to their poor solubility, permeability, and poor stability at a physiological pH of 7.4. That is why lower *in vivo* activities have been found despite good *in vitro* results [33].

9.2.2 Anthocyanins

Anthocyanins are the unique red and pink pigments of strawberries and are considered to be another contributing phytochemical group to the antioxidant activity and health benefits. Anthocyanins play a role in self-protection against biotic and abiotic stresses for the plant itself, and because of the unique pigmentation characteristics, they can also be useful in chemotaxonomy [34]. Although the individual and total anthocyanins contents in dried strawberries are cultivar-dependent, the main aglycones in strawberries are pelargonidin and cyanidin. Strawberry anthocyanins have been shown as strong antioxidants with values of 7156, 4922, and 5514 μM Trolox equivalents/mg for cyanidin-3-glucoside, pelargonidin, and pelargonidin-3-rutinoside, respectively [35].

Anthocyanins of strawberries were the major active components in reducing oxidative stress-induced apoptosis in PC12 cells [17]. Basu *et al.* [36] also reported that whole strawberry (fresh, juice, and freeze-dried) extracts and their purified anthocyanins rendered significant improvements in low-density lipoprotein (LDL) oxidation, lipid peroxidation, total plasma antioxidant activity, dyslipidemia, and glucose metabolism.

Anthocyanins with either a di- or a tri-saccharide were excreted in the urine primarily in the intact form. The urinary recovery of pelargonidin-3-glucoside of strawberries was reported to be 1.8% of intake after a 24-hour postingestion period, which is high for anthocyanins in plants [37]. The limited metabolism of anthocyanins that did occur was via methylation. Wu *et al.* [38] investigated the absorption and metabolism of anthocyanins in three different berries (chokeberries, black currants, or elderberries) by feeding weanling pigs, and suggested that the aglycones and the sugar moieties could alter the absorption and metabolism of anthocyanins via methylation and glucuronidation as well as by the formation of both derivatives on the same anthocyanin molecule. Later, Prior *et al.* [39] found that purified anthocyanins from freeze-dried strawberry powders, but not whole strawberry extracts, could prevent the development of dyslipidemia and obesity in mice via drinking with water. They also found that feeding purified anthocyanins from strawberries instead of the whole freeze-dried blueberry powders could reduce obesity in male mice in terms of body weight gain, body fat, and epididymal fat weight, which may have actually reduced obesity [40].

9.2.3 Flavonols and flavanols

Quercetin and kaempferol derivatives in dried strawberries belong to the subclass of flavonoids called flavonols. Flavanols, such as catechin derivatives, are also found in strawberries. Ripening and environmental factors prior to harvest and during food processing can significantly affect the flavanol content [41]. Wojdyło *et al.* [13] showed that the contents of ellagic acid and flavanol in strawberries were also affected by different drying techniques and cultivars.

Epidemiological studies have found a positive link between quercetin or kaempferol consumption and reduced risk of several diseases [42, 43]. Zhang *et al.* [35] evaluated the antioxidant and antiproliferative activities of 10 phenolic compounds from strawberry extracts (cyanidin-3-glucoside, pelargonidin, pelargonidin-3-glucoside, pelargonidin-3-rutinoside, kaempferol, quercetin, kaempferol-3-(6′-coumaroyl)glucoside), 3,4,5-trihydroxyphenylacrylic acid, glucose ester of (*E*)-*p*-coumaric acid, and ellagic acid) by the Trolox equivalent antioxidant capacity (TEAC) and luminescent adenosine-5′-triphosphate (ATP) cell viability assays. They found that these pure strawberry flavonoids not only had good antioxidant activity at 100 μg/mL but also antiproliferative activity against human oral (CAL-27, KB), colon (HT29, HCT-116), and prostate (LNCaP, DU145) cancer cells. A similar inhibitory activity was found with the crude extracts of strawberry (250 μg/mL) in the same study [35].

Flavanols affect platelet aggregation, vascular inflammation, and endothelial nitric oxide (NO) metabolism and may confer protective effects against neurodegeneration [44]. Nakamura *et al.* [45] found that catechin in green tea could suppress lipopolysaccharide (LPS)-induced bone resorption by inhibiting IL-1β production or osteoclastogenesis. Catechin was also reported to help treatment for men at high risk of prostate cancer and could ameliorate serious obesity and CVD risk factors without raising any safety concerns in Japanese children [46, 47].

The absorbability of flavonols and flavanols depends on the type, chemical binding state (aglycone or glucosides), and food matrix [48]. In general, like other flavonoids, they have a low bioavailability mainly because of the low concentrations detected in plasma and tissues and the instability under digestive conditions.

9.2.4 Hydroxycinnamic acid derivatives

Hydroxycinnamic acid derivatives, such as *p*-coumaric acid, are commonly found in dried strawberries. Among major hydroxycinnamic acids, caffeic acid showed the highest antioxidant activities in 2,2-diphenyl-2-picrylhydrazyl (DPPH) radical and oxygen radical absorbance capacity (ORAC) assays whereas *p*-coumaric acid had the least activity [49]. However, *p*-coumaric acid is the most prevalent cinnamic acid, and was shown to protect against oxidative stress and genotoxicity in cultured mammalian cells (HT-29, EMT6, SW620, LOVO, and HCT-8) [50].

p-Coumaric acid had good antiplatelet activity and was found to be available through diet, suggesting possible applications for primary prevention of vascular disease [51]. This compound was also found to protect rat hearts against doxorubicin-induced oxidative stress, and thus could potentially be used as adjuvant therapy in cancer management [52]. Garrait *et al.* [53] found that *p*-coumaric acid was absorbed by all digestive organs in rats after oral administration. Moreover, urine excretion of *p*-coumaric acid was 23% after oral administration to rats, which showed less metabolized and better health benefits than trans-cinnamic acid (0.3%) [53].

9.3 Factors affecting phytochemicals

Several factors including cultivar, fertilization, mulching, light, temperature, processing, and storage conditions have been reported to affect the contents of phytochemicals in tender fruits such as strawberries [54–56].

9.3.1 Genetics and environment

Genetics and environment are two most important factors that have been found to affect the content and composition of phytochemicals in both fresh and dried strawberries. Tulipani *et al.* [19, 57] evaluated the vitamin C, folate, *p*-coumaric acid, ellagic acid derivatives, pelargonidin, and cyanidin contents of different strawberry varieties, and found that there were significant differences ($P < 0.05$) among genotypes. Others have also found significant cultivar differences for similar phytochemical components in different strawberries [58, 59]. Breeding and biotechnological approaches are currently used to increase the content of phytochemicals in fruits, including strawberries [60].

Studies have also shown that the phytochemical content of a particular strawberry cultivar can vary significantly by other factors. Mineral composition, soil type, temperature, light, and water content are among the frequently reported factors that affect the phytochemical contents in strawberries [61–63]. In addition, agronomic practices may also impact the phytochemical compositions; for example, organically grown strawberry fruits were found to have longer shelf life, greater dry matter, higher vitamin C, antioxidant activity, ascorbic acid, and phenolic compounds than their conventional counterparts [64, 65]. This was attributed to the greater microbial functional capability and resilience to stress of the organic soils [64]. Organically grown strawberries had a higher ratio of ascorbate to dehydroascorbate and antiproliferative activity for both HT29 cells and MCF-7 cells than the conventionally produced ones [66]. Shehata *et al.* [67] showed that strawberries grown from compost fertilized plots exhibited a generally higher total yield, total soluble solid, and anthocyanin content as compared to mineral fertilizer.

9.3.2 Postharvest storage

Postharvest storage may affect the composition of some phytochemicals in strawberries, depending on storage conditions. The main causes of degradation of phytochemicals in fresh strawberries are oxygen, light, and enzymes such as polyphenol oxidase [68]. This is also affected by the biosynthesis of anthocyanins since the synthesis continues in the harvested strawberry (31% increased after 10 days at 5°C) [69]. In addition, strawberries are also dried for an extended storage period for food applications such as breakfast cereal. It is reported that freeze-dried strawberries may be stored in sealed #10 can (with included oxygen absorber packet) for 10 to 15 years under ideal storage conditions (a cool, dry place).

Applications of freeze-dried strawberry pieces in liquid carriers are limited by the high rehydration rate and dissolution of anthocyanins, which lead to collapse of the dried strawberry pieces and loss of color. The rehydration characteristics were influenced by the coating time and different drying conditions [70]. Attempts have been made to tackle this problem by coating the dried strawberry pieces. A unique formula consisting of 10% whey protein, 3% glycerol, and 10% lactose was found to be effective in keeping the integrity of the pieces [70]. By adding 3 mg/mL Na^+ and 0.5 mg/mL β-cyclodextrin in the coating solution, the authors were able to keep the original color of the dried strawberries.

9.3.3 Food processing (drying methods)

Many drying methods such as heated air drying, convective drying, freeze drying, microwave drying, osmotic dehydration, spray drying, and spouted bed drying are commonly used in preserving strawberries. Among them, convective drying is the most popular method for its low cost and easy operation. However, this method is relatively time-consuming, thus it usually results in decrease in the quality of the dried product due to the high temperature and air application [71]. Sublimation drying, for example, freeze drying was reported to be more effective in preserving valuable phytochemicals in food than traditional air drying; however, it is relatively costly [13, 55, 72]. Vacuum-microwave drying has become increasingly popular due to the many advantages such as short drying time, energy saving, and improved product quality in final dried products [73, 74]. All drying methods may reduce the contents of phytochemicals including ascorbic acid, ellagic acid, flavanol, anthocyanins, and antioxidant activity of dried strawberries [13]. However, there are considerable differences in the retention of phytochemical content and antioxidant activity in dried strawberries obtained using different methods.

Yurdugül [75] reported that freeze-dried strawberries had no difference in characteristics such as firmness, sugar content, pH, color, weight loss, dissolved solids, anthocyanin, and vitamin C content when compared with their fresh counterparts. Freeze drying was also found to better preserve total phenolic content than air drying in strawberries [71]. Similarly, freeze-dried strawberries showed greater total phenolic and anthocyanin contents and better antioxidant activity than the air-dried products [72]. Moreover, enzymes such as trehalase have been used and found to be a good drying aid for better aroma retention in freeze-dried strawberry puree [76].

Wojdyło *et al.* [13] found that vacuum-microwave drying (especially at 240 W) could actually produce high-quality strawberries of higher anthocyanins, flavanols, hydroxycinnamic acids, and flavonols contents and greater antioxidant activity, and also reduce processing times compared to freeze drying, convective drying, and vacuum drying. Böhm *et al.* [14] found that the residual phytochemical contents (phenolic compounds and anthocyanins) in vacuum-microwave-dried processing were close to those in freeze-dried products except for ascorbic acid. Attenuation of temperature can increase the stability of ascorbic acid. In the same study, it was found that the soluble phenolic and anthocyanin contents and the antioxidant activity were affected significantly by the pretreatment bath, which could lead to a significant loss during thawing of the materials [14].

Certain drying methods may be more advantageous than others over the quality of dried strawberries. A microwave-assisted hot-air drying method was found to be more effective than hot air drying, although the rehydration parameters and drying rate were dependent on the microwave energy used [55]. Krulis *et al.* [73] found that a low initial moisture content and high microwave power could lead to low density, porous structure, and optimum puffing effects of dried strawberries, but at the expense of slower moisture evaporation rate and higher energy consumption. Osmotic dehydration of strawberries improved dehydration rates than atmospheric pressure operation, resulting in a dried strawberry with intermediate moisture content, which could be used as input material of further processes [77]. A new processing method of sequential infrared and freeze-drying method (SIRFD) has been studied for producing high-quality crispy fruit pieces. Crispy strawberry pieces produced with SIRFD were found to be more desirable in color, crispness, and shrinkage, although lower in rehydration ratio than regular freeze drying [78]. In fact, freeze drying could not produce a high-crispness product as found in the same study [78]. However, more studies are needed to better examine the effects of phytochemical contents on antioxidant activities.

9.4 Health benefits of strawberries

Strawberries have been shown to have strong antioxidant activity *in vitro*, and to play important roles in reducing risks of oxidative stress-related chronic diseases *in vivo*. The majority of published literature has been carried out *in vitro* using chemical and cell culture models. However, a growing number of animal and human intervention studies have recently been conducted. The antioxidant activities and the responsible phytochemicals of strawberries are discussed later in this chapter. The focus of this section is on the *in vitro* and *in vivo* biological evidence shown by dried strawberry powder and its phytochemicals.

9.4.1 *In vitro* biological activities

The main reported *in vitro* activities of strawberries are listed in Table 9.2. Numerous studies have convincingly established the antioxidant potential of strawberries. The antioxidant activity of freeze-dried strawberry powder can be stronger than that of some of the well-known berry fruits such as Saskatoon berries and wild blueberries [23]. One serving (100 g) of strawberries in a DPPH radical scavenging assay was equivalent to 182 mg vitamin C and 483 mg vitamin E [23]. Other *in vitro* models such as the ferric reducing antioxidant power (FRAP) assay and LDL oxidation model also support the high antioxidant potential of strawberry extracts [79].

Table 9.2 Reported major *in vitro* activities of strawberries

Activity	Method	Reference
Antioxidant	DPPH radical scavenging activity	[23]
	FRAP assay	[79]
Anti-inflammatory	Inhibition of LDL oxidation	[79]
Antitumor or anti-cancer	Colon cell (HT-29, HCT116, COX-2)	[7, 16]
	Lung cell (A549)	[7]
	Stomach cell (SNU-638)	[7]
	Fibrosarcoma cell (HT-1080)	[7]
	Oral cell (KB, CAL-27)	[16]
	Breast cell (MCF-7)	[16]
	Leukemia cell	[81, 82]
	Prostate cell (LNCaP)	[16]
	Liver cell (HT-29)	[16]
Enzyme inhibitor	Inhibit α-glucosidase	[29]
	Acetylcholinesterase	[84]
	Activation of PI-3-PKB	[85]
Others	LPS-induced iNOS protein	[7]
	Inhibition of copper-induced oxidation of human LDL	[86]

DPPH, 2,2-diphenyl-1-picrylhydrazyl; FRAP, ferric reducing antioxidant power; LDL, low-density lipoprotein; PI-3, phosphatidylinositol-3; PKB, kinase/protein kinase B; LPS, lipopolysaccharide; iNOS, inducible nitric oxide synthase.

Strawberry extracts with good antioxidant activity were found to inhibit the proliferation of several cancer cells including human colon (HCT-116), lung (A549), stomach (SNU-638), and fibrosarcoma (HT-1080) cancer cell lines [7] and of human oral (KB, CAL-27), breast (MCF-7), colon (HT-29, HCT116), and prostate (LNCaP) tumor cell lines in a concentration-dependent manner (25–200 μg fresh weight/mL) [80]. A significant proapoptotic effect was also found in HT-29 [80]. Strawberry extracts at 0.5% (w/v) inhibited the proliferation of HT-29 and MCF-7 cells by 53 and 43%, respectively [66]. Strawberry extracts also inhibited LPS-stimulated NO production and suppressed LPS-induced inducible nitric oxide synthase (iNOS) protein, and mRNA expressions in mouse macrophage RAW 264.7 cells [7]. Phytochemicals in strawberries may also have high potential in the prevention or treatment of high-risk lymphoblastic leukemia and in reducing oxidative stress-induced apoptosis [81]. Induction of apoptotic cell death in the leukemia cells by strawberry powder extracts was also found in studies by Zunino *et al.* [82], who also found that phytochemicals of strawberries (quercetin, kaempferol, and ellagic acid) could induce apoptotic cell death in pre-B high-risk acute lymphoblastic leukemia via loss of nuclear DNA, loss of mitochondrial membrane potential, and activation of caspase-3 [83].

Extracts of strawberries were a good inhibitor of α-glucosidase but not of α-amylase, suggesting that strawberries may be a potential dietary source of antihyperglycemic agents [29]. The same strawberry cultivars had no significant ACE inhibitory activity, indicating low antihypertensive potential [29]. Screening of the acetylcholinesterase (AChE) inhibitory activity by Ellman's method showed that strawberries (at the concentration of 1.36 g dried powder/L) had strong inhibition, which could be helpful in preventing Alzheimer's disease [84]. Edirisinghe *et al.* [85] demonstrated that extract of freeze-dried strawberries caused endothelium-dependent relaxation (EDR), which is mediated by NO produced by activation of endothelial nitric oxide synthase (eNOS) in human umbilical vein endothelial cells (HUVECs). The phosphorylation of eNOS was caused by activation of the phosphatidylinositol-3 (PI-3)–kinase/protein kinase B (PKB) signaling pathway [85]. This effect makes strawberries a vasodilator to help lower the risk of CVD. It was also reported that strawberry juice [at the concentration of 0.01% fruit juice dissolved in phosphate buffered saline (PBS)] showed complete inhibition of copper-induced oxidation of human LDL *in vitro* [86].

9.4.2 *In vivo* effects in animals

The strawberry-fed rats were better able to retain place information (a hippocampally mediated behavior) and offered better protection against spatial deficits in the maze compared to controls [87]. Supplementation with strawberries (745 g/day) could also reduce the oxidative stress by decreasing malondialdehyde formation and protecting mononuclear blood cells against increased DNA damage in a 11 kg (lean body weight) pig model for 22 days [88]. Purified anthocyanins from freeze-dried strawberry powders were found to be key in preventing the development of dyslipidemia and obesity in mice [39]. The alcohol/water-insoluble fraction of strawberries, as well as that of the black raspberries and blueberries, was found to reduce *N*-nitrosomethylbenzylamine (NMBA)-induced tumorigenesis in rat esophagus [89]. Moreover, Stoner *et al.* [90] proved that the serum levels of interleukin-5 (IL-5) and GRO/KC (interleukin-8) might be predictive of the inhibitory effect of chemopreventive agents on rat esophageal carcinogenesis. Three extracts of strawberry cultivars (40 mg/day/kg body weight) prevented exogenous ethanol-induced damage to rats' gastric mucosa, which confirmed that a diet rich in strawberries could exert a beneficial effect in the prevention of gastric diseases [91]. Freeze-dried strawberries could also cause an EDR in rabbit aorta, which could help to lower the risk of CVD [85]. In another study, hamsters received an atherogenic diet and at the same time strawberry juice at a daily dose corresponding to the consumption of 275 mL by a 70 kg human (0.22 g strawberries/mL). After 12 weeks, the group receiving strawberry juice showed 97% reduction in aortic lipid deposition and lowered activity of hepatic antioxidant enzymes, not accompanied by lowered plasma cholesterol, which suggests that moderate consumption of strawberry juice can help prevent the development of early atherosclerosis [92]. Extracts of strawberries having significant antiplatelet and antioxidant effects *in vitro* were further assessed with a laser-induced thrombosis test in mice, and were found to significantly reduce flow-mediated vasodilation. Both the antiplatelet/antioxidant activities and the antithrombotic effect were considered important mechanisms for the potential health benefits of strawberries in reducing the risk of arterial thrombotic disease [93].

9.4.3 *In vivo* effects in humans

Some *in vivo* studies have shown promising health effects of strawberries in humans. Volunteers (three men and five women) were asked to eat 1 kg of strawberries within a space of 10 min to assess an acute effect of strawberries in healthy subjects, and it was found that antioxidant activity (FRAP values) and the ascorbate concentration in plasma significantly increased after 3 hours [94, 95]. In another study, 12 healthy volunteers were invited to consume 500 g of strawberries per day, for 16 days. This led to increases in serum antioxidants [96, 97]. The short-term or medium-term intake of strawberries could help to improve plasma antioxidant status, which could further prevent the development of risk of several chronic diseases caused by oxidative stress. A strawberry beverage containing 10 g freeze-dried fruit was shown to lower triacylglycerols (TAG) and oxidized LDL after high-fat meals, in a study involving 24 hyperlipidemic men and women (14 women, 10 men; mean age 50.9 ± 15 years) [98]. In another study, 28 hyperlipidemic subjects who had followed the strawberry-supplemented diet showed reduced oxidative damage to LDL and blood lipids [99]. In a study of 27 males and 13 females of age 24 ± 3 years, the whole strawberry juice (300 g in drinking water) was found to reduce 70% of carcinogen N-nitrosodimethylamine (NDMA) formation caused by nitrate intake (400 mg/day) [100].

Despite the above human trials, there is lack of clear epidemiologic support for direct human health benefits arising from strawberry consumption, particularly for lowering the risk of CVD. Sesso *et al.* [33] examined strawberry intake for its potential effect on CVD risk in 38,176 women and on the cross-sectional association with lipids and C-reactive protein in a subset of 26,966 women. No associations between strawberry intake and the risk of CVD incidence, lipids, or C-reactive protein were found in middle-aged and older women, though higher strawberry consumption at 2 servings/week may slightly reduce the likelihood of having elevated C-reactive protein levels. The authors suggested that further epidemiologic data are needed to better understand the health beneficial effect of strawberries in CVD prevention.

9.4.4 Bioavailability

The bioavailability of phytochemicals is defined as the fraction of an ingested nutrient that becomes available to the body for utilization in physiological functions or for storage [101]. It includes bioaccessibility, absorption, metabolism, tissue distribution, and bioactivity [102]. In a study in which concentrations of strawberry phenolic acids were analyzed 30 min after strawberries (750 g fresh weight) were consumed by four healthy human volunteers (32 ± 6 years), all cinnamic acids in the plasma were found to be low and only benzoic acid was detected in the plasma. The major free (gentisic, protocatechuic, and *p*-hydroxybenzoic acids) and conjugated (syringic acid) benzoic acids were 26–27% recovered in the urine within 5 hours. Cinnamic acids escaped absorption early in the gastrointestinal tract and went completely undetected in plasma and only trace amounts were found in the urine [103]. In another study comparing the bioavailability of phytonutrients in fresh strawberries and stored strawberries, Azzini *et al.* [104] found that the plasma vitamin C level increased significantly in the plasma after 2, 3, and 5 hours ($P < 0.05$) in subjects consuming both types of strawberries. In the same study, no quercetin or anthocyanins were found in plasma, while coumaric acid, 4-hydroxybenzoic acid (4HBA) (the major human metabolite of pelargonidin-3-glucoside), and protocatechuic acid (the major human metabolite of cyanidin-3-glucoside) were retrieved in the plasma after 8 hours. Among the urinary metabolites, pelargonidin-3-glucoside represented 91 and 95% of the total compounds after fresh and stored strawberry consumption, respectively [104]. Pelargonidin glucuronide, pelargonidin glucoside, and pelargonidin aglycone peaked in urine excreted after 2 and 24 hours and was approximately 0.9% of the pelargonidin glucoside ingested. In general, most of the phenolic compounds were detected in very small amounts both in plasma and in tissues (they are poorly absorbed from the intestine, highly metabolized, or rapidly eliminated), thereby indicating low bioavailability. The bioavailability of anthocyanins is relatively low compared to that of the other flavonoids [105]. The phytochemicals undergo similar but more complex metabolic degradations *in vivo* as compared to the *in vitro* digestion [106]. During the course of absorption, phenolic compounds are conjugated (usually methylated, sulfated, and glucuronidated) in the small intestine by the colonic microflora and later in the liver by a metabolic detoxification process that facilitates biliary and urinary elimination [6]. The low bioavailability of phytochemicals causes lower *in vivo* activities despite the good *in vitro* results in strawberries [7, 33].

Little information is available on the effects of food matrix on the bioavailability of strawberries. Better knowledge of bioavailability is essential for investigating the health effects of strawberries. Further studies are needed to focus on the absorption and metabolism of the phytochemicals in strawberries *in vivo*, and to gain insight into the mechanisms underlying the interactive effects in complex food matrices.

9.5 Conclusions

Strawberries are a rich source of antioxidative phytochemicals, particularly various phenolic acids, flavonoids, and ellagic acids. The composition of these phytochemicals may be affected by different genetics and environmental factors, and more importantly by postharvest storage and processing conditions. Drying methods can particularly alter the quality of dried strawberries, and the quality and quantity of phytochemicals and their potential health benefits. There is sufficient data showing that strawberry extracts are strong antioxidants and can inhibit proliferation of various cancer cells *in vitro*. Biomarkers related to cancer, inflammation, diabetes, and CVD were found to be positively affected by strawberry extracts; however, there seems to be a lack of strong *in vivo* evidence in support of the *in vitro* findings. Further research should focus on the bioavailability and metabolism of strawberry phytochemicals *in vivo*, and on the regulatory effect of these compounds and their metabolites on biomarkers related to chronic diseases. The strong *in vitro* antiproliferative effect suggests that dried strawberry extracts may have chemopreventive potential. Further studies on consumption of dried strawberry products and health benefits are important as dehydration technology advances.

References

1. Steinbrecher, A., Nimptsch, K., Hüsing, A., Rohrmann, S. & Linseisen, J. (2009) Dietary glucosinolate intake and risk of prostate cancer in the EPIC-Heidelberg cohort study. *International Journal of Cancer*, **125**, 2179–2186.
2. Kontou, N., Psaltopoulou, T., Panagiotakos, D., Dimopoulos, M.A. & Linos, A. (2011) The Mediterranean diet in cancer prevention: a review. *Journal of Medicinal Food*, **14**, 1065–1078.
3. Palozza, P., Mele, M.C., Cittadini, A. & Mastrantoni, M. (2011) Potential interactions of carotenoids with other bioactive food components in the prevention of chronic diseases. *Current Bioactive Compounds*, **7**, 243–261.
4. Gullett, N.P., Ruhul Amin, A.R.M., Bayraktar, S., Pezzuto, J.M., Shin, D.M., Khuri, F.R., Aggarwal, B.B., Surh, Y.J. & Kucuk, O. (2010) Cancer prevention with natural compounds. *Seminars in Oncology*, **37**, 258–281.
5. Tulipani, S., Mezzetti, B., Capocasa, F., Bompadre, S. & Battino, M.A. (2008) Antioxidants in strawberry: from the genotype to the fruit composition. *Progress in Nutrition*, **10**, 224–229.

6. Seeram, N.P. (2008) Berry fruits for cancer prevention: current status and future prospects. *Journal of Agricultural and Food Chemistry*, **56**, 630–635.
7. Hong, J.Y., Song, S.H., Park, H.J., Cho, Y.J., Pyee, J.H. & Lee, S.K. (2008) Antioxidant, antiinflamatory, and antiproliferative activities of strawberry extracts. *Biomolecules and Therapeutics*, **16**, 286–292.
8. Bagchi, D., Sen, C.K., Bagchi, M. & Atalay, M. (2004) Review: antiangiogenic, antioxidant, and anticarcinogenic properties of a novel anthocyanin-rich berry extract formula. *Biokhimiya*, **69**, 95–102.
9. Hannum, S.M. (2004) Potential impact of strawberries on human health: a review of the science. *Critical Reviews in Food Science and Nutrition*, **44**, 1–17.
10. Ali, A., Abrar, M., Sultan, M.T., Din, A. & Niaz, B. (2011) Post-harvest physicochemical changes in full ripe strawberries during cold storage. *Journal of Animal and Plant Sciences*, **21**, 38–41.
11. Goulas, V. & Manganaris, G.A. (2011) The effect of postharvest ripening on strawberry bioactive composition and antioxidant potential. *Journal of the Science of Food and Agriculture*, **91**, 1907–1914.
12. Kerch, G., Sabovics, M., Kruma, Z., Kampuse, S. & Straumite, E. (2011) Effect of chitosan and chitooligosaccharide on vitamin C and polyphenols contents in cherries and strawberries during refrigerated storage. *European Food Research and Technology*, **233**, 351–358.
13. Wojdyło, A., Figiel, A. & Oszmiański, J. (2009) Effect of drying methods with the application of vacuum microwaves on the bioactive compounds, color, and antioxidant activity of strawberry fruits. *Journal of Agricultural and Food Chemistry*, **57**, 1337–1343.
14. Böhm, V., Kühnert, S., Rohm, H. & Scholze, G. (2006) Improving the nutritional quality of microwave-vacuum dried strawberries: a preliminary study. *Food Science and Technology International*, **12**, 67–75.
15. Xu, Y., Zhang, M., Mujumdar, A., Duan, X. & Jin-Cai, S. (2006) A two-stage vacuum freeze and convective air drying method for strawberries. *Drying Technology*, **24**, 1019–1023.
16. Seeram, N.P., Lee, R., Scheuller, H.S. & Heber, D. (2006) Identification of phenolic compounds in strawberries by liquid chromatography electrospray ionization mass spectroscopy. *Food Chemistry*, **97**, 1–11.
17. Heo, H.J. & Lee, C.Y. (2005) Strawberry and its anthocyanins reduce oxidative stress-induced apoptosis in PC12 cells. *Journal of Agricultural and Food Chemistry*, **53**, 1984–1989.
18. Tsao, R., Yang, R., Sockovie, E. & Zhou, T. (2003) Antioxidant phytochemicals in antioxidant phytochemicals in cultivated and wild Canadian strawberries. *Acta Horticulturae*, **626**, 25–35.
19. Tulipani, S., Marzban, G., Herndl, A., Laimer, M., Mezzetti, B. & Battino, M. (2011) Influence of environmental and genetic factors on health-related compounds in strawberry. *Food Chemistry*, **124**, 906–913.
20. Ozcan, G. & Barringer, S. (2011) Effect of enzymes on strawberry volatiles during storage, at different ripeness level, in different cultivars, and during eating. *Journal of Food Science*, **76**, 324–333.
21. Tulipani, S., Mezzetti, B. & Battino, M. (2009) Impact of strawberries on human health: insight into marginally discussed bioactive compounds for the Mediterranean diet. *Public Health Nutrition*, **12**, 1656–1662.
22. Crozier, A., Del Rio, D. & Clifford, M.N. (2010) Bioavailability of dietary flavonoids and phenolic compounds. *Molecular Aspects of Medicine*, **31**, 446–467.
23. Li, W., Hydamaka, A.W., Lowry, L. & Beta, T. (2009) Comparison of antioxidant capacity and phenolic compounds of berries, chokecherry and seabuckthorn. *Central European Journal of Biology*, **4**, 499–506.
24. Patras, A., Brunton, N.P., Da Pieve, S. & Butler, F. (2009) Impact of high pressure processing on total antioxidant activity, phenolic, ascorbic acid, anthocyanin content and colour of strawberry and blackberry purées. *Innovative Food Science and Emerging Technologies*, **10**, 308–313.
25. Aaby, K., Skrede, G. & Wrolstad, R.E. (2005) Phenolic composition and antioxidant activities in flesh and achenes of strawberries (*Fragaria ananassa*). *Journal of Agricultural and Food Chemistry*, **53**, 4032–4040.
26. Larrosa, M., García-Conesa, M.T., Espín, J.C. & Tomás-Barberán, F.A. (2010) Ellagitannins, ellagic acid and vascular health. *Molecular Aspects of Medicine*, **31**, 513–539.
27. Yoshida, T., Amakura, Y. & Yoshimura, M. (2010) Structural features and biological properties of ellagitannins in some plant families of the order myrtales. *International Journal of Molecular Sciences*, **11**, 79–106.
28. Fotie, J. (2010) The potential of ellagic acid as a possible antimalarial drug candidate. *Current Bioactive Compounds*, **6**, 161–177.

29. Pinto, M.D.S., Kwon, Y.I., Apostolidis, E., Lajolo, F.M., Genovese, M.I. & Shetty, K. (2008) Functionality of bioactive compounds in Brazilian strawberry (*Fragaria* x *ananassa* Duch.) cultivars: evaluation of hyperglycemia and hypertension potential using *in vitro* models. *Journal of Agricultural and Food Chemistry*, **56**, 4386–4392.
30. Sharma, M., Li, L., Celver, J., Killian, C., Kovoor, A. & Seeram, N.P. (2010) Effects of fruit ellagitannin extracts, ellagic acid, and their colonic metabolite, urolithin A, on Wnt signaling. *Journal of Agricultural and Food Chemistry*, **58**, 3965–3969.
31. Devipriya, N., Sudheer, A.R. & Menon, V.P. (2007) Dose–response effect of ellagic acid on circulatory antioxidants and lipids during alcohol-induced toxicity in experimental rats. *Fundamental and Clinical Pharmacology*, **21**, 621–630.
32. Pinto, M.S., De Carvalho, J.E., Lajolo, F.M., Genovese, M.I. & Shetty, K. (2010) Evaluation of antiproliferative, anti-type 2 diabetes, and antihypertension potentials of ellagitannins from strawberries (*Fragaria* × *ananassa* Duch.) using *in vitro* models. *Journal of Medicinal Food*, **13**, 1027–1035.
33. Sesso, H.D., Gaziano, J.M., Jenkins, D.J.A. & Buring, J.E. (2007) Strawberry intake, lipids, C-reactive protein, and the risk of cardiovascular disease in women. *Journal of the American College of Nutrition*, **26**, 303–310.
34. Wold, A.-B., Rosenfeld, H.J., Holte, K., Baugerød, H., Blomhoff, R. & Haffner, K. (2004) Colour of post-harvest ripened and vine ripened tomatoes (*Lycopersicon esculentum* Mill.) as related to total antioxidant capacity and chemical composition. *International Journal of Food Science & Technology*, **39**, 295–302.
35. Zhang, Y., Seeram, N.P., Lee, R., Feng, L. & Heber, D. (2008) Isolation and identification of strawberry phenolics with antioxidant and human cancer cell antiproliferative properties. *Journal of Agricultural and Food Chemistry*, **56**, 670–675.
36. Basu, A., Rhone, M. & Lyons, T.J. (2010) Berries: emerging impact on cardiovascular health. *Nutrition Reviews*, **68**, 168–177.
37. Mullen, W., Edwards, C.A., Serafini, M. & Crozier, A. (2008) Bioavailability of pelargonidin-3-*O*-glucoside and its metabolites in humans following the ingestion of strawberries with and without cream. *Journal of Agricultural and Food Chemistry*, **56**, 713–719.
38. Wu, X., Pittman Iii, H.E., Mckay, S. & Prior, R.L. (2005) Aglycones and sugar moieties alter anthocyanin absorption and metabolism after berry consumption in weanling pigs. *Journal of Nutrition*, **135**, 2417–2424.
39. Prior, R.L., Wu, X., Gu, L., Hager, T., Hager, A., Wilkes, S. & Howard, L. (2009) Purified berry anthocyanins but not whole berries normalize lipid parameters in mice fed an obesogenic high fat diet. *Molecular Nutrition & Food Research*, **53**, 1406–1418.
40. Prior, R.L., Wu, X., Gu, L., Hager, T.J., Hager, A. & Howard, L.R. (2008) Whole berries versus berry anthocyanins: interactions with dietary fat levels in the C57BL/6J mouse model of obesity. *Journal of Agricultural and Food Chemistry*, **56**, 647–653.
41. Cermak, R., Durazzo, A., Maiani, G., Böhm, V., Kammerer, D.R., Carle, R., Wiczkowski, W., Piskula, M.K. & Galensa, R. (2009) The influence of postharvest processing and storage of foodstuffs on the bioavailability of flavonoids and phenolic acids. *Molecular Nutrition & Food Research*, **53**, 184–193.
42. Gibellini, L., Pinti, M., Nasi, M., Montagna, J.P., De Biasi, S., Roat, E., Bertoncelli, L., Cooper, E.L. & Cossarizza, A. (2011) Quercetin and cancer chemoprevention. *Evidence-Based Complementary and Alternative Medicine*. Doi: 10.1093/ecam/neq053. Published online at: http://www.hindawi.com/journals/ecam/2011/591356/, last accessed July 5, 2012.
43. Calderón-Montaño, J.M., Burgos-Morón, E., Pérez-Guerrero, C. & López-Lázaro, M. (2011) A review on the dietary flavonoid kaempferol. *Mini-Reviews in Medicinal Chemistry*, **11**, 298–344.
44. Hackman, R.M., Polagruto, J.A., Zhu, Q.Y., Sun, B., Fujii, H. & Keen, C.L. (2008) Flavanols: digestion, absorption and bioactivity. *Phytochemistry Reviews*, **7**, 195–208.
45. Nakamura, H., Ukai, T., Yoshimura, A., Kozuka, Y., Yoshioka, H., Yoshinaga, Y., Abe, Y. & Hara, Y. (2010) Green tea catechin inhibits lipopolysaccharide-induced bone resorption *in vivo*. *Journal of Periodontal Research*, **45**, 23–30.
46. Bettuzzi, S., Rizzi, F. & Belloni, L. (2007) Clinical relevance of the inhibitory effect of green tea catechins (GTCs) on prostate cancer progression in combination with molecular profiling of catechin-resistant tumors: an integrated view. *Polish Journal of Veterinary Sciences*, **10**, 57–60.
47. Matsuyama, T., Tanaka, Y., Kamimaki, I., Nagao, T. & Tokimitsu, I. (2008) Catechin safely improved higher levels of fatness, blood pressure, and cholesterol in children. *Obesity*, **16**, 1338–1348.

48. Bohn, T. (2010) Isoflavone bioavailability from foods and supplements Dietary factors impacting utilization. *Agro Food Industry Hi-Tech*, **21**, 59–62.
49. Shahidi, F. & Chandrasekara, A. (2010) Hydroxycinnamates and their *in vitro* and *in vivo* antioxidant activities. *Phytochemistry Reviews*, **9**, 147–170.
50. Ferguson, L.R., Zhu, S.T. & Harris, P.J. (2005) Antioxidant and antigenotoxic effects of plant cell wall hydroxycinnamic acids in cultured HT-29 cells. *Molecular Nutrition & Food Research*, **49**, 585–593.
51. Luceri, C., Giannini, L., Lodovici, M., Antonucci, E., Abbate, R., Masini, E. & Dolara, P. (2007) *p*-Coumaric acid, a common dietary phenol, inhibits platelet activity *in vitro* and *in vivo*. *British Journal of Nutrition*, **97**, 458–463.
52. Abdel-Wahab, M.H., El-Mahdy, M.A., Abd-Ellah, M.F., Helal, G.K., Khalifa, F. & Hamada, F.M.A. (2003) Influence of *p*-coumaric acid on doxorubicin-induced oxidative stress in rat's heart. *Pharmacological Research*, **48**, 461–465.
53. Garrait, G., Jarrige, J.F., Blanquet, S., Beyssac, E., Cardot, J.M. & Alric, M. (2006) Gastrointestinal absorption and urinary excretion of *trans*-cinnamic and *p*-coumaric acids in rats. *Journal of Agricultural and Food Chemistry*, **54**, 2944–2950.
54. Piljac-Žegarac, J. & Šamec, D. (2011) Antioxidant stability of small fruits in postharvest storage at room and refrigerator temperatures. *Food Research International*, **44**, 345–350.
55. Askari, G.R., Emam-Djomeh, Z. & Mousavi, S.M. (2009) An investigation of the effects of drying methods and conditions on drying characteristics and quality attributes of agricultural products during hot air and hot air/microwave-assisted dehydration. *Drying Technology*, **27**, 831–841.
56. Anttonen, M.J. & Karjalainen, R.O. (2009) Evaluation of means to increase the content of bioactive phenolic compounds in soft fruits. *Acta Horticulturae*, **839**, 309–314.
57. Tulipani, S., Romandini, S., Suarez, J.M.A., Capocasa, F., Mezzetti, B., Busco, F., Bamonti, F., Novembrino, C. & Battino, M. (2008) Folate content in different strawberry genotypes and folate status in healthy subjects after strawberry consumption. *BioFactors*, **34**, 47–55.
58. Alvarengarocha, D., Patto De Abreu, C.M., Correa, A., Dos Santos, C.D. & Nascimento Da Fonseca, E.W. (2008) Comparative analysis of functional nutrients in strawberries of different cultivars in the region of Lavras-MG. *Revista Brasileira de Fruticultura*, **30**, 1124–1128.
59. Wang, S.Y. & Lewers, K.S. (2007) Antioxidant capacity and flavonoid content in wild strawberries. *Journal of the American Society for Horticultural Science*, **132**, 629–637.
60. Battino, M., Beekwilder, J., Denoyes-Rothan, B., Laimer, M., Mcdougall, G.J. & Mezzetti, B. (2009) Bioactive compounds in berries relevant to human health. *Nutrition Reviews*, **67**, 145–150.
61. Mackenzie, S.J., Chandler, C.K., Hasing, T. & Whitaker, V.M. (2011) The role of temperature in the late-season decline in soluble solids content of strawberry fruit in a subtropical production system. *HortScience*, **46**, 1562–1566.
62. Wu, C.C., Hsu, S.T., Chang, M.Y. & Fang, W. (2011) Effect of light environment on runner plant propagation of strawberry. *Acta Horticulturae*, **907**, 297–302.
63. Anttonen, M.J., Hoppula, K.I., Nestby, R., Verheul, M.J. & Karjalainen, R.O. (2006) Influence of fertilization, mulch color, early forcing, fruit order, planting date, shading, growing environment, and genotype on the contents of selected phenolics in strawberry (*Fragaria* × *ananassa* Duch.) fruits. *Journal of Agricultural and Food Chemistry*, **54**, 2614–2620.
64. Johnson, I.T. (2002) Glucosinolates in the human diet. Bioavailability and implications for health. *Phytochemistry Reviews*, **1**, 183–188.
65. Balci, G. & Demirsoy, H. (2008) Effect of organic and conventional growing systems with different mulching on yield and fruit quality in strawberry cvs. Sweet Charlie and Camarosa. *Biological Agriculture and Horticulture*, **26**, 121–129.
66. Olsson, M.E., Andersson, C.S., Oredsson, S., Berglund, R.H. & Gustavsson, K.E. (2006) Antioxidant levels and inhibition of cancer cell proliferation *in vitro* by extracts from organically and conventionally cultivated strawberries. *Journal of Agricultural and Food Chemistry*, **54**, 1248–1255.
67. Shehata, S.A., Gharib, A.A., El-Mogy, M.M., Abdel Gawad, K.F. & Shalaby, E.A. (2011) Influence of compost, amino and humic acids on the growth, yield and chemical parameters of strawberries. *Journal of Medicinal Plant Research*, **5**, 2304–2309.
68. Holcroft, D.M. & Kader, A.A. (1999) Carbon dioxide-induced changes in color and anthocyanin synthesis of stored strawberry fruit. *HortScience*, **34**, 1244–1248.

69. Gil, M.I., Holcroft, D.M. & Kader, A.A. (1997) Changes in strawberry anthocyanins and other polyphenols in response to carbon dioxide treatments. *Journal of Agricultural and Food Chemistry*, **45**, 1662–1667.
70. Huang, L.L., Zhang, M., Yan, W.Q., Mujumdar, A.S. & Sun, D.F. (2009) Effect of coating on post-drying of freeze-dried strawberry pieces. *Journal of Food Engineering*, **92**, 107–111.
71. Asami, D.K., Hong, Y.J., Barrett, D.M. & Mitchell, A.E. (2003) Comparison of the total phenolic and ascorbic acid content of freeze-dried and air-dried marionberry, strawberry, and corn grown using conventional, organic, and sustainable agricultural practices. *Journal of Agricultural and Food Chemistry*, **51**, 1237–1241.
72. Michalczyk, M., Macura, R. & Matuszak, I. (2009) The effect of air-drying, freeze-drying and storage on the quality and antioxidant activity of some selected berries. *Journal of Food Processing and Preservation*, **33**, 11–21.
73. Krulis, M., Kühnert, S., Leiker, M. & Rohm, H. (2005) Influence of energy input and initial moisture on physical properties of microwave-vacuum dried strawberries. *European Food Research and Technology*, **221**, 803–808.
74. Duan, X., Zhang, M., Mujumdar, S. & Wang, R. (2010) Trends in microwave-assisted freeze drying of foods. *Drying Technology*, **28**, 444–453.
75. Yurdugül, S. (2008) An evaluation of the retention of quality characteristics in fresh and freeze-dried alpine strawberries. *International Journal of Food Science & Technology*, **43**, 865–870.
76. Galmarini, M.V., Van Baren, C., Zamora, M.C., Chirife, J., Di Leo Lira, P. & Bandoni, A. (2011) Impact of trehalose, sucrose and/or maltodextrin addition on aroma retention in freeze dried strawberry puree. *International Journal of Food Science & Technology*, **46**, 1337–1345.
77. Nuñez-Mancilla, Y., Perez-Won, M., Vega-Gálvez, A., Arias, V., Tabilo-Munizaga, G., Briones-Labarca, V., Lemus-Mondaca, R. & Di Scala, K. (2011) Modeling mass transfer during osmotic dehydration of strawberries under high hydrostatic pressure conditions. *Innovative Food Science and Emerging Technologies*, **12**, 338–343.
78. Shih, C., Pan, Z., Mchugh, T.H., Wood, D. & Hirschberg, E. (2008) Sequential infrared radiation and freeze-drying method for producing crispy strawberries. *Transactions of the ASABE*, **51**, 205–216.
79. Jiménez-Escrig, A. (2007) Multifunctional *in vitro* antioxidant evaluation of strawberry (*Fragaria virginiana* Dutch.). *International Journal of Food Sciences and Nutrition*, **58**, 629–636.
80. Seeram, N.P., Adams, L.S., Zhang, Y., Lee, R., Sand, D., Scheuller, H.S. & Heber, D. (2006) Blackberry, black raspberry, blueberry, cranberry, red raspberry, and strawberry extracts inhibit growth and stimulate apoptosis of human cancer cells *in vitro*. *Journal of Agricultural and Food Chemistry*, **54**, 9329–9339.
81. Ho, J.H. & Chang, Y.L. (2005) Strawberry and its anthocyanins reduce oxidative stress-induced apoptosis in PC12 cells. *Journal of Agricultural and Food Chemistry*, **53**, 1984–1989.
82. Zunino, S.J., Zhang, Y., Seeram, N.P. & Storms, D.H. (2010) Berry fruit extracts inhibit growth and induce apoptosis of high-risk acute lymphoblastic leukemia cells *in vitro*. *Journal of Functional Foods*, **2**, 187–195.
83. Zunino, S.J., Storms, D.H., Zhang, Y. & Seeram, N.P. (2009) Growth arrest and induction of apoptosis in high-risk leukemia cells by strawberry components *in vitro*. *Journal of Functional Foods*, **1**, 153–160.
84. Kulisic-Bilusic, T., Katalinić, V., Dragović-Uzelac, V., Ljubenkov, I., Kriško, A., Dejanović, B., Jukić, M., Politeo, O., Pifat, G. & Miloš, M. (2008) Antioxidant and acetylcholinesterase inhibiting activity of several aqueous tea infusions *in vitro*. *Food Technology and Biotechnology*, **46**, 368–375.
85. Edirisinghe, I., Burton-Freeman, B., Varelis, P. & Kappagoda, T. (2008) Strawberry extract caused endothelium-dependent relaxation through the activation of PI3 Kinase/Akt. *Journal of Agricultural and Food Chemistry*, **56**, 9383–9390.
86. Kulisic-Bilusic, T., Schnäbele, K., Schmöller, I., Dragovic-Uzelac, V., Krisko, A., Dejanovic, B., Milos, M. & Pifat, G. (2009) Antioxidant activity versus cytotoxic and nuclear factor kappa B regulatory activities on HT-29 cells by natural fruit juices. *European Food Research and Technology*, **228**, 417–424.
87. Shukitt-Hale, B., Carey, A.N., Jenkins, D., Rabin, B.M. & Joseph, J.A. (2007) Beneficial effects of fruit extracts on neuronal function and behavior in a rodent model of accelerated aging. *Neurobiology of Aging*, **28**, 1187–1194.
88. Pajk, T., Rezar, V., Levart, A. & Salobir, J. (2006) Efficiency of apples, strawberries, and tomatoes for reduction of oxidative stress in pigs as a model for humans. *Nutrition*, **22**, 376–384.

89. Wang, L.S., Hecht, S., Carmella, S., Seguin, C., Rocha, C., Yu, N., Stoner, K., Chiu, S. & Stoner, G. (2010) Berry ellagitannins may not be sufficient for prevention of tumors in the rodent esophagus. *Journal of Agricultural and Food Chemistry*, **58**, 3992–3995.
90. Stoner, G.D., Wang, L.S., Seguin, C., Rocha, C., Stoner, K., Chiu, S. & Kinghorn, A.D. (2010) Multiple berry types prevent *N*-nitrosomethylbenzylamine-induced esophageal cancer in rats. *Pharmaceutical Research*, **27**, 1138–1145.
91. Alvarez-Suarez, J.M., Dekanski, D., Ristić, S., Radonjić, N.V., Petronijević, N.D., Giampieri, F., Astolfi, P., González-Paramás, A.M., Santos-Buelga, C., Tulipani, S., Quiles, J.L., Mezzetti, B. & Battino, M. (2011) Strawberry polyphenols attenuate ethanol-induced gastric lesions in rats by activation of antioxidant enzymes and attenuation of MDA increase. *PLoS ONE*, **6**, 1–11.
92. Rouanet, J.M., Décordé, K., Rio, D.D., Auger, C., Borges, G., Cristol, J.P., Lean, M.E.J. & Crozier, A. (2010) Berry juices, teas, antioxidants and the prevention of atherosclerosis in hamsters. *Food Chemistry*, **118**, 266–271.
93. Naemura, A., Mitani, T., Ijiri, Y., Tamura, Y., Yamashita, T., Okimura, M. & Yamamoto, J. (2005) Anti-thrombotic effect of strawberries. *Blood Coagulation and Fibrinolysis*, **16**, 501–509.
94. Tulipani, S., Romandini, S., Alvarez Suarez, J., Busco, F., Mezzetti, B. & Battino, M. (2009) Strawberry consumption and antioxidant status in healthy human subjects. *Progress in Nutrition*, **11**, 178–182.
95. Tulipani, S., Romandini, S., Busco, F., Bompadre, S., Mezzetti, B. & Battino, M. (2009) Ascorbate, not urate, modulates the plasma antioxidant capacity after strawberry intake. *Food Chemistry*, **117**, 181–188.
96. Tulipani, S., Romandini, S., Battino, M., Bompadre, S., Capocasa, F. & Mezzetti, B. (2009) Effects of strawberry consumption on plasma antioxidant status and parameters of resistance to oxidative stress: preliminary evidence from human subjects. *Acta Horticulturae*, **842**, 873–876.
97. Tulipani, S., Alvarez-Suarez, J.M., Busco, F, Busco, F., Bompadre, S., Quiles, J.L., Mezzetti, B. & Battino, M. (2011) Strawberry consumption improves plasma antioxidant status and erythrocyte resistance to oxidative haemolysis in humans. *Food Chemistry*, **128**, 180–186.
98. Burton-Freeman, B., Linares, A., Hyson, D. & Kappagoda, T. (2010) Strawberry modulates LDL oxidation and postprandial lipemia in response to high-fat meal in overweight hyperlipidemic men and women. *Journal of the American College of Nutrition*, **29**, 46–54.
99. Jenkins, D.J.A., Nguyen, T.H., Kendall, C.W.C., Faulkner, D.A., Bashyam, B., Kim, I.J., Ireland, C., Patel, D., Vidgen, E., Josse, A.R., Sesso, H.D., Burton-Freeman, B., Josse, R.G., Leiter, L.A. & Singer, W. (2008) The effect of strawberries in a cholesterol-lowering dietary portfolio. *Metabolism: Clinical and Experimental*, **57**, 1636–1644.
100. Chung, M.J., Lee, S.H. & Sung, N.J. (2002) Inhibitory effect of whole strawberries, garlic juice or kale juice on endogenous formation of *N*-nitrosodimethylamine in humans. *Cancer Letters*, **182**, 1–10.
101. Fraser, P.D. & Bramley, P.M. (2004) The biosynthesis and nutritional uses of carotenoids. *Progress in Lipid Research*, **43**, 228–265.
102. Fernández-García, E., Carvajal-Lérida, I., Jarén-Galán, M., Garrido-Fernández, J., Pérez-Gálvez, A. & Hornero-Méndez, D. (2012) Carotenoids bioavailability from foods: from plant pigments to efficient biological activities. *Food Research International*, **46**, 438–450.
103. Russell, W.R., Scobbie, L., Labat, A. & Duthie, G.G. (2009) Selective bio-availability of phenolic acids from Scottish strawberries. *Molecular Nutrition & Food Research*, **53**, 85–91.
104. Azzini, E., Vitaglione, P., Intorre, F, Napolitano, A., Durazzo, A., Foddai, M.S., Fumagalli, A., Catasta, G., Rossi, L., Venneria, E., Raguzzini, A., Palomba, L., Fogliano, V. & Maiani, G. (2010) Bioavailability of strawberry antioxidants in human subjects. *British Journal of Nutrition*, **104**, 1165–1173.
105. Yang, M., Koo, S.I., Song, W.O. & Chun, O.K. (2011) Food matrix affecting anthocyanin bioavailability: review. *Current Medicinal Chemistry*, **18**, 291–300.
106. Kahle, K., Kempf, M., Schreier, P, Scheppach, W., Schrenk, D., Kautenburger, T., Hecker, D., Huemmer, W., Ackermann, M. & Richling, E. (2011) Intestinal transit and systemic metabolism of apple polyphenols. *European Journal of Nutrition*, **50**, 507–522.

10 Beneficial effects of dried berry fruits in human health and disease prevention

Shirley Zafra-Stone, Manashi Bagchi, and Debasis Bagchi

10.1 Introduction

An overwhelming body of research confirms that the dietary intake of berry fruits has a beneficial and significant impact on human health, function, and disease. Berry fruits, which are commercially cultivated and commonly consumed as fresh and processed in North America, include blueberry (*Vaccinium corymbosum*), cranberry (*Vaccinium macrocarpon* American cranberry and *Vaccinium oxycoccus* European cranberry), red raspberry (*Rubus idaeus*), blackberry (*Rubus* spp.), black raspberry (*Rubus occidentalis*), and strawberry (*Fragaria ananassa*). Other berry fruits, which are less known but consumed in the traditional diets of North American tribal communities, include chokecherry (*Prunus virginiana*), highbush cranberry (*Viburnum trilobum*), serviceberry (*Amelanchier alnifolia*), and silver buffaloberry (*Shepherdia argentea*). In addition, berry fruits such as arctic bramble (*Rubus articus*), and bilberries (*Vaccinuim myrtillus*; also known as bog whortleberries), black currant (*Ribes nigrum*), boysenberries (*Rubus* spp.), cloudberries (*Rubus chamaemorus*), crowberries (*Empetrum nigrum*, *Empetrum hermaphroditum*), elderberries (*Sambucus* spp.), gooseberry (*Ribes uva-crispa*), lingonberries (*Vaccinium vitis-idaea*), loganberry (*Rubus loganobaccus*), marionberries (*Rubus* spp.), Rowan berries (*Sorbus* spp.), as well as sea buckthorn (*Hippophae rhamnoides*), are also popularly consumed in other parts of the world. Recently, the nutraceutical market has shown a surge in the promotion of exotic fruits and berries such as the açaí berry (*Euterpe oleraceae*), mangosteen (*Garcinia mangostana*), pomegranate (*Punica granatum*), goji berries or wolfberry (*Lycium barbarum*), and the maqui berry (*Aristotelia chilensis*).

Berry fruits contain many important dietary components: vitamins, including folate, minerals, and fiber; but it is their polyphenolic content to which their biological action is most often attributed [1]. It is thought that these effects are synergistic as opposed to being due to one constituent alone. Berry phenolics include flavonoids (anthocyanins, flavonols, and flavanols), tannins [condensed tannins (proanthocyanidins) and hydrolyzable tannins (ellagitannins and gallotannins)], stilbenoids, and phenolic acids. Anthocyanins, pigments that account for their attractive colors, are more widely studied and have a broad

Dried Fruits: Phytochemicals and Health Effects, First Edition. Edited by Cesarettin Alasalvar and Fereidoon Shahidi.

range of bioactivities including anti-oxidant, anticancer, and anti-inflammatory properties, among others.

While the majority of research has been carried out on flavonoids and in particular the anthocyanins, there is increasing interest in the tannins as well. The types of tannin molecule vary considerably among different berries. Condensed tannins [also known as oligomeric proanthocyanidin (OPC), pycnogenols, and leukocyanidins] are predominantly found and consumed in blueberries and cranberries. Hydrolyzable tannins or ellagitannins are predominantly found and consumed in blackberries, black raspberries, red raspberries, and strawberries. Thus, the class and specific chemical structures of tannins present in a particular berry type may contribute significantly to its unique biological properties. The OPC found in cranberries hold an A-type structural linkage responsible for its bacterial anti-adhesive properties [2]. Similarly, the biological effects observed in proanthocyanidin-rich blueberries and ellagitannin-rich strawberries on neuronal function and behavior in aging animals may be due to the effects of the individual classes of tannins in different regions of the brain [3]. This chapter highlights the beneficial effects of dried berry fruits in human health and disease risk reduction.

10.2 Antioxidant protection

Anthocyanins are novel anti-oxidants and potent inhibitors of lipid peroxidation as compared to other classic anti-oxidants [4–6]. Endothelial dysfunction has been proposed to play an important role in the initiation and development of vascular diseases, and anthocyanins significantly improve endothelial function [7–9]. The enrichment of endothelial cells with elderberry anthocyanins conferred significant protective effects of endothelial cells against diverse oxidative stressors [9, 10]. Vascular endothelial cells can incorporate anthocyanins into the membrane and cytosol, conferring significant protective effects against oxidative insult, which may have important implications on preserving endothelial cell functions and preventing against vascular diseases [9].

Dietary consumption of blueberry polyphenols has demonstrated significant protection against free radicals and oxidative stress within red blood cells *in vivo* [10]. Black raspberry has also been shown to provide significant anti-oxidant protection in the gut epithelium of weanling pigs due to its high anthocyanin content [11]. A comparative assessment was performed on total anti-oxidant status in serum following consumption of strawberries, spinach, red wine, or vitamin C in eight elderly women. It was concluded that the consumption of strawberries, spinach, or red wine, which are rich in anti-oxidant phenolic compounds, can increase the serum anti-oxidant capacity in humans [12]. Extensive studies were conducted on six edible berry extracts including wild blueberry, wild bilberry, cranberry, elderberry, raspberry seed, and strawberry powder, and accordingly a novel synergistic combination of these six berry extracts was developed, OptiBerry, which demonstrated optimal oxygen radical absorbance capacity (ORAC) value in conjunction with high cellular uptake and low cytotoxicity, as shown by lactate dehydrogenase (LDH) leakage [13].

10.3 Cardiovascular health and metabolic syndrome

Berries show substantial cardioprotective benefits due to their high polyphenol content. Few studies have investigated their efficacy in improving features of metabolic syndrome and related cardiovascular risk factors in obesity. Berry anthocyanins act as cardioprotectant by maintaining vascular permeability, reducing inflammatory responses and platelet aggregation, and offer superior vascular protection as compared to other cardioprotective drugs [8, 9]. Hypertension, atherosclerosis, and arteriosclerosis can reduce the flexibility of arterial walls, which contributes to poor blood flow and plaque formation [8, 9]. Rat aortas exposed to anthocyanin-enriched blueberry extract *in vitro* exhibited relaxation caused by endothelium generated nitric oxide [14]. In another study, treatment of rats with bilberry anthocyanosides (*Vaccinium myrtillus*) for 12 days before the induction of hypertension kept the blood–brain barrier permeability normal and limited the increase in vascular permeability in the skin and the aorta wall [15]. Hamsters given oral doses (10 mg per 10 g body weight) of a commercial product containing 36% bilberry anthocyanosides for 2 or 4 weeks exhibited better capillary perfusion and fewer sticking leukocytes in the capillaries as compared to the untreated controls [16].

10.3.1 Atherosclerosis

Hypercholesterolemia, an integral constituent of the atherosclerotic index and a significant cardiovascular risk factor, is prevalent in the US population [17]. Atherosclerosis is a disease of the arteries in which fatty plaques develop on the inner arterial wall, which eventually obstructs blood flow [17]. Risk factors for atherosclerosis include genetics, diet, lifestyle, smoking, circulating lipid and cholesterol levels, and molecular and circulating signals of chronic vascular inflammation. The protective ability of anthocyanins against atherosclerosis is based partly on their anti-oxidant properties.

In animal studies, hamsters are used due to the similarities in lipid profile to hypercholesterolemic humans when fed a hypercholesterolemic diet of 0.2% cholesterol and 10% coconut oil [17, 18]. The efficacy of OptiBerry supplementation in hamsters against the incidence of atherosclerosis was assessed in male hamsters. The animals were divided into two groups of nine animals with equal weights (about 80 g each) and given powder chow containing 10% coconut oil and 0.2% cholesterol mixed with water and made into a brownie. Treated group animals were fed with chow containing 1% OptiBerry. The animals were weighed after 12-week of feeding the atherogenic diet. Plasma was obtained by cardiac puncture and analysed for triacylglycerols (TAG). All animals in the experimental group gained significantly less weight than the control group (approximately 8% decrease in weight gain) indicating a weight reduction from appetite loss or increased metabolism. Table 10.1 demonstrates the effect of OptiBerry on body weight, lipid profile, and the percentage of aorta covered with foam cells, a biomarker of atherosclerosis *in vivo*. No significant changes ($P > 0.05$) were observed in TAG levels. The atherosclerotic index (% of aorta covered with foam cells) was significantly reduced by 36.6% following supplementation of OptiBerry as compared to the untreated controls. The low-density lipoprotein (LDL) + very low-density lipoprotein (VLDL) oxidation levels in Table 10.2 show that the control oxidized at the greatest rate and OptiBerry group had a significantly lower rate of oxidation of the atherogenic lipoproteins ($P < 0.05$). Thus, OptiBerry which is high in anthocyanins and has a higher ORAC value demonstrated least oxidation of the lipoproteins. Overall, OptiBerry supplementation resulted in less body weight gain as compared to the controls, which might have happened due to its anti-angiogenic potential. Furthermore, OptiBerry may provide significant health benefits by dramatically ameliorating the incidence of atherosclerosis as demonstrated by

Table 10.1 Body weight, lipid, and % atherosclerosis data

Group	Body weight (g)	Plasma TAG (mg/dL)	Atherosclerosis (%)
Control	140 ± 4	163 ± 62	20.5 ± 1.7
OptiBerry	129 ± 3[a]	177 ± 111	13.0 ± 3.1[a]

Source: Adapted with permission from Zafra-Stone *et al.* [18].
Each value represents the mean ± standard deviation of 4–6 animals.
TAG, triacylglycerol.
[a]Significantly different from control ($P < 0.05$).

reducing the formation of foam cells as well as its reduced ability to oxidize functional lipoproteins [17, 18].

In a randomized controlled trial, the effects of blueberry supplementation were examined on features of metabolic syndrome, lipid peroxidation, and inflammation in obese men and women [19]. Forty-eight participants with metabolic syndrome ***4 males and 44 females; BMI: 37.8 ± 2.3 kg/m^2; age: 50.0 ± 3.0 y (mean ± SE)**** consumed freeze-dried blueberry beverage (50 g freeze-dried blueberries, approximately 350 g fresh blueberries) or equivalent amounts of fluids (controls, 960 mL water) daily for 8 weeks. The decreases in systolic and diastolic blood pressures were greater in the blueberry-supplemented group (26 and 24%, respectively) than in controls (21.5 and 21.2%) ($P < 0.05$), whereas the serum glucose concentration and lipid profiles were not affected. The decreases in plasma oxidized LDL, serum malondialdehyde, and hydroxynonenal concentrations were greater in the blueberry group (228 and 217%, respectively) than in the control group (29 and 29%) ($P < 0.01$). This study shows that blueberries may improve selected features of metabolic syndrome and related cardiovascular risk factors at dietary achievable doses [19].

Table 10.2 LDL + VLDL oxidation time

	Conjugate diene formation (234 nm)	
Time (s)	Control	OptiBerry
0	0	0
600	0.314	0.116[a]
900	0.459	0.143[a]
1200	0.518	0.147[a]
1500	0.547	0.142[a]
1800	0.566	0.138[a]
2100	0.569	0.128[a]
2400	0.570	0.130[a]
2700	0.563	0.110[a]
3000	0.560	0.105[a]
3300	0.554	0.093[a]
3600	0.546	0.075[a]

Source: Adapted with permission from Zafra-Stone *et al.* [18].
LDL, low-density lipoprotein; VLDL, very low-density lipoprotein.
[a]Significantly different from control ($P < 0.05$).

10.4 Neuroprotection

Brain functions such as balance, coordination, short-term memory, and information retrieval can be impaired with advancing age. Furthermore, the onset of age-related neurodegenerative diseases such as Alzheimer's or Parkinson's disease, superimposed on a declining nervous system, could exacerbate the motor and cognitive behavioral deficits. Although the mechanisms involved in the behavioral deficits during aging remain to be determined, it is clear that oxidative stress [20] and inflammation [21, 22] are involved. Increased exposure to the long-term effects of oxidative stress and inflammatory insults are thought to be contributing factors to the decline in cognitive and/or motor performance seen in aging and other neurodegenerative diseases. Research has shown that discrepancies in brain functions due to oxidative stress may arise because of degeneration in the endogenous anti-oxidant defense mechanisms [23–26] and the susceptibility of the brain to the deleterious effects of oxidative damage [27]. Research also indicates that not only is the central nervous system particularly vulnerable to oxidative stress but also this vulnerability increases during aging and may also enhance central vulnerability to inflammation [28, 29]. With age, there are increases in inflammatory mediators (e.g., cytokines) [30–32], as well as increased mobilization and infiltration of peripheral inflammatory cells, which have been shown to produce deficits in behavior similar to those observed during aging [33]. Furthermore, age-related changes in brain vulnerability to oxidative stress and inflammation may be the result of membrane changes and differential receptor sensitivity [34].

Research suggests that the polyphenolic compounds found in berry fruits, such as blueberries and strawberries, can reverse age-related and oxidative stress-induced decline in brain functions [35–37]. Berry fruits may exert their beneficial effects either through their ability to lower oxidative stress and inflammation or directly by altering the signaling involved in neuronal communication, calcium buffering ability, neuroprotective stress shock proteins, plasticity, and stress signaling pathways. Berries have been shown to enhance dopamine release in the brain, which improves the ability of brain cells to enhance intracellular communication [35–37]. Strawberry supplementation was shown to enhance striatal muscarinic receptor sensitivity, and this appeared to be reflected in the reversal of cognitive behavioral deficits [36, 37]. Strawberries and blueberries have also been shown to reverse age-induced declines in β-adrenergic receptor function in cerebellar Purkinje neurons, while blueberries prevented and/or reversed age-related declines in cerebellar noradrenergic receptor function [38].

In a series of studies conducted by Joseph *et al.* [39], it was demonstrated that dietary supplementation with berry fruits can prevent or reverse decrements in brain and behavioral aging in the rodent model. In their first study, Fischer 344 (F344) rats underwent long-term feeding from adulthood (6 months) to middle age (15 months) with a control diet or diets supplemented with vitamin E (500 IU/kg of diet) or with extracts of strawberry or spinach that contained identical anti-oxidant level to determine if the feeding would prevent age-related decrements in motor and cognitive behavior as well as brain function [39]. A number of different parameters known to be sensitive to oxidative stress were prevented by the anti-oxidant diets including (1) receptor sensitivity, as measured by oxotremorine-enhanced dopamine release in isolated striatal slices and cerebellar Purkinje cell activity; (2) calcium buffering capacity, or the ability of striatal synaptosomes to extrude calcium following depolarization, deficits of which ultimately result in reduced cellular signaling and eventually cell death; (3) changes in signal transduction assessed by carbachol-stimulated

GTPase coupling/uncoupling in striatal membranes; and (4) cognition (spatial learning and memory) as measured by Morris water maze performance [39]. Spinach-fed rats demonstrated the greatest retardation of age-effects on all parameters except GTPase activity, for which strawberry had the greatest effect; strawberry and vitamin E showed significant but equal protection against these age-induced deficits on the other parameters.

In other studies conducted by Bickford *et al.* [38, 40], they found that dietary supplementations (for 8 weeks) with spinach, strawberry, or blueberry extracts in a control diet were also effective in reversing age-related deficits in brain and behavioral function in aged (19 months) F344 rats. While all of the supplemented diets showed positive effects on cognitive behavior, rats on the blueberry diet showed the greatest increase in motor performance, carbachol-stimulated GTPase activity, and oxotremorine-enhanced dopamine release [36]. In addition, the blueberry-fed group showed no decrements in calcium recovery following exposure to an oxidative stressor [36]. However, though blueberries exhibit the highest anti-oxidant capacity of all fruits and vegetables tested, they were not equally effective in preventing/reversing age-related changes. Therefore, anti-oxidant activity alone was not predictive in assessing the potency of these compounds against certain disorders affected by aging. In fact, oxidative stress markers were only modestly reduced by the diets [36], suggesting that these fruit and vegetable polyphenolics possess a multiplicity of actions, aside from anti-oxidative, and that differences in the polyphenolic composition of these extracts could account for the positive effects observed.

A more recent study has suggested that, in addition to Morris water maze performance, cognitive declines in object recognition were effectively reversed by blueberry supplementation [41]. Furthermore, the beneficial effects of blueberries were seen even when superimposed on an already well-balanced, healthy rodent diet, which was more representative of a balanced human diet [42].

In a separate investigation, rats given intraperitoneal injections of bilberry anthocyanins (200 mg/kg/day) for 5 days had significantly more triiodothyronine (T3) in their brains than rats given only the solvent (26% alcohol). T3 enters the brain by a specific transport in the capillaries; therefore, anthocyanins may mediate T3 transport at the capillary level. Bilberry-treated animals exhibited superior memory, better vision, and better control of sensory input [37].

10.5 Anticancer activity

A large body of evidence suggests that berry fruits may have beneficial effects against several types of human cancers. The anticancer activity of berries has been related to a multitude of bioactive phytochemicals, stilbenoids, lignans, and triterpenoids and shown to reduce and repair damage resulting from oxidative stress and inflammation. In addition, berry bioactives also regulate carcinogen and xenobiotic metabolizing enzymes, various transcription and growth factors, inflammatory cytokines, and subcellular signaling pathways of cancer cell proliferation, apoptosis, and tumor angiogenesis [43]. Berry phytochemicals may also sensitize tumor cells to chemotherapeutic agents by inhibiting pathways that lead to treatment resistance and provide protection from therapy-associated toxicities.

Chemopreventive agents present in berries include vitamins A, C, and E, folic acid, β-carotene, α-carotene, calcium, selenium, lutein, β-sitosterol, stigmasterol, triterpene esters, as well as phenolics—anthocyanins, proanthocyanidins, flavonols, flavanols, ellagitannins, and phenolic acids, among others. As described earlier, the chemistry of berry phenolics directly influences their bioavailability, metabolism, and biological effects *in vivo* [44, 45]. The structural diversity of berry phenolics is observed in several ways including—their measure of oxidation and hydroxylation, their abilities to exist as stereoisomers, glycosylation by sugar moieties, and other substituents, and conjugation to form polymeric molecules, such as tannins and other derived molecules [46].

Blueberry, bilberry, cranberry, strawberry, lingonberry, tart cherry, black raspberry, and red raspberry as such, and their extracts, in either juice or dried form, exhibited potential cancer chemopreventive properties [47–49]. Freeze-dried extracts of strawberries or black raspberries along with ellagic acid displayed potent chemopreventive activity which appears to involve cellular transformation and interference of uptake, activation, detoxification, and/or intervention of DNA binding and DNA repair [47]. Research on black raspberries has shown that they inhibit azoxymethane-induced colon cancer, esophageal tumorigenesis as well as shows antiproliferative effect in liver cancer cells [47–51], while freeze-dried strawberries have been shown to be potent inhibitors of esophageal cancer [48]. An ethanolic extract of black raspberry has been demonstrated to suppress cell proliferation and nitric oxide synthase activity as well as induce both apoptosis and terminal differentiation in human oral squamous cell carcinoma cells [52].

10.5.1 Anti-angiogenic properties

"Angiogenesis" defines the growth of new blood vessels, an important natural process which takes place in the animal or human body contributing to both health and diseased conditions [53]. Angiogenesis occurs in the healthy body for healing wounds and for restoring blood flow to tissues after ischemic injury or insult, while unwanted growth of blood vessels may lead to varicose veins, tumor formation, and cancer metastases, while an anti-angiogenic approach can serve as a form of therapeutic intervention [53, 54]. The healthy body controls angiogenesis through angiogenesis-stimulating growth factors or angiogenesis inhibitors. Black raspberry extract was demonstrated to be anti-angiogenic in human tissue based on *in vitro* angiogenesis assays and can be promising for cancer therapy [55].

Vascular endothelial growth factor (VEGF) is a biomarker of angiogenesis, and plays a crucial role for the vascularization of tumors [56]. The effects of multiple berry extracts on inducible VEGF expression by human HaCaT keratinocytes were evaluated in our laboratory. Six individual berry extracts (wild blueberry, bilberry, cranberry, elderberry, raspberry seed, and strawberry) and OptiBerry a combination of berry extracts were investigated. All of the six berry extracts and OptiBerry demonstrated significant inhibition of both H_2O_2- as well as tumor necrosis factor α (TNFα)-induced VEGF expression by human keratinocytes, while OptiBerry exhibited the greatest effect [57]. In the same experimental setting, anti-oxidants such as grape seed proanthocyanidin extract (GSPE) or α-tocopherol did not influence inducible VEGF expression, while pure flavonoids such as ferulic acid, catechin, and rutin suppressed oxidant-inducible VEGF expression. OptiBerry was also shown to impair angiogenesis in a matrigel assay model [57]. Thus, structural characteristics of berry anthocyanins are responsible for their inhibitory potential on inducible VEGF expression and release [13].

The efficacy of OptiBerry was tested in a model of proliferating hemangioma, a unique model to assess *in vivo* angiogenesis. Macrophages are commonly involved in proliferating hemangiomas. The chemokine monocyte chemotactic protein-1 (MCP-1), a major accessory facilitating angiogenesis, has been shown to be responsible for recruiting macrophages to the infection or inflammation sites, and antagonists to MCP-1 are considered to be anti-angiogenic [58]. MCP-1 transcription in angiogenesis is mediated by several transcription factors among which nuclear factor-κβ (NF-κβ) is a key player. The endothelioma (EOMA) cells derived from spontaneously arising hemangioma were activated with TNFα (400 IU/mL) for 12 hours and elevated levels of basal MCP-1 transcription in these EOMA cells were observed. Pretreatment of these EOMA cells with OptiBerry significantly inhibited basal MCP-1 transcription as well as NF-κβ activity [56]. Subsequently, 8-week-old 129P3/J mice were injected (s.c.) 100 μL of EOMA cell suspension (56,106 cells) with or without OptiBerry pretreatment. OptiBerry pretreatment did not result in hemangioma formation in all the mice and significantly reduced the average mass of tumor growth below 50% [57]. Histological analysis demonstrated that OptiBerry markedly decreased infiltration of macrophages in hemangiomas [57]. Thus, both anti-oxidant and anti-angiogenic properties of edible berry anthocyanins may act synergistically to promote significant health benefits.

10.5.2 Antiproliferative cellular activities and viability

Numerous studies have demonstrated that different berry fruits might act through different mechanisms in their cancer preventive ability. Certain berry extracts and their phenolic constituents inhibit cell proliferation, modulate cell cycle arrest, and induce apoptosis in cancer cells with little or no cytotoxic effects in normal cells. Blackberry, black raspberry, blueberry, cranberry, red raspberry, and strawberry extracts have been shown to inhibit the growth of human oral, breast, colon, and prostate cancer cell lines in a dose-dependent manner [59, 60]. Strongest inhibition of cell growth was observed for the raspberry, lowbush blueberry, and cranberry. Researchers have shown that the inhibition of proliferation by the berry was independent of caspase-dependent apoptosis but appeared to involve cell-cycle arrest, as evidenced by downregulation of the expression of cyclin kinases, cdk 4, cdk 6, cyclin D1, and cyclin D3. Some of the berries also significantly inhibited the TNF-induced activation of the cyclooxygenase-2 (COX-2) enzyme expression and activation of the transcription factor NF-κB.

Researchers have also demonstrated that the berry extracts stimulated apoptosis of a human colon cancer cell line, HT29, which expresses COX-2 [61]. Several berry extracts, including strawberry and raspberry, were recently evaluated for their effects on cell viability and expression of markers of cell proliferation and apoptosis *in vitro* [62]. Researchers concluded that the berry extracts inhibited cancer cell proliferation mainly *via* the p21WAF1 (a member of the cyclin kinase inhibitors) pathway. The pro-apoptosis marker, Bax, was found to be increased in cells treated with the berry extracts in the apoptosis experiments. In addition, researchers demonstrated that in addition to anthocyanins, other phenolic compounds, such as ellagitannins and nonphenolic compounds, might also contribute greatly toward the antiproliferative activity of berries [62].

Red raspberry extract was shown to decrease the population of human HT29 colon cancer cells in the G1 phase of the cell cycle [63]. In addition, the raspberry extract imparted significant protective effects against DNA damage induced by hydrogen peroxide in the colon cancer cells. Researchers also reported that the raspberry extract significantly inhibited the invasion of HT115 colon cancer cells in a matrigel invasion assay [63].

Isolated cell lines from human oral squamous cell carcinoma tumors were recently used to investigate the effects of a freeze-dried black raspberry ethanol extract on cellular growth [52]. The study showed that the black raspberry extract suppressed cell proliferation without perturbing viability, inhibited translation of the complete angiogenic cytokine VEGF, suppressed nitric oxide synthase activity, and induced both apoptosis and terminal differentiation [52]. In another study, red raspberries showed potential in the inhibition of absorption of environmental carcinogens such as polycyclic aromatic hydrocarbons (PAHs) using a Calu-3 cell monolayer model [64]. Researchers demonstrated that the phytochemicals present in red raspberries inhibited PAH absorption across the Calu-3 cell monolayers and are likely to influence the exposure of lung epithelial cells to PAH-induced DNA damage [64].

Three cultivars of blueberries were extracted and assessed for their antiproliferative and pro-apoptotic effects on liver HepG2 cancer cells [65]. The greatest inhibitory effects were observed for the blueberry anthocyanin fractions (ranging from 70 to 150 μg/mL concentrations) with 50% inhibition of cancer cell population growth. Induction of apoptosis was assessed by DNA fragmentation, and the blueberry anthocyanin fraction showed a 2–4-fold increase in apoptosis compared to control [65]. In another study, lowbush blueberries have also been shown to inhibit the activities of enzymes, which play a significant role in cancer metastasis, such as matrix metalloproteinases (MMP) [43]. Researchers investigated the ability of flavonoid-enriched fractions from lowbush blueberry to downregulate MMP activity in DU145 human prostate cancer cells [66]. Differential downregulation of MMP was observed in cells exposed to both anthocyanin- and proanthocyanidin-enriched blueberry fractions. The possible involvement of protein kinase-C and mitogen-activated protein kinase pathways in the flavonoid mediated decreases in MMP activity was observed. Researchers concluded that the downregulation of MMP activities by the blueberry flavonoids might occur through multiple mechanisms [66].

Cranberry extracts were shown to significantly inhibit the growth of human breast cancer MCF7 cells, which was attributed to the ability of the extracts to initiate apoptosis and induce G1 phase arrest in the cell cycle [67]. Researchers isolated and identified 20 pure compounds from cranberries, including ursolic acid, quercetin, and 3,5,7,3′,4′-pentahydroxyflavonol-3-O-β-D-glucopyranoside and showed that these compounds have potent antiproliferative activities against liver HepG2 and breast MCF7 cancer cell growth [68]. Cranberry fruits have also demonstrated the potential for negating drug resistance in cancer cells. Recently, cranberry proanthocyanidin fractions have been reported to show cytotoxicity toward platinum-resistant human ovarian cancer cell lines, neuroblastoma, and prostate cancer cell lines [69]. The cranberry fractions sensitized human ovarian SKOV3 cancer cells to the platinum drug, paraplatin. A significant synergistic effect between cranberry proanthocyanidins and the chemotherapy drug was suggested as being operative [69].

Berry-derived products such as their seed flours have also been evaluated for anticancer properties. Black raspberry, red raspberry, blueberry, and cranberry seed flours were shown to inhibit the proliferation of human HT29 colon cancer cell line [70]. Researchers suggested that berry fruit seed flours might have the potential for the development of value-added products for cancer prevention and optimal health.

10.5.3 *In vivo* anticancer studies

10.5.3.1 *Animal studies*

A rodent model of human esophageal squamous cell carcinoma was used to evaluate the chemopreventive effects of freeze-dried black raspberry powder for this disease and to determine potential mechanisms of action [71]. Researchers showed that dietary freeze-dried black raspberry powder inhibited N-nitrosomethylbenzylamine (NMBA)-induced tumor development in the rat esophagus by inhibiting the formation of DNA adducts and reducing the proliferation rate of preneoplastic cells. On a molecular level, the freeze-dried black raspberry powder downregulated the expression of c-Jun, COX-2, and inducible nitric oxide synthase (iNOS). Researchers also analyzed the effect of freeze-dried black raspberry powder on angiogenesis. Freeze-dried black raspberry powder significantly suppressed VEGF expression from a (2.38 $\pm$ 0.34)-fold increase in animals treated with NMBA alone to a (1.08 $\pm$ 0.22)-fold increase in animals treated with NMBA plus freeze-dried black raspberry powder ($P < 0.005$). In addition, the microvessel density of the esophagus was decreased from 53.7 $\pm$ 5.6 vessels/cm in animals treated with NMBA alone to 22.6 $\pm$ 2.6 vessels/cm in animals treated with NMBA plus freeze-dried black raspberry powder ($P < 0.0001$). This study also showed that the downregulation of VEGF was correlated with suppression of COX-2 and iNOS. It was concluded that because high vascularity is a risk factor for metastasis and tumor recurrence, freeze-dried black raspberry powder might have cancer therapeutic effects in human esophageal cancer [71].

In an earlier study, the same researchers treated F344 rats with NMBA, three times per week for 5 weeks [72]. After one week, the animals were fed a diet containing 5% freeze-dried black raspberry powder for the duration of the bioassay (25 weeks) and were sacrificed at weeks 9, 15, and 25. The expression and enzymatic activities of COX-2 and iNOS, as well as the expression of c-Jun in the esophagi, were evaluated to investigate the potential mechanism(s) by which freeze-dried black raspberry powder modulate tumorigenesis. At week 25, freeze-dried black raspberry powder inhibited tumor multiplicity, from 3.78 $\pm$ 0.41 tumors per rat in NMBA-treated animals to 2.23 $\pm$ 0.21 tumors per rat in animals treated with NMBA plus freeze-dried black raspberry powder ($P < 0.005$). The black raspberry powder reduced mRNA and protein expression levels of COX-2, iNOS, and c-Jun as well as the level of prostaglandin E2 in preneoplastic lesions of the esophagus at week 25. The black raspberry powder inhibited mRNA expression of iNOS and c-Jun, but not COX-2, in papillomatous lesions of the esophagus. Prostaglandin E2 and total nitrite levels were also decreased by black raspberry powder in papillomas. Thus, black raspberry powder was suggested to render a novel tumor suppressive role through the inhibition of COX-2, iNOS, and c-Jun [72].

The mechanistic basis of the cancer anti-initiating effects of freeze-dried black raspberry powder by studying NMBA metabolism in esophageal explant cultures, and in liver microsomes taken from rats fed a control diet versus a control diet containing freeze-dried black raspberry powder (at 5 or 10% concentrations), were investigated [73]. At both test concentrations, dietary black raspberry powder inhibited NMBA metabolism in explants (26 and 20%, respectively) and in microsomes (22 and 28%, respectively). Researchers identified individual active components of black raspberry powder as ellagic acid, and the anthocyanins, cyanidin-3-glucoside and cyanidin-3-rutinoside. NMBA metabolism in explants was inhibited maximally by cyanidin-3-rutinoside (47%) followed by ellagic acid (33%), cyanidin-3-glucoside (23%), and then the black raspberry powder extract (11%). Similarly, in liver microsomes, the inhibition was maximal with cyanidin-3-rutinoside (47%), followed by ellagic acid (33%) and cyanidin-3-glucoside (32%). Dietary freeze-dried black raspberry powder was shown to induce glutathione-S-transferase activity in the liver [73].

In another recent study, diets containing freeze-dried black raspberries suppressed the development of NMBA-induced tumors in the rat esophagus. Using bioassay-directed fractionation, the anthocyanins in freeze-dried black raspberry powder were found to be the most active constituents for downregulation of carcinogen-induced NF-κB and activator protein-1 expression in mouse epidermal cells *in vitro*. The present study was therefore undertaken to determine if the anthocyanins contributed to the chemopreventive activity of freeze-dried black raspberry powder *in vivo*. F344 rats consumed diets containing either (a) 5% whole black raspberry powder, (b) an anthocyanin-rich fraction, (c) an organic solvent-soluble extract (a–c each contained approximately 3.8 μmol anthocyanins/g diet), (d) an organic-insoluble (residue) fraction (containing 0.02 μmol anthocyanins/g diet), (e) a hexane extract, and (f) a sugar fraction (e and f had only trace quantities of anthocyanins), all derived from BRB. Animals were fed diets two weeks before treatment with NMBA and throughout the bioassay. Control rats were treated with NMBA only. Animals were killed at week 30, and esophageal tumors were enumerated. The anthocyanin treatments (diet group a–c) were about equally effective in reducing NMBA tumorigenesis in the esophagus, indicating that the anthocyanins in black raspberry have chemopreventive potential. The organic-insoluble (residue) fraction (d) was also effective, suggesting that components other than berry anthocyanins may be chemopreventive. The hexane and sugar diets were inactive. Diet groups a, b, and d all inhibited cell proliferation, inflammation, and angiogenesis and induced apoptosis in both preneoplastic and papillomatous esophageal tissues, suggesting similar mechanisms of action by the different berry components.

10.5.3.2 *Human studies*

Increased fruit and vegetable consumption has been associated with the decreased risk of a number of cancers of epithelial origin, including esophageal cancer. As discussed above, dietary administration of lyophilized freeze-dried black raspberry powder has been shown to significantly inhibit chemically induced oral, esophageal, and colon carcinogenesis in animal models. A 6-month chemopreventive pilot study conducted by administering 32 or 45 g (female and male, respectively) of black raspberry powder to patients with Barrett's esophagus (BE), a premalignant esophageal condition in which the normal stratified squamous epithelium changes to a metaplastic columnar-lined epithelium, has been reported [74]. BE's importance lies in the fact that it confers a 30–40-fold increased risk for the development of esophageal adenocarcinoma, a rapidly increasing and extremely deadly malignancy. At the time of the publication, interim findings from 10 patients with BE supported the finding that daily consumption of black raspberry powder promoted reductions in the urinary excretion of two markers of oxidative stress, 8-epi-prostaglandin F2R and, to a lesser more variable extent, 8-hydroxy-2′-deoxy-guanosine [74].

It is noteworthy that this group of researchers has also investigated the formulation and characterization of a novel gel formulation for local delivery of the freeze-dried black raspberry powder chemopreventive compounds to human oral mucosal tissues [75]. Anthocyanins contained in mucoadhesive berry gel formulations were readily absorbed into human oral mucosa tissue as evidenced by detectable blood levels within five minutes after gel application. There was a trend for greater penetration of berry anthocyanins into tissue explants for berry gels with a final pH of 6.5 *versus* 3.5. The results from this study showed that the berry anthocyanin stability was dependent upon gel pH and storage temperature and also demonstrated that the gel composition was well suited for absorption and penetration into the target oral mucosal tissue site [75].

10.6 *Helicobacter pylori* and inflammatory response

Approximately 50% of the world's population is infected with *Helicobacter pylori*, which has been demonstrated as a causative factor for diverse gastrointestinal diseases including duodenal ulcer and gastric cancer [76]. *H. pylori* is slowly developing resistance to clarithromycin, a proven antibiotic agent against *H. pylori* infection. In our laboratory, we evaluated the *in vitro* bactericidal activities of various berry extracts including blueberry, bilberry, elderberry, cranberry, strawberry and raspberry seeds, and OptiBerry, with or without clarithromycin on *H. pylori* [76]. All samples tested, at all concentrations, inhibited the growth of *H. pylori*, compared with controls, with maximum inhibition with OptiBerry. At the lowest concentration of 0.25%, significant inhibition of *H. pylori* was observed with elderberry (30%), bilberry (50%), blueberry (50.5%), and OptiBerry (62%). A concentration-dependent increase in inhibition of *H. pylori* with the higher concentrations of 0.5 and 1% of all the berry extracts was observed. Modest increases in bactericidal effect were seen with the 0.5% concentration of strawberry, raspberry, and cranberry extracts, compared with the increases noted for elderberry, bilberry, blueberry, and OptiBerry. At the 1% concentration, all extracts showed >70% inhibition, with cranberry, elderberry, bilberry, and blueberry extracts showing >90% inhibition, and OptiBerry exhibiting 100% inhibition. The addition of clarithromycin to the 0.25% berry concentrations led to a significant increase in the bactericidal effects of the elderberry, bilberry, blueberry, and OptiBerry extracts against *H. pylori* compared with other berry extracts alone [76]. When clarithromycin was added to the 0.5% berry concentrations, a significant increase in the inhibition of *H. pylori* was observed with all the extracts tested. Finally, when clarithromycin was added to the 1% berry concentrations, >90% inhibition was noted for all the extracts, with elderberry, bilberry, blueberry, and OptiBerry exhibiting 100% inhibition [76].

10.6.1 Inhibitory effect of berry anthocyanins on *Helicobacter pylori*-induced IL-8 production in gastric MKN45 cells

In a pilot study, we assessed the effect of OptiBerry on IL-8 (one of the major mediators of the inflammatory response and also a potent angiogenic factor) inhibition in *H. pylori*-treated cultured human gastric cancer cells MKN45, and this is being reported for the first time [76]. *H. pylori* ATCC 49503 cells were used in this study. OptiBerry dramatically inhibited *H. pylori*-induced IL-8 production in MKN45 gastric cells.

Cultured human gastric MKN45 cells (16,106 cells) were grown in Roswell Park Memorial Institute (RPMI) media commonly used in human lymphoid cells culture, and treated with or without 0.5% OptiBerry, 10 ng of TNFα. Supernatants were collected at 6, 12, and 24 hours of treatment, centrifuged and IL-8 levels measured using assay kits.

A significant increase in IL-8 production was observed in cultured MKN45 following treatment with *H. pylori* cells. Approximately, 20, 22, and 25% increases in IL-8 production were observed at 6, 12, and 24 hours of treatment, respectively. Addition of OptiBerry completely inhibited IL-8 production in the *H. pylori*-treated cultured MKN45 cells at all time points. Table 10.3 demonstrates the effect of *H. pylori* on IL-8 production in cultured MKN45 cells and the time-dependent inhibition by OptiBerry [76].

Table 10.3 Effect of *Helicobacter pylori* on IL-8 production in cultured MKN45 cells and the time-dependent inhibition by OptiBerry

	IL-8 (pg/mL) at time points (h)		
	6 h	**12 h**	**24 h**
Control	413.0 ± 21.6	546.0 ± 42.0	600.0 ± 81.0
TNFα	895.3 ± 68.4[a]	1156.3 ± 170.3[a]	1296.0 ± 184.0[a]
TNFα + OptiBerry	546.0 ± 69.0	593.0 ± 74.0	546.0 ± 70.0
TNFα + *H. pylori*	1031.3 ± 126.2[a]	1406.3 ± 218.3[a]	1625.0 ± 195.4[a]
TNFα+ *H. pylori* + OptiBerry	546.0 ± 28.5	562.0 ± 67.0	578.0 ± 80.3

Source: Adapted from Chatterjee *et al.* [76] and Zafra-Stone *et al.* [18].
TNFα, tumor necrosis factor α.
[a]Significantly different from control ($P < 0.05$).

10.7 Diabetes and vision

The leaves and fruits of *Vaccinium myrtillus* (containing 25% anthocyanidins) have been used for centuries in Europe to ameliorate the symptoms of diabetes [77]. Consumption of *V. myrtillus* with breakfast cereals and other whole grain foods could add another level of protection against the onset of diabetes [77]. An aqueous alcoholic extract of *V. myrtillus* leaves produced a 26% reduction in plasma glucose levels in streptozotocin-induced diabetic rats. Plasma TAG decreased in proportion with the amount of bilberry leaf extract given to rats (1.2 or 3.0 g/kg body weight) fed a hyperlipidemic diet [77]. Antihyperglycemic potential of anthocyanins has also been demonstrated by Matsui *et al.* [78].

The berry anthocyanins appear to benefit vision in several ways, including improving night vision by enhanced generation of retinal pigments, increasing circulation within the capillaries of the retina, decreasing macular degeneration and diabetic retinopathy, and improving or preventing glaucoma, retinitis pigmentosa, and cataracts [79]. Bilberry has been demonstrated to improve eyesight particularly night vision. It is worthwhile to mention that bilberry jam was extensively used by British Air Force Pilots during World War I and II before their bombing mission. Since carotenoids with vitamin A activity are found in *Vaccinium* species, some of the benefits pertaining to vision are attributable to these compounds. A double-blind, placebo-controlled study showed that oral doses of anthocyanins are important for generation of visual purple, which helps to convert light into electrical signals for the brain. Adapto-electroretinograms of two sets of six subjects were made before treatment at 1 and 3 hours post administration. Subjects given the bilberry adapted to the light within 6.5 minutes, compared with 9 minutes for the control group. In another study, 50 patients with senile cataracts were given a combination of bilberry extract standardized to contain 25% anthocyanosides (180 mg twice daily) and vitamin E in the form of dl-tocopheryl acetate (100 mg twice daily) administered for 4 months. The progression of cataracts was stopped in 96% of the subjects treated ($n = 25$) compared to 76% in the control group ($n = 25$) [80].

In a double-blind study, 14 diabetic and/or hypertensive outpatients with vascular retinopathy underwent therapy with vanillylmandelic acid (VMA) (160 mg b.i.d.) or placebo ($n = 20$) for 1 month. At the end of the month, placebo-treated patients received the active drug for one additional month. Ophthalmoscopic and fluoroangiographic findings recovered before and after treatments showed an improvement ranging from 77 to 90% of anthocyanosides-treated patients. In another randomized, double-blind, placebo-controlled trial, 50 patients, 21 men and 29 women (mean age 67 year, range 48–81) suffering from mild senile cortical cataract underwent therapy with vitamin E plus VMA (2 tabs b.i.d.) for 4 months. VMA was able to stop lens opacity progress in 97% of the cataracts. No adverse-drug reactions were recorded [81].

Long-term administration of berry supplements including anthocyanins is safe and can inhibit the development of the early stages of diabetic retinopathy, and this finding warrants further investigation to unveil the mechanism [82, 83].

In another study, Tsuda *et al.* [84] explored the gene expression profiles in human adipocytes treated with anthocyanins and demonstrated that anthocyanins can regulate adipocytokine gene expression to positively modulate adipocyte function to control obesity and diabetes.

10.8 Conclusions

There are numerous research studies demonstrating consumption of anthocyanin-rich fruit and vegetables might slow or prevent the onset of chronic diseases. Edible anthocyanin and anthocyanidin phytopharmaceuticals exhibit complementary, pharmacologic and diverse overlapping mechanisms of action, including anti-oxidative, antibacterial, antiviral, induction of detoxification enzymes, stimulation of the immune system, reduction of platelet aggregation, modulation of cholesterol synthesis, antihypertensive, and anti-angiogenic effects [85]. Edible berry anthocyanins have been shown to inhibit cellular transformation and our study exhibited the potent inhibitory effect on inducible VEGF expression [13, 57]. It is worthwhile to mention that the number of novel anti-oxidants did not exhibit anti-angiogenic effect such as berry anthocyanins [13, 56, 57]. Thus, anti-oxidant property alone may not account for the total observed anti-angiogenic effect. OptiBerry, a novel combination of six berry anthocyanins, significantly inhibited basal MCP-1 and inducible NF-κβ transcriptions, and that EOMA cells pretreated with OptiBerry showed a diminished ability to form hemangioma [13]. Histological analysis exhibited markedly decreased infiltration of macrophages in hemangioma of berry formulation-treated mice compared to the control animals [13]. Thus, berry anthocyanins may exert novel chemoprevention, as well as anti-oxidant and anti-angiogenic properties by several mechanistic pathways. These broad spectrum beneficial and mechanistic effects have been examined primarily in *in vivo* models, experimental dietary studies in humans have also demonstrated the capacity of anthocyanin-rich fruit and vegetables and their constituents to modulate some significant benefits and disease-preventive actions. Furthermore, anthocyanins can offer an achievable, safe, and inexpensive adjunct therapy to inhibit the development of several degenerative diseases including obesity, diabetes, cardiovascular dysfunctions, cancer, and retinopathy, among others.

References

1. Seeram, N.P. (2008) Berry fruits: compositional elements, biochemical activities, and the impact of their intake on human health, performance, and disease. *Journal of Agricultural and Food Chemistry*, **56**, 627–629.
2. Howell, A.B. (2007) Bioactive compounds in cranberries and their role in prevention of urinary tract infections. *Molecular Nutrition & Food Research*, **51**, 732–737.

3. Shukitt-Hale, B., Carey, A.N., Jenkins, D., Rabin, B.M. & Joseph, J.A. (2007) Beneficial effects of fruit extracts on neuronal function and behavior in a rodent model of accelerated aging. *Neurobiology and Aging*, **28**, 1187–1194.
4. Kong, J.M., Chia, L.S., Goh, N.K., Chia, T.F. & Brouillard, R. (2003) Analysis and biological activities of anthocyanins. *Phytochemistry*, **64**, 923–933.
5. Prior, R.L. (2003) Fruits and vegetables in the prevention of cellular oxidative damage. *American Journal of Clinical Nutrition*, **78**, 570S–578S.
6. Karakaya, S. (2004) Bioavailability of phenolic compounds. *Critical Reviews in Food Science and Nutrition*, **44**, 453–464.
7. Mullen, W., McGinn, J., Lean, M.E., MacLean, M.R., Gardner, P., Duthie, G.G., Yokota, T. & Crozier, A. (2002) Ellagitannins, flavonoids, and other phenolics in red raspberries and their contribution to antioxidant capacity and vasorelaxation properties. *Journal of Agricultural and Food Chemistry*, **50**, 5191–5196.
8. Mullen, W., Stewart, A.J., Lean, M.E., Gardner, P., Duthie, G.G. & Crozier, A. (2002) Effect of freezing and storage on the phenolics, ellagitannins, flavonoids, and antioxidant capacity of red raspberries. *Journal of Agricultural and Food Chemistry*, **50**, 5197–5201.
9. Youdim, K.A., Martin, A. & Joseph, J.A. (2000) Incorporation of the elderberry anthocyanins by endothelial cells increases protection against oxidative stress. *Free Radical and Biology Medicine*, **29**, 51–60.
10. Youdim, K.A., Shukitt-Hale, B., MacKinnon, S., Kalt, W. & Joseph, J.A. (2000) Polyphenolics enhance red blood cell resistance to oxidative stress: *in vitro* and *in vivo*. *Biochimica Biophysica Acta*, **1523**, 117–122.
11. Wu, X., Pittman, H.E. & Prior, R.L. (2006) Fate of anthocyanins and antioxidant capacity in contents of the gastrointestinal tract of weanling pigs following black raspberry consumption. *Journal of Agricultural and Food Chemistry*, **54**, 583–589.
12. Cao, G., Russel, R.M., Lischner, N. & Prior, R.L. (1998) Serum antioxidant capacity is increased by consumption of strawberries, spinach, red wine or vitamin C in elderly women. *Journal of Nutrition*, **128**, 2383–2390.
13. Roy, S., Khanna, S., Alessio, H.M., Vider, J., Bagchi, D., Bagchi, M. & Sen, C.K (2002) Antiangiogenic property of edible berries. *Free Radical Research*, **36**, 1023–1031.
14. Zaragoza, F., Iglesias, I. & Benedi, J. (1985) Comparative study of the anti-aggregation effects of anthocyanosides and other agents. *Archives of Pharmacological Toxicology*, **11**, 183–188.
15. Detre, Z., Jellinek, H., Miskulin, M. & Robert, A.M. (1986) Studies on vascular permeability in hypertension: action of anthocyanosides. *Clinical Physiological Biochemistry*, **4**, 143–149.
16. Kadar, A., Robert, L., Miskulin, M., Tixier, J.M., Brechemier, D. & Robert, A.M. (1979) Influence of anthocyanoside treatment on the cholesterol-induced atherosclerosis in the rabbit. *Paroi Arterielle*, **5**, 187–205.
17. Bagchi, D., Sen, C.K., Ray, S.D., Das, D.K., Bagchi, M., Preuss, H.G. & Vinson, J.A. (2003) Molecular mechanisms of cardioprotection by a novel grape seed proanthocyanidin extract. *Mutation Research*, **523–524**, 87–97.
18. Zafra-Stone, S., Yasmin, T., Bagchi, M., Chatterjee, A., Vinson, J.A. & Bagchi, D. (2007) Berry anthocyanins as novel antioxidants in human health and disease prevention. *Molecular Nutrition & Food Research*, **51**, 675–683.
19. Basu, A., Du, M., Leyva, M.J., Sanchez, K., Betts, N.M., Wu, M., Aston, C.E. & Lyons, T.J. (2010) Blueberries decrease cardiovascular risk factors in obese men and women with metabolic syndrome. *Journal of Nutrition*, **140**, 1582–1587.
20. Shukitt-Hale, B. (1999) The effects of aging and oxidative stress on psychomotor and cognitive behavior. *Age*, **22**, 9–17.
21. Hauss-Wegrzyniak, B., Vannucchi, M.G. & Wenk, G.L. (2000) Behavioral and ultrastructural changes induced by chronic neuroinflammation in young rats. *Brain Research*, **859**, 157–166.
22. Hauss-Wegrzyniak, B., Vraniak, P. & Wenk, G.L. (1999) The effects of a novel NSAID on chronic neuroinflammation are age dependent. *Neurobiology in Aging*, **20**, 305–313.
23. Halliwell, B. (1994) Free radicals and antioxidants: a personal view. *Nutrition Review*, **52**, 253–265.
24. Harman, D. (1981) The aging process. *Proceedings of the National Academy of Sciences U.S.A.*, **78**, 7124–7128.

25. Yu, B.P. (1994) Cellular defenses against damage from reactive oxygen species. *Physiology Review*, **74**, 139–162.
26. Olanow, C.W. (1992) An introduction to the free radical hypothesis in Parkinson's disease. *Annals of Neurology*, **32**(Suppl.), 2S–9S.
27. Olanow, C.W. (1993) A radical hypothesis for neurodegeneration. *Trends Neuroscience*, **16**, 439–444.
28. Joseph, J.A., Denisova, N.A., Fisher, D., Bickford, P., Prior, R. & Cao, G. (1998) Age-related neurodegeneration and oxidative stress: putative nutritional intervention. *Neurology Clinicals*, **16**, 747–755.
29. Joseph, J.A., Shukitt-Hale, B., Denisova, N.A., Martin, A., Perry, G. & Smith, M.A. (2001) Copernicus revisited: amyloid beta in Alzheimer's disease. *Neurobiology and Aging*, **22**, 131–146.
30. Rozovsky, I., Finch, C.E. & Morgan, T.E. (1998) Age-related activation of microglia and astrocytes: *in vitro* studies show. *Neurobiology and Aging*, **19**, 97–103.
31. McGeer, P.L. & McGeer, E.G. (1995) The inflammatory response system of the brain: implications for therapy of Alzheimer and other neurodegenerative diseases. *Brain Research Reviews*, **21**, 195–218.
32. Volpato, S., Guralnik, J.M., Ferrucci, L., Balfour, J., Chaves, P., Fried, L.P. & Harris, T.B (2001) Cardiovascular disease, interleukin-6, and risk of mortality in older women: the women's health and aging study. *Circulation*, **103**, 947–953.
33. Chen, J., Buchanan, J.B., Sparkman, N.L., Godbout, J.P., Freund, G.G. & Johnson, R.W. (2008) Neuroinflammation and disruption in working memory in aged mice after acute stimulation of the peripheral innate immune system. *Brain, Behavior and Immunity*, **22**, 301–311.
34. Joseph, J.A., Denisova, N.A., Youdim, K.A., Bielinski, D., Fisher, D. & Shukitt-Hale, B. (2001) Neuronal environment and age-related neurodegenerative disease: Nutritional modification. In: *Annual Review of Gerontology and Geriatrics, Focus on Modern Topics in the Biology of Aging*, Vol. 21 (eds V.J. Cristofalo & D. Adelman). Springer Publishing, New York, pp. 195–235.
35. Wei, L., Keogh, C.L., Whitaker, V.R., Theus, M.H. & Yu, S.P. (2005) Angiogenesis and stem cell transplantation as potential treatments of cerebral ischemic stroke. *Pathophysiology*, **12**, 47–62.
36. Joseph, J.A., Shukitt-Hale, B., Denisova, N.A., Bielinski, D., Martin, A., McEwen, J.J., Bickford, P.C. (1999) Reversals of age-related declines in neuronal signal transduction, cognitive, and motor behavioral deficits with blueberry, spinach, or strawberry dietary supplementation. *Journal of Neuroscience*, **19**, 8114–8121.
37. Saija, A., Princi, P., D'Amico, N., De Pasquale, R. & Costa, G. (1990) Effect of *Vaccinium myrtillus* anthocyanins on triiodothyronine transport into brain in the rat. *Pharmacological Research*, **22**, 59–60.
38. Bickford, P.C., Gould, T., Briederick, L., Chadman, K., Pollock, A., Young, D., Shukitt-Hale, B. & Joseph, J. (2000) Antioxidant-rich diets improve cerebellar physiology and motor learning in aged rats. *Brain Research*, **866**, 211–217.
39. Joseph, J.A., Shukitt-Hale, B., Denisova, N.A., Prior, R.L., Cao, G., Martin, A., Taglialatela, G. & Bickford, P.C. (1998) Long-term dietary strawberry, spinach, or vitamin E supplementation retards the onset of age-related neuronal signal-transduction and cognitive behavioral deficits. *Journal of Neuroscience*, **18**, 8047–8055.
40. Bickford, P.C., Shukitt-Hale, B. & Joseph, J.A. (1999) Effects of aging on cerebellar noradrenergic function and motor learning: nutritional interventions. *Mechanisms of Ageing Development*, **111**, 141–154.
41. Goyarzu, P., Malin, D.H., Lau, F.C., Taglialatela, G., Moon, W.D., Jennings, R., Moy, E., Moy, D., Lippold, S., Shukitt-Hale, B. & Joseph, J.A. (2004) Blueberry supplemented diet: effects on object recognition memory and nuclear factor-kappa B levels in aged rats. *Nutrition Neuroscience*, **7**, 75–83.
42. Youdim, K.A., Shukitt-Hale, B., Martin, A., Wang, H., Denisova, N.A., Bickford, P.C. & Joseph, J.A. (2000) Short-term dietary supplementation of blueberry polyphenolics: Beneficial effects on aging brain performance and peripheral tissue function. *Nutrition Neuroscience*, **3**, 383–397.
43. Seeram, N. P. (2006) Berries. In: *Nutritional Oncology*, 2nd edn. (eds D. Heber, G. Blackburn, V.L.W. Go, & J. Milner). Academic Press, London, UK, pp. 615–625.
44. Manach, C., Scalbert, A., Morand, C., Rémésy, C. & Jimenez, L. (2004) Polyphenols: food sources and bioavailability. *American Journal of Clinical Nutrition*, **79**, 727–747.
45. Manach, C., Williamson, G., Morand, C., Scalbert, A. & Rémésy, C. (2005) Bioavailability and bioefficacy of polyphenols in humans. I. Review of 97 bioavailability studies. *American Journal of Clinical Nutrition*, **81**(Suppl. 1), 230S–242S.

46. Seeram, N.P. (2006) Bioactive polyphenols from foods and dietary supplements: challenges and opportunities. In: *Herbs: Challenges in Chemistry and Biology* (eds C.-T. Ho, M. Wang, & S. Sang). ACS Symposium Series 925, American Chemical Society, Washington, DC, pp. 25–38.
47. Xue, H., Aziz, R.M., Sun, N., Cassady, J.M., Kamendulis, L.M., Xu, Y., Stoner, G.D. & Klaunig, J.E. (2001) Inhibition of cellular transformation by berry extracts. *Carcinogenesis*, **22**, 351–356.
48. Stoner, G.D., Kresty, L.A., Carlton, P.S., Siglin, J.C. & Morse, M.A. (1999) Isothiocyanates and freeze-dried strawberries as inhibitors of esophageal cancer. *Toxicological Sciences*, **52**, 95–100.
49. Harris, G.K., Gupta, A., Nines, R.G., Kresty, L.A., Habib, S.G., Frankel, W.L., LaPerle, K., Gallaher, D.D., Schwartz, S.J. & Stoner, G.D. (2001) Effects of lyophilized black raspberries on azoxymethane induced colon cancer and 8-hydroxy-2′-deoxyguanosine levels in the Fischer 344 rat. *Nutrition Cancer*, **40**, 125–133.
50. Kresty, L.A., Morse, M.A., Morgan, C., Carlton, P.S., Lu, J., Gupta, A., Blackwood, M. & Stoner, G.D. (2001) Chemoprevention of esophageal tumorigenesis by dietary administration of lyophilized black raspberries. *Cancer Research*, **61**, 6112–6119.
51. Liu, M., Li, X.Q., Weber, C., Lee, C.Y., Brown, J. & Liu, R.H. (2002) Antioxidant and antiproliferative activities of raspberries. *Journal of Agricultural and Food Chemistry*, **50**, 2926–2930.
52. Rodrigo, K.A., Rawal, Y., Renner, R.J., Schwartz, S.J., Tian, Q., Larsen, P.E. & Mallery, S.R. (2006) Suppression of the tumorigenic phenotype in human oral squamous cell carcinoma cells by an ethanol extract derived from freeze-dried black raspberries. *Nutrition Cancer*, **54**, 58–68.
53. Auerbach, R., Auerbach, W. & Polakowski, L. (1991) Assays for angiogenesis: a review. *Pharmacology Therapeutics*, **51**, 1–11.
54. Griffioen, A.W. & Molema, G. (2000) Angiogenesis: potentials for pharmacologic intervention in the treatment of cancer, cardiovascular diseases, and chronic inflammation. *Pharmacological Reviews*, **52**, 237–268.
55. Liu, Z., Schwimer, J., Liu, D., Greenway, F.L., Anthony, C.T. & Woltering, E.A. (2005) Black raspberry extract and fractions contain angiogenesis inhibitors. *Journal of Agricultural and Food Chemistry*, **53**, 3909–3915.
56. Nguyen, M. (1997) Angiogenic factors as tumor markers. *Investigation into New Drugs*, **15**, 29–37.
57. Bagchi, D., Sen, C.K., Bagchi, M. & Atalay, M. (2004) Anti-angiogenic, antioxidant, and anti-carcinogenic properties of a novel anthocyanin-rich berry extract formula. *Biochemistry (Mosc)*, **69**, 75–80.
58. Salcedo, R., Ponce, M.L., Young, H.A., Wasserman, K., Ward, J.M., Kleinman, H.K., Oppenheim, J.J. & Murphy, W.J. (2000) Human endothelial cells express CCR2 and respond to MCP-1: direct role of MCP-1 in angiogenesis and tumor progression. *Blood*, **96**, 34–40.
59. Seeram, N.P., Adams, L.S., Zhang, Y., Sand, D. & Heber, D. (2006) Blackberry, black raspberry, blueberry, cranberry, red raspberry and strawberry extracts inhibit growth and stimulate apoptosis of human cancer cells *in vitro*. *Journal of Agricultural and Food Chemistry*, **54**, 9329–9339.
60. Boivin, D., Blanchette, M., Barrette, S., Moghrabi, A. & Beliveau, R. (2007) Inhibition of cancer cell proliferation and suppression of TNF-induced activation of NFkappaB by edible berry juice. *Anticancer Research*, **27**, 937–948.
61. Seeram, N.P., Momin, R.A., Bourquin, L.D. & Nair, M.G. (2001) Cyclooxygenase inhibitory and antioxidant cyanidin glycosides from cherries and berries. *Phytomedicine*, **8**, 362–369.
62. Wu, Q.K., Koponen, J.M., Mykkänen, H.M. & Törrönen, A.R. (2007) Berry phenolic extracts modulate the expression of p21(WAF1) and Bax but not Bcl-2 in HT-29 colon cancer cells. *Journal of Agricultural and Food Chemistry*, **55**, 1156–1163.
63. Coates, E.M., Popa, G., Gill, C.I., McCann, M.J., McDougall, G.J., Stewart, D. & Rowland, I. (2007) Colon-available raspberry polyphenols exhibit anti-cancer effects on *in vitro* models of colon cancer. *Journal of Carcinogenesis*, **6**, 4–9.
64. Mahadevan, B., Mata, J.E., Albershardt, D.J., Stevens, J.F., Pereira, C.B., Rodriguez-Proteau, R. & Baird, W.M. (2005) The effects of red raspberry extract on PAH transport across Calu-3 cell monolayer, an *in vitro* cell model. *International Journal Cancer Prevention*, **2**, 129–141.
65. Yi, W., Akoh, C.C., Fischer, J. & Krewer, G. (2006) Effects of phenolic compounds in blueberries and muscadine grapes on HepG2 cell viability and apoptosis. *Food Research International*, **39**, 628–638.
66. Matchett, M.D., MacKinnon, S.L., Sweeney, M.I., Gottschall-Pass, K.T. & Hurta, R.A. (2006) Inhibition of matrix metalloproteinase activity in DU145 human prostate cancer cells by flavonoids from lowbush

blueberry (*Vaccinium angustifolium*): possible roles for protein kinase C and mitogen-activated protein-kinase-mediated events. *Journal of Nutrition Biochemistry*, **2**, 117–125.

67. Sun, J. & Liu, R.H. (2006) Cranberry phytochemical extracts induce cell cycle arrest and apoptosis in human MCF-7 breast cancer cells. *Cancer Letters*, **241**, 124–134.
68. He, X. & Liu, R.H. (2006) Cranberry phytochemicals: isolation, structure elucidation, and their antiproliferative and antioxidant activities. *Journal of Agricultural and Food Chemistry*, **54**, 7069–7074.
69. Singh, A.P., Singh, R.K., Kalkunte, S.S., Nussbaum, R., Kim, K., Jin, H., Torres, M.S., Brard, L., Vorsa, N. (2007) *Cranberry Proanthocyanidins Sensitize Ovarian Cancer Cells to Platinum Drug*. AGFD-140; Presented at the 234th American Chemical Society National Meeting, Boston, MA, 19–23 August 2007.
70. Parry, J., Su, L., Moore, J., Cheng, Z., Luther, M., Rao, J.N., Wang, J.Y. & Yu, L.L. (2006) Chemical compositions, antioxidant capacities, and antiproliferative activities of selected fruit seed flours. *Journal of Agricultural and Food Chemistry*, **54**, 3773–3778.
71. Chen, T., Rose, M.E., Hwang, H., Nines, R.G. & Stoner, G.D. (2006) Black raspberries inhibit N-nitrosomethylbenzylamine (NMBA)-induced angiogenesis in rat esophagus parallel to the suppression of COX-2 and iNOS. *Carcinogenesis*, **27**, 2301–2307.
72. Chen, T., Hwang, H., Rose, M.E., Nines, R.G. & Stoner, G.D. (2006) Chemopreventive properties of black raspberries in N-nitrosomethylbenzylamine-induced rat esophageal tumorigenesis: downregulation of cyclooxygenase-2, inducible nitric oxide synthase, and c-Jun. *Cancer Research*, **66**, 2853–2859.
73. Reen, R.K., Nines, R. & Stoner, G.D. (2006) Modulation of N-nitrosomethylbenzylamine metabolism by black raspberries in the esophagus and liver of Fischer 344 rats. *Nutrition Cancer*, **54**, 47–57.
74. Kresty, L.A., Frankel, W.L., Hammond, C.D., Baird, M.E., Mele, J.M., Stoner, G.D. & Fromkes, J.J. (2006) Transitioning from preclinical to clinical chemopreventive assessments of lyophilized black raspberries: interim results show berries modulate markers of oxidative stress in Barrett's esophagus patients. *Nutrition Cancer*, **54**, 148–156.
75. Mallery, S.R., Stoner, G.D., Larsen, P.E., Fields, H.W., Rodrigo, K.A., Schwartz, S.J., Tian, Q., Dai, J. & Mumper, R.J. (2007) Formulation and *in-vitro* and *in-vivo* evaluation of a mucoadhesive gel containing freeze dried black raspberries: implications for oral cancer chemoprevention. *Pharmacology Research*, **24**, 728–737.
76. Chatterjee, A., Yasmin, T., Bagchi, D. & Stohs, S.J. (2004) Inhibition of *Helicobacter pylori in vitro* by various berry extracts, with enhanced susceptibility to clarithromycin. *Molecular Cellular Biochemistry*, **265**, 19–26.
77. Cignarella, A., Nastasi, M., Cavalli, E. & Puglisi, L. (1996) Novel lipid-lowering properties of *Vaccinium myrtillus* L. leaves, a traditional antidiabetic treatment, in several models of rat dyslipidaemia: a comparison with ciprofibrate. *Thrombosis Research*, **84**, 311–322.
78. Matsui, T., Ogunwande, I.A., Abesundara, K.J. & Matsumoto, K. (2006) Anti-hyperglycemic potential of natural products. *Mini Review in Medicinal Chemistry*, **6**, 349–356.
79. Camire, M.E. (2000) Bilberries and blueberries as functional foods and nutraceuticals. In: *Herbs, Botanicals & Teas* (eds G. Mazza & B.D. Ooomah). Technomic Publishing Company, Lancaster, PA, pp. 289–319.
80. Head, K.A. (2001) Natural therapies for ocular disorders, Part two:cataracts and glaucoma. *Alternatives in Medicinal Reviews*, **6**, 141–166.
81. Bravetti, G.O., Fraboni, E. & Maccolini, E. (1989) Preventive medical treatment of senile cataract with vitamin E and *Vaccinum myrtillus* anthocyanosides: clinical evaluation. *Annali di Ottalmologia e Clinica Oculistica*, **115**, 109–116.
82. Perossini, M., Guidi, G., Chiellini, S. & Siravo, D. (1987) Diabetic and hypertensive retinopathy therapy with *Vaccinum myrtillus* anthocianosides (Tegens) double blind placebo-controlled clinical trial. *Annali di Ottalmologia e Clinica Oculistica*, **12**, 1173–1190.
83. Kowluru, R.A., Tang, J. & Kern, T.S. (2001) Abnormalities of retinal metabolism in diabetes and experimental galactosemia. VII. Effect of long-term administration of antioxidants on the development of retinopathy. *Diabetes*, **50**, 1938–1942.
84. Tsuda, T., Ueno, Y., Yoshikawa, T., Kojo, H. & Osawa, T. (2006) Microarray profiling of gene expression in human adipocytes in response to anthocyanins. *Biochemistry Pharmacology*, **71**, 1184–1197.
85. Seeram, N.P. & Heber, D. (2006) Impact of berry phytochemicals on human health: effects beyond antioxidation. In: *Lipid Oxidation and Antioxidants: Chemistry, Methodologies and Health Effects* (eds C.-T. Ho & F.S. Shahidi). ACS Symposium Series 956, American Chemical Society, Washington, DC, pp. 326–336.

Part 2
Nontropical Dried Fruits

11 Phytochemicals and health benefits of dried apple snacks

H.P. Vasantha Rupasinghe and Ajit P.K. Joshi

11.1 Introduction

Snack foods such as dehydrated chips made from fruits and vegetables make up an important part of consumer's diet in North America who are becoming more concerned about the nutritional value of foods [1]. Recently, with increasing public awareness about dietary antioxidants and biologically active phytochemicals and their associated health benefits, there is a growing demand for convenience snacks with nutritional and health benefits [2, 3]. Hence, the innovative drying processes for producing non-fried snacks could provide the consumers with nutritional benefit, convenience, and taste [3].

Preservation of phytochemicals in fruits during dehydration is important for obtaining their optimum health benefits. Furthermore, fortification of dried fruits with additional nutrients, antioxidants, and flavors can become attractive value-added food products and serve as a source of minerals, vitamins, and indispensable amino acids to meet daily dietary requirement of the public [4–9]. The appropriate use of these technologies can result in the enhancement of both quality and nutritional benefits in fruit-based snacks [4, 10–13]. In this chapter, food industry applications, phytochemicals, potential health benefits, and compositional and nutritional characteristics of dried apple snacks are discussed. Comparison is made with fresh and dried apples where it is needed.

11.2 Food applications of dried apple snacks

The total retail sale of snack foods in Canada was estimated at $981.4 million for the year 2004 and it is predicted that the domestic snack food market will increase in the coming years [14]. Revenues from Canada's snack food industry have increased from $1.2 billion in 1999 to $2.1 billion in 2008 [15]. In 2009, snack food sale in Canada was $1.4 billion [16]. A 20-year study from 1977 to 1996 on snack consumption trends in the United States showed that snacks consumed by most of the children and young people in 1996 were high

Dried Fruits: Phytochemicals and Health Effects, First Edition. Edited by Cesarettin Alasalvar and Fereidoon Shahidi.

in energy density and low in calcium density as compared to those consumed during 1977 and 1988 [17]. In another study, it was determined that some of the available snacks made from potatoes, corn, and wheat had a total fat content in the range of 9–46% [18]. High energy and high fat containing foods are considered as the leading cause of obesity among children [17], and this may further lead to several other adult onset of chronic diseases such as hypertension, diabetes, and cardiovascular disease (CVD) [19]. An increasing incidence of obesity at all age levels of North American population has encouraged introduction of non-traditional chips and snacks containing low amount of oil [14].

The snack food industry is now focusing more on manufacturing fruit- and vegetable-based products because of the associated health benefits and consumer preferences for low caloric healthy snack foods [2, 3]. Some of the health benefits associated with the intake of fruits and vegetables rich in phytochemicals, dietary fiber, and minerals, are the reduced risk of obesity, CVD, diabetes, and other chronic diseases [20]. In addition, the consumption of fruit has been found to inversely correlate with cancers in the esophagus, oral cavity, pancreas, and stomach [21]. Considering the health benefits of apples and their suitability for snack production [3], the promotion of apple-based snack products such as non-fried apple snacks could also be an alternative marketing option for the apple producing and processing industries.

11.3 Effects of drying methods and vacuum impregnation (VI) on apple phytochemicals

11.3.1 Drying methods

Most of the fruit- and vegetable-based snacks currently available in the market are processed by frying in oil. Investigations are in process to develop alternative processing methods to reduce the oil content in these fried snacks. A study was conducted on making non-fried carrot slices by vacuum frying at different temperatures (70, 90, 100, and 110°C) and times (5, 10, 15, 20, 25, and 30 min). The resultant snacks with higher oil content (22.5%) were observed to have more crispiness than other snacks with lower oil content [22]. Another study focused on vacuum frying of apple slices and to achieve the desired crispiness, high-temperature frying and a longer time (110°C for 25 min) was required which not only resulted in an increased oil gain of 39%, but also resulted in non-enzymatic browning in the fried apple slices [23]. The health concerns associated with consumption of high fat foods further emphasizes the need for exploration of non-frying methods for the development of low- or no-fat snack foods with desired crispiness and maximum retention of nutrients [19]. Thus, drying could be a promising alternative process for producing healthy snacks as it does not cause uptake of oil into the food and does allow much of the nutritional value to be maintained.

Drying of fruits is an important method of preservation and production of a wide variety of snack products. To have a desirable shelf-life of fruit products, it is necessary to reduce the moisture content to appropriate levels [24]. In turn, drying has a direct effect on the fruit's quality including nutritional and nutraceutical value, texture, taste, and shelf-life stability [24, 25]. Some of the drying methods that can be used in the fruit processing industry include drying by forced air, convection oven, microwaving, freeze-drying, fluidized-bed drying, osmotic dehydration, and vacuum drying [13, 26, 27].

Air- and oven-drying are the most commonly studied methods for drying apple tissues. However, the occurrence of shrinkage, non-enzymatic browning, and the loss of nutritional quality are recognized as major disadvantages associated with the use of these drying methods [26, 28, 29]. Air drying of apple tissues at 70°C and air velocity 1.5 m/s, resulted in breakage of the cell walls and formation of micro-cavities [29], which yields a lower quality end product. Oven drying of apple slices at 60°C resulted in poor quality in the final product because of uneven drying [30]. In oven drying, the product layer comes in direct contact with the hot surface of the dryer , which results an over-dried food product layer in contact. This may result in oxidation processes and loss of nutrients. Furthermore, high temperature application in air- and oven-drying has also been reported to result in loss of phenolic compounds [31]. The process of freeze-drying can help to retain the shape and nutrients in the final dried product [13, 26, 27]. However, freeze-dried apple slices resulted in the loss of crispiness due to sponginess of apple tissue resulting due to the formation of large cavities during freeze-drying [13, 29]. Microwave energy is another alternative method used to obtain greater quality dehydrated apple chips [32]. A method developed by Yücel *et al.* [33] applied high hydrostatic pressure (HHP) combined with conventional air drying at different pressure-time-temperature combinations (100–300 MPa for 5–45 min at 20 and 35°C) before drying. Significant increase of drying rate occurred when pressure of more than 100 MPa was applied as high pressure caused cell permeabilization. However, these alternative methods need to be explored for their impact on the phytochemicals in dried apple snacks.

The bioactive compounds (phenolic acids, anthocyanins, flavonols, and flavan-3-ols) present in apple are associated with the color, taste, and nutritional quality including their antioxidant activity [34, 35]. The impact of different drying methods on the phenolic compounds in apple tissue is compound dependent (Table 11.1). Phloridzin and quercitin-3-*O*-rhamnoside are well retained under the most commonly used drying process whereas the concentrations of catechin and epicatechin can get significantly reduced ($P < 0.05$) in oven-dried apple slices. Compared to fresh apple slices, concentration of chlorogenic acid is also significantly reduced ($P < 0.05$) in apple slices exposed to all of the drying processes. Air- and oven-drying of apple slices can cause significant loss of cyanidin-3-*O*-galactoside. However, the concentration of quercetin glycosides (with the exception of quercetin-3-*O*-rhamnoside) is well retained and shown to be significantly higher ($P < 0.05$) in vacuum-dried apple slices as compared to fresh, oven-, and air-dried apple slices.

The varying effects of different drying methods on the individual phenolic compounds in dried apple slices can be due to their different thermo-stability properties and their susceptibility to the drying conditions. Phenolic compounds such as phloretin are found to be more stable under high temperature conditions (air- and oven-drying) than chlorogenic acid, catechin, epicatechin, and cyanidin-3-*O*-galactoside. Rupasinghe *et al.* [36] found that baked muffins (at 175°C for 20 min) containing apple peel powder resulted in a 784% increase in free phloretin levels due to the thermohydrolysis of glucosides of phloretin. The loss of phenolic compounds such as catechin and epicatechin during oven drying can be attributed to the application of high temperature which has shown to induce strong decomposition of these compounds [31]. In contrast to high temperature conditions, fresh samples prepared by freeze-drying have shown to minimize the loss of flavan-3-ols as compared to flavonols. Several studies have shown that freeze-drying process can significantly reduce the concentration of flavonoids including quercetin glycosides, whereas vacuum- and oven-dryings at low temperature (48 h at 40°C) have better potential for preserving flavonoids than freeze-drying [37–39]. Cannac *et al.* [37] reported a loss of about 39% of the total flavonols during

Table 11.1 Concentration of phenolic compounds and vitamin C in fresh and dried apple slices (mg/100 g)

Phenolics and vitamin C	Apple flesh				Apple peel
	Fresh	Vacuum-dried (30°C; 15 h)	Oven-dried (70°C; 8 h)	Air-dried (60°C; 0.8 m/s; 7 h)	Oven-dried (70°C; 8 h)
Catechin	0.30 ± 0.02^{a}	0.27 ± 0.02^{ab}	0.26 ± 0.01^{b}	0.29 ± 0.01^{a}	26.53 ± 2.27
Epicatechin	6.09 ± 6.09^{a}	4.71 ± 0.92^{ab}	3.11 ± 0.25^{b}	4.50 ± 1.84^{ab}	116.08 ± 0.74
Chlorogenic acid	191.86 ± 6.95^{a}	153.04 ± 3.71^{b}	138.84 ± 7.25^{b}	136.81 ± 29.31^{b}	51.21 ± 2.67
Cyanidin-3-*O*-galactoside	4.86 ± 0.79^{a}	3.86 ± 0.43^{ab}	2.97 ± 0.43^{b}	2.88 ± 1.18^{b}	49.45 ± 3.12
Phloridzin	46.88 ± 5.57^{a}	47.23 ± 4.94^{a}	35.44 ± 3.86^{a}	43.02 ± 13.44^{a}	11.52 ± 1.34
Phloretin	0.23 ± 0.00^{b}	0.24 ± 0.01^{b}	0.32 ± 0.01^{a}	0.32 ± 0.02^{a}	2.75 ± 0.23
Quercetin-3-*O*-rutinoside	1.37 ± 1.39^{b}	4.55 ± 2.28^{a}	1.71 ± 0.19^{b}	1.84 ± 0.51^{b}	13.11 ± 1.88
Quercitin-3-*O*-galactoside	10.88 ± 3.64^{b}	23.82 ± 6.23^{a}	10.49 ± 1.68^{b}	9.58 ± 2.95^{b}	249.00 ± 9.78
Quercitin-3-*O*-glucoside	5.56 ± 2.61^{b}	11.57 ± 3.28^{a}	5.93 ± 0.22^{b}	5.44 ± 1.40^{b}	25.21 ± 3.49
Quercitin-3-*O*-rhamnoside	8.41 ± 2.21	11.66 ± 1.60	9.21 ± 0.45	8.25 ± 1.79	80.25 ± 0.89
Vitamin C	83.17 ± 9.43^{a}	$65.87 \pm 1.05b^{c}$	57.40 ± 4.15^{c}	77.39 ± 6.65^{ab}	nd

Source: Adapted with permission from Joshi *et al.* [10].
Data are expressed as mean ± SD ($n = 3$) on a dried weight basis.
Means ± SD followed by the same letter, within each row (apple flesh), are not significantly different ($P < 0.05$).
nd, not determined.

pre-freezing process and a loss of 87% of flavonols throughout the entire freeze-drying process. The variation in phenolic compounds during freeze-drying can occur due to the interconversion of compounds by chemical or biochemical decomposition when submerging the samples in liquid nitrogen before freeze-drying [31]. van Sumere *et al.* [40] reported that phenolics can be lost during freeze-drying under high vacuum conditions as some low molecular weight compounds get volatilized and concentrated in the ice trap.

Similarly, drying processes impart significant effects on vitamin C concentration of apple slices (Table 11.1). The vitamin C concentration can be well retained in both air-dried and vacuum-dried apple slices as compared to oven drying. The loss of vitamin C during thermal processing has previously been reported by several other researchers [41, 42]. The oxidation of ascorbic acid under high temperature conditions and depletion of ascorbic acid due to its utilization for oxidative protection of polyphenols are main causes for vitamin C depletion [28, 43]. It has been reported that even drying in the absence of oxygen could not prevent the loss of vitamin C in potatoes [44].

11.3.2 Vacuum impregnation (VI)

Food fortification is a common practice used to further enhance the nutritional value of food products. For producing fortified, value-added fruit products, dipping fruits in fruit juices and sugar solutions containing minerals, vitamins, and other food ingredients is a commonly used processing method [7, 45]. To enhance the efficiency of these additional nutrient uptakes, VI processes can be used. It is a method which introduces controlled quantities of a solution to the porous structure of the fruit matrix [8, 46]. This method involves immersion of high porosity food samples in a solution containing desired solutes, under vacuum conditions [47, 48]. Incorporation of nutritional and functional food ingredients using VI has been found to be more effective in improving product quality than simple immersion treatments [49]. Thus, among numerous other methods, application of VI process can play a significant role in developing functional foods with desired food ingredients as minerals, vitamins, antioxidants, and antimicrobial agents, among others [9, 12, 49]. Incorporation of nutrients using VI process is achieved more quickly as compared to atmospheric impregnation. This technique was applied for incorporating probiotics into apple cylinders using vacuum pressure of 1.5 inch of mercury for a period of 10 minutes [44]. The same amount of calcium impregnated in apple tissues under atmospheric conditions after a period of 10 hours is obtained by 10 minutes exposure to VI [6]. Food products such as apples which have a higher porosity are more suitable for VI methods [47, 48], which allows impregnation with external solution and energy saving while processing. VI can also be studied to modify the compositional, thermal, and physicochemical properties of food products which, in turn, will increase the process efficiency in a manner beneficial for attaining desired product characteristics such as improved taste, texture, and shelf-life [8, 50]. In assessing the beneficial effects of VI, use of VI pre-treatment processes appears to be a good approach to develop value-added apple-based snack products.

11.4 Antioxidant capacity of dried apple snacks

Many *in vitro* methods have been developed to analyze antioxidant activity of food products. These methods each have different merits and limitations and to use any specific one as a

Table 11.2 Total phenolic content and total antioxidant capacity in dried apple slices

Apple slices	Total phenolics (μmol GAE/100 g)	FRAP (mmol TE/100 g)	ORAC (mmol TE/100 g)
Fresh	17.55 ± 1.13	0.99 ± 0.09	9.94 ± 0.37
Vacuum-dried (30°C; 15 h)	19.07 ± 0.33	1.00 ± 0.07	11.02 ± 0.72
Oven-dried (70°C; 8 h)	17.98 ± 1.22	1.00 ± 0.08	9.09 ± 1.64
Air-dried (60°C; 0.8 m/s; 7 h)	17.96 ± 4.98	0.92 ± 0.39	6.91 ± 3.08
Oven-dried apple peel (70°C; 8 h)	17.33 ± 0.09	20.27 ± 2.18	2.64 ± 0.43

Source: Adapted with permission from Joshi *et al.* [10].
Data are expressed as mean ± SD ($n = 3$) on a dried weight basis.
GAE, gallic acid equivalents; FRAP, ferric reducing antioxidant power; TE, trolox equivalents; ORAC, oxygen radical absorbance capacity.

measure of total antioxidant capacity is sometimes misleading. The mechanisms involved in these *in vitro* methods have been generally divided into two types: hydrogen atom transfer (HAT) and single electron transfer (SET) mechanisms. The Folin–Ciocalteu assay and the ferric reducing antioxidant power (FRAP) assay are *in vitro* systems which utilize a SET mechanism. The oxygen radical absorbance capacity (ORAC) assay is an example of an *in vitro* system involving a HAT mechanism.

Dehydration process of dried apple slices had promising effects on the total phenolic content (measured by using the Folin–Ciocalteu assay) and the total antioxidant capacity measured by using both FRAP and ORAC assays as they were well retained after dehydration (Table 11.2). Similar results have been reported where the drying of red grape pomace at 60°C for 9 hours showed no effect on its phenolic content and antioxidant capacity as compared to drying at 100 and 140°C for 5 and 4 hours, respectively [51]. Higher antioxidant activity in air-dried apple cubes of seven different cultivars (100°C for 1 hour followed by 85°C at air velocity 1.75 m/s till final moisture content 7%) was observed by Sacchetti *et al.* [52]. The changes that phenolic compounds undergo during the drying process could increase the content of free phenolic compounds, which could, in turn, act as antioxidants or as new substrates for further oxidation [53, 54]. Thus, even after exposure to high temperature and atmospheric conditions during air- and oven-drying, the dried apple slices can show similar total phenolic content and total antioxidant capacity as compared to fresh apple slices.

Application of thermal treatments such as frying can significantly impact the bioactive compounds and the associated antioxidant capacity of the snacks (Table 11.3). Non-fried apple snacks showed approximately 20 times higher antioxidant capacity as compared to commercially fried potato snacks. The total phenolic content was found to be significantly lower ($P < 0.05$) for fried potato snacks; however, no difference ($P > 0.05$) in the total phenolic content was observed when non-fried apple snacks were compared with fried apple snacks.

A study was carried out to investigate the incorporation of different fruit juices into apple tissues before dehydration (Joshi and Rupasinghe, 2011; Unpublished data, Table 11.4). This resulted in further enhancement of antioxidant capacity and provided different colors to the apple snacks and different fruit juices showed varying abilities to be incorporated into the apple snacks. The FRAP values of dried apple snacks pre-treated with different juices varied considerably [6.3–14.0 mg trolox equivalents (TE)/g dried weight]. The overall trend for antioxidant capacity of the apple snack samples pre-treated with various solutions was

Table 11.3 Total phenolic content and antioxidant capacity of snack products

Snack products	Total phenolics (μmol GAE/100 g)	FRAP (mmol TE/100 g)
Developed non-fried apple snacks	23.52 ± 0.97^{a}	2.05 ± 0.03^{a}
Commercial fried apple snacks	24.03 ± 1.7^{a}	1.60 ± 0.19^{b}
Commercial fried potato snacks	5.31 ± 0.59^{b}	0.11 ± 0.08^{c}

Source: Adapted with permission from Joshi *et al.* [11].
Data are expressed as mean ± SD ($n = 3$) on a dried weight basis.
Means ± SD followed by the same letter, within each column, are not significantly different ($P < 0.05$).
GAE, gallic acid equivalents; FRAP, ferric reducing antioxidant power; TE, trolox equivalents.

as follows: cranberry juice > grape juice > carrot juice > apple juice > apple juice + calcium chloride > pineapple juice > maple syrup. This can be explained by the significant differences of antioxidant capacity of the juices, which were incorporated into the dried apple snacks. Therefore, the high antioxidant capacity of the juice can directly enhance the antioxidant potency of the apple snacks. Apple puree immersion in green tea extracts resulted in substantial levels of epigallocatechin (EGC), epigallocatechin gallate (EGCG), epicatechin gallate (ECG), and gallocatechin gallate (GCG), as well as caffeine [55] in the products. The combination of osmotic pre-treatment mainly in sucrose solutions and microwave-vacuum dehydration of strawberries and apples preserved the cellular structure and vitamin C retention was around 60% with employing the microwave procedure [56].

Table 11.4 ORAC and FRAP values of samples incorporated with different fruit juice

Juice	Treatment	ORAC (mmol TE/100 g)	FRAP (mmol TE/100 g)
Control	control	0.66^{cde}	4.60^{lcd}
Apple	dipping	0.62^{cde}	4.58^{cd}
Apple	vacuum	0.62^{cde}	5.62^{bc}
Apple + calcium chloride	dipping	0.56^{de}	5.00^{cd}
Apple + calcium chloride	vacuum	0.56^{de}	4.88^{cd}
Cranberry	dipping	1.22^{a}	6.21^{ab}
Cranberry	vacuum	0.97^{abc}	6.92^{a}
Carrot	dipping	1.12^{ab}	4.18^{de}
Carrot	vacuum	0.88^{abcde}	4.59^{cd}
Grape	dipping	0.92^{abcde}	4.72^{cd}
Grape	vacuum	0.89^{abcde}	5.05^{cd}
Pineapple	dipping	0.81^{bcde}	4.19^{de}
Pineapple	vacuum	0.93^{abcd}	4.02^{de}
Maple	dipping	0.53^{e}	3.44^{ef}
Maple	vacuum	0.71^{cde}	2.51^{f}

Source: Adapted from Joshi and Rupasinghe [2011, unpublished].
Data are expressed as mean ± SD ($n = 3$) on a dried weight basis.
Means ± SD followed by the same letter, within each column, are not significantly different ($P < 0.05$).
ORAC, oxygen radical absorbance capacity; TE, trolox equivalents; FRAP, ferric reducing antioxidant power.

Hence, similar type of extracts can be used for preparing phytochemical enriched apple snacks using VI processes.

11.5 Compositional and nutritional characteristics of dried apple snacks

Apples are a good source of carbohydrates, ascorbic acid, phenolics, dietary fiber, minerals (such as P, K, and Ca), and vitamins (such as thiamin, niacin, and vitamin A) [57]. The phenolic compounds present in apples are of special interest as these act as a source of dietary antioxidants. The total antioxidant capacity of 100 g of whole apple is equivalent to the antioxidant capacity of 1500 mg of vitamin C [58]. The phenolic compounds, such as protocatechuic acid and chlorogenic acid have shown to alter the activity of enzymes involved in carcinogen activation, inhibit the formation of polycyclic aromatic hydrocarbon-DNA adducts in mouse epidermis, and decrease the level of lipid peroxidation in the epidermal microsomes [59]. Wu *et al.* [57] performed the chemical compositional studies on eight apple cultivars (Delicious, Golden Delicious, Ralls, Fuji, QinGuan, Jonagold, Granny Smith, and Orin); total phenolic content of these cultivars ranged from 26.21 to 88.27 mg/L, of which chlorogenic acid and epicatechin were the most predominant phenolics, followed by coumarin, phloridzin, catechin, and caffeic acid among all apple cultivars.

Application of VI process on apple slices resulted in uptake of calcium (780 mg/100 g) and vitamin E (168 mg/100 g) in the fruit matrix and thus, holds a great potential to prepare fortified apple-based snacks to meet the daily requirement of calcium and vitamin E in the consumer's diet. The calcium concentration in VI-treated apple slices as well as apple slices given anti-browning treatment was 0.78%. Increase in uptake of added calcium chloride by apple slices occurred during both of these pre-treatment processes (Table 11.5). Hence, this increase in calcium content in 100 g of apple snacks obtained from anti-browning and VI treatment was 780 mg of calcium which can help in providing 70% of the daily required calcium in the diet [daily reference intake (RDI): 1100 mg per day] [60]. The concentration of vitamin E in VI-treated apple snacks was 1.81 mg/g, thus 5 g of apple snacks would be sufficient to meet the daily requirements of vitamin E (10 mg per day) [60]. Increase in protein content in VI-treated apple slices was 65% when compared to that of the protein content of untreated apple slices which can be attributed to the addition of whey proteins.

VI process can be used as a successful tool for increasing the nutritional value of the food products [4, 5]. In a study by Xie and Zhao [7], fortification of 200 g of fresh-cut apples using VI methods increased calcium and zinc concentrations equivalent to 15–20 and 40% of RDI, respectively, as compared to fresh apple which provided about 0.84 and 2.30% of RDI of calcium and zinc, respectively. In another study, fortification of fresh-cut apples with vitamin E, calcium, and zinc using VI, resulted in an increased vitamin E (about 100-fold increase), calcium, and zinc contents (about 20-fold increase) as compared to the unfortified apples [12]. Hence, these fortified apple snacks can be introduced to the general public as a choice for delivering the required amount of dietary vitamins, minerals, and other nutritionally significant compounds.

Non-fried apple snacks showed comparable nutritional and compositional characteristics to commercially fried apple and potato snacks (Table 11.6). The amount of oil content was relatively higher in both commercial fried snack products and non-fried apple snacks. Generally, apple contains traces of lipids and for example "Empire" apples reported to have

Table 11.5 Compositional and nutritional characteristics of different apple snacks

		Pre-treatment before drying		
Proximate analysis	**Unit**	**Untreated**	**Anti-browning treatment**	**VI treatment**
Proximate composition				
Dry matter	g/100 g	95.97	91.36	93.42
Crude protein	g/100 g	1.59	1.97	2.63
Crude fat	g/100 g	0.90	1.71	1.39
Ash	g/100 g	1.54	2.18	2.50
Minerals				
Calcium	mg/100 g	<50	780	780
Phosphorus	mg/100 g	70	50	70
Sodium	mg/100 g	50	50	50
Potassium	mg/100 g	680	480	620
Iron	mg/100 g	nd	0.71	1.15
Manganese	mg/100 g	0.26	0.26	0.26
Copper	mg/100 g	0.30	0.33	0.31
Zinc	mg/100 g	0.25	0.16	0.22

Source: Adapted with permission from Joshi *et al.* [10].
Data are expressed as mean $\pm$ SD ($n = 3$) on a dried weight basis.
nd, not determined.

Table 11.6 Compositional and nutritional characteristics of snack products

	Unit	**Developed non-fried apple snacks**	**Commercial fried apple snacks**	**Commercial fried potato snacks**
Water activity	a_w	0.24^a	0.26^a	0.19^b
Proximate composition				
Moisture	g/100 g	2.57^a	1.33^b	0.66^b
Protein	g/100 g	1.04^b	0.95^b	5.26^a
Oil	g/100 g	0.78^c	31.20^b	35.06^a
Ash	g/100 g	2.34^b	0.88^c	3.00^a
Minerals				
Calcium	mg/100 g	560^a	40^a	40^a
Phosphorus	mg/100 g	40^b	40^b	130^a
Sodium	mg/100 g	40^a	40^a	220^a
Potassium	mg/100 g	620^b	390^c	1190^a
Magnesium	mg/100 g	20^a	20^a	50^a
Iron	mg/100 g	1.65^b	2.12^b	23.33^a
Copper	mg/100 g	0.32^a	0.29^a	0.29^a
Manganese	mg/100 g	1.92^a	0.17^b	0.56^b
Zinc	mg/100 g	2.57^a	0.23^c	1.15^b

Source: Adapted with permission from Joshi *et al.* [11].
Data are expressed as mean $\pm$ SD ($n = 3$) on a dried weight basis.
Means $\pm$ SD followed by the same letter, within each row, are not significantly different ($P < 0.05$).

0.9% of lipid content [61]. Similarly, raw peeled potato naturally contains only trace of lipid [61]. Thus, the higher oil content in fried apple and potato snacks is the amount of oil gained during the frying process, as both apple and potato without any processing contain very low amount of oil. Differences in the compositional attributes as crude protein and ash contents can be attributed to the difference in the source and cultivar used for preparing fried and non-fried apple snacks.

11.6 Health benefits of fresh and dried apples

Though the reported literature on health benefits of dried apple products is scarce, scientific evidence for numerous health benefits of consumption of fresh apples is promising. The bioactive phytochemicals in apples contribute to high antioxidant capacity, antiproliferative activity, and cholesterol lowering effects, thereby promoting certain physiological functions in the human body associated with the reduced risk of cancer, heart diseases, asthma, and type-2 diabetes [62].

Apple phytochemicals have been shown to affect cholesterol biosynthesis, metabolism as well as the low-density lipoprotein (LDL) oxidation in different research model systems. A procyanidin-rich fraction isolated from apple polyphenols decreased the esterification of cholesterol and the secretion of apoB containing lipoprotein by human Caco-2/TC7 enterocytes [63]. Freeze-dried apple peels and pulp fed to normal and atherogenic rats showed significant reduction in plasma total and LDL cholesterols [64]. Similar results were reported by Aprikian *et al.* [65] where male Wistar rats fed diets containing 0.3% cholesterol with lyophilized apples showed a reduction in cholesterol in triacylglycerol (TAG) rich lipoproteins, rise in high-density lipoprotein (HDL) fraction, and reduced malondialdehyde excretion in urine. The apparent absorption of dietary cholesterol was also markedly depressed.

Human studies involving apple and their consumption and dietary supplementation of their phytochemicals and health benefits are reported. For example, a 12-week ingestion of apple polyphenol containing capsules by healthy human subjects with moderate obesity significantly reduced the plasma total and LDL cholesterols [66]. Suppression of LDL oxidation *in vitro* was observed for apple extract where 16 mg/mL inhibited the oxidation by 81.9% [67]. Similarly, six commercial apple juices and extracts from "Red Delicious" apples inhibited LDL oxidation in an *in vitro* Cu^{2+}-catalyzed human LDL oxidation system [68]. Interestingly, Seeram *et al.* [69] found that unblended apple juice consumption increased *ex vivo* Cu^{2+}-mediated LDL oxidation lag time by 20% compared with the baseline.

11.7 Conclusions

The phytochemicals present in apples are associated with the color, taste, and nutritional quality including their antioxidant and biological activity. It is believed that growing interest in fruit-based healthy snack foods assures a promising future for the apple-based snack industry. Furthermore, convincing the consumer about the nature and healthiness of dried apple snacks is expected to influence the choice of selecting non-fried snacks over the current commercially available fried snacks. Food technology applications of modern drying processes alone or in combination with other techniques such as VI can increase the market potential of dried fruit products with promises of added health benefits.

References

1. Food Processing (2004) *The Canadian Snack Food Industry*. Published online at: http://www.agr.gc.ca/misb/fb-ba/index_e.php?s1=proc-trans&s2=prof&page=snack-grignot, last accessed 1 June 2011.
2. Zandstra, E.H., Graaf, C.D. & Staveren, W.A.V. (2001) Influence of health and taste attitudes on consumption of low- and high-fat foods. *Food Quality and Preference*, **12**, 75–82.
3. Jack, F.R., O'Neill, J., Piacentini, M.G. & Schroder, M.J.A. (1997) Perception of fruit as a snack: a comparison with manufactured snack foods. *Food Quality and Preference*, **8**, 175–182.
4. Joshi, A.P.K. & Rupasinghe, H.P.V. (2010) Optimization of bioactive fortification in apple snacks through vacuum impregnation using response surface methodology. *Food and Nutrition Sciences*, **2**, 45–52.
5. Joshi, A.P.K., Rupasinghe, H.P.V. & Pitts, N.L. (2010) Sensory and nutritional quality of the apple snacks prepared by vaccum impregnation process. *Journal of Food Quality*, **33**, 758–767.
6. Anino, S.V., Salvatori, D.M. & Alzamora, S.M. (2006) Changes in calcium level and mechanical properties of apple tissue due to impregnation with calcium salts. *Food Research International*, **39**, 154–164.
7. Xie, J. & Zhao, Y. (2003) Nutritional enrichment of fresh apple (Royal Gala) by vacuum impregnation. *International Journal of Food Sciences and Nutrition*, **54**, 387–398.
8. Fito, P., Chiralt, A., Barat, J.M., Andres, A., Martinez-Monzo, J. & Martinez-Navarrete, N. (2001) Vacuum impregnation for development of new dehydrated products. *Journal of Food Engineering*, **49**, 297–302.
9. Fito, P., Chiralt, A., Betoret, N., Gras, M., Cháfe, M., Martínez-Monzó, J., Andrés, A. & Vidal, D. (2001) Vacuum impregnation and osmotic dehydration in matrix engineering application in functional fresh food development. *Journal of Food Engineering*, **49**, 175–183.
10. Joshi, A.P.K., Rupasinghe, H.P.V. & Khanizadeh, S. (2011) Impact of drying processes on bioactive phenolics, vitamin C and antioxidant capacity of red-fleshed apple slices. *Journal of Food Processing and Preservation*, **35**, 453–457.
11. Joshi, A.P.K., Rupasinghe, H.P.V. & Pitts, N.L. (2011) Comparison of non-fried apple snacks with commercially available fried snacks. *Food Science Technology International*, **17**, 249–257.
12. Park, S.I., Zhao, Y. & Koddihalli, I. (2005) Nutritional, sensory, and physicochemical properties of vitamin E and mineral-fortified fresh-cut apples by use of vacuum impregnation. *Journal of Food Science*, **70**, 593–599.
13. Sham, P.W.Y., Scaman, C.H. & Durance, T.D. (2001) Texture of vacuum microwave dehydrated apple chips as affected by calcium pretreatment, vacuum level, and apple variety. *Journal of Food Science*, **66**, 1341–1347.
14. Treidlinger, M. (2007) *The Canadian Snack Food Industry. Food Value Chain Bureau Agriculture and Agri-Food Canada Ottawa, Ontario Canada.* Published online at: http://www4.agr.gc.ca/AAFC-AAC/display-afficher.do?id=1172692863066&lang=e, last accessed 1 June 2011.
15. Canadian Industry Statistics (2010) *Snack Food Manufacturing (NAICS 31191).* Published online at: http://www.ic.gc.ca/cis-sic/cis-sic.nsf/IDE/cis-sic31191prde.html, last accessed 1 June, 2011.
16. Barrett, B. (2010) *Canada: HRI Food Service Annual Report 2010.* Published online at: http://www.linkbc.ca/torc/downs1/Food%20Service%20%20Hotel%20Restaurant%20Institutional_Ottawa_Canada_11-01-2010.pdf, last accessed 1 June, 2011.
17. Jahns, L., Siega-Riz, A.M. & Popkin, M.B. (2001) The increasing prevalence of snacking among US children from 1977 to 1996. *Journal of Pediatrics*, **138**, 493–498.
18. Vardavas, C.I., Yiannopoulos, S., Kiriakakis, M., Poulli, E. & Kafatos, A. (2007) Fatty acid and salt contents of snacks in the Cretan and Cypriot market: a child and adolescent dietary habit. *Food Chemistry*, **101**, 924–931.
19. Kavey, R-E.W., Daniels, S.R., Lauer, R.M., Atkins, D.L., Hayman, L.L. & Taubert, K. (2003) American Heart Association guidelines for primary prevention of atherosclerotic cardiovascular disease beginning in childhood. *Circulation*, **107**, 1562–1566.
20. Ignarro, L.J., Balestrieri, M.L. & Napoli, C. (2007) Nutrition, physical activity and cardiovascular disease: an update. *Cardiovascular Research*, **73**, 326–340.
21. Block, G., Patterson, B. & Subar, A. (1992) Fruit, vegetables and cancer prevention: a review of the epidemiological evidence. *Nutrition and Cancer*, **18**, 1–29.

22. Shyu, S.L., Hau, L.B. & Hwang, L.S. (2005) Effects of processing conditions on the quality of vacuum-fried carrot chips. *Journal of the Science of Food and Agriculture*, **85**, 1903–1908.
23. Shyu, S.L. & Hwang, L.S. (2001) Effect of processing conditions on the quality of vacuum fried apple chips. *Food Research International*, **34**, 133–142.
24. Draudt, H.N. & Huang, Y. (1966) Effect of moisture content of freeze-dried peaches and bananas on changes during storage related to oxidative and carbonyl-amine browning. *Journal of Agricultural and Food chemistry*, **14**, 170–176.
25. Konopacka, D., Plocharski, W. & Beveridge, T. (2002) Water sorption and crispness of fat-free apple chips. *Journal of Food Science*, **67**, 87–92.
26. Lewicki, P.P. (2006) Design of hot air drying for better foods. *Trends in Food Science and Technology*, **17**, 153–163.
27. Bazyma, L.A., Guskov, V.P., Basteev, A.V., Lyashenko, A.M., Lyakhno, V. & Kutovoy, A.V. (2006) The investigation of low temperature vacuum drying processes of agricultural materials. *Journal of Food Engineering*, **74**, 410–415.
28. Toor, R.K. & Savage, G.P. (2006) Effect of semi-drying on the antioxidant components of tomatoes. *Food Chemistry*, **94**, 90–97.
29. Lewicki, P.P. & Pawlak, G. (2003) Effect of drying on microstructure of plant tissue. *Drying Technology*, **21**, 657–683.
30. Sjoholm, I. & Gekas, V. (1995) Apple shrinkage upon drying. *Journal of Food Engineering*, **25**, 123–130.
31. Julkunen-Tiitto, R. & Sorsa, S. (2001) Testing the effects of drying methods on willow flavonoids, tannins, and salicylates. *Journal of Chemical Ecology*, **27**, 779–789.
32. Nijhuis, H.H., Torringa, H.M., Muresan, S., Yuksel, D., Leguijt, C. & Kloek, W. (1998) Approaches to improving the quality of dried fruit and vegetables. *Trends in Food Science and Technology*, **9**, 13–20.
33. Yücel, U., Alpas, H. & Bayindirli, A. (2010) Evaluation of high pressure pre-treatment for enhancing the drying rates of carrot, apple, and green bean. *Journal of Food Engineering*, **98**, 266–272.
34. Ho, C.-T, Chen, Q., Shi, H., Zhang, K.-Q. & Rosen, R.T. (1992) Antioxidative effect of polyphenol extract prepared from various Chinese teas. *Preventive Medicine*, **21**, 520–525.
35. Macheix, J.J., Sapis, J.C. & Fleuriet, A. (1991) Phenolic compounds and polyphenoloxidase in relation to browning in grapes and wines. *Critical Reviews in Food Science and Nutrition*, **30**, 441–486.
36. Rupasinghe, H.P.V., Wang, L., Huber, G.M. & Pitts, N.L. (2008) Effect of baking on dietary fiber and phenolics of muffins incorporated with apple skin powder. *Food Chemistry*, **107**, 1217–1224.
37. Cannac, M., Ferrat, L., Barboni, T., Pergent, G. & Pasqualini, V. (2007) The influence of tissue handling on the flavonoid content of the aquatic plant *Posidonia oceanica*. *Journal of Chemical Ecology*, **33**, 1083–1088.
38. Julkunen-Tiitto, R. & Tahvanainen, J. (1989) The effect of the sample preparation method of extractable phenolics of salicaceae species. *Planta Medica*, **55**, 55–58.
39. Orians, C.M. (1995) Preserving leaves for tannin and phenolic glycoside analyses: a comparison of methods using three willow taxa. *Journal of Chemical Ecology*, **21**, 1235–1243.
40. van Sumere, C., Geiger, H., Bral, D., Fockenier, G., Vande Casteele, K., Martens, M., Hanselaer, R. & Gevaert, L. (1983) Freeze-drying and analysis of plant and other biological material. *Analytical Biochemistry*, **131**, 530–532.
41. Dewanto, V., Wu, X., Adom, K.K. & Liu, R.H. (2002) Thermal processing enhances the nutritional value of tomatoes by increasing total antioxidant activity. *Journal of Agricultural and Food Chemistry*, **50**, 3010–3014.
42. Inyang, U.E. & Ike, C.I. (1998) Effect of blanching, dehydration method temperature on the ascorbic acid, colour, sliminess and other constituents of okra fruit. *International Journal of Food Sciences and Nutrition*, **49**, 125–130.
43. Nicholas, J.J., Richard-Forget, F.C., Goupy, P.M., Amiot, M.J. & Aubert, S.Y. (1994) Enzymatic browning reactions in apple and apple products. *Critical Reviews in Food Science and Nutrition*, **34**, 109–157.
44. Ramesh, M.N., Wolf, W., Tevini, D. & Jung, G. (1999) Studies on inert gas processing of vegetables. *Journal of Food Engineering*, **40**, 199–205.
45. Betoret, N., Puente, L., Diaz, M.J., Pagán, M.J., Garcia, M.J., Gras, M.L., Martínez-Monzó, J. & Fito, P. (2003) Development of probiotic-enriched dried fruits by vacuum impregnation. *Journal of Food Engineering*, **56**, 273–277.
46. Mujica-Paz, H., Valdez-Fragoso, A., Lopez-Malo, A., Palou, E. & Welti-Chanes, J. (2003) Impregnation properties of some fruits at vacuum pressure. *Journal of Food Engineering*, **56**, 307–314.

47. Shi, X.Q. & Fito, P. (1993) Vacuum osmotic dehydration of fruits. *Drying Technology*, **11**, 1429–1442.
48. Shi, X.Q., Fito, P. & Chiralt, A. (1995) Influence of vacuum treatment on mass transfer during the osmotic dehydration of fruits. *Food Research International*, **28**, 445–454.
49. Jeon, M. & Zhao, Y. (2005) Honey in combination with vacuum impregnation to prevent enzymatic browning of fresh-cut apples. *International Journal of Food Sciences and Nutrition*, **56**, 165–176.
50. Martinez-Monzo, J., Barat, J.M., Gonzalez-Martinez, C., Chiralt, A. & Fito, P. (2000) Changes in thermal properties of apple due to vacuum impregnation. *Journal of Food Engineering*, **43**, 213–218.
51. Larrauri, J.A., Ruperez, P. & Saura-Calixto, F. (1997) Effect of drying temperature on the stability of polyphenols and the antioxidant activity of red grape pomace peels. *Journal of Agricultural and Food Chemistry*, **45**, 1390–1393.
52. Sacchetti, G., Cocci, E., Pinnavaia, G., Mastrocola, D. & Rosa, M.D. (2008) Influence of processing and storage on the antioxidant activity of apple derivatives. *International Journal of Food Science & Technology*, **43**, 797–804.
53. Fu, H.Y. (2004) Free radical scavenging and leukemia cell growth inhibitory properties of onion powders treated by different heating processes. *Journal of Food Science*, **69**, 50–54.
54. Manzocco, L., Calligaris, S., Mastrocola, D., Nicoli, M.C. & Lerici, C.R. (2000) Review of non-enzymatic browning and antioxidant capacity in processed foods. *Trends in Food Science and Technology*, **11**, 340–346.
55. Corey, M.E., Kerr, W.L., Mulligan, J.H. & Lavelli, V. (2011) Phytochemical stability in dried apple and green tea functional products as related to moisture properties. *LWT—Food Science & Technology*, **44**, 67–74.
56. Erle, U. & Schubert, H. (2001) Combined osmotic and microwave-vacuum dehydration of apples and strawberries. *Journal of Food Engineering*, **49**, 193–199.
57. Wu, J., Gao, H., Zhao, L., Liao, X., Chen, F., Wang, Z. & Hu, X. (2007) Chemical compositional characterization of some apple cultivars. *Food Chemistry*, **103**, 88–93.
58. Eberhardt, M.V., Lee, C.Y. & Liu, R.H. (2000) Nutrition: antioxidant activity of fresh apples. *Nature*, **405**, 903–904.
59. Szaefera, H., Kaczmarekb, J., Rybczyńskab, M. & Baer-Dubowskaa, W. (2007) The effect of plant phenols on the expression and activity of phorbol ester-induced PKC in mouse epidermis. *Toxicology*, **230**, 1–10.
60. Food and Drugs Act and Regulations (2008) *Vitamins, Minerals and Amino acids.* Published online at: http://laws.justice.gc.ca/en/showdoc/cr/C.R.C.-c.870/bo-ga:l_D//en#anchorbo-ga:l_D, last accessed 2 June 2011.
61. Health Canada (2008) *Fruit and Fruit Juices. In: Nutrient Value of Some Common Foods.* Authority of the Minister of Health. Published online at: http://www.hc-sc.gc.ca/fn-an/nutrition/fiche-nutri-data/nutrient_value-valeurs_nutritives-table5-eng.php, last accessed 1 June 2011.
62. Boyer, J. & Liu, R.H. (2004) Apple phytochemicals and their health benefits. *Nutrition Journal*, **1**, 3–5.
63. Vidal, R., Hernandez-Vallejo, S., Pauquai, T., Texier, O., Rousset, M., Chambaz, J., Demignot, S. & Lacorte, J.-M. (2005) Apple procyanidins decrease cholesterol esterification and lipoprotein secretion in Caco-2/TC7 enterocytes. *Journal of Lipid Research*, **46**, 258–268.
64. Leontowicz, H., Leontowicz, M., Gorinstein, S., Martin-Belloso, O. & Trakhtenberg, S. (2007) Apple peels and pulp as a source of bioactive compounds and their influence on digestibility and lipid profile in normal and atherogenic rats. *Medycyna Weterynaryjna*, **63**, 1434–1436.
65. Aprikian, O., Levrat-Verny, M., Besson, C., Busserolles, J., Rémésy, C. & Demigné, C. (2001) Apple favourably affects parameters of cholesterol metabolism and of anti-oxidative protection in cholesterol-fed rats. *Food Chemistry*, **75**, 445–452.
66. Akazome, N.Y., Kanda, T., Ohtake, Y., Shimasaki, H. & Kobayashi, T. (2007) Apple polyphenols influence cholesterol metabolism in healthy subjects with relatively high body mass index. *Journal of Oleo Science*, **56**, 417–428.
67. Chu, Y. & Liu, R.H. (2004) Novel low-density lipoprotein (LDL) oxidation model: antioxidant capacity for the inhibition of LDL oxidation. *Journal of Agricultural and Food Chemistry*, **52**, 6818–6823.
68. Pearson, D.A., Tan, C.H., German, J.B., Davis, P.A. & Gershwin, M.E. (1999) Apple juice inhibits human low density lipoprotein oxidation. *Life Sciences*, **64**, 1913–1920.
69. Seeram, N.P., Aviram, M., Zhang, Y., Henning, S.M., Feng, L., Dreher, M. & Heber, D. (2008) Comparison of antioxidant potency of commonly consumed polyphenol-rich beverages in the United States. *Journal of Agricultural and Food Chemistry*, **56**, 1415–1422.

12 Phytochemicals and health benefits of dried apricots

Neslihan Göncüoğlu, Burçe Ataç Mogol, and Vural Gökmen

12.1 Introduction

Apricots (*Prunus armeniaca* L.) are classified under the *Prunus* species of the Rosaceae family. This type of fruit is produced by cultivation of wild apricot, called "Zerdali" [1] which has sweet and delicious taste and charming appearance with its yellow color and random reddish overlay.

Apricots have been domesticated well over 5000 years ago in the area of Iran, Turkistan, Afghanistan, Middle Asia, and Western China [2]. Today, the greatest percentage of the world's apricots is produced in the countries around Mediterranean Sea (such as Turkey, Spain, Italy, France, and Greece) [3]. Turkey is the leading apricot producer with 700,000 tonnes production in 2009, followed by Iran, Uzbekistan, Pakistan, Italy, China, and the United States [4].

Apricots have perishable nature and very short shelf-life, about 4–5 days under ambient conditions. Most of the apricots are either dried or consumed as fresh where they are grown [5]. Turkey, Iran, China, the United States, Australia, and South Africa are the major dried apricots producing countries. Turkey itself produces 80% of the dried apricots and exports 70% of the world's dried apricots [2]. This chapter reports on the composition, nutritional value, phytochemicals, and health benefits of dried apricots. Chemical changes occurring during drying are also discussed along with the effect of sulfur dioxide treatment.

12.2 Production

12.2.1 Dehydration methods

Drying is one of the oldest techniques used for food preservation [6, 7]. Drying lowers the water activity of foods and decreases the moisture content to a safe level. For storing and further processing, final moisture content of dried apricots must be less than 16–18% (by weight). It is also possible to produce 5% moisture content apricots with a second step drying [8].

Dried Fruits: Phytochemicals and Health Effects, First Edition. Edited by Cesarettin Alasalvar and Fereidoon Shahidi.

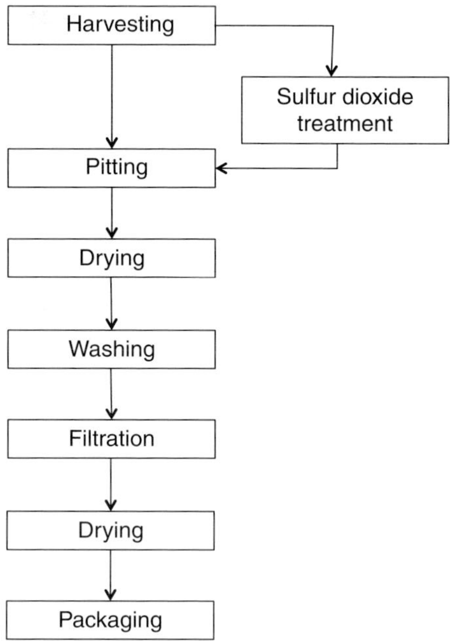

Figure 12.1 Flowchart of dried apricot production.

Sun drying is a widely used technique both in Turkey and elsewhere. Apricots, that are not suitable for fresh consumption or have high fiber content, are generally selected for drying. They are cut into half, pitted, and spread in depths on wooden trays. If sulfur dioxide-treated sun-dried apricots are produced, they can be treated with sulfur dioxide containing solutions or sulfur dioxide in gaseous form and then placed under sun. Unless they are sulfur dioxide-treated, they can directly be placed under sun [9]. Before marketing, they are being washed, filtered, dried, and packed. A flowchart for production of dried apricots is given in Figure 12.1. Drying time may vary between 2 and 8 days, depending upon the moisture content [10]. The dried apricots are stored at room temperature [9] and the yield from this process is approximately 5:1 [10].

After sun drying, it is possible to produce dried apricots having moisture content of less than 20% [10] with rich orange color, translucent appearance, and desirable gummy texture [11, 12]. Besides these advantages, dusts and insects may contaminate apricots when sun drying time is longer. Drying time required for apricots is around 3 to 4 weeks, depends on the thickness of the wet material layer, humidity, and temperature [13]. Longer drying time may cause significant economical loss. Therefore, an alternative technique instead of sun drying is required.

Solar drying is thought to be an alternative technique for apricot drying since it prevents microbial contamination, provides controllable drying parameters, and shortens the drying time. According to energy sources, solar dryers can be classified into three groups such as solar natural dryers (only use ambient energy sources), semi artificial natural dryers (have a fan driven by electrical motors to provide continuous air flow), and solar-assisted artificial dryers (use conventional energy source) [14].

12.3 Compositional and nutritional characteristics of dried apricots

Compositional and nutritional characteristics of fresh apricots and comparison with their dried counterparts are reported in the literature. However, dried apricots attracted much attention because of their mineral and vitamin contents in the last few years [15–17].

12.3.1 Proximate composition

Table 12.1 compares nutrient contents, including proximate composition, minerals, and vitamins of fresh and sulfur dioxide-treated dried apricots [18]. Moisture content of fresh apricots ranges from 75 to 95% [19]. Apricots can be dried until a desired moisture content, which is usually less than 20%, has been reached [1]. Moisture content of sulfur dioxide-treated dried apricots is around 7.5% in the data given by USDA [18]. Its carbohydrate

Table 12.1 Nutrient composition of fresh and sulfur dioxide-treated dried apricots (values in per 100 grams edible portion in fresh weight basis)

	Unit	Fresh apricots	Dried apricots[a]
Proximate composition			
Water	g	86.35	7.50
Energy	kcal	48	320
Protein	g	1.40	4.90
Lipid	g	0.39	0.62
Ash	g	0.75	4.09
Carbohydrate	g	11.12	82.89
Vitamins			
Vitamin C	mg	10	9.5
Thiamin	mg	0.03	0.04
Riboflavin	mg	0.04	0.15
Niacin	mg	0.60	3.58
Pantothenic acid	mg	0.24	1.07
Pyridoxine	mg	0.05	0.52
Folate	μg	9.0	4.0
Vitamin A (RAE)	μg	96	633
Minerals			
Calcium	mg	13	61
Iron	mg	0.39	6.31
Magnesium	mg	10	63
Phosphorus	mg	23	157
Potassium	mg	259	1850
Sodium	mg	1.0	13
Zinc	mg	0.20	1.0
Copper	mg	0.08	0.58
Manganese	mg	0.08	0.37
Selenium	μg	0.1	nd

Source: Adapted from USDA [18].
RAE, retinol activity equivalents; nd, not detected.
[a]Sulfur dioxide-treated.

content is 82.89%, which is 7.5-fold more concentrated than that of fresh apricots. Sucrose is the major sugar in apricots. In fresh apricots, sucrose concentration ranges between 22.96 and 56.83 mg/100 g in dry weight basis [20]. Sorbitol, a sugar alcohol, is also present in apricots in significant amounts. Its concentration ranges between 2.47 and 26.80 mg/100 g (in dry weight basis) among different cultivars [20]. Protein content of sulfur dioxide-treated dried apricots is 4.90%, which is 3.5-fold more concentrated than that of fresh apricots.

12.3.2 Vitamins and minerals

Apricots are an important source of vitamins such as thiamin, riboflavin, niacin, pantothenic acid, pyridoxine, folic acid, vitamin C, and vitamin A [21, 22]. Table 12.1 compares the contents of vitamins and minerals in fresh and sulfur dioxide-treated dried apricots [18]. On a dry weight basis, dried apricots contained lower amounts of vitamin C and folate than their fresh counterparts. Ascorbic acid (vitamin C) is indeed very sensitive to oxidation, hence, its loss may easily be accelerated depending on the oxidation potential of apricots and the drying temperature [22]. Other water-soluble vitamins do not change considerably by sulfur dioxide-treatment of apricots during drying. Sulfur dioxide-treated dried apricots are in fact a rich source of vitamin A and a 40 g portion of them provide 8% of reference dietary intake (RDI) of it [23]. For vitamin C, it also provides 4 and 5.04% of RDI for adult men and women, respectively [24].

Potassium is the most abundant mineral in dried apricots. Its content ranges between 1328 and 2087 mg/100 g on a dry weight basis [3] and this is followed by phosphorus, magnesium, and calcium (Table 12.1). According to the data established by the Food and Nutrition Center of the Institute of Medicine, 40 g edible portion of sulfur dioxide-treated dried apricot provides 15.6, 8.8, 8, and 2.4% of RDI of potassium, phosphorus, magnesium, and calcium, respectively, for adults [25]. Fresh and dried apricots are not good sources of selenium [20].

With regard to the amino acids content (Table 12.2), glutamic acid is the most predominant amino acid found in sulfur dioxide-treated dried apricots [18].

12.4 Phytochemicals in dried apricots

Dried apricots are composed of certain phytochemicals mainly based on polyphenols and carotenoids. Major polyphenols and carotenoids found in dried apricots are shown in Figure 12.2.

12.4.1 Polyphenols

Polyphenols are of much interest because of their high antioxidant capacity and effects in lowering the risk of chronic diseases [26]. In addition, they play a role as anti-carcinogenic, anti-microbial, anti-allergic, anti-mutagenic, and anti-inflammatory agents. Their effects on cardiovascular diseases (CVD) are also well known [27].

Flavonoids detected in fresh and sulfur dioxide-treated dried apricots are listed in Table 12.3 [28–30]. Radi *et al.* [31] identified major classes of phenolics in fresh apricot cultivars as procyanidins, anthocyanins, flavonols, and hydroxycinnamic acid derivatives. They reported chlorogenic acid, neochlorogenic acid, (+)-catechin, (−)-epicatechin, and rutin as principal phenolic compounds. Chlorogenic acid, which is a hydroxycinnamic acid

Table 12.2 Amino acids in sulfur dioxide-treated dried apricots

Amino acids	Concentration (g/100 g)
Alanine	0.24
Arginine	0.19
Aspartic acid	1.12
Cystine	0.02
Glutamic acid	0.50
Glycine	0.15
Histidine[a]	0.08
Isoleucine[a]	0.15
Leucine[a]	0.29
Lysine[a]	0.34
Methionine[a]	0.02
Phenylalanine[a]	0.20
Proline	0.29
Serine	0.28
Threonine[a]	0.18
Tryptophan[a]	0.09
Tyrosine	0.11
Valine[a]	0.18

Source: Adapted from USDA [18].
[a]Indispensable amino acids.

derivative, has been reported as the most dominant phenolic acid found in fresh apricots with 103–390 mg/g dry weight among nine cultivars [32]. Moreover, protocatechuic acid, and prunin along with procyanidins B2, B3, and C1 were characterized in fresh apricots [31]. (+)-Epicatechin, (-)-epicatechin, procyanidins B1, B2, and B4, some procyanidin trimers, quercetin 3-rutinoside, kaempferol 3-rhamnosyl-hexoside, quercetin 3-acetyl-hexoside, and cyanidin 3-rutinoside are the other phenolic compounds detected in fresh apricots by different researchers [20, 29, 32]. Sochor *et al.* [33] identified gallic acid, 4-aminobenzoic acid, chlorogenic acid, ferulic acid, caffeic acid, procatechin, salicylic acid, *p*-coumaric acid, quercetin, quercitrin, rutin, resveratrol, vanillin, epicatechin, and catechin in fresh apricots. Fernandez de Simon *et al.* [27] detected small amounts of coumarins such as aesculetin and scopoletin in fresh apricots and apricot juices.

Dried apricots contain daidzein, genistein, and biochanin A in very small amounts (Table 12.3). These are also known as phytoestrogens [34, 35]. Biochanin A is rarely found in most foods, less than 1 μg/100 g of fresh weight basis. The content of these phytoestrogens depends on a large number of genetic and environmental conditions such as variety, harvest, and processing [35].

12.4.2 Carotenoids

Carotenoids that have some protective effect against cancers and degenerative diseases are naturally occurring pigments in apricots [36]. They are classified into two major subgroups namely xanthophylls (with oxygen) and carotenes (without oxygen). As a source of provitamin A, 250 g of fresh apricots or 30 g of dried apricots provides 100% of RDI for adults [3].

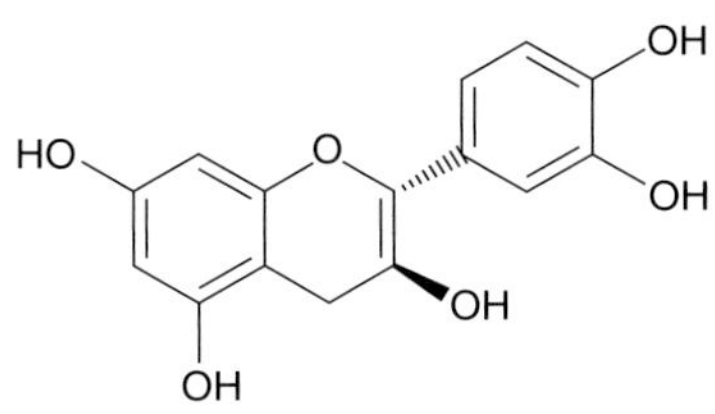

(+)-Catechin

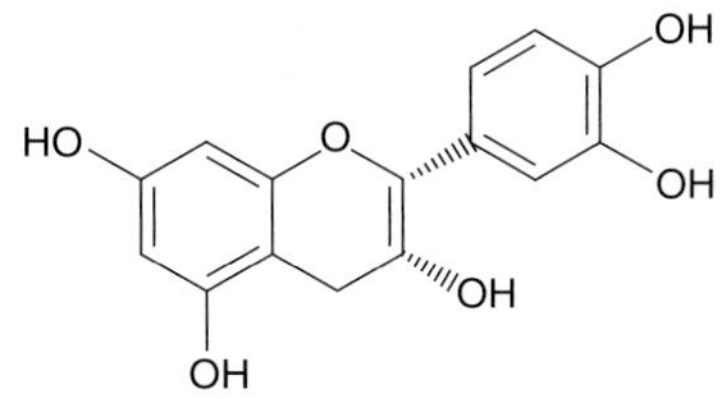

(-)-Epicatechin

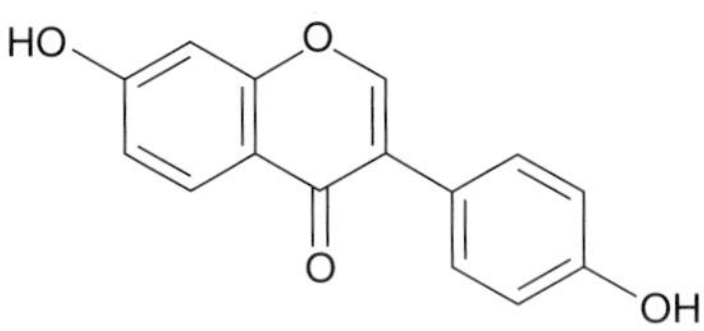

Daidzein

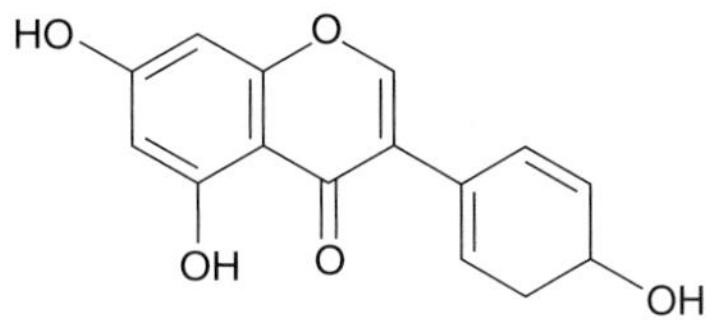

Genistein

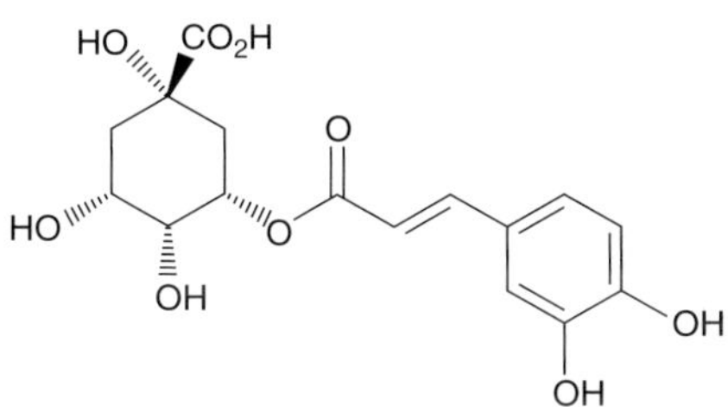

Chlorogenic acid

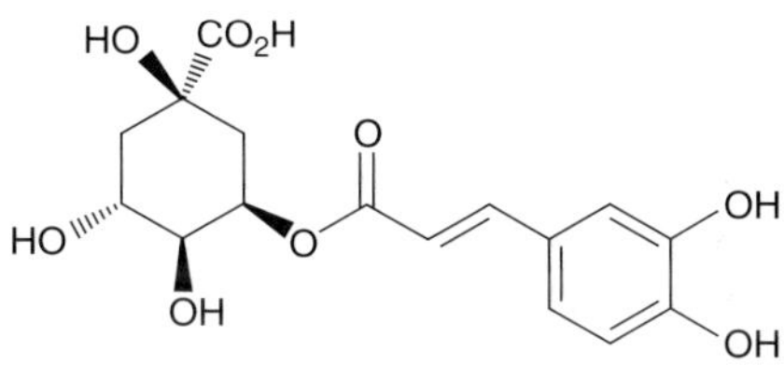

Neochlorogenic acid

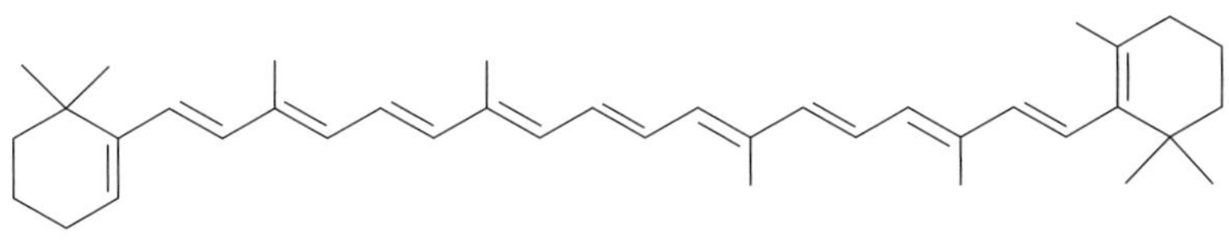

Beta-carotene

Figure 12.2 Chemical structures of some phytochemicals commonly found in dried apricots.

The most abundant carotenoid in apricots is β-carotene that is mainly responsible for the color of both fresh and dried apricots. Besides β-carotene, trace amounts of lutein, zeaxanthin, β-cryptoxanthin, γ-carotene, phytoene, phytofluene, and lycopene are found in fresh and dried apricots (Table 12.4) [36–38]. Several factors such as exposure time of apricots to sunlight, soil, season, region of cultivation, type of apricot, or stage of maturity affect the carotenoid levels [27]. In Turkish apricots, the total carotenoids range between 14.83 and 91.89 mg/100 g (dry weight basis). The percentage of β-carotene in total carotenoids varies

Table 12.3 Flavonoids content of fresh and sulfur dioxide-treated dried apricots (values in mg per 100 grams edible portion)

Apricots	Flavonoids		Min–Max	Reference
Fresh apricots	**Flavan-3-ols**	(−)-Epicatechin	0.02–8.29	[28]
		(+)-Catechin	0.31–7.34	[28]
	Flavonols	Kaempferol	nd −1.32	[28]
		Quercetin	0.38–2.90	[28]
	Proanthocyanidins	Monomers	0.33–2.80	[29]
		Dimers	0.15–3.10	[29]
		Trimers	0.01–1.90	[29]
		4-6mers	4.90–4.90	[29]
		7-10mers	2.20–2.20	[29]
		Polymers	0.80–0.80	[29]
Dried apricots (Sulfur dioxide-treated)	**Isoflavones**	Daidzein	nd–0.01	[30]
		Genistein	nd–0.02	[30]
		Total isoflavones	nd–0.03	[30]
		Biochanin A	0.05 (mean)	[30]

nd, not detected.

from 39 to 65% [20]. Kurz *et al.* [39] reported β-carotene (1.44–39.97 μg/g) as the dominant form of carotenoids in fresh apricots grown in Germany, followed by lutein (0.06–0.36 μg/g) and zeaxanthin (trace-0.46 μg/g).

Karabulut *et al.* [38] reported that apricots dried under hot air (80°C) contained 6.48 mg/100 g β-carotene, whereas sulfur dioxide-treated apricots dried under hot air (80°C) contained 7.17 mg/100 g β-carotene. In addition, sun dried and sulfur dioxide-treated sun-dried apricots had 3.38 and 3.87 mg/100 mg β-carotene, respectively. They reported that conditions of drying may affect the β-carotene content of dried apricots and sulfur dioxide treatment had no significant effect ($P > 0.05$) on β-carotene content upon sun drying under the same conditions [38].

12.5 Antioxidant activity of dried apricots

Dried fruits represent a relatively concentrated form of fresh fruits. As a consequence of concentration, dried fruits have higher total energy, nutrient density, fiber content, and often greater antioxidant activity as compared to their fresh counterparts. Antioxidant activity is a result of the cumulative effect of losses and concentration of polyphenols during drying, and

Table 12.4 Carotenoids found in fresh and dried apricots

Carotenoids	Reference
β-Carotene	[36–38]
β-Cryptoxhantin	[36, 37]
γ-Carotene	[36, 37]
Phytoene	[36, 37]
Phytofluene	[36, 37]
Lutein	[37]
Lycopen	[37]

Table 12.5 ORAC values and total phenolic contents of fresh and partially dried apricots

Apricots	ORAC value (μmol TE/100 g)	Total phenolics (mg GAE/100 g)	Reference
Fresh apricots	1110	79	[40–42]
Partially dried apricots (40% moisture)	3234	248	[40]

ORAC, oxygen radical absorbance capacity; TE, trolox equivalents; GAE, gallic acid equivalents.

generation of certain Maillard reaction products. The antioxidant activity of some fruits and vegetables, including fresh and dried apricots, have been surveyed using the oxygen radical absorbance capacity (ORAC) assay, the ferric reducing antioxidant power (FRAP) assay, and the trolox equivalent antioxidant capacity (TEAC) assay [40–43]. The total antioxidant capacities of fresh and dried apricots are given in Table 12.5 [41–43]. However, lipophilic antioxidants account only for <3% of the total antioxidant activity [41–43].

Bennett *et al.* [44] found that total phenolic content of dried apricots was 19.1 μmol gallic acid equivalents (GAE)/g dried fruit and total antioxidant activity by FRAP assay was 39.6 μmol Fe^{2+} reduced/g dried fruit. Among the dried ready-to-eat fruits, dried apricots were found to have less total phenolics and antioxidant activity than certain berries such as currants, prunes, and raisins [44]. Igual *et al.* [45] determined total phenolic content and 2,2-diphenyl-1-picrylhydrazyl (DPPH) radical scavenging activities of fresh and partially dried apricots. They found that total phenolic content of fresh and partially dried apricots was 16.6 and 64.73 mg GAE/100 g, respectively. Their corresponding DPPH scavenging activities were 2.4 and 3.8% for fresh and partially dried apricots [45].

The total antioxidant activities of different apricot forms are in the descending order of sulfur dioxide-treated sun dried > sun dried > fresh [46]. Sulphites, that are used to prevent flavor and color degradation in beverages and fruits, are reducing agents and weak antioxidants in foods. They react with molecular oxygen to form sulphates and also act as reducing agents that promote the formation of phenols from quinones, thereby preventing browning reactions [47].

12.6 Chemical changes during drying of apricots

Sun drying is the dominant mode of processing for the production of dried apricots, because it requires only free and renewable energy [9]. It is largely practiced in the provinces of Turkey where they are produced. There are also some other techniques such as cabinet/tunnel drying which use solar energy, hot air drying, and microwave drying that shorten drying time and have less effect on nutritional and functional composition of apricots [48].

There are certain chemical changes occurring in apricots during drying depending on the conditions (time, temperature, sulfur dioxide treatment, etc.). The major concern in the drying process of apricots is enzymatic browning catalyzed by polyphenol oxidase (PPO). Meanwhile, non-enzymatic browning such as sugar dehydration and Maillard reaction are also responsible for chemical changes in apricots during drying. The conditions of drying are favorable, especially for the Maillard reaction, a condensation reaction between carbonyls and amines leading to brown nitrogenous melanoidin polymers [49].

12.6.1 Enzymatic browning

PPO catalyzes the oxidation of phenolic compounds into quinones, which further form pigments in wounded tissues [50]. The action of PPO onto phenolic compound requires damage in the cells to let them come into contact [51]. In the presence of oxygen, two reactions begin to occur: the hydroxylation of monophenols (monophenolase activity) and the oxidation of *o*-diphenols to *o*-quinones (diphenolase activity) [52]. These reactions proceed by the polymerization of quinones leading to high molecular weight melanins with dark colors.

Phenolic compounds differ in their ability to act as substrate for PPO. Chlorogenic acid, which is the primary phenolic compound in apricots, is known as a suitable PPO substrate [50, 53]. Browning rate of apricots depends on several factors including concentration of PPO, type and concentration of phenolic compounds, presence and concentration of inhibitors (such as sulfur containing substances), oxygen, pH, and temperature [54]. During drying, PPO activity remains high for longer periods when temperature is around 55–60°C. On the other hand, exposure to 75–80°C in a shorter time inactivates the enzyme [39]. Queiroz *et al.* [50] expressed that optimum pH to PPO ranges from 5.0 to 7.5. Lowering the pH below 5.0 inhibits the enzyme. Several fruit's PPOs including almonds, apricots, peaches, and plums generally have maximal PPO activities at around pH 5.0 [55].

12.6.2 Non-enzymatic browning

Non-enzymatic browning in dried fruits is mainly associated with carbohydrate degradation reactions [56]. The Maillard reaction, which is mainly responsible for browning in foods containing reducing sugars and free amino groups, takes place during processing or storage of many foods [57]. It is controlled by different parameters such as pH, water activity, time, and temperature, as well as the concentration of reactants [49]. Hofmann [58] hypothesized that low molecular weight chromophores rather than high molecular weight melanoidins are mainly responsible for color formation in foods *via* the Maillard reaction. On the other hand, many researchers have related brown color with the formation of high molecular weight melanoidin polymers [59, 60].

5-Hydroxymethylfurfural (HMF) is a furane compound that forms as an intermediate in the Maillard reaction [61]. It is also formed from direct dehydration of hexose sugars under acidic conditions [62]. High temperature is known to accelerate the formation of HMF in sugar rich foods such as fruits and fruit-based products.

Sanz *et al.* [49] specified 2-furoylmethyl amino acids in dates, figs, apricots, and prunes. They observed 2-furoylmethyl-proline, 2-furoylmethyl-γ aminobutyric acid, and furosine in all dried fruits analyzed. Furosine was the main compound found in figs and dried apricots. For dried apricots, the main compound was found to be 2-furoylmethyl-γ-lysine (7.74 mg/100 g), followed by 2-furoylmethyl-γ-proline (5.73 mg/100 g), and 2-furoylmethyl-γ-aminobutyric acid (3.59 mg/100 g). No 2-furoylmethyl-γ-alanine was detected in dried apricots [49].

12.7 Effects of sulfur treatment on phytochemical content of apricots

Sulfurization is a process that apricots are treated with sulfur dioxide gas or sulfur dioxide containing solutions. It is a pre-treatment that can be required to decrease the effect of spoilage reactions, to facilitate the drying process, and to improve the quality [7, 8]. It also reduces fruit darkening rate by preventing enzymatic browning during drying. Sulfur dioxide also helps in stabilizing carotenes with its oxygen-scavenging action [5]. In practice, apricots are placed in fumigation rooms where sulfur is burnt to generate sulfur dioxide gas. They are exposed to sulfur dioxide for about 8–12 hours [50]. It is also possible to treat apricots with solutions of water-soluble sulphite salts such as potassium meta-bisulphite and sodium meta-bisulphite [8, 12].

Sulfur dioxide, sodium sulphite, sodium bisulphite, potassium bisulphite, sodium meta-bisulphite, and potassium meta-bisulphite have been recognized as "Generally Recognized as Safe" in foods and drugs [63]. There is an international maximum value, 2000 ppm, for residual sulfur dioxide in apricots. The residual sulfur is very important from a sensory point of view, because it affects both the color and the taste of dried apricots [8, 12].

Compounds that inhibit enzymatic browning are classified based on the inhibition mechanism as reducing agents, chelating agents, acidulants, enzyme inhibitors, enzyme treatments, and complexing agents [48]. Reducing agents prevent accumulation of *o*-quinones and the formation of melanins. They may also form colorless stable compounds. Sulphiting agents, especially sulfur dioxide, are able to inhibit PPO directly [55]. Therefore, sulfur dioxide-treated sun-dried apricots have golden yellow color, whereas sun-dried apricots have a brown color (Figure 12.3).

Sulfur dioxide-treated sun-dried apricots have some adverse effects on health such as asthma [64]. According to World Health Organization (WHO), the acceptable daily intake of sulfur dioxide in foodstuffs has been established as 0.7 mg/kg of body weight [48, 55]. Most consumers do not prefer sulfur-treated sun-dried apricots nowadays; sun-dried apricots that

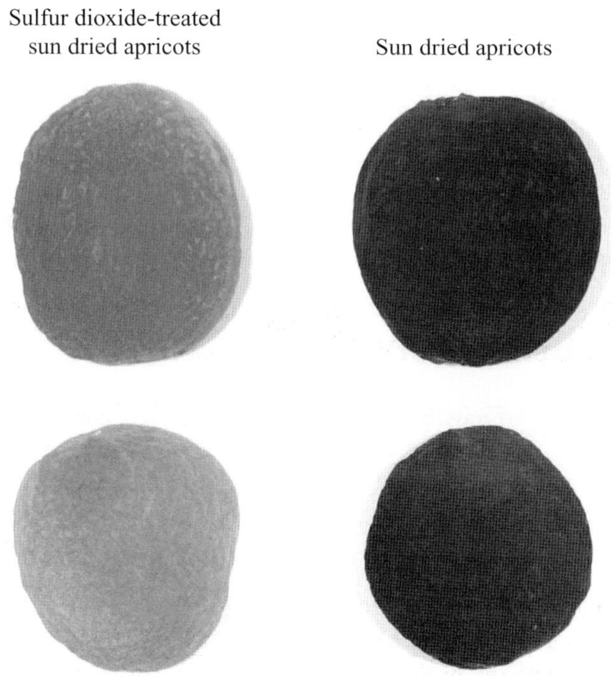

Figure 12.3 Typical images of sun-dried apricots with and without sulfur dioxide treatments. For color detail, see color plate section.

are produced without any treatments are most desirable. Consequently, sun-dried apricots have been produced as an alternative to sulfur dioxide-treated dried apricots in Turkey [38].

Sulfur dioxide treatment affects not only the color, but also phytochemicals, especially phenolic compounds. Even though sulfurization is known as an effective process in preventing the loss of natural color [10], it actually does not protect carotenoids. Dried apricots, sulfur dioxide treated or not, usually have similar contents of β-carotene [38]. However, phenolic profiles differ significantly in the two products. As shown in Figure 12.4, the most dominant phenolic compounds in sulfur dioxide-treated apricots are chlorogenic acid (167 mg/kg dry weight basis) and neochlorogenic acid (152 mg/kg dry weight basis). On the other hand, sun-dried apricots contain neither chlorogenic acid nor neochlorogenic acid [65].

The loss of phenolic compounds with sulfur dioxide treatment minimized the loss of total phenolics and total antioxidant capacity in sun-dried apricots. Thus, sulfur dioxide-treated dried apricots had a total phenolic content of 170.2 GAE/100 g compared to sun-dried apricots with 76.3 mg GAE/100 g dry weight. Dried apricots treated with sulfur dioxide have 2–6-fold more total antioxidant activity, and are measured by different assays [65].

Sulfur dioxide treatment has a significant impact on the pH of dried apricots. It was found that sulfur dioxide-treated dried apricots have an average pH of 4.20, whereas sun-dried apricots have 5.37. As a result of the dissolution of sulfur dioxide in apricot cells, pH of treated sun-dried apricots decreased by approximately one unit [65]. The pH of sun-dried apricots promotes the PPO activity, hence resulting in a complete loss of chlorogenic and neo-chlorogenic acids.

Low pH has an obvious effect on the HMF level of sulfur dioxide-treated sun-dried apricots as the formation of HMF proceeds better under acidic conditions. HMF contents of the sulfur-dioxide treated sun-dried apricots were significantly higher ($P < 0.05$) than those of the sun-dried apricots. Figure 12.5 shows HMF levels of sulfur dioxide-treated sun-dried apricots and sun-dried apricots. The mean HMF levels were 21.48 and 1.19 mg/kg for sulfur dioxide-treated dried apricots and sun-dried apricots, respectively. Figure 12.6 illustrates the chromatograms of standard solution of HMF (1 mg/kg), a sulfur dioxide-treated sun-dried apricot and a sun-dried apricot [65].

The difference in the HMF contents can be considered as a result of high sugar dehydration under acidic conditions and proceeding of Maillard reaction on the route of Schiff base and consequently HMF formation. There is no need of high temperature for HMF formation under acidic conditions as drying temperatures of about 50–60°C is suitable for HMF formation.

12.8 Health benefits of dried apricots

Foods, which are capable of providing additional physiological benefit, such as preventing or delaying the onset of chronic diseases, as well as meeting basic nutritional requirements, are described as functional foods [66]. Fruits and vegetables contain significant amounts of biologically active components and supply basic nutrition along with health benefits [67]. Dried apricots can be classified as functional foods with their concentrated nutritional content and phytochemicals. They are rich in phytochemicals, namely carotenoids and polyphenols. Consumption of plant foods rich in phenolic compounds counteracts the risk of CVD, cancer, and cataract as well as a number of other degenerative diseases [68].

The effect of dried apricots in the prevention of CVD can be explained with different mechanisms. One of these mechanisms is the effect of antioxidants present. Carotenoids and

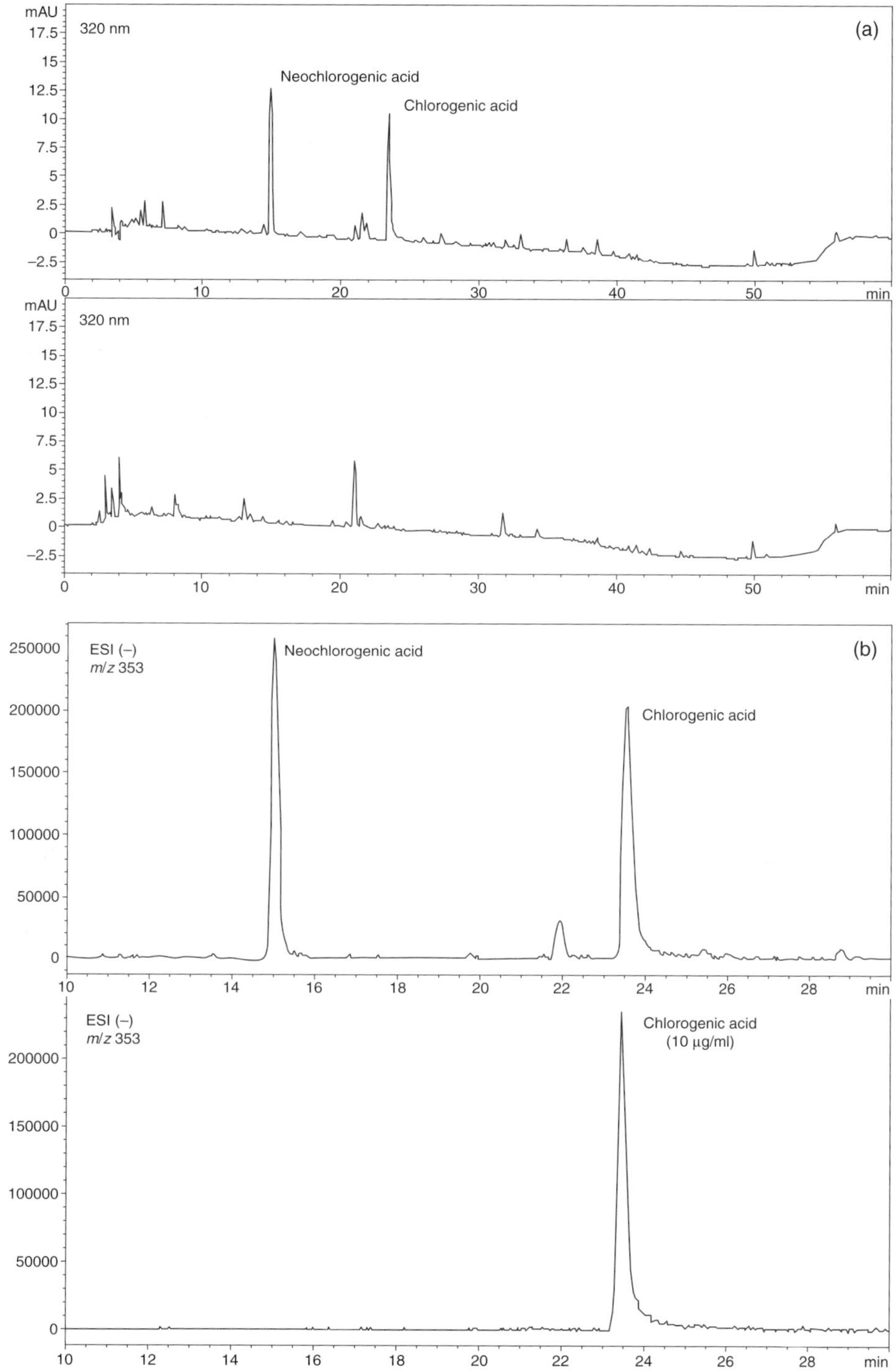

Figure 12.4 (a) HPLC chromatograms (at 320 nm) of sulfur dioxide-treated sun-dried apricots (upper panel) and sun-dried apricots (lower panel) [65]. (b) LC-MS chromatograms (at m/z 353) of sulfur dioxide-treated sun-dried apricots confirming the presence of chlorogenic and neochlorogenic acids as major phenolic compounds [65].

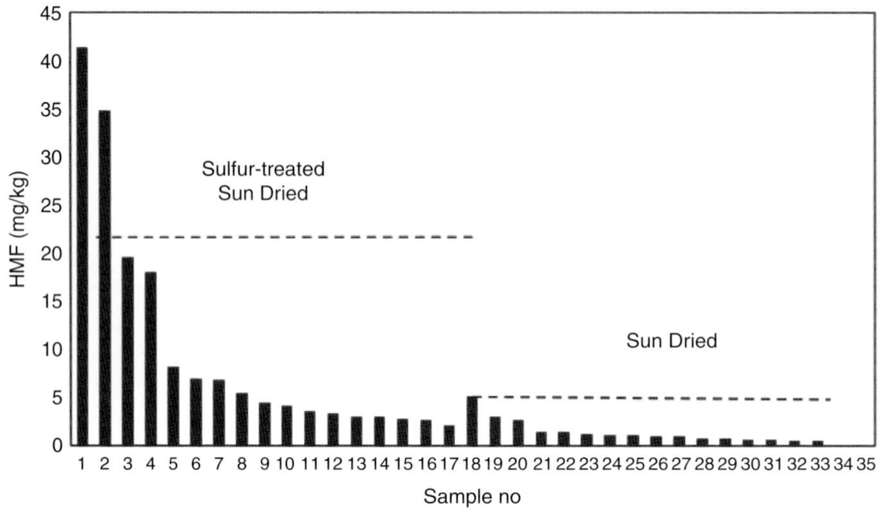

Figure 12.5 Comparison of the occurrence of HMF in sulfur dioxide-treated sun-dried apricots and sun-dried apricots [65].

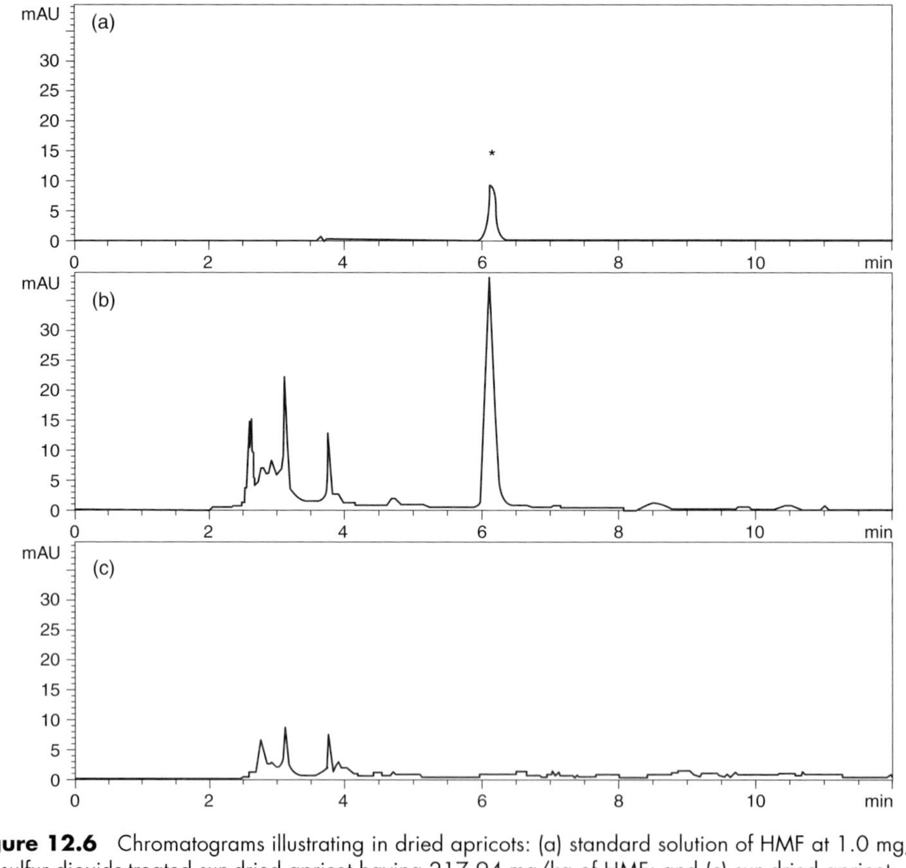

Figure 12.6 Chromatograms illustrating in dried apricots: (a) standard solution of HMF at 1.0 mg/L; (b) sulfur dioxide-treated sun-dried apricot having 217.94 mg/kg of HMF; and (c) sun-dried apricot having 0.67 mg/kg of HMF [65].

flavonols found in dried apricots as phenolic compounds play a major role as antioxidants and take part in reducing the cholesterol oxidation in the arteries. β-carotene, the dominant carotenoid found in dried apricots, is the pioneer substance related to vitamin A. It is necessary for epithelial tissues covering our bodies and organs, eye-health, bone and teeth development, and working of endocrine glades [1]. Another mechanism is, with the help of soluble dietary fiber, the reduction of serum cholesterol level [67]. Dried apricots are a good source of dietary fiber and could play a role in the prevention of obesity [69]. In addition to weight control, dietary fiber consumption helps control serum cholesterol levels and blood sugar [70].

12.9 Conclusions

Dried apricots are appealing due to their health benefits, nutritional composition, and delightful taste. Their health benefits originate from their rich dietary fiber content, polyphenols (particularly chlorogenic and neochlorogenic acids), carotenoids (particularly β-carotene), and other minor components such as vitamins and minerals. Antioxidant activity of naturally occurring polyphenols and carotenoids particularly contributes to health benefits of dried apricots by taking role in the prevention of CVD. Sulfur dioxide-treatment of apricots has both beneficial and adverse effects in terms of food quality and safety. It protects phenolic compounds against oxidation and keeps the antioxidant activity high, but it also accelerates HMF formation due to higher acid conditions arising from the dissociation of sulfur dioxide.

References

1. Haciseferogullari, H., Gezer, I., Ozcan, M.M. & Asma, B.M. (2007) Post harvest chemical and physical-mechanical properties of some apricot varieties cultivated in Turkey. *Journal of Food Engineering*, **79**, 364–373.
2. Ercisli, S. (2009) Apricot culture in Turkey. *Scientific Research and Essay*, **4**, 715–719.
3. Drogoudi, P.D., Vemmos, S., Pantelidis, G., Petri, E., Tzoutzoukou, C. & Karayiannis, I. (2008) Physical characters and antioxidant, sugar, and mineral nutrient contents in fruit from 29 Apricot (*Prunus armeniaca* L.) cultivars and hybrids. *Journal of Agricultural and Food Chemistry*, **56**, 10754–10760.
4. FAO (2009) *FAO Statistical Database*. Published online at: http://faostat.fao.org/site/339/default. aspx, last accessed 30 August 2011.
5. Mir, M.A., Hussain, P.R., Fouzia, S. & Rather, A.H. (2009) Effect of sulphiting and drying methods on physico-chemical and sensorial quality of dried apricots during ambient storage. *International Journal of Food Science & Technology*, **44**, 1157–1166.
6. Rosello, C., Canellas, J., Santiestaban, I. & Mullet, A. (1993) Simulation of absorption process of sulphur dioxide in apricots. *Lebensmittel-Wissenschaft und-Technologie- Food Science and Technology*, **26**, 322–328.
7. Lewicki, P.P. (2006) Design of hot air drying for better foods. *Trends in Food Science & Technology*, **17**, 153–163.
8. Togrul, T.I. & Pehlivan, D. (2003) Modelling of drying kinetics of single apricot. *Journal of Food Engineering*, **58**, 23–32.
9. Sabry, Z.I. (1961) Browning in dried fruit products: non-enzymatic browning and its effects on the carotenoids in Qamareddeen, a dried apricot pulp. *Food Discoloration*, **1**, 53–55.
10. Heid, J.L. & Joslyn, M.A. (1967) *Fundamentals of Food Processing Operations*. AVI Publishing Company Inc., Westport, CT.
11. Vagenas, G.K. & Marinos-Kouris, D. (1991) Drying kinetics of apricots. *Drying Technology*, **9**, 735–752.

12. Kostaropoulos, A.E. & Saravacos, G.D. (1995) Microwave pre-treatment for sun dried raisins. *Journal of Food Science*, **60**, 344–347.
13. Sokhansanj, S. & Jayas, D.S. (1995) Drying of foodstuffs. In: *Handbook of Industrial Drying*, (ed. A.S. Mujumdar). CRC Press, New York, NY, pp. 589–625.
14. Imre, L. (2007) Solar drying. In: *Handbook of Industrial Drying*, 3rd edn. (ed. A.S. Mujumdar). CRC Press, Taylor & Francis Group, Boca Raton, FL, pp. 308–356.
15. Bennett, L.E., Singh, D.P. & Clingeleffer, P.R. (2011) Micronutrient mineral and folate content of Australian and imported dried fruit products. *Critical Reviews in Food Science and Nutrition*, **51**, 38–49.
16. Muller, L., Fröhlich, K. & Böhm, V. (2011) Comparative antioxidant activities of carotenoids measured by ferric reducing antioxidant power (FRAP), ABTS bleaching assay (αTEAC), DPPH assay and peroxyl radical scavenging assay. *Food Chemistry*, **129**, 139–148.
17. Saracoglu, S., Tuzen, M. & Soylak, M. (2009) Evaluation of trace element contents of dried apricot samples from Turkey. *Journal of Hazardous Materials*, **167**, 647–652.
18. USDA (2011) *National Nutrient Database for Standard Reference, Release 24.* Published online at: http://www.nal.usda.gov/fnic/foodcomp/cgi-bin/list_nut_edit.pl, last accessed 28 November 2011.
19. Gezer, I., Haciseferogullari, H. & Demir, F. (2003) Some physical properties of Hacihaliloglu apricot pit and its kernel. *Journal of Food Engineering*, **56**, 49–57.
20. Akin, E.B., Karabulut, I. & Topcu, A. (2008) Some compositional properties of main Malatya apricot (*Prunus armeniaca* L.) varieties. *Food Chemistry*, **107**, 939–948.
21. Munzuroglu, O., Karatas, F. & Geckil, H. (2003) The vitamin and selenium contents of apricot fruit of different varieties cultivated in different geographical regions. *Food Chemistry*, **83**, 205–212.
22. Orhan-Erdogan, I. & Kartal, M. (2011) Insights into research on phytochemistry and biological activities of *Prunus armeniaca* L. (apricot). *Food Research International*, **44**, 1238–1243.
23. Food and Nutrition Board DRI Committee (2001) *Dietary Reference Intakes for Vitamin A, Vitamin K, Arsenic, Boron, Chromium, Copper, Iodine, Iron, Manganese, Molybdenum, Nickel, Silicon, Vanadium, and Zinc*. Food and Nutrition Board Institute of Medicine, Washington, DC.
24. Food and Nutrition Board DRI Committee (2001) *Dietary Reference Intakes for Vitamin C, Vitamin E, Selenium, and Carotenoids*. Food and Nutrition Board Institute of Medicine, Washington DC.
25. US National Institute of Health (2011) *Health Topics.* Published online at: http://www.nlm.nih.gov/medlineplus/minerals.html, last accessed 28 November 2011.
26. Dragovic-Uzelac, V., Levaj, B., Mrkic, V., Bursac, D. & Boras, M. (2007) The content of polyphenols and carotenoids in three apricot cultivars depending on stage of maturity and geographical region. *Food Chemistry*, **102**, 966–975.
27. Fernandez de Simon, B., Perez-Ilzarbe, J., Hernandez, T., Gomez-Cordoves, C. & Estrella, I. (1992) Importance of phenolic compounds for the characterization of fruit juices. *Journal of Agricultural and Food Chemistry*, **40**, 1531–1535.
28. Bhagwat, S., Haytowitz, D.B. & Holden, J.M. (2011) *USDA Database for the Flavonoid Content of Selected Foods*, Release 3. USDA, Beltsville, MD.
29. Nutrient Data Laboratory (2004) *USDA Database for the Proanthocyanidin Content of Selected Foods.* USDA, Beltsville, MD.
30. Nutrient Data Laboratory (2008) *USDA Database for the Isoflavone Content of Selected Foods, Release* 2. USDA, Beltsville, MD.
31. Radi, M., Mahrouz, M. & Jaouad, A. (1997) Phenolic composition, browning, susceptibility, and carotenoid content of several apricot cultivars at maturity. *HortScience*, **32**, 1087–1091.
32. Ruiz, D., Egea, J., Gil, M.I. & Tomas-Barberan, F.A. (2005) Characterization and quantitation of phenolic compounds in new apricot (*Prunus armenica* L.) varieties. *Journal of Agricultural and Food Chemistry*, **53**, 9544–9552.
33. Sochor, J., Zitka, O., Skutkova, H., Pavlik, D., Babula, P., Krska, B., Horna, A., Adam, V., Provaznik, I. & Kizek, R. (2010) Content of phenolic compounds and antioxidant capacity in fruits of apricot genotypes. *Molecules*, **15**, 6285–6305.
34. Liggins, J., Bluck, L.J.C., Runswick, S., Atkinson, C., Coward, W.A. & Bingham, S.A. (2000) Daidzein and genistein content of fruits and nuts. *Journal of Nutritional Biochemistry*, **11**, 326–331.
35. Kuhnle, G.G.C., Dell'Aquila, C., Aspinall, S.M., *et al.* (2009) Phytoestrogen content of fruits and vegetables commonly consumed in the UK based on LC-MS and 13C-labelled standards. *Food Chemistry*, **116**, 542–554.

36. Ruiz, D., Reich, M., Bureau, S., Renard, C.M.G.C. & Audergon, J.M. (2008) Application of reflectance colorimeter measurements and infrared spectroscopy methods to rapid and nondestructive evaluation of carotenoids content in apricot (*Prunus armeniaca* L.). *Journal of Agricultural and Food Chemistry*, **56**, 4916–4922.
37. Curl, A.L. (1960) The carotenoids of apricots. *Journal of Food Science*, **25**, 190–196.
38. Karabulut, I., Topcu, A., Duran, A., Turan, S. & Ozturk. B. (2007) Effect of hot air drying and sun drying on color values and β-carotene content of apricot (*Prunus armenica* L.). *Lebensmittel-Wissenschaft und-Technologie*, **40**, 753–758.
39. Kurz, C., Carle, R. & Chieber, A. (2008) HPLC–DAD–MS characterization of carotenoids from apricots and pumpkins for the evaluation of fruit product authenticity. *Food Chemistry*, **110**, 522–530.
40. Halvorsen, B.L., Holte, K., Myhrstad, M.C.W., Barikmo, I., Hvattum, E., Remberg, S.V., Wold, A.B., Haffner, K., Baugerod, H., Andersen, L.F., Moskaug, O., Jacobs, D.R. & Blomhoff, R. (2002) A systematic screening of total antioxidants in dietary plants. *Journal of Nutrition*, **132**, 461–471.
41. Bacchiocca, M., Biagiotti, E. & Ninfali, P. (2006) Nutritional and technological reasons for evaluating the antioxidant capacity of vegetable products. *Italian Journal of Food Science*, **18**, 209–217.
42. Kevers, C., Falkowski, M., Tabart, J., Defraigne, J-O., Dommes, J. & Pincemail, J. (2007) Evolution of antioxidant capacity during storage of selected fruits and vegetables. *Journal of Agricultural and Food Chemistry*, **55**, 8596–8603.
43. Wu, X., Beecher, G.R., Holden, J.M., Haytowitz, D.B., Gebhardt, S.E. & Prior, R.L. (2004) Lipophilic and hydrophilic antioxidant capacities of common foods in the United States. *Journal of Agricultural and Food Chemistry*, **52**, 4026–4037.
44. Bennett, L.E., Jegasothy, H., Konczak, I., Frank, D., Sudharmarajan, S. & Clingeleffer, P.R. (2011) Total polyphenolics and antioxidant properties of selected dried fruits and relationships to drying conditions. *Journal of Functional Foods*, **3**, 115–124.
45. Igual, M., Garcia-Martinez, E., Camacho, M.M. & Martinez-Navarrete, N. (2010) Effect of thermal treatment and storage on the stability of organic acids and the functional value of grapefruit juice. *Food Chemistry*, **118**, 291–299.
46. Guclu, K., Altun, M., Ozyurek, M., Karademir, S.E. & Apak, R. (2006) Antioxidant capacity of fresh, sun- and sulphited-dried Malatya apricot (*Prunus armeniaca*) assayed by CUPRAC, ABTS/TEAC and folin methods. *International Journal of Food Science & Technology*, **41**, 76–85.
47. Reische, D.W., Lillard, D.A. & Eitenmiller, R.R. (2002) Antioxidant. In: *Food Lipids: Chemistry, Nutrition, and Biotechnology*, 2nd edn. (eds C.C. Akoh & D.B. Min). CRC Press, Taylor and Francis Group, Boca Raton, FL, p. 525.
48. Igual, M., Garcia-Martinez, E., Martin-Esparza, M.E. & Martinez-Navarrete, N. (2012) Effect of processing on the drying kinetics and functional value of dried apricot. *Food Research International*, **47**, 284–290.
49. Sanz, M.L., Castillo, M.D., Corzo, N. & Olano, A. (2001) Formation of amadori compounds in dehydrated fruits. *Journal of Agricultural and Food Chemistry*, **49**, 5228–5231.
50. Queiroz, C., Mendes Lopes, M.L., Fialho, E. & Valente Mesquita, V.L. (2008) Polyphenol oxidase: characteristics and mechanisms of browning control. *Food Reviews International*, **24**, 361–375.
51. Femenia, A., Sanchez, E.S., Simal, S. & Rossello, C. (1998) Development and ripening-related effects on the cell wall of apricot (*Prunus armeniaca*) fruit. *Journal of the Science of Food and Agriculture*, **77**, 487–493.
52. Acar, J. & Gökmen, V. (2005) Meyve ve sebzelerin işlenmelerinde olusan kimyasal değisimler. In: *Meyve ve Sebze Isleme Teknolojisi, Cilt 1- Meyve ve Sebze Suları Uretimi* (eds J. Acar & V. Gökmen). Hacettepe Universitesi Yayınları, Ankara, Turkey [in Turkish], pp. 69–156.
53. Roussos, P.A., Sefferou, V., Denaxa, N.K., Tsantili, E. & Stathis, V. (2011) Apricot (*Prunus armeniaca* L.) fruit quality attributes and phytochemicals under different crop load. *Scientia Horticulturae*, **129**, 472–478.
54. Matheis, G. (1983) Enzymatic browning of foods—Quantitative relationships between browning and food constituents. *Zeitschrift for Lebensmittel-Untersuchung und-Forschung*, **176**, 454–462.
55. Yoruk, R. & Marshall, M.R. (2003) Physicochemical properties and function of plant polyphenol oxidase: a review. *Journal of Food Biochemistry*, **27**, 361–422.
56. BeMiller, J.N. & Whistler, R.L. (1996) Carbohydrates. In: *Food Chemistry* (ed. O.R. Fennema). Marcel Dekker, New York, NY, pp. 157–224.

57. Serpen, A., Capuano, E., Fogliano, V. & Gökmen, V. (2007) A new procedure to measure the antioxidant activity of insoluble food components. *Journal of Agricultural and Food Chemistry*, **55**, 7676–7681.
58. Hofmann, T. (1998) Characterization of the chemical structure of novel colored Maillard reaction products from furan-2-carboxaldehyde and amino acids. *Journal of Agricultural and Food Chemistry*, **46**, 932–940.
59. Bailey, R.G., Ames, J.M. & Monti, S.M. (1996) An analysis of the non-volatile reaction products of aqueous Maillard model system at pH 5, using reversed phase HPLC with diode-array detection. *Journal of the Science of Food and Agriculture*, **72**, 97–103.
60. Caemmerer, B. & Kroh, L.W. (1995) Investigation of the influence of reaction conditions on the elementary composition of melanoidins. *Food Chemistry*, **53**, 55–59.
61. Ames, J.M. (1992) The Maillard reaction. In: *Biochemistry of Food Proteins* (ed. B.J.F. Hudson). Elsevier Applied Science, London, UK, pp. 99–105.
62. Kroh, L.W. (1994) Caramelisation in food and beverages. *Food Chemistry*, **51**, 373–379.
63. Taylor, S.L., Higley, N.A. & Bush, R.K. (1986) Sulfites in foods: uses, analytical methods, residues, fate, exposure assessment, metabolism, toxicity, and hypersensitivity. *Advances in Food Research*, **30**, 1–76.
64. Koksal, N., Hasanoglu, H.C., Gokirmak, M., Yildirim, Z. & Gultek, A. (2003) Apricot sulphurization: an occupation that induces an asthma like syndrome in agricultural environments. *American Journal of Industrial Medicine*, **43**, 447–453.
65. Göncüoğlu, N., Mogol, B.A., Durmaz, G. & Gökmen, V. (2010) HMF formation in dried apricots. Paper given at *5th Central European Congress on Food*, Bratislava, Slovak Republic, 19–22 May, 2010 (Abstract No: 128).
66. Scalzo, J., Politi, A., Pellegrini, N., Mezzetti, B. & Battino, M. (2005) Plant genotype affects total antioxidant capacity and phenolic contents in fruit. *Nutrition*, **21**, 207–213.
67. Oomah, B.D. & Mazza, G. (2000) Functional foods. In: *The Wiley Encyclopedia of Science & Technology*, 2nd edn. (ed. F.J. Francis). Wiley, New York, NY, pp. 1176–1182.
68. Shahidi, F. & Naczk, M. (2003) *Phenolics in Food and Nutraceuticals*. CRC Press, Boca Raton, FL.
69. Van Duyn, M.A.S. & Pivonka, E. (2000) Overview of the health benefits of fruit and vegetable consumption for the dietetics professional: selected literature. *Journal of the American Dietetic Association*, **100**, 1511–1521.
70. Slavin, J.L. (2008) Position of the American Dietetic Association: health implications of dietary fiber. *Journal of the American Dietetic Association*, **108**, 1716–1731.

13 Dried cherries: phytochemicals and health perspectives

Letitia McCune

13.1 Introduction

The United States Department of Agriculture (USDA) has recognized dried cherries as an expanding market for cherries produced in the United States [1]. Increased attention on cherries in general has been a result of cherries labeled as a "super food" and a "super fruit" following the recognition of an increased number of research articles focused on cherries' high antioxidant levels and associated health benefits [2–4]. Considering that increasing consumption of fruits and vegetables is recognized as a method of improving overall health through nutrition, and that the dried fruit consumption in particular is an indicator of better nutrient intake and lower body weight measurements [5], dried cherries are a means of increasing health through consumption. This chapter presents some of the latest data on nutritive values and antioxidant phytochemicals as well as the potential health benefits of dried cherries in relation to cancer, cardiovascular disease (CVD), diabetes, and inflammation.

13.2 Production

Dried cherries are most often made from sour cherries (*Prunus cerasus* L.), for storage and supplementary sugar addition reasons, but they are also made from sweet cherries (*Prunus avium* L.). Both sweet and sour cherries are native to the European and Western Asia regions and currently are grown worldwide, especially in northern temperate regions. The Food and Agriculture Organization (FAO) 2008 data show Turkey as the leading producer of cherries at 338,361 metric tonnes (MT) followed by the United States at 225,073 MT. Under the category of sour cherries, Poland was the leading producer at 201,681 MT followed by Turkey at 185,435 MT with the United States at sixth with 97,250 MT [6]. In the United States, Michigan is the leading state in producing tart cherries while Washington is the leading state in producing sweet cherries.

Dried cherries are recognized as a form of processed cherries usually, but not always, produced from tart or sour cherries. There is limited data that separates out dried from

Dried Fruits: Phytochemicals and Health Effects, First Edition. Edited by Cesarettin Alasalvar and Fereidoon Shahidi.

Table 13.1 US cherry production and utilization amounts (2010 values)

	Tart cherry (in thousand MT)	Sweet cherry (in thousand MT)
Total produced	86.4	283.7
Fresh produced	0.4	225.3
Total processed	82.7	53.3
Frozen	56.2	–
Canned	16.0	2.8
Brined	–	30.9
Others	10.5[a]	19.7[b]

Source: Adapted from USDA [7].
[a]Includes juice, wine, brined, and dried.
[b]Includes California canned and all others processed (dried, frozen, and juice) from other states.

other forms of processed cherries. In 2010, the United States had bearing acreage of sweet cherries in the amount of 88,030 with a total production of 283.7 thousand MT and utilizing 225.3 thousand MT as fresh and 53.3 thousand MT as processed (Table 13.1). Of this, 19.7 thousand MT is for some forms of processed sweet cherries including dried. For tart cherries, the bearing acreage was 35,650 with a total production of 86.4 thousand MT with 0.4 thousand MT used as fresh and 82.7 thousand MT processed. Of those processed, 10.5 thousand MT fall under the category that includes dried cherries [7]. In 2004, 166 firms were recognized as processors of tart cherries in the United States. A survey sent to cherry processor firms in Michigan found that approximately 30% recognized that there had been an increase in dried cherries produced over the last 5 years, while approximately 70% of these firms recorded no change in dried fruit production (none were noted as recognizing a decrease) [8].

In general, the United States utilizes approximately 8% of all its non-citrus fruits, including cherries, in a dried form (2035.3 thousand MT in 2010) [7]. In 2005, the United States documented importing 188 MT of dried cherries. These numbers have gradually increased over the prior of 10 years with a spike in 2002 of 568 MT [9]. Customarily, it is the tart or the sour cherries that are dried considering the percentage of the sweet cherry harvest that is eaten fresh and the potential to add sugar to the dried tart cherry product to improve palatability. The Large Montmorency and Early Richmond are the typical US tart cherries that are dried while in the sweet cherry category it is often the Royal, Napoleon, Bing, and Lambert cultivars.

13.3 Methods of drying

Traditionally, cherries have been dried using open air drying in the sun or solar methods utilizing fans and greenhouses. Commercial drying techniques can also include chemical pretreatments to decrease loss of product and speed drying time. These methods often target the waxy cuticle of the fruit, creating small cracks to speed drying in addition to providing some antibacterial or antifungal properties. Fruit can be dipped in solutions of fatty acid esters (e.g., ethyl oleate), NaOH, Na_2CO_3, K_2CO_3, or citric acid at different concentrations and for different lengths of time. A study on sour cherries found that the best retention of color after drying was produced using a room temperature dipping solution of 2% ethyl oleate for 1 minute followed by drying at 45.2°C. [10]. Sulfur dioxide is sometimes added to prolong shelf life. Sweet cherries consumed fresh, on the other hand, can have post-harvest applications of oils and acids to delay ripening and deterioration before purchase [11]. With the increased demand for organic fruit products, many of these applications are being limited.

13.4 Nutritional characteristics

In contrast to fresh cherries, dried cherries can have the addition of oils, sugars, flavorings, or chocolate to the finished product. These additions can significantly change the nutritive values for dried cherries increasing caloric, fat, and carbohydrate levels as well as affecting glycemic index (GI). The contrast of these nutrient values between fresh and dried cherries can be seen in Table 13.2. Considering that 1 kg of dried cherries is produced from 6 to 8 kg of fresh cherries (cherry marketing board website facts), the amounts of calories and other components are increased in the same 100 g portion. However, a serving of dried fruit is usually considered less than half a cup (60 g) while a serving of fresh fruit is often 1 cup (140 g). Consuming dried cherries over fresh produce, therefore, not only has the potential to increase calories and sugars but also the amounts of protein, fiber, and vitamins.

13.4.1 Protein

The amount of protein in dried cherries is over three times that of fresh cherries. Dried cherries are, therefore, an efficient means to increase plant protein in the diet considering that the increased ratio of plant protein to animal protein in the diet is associated with increased health and decreased obesity [14]. This is also in agreement with the higher level of fruit and vegetable component of the healthy "Mediterranean diet" as compared to the unhealthy "Western diet" [15, 16]. In addition, the consumption of dried fruits has been linked to better quality dietary intakes and lower body weight [5]. Considering the many health benefits of increasing fruit and vegetable consumption, dried cherries is one method of incorporating fruit into the diet year round.

Table 13.2 Some nutrition facts of dried and fresh cherries (tart and sweet) per 100 g

Nutrient	Unit	Dried cherries [12]	Tart cherries [13]	Sweet cherries [13]
Energy	kcal	340	50	63
Protein	g	3.29	1.0	1.06
Fat	g	0.1	0.3	0.2
Carbohydrate	g	82.2	12.2	16.0
Sugars	g	68.3	8.5	12.8
Fiber	g	2.57	1.6	2.1
Vitamin C	mg	<0.5	10	7
Vitamin A	IU	3701	1283	64
Potassium	mg	416	173	222

13.4.2 Fiber

Consuming dried cherries provides a higher amount of fiber in the diet than the same 100 g of fresh cherries (Table 13.2). At a rate of 2.57 g/100 g, this can contribute to the suggested consumption of 14 g/1000 kcal intake. Diets high in fiber have been linked to increased health and weight loss through decreases in cholesterol and blood glucose as well as through increased levels of satiety [17, 18]. While there has been some concern that higher levels of fruit fibers detrimentally affect the bioavailability of dietary antioxidants, it is likely the absorption of the antioxidants occurs in the colon rather than in the small intestine [19].

13.4.3 Vitamins

Vitamin C is decreased in the process of drying of cherries (Table 13.2). In studies of apple juice, it has been shown that UV radiation and higher temperatures lead to vitamin C degradation [20]. Such processing often results in the conversion of L-ascorbic acid (vitamin C) to dehydroascorbic acid and subsequent loss of biological activity.

Vitamin A, on the other hand, is present at a higher level in the dried product as compared to the fresh fruit. Dried cherries constitute a significant source of vitamin A; the amount in 100 g equaling 75% of the recommended daily intake (RDI) in a 2000-calorie diet. Vitamin A benefits many physiological functions including immunity, inflammation, vision, and reproduction. Deficiencies in vitamin A are particularly significant in malnourished populations contributing to infant mortality and blindness. Supplementation of wheat flour with vitamin A in relief projects has been recommended [21] as well as projects to recognize the importance of traditional sources of vitamin A in many Indigenous Peoples' food systems such as in Pohnpei, for the Karen People in Thailand, and the Awajun in Peru, etc. [22, 23].

Dried cherries also serve as a good source of potassium with amounts of 416 mg/100 g [12], almost twice that in fresh sweet cherries (222 mg/100 g) [13]. Higher consumption of potassium rich foods in conjunction with less sodium has been described in many studies as a means to lower high blood pressure and stroke risk [24, 25]. The Dietary Approaches to Stop Hypertension (DASH) intervention trial [26] consisted of a diet high in potassium rich foods including a high intake of fruits and vegetables and a reduced intake of saturated and total fat. High compliance with this diet as well as an increase in calcium and weight control has been shown to reduce hypertension [27]. Low potassium affects the hormones angiotension II and aldosterone linked to sodium and potassium transport and cardiorenal damage as well as the adenosine triphosphate (ATP)-sensitive potassium channels responsible for maintaining glucose homeostasis [28, 29].

13.5 Antioxidant phytochemicals

Cherries are a good source of phenolic antioxidants with many different flavonoids including various anthocyanins [4]. Flavonols have been identified as the major class of polyphenols in tart cherries [30] with 7-dimethoxy-5,8,4′-trihydroxyflavone being the most active one in some antioxidant assays [31]. When tart cherries are converted to dried cherries, many derivatives of the anthocyanins and cyanidins appear [32]. Table 13.3 illustrates the amounts of many of these compounds in dried cherries from two tart cultivars, namely Balaton and Montmorency.

Table 13.3 Antioxidant components of two cultivars of tart cherries

Antioxidant component	Montmorency (dried without sugar)	Montmorency (dried with sugar)	Balaton (dried without sugar)	Balaton (dried with sugar)
Total phenolics[a]	7813 ± 855	5103 ± 455	6343 ± 776	3522 ± 512
Total anthocyanins[b]	173 ± 31	62 ± 5.3	564 ± 65	273 ± 33
Cyanidin 3-sophoroside[c]	4.6 ± 0.8	1.9 ± 0.6	15.7 ± 4.3	13.9 ± 4.2
Cyanidin 3-glucosylrutinoside[c]	33.6 ± 6.4	11.1 ± 4.7	203.6 ± 44.2	64.8 ± 9.2
Cyanidin-3-glucoside[c]	0.7 ± 0.3	nd	7.6 ± 0.9	3.6 ± 0.7
Cyanidin-3-rutinoside[c]	19.5 ± 4.8	6.9 ± 0.9	24.9 ± 6.3	19.5 ± 4.8
Peonidin-3-glucoside[c]	4.5 ± 0.9	1.1 ± 0.6	16.2 ± 4.2	4.2 ± 0.9
Cyanidin[c]	0.3 ± 0.2	nd	1.6 ± 0.7	nd
Pelargonidin[c]	0.8 ± 0.3	0.3 ± 0.2	0.4 ± 0.2	nd
Isorhamnetin rutinoside[c]	383.1 ± 62.1	203 ± 52.1	35.8 ± 8.5	158.7 ± 44.5
Kaempferol[c]	16.9 ± 3.8	12.9 ± 4.1	42.9 ± 6.7	16.9 ± 4.4
Quercetin[c]	7.5 ± 0.9	8.8 ± 0.9	3.1 ± 0.8	1.9 ± 0.8

Source: Adapted with permission from Kirakosyan *et al.* [32].
Data are expressed as means ± standard deviation ($n = 3$) on a dry weight basis.
nd, not detected.
[a] μg of gallic acid equivalents (GAE)/g.
[b] μg of cyanidin 3-glucoside equivalents (C3GE)/g.
[c] μg/g biomass.

Oxygen radical absorption capacity (ORAC) measurements, a standard antioxidant calculation based on μmoles of trolox equivalents (TE) per 100 g sample, indicate that dried cherries have an ORAC value of 6800 versus 2033 for frozen tart cherries and 1700 for canned tart cherries [2]. These high antioxidant values are likely due to the presence of antioxidants in the skins (Table 13.4). When scrutinizing the antioxidants present, one must consider the relative activity of each antioxidant. Kirakosyan *et al.* [32] suggests that cyanidin and its derivatives are the primary contributors to antioxidant activity in tart cherries. Processing of tart cherries into a dried form, however, has the potential of decreasing certain antioxidants when increased temperature is used in the process. Some anthocyanins may be unstable during processing (hence the derivatives of cyanidins seen in Table 13.3) and the

Table 13.4 Montmorency tart cherry antioxidant measures

Portion	Anthocyanins (mg of C3GE/100 g)	Total phenolics (mg of GAE/g)	ORAC (μmoles of TE/g)
Edible portion	8.7 ± 0.8	4.07 ± 0.18	25.57 ± 3.99
Flesh	0.0 ± 0.09	3.01 ± 0.29	15.00 ± 1.00
Pits	0.8 ± 0.08	1.57 ± 0.02	9.78 ± 0.28
Skins	36.5 ± 1.6	5.58 ± 0.33	51.02 ± 1.97

Source: Adapted with permission from Chaovanalikit and Wrolstad [30].
Data are expressed as means ± standard deviation ($n = 2$) on a fresh weight basis.
C3GE, cyanidin 3-glucoside equivalents; GAE, gallic acid equivalents; ORAC, oxygen radical absorbance capacity; TE, trolox equivalents.

antioxidant melatonin may also be affected [32]. To preserve melatonin, Kirakosyan *et al.* [32] indicated that individually quick freezing may provide the best method of processing tart cherries. Brining and canning of the fruit contributes to significant loss of antioxidants into the accompanying solution [30]. In contrast, dried cherries are more antioxidative than frozen, concentrated, or individually quick frozen products of the same cultivars [32]. Adding sugar to dried cherries appears to decrease the antioxidants available but still allows for antioxidant capacities to be substantial [32]. Synergistic actions of the antioxidant compounds in tart cherries contribute to its overall high antioxidant activity [33].

The choice of cultivar as well as elevation and harvesting time can have significant effects on the amount of antioxidants. Differences between the tart cherry cultivars Mortmorency and Balaton can be seen in Table 13.3. Balaton has higher levels of most of the anthocyanins while Mortmorency has higher total phenolics. Significant differences exist between early stage ripening and full ripened stages of harvest for dried sour cherries [34]. In addition, the growing location can affect antioxidant levels. Higher elevations were confirmed to increase phenolic and antioxidant content in four cultivars of sweet cherries in Greece [35]. This is in agreement with theories and studies on ecological conditions related to phytochemical production and antioxidant rates [36]. The stress conditions of the plant (including amounts of UV radiation, drought, and metals in the soil) can increase the antioxidant levels in the harvested plant part. In the conversion of grapes to wine the harvest year, with its associated differences in ecological stresses, can have an effect on phenolics and antioxidants. Further studies are needed in sour cherry cultivars.

Sour cherries have been ranked as one of the top 50 foods in terms of their antioxidant activity, according to the ferric acid in plasma assay (FRAP) analysis of 113 foods [37]. These rates were above that of such high antioxidant producers as grape juice, plums, and coffee. Sour cherries also ranked as 14th in their content of antioxidants present per serving; this is above the antioxidant content of red wine, cranberry juice, and oranges. It remains to be seen what ranking dried cherries would have considering the decrease in some antioxidant rates with processing (Table 13.3). Conversely, some of the antioxidants present in Table 13.3 have been shown to act synergistically, creating higher antioxidant activity in combination, particularly the pairing of cyanidin-3-rutinoside and isorhamnetin-3-rutinoside [33].

13.6 Health benefits

13.6.1 Inflammation

Cherries are a good source of antioxidants, many of which could influence the immune system. The immune system includes the production of pro-inflammatory cytokines such as interleukin (IL)-1, IL-6, and tumor necrosis factor (TNF)α that result in oxidative stress leading to tissue damage and pain. Antioxidants can influence the production of the cytokines, mediate the oxidative stress through the scavenging of free radicals and enhance endogenous antioxidant systems that help in tissue repair and recovery [38]. Natural antioxidants have been studied using the inhibition of the cyclooxygenase (COX-I and COX-II) enzymes that result in the production of prostaglandins leading to inflammation. The inhibition of these enzymes is the basis of the reduction of pain and inflammation by such products as ibuprofen.

In a cell system, studying these cyclooxygenase enzymes, Seeram *et al.* [39] found an inhibition of 25% in a COX-I assay and 38% in a COX-II assay using the anthocyanins from tart cherries. These results were over 50% that of ibuprofen in the COX-I assay and comparable to the results for ibuprofen in the COX-II assay, thus indicating strong anti-inflammatory activity of isolated antioxidants from tart cherries (especially cyanidin). In some cases, the effect was stronger than that of aspirin in a human prostaglandin endoperoxide H synthase-1 (PGHS-1) isozyme assay [31].

In the Zucker obese rat model, Seymour *et al.* [40] found that rats fed a high fat diet in conjunction with whole tart cherry powder (1% weight/weight) for 90 days had a reduced amount of systemic and local inflammation as indicated by lowered levels of plasma and IL-6. Both this study and a previous one [41] on lean rats fed a low fat diet with cherry powder, found an affect on peroxisome proliferator-activated receptors (PPARs) often involved in inflammation.

A number of human trial studies, although of relatively small sample size, have followed these positive studies on anti-inflammatory activity of tart cherries in cellular systems and animal models. Recent studies have focused on exercise recovery using markers associated with inflammation and pain that represent a reduction of the damage caused by the inflammatory response after myofibril initial damage and the resulting oxidative stress. Three human studies have used a tart cherry juice blend with anthocyanin contents that can be extrapolated to dried cherry use. The drink was pasteurized and contained approximately 690 mg phenolics (gallic acid equivalents) and 46 mg anthocyanins (cyanidin-3-glucoside) consumed twice per day. For the same amount of phenolics, about three-fourth cup of unsweetened dried tart cherries (according to Table 13.3) would be required.

When long distance runners ($n = 54$) ingested the cherry blended drink for 7 days before and during running, there was a significant reduction in the pain associated on the day of the race. Results were assessed using a visual analog scale of pain intensity [42]. Howatson *et al.* [43], using the same amount of juice blend product, found marathon runners ($n = 20$) that consumed the cherry blend had significantly lower inflammation [IL-6, C-reactive protein (CRP), uric acid], increased total antioxidants, lower levels of oxidative stress [lipid peroxidation *via* thiobarbituric acid reactive substances (TBARS)] and quicker return of knee muscle strength as compared to the placebo. When healthy older individuals ($n = 12$) consumed the cherry juice blend at a slightly lower dose over 14 days, decreased damage from forearm ischemia/reperfusion was noted as witnessed by the significant difference in integrated plasma F_2- isoprostane produced. The area-under-the-curve (AUC) pre-post as compared to the placebo was significantly lower as was the production of urinary oxidized nucleic acids [44]. Another study using a cherry concentrate (270 mg total anthocyanins twice per day for 10 days) with men ($n = 10$) showed an effect on leg muscle exercise recovery after 7 days of consumption [45]. Force recovery was found to be faster, while the total and absolute increase in protein carbonyl activity (a measurement of oxidized proteins) was lower with cherry consumption.

Another potential use of the anti-inflammatory properties of the anthocyanins in tart cherries is for the alleviation of symptoms of inflammatory diseases such as arthritis and gout. He *et al.* [46] found in a rat model of human rheumatoid arthritis that a reduction of the amounts of prostaglandin E2 associated with inflammation following an injury was achieved with the administration of 40 mg/kg of anthocyanins from sour cherries for 28 days. Unfortunately, in humans (weighing 67.5 kg) the amount of anthocyanins would be 2720 mg/day which is close to 40 cups of dried cherries. In a preliminary study of healthy women ($n = 10$), the levels of plasma urate were decreased with a single dose of 280 g of sweet cherries, more than the other fruits tested [47]. High level of urate is important in the arthritic disease state of gout. A statistically insignificant trend was seen in the decrease of plasma levels of CRP and nitric oxide. In another human feeding study ($n = 18$) using the same amount of sweet cherries for 28 days, there was a reduction of the two inflammatory markers, CRP in serum and nitric oxide in plasma [48]. However, further studies are needed in men (the primary contractors of gout). When these results are evaluated with the above research on inflammatory response and exercise, it appears that cherries have the ability to decrease inflammatory markers that could be significant in arthritis parameters. Larger, more targeted trials in diseased states are still needed.

13.6.2 Melatonin and sleep

Melatonin is a potent free radical scavenger and antioxidant produced in some plants as well as by the vertebrate pineal gland. It is most known for its potential use in improving sleep patterns, but other adjuvant uses have been analyzed [49]. Burkhardt *et al.* [50] has documented the amounts of melatonin in Montmorency and Balaton tart cherries to be in the order of 1350 and 200 ng/100 g of fresh cherry tissue, respectively. This was considered higher than sweet cherries (0–22.4 ng/100 g fresh weight [51]) and other fruits (13–29 ng/100 g for strawberries and pomegranates) as well as selected grains (87–187 ng/100 g) [52]. Although Kirakosyan *et al.* [32] described a loss of melatonin upon drying of tart cherries, they did not find this in freeze dried products. As the drying methods continue to improve with technology, the amount of melatonin in the final product should be monitored in order to maximize the presence of this compound.

Specific human trials of melatonin documentation upon cherry consumption are limited. Pigeon *et al.* [53] found in elderly insomniacs ($n = 15$ in a cross-over design) that upon ingestion of a tart cherry juice blend for 2 weeks, there was a significant reduction of wakefulness after sleep onset as compared to the placebo. Garrido *et al.* [54] found in three age groups of healthy adults ($n = 18$, divided into 3) that consuming 27.85 g of a freeze dried sweet cherry product (equivalent to 142 g fresh) twice per day resulted in a significant increase in sleep time and a significant decrease in nocturnal activity across all three groups. The excretion of urinary 6-sulfatoxymelatonin also increased significantly in all groups. Further studies using cherries in human clinical trials are needed.

13.6.3 Cancer

Not only do cherries have a high level of antioxidants, but the antioxidants present include some that have been studied for their anti-cancer potential, including quercetin, anthocyanins, polyphenols, carotenoids, melatonin, and vitamin C. Oxidative stress is responsible for DNA damage leading to cancer initiation, while antioxidants have the potential to scavenge free radicals and support endogenous antioxidant systems to decrease the oxidative stress load.

Inclusion of cherry in the diet of mice induced with lipopolysaccharide (LPS) resulted in an apparent modulation of liver nuclear factor kappa B (NF-κB) production. NF-κB is important in the transcription of DNA [55]. In mice already with cancerous growths, Bobe *et al.* [56] found a decrease in intestinal tumorigenesis when anthocyanin rich extracts of tart cherry were fed to them. This study showed that when this extract was used in combination with an anti-inflammatory drug (sulindac), much lower doses of the drug could be used to reduce tumor amounts in the intestine. Kang *et al.* [57] in a study on mice and colorectal cancer found those given a cherry diet, anthocyanins or cyanidin had fewer cecal tumors than those on sulindac or a control diet.

These animal model studies followed cancer cell line studies showing protective effects of cyanidin glucosides on improving apoptotic growth cycle arrest [58], protecting DNA cleavage, increasing levels of free radical scavenging, and inhibition of xanthine oxidase [59]. Cyanidin has been shown to increase cellular differentiation and decrease malignant transformation [60] as well as inhibit epidermal growth factor receptor [61]. Cherry anthocyanins caused cell cycle arrest and apoptosis in mutated cells [62, 63] and decreased growth in stomach and colon cancer cells [64]. Yoo *et al.* [65] found in Chinese hamster lung fibroblast cells that both cherry juice as well as individual phenolic cherry components reduced oxidative stress. In this study, cyanidin 3-glucoside was a stronger antioxidant than cyanidin 3-rutinoside.

Often antioxidants can work through synergistic mechanisms demonstrating the importance of the whole plant product rather than its individual chemical components. The antioxidant components in cherries, for example, have been shown to act synergistically [33]. The importance of whole foods is illustrated in the relation of diet and cancer [66] and the protection of DNA *via* the proportion of fruit and vegetables in the Mediterranian diet [67]. Anthocyanins have multifaceted pharmacological activities [68] just as quercetin has cancer pro-apoptotic effects [69], and dietary polyphenols have been reviewed for their anti-cancer properties *via* apoptosis [70]. Melatonin also has many mechanisms related to anti-tumor growth including immune, antioxidant, telomerase, and cell differentiation activity [71]. Given the above research in animal and cell models, human clinical trials with cherries, as with other high antioxidant foods and herbs, are now needed to directly link cherries with chemopreventive effects. In the interest of dried cherries, the aforementioned antioxidant components need to be documented in any product to extrapolate the research to the consumed product as well as the method of drying needs to ensure retention of the maximum amount of these antioxidant components.

13.6.4 Cardiovascular disease

CVD by definition is a group of disorders associated with the heart and blood vessels with risk factors including obesity, high blood pressure, and high cholesterol. The risk of cardiovascular events (such as stroke or heart attack) is often related to the blockage of blood vessels with fatty deposits and associated with high levels of cholesterol and inflammatory markers [72]. As the Mediterranean diet's increase in fruits and vegetables is linked to lower levels of CVD [15], dried fruit consumption has also been linked to reduced rates of obesity [5]. Cherries are high in flavonoids which in general have been linked to lower rates of coronary heart disease (CHD) through the Zutphen Elderly Study [73]. Anthocyanins, one type of flavonoids in cherries, have been described as having cardiovascular effects through angiotensin-converting enzyme (ACE) and protein kinase B/endothelial nitric oxide synthase (Akt/eNOS) pathways (responsible for endothelial effects on blood vessel dilation) as well as vascular cell adhesion molecules [68]. Anthocyanins from some plant extracts have been shown to have vasoactive relaxation effects on excised porcine heart arteries [74].

Studies specific to cherry effects on CVD have focused on the effect of certain cherry anthocyanins on oxidative stress, arterial endothelial and foam cells, vascular inflammation, and ischemic injury to the heart. Bovine arterial endothelial cells exposed to cyanidin-3-glycoside increased nitric oxide production illustrating a potential ability to decrease oxidant stress to the cardiac tissue [75]. In rat hearts with ischemic injury (and therefore irregular heartbeats), tart cherry seed extracts were found to decrease the incidence of irregular heartbeats and heart damage [76]. Cyanidin-3-0-glucoside was found to alter blood lipids in rats by modifying the level of vitamin E [77] and in mice cells to remove cholesterol from macrophage and foam cells [78]. In the Seymour *et al.* [40] study on fat Zucker rats, a tart cherry powder added to the diet resulted in lowered hyperlipidemia, decreased fat mass, lowered plasma expression of IL-6 and TNFα, and an increase in inflammation related peroxisome proliferation-activated receptor (PPAR) mRNA. Studies are now needed in human clinical trials to see if cherries have the same effect of lowering human lipid and inflammation parameters associated with CVD and risk. The use of dried cherries would depend on the level of associated anthocyanins proven in *in vitro* and animal models to have affected these parameters.

13.6.5 Diabetes

Metabolic syndrome is a group of risk factors that can lead to CVD and diabetes. Diagnosis focuses on central obesity and insulin resistance with three or more measurements of high fasting glucose, high blood pressure, large waist circumference, low low-density lipoprotein (LDL) cholesterol or high triacylglycerols (TAG). These health parameters encompass the health issues discussed above such that this section on diabetes also relies on the research on anthocyanins and cherries in the above sections on inflammation and CVD. In addition, recent research is linking many parameters of cancer and diabetes [79].

Oxidative stress is linked to many of the symptoms and complications of diabetes and as such the use of plant products high in antioxidants has the potential to ameliorate these symptoms [80, 81]. Jayaprakasam *et al.* [82] found in a cell culture study, as well as a mouse study [83], that anthocyanins and anthocyanidins from *Cornus* cherry fruit enhanced insulin production and improved glucose levels. The greater activity on insulin production was dependent on the number of hydroxyl groups on the B ring of the anthocyanins tested. In addition, Tsuda *et al.* [84] found that mice fed cyanidin 3-O-β-D-glucoside rich color with their high fat diet for 12 weeks reduced their level of adipose accumulation and normalized hyperglycemic and hyperinsulinemia levels.

In specific animal models associated with diabetes, Seymour *et al.* [40] used the Zucker rat, an animal model of obesity often used in studies of type 2 diabetes, to show that dried tart cherry powder in their diet lowered fat mass, hyperlipidemia, and markers of inflammation. In reducing these parameters of metabolic syndrome, it was concluded that this could influence the onset of diabetes. In their studies in 2008, they used the lean Dahl salt-sensitive rat that has insulin resistance and hyperlipidemia. After similar feeding for 90 days, there was a decrease in fasting blood glucose and hyperinsulinemia, two factors important in diabetes.

Dried cherries often have added sugar which can be detrimental to diabetics. Cherries were initially documented as having a low GI of 22, but dried tart Montmorency cherries have recently been documented at a value of 58. More data are needed on the particular dried cherry product and the amount of sugar present. Jenkins *et al.* [85] found that low GI fruit consumption can reduce HbA1c levels (a measure of blood glucose representative of the past 8–12 weeks). In dried cranberry products, it was found that those with less sugar had significantly improved glycemic and plasma insulin response levels upon consumption [86]. In addition, as seen in Table 13.3, the phenolic and anthocyanin levels can be increased with less sugar in a dried cherry product.

13.7 Conclusions

Dried cherries have been found to be a good source of plant protein, fiber, vitamin A, and potassium, and the level of antioxidant phenolics and derivatives of anthocyanin and cyanidin rank it high in antioxidant foods. Lowered rates of local and systemic inflammation have been documented in cell and rat models, and in small studies on humans consuming tart cherry products. A couple of human pilot trials have also shown an effect of melatonin in tart cherries on sleep patterns. With regard to cancer, cell system studies have documented an improvement using cherry anthocyanins and cherry extracts added to the diet in animal models which were able to reduce the amount of anti-cancer drugs needed. The flavonoids and anthocyanins of cherries have been shown, in experimental systems, to lower the oxidative stress, blood lipids, and heart arrhythmias detrimental in cardiovascular and diabetes disease pathogenesis.

Most of the research presented here relates to dried cherries through the amounts of anthocyanins and use of tart cherry products (chiefly juice blends and dried powder). Improved methods of drying could cause some of the individual anthocyanin levels, as well as vitamin C and melatonin, to increase. Just as the method of drying could improve antioxidant levels (and thereby health parameters), a reduction in the sugar added to a dried cherry product may also improve antioxidant levels and health promotion. Further research is needed specifically using dried cherries over other forms of cherries as well as studies in human clinical trials over cell and animal systems. With designing human clinical trials, researchers must take into account the need for bioavailability measures to assess absorption, function, and synergistic action of antioxidant metabolites [87]. Extrapolation of the results presented here to dried cherries often relies on the amounts of phenolics and anthocyanins present in the product as well as in the blood stream and tissues of consumers.

References

1. USDA (2011) Tart Cherries Grown in the States of Michigan. Final Free and Restricted Percentages for the 2010-2011 Crop Year for Tart Cherries. A Rule by the Agricultural Marketing Service on 2/25/2011, Document No. 2011-4269. *Federal Register* **76**, 10471–10476.
2. Cherry Marketing Institute (2007) *Press Release: The Cherry Nutrition Report: A Close-Up Look at Today's New Antioxidant Super Fruit.* Published online at: www.choosecherries.com, last accessed 13 September 2011.
3. Cherry Marketing Institute (2009) *Press Release: Choose Cherries—America's Super Fruit.* Published online at: http://www.choosecherries.com/news/pressReleases.aspx, last accessed 13 September 2011.
4. McCune, L.M., Kubota, C., Stendell-Hollis, N.R. & Thomson, C.A. (2011) Cherries and health: a review. *Critical Reviews in Food Science and Nutrition*, **51**, 1–12.
5. Keast, D.R., O'Neil, C.E. & Jones, J.M. (2011) Dried fruit consumption is associated with improved diet quality and reduced obesity in US adults: National Health and Nutrition Examination Survey, 1999-2004. *Nutrition Research*, **31**, 460–467.
6. FAOSTAT (2011) *Production*: *Countries by Commodity*. Published online at: http://faostat.fao.org/site/339/default.aspx, last accessed 13 September 2011.
7. USDA (2011) *National Agricultural Statistics Service*: *Non-citrus Fruits and Nuts 2010 Summary (July 2011).* Published online at: http://usda.mannlib.cornell.edu/usda/nass/NoncFruiNu/2010s/2011/NoncFruiNu-07-07-2011.pdf, last accessed 13 September 2011.
8. Martinez, L. & Thornsbury, S. (2006) *Agricultural Economics Report No. 627, September 2006. Michigan Tart Cherry Processors: Issues and Strategy*. Department of Agricultural Economics, Michigan State University, East Lansing, MI.

9. Nagai, T.M., Woods, M. & Thornsbury, S. (2006) *Tariff Intervention in Trade of US and EU Cherry Products: A Guide to Information. Agricultural Economics Report 630*. Department of Agricultural Economics, Michigan State University, East Lansing, MI.
10. Tarhan, S., Ergunes, G. & Taser, O.F. (2006) Selection of chemical and thermal pretreatment combination to reduce the dehydration time of sour cherry (*Prunus cerasus* L.). *Journal of Food Process Engineering*, **29**, 651–663.
11. Valero, D., Díaz-Mula, H.M., Zapata, P.J., Castillo, S., Guillén, F., Martínez-Romero, D. & Serrano, M. (2011) Postharvest treatments with salicylic acid, acetylsalicylic acid or oxalic acid delayed ripening and enhanced bioactive compounds and antioxidant capacity in sweet cherry. *Journal of Agricultural and Food Chemistry*, **59**, 5483–5489.
12. Cherry Marketing Institute (2011) *Dried Tart Cherry*. Published online at: http://choosecherries.com/Uploads/Documents/8590296135698192445.pdf, last accessed 13 September 2011.
13. USDA Agricultural Research Service (2010) *USDA National Nutrient Database for Standard Reference, Release 23*. Published online at: http://www.ars.usda.gov/ba/bhnrc/ndl, last accessed 14 September 2011.
14. Bujnowski, D., Xun, P., Daviglus, M.L., Van Horn, L., He, K. & Stamler, J. (2011) Longitudinal association between animal and vegetable protein intake and obesity among men in the United States: The Chicago Western Electric Study. *Journal of the American Dietetic Association*, **111**, 1150–1155.
15. Tyrovolas, S. & Panagiotakos, D.B. (2010) The role of Mediterranean type of diet on the development of cancer and cardiovascular disease, in the elderly: a systematic review. *Maturitas*, **65**, 122–130.
16. Kapiszewska, M. (2006) A vegetable to meat consumption ration as a relevant factor determining cancer preventive diet. The Mediterranean versus other European countries. *Forum of Nutrition*, **59**, 130–153.
17. Kristensen, M. & Bügel, S. (2011) A diet rich in oat bran improves blood lipids and hemostatic factors, and reduces apparent energy digestibility in young healthy volunteers. *European Journal of Clinical Nutrition*, **65**, 1053–1058.
18. Wolfram, T. & Ismail-Beigi, F. (2011) Efficacy of high-fiber diets in the management of type 2 diabetes mellitus. *Endocrine Practice*, **17**, 132–142.
19. Palafox-Carlos, H., Ayala-Zavala, J.F. & González-Aguilar, G.A. (2011) The role of dietary fiber in the bioaccessibility and bioavailability of fruit and vegetable antioxidants. *Journal of Food Science*, **76**, R6–R15.
20. Tikekar, R.V., Anantheswaran, R.C. & LaBorde, L.F. (2011) Ascorbic acid degradation in a model apple juice system and in apple juice during ultraviolet processing and storage. *Journal of Food Science*, **76**, H62–H71.
21. Klemm, R.D.W., West, K.P., Jr., Palmer, A.C., Johnson, Q., Randall, P., Ranum, P. & Northrop-Clewes, C.(2010) Vitamin A fortification of wheat flour: considerations and current recommendations. *Food & Nutrition Bulletin*, **31**, 47S–61S.
22. Englberger, L., Kuhnlein, H.V., Lorens, A., Pedrus, P., Albert, K., Currie, J., Pretrick, M., Jim, R. & Kaufer, L. (2010) Pohnpei, FSM case study in a global health project documents its local food resources and successfully promotes local food for health. *Pacific Health Dialog*, **16**, 129–136.
23. Kuhnlein, H.V.K., Erasmus, B. & Spigelski, D. (2009) *Indigenous Peoples' Food Systems: The Many Dimensions of Culture, Diversity and Environment for Nutrition and Health*. FAO, Rome, Italy.
24. Ding, E.L. & Mozaffarian, D. (2006) Optimal dietary habits for the prevention of stroke. *Seminars in Neurology*, **26**, 11–23.
25. Wassertheil-Smoller, S. (2010) Stroke in women. *Nutrition, Metabolism & Cardiovascular Diseases*, **20**, 419–425.
26. Appel, L.J., Moore, T.J., Obarzanek, E., Vollmer, W.M., Svetkey, L.P., Sacks, F.M., Bray, G.A., Vogt, T.M., Cutler, J.A., Windhauser, M.M., Lin, P.H. & Karanja, N. (1997). A clinical trial of the effects of dietary patterns on blood pressure. DASH collaborative research group. *New England Journal of Medicine*, **336**, 117–1124.
27. Wexler, R. & Auckerman, G. (2006) Nonpharmacologic strategies for managing hypertension. *American Family Physician*, **73**, 1953–1956.
28. Hoorn, E.J., McCormick, J.A. & Ellison, D.H. (2011) The WNK kinase network regulating sodium, potassium, and blood pressure. *Journal of the American Society of Nephrology*, **22**, 605–614.
29. McTaggart, J.S., Clark, R.H. & Ashcroft, F.M. (2010) The role of the KATP channel in glucose homeostasis in health and disease: more than meets the islet. *Journal of Physiology*, **588**, 3201–3209.
30. Chaovanalikit, A. & Worlstad, R.E. (2004) Total anthocyanins and total phenolics of fresh and processed cherries and their antioxidant properties. *Journal of Food Science*, **69**, FCT67–FCT72.

31. Wang, H., Nair, M.G., Strasburg, G.M., Booren, A.M. & Gray, J.I. (1999) Antioxidant polyphenols from tart cherries (*Prunus cerasus*). *Journal of Agricultural and Food Chemistry*, **47**, 840–844.
32. Kirakosyan, A., Seymour, E.M., Llanes, D.E.U., Kaufman, P.B. & Bolling, S.F. (2009) Chemical profile and antioxidant capacities of tart cherry products. *Food Chemistry*, **115**, 20–25.
33. Kirakosyan, A., Seymour, E.M., Kathleen, R., Noon, K.R., Llanes, D.E.U., Kaufman, P.B., Warber, S.L. & Bolling, S.F. (2010) Interactions of antioxidants isolated from tart cherry (*Prunus cerasus*) fruits. *Food Chemistry*, **122**, 78–83.
34. Ferretti, G., Bacchetti, T., Masciangelo, S. Saturni, L., Antinori, M., Santinelli, A., Neri, D. (2009) *Antioxidant potential of "visciola" dried tart cherry* (*Prunus cerasus*, L.). Published online at: http://nutrimenti.simplicissimus.it/files/2009/06/poster2009foodomics1.pdf, last accessed 13 September 2011.
35. Faniadis, D., Drogoudi, P.D. & Vasilakakis, M. (2010) Effects of cultivar, orchard elevation, and storage on fruit quality characters of sweet cherry (*Prunus avium* L.). *Scientia Horticulturae*, **125**, 301–304.
36. McCune, L.M. & Johns, T. (2007) Antioxidant activity relates to plant part, life form and growing condition in some diabetes remedies. *Journal of Ethnopharmacology*, **112**, 461–469.
37. Halvorsen, B.L., Carlsen, M.H., Phillips, K.M., Bohn, S.K., Holte, K., Jocobs, D.R. & Blomhoff, R. (2006) Content of redox-active compounds (i.e., antioxidants) in foods consumed in the United States. *American Journal of Clinical Nutrition*, **84**, 95–135.
38. Tall, J.M., Seeram, N.P., Zhao, C., Nair, M.G., Meyer, R.A. & Raja, S.N. (2004) Tart cherry anthocyanins suppress inflammation-induced pain behaviour in rat. *Behavioural Brain Research*, **153**, 181–188.
39. Seeram, N.P., Momin, R.A., Nair, M.G. & Bourquin, L.D. (2001) Cyclooxygenase inhibitory and antioxidant cyanidin glycosides in cherries and berries. *Phytomedicine*, **8**, 362–369.
40. Seymour, E.M., Lewis, S.K., Urcyo-Llanes, D.E., Tanone, I.I., Kirakosyan, A., Kaufman, P.B. & Bolling, S.F. (2009) Regular tart cherry intake alters abdominal adiposity, adipose gene transcription, and inflammation in obesity-prone rats fed a high fat diet. *Journal of Medicinal Food*, **12**, 935–942.
41. Seymour, E.M., Singer, A.M., Kirakosyan, A., Urcuyo-Llanes, D.E., Kaufman, P.B. & Bolling, S.F. (2008) Altered hyperlipidemia, hepatic steatosis, and hepatic peroxisome proliferator-activated receptors in rats with intake of tart cherry. *Journal of Medicinal Food*, **11**, 252–259.
42. Kuehl, K.S., Perrier, E.T., Elliot, D.L. & Chesnutt, J.C. (2010) Efficacy of tart cherry juice in reducing muscle pain during running: a randomized controlled trial. *Journal of the International Society of Sports Nutrition*, **7**, 7–17.
43. Howatson, G., McHugh, M.P., Hill, J.A., Brouner, J., Jewell, A.P., van Someren, K.A., Shave, R.E. & Howatson, S.A. (2010) Influence of tart cherry juice on indices of recovery following marathon running. *Scandinavian Journal of Medicine & Science in Sports*, **20**, 843–852.
44. Traustadottir, T., Davies, L.S., Stock, A.A., Su, Y., Heward, C.B., Roberts, L.J. & Harman, S.M. (2009) Tart cherry juice decreases oxidative stress in healthy older men and women. *Journal of Nutrition*, **139**, 1896–1900.
45. Bowtell, J.L., Sumners, D.P., Dyer, A., Fox, P. & Mileva, K.N. (2010) Montmorency cherry juice reduces muscle damage caused by intensive strength exercise. *Medicine & Science in Sports & Exercise*, **43**, 1544–1551.
46. He, Y.-H., Zhou, J., Wang, Y.-S., Xiao, C., Tong, Y., Tang, J.C.-O., Chan, A.S. & Lu, A.P. (2006) Anti-inflammatory and anti-oxidative effects of cherries on Freund's adjuvant-induced arthritis in rats. *Scandinavian Journal of Rheumatology*, **35**, 356–358.
47. Jacob, R.A., Spinozzi, G.M., Simon, V.A., Kelley, D.S., Prior, R.L., Hess-Pierce, B. & Kader, A.A. (2003) Consumption of cherries lowers plasma urate in healthy women. *Journal of Nutrition*, **133**, 1826–1829.
48. Kelley, D.S., Rasooly, R., Jacob, R.A., Kader, A.A. & Mackey, B.E. (2006) Consumption of Bing sweet cherries lowers circulation concentrations of inflammation markers in healthy men and women. *Journal of Nutrition*, **136**, 981–986.
49. Sanchez-Barcelo, E.J., Mediavilla, M.D., Tan, D.X. & Reiter, R.J. (2010) Clinical uses of melatonin: evaluation of human trials. *Current Medicinal Chemistry*, **17**, 2070–2095.
50. Burkhardt, S., Tax, D.X., Manchester, L.C., Hardeland, R. & Reiter, R.J. (2001) Detection and quantification of the antioxidant melatonin in Montmorency and Balaton tart cherries (*Prunus cerasus*). *Journal of Agricultural and Food Chemistry*, **49**, 4898–4902.

51. Gonzalez, C.A. & Riboli, E. (2009) Diet and cancer prevention: contributions from the European Prospective Investigation into Cancer and Nutrition (EPIC) study. *European Journal of Cancer*, **46**, 2555–2562.
52. Badria, F.A. (2002) Melatonin, serotonin, and tryptamine in some Egyptian food and medicinal plants. *Journal of Medicinal Food*, **5**, 153–157.
53. Pigeon, W.R., Carr, M., Gorman, C. & Perlis, M.L. (2010) Effects of a tart cherry juice beverage on the sleep of older adults with insomnia: a pilot study. *Journal of Medicinal Food*, **13**, 579–583.
54. Garrido, M., Espino, J., Gonzalez-Gomez, D., Lozano, M., Cubero, J., Toribio-Delgado, A.F., Maynar-Marino, J.I., Terron, M.P., Munoz, J.L., Pariente, J.A., Barriga, C., Paredes, S.D. & Rodriguez, A.B. (2009) A nutraceutical product based on Jerte Valley cherries improves sleep and augments the antioxidant status in humans. *e-SPEN, European Journal of Clinical Nutrition and Metabolism*, **4**, e321–e323.
55. Balstad, T.R., Paur, I., Poulsen, M., Markowski, J., Kolodziejczyk, K., Dragsted, L.O., Myhrstad, M.C.W. & Blomhoff, R. (2010) Apple, cherry, and black currant increases nuclear factor kappa B activation in liver of transgenic mice. *Nutrition and Cancer*, **62**, 841–848.
56. Bobe, G., Wang, B., Seeram, N.P., Nair, M.G. & Bourquin, L.D. (2006) Dietary anthocyanin-rich tart cherry extract inhibits intestinal tumorigenesis in APC (Min) mice fed suboptimal levels of sulindac. *Journal of Agricultural and Food Chemistry*, **54**, 9322–9328.
57. Kang, S.Y., Seeram, N.P., Nair, M.G. & Bourquin, L.D. (2003) Tart cherry anthocyanins inhibit tumor development in ApcMin mice and reduce proliferation of human colon cancer cells. *Cancer Letters*, **194**, 13–19.
58. Chen, P.N., Chu, S.C., Chiou, H.L., Chiang, C.L., Yang, S.F. & Hsieh, Y.S. (2005) Cyanidin 3-glucoside and peonidin 3-glucoside inhibit tumor cell growth and induce apoptosis *in vitro* and suppress tumor growth *in vivo*. *Nutrition and Cancer*, **53**, 223–243.
59. Acquaviva, R., Russo, A., Galvano, F., Galvano, G., Barcellona, M.L., Li Volti, G. & Vanella, A. (2003) Cyanidin and cyanidin 3-O-β-D-glucoside as DNA cleavage protectors and antioxidants. *Cell Biology and Toxicology*, **19**, 243–252.
60. Serafino, A., Sinibaldi-Vallebona, P., Lazzarino, G., Tavazzi, B., Rasi, G., Pierimarchi, P., Andreola, F., Moroni, G., Galvano, G., Galvano, F. & Garaci, E. (2004) Differentiation of human melanoma cells induced by cyanidin 3-O-β-glucopyranoside. *FASEB Journal*, **18**, 1940–1942.
61. Meiers, S., Kemeny, M., Weyland, U., Gastpar, R., von Angerer, E. & Marko, D. (2001) The anthocyanidins cyanidin and dephinidin are potent inhibitors of the epidermal growth-factor receptor. *Journal of Agricultural and Food Chemistry*, **49**, 958–962.
62. Lazze, M.D., Savio, M., Pizzala, R., Cazzalini, O., Perucca, P., Scovassi, A.I., Stivala, L.A. & Bianchi, L. (2004) Anthocyanins induce cell cycle perturbations and apoptosis in different human cell lines. *Carcinogenesis*, **25**, 1427–1433.
63. Shih, P.H., Yeh, C.T. & Yen, G.C. (2005) Effects of anthocyanidin on the inhibition of proliferation and induction of apoptosis in human gastric adenocarcinoma cells. *Food and Chemical Toxicology*, **43**, 1557–1566.
64. Serra, A.T., Duarte, R.O., Bronze, M.R. & Duarte, C.M.M. (2011) Identification of bioactive response in traditional cherries from Portugal. *Food Chemistry*, **125**, 318–325.
65. Yoo, K.M., Al-Farsi, M., Lee, H., Yoon, H. & Lee, C.Y. (2010) Antiproliferative effects of cherry juice and wine in Chinese hamster lung fibroblast cells and their phenolic constituents and antioxidant activities. *Food Chemistry*, **123**, 734–740.
66. Gonzalez, C.A. & Riboli, E. (2010) Diet and cancer prevention: contributions from the European Prospective Investigation into Cancer and Nutrition (EPIC) study. *European Journal of Cancer*, **46**, 2555–2562.
67. Kapiszewska, M. (2006) A vegetable to meat consumption ratio as a relevant factor determining cancer preventive diet. The Mediterranean versus other European countries. *Forum of Nutrition*, **59**, 130–153.
68. Domitrovic, R. (2011) The molecular basis for the pharmacological activity of anthocyans. *Current Medicinal Chemistry*, **18**, 4454–4469.
69. Gibellini, L., Pinti, M., Nasi, M., Montagna, J.P., De Biasi, S., Roat, E., Bertoncelli, L., Cooper, E.L. & Cossarizza, A. (2012) Quercetin and cancer chemoprevention. *Evidence-Based Complementary and Alternative Medicine* (in press)
70. Fresco, P., Borges, F., Marques, M.P. & Diniz, C. (2010) The anticancer properties of dietary polyphenols and its relation with apoptosis. *Current Pharmaceutical Design*, **16**, 114–134.

71. Mediavilla, M.D., Sanchez-Barcelo, E.J., Tan, D.X., Manchester, L. & Reiter, R.J. (2010) Basic mechanisms involved in the anti-cancer effects of melatonin. *Current Medicinal Chemistry*, **17**, 4462–4481.
72. Ridker, P.M., Hennekens, C.H., Buring, J.E. & Rifai, N. (2000) C-reactive protein and other markers of inflammation in the prediction of cardiovascular disease in women. *New England Journal of Medicine*, **342**, 836–843.
73. Hertog, M.G., Feskens, E.J., Hollman, P.C., Katan, M.B. & Kromhout, D. (1993) Dietary antioxidant flavonoids and risk of coronary heart disease: the Zutphen Elderly Study. *Lancet*, **342**, 1007–1011.
74. Bell, D.R. & Gochenaur, K. (2006) Direct vasoactive and vasoprotective properties of anthocyanin-rich extracts. *Journal of Applied Physiology*, **100**, 1164–1170.
75. Xu, J.W., Ikeda, K. & Yamori, Y. (2004) Upregulation of endothelia nitric oxide synthase by cyanidin-3-glucoside, a typical anthocyanin pigment. *Hypertension*, **44**, 217–222.
76. Bak, I., Lekli, I., Juhasz, B., Nagy, N., Varga, E., Varadi, J., Gesztelyi, R., Szabo, G., Szendrei, L., Bacskay, I., Vecsernyes, M., Antal, M., Fesus, L., Boucher, F., de Leiris, J. & Tosaki, A. (2006) Cardioprotective mechanisms of *Prunus cerasus* (sour cherry) seed extract against ischemia-reperfusion-induced damage in isolated rat hearts. *American Journal of Physiology—Heart and Circulatory Physiology*, **291**, H1329–H1336.
77. Frank, J., Kamal-Eldin, A., Lundh, T., Maatta, K., Torronen, R. & Vessby, B. (2002) Effects of dietary anthocyanins on tocopherols and lipids in rats. *Journal of Agricultural and Food Chemistry*, **50**, 7226–7230.
78. Xia, M., Hou, M., Zhu, H., Ma, J., Tang, Z., Wang, Z., Li, Y., Chi, D., Yu, X., Zhao, T., Han, P., Xi, X. & Ling, W. (2005) Anthocyanins induce cholesterol efflux from mouse peritoneal macrophages. The role of peroxisome proliferators-activated receptor Y-liver X-receptor α-ABCA1 pathway. *The Journal of Biological Chemistry*, **280**, 36792–36801.
79. Mussig, K., Staiger, H., Kantartzis, K., Fritsche, A., Kanz, L. & Haring, H.U. (2011) Type 2 diabetes mellitus and risk of malignancy: is there a strategy to identify a subphenotype of patients with increased susceptibility to endogenous and exogenous hyperinsulinism? *Diabetic Medicine*, **28**, 276–286.
80. McCune, L.M. & Johns, T. (2003) Symptom-specific antioxidant activity of boreal diabetes treatments. *Pharmaceutical Biology*, **41**, 362–370.
81. Montonen, J., Knekt, P., Jarvinen, R. & Reunanen, A. (2004) Dietary antioxidant intake and risk of type 2 diabetes. *Diabetes Care*, **27**, 362–366.
82. Jayaprakasam, B., Vareed, S.H., Olson, L.K. & Nair, M. (2005) Insulin secretion by bioactive anthocyanins and anthocyanidins present in fruits. *Journal of Agricultural and Food Chemistry*, **53**, 28–31.
83. Jayaprakasam, B., Olsen, L.K., Schutzki, E., Tai, M.H. & Nair, M. (2006) Amelioration of obesity and glucose intolerance in high-fat-fed C57BL/6 mice by anthocyanins and ursolic acid in Cornelian cherry (*Cornus mas*). *Journal of Agricultural and Food Chemistry*, **53**, 243–248.
84. Tsuda, T., Horio, F., Uchida, K., Aoki, H. & Osawa, T. (2003) Dietary cyanidin 3-O-β-D-glucoside-rich purple corn color prevents obesity and ameliorates hyperglycemia in mice. *Journal of Nutrition*, **133**, 2125–2130.
85. Jenkins, D.J., Srichikul, K., Kendall, C.W., Sievenpiper, J.L., Abdulnour, S., Mirrahimi, A., Meneses, C., Nishi, S., He, X., Lee, S., So, Y.T., Esfahani, A., Mitchell, S., Parker, T.L., Vidgen, E., Josse, R.G. & Leiter, L.A. (2011) The relation of low glycemix index fruit consumption to glycemix control and risk factors for coronary heart disease in type 2 diabetes. *Diabetologia*, **54**, 271–279.
86. Wilson, T., Luebke, J.L., Morcomb, E.F., Carrell, E.J., Leveranz, M.C., Kobs, L., Schmidt, T.P., Limburg, P.J., Vorsa, N. & Singh, A.P. (2010) Glycemic responses to sweetened dried and raw cranberries in humans with type 2 diabetes. *Journal of Food Science*, **75**, H218–H223.
87. Serafini, M., Peluso, I. & Raguzzini, A. (2010) Antioxidants and the immune system: flavonoids as anti-inflammatory agents. *Proceedings of the Nutrition Society*, **69**, 273–278.

14 Dried citrus fruits: phytochemicals and health beneficial effects

Tzou-Chi Huang and Chi-Tang Ho

14.1 Introduction

Citrus fruits make up the largest sector of the world's fruit production. Their worldwide production is over 88×10^6 tonnes and one-third of the crop is processed. Oranges, lemons, grapefruits, and mandarins represent approximately 98% of the entire industrialized crop; among them oranges occupy approximately 82% of the total. In addition to fresh consumption, majority of citrus fruits are processed to obtain juice [1]. Part of the citrus wastes, including peels, seeds, and fruit pulp, remaining after juice extraction, are subjected to the extraction of flavonoids and essential oils [2]. Different from that of Western world, whole immature and mature citrus fruits and peels are dried, and widely used as remedies to stimulate the appetite, aid digestion, improve menopausal syndromes, alleviate cough, and improve inflammatory syndromes of the respiratory tract such as bronchitis and asthma in oriental countries [3].

According to the morphological system established by Tanaka [4], citrus genus is classified into two subgenera (*Archicitrus* and *Metacitrus*), 8 sections (*Papeda*, *Limonellus*, *Citrophorum*, *Cephacitrus*, *Aurantium*, *Osmocitrus*, *Acrmen*, and *Pseudo Fortunella*) and 16 subsections, involving 149 species. Citrus is a common term and genus of flowering plants in the family *Rutaceae*. Commercially important citrus fruits include oranges, grapefruits, lemons, some limes, and some tangerines. The name "orange" applies primarily to *Citrus sinensis*, which accounts for about 70% of world citrus production. Other citrus species known as oranges include:

(1) *C. reticulata*, which itself has an enormous number of cultivars (most notably the satsuma (*C. unshiu*), tangerine (*C. tangerina*), and clementine (*C. clementina*),
(2) *C. bergamia* Risso, which is grown primarily in Italy and used primarily for the peel, and
(3) *C. aurantium*, also known as Seville orange, sour orange, bitter orange, bigarade orange, and marmalade orange.

Dried Fruits: Phytochemicals and Health Effects, First Edition. Edited by Cesarettin Alasalvar and Fereidoon Shahidi.

C. limon (L.) Burm.f. (Rutaceae) is an important fruit and medicinal plant. Lemon peel is utilized as a stomachic, carminative, diaphoretic, astringent, febrifuge, and diuretic agent in the traditional Indian medicine [5]. California and Arizona are the leading sources of lemons in the Western hemisphere. In recent years, grapefruit has received much attention because of its nutritional and antioxidant properties. It is cultivated principally to obtain the juice with high concentration of naringin. The beneficial effects of the grapefruit juice on human health, such as antioxidant, antiallergic, and anticarcinogenic benefits as well as protection against high blood pressure or cholesterol increase have been reported by Kawaii *et al.* [6]. Although grapefruit peels are the major sources of natural naringin, hesperidin, and other flavonoids, they are currently generally treated as waste by the juice producing industry [7]. This chapter outlines the phytochemicals and health beneficial effects of citrus fruits such as oranges, tangerines, grapefruits, lemons, and limes.

14.2 Compositional and nutritional characteristics of citrus

14.2.1 Structure of citrus fruit

Citrus fruits consist of an epicuticular wax layer. Immediately under epidermis is located the flavedo characterized by a green, yellow, or orange color. Oil glands are interspersed around the flavedo layer and characterized by very thin and fragile walls. Beneath flavedo is the albedo that is composed of loosely packed, many branched, tube-like cells that form a continuous network with the greatest part of tissue volume. Albedo is the major source of citrus phytochemicals which impart a bitter taste to the juice and the health beneficial effects to human body. The edible part of the fruit with segments membrane containing juice sacs is the endocarp.

14.2.2 Nutritional characteristics of citrus fruits

Citrus fruits are the principal source of various important nutrients. They contain vitamin C, folate, dietary fiber, and a number of physiologically functional components such as citric acid, minerals, and phytochemicals which are regarded as chemopreventive agents for various diseases [8]. Citrus soluble solids are on the average made up by a 70% of sugars, whilst pulp solids are made by a 40% of sugars, and by a 50% of polysaccharides. Citrus polysaccharides are mainly pectic acid (polygalacturonic acid). Commercial pectin is mostly derived from citrus peels including lime, grapefruit, and orange [9]. High dietary fiber powder preparations from citrus fruits contain pectin as their major constituent and are thus considered as having a better quality than other sources of dietary fiber due to the presence of associated bioactive compounds which may exert additional health promoting effects than the dietary fiber itself [10].

14.3 Phytochemicals in citrus

At least five types of flavonoids (flavanones, flavones, polymethoxyflavones, flavonols, and anthocyanins) occur in citrus, and more than 60 individual flavonoids have been

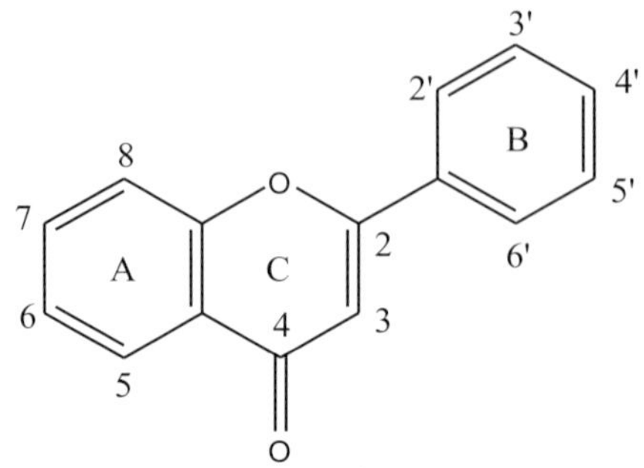

Figure 14.1 The generic structure of flavonoids.

identified. The major flavonoids in orange peel are flavanone glycosides (narirutin 4′-*O*-glucoside, eriocitrin, narirutin, hesperidin, and isosakuranetin rutinoside) [11], flavone glycosides (diosmin, isorhoifolin, and rutin) [12], and C-glycosylated flavones (6,8-di-*C*-glucosylapigenin) [13]. Anthocyanins [cyanidin 3-glucoside and cyanidin 3-(6′′-malonylglucoside)] occur only in blood oranges, whereas polymethoxyflavone (PMF) (sinensetin, hexa-*O*-methylquercetagetin, nobiletin, hexa-*O*-methylgossypetin, 3,5,6,7,8,3′,4′-heptamethoxyflavone, tetra-*O*-methylscutellarein, tangeritin, and 5-hydroxy-3,7,8,3′,4′-pentamethoxyflavone) distributed widely in citrus peels [14]. The highly methoxylated flavones exhibit higher biological activity even though they occur in relatively lower concentrations. Citrus flavonoids characterized with the various combinations of multiple hydroxy, methoxy, and *O*-glycoside group substituents on the basic benzo-γ-pyrone (C6-C3-C6) as shown in Figure 14.1.

14.3.1 Flavanones

Flavanones are the most abundant flavonoids (e.g., 98% in grapefruits, 96% in limes, and 90% in lemons) in citrus [15]. The flavanones in citrus leaves, peels, and fruits are found mainly in glycosylated states. Flavanone glycosides including eriocitrin (eriodictyol 7-*O*-rutinoside), neoeriocitrin (eriodictyol 7-*O*-neohesperidoside), homoeriodictyol 7-*O*-rutinoside, hesperidin (hesperetin 7-*O*-rutinoside), neohesperidin (hesperetin 7-*O*-neohesperidoside), narirutin (naringenin 7-*O*- rutinoside), naringin (Naringenin 7-*O*-neohesperidoside), poncirin (Isosakuranetin 7-*O*-neohesperidoside), and neoponcirin (didymin, isosakuranetin 7-*O*-putinoside) have been identified in citrus fruits [16, 17]. The structures of major flavanone aglycones (naringenin, hesperetin, isosakuranetin, and eriodictyol) are shown in Figure 14.2.

The glycoside flavanone hesperidin is present predominately in the peels of orange (*Citrus sinensis*) and lemon (*Citrus limonium*) [18]. Hesperidin is the primary flavanone glycoside in Valencia, Navel, Temple, and Amber sweet oranges [19], whereas naringin is the dominant flavonoid in grapefruits (*Citrus paradisi*) [18]. Hesperidin accounts for more than 92% of the total flavanone glycosides in the peel extracts of lemon. In *C. reticulata* Blanco fruits, the highest flavanone glycoside contents were found in young fruits and hesperidin represented 49% of the dry weight (DW) of the fruit [20]. Wang *et al.* [21] characterized several rutinosides and neohesperosides of naringenin, eriodictyol, and hesperetin from a less commercialized *C.*

Flavanone

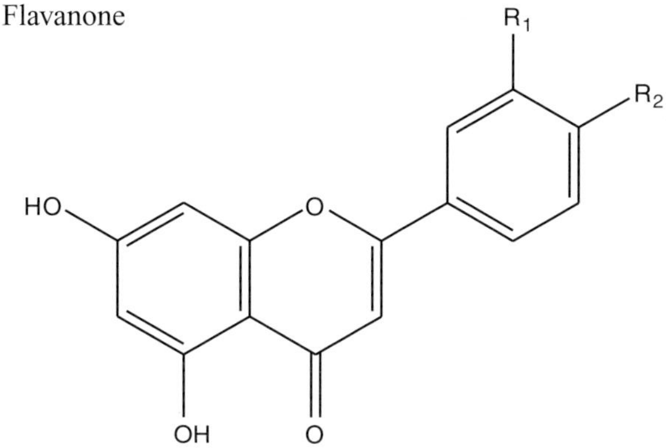

Naringenin	R_1=H;	R_2=OH;
Hesperetin	R_1=OH;	R_2=OCH_3;
Isosakuranetin	R_1=H;	R_2=OCH_3;
Eriodictyol	R_1=H;	R_2=OH;

Figure 14.2 Structures of citrus flavanone aglyocne.

bergamia Risso. Among the neohesperidoside flavanones, including naringin, neohesperidin, and neoeriocitrin, are the most dominant flavonoid found in bergamot, grapefruit, and bitter orange juices, whereas rutinoside flavanones, such as hesperidin, narirutin, and didymin dominated in bergamot, orange, mandarin, and lemon juices [22].

On the other hand, naringin was the most abundant flavanone glycoside (92% of the total flavanone glycosides) in grapefruit peel and narirutin was the major flavanone glycoside in murcott [23]. Vanamala *et al.* [24] reported that naringin, narirutin, and poncirin were the major flavanone glycosides in grapefruits with concentrations of 30.4, 10.1, and 1.24 mg/100 mL of juice, respectively. Naringin was mainly found in lemon peel and seed as well as in mandarin seed, but not in the juices of these fruits [25].

Eriocitrin is particularly abundant in lemons and limes, while they are almost absent in other citrus fruits [26]. Lemon seeds contain large amounts of eriocitrin and hesperidin and a low level of naringin. On the contrary, the lemon peel is rich in neoeriocitrin, neohesperidin, and naringin and has a minor amount of narirutin [27].

14.3.2 Flavones

The flavones in citrus occur mainly in glycosylated states. Flavone glycosides including diosmin (diosmetin-7-*O*-rutinoside), neodiosmin (diosmetin-7-*O*-neohsperidoside), diosmetin 6,8-di-C-β-glucoside, diosmetin-6-*C*-β-glucoside, diosmetin-8-*C*-β-glucoside, vicenin-2 (apigenin 6,8-di-*C*-glucoside), apigenin-7-(malonylapiosyl)-glucoside, isorhoifolin (apigenin-7-*O*-rutinoside), rhoifolin (apigenin-7-*O*-neohesperidoside), lucenin-2 (luteolin 6,8-di-*C*-glucoside), luteolin 7-*O*-rutinoside, scoparin (chrysoeriol 8-*C*-glucoside), and stellarin-2 (chrysoeriol 6,8-di-*C*-glucoside), have been characterized [16, 28]. Figure 14.3 shows the structures of major flavones aglycones such as apigenin, luteolin, diosmetin, chrysoeriol, and chrysin.

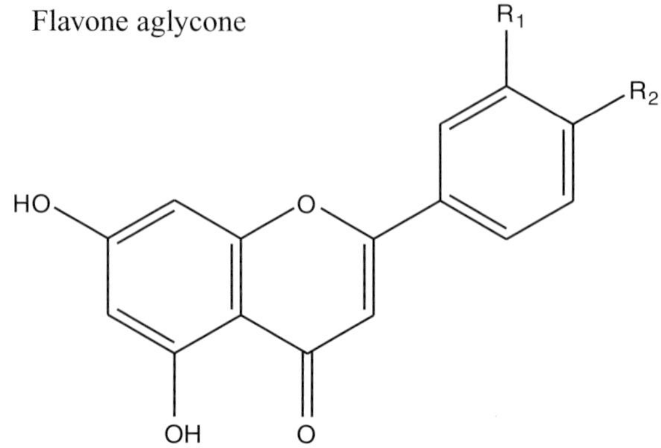

Apigenin	R_1=H;	R_2=OH;
Luteolin	R_1=OH;	R_2=OH;
Diosmetin	R_1=H;	R_2=OCH_3;
Chrysoeriol	R_1=OH;	R_2=OCH_3;
Chrysin	R_1=H;	R_2=H;

Figure 14.3 Structures of citrus flavone aglycone.

Caristi *et al.* [29] reported that orange juice serves as a rich source of 6,8-di-*C*-glucopyranosylapigenin while lemon juice shows an abundance of 6,8-di-*C*-glucopyranosyldiosmetin. They found a 6,8-di-*C*-glucopyranosylapigenin content in the range of 70–76 mg/L in the commercially available Italian orange juices. Apigenin was identified in flower, leaf, and fruit (mesocarp and endocarp) extracts of *C. medica* cv Diamante, with values ranging from 941 mg/kg for flowers to 58 mg/kg for mature fruits endocarp in fresh weight (FW). Besides apigenin, quercetin (580.8 mg/kg) and diosmin (372.5 mg/kg) were also found in significant quantities in extracts of *C. medica* cv Diamante [30]. Recently, a new C-glycosyl flavonoid, lucenin-2, was identified in sour orange (*C. aurantium* L.) juice [28].

Diosmin is the major flavone in navel orange, bergamot, and lemon peels. Lemon peels contain three most abundant flavones: diosmetin 6,8-di-*C*-glucoside, vicenin-2, and diosmin [31]. Miyake *et al.* [32] isolated two *C*-glucosylflavones from the peel of lemon fruits: diosmetin 6,8-di-*C*-glucoside and diosmetin 6-*C*-d-glucoside. These flavones are also present in limes, but not in other citrus fruits. Recently, a new flavonoid possessing strong antioxidant activity [33] was characterized. The major isolated compound from *C. unshiu* peel was quercetagetin (3,5,6,7,3′,4′-hexahydroxyflavone).

14.3.3 Flavonols

Flavonols, including rutin (quercetin 3-*O*-rutinoside), isolimocitrol 3-β-D-glucoside, and limocitrin-3-β-D-glucoside, have been identified [16] in citrus. Dugo *et al.* [34] reported that rutin and myrecetin were the most abundant flavonols identified in lemon juice, while quercetin and kaempferol were present in both peel and juice. Isolimocitrol 3-β-D-glucoside,

Flavonol aglycone

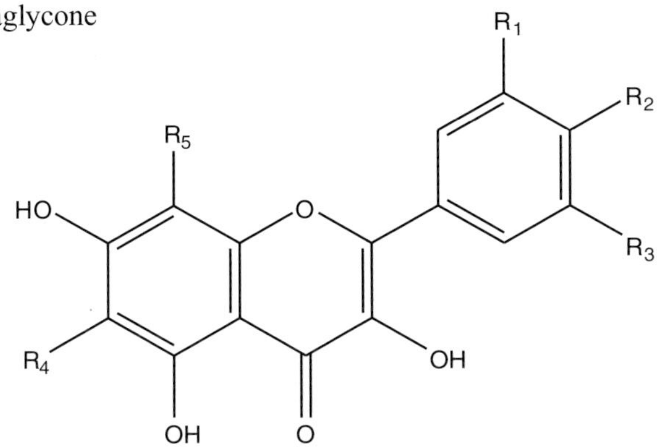

Quercetin	R_1=OH;	R_2=OH;	R_3=H;	R_4=H;	R_5=H;
Kaempferol	R_1=H;	R_2=OH;	R_3=H;	R_4=H;	R_5=H;
Myricetin	R_1=OH;	R_2=OH;	R_3=OH;	R_4=H;	R_5=H;
Limocitrol	R_1=OCH_3;	R_2=OH;	R_3=H;	R_4=OCH_3;	R_5=OCH_3;
Isolimocitrol	R_1=OH;	R_2=OCH_3;	R_3=H;	R_4=OCH_3;	R_5=OCH_3;

Figure 14.4 Structures of citrus flavonol aglycones.

limocitrin 3-β-D-glucoside, and limocitrol were identified as polymethoxylated flavonols, in citrus peels. Figure 14.4 shows the structures of major citrus flavonol aglycones such as quercetin, kaempferol, myricetin, limocitrol, and isolimocitrol.

14.3.4 Polymethoxyflavones

Among the well-known citrus bioactive compounds, PMFs exist in citrus peel deserve special interest for their various biological activities [35]. These polymethoxylated flavones are scutellarein (5,6,7,4′-tetramethoxyflavone), sinensetin (5,6,7,3′,4′-pentamethoxyflavone), tangeretin (5,6,7,8,4′-pentamethoxyflavone), quercetagetin (3,5,6,7,3′,4′-hexamethoxyflavone), nobiletin (5,6,7,8,3′,4′-hexamethoxyflavone), 3,5,6,7,8,3′,4′-heptamethoxyflavone, 7-hydroxy-3,5,6,3′,4′-pentamethoxyflavone, and 7-hydroxy-3,5,6,8,3′,4′-hexamethoxyflavone [16, 36]. Nogata *et al.* [37] reported that nobiletin was the most abundant, while sinensetin was the least abundant of the PMFs. Figure 14.5 shows the structures of citrus PMFs (such as sinensetin, tangeretin, and nobiletin).

Recent studies attributed the beneficial function of chen-pi (dried orange peel) to the PMFs present in the orange peels [38]. Dancy tangerine (*C. tangerina*) and Ponkan, the two most widely used citrus species for the production of chen-pi in Chinese medicine, have the highest amounts of PMFs among citrus species. The typical quantities and distribution of PMFs in the extract of Ponkan peel were found to be in the range of between 6.41 and 12.8 mg/g solid extract. Nogata *et al* [39] analyzed the citrus PMFs in the dried edible parts of fruits. They found that the concentration of tangeretin (9.1 mg/100 g) in Ponkan was highest and that of nobiletin (12.8 mg/100 g) second highest among the 66 citrus species investigated. The level of sinensetin in the peel of Ponkan was determined as 6.6 mg/100 g.

Polymethoxyflavones

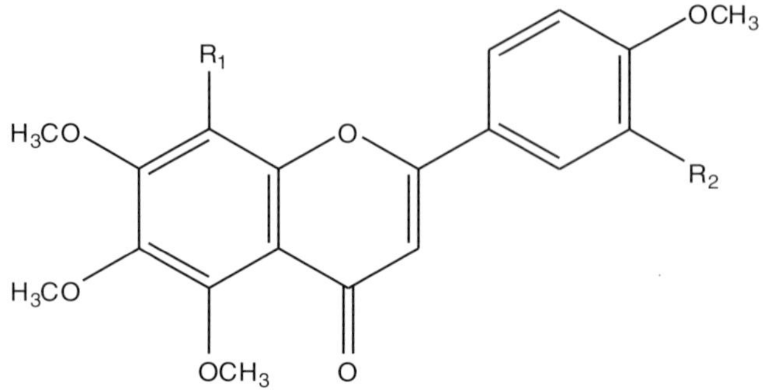

Sinensetin	R_1=H;	R_2=OCH_3;
Tangeretin	R_1=OCH_3;	R_2=H;
Nobiletin	R_1=OCH_3;	R_2=OCH_3;

Figure 14.5 Structures of citrus polymethoxyflavones.

However, the relative non-polar PMFs may also appear in the juice products. The contents of hand-press Ponkan (*C. reticulata*) juice were estimated to be nobiletin (3.56 mg/100 mL), tangeretin (4.10 mg/100 mL), and sinensetin (0.13 mg/100 mL), respectively [39]. Concentrations of PMFs (tangeretin, nobiletin, and sinensetin) in hand-press Ponkan (*C. reticulata*) juice were found to be the highest among the three juices processed by in-line, chopper pulper, and hand-press extractions, respectively. Obviously, the destruction of peel structure leads to the accumulation of PMFs in juice although the level is relatively lower as compared with that in citrus peel.

14.3.5 Hydroxycinnamic acids

Although the main phenolic constituents of citrus peel are flavanone and flavone glycosides, some hydroxycinnamates at relatively low levels were also characterized. Most of the hydroxycinnamates in citrus occur as alkaline hydrolyzable esters [40]. The hydroxycinnamic acids identified were caffeic, ferulic, *p*-coumaric, and sinapinic acids as shown in Figure 14.6 [41]. Manthey and Grohmann [13] reported that the levels of lemon hydroxycinnamic acids were much higher in the peel than in the juice. Unlike other citrus varieties, the main hydroxycinnamic acid resulting after alkaline hydrolysis of lemon molasses was *p*-coumaric acid. In addition, some benzoic acids such as protocatechuic, *p*-hydroxybenzoic, and vanillic acids have also been identified in several citrus varieties cultivated in China [42]. Interestingly, two glycosides of phenolic acid, such as 1-feruloyl-β-D-glucopyranoside and 1-sinapoyl-β-D-glucopyranoside, were identified in lemon juice by Miyake *et al.* [43].

14.3.6 Limonoids

At least 39 limonoid aglycones and 21 glucosides have been isolated and characterized from citrus plants [44]. Five most found limonoid glucosides (limonin, nomilin, deacetylnomilin, nomilinic acid, and obacunone) (structures shown in Figure 14.7) have been identified from

Hydroxycinnamic acid

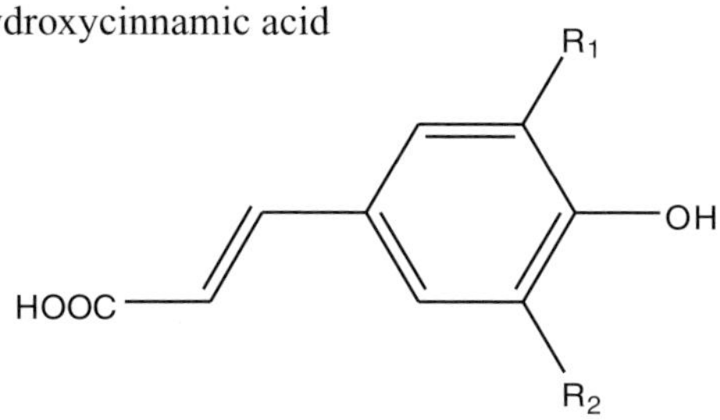

Caffeic acid	R_1=OH;	R_2=H;
Ferulic acid	R_1=OCH_3;	R_2=H;
p-Coumaric acid	R_1=H;	R_2=H;
Sinapic acid	R_1=OCH_3;	R_2=OCH_3;

Figure 14.6 Structures of citrus hydroxycinnamic acids.

the tissues of members of the genus Citrus and related genera in the plant family Rutaceae [45]. Total limonoid glucosides can exceed 300 mg/kg in citrus fruit and exceed 500 mg/kg in citrus peel [46]. In citrus juices, limonin glucoside is present in about twice the amount of the other limonoid glucosides combined [47]. In citrus seeds, nomilin glucoside occurs in higher concentration than limonin glucoside [48].

Among the limonoid aglycones, limonin, and nomilin are most prevalent in citrus [49]. Sun *et al.* [50] investigated the limonin and nomilin contents in the tissues of flavedo, albedo,

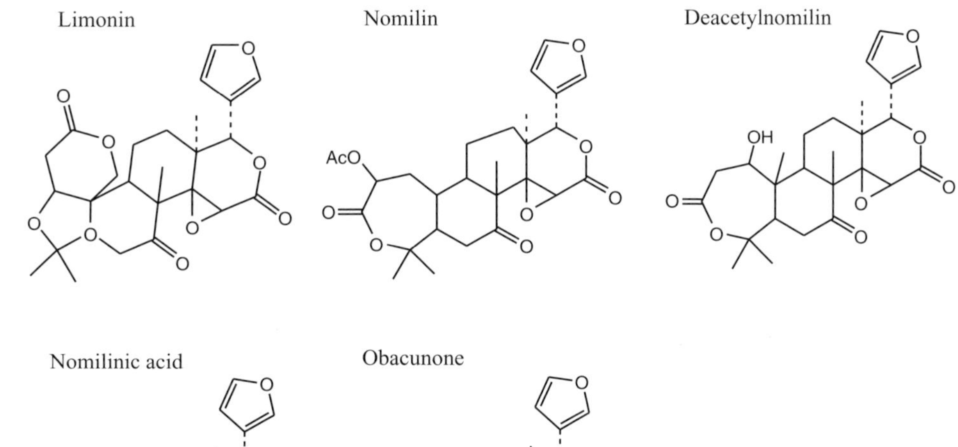

Figure 14.7 Structures of citrus limonin, nomilin, deacetylnomilin, nomilinic acid, and obacunone.

segment membrane, and juice vesicles of four citrus cultivars, including *C. grandis*, *C. unshiu*, *C. reticulata*, and *C. changshanensis* cultivated in China. In albedo of *C. changshanensis,* they found limonin and nomilin contents of 0.578 and 0.635 mg/g DW and in segment membrane of *C. changshanensis* 3.52 and 2.94 mg/g DW, respectively.

14.3.7 Adrenergic amines

Adrenergic amines have been identified and quantified in citrus members of the Rutaceae family for weight management. At least, five adrenergic proto alkaloids including tyramine, *N*-methyltyramine, octopamine, hordenine, and synephrine in *C. aurantium* have been characterized [51]. Synephrine can potentially exist in three different structural or positional isomeric forms (para-*p*, meta-*m*, or ortho-*o*) [52]. Andrade *et al.* [53] reported that only *p*-synephrine can be found in *C. aurantium* fruits. Among the identified adrenergic proto alkaloids, *p*-synephrine has attracted much attention due to its lipolytic and thermogenic effects. Its chemical structure is similar to ephedrine in ephedra, aka, and ma-huang. Figure 14.8 shows the structures of adrenergic amines (ephedrine, *p*-synephrine, and epinephrine) in citrus.

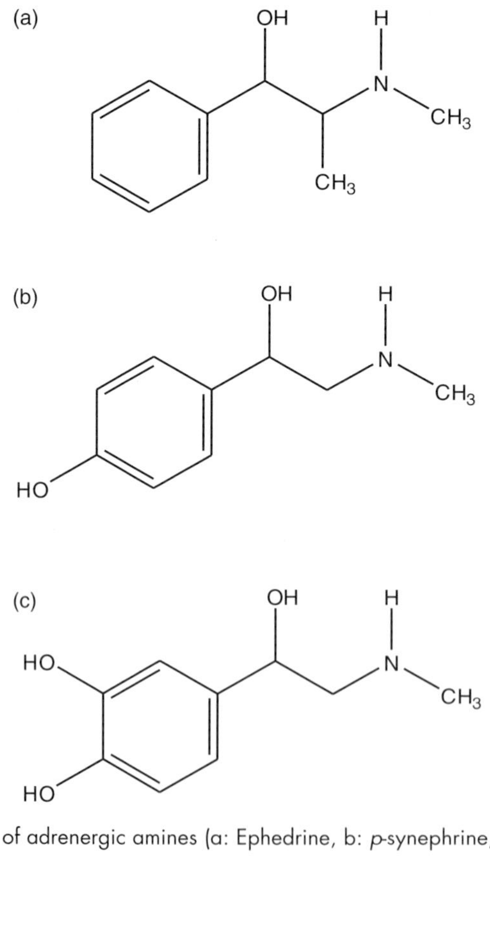

Figure 14.8 Structures of adrenergic amines (a: Ephedrine, b: *p*-synephrine, and c: Epinephrine).

The concentrations of *p*-synephrine in citrus were in the range of 0.001–0.3% in fruits from China [54], Japan [54], Italy [55], United States [56], and Brazil [57]. Mattoli *et al.* [56] analyzed the *p*-synephrine content of *C. aurantium* cultivated in Italy and reported that a high content of *p*-synephrine was found in tarocco orange freshly squeezed juice (26.65–33.22 μg/g) and in commercial red orange juices (29.9–32.07 μg/g). Recently, Arbo *et al.* [57] reported that the peel of *C. aurantium* contained a higher amount of *p*-synephrine than that in the pulp or albedo. The mean *p*-synephrine concentration in Seville orange juice samples was determined to be 56.9 μg/mL. They also compared the concentration of *p*-synephrine in unripe fruits and leaves from *C. aurantium* Lin, *C. sinensis* Osbeck, *C. deliciosa* Ten., *C. limon* Burm, and *C. limonia* Osbeck, collected in Southern Brazil. They found that both the *C. deliciosa* fruits and the leaves contained a higher concentration of *p*-synephrine than that in other citrus species.

14.4 Health effects of dried citrus peels

The health-related properties of phytochemicals in citrus are discussed in this section. The beneficial health effects of citrus flavonoids include antimicrobial, antiviral, antioxidant, anti-inflammatory, anti-allergic, hyperglycemic, antiproliferative, and cholesterol-lowering properties, and effects on capillary fragility and an ability to inhibit human platelet aggregation.

14.4.1 Antimicrobial activity

Natural phytochemical compounds, including phenolics and terpenoids, have been widely used as novel preservatives in the food industry. Apigenin has antibacterial activity over some Gram negative bacterial strains including *Enterobacter cloacae*, *Enterobacter aerogenes*, and *Pseudomonas aeruginosa* [58]. Quercetin also had antibacterial activity against *Bacillus cereus*, *Salmonella enteriditis*, *Listeria monocytogenes*, *and Pseudomonas putida* [59, 60]. Flavonoids, ponciretin, hesperetin, naringenin, and diosmetin extracted from citrus peels were active against *Helicobacter pylori* [63]. Rauha *et al.* [61] reported that naringenin exhibited strong activity against several bacteria including *Bacillus subtilis*, *Escherichia coli*, *Micrococcus luteus*, *Pseudomonas aeruginosa*, *Saccharomyces cerevisiae*, *Staphylococcus aureus*, and *Staphylococcus epidermidis*. Recently, the ethanolic fractions of Bergamot were tested against Gram-negative bacteria (*Escherichia coli*, *Pseudomonas putida*, and *Salmonella enterica*), Gram-positive bacteria (*Listeria innocua*, *Bacillus subtilis*, *Staphylococcus aureus*, and *Lactococcus lactis*) and the yeast *Saccharomyces cerevisiae*. Of the aglycones tested, eriodictyol was the most active and inhibited all the bacteria and the yeast *Saccharomyces cerevisiae* and the minimum inhibitory concentration (MIC) values were 200–800 μg/mL. Naringenin was the next most effective compound [62].

14.4.2 Antiviral activities

An important structure-activity relationship strongly influences the activity of a flavonoid against viruses. Kaul *et al.* [63] reported that quercetin and hesperetin inhibited the infectivity and/or replication of herpes simplex type, polio, parainfluenza type, and syncytial viruses. They postulated that the formation of quercetin-virus complexes may lead to the loss of the ability to induce infection. Recent research demonstrates that nobiletin extracted from *C. unshiu* peel possesses anti-hepatitis C virus effect in MOLT-4 cells [64]. The anti-picornoviral activity of the PMF structure was dependent on the 3′-methoxy and the 4′-hydroxy function in the parent flavone structure.

14.4.3 Antioxidant activity

The 1,1-diphenyl-2-picrylhydrazyl (DPPH) radical-scavenging assay has been used to evaluate antioxidant activities of orange, pomelo, and lemon fruits. Ghasemi *et al.* [65] investigated the antioxidant activity of the extracts of 13 commercially available Citrus spp. peels and edible parts. By using DPPH method, they found the antioxidant activity with the half maximal inhibitory concentration (IC_{50}) values ranged from 0.6 to 3.8 mg/mL. They attributed the DPPH radical-scavenging activities to the phenolics and flavonoids with contents varied from 66.5 to 396.8 mg gallic acid equivalents (GAE)/g of extract and from 0.3 to 31.1 mg quercetin equivalents (QE)/g of extract, respectively. Lee *et al.* [66] investigated relationship between the total polyphenol and flavonoid content and antioxidant activity of the seeds of *C. junos*. The hull and embryo of seeds were extracted with n-hexane and 70% ethanol. The DPPH radical-scavenging activity of hydroalcoholic extract was higher than that of n-hexane extract with IC_{50} values of 3.18-8.43 mg/mL. The scavenging effects of the *C. medica* cv Diamante extracts on DPPH were examined at different concentrations. The best DPPH free radical-scavenging activity was exerted by mesocarp of immature fruits (IC_{50} of 382 lg/mL), followed by flowers and leaves extracts with IC_{50} values of 425 and 502 μg/mL, respectively [67].

14.4.4 Hypoglycemic activity

The hypoglycemic potential of *C. medica* cv Diamante extracts was evaluated by the α-amylase and α-glucosidase inhibition assays [68]. It was found that the leaves extract inhibited α-amylase activity with IC_{50} value of 438.5 μg/mL. Interestingly, extract from endocarp of mature fruits inhibited the α-amylase with IC_{50} value 2-fold higher than that of immature fruits. On the contrary, extract from the immature fruits showed a higher inhibition activity against the α-glucosidase.

14.4.5 Hypolipidemic activity

Accumulated studies indicate that certain citrus flavonoids may have a protective and therapeutic effect in coronary heart disease (CHD). Epidemiological studies show evidence that consumption of citrus fruits is associated with reduced risk of cardiovascular disease (CVD). This cardioprotective effect is thought to be due to the components of citrus phytochemicals. Several studies have shown that flavonoids, limonoids, and PMFs from citrus have cholesterol-lowering properties.

Accumulated clinical data indicate that consumption of citrus juice reduces the risk of CHD in human subjects. The hematological and serum chemical profiles on patients with CHD show significantly higher serum cholesterol. In a clinical experiment [69], sixteen healthy men and nine healthy women with elevated plasma total and low-density lipoprotein (LDL) cholesterols and normal plasma triacylglycerol (TAG) concentrations received

250 mL of orange juice sequentially in their diets over a period of 4 weeks. They reported that orange juice consumption significantly increased high-density lipoprotein (HDL) cholesterol by 21% and reduced the LDL/HDL cholesterol ratio by 16%. Kurowska *et al.* [69] conducted a similar experiment in which hyperlipidemic patients, aged 39–72 years, after coronary bypass surgery, were served with red grapefruit for 30 consecutive days. Diet supplemented with fresh red grapefruits was found to decrease serum lipid levels of total and LDL cholesterols, especially serum TAG. It was hypothesized that the hypocholesterolemic effects of citrus juices were due to the naringenin in grapefruit juice and hesperetin in orange juice.

Several animal model studies have revealed that citrus flavanone glycosides including hesperidin and naringin render cholesterol-lowering properties [70–73]. After 30 days of different feeding, Gorinstein *et al.* [74] found that diets supplemented with red grapefruit juice improved the plasma lipid levels in rats fed cholesterol and increased the plasma antioxidant activity. Recently, Miceli *et al.* [75] reported that chronic administration of *C. bergamia* (1 mL/rat/day) provoked a significant reduction in total and LDL cholesterols as well as TAG levels and an increase in HDL cholesterol level. Analysis showed that *C. bergamia* juice contained naringin (520 mg/kg), neoeriocitrin (370 mg/kg), and neohesperidin (310 mg/kg).

Systematic studies have been conducted to elucidate the mechanism of hypolipidemic activity of citrus flavanones. Borradaile *et al.* [76] examined the effects of citrus flavanones on the secretion of apolipoprotein B (apo B)-containing lipoproteins in the human hepatoma cell line HepG2. They reported that naringenin and hesperetin dose-dependently reduced the accumulation of apo B in the culture media (76–81% at 200 mM) over 24 hours. Reduced hepatic secretion of apo B-containing lipoproteins would be expected to contribute to a hypocholesterolemic effect of naringenin *in vivo*. Lee *et al.* [77] showed that naringenin lowered the plasma and hepatic cholesterol concentrations by suppressing HMG-CoA reductase (HMGR) and acyl CoA: cholesterol acyltransferase (ACAT) in male rats fed a high-cholesterol diet for 42 days. Borradaile *et al.* [78] indicated that naringenin reduced plasma lipids *in vivo* and inhibited apo B secretion, cholesterol esterification, and TAG transfer protein (MTP), a protein known to be essential for the assembly and hepatic secretion of apo B-containing lipoproteins activity in HepG2 human hepatoma cells. Jung *et al.* [79] attributed the plasma and hepatic cholesterol-lowering effect of hesperidin and naringin to the decreased hepatic 3-hydroxy-3-methylglutaryl-coenzyme (HMG-CoA) reductase and ACAT activities. Both enzymes are reported to regulate the extracellular and intracellular cholesterol metabolism [80].

On the other hand, citrus PMFs, especially tangeretin and nobiletin displayed an even stronger potential lowering hepatic very low-density lipoprotein (VLDL), LDL, and total cholesterol than that observed for hesperidin and naringin. Borradaile *et al.* [76] reported that as compared with that of hesperetin, naringenin, and PMFs from tangerines presented a more pronounced reduction of medium apo B in a human liver HepG2 cell line. The IC_{50} concentrations for hesperetin, naringenin, and PMFs were 43.0, 48.5, and 2.5 mg/mL, respectively. Kurowska and Manthey [81] found that diets containing 1% PMFs significantly reduced serum total and VLDL + LDL cholesterols by 19–27% and 32–40%, respectively, in hamsters with diet-induced hypercholesterolemia. Recent clinical trials demonstrated that citrus PMF extract (270 mg/d), in combination with palm tocotrienols (30 mg/d), reduced total cholesterol (20–30%), LDL cholesterol (19–27%), apo B (21%), and TAG (24–34%) [82].

14.4.6 Thermogenic and lipolytic activities

Traditionally, *C. aurantium* L. extracts were used in Chinese medicine to activate vital energy and circulation, eliminate phlegm, and disperse stagnation [3]. The immature dried fruit of *C. aurantium* has been used for the treatment of chest congestion and to stimulate gastrointestinal functions. The growing interest in sour orange products is due to the fact that they are rich in bioactive synephrine alkaloids [83]. Synephrine is a sympathomimetic amine with structural similarities to ephedra. Commercial synephrine (*C. aurantium* extract) is obtained by extraction from immature fruits of *C. aurantium* L. Recently, products containing sour orange extracts have been exploited for the production of ephedrine-free dietary supplements for weight loss and appetite control.

The biological benefits of *p*-synephrine include thermogenic and lipolytic activities [84], weight management [85], and athletic performance [86] as well as the controversial effects on blood pressure and heart rate have been discussed. Several human clinical studies have indicated that *p*-synephrine exibits remarkable thermogenic and lipolytic activities [87–89].

14.4.7 Antiproliferative activity

14.4.7.1 *Antiproliferative effects of citrus flavonones*

The antiproliferative activity of fresh fruit juices containing flavanone glycosides was evaluated. All tested citrus juices exhibited evident antiproliferative activities toward the human chronic myelogenous leukemia K562, human leukemia HL-60, and human breast adenocarcinoma MCF-7. By using high performance liquid chromatography (HPLC), the contents of flavanone glycosides of the tested citrus juices was 7.90–53.57 mg/100 mL for fruits including *C. sinensis*, *C. deliciosa* cv. Avana, *C. clementina* cv. Nules, and *C. aurantium* subsp. Myrtifolia. They attributed the antiproliferative activity to the presence of citrus flavanones [90].

Hesperidin is the most abundant flavanone glycoside in citrus fruits such as lemons and oranges [91]. Hesperidin has a dose-dependent cytotoxic effect on human colon cancer cells, accompanied by DNA fragmentation and caspase-3 activation [92]. Hesperidin has been reported to inhibit cell cycle progression in human pancreatic cells [93]. It has been shown that hesperidin possesses cancer chemopreventive effect on the 1,2-dimethylhydrazine induced colon carcinogenesis in male Wistar rats [94]. On the other hand, naringenin found in grapefruit was associated with cancer prevention [95].

Menon *et al.* [96] found that naringenin possessed an excellent antitumor potential and antioxidant effect. The decreased levels of the superoxide and hydroxyl radicals in cells were attributed to the free radical scavenging effects of naringenin. Kooststra *et al.* [97] also showed the protective effect of naringenin against UV-induced DNA damage. Due to its antioxidant properties and the ability to absorb UV light, naringenin was expected to act in all stages of the carcinogenic process: damage to the DNA (or initiation step), tumor growth (or promotion step), and invasion (or proliferative step).

In addition to the direct radical scavenging activity, the aglycone of hesperidin was reported to be a promoter of cellular antioxidant defense-related enzyme activities as well [98]. Choi [99] reported that the reduced catalase and total superoxide dismutase (SOD) activity by 7,12-dimethylbenz(a)anthracene (DMBA) were restored with the pretreatment of hesperetin a mice model. Recently, Ekambaram *et al.* [100] indicated that administration of naringenin to gastric carcinoma-induced rats could have up-regulated the redox status *via* increasing the expression of SOD, catalase (CAT), and glutathione peroxidase (GPx) to decrease the risk of cancer by antioxidant potential.

Glutathione (GSH), a major nonenzymatic antioxidant in conjugation with GPx and glutathione S-transferase (GST), protects cells against cytotoxic and carcinogenic chemicals [101]. Evidence showed that naringenin enhanced glutathione level and GST expression in rats [102]. Hesperetin treatment significantly increased the GSH/oxidized glutathione (GSSG) ratio in the DMBA-treated group in a dose-dependent manner. Gao *et al.* [103] provided another anti-cancer pathway for citrus flavonoids. They reported that an increased dietary intake of naringenin induced DNA repair enzyme expression in a hormone-responsive human prostate cancer cell line (LNCaP).

14.4.7.2 *Antiproliferative effects of citrus flavones and flavonols*

Many studies have revealed that apigenin exhibited cytotoxicity *in vitro* to human cell lines, including those of colon cancer [104], hepatoma [105], prostate carcinoma [106], and human cervical carcinoma [107]. Other reports indicated that luteolin exhibited cytotoxicity to colon cancer [108] and human cervical carcinoma cells [109]. For flavonol, quercetin was found to show apoptotic effect on both hepatoma [110] and prostate carcinoma cells [111]. A number of reports in different cell lines, animal models and human epidemiological trials have pointed out an association between intake of citrus flavones and flavonols and reduced risk of cancer [112, 113].

Zhang *et al.* [114] studied the molecular mechanisms responsible for the cytotoxic effects of citrus flavones and flavonols and showed that both were able to induce cytotoxicity in KYSE-510 cells in a dose- and time-dependent manner. Their cytotoxic potency was in the order of: luteolin > quercetin > chrysin > kaempferol > apigenin > myricetin. They indicated that all the tested citrus flavonoids caused G2/M arrest through up-regulation of p21waf1 and down-regulation of cyclin B1 at the mRNA and protein levels.

14.4.7.3 *Antiproliferative effects of citrus polymethoxylated flavones*

Manthey and Guthrie [115] have shown that PMFs exhibited a higher antiproliferative activity than other citrus flavonoids. Lipophilic citrus nobiletin and tangeretin inhibited the cell growth of squamous cell carcinoma in a dose-dependent manner [116]. Abe [117] found that a standardized preparation of orange peel extract with 30% PMFs decreased development of an atypical hyperplastic lesion and increased apoptosis in ductal epithelial cells of C57Bl/6 mice. The orange peel extract containing tangeretin (19.0%), heptamethoxyflavone (15.24%), tetramethoxyflavone (13.6%), nobiletin (12.49%), hexamethoxyflavone (11.06%), and sinensitin (9.16%) was used in the experiment. Fan *et al.* [118] reported that after feeding 0.5% orange peel extract in new Western-style diet, the development of tumors markedly decreased, with multiplicity decreasing 49% in the small intestine and 38% in the colon.

The antiproliferative effects of individual citrus PMFs, especially tangeretin and nobiletin have been investigated. Hirano *et al.* [119] showed that tangeretin induced apoptosis in human promyelocytic leukaemia HL-60 cells. Lust *et al.* [120] explored the mechanism of cell death of tangeretin in K562 breakpoint cluster region-abelson murine leukemia (Bcr-Abl+) cells. Zheng *et al.* [121] reported that nobiletin had apoptosis-inducing effects in a concentration- and time-dependent manner in human colon cancer cells. Yoshimizu *et al.*

[122] confirmed the antitumour effects of nobiletin extracted from *C. depressa* Hayata on several human gastric cancer cell lines. Suzuki *et al.* [123] found that nobiletin feeding on 5-week-old male F344 rats reduced the cell-proliferation activity and increased the apoptotic index in colonic adenocarcinoma and/or colonic mucosa. Tang *et al.* [124] investigated the influence of nobiletin on prostate carcinogenesis using transgenic rats developing adenocarcinoma of the prostate (TRAP) and human prostate cancer cells. Nobiletin (powder diet containing 500 mg/kg nobiletin) caused significant reduction in the ventral, lateral, and dorsal prostate lobes.

14.4.7.4 *Antiproliferative effects of citrus limonoids*

Experimental evidence has revealed that citrus limonoids are biologically active, displaying antibacterial, antifungal, antiviral, antimalarial, anticancer, and other pharmacological activities on humans [125]. Citrus limonoids, especially obacunone, limonin, and nomilin were able to inhibit chemically induced carcinogenesis and were effective in a series of human cancer cell lines, with remarkable cytotoxicity against lung, colon, oral, and skin cancer in animal models and human breast cancer cells [126].

In vitro limonin, nomilin, and limonoid glucosides proved to have a significant ability to inhibit proliferation of human breast cancer. Among the tested limonoids, nomilin was the most effective followed by limonin and limonoid glucosides; these compounds showed an IC_{50} of 0.4, 12.5, and 75 μg/mL, respectively [127]. Limonin 17-β-D-glucopyranoside was responsible for the decreased colon tumour-genesis associated with feeding orange juice [128]. Pure limonin glucoside and limonin were also found to possess significant anti-tumor properties in animal models and with human cells [129]. Obacunone and its glucoside inhibit the proliferation of estrogen positive and negative human breast cancer cells [130] as well as colon adenocarcinoma cells [131]. Results from *in vivo* studies suggested that obacunone could inhibit carcinogen-induced colon cancer by blocking the development of a precursor lesion, aberrant crypt foci, and induction of apoptosis [132].

14.4.8 Anti-inflammatory activity

Chronic inflammation is induced by microbial and viral infections; exposure to allergens, radiation, and toxic chemicals; tobacco use and obesity. Accumulated epidemiological and experimental data indicate a close link between inflammation and cancer. The prolonged inflammatory/oxidative environment results in the damage of healthy neighboring cells and leads to carcinogenesis [133]. Overexpression of cyclooxygenase-2 (COX-2) and inducible nitric oxide synthase (iNOS) leads to the accumulation of prostaglandin E2 (PGE2) and nitric oxide (NO), respectively, and plays a major role in oxidative stress-induced inflammation [134]. Inhibitions of COX-2 and iNOS expression are the major targets of anti-inflammatory phytochemicals.

14.4.8.1 *Anti-inflammatory of citrus flavonols*

Rutin has been shown to have significant anti-inflammatory properties in an adjuvant arthritis rat model [135]. Several *in vitro* studies have shown that quercetin, the aglycone moiety of rutin, is capable of inhibiting lipopolysaccharide (LPS)-induced cytokine production. Boots *et al.* [136] found that quercetin inhibits the gene expression of TNFα *via* modulation of NF-kappaB in human peripheral blood mononuclear cells.

14.4.8.2 *Anti-inflammatory of citrus flavones*

Apigenin blocks the development of mouse skin tumours and the proliferation of human breast-cancer cells. It is a potent antioxidant, PGE2 and NO production inhibitor, cell cycle inhibitor, protein kinase C inhibitor, and apoptosis inducer. Apigenin is also found to inhibit the adhesion of monocytes to human umbilical vein endothelial cells and the expression of cellular adhesion molecules [137].

14.4.8.3 *Anti-inflammatory of citrus flavanones*

Da Silva *et al.* [138] tested the functionality of citrus flavonoids in rats and mice models and found that hesperidin may have potential therapeutic use as mild anti-inflammatory agents. Benavente-Garcia *et al.* [10] reported that several citrus flavanones exhibited potent inhibitory activity on the catalytic function of lipoxygenase and cyclooxygenase. Sakata *et al.* [139] indicated that hesperidin inactivated the LPS-induced overexpression of cyclooxygenase-2 and iNOS in a mouse macrophage cell model. Kim *et al.* [140] showed that hesperetin suppressed NF-kappaB activation and related gene expressions in kidneys of 6- and 24-monthold rats. They postulated that hesperetin suppresses NF-kappaB activity through four signal transduction pathways, NIK/IKK, ERK, p38, and JNK.

14.4.8.4 *Anti-inflammatory of citrus polymethoxyflavones*

Murakami *et al.* [141] studied the antiinflammatory activities of 31 citrus fruits. The suppressive activity of peel and juice extracts toward NO generation in murine macrophage RAW264.7 cells stimulated with LPS. They found that the inhibitory activities of peel parts containing nobiletin as the major PMFs were largely higher than those of the corresponding juice sac parts. The effects of citrus phytochemicals on a variety of inflammatory processes have been extensively studied and it has been demonstrated to inhibit several enzymes that are activated in certain inflammatory conditions [142]. The content of nobiletin in the peel extracts of 20 citrus fruits has been shown to correlate well with their inhibitory activities on NO production in the LPS-activated RAW264.7 cells [143]. Ho and Lin [144] showed that PMFs but not flavanone glycosides, were the major contributors to the NO-suppressing activity of heat-treated chen-pi. The content of PMFs (nobiletin and tangeretin) was found to be highly correlated with anti-inflammatory activity.

Wu *et al.* [145] reported that nobiletin could alleviate the airway inflammation of asthmatic rats. Nobiletin significantly reduced ovalbumin (OVA)-induced increases in eosinophils, remarkably lowered the level of eotaxin in blood and broncho-alveolar lavage fluid (BALF) of asthmatic rats. Lin *et al.* [146] indicated that nobiletin downregulated the interleukin (IL)-1-induced gene expression and production of proMMP-1/procollagenase-1 as well as proMMP-3/prostromelysin-1 in human synovial fibroblasts. They claimed that nobiletin has anti-inflammatory actions similar to those of dexamethasone.

On the other hand, tangeretin suppresses IL-1-β-induced COX-2 expression [147]. The effects of tangeretin on the expression of COX-2 in human lung epithelial carcinoma cells, A549, and human non-small cell lung carcinoma cells, H1299 were examined. Tangeretin was found to exert a much better inhibitory activity than nobiletin against IL-1-β-induced production of COX-2 in A549 cells. Li *et al.* [148] interpreted that the hydrophobic nature of the PMFs enables them to cross the intestinal membrane easily leading to the remarkable anti-inflammatory activities. They regarded citrus PMFs as promising therapeutic phytochemicals.

14.5 Food application of citrus and their by-products

At least one third of citrus fruits are used for juice production. During juice processing, peels are the major by product. It is well established that phytochemicals beneficial to human health are more abundant in citrus peels rather than the juice. As manufacturing processes can modify the level of the biologically active phytochemicals present in processed products, various processing methods to increase the biological activities of citrus were developed basically by trial and error in ancient China. Table 14.1 lists citrus varieties, especially those dried to produce value-added products.

14.5.1 Dried citrus peels

Chen-pi (dried citrus peels) has long been used as a remedy to help digestion and to treat inflammatory syndromes including bronchitis and asthma in China, Japan, Korea, and Taiwan. Varieties of *C. unshiu* Markovich, *C. reticulate* Blanco, and *C. tachibana* Makino Tanaka, are most frequently used. Some other citrus immature peels, such as *C. reticulata* Blanco ("Qing-Pi"), *C. aurantium* L., and *C. wilsonii* Tanaka ("Zhi-Ke"), are used in Chinese folk medicine preparation as well [149].

A modified intermittent drying process was developed by ancient Chinese to produce Chen-pi. Modified intermittent drying techniques for Chen-pi manufacturing are characterized with several tempering periods which allow moisture diffusion from the interior to the external surface to achieve a concentrating effect. The interaction between enzyme and substrate leads to the brown or dark brown appearance of the finished dried products.

Although sun drying is a common dehydration process used in the traditional Chinese medicine industry, high temperature treatment is found to enhance the antioxidative and antiinflamatory activity [146]. Heat treatment of citrus peels was found to enhance the antioxidant activity, perhaps due to the release of phenolic compounds from an unextractable form covalently bound to insoluble polymers to an extractable free form [150]. Interestingly, heat treatment at 100°C for 2 h also significantly increased the anti-inflammatory activity of

Table 14.1 Commercial dried citrus fruits

Common name	Scientific name	Part used	Products
Ponkan	*Citrus reticulate*	Peel	Folk medicine
Dancy tangerine	*Citrus tangerina*	Peel	Folk medicine
Satsuma	*Citrus unshiu*	Peel	Weight loss formula
Kumquat	*Fortunella margarita* S.	Whole	Preserved fruit
Sour orange	*Citrus aurantium*	Whole	Weight loss formula
Lemon	*Citrus lemon* L. *Citrus aurantifolia*	Whole	Dried slice, whole

citrus peels as evaluated by their ability to inhibit NO production in LPS-activated RAW264.7 macrophages. Jung *et al.* [151] proposed that PMFs were released by heat treatment and act as the key determinants of the anti-inflammatory activity of Chen-pi.

14.5.2 Dried sour orange

Sour orange (*C. aurantium*) is dried in Taiwan to make suan-gang tea. The entire production process of suan-gang tea is quite complicated and time-consuming. Similar to that of Chen-Pi, a modified intermittent drying process is applied to process dried sour orange. The aged suan-gang tea is charcoal black in color and has a pleasant fragrance when brewed. Before brewing, it is required to chop the product into smaller pieces. Some of the tea shops would provide the aforementioned service. A steaming hot cup of suan-gang tea tastes amazingly soothing and refreshing.

14.5.3 Preserved kumquat

Kumquats (*Fortunella margarita* S.) have a sweet outer coat and a tart, juicy center, and they can be dried and processed into preserves. Dried whole kumquats are often used as folk medicine to cure inflammatory syndromes of the respiratory tract, such as coughing, hoarseness, and sore throats [152]. Preserved kumquat is processed following several steps of material selection, washing, puncturing, color protection, blanching, vacuum sugar-dipping, draining, and drying in Taiwan.

14.5.4 Dried lemon

14.5.4.1 *Dried lemon slices*

Dried lemon slices are now commercially available in Taiwan. Traditionally, lemon slices are dried by conventional hot air drier. Recently, a closed-type dryer associated with a photovoltaic system (PV) was developed by Chen *et al.* [153]. Dried lemon slices using a closed-type dryer have better quality in terms of their sensory parameters. There are significant color differences resulting from the browning reaction experienced under the two drying methods. The constant (at 60°C) drying using hot air dryer resulted in a deeper brown color than did the closed-type dryer.

14.5.4.2 *Dried black lemon*

For decades, people in Eastern Guatemala dried native lemon (*C. aurantifolia* L.) and this practice dates back to the ancient Mayan civilization. *C. aurantifolia* originated in Asia but has adapted well to Guatemala and can be cultivated in warm temperate climates. The lemons are small and round measuring from 30 to 50 mm in diameter.

The lemon dehydration process developed in Guatemala involves the complete drying of lemons in direct sunlight [154]. Ripe fruit is picked, spread uniformly on raised earth beds (1.5 m wide) covered with black plastic held in place by stones or other heavy objects at the sides. Lemons are sun-dried on the black plastic for 3–4 months until 80–90% water is removed. Lemons are constantly turned and moved around to ensure homogenous drying. They are daily covered with the black plastic in the late evening and are uncovered to expose them to direct sunlight in the morning. Optimal conditions are reached when the lemons turn brown or black and have a slightly burnt look. Products include whole sun-dried lemons, available in different sizes (Jumbo 40 mm, 30–40 mm, 20–30 mm, and Baby) and ground lemon. Most of the dried lemon is exported to the Middle Eastern countries.

14.6 Conclusions

Among world fruit production, citrus is undoubtedly the largest one. Different from that of Western world in which citrus fruits are processed for juice production, in Oriental countries the whole immature and mature citrus fruits and peels are dried, and widely used as remedies for health promotion. Citrus phytochemicals, particularly flavonoids have wide ranges of health beneficial properties such as antioxidant, hypoglycaemic, hypolipidemic, and anticancer activities, among others.

References

1. Li, S., Lo, C.Y. & Ho, C.-T. (2006) Hydroxylated polymethoxyflavones and methylated flavonoids in sweet orange (*Citrus sinensis*) peel. *Journal of Agricultural and Food Chemistry*, **54**, 4176–4185.
2. Lario, Y., Sendra, E., Garcia-Perez, J., Fuentes, C., Sayas-Barbera, E., Fernandez-Lopez, J. & Perez-Alvarez, J.A. (2004) Preparation of high dietary fiber powder from lemon juice by-products. *Innovative Food Science and Emerging Technologies*, **5**, 113–117.
3. Ou, M. (1999) *Regular Chinese Medicine Handbook*. Warmth Publishing Ltd., Tapei, Taiwan.
4. Tanaka, T. (1969) Misunderstanding with regards citrus classification and nomenclature. *Bulletin of the University of Osaka Prefecture, Series B*, **21**, 139–145.
5. Sultana, S., Ali, M., Ansari, S.H. & Bagri, P. (2008) New 40-substituted flavones from the fruit peels of *Citrus limon* (L.) Burm.f. *Journal of Asian Natural Products Research*, **10**, 1123–1127.
6. Kawaii, S., Tomono, Y., Katase, E. Ogawa, K. & Yano, M. (1999) Quantitation of flavonoid constituents in citrus fruits. *Journal of Agricultural and Food Chemistry*, **47**, 3565–3571.
7. Gorinstein, S., Zachwieja, Z., Katrich, E., Pawelzikc, E., Haruenkitd, R., Trakhtenberge, S. & Martin-Belloso, O. (2004) Comparison of the contents of the main antioxidant compounds and the antioxidant activity of white grapefruit and his new hybrid. *Lebensmittel-Wissenschaft und-Technologie*, **37**, 337–343.
8. Ejaz, S., Ejaz, A., Matsuda, K. & Lim, C.W. (2006) Limonoids as cancer chemopreventive agents. *Journal of the Science of Food and Agriculture*, **86**, 339–345.
9. Liu, Y., Ahmad, H., Luo, Y., Gardinar, D.T., Gunasekera, R.S., McKeehan, W.L. & Patil, B.S. (2001) Citrus pectin: Characterization and inhibitory effect on fibroblast growth factor-receptor interaction. *Journal of Agricultural and Food Chemistry*, **49**, 3051–3057.
10. Benavente-Garcia, O., Castillo, J., Marin, F.R., Ortuño, A. & Del Río, J.A. (1997) Uses and properties of citrus flavonoids. *Journal of Agricultural and Food Chemistry*, **45**, 4505–4515.
11. Horowitz, R. & Gentili, B. (1977) Flavonoids constituents of citrus. In: *Citrus Science and Technology* (eds S. Nagy, P.E. Shaw & M.K. Vedhuis). AVI Publishing, Westport, CT, pp. 397–426.
12. Kanes, K., Tisserat, B., Berhow, M. & Vandercook, C. (1993) Phenolic composition of various tissues of Rutaceae species. *Phytochemistry*, **32**, 967–974.
13. Manthey, J.A. & Grohmann, K. (2001) Phenols in citrus peel byproducts. Concentrations of hydroxycinnamates and polymethoxylated flavones in citrus peel molasses. *Journal of Agricultural and Food Chemistry*, **49**, 3268–3273.
14. Hillebrand, S., Schwarz, M. & Winterhalter, P. (2004) Characterization of anthocyanins and pyranoanthocyanins from blood orange (*Citrus sinensis* (L.) Osbeck) juice. *Journal of Agricultural and Food Chemistry*, **52**, 7331–7338.
15. Peterson, J.J., Dwyer, J.T., Beecher, G.R., Bhagwat, S.A., Gebhardt, S.E., Haytowitz, D.B. & Holden, J.M. (2006) Flavanones in oranges, tangerines (*mandarins*), tangors, and tangelos: A compilation

and review of the data from the analytical literature. *Journal of Food Composition and Analysis*, **19**, S66–S73.

16. Gonzalez-Molina, E., Dominguez-Perles, R., Moreno, D.A. & García-Viguera, C. (2010) Natural bioactive compounds of *Citrus limon* for food and health. *Journal of Pharmaceutical and Biomedical Analysis*, **51**, 327–345.
17. Ramful, D., Bahorun, T., Bourdon, E., Tarnus, E. & Aruoma, O.I. (2010) Bioactive phenolics and antioxidant propensity of flavedo extracts of Mauritian citrus fruits: potential prophylactic ingredients for functional foods application. *Toxicology*, **278**, 75–87.
18. Kanaze, F.I., Gabrieli, C., Kokkalou, E., Georgarakis, M. & Niopas, I. (2003) Simultaneous reversed-phase high-performance liquid chromatographic method for the determination of diosmin, hesperidin and naringin in different citrus fruit juices and pharmaceutical formulations. *Journal of Pharmaceutical and Biomedical Analysis*, **33**, 243–249.
19. Manthey, J.A. & Grohmann, K. (1996) Concentrations of hesperidin and other orange peel flavonoids in citrus processing byproducts. *Journal of Agricultural and Food Chemistry*, **44**, 811–814.
20. Ortuno, A., Reynaldo, I., Fustera, M.D., Botía, J., Puig, D.G., Sabater, F., Lidón, A.G., Porras, I. & Del Río, J.A. (1997) Citrus cultivars with high flavonoid contents in the fruits. *Science Horticulture*, **68**, 231–236.
21. Wang, Y.C., Chuang, Y.C. & Hsu, H.W. (2008) The flavonoid, carotenoid and pectin content in peels of citrus cultivated in Taiwan. *Food Chemistry*, **106**, 277–284.
22. Horowitz, R.M. (1986) Taste effects of flavonoids. *Progress in Clinical and Biological Research*, **213**, 163–175.
23. Wu, T., Guan, Y. & Ye, J. (2007) Determination of flavonoids and ascorbic acid in grapefruit peel and juice by capillary electrophoresis with electrochemical detection. *Food Chemistry*, **100**, 1573–1579.
24. Vanamala, J.L., Reddivari, K.S., Yoo, L.M., Pike, L.M. & Patil, B.S. (2006) Variation in the content of bioactive flavonoids in different brands of orange and grapefruit juice. *Journal of Food Composition and Analysis*, **19**, 157–166.
25. Mouly, P., Gaydou, E.M. & Auffray, A. (1998) Simultaneous separation of flavanone glycosides and polymethoxylated flavones in citrus juices using liquid chromatography. *Journal of Chromatography A*, **800**, 171–179.
26. Miyake, Y., Yamamoto, K., Morimitsu, Y. & Osawa, T. (1998) Characteristics of antioxidative flavonoid glycosides in lemon fruit. *Food Science and Technology International*, **4**, 48–53.
27. Bocco, A., Cuvelier, M.E., Richard, H. & Berset, C. (1998) Antioxidant activity and phenolic composition of citrus peel and seed extract. *Journal of Agricultural and Food Chemistry*, **46**, 2123–2129.
28. Barreca, D., Bellocco, E., Caristi, C., Leuzzi, U. & Giuseppe Gattuso, G. (2011) Distribution of *C*- and *O*-glycosyl flavonoids, (3-hydroxy-3-methylglutaryl)glycosyl flavanones and furocoumarins in *Citrus aurantium* L. juice. *Food Chemistry*, **124**, 576–582.
29. Caristi, C., Bellocco, E., Gargiulli, C., Toscano, G. & Leuzzi, U. (2006) Flavone-di-*C*-glycosides in citrus juices from Southern Italy. *Food Chemistry*, **95**, 431–437.
30. Menichini, F., Loizzo, M.R., Bonesi, M., Conforti, F. & De Luca, D. (2011) Phytochemical profile, antioxidant, anti-inflammatory and hypoglycemic potential of hydroalcoholic extracts from *Citrus medica* L. cv Diamante flowers, leaves and fruits at two maturity stages. *Food and Chemical Toxicology*, **49**, 1549–1555.
31. Mandalari, G., Bennett, R.N., Bisignano, G., Saija, A., Dugo, G., Lo Curto, R.B., Faulds, C.B. & Waldron, K.W. (2006) Characterization of flavonoids and pectins from bergamot (*Citrus bergamia* Risso) peel, a major byproduct of essential oil extraction. *Journal of Agricultural and Food Chemistry*, **54**, 197–203.
32. Miyake, Y., Yamamoto, K., Morimitsu, Y. & Osawa, T. (1997) Isolation of *C*-glucosylflavone from lemon peel and antioxidative activity of flavonoid compounds in lemon fruit. *Journal of Agricultural and Food Chemistry*, **45**, 4619–4623.
33. Yang, X., Kang, S.M., Jeon, B.T., Kim, Y.D., Ha, J.H., Kim, Y.T. & Jeon, Y.J. (2011) Isolation and identification of an antioxidant flavonoid compound from citrus-processing by-product. *Journal of the Science of Food and Agriculture*, **91**, 1925–1927.
34. Dugo, P., Presti, M.L., Öhman, M., Fazio, A., Dugo, G. & Mondello, L. (2005) Determination of flavonoids in citrus juices by micro-HPLC-ESI/MS. *Journal of Separation Science*, **28**, 1149–1156.
35. Tripoli, E., Guardia, M.L., Giammanco, S., Majo, D.D. & Giammanco, M. (2007) Citrus flavonoids: molecular structure, biological activity and nutritional properties: a review. *Food Chemistry*, **104**, 466–479.

36. Horowitz, R. & Gentili, B. (1977) Flavonoids constituents of citrus. In: *Citrus Science and Technology Westport* (eds S. Nagy, P.E. Shaw & M.K. Vedhuis). AVI Publishing, Westport, CT, pp. 397–426.
37. Nogata, Y., Sakamoto, K., Shiratsuchi, H., Ishii, T., Yano, M. & Ohta, H. (2006) Flavonoid composition of fruit tissues of citrus species. *Bioscience, Biotechnology and Biochemistry*, **70**, 178–192.
38. Huang, Y.S. & Ho, S.C. (2010) Polymethoxy flavones are responsible for the anti-inflammatory activity of citrus fruit peel. *Food Chemistry*, **119**, 868–873.
39. Nogata, Y., Ohta, H., Sumida, T. & Sekiya, K. (2003) Effect of extraction method on the concentrations of selected bioactive compounds in mandarin juice. *Journal of Agricultural and Food Chemistry*, **51**, 7346–7351.
40. Peleg, H., Naim, M., Rouseff, R.L. & Zehavi, U. (1991) Distribution of bound and free phenolic acid in oranges (*Citrus sinensis*) and grapefruits (*Citrus paradisi*). *Journal of Science of Food and Agriculture*, **57**, 417–426.
41. Robbins, R.J. (2003) Phenolic acids in foods: An overview of analytical methodology. *Journal of Agricultural and Food Chemistry*, **51**, 2866–2887.
42. Xu, G., Liu, D., Chen, J., Ye, X., Ma, Y. & Shi, J. (2008) Juice components and antioxidant capacity of citrus varieties cultivated in China. *Food Chemistry*, **106**, 545–551.
43. Miyake, Y., Mochizuki, M., Okada, M., Hiramitsu, M., Morimitsu, Y. & Osawa, T. (2007) Isolation of antioxidative phenolic glucosides from lemon juice and their suppressive effect on the expression of blood adhesion molecules. *Bioscience, Biotechnology & Biochemistry*, **71**, 1911–1199.
44. Jayaprakasha, G.K., Mandadi, K.K. & Poulose, S.M. (2008) Novel triterpenoid from *Citrus aurantium* L. possesses chemopreventive properties against human colon cancer cells. *Bioorganic & Medicinal Chemistry*, **16**, 5939–5951.
45. Hasegawa, S., Fong Chi, H., Herman, Z. & Miyake, M. (1992) Glucosides of limonoids. In: *Flavor Precursors* (eds R. Teranish, G.R. Takeoka & G. Matthias). American Chemical Society, Washington, DC, pp. 88–97.
46. Shimoda, M.M., Osajima, Y., Inaba, N., Inaba, N., Ayano, S., Ozaki, Y. & Hasegawa, S. (2000) Extraction and recovery of limonoids with the supercritical carbon dioxide microbubble method. In: *Citrus Limonoids: Functional Chemicals in Agriculture and Foods* (eds M.A. Berhow, S. Hasegawa & G.D. Manners). ACS Symposium Series 757, American Chemical Society, Washington, DC, pp. 96–106.
47. Herman, Z., Fong Chi, H., Ou, P. & Hasegawa, S. (1990) Limonoid glucosides in orange juice by HPLC. *Journal of Agricultural and Food Chemistry*, **38**, 1860–1861.
48. Ozaki, Y., Fong, C.H., Herman, Z., Maeda, H., Miyake, M., Ifuku, Y. & Hasegawa, S. (1991) Limonoid glucosides in citrus seeds. *Agricultural and Biological Chemistry*, **55**, 137–141.
49. Ohta, H., Berhow, M., Bennett, R.D. & Hasegawa, S. (1992) Limonoids in seeds of *citrus hanaju*. *Phytochemistry*, **31**, 3905–3907.
50. Sun, C.D., Chen, K.S., Chen, Y. & Chen, Q.J. (2005) Contents and antioxidant capacity of limonin and nomilin in different tissues of citrus fruit of four cultivars during fruit growth and maturation. *Food Chemistry*, **93**, 599–605.
51. Nelson, B.C., Putzbach, K., Sharpless, K.E. & Sander, L.C. (2007) Mass spectrometric determination of the predominant adrenergic protoalkaloids in bitter orange (*Citrus aurantium*). *Journal of Agricultural and Food Chemistry*, **55**, 9769–9775.
52. James, M.I., Midgley, J.M. & Williams, C.M. (1983) The metabolism and biosynthesis of +/-o-octopamine and +/-o-synephrine in the rat. *Journal of Pharmacy and Pharmacology*, **35**, 559–565.
53. Andrade, A.S., Schmitt, G.C., Rossato, L.G., Russowsky, D. & Limberger, R.P. (2009) Gas chromatographic method for analysis of *p*-synephrine in *Citrus aurantium* L. products. *Chromatographia*, **69** (Supplement), 225–229.
54. Takei, H., Hirabuki, M. & Yoshisaki, F. (1999) Analysis of synephrine in the peel of citrus fruit, immature citrus fruit and decoctions of Chinese medicinal prescriptions containing these crude drugs by capillary electrophoresis. *Analytical Science*, **15**, 1017–1020.
55. Dragull, K., Breksa, A.P., III & Cain, B. (2008) Synephrine content of juice from satsuma mandarins (*Citrus unshiu* Marcovitch). *Journal of Agricultural and Food Chemistry*, **56**, 8874–8878.
56. Mattoli, L., Cangi, F., Maidecchi, A., Ghiara, C., Tubaro, M. & Pietro, T. (2005) A rapid liquid chromatography electrospray ionization mass spectrometry method for evaluation of synephrine in *Citrus aurantium* L. samples. *Journal of Agricultural and Food Chemistry*, **53**, 9860–9866.

57. Arbo, M.D., Larentis, E.R., Linck, V.M., Aboy, A.L., Pimentel, A.L., Henriques, A.T., Dallegrave, E., Garcia, S.C., Leal, M.B. & Limberger, R.P. (2008) Concentrations of *p*-synephrine in fruits and leaves of Citrus species (Rutaceae) and the acute toxicity testing of *Citrus aurantium* extract and *p*-synephrine. *Food and Chemical Toxicology*, **46**, 2770–2775.
58. Basile, A., Giordano, S., Lopez-Saez, J.A. & Cobianchi, R.C. (1999) Antibacterial activity of pure flavonoids isolated from mosses. *Phytochemistry*, **52**, 1479–1482.
59. Arima, H., Ashida, H. & Danno, G. (2002) Rutin-enhanced antibacterial activities of flavonoids against *Bacillus cereus* and *Salmonella enteritidis*. *Bioscience, Biotechnology and Biochemistry*, **66**, 1009–1014.
60. Bae, E.A., Han, M.J. & Kim, D.H. (1999) *In vitro* Anti-*Helicobacter pylori* activity of some flavonoids and their metabolites. *Planta Medica*, **655**, 442–443.
61. Rauha, J.P., Remes, S., Heinonen, M., Hopia, A., Kähkönen, M., Kujala, T., Pihlaja, K., Vuorela, H. & Vuorela, P. (2000) Antimicrobial effects of Finnish plant extracts containing flavonoids and other phenolic compounds. *International Journal of Food Microbiology*, **56**, 3–12.
62. Mandalari, G., Bennett, R.N., Bisignano, G., Trombetta, D., Saija, A., Faulds, C.B., Gasson, M.J. & Narbad, A. (2007) Antimicrobial activity of flavonoids extracted from bergamot (*Citrus bergamia* Risso) peel, a byproduct of the essential oil industry. *Journal of Applied Microbiology*, **103**, 2056–2064.
63. Kaul, T.N., Middleton, E. & Ogra, P.L. (1985) Antiviral effect of flavonoids on human viruses. *Journal of Medical Virology*, **15**, 71–79.
64. Suzuki, M., Sasaki, K., Yoshizaki, F., Fujisawa, M., Katsuji Oguchi, K. & Cyong, J.C. (2005) Anti-hepatitis C virus effect of *Citrusunshiu* peel and its active ingredient nobiletin. *American Journal of Chinese Medicine*, **33**, 87–94.
65. Ghasemi, K., Ghasemi, Y. & Ebrahimzadeh, M.A. (2009) Antioxidant activity, phenol and flavonoid contents of 13 citrus species peels and tissues. *Pakistan Journal of Pharmacological Science*, **22**, 277–281.
66. Lee, Y.J., Hwang, I.G., Joung, E.M., Kim. H.Y., Park, E.S., Woo, K.S. & Jeong, H.S. (2009) Physiological activity and antiproliferation effects of citron seed extracts on cancer cells. *Journal of the Korean Society of Food Science and Nutrition*, **38**, 1672–1678.
67. Menichini, F., Loizzo, M.R. & Bonesi, M. (2011) Phytochemical profile, antioxidant, anti-inflammatory and hypoglycemic potential of hydroalcoholic extracts from *Citrus medica* L. cv Diamante flowers, leaves and fruits at two maturity stages, *Food and Chemical Toxicology*, **49**, 1549–1555.
68. Tundis, R., Loizzo, M.R., Statti, G.A. & Menichini, F. (2007) Inhibitory effects on the digestive enzyme a-amylase of three Salsola species (Chenopodiaceae) *in vitro*. *Pharmazie*, **62**, 473–475.
69. Kurowska, E.M., Spence, J.D., Jordan, J., Wetmore, S., Freeman, D.J., Piché, L.A. & Serratore, P. (2000) HDL cholesterol raising effect of dietary orange juice in subjects with hypercholesterolemia. *American Journal of Clinical Nutrition*, **72**, 1095–1100.
70. Shin, Y.W., Bok, S.H., Jeong, T.S., Bae, K.H., Jeoung, N.H., Choi, M.S. Lee, S.H. & Park, Y.B. (1999) Hypocholesterolemic effect of naringin associated with hepatic cholesterol regulating enzyme changes in rats. *International Journal of Vitamin & Nutrition Research*, **69**, 341–347.
71. Monforte, M.T., Trovato, A., Kirjavainen, S., Forestieri, A.M., Galati, E.M. & Lo Curto, R.B. (1995) Biological effects of hesperidin, a citrus flavonoid. (note II): hypolipidemic activity on experimental hypercholesterolemia in rat. *Farmaco*, **50**, 595–599.
72. Kurowska, E.M., Borradaile, N.M., Spence, J.D. & Carroll, K.K. (2000) Hypocholesterolemic effects of dietary citrus juices in rabbits. *Nutrition Research*, **20**, 121–129.
73. Cha, J.Y., Cho, Y.S., Kim, I., Anno, T., Rahman, S.M. & Yanagita, T. (2001) Effect of hesperetin, a citrus flavonoid, on the liver triacylglycerol content and phosphatidate phosphohydrolase activity in orotic acid-fed rats. *Plant Foods for Human Nutrition*, **56**, 349–358.
74. Gorinstein, S., Leontowicz, H. & Leontowicz, M. (2005) Changes in plasma lipid and antioxidant activity in rats as a result of naringin and red grapefruit supplementation. *Journal of Agricultural and Food Chemistry*, **53**, 3223–3228.
75. Miceli, N., Mondello, M.R., Monforte, M.T., Sdrafkakis, V., Dugo, P., Crupi, M.L., Taviano, M.F., Pasquale, R.D. & Trovato, A. (2007) Hypolipidemic effects of *Citrus bergamia* Risso et Poiteau juice in rats fed a hypercholesterolemic diet. *Journal of Agricultural and Food Chemistry*, **55**, 10671–10677.
76. Borradaile, N.M., Carroll, K.K. & Kurowska, E.M. (1999) Regulation of HepG2 cell apolipoprotein B metabolism by the citrus flavanones hesperetin and naringenin. *Lipids*, **34**, 591–598.

77. Lee, S.H., Park, Y.B., Bae, K.H., Bok, S.H., Kwon, Y.K., Lee, E.S. & Choi, M.S. (1999) Cholesterol-lowering activity of naringenin *via* inhibition of 3-hydroxy-3-methylglutaryl coenzyme A reductase and acyl coenzyme A: cholesterol acyltransferase in rats. *Annual Nutrition Metabolism*, **43**, 173–180.
78. Borradaile, N.M., de Dreu, L.E., Barrett, P.H., Behrsin, C.D. & Huff, M.W. (2003) Hepatocyte apoB-containing lipoprotein secretion is decreased by the grapefruit flavonoid, naringenin, via inhibition of MTP-mediated microsomal triglyceride accumulation. *Biochemistry*, **42**, 1283–1291.
79. Jung, U.J., Lee, M.K. & Park, Y.B. (2006) Effect of citrus flavonoids on lipid metabolism and glucose-regulating enzyme mRNA levels in type-2 diabetic mice. *International Journal of Biochemistry & Cell Biology*, **38**, 1134–1145.
80. Bok, S.H., Lee, S.H. & Park, Y.B. (1999) Plasma and hepatic cholesterol and hepatic activities of 3-hydroxy-3-methyl-glutaryl-CoA reductase and acyl CoA: cholesterol transferases are lower in rats fed citrus peel extract or a mixture of citrus bioflavonoids. *Journal of Nutrition*, **129**, 1182–1185.
81. Kurowska, E.M. & Manthey, J.A. (2004) Hypolipidemic effects and absorption of citrus polymethoxylated flavones in hamsters with diet-induced hypercholesterolemia. *Journal of Agricultural and Food Chemistry*, **52**, 2879–2886.
82. Roza, J.M., Zheng, X.L. & Guthrie, N. (2007) Effect of citrus flavonoids and tocotrienols on serum cholesterol levels in hypercholesterolemic subjects. *Alternative Therapies Health Medicine*, **13**, 44–48.
83. Santana, J., Sharpless, K.E. & Bryant, C. (2008) Determination of para-synephrine and meta-synephrine positional isomers in bitter orange-containing dietary supplements by LC/UV and LC/MS/MS. *Food Chemistry*, **109**, 675–682.
84. Arch, J.R. (2002) 3-Adrenoceptor agonists: potential, pitfalls and progress. *European Journal of Pharmacology*, **440**, 99–107.
85. Haaz, S., Fontaine, K.R., Cutter, G., Limdi, N., Perumean-Chaney, S. & Allison, D.B. (2006) *Citrus aurantium* and synephrine alkaloids in the treatment of overweight and obesity: an update. *Obesity Reviews*, **7**, 79–88.
86. Haller, C.A., Benowitz, N.L. & Jacob, P., III. (2005) Hemodynamic effects of ephedra-free weight-loss supplements in humans. *American Journal of Medicine*, **118**, 998–1003.
87. Colker, C.M., Kalman, D.S., Torina, G.C., Perlis, T. & Street, C. (1999) Effects of *Citrus aurantium* extract, caffeine and St-John's wort on body fat loss, lipid levels and mood states in overweight healthy adults. *Current Therapy Research*, **60**, 145–153.
88. Gougeon, R., Harrigan, K. & Tremblay, J.F. (2005) Increase in the thermic effect of food in women by adrenergic amines extracted from *Citrus aurantium*. *Obesity Research*, **13**, 1187–1194.
89. Stohs, S.J., Preuss, H.G., Keith, S.C., Keith, P.L., Miller, H. & Kaats, G.R. (2011) Effects of *p*-synephrine alone and in combination with selected bioflavonoids on resting metabolism, blood pressure, heart rate and self-reported mood changes. *International Journal of Medical Science*, **8**, 295–301.
90. Camarda, L., Stefano, V.D., Del Bosco, S.F. & Schillaci, D. (2007) Antiproliferative activity of citrus juices and HPLC evaluation of their flavonoid composition. *Fitoterapia*, **78**, 426–429.
91. Ross, J.A. & Kasum, C.M. (2002) Dietary flavonoids: bioavailability, metabolic effects, and safety. *Annual Review of Nutrition*, **22**, 19–34.
92. Park, H.J., Kim, M.J., Ha, E. & Chung, J.H. (2008) Apoptotic effect of hesperidin through caspase 3 activation in human colon cancer cells, *SNU-C4. Phytomedicine*, **15**, 147–151.
93. Patil, J.R., Chidambara Murthy, K.N., Jayaprakasha, G.K., Chetti, M.B. & Patil, B.S. (2009) Bioactive compounds from Mexican lime (*Citrus aurantifolia*) juice induce apoptosis in human pancreatic cells. *Journal of Agricultural and Food Chemistry*, **57**, 10933–10942.
94. Aranganathan, S. & Nalini, N. (2009) Efficacy of the potential chemopreventive agent, hesperetin (citrus flavanone), on 1,2-dimethylhydrazine induced colon carcinogenesis. *Food and Chemical Toxicology*, **47**, 2594–2600.
95. Liu, R.H. (2004) Potential synergy of phytochemicals in cancer prevention: mechanism of action. *Journal of Nutrition*, **134**, 3479S–3485S.
96. Menon, L.G., Kuttan, R. & Kuttan, G. (1995) Inhibition of lung metastasis in mice induced by B16F10 melanoma cells by polyphenolic compounds. *Cancer Letters*, **95**, 221–225.
97. Kooststra, M. (1994) Protection from UV-B induced DNA damage by flavonoids. *Plant Molecular Biology*, **26**, 771–774.
98. Kim, D.O. & Lee, C.Y. (2004) Comprehensive study on vitamin C equivalent antioxidant capacity (VCEAC) of various polyphenolics in scavenging a free radical and its structural relationship. *Critical Reviews in Food Science and Nutrition*, **44**, 253–273.

99. Choi, E.J. (2008) Antioxidative effects of hesperetin against 7,12-dimethylbenz(a)anthracene-induced oxidative stress in mice. *Life Sciences*, **82**, 1059–1064.
100. Ekambaram, G., Rajendran, P., Magesh, V. & Sakthisekaran, D. (2008) Naringenin reduces tumor size and weight lost in N-methyl-N′-nitro-nitrosoguanidine-induced gastric carcinogenesis in rats. *Nutrition Research*, **28**, 106–112.
101. Obrador, E., Navarro, J., Mompo, J., Asensi, M., Pellicer, J.A. & Estrela, J.M. (1997) Glutathione and the rate of cellular proliferation determine tumour cell sensitivity to tumour necrosis factor *in vivo*. *Biochemical Journal*, **325**, 183–189.
102. Lee, M.K., Bok, H.S., Jeong, T.S., Moon, S.S., Lee, S.E., Park, Y.B. & Choi, M.S. (2002) Supplementation of naringenin and its synthetic derivative alters antioxidant enzyme activities of erythrocyte and liver in high cholesterol-fed rats. *Bioorganic and Medicinal Chemistry*, **10**, 2239–2244.
103. Gao, K., Henning, S.M., Niu, Y., Youssefian, A.A., Seeram, N.P., Xu, A. & Heber, D. (2006) The citrus flavonoid naringenin stimulates DNA repair in prostate cancer cells. *Journal of Nutritional Biochemistry*, **17**, 89–95.
104. Takagaki, N., Sowa, Y., Oki, T., Nakanishi, R., Yogosawa, S. & Sakai, T. (2005) Apigenin induces cell cycle arrest and p21/WAF1 expression in a p53-independent pathway. *International Journal of Oncology*, **26**, 185–189.
105. Chiang, L.C., Ng, L.T., Lin, I.C., Kuo, P.L. & Lin, C.C. (2006) Anti-proliferative effect of apigenin and its apoptotic induction in human Hep G2 cells. *Cancer Letters*, **237**, 207–214.
106. Shukla, S. & Gupta, S. (2006) Molecular targets for apigenin-induced cell cycle arrest and apoptosis in prostate cancer cell xenograft. *Molecular Cancer Therapeutics*, **5**, 843–852.
107. Zheng, P.W., Chiang, L.C. & Lin, C.C. (2005) Apigenin induced apoptosis through p53-dependent pathway in human cervical carcinoma cells. *Life Science*, **76**, 1367–1379.
108. Lim, D.Y., Jeong, Y., Tyner, A.L. & Park, J.H.Y. (2007) Induction of cell cycle arrest and apoptosis in HT-29 human colon cancer cells by the dietary compound luteolin. *American Journal of Physiology, Gastrointestinal and Liver Physiology*, **292**, G66–G75.
109. Horinaka, M., Yoshida, T., Shiraishi, T., Nakata, S., Wakada, M., Nakanishi, R., Nishino, H., Matsui, H. & Sakai, T. (2005) Luteolin induces apoptosis via death receptor 5 upregulation in human malignant tumor cells. *Oncogene*, **24**, 7180–7189.
110. Granado-Serrano, A.B., Martin, M.A., Bravo, L., Goya, L. & Ramos, S. (2006) Quercetin induces apoptosis via caspase activation, regulation of Bcl-2, and inhibition of PI-3-kinase/Akt and ERK pathways in a human hepatoma cell line (HepG2). *Journal of Nutrition*, **136**, 2715–2721.
111. Vijayababu, M.R., Kanagaraj, P., Arunkumar, A. & Ilangovan, R. (2006) Quercetin induces p53-independent apoptosis in human prostate cancer cells by modulating Bcl-2-related proteins: a possible mediation by IGFBP-3. *Oncology Research*, **16**, 67–74.
112. Deschner, E.E., Ruperto, J., Wong, G. & Newmark, H.L. (1991) Quercetin and rutin as inhibitors of azoxymethanol-induced colonic neoplasia. *Carcinogenesis*, **12**, 1193–1196.
113. Elangovan, V., Sekar, N. & Govindasamy, S. (1994) Chemopreventive potential of dietary bioflavonoids against 20-methylcholanthrene-induced tumorigenesis. *Cancer Letters*, **87**, 107–113.
114. Zhang, Q., Zhao, X.H. & Wang, Z.J. (2009) Cytotoxicity of flavones and flavonols to a human esophageal squamous cell carcinoma cell line (KYSE-510) by induction of G2/M arrest and apoptosis. *Toxicology in Vitro*, **23**, 797–807.
115. Manthey, J.A. & Guthrie, N. (2002) Antiproliferative activities of citrus flavonoids against six human cancer cell lines. *Journal of Agricultural and Food Chemistry*, **50**, 5837–5843.
116. Kandaswami, C., Perkins, E., Soloniuk, D.S., Drzewiecki, G. & Middleton, E., Jr. (1991) Antiproliferative effects of citrus flavonoids on a human squamous cell carcinoma *in vitro*. *Cancer Letters*, **56**, 147–152.
117. Abe, S., Fan, K., Ho, C.-T., Ghai, G. & Yang, K. (2007) Chemopreventive effects of orange peel extract (OPE). II: OPE inhibits atypical hyperplastic lesions in rodent mammary gland. *Journal of Medicinal Food*, **10**, 18–24.
118. Fan, K., Kurihara, N., Abe, S., Ho, C.-T., Ghai, G. & Yang, K. (2007) Chemopreventive effects of orange peel extract (OPE). I: OPE inhibits intestinal tumor growth in ApcMin/+ mice. *Journal of Medicinal Food*, **10**, 11–17.
119. Hirano, T., Abe, K., Gotoh, M. & Oka, K. (1995) Citrus flavone tangeretin inhibits leukaemic HL-60 cell growth partially through induction of apoptosis with less cytotoxicity on normal lymphocytes. *British Journal of Cancer*, **72**, 1380–1388.

120. Lust, S., Vanhoecke, B., Van Gele, M., Philippé, J., Bracke, M. & Offner, F. (2010) The flavonoid tangeretin activates the unfolded protein response and synergizes with imatinib in the erythroleukemia cell line K562. *Molecular Nutrition & Food Research*, **54**, 823–832.
121. Zheng, Q., Hirose, Y., Yoshimi, N., Murakami, A., Koshimizu, K., Ohigashi, H., Sakata, K., Matsumoto, Y., Sayama, Y. & Mori, H. (2002) Further investigation of the modifying effect of various chemopreventive agents on apoptosis and cell proliferation in human colon cancer cells. *Journal of Cancer Research & Clinical Oncology*, **128**, 539–546.
122. Yoshimizu, N., Otani, Y., Saikawa, Y., Kobota, T., Yoshida, M., Furukawa, T., Kumai, K., Kameyama, K., Fujii, M., Yano, M., Sato, T., Ito, A. & Kitajima, M. (2004) Anti-tumour effects of nobiletin, a citrus flavonoid, on gastric cancer include: antiproliferative effects, induction of apoptosis and cell cycle deregulation. *Aliment Pharmacology Therapy*, **20**, 95S–101S.
123. Suzuki, R., Kohno, H., Murakami, A., Koshimizu, K., Ohigashi, H., Yano, M., Tokuda, H., Nishino, H. & Tanaka, T. (2004) Citrus nobiletin inhibits azoxymethane-induced large bowel carcinogenesis in rats. *BioFactors*, **22**, 111–114.
124. Tang, M., Ogawa, K., Asamoto, M., Hokaiwado, N., Seeni, A., Suzuki, S., Takahashi, S., Tanaka, T., Ichikawa, K. & Shirai, T. (2007) Protective effects of citrus nobiletin and auraptene in transgenic rats developing adenocarcinoma of the prostate (TRAP) and human prostate carcinoma cells. *Cancer Science*, **98**, 471–477.
125. Manners, G.D. (2007) Citrus limonoids: analysis, bioactivity, and biomedical prospects. *Journal of Agricultural and Food Chemistry*, **55**, 8285–8294.
126. Roy, A. & Saraf, S. (2006) Limonoids: overview of significant bioactive triterpenes distributed in plants kingdom. *Biology Pharmacology Bulletin*, **29**, 191–201.
127. Guthrie, N., Chambers, A.F. & Carroll, K.K. (1997) Inhibition of MDA-MB-435 estrogen receptor-negative human breast cancer cells by citrus limonoids. *Proceeding American Association of Cancer Research*, **38**, 113.
128. Miyagi, Y., Om, A.S., Chee, K.M. & Bennink, M.R. (2000) Inhibition of azoxymethane-induced colon cancer by orange juice. *Nutrition and Cancer*, **36**, 224–229.
129. Manners, G.D., Jacob, R.A., Breksa, A.P., Schoch, T.K. & Hasegawa, S. (2003) Bioavailability of citrus limonoids in humans. *Journal of Agricultural and Food Chemistry*, **51**, 4156–4161.
130. Guthrie, N., Morley, K., Hasegawa, S., Manners, G.D. & Vandenberg, T. (2000) Inhibition of human breast cancer cells by citrus limonoids. In: *Citrus Limonoids: Functional Chemicals in Agriculture and Foods* (eds M.A. Berhow, S. Hasegawa & G.D. Manners). ACS Symposium Series 758, American Chemical Society, Washington, DC., pp. 164–174.
131. Poulose, S.M., Harris, E.D. & Patil, B.S. (2006) Antiproliferative effects of citrus limonoids against human neuroblastoma and colonic adenocarcinoma cells. *Nutrition and Cancer*, **56**, 103–112.
132. Tanaka, T., Kohno, H., Tsukio, Y. & Honjo, S. (2000) Citrus limonoids obacunone and limonin inhibit azoxymethane-induced colon cancer in rats. *BioFactors*, **13**, 213–218.
133. Federico, A., Morgillo, F., Tuccillo, C., Ciardiello, F. & Loguercio, C. (2007) Chronic inflammation and oxidative stress in human carcinogenesis. *International Journal of Cancer*, **121**, 2381–2386.
134. Lin, W.W. & Karin, M. (2007) A cytokine-mediated link between innate immunity, inflammation, and cancer. *Journal of Clinical Investigation*, **117**, 1175–1183.
135. Guardia, T., Rotelli, A.E., Juarez, A.O. & Pelzer, L.E. (2001) Anti-inflammatory properties of plant flavonoids. Effects of rutin, quercetin and hesperidin on adjuvant arthritis in rat. *Pharmacology*, **56**, 683–687.
136. Boots, A.W., Haenen, G.R.M.M. & Bast, A. (2008) Health effects of quercetin: from antioxidant to nutraceutical. *European Journal of Pharmacology*, **585**, 325–337.
137. Lee, J.H., Zhou, H.Y., Cho, S.Y., Kim, Y.S., Lee, Y.S. & Jeong, C.S. (2007) Anti-inflammatory mechanisms of apigenin: inhibition of cyclooxygenase-2 expression, adhesion of monocytes to human umbilical vein endothelial cells, and expression of cellular adhesion molecules. *Archives of Pharmacal Research*, **30**, 1318–1327.
138. Da Silva, E.J.A., Oliveira, A.S. & Lapa, A.J. (1994) Pharmacological evaluation of the anti-inflammatory activity of a citrus bioflavonoid, hesperidin, and the isoflavonoids, duartin and claussequinone, in rats and mice. *Journal of Pharmacy and Pharmacology*, **46**, 118–122.
139. Sakata, K., Hirose, Y., Qiao, Z., Tanaka, T. & Mori, H. (2003) Inhibition of inducible isoforms of cyclooxygenase and nitric oxide synthase by flavonoid hesperidin in mouse macrophage cell line. *Cancer Letters*, **199**, 139–145.

140. Kim, J.Y., Jung, K.J., Choi, J.S. & Chung, H.Y. (2006) Modulation of the age-related nuclear factor-kappaB (NF-kappaB) pathway by hesperetin. *Aging Cell*, **5**, 401–411.
141. Murakami, A., Koshiba, T., Koshimizu, K., Nakamura, Y., Ohto, Y., Yano, M., Tokuda, H., Nishino, H. & Ohigashi, H. (2000) Suppressive effects of citrus fruits on free radical generation and nobiletin, an anti-inflammatory polymethoxyflavonoid. *Biofactors*, **12**, 187–192.
142. Choi, S.Y., Hwang, J.H., Ko, H.C., Park, J.G. & Kim, S.J. (2007) Nobiletin form citrus fruit peel inhibits the DNA-binding activity of NF-kB and ROS production in LPS-activated RAW264.7 cells. *Journal of Ethnopharmacology*, **113**, 149–155.
143. Choi, S.Y., Ko, H.C., Ko, S.Y., Hwang, J.H., Park, J.G., Kang, S.H., Han, S.H., Yun, S.H. & Kim, S.J. (2007) Correlation between flavonoid content and the NO production inhibitory activity of peel extracts from various citrus fruits. *Biological Pharmacology Bulletin*, **30**, 772–778.
144. Ho, S.C. & Lin, C.C. (2008) Investigation of heat treating conditions for enhancing the anti-inflammatory activity of citrus fruit (*Citrus reticulata*) peels. *Journal of Agricultural and Food Chemistry*, **56**, 7976–7982.
145. Wu, Y.Q., Zhou, C.H., Tao, J. & Li, S.N. (2006) Antagonistic effects of nobiletin, a polymethoxyflavonoid, on eosinophilic airway inflammation of asthmatic rats and relevant mechanisms. *Life Science*, **78**, 2689–2696.
146. Lin, C.C., Hung, P.F. & Ho, S.C. (2008) Heat treatment enhances the NO suppressing and peroxynitrite-intercepting activities of kumquat's (*Fortunella margarita* Swingle) peel. *Food Chemistry*, **109**, 95–103.
147. Chen, K.H., Weng, M.S. & Lin, J.K. (2007) Tangeretin suppresses IL-1beta-induced cyclooxygenase (COX)-2 expressions through inhibition of p38 MAPK, JNK, and AKT activation in human lung carcinoma cells. *Biochemical Pharmacology*, **73**, 215–227.
148. Li, S., Wang, Z., Sang, S., Huang, M.T. & Ho, C.-T. (2006) Identification of nobiletin metabolites in mouse urine. *Molecular Nutrition & Food Research*, **50**, 291–299.
149. Dan, B. & Andrew, G. (1986) *Chinese Herbal Medicine*. Eastland Press, Seattle, WA.
150. Jeong, S.M., Kim, S.Y., Kim, D.R., Jo, S.C., Nam, K.C., Ahn, D.U. & Lee, S.C. (2004) Effect of heat treatment on the antioxidant activity of extracts from citrus peels. *Journal of Agricultural and Food Chemistry*, **52**, 3389–3393.
151. Jung, U.J., Lee, M.K., Park, Y.B., Kang, M.A. & Choi, M.S. (2006) Effect of citrus flavonoids on lipid metabolism and glucose-regulating enzyme mRNA levels in type-2 diabetic mice. *International Journal of Biochemical Cell Biology*, **38**, 1134–1145.
152. Chiu, N.C. & Chang, K.S. (1998) *The Illustrated Medicinal Plants of Taiwan*. SMC Publishing Ltd., Tapei, Taiwan.
153. Chen, H.H., Hernandez, C.E. & Huang, T.C. (2005) A study of the drying effect on lemon slices using a closed-type solar dryer. *Journal of Solar Energy Engineering*, **78**, 97–103.
154. Hoyos, N.E. (2011) *Natural Dehydration of Lemons*. Published online at: http://www.ideassonline.org, last accessed 28 July 2011.

15 Functional characteristics of dried figs

Cesarettin Alasalvar

15.1 Introduction

Figs (*Ficus carica* L.) are native to southwest Asia and the eastern Mediterranean region, belonging to the Moraceae family [1, 2]. Figs are a widespread species commonly grown, especially in warm and dry climates and an important crop worldwide for fresh and dried consumption. The world production of figs was around 1,024,138 metric tonnes (MT) in 2008 [3], 70% of which was grown in the countries of the Mediterranean coast. Egypt is the world's largest fig producer, followed by Turkey, Algeria, and Morocco. The majority of fig production is for dry consumption. Since figs are highly perishable fruits, they have to be preserved in the dried form [4]. World dried fig production was approximately 105,453 MT in 2011 [5].

Fresh fig colors vary from dark purple to green. They can be eaten whole and raw, but are often peeled; the flesh is eaten and the skin often discarded [6]. Whereas, dried figs are available in many forms: whole for the consumers and, in industrial products, as paste, concentrate, nuggets, powder, and diced forms [7]. Potassium sorbate is added to dried figs to inhibit yeast fermentation and mould growth. Dried figs are processed to bring their moisture content up to 14–20% to as high as 30% [7]. Compared to other dried fruits such as apricots and apples, color in certain varieties may be stabilized by the addition of low levels of sulfur dioxide.

Some important aspects such as compositional/nutritional characteristics, phytochemicals, and health benefits of dried figs are considered in this chapter. Where available, comparison is made with fresh figs.

15.2 Compositional and nutritional characteristics of fresh and dried figs

Table 15.1 lists the compositional and nutritional characteristics of fresh and dried figs [8]. Figs are low in fat (0.30 g/100 g for fresh and 0.93 g/100 g for dried) and high in fiber (2.9 g/100 g for fresh and 9.8 g/100 g for dried). Dried figs are excellent sources of sugar

Dried Fruits: Phytochemicals and Health Effects, First Edition. Edited by Cesarettin Alasalvar and Fereidoon Shahidi.

Table 15.1 Compositional and nutritional characteristics of fresh and dried figs (values in per 100 g edible portion)

Nutrient	Units	Fresh figs	Dried figs
Proximate composition			
Water	g	79.11	30.05
Energy	kcal	74	249
Protein	g	0.75	3.30
Lipid	g	0.30	0.93
Ash	g	0.66	1.86
Carbohydrate	g	19.18	63.87
Dietary fiber	g	2.9	9.8
Sugars	g	16.26	47.92
Minerals			
Calcium	mg	35	162
Copper	mg	0.07	0.29
Iron	mg	0.37	2.03
Magnesium	mg	17	68
Manganese	mg	0.13	0.51
Phosphorus	mg	14	67
Potassium	mg	232	680
Selenium	μg	0.2	0.6
Sodium	mg	1.0	10
Zinc	mg	0.15	0.55
Vitamins			
Betaine	mg	nd	0.7
Choline	mg	4.7	15.8
Folate	μg	6.0	9.0
Niacin	mg	0.4	0.62
Pantothenic acid	mg	0.3	0.43
Pyridoxine	mg	0.11	0.11
Riboflavin	mg	0.05	0.08
Thiamin	mg	0.06	0.09
Vitamin A (RAE)	μg	7.0	tr
Vitamin C	mg	2.0	1.2
Vitamin E (ATE)	mg	0.11	0.35
Vitamin K	μg	4.7	15.6
Amino acids			
Alanine	g	0.045	0.134
Arginine	g	0.017	0.077
Aspartic acid	g	0.176	0.645
Cystine	g	0.012	0.036
Glutamic acid	g	0.072	0.295
Glycine	g	0.025	0.108
Histidine[a]	g	0.011	0.037
Isoleucine[a]	g	0.023	0.089
Leucine[a]	g	0.033	0.128
Lysine[a]	g	0.030	0.088
Methionine[a]	g	0.006	0.034
Phenylalanine[a]	g	0.018	0.076
Proline	g	0.049	0.610
Serine	g	0.037	0.128
Threonine[a]	g	0.024	0.085
Tryptophan[a]	g	0.006	0.020
Tyrosine	g	0.032	0.041
Valine[a]	g	0.028	0.122

Source: Adapted from USDA [8].
Note: Some numbers are rounded to the second digit after decimal point.
RAE, retinol activity equivalents; ATE, alpha-tocopherol equivalents; nd, not detected; tr, trace.
[a]Indispensable amino acids.

(47.92 g/100 g); fructose and glucose are the main sugars. Trace amounts of sucrose and galactose are also available [8]. Levels may differ according to drying method, regional, and varietal factors.

Based on USDA National Nutrient Database [8], dried figs provide more fiber (9.8 g/100 g) than other dried fruits (Chapter 1, Table 1.2). The recommended dietary fiber intake is 14 g/1000 calories of food consumed each day. This means 25–38 g of fiber per day depending on age and gender [9]. On a per serving basis (40 g dried figs or one-fourth cup and 150 g fresh figs or 1 cup or 3 medium size figs), dried figs deliver between 11.2 and 15.7% of the recommended daily intake of fiber as compared to their corresponding fresh fig counterparts (between 12.5 and 17.4%).

With respect to nutritional aspects, percentage of recommended dietary allowances (RDA) or adequate intake (AI) for minerals for adult males and females (aged 15–50 years) are also given in Table 15.2. Dried figs compare favorably in their daily mineral requirement values to fresh figs. Dried figs, in general, serve as a good source of calcium, copper, iron, magnesium, manganese, phosphorus, and potassium. Consuming 40 g (on a per serving basis) of dried figs (Table 15.2) supplies 6.5% of calcium, 12.9% of copper, 4.5–10.2% of iron, 6.6–8.6% of magnesium, 8.9–11.3% of manganese, 3.8% of phosphorus, and 5.8% of potassium for RDA or AI for adults [8, 10–12]. These values are comparable with their corresponding fresh

Table 15.2 Percentage of RDA values for adults (aged 19–50) in fresh and dried figs

Mineral	RDA or AI*	Unit	Fresh figs (150 g serving basis)[a]	Dried figs (40 g serving basis)[b]	Reference
Males					
Calcium	1000 mg/day*	mg	5.3	6.5	[8, 10]
Copper	0.9 mg/day	mg	11.7	12.9	[8, 11]
Iron	8 mg/day	mg	6.9	10.2	[8, 11]
Magnesium	400–420 mg/day	mg	6.2	6.6	[8, 10]
Manganese	2.3 mg/day*	mg	8.5	8.9	[8, 11]
Phosphorus	700 mg/day	mg	3.0	3.8	[8, 10]
Potassium	4700 mg/day	mg	7.4	5.8	[8, 12]
Selenium	55 μg/day	μg	0.5	0.4	[8, 13]
Sodium	1500 mg/day	mg	0.1	0.3	[8, 12]
Zinc	11 mg/day	mg	2.0	2.0	[8, 11]
Females					
Calcium	1000 mg/day*	mg	5.3	6.5	[8, 10]
Copper	0.9 mg/day	mg	11.7	12.9	[8, 11]
Iron	18 mg/day	mg	3.1	4.5	[8, 11]
Magnesium	310–320 mg/day	mg	8.1	8.6	[8, 10]
Manganese	1.8 mg/day*	mg	10.8	11.3	[8, 11]
Phosphorus	700 mg/day	mg	3.0	3.8	[8, 10]
Potassium	4700 mg/day	mg	7.4	5.8	[8, 12]
Selenium	55 μg/day	μg	0.5	0.4	[8, 13]
Sodium	1500 mg/day	mg	0.1	0.3	[8, 12]
Zinc	8 mg/day	mg	2.8	2.8	[8, 11]

RDA, recommended dietary allowances; AI*, adequate intake.
[a]150 g serving basis (1 cup): Around 3 medium size figs.
[b]40 g serving basis (one-fourth cup): Around 4–5 medium size figs.

fig counterparts (on a 150 g serving or one-fourth cup or 3 medium size figs) and dependent upon variety. Based on RDA and AI values, dried figs contain more calcium, magnesium, and manganese than any other dried fruits (Chapter 1, Table 1.3). O'Brien *et al.* [14] found that fruit-eating animals regularly prefer to eat figs even when other food is abundant. They proposed that high calcium levels contribute to the desirability of figs as food for many forest animals. In addition, unlike most fruits figs are considered a "keystone" plant resource for fruit-eating birds and mammals throughout the tropics [15].

Dried figs contain both water-soluble (choline, folate, niacin, pantothenic acid, pyridoxine, riboflavin, thiamine, and vitamin C) and fat-soluble vitamins (A, E, and K) (Tables 15.1 and 15.3). With regard to RDA of vitamins, 40 g dried figs provide up to 3.4% of pantothenic acid and pyridoxine, 2.5–2.9% of riboflavin, 3.0–3.3% of thiamine, and 5.2–6.9% of vitamin K for RDA or AI for adults [8, 11, 13, 16]. These values are lower than that of their corresponding fresh figs (on a 150 g serving or one-fourth cup or 3 medium size figs). Among eight dried fruits, prunes, apricot, and peaches contain higher amounts of vitamins than other dried fruits including figs (Chapter 1, Tables 1.2 and 1.4).

Table 15.3 Percentage of RDA values for adults (aged 19–50 years) in fresh and dried figs

Vitamin	RDA or AI*	Unit	Fresh figs (150 g serving basis)[a]	Dried figs (40 g serving basis)[b]	Reference
Males					
Choline	550 mg/day*	mg	1.3	1.1	[8, 16]
Folate	400 μg/day	μg	2.3	0.9	[8, 16]
Niacin	16 mg/day	mg	3.8	1.6	[8, 16]
Pantothenic acid	5 mg/day*	mg	9.0	3.4	[8, 16]
Pyridoxine	1.3 mg/day	mg	12.7	3.4	[8, 16]
Riboflavin	1.3 mg/day	mg	5.8	2.5	[8, 16]
Thiamin	1.2 mg/day	mg	7.5	3.0	[8, 16]
Vitamin A (RAE)	900 μg/day	μg	1.2	tr	[8, 11]
Vitamin C	90 mg/day	mg	3.3	0.5	[8, 13]
Vitamin E (ATE)	15 mg/day	mg	1.1	0.9	[8, 13]
Vitamin K	120 μg/day*	μg	5.9	5.2	[8, 11]
Females					
Choline	425 mg/day*	mg	1.7	1.5	[8, 16]
Folate	400 μg/day	μg	2.3	0.9	[8, 16]
Niacin	14 mg/day	mg	4.3	1.8	[8, 16]
Pantothenic acid	5 mg/day*	mg	9.0	3.4	[8, 16]
Pyridoxine	1.3 mg/day	mg	12.7	3.4	[8, 16]
Riboflavin	1.1 mg/day	mg	6.8	2.9	[8, 16]
Thiamin	1.1 mg/day	mg	8.2	3.3	[8, 16]
Vitamin A (RAE)	700 μg/day	μg	1.5	tr	[8, 11]
Vitamin C	75 mg/day	mg	4.0	0.6	[8, 13]
Vitamin E (ATE)	15 mg/day	mg	1.1	0.9	[8, 13]
Vitamin K	90 μg/day*	μg	7.8	6.9	[8, 11]

RDA, recommended dietary allowances; AI*, adequate intake; RAE, retinol activity equivalents; ATE, alpha-tocopherol equivalents; tr, trace.
[a]150 g serving basis (1 cup): Around 3 medium size figs.
[b]40 g serving basis (one-fourth cup): Around 4–5 medium size figs.

Vinson *et al.* [17] compared six fresh fruits (apricots, cranberries, dates, figs, grapes, and plums) with their corresponding dried versions with respect to nutrients for a serving size of the fruit. Dried figs were ranked number 2 in two nutrients, dietary fiber and calcium, and had the best nutrient score among the dried fruits (tied with dried apricots). The nutrient score was significantly ($P < 0.01$) better for dried fruits compared to their fresh counterparts. The best nutrient score of the fresh fruits belonged to dates, which in fact had the best nutrient score of all the fruits.

Despite the fact that fresh and dried fruits contain all indispensable amino acids, in general, they are not good sources of amino acids due to their low content of protein (Table 15.1). Aspartic acid, proline, and glutamic acids are the major amino acids in fresh and dried figs.

15.3 Phytochemicals in dried figs

15.3.1 Antioxidant activity and quality of antioxidants

The oxygen radical absorbance capacity (ORAC) of a selection of dried fruits is shown in Chapter 1, Table 1.5. Raisins (golden seedless) have the highest ORAC value (10,450 μmol trolox equivalents [TE]/100 g), whereas dates (Deglet noor) have the lowest ORAC value (2387 μmol of TE/100 g). Dried figs contain 3383 μmol of TE/100 g [18, 19]. Values are much higher for dried figs than their corresponding ORAC values for fresh, since antioxidants are concentrated during the drying process.

A comparison of antioxidant quality of vitamins and dried fruits is shown in Figure 15.1 [17]. The higher the value of $1/IC_{50}$, the better the quality. Among dried fruits, figs have the highest value and approximately 10-fold higher than vitamin C, vitamin E, and β-carotene. Therefore, polyphenols from dried fruits, including figs, can serve as potent antioxidants at physiological concentrations [17].

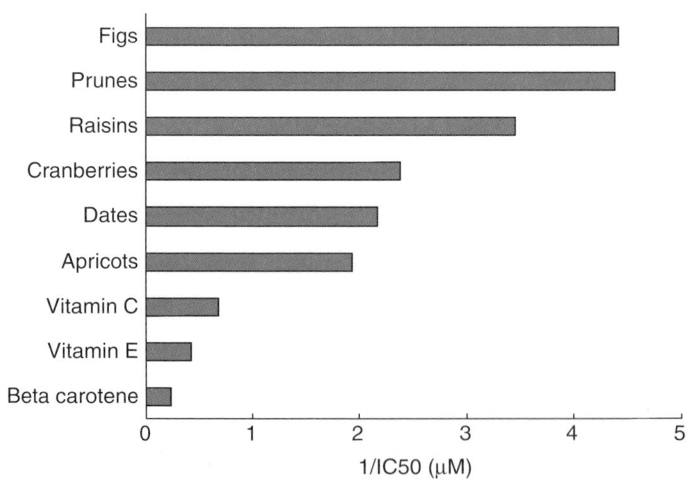

Figure 15.1 Comparison of quality of antioxidants of vitamins and dried fruits. (Adapted with permission from Vinson *et al.* [17]).

In a human study, plasma antioxidant capacity is measured by trolox equivalent antioxidant capacity (TEAC). The plasma TEAC value has been reported to be 9% after eating dried figs (40 g serving) [17]. This value after drinking one serving of green tea by normal subjects compared to baseline is 4% [20]. This shows that dried figs produce a much greater increase in plasma TEAC than green tea. Besides figs, the only other dried fruit to demonstrate human *in vivo* antioxidant activity is raisins from white grapes. In this study, subjects consumed 3 g/kg (about 5 serving) of raisins for a week and the fasting TEAC increased 8% [21].

Slatnar *et al.* [22] measured antioxidant activity of fresh and dried figs and found that antioxidant activity measured by 1,1-diphenyl-2-picrylhydrazyl (DPPH) radical scavenging activity was significantly 2-fold higher in dried figs that fresh ones.

15.3.2 Phenolics

Flavonoids are another group of phenolic compounds that can be classified into flavanones, flavones, isoflavones, anthocyanidins, flavonols, and flavanols or flavan-3-ols [23]. Literature on flavonoid profiles, particularly anthocyanins, in dried figs are scarce. Solomon *et al.* [6] analyzed total phenolics, total flavonoids, total anthocyanins, and antioxidant capacity of six commercial varieties of figs, differing in color (black, red, yellow, and green) (Table 15.4). The dark-purple mission variety contained the highest levels of phenolics, flavonoids, and anthocyanins and exhibited the highest antioxidant capacity among the tested varieties. The order of total phenolics, flavonoids, anthocyanins, and antioxidant capacity of figs were in the order of skin > fruit > pulp. Recently, Çalişkan *et al.* [24] analyzed phytochemical and antioxidant properties of 76 fig cultivars (green, yellow, purple, black, and brown) from the eastern Mediterranean region of Turkey. Black fig cultivars had the highest total antioxidant capacity (range of 7.9–16.1, mean 1.2 Fe^{2+} mmol/kg fresh weight), total anthocyanins (range of 32.3–356, mean 128.4 μg cyaniding-3-rutinoside/g fresh weight), and total phenols (range of 69.1–220, mean 118.9 mg gallic acid equivalents [GAE]/100 g fresh weight). These black-fruited cultivars had 2-fold greater total antioxidant capacity, 15-fold greater total anthocyanin, and 2.5-fold greater total phenolics than green and yellow fig cultivars. Similar results were obtained by Del Caro and Piga [25] who found that black figs were significantly richer in polyphenols (phenolic acids, flavonols, and anthocyanins) than the green ones. They also found that polyphenols were mainly concentrated in the peel rather than pulp of fresh figs. Slatnar *et al.* [22] reported that total phenolics were in the range from 7.49 mg GAE/100 g in fresh figs to 53.02 mg GAE/100 g in dried figs.

Solomon *et al.* [6] measured the total anthocyanins of various colored fresh fig fruits, skins, and pulps (Table 15.4). In all figs tested, anthocyanins concentrated in the fruit skin and constituted the main coloring compounds. Vallejo *et al.* [26] studied phenolic compounds of fresh figs and found that high concentrations of phenolic compounds are present either in the skin (mainly anthocyanins) and pulp (mainly proanthocyanidins). The average daily intake of anthocyanins per person has been estimated to reach 200 mg [27]. Fig skins are major sources of anthocyanins and polyphenols; therefore consumption of whole ripe fruit is recommended [6].

Table 15.5 lists the anthocyanins found in different colored fresh figs, skins, and pulps [6, 25, 26, 28–32]. In total, 19 different anthocyanins have been reported in different varieties and colored figs, three of which are major pigments found in figs and their skin (cyanidin

Table 15.4 Total phenolics, flavonoids, anthocyanins, and TEAC values of 100 g of fresh figs, skins, and pulps

Fig type & color	Total phenolics (mg of GAE/100 g)	Total flavonoids [mg of (+)-catechin/100 g]	Total anthocyanins (mg of cyn-3-glu/100 g)	TEAC (μmol of TE/100 g)
Fig fruits				
Mission (dark purple)	281.1 ± 3.0^{a}	21.5 ± 2.7^{a}	10.9 ± 1.3^{a}	716.3 ± 52.6^{a}
Chechick (dark purple)	80.6 ± 7.2^{b}	15.9 ± 1.7^{b}	1.8 ± 0.2^{b}	192.1 ± 16.1^{b}
Brown Turkey (purple-red)	58.1 ± 6.3^{bc}	3.6 ± 0.4^{c}	1.3 ± 0.1^{b}	120.7 ± 11.3^{c}
Bursa (purple-red)	56.0 ± 6.2^{c}	2.7 ± 0.3^{c}	0.3 ± 0.1^{c}	100.3 ± 8.6^{c}
Brunswick (yellow-green)	50.0 ± 4.6^{c}	2.3 ± 0.3^{c}	nd	69.5 ± 7.8^{cd}
Kodata (yellow)	48.6 ± 3.8^{c}	2.1 ± 0.2^{c}	nd	25.0 ± 3.1^{d}
Fig skins				
Mission (dark purple)	463.0 ± 44.3^{a}	45.6 ± 3.7^{a}	27.3 ± 2.3^{a}	1987.0 ± 175.4^{a}
Chechick (dark purple)	164.2 ± 10.5^{b}	42.9 ± 4.1^{a}	7.7 ± 0.5^{b}	602.5 ± 58.1^{b}
Brown Turkey (purple-red)	141.1 ± 12.4^{bc}	13.4 ± 1.4^{b}	6.5 ± 0.7^{b}	302.2 ± 20.3^{c}
Bursa (purple-red)	123.0 ± 13.4^{c}	10.1 ± 1.3^{b}	4.1 ± 0.3^{c}	292.5 ± 24.6^{c}
Brunswick (yellow-green)	65.5 ± 5.6^{d}	3.8 ± 0.5^{c}	0.7 ± 0.1^{d}	101.2 ± 7.8^{d}
Kodata (yellow)	41.7 ± 3.8^{d}	2.2 ± 0.3^{c}	nd	82.0 ± 5.1^{d}
Fig pulps				
Mission (dark purple)	100.6 ± 8.6^{a}	5.7 ± 0.5^{a}	0.3 ± 1.3^{a}	357.5 ± 30.6^{a}
Chechick (dark purple)	36.5 ± 4.2^{d}	4.5 ± 0.4^{b}	0.1 ± 0.2^{b}	88.4 ± 7.1^{b}
Brown Turkey (purple-red)	42.9 ± 4.7^{d}	1.6 ± 0.2^{d}	0.1 ± 0.1^{b}	79.3 ± 6.3^{b}
Bursa (purple-red)	73.7 ± 6.2^{b}	3.2 ± 0.3^{c}	0.1 ± 0.1^{c}	107.8 ± 9.6^{b}
Brunswick (yellow-green)	37.0 ± 4.6^{d}	1.6 ± 0.3^{d}	nd	36.5 ± 2.8^{c}
Kodata (yellow)	59.1 ± 6.8^{c}	2.1 ± 0.2^{d}	nd	20.8 ± 3.1^{c}

Source: Adapted with permission from Solomon *et al.* [6].
TEAC, trolox equivalent antioxidant capacity; GAE, gallic acid equivalents; cyn-3-glu, cyanidin-3-glucoside; TE, trolox equivalents; nd, not detected.
Different letters in the same column indicate significant differences ($P < 0.05$).
Data expressed as means ± standard deviation ($n = 3$) on a fresh weight basis.

Table 15.5 Anthocyanins found in different colored fresh figs, skins, and pulps

Anthocyanin	Reference
Cyanidin 3-glucoside	[28]
Cyanidin 3-rhamnoglucoside	[29]
Cyanidin 3,5-diglucoside	
Pelargonidin 3-rhamnoglucoside	
Cyanidin 3-glucoside	[6, 30]
Cyanidin 3-rhamnoglucoside	
Cyanidin 3-rutinoside dimer	[31]
(Epi)catechin-(4–8)-cyanidin 3-glucoside	
(Epi)catechin-(4–8)-cyanidin 3-rutinoside	
Cyanidin 3,5-diglucoside	
(Epi)catechin-(4–8)-cyanidin 3-rutinoside	
(Epi)catechin-(4–8)-pelargonidin 3-rutinoside	
(Epi)catechin-(4–8)-pelargonidin 3-rutinoside	
5-Carboxypyranocyanidin-3-rutinoside	
Cyanidin 3-malonylglicosyl-5-glucoside	
Cyanidin 3-glucoside	
Cyanidin 3-rutinoside	
Pelargonidin 3-glucoside	
Pelargonidin 3-rutinoside	
Peonidin 3-rutinoside	
Cyanidin 3-malonylglucoside	
Quercetin-3-rutinoside	[25]
Cyanidin 3-glucoside	
Cyanidin 3-rutinoside	
Quercetin-3-rutinoside	[32]
Cyanidin 3-glucoside	[26]
Cyanidin 3-rutinoside	

3-glucoside, cyanidin 3-rutinoside, and cyanidin 3-rhamnoglucoside) (Figure 15.2). Dark-purple and purple colored fig varieties contain much higher anthocyanins than their yellow- and red-fruited varieties [6, 32].

Phytoestrogens comprise three major classes: isoflavones, lignans, and coumestans [33]. The content of daidzein and genistein, which are classes of isoflavones, in some dried fruits (such as apricots, currants, dates, figs, prunes, and raisins) has been reported [33, 34]. Raisins are the richest source of daidzein and genistein (183.64 μg/100 g), primarily genistein, among six dried fruits (Table 15.6). Dried figs contain relatively low amount of isoflavones (5.97 μg/100 g) as compared to other dried fruits, except dried apricots (4.27 μg/100 g). More detailed information about phytoestrogen content of some dried fruits (apricots, currants, dates, prunes, and raisins) is given in Chapter 1, Table 1.7.

Vallejo *et al.* [26] reported that dried figs contained higher amounts of phenolics than the pulp of fresh figs, due to the contribution of dry skin. Four flavonols (such as kaempferol-rutinoside, quercetin-acetylglucoside, quercetin-rutinoside, and quercetin-glucoside) have been reported in three dried fig cultivars [26] and are given in Figure 15.3. Among them, quercetin-rutinoside is the major one (Table 15.7). All dried fig cultivars showed similar amounts of flavonols to fresh ones. Same authors have also reported that no anthocyanins and proanthocyanidins were found in green colors of dried fig cultivars.

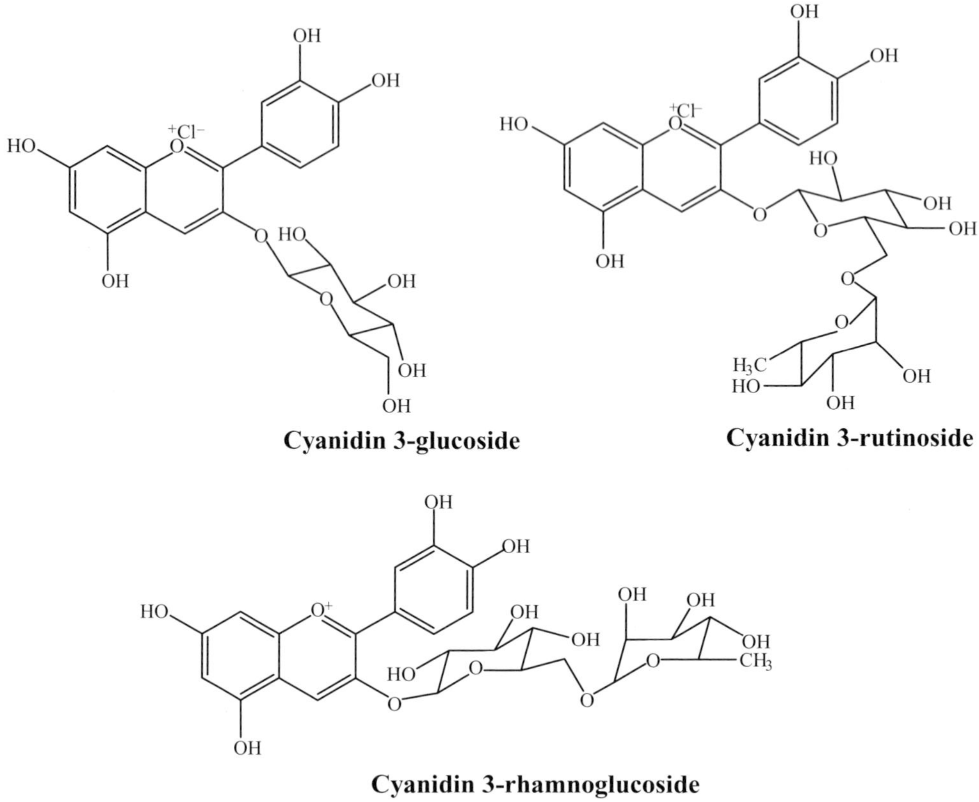

Figure 15.2 Chemical structures of major anthocyanins found in figs.

Recently, Slatnar *et al.* [22] measured the effect of different drying on fig phenolics (Table 15.8). Eight phenolics in fresh and dried figs, belonging to four groups of hydroxycinnamic acids, flavan-3-ols, flavonols, and anthocyanins were identified. The predominant phenolic compound was epicatechin; in very small amounts of luteolin-8-*C*-glucoside. Analysis of individual phenolic compounds revealed a higher content of all phenolic groups determined after the oven-drying process, with the exception of cyaniding-3-*O*-rutinoside.

Table 15.6 Daidzein and genistein content of some dried fruits (μg/100 g edible portion)

Dried fruits	Daidzein	Genistein	Total	Reference
Apricots	4.27	nd	4.27	[34]
Currants	2.20	10.00	12.20	[33]
Dates	1.72	5.16	6.88	[34]
Figs	1.77	4.2	5.97	[34]
Prunes	4.26	8.53	12.79	[34]
Raisins	58.99	124.65	183.64	[34]

nd, not detected.

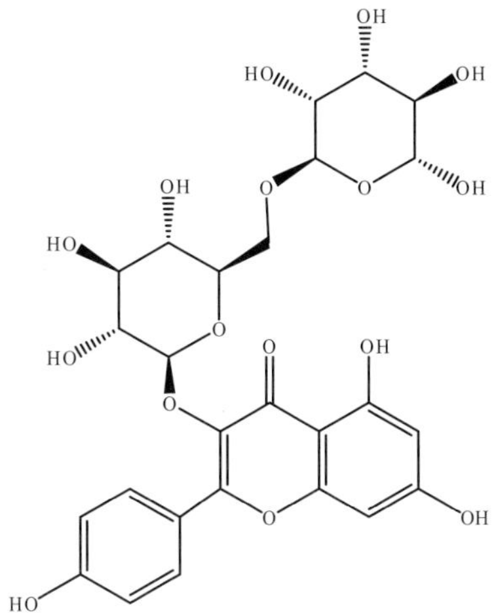

Kaempferol 3-*O*-rutinoside

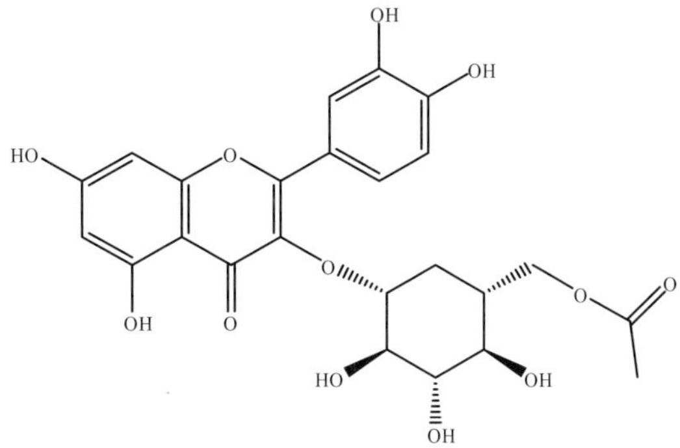

Quercetin 3-(6'' acetylglucoside)

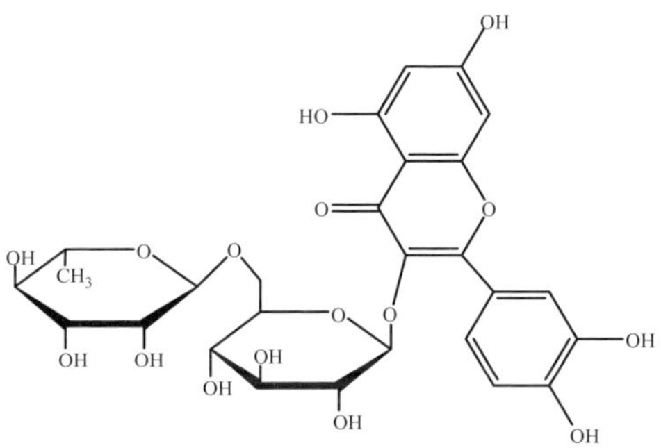

Quercetin 3-rutinoside

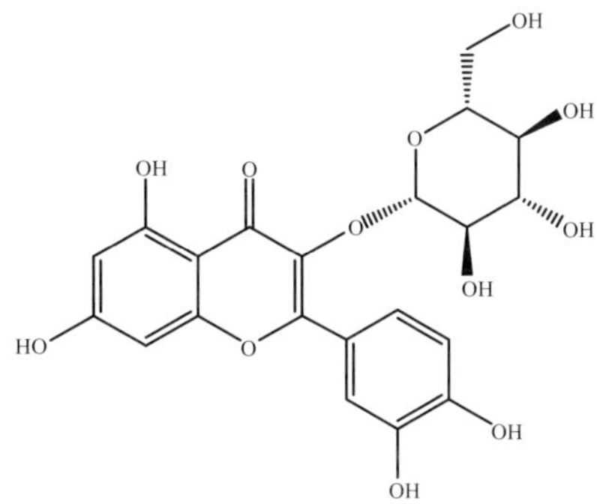

Quercetin 3-glucoside

Figure 15.3 Chemical structures of flavonols found in dried figs.

Similarly, higher total phenolic content and antioxidant activity were determined after the drying process [22].

Finally, two flavones (luteolin 6C-hexose-8C-pentose and apigenin-rutinoside) have been reported in fresh fig skins [26]. These compounds were not detected in fig pulps and dried figs.

Table 15.7 Flavonol content of three dried figs (mg/100 g edible portion)

Dried fig variety (green color)	Kaempferol-rutinoside	Quercetin-acetylglucoside	Quercetin-rutinoside	Quercetin-glucoside
Unknown Turkey	2.0[a]	2.6[a]	10.2[a]	2.5[a]
Cuello Dama	1.0[b]	2.1[b]	13.0[b]	1.0[b]
Unknown Spain	2.0[a]	2.8[a]	10.9[c]	0.7[c]

Source: Adapted from Vallejo *et al.* [26].
Different letters in the same column indicate significant differences ($P < 0.05$).

Table 15.8 Content of phenolic compounds (mg/100 g) in fresh and dried figs of two drying methods at different sampling dates

	Sampling date		
Drying methods	**July 9**	**July 15**	**September 11**
	Chlorogenic acid		
Fresh	1.33 ± 0.15[a]	2.78 ± 0.46[a]	4.91 ± 1.00[a]
Sun drying	9.84 ± 1.41[b]	15.88 ± 1.07[b]	3.42 ± 0.54[a]
Oven drying	13.96 ± 1.48[c]	32.42 ± 0.89[c]	19.92 ± 2.56[b]
	Catechin		
Fresh	1.36 ± 0.24[a]	2.67 ± 0.17[a]	2.88 ± 0.18[a]
Sun drying	11.46 ± 2.45[b]	5.88 ± 0.60[b]	6.60 ± 1.18[b]
Oven drying	16.16 ± 1.32[b]	15.57 ± 2.04[c]	19.75 ± 0.68[c]
	Epicatechin		
Fresh	7.58 ± 1.64[a]	8.67 ± 1.12[a]	7.11 ± 0.54[a]
Sun drying	23.30 ± 3.12[b]	20.37 ± 0.70[b]	10.44 ± 0.86[b]
Oven drying	34.65 ± 2.63[c]	36.65 ± 2.46[c]	26.66 ± 1.85[c]
	Kaempferol-3-*O*-glucoside		
Fresh	0.04 ± 0.00[a]	0.10 ± 0.00[a]	0.13 ± 0.01[a]
Sun drying	0.46 ± 0.04[b]	0.31 ± 0.04[b]	0.59 ± 0.06[b]
Oven drying	0.99 ± 0.09[c]	0.56 ± 0.05[c]	1.43 ± 0.07[c]
	Luteolin-8-*C*-glucoside		
Fresh	nd	nd	nd
Sun drying	0.15 ± 0.02	0.13 ± 0.01	0.16 ± 0.02
Oven drying	0.39 ± 0.03	0.21 ± 0.02	0.45 ± 0.04
	Rutin		
Fresh	0.61 ± 0.14[a]	1.86 ± 0.63[a]	0.89 ± 0.20[a]
Sun drying	6.66 ± 1.39[b]	12.06 ± 1.00[b]	1.38 ± 0.37[a]
Oven drying	7.03 ± 1.03[b]	14.62 ± 1.81[b]	3.75 ± 0.29[b]
	Quercetin-3-*O*-glucoside		
Fresh	0.18 ± 0.04[a]	0.60 ± 0.17[a]	0.41 ± 0.09[a]
Sun drying	2.40 ± 0.46[b]	3.35 ± 0.19[b]	0.56 ± 0.12[a]
Oven drying	2.23 ± 0.24[b]	2.98 ± 0.27[b]	1.10 ± 0.06[b]
	Cyanidin-3-*O*-rutinoside		
Fresh	0.21 ± 0.05[a]	0.31 ± 0.05[b]	0.62 ± 0.04[c]
Sun drying	0.26 ± 0.06[a]	0.12 ± 0.01[a]	0.13 ± 0.05[a]
Oven drying	0.16 ± 0.02[a]	0.12 ± 0.01[a]	0.31 ± 0.05[b]

Source: Adapted with permission from Slatnar *et al.* [22].
Data are expressed as mean ± standard deviation ($n = 5$) on a fresh weight basis.
Different letters in columns indicate significant differences ($P < 0.05$) in the contents of individual compounds between the treatments for each set of sampling dates.
nd, not detected.

Table 15.9 Contents of free and bound phenolic acids in dried figs (mg/100 g)

Phenolic acid	Free phenolic acids	Bound phenolic acids	Total
Chlorogenic	0.40 ± 0.06^a	2.60 ± 1.01^b	3.00
Ferulic	0.31 ± 0.01^a	6.88 ± 0.01^b	7.19
Gallic	tr	tr	tr
p-Coumaric	0.25 ± 0.01^a	9.64 ± 0.01^b	9.89
Protocatechuic	0.43 ± 0.35^a	1.53 ± 2.16^b	1.96
Syringic	0.51 ± 0.60^a	0.93 ± 0.94^b	1.44
Vanillic	3.47 ± 0.05^a	29.88 ± 2.61^b	33.35

Source: Adapted from Aegean Exporter's Associations [35].
Data are expressed as mean ± standard deviation ($n = 3$) on a fresh weight basis.
Different letters in the same row indicate significant differences ($P < 0.05$).
tr, trace.

15.3.3 Phenolic acids

Phenolic acids in foods occur in the free and bound forms. Free forms are known to contribute to the taste of foods. The bound forms contribute to the antioxidant activity. Seven phenolic acids (free and bound), four of which are hydroxylated derivatives of benzoic acid (gallic, protocatechuic, syringic, and vanillic acids) and three are cinnamic acid derivatives (chlorogenic, ferulic, and *p*-coumaric acids), have been reported and tentatively identified in dried figs [35]. Bound phenolics were most abundant in dried figs and are significantly higher ($P < 0.01$) than free phenolics (Table 15.9). Among the identified phenolic acids, vanillic acid was most abundant in dried figs, followed by *p*-coumaric acid, and ferulic acid. Recently, Vallejo *et al.* [26] have found only one phenolic acid (chlorogenic acid) in dried fig cultivars, ranging from 1.4 to 2.0 mg/100 g edible portion. Veberic *et al.* [32] have identified three phenolic acids in fresh fig cultivars, chlorogenic acid (1.71 mg/100 g fresh weight) being the most abundant, followed by gallic acid (0.38 mg/100 g fresh weight), and trace amounts of syringic acid (0.10 mg/100 g fresh weight). Dark cultivars exhibited a higher level of phenolic acids in comparison to the white fruit cultivar. Ferulic and *p*-coumaric acids have been reported as being more antioxidative than syringic, vanillic, and protocatechuic acids [36].

15.3.4 Carotenoids

Five carotenoids in an unknown variety of fresh figs (α-carotene, β-carotene, β-cryptoxanthin, lutein, and lycopene) have been reported, with lycopene being the most abundant one, followed by lutein and β-carotene [37]. Based on USDA [8] data, β-carotene and lutein + zeaxanthin are present in dried figs. Both fresh and dried figs contain most carotenoids appearing in plasma, albeit at relatively low concentrations (470 μg/100 g in fresh figs and 38 μg/100 g in dried figs). The low level of carotenoids in dried figs may be due to the drying process since carotenoids are heat sensitive. As compared to other dried fruits, dried apricots contain the highest carotenoids (2163 μg/100 g), followed by dried peaches (2080 μg/100 g), and prunes (692 μg/100 g) [8].

15.4 Health benefits of dried figs

Dried figs are an excellent source of nutrients such as minerals (calcium), fiber (soluble), and sugars (fructose and glucose) and thus a good choice for snacking. Besides nutrients, they contain an array of health promoting bioactive compounds [2].

It has been estimated that up to 70% of incidences of cancer are related to diet. Previously, it was thought that the antioxidant vitamins, such as vitamins C, and E, and β-carotene, were responsible for the beneficial effects of fruits and vegetables as shown by epidemiological studies across many cultures. However, recent supplement studies have not supported this hypothesis. More than 200 epidemiological studies now support the fact that phytochemicals in the fruits and vegetables (including dried fruits) are responsible for the reduced cancer-risk benefits offered by fruits and vegetables [7]. Major antioxidant activity of fruits including dried fruits is due to the presence of flavonoids and phenolic acids. The antioxidant activity of several flavonoids, measured as scavenging of peroxyl radicals, were found to be higher than that of vitamins E and C, and glutathione [38].

The Mediterranean diet has been reported to promote health and quality of life, specifically by preventing pathophysiological conditions related to coronary heart disease (CHD) and cancer. The high consumption of natural antioxidants and health promoting phytochemicals, achieved by consuming salads, vegetables, fruits, and their derived products, is generally considered to be a major beneficial contributor to the Mediterranean diet. Among them, olives and figs are characteristic and abundant fruits in this diet [6, 7, 17].

Dried fig antioxidants can enrich lipoproteins in plasma and protect them from subsequent oxidation. In addition, they produce a significant increase in plasma antioxidant capacity for 4 hours after consumption, and overcome the oxidative stress of consuming high fructose corn syrup in a carbonated soft drink [17].

Besides their ubiquitous polyphenols, figs have other compounds with anticancer activity, specifically benzaldehyde and coumarins [7]. Benzaldehyde has been used successfully to treat terminal human carcinomas [39]. Coumarins are the major compounds isolated from volatile extracts of figs [40]. The furanocoumarins identified in figs include angelicin, marmesin, psoralen, umbelliferone, and bergapten [41]. Coumarins have also been used for the treatment of prostate cancer [42, 43]. Psoralens, such as in the fig compound angelicin, are currently being investigated for the treatment of skin cancers and have been recommended for clinical trials because they have low skin phototoxicity [44]. These compounds produce free radicals and photoadducts of DNA that inhibit proliferation of the cancer cells [45]. Moreover, potent *in vitro* suppressors of cancer cell proliferation have been found in a fraction isolated from figs [46].

Dietary phytoestrogens present in dried figs have attracted much interest in recent years because of their potential protective effects against many diseases and conditions including cancer, cardiovascular disease (CVD), osteoporosis, and menopausal symptoms [47–53].

15.5 Conclusions

Dried figs are an excellent source of fiber, minerals, and certain phytochemicals and possess strong antioxidant activities. They are low in fat and sodium. Dried figs retain most of the nutritional benefits and health protective bioactive compounds of fresh figs. When fresh figs are not available, properly dried figs could thus be used as a valuable substitute in diets that aim to prevent certain diseases.

References

1. Flaishman, M.A., Rodov, V. & Stover, E. (2008) The fig: botany, horticulture and breeding. *Horticulture Reviews*, **34**, 132–196.
2. Lansky, E.P. & Paavilainen, H.M. (2010) *Figs: The Genus Ficus*. CRC Press, Taylor & Francis Group, Boca Raton, FL.
3. FAO (2011) *Food and Agricultural Commodities Production*. Published online at: http://faostat.fao .org/site/339/default.aspx, last accessed June 1, 2011.
4. Piga, A., Pinna, I., Özer, K.B., Agabbio, M. & Aksoy, U. (2004) Hot air dehydration of figs (*Ficus carica* L.): drying kinetics and quality loss. *International Journal of Food Science & Technology*, **39**, 793–799.
5. INC. (2012) *2011 Dried Fruits Production*. International Nut and Dried Fruit Council Foundation, Reus, Spain.
6. Solomon, A., Golubowicz, S., Yablowicz, Z., Grossman, S., Bergman, M., Gottlieb, H.E., Kerem, Z. & Flaishman, M.A. (2006) Antioxidant activities and anthocyanin content of fresh fruits of common fig (*Ficus carica* L.). *Journal of Agricultural and Food Chemistry*, **54**, 7717–7723.
7. Vinson, J.A. (1999) The functional food properties of figs. *Cereal Foods World*, **44**, 82–87.
8. USDA. (2011) *National Nutrient Database for Standard Reference, Release 24*. Published online at: http://www.nal.usda.gov/fnic/foodcomp/search/, last accessed June 20, 2011.
9. USDA. (2005) *Carbohydrates. Dietary Guidelines for Americans*. US Department of Agriculture and US Department of Health and Human Services. Government Printing Office, Washington, DC.
10. DRIs. (1997) *Dietary Reference Intakes for Calcium, Phosphorus, Magnesium, Vitamin D, and Fluoride*. The National Academies Press, Washington, DC.
11. DRIs. (2001) *Dietary Reference Intakes for Vitamin A, Vitamin K, Arsenic, Boron, Chromium, Copper, Iodine, Iron, Manganese, Molybdenum, Nickel, Silicon, Vanadium, and Zinc*. The National Academies Press, Washington, DC.
12. DRIs. (2004) *Dietary Reference Intakes for Water, Potassium, Sodium, Chloride, and Sulfate*. The National Academies Press, Washington, DC.
13. DRIs. (2000) *Dietary Reference Intakes for Vitamin C, Vitamin E, Selenium, and Carotenoids*. The National Academies Press, Washington, DC.
14. O'Brien, T.G., Kinnaird, M.F., Dierenfeld, E.S., Conklin-Brittain, N.L., Wrangham, R.W. & Silver, S.C. (1998) What's so special about figs. *Nature*, **392**, 668.
15. Slavin, J.L. (2006) Figs: Past, present and future. *Nutrition Today*, **41**, 180–184.
16. DRIs. (1998) *Dietary Reference Intakes for Thiamin, Riboflavin, Niacin, Vitamin* B_6*, Folate, Vitamin* B_{12}*, Pantothenic Acid, Biotin, and Choline*. The National Academies Press, Washington, DC.
17. Vinson, J.A., Zubik, L., Bose, P., Samman, N. & Proch, J. (2005) Dried fruits: excellent *in vitro* and *in vivo* antioxidants. *Journal of the American College of Nutrition*, **24**, 44–50.
18. Wu, X., Beecher, G.R., Holden, J.M., Haytowitz, D.B., Gebhardt, S.E. & Prior, R.L. (2004) Lipophilic and hydrophilic antioxidant capacities of common foods in the United States. *Journal of Agricultural and Food Chemistry*, **52**, 4026–4037.
19. USDA. (2010) *USDA Database for the Oxygen Radical Absorbance Capacity (ORAC) of Selected Foods, Release 2.0*. Published online at: http://www.ars.usda.gov/nutrientdata, last accessed March 9, 2011.
20. Serafini, M., Ghiselli, A. & Ferro-Luzzi, A. (1996) *In vivo* antioxidant effect of green and black tea in man. *European Journal of Clinical Nutrition*, **50**, 28–32.
21. Durak, I., Karaca, L., Cimen, M.B., Kaçmaz, M., Avci, A., Gubat, G. & Oztürk, H.S. (2002) Dried white grapes enhance blood antioxidant potential. *Nutrition, Metabolism & Cardiovascular Diseases*, **12**, 204–205.
22. Slatnar, A., Klancar, U., Stampar, F. & Veberic, R. (2011) Effect of drying of figs (*Ficus carica* L.) on the contents of sugars, organic acids, and phenolic compounds. *Journal of Agricultural and Food Chemistry*, **59**, 11696–11702.

23. Shahidi, F. & Ho, C.-T. (2005) Phenolics in food and natural health products: an overview. In: *Phenolic Compounds in Foods and Natural Products* (eds F. Shahidi & C.-T. Ho). ACS Symposium Series 909, American Chemical Society, Washington, DC, pp. 1–8.
24. Çalişkan, O. & Polat, A.A. (2011) Phytochemicals and antioxidant properties of selected fig (*Ficus carica* L.) accessions from the eastern Mediterranean region of Turkey. *Scientia Horticulture*, **128**, 473–478.
25. Del Caro, A. & Piga, A. (2008) Polyphenol composition of peel and pulp of two Italian fresh fig fruits cultivars (*Ficus carica* L.). *European Food Research & Technology*, **226**, 715–719.
26. Vallejo, F., Marín, J.G. & Tomás-Barberán, F.A. (2012) Phenolic compound content of fresh and dried figs (*Ficus carica* L.). *Food Chemistry*, **130**, 485–492.
27. Kuhnau, J. (1974) The flavonoids. A class of semi-essential food components: their role in human nutrition. *World Review of Nutrition and Dietetics*, **24**, 117–191.
28. Robinson, G.M. & Robinson, R. (1932) A survey of anthocyanins. *Biochemical Journal*, **26**, 1647–1664.
29. Puech, A.A., Rebeiz, C.A., Catlin, P.B. & Crane, J.C. (1975) Characterization of anthocyanins in fig (*Ficus carica* L.) fruits. *Journal of Food Science*, **40**, 775–779.
30. Solomon, A., Golubowicz, S., Yablowicz, Z., Bergman, M., Grossman, S., Altman, A., Kerem, Z. & Flaishman, M.A. (2010) Protection of fibroblasts (NIH-3T3) against oxidative damage by cyanidin-2-rhamnoglucoside isolated from fig fruits (*Ficus carica* L.). *Journal of Agricultural and Food Chemistry*, **58**, 6660–6665.
31. Dueñas, M., Pérez-Alonso, J.J., Santos-Buelga, C. & Escribano-Bailón, T. (2008) Anthocyanin composition in fig (*Ficus carica* L.). *Journal of Food Composition and Analysis*, **21**, 107–115.
32. Veberic, R., Colaric, M. & Stampar, F. (2008) Phenolic acids and flavonoids of fig fruit (*Ficus carica* L.) in the northern Mediterranean region. *Food Chemistry*, **106**, 153–157.
33. Thompson, L.U., Boucher, B.A., Liu, Z., Cotterchio, M. & Kreiger, N. (2006) Phytoestrogen content of foods consumed in Canada, including isoflavones, lignans, and coumestan. *Nutrition and Cancer*, **54**, 184–201.
34. Liggins, J., Bluck, L.J.C., Runswick, S., Atkinson, C., Coward, W.A. & Bingham, S.A (2000) Daidzein and genistein content fruits and nuts. *Journal of Nutritional Biochemistry*, **11**, 326–331.
35. Aegean Exporter's Associations. (2009) *Report on Food Composition and Clinical Research on Seedless Raisins and Dried Figs*. Alsancak, İzmir, Turkey.
36. Cuvelier, M.E., Richard, H. & Berset, C. (1992) Comparison of the antioxidant activity of some acid-phenols: Structure–activity relationship. *Bioscience, Biotechnology, and Biochemistry*, **56**, 324–325.
37. Su, Q., Rowley, K.G., Itsiopoulos, C. & O'Dea, K. (2002) Identification and quantification of major carotenoids in selected components of the Mediterranean diet: green leafy vegetables, figs and olive oil. *European Journal of Clinical Nutrition*, **56**, 1149–1154.
38. Cao, G., Alessio, H.M. & Cutler, R.G. (1993) Oxygen-radical absorbance capacity assay for antioxidants. *Free Radical Biology & Medicine*, **14**, 303–311.
39. Kochi, M., Takeuchi, S., Mizutani, T., Mochizutki, K., Matsumoto, Y. & Saito, Y. (1980) Antitumor activity of benzaldehyde. *Cancer Treatment Reports*, **64**, 21–23.
40. Gibernau, M., Buser, H.R., Frey, J.E. & Hossaert-McKey, M. (1997) Volatile compounds from extracts of figs of *Ficus carica*. *Phytochemistry*, **46**, 241–244.
41. Innocenti, G., Bettero, A. & Caporale, G. (1982) Determination of the coumarinic constituents of *Ficus carica* leaves by HPLC. *Farmaco Sci*, **37**, 475–485 (in Italian).
42. Berkarda, B. (1993) Coumarin derivatives and cancer: a review of my work in this field. *Journal of the Irish Colleges of Physicians and Surgeons*, **22**, 69.
43. Maucher, A., Kager, M. & von Angerer, E. (1993) Evaluation of the antitumour activity of coumarin in prostate cancer models. *Journal of Cancer Research and Clinical Oncology*, **119**, 150–154.
44. Bordin, F., Call'acqua, G., Guiotto, A. (1991) Angelicins, angular analogs of psoralens: chemistry, photochemical, photobiological and phototherapeutic properties. *Pharmacology & Therapeutics*, **52**, 331–363.
45. Conconi, M.T., Montesi, F. & Parnigotto, P.P. (1998) Antiproliferative activity and phototoxicity of some methyl derivatives of 5-methoxypsoralen and 5-methoxyangelicin. *Pharmacology & Toxicology*, **82**, 193–198.
46. Rubnov, S., Kashman, Y., Rabinowitz, R., Schlesinger, M. & Mechoulam, R. (2001) Suppressors of cancer cell proliferation from fig (*Ficus carica*) resin: isolation and structure elucidation. *Journal of Natural Products*, **64**, 993–996.

47. Gültekin, E. & Yıldız, F. (2006) Introduction to phytoestrogens. In: *Phytoestrogens in Functional Foods* (ed. F. Yıldız). CRC Press Taylor & Francis Group, Boca Raton, FL, pp. 3–18.
48. Adlercreutz, H. (2002) Phytoestrogens and breast cancer. *Journal of Steroid Biochemistry and Molecular Biology*, **83**, 113–118.
49. Adlercreutz, H., Heinonen, S.M. & Penalvo-Garcia, J. (2004) Phytoestrogens, cancer, and coronary heart disease. *BioFactors*, **22**, 229–236.
50. Cornwell, T., Cohick, W. & Raskin, I. (2004) Dietary phytoestrogens and health. *Phytochemistry*, **65**, 995–1016.
51. Magee, P.J. & Rowland, I.R. (2004) Phyto-estrogens, their mechanism of action: current evidence for a role in breast and prostate cancer. *British Journal of Nutrition*, **91**, 513–531.
52. Webb, A.L. & McCullough, M.L. (2005) Dietary lignans: potential role in cancer prevention. *Nutrition and Cancer*, **51**, 117–131.
53. Thompson, L.U. (2003) Flaxseed, lignans, and cancer. In: *Flaxseed in Human Nutrition*, 2nd ed. (eds L.U. Thompson & S.C. Cunnane). AOCS Press, Champaign, IL, pp. 194–222.

16 Drying nectarines: functional compounds and antioxidant potential

Daniel Valero, Huertas María Díaz-Mula, and María Serrano

16.1 Introduction

Nectarines are delicious summertime fruits, mainly consumed fresh. Genetically, nectarine fruits are related to peaches since they belong to the same species, *Prunus persica*, and nectarines being *P. persica* var. nectarines. Peaches and nectarines can be differentiated by skin texture; nectarines have smooth skin, while peaches have fuzzy or velvety skin. Both fruits are available in clingstone and freestone types. Nectarines, as other stone fruits, are characterized by highly lignified endocarp (pit or stone), fleshy mesocarp (pulp), and thin epicarp (skin). Production of peaches and nectarines worldwide was about 20,000 million tonnes in 2009, these fruits having the highest production among stone fruits. The main producer is China, with more than 10,000 million tonnes, followed by Italy, the United States, and Spain [1].

In general, nectarine cultivars are grouped into two types: soft flesh and tough flesh. The latter has a longer shelf life and is most preferred by consumers. Also, the flesh of nectarines can be white or yellow, depending on the variety, while all of them have smooth and red skin. Nectarines are not only sweet and tasty, but they are loaded with innumerable health benefits [2–4].

From the point of view of nectarine fruit quality and consumer acceptance, several authors have found a linear relationship between total soluble solids (TSS) and consumer acceptance, showing that TSS below 11 to 12% is considered as being unacceptable, although for low-acid cultivars the acceptance by consumers depends more on the TSS/total acidity (TA) ratio than TSS alone [5, 6]. In addition, consumer acceptability is also dependent on fruit firmness, since too soft or too firm flesh impacts negatively on quality attributes [7]. Color changes that are associated with ripening strongly influence both visual and eating quality of nectarines. In this sense, genotype differences markedly affect color intensity, the main pigments responsible for color (both skin and flesh) being carotenoids [8].

Fruits are an important part of our daily diet and their increased intake has been associated with reduced incidence of human diseases due to their antioxidant potential attributed to several compounds which vary widely in chemical structure and function in plant tissues and are grouped into vitamins (C and E), carotenoids, phenolics, and thiol compounds [9–11].

Dried Fruits: Phytochemicals and Health Effects, First Edition. Edited by Cesarettin Alasalvar and Fereidoon Shahidi.

Nectarines are mainly consumed fresh worldwide. However, as in other climacteric fruits the ripening process evolves quickly after harvesting, leading to quality losses and reducing the fruit shelf life. Therefore, canning and freezing have traditionally been used in the fruit processing industry to extend their shelf life. The drying process is another mode of fruit preservation consisting of the removal of water vapors from the system after they have separated from the fruit tissues. This process tends to extend the storage life by months or even years. In spite of this, there are some chemical and physical changes that occur during and/or after processing, which leads to a gradual loss of quality, such as discoloration and off-flavor development [12]. This chapter reports the available information about the nutritive properties of dried nectarines, with special emphasis on their antioxidant properties due to the content of bioactive compounds such as carotenoids and phenolics, as compared with those present in fresh nectarines.

16.2 How to dry nectarines

The earliest recorded mention of dried fruits can be found in Mesopotamian tablets dating back to about 1700 BC, which contain what are probably the oldest known written recipes. Traditional dried fruits were either sun-dried or dehydrated in wind tunnels and they have a long history of food safety. The high drying and processing temperatures, the intrinsic low pH of the fruit, the low water activity, and the presence of natural antimicrobial compounds in dried fruits make them a remarkable stable food. There is no known incident of a food-borne illness related to dried fruits. Dried nectarines are not a common food as other fruits such as prunes or raisins, although in the recent years some dried nectarine types can be found in the market.

Traditionally, nectarines were sun-dried, although this process was dependent on weather conditions of each year and had risks due to attack by insects and fungi that produce phytotoxins. Therefore, various methods of drying chambers or tunnels under controlled conditions have been developed in recent years for this purpose, which show some interesting advantages against the traditional sun-drying. Principally, the drying time is shortened and different combinations of air velocity, temperature, and time can be used, leading to lower changes in color, taste, and nutritional values than those occurring when high drying temperatures and times are used [13]. In addition, the drying process is under better sanitary conditions, because of a reduction in contamination by dust and other forcing matter and dehydration is not conditioned by rain or weather changes. Thus, dried nectarines are prepared from sound, mature fresh nectarines which have been washed, peeled, pitted, cut into slices, sulfured for color retention, and dried to remove the greater portion of moisture. Nectarine slices should be dried to a moisture content of 18 to 20%, until their water activity falls below 0.6, which is the industry standard for storage of dried fruits. After the drying process, nectarines have a pronounced sweet flavor, chewy texture, and a deep red-orange blush.

16.3 Compositional and nutritional characteristics of dried nectarines

Compositional and nutritional characteristics of dried nectarines as compared to that of fresh ones are shown in Table 16.1 [14–16]. Moisture content of fresh and dried nectarines is

Table 16.1 Compositional and nutritional characteristics of fresh and dried nectarines (values in per 100 g edible portion)

Nutrient	Units	Fresh nectarines [14]	Dried nectarines [15, 16]
Proximate composition			
Water	g	87.6	20
Energy	kcal	44	220–280
Protein	g	1.08	4–6
Lipid	g	0.32	0.6
Ash	g	0.48	
Carbohydrate	g	10.55	46–66
Dietary fiber	g	1.7	5–7
Sugars	g	7.89	
Minerals			
Calcium	mg	6	26
Copper	mg	0.09	–
Iron	mg	0.28	3
Magnesium	mg	9	–
Manganese	mg	0.05	–
Phosphorus	mg	26	–
Potassium	mg	201	960
Selenium	μg	nd	–
Sodium	mg	nd	5
Zinc	mg	0.17	–
Vitamins			
Betaine	mg	0.2	–
Choline	mg	6.2	–
Folate	μg	5	–
Niacin	mg	1.13	–
Pantothenic acid	mg	0.19	–
Pyridoxine	mg	0.03	–
Riboflavin	mg	0.03	–
Thiamin	mg	0.03	–
Vitamin A (RAE)	μg	17	21
Vitamin C	mg	5.4	10
Vitamin E (ATE)	mg	0.77	–
Vitamin K	μg	2.2	–

RAE, retinol activity equivalents; ATE, α-tocopherol equivalents; nd, not detected.
Some numbers are rounded to the second digit after decimal point.

around 88 and 20%, respectively. Dried nectarines tend to be higher in calories than fresh nectarines, as the reduction in water condenses the fruit. Thus, dried nectarines are relatively calorie-dense, since 100 g provides 220 to 280 kcal, which is 12% of the suggested daily intake of 2000 calories, while fresh nectarines would contain just 44 kcal. Dried nectarines, similar to other fruits, are a rich source of carbohydrates, with concentration of 46 to 66 g/100 g, while fresh nectarines contain 10.55 g/100 g. Dried nectarines provide more than threefold higher dietary fiber than fresh ones. Both fresh and dried nectarines are poor sources of lipid and protein. Information about vitamin content of dried nectarines is scarce, and only vitamins A and C have been found in dried nectarines. Finally, dried nectarines are a good source of mineral, especially potassium, with concentration of approximately 1 g/100 g (Table 16.1).

16.4 Phytochemicals in dried nectarines

The health benefits associated with fresh fruit consumption are broadly known and linked with convincing evidence for lowered risk of cardiovascular disease (CVD) [17], cancer [18], and antioxidant activity [19].

It has been reported that nectarines contain different bioactive compounds with antioxidant activity such as polyphenols and carotenoids with large variation among round and flat nectarine cultivars [2–4, 20, 21]. In particular, polyphenols and their associated antioxidant properties are increasingly linked with health and lowering of the risk of chronic diseases. Phenolic compounds are not uniformly distributed within the tissue of fruits, and more of them are concentrated in the epidermal and subepidermal tissues of the fruit as has been reported for peaches, nectarines, and plums [2, 4, 22]. However, the effects of drying practices on polyphenols and antioxidant activities have not been systematically studied in fruits in general and no evidences exist in the literature about this issue in dried nectarines.

We have analyzed the effect of drying process on total phenolics, total carotenoids, and total antioxidant activity in both hydrophilic and lipophilic extracts (H-TAA and L-TAA, respectively) from nectarines. Nectarines were first dried till they had 50% of water content (partial drying) and then totally dried in an oven at 60°C for 24 and 72 hours, respectively. Total phenolic content in fresh nectarines was 121.9 mg gallic acid equivalents (GAE)/ 100 g dry weight basis (DW), which is within the range of nectarines in the literature (10 to 80 mg GAE/100 g of fresh weight depending on cultivars) [2, 3, 20]. Results showed an increase in total phenolic content and H-TAA upon drying of nectarines (Figure 16.1). It has recently been reported that total phenolic content increases during the drying process of two grape cultivars due to increases in phenolic acids, flavans, flavonols, and anthocyanins [23]. Rababah *et al.* [24] reported that the levels of total phenolics were higher in dried apples, strawberries, and peaches than in fresh fruits. Slatnar *et al.* [25] have recently shown that individual phenolic compounds and total phenolics were at higher concentration in dried figs than their fresh counterparts. In addition, total phenolic concentration in apricots increased after hot drying treatments, combined or without microwave energy, although some individual phenolic acids, such as caffeic and gallic acids, decreased [26]. Phenolic acids such as chlorogenic and neochlorogenic acids have been reported to be the dominant phenolic compounds in nectarines [20].

The increment on phenolic concentration upon drying process could be due to greater ease in the extraction of these compounds as a result of cell structure alteration and breakage of the skin during the drying process. Thus, dried fruits represent a shelf-stable and rich source of dietary polyphenols, compared with polyphenolic-rich fresh fruits. Moreover, phenolic content and antioxidant activity in dried fruits remained unchanged after long-term storage [27]. Dried nectarines could fulfill consumer's wishes for a diet with health benefits [28]. However, the drying process can also lead to loss in total phenolic compounds or changes in the ratio of free to total polyphenols [29]. Hot drying of raspberries at 77°C for 4.5 hours significantly reduced the total phenolic content and antioxidant activity, mainly due to the degradation of anthocyanins, which are less stable to temperature treatments than

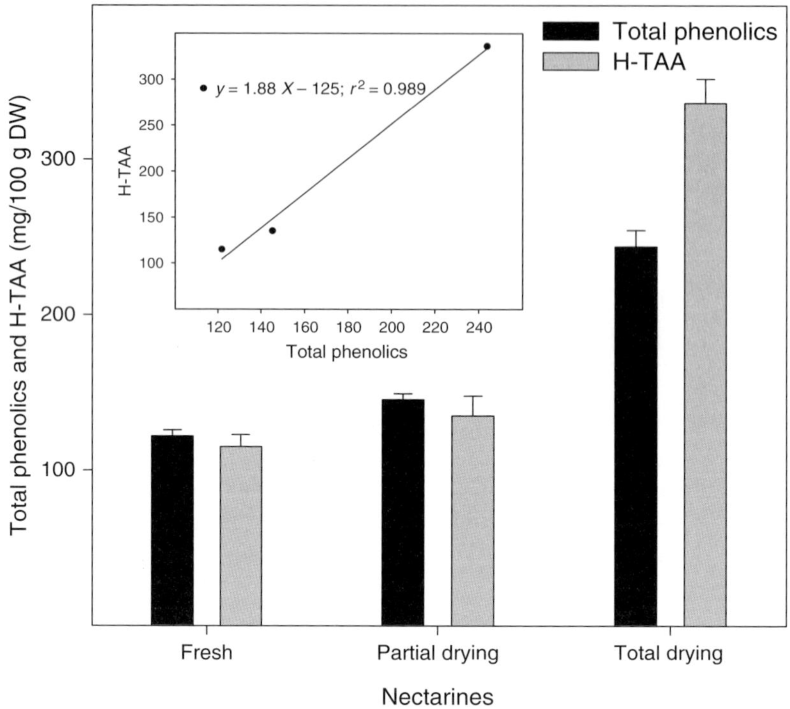

Figure 16.1 Total phenolic concentration (milligrams of gallic acid equivalents per 100 g of dry weight, DW) and hydrophilic total antioxidant activity (H-TAA, milligrams of trolox equivalents per 100 g of DW) in fresh nectarines and after partial and total drying (unpublished data).

polyphenols [30]. Thus, high-temperature treatments of drying fruits could lead to loss of bioactive compounds, especially in those fruits with a high anthocyanin concentration.

H-TAA also increased in both partially and totally dried nectarines with respect to fresh ones (Figure 16.1). It has been reported that the antioxidant activity was significantly increased in dried apricots and raisins with respect to their fresh counterparts [26, 31]. Moreover, the antioxidant activity was higher in dried strawberries, apples, and peaches than those of fresh fruits. This increase was higher if ascorbic acid was added before drying [24]. H-TAA was highly correlated with total phenolic concentration in other dried fruits [23, 24, 27] as well as in a wide range of fresh nectarine cultivars [2–4], showing that, as for fresh fruits, the antioxidant activity in the hydrophilic fraction of dried fruit extracts is mainly due to phenolic compounds [22, 32, 33]. Thus, total phenolics could be considered the main bioactive compounds contributing to H-TAA, since it has been reported that vitamin C (another hydrophilic antioxidant compound) makes a low contribution to H-TAA. This compound was found at low concentrations in peach and nectarine cultivars [21, 34]. Nevertheless, loss of vitamin C as a consequence of thermal processing has been reported in fruits, probably due to its utilization for protection against the oxidation of polyphenols in fruit tissues [35]. The elevated antioxidant activity in dried nectarine could also be due to the generation of Maillard reaction products (MRPs) during the drying process, which can enhance antioxidant potential in dried fruits [36]. The contribution of the MRPs to the antioxidant capacity of dried fruits is supported by the findings of Madrau *et al.* [37], who reported

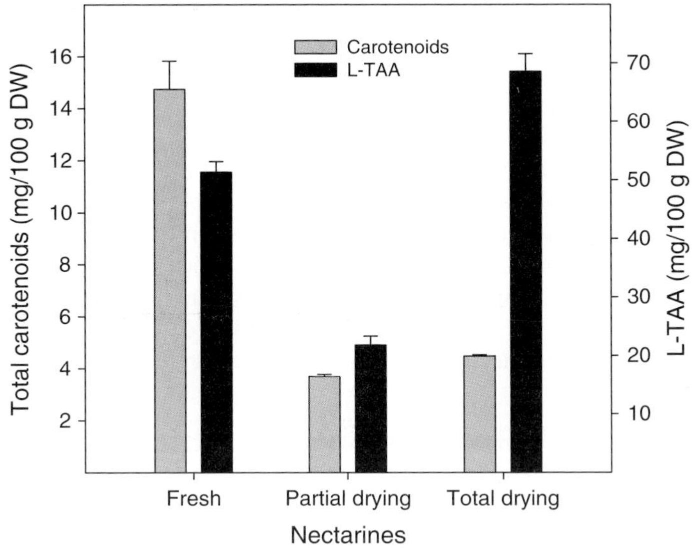

Figure 16.2 Total carotenoid concentration (milligrams of β-carotene per 100 g of dry weight, DW) and lipophilic total antioxidant activity (L-TAA, milligrams of trolox equivalents per 100 g of DW) in fresh nectarines and after partial and total drying (unpublished data).

an increase in antioxidant activity after the drying process of apricots in spite of a decrease in phenolics present. Moreover, the contemporary decrease in polyphenols and increase in antioxidant activity could also be due to an increased antioxidant power of polyphenols in an intermediate state of oxidation.

Total carotenoid content decreased with drying, from 14.74 mg/100 g (DW) in fresh nectarines to 4.48 mg/100 g (DW) in dried nectarines (Figure 16.2). These carotenoid contents are within the range of reported values for other round white-fleshed peaches and nectarines in which the major carotenoid is β-carotene followed by β-cryptoxanthin and lutein + zeaxanthin [2, 4, 14, 21, 34, 38]. Lavelli *et al.* [39] reported a decrease in the content of β-carotene after thermal processing of nectarine purées at 100°C for 5 minutes. However, García-Parra *et al.* [40] have recently shown that less intense stabilization treatment, as 80°C for 5 minutes, does not affect the total carotenoid content or the individual carotenoid compounds. Very few reports have examined the antioxidant activity in the hydrophilic and lipophilic fractions separately, although evidences exist for some fresh peaches, nectarines, and plums, in which correlations exist between carotenoids concentration and L-TAA [4, 22, 38, 39, 41]. In other stone fruits (such as sweet cherries, apricots, peaches, and plums), L-TAA as well as the content of tocopherols (α, β + γ, and δ) in the skin was determined, although no correlations were established [42]. Thus, tocopherols as other lipophilic substances might contribute to the L-TAA.

16.5 Health benefits of dried nectarines

The growing demand for healthy and nutritive foods in the world today has made nutrient analyses a major area in quality control studies. The antioxidant capacity of fruits varies in relation to antioxidant compounds present in different species, although variations can occur among cultivars within a single species [2, 3, 22, 32, 34]. Dietary phenolic intakes, in particular, are known to reduce coronary heart diseases (CHD) and cancer, as well as to act as antimicrobial, antiallergic, antimutagenic, and anti-inflammatory agents due to their antioxidant properties [43]. The antioxidant activity of phenolic compounds is based on their ability to scavenge free radicals, chelate pro-oxidant metal ions, and inhibit some oxidant enzymes. In addition, polyphenols of fruits have an inhibitory effect in fat cell development and intracellular fat deposition, that is, they have antiadipogenic activity leading to reduction of the risk for the development of obesity [30]. Thus, since dried nectarines have even greater content of phenolic compounds and higher antioxidant potential than fresh ones, the consumption of this commodity could be a good alternative to the intake of health beneficial compounds when fresh nectarines are not available in the market.

16.6 Commercial products and industrial applications of dried nectarines

The commercialization of dried nectarines around the world is not as high as that of other dried fruits such as plums or raisins, and only a few companies make dried nectarines, especially in the United States (such as Sunsweet Nutrition, Sid Wainer & Son, Barry Farm Foods, California Gourmet Company, Bella Viva Orchards, and Steward and Jasper Orchards, among others). However, this is an interesting product, with possibilities of reaching new markets, due to their high-quality attributes, with special emphasis on high phenolic content and antioxidant activity upon drying process, leading to health beneficial effects when consumed.

16.7 Conclusions

Nowadays, there is an increasing demand for dried fruits in several parts of the world, such as the United States, the United Kingdom, Germany, and Australia, partly due to the presence of bioactive compounds with beneficial health effects. The production of nectarines as dried fruits is scarce around the world, although it is an interesting fruit to be used for drying process, leading to new market opportunities to solve the problem of their harvest in a very short season. In addition, the content of bioactive compounds, such as polyphenols, and the antioxidant activity increase upon drying process and then the consumption of dried nectarines could have even more health beneficial effect for consumers than those of fresh nectarines.

References

1. FAOSTAT (2009) *Peaches and Nectarines*. Published online at: http://faostat.fao.org/site/339/default.aspx, last accessed January 1, 2012.
2. Gil, M.I., Tomás-Barberán, F.A., Hess-Pierce, B. & Kader, A.A. (2002) Antioxidant capacities, phenolic compounds, carotenoids, and vitamin C contents of nectarine, peach, and plum cultivars from California. *Journal of Agricultural and Food Chemistry*, **50**, 4976–4982.
3. Cantín, C.M., Moreno, M.A. & Gorgocena, Y. (2009) Evaluation of the antioxidant capacity, phenolic compounds, and vitamin C content of different peach and nectarine [*Prunus persica* (L.) Batsch] breeding progenies. *Journal of Agricultural and Food Chemistry*, **57**, 4586–4592.

4. Legua, P., Hernández, F., Díaz-Mula, H.M., Valero, D. & Serrano, M. (2011) Quality, bioactive compounds, and antioxidant activity of new flat-type peach and nectarine cultivars: a comparative study. *Journal of Food Science*, **76**, C729–C735.
5. Crisosto, C. & Crisosto, G. (2005) Relationship between ripe soluble solids concentration (RSSC) and consumer acceptance of high and low acid melting flesh peach and nectarine [*Prunus persica* (L.) Batsch] cultivars. *Postharvest Biology and Technology*, **38**, 239–246.
6. Iglesias, I. & Echeverría, G. (2009) Differential effect of cultivar and harvest date on nectarine colour, quality and consumer acceptance. *Scientia Horticulturae*, **120**, 41–50.
7. Crisosto, C.H., Crisosto, G.M., Echeverría, G. & Puy, J. (2006) Segregation of peach and nectarine [*Prunus persica* (L.) Batsch] cultivars according to their organoleptic characteristics. *Postharvest Biology and Technology*, **39**, 10–18.
8. Valero, D. & Serrano, M. (2010) *Postharvest Biology and Technology for Preserving Fruit Quality*. CRC Press, Boca Raton, FL.
9. Nichenametla, S.N., Taruscio, T.G., Barney, D.L. & Exon, J.H. (2006) A review of the effects and mechanisms of polyphenolics in cancer. *Critical Reviews in Food Science and Nutrition*, **46**, 161–183.
10. Schreiner, M. & Huyskens-Keil, S. (2006) Phytochemicals in fruit and vegetables: health promotion and postharvest elicitors. *Critical Reviews in Plant Science*, **25**, 267–278.
11. Saura-Calixto, F. & Goñi, I. (2009) Definition of the Mediterranean diet based on bioactive compounds. *Critical Reviews in Food Science and Nutrition*, **49**, 145–152.
12. Arthey, D. & Ashurst, P.R. (1996) *Fruit Processing*. The University Press, Cambridge, UK.
13. Tarhan, S. (2006) Selection of chemical and thermal pretreatment combination for plum drying at low moderate drying air temperatures. *Journal of Food Engineering*, **79**, 255–260.
14. USDA (2011) *National Nutrient Database for Standard Reference, Release 24*. Published online at: http://www.nal.usda.gov/fnic/foodcomp/search/, last accessed December 16, 2011.
15. Calories in Dried Nectarines (2012) *Calories in Dried Nectarines*. Published online at: http://caloriecount.about.com/calories-sunsweet-dried-nectarines-i113642, last accessed January 1, 2012.
16. Nutritional Values of Dried Nectarines (2012) *Nutritional Values of Dried Nectarines*. Published online at: http://www.livestrong.com/article/519048-nutritional-value-of-dried-nectarines, last accessed January 1, 2012.
17. He, F.J., Nowson, C.A., Lucas, M. & MacGregor, G.A. (2007) Increased consumption of fruit and vegetables is related to a reduced risk of coronary heart disease: meta-analysis of cohort studies. *Journal of Human Hypertension*, **21**, 717–728.
18. Duthie, S.J. (2007) Berry phytochemicals, genomic stability and cancer: evidence for chemoprotection at several stages in the carcinogenic process. *Molecular Nutrition & Food Research*, **51**, 665–674.
19. Wolfe, K.L., Kang, X.M., He, X.J., Dong, M., Zhang, Q.Y. & Liu, R.H. (2008) Cellular antioxidant activity of common fruits. *Journal of Agricultural and Food Chemistry*, **56**, 8418–8426.
20. Tomás-Barberán, F.A., Gil, M.I., Cremin, P., Waterhouse, A.L., Hess-Pierce, B. & Kader, A.A. (2001) HPLC-DAD-ESIMS analysis of phenolic compounds in nectarines, peaches, and plums. *Journal of Agricultural and Food Chemistry*, **49**, 4748–4760.
21. Vizzotto, M., Cisneros-Zevallos, L. & Byrne, D.H. (2007) Large variation found in the phytochemical and antioxidant activity of peach and plum germplasm. *Journal of the American Society for Horticultural Science*, **132**, 334–340.
22. Díaz-Mula, H.M., Zapata, P.J., Guillén, F., Castillo, S., Martínez-Romero, D., Valero, D. & Serrano, M. (2008) Changes in physicochemical and nutritive parameters and bioactive compounds during development and on-tree ripening of eight plum cultivars. *Journal of the Science of Food and Agriculture*, **88**, 2499–2507.
23. Serratosa, M.P., Marquez, A., Lopez-Toledano, A., Medina, M. & Merida, J. (2011) Changes in hydrophilic and lipophilic antioxidant activity in relation to their phenolic composition during the chamber drying of red grapes at a controlled temperature. *Journal of Agricultural and Food Chemistry*, **59**, 1882–1892.
24. Rababah, T.M., Ereifej, K.I. & Howard, L. (2005) Effect of ascorbic acid and dehydration on concentrations of total phenolics, antioxidant capacity, anthocyanins, and color in fruits. *Journal of Agricultural and Food Chemistry*, **53**, 4444–4447.
25. Slatnar, A., Klancar, U., Stampar, F. & Veberic, R. (2011) Effect of drying of figs (*Ficus carica* L.) on the contents of sugars, organic acids, and phenolic compounds. *Journal of Agricultural and Food Chemistry*, **59**, 11696–11702.

26. Igual, M., García-Martínez, E., Camacho, M.M. & Martínez-Navarrete, N. (2011) Changes in flavonoid content of grapefruit juice caused by thermal treatment and storage. *Innovative Food Science and Emerging Technologies*, **12**, 153–162.
27. Bennett, L.A., Jegasothy, H., Konczak, I., Frank, D., Sudharmarajan, S. & Clingeleffer, P.R. (2011) Total polyphenolics and anti-oxidant properties of selected dried fruits and relationships to drying conditions. *Journal of Functional Foods*, **3**, 115–124.
28. Noratto, G., Porter, W., Byrne, D.H. & Cisneros-Zevallos, L. (2009) Identifying peach and plum polyphenols with chemopreventive potential against estrogen-independent breast cancer cells. *Journal of Agricultural and Food Chemistry*, **57**, 5219–5226.
29. Vinson, J.A., Zubik, L., Bose, P., Samman, N. & Proch, J. (2005) Dried fruits: excellent *in vitro* and *in vivo* antioxidants. *Journal of the American College of Nutrition*, **24**, 44–50.
30. Mejia-Meza, E.I., Yáñez, J.A., Remsberg, C.M., Takemoto, J.K., Davies, N.M., Rasco, B. & Clary, C. (2010) Effect of dehydration on raspberries: polyphenol and anthocyanin retention, antioxidant capacity, and antiadipogenic activity. *Journal of Food Science*, **75**, H5-H12.
31. Igual, M., García-Martínez, E., Camacho, M.M. & Martínez-Navarrete, N. (2010) Effect of thermal treatment and storage on the stability of organic acids and the functional value of grapefruit juice. *Food Chemistry*, **118**, 291–299.
32. Díaz-Mula, H.M., Castillo, S., Martínez-Romero, D., Valero, D., Zapata, P.J., Guillén, F. & Serrano, M. (2009) Organoleptic, nutritive and functional properties of sweet cherry as affected by cultivar and ripening stage. *Food Science and Technology International*, **15**, 535–543.
33. Jacobo-Velázquez, D.A. & Cisneros-Zevallos, L. (2009) Correlations of antioxidant activity against phenolic content revisited: a new approach in data analysis for food and medicinal plants. *Journal of Food Science*, **74**, R107–R113.
34. Tavarini, S., Degl'Innocent, E., Remorini, D., Massai, R. & Guidi, L. (2008) Preliminary characterisation of peach cultivars for their antioxidant capacity. *International Journal of Food Science & Technology*, **43**, 810–815.
35. Joshi, A.P.K., Rupasinghe, H.P.V. & Khanizadeh, S. (2011) Impact of drying processes on bioactive phenolics, vitamin C and antioxidant capacity of red-fleshed apple slices. *Journal of Food Processing and Preservation*, **35**, 453–457.
36. Yilmaz, Y. & Toledo, R. (2005) Antioxidant activity of water soluble Maillard reaction products. *Food Chemistry*, **93**, 273–278.
37. Madrau, M.A., Piscopo, A., Sanguinetti, A.M., Del Caro, A., Poiana, M., Romeo, F.V. & Piga, A. (2009) Effect of drying temperature on polyphenolic content and antioxidant activity in apricots. *European Food Research and Technology*, **228**, 441–448.
38. Di Vaio, C., Graziani, G., Marra, L., Cascone, A. & Ritieni, A. (2008) Antioxidant capacities, carotenoids, and polyphenols evaluation of fresh and refrigerated peach and nectarine cultivars from Italy. *European Food Research and Technology*, **227**, 1225–1231.
39. Lavelli, V., Pompei, C. & Casadei, M.A. (2008) Optimization of color and antioxidant activity of peach and nectarine puree: scale-up study from pilot plant. *Journal of Agricultural and Food Chemistry*, **56**, 7091–7099.
40. García-Parra, J., González-Cebrino, F., Delgado, J., Lozano, M., Hernández, T. & Ramírez, R. (2011) Effect of thermal and high-pressure processing on the nutritional value and quality attributes of a nectarine purée with industrial origin during the refrigerated storage. *Journal of Food* Science, **76**, C618–C625.
41. Wu, X., Beecher, G.R., Holden, J.M., Haytowitz, D.B., Gebhardt, S.E. & Prior, R.L. (2004) Lipophilic and hydrophilic antioxidant capacities of common food in the United States. *Journal of Agricultural and Food Chemistry*, **52**, 4026–4037.
42. Mohamed, R., Pineda, M. & Aguilar, M. (2007) Antioxidant capacity of extracts from wild and crop plants of the Mediterranean region. *Journal of Food Science*, **72**, S59–S63.
43. Kim, D.O., Jeong, S.W. & Lee, C.Y. (2003) Antioxidant capacity of phenolic phytochemicals from various cultivars of plums. *Journal of Agricultural and Food Chemistry*, **51**, 7292–7295.

17 Phytochemical composition and health aspects of peach products

Emilio Alvarez-Parrilla, Laura A. de la Rosa, Gustavo A. González-Aguilar, and Jesús F. Ayala-Zavala

17.1 Introduction

Peaches (*Prunus persica*) are juicy, sweet drupe fruits belonging to the Rosaceae family, which are categorized as "stone fruits" because their seeds are enclosed in a hard, stone-like endocarp. They originated in China and were introduced to Persia and then to Europe by the Romans, and later on to the rest of the world by the Spaniards. Commercially, they can be classified as clingstone and freestone, depending on whether the flesh sticks to the stone or not. They can also be classified as white or yellow, according to their flesh color. Clingstone are especially used in the processing industry, while freestone are used mainly as fresh fruits [1].

World production of peaches has shown a 52% increase since 1999, with a total production of more than 20 million tonnes in 2009. In a recent study, Konopacka *et al.* [2] reported that among 4271 European consumers, 39% consumed three to five peaches per week, and 23.8% more than five fruits per week. China is the main producer with a 50% share of the total production in 2009, followed by Italy, United States, Spain, Greece, and Turkey, with an overall share of 26.5% of the world production [3]. In 2008, 48% of the total production of peaches in the United States was consumed fresh. Canned (37.6%) and frozen (9.8%) are the main processed peach products consumed in the United States. Dried peaches represent approximately 1% of the total production, while other products such as juices, jams, and jellies represent around 3%. Due to the increase in scientific information on the health benefits of consumption of fresh fruits, there has been a decrease (~10%) in the consumption of canned peaches, due to an increase of fresh (~2%), frozen (~7%), and juice (~2%) use compared to the 1980. In the same period of time, the consumption of dried peaches has shown only a slight decrease (0.27%) [4].

Dried peaches are defined by the US standard as "the halved and pitted fruit of the peach tree (*Prunus persica*) from which the greater portion of moisture has been removed. Before packing, the dried fruit is processed to cleanse the fruit and may be treated with sulfur dioxide in order to retain its characteristic color." Federal inspection certificates indicate the moisture content of the finished product not more than 25% by weight [5]. Production of dried peaches

Dried Fruits: Phytochemicals and Health Effects, First Edition. Edited by Cesarettin Alasalvar and Fereidoon Shahidi.

involves washing, peeling, cutting, blanching with hot water, and/or ascorbic acid, or other antioxidants to minimize discoloration, and finally sun or oven drying. It is generally known that during processing, there is a decrease in the phytochemical content of dried products. This chapter summarizes the effect of drying on the phytochemical content and antioxidant capacity of fresh and dried peaches, where available the possible health benefits of dried peaches and other minor products are discussed.

17.2 Compositional and nutritional changes of peaches during dehydration

Among the most complete compositional database of peaches, the USDA [6] proposes the nutrient database information for the edible portion of raw peaches. Table 17.1 summarizes the compositional and nutritional characteristics of raw, dried, and dehydrated peaches [6]. Water represents 88.87% of peaches, hence an energy value of 39 kcal/100 g of edible portion. However, after drying the water content is 31.8 and 7.5% for dried (uncooked, sulfur dioxide-treated dried) peaches and dehydrated peaches, respectively, affecting the contents of all nutritional components [6]. Energy value increases to 239 and 325 kcal/100 g for dried and dehydrated peaches, respectively (Table 17.1).

Protein content of fresh peaches is approximately of 0.91 g/100 g, aspartic acid being the most abundant amino acid. After drying or dehydration, the protein content increases to 3.61 and 4.89 g/100 g, respectively. However, if the analysis is done on a dry weight basis, it is possible to observe a reduction in the protein content from 8.18 g/100 g for fresh peaches to 5.29 g/100 g for dried peaches. This reduction could be due to deteriorative reactions or to lixiviation during the drying process. In all cases, aspartic and glutamic acids remain the major amino acids present. Lipid content in fresh peaches is 0.25 g/100 g, of which 7.6% are saturated fatty acids (SFA), 26.8% monounsaturated fatty acids (MUFA), and 34.4% polyunsaturated fatty acids (PUFA). This composition remains practically unchanged during the drying process. These values suggest that peaches are a good source of PUFA. As in the case of protein, lipid content increases to 0.76 and 1.03 g/100 g for dried and dehydrated peaches, respectively. The total carbohydrate and dietary fiber of raw peaches are 9.54 and 1.5 g/100 g, respectively. Sucrose is the most abundant sugar, followed by glucose and fructose. These three sugars also remain the most abundant in dried peaches (Table 17.1).

The main minerals found in peaches are K, P, Mg, Ca, Fe, Zn, Cu, and Mn. Concentration of all of these minerals increases in dried and dehydrated peaches. The edible portion of raw peaches possesses a variety of fat- and water-soluble vitamins. Most of these compounds can be classified as labile structures, sensitive to light and high temperature, among other factors, that can be involved in the dehydration procedure. Even when no drying conditions are reported in the USDA database, it is evident that drying drastically reduces vitamin E content (Table 17.1). Thiamine, folic acid, and ascorbic acid (vitamin C) are also reduced in dried peaches, but show a recovery in the dehydrated form, probably due to water loss. Reduction in thiamine is associated with the use of sulfur dioxide, which is known to destroy this vitamin [7].

Considering that the drying process can have very drastic effects on the content of natural bioactive compounds of raw peaches, several mechanisms can be contemplated to prevent this loss [8]. For many years, sulfur dioxide has been used as antioxidant in the drying process of peaches, in order to prevent the loss of carotenoids and ascorbic acid. Packaging material and

Table 17.1 Compositional and nutritional characteristics of fresh, dried, and dehydrated peaches (values in per 100 g edible portion)

Nutrient	Units	Fresh	Dried	Dehydrated
Proximate composition				
Water	g	88.87	31.8	7.5
Energy	kcal	39	239	325
Protein	g	0.91	3.61	4.89
Lipid	g	0.25	0.76	1.03
SFA	g	0.02	0.08	0.11
MFA	g	0.07	0.28	0.38
PUFA	g	0.09	0.37	0.50
Ash	g	0.43	2.50	3.39
Carbohydrate	g	9.54	61.33	83.18
Dietary fiber	g	1.5	8.2	nr
Sugars	g	8.39	41.74	nr
Fructose	g	1.53	13.49	nr
Glucose	g	1.95	12.83	nr
Sucrose	g	4.76	15.42	nr
Minerals				
Calcium	mg	6	28	38
Copper	mg	0.07	0.36	0.493
Fluoride	μg	4.0	nr	nr
Iron	mg	0.25	4.06	5.51
Magnesium	mg	9	42	57
Manganese	mg	0.06	0.30	0.41
Phosphorus	mg	20	119	162
Potassium	mg	190	996	1351
Selenium	μg	0.1	0.5	nr
Sodium	mg	nd	7	10
Zinc	mg	0.17	0.57	0.78
Vitamins				
Betaine	mg	0.3	nr	nr
Choline	mg	6.1	12.7	nr
Folate	μg	4	nd	7
Niacin	mg	0.81	4.38	4.82
Pantothenic acid	mg	0.15	0.56	0.52
Pyridoxine	mg	0.03	0.07	0.16
Riboflavin	mg	0.03	0.21	0.07
Thiamin	mg	0.02	tr	0.04
Vitamin A (RAE)	μg	16	108	71
Vitamin C	mg	6.6	4.8	10.6
Vitamin E (ATE)	mg	0.73	0.19	nr
Vitamin K	μg	2.6	15.7	nr
Amino acids				
Alanine	g	0.028	0.215	0.292
Arginine	g	0.018	0.092	0.124
Aspartic acid	g	0.418	0.602	0.817
Cystine	g	0.012	0.029	0.040
Glutamic acid	g	0.056	0.548	0.743
Glycine	g	0.021	0.126	0.171
Histidine[a]	g	0.013	0.067	0.091

(continued)

Table 17.1 *(Continued)*

Nutrient	Units	Fresh	Dried	Dehydrated
Isoleucine[a]	g	0.017	0.104	0.141
Leucine[a]	g	0.027	0.204	0.277
Lysine[a]	g	0.030	0.116	0.157
Methionine[a]	g	0.010	0.087	0.118
Phenylalanine[a]	g	0.019	0.114	0.154
Proline	g	0.018	0.152	0.206
Serine	g	0.038	0.167	0.226
Threonine[a]	g	0.016	0.141	0.192
Tryptophan[a]	g	0.010	0.010	0.014
Tyrosine	g	0.014	0.094	0.128
Valine[a]	g	0.022	0.197	0.267

Source: Adapted from USDA [6].
Some numbers are rounded to the second digit after decimal point.
SFA, saturated fatty acids; MUFA, monounsaturated fatty acids; PUFA, polyunsaturated fatty acids; RAE, retinol activity equivalents; ATE, α-tocopherol equivalents; nr, not reported; nd, not detected; tr, trace.
[a]Indispensable amino acids.

packaging atmosphere can be used to control sulfur dioxide loss from dried peaches during extended storage; nitrogen packaging also reduces the loss of sulfur dioxide from dried peaches and helps to maintain higher levels of antioxidant compounds. This information reveals that due to their composition, dry peaches represent a healthy consumption option, rich in fiber and bioactive compounds, being also an interesting option for the agribusiness to develop functional foods.

17.3 Phytochemicals in fresh and processed peaches

Phytochemicals are defined as bioactive nonnutrient plant compounds found in fruits, vegetables, grains, and other plant-derived foods that have been linked to reducing the risk of major chronic diseases [9]. The major phytochemicals found in peaches are polyphenolic compounds, carotenoids, and nondigestible carbohydrates such as pectin and dietary fiber.

17.3.1 Polyphenolic compounds and antioxidant capacity

Polyphenolic compounds are secondary metabolites of plants with an aromatic ring bearing one or more hydroxyl groups in their structure. They are usually classified into two main groups of flavonoids and non-flavonoids, and each group can be further classified according to the number of carbon atoms, phenol rings, and structural elements that bind these rings [10]. The main classes of flavonoids are flavones, flavonols, flavanones, flavan-3-ols, anthocyanidins (anthocyanins when glycosylated), and isoflavones, among others. Meanwhile, the main classes of non-flavonoids are hydroxybenzoic acid and hydroxycinnamic acid derivatives. Polyphenolic compounds are recognized for their high antioxidant capacity; the antioxidant activity of individual phenolics depends on several structural features, including position and number of hydroxyl groups [11]. The total antioxidant capacity of fruits is usually highly correlated with concentrations of total or individual polyphenols, indicating that these compounds are the main contributors to the antioxidant capacity of fruits [9].

Concentrations of total phenolics, total antioxidant capacity, major flavonoids, and carotenoids reported in fresh (raw) and processed peaches are shown in Table 17.2 [12]. Total phenolic compounds in fresh peaches range from 38 [13] to over 300 mg gallic acid equivalents (GAE)/100 g [14] of fresh fruit, with an average concentration of 133 mg GAE/100 g [15]. These wide variations may be due to natural causes such as differences among cultivars, geographical area, maturity stage, agronomic, and postharvest practices; but they are also caused by differences in the analytical methods, especially in the extraction procedures and solvents. The major classes of phenolic compounds found in peaches are flavan-3-ols (catechin, procyanidin B1, and epicatechin) followed by hydroxycinnamic acid derivatives (such as chlorogenic and neochlorogenic acids). Structures of some phytochemicals found in peaches are given in Figure 17.1. The peels of some red-colored peach cultivars are also rich in anthocyanins [16]. Peaches contain relatively low concentrations of flavonols (quercetin glucosides). Total antioxidant capacity of raw peaches is 1922 μmol of trolox equivalents (TE)/100 g of fresh sample [15].

Processing of peaches may cause some alterations in the contents of phenolic compounds and antioxidant activity in peach products, depending on the type and intensity of the process. Canning is the most common form of peach processing; however, few studies exist on the effect of canning on the content of phenolic compounds in peaches. Thermal processing during canning induces the loss of total phenolic compounds and procyanidins (polymeric flavan-3-ols) in peaches [17, 18]. However, these losses are dependent on the processing temperature. Processing at 90°C for 40 minutes did not result in any significant loss in total polyphenolic compounds [17]. For procyanidins, these losses were also dependent on the degree of polymerization of the compound. Thus, octamers were completely lost in the process while monomers to heptamers were reduced by 6–30% and could be detected in the syrup in quantities that accounted for the loss in the peach tissue [18]. Cold storage at 4°C for 14 days or freezing and storing at −12°C for 3 months produced no loss in total polyphenolic compounds in peaches [17].

Peach juices are common peach products; fresh peach juices contain between 68.8 and 75.4 mg GAE/100 g total phenolic compounds, the main compounds quantified are catechin (2.0–3.4 mg/100 g juice) and chlorogenic acid (1.2–1.9 mg/100 g juice), followed by isoquercetin (quercetin-3-glucoside, 0.52–0.71 mg/100 g juice) [19]. Purees and concentrates are intermediate products in the elaboration of commercial fruit juices. The contents and types of phenolic compounds in these products have been determined by several authors. Bengoechea *et al.* [20] found that chlorogenic acid and catechin were the major polyphenolic compounds in peach purees, while chlorogenic and neochlorogenic acids were major phenolics in peach concentrates. No flavonols were found in these products. The authors concluded that phenolic compounds can be helpful in the characterization of fruit purees and concentrates as well as in the detection of adulteration in the manufacturing of commercial fruit juices. Talcott *et al.* [21] demonstrated that the production of peach purees without periderm removal increased the processing yield and levels of polyphenolic compounds and antioxidant capacity, without significantly impacting product quality attributes. The major polyphenolic compounds found were chlorogenic and neochlorogenic acids, followed by catechin. They also found that blanching time impacted the antioxidant capacity and contents of phenolic compounds in purees. However, the losses during processing (steam-blanching, blending, and pasteurization) or storage were low, in fact a long blanching time (20 minutes) enhanced

Table 17.2 Phenolic compounds, antioxidant capacity, and carotenoids of raw and processed peaches (values in per 100 g edible portion)

	Fresh peaches	Fresh peaches (white flesh)	Canned peaches	Dry peaches (40% moisture)	Reference
Total phenolics (mg of GAE)	133 ($n = 15$)	nr	73 ($n = 2$)	283 ($n = 4$)	[15]
Total antioxidant capacity (μmol of TE)	1922 ($n = 15$)	nr	436 ($n = 2$)	4222 ($n = 4$)	[15]
Individual flavonoids (mg)					
Catechin	4.92 ($n = 49$)	12.25 ($n = 30$)	1.87 ($n = 1$)	nr	[12]
Epicatechin	2.34 ($n = 49$)	4.09 ($n = 30$)	nd	nr	[12]
Epigallocatechin	1.04 ($n = 14$)	nr	nr	nr	[12]
Cyanidin	1.61 ($n = 45$)	0.97 ($n = 30$)	nd	nr	[12]
Quercetin	0.66 ($n = 40$)	0.45 ($n = 30$)	nd	nr	[12]
Kaempferol	0.22 ($n = 3$)	nr	nd	nr	[12]
Carotenoids (μg)					
α-Carotene	nd	nr	nd	3 ($n = 0$)	
β-Carotene	162 ($n = 32$)	nr	333 ($n = 6$)	1074 ($n = 0$)	[6]
β-Cryptoxanthin	67 ($n = 32$)	nr	85 ($n = 6$)	444 ($n = 0$)	[6]
Lutein + zeaxanthin	91 ($n = 8$)	nr	33 ($n = 1$)	559 ($n = 0$)	[6]

Some numbers are rounded to the second digit after decimal point.
Data in parentheses show the number of data points.
GAE, gallic acid equivalents; TE, trolox equivalents; nr, not reported; nd, not detected.

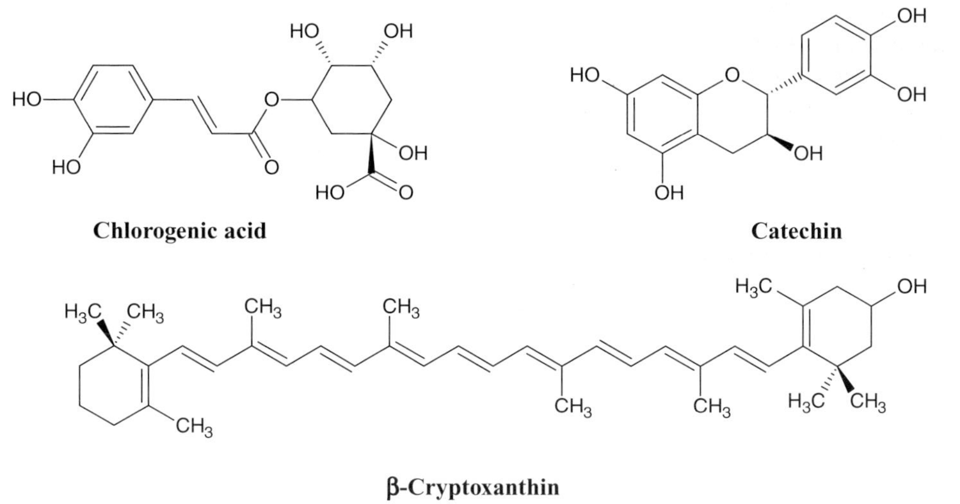

Figure 17.1 Structures of some phytochemicals found in peaches.

the yield of some phenolic compounds in the purees. Antioxidant capacity of peach purees was highly correlated with the concentration of chlorogenic acid [21]. Production of peach purees at room temperature, followed by heat treatment, reduced the contents of polyphenolic compounds in the purees in comparison with a traditional process of hot pulping/finishing, probably due to a lower extraction yield [22]. Loss of individual polyphenols calculated with respect to the amounts present in fresh fruits varied from 5 to 82% depending on the individual compound, peach variety, and type of process. Loss of cyanidin-3-glucoside was considered favorable, since the presence of this anthocyanin in purees was correlated with an undesirable color [22].

Very few studies have analyzed the contents of phenolic phytochemicals and antioxidant capacity of dried peaches or the effect of dehydration on them (Table 17.2). Rababah *et al.* [23] studied the effect of ascorbic acid addition on the concentrations of total phenolics, antioxidant capacity, and anthocyanins of fresh and dried peaches. They found that addition of ascorbic acid had a minor impact while dehydration significantly increased the levels of total phenolics, antioxidant capacity, and anthocyanins (Table 17.3). This is expected, since water elimination from the fresh fruits concentrated all the compounds in the peach tissues.

Table 17.3 Effect of drying on phenolic compounds and antioxidant capacity of peaches

	Fresh peaches	**Dry peaches**
Total phenolics (mg of GAE/100 g)	197.3	759.0
Total antioxidant capacity (μmol of TE/100 g)	1140	6760
Total anthocyanins (mg of cyn-3-glu/100 g)	1.9	5.1

Source: Adapted with permission from Rababah [23].
Data expressed as means ($n = 3$) on a fresh weight basis.
GAE, gallic acid equivalents; TE, trolox equivalents; cyn-3-glu, cyanidin-3-glucoside.

Threlfall *et al.* [24] also found that polyphenolic compounds, antioxidant capacity, and anthocyanins were higher in dry peaches than those of their fresh counterparts. More studies on the effect of drying on peach polyphenolic compounds and antioxidant activity are needed. Nevertheless, it can be said that drying is a convenient method of peach processing, which effectively concentrates phenolic compounds and therefore increases antioxidant capacity in the dried products.

17.3.2 Carotenoids

There are over 600 known carotenoids that are synthesized by fruits and vegetables [25]. Carotenoids along with anthocyanins are the major pigments in peaches. As in the case of other phytochemicals such as polyphenols, carotenoid contents in peaches vary greatly among cultivars due to genotype, stage of maturity, climatic conditions, and postharvest practices, among others. Maturity is one of the main factors that control the amount of carotenoids in peaches, since during ripening there is an increase of carotenogenesis [26]. It has been reported that carotenoid content in yellow-flesh peaches (0.1–2.8 mg β-carotene equivalents/100 g) is higher than in white-fleshed cultivars (0.025–0.08 mg β-carotene equivalents/100 g) [27, 28]. β-Carotene (0.004–0.31 mg/100 g) and β-cryptoxanthin (0.04–0.34 mg/100 g) [22, 29–32] are the main provitamin A carotenoids identified in peaches. Other carotenoids include α-carotene (0.0082 mg/100 g), lutein (0.075 mg/100 g), and zeaxanthin (0.025 mg/100 g) [29]. Khachik *et al.* [33] reported a large difference between fresh and dried peaches (0.122 versus 4.47 mg/100 g). According to the USDA nutrient database [6], the main carotenoids in fresh and dried peaches are β-carotene, β-cryptoxanthin, and lutein + zeaxanthin. Their contents are 162, 67, and 91 μg/100 g for fresh (raw) peaches and 1074, 444, and 559 μg/100 g for dried peaches, respectively (Table 17.2). These results suggest that drying increases carotenoids concentration because of removal of water that concentrates the phytochemicals.

Gil *et al.* [30] reported that carotenoids were concentrated mainly in the peel, and β-carotene in the peel was two to six times higher than that in the flesh. Caprioli *et al.* [34] found that concentration of total carotenoids in peaches was higher than the sum of individual carotenoids determined by high-performance liquid chromatography (HPLC) and suggested that this may be due to different maximum absorption wavelength for each individual carotenoid, which is not taken into account with the spectroscopic method. Tavarini *et al.* [27] found no correlation between peach carotenoids and antioxidant capacity measured by the ferric-reducing antioxidant power (FRAP) assay.

Peach processing has been associated with a decrease in carotenoid content, which has also been correlated with a decrease in the chroma values of the processed peach products [32]. McHugh and Huxsoll [35] reported that color changes in extruded peaches were attributed to carotenoid degradation and nonenzymatic browning. This behavior has also been observed in fresh-cut peaches treated with a precutting heating at 50°C for 10 minutes [36], decreasing both the chroma value and the β-carotene content.

Canned peaches are by far the most commonly consumed form of processed peaches in the United States. There are few studies that report the effect of canning on the content of carotenoids in peaches. Khachik *et al.* [33] reported an increase in the concentration of total carotenoids in canned peaches (as per 100 g edible portion); however, they argued that these results may not reflect the expected carotenoid losses during processing since no information about the moisture content of fruits or initial carotenoid concentration of peaches before processing was available. When the variations in carotenoid content of fresh and canned peaches were analyzed, a 50% reduction in both β-carotene and β-cryptoxanthin was observed [37, 38]. In a later study, a 10% increase in the content of *cis*-isomers was observed during processing, due to light, heat, and acid conditions [25]. It has been reported that this isomerization process changes the bioavailability and biological activity of carotenoids, because *cis*-isomers are less absorbed [25]. Other factors such as canning material and storage can also change the carotenoid content. Kaushal *et al.* [39] observed less carotenoid concentration in jars compared to cans under the same processing conditions. This reduction in the content of carotenoids during processing can be due to their decomposition which depends on several factors such as carotenoid structure, nature of the system, available oxygen, exposure to light, water content, temperature, atmosphere, and presence of antioxidants, prooxidants, and free radicals [26].

Other commonly consumed peach products are purees and nectars. Purees are intermediate moisture products that are used in the formulation of nectars. Lavelli *et al.* [22] reported a 48–65% reduction in carotenoid concentration, during puree processing, depending on peach variety and processing procedures. They observed a larger reduction in carotenoid concentration in Elegant Lady variety, which is more susceptible to enzymatic browning, than in Redhaven, which is a variety normally used for puree and nectar production. No further decrease in carotenoid content with respect to purees was observed during nectar production and storage for as long as 3 months. However, carotenoid concentration in lye-peeled nectars was lower than the whole peach nectars [22, 31]. These results are in agreement with the higher carotenoid concentration in peel than in flesh observed by Gil *et al.* [30].

It is well known that air-dried products undergo a significant reduction in carotenoids concentration. Higher carotenoid concentrations are reported in cabinet-dried vegetables than their solar-dried counterparts [25]. However, there is little information on carotenoid content in dried peaches. Two early studies suggested that peaches retained only 42% of their carotenoid content during dehydration [37, 40]. Mackinney *et al.* [40] observed that sulfur dioxide treatment reduced carotenoid losses during drying.

The main concern when analyzing carotenoids in dried fruits is to prevent their loss; however, in the several last years there has been an increased interest in analyzing the effect of processing on the bioavailability of carotenoids which suggest that their bioavailability is higher compared to the unprocessed samples [26].

17.3.3 Pectin and other nondigestible carbohydrates

Pectin or pectic substances refer to a group of heteropolysaccharides that consist predominantly of partially methylated galacturonic acid residues [41]. Together with cellulose and hemicelluloses, they are the main components of the vegetable cell walls, and their molecular configuration is determinant of the textural properties of fresh and processed peaches [42]. Pectins are not digested and absorbed in the human small intestine; therefore they are part of the dietary fiber fraction of peaches and other fruits. Total dietary fiber constitutes 1.5 g/100 g of the edible portion of fresh peaches (Table 17.1), while in dried peaches it represents 8.2 g/100 g [6]. Total dietary fiber components are usually classified as insoluble (cellulose, hemicellulose, and the non-polysaccharide lignin) and soluble (pectins, β-glucans, and arabinoxylans) dietary fibers. Soluble dietary fiber content is high in fruits, vegetables, and legumes and is associated with colonic degradation and high fermentability, prebiotic properties, slow glucose absorption, lower serum cholesterol levels, and enhanced immune function

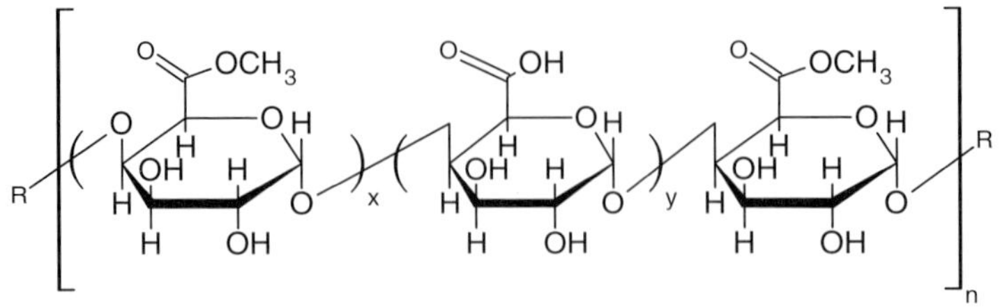

Figure 17.2 Structure of pectin backbone found in peaches. Solubility and functional properties of pectins depend on the chain length (*n*) and relation between methylated (x) and free (y) galacturonic acid residues.

[43]. Fresh and processed peaches are good sources of soluble dietary fiber (mainly pectins). Fresh and canned peaches contain 0.60 and 0.49 g of soluble fiber/100 g of edible portion, respectively [44]. Peach concentrate is rich in total dietary fiber (33.2 g/100 g), consisting of 11.3 g/100 g soluble and 21.9 g/100 g insoluble fibers [45].

Pectin is the most studied component of peach dietary fiber (Figure 17.2), mainly because of its profound effects on the textural properties of food products. The changes in the physicochemical characteristics of pectins during peach dehydration, and their relationship to changes in the texture of the dried products, have been studied [46]. It was found that fresh peaches (Carson variety) contain 2.63% pectin (on a dry weight basis) and that osmotic drying increased total and oxalate-soluble pectins (extracted in 0.005 M sodium oxalate solution, pH 5.6) while decreasing the content of water-soluble pectin and protopectin. However, the decrease in protopectin was minimal, and therefore the firmness and organoleptic properties of the dried peaches were not affected. Forni *et al.* [46] found that peaches behaved better than apricots when processed by osmotic dehydration. Later on, Levi *et al.* [42] studied the effect of blanching and thermal (oven) drying on the pectic constituents of peaches and found that adequate blanching stabilized pectins of the dehydrated peaches, so the higher the rehydration capacity, the lower the rehydration losses after 5 minutes of blanching. Total pectin in fresh peaches was 509 mg/100 g, of which 60% was soluble pectin, 25% protopectin, and 15% calcium pectate. After dehydration, total and soluble pectins were reduced, but protopectin was more stable. The stability of protopectin was also important for the rehydration properties of the peaches [42].

The health effects of pectin are receiving increased attention. Evidence exists that pectin can lower cholesterol and serum glucose levels and may also have anticancer activities [47]. However, to the best of our knowledge, no studies exist on the health benefits or biological activity of peach pectins.

17.4 Health effects of peaches

Fresh and processed peaches are good sources of nutrients such as carbohydrates, proteins, amino acids, vitamins, and minerals as well as health-promoting bioactive compounds such as phytochemicals and soluble dietary fiber. Therefore, peach products contribute to the intake of several types of phytochemicals having a positive influence on human health. Peaches may represent a significant contribution to the daily uptake of carotenoids and polyphenolic compounds. They are highly consumed [2] and their serving size is often larger than that for other fruits [48].

It is well known that high intakes of carotenoids and polyphenolic compounds are associated with the prevention or reduction of the risk of some diseases, such as cardiovascular disease (CVD) [49] and certain types of cancer [50]. Therefore, peach consumption may aid in the prevention of such disorders. However, studies dealing with the healthful properties of fresh and processed peaches are very scarce.

Phenolic compounds extracted from peel and flesh of fresh peaches inhibited the copper-catalyzed human low-density lipoprotein (LDL) oxidation *in vitro* [51] and the proliferation of HepG2 human liver cancer cells [52]. Polyphenol-rich extracts of yellow-fleshed peaches also inhibit the proliferation of an estrogen-independent breast cancer cell line (MDA-MB-435) with a 50% inhibitory concentration (IC_{50}), three-fold lower than for the inhibition of a noncancerous breast cell line (MCF-10A) [53].

One *in vivo* study was carried out in rats fed on a diet with or without 1% cholesterol and 10% whole dry peaches, apples, or pears. Peach supplementation did not affect serum or liver lipid levels in rats fed a cholesterol-free diet. Addition of cholesterol to the diet significantly increased total cholesterol, LDL-cholesterol, and triacylglycerols (TAG) in the serum and total cholesterol in the liver. Peach supplementation to the cholesterol-containing diet significantly hindered the rise in plasma and liver lipids, although the values were still higher than those of rats fed cholesterol-free diets [54]. Oxidation levels of serum lipids were evaluated by means of the thiobarbituric acid reactive substances (TBARS) assay and plasma antioxidant capacity was also measured as an indicator of *in vivo* oxidative status. Peach supplementation had no effect on neither of these two variables independent of the presence or absence of cholesterol in the diet, suggesting that peaches have no effect in regulating oxidative status *in vivo* [54]. Apples were found to be more effective than peaches and pears in regulating lipid levels and markers of oxidative status [54]. The authors argue that this may be due to the higher concentrations of phenolic compounds found in apples and explain that the effects on lipid levels may also be dependent on the fiber content, which was high in all fruits tested [54]. As mentioned earlier, peaches are good sources of soluble dietary fiber [45], and it is known that dietary fiber intake provides many health benefits including reduction in the risk of developing coronary heart disease (CHD), stroke, and gastrointestinal disorders, among others [55]. In a different *in vivo* study, peach juice consumption in humans had a short-term effect on the antioxidant status of human plasma, evaluated using dichlorofluorescein. Consumption of 120 mL of peach juice (and other fruit juices) induced an increment in the plasma antioxidant status within 30 minutes after juice consumption; but the effect was lost after 2 hours [56].

A specific health benefit of peach consumption was found by an epidemiological cross-sectional cohort study [57]. The association between glaucoma and the consumption of specific fruits and vegetables was studied in a cohort of women aged 65 and older, participating in the study of osteoporotic fractures. The odds of glaucoma risk were found to decrease by 47% in women who consumed at least one serving per week of canned or dried peaches compared to those who consumed less than one serving per month. Consumption of one, two, or more servings of fresh peaches per week did not have any effect compared to less than one serving per week [57]. The protective effect of canned/dried peaches may be due to their content of vitamin A or provitamin A carotenoids, especially cryptoxanthin, which may become more bioavailable due to processing [25, 57].

Peach consumption may also have some potentially negative health effects. Fresh peaches may cause food allergies, due to the presence of the lipid transfer protein (Pru p3), which is the major allergen of peach in some populations [58]. Production of dried peaches is usually achieved by thermal drying procedures such as sun drying [59] or oven drying [60], and sulfites (sulfur dioxide or potassium meta-bisulfite) are used as preservatives and antioxidants [59, 60]. However, ingested sulfites may cause several mild to severe, and even fatal, adverse effects to the asthmatic population, including broncho-constriction, urticaria, and anaphylaxis, among others [61]. Therefore, consumption of dried peaches (and other dried fruits) containing high levels of sulfites may pose a threat to sensitive individuals, and use of alternative treatments to help preserve the quality of fruits throughout dehydration and storage is advisable [62, 63].

17.5 Dry peaches and their by-products

Drying is one of the less used processing options for peaches. This fruit is principally canned, and a small proportion is frozen, dried, or used to prepare other products such as jams, preserves, or beverages. The steps involved in common pre-drying of peaches are selection and sorting for size, maturity, and firmness; washing; peeling by hand, lie solution, or abrasion; cutting into halves, slices, cubes, or segments; and sulfating [8]. The quality of dehydrated peaches depends on many factors, such as raw material, drying temperature, process time, moisture content, and sulfur dioxide concentration. Dried peaches are utilized as pie, tart, and turnover filling, while their powder provides excellent purees, spreads, or glazes, after proper dehydration and preparation [8].

The processing of peaches results in the generation of waste in the form of peel, seeds, and trimmings. In the past, this costly problem has been mitigated to some extent by processing the by-products further to yield a product that presents less of disposal problem or that has some marginal economic value [64]. The processing of dried peaches could be improved by developing higher value use for the by-products. For instance, several patents have been published relating the use of peaches as a source of nutraceutical and cosmetic compounds [65, 66]. It has been found that fruit by-products contain high levels of various health-enhancing substances that can be extracted from the by-products to provide nutraceuticals [54, 67, 68].

Peach seeds are considered low-value agro-industrial residue. To achieve the most economical and efficient utilization of these seeds, more information on their varieties, properties, and composition is required [69]. Cyanoglycoside in peach kernel, also called amygdaline, can represent about 71–72 mg/100 g of HCN [70]. In Chinese medicine, peach kernel is one of the nine plant ingredients used in a cocktail for CVD [71]. Peach kernel contains 63.8% oleic acid, 15.4% linoleic acid, 20.7% SFA, and 27.5% protein [72]. Essential amino acids make up 32–34 g/100 g of total amino acids found in apricot kernels [69]. Protein values, which range from 17.11 to 21.33 g/100 g, are comparable with those of black beans, soy beans, and peanuts [73]. Peach seed is an agro-industrial residue that corresponds to approximately 10% of the fruit weight. Peach kernel flour was produced by drying at 45, 55, and 65ºC, with or without maltodextrin.

Several potential uses can be considered for dry peach by-products, one of the most interesting options can be as food additives (antioxidants, antimicrobials, colorants, flavorings, and thickeners). It can be considered that the high content of bioactive compounds present in peach by-products can be used as natural food additives. If this approach is realized, it would be feasible to fulfill the requirements of consumers for natural and preserved healthy food. In addition, the full utilization of peach fruits could lead the industry to a lower waste agribusiness and increasing industrial profitability.

17.6 Conclusions

Peaches are good sources of phytochemicals, mainly polyphenolic compounds and carotenoids. They can be consumed as both fresh and processed products. Among the existing processing options, drying or dehydration is not a very common form of peach preservation. However, this process has the advantage of concentrating the phytochemicals in the final product. Generally speaking, processing induces degradation of phytochemicals, depending on the fruit variety, processing technique, and type of phytochemicals present. However, very few studies have analyzed the effect of drying on the degradation of phenolic compounds and carotenoids in peaches. There are also few studies on the health effects of peach products. One of the most relevant studies showed that consumption of dried or canned peaches, but not fresh peaches, may protect individuals from glaucoma. This may bring about interest for studying, not only the concentration, but also the bioavailability of phytochemicals in dried peaches and their effect on different human and animal models. Finally, production of dry peaches generates large amounts of by-products (mainly kernels) that can be used as sources of economically important natural products.

Acknowledgments

Financial support from Universidad Autónoma de Ciudad Juárez (internal project: PIFI-2011-CA-03) is gratefully acknowledged.

References

1. Siddiq, M. (2006) Peach and nectarine. In: *Handbook of Fruits and Fruit Processing* (eds Y.H. Hui, J. Barta, M.P. Cano, T. Gusek, J.S. Sidhu & N. Sinha). Wiley-Blackwell, Oxford, UK, pp. 519–531.
2. Konopacka, D., Jesionkowska, K., Kruczyńska, D., Stehr, R., Schoorl, F., Buehler, A., Egger, S., Codarin, S., Hilaire, C., Höller, I., Guerra, W., Liverani, A., Donati, F., Sansavini, S., Martinelli, A., Petiot, C., Carbó, J., Echeverria, G., Iglesias, I. & Bonany, J. (2010) Apple and peach consumption habits across European countries. *Appetite*, **55**, 478–483.
3. FAO (2009) *FAOSTAT*. Published online at: http://faostat.fao.org/site/567/default.aspx#ancor, last accessed October 28, 2011.
4. USDA (2009) *USDA Economics, Statistics, and Market Information System*. Published online at: http://usda.mannlib.cornell.edu/MannUsda/viewStaticPage.do?url=http://usda.mannlib.cornell.edu/usda/ers/./89022/2009/index.html, last accessed October 28, 2011.
5. USDA (1967) *United States Standards for Grades of Dried Peaches*. USDA, Washington, DC.
6. USDA (2011) *National Nutrient Database for Standard Reference, Release 24*. Published online at: http://www.nal.usda.gov/fnic/foodcomp/search/, last accessed October 28, 2011.
7. Komarnisky, L.A., Christophenrson, R.J. & Basu, T.K. (2003) Sulfur: its clinical and toxicologic aspects. Review article. *Nutrition*, **19**, 54–61.
8. Salunkhe, D.K. & Kadam, S. (1995) *Handbook of Fruit Science and Technology: Production, Composition, Storage, and Processing*. CRC Press Taylor & Francis Group, Boca Raton, FL.

9. Liu, R.H. (2004) Potential synergy of phytochemicals in cancer prevention: mechanism of action. *Journal of Nutrition*, **134**, 3479S–3485S.
10. Andres-Lacueva, C., Medina-Remon, A., Llorach, R., Urpi-Sarda, M., Khan, N., Chiva-Blanch, G., Zamora-Ros, R., Rotches-Ribalta, M. & Lamuela-Raventós, R.M. (2010) Phenolic compounds: chemistry and occurrence in fruits and vegetables. In: *Fruit and Vegetables Phytochemicals: Chemistry, Nutritional Value and Stability* (eds L.A. De la Rosa, E. Alvarez-Parrilla & G.A. González-Aguilar). Willey-Blackwell, Oxford, UK, pp. 53–88.
11. Rice-Evans, C.A., Miller, N.J. & Paganga, G. (1996) Structure-antioxidant activity relationship of flavonoids and phenolic acids. *Free Radical Biology and Medicine*, **20**, 933–956.
12. USDA (2011) *USDA Database for the Flavonoid Content of Selected Foods, Release 3.* Published online at: http://www.ars.usda.gov/nutrientdata, last accessed October 28, 2011.
13. Proteggente, A.R., Pannala, A.S., Paganga, G., Van Buren, L., Wagner, E., Wiseman, S., Van de Put, F., Dacombe, C. & Rice-Evans, C.A. (2002) The antioxidant activity of regularly consumed fruit and vegetables reflects their phenolic and vitamin C composition. *Free Radicals Research*, **36**, 217–233.
14. Imeh, U. & Khokhar, S. (2002) Distribution of conjugated and free phenols in fruits: antioxidant activity and cultivar variations. *Journal of Agricultural and Food Chemistry*, **50**, 6301–6306.
15. USDA (2010) *USDA Database for the Oxygen Radical Absorbance Capacity (ORAC) of Selected Foods, Release 2.0.* Published online at: http://www.ars.usda.gov/nutrientdata, last accessed October 28, 2011.
16. Tomás-Barberán, F.A., Gil, M.I., Cremin, P., Waterhouse, A.L., Hess-Pierce, B. & Kader, A.A. (2001) HPLC-DAD-ESIMS analysis of phenolic compounds in nectarines, peaches, and plums. *Journal of Agricultural and Food Chemistry*, **49**, 4748–4760.
17. Asami, D.K., Hong, Y.-J., Barrett, D.M. & Mitchell, A.E. (2003) Processing-induced changes in total phenolics and procyanidins in clingstone peaches. *Journal of the Science of Food and Agriculture*, **83**, 56–63.
18. Hong, Y.-J., Barrett, D.M. & Mitchell, A.E. (2004) Liquid chromatography/mass spectrometry investigation of the impact of thermal processing and storage on peach procyanidins. *Journal of Agricultural and Food Chemistry*, **52**, 2366–2371.
19. Versari, A., Castellari, M., Parpinello, G.P., Riponi, C. & Galassi, S. (2002) Characterisation of peach juices obtained from cultivars Redhaven, Suncrest and Maria Marta grown in Italy. *Food Chemistry*, **76**, 181–185.
20. Bengoechea, M.L., Sancho, A.I., Bartolomé, B., Estrella, I., Gómez-Cordovés, C. & Hernández, M.T. (1997) Phenolic composition of industrially manufactured purees and concentrates from peach and apple fruits. *Journal of Agricultural and Food Chemistry*, **45**, 4071–4075.
21. Talcott, S.T., Howard, L.R. & Brenes, C.H. (2000) Contribution of periderm material and blanching time to the quality of pasteurized peach puree. *Journal of Agricultural and Food Chemistry*, **48**, 4590–4596.
22. Lavelli, V., Pompei, C. & Casedei, M.A. (2008) Optimization of color and antioxidant activity of peach and nectarine puree: scale-up study from pilot to industrial plant. *Journal of Agricultural and Food Chemistry*, **56**, 7091–7099.
23. Rababah, T.M., Ereivej, K.I. & Howard, L.R. (2005) Effect of ascorbic acid and dehydration on concentrations of total phenolics, antioxidant capacity, anthocyanins, and color in fruits. *Journal of Agricultural and Food Chemistry*, **53**, 4444–4447.
24. Threlfall, R., Morris, J. & Meullenet, J.-F. (2007) Product development and nutraceutical analysis to enhance the value of dried fruit. *Journal of Food Quality*, **30**, 552–566.
25. Namitha, K.K. & Negi, P.S. (2010) Chemistry and biotechnology of carotenoids. *Critical Reviews in Food Science and Nutrition*, **50**, 728–760.
26. Rodriguez-Amaya, D.B. (1999) Changes in carotenoids during processing and storage of foods. *Archivos Latinoamericanos de Nutrición*, **49**, 38S–47S.
27. Tavarini, S., Degl'Innocenti, E., Remorini, D., Massai, R. & Guidi, L. (2008) Preliminary characterisation of peach cultivars for their antioxidant capacity. *International Journal of Food Science & Technology*, **43**, 810–815.
28. Vizzotto, M., Cisneros-Zevallos, L. & Byrne, D. (2007) Large variation found in the phytochemical and antioxidant activity of peach and plum germplasm. *Journal of the American Society of Horticultural Science*, **132**, 334–340.
29. Dias, M.G., Camões, M.F.G.F.C. & Oliveira, L. (2009) Carotenoids in traditional Portuguese fruits and vegetables. *Food Chemistry*, **113**, 808–815.

30. Gil, M.I., Tomás-Barberán, F.A., Hess-Pierce, B. & Kader, A.A. (2002) Antioxidant capacities, phenolic compounds, carotenoids, and vitamin C contents of nectarine, peach, and plum cultivars from California. *Journal of Agricultural and Food Chemistry*, **50**, 4976–4982.
31. Lavelli, V., Pompei, C. & Casedei, M.A. (2009) Quality of nectarine and peach nectars as affected by lye-peeling and storage. *Food Chemistry*, **115**, 1291–1298.
32. Tourjee, K.R., Barrett, D.M. & Romero, M.V. (1998) Measuring flesh color variability among processing clingstone peach genotypes differing in carotenoid composition. *Journal of the American Society of Horticultural Science*, **123**, 433–437.
33. Khachik, F., Beecher, G.R. & Lusby, W.R. (1989) Separation, identification, and quantification of the major carotenoids in extracts of apricots, peaches, cantaloupe, and pink grapefruit by liquid chromatography. *Journal of Agricultural and Food Chemistry*, **37**, 1465–1473.
34. Caprioli, I., Lafuente, M.T., Rodrigo, M.J. & Mencarelli, F. (2009) Influence of postharvest treatments on quality, carotenoids, and abscisic acid content of stored "Spring Belle" peach (*Prunus persica*) fruit. *Journal of Agricultural and Food Chemistry*, **57**, 7056–7063.
35. McHugh, T.H. & Huxsoll, C.C. (1999) Extrusion processing of restructured peach and peach/starch gels. *LWT-Food Science and Technology*, **32**, 513–520.
36. Koukounaras, A., Diamantidis, G. & Stakiotakis, E. (2008) The effect of heat treatment on quality retention of fresh-cut peach. *Postharvest Biology and Technology*, **48**, 30–36.
37. Eheart, M.S. & Sholes, M.L. (1946) Effects of method of sulfuring, dehydration, and temperature of storage on ascorbic and carotene content of dehydrated peaches. *Virginia Agricultural Experiment Station Scientific Paper*, **8**, 332–340.
38. Lessin, W.J., Catigani, G.L. & Schwartz, S.J. (1997) Quantification of *cis-trans* isomers of provitamin A carotenoids in fresh and processed fruits and vegetables. *Journal of Agricultural and Food Chemistry*, **45**, 3728–3732.
39. Kaushal, M., Sharma, K.D., Kaushal, B.B.L. & Vaidya, D. (2003) Effect of peach juice as covering medium on the quality of canned peach halves. *Acta Horticulturae (ISHS)*, **696**, 503–508.
40. Mackinney, G., Aronoff, S. & Bornstein, B.T. (1940) Some assays of provitamin A carotenoids. *Industrial and Engineering Chemistry*, **14**, 391–395.
41. Faravash, R.S. & Ashtiani, F.Z. (2008) The influence of acid volume, ethanol-to-extract ratio and acid-washing time on the yield of pectic substances extraction from peach pomace. *Food Hydrocolloids*, **22**, 196–202.
42. Levi, A., Ben-Shalom, N., Plat, D. & Reid, D.S. (1988) Effect of blanching and drying on pectin constituents and related characteristics of dehydrated peaches. *Journal of Food Science*, **53**, 1187–1190.
43. Saura-Calixto, F., Pérez-Jiménez, J. & Goñi, I. (2009) Dietary fiber and associated antioxidants in fruits and vegetables. In: *Fruit and Vegetable Phytochemicals: Chemistry, Nutritional Value, and Stability* (eds L.A. De la Rosa, E. Alvarez-Parrilla & G.A. González-Aguilar). Wiley-Blackwell, Oxford, UK, pp. 223–234.
44. Marlett, J.A. & Vollendorf, N.W. (1994) Dietary fiber content and composition of different forms of fruits. *Food Chemistry*, **51**, 39–44.
45. Grigelmo-Miguel, N., Gorinstein, S. & Martin-Belloso, O. (1999) Characterisation of peach dietary fibre concentrate as a food ingredient. *Food Chemistry*, **65**, 175–181.
46. Forni, E., Torreggiani, D., Battiston, P. & Polesello, A. (1986) Research into changes of pectic substances in apricots and peaches processed by osmotic dehydration. *Carbohydrate Polymers*, **6**, 379–393.
47. Willats, W.G.T., Knox, P. & Mikkelsen, D. (2006) Pectin: new insights into an old polymer are starting to gel. *Trends in Food Science & Technology*, **17**, 97–104.
48. Remorini, D., Tavarini, S., Degl'Innocenti, E., Loreti, F., Massai, R. & Guidi, L. (2008) Effect of rootstocks and harvesting time on the nutritional quality of peel and flesh of peach fruits. *Food Chemistry*, **110**, 361–367.
49. Erdman, J.W., Balentine, D., Arab, L., Beecher, G., Dwyer, J.T., Folts, J., Harnly, J., Hollman, P., Keen, C.L., Mazza, G., Messina, M., Scalbert, A., Vita, J., Williamson, G. & Burrowes, J. (2007) Flavonoids and heart health. *Journal of Nutrition*, **137**, 718S–737S.
50. Nishino, H., Murakoshi, M., Tokuda, H. & Satomi, Y. (2009) Review: cancer prevention by carotenoids. *Archives of Biochemistry and Biophysics*, **483**, 165–168.
51. Chang, S., Tan, C., Frankel, E.N. & Barrett, D.M. (2000) Low-density lipoprotein antioxidant activity of phenolic compounds and polyphenol oxidase activity in selected clingstone peach cultivars. *Journal of Agricultural and Food Chemistry*, **48**, 147–151.

52. Sun, J., Chu, Y.-F., Wu, X. & Liu, R.H. (2002) Antioxidant and antiproliferative activities of common fruits. *Journal of Agricultural and Food Chemistry*, **50**, 7449–7454.
53. Noratto, G., Porter, W., Byrne, D. & Cisneros-Zevallos, L. (2009) Identifying peach and plum polyphenols with chemopreventive potential against estrogen-independent breast cancer cells. *Journal of Agricultural and Food Chemistry*, **57**, 5219–5226.
54. Gorinstein, S., Martin-Belloso, O., Lojek, A., Ciz, M., Soliva-Fortuny, R., Park, Y.S., Caspi, A., Libman, I. & Trakhtenberg, S. (2002) Comparative content of some phytochemicals in Spanish apples, peaches and pears. *Journal of the Science of Food and Agriculture*, **82**, 1166–1170.
55. Anderson, J.W., Baird, P., Davis, R.H., Ferreri, S., Knudtson, M., Koraym, A., Waters, V. & Williams, C.L. (2009) Health benefits of dietary fiber. *Nutrition Reviews*, **67**, 188–205.
56. Ko, S.H., Choi, S.-W., Ye, S.K., Kim, H.S. & Chung, M.H. (2005) Comparison of the antioxidant activities of nine different fruits in human plasma. *Journal of Medicinal Food*, **8**, 41–46.
57. Coleman, A.L., Stone, K.L., Kodjebacheva, G., Yu, F., Pedula, K.L., Ensrud, K.E., Cauley, J.A., Hochberg, M.C., Topouzis, F., Badala, F. & Mangione, C.M. (2008) Glaucoma risk and the consumption of fruits and vegetables among older women in the study of osteoporotic fractures. *American Journal of Ophthalmology*, **145**, 1081–1089.
58. Fernandez-Rivas, M. & Cuevas, M. (1999) Peels of rosaceae fruits have a higher allergenicity than pulps. *Clinical & Experimental Allergy*, **29**, 1239–1247.
59. Joubert, E., Wium, G.L. & Sadie, A. (2003) Discoloration of sun-dried and processed Elderta peaches during storage. *Journal of Food Processing and Preservation*, **27**, 351–364.
60. DiPersio, P.A., Kendall, P.A. & Sofos, J.N. (2004) Inactivation of *Listeria monocytogenes* during drying and storage of peach slices treated with acidic or sodium metabisulfite solutions. *Food Microbiology*, **21**, 641–648.
61. Lester, M.R. (1995) Sulfite sensitivity: significance in human health. *Journal of the American College of Nutrition*, **14**, 229–232.
62. DiPersio, P.A., Kendall, P.A. & Sofos, J.N. (2005) Sensory evaluation of home dried fruit prepared using treatments that enhance destruction of pathogenic bacteria. *Journal of Food Quality*, **29**, 47–64.
63. Urfalino, D.P. & Quiroga, A. (2011) Desarrollo de técnicas combinadas de secado para la obtención de duraznos deshidratados con bajo contenido de sulfitos. *Revista de Investigaciones Agropecuarias*. Published online at: http://ria.inta.gov.ar/?p=914 (online journal).
64. Sun-Waterhouse, D., Wen, I., Wibisono, R., Melton, L.D. & Wadhwa, S. (2009) Evaluation of the extraction efficiency for polyphenol extracts from by-products of green kiwifruit juicing. *International Journal of Food Science & Technology*, **44**, 2644–2652.
65. Juillerat, M.A. & Perrinjaquet, J. (2002) *Fruit Kernel Protein and Lipid Extract Compositions*. USA Patent, US 6383550 B1. Nestec S.A., Vevey, Switzerland.
66. Tanaka, K., Matsukuma, S. & Suzuki, T. (2006) *Cosmetics*. USA Patent, US 2006/0165644 A1. Fanel Corporation, Washington, DC.
67. Ayala-Zavala, J.F., Rosas-Dominguez, C., Vega-Vega, V. & Gonzalez-Aguilar, G.A. (2010) Antioxidant enrichment and antimicrobial protection of fresh-cut fruits using their own byproducts: looking for integral exploitation. *Journal of Food Science*, **75**, R175–R181.
68. Ayala-Zavala, J.F., Vega-Vega, V., Rosas-Dominguez, C., Palafox-Carlos, H., Villa-Rodriguez, J.A., Siddiqui, M.W., Davila-Avina, J.E. & Gonzalez-Aguilar, G.A. (2011) Agro-industrial potential of exotic fruit byproducts as a source of food additives. *Food Research International*, **44**, 1866–1874.
69. Femenia, A., Rossello, C., Mulet, A. & Canellas, J. (1995) Chemical composition of bitter and sweet apricot kernels. *Journal of Agricultural and Food Chemistry*, **43**, 356–361.
70. Haque, M.R. & Bradburry, J.H. (2002) Total cyanide determination of plants and foods using the picrate and acid hydrolysis methods. *Food Chemistry*, **77**, 107–114.
71. Tu, Z., Han, X., Wang, X., Hou, Y., Shao, B., Zhou, Q. & Fan, Q. (2003) Protective effects of CVPM on vascular endothelium in rats fed cholesterol diet. *Clinica Chimica Acta*, **333**, 85–90.
72. Rahma, E. & El-Aal, M. (1988) Chemical characterization of peach kernel oil and protein: functional properties, *in vitro* digestibility and amino acids profile of the flour. *Food Chemistry*, **28**, 31–43.
73. Pelentir, N., Block, J.M., Monteiro, F., Reginatto, V. & Amante, E.R. (2009) Production and chemical characterization of peach (*Prunus persica*) kernel flour. *Journal of Food Process Engineering*, **34**, 1253–1265.

18 Dried pears: phytochemicals and potential health effects

Lisete Silva, Fereidoon Shahidi, and Manuel A. Coimbra

18.1 Introduction

The pears (*Pyrus communis* L.) are the pomaceous fruits of the trees within the same genus, *Pyrus*. They grow in temperate zones of Western Europe, among other regions. There are over 2000 varieties of pears, but only a few of them are commercialized [1]. They are produced from the flowers of their trees that often grow in cooler climates. They range in color from red, green, yellow, and white, and in shape, from oblate to globule to the most recognizable cultivar shape: pyriform, with an elongated portion ending in a bulbous bottom [2]. On the average, pears contain carbohydrates (11%), which include reducing sugars and dietary fiber (cellulose, hemicellulose, and pectins), proteins (1.5%), and lipids (0.1–0.5%), making them an important part of the human diet [1, 3–5]. The health benefits of dietary fibers are well known, as recently reviewed by Anderson *et al.* [6]. In addition, pears contain vitamins, minerals, organic acids, and phenolic compounds that contribute to the human health promotion. Pears are widely consumed in the fresh form. However, they can also be consumed, among others, in yogurt, juice, frozen, and dried forms, in order to provide a more stable and good quality dehydrated fruit.

The fruits of *Opuntia matudae*, a cactus variety, is also named "pear," specifically, "prickly pear" and have been shown many human health benefits. Its origin, physical aspect, and composition have no resemblance to *P. communis* L. pear fruit. This plant grows in desert climates and its fruits can be consumed in fresh form, as well as mashed into dips and salsas for certain sour species [7]. They are a good source of fiber and minerals, such as iron, zinc, calcium, potassium, and magnesium [8, 9]. The peel or skin of *O. matudae* fruit is rich in phenolic compounds, namely gallic acid, vanillic acid, 4-hydroxybenzoic acid, vanillin, and catechin. Also, the indole-derived pigments betacyanins and betaxanthins are present in these fruits. In this way, these fruits, derived from *Opuntia* genus, should not be confused with those from *Pyrus* genus, the subject of the present chapter.

One of the traditional and well-appreciated Portuguese food products is a sun-dried small pear (S. Bartolomeu) of reddish brown color which has unique elastic properties [4, 5]. In Portugal, the tradition of drying the S. Bartolomeu pear by direct sun exposure is still used,

Dried Fruits: Phytochemicals and Health Effects, First Edition. Edited by Cesarettin Alasalvar and Fereidoon Shahidi.

Figure 18.1 The S. Bartolomeu pears: (**a**) fresh pears and pears dried by different technologies, (**b**) traditional, (**c**) large glass greenhouse with air convection (GH1), (**d**) small greenhouse with natural convection (GH2), and (**e**) hot air tunnel in the absence of light (HAT). (Adapted with permission from Coimbra *et al.* [14]). For color detail, see color plate section.

although the development of new drying methods, such as greenhouses with or without air convection and a hot air tunnel have recently been tested [10] (Figure 18.1). Pears are rich in phenolic compounds, which are the main phytochemicals present in this fruit, in both the fresh and dried forms [11]. The presence of these compounds has been associated with the high antioxidant activity of pears [12]. However, the content of total phytochemicals present in this fruit may vary, depending on a number of factors, such as cultivar type, environmental growth conditions, harvest, ripening, drying, and storage conditions [1, 11, 13]. Furthermore, during the drying process, both enzymatic and nonenzymatic reactions between reducing carbohydrates and proteins/amines promote the formation of volatile flavor components and brown compounds that change the color, flavor, and texture of the dried product [14]. Phenolic compounds, namely caffeoylquinic acid, (+)-catechin, (–)-epicatechin, and proanthocyanidins, are also responsible for the changes in organoleptic properties (color and specific flavors) during the drying process of pears [4, 11, 15]. The texture of dried pears is also altered due to the modifications in the chemistry of cell wall polysaccharides [1, 4, 15].

Chemical and sensory transformations that occur during the drying process influence the quality of the end-product, having a decisive relevance toward its attractiveness to the consumer, in both physical and nutritional aspects. Thus, this chapter seeks to clarify the role of the main phytochemicals present in dried pears and to understand their relevance as potential benefits to human health.

18.2 Phytochemicals in pears

Much attention has recently been paid to phytochemicals of pears due to their positive association with human health [16]. In the following sections, the amount and structure of carbohydrates, proteins, and phenolic compounds present in pears are described.

18.2.1 Carbohydrates

Pear fruit is composed of pectic polysaccharides (36%), cellulose (31%), and hemicellulose, which includes glucuronoxylans (27%), xyloglucans (4%), and mannans (2%) (Figure 18.2)

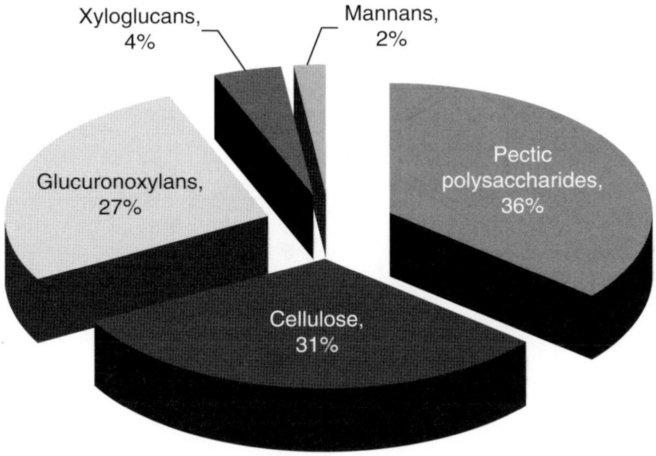

Figure 18.2 Relative amount of cell wall polysaccharides of S. Bartolomeu fresh pulp pears. (Adapted from Ferreira [17]).

[17]. Pectic polysaccharides are a complex class of polysaccharides rich in galacturonic acid (GalA). Rhamnose (Rha), arabinose (Ara), and galactose (Gal) residues are also present. The occurrence of glucose (Glc) residues glycosidically linked to GalA backbone has been found in pectic polysaccharides of different fruits, including pear. Thus, this structural feature could also be common to other pectic polysaccharides.

The composition of pectic polysaccharides includes different associated polysaccharides, such as homogalacturonans (HG), xylogalacturonans (XGA), type I rhamnogalacturonans (RG-I), type II rhamnogalacturonans (RG-II), arabinans, and arabinogalactans [18, 19] and all these types of structures can be found in pear fruit [4, 20]. The most abundant pectic polysaccharides in pears are HG. They are linear homopolymers of α-1,4-linked galacturonic acid in the pyranose form (Gal*p*A), comprising about 60–70% of total amount of pectic polysaccharides. HG could partially be methylesterified at C-6 carboxyl group and/or O-acetylated at O-2 and/or O-3 (Figure 18.3), according to the source [21–23]. Pectic polysaccharides are also composed of xylose (β-D-Xyl*p*) residues substituted at C-3 of Gal*p*A backbone, forming XGA (Figure 18.3). Although xylose is mainly present as single residues, it can occur, occasionally, as an additional β-2-linked or β-4-linked Xyl*p*, to form a disaccharide [21–24]. Moreover, the Gal*p*A residues comprising the XGA backbone can be partially methylesterified and the methyl esters seem to be equally distributed among the substituted and unsubstituted Gal*p*A residues [25]. RG-I like polysaccharide has a backbone composed of repeating units of α-2-linked Rha*p* and α-4-linked Gal*p*A, [→2)-α-L-Rha*p*-(1→4)-α-D-Gal*p*A-(1→]. The residues of Rha can be substituted (20–80%) at O-4 with Gal and/or Ara residues. In these side chains, single residues of β-D-Gal*p* as well as linear or branched polymers of arabinogalactans and/or arabinans have been identified [4, 20]. Furthermore, the Gal*p*A residues in the RG-I backbone may be highly acetylated in O-2 and/or O-3 positions (Figure 18.3) [26]. RG-II are the most complex pectic polysaccharides, with an HG backbone constituted of, at least, 8 1→4-linked α-D-Gal*p*A residues with branching chains consisting of 12 different types of sugars in over 20 different linkages. Four different side chains have been described in the RG-II structure, with peculiar sugar residues, namely apiose (Api), aceric

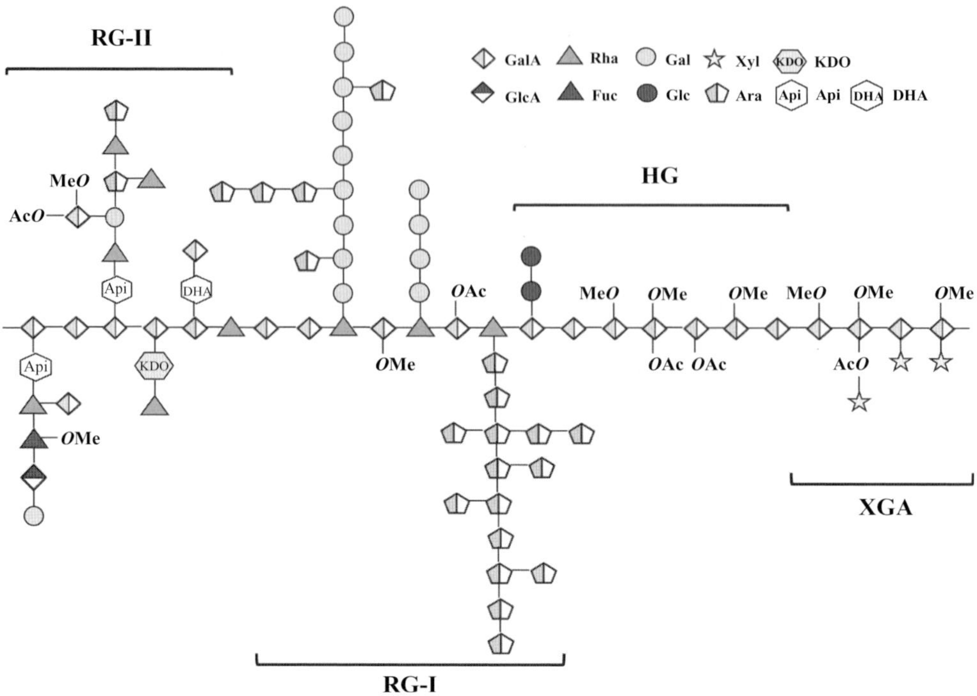

Figure 18.3 Schematic structure of pectic polysaccharides. HG, homogalacturonan; XGA, xylogalacturonan; RG-I, type I rhamnogalacturonan; RG-II, type II rhamnogalacturonan; GalA, galacturonic acid; GlcA, glucuronic acid; Rha, rhamnose; Xyl, xylose; Ara, arabinose; Fuc, fucose; Gal, galactose; KDO, 3-deoxy-D-*manno*-octulosonic acid; DHA, 3-deoxy-*lyxo*-2-heptulosaric acid; Api, apiose. (Adapted from Ferreira [17]).

acid (AceA), 3-deoxy-*lyxo*-2-heptulosaric acid (DHA), and 3-deoxy-*manno*-2-octulosonic acid (KDO) (Figure 18.3) [18, 19, 26].

Cellulose is the second major carbohydrate found in pears. This polysaccharide is constituted by β-4-linked Glc residues, [→4)-β-D-Glc*p*-(→1], that associate with other cellulose chains by hydrogen bonding and van der Waals forces, yielding microfibrils [19].

Glucuronoxylans, xyloglucans, and mannans are regarded as hemicellulosic compounds and are present in pear fruit, although in lesser amount than pectic polysaccharides and cellulose. Glucuronoxylans are about 27% of polysaccharides of fresh S. Bartolomeu pulp cell wall. They are composed of β-(1→4)-D-Xyl*p* backbone with α-D-Gal*p*A residues in O-2 position of the backbone and/or their derivative 4-O-methylated. The high content of these compounds is also observed in Branquilla pear, and probably they are derived from secondary cell wall of sclereids [27]. Xyloglucans have been found comprising a core backbone structure composed of 4-linked β-D-Glc*p* residues, with regular branches at O-6 of α-D-Xyl*p* residues. In S. Bartolomeu pears, decoration with Gal and fucose (Fuc) also occurs, yielding XXFG- and XLFG-type blocks, in which G = Glc*p*; X = Xyl-(1→6)-Glc*p*-; L = Gal*p*-(1→2)-Xyl*p*-(1→6)-Glc*p*-; and F = Fuc*p*-(1→2)-Gal*p*-(1→2)-Xyl*p*-(1→6)-Glc*p*- (Figure 18.4) [17]. The occurrence of mannans (1→4-linked α-D-Man*p* residues) in pears is probably related to the presence of glucomannans, which are present in the secondary cell wall of sclerotic cells, although in small amounts [17].

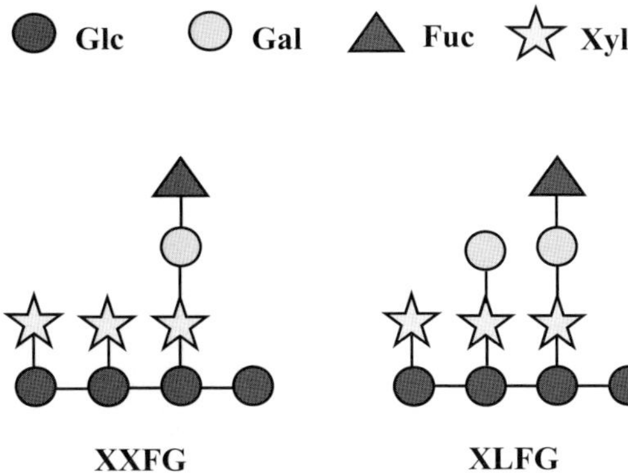

Figure 18.4 Schematic structures of xyloglucans found in S. Bartolomeu pears. XXFG, G = Glc; X = Xyl-(1→6)-Glc-; L = Gal-(1→2)-Xyl-(1→6)-Glc-; F = Fuc-(1→2)-Gal-(1→2)-Xyl-(1→6)-Glc-. Glc, glucose; Gal, galactose; Fuc, fucose; Xyl, xylose. (Adapted from Ferreira [17]).

18.2.2 Proteins

The protein content of pears may range from 1.5 to 2.6% of dry mass, according to their variety [5]. Pears are composed mainly of glutamic acid (Glu)/glutamine (Gln), determined as Glx, and leucine (Leu), accounting for 20–19 mol.%, each, followed by aspartic acid (Asp)/asparagine (Asn), determined as Asx, contributing 17–13 mol.% for S. Bartolomeu pear variety [14]. The designations Asx and Glx derive from the fact that the methodology did not allow the distinction between the amide and carboxylic acid functions for Asn and Asp, as well as for Gln and Glu. In addition, isoleucine (Ile), alanine (Ala), and valine (Val) are present, accounting for 10–7, 8–7, and 7 mol.%, respectively [14, 28]. Despite these protein amino acids, 4-hydroxyproline (Hyp) was found in Beurré variety, but not in S. Bartolomeu variety [14]. Hyp occurs in extensins (hydroxyproline-rich proteins) and arabinogalactan-rich glycoproteins of *P. communis* L. [10]. Finally, serine (Ser) is found in some Chinese pear varieties [29].

Amino acids can also occur in the free form. The most abundant free amino acids in pears are Asx and Glx, accounting for 45–20 and 38–29 mol.%, respectively. Ala and Val are also found, although, in smaller quantities (17–9 and 9–7 mol.%, respectively), as well as proline (Pro) and glycine (Gly) [14, 28].

18.2.3 Phenolic compounds

Phenolic compounds are widely distributed in pear fruit [11]. The pulp of fresh pears contains 3.7 g/kg phenolics, mainly composed of procyanidins (96%), with a mean degree of polymerization (mDP) of about 13–14. (–)-Epicatechin (99%) is the most abundant monomer in the procyanidin structure with only 1% (+)-catechin. Hydroxycinnamic acids, arbutin, and monomeric catechins are also present, accounting for 2.0, 0.8, and 0.7%, respectively [11].

18.2.3.1 *Arbutin*

The simple phenolic arbutin, 4-hydroxyphenyl-β-D-glucopyranoside (Figure 18.5), is a hydroquinone glucoside characteristic of *P. communis* L. pear, including the varieties of

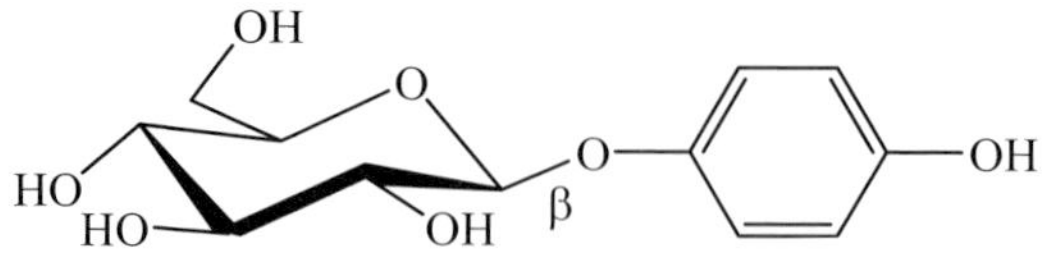

Figure 18.5 Structure of arbutin.

S. Bartolomeu [11], d'Anjou, Red Williams, Alexander Lucas [30], and Decana [31]. This phytochemical can occur in the leaves of *P. communis* L. [32], as well as in pear juice [33]. It is diagnostic of the occurrence of pear pulp in fruit purees (30 mg/kg of fresh pulp), allowing their distinction from other pome fruits, such as apples [34, 35]. There was little change in arbutin content between fresh and dried pears [14].

18.2.3.2 *Phenolic acids*

Phenolic acids comprise the broadly distributed hydroxybenzoic and hydroxycinnamic acids. In pears, hydroxycinnamic acids are the most abundant, accounting for about 85 mg/kg of fresh pulp [11]. These phytochemicals have a C6–C3 structure, with *p*-coumaric or caffeic acids (Figure 18.6) being most abundant. Caffeic acid has been found in the pulp of some pear varieties [12, 31, 36], as well as in pear juice, together with *p*-coumaric acid [33]. Hydroxycinnamic acids are present mainly as derivatives, as esters of caffeic and coumaric acids linked to the hydroxyl groups of quinic or malic acids forming coumaroylquinic, caffeoylquinic, and coumaroylmalic acids. 5-Caffeoylquinic acid (40–141 mg/kg of fresh pulp), a chlorogenic acid (Figure 18.7), is the most abundant hydroxycinnamic acid in pears, as reported for varieties such as S. Bartolomeu [11], d'Anjou, Bosc [37], Williams, Guyot, and Conference [38], Comice and Abbe Fetel [39], Decana, Prassagrana [31], Alexander Lucas [30], as well as in some Chinese pear varieties [12]. In addition, chlorogenic acid occurs in pears juice [40]. 5-*p*-Coumaroylquinic acid (Figure 18.7) has been found in S. Bartolomeu variety [11] and, together with 2-*p*-coumaroylmalic acid (Figure 18.7), is present in Guyot pear variety [41]. 2-*p*-Coumaroylmalic acid (Figure 18.7) (about 15 g/kg of fresh pulp) occurs in S. Bartolomeu [11] and Conference pear varieties [38], as well as in Comice and d'Anjou pear juice [33]. All of these hydroxycinnamic acid esters also occur in other varieties of pear [41].

18.2.3.3 *Flavonoids and procyanidins*

Flavan-3-ols, which are included in the flavonoid family, can be present as monomeric (named catechins), oligomeric, and polymeric forms. Monomeric catechins, namely (+)-catechin and (–)-epicatechin (Figure 18.8), have been reported in several varieties of pear,

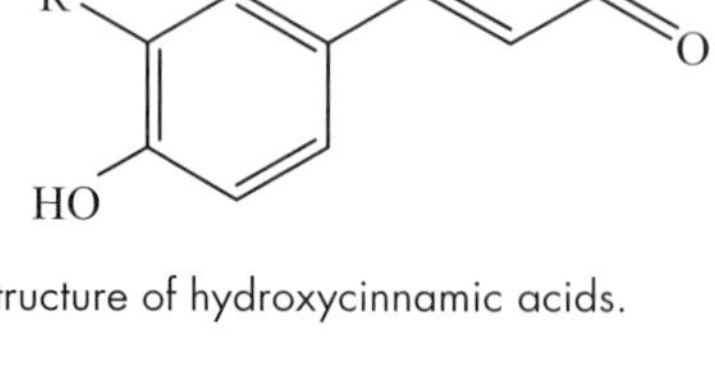

R = H, *p*-Coumaric acid
R = OH, Caffeic acid

Figure 18.6 Structure of hydroxycinnamic acids.

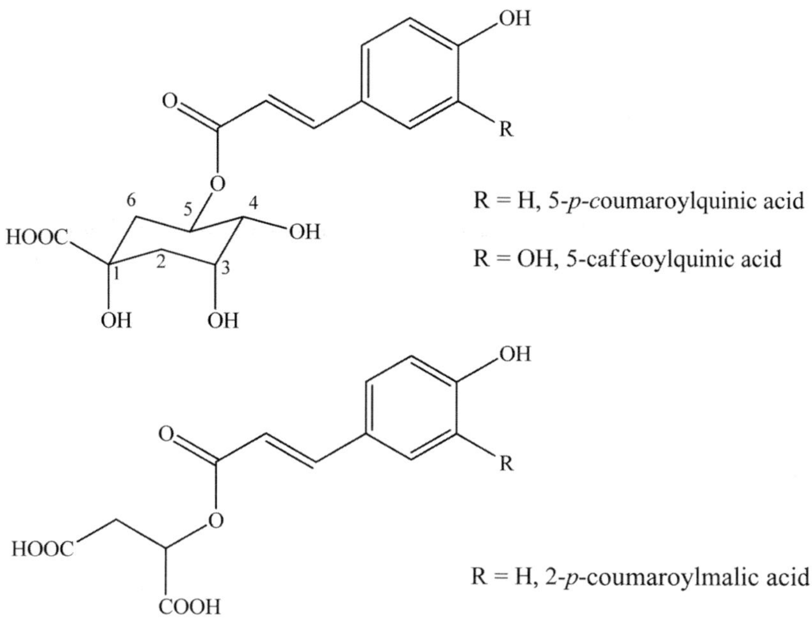

Figure 18.7 Structure of hydroxycinnamic esters.

such as S. Bartolomeu [11], Guyot, Comice, Conference, and Williams (13–26 mg/kg of fresh pulp) [39], among others [30, 42], in which (–)-epicatechin is the main monomer (about 88%). On another study, Schieber *et al.* [30] reported the highest epicatechin content in Red Williams cultivar.

Oligomeric as well as polymeric forms of flavan-3-ols are known as proanthocyanidins (condensed tannins). Proanthocyanidins are colorless precursors of anthocyanidins, the aglycone form of anthocyanins (pigments with red, blue, or purple color), formed upon the rupture of C–C bond, which is cleaved after being heated in acidic solutions. In this way, the terminal flavan unit is released from the oligomers as carbocations that are then oxidized to colored anthocyanidins by atmospheric oxygen [3]. Proanthocyanidins, which are composed

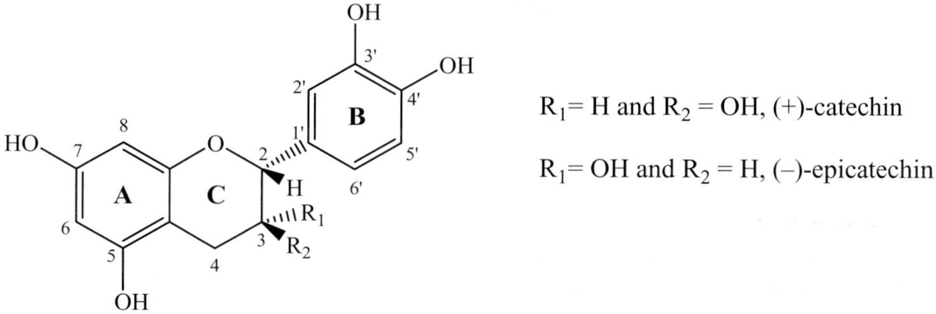

Figure 18.8 Structure of monomeric catechins.

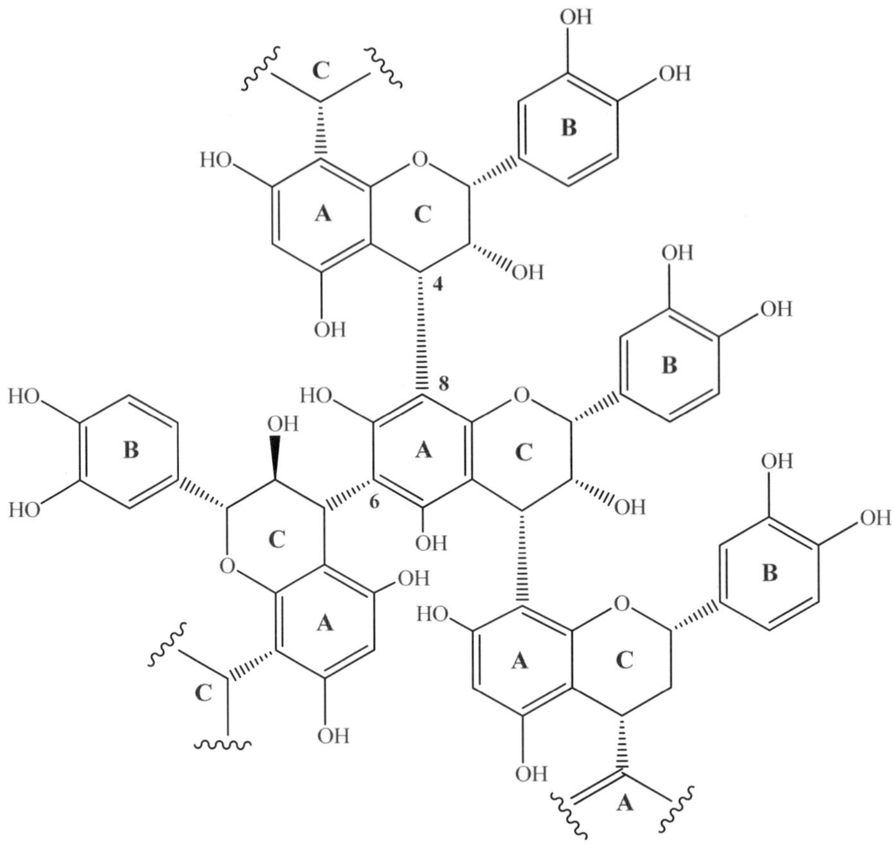

Figure 18.9 Partial structure of procyanidins group B.

by (+)-catechin and/or (–)-epicatechin units, are known as procyanidins due to the release of cyanidin units after being heated in acidic solutions. In the procyanidin structures, both C–C and C–O bonds are possible to occur (procyanidins of group A), but usually a C–C bond from the C-4 of one unit to the C-6 or C-8 of the other (procyanidins of group B) is found in pears (Figure 18.9). Procyanidin dimers of group B with C_4–C_8 bond, such as B1 [43], B2 [33, 43], and B4 [33] have been reported in different pear varieties, as well as polymeric procyanidins [11]. For example in fresh pears, procyanidins B2 account for about 3.6 g/kg of the fresh pulp, where (–)-epicatechin (99%) is in higher amount, rather than (+)-catechin, which account only for 1% of the total procyanidin monomers [11]. The high degree of polymerization of procyanidins seems to affect the sensory properties such as the enhancement of astringency [44, 45], as is observed for fresh S. Bartolomeu pear [11]. The sensation of astringency occurs due to the specific interaction between phenolic compounds and proteins [46], leading to the precipitation of salivary proteins [47]. Furthermore, oligomeric forms seem to be related to the bitterness of pears [44]. Flavonols and their glycosidic forms are also identified in pears. Quercetin, in both aglycone and glycoside forms (Figure 18.10), is the main flavonol found in pears, as well as isorhamnetin glycosides [41, 48]. Quercetin 3-*O*-glycosides, identified as rutinoside, glucoside, and malonyl glucoside, and the isorhamnetin

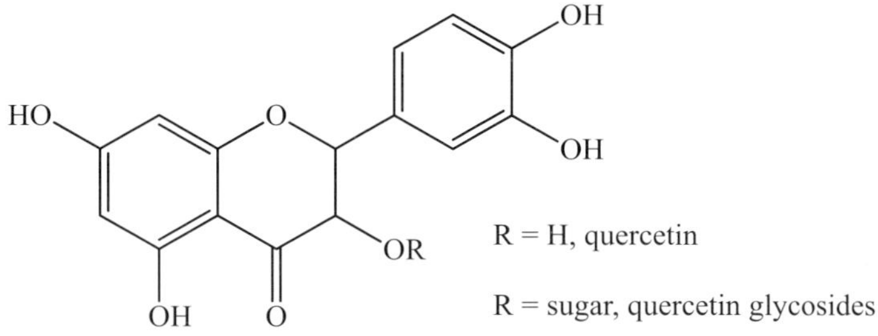

Figure 18.10 Structure of quercetin and its O-3-derivatives.

3-*O*-glycosides determined as rutinoside, galactorhamnoside, glucoside, malonyl galactoside, and malonyl glucoside have been reported in Bartlett and Bon Chretien pears, as well as in Packingham variety [41, 48]. In addition, isorhamnetin-3-rhamnogalactoside and a derivative of isorhamnetin-3-glucoside as well as isoquercitrin (quercetin-3-glucoside) are present [48, 49]. Kaempferol, another flavonol, has been found in pears, although in small quantities [12, 48].

18.3 Changes in phytochemical compounds during drying of pears

Drying of fruits is probably the oldest practice for food preservation. In Portugal, the tradition of drying of S. Bartolomeu pear by direct sun exposure is still in use. After harvesting, the fruits are peeled and dried in sunlight and in open air for about 5 days. Then, the pears are laid in baskets and covered with a cloth for 2 days, in order to obtain pears with the needed elasticity to flatten, without breakage. After this, the fruits are submitted to a second sun-drying process [4]. The final product is a much appreciated sun-dried small pear with singular organoleptic characteristics, such as reddish brown color (Figure 18.1) and unique elastic properties [4, 5]. During the drying process, the astringency is lost and the fruit acquires a typical sweet flavor and chewy texture [5].

This section reviews the changes occurring in cell wall polysaccharides and in protein and free amino acid profiles, as well as the changes associated with the nonenzymatic browning reactions (Maillard reaction). The changes of phenolic compounds during the drying process of pears are also presented.

18.3.1 Carbohydrates

After drying, pears have a moisture content of about 20%, representing a 75% decrease of moisture content compared to the fresh pear [17, 50]. Dried pear is mainly composed of cellulose (33%), pectic polysaccharides (32%), and glucuronoxylans (30%), followed by xyloglucans (3%) and mannans (1%) [17]. Compared with fresh pears, a slight decrease in relative amounts of pectic polysaccharides, xyloglucans, and mannans may be observed. Contrarily, the relative amount of cellulose and glucuronoxylans increased slightly with the sun-drying process. The increase in the relative amount of glucuronoxylans can be explained

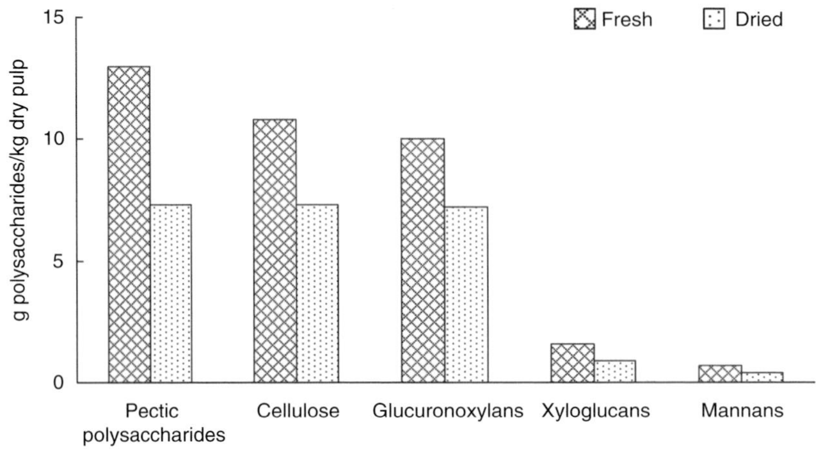

Figure 18.11 Composition of cell wall polysaccharides (g/kg of dry pulp) of S. Bartolomeu fresh and dried pears. (Adapted from Ferreira [17]).

with the rise of polymerization degree in these structures, possibly due to the continuing development of sclereid cells during the drying in the sun, as was observed during Blanquila pear ripeness [27]. In terms of dry weight of pulp, sun-drying of pears accounts for a 36% decrease in total cell wall polysaccharides, in which xyloglucans are most affected (49%; Figure 18.11) [17]. This fact is probably related to the depolymerization of backbone and/or loss of residues from xyloglucan side chains, affecting the solubility of these compounds. Furthermore, cellulose content decreased 32% in dry weight of pulp, compared to the fresh form, suggesting the degradation of cellulose microfibrils and solubilization of glucans. The amount of glucuronoxylans and mannans decreased by 28 and 43%, respectively. Pectic polysaccharides in dried pears showed a decrease of 43%, which can be explained by the degradation of the galacturonan backbone, mainly of arabinan and/or galactan and/or arabinogalactan side chains, which in turn were also reduced (52%). The changes in the composition of sugars of these polysaccharides may contribute to the changes in their solubility in the cell walls of dried pears [17]. The same results were found for sun-dried Bartlett pears [1], where a positive correlation with the decrease of RG-I–arabinan side chains and the solubility of pectic polysaccharides was observed.

18.3.2 Proteins

The drying of pears with different methods, such as Portuguese traditional sun-drying method, in a large glass greenhouse with air convection (GH1), in a small greenhouse with natural convection (GH2), or using a hot air tunnel in the absence of light (HAT), did not seem to significantly affect the proteic amino acid profile of the fruit, when compared to its fresh counterpart [14]. For all dried pears protein, Glx (26–22%), Leu (19–17%), and Asx (14–13%) were the predominant amino acids, contrasting with the high content Asx (35%) and low Glx (8%) and Leu (5%), found in the literature for fresh pear (*P. communis* L.) [28]. The occurrence of high amount of Glx in dried S. Bartolomeu pear seems to be related to the presence of arabinogalactan-rich glycoproteins in the fruit [10]. In addition, other amino acid proteins, namely Ile, Ala, and Val, were present in higher abundance in dried pears, accounting for 11–8, 9–5, and 8–7%, respectively. The higher differences observed between fresh and traditional dried pears are the decrease of the content of Pro and the increase of the content in Glx. Also, a loss in relative content of phenylalanine (Phe), lysine (Lys), and threonine (Thr) also occurred [14]. The decrease of Pro and Lys contents also occurs in the other drying methods, although to different extent, depending on the type of processing. The highest loss of Lys was observed in GH1 and traditional processed pears, followed by GH2 and no loss of Lys was noted in HAT processed fruits. Also, the relative content of Glx increased in all processed S. Bartolomeu pears [14].

Similar to that observed for S. Bartolomeu fresh pear, dried pears processed according to different methodologies also contain Ala, Gly, Val, Pro, Asx, and Glx, as free amino acids, in quantifiable amounts. Thr, Leu, Ile, Phe, and Lys also occur, but in lesser amounts. In dried pears, Pro is the most abundant free amino acid, except for the pears obtained with GH1 drying process, where Glx was most abundant [14]. For all dried fruits, Pro accounted for 48–21% of the free amino acids and Glx contributed 24–12%. Ala also accounted for 17–8%. Val, Asx, and Gly were present in all dried pears, accounting for 15–9, 15–8, and 12–3%, respectively. In general, all other free amino acids decreased upon all drying processes. The decrease in the total free amino acid content was related to the low water activity of fruits, promoting the Maillard reaction at room temperature. Compared with free amino acid profile from fresh S. Bartolomeu pears, all drying processes increased the content of Pro. In fact, the increase in the synthesis of Pro has been reported as a response of fruits to stress conditions [51–53], namely due to the modulation of the expression of proline-rich glycoproteins (extensins) [14]. Also, the high increase of Pro is in accordance with the higher decrease in Pro as integrated protein, as well as the lower amounts in Glx found as free amino acid, in agreement with its high content in the protein.

18.3.3 Phenolic compounds

Similar to that observed for polysaccharides and proteins, sun-drying of pears leads to a slight modification in the profile of their phenolic compounds [11]. Dried S. Bartolomeu pear seems to have a loss of about 64% (in dry weight pulp) of the total amount of phenolic compounds, compared with the fresh form. With the exception of arbutin, all other phenolics appear to be affected by the drying process [11]. Procyanidins were the most abundant phenolic compounds in dried pears, although a decrease of 68% occurred. However, its degree of polymerization was not affected. As observed for fresh pears, (–)-epicatechin was the most abundant flavan-3-ol monomer, accounting for 77%. In sun-dried pear, B2 dimer of procyanidin was not present. The hydroxycinnamic acids, namely chlorogenic acid, as well as monomeric catechins seem to be most affected (decreased 96 and 91%, on a dry weight basis, respectively) [11].

Arbutin was not affected by the sun-drying process [11]. A similar result was observed in pear juice [33]. In fact, it has been postulated that polyphenoloxidase (PPO) might have less influence on the concentration of arbutin than on chlorogenic acid and catechins, as was observed in d'Anjou and Bosc pears [37].

The decrease of polymeric procyanidins and the absence of procyanidin B2 in sun-dried pear, as well as their high insolubility [11], are probably due to the involvement of these compounds in coupled oxidation reactions, as also observed in apple juice [54]. For example, in apple juice, procyanidins are oxidized and highly reactive oxidized intermediates react irreversibly with each other and with proteins to form insoluble complexes [54]. Polymeric procyanidins have been shown to affect sensory properties, such as astringency [44]. The sensation of astringency provided by some fresh pears, such as S. Bartolomeu pear, seems to be lost during the drying process, probably due the decrease of polymeric procyanidins and their insolubility [11]. Furthermore, the reaction of these compounds with other components present in cell wall, such as polysaccharides, could explain the physical characteristics of sun-dried pears, since phenolic compounds form an additional, less hydrophilic, network in these fruits [11].

During processing of fruits, changes in color also occur. Chlorogenic acid and catechins (*o*-diphenols) are involved in enzymatic browning of pears [38, 55]. In this way, the browning of pears, observed during sun-drying process seems to be related with the oxidation of these compounds catalyzed by cathecolase of pear PPO. Consequently, the oxidation of these *o*-diphenols leads to the formation of *o*-quinones, which are rapidly polymerized, yielding brown pigments [34, 56]. The oxidation of these phenolic compounds by PPO could be the cause of their decrease in the pulp-dried pear. Moreover, since the PPO was shown to have higher affinity for chlorogenic acid than for catechins this may explain the higher reduction in caffeoylquinic acid in dried S. Bartolomeu pear [57, 58].

18.3.4 Maillard reactions

Apart from enzymatic reactions involving phenolic compounds and PPOs, the browning of sun-dried pears was recently associated with the presence of Maillard reaction products (MRPs) [14]. Maillard reaction, also called nonenzymatic browning or glycation, comprises a reaction between reducing sugars and the free amino groups of amino acids, peptides, or proteins [59] and has been shown to play an important role in the browning and loss of nutritional value of proteins (e.g., loss of available Lys) during food drying [14, 60]. In addition, MRPs account for both desirable [61, 62] and unpleasant [56, 63, 64] flavors and aroma of processed or stored food. It is known that browning reactions are favored during the dehydration process [65], under conditions of pH 5–7, with intermediate moisture content and temperatures over 50°C, using long processing times [66]. After the immersion of pieces of freeze-dried pulp of S. Bartolomeu fresh pear in buffer solutions with a large range of pH, a fast darkening of the tissues immersed in the solutions with higher pH than the other ones [14] was observed. Major color variations were observed in the pH interval of 4–6, whereas at pH 5.0 the color of the tissue approached the reddish brown color characteristic of the traditional product (Figure 18.12a). In addition, the intensity of color in the pear tissues exposed to sunlight was higher when compared with the same samples placed in the dark (Figure 18.12b). In addition, high temperature under high moisture conditions also promoted the formation of the reddish brown color of the pear tissues (Figure 18.12c) [14].

Maillard reaction comprises three stages: the early, the advanced, and the final Maillard reaction steps, which can occur simultaneously [67]. Amadori compounds, formed during the early stage of Maillard reactions [68, 69], have been detected in dried pears as furosine, 2-furoylmethyllysine (Figure 18.13) [14], a good indicator of the first step of these reactions in dairy products [70, 71]. Taking into account the different drying methodologies of pears (see Section 18.2.2), the content of furosine was higher for S. Bartolomeu pears dried by the traditional sun-drying process (247 mg/100 g protein) and lower for pears dried in a

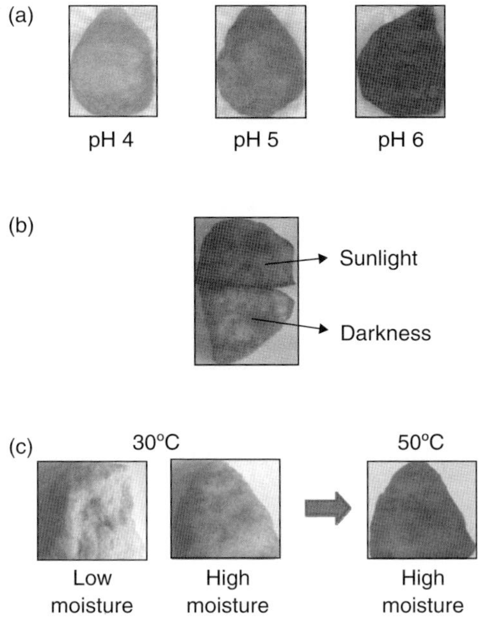

Figure 18.12 Effect of (**a**) pH variation, (**b**) sunlight exposure, and (**c**) moisture and temperature on the development of color of pear tissues. (Adapted from Coimbra *et al.* [14]). For color detail, see color plate section.

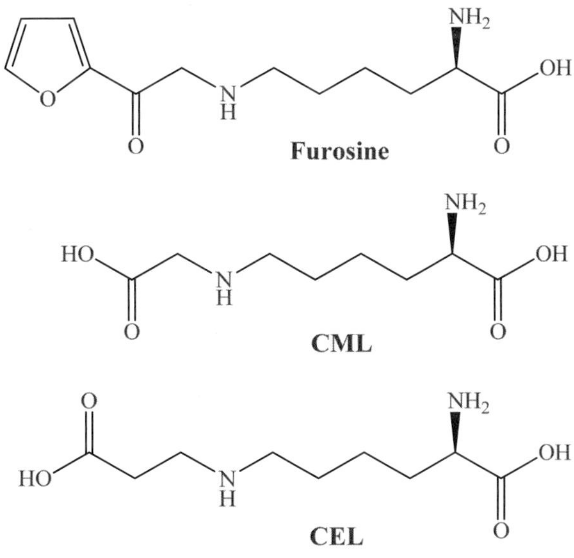

Figure 18.13 Maillard reaction products identified in dried pears. CML, carboxymethyllysine; CEL, carboxyethyllysine.

HAT (80 mg/100 g protein). Pears dried in a GH1 and GH2 showed intermediate levels of furosine (182 and 136 mg/100 g protein, respectively). These amounts of furosine are similar to those observed for products to which moderate heat treatment is applied, such as ultra-high-temperature (UHT) milk products [72, 73], jams, and fruit-based infant foods [71] and cooked salmon [74]. During intense heating or prolonged storage of foods, Amadori compounds may undergo several rearrangements and degradation reactions, yielding the so-called advanced glycation end-products, AGEs (advanced stage) [67].

Carboxymethyllysine (CML) and carboxyethyllysine (CEL) have been shown to be useful markers in this step of browning reactions (Figure 18.13) [75–77]. Both of these AGEs (CML and CEL) were found in dried pears, accounting for 96 mg/g protein for traditional sun-drying method and 37 mg/g protein for HAT. Dried pears in GH1 and GH2 (94 and 71 mg/g protein, respectively) showed intermediate amounts of CML and CEL, which suggests that the sun-drying process of pears is a more severe treatment than HAT. The content of MRPs present in dried pears seem to influence their color, since the reddish brown color observed in traditional dried fruits (Figure 18.1b) is in accordance with the high content of these compounds. On the other hand, HAT-dried pears showed a less intense color (yellow-orange color, Figure 18.1e), which is probably related to their low content of MRPs [14]. Taking into account the decrease in the amount of each amino acid after drying pears with different methodologies and the content of MRPs formed, a positive correlation between the amount of these compounds and a decrease in the relative amount of Lys can be observed.

The loss of Lys can be used as a marker of Maillard reactions [78], since it is the most reactive residue undergoing chemical changes under these conditions [68]. Contrarily, the nutritional value of S. Bartolomeu pear was not affected when pears were dried in HAT process, since the loss of Lys does not occur [14]. The final stage of Maillard reactions is marked by the presence of melanoidins [68]. These compounds, resulting from numerous rearrangements and polymerization reactions of Amadori compounds, as well as AGEs, are commonly described as brown-colored high-molecular-weight peptide-bound MRPs, since the information of their exact chemical structure is still scarce [68]. Melanoidins have been found in several heat-processed dairy foods such as coffee, bread [79, 80], meat [81], and tomato sauce [82]. However, as far as we know, these compounds have not been reported in dried pears.

18.4 Bioavailability and potential health effects

Epidemiological and clinical investigations strongly suggest that the consumption of fruits, including pears, is associated with a decrease of developing chronic diseases, such as cardiovascular diseases (CVD), cancer, and other pathologies [83–86]. The lower incidence of human pathologies has been mainly ascribed to phenolic compounds present in fruits, due to their high antioxidant activity [16, 87]. Furthermore, positive effects in the intake of dietary fiber, widely distributed in pears, are also known [88, 89]. The potential health effects of phytochemicals depend on their bioavailability, which includes absorption in human gastrointestinal tract, metabolism, and excretion [87].

Pear fruit is consumed globally, in both the fresh and dried forms. Few studies have been performed in order to understand the influence of the sun-drying process in the chemical and sensory properties of pears [11, 14]. However, as far as we know, no data are available from *in vitro* and/or *in vivo* experiments with phytochemicals of dried pears in order to evaluate their impact on human health. As the compounds present in pears also occur in other matrices, and some of them have also been isolated and used for the evaluation of bioavailability and potential health effects, the information available in the literature concerning the potential health benefits of these compounds are discussed and related to dried pears with respect to phenolic compounds (Table 18.1) and dietary fiber (Table 18.2).

18.4.1 Phenolic compounds

18.4.1.1 *Bioavailability*

Phenolic compounds, as the major antioxidants present in pears, cannot exert any biological effects unless they are absorbed, metabolized, and distributed in the body [90]. However, little information exists about the bioavailability of these phytochemicals from whole foods, including pears. Thus, as far as we know, information concerning the bioavailability of arbutin, the phytochemical characteristic of pears, is not known.

It is estimated that the total intake of phenolic compounds from fruits and beverages (fruit juice, wine, tea, coffee, chocolate, and beer) is about 1 g/day [91]. Among flavonoids (total 650 mg/day), flavonols account for about 23 mg/day in human dietary, in which about 70% (16 mg/day) is attributed to quercetin and 17% (4 mg/day) to kaempferol [90]. The procyanidin content in food may vary according to the source. Apple fruit is one major source of procyanidins, accounting for 12.3–252.4 mg/serving, depending on its variety. On average, procyanidins from apple are about 147.1 mg/day [92]. The daily intake of flavan-3-ol (catechin and epicatechin) may range between 100 and 500 mg/day, depending on the fruits and beverages ingested [43, 91]. Despite the amount of intake, only about 5–10% of the total phenolic compounds are absorbed in the small intestine [91, 93]. Unabsorbed phenolic compounds reach the large intestine, where they become fermentable substrates for bacteria. Those that are not fermented remain in the colonic microflora, where they may contribute to a healthy antioxidant environment by scavenging free radicals and counteracting the effects of dietary pro-oxidants [94]. Several studies have reported the absorption of these phytochemicals in the small intestines in humans and animals. One *in vitro* study performed in an isolate rat intestine model showed the absorption of chlorogenic and caffeic acids [95]. In another study, injected chlorogenic and caffeic acids showed 33 and 95%, respectively, of absorption in the small intestines of ileostomized humans [96]. Traces of chlorogenic acids and 11% of caffeic acid were recovered in the urine after 24 hours, suggesting that part of the chlorogenic acids from foods will enter the bloodstream, but most will reach the large intestine and may be metabolized by the gut microflora [96]. Furthermore, rats fed with chlorogenic acids excreted very low amount of chlorogenic acids in their urine, but instead they excreted mainly microbially produced metabolites of chlorogenic acids such as hippuric acid and m-coumaric acid [97]. Several studies have demonstrated the absorption of flavonols in humans [98, 99], but in which form they are absorbed, for example, as an aglycone, glycoside, or both, remains to be answered. A study reported that quercetin-3-rutinoside (rutin) is hydrolyzed in the gastrointestinal tract to quercetin aglycone by glycosidases released from gut microflora [100]. Others have suggested that quercetin glucoside is better hydrolyzed to quercetin in small intestines by glucosidases and is then absorbed [101]. A previous study on ileostomized patients on different forms of quercetin reported the absorption as 52% for glucosides, its major dietary form, 24% for aglycone, and 17% for quercetin-3-rutinoside [98]. In addition, another study involving humans suggested the quercetin bioavailability from apples (rich in

Table 18.1 Summary table of some potential health effects of phenolic compounds present in dried pears

Biological activity	Phytochemicals	Biological assays	Findings	Reference
Antioxidant activity	Phenolic acids and flavonoids	*In vitro*	Phenolic extracts of pears have antioxidant activity	[12]
		In vivo	Positive influence on plasma lipid levels and on plasma antioxidant capacity of rats	[112]
Anti-inflammatory activity	Quercetin	*In vitro*	Inhibition at 20 μM for LPS-induced NO and 50 μM for TNF-α production in RAW cells	[124]
			Inhibition of collagen-induced platelet aggregation at a concentration that can be achieved by the diet	[126]
	Phenolic acids and flavonoids	*In vivo*	No correlation between the high content of phytochemicals and anti-inflammatory activity.	[12]
Prevention of CVD	Flavonoids	*In vitro*	Inhibition of LDL oxidation (quercetin was the most effective inhibitor)	[119]
	Hydroxycinnamates	*In vitro*	Inhibition of platelet aggregation (IC_{50} 556 μM; human plasma ≈ 3 μM)	[126]
	Procyanidins	*In vitro* and/or *in vivo*	No direct effect on blood vessel function	[107, 133]
Cancer effect	Caffeic acid	*In vivo*	Carcinogenic at 2%	[136]
			Tumor promoter at 0.5–1.0%	[137, 138]
			Anticarcinogenic at 0.05–0.5%	[135, 139]
	Chlorogenic acid	*In vivo*	Chemopreventive potential against colon carcinogenesis in rats	[140]
			No significant anticarcinogenic effects	[86, 141]
	Arbutin	*In vivo*	At 50 μM showed inhibitory effect; reduction of melanin per cell (39%); and reduction of tyrosinase activity.	[145]
			Reduction on TCCSUP human bladder carcinoma cell proliferation concentrations of <500 μg/mL.	[146]
	Procyanidins	*In vivo*	High protection of gastric lesions (antiulcer effect).	[86]

CVD, cardiovascular disease; LPS, lipopolysaccharide; NO, nitric oxide; TNF-α, tumor necrosis factor-α; LDL, low-density lipoprotein.

Table 18.2 Summary table of some potential health effects of dietary fiber present in dried pears

Biological activity	Phytochemicals	Biological assays	Findings	Reference
Prevention of CVD	Soluble fiber	*In vivo*	Reduction of total plasma and LDL-cholesterol levels with daily intake of 2–10 g; no changes in HDL-cholesterol or TAG in blood	[162]
			Intake ≈ 6 g/day reduced on serum LDL-cholesterol levels of ≈ 5.4% and risk of CHD of ≈ 9%	[6]
			Improve blood glucose control in type 1 diabetes	[169]
	Pectins (HMP and LMP) and cellulose	*In vivo*	10 and 15% of HMP decreased the body weight; probably HMP is associated with high viscosity	[163]
	Soluble fiber of fruits	Prospective study	No correlation with reduction risk of type 2 diabetes	[173]
	Soluble and insoluble fibers	*In vivo*	26 g/day of dietary fiber (≈ 6 sun-dried pears) leads to a 22% lower risk of developing diabetes	[172]
		In vivo	Enhanced weight loss (1.22 kg) compared with fibers from oat cookies (0.88 kg)	[175]
		Meta-analysis	Reduction of blood pressure	[179]
		Randomized trial	Reduction of CRP levels with ≈ 30 g/day fiber intake	[186]
	Propionate (SCFA)	*In vivo*	Reduction in blood cholesterol level through inhibition cholesterol synthesis	[177]
Anti-inflammatory activity	Butyrate (SCFA)	*In vitro*	Anti-inflammatory effect by influencing NF-kB activation in a human colonic epithelial cell line	[178]
Cancer effect	Pectin	*In vivo*	Reduction of tumor growth and cancerous cell migration in rats fed modified citrus pectin	[192]
	Soluble and insoluble fibers	*In vivo*	Improve the stool weight, thereby promoting normal laxation	[196]
		Case–control studies	Positive correlation with lower pancreatic and breast cancer risk	[193, 194]
		Cohort studies	No correlation with lower breast and colorectal cancer risks	[191, 195]

CVD, cardiovascular disease; HMP, high-methoxy pectins; LMP, low-methoxy pectins; SCFA, short-chain fatty acids; LDL, low-density lipoprotein; HDL, high-density lipoprotein; TAG, triacylglycerol; CRP, C-reactive protein; NF-kB, nuclear factor-kappaB.

both quercetin glucosides and aglycone) and pure rutin to be only 30% of that of quercetin from onions (rich in quercetin glucosides) [99]. Thus, it may be inferred that humans absorb quercetin and that its absorption may be enhanced in the form of quercetin glucosides.

Procyanidins, the major phenolic compounds present in pears, as well as flavan-3-ol monomers (catechin and epicatechin), were absorbed by the small intestinal epithelial cells, although in limited concentrations [102, 103]. It has been suggested that the major bioactive forms of flavan-3-ol monomers and procyanidins *in vivo* are probably in the form of their metabolites and/or conjugates of epicatechin [104]. In fact, (–)-epicatechin monomers were strongly metabolized to O-methylated forms and/or conjugated to glucuronides and sulfates during absorption into the circulation, which were detected in the plasma, liver, kidneys, and urine of grape seed extract-fed rats [103]. The amounts of (+)-catechin and (–)-epicatechin metabolites excreted in urine relative to the quantity of the monomers ingested were 27 and 36%, respectively, after 24 hours. Low amounts of procyanidins were also detected in the urine of rats, providing evidence that depolymerization to monomers does not occur in the gastrointestinal tract and these have high stability in the stomach [105]. These results are in accordance with the study conducted by Donovan *et al.* [106], which detected 37% urinary excretion after feeding rats with (–)-epicatechin, supporting the hypothesis that procyanidins are not depolymerized after ingestion. The same findings were also reported in both *in vitro* and *in vivo* models [107].

There are evidences indicating that dietary fiber [107–109], as well as proteins [110, 111], interact with food antioxidants, influencing their absorption and consequently their bioavailability in human bloodstream.

18.4.1.2 *Antioxidant activity*

The antioxidant capacity of phenolic compounds of fruits has been extensively studied over the years and their potential effects in lowering human pathologies have been suggested [16, 87, 112–114]. In addition, compared to vitamins C and E, phenolic compounds are more effective as antioxidants *in vitro*, suggesting their potential protective effects *in vivo* [115, 116]. The mechanism of antioxidant action of these phytochemicals is based on their radical-scavenging activity, donation of hydrogen atoms/electrons, or metal-chelating capacity, which in turn is related to their chemical structure [87]. Recently, the antioxidant capacity of five commercial Chinese pear cultivars was evaluated and showed that the extracts obtained from different pear cultivars exhibited varying degrees of antioxidant activity [12]. This antioxidant activity was highly correlated with the contents of phenolic acids and flavonoids. The antioxidant activity of fruit phenolics appears to be higher than common synthetic antioxidants. A study involving Mediterranean fruits, including four varieties of pears, and tropical fruits was performed in order to assess their antioxidant activity compared with that of common food additives [such as butylated hydroxyanisole (BHA), butylated hydroxytoluene (BHT), and propyl gallate, (PG)] [117]. The results showed very good scavenging activity against hydroxyl radicals ($HO^{\bullet}$) for all Mediterranean and tropical fruits, protecting deoxyribose better than BHA and BHT. However, hypochlorous acid (HOCl)-scavenging activity of all pear varieties was weak compared to other fruits. The inhibition of oxidation and prevention of DNA damage are crucial in lowering human diseases risk, such as CVD (e.g., atherosclerosis) [118–120] and certain types of cancer [84]. A positive influence on plasma lipid levels and on plasma antioxidant capacity of rats by phenolic acids and flavonoids from pears and apples (peel and pulp) was observed [112].

18.4.1.3 *Anti-inflammatory activity*

Numerous studies have suggested a positive correlation between the radical-scavenging activity of phenolic compounds from fruits and their anti-inflammatory activities [121, 122]. However, the opposite has been reported by Li *et al.* [12] who found no correlation between the total phenolic content (mainly phenolic acids and flavonoids) of five pear cultivars and their anti-inflammatory capacity in *in vivo* models. The authors suggested that the observed anti-inflammatory effect in those extracts may be due to the presence of other phytochemicals such as sterols and triterpenes (phytosterols) [123]. Quercetin has been shown to possess anti-inflammatory activities by inhibiting, *in vitro*, lipopolysaccharide (LPS)-induced nitric oxide (NO) and tumor necrosis factor-α (TNF-α) production in a dose-dependent manner [124].

Quercetin accounted for more than 75 and 80% of inhibition at a concentration of 20 μM for LPS-induced NO and 50 μM for TNF-α production, respectively, in RAW cells. TNF-α, an antitumor cytokine, and NO (cytokine mediator), has been shown to play an important role in the pathogenesis of diseases during acute inflammation [124]. However, their overproduction by activated macrophages may lead to several pathophysiological conditions during acute and chronic inflammation [124].

18.4.1.4 *Prevention of cardiovascular diseases*

CVD are the main cause of death and disability in the world [125]. Both *in vitro* [126–128] and *in vivo* [129–131] studies have demonstrated that some phenolic compounds (e.g., procyanidins, flavan-3-ol, and quercetin) are able to inhibit platelet aggregation, suggesting a reduction in thrombosis risk. The effects of hydroxycinnamates and quercetin on platelet activation and cell signaling *in vitro* were evaluated [126].

Hydroxycinnamates inhibited platelet function with collagen as agonist, although not at levels that can be achieved in human plasma by diet (IC_{50} 556 μM; human plasma: ≈ 3 μM [132]). Therefore, these compounds seem to be inefficient as inhibitors of platelet aggregation *in vivo*. On the contrary, quercetin inhibited collagen-induced platelet aggregation *in vitro* at a concentration that can be achieved by diet. This shows that quercetin may have potential effect as inhibitor of platelet aggregation *in vivo* [126]. As procyanidins do not break down in the gastrointestinal tract into their flavanol monomers [107, 133], dietary procyanidins do not make any marked contribution to the systemic pool of flavanols in humans, since they are not absorbed and consequently, their direct effect on blood vessel function is limited [133]. Many studies have shown the capability of flavonoids in inhibiting low-density lipoprotein (LDL) oxidation, which is associated with cell aging and chronic diseases, such as atherosclerosis [134].

18.4.1.5 *Cancer effects*

Most investigations, both *in vitro* and *in vivo*, have suggested the anticancer action of some phenolic compounds present in fruits. However, there are other studies that show no effect or even cancer-inducing effect of these phytochemicals [135]. For example, caffeic acid has been demonstrated to have both carcinogenic and anticarcinogenic effects in *in vivo* studies [135]. It can be carcinogenic at 2% [136], tumor promoter at 0.5–1% [137, 138], and anticarcinogenic at 0.05–0.5% [135, 139]. The lower quantities found in dried pears may infer that these compounds should have a protective effect.

Chlorogenic acid has been shown to have chemopreventive potential against colon carcinogenesis in rats [140]. However, no significant anticarcinogenic effects were observed in other studies [86, 141]. The effect of procyanidins and chlorogenic acid, extracted from Winter Nelis pear fruit, was investigated in ethanol-induced gastric ulcer in rats [86]. After oral administration of 20 mg/rat before 60% ethanol treatment, procyanidins showed a high level of antiulcer capacity, probably due to their strong antioxidant activity. Interestingly, chlorogenic acid, which acts as an antioxidant, had no protective effect on ulcers induced by ethanol [86]. In addition, chlorogenic acid is thought to inhibit human immunodeficiency virus (HIV) activity [142].

Numerous studies have demonstrated that arbutin inhibits tyrosinase, a key enzyme to melanin biosynthesis in melanocytes [143–145]. Melanin is the major pigment responsible for skin color. In this way, arbutin has been used as a whitening agent in cosmetics. Arbutin can also be a potential drug in the treatment of melanin-associated hyperpigmentation disorders. A recent study reported the inhibitory effects of arbutin on melanin production in B16 cells of brownish guinea pig and human skin tissues induced with alpha-melanocyte stimulating hormone (α-MSH) through the reduction of tyrosinase activity in a cell-free system [143]. The maximum arbutin concentration without inhibitory effect on B16 melanoma cell growth was 50 μM [145]. At this concentration, a significant decrease (about 39%) on melanin content per cell was observed, accompanied by a reduction of tyrosinase activity [145]. Arbutin showed significant reduction in TCCSUP human bladder carcinoma cell proliferation in dose–response and time–response at concentrations of <500 μg/mL [146]. Based on the content of arbutin in dried pears and on the serving dose, it is not probable that arbutin could have any anticancer effect by ingestion but its topical use may be beneficial. In addition, arbutin has diuretic and urinary anti-infective properties [147].

18.4.2 Dietary fiber

18.4.2.1 *Bioavailability*

Epidemiologic studies have estimated that the recommended acceptance intake of dietary fiber for adults is 25 g/day for women and 38 g/day for men to ensure optimum CVD protection [148]. Dietary fiber of pears (1 pear = 4 g of total dietary fiber) includes pectic polysaccharides (pectins) as soluble fiber and cellulose, glucuronoxylans, and xyloglucans as insoluble fiber [4, 148, 149]. Their bioavailability defines the utilization and potential biological effect of dietary carbohydrates [150]. Due to the nature of their glycosidic linkages, these compounds cannot be hydrolyzed by the endogenous enzymes of the small intestine, flowing directly into the large intestine. In the colon, the soluble dietary fiber is a potential substrate for the fermentation of the gut microflora, yielding short-chain fatty acids (SCFA), as by-products, that can be absorbed and utilized by the host as an energy source [150]. In a recent study, human subjects with short bowel syndrome were fed with a custom pectin-based supplement (4 g) to evaluate their small intestinal absorption capacity. Evidences showed that 80% of pectin was fermented by the gut microflora, associated with the increase of SCFA detected in the feces, suggesting that soluble fiber intake enhances colonic SCFA production [151]. SCFA have been shown to be involved in numerous physiological processes promoting health [152] and some of them are described below. Insoluble fibers, which are more difficult to ferment, have a bulking action by absorbing water as it moves through the digestive system leading to an easier defecation.

18.4.2.2 *Prevention of cardiovascular diseases*

Hypertension, hypercholesterolemia, obesity [153], and diabetes [153, 154] are the major risk factors associated with CVD. Higher intake of dietary fiber, particularly soluble fiber, reduced the risk of CVD and coronary heart disease (CHD) [155]. Also, an inverse relation between the intake of dietary fiber from fruits and risk of CHD has been reported [156]. Several mechanisms have been postulated to elucidate protective effects of dietary fiber intake against CVD. Viscous fibers, particularly the soluble ones (e.g., pectins), seem to improve blood lipid profiles [157–160], thus avoiding other inflammation pathologies such as atherosclerosis [160]. Dietary fiber intake has been associated with their ability in lowering serum LDL-cholesterol levels, which is the major risk factor for CVD [161]. Brown *et al.* [162] evaluated the cholesterol-lowering effects of various soluble fibers, including pectin and oat bran, and their influence on blood lipid changes in adults subjects. The authors showed a good correlation between the daily intake of 2–10 g different soluble fibers and significant decrease of serum total cholesterol and LDL-cholesterol concentrations by similar amounts. Furthermore, they also observed no changes in high-density lipoprotein (HDL)-cholesterol or triacylglycerol (TAG) concentrations in blood with soluble fiber. Some random studies estimated that the consumption of 12–24 g pectin/day (about three pears) in divided amounts was associated with 13% reduction in LDL-cholesterol values [6, 162]. Soluble fiber intake of about 6 g/day is accompanied by reductions in serum LDL-cholesterol levels of around 5.4% and estimated risk of CHD of about 9% [6]. Viscosity is thought to be an essential feature in the lowering of cholesterol, although solubility and molecular weight of dietary fibers also determine the cholesterol-lowering ability. Fietz *et al.* [163] evaluated the effects of high- and low-methoxy pectins (HMP and LMP, respectively) and cellulose on serum cholesterol levels, as well as TAG in hyperlipidemic rats over 30 days. HMP-fed rats (10 and 15%) significantly decreased the body weight with increased dietary fiber concentrations, suggesting that higher molecular weight fibers are associated with increased viscosity. Soluble and viscous fibers tend to bind to bile acids in the small intestine, making them less likely to enter the body [6]. *In vivo* experiments showed in response to this fact that LDL-cholesterol is removed from blood and converted into bile acids/salts by the liver to replace the bile acids lost in the feces (35–65%) [164, 165]. Therefore, total plasma and LDL-cholesterol concentrations are reduced, but no changes in HDL-cholesterol were observed [166, 167]. The same viscosity seems to improve blood glucose control in type 1 diabetes [168, 169], either by delaying [170] or reducing [171] glucose intestinal absorption, thus lowering postprandial glucose and attenuating insulin responses. The consumption, on an average of 26 g/day, of dietary fiber (equivalent to six sun-dried pears) leads to a 22% lower risk of developing diabetes [172]. On the contrary, except for fiber from cereals, some recent studies have reported no correlation between the consumption of soluble fiber from fruits and vegetables and a reduced risk of type 2 (non-insulin-dependent) diabetes [173, 174]. Dietary fiber has also been shown to regulate energy intake [171], thereby enhancing loss of weight [175] or preserving a healthier body weight. In one study, women (30–50 years of age) were randomized to receive one of three dietary supplements: apples, pears, or oat cookies, three times a day in a total of six meals a day, in order to evaluate the effect of fruit intake on body weight change [175]. After 12 weeks of follow-up, the fruit group lost 1.22 kg, whereas the oat group had an insignificant weight loss of 0.88 kg. Furthermore, a significantly greater decrease of blood glucose was observed among those who had eaten fruits compared with those who had eaten oat cookies, but the glucose to insulin ratio was not statistically different from baseline to follow-up. The mechanisms of how dietary fiber enhances the weight loss are still unclear. Nevertheless, it has been suggested that glucagon-like peptide, a hormone produced as a result of soluble fiber fermentation in the large intestine, may be responsible for the satiety, which refers to the state in which further eating is inhibited and occurs as a consequence of having eaten [148].

Fermented dietary fiber leads to an increase in SCFAs such as acetate, propionate, and butyrate as products by gut microflora. These SCFAs, particularly propionate, have been shown to contribute to a decrease in blood cholesterol level through inhibition of cholesterol synthesis [176, 177]. In addition, butyrate has been associated with anti-inflammatory effects [178]. Dietary fibers also affect blood pressure, as well as C-reactive protein (CRP). Many studies have demonstrated that the prevention of hypertension could be achieved with the consumption of dietary fiber [179–181]. Moreover, it has been observed that the decrease in blood pressure is higher in older humans and in hypertensive populations [179]. CRP is a sensitive marker of inflammation, and thus, CRP has been shown to be a strong predictor of CVD [182, 183]. It is produced by the liver and is present in low amounts ($\leq$1 mg/L) in the blood of healthy subjects. Investigations have shown an inverse association between the intake of dietary fiber and CRP [184, 185]. Moreover, it has also been suggested that the levels of CRP can be reduced with 30 g/day fiber intake from a diet naturally rich in fiber or from a fiber supplement [186].

18.4.2.3 *Cancer effects*

The relationship between dietary fiber intake from fruits and risk in developing cancer is still inconclusive. Although some epidemiological studies have suggested an inverse association between the consumption of dietary fiber from fruits and cancer risk [187–189], others have not found this correlation [190, 191]. For example, Terry *et al.* [188] concluded that individuals who consumed less than 1.5 servings of fruits and vegetables per day had a higher risk for developing colorectal cancer than individuals who consumed more than 2.5 servings. Pectins, the main soluble fiber occurring in dried pears, may play a role in cancer prevention. Meanwhile, Nangia-Makker *et al.* [192] demonstrated the ability of pectin in binding to galectin-3, thus decreasing tumor growth and cancerous cell migration in rats fed modified citrus pectin, suggesting the importance of dietary fibers as cancer-preventive and/or -therapeutic agents. The reduced risk of breast cancer associated with high consumption of dietary fiber has also been shown [193]. The same findings were reported for pancreatic cancer, in which fiber intake from fruits led to a low risk in developing the pathology [194].

However, several studies disagree with these promising results. A pooled analysis of cohort studies of breast cancer found no association between the intake of fruits and vegetables during adulthood and lower breast cancer risk [191]. Similarly, a recent investigation involving 58,279 men found no relationship between dietary fiber and colorectal cancer [195]. Despite these controversial results concerning the potential effects of dietary fibers on cancer, these phytochemicals have been demonstrated to offer potential benefits, such as improving the stool weight, thereby promoting normal laxation [196].

18.5 Conclusions

Pears (*P. communis* L.) are an excellent source of nutrients and phytochemicals for maintaining good health and defending against a large number of diseases. Due to their high fiber content they help against constipation and lower cholesterol. The presence of phenolics also helps ameliorating cardiovascular and neurodegenerative diseases and certain types of cancer.

The traditional sun-drying of S. Bartolomeu pear is still practiced to obtain well-appreciated small sun-dried edible products with reddish brown color and unique elastic properties. The several chemical and sensory changes, responsible for interesting features of dried pear, have been poorly explored. In general, a decrease in the amount of all phytochemicals is observed when the fruit is dried. No reports are available for *in vitro* and/or *in vivo* studies concerning phytochemicals of dried pears. However, we did provide an analogical approach of bioavailability and potential health benefits of dried pears phytochemicals based on the information available in the literature, concerning the observed human health effects of these compounds found in other matrices. Thus, it is possible to infer that some phenolic compounds, as well as dietary fibers that are described to exist in dried pear, should be able to exert human health benefits. In order to support the real implications of dried pear components as health protectors, an extensive investigation in this area is crucial.

References

1. Raffo, M.D., Ponce, N.M., Sozzi, G.O., Vicente, A.R. & Stortz, C.A. (2011) Compositional changes in 'Bartlett' pear (*Pyrus communis L.*) cell wall polysaccharides as affected by sunlight conditions. *Journal of Agricultural and Food Chemistry*, **59**, 12155–12162.
2. Pear Bureau Northwest (2012) *Pear Varieties, USA Pears*. Published on-line at: http://usapears.com/Recipes%20And%20Lifestyle/Now%20Serving/Pears%20and%20Varieties.aspx, last accessed March 26, 2012.
3. Belitz, H.-D., Grosch, W. & Schieberle, P. (2004) *Food Chemistry*, 3rd edn. Springer, Berlin, Germany.
4. Ferreira, D., Barros, A., Coimbra, M.A. & Delgadillo, I. (2001) Use of FT-IR spectroscopy to follow the effect of processing in cell wall polysaccharide extracts of a sun-dried pear. *Carbohydrate Polymers*, **45**, 175–182.
5. Barroca, M.J., Guiné, R.P.F., Pinto, A., Gonçalves, F.M. & Ferreira, D.M.S. (2006) Chemical and microbiological characterization of Portuguese varieties of pears. *Food and Bioproducts Processing*, **84**, 109–113.
6. Anderson, J.W., Baird, P., Davis, R. Jr., Ferreri, S., Knudtson, M., Koraym, A., Waters, V. & Williams, C. (2009) Health benefits of dietary fiber. *Nutrition Reviews*, **67**, 188–205.
7. What's Cooking, Mexico (2012) *Xoconostle and Tomatillo Salsa, What's Cooking, Mexico?* Published on-line at: http://whatscookingmexico.com/2011/03/29/xoconostle-and-tomatillo-salsa/, last accessed March 26, 2012.
8. Guzmán-Maldonado, S., Morales-Montelongo, A., Mondragón-Jacobo, C., Herrera-Hernández, G., Guevara-Lara, F. & Reynoso-camacho, R. (2010) Physicochemical, nutritional, and functional characterization of fruits xoconostle (*Opuntia matudae*) pears from central-México region. *Journal of Food Science*, **75**, C485–C492.
9. Health Canada (2012) *The % Daily Value, Health Canada*. Published on-line at: http://www.hc-sc.gc.ca, last accessed March 29, 2012.
10. Webster, J.M., Oxley, D., Pettolino, F.A. & Bacic, A. (2008) Characterisation of secreted polysaccharides and (glyco)proteins from suspension cultures of *Pyrus communis*. *Phytochemistry*, **69**, 873–881.
11. Ferreira, D., Guyot, S., Marnet, N., Delgadillo, I., Renard, C.M. & Coimbra, M.A. (2002) Composition of phenolic compounds in a Portuguese pear (*Pyrus communis* L. var. S. Bartolomeu) and changes after sun-drying. *Journal of Agricultural and Food Chemistry*, **50**, 4537–4544.
12. Li, X., Gao, W.Y., Huang, L.J., Zhang, J.Y. & Guo, X.H. (2011) Antioxidant and antiinflammation capacities of some pear cultivars. *Journal of Food Science*, **76**, C985–C990.

13. Tiwari, U. & Cummins, E. (2012) Factors influencing levels of phytochemicals in selected fruit and vegetables during pre- and post-harvest food processing operations. *Food Research International* (in press).
14. Coimbra, M.A., Nunes, C., Cunha, P. & Guiné, R. (2011) Amino acid profile and Maillard compounds of sun-dried pears. Relation with the reddish brown colour of the dried fruits. *European Food Research and Technology*, **233**, 637–646.
15. Ferreira, D., Silva, J., Pinto, G., Santos, C., Delgadillo, I. & Coimbra, M.A. (2008) Effect of sun-drying on microstructure and texture of S. Bartolomeu pears (*Pyrus communis* L.). *European Food Research and Technology A*, **226**, 1545–1552.
16. Sun, J., Chu, Y.F., Wu, X. & Liu, R.H. (2002) Antioxidant and antiproliferative activities of common fruits. *Journal of Agricultural and Food Chemistry*, **50**, 7449–7454.
17. Ferreira, D.S.M. (2003) *Estudo das Transformações Bioquímicas e Químicas da Pêra de S. Bartolomeu Durante o Processo de Secagem—Recurso Endógeno da Região de Viseu*, PhD Thesis, University of Aveiro, Aveiro, Portugal (in Portuguese).
18. Voragen, A.G.J., Coenen, G.J., Verhoef, R.P. & Schols, H.A. (2009) Pectin, a versatile polysaccharide present in plant cell walls. *Structural Chemistry*, **20**, 263–275.
19. Caffall, K.H. & Mohnen, D. (2009) The structure, function, and biosynthesis of plant cell wall pectic polysaccharides. *Carbohydrate Research*, **344**, 1879–1900.
20. Dick, A.J. & Labavitch, J.M. (1989) Cell-wall metabolism in ripening fruit. 4. Characterization of the pectic polysaccharides solubilized during softening of bartlett pear fruit. *Plant Physiology*, **89**, 1394–1400.
21. Le Goff, A., Renard, C.M.G.C., Bonnin, E. & Thibault, J.F. (2001) Extraction, purification and chemical characterisation of xylogalacturonans from pea hulls. *Carbohydrate Polymers*, **45**, 325–334.
22. Oechslin, R., Lutz, M.V. & Amado, R. (2003) Pectic substances isolated from apple cellulosic residue: structural characterisation of a new type of rhamnogalacturonan I. *Carbohydrate Polymers*, **51**, 301–310.
23. Zandleven, J., Beldman, G., Bosveld, M., Schols, H.A. & Voragen, A.G.J. (2006) Enzymatic degradation studies of xylogalacturonans from apple and potato, using xylogalacturonan hydrolase. *Carbohydrate Polymers*, **65**, 495–503.
24. Nunes, C., Silva, L., Fernandes, A.P., Guiné, R.P.F., Domingues, M.R.M. & Coimbra, M.A. (2012) Occurrence of cellobiose residues directly linked to galacturonic acid in pectic polysaccharides. *Carbohydrate Polymers*, **87**, 620–626.
25. Strasser, G.R. & Amado, R. (2001) Pectic substances from red beet (*Beta vulgaris conditiva*). Part I. Structural analysis of rhamnogalacturonan I using enzymic degradation and methylation analysis. *Carbohydrate Polymers*, **44**, 63–70.
26. Ridley, B.L., O'Neill, M.A. & Mohnen, D.A. (2001) Pectins: structure, biosynthesis, and oligogalacturonide-related signaling. *Phytochemistry*, **57**, 929–967.
27. Martin Cabrejas, M.A., Waldron, K.W., Selvendran, R.R., Parker, M.L. & Moates, G.K. (1994) Ripening-related changes in the cell-walls of Spanish pear (*Pyrus communis*). *Physiologia Plantarum*, **91**, 671–679.
28. Pilipenko, L., Kalinkov, A. & Spektor, A. (1999) Amino acid composition of fruit in the manufacture of sedimentation-stabilized dispersed products. *Chemistry of Natural Compounds*, **35**, 208–211.
29. Chen, J.L., Wang, Z.F., Wu, J.H., Wang, Q. & Hu, X.S. (2007) Chemical compositional characterization of eight pear cultivars grown in China. *Food Chemistry*, **104**, 268–275.
30. Schieber, A., Keller, P. & Carle, R. (2001) Determination of phenolic acids and flavonoids of apple and pear by high-performance liquid chromatography. *Journal of Chromatography A*, **910**, 265–273.
31. Escarpa, A. & Gonzalez, M.C. (1999) Fast separation of (poly)phenolic compounds from apples and pears by high-performance liquid chromatography with diode-array detection. *Journal of Chromatography A*, **830**, 301–309.
32. Challice, J.S. & Westwood, M.N. (1972) Phenolic compounds of the genus *Pyrus*. *Phytochemistry*, **11**, 37–44.
33. Spanos, G.A. & Wrolstad, R.E. (1990) Influence of variety, maturity, processing, and storage on the phenolic composition of pear juice. *Journal of Agricultural and Food Chemistry*, **38**, 817–824.
34. Spanos, G.A. & Wrolstad, R.E. (1992) Phenolics of apple, pear and white grape juices and their changes with processing and storage—a review. *Journal of Agricultural and Food Chemistry*, **40**, 1478–1487.

35. Williams, A.H. (1957) The simpler phenolic substances of plants. *Journal of the Science of Food and Agriculture*, **8**, 385–389.
36. Mosel, H.D. & Herrmann, K. (1974) Changes in catechins and hydroxycinnamic acid-derivatives during development of apples and pears. *Journal of the Science of Food and Agriculture*, **25**, 251–256.
37. Blankenship, S.M. & Richardson, D.G. (1985) Changes in phenolic-acids and internal ethylene during long-term cold-storage of pears. *Journal of the American Society for Horticultural Science*, **110**, 336–339.
38. Amiot, M.J., Aubert, S. & Nicolas, J. (1992) Phenolic composition and browning susceptibility of various apple and pear cultivars at maturity. In: *International Symposium on the Physiological Basis of Postharvest Technologies* (ed. M.E. Saltveit). International Society for Horticultural Science, Wageningen, The Netherlands, pp. 67–69.
39. Amiot, M.J., Tacchini, M., Aubert, S.Y. & Oleszek, W. (1995) Influence of cultivar, maturity stage, and storage-conditions on phenolic composition and enzymatic browning of pear fruits. *Journal of Agricultural and Food Chemistry*, **43**, 1132–1137.
40. Tanrioven, D. & Eksi, A. (2005) Phenolic compounds in pear juice from different cultivars. *Food Chemistry*, **93**, 89–93.
41. Oleszek, W., Amiot, M.J. & Aubert, S.Y. (1994) Identification of some phenolics in pear fruit. *Journal of Agricultural and Food Chemistry*, **42**, 1261–1265.
42. Escarpa, A. & González, M.C. (2001) Approach to the content of total extractable phenolic compounds from different food samples by comparison of chromatographic and spectrophotometric methods. *Analytica Chimica Acta*, **427**, 119–127.
43. de Pascual-Teresa, S., Santos-Buelga, C. & Rivas-Gonzalo, J.C. (2000) Quantitative analysis of flavan-3-ols in Spanish foodstuffs and beverages. *Journal of Agricultural and Food Chemistry*, **48**, 5331–5337.
44. Lea, A.G. (1978) The phenolics of ciders: oligomeric and polymeric procyanidins. *Journal of the Science of Food and Agriculture*, **29**, 471–477.
45. Sarni-Manchado, P., Cheynier, V. & Moutounet, M. (1999) Interactions of grape seed tannins with salivary proteins. *Journal of Agricultural and Food Chemistry*, **47**, 42–47.
46. Mcmanus, J.P., Davis, K.G., Beart, J.E., Gaffney, S.H., Lilley, T.H. & Haslam, E. (1985) Polyphenol interactions. 1. Introduction—some observations on the reversible complexation of polyphenols with proteins and polysaccharides. *Journal of the Chemical Society, Perkin Transactions*, **2**, 1429–1438.
47. Luck, G., Liao, H., Murray, N.J., Grimmer, H.R., Warminski, E.E., Williamson, M.P., Lilley, T.H. & Haslam, E. (1994) Polyphenols, astringency and proline-rich proteins. *Phytochemistry*, **37**, 357–371.
48. Duggan, M.B. (1969) Identity and occurrence of certain flavonol glycosides in four varieties of pears. *Journal of Agricultural and Food Chemistry*, **17**, 1098–1101.
49. Nortje, B.K. (1965) The flavonol glycosides in the fruit of *Pyrus communis* L. cultivar Bon Chretien. *Biochemical Journal*, **97**, 209–213.
50. McCance & Widdowson's (1992) *Fruits and Nuts. The Composition of Foods*. Royal Society of Chemistry, Cambridge, UK.
51. Shirsat, A.H., Bell, A., Spence, J. & Harris, J.N. (1996) The Brassica napus extA extensin gene is expressed in regions of the plant subject to tensile stresses. *Planta*, **199**, 618–624.
52. Huxham, I.M., Jarvis, M.C., Shakespeare, L., Dover, C.J., Johnson, D., Knox, J.P. & Seymour, G.B. (1999) Electron-energy-loss spectroscopic imaging of calcium and nitrogen in the cell walls of apple fruits. *Planta*, **208**, 438–443.
53. Josè, M. & Puigdomènech, P. (1993) Structure and expression of genes coding for structural proteins of the plant cell wall. *New Phytologist*, **125**, 259–282.
54. Lea, A.G.H. (1984) Tannin and colour in English apple cider. *Flüssiges Obst*, **8**, 356–361.
55. Vamosvigyazo, L. & Nadudvarimarkus, V. (1982) Enzymatic browning, polyphenol content, polyphenol oxidase and peroxidase-activities in pear cultivars. *Acta Alimentaria Hungarica*, **11**, 157–168.
56. Nicolas, J.J., Richard-Forget, F.C., Goupy, P.M., Amiot, M.J. & Aubert, S.Y. (1994) Enzymatic browning reactions in apple and apple products. *Critical Reviews in Food Science and Nutrition*, **34**, 109–157.
57. Gauillard, F. & Richard-Forget, F. (1997) Polyphenoloxidases from Williams pear (*Pyrus communis* L, cv Williams): activation, purification and some properties. *Journal of the Science of Food and Agriculture*, **74**, 49–56.
58. Richard-Forget, F.C. & Gauillard, F.A. (1997) Oxidation of chlorogenic acid, catechins, and 4-methylcatechol in model solutions by combinations of pear (*Pyrus communis* Cv Williams) polyphenol

oxidase and peroxidase: a possible involvement of peroxidase in enzymatic browning. *Journal of Agricultural and Food Chemistry*, **45**, 2472–2476.
59. Hodge, J.E. (1953) Dehydrated foods—chemistry of browning reactions in model systems. *Journal of Agricultural and Food Chemistry*, **1**, 928–943.
60. Moreno, J., Peinado, J. & Peinado, R.A. (2007) Antioxidant activity of musts from Pedro Ximenez grapes subjected to off-vine drying process. *Food Chemistry*, **104**, 224–228.
61. Arnoldi, A., Arnoldi, C., Baldi, O. & Griffini, A. (1988) Flavor components in the Maillard reaction of different amino acids with fructose in cocoa butter water—qualitative and quantitative analysis of pyrazines. *Journal of Agricultural and Food Chemistry*, **36**, 988–992.
62. Umano, K., Hagi, Y., Nakahara, K., Shyoji, A. & Shibamoto, T. (1995) Volatile chemicals formed in the headspace of a heated D-glucose/L-cysteine Maillard model system. *Journal of Agricultural and Food Chemistry*, **43**, 2212–2218.
63. Cornwell, C.J. & Wrolstad, R.E. (1981) Causes of browning in pear juice concentrate during storage. *Journal of Food Science*, **46**, 515–518.
64. Toribio, J.L. & Lozano, J.E. (1984) Nonenzymatic browning in apple juice concentrate during storage. *Journal of Food Science*, **49**, 889–892.
65. Rufian-Henares, J.A., Garcia-Villanova, B. & Guerra-Hernandez, E. (2008) Occurrence of furosine and hydroxymethylfurfural as markers of thermal damage in dehydrated vegetables. *European Food Research and Technology*, **228**, 249–256.
66. Hidalgo, F.J., Alaiz, M. & Zamora, R. (1999) Effect of pH and temperature on comparative nonenzymatic browning of proteins produced by oxidized lipids and carbohydrates. *Journal of Agricultural and Food Chemistry*, **47**, 742–747.
67. Silvan, J.M., van de Lagemaat, J., Olano, A. & Del Castillo, M.D. (2006) Analysis and biological properties of amino acid derivates formed by Maillard reaction in foods. *Journal of Pharmaceutical and Biomedical Analysis*, **41**, 1543–1551.
68. Ledl, F. & Schleicher, E. (1990) New aspects of the Maillard reaction in foods and in the human body. *Angewandte Chemie International Edition*, **29**, 565–594.
69. John, E.H. (1955) The amadori rearrangement. In: *Advances in Carbohydrate Chemistry* (ed. L.W. Melville). Academic Press, New York, NY, pp. 169–205.
70. Erbersdobler, H.F. & Somoza, V. (2007) Forty years of furosine—forty years of using Maillard reaction products as indicators of the nutritional quality of foods. *Molecular Nutrition & Food Research*, **51**, 423–430.
71. Rada-Mendoza, M., Olano, A. & Villamiel, M. (2002) Furosine as indicator of Maillard reaction in jams and fruit-based infant foods. *Journal of Agricultural and Food Chemistry*, **50**, 4141–4145.
72. Vallejo-Cordoba, B., Mazorra-Manzano, M.A. & Gonzalez-Cordova, A.F. (2004) New capillary electrophoresis method for the determination of furosine in dairy products. *Journal of Agricultural and Food Chemistry*, **52**, 5787–5790.
73. VanRenterghem, R. & DeBlock, J. (1996) Furosine in consumption milk and milk powders. *International Dairy Journal*, **6**, 371–382.
74. Charissou, A., Ait-Ameur, L. & Birlouez-Aragon, I. (2007) Evaluation of a gas chromatography/mass spectrometry method for the quantification of carboxymethyllysine in food samples. *Journal of Chromatography A*, **1140**, 189–194.
75. Hartkopf, J. & Erbersdobler, H.F. (1995) Experiments with sausage meat on the formation of *N* epsilon-carboxymethyllysine. *Zeitschrift fur Lebensmittel-Untersuchung und -Forschung*, **201**, 27–29.
76. Hartkopf, J. & Erbersdobler, H.F. (1994) Model experiments of the formation of *N* epsilon-carboxymethyllysine in food products. *Zeitschrift fur Lebensmittel-Untersuchung und -Forschung*, **198**, 15–19.
77. Srey, C., Hull, G.L., Connolly, L., Elliott, C.T., del Castillo, M.D. & Ames, J.M. (2010) Effect of inhibitor compounds on Nepsilon-(carboxymethyl)lysine (CML) and Nepsilon-(carboxyethyl)lysine (CEL) formation in model foods. *Journal of Agricultural and Food Chemistry*, **58**, 12036–12041.
78. Ferrer, E., Alegria, A., Farre, R., Abellan, P. & Romero, F. (1999) Review: indicators of damage of protein quality and nutritional value of milk. *Food Science and Technology International*, **5**, 447–461.
79. Fogliano, V. & Morales, F.J. (2011) Estimation of dietary intake of melanoidins from coffee and bread. *Food & Function*, **2**, 117–123.

80. Nunes, F.M. & Coimbra, M.A. (2007) Melanoidins from coffee infusions. Fractionation, chemical characterization, and effect of the degree of roast. *Journal of Agricultural and Food Chemistry*, **55**, 3967–3977.
81. Obretenov, T.D., Ivanova, S.D., Kuntcheva, M.J. & Somov, G.T. (1993) Melanoidin formation in cooked meat products. *Journal of Agricultural and Food Chemistry*, **41**, 653–656.
82. Adams, A., Borrelli, R.C., Fogliano, V. & De Kimpe, N. (2005) Thermal degradation studies of food melanoidins. *Journal of Agricultural and Food Chemistry*, **53**, 4136–4142.
83. Hertog, M.G., Bueno-de-Mesquita, H.B., Fehily, A.M., Sweetnam, P.M., Elwood, P.C. & Kromhout, D. (1996) Fruit and vegetable consumption and cancer mortality in the Caerphilly Study. *Cancer Epidemiology, Biomarkers & Prevention: a Publication of the American Association for Cancer Research, Cosponsored by the American Society of Preventive Oncology*, **5**, 673–677.
84. Block, G., Patterson, B. & Subar, A. (1992) Fruit, vegetables, and cancer prevention: a review of the epidemiological evidence. *Nutrition and Cancer*, **18**, 1–29.
85. Ko, S.H., Choi, S.W., Ye, S.K., Cho, B.L., Kim, H.S. & Chung, M.H. (2005) Comparison of the antioxidant activities of nine different fruits in human plasma. *Journal of Medicinal Food*, **8**, 41–46.
86. Hamauzu, Y., Forest, F., Hiramatsu, K. & Sugimoto, M. (2007) Effect of pear (*Pyrus communis* L.) procyanidins on gastric lesions induced by HCl/ethanol in rats. *Food Chemistry*, **100**, 255–263.
87. Rice-Evans, C., Miller, N. & Paganga, G. (1997) Antioxidant properties of phenolic compounds. *Trends in Plant Science*, **2**, 152–159.
88. Rimm, E.B., Ascherio, A., Giovannucci, E., Spiegelman, D., Stampfer, M.J. & Willett, W.C. (1996) Vegetable, fruit, and cereal fiber intake and risk of coronary heart disease among men. *Journal of the American Medical Association*, **275**, 447–451.
89. Anderson, J. & Bridges, S. (1988) Dietary fiber content of selected foods. *American Journal of Clinical Nutrition*, **47**, 440–447.
90. Watkins, C.B. & Liu, R.H. (2011) Pome fruit. In: *Health-Promoting Properties of Fruits and Vegetables* (ed. L.A. Terry). CABI Publishing, Oxford, UK, pp. 196–217.
91. Scalbert, A. & Williamson, G. (2000) Dietary intake and bioavailability of polyphenols. *Journal of Nutrition*, **130**, 2073S–2085S.
92. Hammerstone, J.F., Lazarus, S.A. & Schmitz, H.H. (2000) Procyanidin content and variation in some commonly consumed foods. *Journal of Nutrition*, **130**, 2086S–2092S.
93. Palafox-Carlos, H., Ayala-Zavala, J.F. & Gonzalez-Aguilar, G.A. (2011) The role of dietary fiber in the bioaccessibility and bioavailability of fruit and vegetable antioxidants. *Journal of Food Science*, **76**, R6–R15.
94. Goni, I. & Serrano, J. (2005) The intake of dietary fiber from grape seeds modifies the antioxidant status in rat cecum. *Journal of the Science of Food and Agriculture*, **85**, 1877–1881.
95. Spencer, J.P., Chowrimootoo, G., Choudhury, R., Debnam, E.S., Srai, S.K. & Rice-Evans, C. (1999) The small intestine can both absorb and glucuronidate luminal flavonoids. *FEBS Letters*, **458**, 224–230.
96. Olthof, M.R., Hollman, P.C. & Katan, M.B. (2001) Chlorogenic acid and caffeic acid are absorbed in humans. *Journal of Nutrition*, **131**, 66–71.
97. Gonthier, M.P., Verny, M.A., Besson, C., Remesy, C. & Scalbert, A. (2003) Chlorogenic acid bioavailability largely depends on its metabolism by the gut microflora in rats. *Journal of Nutrition*, **133**, 1853–1859.
98. Hollman, P.C., de Vries, J.H., van Leeuwen, S.D., Mengelers, M.J. & Katan, M.B. (1995) Absorption of dietary quercetin glycosides and quercetin in healthy ileostomy volunteers. *American Journal of Clinical Nutrition*, **62**, 1276–1282.
99. Hollman, P.C., van Trijp, J.M., Buysman, M.N., van der Gaag, M.S., Mengelers, M.J., de Vries, J.H. & Katan, M.B. (1997) Relative bioavailability of the antioxidant flavonoid quercetin from various foods in man. *FEBS Letters*, **418**, 152–156.
100. Hawkswor, G., Drasar, B.S. & Hill, M.J. (1971) Intestinal bacteria and hydrolysis of glycosidic bonds. *Journal of Medical Microbiology*, **4**, 45–459.
101. Walle, T., Otake, Y., Walle, U.K. & Wilson, F.A. (2000) Quercetin glucosides are completely hydrolyzed in ileostomy patients before absorption. *Journal of Nutrition*, **130**, 2658–2661.
102. Spencer, J.P. (2003) Metabolism of tea flavonoids in the gastrointestinal tract. *Journal of Nutrition*, **133**, 3255S–3261S.

103. Tsang, C., Auger, C., Mullen, W., Bornet, A., Rouanet, J.M., Crozier, A. & Teissedre, P.L. (2005) The absorption, metabolism and excretion of flavan-3-ols and procyanidins following the ingestion of a grape seed extract by rats. *British Journal of Nutrition*, **94**, 170–181.
104. Spencer, J.P.E., Schroeter, H., Rechner, A.R. & Rice-Evans, C. (2001) Bioavailability of flavan-3-ols and procyanidins: gastrointestinal tract influences and their relevance to bioactive forms *in vivo*. *Antioxidants & Redox Signaling*, **3**, 1023–1039.
105. Rios, L.Y., Bennett, R.N., Lazarus, S.A., Remesy, C., Scalbert, A. & Williamson, G. (2002) Cocoa procyanidins are stable during gastric transit in humans. *American Journal of Clinical Nutrition*, **76**, 1106–1110.
106. Donovan, J.L., Manach, C., Rios, L., Morand, C., Scalbert, A. & Remesy, C. (2002) Procyanidins are not bioavailable in rats fed a single meal containing a grapeseed extract or the procyanidin dimer B3. *British Journal of Nutrition*, **87**, 299–306.
107. Serra, A., Macia, A., Romero, M.P., Valls, J., Blade, C., Arola, L. & Motilva, M.J. (2010) Bioavailability of procyanidin dimers and trimers and matrix food effects in *in vitro* and *in vivo* models. *British Journal of Nutrition*, **103**, 944–952.
108. Parada, J. & Aguilera, J.M. (2007) Food microstructure affects the bioavailability of several nutrients. *Journal of Food Science*, **72**, R21–R32.
109. Perez-Jimenez, J., Serrano, J., Tabernero, M., Arranz, S., Diaz-Rubio, M.E., Garcia-Diz, L., Goni, I. & Saura-Calixto, F. (2009) Bioavailability of phenolic antioxidants associated with dietary fiber: plasma antioxidant capacity after acute and long-term intake in humans. *Plant Foods for Human Nutrition*, **64**, 102–107.
110. Carbonaro, M., Grant, G. & Pusztai, A. (2001) Evaluation of polyphenol bioavailability in rat small intestine. *European Journal of Nutrition*, **40**, 84–90.
111. Kosinska, A., Karamac, M. & Amarowicz, R. (2009) Study on interaction between broad bean phenolic compounds and proteins using fluorescence method. *Czech Journal of Food Science*, **27**, S119–S119.
112. Leontowicz, M., Gorinstein, S., Leontowicz, H., Krzeminski, R., Lojek, A., Katrich, E., Ciz, M., Martin-Belloso, O., Soliva-Fortuny, R., Haruenkit, R. & Trakhtenberg, S. (2003) Apple and pear peel and pulp and their influence on plasma lipids and antioxidant potentials in rats fed cholesterol-containing diets. *Journal of Agricultural and Food Chemistry*, **51**, 5780–5785.
113. Adisakwattana, S., Moonsan, P. & Yibchok-Anun, S. (2008) Insulin-releasing properties of a series of cinnamic acid derivatives *in vitro* and *in vivo*. *Journal of Agricultural and Food Chemistry*, **56**, 7838–7844.
114. Graziani, G., D'Argenio, G., Tuccillo, C., Loguercio, C., Ritieni, A., Morisco, F., Blanco, C. D., Fogliano, V. & Romano, M. (2005) Apple polyphenol extracts prevent damage to human gastric epithelial cells *in vitro* and to rat gastric mucosa *in vivo*. *Gut*, **54**, 193–200.
115. Rice-Evans, C.A., Miller, N.J. & Paganga, G. (1996) Structure-antioxidant activity relationships of flavonoids and phenolic acids. *Free Radical Biology and Medicine*, **20**, 933–956.
116. Rice-Evans, C.A., Miller, N.J., Bolwell, P.G., Bramley, P.M. & Pridham, J.B. (1995) The relative antioxidant activities of plant-derived polyphenolic flavonoids. *Free Radical Research*, **22**, 375–383.
117. Murcia, M.A., Jimenez, A.M. & Martinez-Tome, M. (2001) Evaluation of the antioxidant properties of Mediterranean and tropical fruits compared with common food additives. *Journal of Food Protection*, **64**, 2037–2046.
118. Auger, C., Al-Awwadi, N., Bornet, A., Rouanet, J.M., Gasc, F., Cros, G. & Teissedre, P.L. (2004) Catechins and procyanidins in Mediterranean diets. *Food Research International*, **37**, 233–245.
119. Suda, I., Oki, T., Nishiba, Y., Masuda, M., Kobayashi, M., Nagai, S., Hiyane, R. & Miyashige, T. (2005) Polyphenol contents and radical-scavenging activity of extracts from fruits and vegetables in cultivated in Okinawa, Japan. *Journal of Japan Society of Nutrition and Food Sciences*, **52**, 462–471.
120. Maxwell, S.R.J. (1995) Prospects for the use of antioxidant therapies. *Drugs*, **49**, 345–361.
121. Conforti, F., Sosa, S., Marrelli, M., Menichini, F., Statti, G.A., Uzunov, D., Tubaro, A. & Loggia, R.D. (2008) *In vivo* anti-inflammatory and *in vitro* antioxidant activities of Mediterranean dietary plants. *Journal of Ethnopharmacology*, **116**, 144–151.
122. Rahman, I., Biswas, S.K. & Kirkham, P.A. (2006) Regulation of inflammation and redox signaling by dietary polyphenols. *Biochemical Pharmacology*, **72**, 1439–1452.
123. Huang, L.-J., Gao, W.-Y., Li, X., Zhao, W.-S., Huang, L.-Q. & Liu, C.-X. (2010) Evaluation of the *in vivo* anti-inflammatory effects of extracts from *Pyrus bretschneideri* Rehd. *Journal of Agricultural and Food Chemistry*, **58**, 8983–8987.

124. Nathan, C. (1992) Nitric oxide as a secretory product of mammalian cells. *FASEB Journal*, **6**, 3051–3064.
125. WHO (2011) *Cardiovascular Disease*. Published on-line at: http://www.who.int/cardiovascular_diseases/en/, last accessed March 20, 2012.
126. Hubbard, G.P., Wolffram, S., Lovegrove, J.A. & Gibbins, J.M. (2003) The role of polyphenolic compounds in the diet as inhibitors of platelet function. *Proceedings of the Nutrition Society*, **62**, 469–478.
127. Landolfi, R., Mower, R.L. & Steiner, M. (1984) Modification of platelet function and arachidonic acid metabolism by bioflavonoids. Structure-activity relations. *Biochemical Pharmacology*, **33**, 1525–1530.
128. Chang, W.-C. & Hsu, F.-L. (1989) Inhibition of platelet aggregation and arachidonate metabolism in platelets by procyanidins. *Prostaglandins, Leukotrienes and Essential Fatty Acids*, **38**, 181–188.
129. Keevil, J.G., Osman, H.E., Reed, J.D. & Folts, J.D. (2000) Grape juice, but not orange juice or grapefruit juice, inhibits human platelet aggregation. *Journal of Nutrition*, **130**, 53–56.
130. Tzeng, S.H., Ko, W.C., Ko, F.N. & Teng, C.M. (1991) Inhibition of platelet aggregation by some flavonoids. *Thrombosis Research*, **64**, 91–100.
131. Heiss, C., Jahn, S., Taylor, M., Real, W.M., Angeli, F.S., Wong, M.L., Amabile, N., Prasad, M., Rassaf, T., Ottaviani, J.I., Mihardja, S., Keen, C.L., Springer, M.L., Boyle, A., Grossman, W., Glantz, S.A., Schroeter, H. & Yeghiazarians, Y. (2010) Improvement of endothelial function with dietary flavanols is associated with mobilization of circulating angiogenic cells in patients with coronary artery disease. *Journal of the American College of Cardiology*, **56**, 218–224.
132. Bourne, L.C. & Rice-Evans, C.A. (1998) Urinary detection of hydroxycinnamates and flavonoids in humans after high dietary intake of fruit. *Free Radical Research*, **28**, 429–438.
133. Ottaviani, J.I., Kwik-Uribe, C., Keen, C.L. & Schroeter, H. (2012) Intake of dietary procyanidins does not contribute to the pool of circulating flavanols in humans. *American Journal of Clinical Nutrition*, **95**, 851–858.
134. Cook, N.C. & Samman, S. (1996) Flavonoids—chemistry, metabolism, cardioprotective effects, and dietary sources. *Journal of Nutritional Biochemistry*, **7**, 66–76.
135. Nichenametla, S.N., Taruscio, T.G., Barney, D.L. & Exon, J.H. (2006) A review of the effects and mechanisms of polyphenolics in cancer. *Critical Reviews in Food Science and Nutrition*, **46**, 161–183.
136. Hagiwara, A., Hirose, M., Takahashi, S., Ogawa, K., Shirai, T. & Ito, N. (1991) Forestomach and kidney carcinogenicity of caffeic acid in F344 rats and C57BL/6N x C3H/HeN F1 mice. *Cancer Research*, **51**, 5655–5660.
137. Hirose, M., Mutai, M., Takahashi, S., Yamada, M., Fukushima, S. & Ito, N. (1991) Effects of phenolic antioxidants in low dose combination on forestomach carcinogenesis in rats pretreated with N-methyl-N′-nitro-N-nitrosoguanidine. *Cancer Research*, **51**, 824–827.
138. Hirose, M., Kawabe, M., Shibata, M., Takahashi, S., Okazaki, S. & Ito, N. (1992) Influence of caffeic acid and other o-dihydroxybenzene derivatives on N-methyl-N′-nitro-N-nitrosoguanidine-initiated rat forestomach carcinogenesis. *Carcinogenesis*, **13**, 1825–1828.
139. Tanaka, T., Kojima, T., Kawamori, T., Wang, A., Suzui, M., Okamoto, K. & Mori, H. (1993) Inhibition of 4-nitroquinoline-1-oxide-induced rat tongue carcinogenesis by the naturally occurring plant phenolics caffeic, ellagic, chlorogenic and ferulic acids. *Carcinogenesis*, **14**, 1321–1325.
140. Matsunaga, K., Katayama, M., Sakata, K., Kuno, T., Yoshida, K., Yamada, Y., Hirose, Y., Yoshimi, N. & Mori, H. (2002) Inhibitory effects of chlorogenic acid on azoxymethane-induced colon carcinogenesis in male F344 rats. *Asian Pacific Journal of Cancer Prevention*, **3**, 163–166.
141. Exon, J.H., Magnuson, B.A., South, E.H. & Hendrix, K. (1998) Effect of dietary chlorogenic acid on multiple immune functions and formation of aberrant crypt foci in rats. *Journal of Toxicology and Environmental Health Part A*, **53**, 375–384.
142. Soh, Y., Kim, J.A., Sohn, N.W., Lee, K.R. & Kim, S.Y. (2003) Protective effects of quinic acid derivatives on tetrahydropapaveroline-induced cell death in C6 glioma cells. *Biological & Pharmaceutical Bulletin*, **26**, 803–807.
143. Lim, Y.J., Lee, E.H., Kang, T.H., Ha, S.K., Oh, M.S., Kim, S.M., Yoon, T.J., Kang, C., Park, J.H. & Kim, S.Y. (2009) Inhibitory effects of arbutin on melanin biosynthesis of alpha-melanocyte stimulating hormone-induced hyperpigmentation in cultured brownish guinea pig skin tissues. *Archives of Pharmacal Research*, **32**, 367–373.
144. Chakraborty, A.K., Funasaka, Y., Komoto, M. & Ichihashi, M. (1998) Effect of arbutin on melanogenic proteins in human melanocytes. *Pigment Cell Research*, **11**, 206–212.

145. Akiu, S., Suzuki, Y., Asahara, T., Fujinuma, Y. & Fukuda, M. (1991) Inhibitory effect of arbutin on melanogenesis–biochemical study using cultured B16 melanoma cells. *Nihon Hifuka Gakkai Zasshi*, **101**, 609–613.
146. Li, H., Jeong, Y.M., Kim, S.Y., Kim, M.K. & Kim, D.S. (2011) Arbutin inhibits TCCSUP human bladder cancer cell proliferation via up-regulation of p21. *Pharmazie*, **66**, 306–309.
147. Lukas, B., Schmiderer, C., Mitteregger, U. & Novak, J. (2010) Arbutin in marjoram and oregano. *Food Chemistry*, **121**, 185–190.
148. Slavin, J.L. (2008) Position of the American Dietetic Association: health implications of dietary fiber. *Journal of the American Dietetic Association*, **108**, 1716–1731.
149. Tungland, B.C. & Meyer, D. (2002) Nondigestible oligo- and polysaccharides (dietary fiber): their physiology and role in human health and food. *Comprehensive Reviews in Food Science and Food Safety*, **1**, 90–109.
150. Englyst, K.N. & Englyst, H.N. (2005) Carbohydrate bioavailability. *British Journal of Nutrition*, **94**, 1–11.
151. Atia, A., Girard-Pipau, F., Hebuterne, X., Spies, W.G., Guardiola, A., Ahn, C.W., Fryer, J., Xue, F., Rammohan, M., Sumague, M., Englyst, K. & Buchman, A.L. (2011) Macronutrient absorption characteristics in humans with short bowel syndrome and jejunocolonic anastomosis: starch is the most important carbohydrate substrate, although pectin supplementation may modestly enhance short chain fatty acid production and fluid absorption. *Journal of Parenteral and Enteral Nutrition*, **35**, 229–240.
152. Wong, J.M., de Souza, R., Kendall, C.W., Emam, A. & Jenkins, D.J. (2006) Colonic health: fermentation and short chain fatty acids. *Journal of Clinical Gastroenterology*, **40**, 235–243.
153. Ogden, C.L., Carroll, M.D., Curtin, L.R., McDowell, M.A., Tabak, C.J. & Flegal, K.M. (2006) Prevalence of overweight and obesity in the United States, 1999-2004. *Journal of the American Medical Association*, **295**, 1549–1555.
154. Mokdad, A.H., Ford, E.S., Bowman, B.A., Dietz, W.H., Vinicor, F., Bales, V.S. & Marks, J.S. (2003) Prevalence of obesity, diabetes, and obesity-related health risk factors, 2001. *Journal of the American Medical Association*, **289**, 76–79.
155. Bazzano, L.A., He, J., Ogden, L.G., Loria, C.M. & Whelton, P.K. (2003) Dietary fiber intake and reduced risk of coronary heart disease in US men and women—the National Health and Nutrition Examination Survey I Epidemiologic Follow-up Study. *Archives of Internal Medicine*, **163**, 1897–1904.
156. Pereira, M.A., O'Reilly, E., Augustsson, K., Fraser, G.E., Goldbourt, U., Heitmann, B.L., Hallmans, G., Knekt, P., Liu, S., Pietinen, P., Spiegelman, D., Stevens, J., Virtamo, J., Willett, W.C. & Ascherio, A. (2004) Dietary fiber and risk of coronary heart disease: a pooled analysis of cohort studies. *Archives of Internal Medicine*, **164**, 370–376.
157. Babio, N., Balanza, R., Basulto, J., Bullo, M. & Salas-Salvado, J. (2010) Dietary fibre: influence on body weight, glycemic control and plasma cholesterol profile. *Nutrición Hospitalaria*, **25**, 327–340.
158. Jenkins, D.J.A., Kendall, C.W.C., Vuksan, V., Vidgen, E., Parker, T., Faulkner, D., Mehling, C.C., Garsetti, M., Testolin, G., Cunnane, S.C., Ryan, M.A. & Corey, P.N. (2002) Soluble fiber intake at a dose approved by the US Food and Drug Administration for a claim of health benefits: serum lipid risk factors for cardiovascular disease assessed in a randomized controlled crossover trial. *American Journal of Clinical Nutrition*, **75**, 834–839.
159. Jenkins, D.J.A., Newton, C., Leeds, A.R. & Cummings, J.H. (1975) Effect of pectin, guar gum, and wheat fiber on serum-cholesterol. *Lancet*, **1**, 1116–1117.
160. Fisher, H., Griminger, P. & Siller, W.G. (1967) Effect of pectin on atherosclerosis in the cholesterol-fed rabbit. *Journal of Atherosclerosis Research*, **7**, 381–386.
161. Pearson, T.A., Blair, S.N., Daniels, S.R., Eckel, R.H., Fair, J.M., Fortmann, S.P., Franklin, B.A., Goldstein, L.B., Greenland, P., Grundy, S.M., Hong, Y.L., Miller, N.H., Lauer, R.M., Ockene, I.S., Sacco, R.L., Sallis, J.F., Smith, S.C., Stone, N.J. & Taubert, K.A. (2002) AHA guidelines for primary prevention of cardiovascular disease and stroke: 2002 update—consensus panel guide to comprehensive risk reduction for adult patients without coronary or other atherosclerotic vascular diseases. *Circulation*, **106**, 388–391.
162. Brown, L., Rosner, B., Willett, W.W. & Sacks, F.M. (1999) Cholesterol-lowering effects of dietary fiber: a meta-analysis. *American Journal of Clinical Nutrition*, **69**, 30–42.
163. Fietz, V.R. & Salgado, J.M. (1999) Efeito da pectina e da celulose nos níveis séricos de colesterol e triglicerídeos em ratos hiperlipidêmicos. *Ciência e Tecnologia de Alimentos*, **19**, 318–321.

164. Everson, G.T., Daggy, B.P., Mckinley, C. & Story, J.A. (1992) Effects of psyllium hydrophilic mucilloid on LDL-cholesterol and bile-acid synthesis in hypercholesterolemic men. *Journal of Lipid Research*, **33**, 1183–1192.
165. Lia, A., Hallmans, G., Sandberg, A.S., Sundberg, B., Aman, P. & Andersson, H. (1995) Oat beta-glucan increases bile-acid excretion and a fiber-rich barley fraction increases cholesterol excretion in ileostomy subjects. *American Journal of Clinical Nutrition*, **62**, 1245–1251.
166. Arjmandi, B.H., Craig, J., Nathani, S. & Reeves, R.D. (1992) Soluble dietary fiber and cholesterol influence *in vivo* hepatic and intestinal cholesterol biosynthesis in rats. *Journal of Nutrition*, **122**, 1559–1565.
167. Moundras, C., Behr, S.R., Remesy, C. & Demigne, C. (1997) Fecal losses of sterols and bile acids induced by feeding rats guar gum are due to greater pool size and liver bile acid secretion. *Journal of Nutrition*, **127**, 1068–1076.
168. Anderson, J.W., Randles, K.M., Kendall, C.W. & Jenkins, D.J. (2004) Carbohydrate and fiber recommendations for individuals with diabetes: a quantitative assessment and meta-analysis of the evidence. *Journal of the American College of Nutrition*, **23**, 5–17.
169. Giacco, R., Parillo, M., Rivellese, A.A., Lasorella, G., Giacco, A., D'Episcopo, L. & Riccardi, G. (2000) Long-term dietary treatment with increased amounts of fiber-rich low-glycemic index natural foods improves blood glucose control and reduces the number of hypoglycemic events in type 1 diabetic patients. *Diabetes Care*, **23**, 1461–1466.
170. Weickert, M.O. & Pfeiffer, A.F.H. (2008) Metabolic effects of dietary fiber consumption and prevention of diabetes. *Journal of Nutrition*, **138**, 439–442.
171. Tucker, L.A. & Thomas, K.S. (2009) Increasing total fiber intake reduces risk of weight and fat gains in women. *Journal of Nutrition*, **139**, 576–581.
172. Lattimer, J.M. & Haub, M.D. (2010) Effects of dietary fiber and its components on metabolic health. *Nutrients*, **2**, 1266–1289.
173. Meyer, K.A., Kushi, L.H., Jacobs, D.R., Slavin, J., Sellers, T.A. & Folsom, A.R. (2000) Carbohydrates, dietary fiber, and incident type 2 diabetes in older women. *American Journal of Clinical Nutrition*, **71**, 921–930.
174. Salmeron, J., Ascherio, A., Rimm, E.B., Colditz, G.A., Spiegelman, D., Jenkins, D.J., Stampfer, M.J., Wing, A.L. & Willett, W.C. (1997) Dietary fiber, glycemic load, and risk of NIDDM in men. *Diabetes Care*, **20**, 545–550.
175. Conceição de Oliveira, M., Sichieri, R. & Sanchez Moura, A. (2003) Weight loss associated with a daily intake of three apples or three pears among overweight women. *Nutrition*, **19**, 253–256.
176. Bridges, S.R., Anderson, J.W., Deakins, D.A., Dillon, D.W. & Wood, C.L. (1992) Oat bran increases serum acetate of hypercholesterolemic men. *American Journal of Clinical Nutrition*, **56**, 455–459.
177. Wright, R.S., Anderson, J.W. & Bridges, S.R. (1990) Propionate inhibits hepatocyte lipid synthesis. *Proceedings of the Society for Experimental Biology and Medicine*, **195**, 26–29.
178. Inan, M.S., Rasoulpour, R.J., Yin, L., Hubbard, A.K., Rosenberg, D.W. & Giardina, C. (2000) The luminal short-chain fatty acid butyrate modulates NF-kappaB activity in a human colonic epithelial cell line. *Gastroenterology*, **118**, 724–734.
179. Streppel, M.T., Arends, L.R., van 't Veer, P., Grobbee, D.E. & Geleijnse, J.M. (2005) Dietary fiber and blood pressure: a meta-analysis of randomized placebo-controlled trials. *Archives of Internal Medicine*, **165**, 150–156.
180. Brussaard, J.H., Vanraaij, J.M.A., Stassewolthuis, M., Katan, M.B. & Hautvast, J.G.A.J. (1981) Blood-pressure and diet in normotensive volunteers—absence of an effect of dietary fiber, protein, or fat. *American Journal of Clinical Nutrition*, **34**, 2023–2029.
181. Kelsay, J.L., Behall, K.M. & Prather, E.S. (1978) Effect of fiber from fruits and vegetables on metabolic responses of human subjects 1. Bowel transit-time, number of defecations, fecal weight, urinary excretions of energy and nitrogen and apparent digestibilities of energy, nitrogen, and fat. *American Journal of Clinical Nutrition*, **31**, 1149–1153.
182. Kuller, L.H., Tracy, R.P., Shaten, J. & Meilahn, E.N. (1996) Relation of C-reactive protein and coronary heart disease in the MRFIT nested case-control study. Multiple Risk Factor Intervention Trial. *American Journal of Epidemiology*, **144**, 537–547.
183. Ridker, P.M., Hennekens, C.H., Buring, J.E. & Rifai, N. (2000) C-reactive protein and other markers of inflammation in the prediction of cardiovascular disease in women. *New England Journal of Medicine*, **342**, 836–843.

184. Ajani, U.A., Ford, E.S. & Mokdad, A.H. (2004) Dietary fiber and C-reactive protein: findings from National Health and Nutrition Examination Survey Data. *Journal of Nutrition*, **134**, 1181–1185.
185. Ma, Y., Griffith, J.A., Chasan-Taber, L., Olendzki, B.C., Jackson, E., Stanek, E.J., Li, W., Pagoto, S.L., Hafner, A.R. & Ockene, I.S. (2006) Association between dietary fiber and serum C-reactive protein. *American Journal of Clinical Nutrition*, **83**, 760–766.
186. King, D.E., Egan, B.M., Woolson, R.F., Mainous, A.G., Al-Solaiman, Y. & Jesri, A. (2007) Effect of a high-fiber diet vs a fiber-supplemented diet on C-reactive protein level. *Archives of Internal Medicine*, **167**, 502–506.
187. Cummings, J.H., Bingham, S.A., Heaton, K.W. & Eastwood, M.A. (1992) Fecal weight, colon cancer risk, and dietary intake of nonstarch polysaccharides (dietary fiber). *Gastroenterology*, **103**, 1783–1789.
188. Terry, P., Giovannucci, E., Michels, K.B., Bergkvist, L., Hansen, H., Holmberg, L. & Wolk, A. (2001) Fruit, vegetables, dietary fiber, and risk of colorectal cancer. *Journal of the National Cancer Institute*, **93**, 525–533.
189. Prentice, R.L. (2000) Future possibilities in the prevention of breast cancer: fat and fiber and breast cancer research. *Breast Cancer Research*, **2**, 268–276.
190. Schatzkin, A., Lanza, E., Corle, D., Lance, P., Iber, F., Caan, B., Shike, M., Weissfeld, J., Burt, R., Cooper, M.R., Kikendall, J.W. & Cahill, J. (2000) Lack of effect of a low-fat, high-fiber diet on the recurrence of colorectal adenomas. Polyp Prevention Trial Study Group. *New England Journal of Medicine*, **342**, 1149–1155.
191. Smith-Warner, S.A., Spiegelman, D., Yaun, S.S., Adami, H.O., Beeson, W.L., van den Brandt, P.A., Folsom, A.R., Fraser, G.E., Freudenheim, J.L., Goldbohm, R.A., Graham, S., Miller, A.B., Potter, J.D., Rohan, T.E., Speizer, F.E., Toniolo, P., Willett, W.C., Wolk, A., Zeleniuch-Jacquotte, A. & Hunter, D.J. (2001) Intake of fruits and vegetables and risk of breast cancer: a pooled analysis of cohort studies. *Journal of the American Medical Association*, **285**, 769–776.
192. Nangia-Makker, P., Hogan, V., Honjo, Y., Baccarini, S., Tait, L., Bresalier, R. & Raz, A. (2002) Inhibition of human cancer cell growth and metastasis in nude mice by oral intake of modified citrus pectin. *Journal of the National Cancer Institute*, **94**, 1854–1862.
193. Howe, G.R., Hirohata, T., Hislop, T.G., Iscovich, J.M., Yuan, J.M., Katsouyanni, K., Lubin, F., Marubini, E., Modan, B. & Rohan, T. (1990) Dietary factors and risk of breast cancer: combined analysis of 12 case-control studies. *Journal of the National Cancer Institute*, **82**, 561–569.
194. Howe, G.R., Jain, M. & Miller, A.B. (1990) Dietary factors and risk of pancreatic cancer: results of a Canadian population-based case-control study. *International Journal of Cancer*, **45**, 604–608.
195. Simons, C.C.J.M., Schouten, L.J., Weijenberg, M.P., Goldbohm, R.A. & van den Brandt, P.A. (2010) Bowel movement and constipation frequencies and the risk of colorectal cancer among men in the Netherlands Cohort Study on Diet and Cancer. *American Journal of Epidemiology*, **172**, 1404–1414.
196. Haack, V.S., Chesters, J.G., Vollendorf, N.W., Story, J.A. & Marlett, J.A. (1998) Increasing amounts of dietary fiber provided by foods normalizes physiologic response of the large bowel without altering calcium balance or fecal steroid excretion. *American Journal of Clinical Nutrition*, **68**, 615–622.

19 Prunes: are they functional foods?

Alessandra Del Caro and Antonio Piga

19.1 Introduction

It is widely recognized that regular consumption of different foods, particularly fruit and vegetables, may help to promote an optimal health and lower or reduce the risk of different chronic diseases. The above-cited effects are associated to naturally present or process-induced biologically active compounds. Foods with these characteristics are often classified as functional foods. Prunes have long been recognized as having beneficial effects on human health, thus they can surely be included in the functional foods [1]. Prunes are obtained from drying fresh plum fruits (*Prunus domestica* L.) that are thought to originate in the Caucasus Mountains. The actual polyploidy species are the results from two wild plums, the diploid *Prunus cerasifera* Ehrh (cherry plum) and the tetraploid *Prunus spinosa* L. (blackthorn) [2]. The cross-breeding generated an appreciated fruit species that was propagated worldwide during the centuries. Prunes are industrially obtained by tunnel air drying up to 20–24% in moisture at temperatures ranging from 60 to 85°C [3]. They can also be rehydrated up to 35% in moisture. Prune juice may be obtained by boiling prunes in water up to 18.5% solid content. Other products can be obtained from prunes, such as purees and powders. Most prunes are obtained from cv. D'Agen plums and until 2000 California was the dominant world prunes supplier with a market share of over 70%. Nowadays, other countries in the world are emerging and can become rivals with regard to prune production [3]. One of these countries is Chile, which ranks second in prune production, with an estimated 55,000 metric tonnes (MT) produced in 2009, compared to US production of 139,000 MT. Argentina is third in the world prune production with a 25% increase in production to an expected 50,000 MT over the next 5 years.

A certain number of articles have been published on biological effect of prunes on human health. The main compounds claimed to have beneficial effects are mainly phytochemicals. This chapter outlines the present state of knowledge about composition, nutritional characteristics, phytochemicals, health effects, and food applications of prunes and their by-products.

Dried Fruits: Phytochemicals and Health Effects, First Edition. Edited by Cesarettin Alasalvar and Fereidoon Shahidi.

19.2 Compositional and nutritional characteristics of prunes

19.2.1 Proximate composition

There is extensive literature that summarizes the current knowledge of the proximate composition of prunes [1], but most of the scientific papers show mainly data related to fresh plum cultivars [4–13]; some refer to prune juice [14–16] and others to dried prunes [17–19].

Table 19.1 reports the nutrient composition of prunes. Their content vary considerably, even in the same cultivar, surely due to the growing conditions, geographic area, and degree of ripeness. Prunes contain a considerable amount of carbohydrates compared to the fresh fruit due the dehydration process, so they can be considered a good source of energy.

19.2.2 Dietary fiber

Dietary fiber has been reported to have some biological activities, such as lowering of serum cholesterol and postprandial glycemia [20, 21] and reductions in chronic disease, such as coronary heart disease (CHD), diabetes mellitus, cancer [22–25], and irritable bowel syndrome [26, 27].

Prunes have been reported to contain from 6.0 to 16.1 g/100 g of dietary fiber (Table 19.1). This wide range could be due to the different analytical methods employed and variety. USDA Food Composition Tables [28] report 7.0 g/100 g of dietary fiber in prunes. Tinker *et al.* [29] reported 6.0 g/100 g of dietary fiber in prunes, while Labavitch *et al.* [30] found 6.2 g/100 g of dietary fiber. Recently, Cheryl *et al.* [31] analyzed the dietary fiber content of prune and prune preparations and expressed the content as percent of dry weight. After converting the values from percentage dry weight to fresh weight, the following amounts were obtained: 10.8, 8.4, 8.0, and 5.3 g/100 g in prune powders, prunes, pitted prunes, and prune puree, respectively. These values are lower than those reported by the European Food Composition Tables [32], which give a value of 16.1 g/100 g. The French Prune Association [33] reports similar values of 13–16 g/100 g. These data show that prunes, along with dried figs, are probably the fruits with the highest dietary fiber content.

Data on the composition of the dietary fiber show different values for soluble and insoluble fibers. An ethanol extract of prune has been reported to have 49% of soluble and 51% of insoluble dietary fiber [34], while a methanol extract afforded 42% of insoluble and 58% of soluble dietary fiber [35].

With respect to prune juice which is lower in fiber, values ranging from 0.01 to 1.1 g/100 g have been reported [31, 36]. Some prune by-products such as waste cake and prune pits contain 6.0 and 70.4 g/100 g of dietary fiber, respectively [31].

19.2.3 Sugars

Individual sugar contents (Table 19.1) reported in several studies [11, 37–39] show wide ranges, probably due to the variations in drying conditions employed. The sugar content is different for each species, for example, apricots contain more sucrose and much less sorbitol than prunes [39]. These individual sugars are also present in the fresh fruit but their percentages change upon drying [17]. During processing, sucrose is hydrolyzed to fructose

Table 19.1 Nutrient composition of prunes (values per 100 g edible portion)

Nutrient	Unit	Prunes	References
Proximate composition			
Water	g	30.9–32.4[a]	[37, 47]
Energy	kcal	240	[47]
Protein	g	2.2–2.6	[37, 47]
Lipid	g	0.4–0.5	[37, 47]
Carbohydrate	g	62.7–63.9	[37, 47]
Dietary fiber	g	6.0–16.1	[32, 47, 48]
Sugars			
Fructose	g	13.1	[11, 38, 39]
Glucose	g	23.1	[11, 38, 39]
Sucrose	g	0.6	[11, 38, 39]
Sorbitol	g	14.7	[11, 38, 39]
Minerals			
Boron	mg	2.2	[48]
Calcium	mg	43–78	[42, 43, 47, 48]
Copper	mg	0.3–0.4	[47, 48]
Fluoride	μg	4.0	[47]
Iron	mg	0.9–3.9	[42, 47, 48]
Magnesium	mg	41–45	[47, 48]
Manganese	mg	0.2–0.3	[47, 48]
Phosphorus	mg	69–85	[42, 47, 48]
Potassium	mg	732–990	[42, 43, 47, 48]
Selenium	μg	0.3	[47]
Sodium	mg	2–8	[42, 47, 48]
Zinc	mg	0.4–0.5	[42, 47, 48]
Vitamins			
Betaine	mg	0.4	[47]
Choline	mg	10.1	[47]
Folate	μg	3.7–4.0	[47, 50]
Niacin	mg	2.0	[50]
Pantothenic acid	mg	0.46	[50]
Pyridoxine	mg	0.21–0.28	[47, 50]
Riboflavin	mg	0.16–0.19	[47, 50]
Thiamin	mg	0.05–0.08	[47, 50]
Vitamin A (RAE)	μg	39	[47]
Vitamin C	mg	0.6–4.0	[47, 48, 50]
Vitamin E (ATE)	mg	0.43	[47]
Vitamin K	μg	5.5–59.5	[47, 50]

RAE, retinol activity equivalents; ATE, α-tocopherol equivalents.
[a]Range (minimum–maximum).

and glucose, which can undergo Maillard reaction as a result of drying up to their total loss. Sorbitol is extremely important since it represents one of the criteria used to select the plum cultivar for drying [40]. In fact, this sugar is responsible for the laxative effect of dried prunes at low doses (70 g/day) and prevents excessive browning in prunes due to its resistance to caramelization and to the fact that it is not a reactant molecule in the Maillard reaction [41].

19.2.4 Minerals

Table 19.1 shows the content of minerals present in prunes. Prune-making plums, once subjected to the dehydration process, maintain the same mineral content as the fresh fruit [1, 42]. A study on California prunes with 19% of moisture showed an iron content of 3.2 mg/100 g of dry weight, while potassium was 990 mg/100 g and calcium 78 mg/100 g [43]. Rehydration does not cause any significant loss of minerals. For this reason, a 100 g serving of prunes can supply 20% of the daily reference value (DRV) for potassium, 20% of the reference daily intake (RDI) for copper, 14% of RDI for iron, about 10% of the RDI for zinc and magnesium, and 10% of intake of manganese. Moreover, prunes contain significant amounts of boron [44]. A 100 g serving of prunes is equal to an average daily intake for adult males of about 2.2 mg of boron [45], where the dietary requirement could be 1 mg/day.

19.2.5 Vitamins

The content of vitamins in prunes is reported in Table 19.1. Some vitamins are perhaps lost during the dehydration process, due to the high temperature used, especially ascorbic acid, as reported in the literature [19, 46]. There is a wide difference in the content of vitamin C reported in the United States (0.6 mg/100 g) [47] with respect to Europe (4.0 mg/100 g) [48]. A large serving of dried prunes could provide 6–10% of the RDI of vitamin C, whereas 9% of RDI for vitamin E. Other tocopherols are present, but only in trace amounts as β-tocopherol and γ-tocopherol. Dried plum also contains a high amount of vitamin K (up to 59.5 mg phylloquinone/100 g) among commonly consumed foods [49].

19.3 Phytochemicals in prunes and their by-products

Phytochemicals are generally referred to as nonnutritive plant compounds, which may affect health through protective or disease-preventive properties. More than several thousand phytochemicals have so far been reported. Some of the well-known phytochemicals belong to the classes of phenolics, carotenoids, organosulfur compounds, nitrogen-containing compounds, and alkaloids. Prunes and their by-products are rich in phytochemicals, mainly polyphenols, but minor amounts of carotenoids and other compounds are also present.

19.3.1 Polyphenols

This class of compounds is well represented in fresh-prune-making plums and, in spite of the known heat damage caused by the drying process, they are still present in prunes. Dried prunes have a high content of polyphenols. The main compounds are hydroxycinnamic acids, namely 3-*O*-caffeoylquinic acid (neochlorogenic acid) and 5-*O*-caffeoylquinic acid (chlorogenic acid). Table 19.2 reports the range of polyphenols found in prunes and prune juices.

Hydroxycinnamates found in prunes have been reported by different authors [15, 19, 50–53]. All of these authors found a higher amount of neochlorogenic acid than chlorogenic acid in the products. Both acids account for 95–98%, but their content declined to 75% when another isomer, 4-*O*-caffeoylquinic acid (cryptochlorogenic acid), was present [51]. Higher amount of neochlorogenic and chlorogenic acid was found by Piga *et al.* [19] in

Table 19.2 Content (mg/kg dry weight basis) of phenolic compounds in prunes and prune juices

Phenolics	Prunes	Prune juices	References
Neochlorogenic acid	928–3045[a]	225	[15, 19, 51, 53]
Chlorogenic acid	67–562	193–335	[14, 15, 19, 51, 53]
Caffeic acid	1–35	3–45	[14, 15, 19, 51–53]
p-Coumaric acid	2–43	4	[15, 19, 51–53]
Protocatechuic acid	0.5–2	–	[52]
Cyanidin 3-rutinoside	15	–	[19, 53]
Rutin	3–89	4	[15, 19, 51–53]
(–)-Epicatechin	7.2	–	[52]
Catechin	–	126–169	[14]

[a]Range (minimum–maximum).

President prunes obtained by high-temperature drying. According to the literature, prunes are the richest fruits in neochlorogenic acid. It is important to highlight that the use of drying temperatures higher than 75°C reduces considerably the depletion of these two phenolic acids, due to the fact that the degradation process is mediated by polyphenoloxidase [19, 50]. Kayano *et al.* [54] reported different caffeoylquinic acid isomers in prunes for the first time. Caffeic and coumaric acids, which are absent in fresh fruits, appear in prunes [15, 19, 51–53], probably due to the hydrolysis of cinnamic acid during drying.

Other compounds found in prunes are rutin, (–)-epicatechin, protocatechuic acid, coniferin, vanillic acid, scopoletin, magnolioside, anthocyanidins, and proanthocyanidins [15, 19, 51–55]. Catechin has been found in dried fruits by Raynal *et al.* [50], but they provided separate data for pulp and exocarp, thus it is not simple to report the exact amount of phenolic content in whole dried fruits. Rutin accounted for about 2–4% of the total polyphenol content. Anthocyanins have been found in prunes at very low concentrations due to their rapid degradation during the first phase of drying [55]. Literature reports the presence of peonidin 3-rutinoside, peonidin 3-glucoside, cyanidin 3-rutinoside, and cyanidin 3-glucoside [19, 55], but data are available only for cyanidin 3-rutinoside [19, 53], for the reason given earlier as noted by Raynal *et al.* [50]. Coniferin, vanillic acid, scopoletin, and magnolioside have been reported for the first time by Kayano *et al.* [56]. Recently, Kimura *et al.* [57] found an oligomeric proanthocyanidin in a prune extract. The oligomer is composed mainly of epicatechin units with catechin as a terminal unit. Other important components found in prunes include two lignans [58].

Prune juices are the main product derived from prunes. They are obtained using different processing technologies, with respect to the other fruit juices. In fact, there is no pressing and refining operation, but only water extraction by boiling dried prunes until 18.5% sugar content is achieved. This process surely results in further depletion of polyphenols. Donovan *et al.* [15] found significantly lower content of polyphenols in prune juices than in prunes. They also noticed a decreased ratio of neochlorogenic acid to chlorogenic acid. Apart from these two compounds, they detected low amounts of caffeic and coumaric acids as well as rutin. It is important to note that the boiling treatment increases the concentration of hydroxymethylfurfural, an intermediate compound in the Maillard reaction. Catechin is also reported in prune juices [14]. To best of our knowledge, data on other prune derivatives or by-products have not been published.

Table 19.3 Carotenoid composition of prunes (values per 100 g edible portion)

Nutrient	Unit	Prunes	References
α-Carotene	μg	31–57	[47, 59]
β-Carotene	μg	140–394	[47, 59]
β-Cryptoxanthin	μg	93	[47]
Lutein + zeaxanthin	μg	148	[47]

19.3.2 Carotenoids

Carotenoids are yellow to red fat-soluble pigments that may be converted to vitamin A in the body, provided they have a β-ionone ring. Table 19.3 shows carotenoid and xanthophyll profiles (α-carotene, β-carotene, β-cryptoxanthin, and lutein + zeaxanthin) of prunes. Among these, β-carotene acts as provitamin A. Finnish researchers found very low amounts of α-carotene, β-carotene, and lutein, in particular, with respect to the content of the fresh plums [59]. One hundred grams of prunes may provide only 5% of RDI. This low RDI contrasts with the European Tables and US claim of 22 and 25% RDI, respectively [1, 32]. Scientific literature also includes a study by Bolin [43] who reported that steam-rehydrated Californian prunes contained 0.56 mg β-carotene/100 g of fresh weight, which is less than half of the provitamin A activity reported by the USDA Food Composition Tables [28]. Similar results have also been obtained by Korobkina [60] with four cultivars of Uzbek prunes.

When looking for changes in carotenoid content after drying of fresh plums, a detailed study was carried out by Moutounet [61]. In particular, the author demonstrated that drying results in a 45–75% loss of total carotenoids, β-carotene being less susceptible. It is interesting to note that the paper reports that prunes dehydrated in two harvesting seasons had very different carotenoid contents.

Prune juice contains only traces of vitamin A activity [47], probably due to the hot water extraction used to obtain the prune juice.

19.4 Natural antioxidant in prunes

A large group of phytochemicals, which have protective effects against cell oxidation, have been discovered. These naturally occurring compounds act as antioxidants in the body by scavenging harmful free radicals, which are implicated in most degenerative diseases. Thus, they may well be defined as the substances that are capable of quenching or stabilizing free radicals. Epidemiological studies have established a positive correlation between the intake of fruits and vegetables and prevention of diseases such as atherosclerosis, cancer, diabetes, arthritis, and ageing. Vitamins (A, C, and E) present in dried prunes are all considered as natural antioxidants. The other natural antioxidants in prunes can be phenolic compounds, such as flavonoids and phenolic acids, among others. The antioxidant properties of phenolics are mainly due to their redox properties, which allow them to act as reducing agents, hydrogen donators, and singlet oxygen quenchers. Prunes and prune juices are an excellent source of dietary antioxidants [15]; in fact they are characterized by high concentrations of hydroxycinnamic acids, especially neochlorogenic and chlorogenic acids that are predominant. The prune extracts, as well as neochlorogenic and chlorogenic acids, exhibit antioxidant activity by inhibiting low-density lipoprotein (LDL) oxidation *in vitro*. Other phenolic compounds present in prunes, such as rutin and caffeic acid, have been reported in the literature as active inhibitors of human LDL oxidation [62].

19.5 Health effects of prunes

It is well known that certain foods help to keep us healthy and prevent chronic disease [63]. Above all, consumption of fruits and vegetables has shown to reduce the risk of cancer, heart disease, and stroke [64]. These foods are often classified as functional foods, and dried prunes fall into this category. The literature reports many biological effects of prunes, whereas the most known is the laxative effect.

Prunes alleviate constipation with an effect on the bowel actually, which is not well understood. It is reasonable to ascribe this action to the prune fiber content. Probably, the combination of soluble and insoluble fibers acts in a gentle way in the lower intestine, softening the stool, increasing its bulk, and promoting in this way the intestinal mobility. In addition, prunes contain sorbitol that is known to cause a laxative action in animals and humans [65]. Phenolic compounds can also contribute to the laxative effect.

Prunes have the ability to lower the glycemic index (GI) in humans [66, 67]. They provide reasonable amount of energy (239 kcal/100 g), due to their high sugar content, but do not increase the blood level of sugar and their GI is in the moderate range (54 against 100 for the GI of glucose) [67]. Besides, prunes reduce the plasma insulin concentration and C-peptide, a product of insulin breakdown. This action reduces the level of insulin secretion in humans who consume dried prunes. This behavior is linked to the sugar and fiber content of prunes; in fact, the combination of glucose, fructose, and sorbitol with significant amounts of dietary fiber has a beneficial influence on glucose metabolism and diabetes management [66].

Phenolic compounds appear to have some effect on glucose absorption and metabolism, as reported by Welsch *et al.* [68], where the responsible compound is chlorogenic acid, and can decrease 80% of glucose active transport capacity. Phenolic compounds also have the ability to inhibit the formation of endogenous glucose in the liver, where chlorogenic acid is responsible for inhibition of glucose-6-phosphate translocase in the enzyme system of microsomes in rat liver [69].

Prunes have a very important role in bone metabolism. Several studies are reported in the literature that evidence the important role of prune consumption in both preventing and reversing bone loss in male and female animal models of osteoporosis [70]. In addition to the animal studies, the results of a short-term clinical trial indicated that the consumption of prunes (100 g/day) by postmenopausal women significantly increased markers of bone formation such as serum total alkaline phosphatase (ALP) activity, bone-specific ALP activity, and insulin-like growth factor-I (IGF-I) that is reported to correlate with bone formation and bone mass in women [71]. Prunes contain about 50 mg of calcium, the same quantity of magnesium, and 80 mg of phosphorus per 100 g of fruit. Even if these minerals are not high in concentration, compared with the RDI for these nutrients, the absorption of bone-building minerals can be improved due to the high organic acid content of prunes. Besides, they contain boron and vitamin K [49], which are responsible for influencing bone-building process. Boron has the ability to increase plasma steroid hormone concentrations and reduce the urinary excretion of calcium in both humans and laboratory animals [72, 73]. Therefore, boron can be considered as an important nutritional factor in the prevention of the incidence of osteoporosis. Regarding vitamin K, studies in the literature report that it influences bone health by improving calcium balance. It is also considered as a cofactor needed for γ-carboxylation of osteocalcin. γ-Carboxylated osteocalcin promotes normal bone mineralization by regulating the growth of hydroxyapatite crystals [74].

Many studies in the literature have reported the role of bioactive phenolic compounds (isoflavones and lignans) on bone health [75, 76]. Prunes are an important source of phenolic compounds, such as neochlorogenic acid and chlorogenic acid, which act as antioxidants and inhibit bone resorption and stimulate bone formation [77]. Therefore, the beneficial effects of dried prunes on bone may be mediated, in part, through their antioxidant properties. Dried prunes polyphenols seem to be able to suppress osteoclast differentiation and activity under normal, oxidative stress, and inflammatory conditions *in vitro* [78]. Another study on ovariectomized rats evidenced the role of prunes in bone metabolism [79] as linked to the high content of phenolic compounds. However, some researchers suggest the presence of boron as an agent which prevents bone loss [1].

Prunes are also responsible for beneficial influences on cardiovascular health. They have a role on three major risk factors involved in the cardiovascular diseases (CVD) such as hypertension, dyslipidemia, and oxidative stress [1]. With regard to hypertension, prunes with their high potassium and low sodium content can protect against hypertension, especially by preventing the adverse effects of high sodium diets [80].

Prunes are able to control dyslipidemia due to the significant quantities of soluble fiber in the form of pectin responsible for decreasing the plasma cholesterol [29]. Studies on ovariectomized female rats fed with dried powdered prunes [81] indicated that their cholesterol-lowering effect was probably due to both their high fiber content and some estrogen-like actions [82]. Prunes also have the capacity of binding bile acids. Binding of bile acids and increasing their fecal excretion has been hypothesized as a possible mechanism by which dietary fibers lower cholesterol [83].

Prunes are responsible for inhibiting human LDL oxidation *in vitro* [84, 85]. In particular, the antioxidant activity of prunes is very high in comparison to the antioxidant activities of other fruits and vegetables on the basis of the oxygen radical absorbance capacity (ORAC) [86]. It is well known that neochlorogenic, chlorogenic, and cryptochlorogenic acids exhibit antioxidant activity against human LDL oxidation [62], scavenge reactive oxygen and nitrogen species [51], and inhibit the formation of conjugated dienes from linoleic acid oxidation. It is reported that the antioxidant activity of prunes is highly dependent on caffeoylquinic acid isomers [1, 87]. However, the contribution of caffeoylquinic acid isomers to the antioxidant activity of prunes has not yet been proven, as reported by Kayano *et al.* [54]. Another recent study [57] shows proanthocyanidin oligomers as contributing to the antioxidative activity of prunes. The results of this study suggest that the high antioxidative activity of prunes depends on the effects of both proanthocyanidin oligomers and caffeoylquinic acids. Prunes also contain small amounts of flavonoids, especially quercetin, which influences the enzyme system involved in immune response and generation of inflammatory process by decreasing the synthesis of prostaglandins, inhibiting histamine release, and reducing cell aggregation or adhesion in various types of cells. In addition, flavonoids can exert an antithrombotic effect, causing a decrease in the incidence of heart attack and stroke [88]. The presence of copper in prunes can be considered as beneficial for cardiovascular health, since copper is needed for the integrity of blood vessels and hemoglobin formation. Many oxidative and antioxidative enzymes are copper dependent, so it can be useful for maintaining the oxidant/antioxidant balance.

Prunes also have a role against cancer. It is known that phenolic compounds are responsible for the inhibitory effect on the carcinogenic action of many chemical carcinogens. In particular, caffeic and chlorogenic acids are able to suppress the mutagenic activity of *N*-methyl-*N*′-nitro-nitrosoguanidine (MNNG) in *Salmonella typhimurium* test [89]. Chlorogenic acid is responsible for the reduction of tumors such as adenocarcinomas, hemangiomas, and liver cell adenomas in methylazoxymethanol-injected hamsters [90]. Consumption of foods rich in fiber diminishes the risk of colon and rectal cancer [91]. Prune extracts have been evaluated *in vitro* for their antiviral, antibacterial, and antifungal activities, but they showed a very low level of activity, as compared to other fruits [92–95].

19.6 Food application of prunes and their by-products

Food applications of prunes include the use of prune itself as well as its derived products.

19.6.1 Whole prunes

Whole prunes may be sold intact or be processed as juice or juice concentrate. Prune juice is obtained from extraction in boiling water, pressing or centrifugation, clarification, concentration up to a minimum of 18.5% soluble solids, pasteurization, and bottling. This juice after depectinization may be vacuum evaporated at low temperature up to 60°Brix (48°C), then packed in cans, and frozen [96]. A more concentrated juice, up to 72°Brix, may be produced for industrial markets. Concentrated juice is used to make prune juice after reconstitution. Apart from its direct use, prune juice has been reported for a number of applications, as listed below:

(a) in bread preparation [97];
(b) for preparing sauces [98];
(c) for extending the shelf life in baked goods [99] due to the mold control exerted by malic acid and the humectant function of sorbitol, fibers, and monosaccharides that result in maintaining a soft and moist texture;
(d) as antioxidant in precooked roast beef [100];
(e) as natural preservative in whole grain bread [101]; and
(f) for controlling pathogen growth [102].

19.6.2 Pitted prunes

Pitted prunes may be sold as whole, canned, chopped, and diced and used to make pastes and purees, granules and powder (low moisture content), fiber, fillings, and toppings.

Significant amounts of prune products are bought by institutional users. Derived products may have different food applications. Pitted canned prunes are produced in three types: (a) regular, packed in sugar syrup; (b) nectarized, water-packed; and (c) moist pack, regular pitted prunes with higher moisture content. Their uses are for direct consumption. The main food applications come from puree, fiber, fillings, and toppings. Prune puree, obtained from whole prunes by extrusion, is in the form of homogenous paste and has a number of applications, such as fat replacer for reducing fat content in a series of goods. The main segments of the food industry in which prune puree has been experimented are bakery, confectionery, and meat. In fact, the unique multifunctional "fruit system" of prunes confers prunes the particular fat-sparing capacity. This characteristic is due to the presence of

(a) pectins, with the properties of creating a stable film during mixing and of flavor entrapping and gradual releasing during mastication;
(b) very high content of sorbitol and reducing sugars, providing humectancy; and
(c) malic acid, with acts as flavor enhancer.

All of these characteristics match the need to lower the fat content of food while imparting delicious flavor. Thus, fruit purees have been proposed as moisture enhancer in precooked beef, pork, lamb, turkey, and chicken products to retain moisture particularly after freezing, reheating, and holding [103] as well as on hot dogs [104], as a fat replacer and moisture enhancer in ground beef for school lunch hamburgers [105], for replacing shortening and thus reducing trans fat in baked goods [106], for salt reduction in bakery and meat production [107], and for adding protein value in underutilized meats [108], as well as for all the features cited for juice. Moreover, dried prune powder has been demonstrated to reduce lipid oxidation in meat [109]. All of these literature reports confirm the important technological functionality of prune derivatives.

19.7 Conclusions

Prunes have attracted the attention of consumers mainly for their bowel function, but nowadays there is increasing scientific evidence for their nutritional and technological functionality. Prunes' high amounts of dietary fiber, sorbitol, boron, potassium, and phenolics may be beneficial in regulating digestion and sugar metabolism, in lowering plasma cholesterol concentration and reducing the risk of bowel cancer, in maintaining cardiovascular health, and in improving bone metabolism. Phenolic compounds, on the other hand, due to their antioxidant properties, may have a role in preventing or decreasing the risk of various diseases.

The particular composition of prunes confers them technological functionality, such as fat replacer in meat and bakery products, antioxidant effect in roast beef, and preservative in bread, or for inhibiting pathogen growth. Despite the above-cited functions, more research is needed on the composition, in particular on the amounts of carotenoids, vitamin E, and other vitamins and on the relationship between prune consumption and disease risks. Thus, only a multidisciplinary research program could reveal the exact potential related to the functionality of prunes.

References

1. Stacewicz-Sapuntzakis, M., Bowen, P.E., Hussain, E.A., Damayanti-Wood, B.I. & Farnsworth, N.R. (2001) Chemical composition and potential health effects of prunes: a functional food? *Critical Reviews in Food Science*, **41**, 251–286.
2. Groh, B., Bauer, H. & Treutter, D (1994) Chemotaxonomical investigations of *Prunus domestica* by isoenzyme markers and phenolic compounds. *Scientia Horticulturae*, **58**, 41–55.
3. Sabarez, H.T. & Price, W.E. (1999) A diffusion model for prune dehydration. *Journal of Food Engineering*, **42**, 167–172.

4. Wrolstad, R.E. & Shallenberger, R.S. (1981) Free sugars and sorbitol in fruits—a compilation from the literature. *Journal of Association of Official Analytical Chemistry*, **64**, 91–103.
5. Wehmeyer, A.S. & Nortje, B.K. (1980) Nutrient composition of single soft fruit cultivars. *South African Food Review*, **7**, 98–99.
6. Nergiz, C. & Yildiz, H. (1997) Research on chemical composition of some varieties of European plums (*Prunus domestica*) adapted to the Aegean district of Turkey. *Journal of Agricultural and Food Chemistry*, **45**, 2820–2823.
7. Renard, C.M.G.C. & Ginies, C. (2009) Comparison of the cell wall composition for flesh and skin from five different plums. *Food Chemistry*, **114**, 1042–1049.
8. Fernandez-Flores, E., Kline, D., Johnson, A.R. & Leber, B.L. (1970) GLC determination of free amino acids in fruits and fruit juice. *Journal of the Association of Official Analytical Chemists*, **53**, 1203–1208.
9. Usenik, V., Kastelec, D., Veberic, R. & Stampar, F. (2008) Quality changes during ripening of plums (*Prunus domestica* L.). *Food Chemistry*, **111**, 830–836.
10. Kristl, J., Slekovec, M., Tojinko, S. & Unuk, T. (2011) Extractable antioxidants and non-extractable phenolics in the total antioxidant activity of selected plum cultivars (*Prunus domestica* L.): evolution during on-tree ripening. *Food Chemistry*, **125**, 29–34.
11. Hansmann, C.F. & Nortje, B.K. (1980) Sugars in stone fruits. *South African Food Review*, **7**, 96–97.
12. Jouret, C., Maugenet, J. & Mesnier, Y. (1969) Maturation of plums d'Ente: chemical composition changes in fruits. *Industrie Alimentaire et Agro-industrielle*, **86**, 795–799 (in French).
13. Fusi, P., Bosetto, F.M. & Cecconi, S. (1981) Study of the dynamics of sugars in fruit and leaves of prunes d'Ente P707 during the maturation process. *Agrochimica*, **25**, 492–500 (in Italian).
14. van Gorsel, H., Li, C., Kerbel, E.L., Smits, M. & Kader, A.A. (1992) Compositional characterization of prune juice. *Journal of Agricultural and Food Chemistry*, **40**, 784–789.
15. Donovan, J.L., Meyer, A.S. & Waterhouse, A.L. (1998) Phenolic composition and antioxidant activity of prunes and prune juice (*Prunus domestica*). *Journal of Agricultural and Food Chemistry*, **46**, 1247–1252.
16. Dikeman, C.L., Bauer, L.L. & Fahey Jr, G.C. (2004) Carbohydrate composition of selected plum/prune preparations. *Journal of Agricultural and Food Chemistry*, **52**, 853–859.
17. Wilford, L.G., Sabarez, H. & Price, W.E. (1997) Kinetics of carbohydrates change during dehydration of d'Agen prunes. *Food Chemistry*, **59**, 149–155.
18. Fang, N., Yu, S. & Prior, R.L. (2002) LC/MS/MS characterization of phenolic constituents in dried plums. *Journal of Agricultural and Food Chemistry*, **50**, 3579–3585.
19. Piga, A., Del Caro, A. & Corda, G. (2003) From plums to prunes: influence of drying parameters on polyphenols and antioxidant activity. *Journal of Agricultural and Food Chemistry*, **51**, 3675–3681.
20. Schafer, G., Schenk, U., Ritzel, U., Ramadori, G. & Leonhardt, U. (2003) Comparison of the effects of dried peas with those of potatoes in mixed meals on postprandial glucose and insulin concentrations in patients with type 2 diabetes. *American Journal of Clinical Nutrition*, **78**, 99–103.
21. Naumann, E., van Rees, A.B., Önning, G, Öste, R., Wydra, M. & Mensink, R.P. (2006) ß-Glucan incorporated into a fruit drink effectively lowers serum LDL-cholesterol concentrations. *American Journal of Clinical Nutrition*, **83**, 601–606.
22. Mozaffarian, D., Micha, R. & Wallace, S. (2003) Effects on coronary heart disease of increasing polyunsaturated fat in place of saturated fat: a systematic review and meta-analysis of randomized controlled trials. *PLoS Medicine*, **7**, 1–10.
23. Gross, L.S., Li, L., Ford, E.S & Liu, S. (2004) Increased consumption of refined carbohydrates and the epidemic of type 2 diabetes in the United States: an ecologic assessment. *American Journal of Clinical Nutrition*, **79**, 774–779.
24. Shin, A., Li, H., Shu, X.O., Yang, G., Gao, Y.T. & Zheng, W. (2006) Dietary intake of calcium, fiber and other micronutrients in relation to colorectal cancer risk: results from the Shanghai Women's Health Study. *International Journal of Cancer*, **119**, 2938–2942.
25. Holmes, M.D., Liu, S., Hankinson, S.E., Colditz, G.A., Hunter, D.J. & Willett, W.C. (2004) Dietary carbohydrates, fiber, and breast cancer risk. *American Journal of Epidemiology*, **159**, 732–739.
26. Giacco, R., Parillo, M., Rivellese, A.A., Lasorella, G., Giacco, A., D'Episcopo, L. & Riccardi, G. (2000) Long-term dietary treatment with increased amounts of fiber-rich low-glycemic index natural foods improves blood glucose control and reduces the number of hypoglycemic events in type 1 diabetic patients. *Diabetes Care*, **23**, 1461–1466.

27. Ford, A.C., Forman, D., Bailey, A.G., Axon, A.T.R. & Moayyedi, P. (2008) Irritable bowel syndrome: a 10-year natural history of symptoms, and factors that influence consultation behavior. *American Journal of Gastroenterology*, **103**, 1229–39.
28. USDA (1988) *Provisional Table on the Dietary Fiber Content of Selected Foods*. US Department of Agriculture, Washington, DC.
29. Tinker, L.F., Davis, P.A., Schneeman, B.O., Gallaher, D.D. & Waggoner, C.R. (1991) Consumption of prunes as a source of dietary fiber in men with mild hypercholesterolemia. *American Journal of Clinical Nutrition*, **53**, 1259–1265.
30. Labavitch, J.M., Rae, H.L. & Sessoms, D. (1986) *Dietary Fiber of Prunes*. California Prune Board Annual Report, Pleasanton, CA.
31. Cheryl, L., Dikeman, L., Bauer, L. & Fahey Jr, G.C. (2004) Carbohydrate composition of selected plum/prune preparations. *Journal of Agricultural and Food Chemistry*, **52**, 853–859.
32. California Prune Board (1993) *Technical Bulletin # 5*. California Prune Board, Pleasanton, CA.
33. Bureau National Interprofessionnel du Pruneau (BIP) (1998) *An Outstanding Profile*. Villeneuve sur Lot, France.
34. Tinker, L.F., Davis, P.A. & Schneeman, B.O. (1994) Prune fiber or pectin compared with cellulose lowers plasma and liver lipids in rats with diet-induced hyperlipidemia. *Journal of Nutrition*, **124**, 31–40.
35. Prosky, L. (1990) Collaborative study of a method for soluble and insoluble dietary fiber. *Advances in Experimental and Medical Biology*, **270**, 193–203.
36. Dreher, M. (1999) Food sources and uses of dietary fiber. In: *Complex Carbohydrates in Foods* (eds S.S. Cho, L. Prosky & M. Dreher). Marcel Dekker, New York, pp. 334–335.
37. California Prune Board (1997) *Buyer's Guide*. California Prune Board, Pleasanton, CA.
38. Kline, D.A., Fernandez-Flores, E. & Johnson, A.R. (1970) Quantitative determination of sugars in fruits by GLC separation of TMS derivatives. *Journal of the Association of Official Analytical Chemists*, **53**, 1198–1202.
39. Ishii, Y. (1983) Sugar components of some dry fruits. *Nippon Eiyo, Shokuryo Gakkaishi*, **36**, 53–55 (in Japanese).
40. Forni, E., Erba, M.L., Maestrelli, A. & Poleselli, A. (1992) Sorbitol and free sugar content in plums. *Food Chemistry*, **44**, 269–275.
41. Cinquanta, L., Di Matteo, M. & Esti, M. (2002) Physical pre-treatment of plums (*Prunus domestica*). Part 2. Effect on the quality characteristics of different prune cultivars. *Food Chemistry*, **79**, 233–238.
42. Salvini, S., Parpinel, M., Gnagnarella, P., Maisonneuve, P. & Turrini, A. (1998) *Banca Dati di Composizione Degli Alimenti per Studi Epidemiologici in Italia*. Istituto Europeo di Oncologia, Milan, Italy (in Italian).
43. Bolin, H.R. (1977) Effects of processing on nutrient composition and texture of prunes. *Journal of Food Quality*, **1**, 123–133.
44. Naghii, M.R., Lyons Wall, P.M. & Samman, S. (1996) The boron content of selected foods and the estimation of its daily intake among free-living subjects. *Journal of the American College of Nutrition*, **15**, 614–619.
45. Nielsen, F.H. (1996) Other trace elements. In: *Present Knowledge in Nutrition*, 7th edn. (eds E.E. Ziegler & L.J. Filer Jr). ILSI Press, Washington, DC, pp. 355–358.
46. Ryley, J. & Kayda, P. (1993) Vitamins in thermal processing. *Food Chemistry*, **49**, 119–129.
47. USDA (2010) *National Nutrient Database for Standard Reference, Release 23*. Published on-line at: http://www.nal.usda.gov/fnic/foodcomp/search/, last accessed March 7, 2011.
48. Souci, S.W., Fachmann, W. & Kraut, H. (1982) *Food Composition and Nutrition Tables (1981–1982)*. Wissenschaftliche Verlagsgesellschaft mbH, Stuttgart, Germany.
49. Dismore, M.L., Haytowitz, D.B., Gebhardt, S.E., Peterson, J.W. & Booth, S.W. (2003) Vitamin K content of nuts and fruits in the US diet. *Journal of the American Dietetics Association*, **103**, 1650–1652.
50. Raynal, J., Moutounet, M. & Souquet, J.M. (1989) Intervention of phenolic compounds in plum technology. 1. Changes during drying. *Journal of Agricultural and Food Chemistry*, **37**, 1046–1050.
51. Nakatani, N., Kayano, S., Kikuzaki, H., Sumino, K., Katagiri, K. & Mitani, T. (2000) Identification, quantitative determination, and antioxidative activities of chlorogenic acid isomers in prune (*Prunus domestica* L.). *Journal of Agricultural and Food Chemistry*, **48**, 5512–5516.

52. Kayano, S., Yamada, N.F., Suzuki, T., Ikami, T., Shioaki, K., Kikuzaki, H., Mitani, T. & Nakatani, N. (2003) Quantitative evaluation of antioxidant components in prunes (*Prunus domestica* L.). *Journal of Agricultural and Food Chemistry*, **51**, 1480–1485.
53. Del Caro, A., Piga, A., Pinna, I., Fenu, P.M. & Agabbio, M. (2004) Effect of drying conditions and storage period on polyphenolic content, antioxidant capacity, and ascorbic acid of prunes. *Journal of Agricultural and Food Chemistry*, **52**, 4780–4784.
54. Kayano, S., Kikuizaki, H., Fukutsuka, N., Mitani, T. & Nakatani, N. (2002) Antioxidant activity of prune (*Prunus domestica* L.) constituents and a new synergist. *Journal of Agricultural and Food Chemistry*, **50**, 3708–3712.
55. Raynal, J. & Moutounet, M. (1989) Intervention of phenolic compounds in plum technology: II. Mechanisms of anthocyanin degradation. *Journal of Agricultural and Food Chemistry*, **37**, 1051–1053.
56. Kayano, S., Kikuizaki, H., Ikami, T., Suzuki, T., Mitani, T. & Nakatani, N. (2004) A new bipyrrole and some phenolics constituents in prunes (*Prunus domestica* L.) and their oxygen radical absorbance capacity (ORAC). *Bioscience, Biotechnology and Biochemistry*, **68**, 942–944.
57. Kimura, Y., Ito, H., Kawaji, M., Ikami, T. & Hatano, T. (2008) Characterization and antioxidative properties of oligomeric proanthocyanidin from prunes, dried fruit of Prunus domestica L. *Bioscience, Biotechnology and Biochemistry*, **72**, 1615–1618.
58. Kikuzaki, H., Kayano, S., Fuktsuka, N., Aoki, A., Kasamatsus, K., Yamasaki, Y., Mitani, D. & Nakatani, N. (2004) Abscisic acid related compounds and lignans in prunes (*Prunus domestica* L.) and their oxygen radical absorbance capacity (ORAC). *Journal of Agricultural and Food Chemistry*, **52**, 344–349.
59. Heinonen, M.I., Ollilainen, V., Linkola, E.K., Varo, P.T. & Koivistoinen, P.E. (1989) Carotenoids in Finnish foods: vegetables, fruits, and berries. *Journal of Agricultural and Food Chemistry*, **37**, 655–659.
60. Korobkina, Z.V. (1968) Food and biological value of sun dried fruits and grapes. *Konserv Ovoshchesush Prom*, **23**, 14–16.
61. Moutounet, M. (1976) Loss of carotenoids during processing of plums to make prunes. *Annals of Technology and Agriculture*, **25**, 73–84.
62. Nardini, M., D'Aquino, M., Tomassi, G., Gentili, V., Di Felice, M. & Scaccini, C. (1995) Inhibition of human low-density lipoprotein oxidation by caffeic acid and other hydroxycinnamic acid derivatives. *Free Radical Biology & Medicine*, **19**, 541–552.
63. Ames, B.N. (1983) Dietary carcinogens and anticarcinogens: oxygen radicals and degenerative diseases. *Science*, **221**, 1256–1264.
64. Vinson, J.A., Zubik, L., Bose, P., Samman, N. & Proch, J. (2005) Dried fruits: excellent *in vitro* and *in vivo* antioxidants. *Journal of the American College of Nutrition*, **24**, 1, 44–50.
65. Ellis, F.W. & Krantz, J.C. (1941) Sugar alcohols. XII. Metabolism and toxicity studies with mannitol and sorbitol in man and animals. *The Journal of Biological Chemistry*, **141**, 147–154.
66. Brand-Miller, J.C. (1994) Importance of glycemic index in diabetes. *American Journal of Clinical Nutrition*, **59**, 747–752.
67. Iron, A., Rigalleau, V., Bignon, J., Dubroca, H., Aubertin, J. & Gin, H. (1998) *Effect of prunes on insulin secretion in healthy young men. An Outstanding Profile*. Villeneuve sur Lot, France.
68. Welsch, C.A, Lachance, P.A. & Wasserman, B.P. (1989) Dietary phenolic compounds: inhibition of Na^+-dependent D-glucose uptake in rat intestinal brush border membrane vescicles. *Journal of Nutrition*, **119**, 1698–1704.
69. Hemmerle, H., Burger, H.J., Below, P., Schubert, G., Rippel, R., Schindler, P.W., Paulus, E. & Herling, A.W. (1997) Chlorogenic acid and synthetic chlorogenic acid derivatives: novel inhibitors of hepatic glucose-6-phosphate translocase. *Journal of Medicinal Chemistry*, **40**, 137–145.
70. Hooshmand, S. & Arjmandi, B.H. (2009) Viewpoint: dried plum, an emerging functional food that may effectively improve bone health. *Ageing Research Reviews*, **8**, 122–127.
71. Arjmandi, B.H., Khalil, D.A., Lucas, E.A., Georgis, A., Stoecker, B.J., Hardin, C., Payton, M.E. & Wild, R.A. (2002) Dried plums improve indices of bone formation in postmenopausal women. *Journal of Womens Health Gender Based Medicine*, **11**, 61–68.
72. Chapin, R.E., Ku, W.W., Kenny, M.A., McCoy, H., Gladen, B., Wine, R.N., Wilson, R. & Elwell, M.R. (1997) The effects of dietary boron on bone strength in rats. *Fundamental Applied Toxicology*, **35**, 205–215.
73. Naghii, M.R., Torkaman, G. & Mofid, M. (2006) Effects of boron and calcium supplementation on mechanical properties of bone in rats. *Biofactors*, **28**, 195–201.

74. Iwamoto, J., Takeda, T. & Sato, Y. (2004) Effects of vitamin K2 on osteoporosis. *Current Pharmaceutical Design*, **10**, 2557–2576.
75. Arjmandi, B.H., Alekel, L., Hollis, B.W., Amin, D., Stacewicz-Sapuntzakis, M., Guo, P. & Kukreja, S.C. (1996) Dietary soybean protein prevents bone loss in an ovariectomized rat model of osteoporosis. *Journal of Nutrition*, **126**, 161–167.
76. Arjmandi, B.H., Khalil, D.A., Smith, B.J., Lucas, E.A., Juma, S., Payton, M.E. & Wild, R.A. (2003) Soy protein has a greater effect on bone in postmenopausal women not on hormone replacement therapy, as evidenced by reducing bone resorption and urinary calcium excretion. *Journal of Clinical Endocrinology & Metabolism*, **88**, 1048–1054.
77. Basu, S., Michaelsson, K., Olofsson, H., Johansson, S. & Melhus, H. (2001) Association between oxidative stress and bone mineral density. *Biochemical and Biophysical Research Communication*, **288**, 275–279.
78. Bu, S.Y., Lerner, M., Stoecker, B.J., Boldrin, E., Brackett, D.J., Lucas, E.A. & Smith, B.J. (2008) Dried plum polyphenols inhibit osteoclastogenesis by downregulating NFATc1 and inflammatory mediators. *Calcified Tissue International*, **82**, 475–488.
79. Arjmandi, B.H., Lucas, E.A., Juma, S. & Soliman, A. (2001) Prune prevents ovariectomy-induced bone loss in rats. *JANA*, **4**, 50–56.
80. Luft, F.C. (1996) Potassium and its regulation. In: *Present Knowledge in Nutrition* (eds E.E. Ziegler & L.J. Filer Jr). ILSI Press, Washington, DC, pp. 272–276.
81. Lucas, E.A., Juma, S., Stoecker, B.J. & Arjmandi, B.H. (2000) Prune suppresses ovariectomy-induced hypercholesterolemia in rats. *Journal of Nutritional Biochemistry*, **11**, 255–259.
82. Bradbury, R.B. & White, DE. (1954) Estrogens and related substances in plants. *Vitamins and Hormones*, **12**, 207–233.
83. Anderson, J.W. & Siesel, A.E. (1990) Hypocholesterolemic effects of oat products. In: *New Developments in Dietary Fiber: Physiological, Physiochemical, and Analytical Aspects* (eds I. Furda & C.J. Brine). Plenum Press, New York, pp. 17–36.
84. Vinson, J.A., Jang, J., Dabbagh, Y.A. & Serry, M.M. (1995) Plant flavonoids, especially tea flavonoids, are powerful antioxidants using an *in vitro* oxidation model for hearth disease. *Journal of Agricultural and Food Chemistry*, **43**, 2800–2802.
85. Meyer, A., Donovan, J.L., Pearson, D.A., Waterhouse, A.L. & Frankel, E.N. (1998) Fruit hydroxycinnamic acids inhibit human low-density lipoprotein oxidation *in vitro*. *Journal of Agricultural and Food Chemistry*, **46**, 1783–1787.
86. Mc Bride, J. (1999) Can foods forestall aging? *Agricultural Research*, **47**, 15–17.
87. Gallaher, C.M. & Gallaher, D.D. (2009) Dried plums (prunes) reduce atherosclerosis lesion area in apolipoprotein E-deficient mice. *British Journal of Nutrition*, **101**, 233–239.
88. Gryglewsky, R.J., Korbut, R., Robak, J. & Swies, J. (1987) On the mechanism of antithrombotic action of flavonoids. *Biochemical Pharmacology*, **36**, 317–322.
89. Chan, R.I.M., San, R.C.H. & Stich, H.F. (1986) Mechanism of inhibition of N-methyl-N′-nitro-nitrosoguanidine induced mutagenesis by phenolic compounds. *Cancer Letters*, **31**, 27–34.
90. Mori, H., Tanaka, T., Shima, H., Kuniyasu, T. & Takahashi, M. (1986) Inhibitory effect of chlorogenic acid on methylazoxymethanol acetate-induced carcinogenesis in large intestine and liver of hamsters. *Cancer Letters*, **30**, 49–54.
91. Mathew, A., Peters, U., Chatterjee, N., Kulldorff, M. & Sinha, R. (2004) Fat, fiber, fruits, vegetables, and risk of colorectal adenomas. *International Journal of Cancer*, **108**, 287–292.
92. Dornberger, K. & Lich, H. (1982) Screening for antimicrobial and presumed cancerostatic plant metabolites. *Pharmazie*, **37**, 215–221 (in German).
93. Gaworski, C.L., Vollmuth, T.A., Dozier, M.M., Heck, J.D., Dunn, L.T., Ratajczak, H.V. & Thomas, P.T. (1994) An immunotoxicity assessment of food flavoring ingredients. *Food and Chemical Toxicology*, **32**, 409–415.
94. Gottshall, R.Y., Lucas, E.H., Lickfeldt, A. & Roberts, J.M. (1949) The occurrence of antibacterial substances active against *Mycobacterium tuberculosis* in seed plants. *The Journal of Clinical Investigation*, **28**, 920–923.
95. Sauter, C. & Wolfensberger, C. (1989) Anticancer activities as well as antiviral and virus enhancing properties of aqueous fruit extracts from 56 European plant species. *European Journal of Cancer and Clinical Oncology*, **25**, 987–990.

96. Luh, B.S., Kean, C.E. & Woodroof, J.G. (1986) Canning of fruits. In: *Commercial Fruit Processing*, 2nd edn. (eds J.G. Woodroof & B.S. Luh). AVI Publishing, Westport, CT, pp. 163–262.
97. Anonymous (1993) Prune preparation as raw material for the food industry. *Gordian*, **93**, 169.
98. California Prune Board (2009) *Dried Plum Rubs, Marinades and Sauces Achieve Deeper Flavor, Added Moisture, Extended Shelf Life . . . Naturally*. Technical Bulletin, California Prune Board, Pleasanton, CA.
99. California Prune Board (2005) *California Dried Plums Offer Bakers Innovative Solutions to Control and Reduce Fat, Calories, Sugar and Carbohydrates While Extending Shelf Life*. Technical Bulletin, California Prune Board, Pleasanton, CA.
100. Nuñez de Gonzalez, M.T., Hafley, B.S., Boleman, R.M., Miller, R.K., Rhee, K.S. & Keeton, J.T. (2008) Antioxidant properties of plum concentrates and powder in precooked roast beef to reduce lipid oxidation. *Meat Science*, **80**, 997–1004.
101. Sanders, S.W. (1993) *Dried Plums: A Multi-functional Bakery Ingredient*. Bulletin No. 228, American Society of Bakery Engineers, Swedesboro, NJ.
102. Fung, D.Y.C. & Thompson, L.K. (2009) "Natural" suppression of the growth of foodborne pathogens in meat products. *International Review of Food Science and Technology*, **2**, 80–81.
103. California Prune Board (2007) *USDA Puree/Plum Plum Juicy™ 100 Meat Moisture Enhancer*. Technical Bulletin, California Prune Board, Pleasanton, CA.
104. California Prune Board (2007) *Dried Plum Puree Helps Retain the Moisture and Flavor of Hot Dogs While Lowering Cost*. Technical Bulletin, California Prune Board, Pleasanton, CA.
105. California Prune Board (2007) *Dried Plum Puree Increases the Moisture and Yield of School Lunch Hamburgers Thus Lowering Cost- and They Taste Fast Food Good*. Technical Bulletin, California Prune Board, Pleasanton, CA.
106. California Prune Board (2006) *California Dried Plums. Natural Alternative to Reduce Trans Fat in Baked Goods*. Technical Bulletin, California Prune Board, Pleasanton, CA.
107. California Prune Board (2010) *Limit Added Salt in Formulas Naturally with Dried Plums*. Technical Bulletin, California Prune Board, Pleasanton, CA.
108. California Prune Board (2008) *Dried Plums Naturally Raise the Value of Underutilized Meats*. Technical Bulletin, California Prune Board, Pleasanton, CA.
109. Osburn, W.N. (2009) *Evaluation of Dried Plum Powder in Meat Products Destined for Convenience and Foodservice Outlets*. Final Report, California Prune Board, Pleasanton, CA.

20 Raisins: processing, phytochemicals, and health benefits

Fereidoon Shahidi and Zhuliang Tan

20.1 Introduction

Raisins are defined as grapes that have been dried and have a sweet taste and a wrinkly texture. The word raisin comes from the French "*raisin sec*," which literally means "dried grapes." The French word itself comes from the Latin term *racemes*, meaning "a bunch of grapes" [1]. Raisins have been consumed as food since 1490 BC due to their high nutritional value and the presence of high levels of micronutrients [2]. They are known as nature's candy and counted among the most nutritious dried fruits globally. Popularity of raisins as healthy food for millennia is due to their nutritional value and the enjoyable taste. They may be eaten as such or included in breakfast cereals, dairy, and bakery as well as confectionery products and nutrition bars [3].

Raisins are of major commercial interest as they are the most important horticultural products. They are the second most important product of grapevine after wine [1]. The world production of raisins from 2005 to 2010 by different countries is given in Table 20.1 [4]. The United States is the largest raisin producer in the world, followed by Turkey. Both countries typically account for more than 95% of production among the major northern hemisphere producing countries and, generally, about 80% of global production. The other major raisin producers include Iran, China, Chile, South Africa, and Afghanistan, among others.

Studies on raisins have shown that they are rich sources of natural bioactive compounds [5, 6]. In this chapter, the processing of raisins will be presented and the characteristic phytochemicals are summarized. The bioactivities and health benefits of raisin consumption are also discussed.

20.2 Types of raisins

Four main grape varieties used for raisin production are Thompson seedless, Muscat, Sultana, and black Corinth. Thompson seedless is the most common variety that constitutes 90% of the total raisin supply in the United States [7]. Dark raisins, golden raisins, Sultanas, Zante

Dried Fruits: Phytochemicals and Health Effects, First Edition. Edited by Cesarettin Alasalvar and Fereidoon Shahidi.

Table 20.1 World raisin production by countries (metric tonnes, dry weight basis)

Production	2005/2006	2006/2007	2007/2008	2008/2009	2009/2010
United States	316,245	274,877	321,143	347,452	294,835
Turkey	250,000	280,000	250,000	310,000	260,000
Iran	155,000	130,000	150,000	60,000	100,000
China	105,000	125,000	150,000	135,000	150,000
Chile	65,500	61,500	67,350	80,000	67,000
South Africa	36,000	41,800	40,200	28,000	43,000
Afghanistan	31,000	20,000	25,000	25,300	29,000
Australia	30,400	15,000	12,000	16,000	14,000
EU-27	30,000	20,000	10,000	10,000	12,000
Uzbekistan	26,000	30,000	37,000	25,700	26,000
Argentina	25,000	36,000	33,000	30,000	36,000
Mexico	8,200	8,500	8,500	8,500	8,300
Total	1,078,345	1,042,677	1,104,193	1,075,952	1,040,135

Source: Adapted from USDA [4].

currants, and Monukka raisins are among the most popular types of raisins [7]. Dark raisins are most commonly found in the marketplace, usually made from Thompson seedless grapes. Although they start as green grapes, the fruit naturally darkens as it dries. Golden raisins, also known as Muscats, are generally made of white Muscat grapes. They are oven-dried rather than being dried by sun and are treated with sulfur dioxide in order to retain their light color. Some of the golden raisins may also be made from Thompson seedless. Sultanas are more popular in Europe and originate from seedless yellow grapes; they are usually softer and sweeter than other varieties. The American variety of Sultana grape is the Thompson seedless. Zante currants are made from black Corinth grape. Tiny dried currants are very sweet and aromatic [7]. Other criteria can also be used to classify raisin products or distinguish them from one another; these include the method of drying (natural, golden-bleached, sulfur-bleached, and lexia), the place of origin (Vostizza, Patras, Pyrgos, Smyrne, Malaga, and Valencia), the conditions under which the product is offered for sale (layers, loose, and seeded), the size grades (4 crown, 3 crown, and 2 crown), the US maturity grades (B or better, C, and substandard), and the quality grades (extra standard, standard, substandard, extra fancy, choice, etc.) [1].

20.3 Processing of raisins

Drying is the process of removing moisture through simultaneous heat and mass transfer and is a classical method of food preservation that provides longer shelf life and reduces weight and volume [8]. It is probably the oldest and one of the most cost-effective methods for preserving fruits, including grapes. The preservation of grapes by drying is a major industry in many parts of the world where grapes are grown. The technique for drying of grapes into raisins has been in use since ancient times. Raisins were produced in Persia and Egypt as early as 2000 BC with one of their first mention being in the Old Testament [9]. A typical block flow diagram for modern raisin production is given in Figure 20.1.

Figure 20.1 Raisin production by drying.

The grape with an outer layer waxy cuticle and a pulpy material inside is a complex product for dehydration as the outer layer cuticle controls the moisture diffusion rates during drying [1]. The low diffusivity implies that there is a formidable barrier that gives consistency to the skin of grape, which may lead to a time-consuming drying process [10]. Thus, a chemical or physical pretreatment is generally applied to decrease skin resistance and, hence, improving moisture diffusion through the waxy cuticle [11]. The commonly used chemical pretreatment in grape drying includes dipping the grapes in hot water and using chemicals such as sulfur dioxide, caustic, and ethyl oleate or methyl oleate emulsions [1]. Different physical treatments have also been reported; for example, superficial abrasion [12], microwave heating [13], and ohmic heating have been used [14].

Drying of grapes is largely traditional and varies in different parts of the world, depending on the cultivation conditions. There are three main methods that are widely used in raisin production, namely, sun drying, solar drying, and mechanical drying [15]. Comparison of the different drying techniques is given in Table 20.2. Historically, the production of raisin from grapes by open sun drying can be traced back to 1490 BC in Greece since it is a simple and cost-effective method [16]. Some of the traditional sun drying methods include open sun drying without cover, open sun drying with cover, and natural rack dryer [11]. Traditionally, the natural sun drying of grapes is carried out in trays. More recently, a new natural sun drying method named within-row-alternate-bearing drying-on-vine (WRAB DOV) has been developed. Compared with the traditional method of hand-harvest, the dried-on-vine raisins are machine harvested, thus reducing human contact and production costs. This novel method also eliminates the need for intensive cultivation to prepare terraces down

Table 20.2 Comparison of different drying techniques employed in raisin production

Drying method	Advantages	Disadvantages
Sun drying	Simple Cost effective Most efficient use of solar energy	Long drying time (8–10 d) Possibility of contamination Uncontrolled drying conditions
Solar drying	Renewable energy source Controllable process Short drying time (within 4 d)	Specially designed dehydrator Moderate equipment investments Small drying capacity
Mechanical drying	Rapid drying rate (within hours) Controllable processing Minimizing the composition, texture, and color changes Slightly higher air-stream grade	High initial investment Significantly higher operational cost High energy cost Only cost effective for specialty, high-valued products such as sulfured raisins or reconditioned rain-damaged fruit

More than any other factor, raisin quality is directly influenced by fruit maturity which is determined by the soluble solid content of the grapes.

row middles [17]. Although sun drying is a popular and inexpensive method, a considerable risk of deterioration of quality due to dust, insect infection, undesirable changes in color, and the presence of foreign matters exists. The other main disadvantage of sun drying is the uncontrollable drying conditions such as weather, intensity of radiation, and ventilation, among others. The basic principle of a solar dryer is that air is heated by the sun in a collector and then passed over the grapes, which are to be dried. It has been established that solar drying is the most technically and economically viable option for most developing countries, especially for those countries within the belt of good solar radiation [18]. Solar dryers for raisin production can be broadly classified into three types, namely, direct type, indirect type, and mixed type. Grapes are exposed directly to solar radiation in the direct type of solar dryer. However, in the indirect type of solar dryer, air heated by solar radiation is made to flow through grapes. Depending on the air circulation, the indirect type solar dryers can be further classified into natural circulated or forced circulated solar dryers. The mixed type of solar dryer combines the advantages of the above two types of solar dryers, in which grapes are exposed directly to solar radiation and hot air is also allowed to flow through them [19]. Advantages and disadvantages of various solar dryer systems that have been developed, fabricated, and tested for effective grape drying are discussed by Pangavhane and Sawhney [15] and Jairaj *et al.* [19]. Mechanical drying, which is safe, rapid, and controllable, is attractive to raisin production, when high throughput is needed. Freeze and microwave vacuum dehydration have also been employed to minimize the compositional changes in grapes during the drying process [20–22]. The industrial application of microwave heating for the preparation of dried grapes has been reported in the literature [23]. The main disadvantages of mechanical drying are the high initial investment and additional operation and running energy cost.

The dried grapes, by either sun drying or other drying techniques, are then delivered to an appropriate unit for further processing. Water is first used to wash and eliminate foreign materials such as dust and soil. Cleaning involves individualizing the dried grapes, removal of stems and foreign materials, and removal of off-grade raisins. During the above two processing steps, rehydration, leakage of some of the dissolved sugars of the dried grapes into the washing water, and extension of micro-cracks and skin wounds are the main changes that may affect dried raisins [11]. To control the amount of moisture in the final products, one more step, finishing drying, is needed, especially for rehydrated dried raisins during the washing or the grapes that have been partially dried or washing off the grapes before packaging. The efficiency and quality of the finishing drying operations are significantly influenced by biomechanical and physical properties of the dried grapes [24].

The criteria for quality determination of raisins have been summarized by Winkler *et al.* [25] based on color, appearance, texture, free flow (having nonsticky surface), flavor, and nutritional value of raisins, which are the main attributes for quality evaluation of raisins and their consumer acceptance. As detailed in the US Standards for Grades of Processed Raisins [26], which became effective December 1, 1978, several factors may affect the grading and quality of raisins. These include maturity and size of grapes, hue, uniformity, and brilliance of the color, condition of the grape surfaces, texture of the skin and pulp, moisture content, chemical composition, presence of decay or rot, mould, yeast, and foreign matters, insect infection, and pretreatment methods as well as drying and storage conditions [1, 11].

20.4 Composition of raisins

Raisins, similar to other dried fruits, have an appealing, sweet taste, and high nutritional value. They provide essential nutrients, dietary fiber, and other bioactive compounds. The nutrient composition, including proximate composition, minerals, and vitamins, of grapes and raisins is given in Table 20.3 [27]. Grapes and raisins are essentially the same, as raisins are dehydrated grapes. The fresh grapes normally contain 80% water. However, the moisture content in raisins is decreased to 15% after the drying process. Consequently, macronutrients such as total protein, carbohydrates, sugars, and fiber become concentrated in raisins compared with fresh grapes. The high sugar and low moisture content of raisins makes them naturally resistant to spoilage during storage. It is important to note that the high content of dietary fiber (3.7–4.0 g/100 g raisins) in raisins helps meet dietary fiber recommendations (14 g of fiber for every 1000 calories of food) [28]. Higher fiber content of 5.05–5.37 g/100 g raisins has also been reported [29]. High-fiber diets help reduce the risk of developing various conditions, including constipation, heart disease, diabetes, colon cancer, and obesity [30]. As indicated in Table 20.3, the content of minerals in raisins is generally 3–7 times higher than those in fresh grapes. Raisins are rich in potassium, phosphorus, magnesium, and calcium, but very low in sodium. It is known that the drying process induces compositional changes in the type of sugar, enzymes, and certain pigments. Grapes contain pentose and furanose, as compared to hexose in raisins as a main sugar group. The polyphenol oxidase, which is present in grapes, does not occur in raisins. Chlorophyll *a*, chlorophyll *b*, xanthophylls, and carotenes are the main pigments identified in grapes, but these are not present in raisins [31]. The content of vitamins C, E, and K is deceased in raisins, while that of other vitamins, namely, thiamine, riboflavin, niacin, vitamin B_6, folate, and choline, remains at the same level or is increased compared with fresh grapes. The decrease in vitamins may be caused by the drying process or the chemical pretreatment [32]. Raisins, like other fruits, contain very low levels of fat and are devoid of cholesterol. Raisins are also an important source of dietary boron, which is not listed in Table 20.3. According to a study by Rainey *et al.* [33],

Table 20.3 Nutrient composition of grapes and raisins (values per 100 g grapes or raisins)

Nutrient	Units	Grapes (American type)	Grapes (European type)	Raisins (golden seedless)	Raisins (seedless)
Proximate composition					
Water	g	81.30	80.54	14.97	15.43
Protein	g	0.63	0.72	3.39	3.07
Total lipid	g	0.35	0.16	0.46	0.46
Ash	g	0.57	0.48	1.66	1.85
Carbohydrate	g	17.15	18.10	79.52	79.18
Fiber	g	0.90	0.90	4.00	3.70
Sugars	g	16.25	15.48	59.19	59.19
Minerals					
Calcium	mg	14.00	10.00	53.00	50.00
Iron	mg	0.29	0.36	1.79	1.88
Magnesium	mg	5.00	7.00	35.00	32.00
Phosphorus	mg	10.00	20.00	115.00	101.00
Potassium	mg	191.00	191.00	746.00	749.00
Sodium	mg	2.00	2.00	12.00	11.00
Zinc	mg	0.04	0.07	0.32	0.22
Copper	mg	0.04	0.13	0.36	0.32
Manganese	mg	0.72	0.07	0.31	0.30
Selenium	μg	0.10	0.10	0.70	0.60
Vitamins					
Vitamin C	mg	4.00	10.80	3.20	2.30
Thiamin	mg	0.09	0.07	0.01	0.11
Riboflavin	mg	0.06	0.07	0.19	0.13
Niacin	mg	0.30	0.19	1.14	0.77
Vitamin B6	mg	0.11	0.09	0.32	0.17
Folate	μg	4.00	2.00	3.00	5.00
Choline	mg	5.60	5.60	11.10	11.10
Vitamin E	mg	0.19	0.19	0.12	0.12
Vitamin K	μg	14.60	14.60	3.50	3.50

Source: Adapted from USDA [27].
The numbers are rounded to the second digit after decimal point.

raisins have the highest concentration of boron at 2.2 mg/100 g among the 50 major food contributors of boron in the American diet.

20.5 Phytochemicals in raisins

Phytochemicals or plant chemicals are known to possess certain biological and physiological activities. These include antioxidant activity, inhibiting cholesterol absorption, blocking the activity of bacteria or viral toxins, decreasing platelet aggregation, or destroying the harmful gastrointestinal bacteria [34]. In recent years, phytochemicals originating from fruits, vegetables, tree nuts, cereals, and other plant food products have attracted much attention by the functional food and nutraceutical market due to their potential health benefits and therapeutic value [35]. In addition to providing essential nutrients, raisins are also a rich source of phytochemicals. The known phytochemicals in raisins include phenolics, prebiotics, tartaric acid, and other minor phytochemicals which may positively affect human health.

20.5.1 Phenolics

Phenolic compounds are ubiquitous and widely distributed secondary metabolites in the plant kingdom. They contain at least one aromatic ring with one or more hydroxyl groups attached to it and play a very important role in plants and plant-derived foods [36]. A wide range of beneficial physiological properties, such as anti-allergic, anti-atherogenic, anti-inflammatory, antimicrobial, anti-thrombotic, cardioprotective, and vasodilatory effects, have been attributed to phenolics [37]. Phenolic compounds may serve as a major determinant of antioxidant potential of foods because of their ability to donate a hydrogen atom or an electron in order to form stable radical intermediates. Different mechanisms may be used to explain their antioxidant activity including scavenging of free radicals, quenching of reactive oxygen species, inhibition of oxidative enzymes, chelating of transition metals, or interaction with biomembranes [38]. The major phenolics found in raisins are phenolic acids, flavonols, isoflavones, and their derivatives. The chemical structures of representative phenolics found in raisins are given in Figure 20.2.

While the phenolics composition of grapes, grape juices, and wines has been extensively investigated, few studies have explored the phenolics composition of raisins. The total phenolics content (TPC) and antioxidant capacity of selected dried fruits and grapes are presented in Table 20.4. The TPC of raisins is higher than fresh grapes since the phenolics become concentrated and possibly be partially modified during the dehydration process [39].

The phenolic profiles of raisin as reported by Karadeniz *et al.* [40] and Parker *et al.* [41] are summarized in Table 20.5. The difference in phenolics content between raisins and fresh grapes is striking, partially due to the enzymatic oxidation and nonenzymatic browning reactions happening during the drying process. Similar trends have been reported in both studies; for example, raisins possess higher levels of phenolics than grapes in most cases. Phenolic acids and flavonols are the two major types of phenolics in both raisins and grapes. Phenolic acids, namely, caftaric and coutaric, exist mainly as *trans* isomers. Due to sulfur dioxide pretreatment, higher levels of *trans*-caftaric and *trans*-coutaric acids are present in golden raisins. Protocatechuic acid was detected only in sun-dried raisins and dipped raisins. Among the flavonols, quercetin glycosides have the highest concentration, followed by kaempferol glycosides and rutin. Except the similar trends mentioned above, there are also certain inconsistencies between these two reports. The content of some phenolics given in the two reports varied considerably. The difference was up to 12 times as indicated by the content of *trans*-caftaric acid in fresh grapes. Ninety percent of caftaric and coutaric acids were lost when fresh grapes were processed into sun-dried, dipped, and golden raisins as reported by Karadeniz *et al.* [40]. However, in the report by Parker *et al.* [41], the content of these two phenolic acids in raisins was higher than that in their corresponding fresh grapes, According to the study of Karadeniz *et al.* [40], procyanidins and flavanols were completely degraded during raisin formation. This is also inconsistent with the USDA database [42], which reported that raisins contain catechin (0.42 mg/100 g), epicatechins (0.10 mg/100 g), and cyanidin (0.03 mg/100 g). The inconsistency may be explained by the use of different samples of grapes/raisins, highly complex structures produced during the drying process, difficulty in working with the raisins matrix as well as interference with non-phenolic

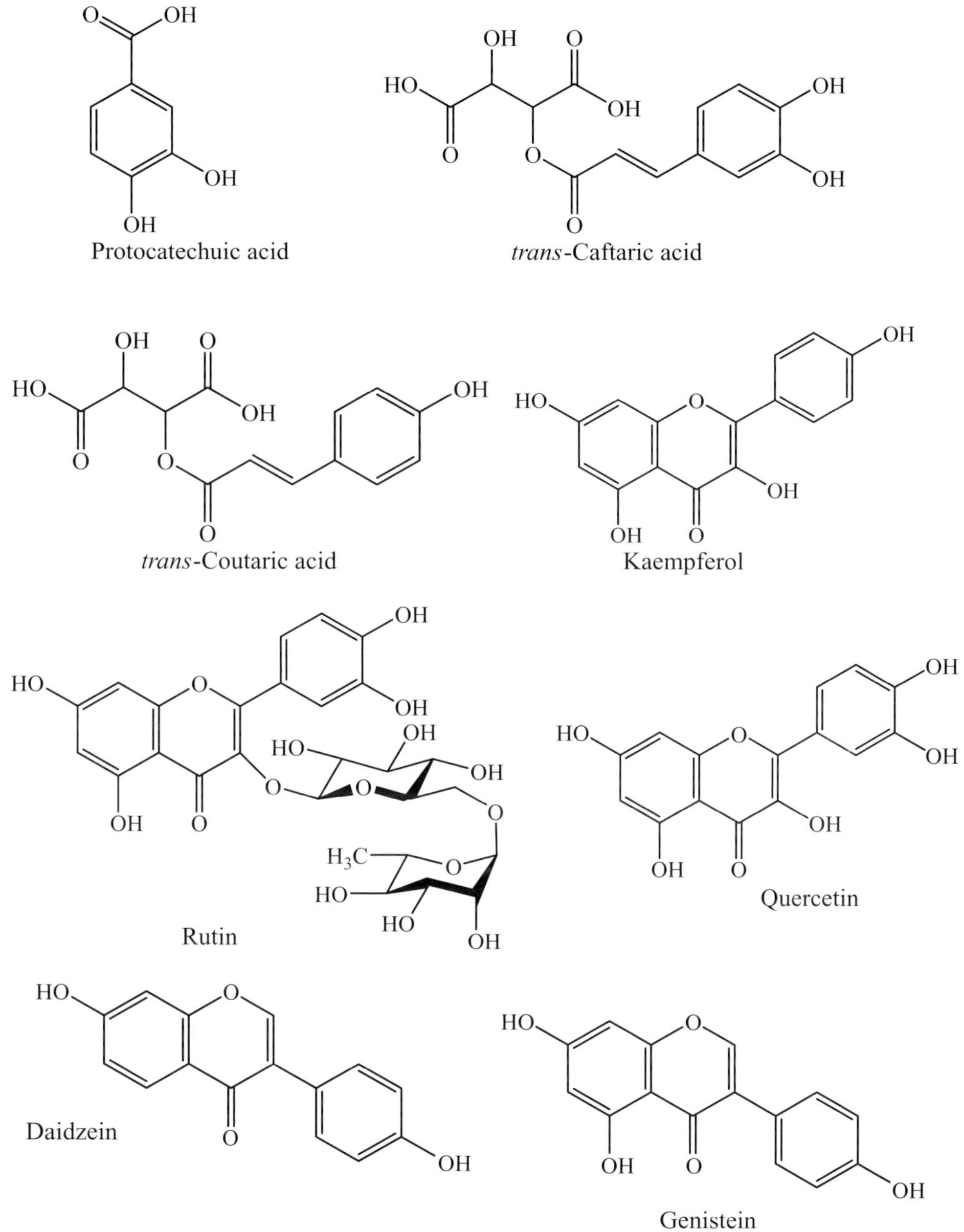

Figure 20.2 Structures of representative phenolics found in raisins.

Table 20.4 Total phenolics and antioxidant capacity of dried fruits and grapes

Food	TPC (mg of GAE/g)	TAC (μmol of TE/g)	L-ORAC$_{FL}$ (μmol of TE/g)	H-ORAC$_{FL}$ (μmol of TE/g)
Medjool dates	5.72	23.87	0.27	23.60
Figs	9.60	33.83	1.83	32.00
Prunes	11.95	85.78	1.79	83.99
Raisins	10.65	30.37	0.35	30.02
Green grapes	1.45	11.18	na	nc
Red grapes	1.75	12.60	na	nc

Source: Adapted from Wu *et al.* [39].
TPC, total phenolics content; TAC, total antioxidant capacity; L-ORAC$_{FL}$, lipophilic-oxygen radical absorbance capacity; H-ORAC$_{FL}$, hydrophilic-oxygen radical absorbance capacity; GAE, gallic acid equivalents; TE, trolox equivalents; na, not available; nc, not calculated.

compounds [43]. The discrepancy may also be explained by the different extraction conditions employed. Zhao and Hall [44] studied the optimization of solvent extraction of raisins phenolics. Their results indicated that the extraction method, such as the types of solvent, pH, temperature, and other factors, plays a major role in the yield of phenolics and hence the apparent composition of products. In their study, except the phenolics mentioned above, they also reported the presence of gallic acid, ferulic acid, chlorogenic acid, and (+)-catechin in raisin extracts. However, Ong and Nagel [45] reported that chlorogenic acids are absent in grapes, and phenolic acids were present mainly as tartaric and not quinic acid esters in grapes. Thus, the presence and source of chlorogenic acid in raisins may still remain unclear and more research is needed to shed light on the composition of phenolics in raisins.

Other minor phenolics reported in raisins include isoflavones, 2-*S*-glutathionylcaftaric acid, and other oxidized cinnamics. Isoflavones such as daidzein and genistein are thought to protect against chronic diseases that are common in the Western countries, such as cancer,

Table 20.5 Phenolics composition in raisins and fresh grapes (mg/kg of sample)

		CAA		COA					QG		KG	
Food	**Reference**	***trans***	***cis***	***trans***	***cis***	**2-S-GCA**	**PA**	**RU**	**A**	**B**	**A**	**B**
Sun-dried raisins	[40]	39.6	nd	6.7	nd	nd	6.8	5.2	7.3	34.7	11.2	23.7
	[41]	41.4	nr	nd	nr	nd	4.4	8.3	15.6	6.5	7.0	9.3
Golden raisins	[40]	84.3	nd	27.3	nd	nd	nd	3.5	41.5	37.1	6.5	7.6
	[41]	130.4	nr	31.4	nr	nd	nd	14.4	65.7	43.4	9.8	14.3
Dipped raisins	[40]	45.2	nd	7.7	nd	8.1	2.8	6.5	20.6	39.0	16.7	29.5
Fresh grapes	[40]	100.7	2.7	31.8	8.0	nd	nd	0.9	21.9	3.9	nd	19.4
	[41]	7.9	nr	14.8	nr	8.8	nd	tr	15.2	25.6	tr	tr
Frozen grapes	[40]	18.2	nd	1.2	nd	nd	nd	0.9	21.9	3.9	nd	19.4
	[41]	7.9	nr	nd	nr	12.1	nd	tr	21.5	27.7	tr	tr

CAA, caftaric acid; COA, coutaric acid; 2-S-GCA, 2-*S*-glutathionylcaftaric acid; PA, protocatechuic acid; RU, rutin; QG, quercetin glycoside; KG, kaempferol glycoside; nd, not detected; nr, not reported; tr, trace.

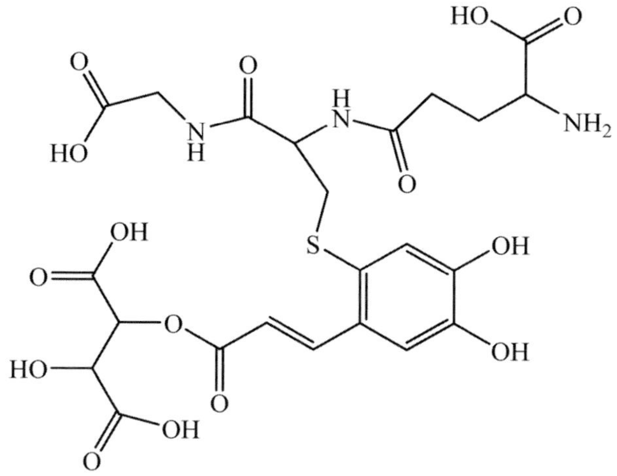

Figure 20.3 Chemical structure of 2-*S*-glutathionylcaftaric acid as oxidation product of glutathione and caftaric acid.

osteoporosis, and coronary heart diseases (CHD) [46]. California raisins are the richest sources of daidzein and genistein among 36 samples of fruits and nuts that contain daidzein or genistein, being 2.25 and 1.84 mg/kg, respectively [47]. 2-*S*-Glutathionylcaftaric acid (also known as grape reaction product, its structure is shown in Figure 20.3) is formed when glutathione and caftaric acids are decompartmentalized upon processing [48]. The amount of 2-*S*-glutathionylcaftaric acid present in a wine provides information on the oxidation history of the wine over its elaboration and aging [49]. The content of 2-*S*-glutathionylcaftaric acid in different raisin and grape samples from two sources is shown in Table 20.5. Karadeniz *et al.* [40] reported that 2-*S*-glutathionylcaftaric acid exists only in dipped raisins. This may indicate that enzymatic oxidation reactions probably occur during the drying process as the polyphenol oxidase is only partially inactivated in the dipped raisins during production. In contrast, in sun-dried raisins, 2-*S*-glutathionylcaftaric acid does not normally form due to the compartmentalization of glutathione and caftaric acid, and in golden raisins, pretreatment with sulfur dioxide inhibits phenolic oxidation. However, different results about the content of 2-*S*-glutathionylcaftaric acid have been reported by Parker *et al.* [41]. A similar level of 2-*S*-glutathionylcaftaric acid is present in grapes, with a higher level in frozen grapes than in fresh ones. The different sample preparation methods may contribute to such difference as explained by the authors [41].

High level of resveratrol has been reported in both grapes and wine [50–52]; however, no resveratrol has been reported in raisins by Karadeniz *et al.* [40] and Parker *et al.* [41]. This may be partially explained by the fact that grapes used for raisin production are normally fully ripened and grapes are known to lose their capacity to produce resveratrol during ripening even in the face of fungal infection [53, 54] and have a reduction in their content of resveratrol. However, resveratrol (40–1088 μg/g) in raisin extracts was reported by Zhao and Hall [44] who were using different extraction solvents and various concentrations and combination.

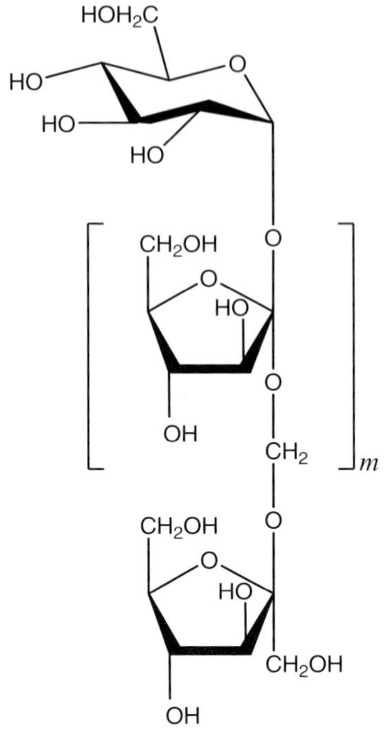

Figure 20.4 Structure of fructan.

20.5.2 Prebiotics

A prebiotics is defined as a nonviable food component that confers a health benefit on the host associated with modulation of the microbiota [55]. Inulin, fructooligosaccharides (FOS), galactooligosaccharides (GOS), soya oligosaccharides, xylooligosaccharides, pyrodextrins, isomalto-oligosaccharides, and lactulose are the commonly used prebiotics. The majority of studies have, so far, focused on inulin, FOS, and GOS [56, 57]. Inulin and FOS are two subgroups of fructans, also known as polyfructoses (Figure 20.4), and differ in their molecular size. The larger sized polyfructoses (more than 10 fructose units, $m > 10$) are known as inulin while the smaller ones (chain length of 3–10 fructose units, $m = 3$–10) are known as FOS [58]. Fructans can reach the large intestine practically intact and are then fermented by bacteria that can cleave the β(2–1)-glycosidic bonds between fructosyl units and act as prebiotic compounds by selectively stimulating the growth of beneficial intestinal microflora, namely, *Bifidobacteria* and *Lactobacilli*. Clinical studies have shown that fructans increase the number of these health-promoting bacteria in the colon mucosa when taken in the diet [59]. Prebiotics, such as fructans, not only help keep microbial balance in the intestinal track, but also offer cardiovascular benefits through a triacylglycerol (TAG)- and cholesterol-lowering effect. They may also increase the absorption of calcium and magnesium and so enhance bone mineralization during growth and prevent osteoporosis [60].

While fresh grapes have no detectable fructans, the dehydration process converts grape sugars into fructans. Thus raisins, unlike fresh grapes, are a rich source of prebiotics in the diet. Sun-dried raisins contain 5.7 g fructans per 100 g of fruit [29], which is the highest among all commonly consumed fruits [61]. Carughi [62] reported that the inulin content

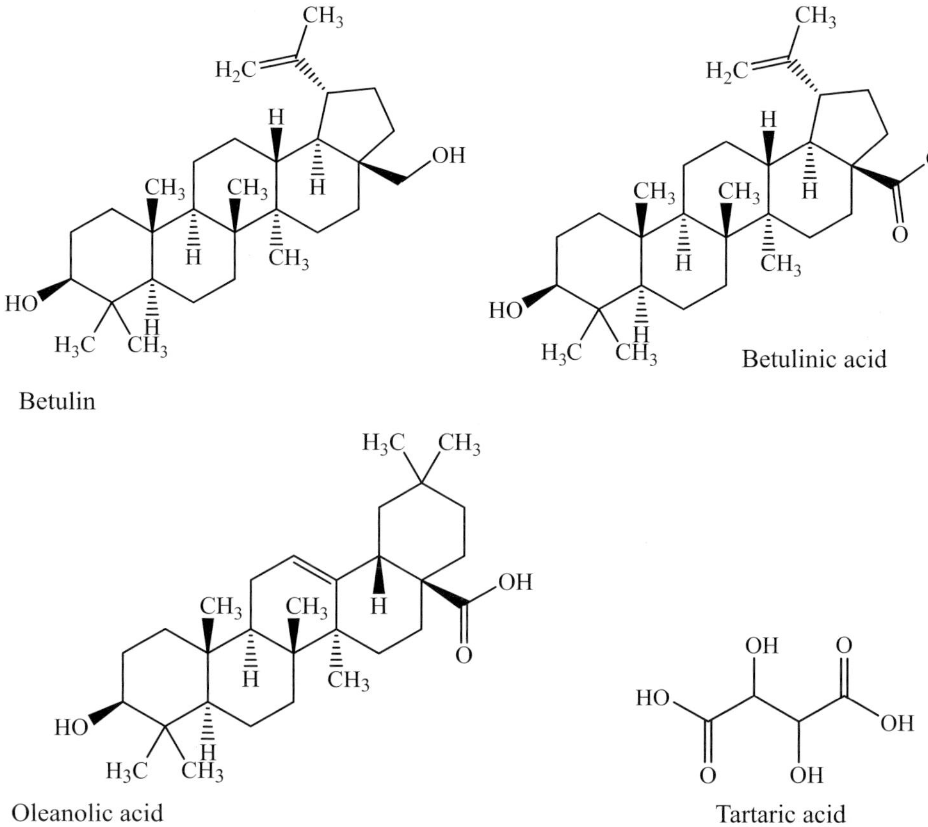

Figure 20.5 Structures of some minor phytochemicals in raisins (betulin, betulinic acid, oleanolic acid, and tartaric acid).

in raisins ranged between 0.7 and 1.2% and FOS between 0.78 and 0.96%, thus having the highest content of prebiotics in commonly consumed fruits. The type of dehydration, storage, and grape variety also influenced the content of fructans in raisins.

20.5.3 Other minor phytochemicals

Triterpenoids are terpenoid derivatives with 30 carbon atoms. The structures of betulin, betulinic acid, and oleanolic acid belonging to this group are given in Figure 20.5. This group of compounds exhibits a wide range of bioactivities including antiretroviral, antimalarial, anti-inflammatory, and anticancer properties. The main plant source for triterpenoids includes birch bark, rosemary leaves, apple peel, and mistletoe shoots [63–65]. Triterpenoids, namely oleanolic acid, oleanolic aldehyde, betulin, and betulinic acid, have been isolated and indentified in raisins. However, there is no quantitative report of their presence in raisins [66, 67].

Tartaric acid (Figure 20.5) is a diprotic organic acid and occurs naturally in many plants, especially grapes, raisins, bananas, and tamarinds, and is used as both a flavorant and an antioxidant in the food industry. Tartaric acid is of interest in studies of fiber and health as it is known to affect colon function. Tartaric acid and fiber in raisins work synergistically to maintain a healthy digestive system [68]. The content of tartaric acid in grapes and raisins is 0.6–0.9 and 2.0–3.5 g/100 g, respectively [69].

The high content of phytosterols (200 mg/kg of fresh weight), with dominance of β-sitosterol, has been reported in grapes [70]. However, there is no report on the composition of phytosterols in raisins, a subject that deserves more attention.

20.6 Bioactivities and health benefits of raisins

20.6.1 Antioxidant activity of raisins

Phenolic compounds are believed to account for a major portion of the antioxidant capacity in many plants [71]. Antioxidant activity of raisins has been studied by several groups of researchers as it is the most notable bioactivity of phenolics in raisin. Antioxidant activity assays such as oxygen radical absorbance capacity (ORAC), total radical-trapping antioxidant parameter (TRAP), trolox equivalent antioxidant capacity (TEAC), ferric-reducing antioxidant power (FRAP), and cellular antioxidant activity (CAA) have been employed to evaluate the antioxidant potential of raisins [39, 41, 72–74]. Comparison of the antioxidant activity of two types of raisins and Thomson seedless grapes is shown in Table 20.6. As shown, raisins generally possess a higher antioxidant activity than fresh grapes because of the removal of water and concentration of antioxidants during the drying process. Parker *et al.* [41] indicated that the ORAC value of golden raisins was much higher than that of the sun-dried raisins, possibly because the pretreatment with sulfur dioxide in the production of golden raisins may inactivate polyphenol oxidase and minimize nonenzymatic browning. Thus, more phenolics are retained during the drying process. The lipophilic and hydrophilic antioxidants are shown in Table 20.4 for dried fruits such as dates, prunes, and raisins. For all those dried fruits, the hydrophilic antioxidants contributed more than 94% to the total antioxidant activity [39]. The antioxidant activity measured by FRAP, TRAP, and TEAC is reported by Pellegrini *et al.* [73]. In their study, raisins showed a moderate antioxidant activity in all three assays employed among several dried fruits examined, including prune, apricot, chestnut, and fig.

Antioxidants can inhibit copper-induced oxidation of human low-density lipoprotein (LDL) and very low density lipoprotein (VLDL) particles as estimated using the thiobarbituric acid reactive substances (TBARS) [75]. The quality of antioxidants in raisin extracts has

Table 20.6 Antioxidant activity of raisins and grapes using different methods

Food	ORAC (μmol of TE/g) [39]	ORAC (μmol of TE/g) [41]	FRAP (mmol antioxidant content/100 g) [72]
Sun-dried raisins	na	37.4	0.78
Golden raisins	30.37	104.5	na
Thompson seedless grapes	11.80	10.8	0.13

ORAC, oxygen radical absorbance capacity; FRAP, ferric-reducing antioxidant power; TE, trolox equivalents; na, not available.

also been reported as IC_{50} (the concentration to inhibit *in vitro* oxidation of LDL particles by 50%) using this method. Values of $1/IC_{50}$ (reciprocal of the amount needed to inhibit oxidation by 50%) for fresh cranberries, green grapes, and plums are 1.16, 1.32, and 1.42 [76], respectively, compared to values for dried cranberries, raisins, and dried plums of 2.38, 3.45, and 4.38, respectively [77]. This is in agreement with the conclusions reached by Wu *et al.* [39] that antioxidant activity improved during the drying process.

It is reported that consumption of phenolic-rich foods can increase serum antioxidant capacity in humans [78]. Study of serum antioxidant capacity after consumption of raisins/grapes was carried out by Parker *et al.* [41]. Long-term consumption of raisins/grapes (4 weeks) increased serum antioxidant capacity by the second and third week of sample consumption. However, values fell again in the fourth week. They speculated that there may be a physiological plateau approximately 2 or 3 weeks after consistent consumption; Cao *et al.* [79] reported similar results. However, serum ORAC values after 1 and 2 h of raisins/grapes consumption weekly during each of 4-week consumption indicated increased oxidation. The authors speculated that the high sugar content of the grapes/raisins may have affected the antioxidant capacity of blood by causing postprandial oxidative stress [41]. A more recent study found that serum antioxidant capacity was modestly increased by daily consumption of raisin, but this did not alter fasted or postprandial inflammatory response in these relatively healthy but overweight individuals [80].

Foods high in phenolic antioxidants can protect against DNA damage during intense physical activity by counteracting oxidative stress. Presence of 8-hydroxy-2-deoxyguanosine (8OHdG) in urine or in leukocytes is an indication of free radical oxidative damage to DNA [81]. Feeding 170 g of sun-dried raisins prior to and during a triathlon to trained athletes significantly lowered the level of 8OHdG in urine compared to feeding of a glucose drink with the same amount of calories. This suggests that raisins can protect against DNA damage caused by oxidative stress during strenuous exercise [82]. Thus, raisins exhibit excellent antioxidant activity in both *in vitro* and *in vivo* tests.

20.6.2 Cardioprotection

Cardiovascular disease (CVD) is a main cause of mortality and morbidity in the world and ranks first in this respect. In Canada, CVD is the leading cause of death and accounts for at least 36% of all deaths [83]. The cause of most CVD is atherosclerosis. High levels of total and LDL cholesterol concentrations are the major atherogenic components of plasma.

Raisins are a significant source of dietary fiber and other phytochemicals which may reduce the risk of CVD by affecting lipoprotein metabolism and inflammation. Phytochemicals, especially phenolics, can protect from atherosclerosis because of their antioxidant potential and through their anti-inflammatory activity [84]. The antioxidant activity of raisins has been discussed in the previous section. The effect of diets containing raisins alone or together with other plant material on blood lipids has been reported by different research groups. In adults with hypercholesterolemia, consumption of Mediterranean-style diets high in whole grains, nuts, and 84 g raisins daily, for 4 weeks, lowered the total and LDL cholesterol concentrations by 9 and 15%, respectively, but there was no significant change in high-density lipoprotein (HDL) cholesterol [85]. A similar result was reported in a crossover study [86] in which hyperlipidemic volunteers were fed a diet with unrefined food, which contained 126 g raisins daily. By the end of the study, total and LDL cholesterol concentrations were lowered by 13 and 16%, respectively, compared to the baselines.

More recently, Puglisi *et al.* [87] indicated that simple lifestyle modifications such as adding raisins to the diet or increasing walking steps have distinct beneficial effects on CVD risk. Volunteers were assigned to consume 160 g raisins per day, increase the number of walking steps per day, or a combination of both interventions for 6 weeks. Results indicated that the total and LDL cholesterol concentrations were decreased by 9.4 and 13.7%, respectively, for all subjects. Subjects in the raisin group had significantly lower levels of α-tumor necrosis factor (TNF-α) and soluble intercellular adhesion molecule-1 (sICAM-1). TNF-α is a powerful pro-inflammatory cytokine. Reducing TNF-α could potentially prevent progression of inflammatory damage. Lower levels of sICAM-1 could potentially prevent progression of atherosclerosis by decreasing adhesion of monocytes to the vascular endothelium.

Addition of raisins to the daily diet could reduce the risk of CVD by increasing the plasma antioxidant capacity, lowering the total and LDL cholesterol concentrations, and reducing inflammation.

20.6.3 Diabetes

Diabetes is a chronic disease marked by high levels of sugar in the blood. It not only increases the risk of death, but also results in a variety of complications including heart disease, kidney disease, eye disease, problems with erection (impotence), and nerve damage [88]. Controlling blood glucose levels closer to normal values through different strategies may lower the risk of diseases and death from these complications.

Potential concerns about raisins causing a high postprandial glycemic response, especially in diabetic or prediabetic individuals, may arise as raisins are a concentrated source of natural sugars. However, except the absolute amount of carbohydrate, other factors in foods may also play important roles in influencing the postprandial glycemic response, such as the glycemic index (GI) of a carbohydrate source and the presence of other antidiabetic bioactive compounds [89]. Although raisins are a concentrated source of carbohydrates, roughly half of their available carbohydrates is fructose, which has a low GI value of 19 (glucose = 100) [90]. The GI values for raisins in healthy adults and women with gestational diabetes were 64 and 65.7, respectively [91, 92]. More recently, GI of raisin from 49.4 to 69 with corresponding insulin index of 47.3–54.4 has also been reported [89].

In addition to its low GI and high dietary fiber content, the high content of antioxidants, especially phenolics such as flavonoids, may also serve as important factors in the management of diabetes. It has been hypothesized that flavonoids, functioning as free radical scavengers and metal chelator, may preserve β-cell function by reducing oxidative stress-induced tissue damage and protect against the progression of insulin resistance to type-2 diabetes. However, results from the available epidemiological studies on the relationship between dietary flavonoids and development of type-2 diabetes are controversial [93–95]. More research is needed to confidently conclude the role of dietary flavonoids in the control of type-2 diabetes.

20.6.4 Anticancer activities

Cancer is the second leading cause of death in the world after heart disease. Epidemiological studies have consistently shown that diets rich in fruits and vegetables help prevent many types of cancer [96]. Although there is no study on the anticancer effect of raisin-containing diets, raisins could serve as important ingredients of a healthy diet to help prevent cancer due to their high content of fiber (fructans) and phenolics (flavonols). The cancer-protective effect of fructans and flavonols is well documented. Animal model studies indicated that fructans possessed anticarcinogenic properties. For example, dietary fructans inhibit the formation of chemically induced aberrant crypt foci in rats. These are neoplastic lesions in the colon from which adenomas and carcinoma may develop [97–99]. High intakes of selected flavonoids, including flavonols, have been found to reduce the risk of developing several types of cancer, such as ovarian cancer [100], breast cancer [101], oral and pharyngeal cancers [102], colorectal cancer [103], and pancreatic cancer [104]. More research is needed to explore the cancer-protective effects of raisins.

20.6.5 Dental health

Traditionally, raisins have been thought of as cariogenic foods because of their high content of sugar and sticky texture. However, studies demonstrated that the perceived "stickiness" bears little relationship to the retention of food particles on tooth surface. Actually, raisins are one of the least retentive foods among the 21 tested commercial snack foods due to their rapid clearing rates after consumption [105]. The sugar composition of raisins may also benefit the dental health. Compared with most commercial sweets, raisins contain a high amount of fructose and glucose instead of sucrose. Sucrose normally induces the most smooth-surface-type and fissure-type caries among the five tested sugars, namely, sucrose, maltose, lactose, fructose, and glucose [74].

Triterpenoids from raisins such as oleanolic acid, oleanolic aldehyde, betulin, and betulinic acid have been shown to inhibit the growth of two species of oral bacteria, *Streptococcus mutans* and *Porphyromonas gingivitis*, which can cause tooth cavities and periodontal disease, respectively [66]. Oleanolic acid also blocks the adherence of *S. mutans* to experimental surfaces, which reduces the risk of forming dental plaque [67]. Thus, raisins may protect teeth and gums through different means as mentioned above.

20.7 Conclusions

Raisins are important dried fruits that enhance diet quality due to their pleasing flavor (taste and aroma) as well as offering nutrition and physiological benefits. They are a rich source of phytochemicals, including phenolics, prebiotics, tartaric acid, and other minor components, which contribute to the different bioactivities and health benefits of raisins.

Although various drying techniques for raisin production have been developed, more research is needed to further refine and optimize the drying process. Pretreatment with chemicals such as sulfur dioxide should be avoided. The undesirable changes in color, flavor, and composition of raisins should also be minimized. In addition, the development of new raisin grape cultivars with high contents of total phenolics and low polyphenol oxidase activity would be valuable to growers and consumers.

More research is particularly needed in the areas related to raisin phytochemicals and in the extraction and analyses of minor phytochemicals and composition of flavonoids other than flavonols as well as the bioavailability of certain raisins phytochemicals, especially the absorption and metabolism of tartaric acid esters.

References

1. Bhat, N.R., Desai, B.B. & Suleiman,M.K. (2006) Grapes and raisins. In: *Handbook of Fruits and Fruit Processing* (ed. Y.H. Hui). Blackwell Publishing, Ames, IA, pp. 439–452.
2. Witherspoon, B. (2000) Raisins to the rescue. *School Food Service and Nutrition*, **54**, 60–63.
3. Ramos, I.N., Silva, C.L.M., Sereno, A.M. & Aguilera, J.M. (2004) Quantification of microstructural changes during first stage air drying of grape tissue. *Journal of Food Engineering*, **62**, 159–164.
4. USDA (2010) *Raisin Situation and Outlook in Selected Countries*. Published on-line at: http://www.fas.usda.gov/psdonline/circulars/raisins.pdf, last accessed December 12, 2010.
5. Karakaya, S., El, S.N. & Tas, A.A. (2001) Antioxidant activity of some foods containing phenolic compounds. *International Journal of Food Sciences and Nutrition*, **52**, 501–508.
6. Yeung, C.K., Glahn, R., Wu, X., Liu, R.H. & Miller, D. (2003) *In vitro* iron bioavailabilty and antioxidant activity of raisins. *Journal of Food Science*, **68**, 701–705.
7. Christensen, L.P. (2000) Raisin grape varieties. In: *Raisin Production Manual*. University of California, Agricultural, and Natural Resources Publication 3393, Oakland, CA, pp. 38–47.
8. Yaldiz, I. (2004) Drying characteristics and kinetics of okra. *Journal of Food Engineering*, **69**, 275–279.
9. Savoy, E.F., Morris, J.R. & Petrucci, V.E. (1983) Processing muscadine grapes into raisins. *Proceedings for Florida State Horticultural Society*, **96**, 355–357.
10. Casado, C.G. & Heredia, A. (1999) Structure and dynamic of reconstituted cuticular waxes of grape berry cuticle (*Vitis vinifera L.*). *Journal of Experimental Botany*, **50**, 175–182.
11. Esmaiili, M., Sotudeh-Gharebagh, R., Cronin, K., Mousavi, M.A.E. & Rezazadeh, G. (2007) Grape drying: a review. *Food Reviews International*, **23**, 257–280.
12. Di Matteo, M., Cinquanta, L., Galiero, G. & Crescitelli, S. (2000) Effect of a novel physical pretreatment process on the drying kinetics of seedless grapes. *Journal of Food Engineering*, **46**, 83–89.
13. Kostaropoulos, A.E. & Saravacos, G.D. (1995) Microwave pretreatment for sun-dried raisins. *Journal of Food Science*, **60**, 344–347.
14. Salengke, S. & Sastry, S.K. (2005) Effect of ohmic pretreatment on drying rate of grapes and adsorption isotherm of raisins. *Drying Technology*, **23**, 551–564.
15. Pangavhane, D.R. & Sawhney, R.L. (2002) Review of research and development work on solar dryers for grape drying. *Energy Conversion and Management*, **43**, 45–61.
16. Sharma, A.K. & Adulse, P.G. (2007) *Raisin Production in India*. National Research Centre (NRC) for Grapes, Pune, India.
17. Peacock, W.L. & Swanson, F.H. (2005) The future of California raisins is drying on the vine. *California Agriculture*, **59**, 70–74.
18. Wakjira, M. (2010) Solar drying of fruits and windows of opportunities in Ethiopia. *African Journal of Food Science*, **4**, 790–802.
19. Jairaj, K.S., Singh, S.P. & Srikand, K. (2009) A review of solar dryer developed for grape drying. *Solar Energy*, **83**, 1698–1712.
20. Clary, C.D. & Ostrom, G.A. (1995) *Use of Microwave Vacuum for Dehydration of Thompson Seedless Grapes*. Research Bulletin, California Agricultural Technology Institute (CATI) Publication No. 950405, Fresno, CA.
21. King, C.J. (1973) Freeze drying. In: *Food Dehydration*, 2nd edn. (eds W.B. Van Arsdel, M.J. Copley & A.I. Morgan, Jr). AVI Publication Company Inc., Westport, CT, pp. 161–200.
22. Tulasidast, T.N., Ratti, C. & Raghavan, G.S.V. (1997) Modelling of microwave drying of grapes. *Canadian Agricultural Engineering*, **39**, 57–67.
23. Yaldiz, O., Ertekin, C. & Uzun, H.I. (2001) Mathematical modelling of thin layer solar drying of Sultana grapes. *Energy*, **26**, 457–465.
24. Kostaropoulos, A.E., Mandala, J., Spiess, W.E.L. & Saravacos, G.D. (1997) Factors influencing the friction of raisins during processing and handling. *Journal of Food Engineering*, **33**, 385–393.
25. Winkler, A.J., Cook, J.A., Kliewer, J.A. & Lider, L.A. (1974) *General Viticulture*. University of California Press, Berkeley, CA.
26. USDA (1978) *United States Standards for Grades of Processed Raisins*. Agricultural Marketing Service, Fruit and Vegetable Division, Processed Products Branch, Washington, DC.
27. USDA (2010) *National Nutrient Database for Standard Reference, Release 23*. Published on-line at: http://www.nal.usda.gov/fnic/foodcomp/search/, last accessed January 14, 2011.

28. USDA (2005) *Carbohydrates. Dietary Guidelines for Americans*. US Department of Agriculture and US Department of Health and Human Services, Government Printing Office, Washington, DC.
29. Camire, M.E. & Dougherty, M.P. (2003) Raisin dietary fiber composition and *in vitro* bile acid binding. *Journal of Agricultural and Food Chemistry*, **51**, 834–837.
30. Kay, R.M. (1982) Dietary fiber. *Journal of Lipid Research*, **23**, 221–242.
31. Krueger, J., Kuelbs, M., May, L. & Wolcott, J. (2003) *Carbohydrate, Paper Chromatography, and Enzyme Tests Distinguish Chemical Differences Between Grapes and Raisins*. Published on-line at: https://www.msu.edu/course/lbs/145/luckie/inquiriesf2003/cellular4.html, last accessed January 16, 2011.
32. Moegan, A.F., Kimmel, L., Field, A. & Nichols, P.F. (1934) The vitamin content of sultanina (Thompson seedless) grapes and raisins. *Journal of Nutrition*, **9**, 369–382.
33. Rainey, C.J., Nyquist, L.A., Christensen, R.E., Strong, P.L., Culver, B.D. & Coughlin, J.R. (1999) Daily boron intake from the American diet. *Journal of American Dietetic Association*, **99**, 335–340.
34. Shahidi, F. & Tan, Z. (2010) Physiological effects of sesame: bioactive and antioxidant compounds. In: *Sesame: The Genus Sesamum* (ed. B. Dorothea). CRC Press, Taylor & Francis Group, Boca Raton, FL, pp. 139–154.
35. Shahidi, F. & Tan, Z. (2009) Bioactives and health benefits of Brazil nut. In: *Tree Nuts: Composition, Phytochemicals, and Health Effects* (eds C. Alasalvar & F. Shahidi). CRC Press, Taylor & Francis Group, Boca Raton, FL, pp. 143–156.
36. Shahidi, F. & Wanasundara, P.K.J.P.D. (1992) Phenolic antioxidants. *Critical Reviews in Food Science and Nutrition*, **32**, 67–103.
37. Shahidi, F. & Naczk, M. (2004) *Phenolics in Food and Nutraceuticals*. CRC Press, Boca Raton, FL.
38. Antolovish, M., Prenzler, P.D., Patsalides, E., Mcdonald, S. & Robards, K. (2002) Methods for testing antioxidant activity. *Analysis*, **127**, 183–197.
39. Wu, X., Beecher, G.R., Holden, J.M., Haytowitz, D.B., Gebhardt, S.E. & Prior, R. (2004) Lipophilic and hydrophilic antioxidant capacities of common foods in the United States. *Journal of Agriculture and Food Chemistry*, **52**, 4026–4037.
40. Karadeniz, F., Durst, R.W. & Wrolstad, R.E. (2000) Polyphenolic composition of raisins. *Journal of Agricultural and Food Chemistry*, **48**, 5343–5350.
41. Parker, T.L., Wang, X.H., Pazmino, J. & Engeseth, N.J. (2007) Antioxidant capacity and phenolic content of grapes, sun-dried raisins & golden raisins & their effect on *ex vivo* serum antioxidant capacity. *Journal of Agricultural and Food Chemistry*, **55**, 8472–8477.
42. USDA (2007) *United States Department of Agriculture Database for the Flavonoids Content of Selected Foods, Release 2.1*. Published on-line at: http://www.nal.usda.gov/fnic/foodcomp/data/flav/flav02–1.pdf, last accessed January 16, 2011.
43. Williamson, G. & Carughi, A. (2010) Polyphenol content and health benefits of raisins. *Nutrition Research*, **30**, 511–519.
44. Zhao, B. & Hall III, C.A. (2008) Composition and antioxidant activity of raisin extracts obtained from various solvents. *Food Chemistry*, **108**, 511–518.
45. Ong, B.Y. & Nagel, C.W. (1978) High-pressure liquid chromatographic analysis of hydroxycinnamic acid-tartaric acid esters and their glucose esters in *Vitis vinifera*. *Journal of Chromatography A*, **157**, 345–355.
46. Reinli, K. & Block, G. (1996) Phytoestrogen content of foods: a compendium. *Nutrition and Cancer*, **26**, 123–148.
47. Liggins, J., Bluck, L.J.C. & Runswick, S. (2000) Daidzein and genistein content of fruits and nuts. *Journal of Biochemistry*, **11**, 326–331.
48. Singleton, V.L. & Trousdale, E. (1983) White wine phenolics: varietal and processing differences as shown by HPLC. *American Journal of Enology and Viticulture*, **34**, 27–34.
49. Cejudo-Bastante, M.J., Perzcoello, M.S. & Hermosin-Guntierrez, I. (2010) Identification of new derivatives of 2-S-glutathionylcaftaric acid in aged white wines by HPLC-DAD-ESI-MS^n. *Journal of Agricultural and Food Chemistry*, **58**, 11483–11492.
50. Burns, J., Yokota, T., Ashihara, H., Lean, M.E. & Crozier, A. (2002) Plant foods and herbal sources of resveratrol. *Journal of Agricultural and Food Chemistry*, **50**, 3337–3340.
51. Pezet, R. & Cuenat, P.H. (1996) Resveratrol in wine: extraction from skin during fermentation and post-fermentation standing of must from Gamay grapes. *American Journal of Enology and Viticulture*, **47**, 287–290.

52. Siemann, E.H. & Creasy, L.L. (1992) Concentration of the phytoalexin resveratrol in wine. *American Journal of Enology and Viticulture*, **43**, 49–51.
53. Jeandet, P., Bessis, R. & Gautheron, B. (1991) The production of resveratrol (3,5,4'-trihydroxystilbene) by grape berries in different developmental stages. *American Journal of Enology and Viticulture*, **42**, 41–46.
54. Lamuela-Raventos,R.M. & Waterhouse,A.L. (1993) Occurrence of resveratrol in selected California wines by a new HPLC method. *Journal of Agricultural and Food Chemistry*, **41**, 521–523.
55. FAO (2007) *FAO Technical Meeting on Prebiotics*. Published on-line at: http://www.fao.org/ag/agn/agns/files/prebiotics_tech_meeting_report.pdf, last accessed January 10, 2011.
56. Hu, B. (2003) Vegetarian diets and risk of selected chronic diseases: plant based foods and prevention of cardiovascular disease: an overview. *American Journal of Clinical Nutrition*, **78**, 544–551.
57. Ness, A.R. & Powles, J.W. (1997) Fruit and vegetables, and cardiovascular disease: a review. *International Journal of Epidemiology*, **269**, 1–12.
58. Gibson, G.R., Probert, H.M., Van Loo, J., Rastall, R.A. & Roberfroid, M. (2004) Dietary modulation of the human colonic microbiota: updating the concept of prebiotics. *Nutrition Research Reviews*, **17**, 259–275.
59. Langlands, S.J., Hopkins, M.J., Coleman, N. & Cummings, J.H. (2004) Prebiotic carbohydrates modify the mucosa-associated microflora of the human large bowel. *Gut*, **53**, 1610–1616.
60. Roberfroid, M.B. & Delzenne, N.M. (1998) Dietary fructans. *Annual Review of Nutrition*, **18**, 117–143.
61. Gibson, G.R. & Roberfroid, M.B. (1995) Dietary modulation of the human colonic microbiota: introducing the concept of prebiotics. *Journal of Nutrition*, **125**, 1401–1412.
62. Carughi, A. (2009) Raisins as source of prebiotic compounds in the diet. *Journal of Federation of American Societies for Experimental Biology*, **23**, 716.9.
63. Laszczyk, M.N. (2009) Pentacyclic triterpenes of the lupane, oleanane, and ursane group as tools in cancer therapy. *Planta Medica*, **75**, 1549–1560.
64. Liu, J. (1995) Pharmacology of oleanolic acid and ursolic acid. *Journal of Ethnopharmacology*, **49**, 57–68.
65. Wick, W., Grimmel, C., Wagenknecht, B., Dichgans, J. & Weller, M. (1999) Betulinic acid-induced apoptosis in glioma cells: a sequential requirement for new protein synthesis, formation of reactive oxygen species, and caspase processing. *Journal of Pharmacology and Experimental Therapeutics*, **289**, 1306–1312.
66. Rivera-Cruz, J.F., Zhu, M., Su, B., Kinghorn, A.D. & Wu, C.D. (2005) Antimicrobial phytochemicals in Thompson seedless raisins (*Vitis vinifera* L.) inhibit dental plaque bacteria. Paper given at *the American Society of Microbiology Meeting*, Atlanta, GA, 5–9 June 2005 (Abstract No: O-068).
67. Wu, C.D., Zhu, M., Su, B.N., Workman, J.T. & Kinghorn, A.D. (2003) Phytochemicals in raisins inhibit growth and adherence of plaque bacteria. Paper given at *81st General Session of the International Association for Dental Research*, Guthenburg, Sweden, 25–28 June 2003 (Abstract No: 2053).
68. Spiller, G.A., Story, J.A., Furumoto, E.J., Chezem, J.C. & Spiller, M. (2003) Effect of tartaric acids and dietary fiber from sun-dried raisins on colonic function and on bile acid and volatile fatty acid excretion in healthy adults. *British Journal of Nutrition*, **90**, 803–807.
69. Spiller, G.A. & Spiller, M. (2001) Tartaric acid content of foods. In: *Handbook of Dietary Fiber in Human Nutrition*, 3rd edn. (ed. G.A. Spiller). CRC Press, Boca Raton, FL, pp. 681–882.
70. Piironen, V., Toivo, J., Puupponen-Pimia, R. & Lampi, A.M. (2003) Plant sterols in vegetables, fruits, and berries. *Journal of the Science of Food and Agriculture*, **83**, 330–337.
71. Duthie, G. & Crozier, A. (2000) Plant-derived phenolic antioxidants. *Current Opinion in Lipidology*, **11**, 43–47.
72. Halvorsen, B.E., Carlsen, M.H., Phillips, K.M., Bohn, S.K., Holte, K., Jacobs Jr, D.R. & Blomhoff, R. (2006) Content of redox active compounds (ie, antioxidants) in foods consumed in the United States. *American Journal of Clinical Nutrition*, **84**, 95–135.
73. Pellegrini, N., Serafini, M., Salvatore, S., Del Rio, D., Bianchi, M. & Brighenti, F. (2006) Total antioxidant capacity of spices, dried fruits, nuts, pulses, cereals and sweets consumed in Italy assessed by three different *in vitro* assays. *Molecular Nutrition & Food Research*, **50**, 1030–1038.
74. Wolfe, K.L. & Liu, R.H. (2007) Cellular antioxidant activity (CAA) assay for assessing antioxidants, foods, and dietary supplements. *Journal of Agricultural and Food Chemistry*, **55**, 8896–8907.
75. Vinson, J.A., Proch, J. & Bose, P. (2001) Determination of the quantity and quality of polyphenol antioxidants in foods and beverages. *Methods in Enzymology*, **335**, 103–114.

76. Vinson, J.A., Xuehui, S., Zubik, L. & Bose, P. (2001) Phenol antioxidant quantity and quality in foods: fruits. *Journal of Agricultural and Food Chemistry*, **49**, 5315–5321.
77. Vinson, J.A., Zubik, L., Bose, P., Samman, N. & Proch, J. (2005) Dried fruits: excellent *in vitro* and *in vivo* antioxidants. *Journal of the American College of Nutrition*, **24**, 44–50.
78. Cao, G., Russell, R.M., Lischener, N. & Prior, R.L. (1998) Serum antioxidant capacity is increased by consumption of strawberries, spinach, red wine or vitamin C in elderly women. *Journal of Nutrition*, **128**, 2383–2390.
79. Cao, G., Booth, S.L., Sadowski, J.A. & Prior, R.L. (1998) Increases in human plasma antioxidant capacity after consumption of controlled diets high in fruit and vegetables. *American Journal of Clinical Nutrition*, **68**, 1081–1087.
80. Rankin, J.W., Andreae, M.C., Chen, O. & O'Keefe, S.F. (2008) Effect of raisin consumption on oxidative stress and inflammation in obesity. *Diabetes, Obesity and Metabolism*, **10**, 1086–1096.
81. Halliwell, B. (2002) Effect of diet on cancer development: is oxidative DNA damage a biomarker? *Free Radical Biology & Medicine*, **32**, 968–974.
82. Spiller, G.A., Schultz, L., Spiller, M. & Ou, B. (2002) Sun-dried raisins help prevent oxidative DNA damage during intense athletic activity. *Journal of the American College of Nutrition*, **21**, 482 (Abstract 64).
83. Statistics Canada. (2007) *Table 102–0529—Deaths by Cause: Diseases of the Circulatory System (I00 to I99), Age Group, and Sex. CANSIM (Database)*. Published on-line at: http://www5.statcan.gc.ca/cansim/a01?lang=eng, last accessed March 7, 2011.
84. Graf, B.A., Milbury, P.E. & Blumberg, J.B. (2005) Flavonols, flavones, flavanones, and human health: epidemiological evidence. *Journal of Medicinal Food*, **8**, 281–290.
85. Bruce, B., Spiller, G.A. & Farquhar, J.W. (1997) Effects of a plant-based diet rich in whole grains, sun-dried raisins, and nuts on serum lipoproteins. *Vegetarian Nutrition: An International Journal*, **1**, 58–63.
86. Bruce, B., Spiller, G.A., Klevay, L.M. & Gallagher, S.K. (2000) A diet high in whole and unrefined foods favorably alters lipids, antioxidant defences, and colon function. *Journal of the American College Nutrition*, **19**, 61–67.
87. Puglisi, M.J., Vaishnav, U., Shrestha, S., Torres-Gonzalez, M., Wood, R.J., Volek, J.S. & Fernandez, M.L. (2008) Raisins and additional walking have distinct effects on plasma lipids and inflammatory cytokines. *Lipids in Health and Disease*, **7**, 14–18.
88. The Canadian Diabetes Association. (2010) *Diabetes Facts*. Published on-line at: http://www.diabetes.ca/files/diabetes_fact_sheet.pdf, last accessed February 8, 2011.
89. Kim, Y., Hertzler, S.R., Byrne, H.K. & Mattern, C.O. (2008) Raisins are a low to moderate glycemic index food with a correspondingly low insulin index. *Nutrition Research*, **28**, 304–308.
90. Foster-Powell, K., Holt, S.H.A. & Brand-Miller, J.C. (2002) International table of glycemic index and glycemic load values. *American Journal of Clinical Nutrition*, **76**, 55–56.
91. Jenkins, D.J., Wolever, T.M., Taylor, R.H., Barker, H., Fielden, H., Baldwin, J.M., Bowling, A.C., Newman, H.C., Jenkins, A.L. & Goff, D.V. (1981) Glycemic index of foods: a physiological basis for carbohydrate exchange. *American Journal of Clinical Nutrition*, **34**, 362–366.
92. Lock, D.R., Bar-Eyal, A., Voet, H. & Madar, Z. (1988) Glycemic indices of various foods given to pregnant diabetic subjects. *Obstetrics and Gynaecology*, **71**, 180–183.
93. Knekt, P., Kumpulainen, J., Jarvinen, R., Rissanen, H., Heliovaara, M., Reunanen, A., Hakulinen, T. & Aromaa, A. (2002) Flavonoid intake and risk of chronic diseases. *American Journal of Clinical Nutrition*, **76**, 560–568.
94. Nettleton, J.A., Harnack, L.J., Scrafford, C.G., Mink, P.J., Barraj, L.M. & Jacobs Jr,D.R. (2006) Dietary flavonoids and flavonoid-rich foods are not associated with risk of type 2 diabetes in postmenopausal women. *Journal of Nutrition*, **136**, 3039–3045.
95. Song, Y., Manson, J.E., Buring, J.E., Sesso, H.D. & Liu, S. (2005) Associations of dietary flavonoids with risk of type 2 diabetes, and markers of insulin resistance and systemic inflammation in women: a prospective study and cross-sectional analysis. *Journal of the American College of Nutrition*, **24**, 376–384.
96. Steinmetz, K.A. & Potter, J.D. (1996) Vegetables, fruit, and cancer prevention: a review. *Journal of the American Dietetic Association*, **96**, 1027–1039.
97. Reddy, B.S. (1999) Possible mechanisms by which pro- and prebiotics influence colon carcinogenesis and tumor growth. *Journal of Nutrition*, **129** (Suppl.), 1478S–1482S.

98. Reddy, B.S., Hamid, R. & Rao, C.V. (1997) Effect of dietary oligofructose and inulin on colonic preneoplastic aberrant crypt foci inhibition. *Carcinogenesis*, **18**, 1371–1374.
99. Poulsen, M., Molck, A.M. & Jacobsen, B.L. (2002) Different effects of short and long chained fructans on large intestinal physiology and carcinogen-induced aberrant crypt foci in rats. *Nutrition and Cancer*, **42**, 194–205.
100. Rossi, M., Negri, E., Lagiou, P., Talamini, R., Dal Maso, L., Montella, M., Franceschi, S. & La Vecchia, C. (2008) Flavonoids and ovarian cancer risk: a case control study in Italy. *Epidemiology*, **123**, 895–898.
101. Bosetti, C., Spertini, L., Parpinel, M., Gnagnarella, P., Lagiou, P., Negri, E., Franceschi, S., Montella, M., Peterson, J., Dwyer, J., Giacosa, A. & La Vecchia, C. (2005) Flavonoids and breast cancer risk in Italy. *Cancer Epidemiology, Biomarkers & Prevention*, **14**, 805–808.
102. Rossi, M., Garavello, W., Talamini, R., Negri, E., Bosetti, C., Dal Maso, L., Lagiou, P., Tavani, A., Polesel, J., Barzan, L., Ramazzotti, V., Franceschi, S. & La Vecchia, C. (2007) Flavonoids and the risk of oral and pharyngeal cancer: a case control study in Italy. *Cancer Epidemiology, Biomarkers & Prevention*, **16**, 1621–1625.
103. Rossi, M., Negril, E., Talamini, R., Bosetti, C., Parpinel, M., Gnagnarella, P., Franceschi, S., Dal Maso, L., Montilla, M., Giacosa, A. & La Vecchia, C. (2006) Flavonoids and colorectal cancer in Italy. *Cancer Epidemiology, Biomarkers & Prevention*, **15**, 1555–1558.
104. Nothlings, U., Murphy, S.P., Wilkens, L.R., Henderson, B.R. & Kolonel, L.N. (2007) Flavonols and pancreatic cancer risk: the multiethnic cohort study. *American Journal of Epidemiology*, **166**, 924–931.
105. Kahket, S., Van Houte, L.R. & Stocks, S. (1991) Lack of correlation between food retention on the human dentition and consumer perception of food stickiness. *Journal of Dental Research*, **70**, 1314–1319.
106. Newbrun, E. (1991) Dental effects of sugars and sweeteners. In: *Sugars and Sweeteners* (eds N. Kretchmer & C.B. Hollenbeck). CRC Press, Boca Raton, FL, pp. 175–202.

Part 3
Tropical Dried Fruits

21 Açai fruits: potent antioxidant and anti-inflammatory superfruits with potential health benefits

Alexander G. Schauss

21.1 Introduction

The terms "superfood" and "superfruit" first appeared in food advertising at the beginning of the twenty-first century to describe a food that possessed functional health properties beyond its nutritive value. Familiar foods such as blueberries [1], strawberries [2, 3], cranberries [4, 5], walnuts [6], and pomegranates [7] were the first to be categorized as superfoods, which resulted in a significant increase in their consumption based on the assumption that food sources containing exogenous nutrient and/or phytochemical antioxidants could protect the body from damage arising from chronic oxidative stress—caused by excessive production of free radicals. However, initial clinical evidence for any potential health benefits of superfoods was open to question based on clinical trial outcomes. For example, a walnut consumption randomized crossover trial did not see a significant change in plasma antioxidant capacity in healthy, well-nourished older adults [8], while a cranberry juice randomized, double-blind, and placebo-controlled trial of female subjects with metabolic syndrome observed a significant reduction in lipid oxidation and an increase in plasma antioxidant capacity [9]. Evidence is accumulating that certain superfoods, such as the subject of this chapter, not only have antioxidant bioactivity but also possess anti-inflammatory properties.

There is a growing consensus in the medical community that chronic non-resolved inflammation contributes to numerous diseases including atherosclerosis [10–12], ischemic heart disease [13, 14], hypertension [15], cancer [16–18], obesity [19], inflammatory bowel disease [20, 21], Crohn's disease [19, 22], type 2 diabetes [23, 24], end-stage renal disease [25], and autoimmune disorders [26], among others.

Of all fruits that quickly became identified as superfruits based on growing and increased year-to-year consumer interest through 2010, none captured as much attention and interest among food scientists as that of a small Amazonian palm fruit known as "açai" (pronounced, "ah-sigh-ee") [27]. From virtual obscurity in the 1990s to superfruit status a decade later [28], this small and tasteless nutritionally dense fruit with remarkable antioxidant and anti-inflammatory properties *in vitro* and *in vivo* came to define what a superfruit is.

Dried Fruits: Phytochemicals and Health Effects, First Edition. Edited by Cesarettin Alasalvar and Fereidoon Shahidi.

21.1.1 Açai palm trees

Palms belong to the plant family known as Arecaceae, which consists of nearly 4000 species that grow primarily in the tropics and subtropics. Within the palm family there are more than 200 genera. Nearly 220 different palm species have been identified as growing in the Amazon, of which only 40 species of palms are within the genera *Euterpe*. Three commercial species of açai fruits, *Euterpe oleracea*, *Euterpe precatoria*, and *Euterpe edulis*, are only found growing at various altitudes in the southern hemisphere within a geographical area commonly referred to as the Amazon.

The origin of the word, *Euterpe*, comes from the Greek words, "forest grace," owing to its elegant appearance. Its dropping leaflets give the long pendulous fronds the appearance that rain just fell on them. Among the three açai palms of interest, *E. edulis* is primarily harvested for its heart-of-palm, not its fruit. Of the remaining two *Euterpe* species, *E. oleracea* has received the most research attention due to its ample availability and documented traditional use as a food in and between the Amazon River and its tributaries and estuaries in the Amazon floodplains of Brazil. In the State of Para, Brazil, during any 8-month açai fruits harvest season, more than 35,000 people are employed at various occupations in the state's largest city, Belem, gathering, wholesaling, retailing or processing the fruit into frozen pulp or either spray-dried or freeze-dried powders for domestic consumption and/or export. More recently, *E. precatoria* has drawn attention with the discovery of remarkably stronger antioxidant and anti-inflammatory activity than even *E. oleracea*, which had already gained recognition for having the highest antioxidant activity on the oxygen radical absorbance capacity (ORAC) assay of almost any food tested by the US Department of Agriculture (USDA) [29].

21.2 Compositional and nutrition characteristics of açai fruits

21.2.1 *Euterpe oleracea* fruits

Euterpe oleracea fruits have been found to have a remarkable range of nutrients, including unusually high content of mono- and polyunsaturated fatty acids (MUFA and PUFA) approximating that of olives and avocadoes, phytosterols, amino acids, soluble and insoluble fibers, vitamins, minerals, and trace elements, among others [30, 31]. The compositional and nutritional characteristics of freeze-dried açai fruits are shown in Table 21.1.

The phytochemistry of *E. oleracea* fruits has proven to have similarities to other antioxidant-rich berries. Anthocyanins are believed to be the major compounds that contribute to the *E. oleracea*'s antioxidant activity. *E. oleracea* fruits contain two major anthocyanins, namely cyanidin 3-glucoside and cyanidin 3-rutinoside, and three minor anthocyanins, cyanidin 3-sambubioside, peonidin 3-glucoside, and peonidin 3-rutinoside [30]. However, quantitatively, total anthocyanin content is lower than other dark-colored berries, such as blackberries, blueberries, and cranberries, which suggest that açai fruits' extraordinary antioxidant activity may be due to other compounds or a combination of anthocyanins and other compounds. Proanthocyanidin content has also been characterized and quantified, as proanthocyanidin have been found to possess strong antioxidant capacity. Açai fruits contain monomers (catechin and epicatechin) and B type procyanidins from dimers to polymers,

Table 21.1 Compositional and nutritional characteristics of freeze-dried açai fruits

Nutrient	Units	Freeze-dried[a] [31]
Proximate composition		
Water	g	3.4
Energy	kcal	534
Protein	g	8.1
Lipid	g	32.5
Ash	g	3.8
Carbohydrate	g	52.2
Dietary fiber	g	44.2
Sugars	g	1.3
Minerals		
Calcium	mg	260
Iron	mg	4.4
Potassium	mg	na
Sodium	mg	30.4
Vitamins		
β-Carotene	IU	<5.0
Vitamin A	IU	1002
Vitamin C	mg	<0.1
Vitamin E (ATE)	mg	na
Amino acids		
Alanine	g	0.46
Arginine	g	0.42
Aspartic acid	g	0.83
Cystine	g	0.18
Glutamic acid	g	0.80
Glycine	g	0.39
Histidine[b]	g	0.17
Hydroxyproline	g	<0.01
Isoleucine[b]	g	0.38
Leucine[b]	g	0.65
Lysine[b]	g	0.66
Methionine[b]	g	0.12
Phenylalanine[b]	g	0.43
Proline	g	0.53
Serine	g	0.32
Threonine[b]	g	0.31
Tryptophan[b]	g	0.13
Tyrosine	g	0.29
Valine[b]	g	0.51
Fatty acids		
SFA	g/100 g oil	26.1
MUFA	g/100 g oil	60.6
PUFA	g/100 g oil	13.3
Phytosterols		
β-Sitosterol	mg/g	0.44
Campesterol	mg/g	<0.03
Stigmasterol	mg/g	0.04

Note: Some numbers are rounded to the second digit after decimal point.
ATE, alpha-tocopherol equivalents; na, not available; SFA, saturated fatty acids; MUFA, monounsaturated fatty acids; PUFA, polyunsaturated fatty acids.
[a]Values per 100 g in dry weight unless otherwise stated.
[b]Indispensable amino acids.

with the latter being the major proanthocyanidins accounting for 12.89 mg/g dry weight (DW) [30]. Interestingly, the proanthocyanidin profiles of açai fruits are similar to that of blueberries.

21.3 Antioxidant and anti-inflammatory activities of açai fruits

The ORAC assay was initially developed to measure antioxidant inhibition of hydrophilic and lipophilic compounds and components in foods against peroxyl radical–induced oxidation [32]. Over time, a series of additional assays have been developed based on the original ORAC peroxyl radical scavenging assay that measures antioxidant inhibition against other reactive oxygen species/reactive nitrogen species (ROS/RNS) moieties, including $^{\bullet}OH$, $O_2^{\bullet-}$, $ONOO^-$, and $^1O_2$ [33], and which have been given the corresponding names of hydroxyl-ORAC (HORAC), superoxide-ORAC (SORAC), peroxynitrite-ORAC (NORAC), and singlet-oxygen-ORAC (SOAC) to distinguish them, as there are various reactive nitrogen or oxygen free radical species involved in the pathophysiology of human diseases. The antioxidant and anti-inflammatory activities of açai fruits have been evaluated by numerous assays. Besides the ORAC assay, 2,2′-diphenyl-1-picrylhydrazyl (DPPH) assay has also been widely employed [34].

Characterization of *E. precatoria* fruits has only recently focused on their phytochemical compositions. As was the case with *E. oleracea* fruits, anthocyanins were the predominant polyphenolics. Various flavones, flavanol derivatives including procyanidin dimers and trimers, and phenolic acids are present in *E. precatoria* fruits, leading to the conclusion that both açai species are characterized by similar polyphenolic profiles and antioxidant activities [35]. However, later investigations found significantly higher antioxidant activity in *E. precatoria* fruits than *E. oleracea* fruits [36]. As shown in Table 21.2, where the combination of hydrophilic ORAC ($ORAC_{hydro}$) and lipophilic ORAC ($ORAC_{lipo}$) scavenging capacity yields an ORAC value of 1027 μmol trolox equivalents (TE)/g for *E. oleracea* fruits, the combined ORAC result for *E. precatoria* fruits yield a value of 1824 μmol TE/g, a significant difference in scavenging capacity. The antioxidant activity of *E. precatoria* fruits in scavenging the peroxyl radical is superior to *E. oleracea* fruits, and far superior to other antioxidant-rich fruits and berries, nuts, vegetables, and grains, with the exception of sumac bran, based on USDA database of ORAC values for foods [29].

A further comparison between *E. oleracea* and *E. precatoria* fruits' free radical scavenging activities that include the ORAC, HORAC, SORAC, NORAC, and SOAC assays, to provide a more comprehensive understanding of each açai fruits' radical scavenging activities, has been carried out, given the complexity of the body's antioxidant defense system. This is now becoming the standard method of *in vitro* antioxidant scavenging assessment since a single antioxidant assay cannot reflect all aspects of scavenging activities of foods otherwise. As seen in Table 21.3, *E. precatoria* fruits have nearly three-fold higher total ORAC value against multiple radicals of *E. oleracea* fruits *in vitro*.

In evaluating the antioxidant activities of a given food or natural product, combining both chemical and cell-based assays provides a useful approach toward understanding the antioxidant effects of antioxidants found in food and their biological relevance to any health benefits that may be observed *in vivo* [37]. The advantages of a cell-based assay over current

Table 21.2 ORAC values of açai species' pulp compared to other antioxidant-rich foods

Food	μmol TE/g DW	Reference
Sumac, bran	3124	[29]
Açai pulp (*E. precatoria*)	1824	[38]
Açai pulp (*E. edulis*)	1193	[a]
Açai pulp (*E. oleracea*)	1027	[29, 30]
Sorghum, bran, black	1008	[29]
Cocoa, dry powder, unsweetened	557	[29]
Chocolate, unsweetened	499	[29]
Black raspberry	192	[29]
Pecan	179	[29]
Chokeberry	160	[29]
Elderberry	147	[29]
Walnut	135	[29]
Cranberry	91	[29]
Blueberry (wild)	96	[29]
Plums	76	[29]
Blackberries	59	[29]
Pomegranate	45	[29]
Strawberry	43	[29]

DW, dry weight; TE, trolox equivalents.
[a]Unpublished data.

chemical assays such as the ORAC and DPPH assays are numerous. Whereas chemical assays depend on chemical reactions using a limited number of reagents, cell-based assays take into consideration complex enzymatic reactions in biological systems, providing qualitative rather than just quantitative assessment. In this way, cell-based assays incorporate those enzymes involved in redox reactions that are occurring in red blood cells (RBCs) as well as polymorphonuclear (PMN) cells or other living cells.

Table 21.3 ORAC values of two different freeze-dried açai fruit species (*E. precatoria* and *E. oleracea*)

ORAC-based assays	***E. precatoria*** **(μmol of TE/g DW)**	***E. oleracea*** **(μmol of TE/g DW)**
ORAC (against peroxyl radical)	1828	1014
H-ORAC (hydrophilic fraction)	1792 ± 90	986 ± 57
L-ORAC (lipophilic fraction)	36 ± 2.6	28 ± 2.1
HORAC (against hydroxyl radical)	4114 ± 313	1357 ± 68
SORAC (against superoxide anion radical)	1040 ± 55	169 ± 12
NORAC (against peroxynitrite anion)	87 ± 5.9	37.2 ± 2.6
SOAC (against singlet oxygen)	629 ± 33.4	71.6 ± 8.8
Total ORAC	7698	2649

Source: Adapted with permission from Kang *et al.* [38].
Note: Some numbers are rounded to the second digit after decimal point.
DW, dry weight; TE, trolox equivalents.

The comparison of the antioxidant activities of *E. oleracea* and *E. precatoria* fruits has been evaluated by multiple assays, including ORAC, DPPH, and cell-based antioxidant protection in erythrocyte (CAP-e). The CAP-e assay showed that both *E. oleracea* and *E. precatoria* fruits protect human cells from oxidative damage [38]. In addition, potential anti-inflammatory activity was evaluated by the lipopolysaccharide (LPS)-induced secreted embryonic alkaline phosphatase (SEAP) reporter assay, designed to measure nuclear factor-kappa B (NF-κB) activation [39, 40]. *E. precatoria* fruits contained much stronger water-soluble antioxidants that can enter live cells and effectively inhibit ROS formation compared to those from *E. oleracea* fruits. *E. precatoria* fruits, but not *E. oleracea* fruits, inhibited NF-κB activation as assessed by the SEAP reporter assay, suggesting that *E. precatoria* fruits have potential anti-inflammatory effects. NF-κB is a major mechanism for induction of inflammatory and immune responses.

Considering the extremely high SOAC value observed (Table 21.3), carotenoids were analyzed in *E. precatoria* fruits and five carotenoids, namely β-carotene, lycopene, astaxanthin, lutein, and zeaxanthin, were detected and quantified. Their levels ranged from 18.7 μg/g DW of astaxanthin to 483.0 μg/g DW of lutein for *E. precatoria* fruits; total cartenoid in *E. precatoria* fruits was calculated as 963.7 μg/g DW. Only β-carotene reached a detectable level, quantified at 10.8 μg/g DW in *E. oleracea* fruits [38]. These differences may explain in part the significant difference in SOAC values between both açai fruit species and the role that carotenoids play in singlet-oxygen scavenging.

Açai fruits have been shown to improve the lifespan and survival of *Drosophila melanogaster* females fed a high-fat diet through activation of stress response pathways and the suppression of phosphoenolpyruvate carboxykinase (*Pepck*) expression, increased transcription of a small heat-shock-related protein (*l(2)efl*), and two detoxification genes, while decreasing the transcript level of *Pepck* [41]. The study demonstrated a potential benefit in alleviating the effect of oxidative stress on aging, as oxidative damage that accumulates in cells with advancing age is thought to be one of the causative contributions to aging [42].

Studies on the cardioprotective properties of açai fruits have been demonstrated. Animals were fed high-fat hypercholesterolemic diets, while controls were fed standard diets. De Souza *et al.* [43] fed rats a hypercholesterolemic diet for 6 weeks. Serum levels of carbonyl proteins and total, free, and protein sulfhydryl groups, and superoxide dismutase activity, at the end of the period were reduced. Feio *et al.* [44] hypothesized that feeding of açai fruits to New Zealand rabbits on a cholesterol-enriched diet for 12 weeks followed by 12 weeks of a diet containing an açai fruit extract would attenuate the development of atherosclerosis by decreasing cholesterol absorption and synthesis. Rabbits fed açai fruits in the last half of the feeding study had marked improvement in their lipid profile, smaller atherosclerotic plaque area in aortas, smaller intima/media ratio, and differences in plaque composition, than controls. The attenuation of atherosclerosis was marked. These improvements in cardiovascular status may partially be due to the endothelial anti-inflammatory effects of açai fruit polyphenolics seen during oxidative stress in another study as açai fruits inhibited gene expression of adhesion molecules and NF-κB activation [45].

An açai-enriched fruit/berry juice attenuated atherosclerosis in apolipoprotein E deficient ($apoE^-$) mice through antioxidant and anti-inflammatory activities [46]. The treatment group fed the juice experienced a reduction in mean lesion area of 58% in the aorta compared to the controls ($P < 0.001$). Treated mice also had higher high-density lipoprotein (HDL)-cholesterol and significantly lower levels of lipid peroxidation, including F_2-isoprostanes and isomers of hydroxyoctadecadienoic acids (HODEs), and hydroxyeicosatetraenoic acids

(HETEs) in serum and in liver. The expression of two antioxidant enzyme genes was significantly upregulated in the aorta, while glutathione peroxidase (GPX), glutathione reductase (GSR), and serum paraoxonase-1 (PON1) activity was increased in serum and/or liver of animals fed the açai juice. Tumor necrosis factor-alpha (TNF-α) and interleukin-6 (IL-6) levels were lower in resident macrophages with or without LPS stimulation, as well. The açai-enriched juice has assembled a body of experimental evidence of safety [47], allowing for future studies on the juice's impact on various disease processes and their progression.

A number of studies have also examined the effect of açai fruits fractions on brain cells *in vitro*. Poulose *et al.* [48] carried out experiments to assess the effects of ethanol, methanol, acetone, and ethyl acetate fractions of açai fruits on dopamine-induced changes in recovery, reactive ROS production, and viability, in primary rat hippocampal cells. The results indicate that a 1-hour pretreatment of hippocampal cells with ethanol (50 μg/mL) and acetone (10 μg/mL) fractions were effective in preventing decreases in calcium recovery. In BV-2 microglial cells, all of the fractions were effective in preventing LPS-induced increases in nitrite. The protection of microglial cells by ethanol, methanol, and acetone açai fruit fractions was accompanied by a significant concentration-dependent reduction in NF-κB, TNF-α, cyclooxygenase-2 (COX-2), and p38 mitogen-activated protein kinase (p38-MAPK). These results suggest that açai fruits might contribute to mitigating the adverse effects of aging brain cells by combating the inflammatory and oxidative mediators of aging at the cellular level.

In another study by Poulose *et al.* [48], the effects of açai fruit fractions on autophagy in BV-2 microglial and HT-22 hippocampal cells were examined. Autophagy is critical for intracellular degradation or recycling of toxic proteins and cellular debris. As the brain ages, this function declines leading to accumulation of toxic debris, a known contributor to age-related neurodegenerative diseases. Cells were pretreated with 6-hydroxydopamine (HDA) causing substantial accumulation of ubiquinated protein aggregates. Açai fruit fractions induced activation of neuronal housekeeping (autophagy) and inhibited the action of a protein that shuts down the autophagic process in both microglial and hippocampal cells. Microglial cells can be activated during chronic inflammation and lead cells to leak toxic proteins. During activation inflammatory gene production is stimulated which increased oxidative stress and promotes cell death.

The findings from these studies offer valuable insight into the neuroprotective effects of açai fruit fractions on certain brain cells, which could have implications for the mitigation of neurodegenerative diseases as well as maintenance of cognitive and motor functions.

Components other than anthocyanins in açai fruits have been shown to have "remarkably suppressive" antiproliferative activity against C-6 rat brain glioma cells (C-6) [49]. The antioxidant and antiproliferative activity of an anthocyanin-rich açai fruit extract against C-6 was studied leading to the discovery that only an açai fruit extract exhibited antiproliferative activity against C-6 in a dose-dependent manner with a half-maximal inhibitory concentration (IC_{50}) of 121 μg/mL, whereas other anthocyanin-rich extracts, including blueberry, strawberry, raspberry, blackberry, and wolfberry, did not exhibit any antiproliferative activity. This suggested that the active antiproliferative components in the extract were not due to anthocyanins. The contributive moieties' chemical structures remain to be characterized, as does the apoptotic pathways, warranting additional research.

A comprehensive review of all research on açai fruits is not within the scope of this chapter. One area worth adding to the discussion is recent reports on potential benefit of açai's seeds that fits well within the concept of novel foods.

21.3.1 Antioxidant and anti-inflammatory properties of açai seeds

There is anticipation that interest in the commercial value of açai seeds, currently considered a "waste product" by açai fruit producers, will change given recent studies that demonstrate surprising evidence that a hydroalcoholic extract of açai seeds can attenuate damage in the lungs of mice exposed to short-term or chronic cigarette smoking.

De Moura *et al.* [50] have described experiments conducted in Brazil medical school laboratories in which short-term (5 days) inhalation of tobacco smoke by mice orally administered a hydroalcoholic açai seed extract reduced the inflammatory and pro-oxidant damage associated with cigarette smoking. It has been shown that short-term inhalation of tobacco smoke leads to acute lung inflammation due to oxidative stress [51].

In a 60-day chronic inhalation of tobacco smoke study, oral administration of the açai seed extract resulted in a protective effect against the development of emphysema in mice [52]. Oxidative stress based on differences in antioxidant enzyme activities, and the observed reduction in macrophage and neutrophil elastase levels, provided mechanisms that partially explain the protective effect of the extract against the damage from tobacco seen in control mice not fed the extract. The possibility that compounds in the seed of açai might reduce the harmful effects associated with cigarette smoking and possibly other tobacco products, and possibly prevent the development of such tobacco-related diseases as chronic obstructive pulmonary disease (COPD), warrants further research and independent confirmation.

21.4 Phytochemicals in açai fruits

Numerous analytical efforts have been made to identify and quantify the anthocyanins and other flavonoids in açai fruits. Early attempts disagreed on what the major anthocyanins were in the fruit. While Bobbio *et al.* [53] reported that the major anthocyanins were cyanidin 3-arabinoside and cyanidin 3-arabinosylarabinoside, Lichtenthaler *et al.* [54] reported cyanidin 3-glucoside and cyanidin 3-rutinoside as the main anthocyanins. These disagreements mirrored much of the literature related to açai's nutritional composition up to 2005; hence, we attempted not only to elucidate the fruits' nutritional composition but also to establish definitively what the major and minor anthocyanins were in the fruits, a project that began in 2000 to identify and elucidate all of the bioactive compounds in the fruit.

Over a span of several years, characterization of açai fruits determined that açai mesocarp/epicarp contain numerous phytochemicals, including the following:

- Homoorientin, taxifolin deoxyhexose, isovitexin, cyanidin 3-glucoside, and cyanidin 3-rutinoside [55, 56]
- Cyanidin 3-sambubioside, peonidin 3-glucoside, peonidin 3-rutinoside, trans-resveratrol, hydroxybenzoic acids, isoquercitin, apigenin, eriodictyol, eriodictyol 7-glucoside, luteolin 4′-glucoside, eupatorin, catechin, protocatechuic acid, quercitin-3-arabinoside, kaempferol, and chrysoeriol [30]
- Luteolin, quercitin, orientin, vitexin, and dihydrokaempferol [57]
- The novel dihydroflavone glucoside, (2*S*,3*S*)-dihydrokaempferol 3-*O*-β-D-glucoside, and its isomer (2*R*,3*R*)-dihydrokaempferol 3-*O*-β-D-glucoside, velutin, and 5,4′-dihydroxy-7,3′,5′-trimethoxyflavone [58]

Table 21.4 ORAC antioxidant activities of açai (*E. oleracea*) fruit flavones

Compound	ORAC (μmol of TE/g) [57, 58]
Isovitexin	22,404 ± 1322
(2*S*,3*S*)-dihydrokaempferol 3-*O*-β-D-glucoside	15,199 ± 503
(2*R*,3*R*)-dihydrokaempferol 3-*O*-β-D-glucoside	11,219 ± 433
Vitexin	14,800 ± 451
Velutin	13,643 ± 1119
Quercetin	12,300 ± 1070
Dihydrokaempferol	8390 ± 94
Luteolin	7870 ± 350
5,4′-dihydroxy-7,3′,5′-trimethoxyflavone	4458 ± 409
Chrysoeriol	4400 ± 189
Orientin	1700 ± 79
Homoorientin	1420 ± 63

Note: Some numbers are rounded to the second digit after decimal point.
TE, trolox equivalents.

Agawa *et al.* [59] determined the concentration of various anthocyanins in the mesocarp, epicarp, and endocarp, of fresh açai, and had each component's antioxidant activities measured, using the ORAC, DPPH, 2,2′-azino-bis(3-ethylbenzothiazoline-6-sulphonic acid) (ABTS), ferric reducing antioxidant power (FRAP), and superoxide dismutase (SOD) assays. Açai mesocarp/epicarp extracts were found to have potent antioxidant activity compared to blueberry extract on every assay. The major anthocyanins, namely cyanidin 3-glucoside and cyanidin 3-rutinoside, were found to provide strong antioxidant potency if derived from the mesocarp/epicarp portion. These findings, in general, support previous work reported on the polyphenolic constituents of the *E. oleracea* açai fruit [55].

Isolation and structure identification studies by Kang *et al.* [57, 58] confirmed that flavonoids are the major polyphenols in açai fruits. The antioxidant activities, bioactivities, and bioavailability of these flavonoids were evaluated by ORAC, CAP-e, ROS, and PMN assays. The assays determined that the antioxidant activities of the flavonoid compounds in açai fruits varied significantly; aglycones had much higher antioxidant activities than their corresponding C-glycosides. The position of the methoxy group exerted a significant effect on antioxidant activity, while the CAP-e assay demonstrated that the polyphenols were able to penetrate into living cells.

The difference in ORAC antioxidant activities between the pure flavones is shown in Table 21.4. The difference in antioxidant activities is based on the number and position of the hydroxyl groups and/or other substitute groups. For example, in terms of the difference in ORAC values between chrysoeriol and luteolin, the *O*-methylation of the hydroxyl groups reduced the ORAC value of chrysoeriol by nearly 50%.

The CAP-e assays of the flavones revealed that of the compounds able to get into RBCs all were aglycones, with the exception of chrysoeriol, which bears a methoxy group, possibly due to its structural properties.

The isolation and identification of velutin in açai fruits is particularly significant since it exhibits the most potent anti-inflammatory bioactivity of any flavonoid so far reported in the literature (IC_{50} calculated at 2.0 μM) [36]. In addition, four other flavones identified in the fruits were found to exhibit strong anti-inflammatory properties, including

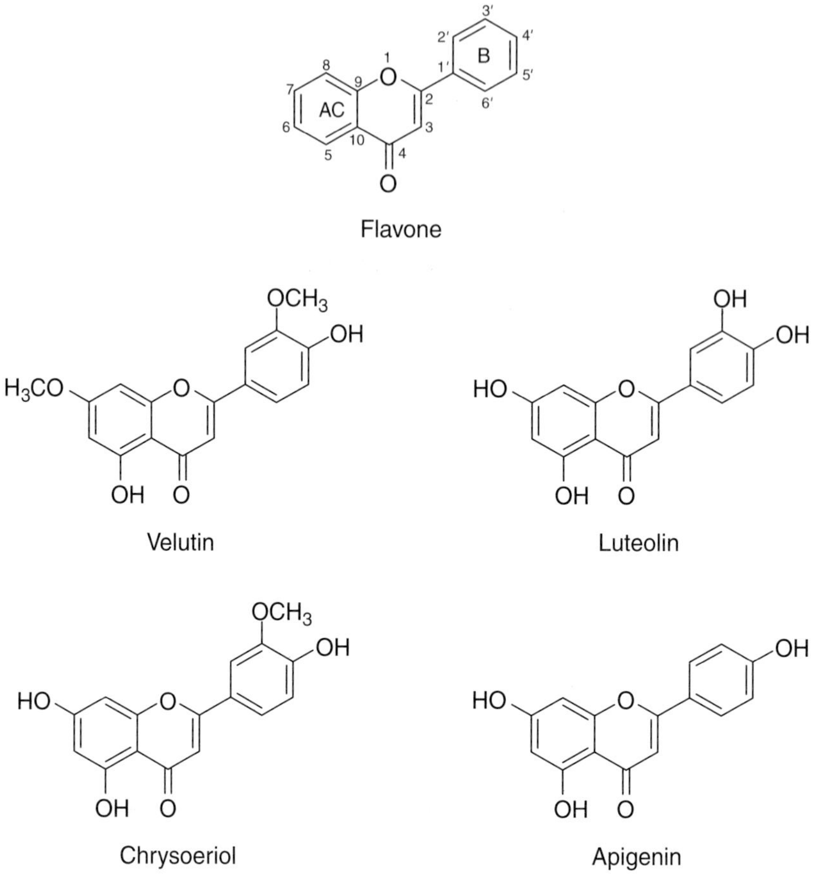

Figure 21.1 Major flavones found in açai fruits.

luteolin (12.4 μM), whose only difference compared to velutin is in the position of the hydroxy/methoxy groups. Velutin bears two methoxy groups at the 7′- and 3′-positions, whereas luteolin has two hydroxy groups at the 7′- and 3′-positions. It was determined that the substitution of the methoxy groups is the significant difference that contributes to its more pronounced antioxidant activity and anti-inflammatory properties. Isovitexin also exhibited anti-inflammatory bioactivity (IC_{50} calculated at 58.5 μM). Isovitexin has been shown to inhibit the translocation of NF-κB from the cytoplasm to the nucleus [60]. Velutin's potent anti-inflammatory bioactivity has been shown to reduce NF-κB activation. NF-κB plays a crucial role as a transcription factor in regulating many pro-inflammatory cytokine genes [61, 62]. It is likely that the combination of the açai flavone's antioxidant and anti-inflammatory properties work synergistically, a theory being researched to determine whether they act synergistically or are additive in nature. Exactly to what extent the methoxy groups and stereo configurations modulate the inhibitory properties of the flavone compounds has not been determined. Figure 21.1 shows the chemical structures of flavones found in açai fruits.

The results seen for velutin and other flavones found in açai fruits as observed in the SEAP reporter gene assays implicate potential protective actions of these compounds in

Table 21.5 Content of proanthocyanidins in freeze-dried açai fruits

Proanthocyanidins	mg/g DW
Monomers	0.21
Dimers	0.30
Trimers	0.25
Tetramers	0.32
Pentamers	0.31
Hexamers	0.52
Heptamers	0.32
Octamers	0.39
Nonamers	0.64
Decamers	0.34
Polymers	9.28
Total	12.98

Source: Adapted with permission from Schauss *et al.* [31].
DW, dry weight.

attenuating the progression of atherosclerosis. As part of the study, the inhibitory effect of velutin in oxidized low-density lipoprotein (oxLDL) induction of NF-κB activation was examined. Velutin was found to inhibit NF-κB activation induced by oxLDL. It is known that oxLDL plays a crucial role in the initiation and progression of atherosclerosis [63]. These findings, combined with the growing body of *in vivo* evidence in animals fed a high-fat/high-cholesterol diet with açai fruits, suggest that the flavones in açai fruits may play an important role in attenuating the progressive development of atherosclerosis.

Besides anthocyanins, other polyphenols in açai fruits also contain proanthocyanidins that have strong antioxidant properties [31], such as flavan-3-ol (epicatechin), as shown by the concentration of proanthocyanidins in Table 21.5. They have also been shown to counteract the negative effects of high-fat diets on the arteries [64].

Polyphenols in açai fruits are associated with enhanced antioxidant protection in cells and generalized vascular protection. In a pharmacokinetic study in humans, anthocyanins from açai fruit were shown to be bioavailable in healthy volunteers [65]. Açai fruits caused a significant increase in antioxidant activity of plasma, in peripheral blood mononuclear cells (PBMC) collected from patients hourly for 4 hours. However, the investigators used an açai juice drink that contained frozen açai fruits imported from Brazil. The beverage was sweetened with sucrose. Although the antioxidant activity in PBMC increased, the concentration of antioxidant polyphenolics in plasmas did not rise sufficiently to reduce the hydrogen peroxide–induced generation of ROS in any of the volunteers, which may be due to the use of frozen fruits and added sugar. Stress situations, where soluble sugars are involved such as chilling, are related to important changes in ROS balance [66]. A high-sucrose diet has been shown to increase ROS generation [67]. In contrast, in two studies using another açai-rich beverage fed to volunteers placed under oxidative stress, an increase in antioxidant activity of both plasma in RBCs and PMN cells was observed [68, 69], as determined by the antioxidant protection in erythrocytes (CAP-e) bioassay and evaluation in PMN cells of the inhibition of ROS production [37]. The difference between the two studied beverages was that the MonaVie açai beverage contained a combination of freeze-dried and frozen açai fruits, containing more of the antioxidant polyphenolics needed either to inhibit ROS production or to indirectly increase endogenous antioxidant activity. The practical benefit of incorporating such a juice in the diet was seen in both the randomized, double-blind, placebo-controlled, and crossover clinical trial, as well as in the range of motion (ROM) study involving 48- to 84-year-old adults, who started the study with limited ROM affecting the quality of life, and who within weeks and through to the end of a 12-week study had significant improvements in mobility and reduction of pain in key joints [68].

Given how many compounds are found in foods, many of which exhibit bioactivity and contribute to health benefits, their identification and characterization can determine which foods may play the greatest role in the prevention, mitigation, and treatment of diseases. It is also important to know how many compounds are there in the açai fruits, besides their contents of vitamins, minerals, protein, fat, carbohydrates, and soluble and insoluble fibers. This is because any number of compounds in a food, such as açai fruits, might be of therapeutic value, depending on the degree of evidence. For example, proteins in açai fruits have been found to show high-antitryptic activity, while some may inhibit human salivary α-amylase overproduction [70]. The effect of an açai extract on the antiproliferation and induction of apoptosis in human leukemia-60 (HL-60) cells has been reported [56].

To determine the phytochemical compounds in açai fruits, a semiquantitative matrix-assisted laser desorption time-of-flight mass spectrometry (MALDI-TOF MS) analysis was performed at the University of Wisconsin–Madison (unpublished data). MALDI-TOF MS characterizes complex mixtures of oligomeric polyphenolic compounds in various foods, including fruits, berries, and grains [71]. MS does not distinguish between isomers (although various stereoisomers and structural isomers can be described of dimers and trimers). The number of possible structures significantly increases by isomerization of each of the masses detected by MALDI-TOF MS. Analysis of açai fruits by MALDI-TOF MS determined that classical proanthocyanidins (homopolymers of repeating catechin/epicatechin units) were not present in appreciable levels. The açai spectra showed more than 3000 compounds, indicating that açai is far more complex than fruits such as cranberries and blueberries, also suggesting that there is greater structural heterogeneity in the fruits' polyphenols waiting to be elucidated [72].

21.5 Processing of açai fruits for value-added products

Açai palm trees are found in dense grooves and virtually every island among the Amazon River's estuaries and riverines. In Para state, it is estimated that more than 11 million hectares of the floodplains contain dense groves of açai palm trees. Each hectare can have 2000 to more than 7000 such palms. Government agricultural surveys have estimated that a palm tree can produce more than 1000 kg in a 5- to 7-year period (depending on rain levels) upon reaching maturity. In essence, there is a seemingly limitless quantity of açai fruits available in the floodplains. Harvesting occurs over an 8-month period as the trees produce up to four crops of fruit during the "drier" months, even though some rain may fall nearly every day during the dry season that runs from June to late December. Hence, most natives have access to the fruits as açai palm trees abound, allowing for daily harvesting of the fruits that ends when the rainy season begins in January.

21.5.1 Traditional preparation

Açai fruits are obtained from mature infructescences containing from 200 to 500 fruits removed from near the crown of the palm tree. The fruits are then stripped from the infructescence and sorted. Once filled, the basket is taken to one's home where it is placed in a large bowl. Warm water is added and the fruits soaked for half an hour to an hour. This softens the exocarp, epicarp, and mesocarp and removes dirt or debris that either sinks to the bottom of the bowl or floats to the top. In the most remote areas of the Amazon floodplains, a square device called a *caroceiro* is used which is made from strips of the Aruma tree. Fruit is pressed against the interweaving strips that force the seed to drop through weaving. The remaining pulp and skin is then rubbed repeatedly against a large mesh sieve under which is a finer screen that separates approximately 80–90% of the solid matter from the juice. In recent years, families have acquired mechanically handheld devices that far more easily remove the seed. The seeds are either thrown into the garden to turn into mulch within 2–3 months, to provide a soil amendment in the garden, or fed to pigs.

Most families consume the juice within 2 hours of harvesting to produce a soup eaten at breakfast with a spoon, or as a beverage, much like one would drink unfiltered pulp-rich orange juice. Some prefer the açai fruit to be more viscous. It is estimated that in cities and towns in açai tree–rich areas in the floodplains people consume an average of 2 L of fresh açai juice a day during the harvest season. Açai forms a major part of the diet of inhabitants of the Amazon floodplains along with farinha and whatever other food sources are available. Açai and farinha constitute approximately 65% of the diet of natives [27, 73], which confirms its nutritional density, as determined by nutrient composition analysis [30].

If produced for local consumption, only the fruits are sold at wholesale markets to local retail vendors or residents who take them home to make a beverage according to traditional methods of preparation.

21.5.2 Commercial production for value-added products

Large processing operations, located in larger cities in the Amazon floodplains, purchase açai fruits from large wholesalers. Fruits are ranked according to a quality scale. To produce pulp for sale in tonnage quantities, the fruits are sorted, washed, and softened by immersion in large tanks filled with water. After removing any debris, damaged fruits, or unripe or green fruits, the fruits are placed into mechanical mills that strip the seed and some of the exocarp, leaving the remaining mesocarp (pulp) with some skin segments for further processing. Ultraviolet (UV) sterilized water is added to the pulp, homogenized, during which time citric acid is added to stabilize acidity and inhibit microbial growth (usually, the pulp is standardized to 10–12% total solids, depending on customer's preference). Thereafter, the pulp is pasteurized at around 85°C for approximately 1 minute then transferred into polypropylene bags that line steel drums with removable lids. The drums are then frozen in a chamber, where after moved into refrigerated cold storage, maintained at around 17°C. Facilities producing açai pulp for export must be government certified, meet hazard analysis critical control point (HACCP), sanitation operating procedures, guided by good manufacturing practice (GMP) procedures and practices.

Frozen pulp can be spray- or freeze-dried into a powder that is stored in sealed aseptic polypropylene to prevent reabsorption of moisture. Spray-dried or freeze-dried powders are used to primarily produce açai juice beverages. Because of the relatively bland taste of the fruits, other fruits or berries are typically mixed with the pulp powder to produce an organoleptically acceptable beverage for consumers. Almost all of the research on açai has been performed on freeze-dried açai or beverages that include freeze-dried açai due to the powders' shelf life and retention of the pulp's nutritional composition and bioactivities. ORAC assays have shown that freeze-dried açai powders are superior in terms of antioxidant activity compared to the spray-dried powders [31].

Nindo *et al.* [74] have studied the rheology of fruit pulps and have pointed out that extended application of heat coupled with continuous shearing during component fruit separation and processing can cause irreversible structural breakdown of the products. Tonon *et al.* [75] have shown that at a temperature range of 10–70°C, açai pulp shows shear-thinning behavior with yield stress in the steady-shear measurements and other characteristics unlike that of most fruit pulps. This rheological information about açai pulp is partially due to the unusual chemical composition of the fruits, such as their high levels of lipids, proteins, and fibers.

21.6 Conclusions

Research interest in the fruits of the Amazonian açai has been of relatively recent but intensive interest following the discovery of the fruit's pronounced potent antioxidant and anti-inflammatory properties. Two of the species that have earned particular interest due to their superior antioxidant activities compared to almost all other foods and commercial availability are that of *E. precatoria* and *E. oleracea*, both of which are commonly called "açai" by natives living in the Amazon.

Years of characterization of the fruits has lead to a much better qualitative and quantitative understanding of their chemistry and bioactive properties. The three species of açai fruits occupy the top four out of five ORAC assay values of all foods assayed for the peroxyl radical scavenging capacity *in vitro*. The fruits contain a class of flavones that includes the most potent anti-inflammatory flavonoid found. Most of these flavones are not only bioavailable and protect cells from oxidative damage and chronic inflammation, but they may act synergistically, suggesting that the fruits as a whole may have therapeutic benefits beyond their considerable nutritive values. It also suggests that there may be a synergy between many compounds in the fruits, from anthocyanins, proanthocyanidins, fatty acids (to facilitate gut absorption), to numerous polyphenolics, some unique to the fruits.

Preliminary evidence shows açai fruits decrease production or signaling of pro-inflammatory cytokines and/or inflammatory mediators. Isolation of all the compounds in açai that may act to inhibit or reduce inflammation and oxidative stress and/or enhance protective mechanisms such as has been shown by the induction of autophagy will take time.

The evidence in support of a protective effect of açai compounds in providing cardiovascular protection is more compelling based on a number of well-controlled animal studies that compared the effect of feeding hypercholesterolemic and/or high-fat diets. Downregulation of pro-inflammatory moieties such as NF-κB has been consistently seen in each experiment. Given the prevalence and incidence of atherosclerosis, it seems warranted at this time to include açai fruits in the diet.

An intriguing finding has been the unexpected emergence of evidence that what was once considered a waste product, namely, the seeds of the açai fruits, may have potential benefit in mitigating the harmful effects of cigarette smoking. Considerable funding should be committed toward a greater understanding of the mechanisms of action to account for the observed protection seen in animal lung tissue following short-term and long-term inhalation of cigarettes either laced with a hydroalcoholic extract of the seed or fed orally to rodents unable to prevent the inhalation of the smoke.

Fortunately, there are ample year-to-year supplies of açai fruits available to support world-wide interest and consumption. Millions of hectares of açai palms can be found throughout the floodplains of the Amazon, in addition to the availability of equally as much of the other açai species at higher elevations, including the Bolivian Amazon.

The rate of publications on açai fruits related to its potential health benefits is mirrored by the exponential growth of papers appearing in databases such as PubMed [managed by the US National Library of Medicine, National Institutes of Health (NIH)]. Whereas in 2005, there were only two papers on açai in PubMed, by early 2012 there were more than 100 published papers, including a 2012 report that in hypercholesterolemic zebrafish fed for 4 weeks to evaluate the fruit's hypolipidemic activity, açai inhibited cholesteryl ester transfer protein (CETP) production [76]. The fish fed açai demonstrated lower serum triglycerides levels, total cholesterol, attenuation of fatty liver, reduced content of oxidized species, and significantly less hepatic inflammation. This study as well as others discussed earlier providing the growing body of data showing açai to have antioxidant, anti-inflammatory, and anti-atherosclerotic activity in cellular assays as well as *in vivo* models. The pace of research is continuing given the interest in açai generated by scientists studying their properties and attributes at the NIH's National Institute on Aging, various USDA research centers including the USDA Human Nutrition Research Center on Aging at Tufts University, USDA Arkansas Children's Nutrition Center, FDA's Center for Biologics Evaluation and Research, and numerous universities and research foundations.

Given the exponential research interest in the attributes, properties, and applications of each species of açai fruits, consumer interest would be expected to continue to support their status as one of the true "superfoods" worth considering as part of a nutritious human diet; a diet that should include foods that promote healthy aging and is able to mitigate damage associated with acute and/or chronic oxidative stress.

References

1. Giacalone, M., Di Sacco, F., Traupe, I., Topini, R., Forfori, F. & Giunta, F. (2011) Antioxidant and neuroprotective properties of blueberry polyphenols: a critical review. *Nutrition and Neuroscience*, **14**, 119–125.
2. Giampieri, F., Tulipani, S., Alvarez-Suarez, J.M., Quiles, J.L., Mezzetti, B. & Battino, M. (2012) The strawberry: composition, nutritional quality, and impact on human health. *Nutrition*, **28**, 9–19.
3. Azzini, E., Vitaglione, P., Intorre, F., Napolitano, A., Durazzo, A., Foddal, M.S., Fumagalli, A., Catasta, G., Rossi, L., Venneria, E., Raguzzini, A., Palomba, L., Fogliano, V. & Maiani, G. (2010) Bioavailability of strawberry antioxidants in human subjects. *British Journal of Nutrition*, **104**, 1165–1173.
4. Brown, J. (2011) Cranberries: this superfood is not just for thanksgiving. *CDS Review*, **104**, 24.
5. Cote, J., Caillet, S., Doyon, G., Sylvain, J.F. & Lacroix, M. (2010) Bioactive compounds in cranberries and their biological properties. *Critical Reviews in Food Science and Nutrition*, **50**, 666–679.
6. Vinson, J.A. & Cai, Y. (2011) Nuts, especially walnuts, have both antioxidant quantity and efficacy and exhibit significant potential health benefits. *Food & Function*, **3**, 134–140.
7. Johanningsmeier, S.D. & Harris, G.K. (2011) Pomegranate as a functional food and nutraceutical source. *Annual Review of Food Science and Technology*, **2**, 181–201.
8. McKay, D.L., Chen, C.Y., Yeum, K.J., Matthan, N.R., Lichtenstein, A.H. & Blumberg, J.B. (2010) Chronic and acute effects of walnuts on antioxidant capacity and nutritional status in humans: a randomized, cross-over pilot study. *Nutrition Journal*, **9**, 21.

9. Basu, A., Betts, N.M., Ortiz, J., Simmons, B., Wu, M. & Lyons, T.J. (2011) Low-energy cranberry juice decreases lipid oxidation and increases plasma antioxidant capacity in women with metabolic syndrome. *Nutrition Research*, **31**, 190–196.
10. Tousoulis, D., Kampoli, A.M., Papageorgiou, N., Androulakis, E., Antoniades, C., Toutouzas, K. & Stefanadis, C. (2011) Pathophysiology of atherosclerosis: the role of inflammation. *Current Pharmaceutical Design*, **17**, 4089–4110.
11. Rull, A., Garcia, R., Fernandez-Sender, L., Garcia-Heredia, A., Aragones, G., Beltran-Debon, R., Marsillach, J., Alegret, J.M., Martin-Paredero, V., Mackness, B., Mackness, M., Joven, J. & Camps, J. (2012) Serum paraoxonase-3 concentration is associated with insulin sensitivity in peripheral artery disease and with inflammation in coronary artery disease. *Atherosclerosis*, **220**, 545–551.
12. Steffen, B.T., Steffen, L.M., Tracy, R., Siscovick, D., Jacobs, D., Liu, K., He, K., Hanson, N.Q., Nettleton, J.A. & Tsai, M.Y. (2012) Ethnicity, plasma phospholipid fatty acid composition and inflammatory/endothelial activation biomarkers in the Multi-Ethnic Study of Atherosclerosis (MESA). *European Journal of Clinical Nutrition*, **66**, 600–605.
13. Kalogeropoulos, A.P., Georgiopoulou, V.V. & Butler, J. (2012) From risk factors to structural heart disease: the role of inflammation. *Heart Failure Clinic*, **8**, 113–123.
14. Raaz-Schrauder, D., Klinghammer, L., Baum, C., Frank, T., Lewczuk, P., Achenbach, S., Cicha, I., Stumpf, C., Wiltfang, J., Kornuber, J., Daniel, W.G. & Garlichs, C.D. (2012) Association of systemic inflammation markers with the presence and extent of coronary artery calcification. *Cytokine*, **57**, 251–257.
15. Androulakis, E., Tousoulis, D., Papageorgiou, N., Latsios, G., Siasos, G., Tsioufis, C., Giolis, A. & Stefanadis, C. (2011) Inflammation in hypertension: current therapeutic approaches. *Current Pharmaceutical Design*, **17**, 4121–4131.
16. Sfanos, K.S. & De Marzo, A.M. (2012) Prostate cancer and inflammation: the evidence. *Histopathology*, **60**, 199–215.
17. Chow, M.T., Moller, A. & Smyth, M.J. (2012) Inflammation and immune surveillance in cancer. *Seminar in Cancer Biology*, **22**, 23–32.
18. Prasad, S., Sung, B. & Aggarwal, B.B. (2011) Age-associated chronic diseases require age-old medicine: role of chronic inflammation. *Preventive Medicine*, **54**, S29–S37.
19. Drouet, M., Dubuquoy, L., Desreumaux, P. & Bertin, B. (2012) Visceral fat and gut inflammation. *Nutrition*, **28**, 113–117.
20. Maher, M.M. (2012) Inflammatory bowel disease: review and future view. *Frontiers in Bioscience*, **4**, 1638–1647.
21. Perrier, C. & Rutgeerts, P. (2011) Cytokine blockade in inflammatory bowel diseases. *Immunotherapy*, **3**, 1341–1352.
22. Marcuzzi, A., Girardelli, M., Bianco, A.M., Martelossi, S., Magnolato, A., Tommasini, A. & Crovella, S. (2012) Inflammation profile of four early onset Crohn patients. *Gene*, **493**, 282–285.
23. Brooks-Worrell, B. & Palmer, J.P. (2012) Immunology in the clinic review series; focus on metabolic diseases: development of islet autoimmune disease in type 2 diabetes patients: potential sequelae of chronic inflammation. *Clinical and Experimental Immunology*, **167**, 40–46.
24. Hyun, E., Ramachandran, R., Hollenberg, M.D. & Vergnolle, N. (2011) Mechanisms behind the anti-inflammatory actions of insulin. *Critical Reviews in Immunology*, **31**, 307–340.
25. Turkmen, K., Guney, I., Yerlikaya, F.H. & Tonbul, H.Z. (2012) The relationship between neutrophil-to-lympocyte ratio and inflammation in end-stage renal disease patients. *Renal Failure*, **34**, 155–159.
26. Saccucci, P., Banci, M., Amante, A., Bottini, E. & Gloria-Bottini, F. (2011) Coronary artery disease: evidence of interaction between PTPN22 and p53 genetic polymorphisms. *Cardiology*, **120**, 166–168.
27. Schauss, A.G. (2011) *Açai: An Extraordinary Antioxidant-Rich Palm Fruit*, 3rd edn. Biosocial Publications, Tacoma, WA.
28. Heinrich, M., Dhanji, T. & Casselman, I. (2011) Açai (*Euterpe oleracea* Mart.): a phytochemical and pharmacological assessment of the species' health claims. *Phytochemistry Letters*, **4**, 10–21.
29. USDA (2010) *USDA Database for the Oxygen Radical Absorbance Capacity (ORAC of Selected Foods, Release 2.0.* Published online at: http://www.ars.usda.gov/nutrientdata, last accessed March 9, 2011.
30. Schauss, A.G., Wu, X., Prior, R.L., Ou, B., Huang, D., Owens, J., Agarwal, A., Jensen, G.S., Hart, A.N. & Shambrom, E. (2006) Antioxidant capacity and other bioactivities of the freeze-dried Amazonian palm berry, *Euterpe oleraceae* Mart. (Açai). *Journal of Agricultural and Food Chemistry*, **54**, 8604–8610.

31. Schauss, A.G., Wu, X., Prior, R.L., Ou, B., Patel, D., Huang, D. & Kababick, J.P. (2006) Phytochemical and nutrient composition of the freeze-dried Amazonian palm berry, *Euterpe oleraceae* Mart. (Açai). *Journal of Agricultural and Food Chemistry*, **54**, 8598–8603.
32. Ou, B., Hampsch-Woodhill, M., Flanagan, J., Deemer, E.K., Prior, R.L. & Huang, D. (2002) Novel fluorometric assay for hydroxyl radical prevention capacity using fluorescein as the probe. *Journal of Agricultural and Food Chemistry*, **50**, 2772–2777.
33. Zhang, L., Huang, D., Kondo, M., Fan, E., Ji, H., Kou, Y. & Ou, B. (2009) Novel high-throughput assay for antioxidant capacity against superoxide anion. *Journal of Agricultural and Food Chemistry*, **57**, 2661–2667.
34. Ozgen, M., Reese, R.N., Tulio, A.Z., Scheerens, J.C. & Miller, A.R. (2006) Modified 2,2-azino-bis-3-ethylbenzothiazoline-6-sulfonic acid (ABTS) method to measure antioxidant capacity of selected small fruits and comparison to ferric reducing antioxidant power (FRAP) and 2,2′-diphenyl-1-picylhydrazyl (DPPH) methods. *Journal of Agricultural and Food Chemistry*, **54**, 1151–1157.
35. Pacheco-Palencia, L.A., Duncan, C.E. & Talcott, S.T. (2009) Phytochemical composition and thermal stability of two commercial açai species, *Euterpe oleracea* and *Euterpe precatoria*. *Food Chemistry*, **115**, 1199–1205.
36. Xie, C., Kang, J., Li, Z., Schauss, A.G., Badger, T.M., Nagarajan, S., Wu, T. & Wu, X. (2012) The açai flavonoid velutin is a potent anti-inflammatory agent: blockade of LPS-mediated TNF-α and IL-6 production through inhibiting NF-κB activation and MAPK pathway. *Journal of Nutritional Biochemistry*, **23**, 1184–1191.
37. Honzel, D., Carter, S.G., Redman, K.A., Schauss, A.G., Endres, J.R. & Jensen, G.S. (2008) Comparison of chemical and cell-based antioxidant methods for evaluation of foods and natural products: generating multifaceted data by parallel testing using erythrocytes and polymorphonuclear cells. *Journal of Agricultural and Food Chemistry*, **56**, 8319–8325.
38. Kang, J., Thakali, K.M., Xie, C., Kondo, M., Tong, Y., Ou, B., Jensen, G., Medina, M.B., Schauss, A.G. & Wu, X. (2012) Bioactivities of açaí (*Euterpe precatoria* Mart.) fruit pulp, superior antioxidant and anti-inflammatory properties to *Euterpe oleracea* Mart. *Food Chemistry*, **133**, 671–677.
39. Berger, J., Hauber, J., Hauber, R., Geiger, R. & Cullen, B.R. (1988) Secreted placental alkaline phosphatase: a powerful new quantitative indicator of gene expression in eukaryotic cells. *Gene*, **66**, 1–10.
40. Moon, K.Y., Hahn, B.S., Lee, J. & Kim, Y.S. (2001) A cell-based assay system for monitoring NF-kappaB activity in human HaCat transfectant cells. *Analytical Biochemistry*, **292**, 17–21.
41. Sun, X., Seeberger, J., Albericol, T., Schauss, A.G. & Zou, S. (2010) Açai palm fruit (*Euterpe oleracea* Mart.) pulp improves survival of flies on a high fat diet. *Experimental Gerontology*, **45**, 243–251.
42. Perez, V.I., Bokov, A., Remmen, H.V., Mele, J., Ran, Q., Ikeno, Y. & Richardson, A. (2009) Is the oxidative stress theory of aging dead? *Biochimica Biophysica Acta*, **1790**, 1005–1014.
43. De Souza, O., Silva, M., Silva., M.E., de Paula Oiveira, R. & Pedrosa, M.L. (2010) Diet supplementation with açai (*Euterpe oleracea* Mart.) pulp improves biomarkers of oxidative stress and the serum lipid profile in rats. *Nutrition*, **26**, 804–810.
44. Feio, C.A., Iaar, M.C., Ihara, S.S., Kasmas, S.H., Martins, C.M., Feio, M.N., Maués, L.A., Borges, N.C., Moreno, R.A., Póvoa, R.M. & Fonseca, F.A. (2012) *Euterpe oleracea* (açai) modifies sterol metabolism and attenuates experimentally-induced atherosclerosis. *Journal of Atherosclerosis and Thrombosis*, **19**, 237–245.
45. Noratto, G.D., Angel-Morales, G., Talcott, S.T. & Mertens-Talcott, S.U. (2011) Polyphenolics from açai (*Euterpe oleracea* Mart.) and red muscadine grape (*Vitis rotundifolia*) protect human umbilical vascular endothelial cells (HUVEC) from glucose- and lipopolysaccharide (LPS)-induced inflammation and target microRNA-126. *Journal of Agricultural and Food Chemistry*, **59**, 7999–8012.
46. Xie, C., Kang, J., Burris, R., Ferguson, M.E., Schauss, A.G., Nagarajan, S. & Wu, X. (2011) Açaí juice attenuates atherosclerosis in ApoE deficient mice through antioxidant and anti-inflammatory activities. *Atherosclerosis*, **216**, 327–333.
47. Schauss, A.G., Clewell, A., Balogh, L., Szakonyl, I.P., Financsek, I., Horvath, J., Thuroczy, J., Beres, E., Vertesi, A. & Hirka, G. (2010) Safety evaluation of an açai-fortified fruit and berry functional juice beverage (MonaVie Active.5®). *Toxicology*, **278**, 46–54.
48. Poulose, S.M., Fischer, D.R., Larson, J.A., Bielinski, D.F., Rimando, A.M., Carey, A.N., Schauss, A.G. & Shuckitt-Hale, B. (2012) Anthocyanin-rich açai (*Euterpe oleracea* Mart.) fruit pulp fractions attenuate inflammatory stress signaling in mouse brain BV-2 microglial cells. *Journal of Agricultural and Food Chemistry*, **60**, 1084–1093.

49. Hogan, S., Chung, G.H., Zhang, L., Li, J., Lee, Y., Dai, Y. & Zhou, K. (2010) Antiproliferative and antioxidant properties of anthocyanin-rich extract from açai. *Food Chemistry*, **118**, 208–214.
50. De Moura, R.S., Ferreira, T.S., Lopes, A.A., Pires, K.M., Nesi, R.T., Resende, A.C., Souza, P.J., da Silva, A.J., Borges, R.M., Porto, L.C. & Valenca, S.S. (2012) Effects of *Euterpe oleracea* Mart. (Açai) extract in acute lung inflammation induced by cigarette smoke in the mouse. *Phytomedicine*, **19**, 262–269.
51. Valenca, S.S., Pimenta, W.A., Ruef-Barroso, C.R., Ferreira, T.S., Resende, A.C., Moura, R.S. & Porto, L.C. (2009) Involvement of nitric oxide in acute lung inflammation induced by cigarette smoke in the mouse. *Nitric Oxide*, **20**, 175–181.
52. De Moura, R.S., Pires, K.M., Santos Ferreira, T., Lopes, A.A., Nesi, R.T., Resende, A.C., Sousa, P.J., da Silva, A.J., Porto, L.C. & Valenca, S.S. (2011) Addition of açai (*Euterpe oleracea*) to cigarettes has a protective effect against emphysema in mice. *Food and Chemical Toxicology*, **49**, 855–863.
53. Bobbio, F.O., Bibbio, P.A., Oliveira, P.A. & Fadelli, S. (2002) Stability and stabilization of the anthocyanins from *Euterpe oleracea* Mart. *Acta Alimentaria*, **31**, 371–377.
54. Lichtenthaler, R., Rodriques, R.B., Maia, J.G., Papgiannipoulos, M., Fabricius, H. & Marx, F. (2005) Total oxidant scavenging capacities of *Euterpe oleracea* Mart. (açai) fruit. *International Journal of Food Sciences and Nutrition*, **56**, 53–64.
55. Gallori, S., Bilia, A.R., Bergonzi, M.C., Barbosa, W.L.R. & Vincieri, F.F. (2004) Polyphenolic constituents of fruit pulp of *Euterpe oleracea* Mart. (Açai palm). *Chromatographia*, **59**, 739–743.
56. Del Pozo-Insfran, D., Percival, S.S. & Talcott, S.T. (2006) Açai (*Euterpe oleracea* Mart.) polyphenolics in their glycoside and aglycones forms induce apoptosis of HL-60 leukemia cells. *Journal of Agricultural and Food Chemistry*, **54**, 1222–1229.
57. Kang, J., Li, Z., Wu, T., Jensen, G.S., Schauss, A.G. & Wu, X. (2010) Antioxidant capacities of flavonoid compounds isolated from açai pulp (*Euterpe oleracea* Mart.). *Food Chemistry*, **122**, 610–617.
58. Kang, J., Xie, C., Li, Z., Wu, T., Nagaraja, S., Schauss, A.G. & Wu, X. (2011) Flavonoids from açai (*Euterpe oleracea* Mart.) pulp and their antioxidant and anti-inflammatory activities. *Food Chemistry*, **128**, 152–157.
59. Agawa, S., Sakakibara, H., Iwata, R., Shimoi, K., Hergesheimer, A. & Kumazawa, S. (2011) Anthocyanins in mesocarp/epicarp and endocarp of fresh açai (*Euterpe oleracea* Mart.) and their antioxidant activities and bioavailability. *Food Science and Technology Research*, **17**, 327–334.
60. Lin, C.M., Huang, S.T., Liang, Y.C., Lin, M.S., Shih, C.M., Chang, T.C., Chen, T.Y. & Chen, C.T. (2005) Isovitexin suppresses lipopolysaccharide-mediated inducible nitric oxide synthase through inhibition of NF-kappa B in mouse macrophages. *Planta Medica*, **71**, 748–753.
61. De Winther, M.P., Kanters, E., Kraal, G. & Hofker, M.H. (2005) Nuclear factor kappa B signaling in atherogenesis. *Arteriosclerosis and Thrombosis in Vascular Biology*, **25**, 904–914.
62. Xanthoulea, S., Curfs, M.D., Hofker, M.H. & de Winther, M.P. (2005) Nuclear factor kappa B signaling in macrophage function and atherogenesis. *Current Opinions in Lipidology*, **16**, 536–542.
63. Pietta, P.G. (2000) Flavonoids as antioxidants. *Journal of Natural Products*, **63**, 1035–1042.
64. Schroeter, H., Heiss, C., Spencer, J.P., Keen, C.L., Lupton, J.R. & Schmitz, H.H. (2010) Recommending flavanols and procyanidins for cardiovascular health: current knowledge and future needs. *Molecular Aspects of Medicine*, **31**, 546–557.
65. Mertens-Talcott, S.U., Rios, J., Jilma-Stohlawetz, P., Pachecho-Palencia, L.A., Meibohm, B., Talcott, S.T. & Derendorf, H. (2008) Pharmacokinetics of anthocyanins and antioxidant effects after the consumption of anthocyanin-rich açai juice and pulp (*Euterpe oleracea* Mart.) in human healthy volunteers. *Journal of Agricultural and Food Chemistry*, **56**, 7796–7802.
66. Couee, I., Sulmon, C., Gouesbet, G. & El Amrani, A. (2005) Involvement of soluble sugars in reactive oxygen species balance and responses to oxidative stress in plants. *Journal of Experimental Botany*, **57**, 449–459.
67. Ruiz-Ramirez, A., Chavez-Salgado, M., Peneda-Flores, J.A., Zapata, E., Masso, F. & El-Hafidi, M. (2011) High-sucrose diet increases ROS generation, FFA accumulation, UCP2 level, and proton leak in liver mitochondria. *American Journal of Physiology, Endocrinology and Metabolism*, **301**, E1198–E1207.
68. Jensen, G.S., Wu, X., Patterson, K.M., Barnes, J., Carter, S.G., Scherwitz, L., Beaman, R., Endres, J.R. & Schauss, A.G. (2008) *In vitro* and *in vivo* antioxidant and anti-inflammatory capacity of an antioxidant-rich fruit and berry juice blend. Results of a pilot and randomized, double-blind, placebo-controlled, crossover study. *Journal of Agricultural and Food Chemistry*, **56**, 8326–8333.

69. Jensen, G.S., Ager, D.M., Redman, K.A., Mitzner, M.A., Benson, K.F. & Schauss, A.G. (2011) Pain reduction and improvement in range of motion after daily consumption of açai (*Euterpe oleracea* Mart.) pulp-fortified polyphenolic-rich fruit and berry juice blend. *Journal of Medicinal Food*, **14**, 702–711.
70. Araújo, C.L., Bezerra, I.W.L., Dantas, I.C., Lima, T.V.S., Oliveira, A.S., Miranda, M.R.A., Leite, E.L. & Sales, M.P. (2004) Biological activity of proteins from pulps of tropical fruits. *Food Chemistry*, **85**, 107–110.
71. Reed, J.D., Krueger, C.G. & Vestling, M.M. (2005) MALDI-TOF mass spectrometry of oligomeric food polyphenols. *Phytochemistry*, **66**, 2248–2263.
72. Krueger, C.G., Vestling, M.M. & Reed, J.D. (2004) Matrix-assisted laser desorption/ionization time-of-flight mass spectrometry of anthocyanin-polyflavan-3-ol oligomers in cranberry fruit (*Vaccinium macrocarpon* Ait.) and spray dried cranberry juice. In: *Red Wine Color: Revealing the Mysteries* (eds A.L. Waterhouse & J.A. Kenedy). ACS Sympoiusm Series 886, American Chemical Society, Washington, DC, pp. 232–246.
73. Balick, M.J. (1988) The palm, tree of life: biology, utilization and conservation. *Advances in Economic Botany*, **6**, 224–253.
74. Nindo, C.I., Tang, J., Powers, J.R. & Takhar, P.S. (2007) Rheological properties of blueberry puree for processing applications. *Food Science and Technology*, **40**, 292–299.
75. Tonon, R.V., Alexandre, D., Hubinger, M.D. & Cunha, R.L. (2009) Steady and dynamic shear rheological properties of açai pulp. *Journal of Food Engineering*, **92**, 425–431.
76. Kim, J.Y., Hong, J.H., Jung, H.K., Jeong, Y.S. & Cho, K.H. (2012). Grape skin and loquat leaf extracts and acai puree have potent anti-atherosclerotic and anti-diabetic activity *in vitro* and *in vivo* in hypercholesterolemic zebrafish. *International Journal of Molecular Medicine*, **30**, 606–614.

22 Bananas, dried bananas, and banana chips: nutritional characteristics, phytochemicals, and health effects

Arianna Carughi

22.1 Introduction

The word "banana" encompasses a variety of species or hybrids derived from two wild species of *Musa acuminata* and *Musa balbisiana*. Bananas are often categorized as "dessert" sweet bananas, which are ripened and eaten raw, and starchy or "cooking" bananas and plantains intended for cooking or processing, but also edible when fully ripe. Bananas have their origin in the Asia Pacific region and are now cultivated throughout the humid tropical and subtropical areas of the world [1]. They are perennial, giant herbs, seedless and parthenocarpic—the fruit develops without fertilization or pollination—and produce fruit throughout the year [2]. More than 300 types of bananas are cultivated, but little is known about the variation in nutrient content among them and their contribution to dietary requirements of different populations. Part of the reason for this may be the commercial popularity of only a few varieties, exports almost exclusively dominated by the Cavendish banana. Yet this variety accounts for little more than about 13% of the world's banana and plantain production. The other 87% is made up of a broad range of varieties adapted to specific climate/growing conditions and selected for specific eating/cooking qualities [3]. This chapter is an overview of the nutritional properties of fresh bananas, dried bananas, and banana chips highlighting characteristic phytochemicals in the fresh fruit as a potential source in dried fruit products.

22.2 Production and consumption

Bananas and plantains represent the fourth most important crop in the world after rice, wheat, and maize. They are grown in more than 130 countries worldwide and harvested over approximately 11 million hectares, with an annual production that surpassed 132 million tonnes in 2009 [4]. Table 22.1 shows banana and plantain production data. These figures are an approximation since most of the world's banana production comes from small plots and home gardens and so may have escaped reporting.

Bananas have been a staple food for centuries and one of the most important sources of energy in the diet of millions of people, in regions in Africa, Asia, Central and South America,

Dried Fruits: Phytochemicals and Health Effects, First Edition. Edited by Cesarettin Alasalvar and Fereidoon Shahidi.

Table 22.1 Top banana and plantain producers and production in 2009

Rank	Country	(MT × 1000)
Bananas		
1	India	26,470
2	Philippines	9013
3	China	9006
4	Ecuador	7637
5	Brazil	6783
6	Indonesia	6374
7	United Republic of Tanzania	3219
8	Guatemala	2544
9	Mexico	2232
10	Colombia	2020
Plantains		
1	Uganda	9512
2	Ghana	3563
3	Colombia	3012
4	Rwanda	2993
5	Nigeria	2911
6	Cameroon	2450
7	Peru	1867
8	Côte d'Ivoire	1497
9	Democratic Republic of the Congo	1200
10	Kenya	843

Source: Adapted from FAO [4].
MT, metric tonnes.

and the Pacific. Today, in addition to being a major cash crop around the world, more than 85% of bananas are produced for local consumption by small scale farmers growing the crop either for home consumption or for trading at local markets. They are, therefore, very important for income and food security. Bananas are the most popular fresh fruit in the United States where per annual capita consumption is about 11 kg [5], slightly above the European average of about 9 kg in 2009 [6].

Most bananas are eaten raw or cooked (baked, boiled, roasted, or fried) and only a very minor proportion is processed (5% dessert bananas and 24% plantains) or dried [7]. This is true both at the village level and in the international trade of "dessert" bananas. Barriers to processing include a susceptibility to enzymatic browning, discoloration, flavor deterioration, and lack of demand since the fresh fruit is available year round [8]. However, high volumes of banana that is lost after harvest, and those that are rejected for the export marked has stimulated finding solutions to these problems [9]. Also, in many areas dried bananas are important in periods where food is scarce [10].

22.3 Dried bananas or banana figs

Banana figs are dried or dehydrated bananas, usually whole, but sometimes sliced lengthwise. They are very sweet and have a sticky, fig-like consistency. Traditionally, sun drying was the most widespread method of producing banana figs but now hot-air circulation in tunnel or cabinet dryers are more popular. Banana figs are made from fully ripe fruits from varieties with high total solid content and a sugar content of about 19.5%. They are peeled, sulfured either by dipping in sulfurous acid solution or by burning sulfur in an enclosed room and dried at temperatures ranging from 50 to 80°C for 5–24 hours, depending if bananas are whole or sliced, to reach a final moisture content of 20–25% [8]. Bananas can also be dipped in a boiling sugar solution before drying. Osmotic dehydration has recently gained popularity. It requires less energy and the nutrient loss is minimized. In this process, banana slices are dipped into an isotonic sugar solution and a sweeter, stable banana fig of intermediate moisture content is produced [11]. Ecuador is one of the main producers of banana figs, but India, Brazil, Philippines, and Costa Rica also produce considerable amounts of it. Europe is the main importer, especially France and Germany [12].

22.4 Dried and fried banana chips (crisps)

Typically green (unripe) bananas are used for producing dehydrated chips. They are thinly sliced, blanched in hot water or steam, or dipped in a solution of a browning inhibitor (SO_2 and citric acid) and dried by different methods to a very low moisture content (2.5–9%) [8]. For banana crisps, the slices are deep-fried in cooking oil (usually coconut, palm or cottonseed oil, depending on the location) at 180–200°C and dusted with salt and occasionally an antioxidant [12]. Sometimes, the slices are partially dried or dipped in salt or sugar solutions before frying. In Philippines, they are fried to intermediate moisture content, soaked in a sugar solution, and fried again in hot oil [9]. Many variations of banana chips exist with a concomitant variation in caloric content and nutrient composition. Philippines is the main exporting country of banana chips [12].

22.5 Nutritional content of bananas, dried bananas, and banana chips

Tables 22.2 and 22.3 show compositional and nutritional characteristics of fresh bananas, plantains, chips, and banana figs [13–17]. Table 22.4 depicts the contributions that one serving of these foods makes toward meeting daily requirements of essential nutrients, expressed as percentage of the Daily Value (DV) [18]. It is important to note that these tables reflect the fruit and dried fruit products available in the United States and Europe. They represent only a fraction of the varieties and products consumed by people in tropical and subtropical areas of the world. Significant compositional differences exist among banana varieties/cultivars particularly in provitamin A activity, but also in fat, iron, calcium, and zinc contents, among others [19]. Data on nutritional content of locally consumed dried banana figs/chips are lacking or are incomplete.

Bananas are a major source of carbohydrate in the diet of people worldwide. In the green fruit, they provide 12–30 g/100 g of starch, and as low as 1% in some fully ripe varieties [20]. The ripe Dwarf Cavendish has 5% starch and provides 12 g/100 g sugar (glucose, fructose, and sucrose). Bananas and plantains are a particularly good source of fiber, one cup (about one fruit) providing 16 and 14%, respectively, of the Dietary Reference Value (DRV). They

Table 22.2 Compositional and nutritional characteristics of bananas and plantains, and their chips (values in per 100 g edible portion)

Nutrient	Unit	Fresh bananas	Fresh plantains	Banana chips (snacks)	Plantain chips (salted snacks)
Proximate composition					
Water	g	74.91	65.28	4.30	2.09
Energy	kcal	89	122	519	531
Protein	g	1.09	1.30	2.30	2.28
Lipid	g	0.33	0.37	33.60	29.59
Ash	g	0.82	1.19	1.4	2.19
Carbohydrate	g	22.84	31.89	58.40	63.84
Dietary fiber	g	2.6	2.3	7.7	3.5
Sugars	g	12.23	15.00	35.34	0.92
Minerals					
Calcium	mg	5	3	18	9
Copper	mg	0.08	0.081	0.21	0.20
Fluoride	μg	2.2	na	na	na
Iron	mg	0.26	0.60	1.25	0.97
Magnesium	mg	27	37	76	71
Manganese	mg	0.27	na	1.56	0.28
Phosphorus	mg	22	34	56	78
Potassium	mg	358	499	536	786
Selenium	μg	1.0	1.5	1.5	0.4
Sodium	mg	1.0	4	6	202
Zinc	mg	0.15	0.14	0.75	0.37
Vitamins					
Choline	mg	9.8	13.5	21.3	na
Folate	μg	20	22	14	35
Niacin	mg	0.67	0.69	0.71	0.80
Pantothenic acid	mg	0.33	0.26	0.62	1.1
Pyridoxine	mg	0.37	0.30	0.26	0.46
Riboflavin	mg	0.07	0.05	0.02	0.04
Thiamine	mg	0.03	0.05	0.09	0.07
Vitamin A (RAE)	μg	3.0	56	4.0	69
Vitamin C	mg	8.7	18.4	6.3	32.1
Vitamin E (ATE)	mg	0.10	0.14	0.24	5.1
Vitamin K	μg	0.5	0.7	1.3	28.6
Amino acids					
Alanine	g	0.040	0.051	na	na
Arginine	g	0.049	0.108	na	na
Aspartic acid	g	0.124	0.645	na	na
Cystine	g	0.009	0.108	na	na
Glutamic acid	g	0.152	0.116	na	na
Glycine	g	0.038	0.045	na	na
Histidine[a]	g	0.077	0.064	na	na
Isoleucine[a]	g	0.028	0.036	na	na
Leucine[a]	g	0.068	0.059	na	na
Lysine[a]	g	0.050	0.060	na	na
Methionine[a]	g	0.008	0.017	na	na

(continued)

Table 22.2 (*Continued*)

Nutrient	Unit	Fresh bananas	Fresh plantains	Banana chips (snacks)	Plantain chips (salted snacks)
Phenylalanine[a]	g	0.049	0.044	na	na
Proline	g	0.028	0.050	na	na
Serine	g	0.040	0.014	na	na
Threonine[a]	g	0.028	0.034	na	na
Tryptophan[a]	g	0.009	0.015	na	na
Tyrosine	g	0.009	0.032	na	na
Valine[a]	g	0.047	0.046	na	na
Lipids					
SFA	g	0.11	0.14	28.97	8.34
MUFA	g	0.03	0.03	1.95	5.63
PUFA	g	0.07	0.07	0.63	11.74

Source: Adapted from USDA [13].
Note: Some numbers are rounded to the second digit after decimal point.
RAE, retinol activity equivalents; ATE, alpha-tocopherol equivalents; na, not available; SFA, saturated fatty acids; MUFA, monounsaturated fatty acids; PUFA, polyunsaturated fatty acids.
[a]Indispensable amino acids.

are excellent sources of potassium and also provide 10 and 14% of the recommended intake of magnesium and 20% of that of manganese. Bananas and plantains are rich in vitamins C and pyridoxine. They also provide significant amounts of the daily requirement of folic acid, niacin, riboflavin, and copper. The yellow- and orange-fleshed bananas are rich in vitamin A, providing as much as 45% of the daily requirement [21]. As is the case of most fruits, they are low in sodium and fat.

Banana figs have comparable proximate compositions to other dried fruits. They are an excellent source of sugar (57–79 g/100 g) and provide fair amounts of fiber (2.1–3.0 g/100 g) [14]. Since drying concentrates the minerals, dried banana and plantain chips are excellent sources of potassium, magnesium, manganese, and copper and provide a significant amount of iron (Table 22.3). Depending on the drying method and the variety of the fruit, they can provide appreciable amounts of the daily intake of pyridoxine, niacin, thiamine, vitamin C, and vitamin A. Dried bananas and banana chips could play an important role in alleviating micronutrient deficiencies particularly in tropical and subtropical areas where they are most severe and where bananas are a staple component of the diet.

Fried banana and plantain chips or crisps, in terms of their composition, could be more accurately compared to potato chips than dried fruits. In addition to being high in carbohydrate, they provide more than 50% of the calories from fat. Since they are highly caloric, vitamin and mineral content is diluted and consequently is lower than that of their fresh counterparts. Nonetheless, in populations at risk of micronutrient deficiency, banana crisps can be a source of essential minerals and vitamin A, if high β-carotene varieties are used [22]. Techniques are being investigated to reduce the oil uptake and preserve nutritional properties [23]. Vitamin C is particularly susceptible to drying and deep frying [24, 25].

Table 22.3 Compositional and nutritional characteristics of dried bananas, plantains, and banana figs (values in per 100 g edible portion)

Nutrient	Unit	Dried bananas (banana figs) [14]	Smoked/ dried (banana figs) [15]	Hawaiian dried bananas [16]	Dried plantains (green)[a][14]	Dried bananas (ripe)[a][17]
Proximate composition						
Moisture	g	19.5–27.7	12.5	29.2	9.0	4
Energy	kcal	300	337	–	359	346
Protein	g	2.8–3.5	5.1	5.3	3.3	4
Lipid	g	0.8–3.5	0.2	2.3	1.4	2
Ash	g	2.1–2.8	–	5.3	2.4	
Carbohydrate	g	60–69	79	57.9	83.9	88
Dietary fiber	g	2.1–3.0	–	–	1 (4%)	10 (40%)
Other nutrients						
Calcium	mg				50 (5%)	22 (2%)
Copper	mg				–	0.4 (20%)
Iron	mg				1.1 (6%)	1.2 (6%)
Magnesium	mg				–	108 (27%)
Manganese	mg					0.6 (29%)
Phosphorus	mg				65 (7%)	74 (7%)
Potassium	mg				–	1491 (43%)
Vitamin C	mg				1 (2%)	7 (12%)
β-Carotene	μg				45 (2%)	91 (5%)
Niacin	mg				1.9 (10%)	2.8 (14%)
Riboflavin	mg				0.16 (9%)	0.2 (10%)
Thiamin	mg				0.10 (7%)	0.2 (8%)
Folic acid	μg				–	14 3%)

[a]Value (Percentage RDI for vitamins and minerals; percentage DRV for fiber and potassium).

Table 22.4 Percentage of the daily value for fiber, vitamins, and minerals met by a serving of fresh bananas and plantains, and their chips

Nutrient	DV[a][18]	Unit	Bananas (1 cup sliced serving basis or 150 g)[b]	Plantains (1 cup sliced serving basis or 148 g)[c]	Fried banana chips (1 oz or 28.5 g serving basis)[d]	Fried plantain chips (1 oz or 28.5 g serving basis)[d]
Fiber	25 g/day	g	16	14	8	4
Vitamins						
Folate	400 μg/day	μg	8	8	1	2
Niacin	20 mg/day	mg	5	5	1	1
Pantothenic acid	10 mg/day	mg	5	4	2	3
Pyridoxine	2.0 mg/day	mg	28	22	4	7
Riboflavin	1.7 mg/day	mg	6	5	0	1
Thiamine	1.5 mg/day	mg	3	5	1	1
Vitamin A (RAE)	1500 μg /day	μg	2	33	0	8
Vitamin C	60 mg/day	mg	22	45	3	15
Vitamin E (ATE)	20 mg/day	mg	1	1	0	7
Vitamin K	80 μg /day	μg	0	1	0	10
Minerals						
Calcium	1000 mg/day	mg	1	0	1	0
Copper	2 mg/day	mg	6	6	3	3
Iron	18 mg/day	mg	2	5	2	2
Magnesium	400 mg/day	mg	10	14	5	5
Manganese	2 mg/day	mg	20	na	22	4
Phosphorus	1000 mg/day	mg	3	5	2	2
Potassium	3500 mg/day	mg	15	21	4	5
Selenium	70 μg/day	μg	2	3	1	0
Sodium	2500 mg/day	mg	0	0	0	2
Zinc	15 mg/day	mg	2	1	1	1

DV, Daily Value; RAE, retinol activity equivalents; ATE, alpha-tocopherol equivalents; na, not available.

[a] DV is derived from the Daily Reference Values and Reference Daily Intakes and is based on a caloric intake of 2000 calories, for adults and children 4 or more years of age.

[b] Serving size is about one large banana.

[c] Serving size is about one small plantain.

[d] Serving size is defined by USDA.

22.6 Phytochemicals in bananas and dried fruit products

Phytochemicals are bioactive compounds in plants that appear to play a role in health and longevity and have been associated with a reduction in the risk of major chronic degenerative diseases. Like other fruits, bananas have a characteristic array of phytonutrients (Table 22.5). Those that have received most attention in the raw fruit are carotenoids, flavonoids, and phenolic acids. Phytosterols, particularly β-sitosterol, have been found at low levels in banana pulp (530 mg/kg dry weight) and peel and are being investigated as a means to utilize banana refuse [26].

22.6.1 Phenolics in fresh and dried bananas

Phenolic compounds are ubiquitous in plants. They not only contribute to the taste and color of plant-derived foods and beverages but also to the health benefits associated with consumption of diets high in fruits and vegetables. Phenolic acids and flavonoids are major classes of phenolic compounds. They are considered to be potent antioxidants and exhibit a wide range of beneficial physiological properties. Many studies have measured total phenolic content of different varieties of banana using the Folin–Ciocalteu colorimetric method (Table 22.5). Values range from 14 to 518 mg of gallic acid equivalents (GAE)/100 g [27–32]. The wide variation of total phenolics not only reflects the intrinsic (e.g., cultivar and degree of ripeness) and extrinsic factors (e.g., agronomic and storage), but also the method of extraction. Phenolic compounds in plants occur in both bound and free forms. As the former may be excluded from the analysis, total phenolic content is sometimes underestimated.

Literature on flavonoid profiles of bananas is scarce and inconsistent. Most data available come from studies surveying flavonoid content in a range of fruits/vegetables (Table 22.5). The main classes of flavonoids detected in bananas are flavan-3-ols and flavones. High levels of +(-catechin) have been found in two varieties of bananas from Tenerife and one from Ecuador [33]. Epicatechin, epigallocatechin, and gallocatechin have been detected in Dwarf Cavendish bananas [34–36] as well the flavonols quercetin, myricetin, and kaempferol [27, 32]. The only anthocyanin that has been detected is delphinidin at 7.4 mg/100 g edible portion [34] and in banana pulp cell walls [37].

Total content of phenolic acids in bananas has been reported at 7 mg/100 g fresh weight [38]. Phenolic acids bound to other plant components such as polysaccharides and lignin in cell walls, predominate. Ferulic acid is particularly high in bananas accounting for 69% of the total [39]. Ferulic, gallic, hydroxybenzoic, *p*-coumaric, protacatechuic, salicylic, sinapic, syringic, and vanillic acids have all been detected in bananas (Table 22.5).

No data are yet available on the flavonoid and phenolic content of dried bananas. It is known that there is a loss and modification of phenolic acids of up to 90% and approximately 60% loss of flavonols due to sun drying [40]. However, total phenolic content remains relatively unchanged, which implies that many of these modified compounds are yet to be identified and could include oligomeric or polymeric products that are difficult to categorize. Treatment with sulfiting agents inactivates polyphenol oxidase and inhibits non-enzymatic browning and so pre-treated dried bananas would better retain the phenolic compounds than their fresh counterparts [40].

Table 22.5 Major phytochemicals detected in fresh bananas

	Unit	Fresh bananas	Reference
Total phenolics	mg of GAE/100 g fw	155	[27]
	mg of GAE/100 g fw	14	[27]
	mg of GAE/100 g fw	51	[28]
	mg of GAE/100 g fw	321–518	[29]
	mg of GAE/100 g fw	68–263	[30]
	mg of GAE/100 g fw	90.4	[31]
	mg of CAE/100 g fw	475	[32]
Flavan-3-ols			
+(-catechin)	mg/100 g fw	6.23–6.30	[33]
	mg/100 g fw	nd	[34]
	mg/100 g fw	10.29	[33]
Epicatechin	mg/100 g fw	0.03–0.2	[34, 35]
Epigallocatechin	mg/100 g fw	0.01	[35]
Gallocatechin	mg/100 g fw	29.6	[36]
Anthocyanins	mg/100 g fw	nd	[32]
Delphinidin	mg/100 g fw	7.4	[34]
Flavonols (aglycones)			
Kaempferol	mg/100 g fw	0.012–0.11	[27, 32]
Quercetin	mg/100 g fw	0.06–0.292	[27, 32]
Myricetin	mg/100 g fw	0.01–0.143	[27, 32]
Flavones	mg/100 g fw	nd	[34]
Flavanols	mg/100 g fw	nd	[34]
Phenolic acids	mg/100 g fw	7	[38]
Gallic acid (aglycones)	mg/100 g fw	0.87–1.06	[33]
Bound			
Ferulic	mg/100 g dw	40.07	[39]
Gallic acid	mg/100 g dw	3.22	[39]
Hydroxybenzoic	mg/100 g dw	1.64	[39]
p-Coumaric	mg/100 g dw	3.47	[39]
Salicylic	mg/100 g dw	6.73	[39]
Sinapic	mg/100 g dw	0.27	[39]
Syringic	mg/100 g dw	3.54	[39]
Vanillic	mg/100 g dw	1.04	[39]
Free			
Protocatechuic	mg/100 g dw	0.16	[39]
Syringic	mg/100 g dw	0.21	[39]
Vanillic	mg/100 g dw	0.30	[39]
Fructooligosaccharides	g/100 g fw	nd–0.7	[42, 43]
Fructans 1-kestose, nystose and 1-β-fructafuranosyl-nystone measured only	g/100 g fw	0.02	[44]
	g/100 g fw	0.04–0.13	[46]
Carotenoids			
β-Carotene	μg/100 g fw	25.2–58.8	[51]
	μg/100 g fw	26–457	[13]
	μg/100 g fw	50.6–61.6	[52]
	μg/100 g fw	42.8–131.4	[53]
α-Carotene	μg/100 g fw	25–438	[13]
	μg/100 g fw	61–1055	[21]
	μg/100 g fw	67.9–155.6	[53]
Lutein	μg/100 g fw	22–30	[13]
	μg/100 g fw	86.3–192.2	[53]
Cryptoxanthin	μg/100 g fw	nd	[13, 53]
Lycopene	μg/100 g fw	nd	[51]

nd, not detected; dw, dry weight; fw, fresh weight; GAE, gallic acid equivalents; CAE, chlorogenic acid equivalents.

22.6.2 Fructooligosaccharides (FOS)

FOS are oligosaccharides consisting of short chains of fructose units linked by beta (2–1) bonds. They occur naturally in plant foods or are added to foods as functional ingredients. Since they are not digested in the human small intestine, they are considered dietary fibers. Fructans have important physiological effects. They appear to promote the growth of beneficial intestinal flora and suppress the growth of potential pathogens in the colon. They increase stool-bulking capacity, enhance mineral absorption, maintain the integrity of the intestinal mucosal barrier, and may stimulate the gastrointestinal immune system [41]. Bananas are among the few fruits where FOS have been detected (Table 22.5) at levels ranging from 0.02 [42] to 0.7 mg/100 g [43, 44]. Levels of FOS in dried bananas have not been measured, but it is possible that they are present and become concentrated. Raisins have relatively high levels of fructans while they are absent in the grapes they are made from [45]. It is possible that fructans are formed from the sugars in the grapes during sun drying. There are no detectable levels of FOS in banana chips [46].

22.6.3 Carotenoids

Other than the provitamin A activity of a number of carotenoids, these compounds may have a protective effect against heart disease and certain types of cancer. In addition, epidemiological studies suggest that diets rich in lutein and zeaxanthin may prevent age-related macular degeneration and cataracts [47]. A broad survey of banana varieties indicate that many have very high levels of provitamin A and total carotenoids [21, 48, 49]. More than 90% of the carotenes in bananas consist of all trans β-carotene and α-carotene with only very small amounts of the 13-cis-isomer of μg/100 g [5, 50–53]. Levels for β- and α-carotene range from 25 μg/100 g in creamy fleshed bananas to more than 1000 μg/100 g in deep orange- or yellow-fleshed fruits (Table 22.5). The only other major carotenoid present is lutein, detected at levels ranging from 22 to 192 μg/100 g [5, 53]. Proportion of carotenoids in bananas is characteristic of the variety [50].

Dried bananas, banana figs, and banana chips do retain part of the carotenoid content of the fresh fruit, the proportion that is retained depends mostly on the duration and the severity of the drying process [22, 54]. While sun drying is the most affordable method, it is slow and can lead to considerable losses. With hot-air drying, significantly higher retention of carotenoids is achieved in the final product. Other than destruction by oxidation, drying causes isomerization of all-trans carotenoids into the cis-form, thus affecting provitamin A activity, bioavailability, and antioxidant activity [23]. Pre-treatments such as dipping in a salt or ascorbic acid solution, sodium sulfiting increase carotenoid retention during drying [54]. Conventional deep frying also leads to significant losses which can range from 20 to 50% of total carotenoids and 16 to 60% of β-carotene, depending on the duration, temperature, type, and freshness of the oil [25, 55]. Double frying of plantain chips leads to further degradation of carotenoids [24].

22.7 Potential health benefits of dried bananas

Dried bananas, banana figs, and banana chips (non-fried) are low in fat and sodium and high in potassium, magnesium, and fiber. This makes them desirable snacks for a diet aimed at cardiovascular health. It is recognized that increasing dietary potassium intake can lower blood pressure and attenuates the adverse effects of sodium [56]. Potassium intake is low among both adults and children and has become a public health concern [57]. High fiber diets are recommended to reduce the risk of developing cardiovascular disease (CVD) and various other conditions including constipation, diverticulitis, diabetes, and colon cancer.

Dried bananas are also a rich source of carbohydrate and depending on the drying process and the variety of the fruit; they can provide appreciable amounts of the daily intake of pyridoxine, niacin, thiamine, vitamin C, and vitamin A. They could play an important role in alleviating micronutrient deficiency, particularly in tropical and subtropical areas where they are most severe and where bananas are a staple component of the diet. Green banana starch has a low glycemic index (GI) as it resists digestive enzyme's catalytic hydrolysis [58]. Green bananas have been found to have a protective effect against experimentally induced gastric mucosal lesions [59]. It is not clear how these effects would transfer to dried chips made with green bananas.

Total antioxidant capacity of bananas, as measured by the oxygen radical absorbance capacity (ORAC), has been reported as 879 μmol of trolox equivalents (TE)/100 g, which is intermediate among fruits but equal or higher than most vegetables [60]. This is due to the presence of micronutrients such as vitamin C and phytonutrients such as phenolic acids, flavonoids, and carotenoids with high antioxidant activity. Current scientific thought regards oxidative stress as an important contributing factor in the development of chronic diseases such as heart disease, diabetes, cancer, neurodegenerative diseases such as in Alzheimer's and Parkinson's disease as well as in the aging process itself. Dietary antioxidants are part of the body's antioxidant defence system. Epidemiological, case-control, and dietary intervention studies suggest or show the importance of plant foods rich in antioxidants in the prevention against chronic disease. Depending on the processing method, it is possible that dried bananas retain their antioxidant potential as other dried fruits do.

22.8 Conclusions

While fresh bananas are very nutritious, high in fiber and potassium, excellent sources of many micronutrients and phytonutrients, more nutritional information is needed to fully appreciate the potential health benefits of dried bananas.

References

1. Perrier, X., De Langhe, E., Donohue, M., Lentfer, C., Vrydaghs, L., Bakry, F., Carreel, F., Hippolyte, I., Lebot, V., Risterucci, A., Tomekpe, K., Doutrelepont, H., Ball, T., Manwaring, J., Maret, P. & Denham, T. (2011) Multidisciplinary perspectives on banana (Musa spp.) domestication. *Proceedings of the National Academy of Sciences*, **108**, 11311–11318.
2. Helop-Harrison, J.S. & Schwarzacher, T. (2007) Domestication, genomics and the future for banana. *Annals of Botany*, **100**, 1073–1084.
3. FAO (2003) *The World Banana Economy (1985–2005). FAO Corporate Document Repository*. Published online at: http://www.fao.org/docrep/007/y5102e/y5102e0a.htm#bm10, last accessed March 5, 2012.
4. FAO (2012) *FAOSTAT Agricultural Production Data*. Published online at: http://faostat.fao.org/site/339/default.aspx, last accessed March 5, 2012.

5. USDA/ERS (2011) *Food Availability (per capita) Data System.* Published online at: http://www.ers.usda.gov/Data/FoodConsumption/FoodAvailSpreadsheets.htm#fruitfr, last accessed March 5, 2012.
6. CBI (2011) *Centre for the Promotion of Imports from Developing Countries, Netherlands Ministry of Foreign Affairs, Fresh Bananas in the United Kingdom.* Published online at: http://www.cbi.eu/?pag=85&doc=6102&typ=mid_document, last accessed March 5, 2012.
7. INIBAP (2006) *International Network for the Improvement of Banana and Plantain Progress Report to the Rockefeller Foundation: Adding Value to Bananas.* Published online at: http://bananas.bioversityinternational.org/files/files/pdf/publications/addingvalue-june2006.pdf, last accessed March 5, 2012.
8. Narayana, C.K. & Pillay, M. (2011) Postharvest products from bananas. In: *Banana Breeding: Progress and Challenges* (eds M. Pillay & A. Tenkouano). CRC Press, Taylor & Francis Group, Boca Raton, FL, pp. 269–284.
9. Ocena-Po, L.G. (2006) Banana, mango and passion fruit. In: *Handbook of Fruit and Fruit Processing* (ed. Y.H. Hui). Wiley-Blackwell, Oxford, UK.
10. Eyabe, G., Salamang, P., Ehabe, E. & Numfor, F.A. (2000) Chemical and organoleptic figs from selected sweet bananas (Musa Spp.). *Journal of Food Technology Africa*, **5**, 70–72.
11. Mohapatra, D., Mishra, S., Singh, C.B. & Jayas, D.S. (2011) Post-harvest processing of banana: opportunities and challenges. *Food Bioprocess Technology*, **4**, 327–339.
12. Sole, P. (2005) Bananas (processed). In: *Processing Fruits* (eds D.M. Barrett, L. Somogyi & H. Ramaswamy). CRC Press, Taylor and Francis Group, Boca Raton, FL, pp. 657–678.
13. USDA (2011) *National Nutrient Database for Standard Reference, Release 24.* Published online at: http://www.nal.usda.gov/fnic/foodcomp/search/, last accessed February 7, 2012.
14. Morton, J. (1987) Banana. In: *Fruits of Warm Climates* (ed. J.F. Morton). Creative Resource Systems, Miami, FL, pp. 29–46.
15. Zaidah, I. (2004) Snacks made from bananas. Paper given at the *International Congress on Musa: Harnessing Research to Improve Livelihoods*, Penang, Malaysia, 6–9 July 2004 (Abstract No. P-85).
16. Aalbersberg, B (1990) Food preparation and preservation. In: *Food and Nutrition in Fiji: Food Production, Composition and Intake* (eds A.J. Jansen, S. Parkinson & A.F.S. Robertson). The University of the South Pacific, Suva, Fiji, pp.156–171.
17. Dried Fruit Resources and Dried Bananas (2012) *Nutritional Information.* Published online at: http://www.triedtastedserved.com/dried-fruits/dried-bananas.php, last accessed February 20, 2012.
18. Food and Drug Administration Guidance for Industry (2012) *A Food Labelling Guide, Appendix 14.* Published online at: http://www.fda.gov/Food/GuidanceComplianceRegulatoryInformation/GuidanceDocuments/FoodLabelingNutrition/FoodLabelingGuide/ucm064928.htm, last accessed February 7, 2012.
19. Stadlmayer, B., Nilsson, E., Mouille, B., Methammer, E., Burlingame, B. & Charrondiere, U.R. (2011) Nutrition indicator for biodiversity on food composition—a report on the progress of data availability. *Journal of Food Composition and Analysis*, **24**, 692–696.
20. Soares, C.A., Goncalves Peroni-Okita, F.H., Borba Cardoso, M., Shitakubo, R., Lajolo, F.M. & Cordenunsi, B.R. (2011) Plantain and banana starches: granule structural characteristics explain the differences in their starch degradation patterns. *Journal of Agricultural and Food Chemistry*, **59**, 6672–6681.
21. Blades, B.L., Dufficy, L., Englberger, L., Daniells, J.W., Coyne, T., Hamill, S. & WillS, R.B.H. (2003) Bananas and plantains as a source of provitamin A. *Asia Pacific Journal of Clinical Nutrition*, **12**, 36–43.
22. Adeniji, T.A. & Tenkouano, A. (2007) Effect of processing on micronutrient content of chips produced from some plantain and banana hybrids. *Fruits*, **63**, 345–352.
23. Dueik, V., Robert, P. & Boudin, P. (2010) Vacuum frying reduces oil uptake and improves the quality parameters or carrot crisps. *Food Chemistry*, **119**, 114–1149.
24. Avallone, S., Rojas-Gonzales, J.A., Trystam, G. & Bohuon, P. (2009) Thermal sensitivity of some plantain micronutrients during deep-fat frying. *Journal of Food Chemistry*, **74**, C339–C347.
25. Rojas-Gonzales, J.A., Avallone, S., Brat, P., Trystam, G. & Bohuo, P. (2006) Effect of deep-fat frying on ascorbic acid, carotenoids and potassium content of plantain cylinders. *International Journal of Food Science and Nutrition*, **57**, 123–136.

26. Oliveira, L., Freire, C.S.R., Silvestre, A.J. & Cordero, N. (2008) Lipophilic extracts from banana fruit residues: a source of valuable phytosterols. *Journal of Agricultural Food Chemistry*, **56**, 9520–9525.
27. USDA (2011) *USDA Database for the Flavonoid Content of Selected Foods, Release 3*. Published online at: http://www.ars.usda.gov/nutrientdata, last accessed January 9, 2012.
28. Lim, Y.Y., Lim, T.T. & Tee, J.J. (2007) Antioxidant properties of several tropical fruits: a comparative study. *Food Chemistry*, **103**, 1003–1008.
29. Shian, T.E., Abdullah, A., Musa, K.H., Maskat, M.Y. & Ghani, M.A. (2012) Antioxidant properties of three banana cultivars (*Musa acuminata* 'Berangan,' 'Mas' and 'Raja') extracts. *Sains Malaysiana*, **41**, 319–324.
30. Sulaiman, S.F., Yusoff, M.N., Eldeen, I.M., Seow, E.M., Sajak, A.A.B. & Ooi, K.L (2011) Correlation between total phenolic and mineral content with antioxidant activity of eight Malaysian bananas (Musa sp.). *Journal of Food Composition and Analysis*, **24**, 1–10.
31. Sun, J., Chu, Y.-F. & Liu, R.H. (2002) Antioxidant and antiproliferative activities of common fruits. *Journal of Agricultural and Food Chemistry*, **50**, 7449–7454.
32. Kevers, C., Falkowski, M., Tabart, J., Defraigne, J.O., Dommes, J. & Pincemail, J. (2007) Evolution of antioxidant capacity during storage of selected fruits and vegetables. *Journal of Agricultural and Food Chemistry*, **55**, 8596–8603.
33. Mendez, C.M.V., Foster, M.P., Rodriguez-Delgado, M.A., Rodriguez-Rodriguez, E.M. & Romero, C.D. (2003) Content of free phenolic compounds in bananas from Tenerife (Canary Island) and Ecuador. *European Food Research and Technology*, **217**, 287–290.
34. Harnly, J., Doherty, R.F., Beecher, G.R., Holden, J.M., Haytowitz, D.B., Bhagwat, S. & Gephart, S. (2006) Flavonoid content of U.S. fruits, vegetables, and nuts. *Journal of Agricultural and Food Chemistry*, **54**, 9966–9977.
35. de Pascual-Teresa, S., Santos-Buelgas, C. & Rives-Gonzalo, J.C. (2000) Quantitative analysis of flavan-3-ols in Spanish foodstuff and beverages. *Journal of Agricultural and Food Chemistry*, **48**, 5331–5337.
36. Someya, S., Yoshiki, Y. & Okubo, K. (2002) Antioxidant compounds from bananas. *Food Chemistry*, **79**, 351–354.
37. Bennett, R., Shiga, T.M., Hassimotto, N.M., Rosa., E.A., Lajolo, F.M. & Cordenunsi, B.R. (2010) Phenolics and antioxidant properties of fruit pulp and cell wall fractions of postharvest banana (*Musa acuminata* Juss.) cultivars. *Journal of Agricultural and Food Chemistry*, **58**, 7991–8003
38. Matilla, P., Hellstrom, J. & Torronen, R. (2006) Phenolic acid in berries, fruits, and beverages. *Journal of Agricultural and Food Chemistry*, **54**, 7193–7199.
39. Russell, W.R., Labat, A., Scobbie, L., Duncan, G.J. & Duthie, G.G. (2009) Phenolic acid content of fruits commonly consumed and locally produced in Scotland. *Food Chemistry*, **115**, 100–104.
40. Williamson, G. & Carughi, A. (2010) Polyphenol content and health benefits of raisins. *Nutrition Research*, **30**, 511–519.
41. Sabater-Molina, M., Larque, E., Torrella, F. & Zamora, S. (2009) Dietary fructooligosaccharides and potential benefits on health. *Journal of Physiological Biochemistry*, **65**, 315–328.
42. Muir, J., Shepherd, S.J., Rosella, O., Barrett, J.S. & Gibson, P.R. (2007) Fructan and free fructose content of common Australian vegetables and fruit. *Journal of Agricultural and Food Chemistry*, **55**, 6619–6627.
43. Van Loo, J., Coussement, P., De Leenheer, L., Hoebregs, H. & Smits, G. (1995) On the presence of inulin and oligofructose as natural ingredients in the Western diet. *Critical Reviews in Food Science and Nutrition*, **35**, 525–552.
44. Campbell, J.M, Bauer, L.L., Fahey, G.C., Hogarth, A.J.C.L., Wolf, B.W. & Hunter, D.E. (1997) Selected fructooligosaccharide (1-kestose, nystose, and 1^F-b-fructofuranosylnystose) composition of foods and feeds. *Journal of Agricultural and Food Chemistry*, **45**, 3076–3082.
45. Camire, M.E. & Dougherty, M.P. (2003) Raisin dietary fiber composition and *in vitro* bile acid binding. *Journal of Agricultural and Food Chemistry*, **51**, 834–837.
46. Hogarth, A.J.C.L., Hunter, D.E., Jacobs, W.A., Garleb, K.A. & Wolf, B.W. (2000) Ion chromatographic determination of three fructooligosaccharide oligomers in prepared and preserved foods. *Journal of Agricultural and Food Chemistry*, **48**, 5326–5330.
47. Rhone, M. & Basu, A. (2008) Phytochemicals and age related eye diseases. *Nutrition Reviews*, **66**, 465–472.
48. Englberger, L. (2003) Carotenoid-rich bananas in Micronesia. *Info Musa*, **12**, 2–5.

49. Englberger, L., Wills, R.B., Blades, B., Dufficy, L., Daniells, J.W. & Coyn, T. (2006) Carotenoid content and flesh color of selected banana cultivars in Australia. *Food and Nutrition Bulletin*, **27**, 281–291.
50. Davey, M.W., Saeys, W., Hof, E., Ramon, H., Swennen, R.L. & Keulemans, J. (2009) Application of visible and near infrared reflectance spectroscopy (Vis/NIRS) to determine carotenoid content in banana (Musa spp.) fruit pulp. *Journal of Agricultural and Food Chemistry*, **57**, 1742–1751.
51. Kongkachuichai, R., Charoensiri, R. & Sungpuag, P. (2010) Carotenoid, flavonoid profiles and dietary fiber contents of fruits commonly consumed in Thailand. *International Journal of Food Science and Nutrition*, **61**, 536–548.
52. Fungo, R. & Pillay, M. (2011) Beta carotene content of selected banana genotypes from Uganda. *African Journal of Biotechnology*, **10**, 5423–5430.
53. Wall, M. (2006) Ascorbic acid, provitamin A and mineral composition of banana (Musa sp.) and papaya (*Carica papaya*) cultivars grown in Hawaii. *Journal of Food Composition and Analysis*, **19**, 432–455.
54. Bechoff, A. (2010) *Investigating Carotenoid Loss After Drying and Storage of Orange-Fleshed Sweet Potato*. PhD thesis, University of Greenwhich, Greenwich, UK.
55. Vimala, B., Thushara, R., Nambisan, B. & Sreekumar, J. (2011) Effect of processing on the retention of carotenoids in yellow fleshed cassava (*Manihot escueta* Crantz) roots. *International Journal of Food Science & Technology*, **46**, 166–169.
56. Appel, L.J. (2003) Lifestyle modifications as a means to prevent and treat high blood pressure. *Journal of the American Society of Nephrology*, **14**, S99–S102.
57. van Mierlo, L.A., Greylin, A., Zock, P.L., Kok, F.J. & Geleijmse, J.M. (2010) Suboptimal potassium intake and potential impact on population blood pressure. *Archives of Internal Medicine*, **170**, 1501–1502.
58. Zhang, S., Whistler, R.L., BeMiller, J.N. & Hammaker, B.R. (2005) Banana starch: production, physicochemical properties and digestibility-a review. *Carbohydrate Polymers*, **50**, 443–458.
59. Dunjic, B.S., Svensson, I., Axelson, J., Adlercreutz, P., Ar'Rajab, A., Larsson, K. & Bengmark, S. (1993) Green banana protection of gastric mucosa against experimentally induced injuries in rats. *Scandinavian Journal of Gastroenterology*, **28**, 894–898.
60. Wu, X., Beecher, G.R., Holden, J.M., Haytowitz, D.B., Gebhardt, S.E. & Prior, R.L. (2004) Lipophilic and hydrophilic antioxidant capacities of common foods in the United States. *Journal of Agricultural and Food Chemistry*, **52**, 4026–4037.

23 Nutritional composition, phytochemicals, and health benefits of dates

Cesarettin Alasalvar and Fereidoon Shahidi

23.1 Introduction

Dates are produced largely in the hot desert regions of the Middle East and North Africa and are marketed worldwide as a high-value crop. The world production of dates has increased from 4,600,000 metric tonnes (MT) in 1994 to 7,321,090 MT in 2009 [1]. Egypt is the largest producer of dates, contributing 17.4% to the total global production, followed by Iran (14.0%), Saudi Arabia (13.5%), United Arab Emirates (10.4%), and Pakistan (10.0%). Other countries (Algeria, Iraq, Sudan, Oman, Tunisia, Libya, China, Morocco, Yemen, Niger, Turkey, Israel, the United States, and Mauritania) contribute 34.7% to the total global production [1]. There are more than 600 known date varieties [2].

Sun drying is the common method of drying dates; taking around 7–10 days depending on the daytime temperature which varies from 30 to 50°C and the humidity (60–85%) [3, 4]. Several factors affect the quality of dates during sun drying. These include insect infestation, microbial infestation, and browning by enzymatic and non-enzymatic reactions [5]. The drying of dates can also be achieved by artificial heat treatment in circumstances where early rains threaten to damage the crop. The process requires rooms in which temperature, humidity, and air ventilation can be controlled. Drying rate is a function of temperature, relative humidity, and velocity of the air. For drying of soft dates, a temperature of 65°C is recommended, which ensures a reasonable drying rate with a minimal effect on the basic qualities. Relative humidity should be maintained at over 40%, but should not exceed 60% to avoid case-hardening and also for fuel economy [5].

Dates are considered to be an important component of daily diet in the Middle East and North Africa. They are consumed either fresh (30–40%) or sun-dried (60–70%). Sun-dried dates are consumed throughout the year, but their use reaches peak during the month of Ramadan, for breaking the fast before eating. The average daily consumption of dates per capita was estimated in Oman and United Arab Emirates at 164 and 114 g, respectively [6, 7]. Socio-economic changes have reduced date consumption due to improvement in living standards, changes in eating habits, continued urban drift, and the tendency toward a smaller sized family. The year round availability of alternative competitive confectioneries and other fruits have aggravated the problem [4].

Dried Fruits: Phytochemicals and Health Effects, First Edition. Edited by Cesarettin Alasalvar and Fereidoon Shahidi.

Because of their nutritional value and potential health promoting activities, dates may be considered as an emerging and potential candidate for the development of foods with disease risk reduction potential [2]. This chapter reviews the latest research on fresh and dried dates' nutrition, phytochemicals, and their roles in antioxidant status and other human health potentials. Factors affecting the phytochemical composition and antioxidant activity during drying/processing are also discussed. Readers who are interested in detailed information and in-depth discussions about the specific nutritional, phytochemicals or functions of dates, are referred to comprehensive reviews by others [2, 4]. In addition, final section of a recent book entitled "*Dates: Production, Processing, Food, and Medicinal Values*" focuses on the potential medicinal values of dates along with the latest research findings [8].

23.2 Compositional and nutritional characteristics of fresh and dried dates

Dates have provided nutrition to millions of people around the world for thousands of years. Therefore, it is worthwhile to assess their nutritional significance as a food for regular consumption and as a good source of nutritional components for maintaining health and for reducing the risk of various diseases [2]. Dates contain many nutrients, at 3–10-fold higher than the other commonly consumed fruits (such as apples, grapes, oranges, cranberries, and blueberries), and may be considered as one of the highly nutritious fruits available on earth for human consumption [2]. Vinson *et al.* [9] have indicated that fresh dates have the best nutrient score among the fresh apricots, cranberries, figs, green grapes, and plums.

23.2.1 Proximate composition

The average proximate composition and caloric value of fresh and dried dates are summarized in Table 23.1. The detailed information on nutritional and functional properties of 10 fresh and 16 dried date varieties have been extensively reviewed by Al-Farsi and Lee [4]. In addition, Table 1.2 (see Chapter 1) compares compositional and nutritional characteristics of some dried fruits (apples, apricots, dates, figs, peaches, pears, prunes, and raisins). Here, comparison between fresh and dried dates is discussed briefly. Carbohydrate is the predominant component in both fresh and dried dates, ranging in concentration from 54.9 g/100 g (in fresh dates) to 80.6 g/100 g (in dried dates), followed by moisture (42.4 g/100 g in fresh dates and 15.2 g/100 g in dried dates), along with small amounts of protein, lipid, and ash (Table 23.1). The significant reduction in the moisture content of dried dates is mainly due to sun drying, which is a traditional way of preserving dates. The energy values range from 213 kcal/100 g in fresh dates to 314 kcal/100 g in dried dates, due to their high sugar content.

23.2.2 Dietary fiber

Table 23.1 presents the content of dietary fiber in fresh and dried dates. Dietary fiber is divided into water-soluble (pectin and hydrocolloids) and water-insoluble (cellulose, hemicellulose, and lignin). Dietary fiber plays an important role in human health as its low intake has been linked to several diseases in the industrialized nations [2]. The average total fiber content in fresh and dried dates ranges from 7.5 to 8.0 g/100 g, respectively; due to moisture reduction

Table 23.1 Compositional and nutritional characteristics of fresh and dried dates (average values in per 100 g edible portion)

Nutrient	Units	Fresh dates	Dried dates
Proximate composition			
Water	g	42.4	15.2
Energy	kcal	213	314
Protein	g	1.5	2.14
Lipid	g	0.14	0.38
Ash	g	1.16	1.67
Carbohydrate	g	54.9	80.6
Dietary fiber	g	7.5	8.00
Soluble	g	0.96	0.84
Insoluble	g	5.89	5.76
Sugars	g	43.4	64.1
Fructose	g	19.4	29.4
Glucose	g	22.8	30.4
Sucrose	g	4.03	11.6
Minerals			
Calcium	mg	20.2	70.7
Copper	mg	0.21	0.24
Iron	mg	0.64	0.83
Magnesium	mg	43.3	64.2
Manganese	mg	0.29	0.27
Phosphorus	mg	41.0	58.1
Potassium	mg	486	713
Selenium	mg	0.24	0.31
Sodium	mg	90.9	32.9
Zinc	mg	0.24	0.27
Vitamins			
Folate	μg	na	53.75
Niacin	μg	na	1442
Pyridoxine	μg	na	207
Riboflavin	μg	na	116.5
Thiamin	μg	na	78.67
Vitamin A (RAE)	μg	na	23.85
Vitamin C	μg	na	3900
Amino acids			
Alanine	mg	50.0	93.2
Arginine	mg	52.0	80.9
Aspartic acid	mg	118	152
Cystine	mg	27.3	46.0
Glutamic acid	mg	147	244
Glycine	mg	60.3	107
Histidine[a]	mg	23.7	27.7
Isoleucine[a]	mg	21.3	46.2
Leucine[a]	mg	68.0	98.7
Lysine[a]	mg	86.3	66.9
Methionine[a]	mg	7.7	22.9
Phenylalanine[a]	mg	37.0	53.2
Proline	mg	54.0	105
Serine	mg	48.7	67.4
Threonine[a]	mg	39.3	52.6
Tryptophan[a]	mg	nd	40.6
Tyrosine	mg	26.3	41.2
Valine[a]	mg	na	na

Source: Adapted with permission from Al-Farsi and Lee [4].
Average values are composed of between three and ten date varieties.
RAE, retinol activity equivalents; na, not available; nd, not detected
[a]Indispensable amino acids.

and to the ripening process in which enzymes gradually break down these substances to the more soluble compounds which soften the fruit [10]. It is important to note that high content of dietary fiber in dates helps meeting dietary recommendations (14 g of fiber for every 1000 calories of food consumed each day). This becomes 25–38 g of fiber per day depending on the age and gender [11]. On a per serving basis (40 g), fresh and dried dates deliver between 7.9 and 12.8% of the daily required value of fiber [12]. The high content of the insoluble fiber induces satiety and has a laxative effect due to increased stool weight. It, therefore, may reduce the risk of serious conditions such as bowel cancer and diverticular disease [13, 14]. In comparison with other dried fruits (see Table 1.2 in Chapter 1), dried dates are a good source of dietary fiber.

Based on the published reports, the total fiber content in dates varies from 1.7 to 11.4% (w/w) depending on the date variety and the method of analysis [15]. However, in a recent study [16], *Deglet Noor* and *Allig* cultivars were analyzed and their total dietary fiber content was reported as 14.4 and 18.4% of the dry matter, respectively.

23.2.3 Sugars

The predominant sugars in dates are simple reducing sugars such as fructose and glucose, and small amounts of non-reducing sugar sucrose, depending on the stage of the fruit development and the date palm varieties [2]. The average content of fructose, glucose, and sucrose in fresh dates are 19.4, 22.8, and 4.03 g/100 g, respectively, with an average total of 43.4 g/100 g. The content of sugars in dried dates increases to 29.4, 30.4, and 11.6 g/100 g for fructose, glucose, and sucrose, respectively, with a total of 64.1 g/100 g [4]. Both fructose and glucose are the major sugars in most date varieties and are present almost in equal amounts [3]. *Deglet Noor* variety of dates, however, presented an exception, where sucrose constituted 38% of the total carbohydrates in the fruit, possibly due to a lower invertase activity compared to other varieties [2, 17].

Fully ripened dates contain more than twice the amount of fructose compared to any other commonly consumed fruits (such as apples, grapes, oranges, cranberries, and blueberries). Occurrence of high amount of fructose in dates may provide several beneficial effects on human health and delay or prevent the development of chronic diseases [2, 18–20].

23.2.4 Minerals

Fresh and dried dates, in general, serve as a reasonable source of copper, iron, magnesium, manganese, phosphorus, and potassium (Table 23.1). Based on daily 40 g consumption of dried dates (on a per serving basis), recommended dietary allowances (RDA) or adequate intake (AI) of minerals for adult males and females are given in Table 1.3 (see Chapter 1). Consuming 40 g of dried dates (Table 1.3, see Chapter 1) supplies 1.6% of calcium, 9.3% of copper, 2.3–5.1% of iron, 4.2–5.5% of magnesium, 4.5–5.8% of manganese, 3.5% of phosphorus, and 5.6% of potassium for RDA or AI for adults [12, 21–23]. The values in Table 23.1 are within the range of the above given RDA or AI. The high potassium and low sodium content in dates are desirable for individuals suffering from hypertension [24]. Dates contain the highest amount of selenium among eight dried fruits (Tables 1.2 and 1.3, see Chapter 1) [12].

23.2.5 Vitamins

Dried fruits, in general, contain a small amount of vitamins (Table 23.1). With regard to RDA or AI of vitamins, 40 g of dried dates provide 5.4% of folate, 3.6–4.1% of niacin, 6.4% of pyridoxine, 3.6–4.2% of riboflavin, 2.6–2.9% of thiamine, 1.1–1.4% of vitamin A, and 1.7–2.1% of vitamin C [4, 22, 25, 26]. Among eight dried fruits, dates contain higher amount of folate and pantothenic acid than other dried fruits based on USDA database (Tables 1.2 and 1.4, see Chapter 1) [12].

23.2.6 Amino acids

Despite the fact that fresh and dried dates contain all indispensable amino acids (except tryptophan in fresh and valine in both fresh and dried dates), in general, they are not good sources of amino acids due to their low content of protein (Table 23.1). Glutamic acid, aspartic acid, lysine, leucine, and glycine are the predominant amino acids in fresh dates, whereas glutamic acid, aspartic acid, glycine, proline, and leucine are the predominant amino acids in dried dates [4, 27]. Among indispensable amino acids, only lysine was lost as a result of drying.

23.3 Phytochemicals in fresh and dried dates

Phytochemicals (non-nutritive bioactive compounds) have gained increased interest due to their antioxidant activity, cholesterol-lowering properties, and other potential health benefits such as cancer chemoprevention, prevention of diabetes, and cardiovascular diseases (CVD). Dates are renowned for the presence of many classes of phytochemicals such as phenolics (phenolic acids, tannins, and certain flavonoids such as flavonols, flavones, anthocyanins, and phytoestrogens) and carotenoids [2, 4, 28, 29]. These phytochemicals together with other minor components and certain antioxidant vitamins render antioxidant properties to dates. However, studies pertaining to the detailed identification, characterization, and quantification of phytochemicals in different date varieties at different stages of fruit ripening and processing are still insufficient. The effect of sun drying on antioxidant status and phytochemicals of dates are discussed below.

23.3.1 Antioxidant activity

The antioxidant activity of dates has been evaluated *in vitro* by different methods, such as oxygen radical absorbance capacity (ORAC), trolox equivalent antioxidant capacity (TEAC), ferric-reducing ability of plasma (FRAP), and 1,1-diphenyl-2-picrylhydrazyl (DPPH) radical scavenging assays [9, 30–38]. The antioxidant activities of dates using these methods have been discussed extensively by Vayalil [2]. Therefore, the effect of sun drying on antioxidant activity of dates is discussed here.

Most available data on antioxidants in dates are reported using the ORAC assay. Significant differences ($P < 0.05$) in ORAC values exist among fresh and sun-dried date varieties (Table 23.2). Comparison of ORAC values of some dried fruits, including dates, is given in Table 1.5 (see Chapter 1).

Table 23.2 Antioxidant activities and phenolics of fresh and dried dates

	Fresh dates	Dried dates	Reference
ORAC values (μmol of TE/100 g)	11,687–20,604	2387–12,543	[35, 36]
Total phenolics(mg of GAE/100 g)	134–280	172–661	[35, 36, 66]
Total anthocyanins (mg of C3GE/100 g)	0.24–1.52	nd	[36]
Total carotenoids (mg/100 g)	1.31–3.03	0.92–2.91	[36]
Total phytoestrogens (μg/100 g)	na	329.5	[58]

The values are given as a range from minimum to maximum (three varieties of fresh dates and eight varieties of dried dates).
ORAC, oxygen radical absorbance capacity; TE, trolox equivalents; GAE, gallic acid equivalents; C3GE, cyanidin-3-glucoside equivalents; na, not available; nd, not detected.

Al-Farsi *et al.* [36] found that sun drying caused a significant loss (ranging from 29.7 to 42.5%) of antioxidant activity in date varieties (Figure 23.1). This loss could be due to the decomposition of natural antioxidants in dates after drying. The reduction in antioxidant activity after drying has also been reported for other fruits. Larrauri *et al.* [39] reported decreases of 28 and 50% in the antioxidant activity of red grape pomace peel after drying at temperatures of 100 and 140°C, respectively. The antioxidant activity of blueberry decreased after drying (52%) and canning (65%) [40]. The difference between antioxidants in fresh and sun-dried dates observed by Al-Farsi *et al.* [36] was of similar magnitude. In contrast, there is only a few studies in which either an increase in antioxidant activity or no change

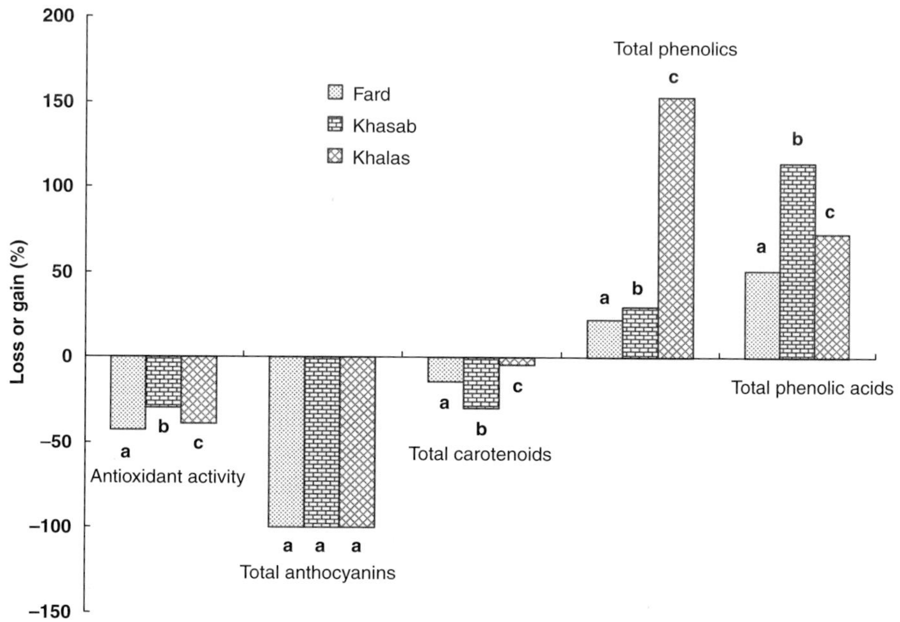

Figure 23.1 Effect of sun drying on different varieties of date antioxidants. Different letters in the same column (Fard, Khasab, and Khalas) indicate significant differences ($P < 0.05$). (Adapted with permission from Al-Farsi *et al.* [36].

after drying has been observed. Piga *et al.* [41] reported an increase in antioxidant activity in plums and prunes after drying at 85°C for 40 hours. They explained that this increase was due to the formation of Maillard reaction products during drying process. Thus, chemical and biochemical changes that affect the antioxidant activity may occur during drying.

23.3.2 Total phenolics

The average content of total phenolics ranges from 134 to 280 mg of gallic acid equivalents (GAE)/100 g and from 172 to 661 mg of GAE/100 g in fresh and dried date varieties, respectively (Table 23.2) [35, 36]. Vinson *et al.* [9] compared the amount and quality of phenolic antioxidants among dried and fresh fruits. Among seven fresh and dried fruits studied (apricots, cranberries, dates, figs, green grapes, and plums), it was observed that dates contained the highest concentration of total phenolics in both fresh and dried states when compared to other fruits.

Al-Farsi *et al.* [36] found that total phenolic contents of sun-dried dates were significantly ($P < 0.05$) higher than those of fresh dates in Fard, Khasab, and Khalas by 22.5, 29.9, and 153%, respectively (Figure 23.1). It was also observed that drying increased the amount of total phenols in dates which was statistically significant in Iranian dates [34] and not significant in California dates [9]. Shahidi and Naczk [42] reported that drying, in general, is regarded as being unfavorable due to the possibility of inducing oxidative decomposition either enzymatically by polyphenol oxidase and glycosidase or by thermal degradation of phenolic compounds. However, total phenolic content in dates showed increases after sun drying, possibly due to the release of phenolic compounds [43]. Thus, the linkages between *p*-coumaric acid and lignin, and between ferulic acid and arabinoxylans could be broken down at high temperatures.

23.3.3 Anthocyanins

Anthocyanins are a group of phenolic phytochemicals that are found in small amounts in fresh dates. Their presence has been reported only in red- and dark-colored fresh dates, from 0.24 to 1.52 mg of cyanidin-3-glucoside equivalents (C3GE)/100 g in three fresh date varieties (Fard, Khasab, and Khalas) (Table 23.2). These differences are related to the color of these varieties (both Khasab and Fard varieties are red, whereas Khalas variety is yellow due to the presence of carotenoids).

The absence of anthocyanins in sun-dried dates is probably due to their destruction upon drying (Figure 23.1) [36, 42, 44]. In a previous study by Alasalvar *et al.* [45], a significant loss of anthocyanins (92.5%) during heat processing was reported in the production of pekmez. Raynal [46] found that only 14.6% of the initial anthocyanins in plums remained after 1 hour of drying at 95°C, whereas 45.6% remained upon drying at 55°C. Studies have shown that anthocyanins are readily destroyed by heat during food processing [44, 47]. Apart from heat, many other factors such as light, temperature, agronomics, and storage conditions, among other variables, are also responsible for the degradation of anthocyanins during drying and processing of fruits [42, 47]. Wrolstad [48] stated that the degradation of anthocyanins during drying and storage is due to enzymatic and non-enzymatic browning reactions. Several enzymes have been found to be involved in the degradation of anthocyanins; these include glycosidase and polyphenol oxidase [42].

Table 23.3 Carotenoid content of fresh and dried dates (μg/100 g)

Carotenoid	Fresh dates	Dried dates
α-Carotene	3.0	nd
β-Carotene	34.4	59.8
Zeaxanthin	33.0	nd
β-Zeaxanthin	9.0	nd
Lutein	244	289
Neoxanthin	306	252
Total	913	973

Source: Adapted with permission from Al-Farsi and Lee [4].
Average values are composed of nine date varieties.
nd, not detected.

23.3.4 Carotenoids

The major class of phytochemicals found in the lipid fractions of dates is carotenoids [2]. Six carotenoids, namely α-carotene, β-carotene, lutein, neoxanthin, zeaxanthin, and β-zeaxanthin, are present, albeit at different levels, in both fresh and dried dates (Table 23.3). The average total carotenoids content of fresh and dried dates are 913 and 973 μg/100 g, respectively. Of these, lutein, neoxanthin, and β-carotene were the major carotenoids (Figure 23.2) among nine fresh (244, 306, and 34.4 μg/100 g, respectively) and dried dates (289, 252, and 59.8 μg/100 g, respectively) [4]. The total carotenoid content in date varieties differs between yellow and red colors. The high content of carotenoids is expected to be in the yellow variety. Al-Farsi *et al.* [36] measured total carotenoid content in three date varieties (Fard, Khasab, and Khalas) of fresh (ranged from 1.31 to 3.03 mg/100 g fresh weight) and dried (ranged from 0.92 to 2.91 mg/100 g fresh weight) products (Table 23.2). The highest total carotenoids were expected in Khalas, as this variety has a yellow color, whereas the other two varieties are red. A detailed analysis of total carotenoids in few varieties of Algerian dates has shown that lutein (89–94%) and β-carotene (3–10%) were the major carotenoids present in dates and another 2–8% consisted of unidentified minor carotenoids [49]. Ben-Amotz and Fishier [50] analyzed different carotenoids in freeze-dried dates. They estimated 220 μg/100 g of total carotenoids as dry weight and the major carotenoids identified were cis-violaxanthin (10 μg/100 g), zeaxanthine (110 μg/100 g), β-zeacarotene (30 μg/100 g), α-carotene (10 μg/100 g), and β-carotene (60 μg/100 g). However, details of the dates used for the analysis are unavailable. Among eight dried fruits (apples, apricots, dates, figs, peaches, pears, prunes, and raisins), dates are the fourth richest source of carotenoids after apricots, peaches, and prunes [12] (Figure 1.2, see Chapter 1).Therefore, dates can be considered a moderate source of carotenoids.

Similar to anthocyanins and antioxidant activity, commercial sun drying caused a significant ($P < 0.05$) loss (ranging from 4 to 29.8%) of carotenoids in date varieties except in Khalas, which was insignificant ($P > 0.05$) (Figure 23.1). Several studies reported the loss of carotenoids in carrot and orange juices during processing and storage [51, 52]. Mahanom *et al.* [53] reported that 64% of carotenoids were lost after drying (at 50°C for 9 hours) in the herbal preparation of eight medicinal plant leaves. In addition, lycopene losses during processing of tomato paste varied from 9 to 28% [54]. A number of mechanisms for the

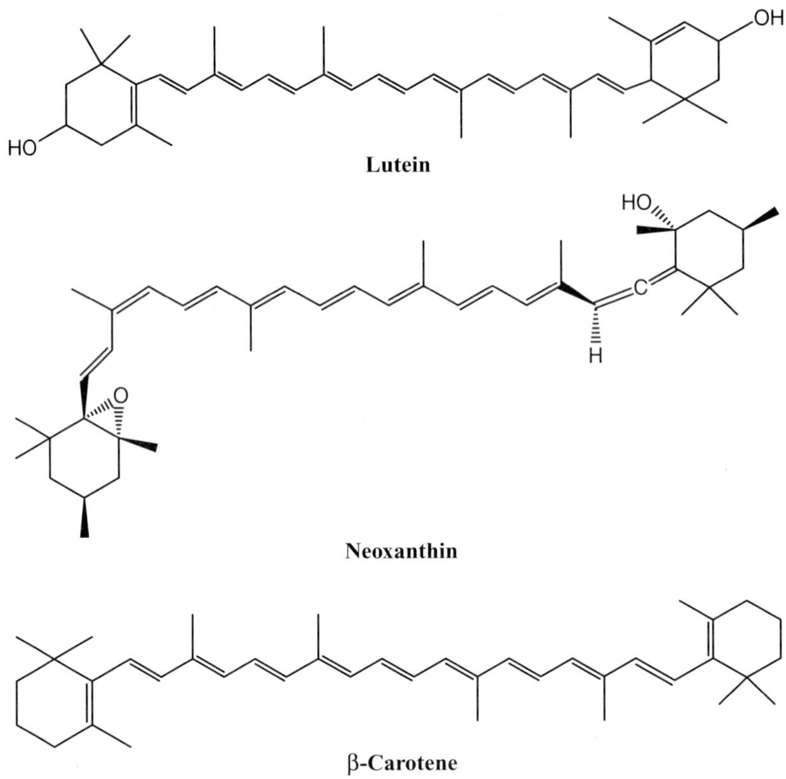

Figure 23.2 Structures of major carotenoids in dates.

reaction and decomposition of carotenoids in plant materials have been reported. These include enzymatic processes, autoxidation, and thermal degradation [55–57].

23.3.5 Phytoestrogens

There are three major classes of phytoestrogens such as isoflavones, coumestans, and lignans that occur in plants, fruits, and their byproducts. Total phytoestrogens in dried dates has been reported to be 329.5 μg/100 g (Table 23.2), lariciresinol being a major one (Figure 23.3) [58]. Among five dried fruits (apricots, currants, dates, prunes, and raisins), dates rank the second richest source of phytoestrogens after apricots (444.5 μg/100 g) (Table 1.7, see Chapter 1). Despite being complex, detailed quantitative analysis on the different classes of phytoestrogens present in different forms and varieties of dates at different stages of ripening remains unexplored [2].

23.3.6 Phenolic acids

Based on the available data [3, 28, 59], dates may be considered as a rich source of phenolic acids when compared with other fruits and berries [60]. Al-Farsi *et al.* [36] identified a total

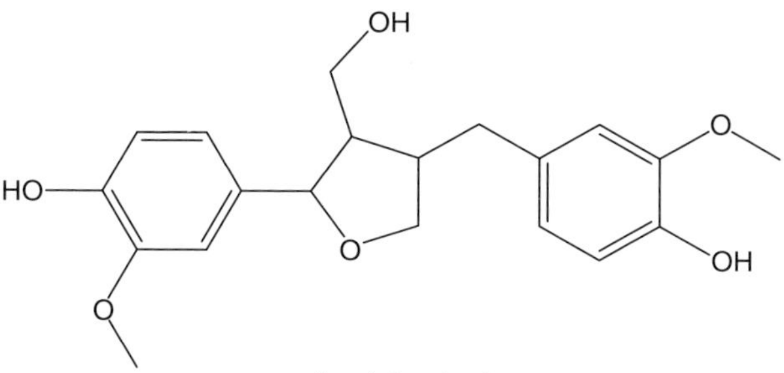

Lariciresinol

Figure 23.3 Structure of major phytoestrogen (lignan) in dates.

of nine phenolic acids (caffeic, ferulic, gallic, *o*-coumaric, *p*-coumaric, *p*-hydroxybenzoic, protocatechuic, syringic, and vanillic) in fresh and sun-dried dates (Table 23.4). Four free phenolic acids (protocatechuic, ferulic, syringic, vanillic) were present in dates, with total concentrations varying between 2.61 and 12.27 mg/100 g in fresh dates, and between 6.06 and 14.77 mg/100 g in their sun-dried counterparts (Table 23.4). The total bound phenolic contents were in the range of 6.84–30.25 mg/100 g in fresh dates and 14.18–49.67 mg/100 g in sun-dried dates. Ferulic and *p*-coumaric were major phenolic acids in the bound form in dried dates (Figure 23.4). Sun drying significantly ($P < 0.05$) increased the total concentration of bound phenolic acids in all date varieties (Figure 23.1). This was probably due to degradation of complex, high molecular weight polymers such as tannins [4, 36].

Detailed phenolics profile of dates was first published in the late 1980s by Regnault-Roger *et al.* [28]. They identified eight phenolic acids in dried Tunisian dates (namely caffeic, ferulic, gallic, *p*-coumaric, *p*-hydroxybenzoic, protocatechuic, syringic, and vanillic). In

Table 23.4 Composition of free and bound phenolic acids (mg/100 g) in fresh and sun-dried dates

	Free		Bound	
Phenolic acids	**Fresh**	**Dried**	**Fresh**	**Dried**
Benzoic acids				
Gallic	nd	nd	nd–0.48	nd–3.09
p-Hydroxybenzoic	nd	nd	nd–0.49	nd
Protocatechuic	nd	nd–2.04	nd–4.49	nd–8.34
Syringic	nd–5.49	nd–6.02	nd–1.70	nd–3.22
Vanillic	1.45–2.13	2.18–4.14	nd	nd–2.26
Cinnamic acids				
Caffeic	nd	nd	nd–10.10	nd–7.57
o-Coumaric	nd	nd	nd–1.46	nd–6.71
p-Coumaric	nd	nd	0.49–6.25	1.41–14.19
Ferulic	1.16–4.71	1.84–5.08	2.06–12.27	6.08–13.28
Total	2.61–12.27	6.06–14.77	6.84–30.25	14.18–49.67

Source: Adapted with permission from Al-Farsi *et al.* [36].
The values are given as a range from minimum to maximum (three varieties of dried dates).
nd, not detected.

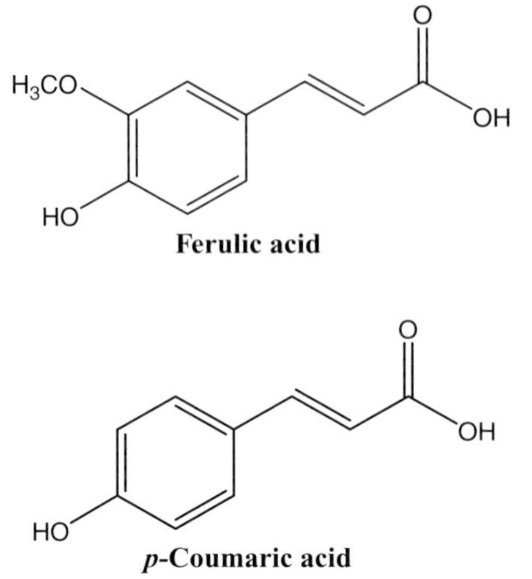

Figure 23.4 Structures of major phenolic acids in dates.

addition to these phenolic acids, recent studies have demonstrated the presence of sinapic acid, 5-*o*-caffeoylshikimic acid and its three isomers, xantoxylin, hydrocaffeic acid, and coumaroylquinic acid in seven varieties of ripe Algerian dates [59].

It has been reported that caffeic, sinapic, ferulic, and *p*-coumaric acids are more antioxidative than protocatechuic, syringic, vanillic, and protocatechuic acids [61]. Since dates are a good source of the more active phenolic acids, they may be considered as good sources of natural antioxidant. Besides their naturally occurring antioxidant properties, phenolic acids can also influence product flavor and color.

23.3.7 Other phenolics

Some flavonoids including flavonols and proanthocyanidins, especially procyanidins and flavanoid glycosides, have been reported [29, 37, 59, 62–64]. Gu *et al.* [37] demonstrated that *Deglet Noor* contained proanthocyanidins of type B consisting exclusively of (epi) catechin while in *Medjool* proanthocyanidins were not detected. Mansouri *et al.* [59] indicated the presence of flavonoid glycosides in many Algerian dates. Hong *et al.* [63] identified that procyanidins in dates are homogeneous B type oligomers and these oligomers were identified through decamers. Higher molecular weight polymers were also well documented and composed mainly of undecamers through heptadecamers. Moreover, in their study, 19 flavonoid glycosides of luteolin, quercetin, and apigenin were identified together with their isomeric forms. Luteolin, quercetin, and apigenin glycosides also existed as methylated and sulfated forms in dates [2].

23.4 Health benefits of dates

A recent review article on *date fruits: an emerging medicinal foods* has been published by Vayalil [2]. This review provides an account of health benefits of dates, hence only a summary of health effects of dates is given here. Dates are considered to have several health benefits due to their compositional and nutritional characteristics as well as their non-nutritive phytochemicals. In certain countries (such as the Middle East and North Africa), dates and their byproducts are used as folk medicine (such as diabetes and hypertension) for centuries [65]. Although both fresh and dried dates are sugar-packed fruits, consumption of certain varieties of dates do not induce any metabolic and inflammatory markers that are associated with metabolic syndrome and related diseases and the effect is comparable to that of other fruits. The main health benefits of date consumption are detailed by Vayalil [2], as shown below:

- a potential medical nutritional therapy for the prevention and control of diabetes mellitus,
- a potential medical nutritional therapy for cardio-and cerebrovascular diseases,
- a potential cancer preventive diet,
- a potential medicinal food against bacterial and fungal infections,
- other potential health benefits (laxative, anti-inflammatory activity in adjuvant arthritis model, protective effects against chemical-induced toxicity, neuro-protective effects, anti-ulcer effect, and dates for deficiency diseases, among others).

23.5 Food application of dates, syrups, and their byproducts

Limited data is available regarding the compositional and functional characteristics of date syrups and byproducts (such as press cakes and seeds) [66]. Table 23.5 shows the nutritional composition, total phenolics, and antioxidant activity of date flesh, syrup, press cake, and seed. Date seeds contain a relatively high amount of protein (2.29–5.40 g/100 g) as compared

Table 23.5 Nutritional composition, total phenolics, and antioxidant activity of date flesh, syrup, and their byproducts

			Byproducts	
	Date flesh	**Syrup**	**Press cake**	**Seed**
Proximate composition				
Protein	1.10–1.79	0.95–1.09	3.62–5.23	2.29–5.40
Fat	2.04–3.25	0.62–2.84	1.40–2.20	5.02–5.90
Moisture	9.73–17.52	20.56–34.33	8.30–10.59	3.14–5.19
Ash	1.41–1.99	1.23–1.76	1.68–2.46	0.89–1.16
Carbohydrates	77.34–84.45	62.73–74.24	81.86–83.33	83.14–86.89
Dietary fiber	5.94–8.72	0.01–0.18	25.39–33.81	77.75–80.15
Total phenolics	172–246	96–162	165–435	3102–4430
Total antioxidant activity	14,600–16,200	8400–17,400	13,400–35,700	58,000–92,900

Source: Adapted with permission from Al-Farsi *et al.* [66].
The values are given as a range from minimum to maximum (three varieties of dried dates).
Proximate composition is expressed as g/100 g fruits.
Total phenolics are expressed as milligrams of gallic acid equivalents (GAE) per 100 g.
Total antioxidant activities are expressed as micromoles of trolox equivalents (TE) per 100 g.

to date flesh (1.10–1.79 g/100 g), syrups (0.95–1.09 g/100 g), and press cakes (3.62–5.23 g/100 g). Date seeds are also the richest sources of dietary fiber (77.75–80.15 g/100 g), total phenolics (3102–4430 mg of GAE/100 g), and antioxidant activity (58,000–92,900 μmol of TE/100 g) among date flesh and byproducts (Table 23.5).

At present, date seeds are used mainly for animal feed, whereas most is regarded as waste. Utilization of such waste is very important to date cultivation and to increase the income to this sector. Date byproducts (particularly seeds) can be used in a variety of food and specialty products, including functional foods and ingredients in nutraceuticals, pharmaceuticals, and medicinal products. In addition, date seeds could potentially be considered as an inexpensive source of dietary fiber and natural antioxidants. Date seeds, roasted and ground into powder make a beverage like coffee, called "date coffee." Date syrup is a popular product world-wide and can be used for several purposes, including a sweetener in tea and hot-chocolate, topping on ice creams, breadmaking, spreading in breads, and mixing with cold and hot milk, among others.

23.6 Conclusions

Dates can be stored for months after harvest without need for using preservatives. They are an energy-dense food containing high concentrations of carbohydrates (up to 80% in dried dates), several essential nutrients (minerals, amino acids, and vitamins), and health-promoting phytochemicals (such as phenolic acids, tannins, flavonols, flavones, anthocyanins, phytoestrogens, and carotenoids) in maintaining human health. They serve as a good source of natural antioxidants and could potentially be considered as a functional food or functional food ingredient.

Date consumption would be a good alternative for the malnourished infants and adults for their basic nutritional requirements and to fight against nutrient deficiency-related diseases and infections which is an endemic in the poor nations of the world. Therefore, dates may be considered as a gift of nature to the people not only for those living in the hot arid regions, but also other parts of the world for general nutrition and a potential emerging functional food [2].

A significant ($P < 0.05$) amount of antioxidants, anthocyanins, and carotenoids is lost during sun drying of dates. This is an important issue that needs to be addressed in order to maximize their retention during the drying process, perhaps by devising innovative and alternative techniques.

References

1. FAO (2009) *FAOSTAT-Date Production 2009*. Published online at: http://faostat.fao.org/site/339/default.aspx, last accessed January 9, 2012.
2. Vayalil, P.K. (2012) Date fruits (*Phoenix dactylifera* Linn): an emerging medicinal food. *Critical Reviews in Food Science and Nutrition*, **52**, 249–271.
3. Al-Farsi, M., Alasalvar, C., Morris, A., Barron, M. & Shahidi, F. (2005) Compositional and sensory characteristics of three native sun-dried date (*Phoenix dactylifera* L.) varieties grown in Oman. *Journal of Agricultural and Food Chemistry*, **53**, 7586–7591.
4. Al-Farsi, M.A. & Lee, C.H. (2008) Nutritional and functional properties of dates: a review. *Critical Reviews in Food Science and Nutrition*, **48**, 877–887.

5. Barreveld, W.H. (1993) *Date Palm Products—Agricultural Services Bulletin No.101*. FAO, Rome, Italy.
6. MAF (2005) *Agricultural and Fisheries Research Report*. Ministry of Agriculture and Fisheries, Muscat, Oman.
7. Ismail, B., Henry, J., Haffar, I. & Baalbaki, R. (2006) Date consumption and dietary significance in the United Arab Emirates. *Journal of the Science of Food and Agriculture*, **86**, 1196–1201.
8. Manickavasagan, A., Mohamed Essa, M. & Sukumar, E. (eds) (2012) *Dates: The Genus* Phoenix*: Production, Processing, Food, and Medicinal Values*. CRC Press, Taylor & Francis Group, Boca Raton, FL.
9. Vinson, J.A., Zubik, L., Bose, P., Samman, N. & Proch, J. (2005) Dried fruits: excellent *in vitro* and *in vivo* antioxidants. *Journal of the American College of Nutrition*, **24**, 44–50.
10. Fennema, O.R. (1996) *Food Chemistry*, 3rd ed. Dekker, New York, NY.
11. USDA (2005) *Carbohydrates. Dietary Guidelines for Americans*. US Department of Agriculture and US Department of Health and Human Services. Government Printing Office, Washington, DC.
12. USDA (2010) *National Nutrient Database for Standard Reference, Release 23*. Published online at: http://www.nal.usda.gov/fnic/foodcomp/search/, last accessed March 7, 2011.
13. Marlett, J.A., McBurney, M.I. & Slavin, J.L. (2002) Position of the American Dietetic Association: health implications of dietary fiber. *Journal of the American Dietetic Association*, **102**, 993–1000.
14. Cummings, J.H., Bingham, S., Heaton, K.W. & Eastwood, M.A. (1992) Fecal weight, colon cancer risk, and dietary intake of nonstarch polysaccharides (dietary fibre). *Gastroenterology*, **103**, 1783–1789.
15. Al-Shahib, W. & Marshall, R.J. (2003) The fruit of the date palm: its possible use as the best food for the future? *International Journal of Food Sciences and Nutrition*, **54**, 247–259.
16. Elleuch, M., Besbes, S., Roiseux, O., Blecker, C., Deroanne, C., Drira, N.E. & Attia, H. (2008) Date flesh: chemical composition and characteristics of the dietary fibre. *Food Chemistry*, **111**, 676–682.
17. USDA (2011) *National Nutrient Database for Standard Reference, Release 24*. Published online at: http://www.nal.usda.gov/fnic/foodcomp/search/, last accessed January 9, 2012.
18. Watford, M. (2002) Small amounts of dietary fructose dramatically increase hepatic glucose uptake through a novel mechanism of glucokinase activation. *Nutrition Reviews*, **60**, 253–257.
19. Shiota, M., Moore, M.C., Galassetti, P., Monohan, M., Neal, D.W., Shulman, G.I. & Cherrington, A.D. (2002) Inclusion of low amounts of fructose with an intraduodenal glucose load markedly reduces postprandial hyperglycemia and hyperinsulinemia in the conscious dog. *Diabetes*, **51**, 469–478.
20. Ali, A., Al-Kindi, Y.S.M. & Al-Said, F. (2009) Chemical composition and glycemic index of three varieties of Omani dates. *International Journal of Food Sciences and Nutrition*, **60**, 51–62.
21. DRIs (1997) *Dietary Reference Intakes for Calcium, Phosphorus, Magnesium, Vitamin D, and Fluoride*. The National Academies Press, Washington, DC.
22. DRIs (2001) *Dietary Reference Intakes for Vitamin A, Vitamin K, Arsenic, Boron, Chromium, Copper, Iodine, Iron, Manganese, Molybdenum, Nickel, Silicon, Vanadium, and Zinc*. The National Academies Press, Washington, DC.
23. DRIs (2004) *Dietary Reference Intakes for Water, Potassium, Sodium, Chloride, and Sulfate*. The National Academies Press, Washington, DC.
24. Appel, L.J., Moore, T.J., Obarzanek, E., Vollmer, W.M., Svetkey, L.P., Sacks, F.M., Bray, G.A., Vogt, T.M., Cutler, J.A., Windhauser, M.M., Lin, P.H., Karanja, N., Simons-Morton, D., McCullough, M., Swain, J., Steele, P., Evans, M.A., Miller, E.R. & Harsha, D.W. (1997) A clinical trial of the effects of dietary patterns on blood pressure. *New England Journal of Medicine*, **336**, 1117–1124.
25. DRIs (1998) *Dietary Reference Intakes for Thiamin, Riboflavin, Niacin, Vitamin B_6, Folate, Vitamin B_{12}, Pantothenic Acid, Biotin, and Choline*. The National Academies Press, Washington, DC.
26. DRIs (2000) *Dietary Reference Intakes for Vitamin C, Vitamin E, Selenium, and Carotenoids*. The National Academies Press, Washington, DC.
27. Ishurd, O., Zahid, M., Xiao, P. & Pan, Y. (2004) Protein and amino acid contents of Libyan dates at three stages of development. *Journal of the Science of Food and Agriculture*, **84**, 481–484.
28. Regnault-Roger, C., Hadidane, R., Biard, J.F. & Boukef, K. (1987) High performance liquid and thin-layer chromatographic determination of phenolic acids in palm (*Phoenix dactilifera*) products. *Food Chemistry*, **25**, 61–67.
29. USDA (2011) *USDA Database for the Flavonoid Content of Selected Foods, Release 3*. Published online at: http://www.ars.usda.gov/nutrientdata, last accessed November 2, 2011.

30. Chaira, N., Ferchichi, A., Mrabet, A. & Sghairoun, M. (2007) Chemical composition of the flesh and the pit of date palm fruit and radical scavenging activity of their extracts. *Pakistan Journal of Biological Sciences*, **10**, 2202–2207.
31. Ranilla, L.G., Kwon, Y.I., Genovese, M.I., Lajolo, F.M. & Shetty, K. (2008) Antidiabetes and antihypertension potential of commonly consumed carbohydrate sweeteners using *in vitro* models. *Journal of Medicinal Food*, **11**, 337–348.
32. Rock, W., Rosenblat, M., Borochov-Neori, H., Volkova, N., Judeinstein, S., Elias, M. & Aviram, M. (2009) Effects of date (*Phoenix dactylifera* L., Medjool or Hallawi Variety) consumption by healthy subjects on serum glucose and lipid levels and on serum oxidative status: a pilot study. *Journal of Agricultural and Food Chemistry*, **57**, 8010–8017.
33. Cook, J.A., VanderJagt, D.J., Dasgupta, A., Mounkaila, G., Glew, R.S., Blackwell, W. & Glew, R.H. (1998) Use of the Trolox assay to estimate the antioxidant content of seventeen edible wild plants of Niger. *Life Sciences*, **63**, 105–110.
34. Biglari, F., AlKarkhi, A.F.M. & Easa, A.M. (2008) Antioxidant activity and phenolic content of various date palm (*Phoenix dactylifera*) fruits from Iran. *Food Chemistry*, **107**, 1636–1641.
35. Wu, X., Beecher, G.R., Holden, J.M., Haytowitz, D.B., Gebhardt, S.E. & Prior, R.L. (2004) Lipophilic and hydrophilic antioxidant capacities of common foods in the United States. *Journal of Agricultural and Food Chemistry*, **52**, 4026–4037.
36. Al-Farsi, M., Alasalvar, C., Morris, A., Barron, M. & Shahidi, F. (2005) Comparison of antioxidant activity, anthocyanins, carotenoids, and phenolics of three native fresh and sun-dried date (*Phoenix dactylifera* L.) varieties grown in Oman. *Journal of Agricultural and Food Chemistry*, **53**, 7592–7599.
37. Gu, L.W., Kelm, M.A., Hammerstone, J.F., Beecher, G., Holden, J., Haytowitz, D. & Prior, R.L. (2003) Screening of foods containing proanthocyanidines and their structural characterization using LC-MS/MS and thiolytic degradation. *Journal of Agricultural and Food Chemistry*, **51**, 7513–7521.
38. Abdul Ameer, A.A. (2008) Antioxidant activity of Bahraini date palm (*Phoenix dactylifera* L.) fruit of various cultivars. *International Journal of Food Science & Technology*, **43**, 1033–1040.
39. Larrauri, J.A., Rupérez, P. & Saura-Calixto, F. (1997) Effect of drying temperature on the stability of polyphenols and antioxidant activity of red grape pomace peels. *Journal of Agricultural and Food Chemistry*, **45**, 1390–1393.
40. Kalt, W., McDonald, J.E. & Donner, H. (2000) Anthocyanins, phenolics, and antioxidant capacity of processed lowbush blueberry products. *Journal of Food Science*, **65**, 390–393.
41. Piga, A., Del Caro, A. & Corda, G. (2003) From plums to prunes: influence of drying parameters on polyphenols and antioxidant activity. *Journal of Agricultural and Food Chemistry*, **51**, 3675–3681.
42. Shahidi, F. & Naczk, M. (2004) *Phenolics in Food and Nutraceuticals*. CRC Press, Boca Raton, FL.
43. Maillard, M.-N. & Berset, C. (1995) Evolution of antioxidant activity during kilning: role of insoluble bound phenolic acids of barley and malt. *Journal of Agricultural and Food Chemistry*, **43**, 1789–1793.
44. Markakis, P. (1982) *Anthocyanins as Food Colors*. Academic Press, New York, NY.
45. Alasalvar, C., Al-Farsi, M. & Shahidi, F. (2005) Compositional characteristics and antioxidant components of cherry laurel varieties and pekmez. *Journal of Food Science*, **70**, S47–S52 [errata: **2005**, 70 (5), pp. ix].
46. Raynal, J. (1987) *Modification Stucturales et Biochimiques de la Prune D'ente au Début du Séchage. Effets de Diverses Températures sur la Structure Cellulaire et Ses Incidences sur les Composés Pectiques et Phénoliques*. DPhil Thesis (in French), University of Montpellier, Montpellier, France.
47. Mazza, G. & Miniati, E. (1993) *Anthocyanins in Fruits, Vegetables, and Grains*. CRC Press, Boca Raton, FL.
48. Wrolstad, R.E. (2004) Interaction of natural colors with other ingredients: anthocyanin pigments-bioactivity and coloring properties. *Journal of Food Science*, **69**, C419–C425.
49. Boudries, H., Kefalas, P. & Hornero-Mendez, D. (2007) Carotenoid composition of Algerian date varieties (*Phoenix dactylifera*) at different edible maturation stages. *Food Chemistry*, **101**, 1372–1377.
50. Ben-Amotz, A. & Fishier, R. (1998) Analysis of carotenoids with emphasis on 9-cis [beta]-carotene in vegetables and fruits commonly consumed in Israel. *Food Chemistry*, **62**, 515.
51. Chen, B.H., Peng, H.Y. & Chen, H.E. (1995) Changes of carotenoids, color, and vitamin A contents during processing of carrot juice. *Journal of Agricultural and Food Chemistry*, **43**, 1912–1918.
52. Sánchez-Moreno, C., Plaza, L., De Ancos, B. & Cano, M.P. (2003) Vitamin C, provitamin A carotenoids, and other carotenoids in high-pressurized orange juice during refrigerated storage. *Journal of Agricultural and Food Chemistry*, **51**, 647–653.

53. Mahanom, H., Azizah, A.H. & Dzulkifly, M.H. (1999) Effect of different drying methods on concentrations of several phytochemicals in herbal preparation of 8 medicinal plants leaves. *Malaysian Journal of Nutrition*, **5**, 47–54.
54. Takeoka, G.R., Dao, L., Flessa, S., Gillespie, D.M., Jewell, W.T., Huebner, B., Bertow, D. & Ebeler, S.E. (2001) Processing effects on lycopene content and antioxidant activity of tomatoes. *Journal of Agricultural and Food Chemistry*, **49**, 3713–3717.
55. Talcott, S.T. & Howard, L.R. (1999) Phenolic autoxidation is responsible for color degradation in processed carrot puree. *Journal of Agricultural and Food Chemistry*, **47**, 2109–2115.
56. Kanasawud, P. & Crouzet, J.C. (1990) Mechanism of formation of volatile compounds by thermal degradation of carotenoids in aqueous medium. 1. β-carotene degradation. *Journal of Agricultural and Food Chemistry*, **83**, 237–243.
57. Bonnie, T.P. & Choo, Y.M. (1999) Oxidation and thermal degradation of carotenoids. *Journal of Oil Palm Research*, **2**, 62–78.
58. Thompson, L.U., Boucher, B.A., Liu, Z., Cotterchio, M. & Kreiger, N. (2006) Phytoestrogen content of foods consumed in Canada, including isoflavones, lignans, and coumestan. *Nutrition and Cancer*, **54**, 184–201.
59. Mansouri, A., Embarek, G., Kokkalou, E. & Kefalas, P. (2005) Phenolic profile and antioxidant activity of the Algerian ripe date palm fruit (*Phoenix dactylifera*). *Food Chemistry*, **89**, 411–420.
60. Mattila, P., Hellstrom, J. & Torronen, R. (2006) Phenolic acids in berries, fruits, and beverages. *Journal of Agricultural and Food Chemistry*, **54**, 7193–7199.
61. Cuvelier, M.E., Richard, H. & Berset, C. (1992) Comparison of the antioxidant activity of some acid-phenols: structure-activity relationship. *Bioscience, Biotechnology, and Biochemistry*, **56**, 324–325.
62. Tomás Lorente, F. & Ferreres, F. (1988) Sulfatos de flavonoides en frutos de phoenix dactilifera. *Revista de Agroqímica y Tecnología de Alimentos*, **28**, 581–585.
63. Hong, Y.J., Tomas-Barberan, F.A., Kader, A.A. & Mitchell, A.E. (2006) The flavonoid glycosides and procyanidine composition of Deglet Noor dates (*Phoenix dactylifera*). *Journal of Agricultural and Food Chemistry*, **54**, 2405–2411.
64. Harnly, J.M., Doherty, R.F, Beecher, G.R., Holden, J.M., Haytowitz, D.B., Bhagwat, S. & Gebhardt, S. (2006) Flavonoid content of U.S. fruits, vegetables, and nuts. *Journal of Agricultural and Food Chemistry*, **54**, 9966–9977.
65. Tahraoui, A., El-Hilaly, J., Israili, Z.H. & Lyoussi, B. (2007) Ethnopharmacological survey of plants used in the traditional treatment of hypertension and diabetes in south-eastern Morocco (Errachidia province). *Journal of Ethnopharmacology*, **110**, 105–117.
66. Al-Farsi, M., Alasalvar, C., Al-Abid, M., Al-Shoaily, K., Al-Amry, M. & Al-Rawahy, F. (2007) Compositional and functional characteristics of dates, syrups, and their by-products. *Food Chemistry*, **104**, 943–947.

24 Neutraceutical properties of dried tropical fruits: guavas and papayas

K. Nagendra Prasad, Azrina Azlan, and
Barakatun Nisak Mohd Yusof

24.1 Introduction

Fruits and vegetables are of increasing economic importance due to their contribution to health promotion as "functional" foods. Thus, they are capable of not only providing basic nutritional requirements, but also an array of phytochemicals that may prevent or delay onset of chronic diseases [1, 2]. Inadequate intake of fruits and vegetables in the diet has now been recognized as one of the top 10 leading global disease burden risk factors [3]. Epidemiological studies have shown that a high intake of fruits can lower the occurrence of cardiovascular diseases (CVD), stroke, and cancer [4].

Fruits are broadly classified as temperate, Mediterranean-subtropical, and tropical. Tropical fruits originate from the warm climate of the tropics where the temperatures remain relatively constant throughout the year. It is predicted that world production of tropical fruits is likely to expand over the next 10 years. More than 3000 types of tropical fruits with production of approximately 140 million tonnes are produced annually worldwide [5]. Amongst them, avocados, bananas, mangos, papayas, and pineapples have the largest global production. Worldwide tropical fruit utilization per capita has improved by 33% in the last two decades [6].

Many of the tropical fruits have high moisture content (>80%) with short storage life, leading to their easy spoilage and infection [7]. It is estimated that about 40% of fruit production is wasted due to poor post-harvest management [8, 9]. To overcome this wastage, processing of fruits to increase their shelf life is an alternative. Worldwide markets of processed fruits have constantly increased due to the improved quality, convenience of ready-to-serve products, and availability of seasonal commodities, which has been expanded by processing [10].

Drying is the most common, cost effective, and viable processing method. It provides an alternative to expensive preservation methods such as cold storage. It is the oldest method to preserve fruits by selectively removing water, thus preventing microbial spoilage and quality deterioration due to undesirable biochemical reactions [11]. Dried fruits are being used traditionally and appreciated for their convenience and usefulness [12]. Dried fruits

Dried Fruits: Phytochemicals and Health Effects, First Edition. Edited by Cesarettin Alasalvar and Fereidoon Shahidi.

also have wider application as they are included in many breakfast cereal-based packages to offer more nutritious products besides providing an appealing color. In addition, dried fruits have been regarded as alternative fat-free snacks for health conscious consumers. Among popular dried fruits, raisins, currants, plums, figs, and prunes may be included. Dried fruits are more concentrated with nutrients and are easily available throughout the year, retain the characteristics as natural products, easy to handle and stored with reduced transportation cost [10, 11]. Moreover, drying reduces fruit wastes and post-harvest losses, and allows the whole fruit to be used in production by the food industry. The dried fruit industry maintains the potential to contribute to the well-being of the country through substituting a natural product in place of the highly refined foods [13].

During drying, increase in solid concentration due to the removal of water and partial inversion of sucrose occurs in fruits having a high acid content. This leads to more hygroscopic dried fruits with altered texture, taste, and appearance. Reaction of reducing sugars with nitrogen compounds (Millard reaction) imparts a brown color and caramel-like flavor to the dehydrated products. In addition, due to loss in some volatile compounds, a difference in flavor between fresh and dried fruits is noticed [10]. However, the existing literature has focused more on fresh products, their nutritional qualities and health benefits than their dried counterparts. Therefore, information on some dried fruits is scarce. Hence, in the present chapter we have focused to review the nutritional qualities, bioactive compounds, and health benefits of dried tropical fruits with special emphasis on dried guavas and papayas. This information will be useful for researchers, food industries, and nutritionists. A comparison with fresh fruits is also provided, when possible.

24.2 Guavas

Guavas (*Psidium guajava* L.) belong to the family *Myrtaceae*. The tree is indigenous to Central America, South America, Mexico, India, and Caribbean. Guava pulps are sweet and have a white or deep pink color with many seeds of variable number and their texture depends on the species [14, 15]. Guavas are commercially used in the production of juice, jam, puree, marmalade, ice cream, cookies, and several bakery products, among others.

24.2.1 Nutritional composition of dried guavas

Nutritional compositions of dried guavas are given in Table 24.1. Osorio *et al.* [16] developed a guava powder by hot-air drying and freeze drying. Both powders were rich in pectin and had a good retention of guava aroma. Guava powders obtained from hot-air drying had moisture and crude fiber content of 5.3 and 47%, respectively. In addition, guava powders obtained by freeze drying had lower moisture (2.75%), but a higher crude fiber (54.5%) content. Munhoz *et al.* [17] extracted pectin from dried pulp and peel of guavas. They developed an optimized method for extraction of pectin using 5% citric acid, 60 minute extraction time, and a temperature of 97°C. Under these conditions, the yield of pectin was more than 11%. Pectins from dried guava slices were analyzed by Tsai *et al.* [18]. Dried guavas were rich in total pectin (6108 ppm), cold water soluble pectin (3384 ppm), hot water soluble pectin (841 ppm), and hot acid soluble pectin (111 ppm).

Nutritional composition of sun dried by-product of guavas, consisting of seed, pulp, and fruits, was analyzed by El-Deek *et al.* [19]. The products had a moisture content of 6.94%,

Table 24.1 Nutritional composition of dried guavas

Nutritional composition	Product	g/100 g	Reference
Moisture	Guava powder	2.7–5.3	[16]
	Dried fruit	6.9	[19]
	Dried guavas	13.8	[21]
	Freeze-dried pulp	7.0	[10]
Protein	Dried fruit	9.7	[19]
Carbohydrate	Dried fruit	33.0	[19]
Fat	Dried fruit	4.5	[19]
Crude fiber	Guava powder	47.0–54.5	[16]
	Dried fruit	40.0	[19]
Ash	Dried fruit	5.6	[19]
Pectin	Dried pulp and peel	11.0	[17]
	Guava slices	61.0	[18]
Vitamin C	Freeze-dried pulp	3.0	[10]

while their crude protein, carbohydrate, and lipid were at 9.78, 33, and 4.52%, respectively. Meanwhile, their crude fiber, ash, and calcium were 40, 5.62, and 0.37%, respectively. Chemical characteristics such as pH, soluble solids, total acidity, protein, lipids, fiber, ash, moisture, and total and reducing sugars of cookies enriched with guava fibers were evaluated by Matias *et al.* [20]. Marques *et al.* [10] estimated some nutritional content (moisture, vitamin C, phosphorus, and calcium) of fresh and freeze-dried guavas pulps. The nutritional quality of freeze-dried guava pulps was reported as follows: moisture (7%), vitamin C (30 mg/100 g), phosphorus (9.69 mg/100 g), and calcium (12.49 mg/100g). Dried guavas had a moisture content of 13.8%, total sugar (16%), and titratable total acidity (5.7%), as reported by Sanjinez-Argandona *et al.* [21].

Queiroz *et al.* [22] reported the loss of vitamin C and other minerals (sodium, potassium, calcium, magnesium, zinc, and manganese) during the dehydration process of dried guavas. A reduction in mineral content (20–64%) and vitamin C (32–68%) was noticed. Retention of vitamin C in freeze-dried guavas was more (63%) than air-dried (25%) and vacuum-dried (58%) products [23]. Degradation of ascorbic acid and carotenoids in osmotic pre-treated and convective-dried guavas was investigated by Sanjinez-Argandona *et al.* [21]. The results showed that the degradation values for carotenoid varied from 66 to 70% and ascorbic acid from 20 to 35% of the original values in osmotic pre-treated and dried guavas, respectively. Effects of cyclic air temperature on moisture and ascorbic acid content of dried guava pieces were also evaluated by Chua *et al.* [24].

The color change of dried guavas during blanching increased with increasing temperature and immersion time. For a 90 second immersion, the total color change increased by 400% between 80 and 95°C. Red color of guavas was also deepened after air drying when osmotic dehydration was applied as pre-treatment. The L and a values were increased during vacuum drying while no significant changes were noticed for the b value compared to the fresh samples. This could probably be due to the browning reaction and pigment destruction [14]. Chemical [total soluble solids (TSS), acidity, color, pH, aroma, appearance, flavor, and texture] and sensory properties of fresh and osmotically dehydrated guavas were analyzed by Queiroz *et al.* [25]. Dried samples showed alteration in color values compared to the osmotic dehydration. Sensory qualities of osmotically dehydrated samples were also most preferred to those of air-dried guavas. Guava tea sample was found to have a higher color retention (*L*, *a*, *b*) as investigated by Lee *et al.* [26].

With regard to texture, penetration tests show that osmotically dehydrated guavas require 109% more force than air-dried guavas. Blanching prior to osmotic dehydration could decrease the penetration force by 15%, while ascorbic acid and calcium lactate increased the penetration force by 14.6 and 87.9%, respectively [14]. Dried guava slices treated with phenolic compounds such as gallic acid, ferulic acid, and caffeic acid retained a better hardness (1.25–1.60 folds) when compared to the control [18]. Sensory analysis of osmotically dehydrated guavas was more acceptable by consumers than air-dried guavas. Rating for osmotically dehydrated guavas was more than double that of the air-dried guavas for appearance, flavor, and texture [14].

24.2.2 Phytochemicals in dried guavas

Research on phytochemicals from dried guavas is very limited. Tea prepared from guavas was shown to have the highest phenolic (144.7 mg/g) and flavonoid contents (3.02 mg/g) [26] which was higher than that of the dried fruit. Gallic acid was the predominant phenolic acid present in guava tea (Table 24.2).

24.2.3 Antioxidant activity of dried guavas

Guava tea prepared from dried guavas exhibited a weaker antioxidant activity when compared to the extract of guava leaves [27, 28]. However, dried guava tea still exhibited a stronger antioxidant activity than the other commercial products, as evaluated by the linoleic acid method, especially at higher concentrations (Table 24.3). Although the exact mechanism of action as an antioxidant is not clear, it is believed that dried guava fruits may possess strong free-radical scavenging ability [28]. Table 24.4 summarizes the scavenging effects of dried guava fruits as compared to guava leaves and commercial tea. The inhibitory effect of dried guava on peroxyl radicals was increased with increase in the concentration, reaching 91% at higher concentrations.

Table 24.2 Total phenolic and phenolic acid content of extracts from guava tea and guava dried fruit

	Total phenolic (mg/g)		Phenolic acid (μg/g)	
Samples	**Catechin equivalent**	**Gallic acid equivalent**	**Gallic acid**	**Ferulic acid**
Guava tea	177 ± 9.2	279 ± 14.0	1278 ± 92.7	234 ± 27.5
Guava dried fruit	69.6 ± 2.8	115 ± 4.2	266 ± 15.4	–
Tea polyphenon 60	643 ± 8.5	985 ± 12.8	–	–

Source: Adapted with permission from Chen and Yen [28].
Data expressed as means ± standard deviation ($n = 3$) on an extract.

Table 24.3 Antioxidant activity (by linoleic acid method) of extracts from guava tea and guava dried fruit

Sample concentration (μg/mL)	Peroxidation (%)[a]		
	Guava tea	Guava dried fruit	Tea polyphenon 60
50	17.8 ± 2.6	49.9 ± 2.7	17.8 ± 1.8
100	4.7 ± 0.2	11.6 ± 0.8	10.1 ± 1.1
150	3.7 ± 0.2	7.3 ± 0.3	11.3 ± 0.4
200	4.5 ± 0.1	4.8 ± 0.2	13.1 ± 0.4
500	6.2 ± 0.2	0.7 ± 0.1	19.5 ± 1.2

Source: Adapted with permission from Chen and Yen [28].
Data expressed as means ± standard deviation ($n = 3$) on an extract.
[a]A lower peroxidation indicates higher antioxidant activity.

24.2.4 Health benefits of dried guavas

There is a growing interest in guavas and their effects on human health. They have been used in traditional medicine around the world. They are claimed to be beneficial in the treatment of hypertension and diabetes, act as anti-inflammatory, pain relieving, and wound-healing agents and improve diarrhea, and reduce fever [27]. Other parts of guava that have been extensively studied are dried leaves, root barks, fresh ripe fruits or extracts of the whole fruits, or plants, but limited studies have been carried out on dried fruit components [27].

Much of the health benefits of guavas in any form of the plant have been attributed to their phytochemical components such as flavonoids and phenolics that possess strong antioxidant activity [28]. This is also governed by the content of phytochemicals present.

Increased production of free radicals causes oxidation in the cells which play a key role in CVD, cancer initiation, cataract formation, aging process, inflammatory diseases, and a variety of neurological disorders [29]. The antioxidant defence system in dried guavas may potentially protect against oxidative damage by preventing radical formation, repair oxidative damage, eliminate damaged molecules, and prevent mutation [30].

The hypothetical protective effect of dried guavas against oxidation is consistent with an experimental study conducted in healthy young adults [31] and in an animal model [32]

Table 24.4 Scavenging effects (by peroxyl radicals) of extracts from guava tea and guava dried fruit

Sample concentration (μg/mL)	Scavenging effects (%)		
	Guava tea	Guava dried fruit	Tea polyphenon 60
2.5	37.9 ± 2.4	16.6 ± 7.3	90.6 ± 6.7
5	66.8 ± 1.7	44.0 ± 4.3	89.2 ± 4.0
10	90.5 ± 3.3	79.3 ± 6.6	79.0 ± 4.6
25	89.0 ± 4.9	89.2 ± 2.3	62.9 ± 5.7
50	78.7 ± 8.0	91.2 ± 4.2	39.8 ± 5.7

Source: Adapted with permission from Chen and Yen [28].
Data expressed as means ± standard deviation ($n = 3$) on an extract.

using either fresh guava fruits or guava purees. In both studies, consumption of fresh guavas [31] or guava purees [32] resulted in a significant increase in the plasma antioxidant levels and reduced oxidative stress as indicated by decreased level of glutathione peroxidase and glutathione reductase [31].

Aside from their phytochemical properties, dried guavas are also high in fiber. Isolated dietary fiber, particularly pectins in fresh guavas have been shown to have hypocholesterolemic effect in humans [33]. However, the contribution of dried guava fibers in their specific hypolipidemic effect has not yet been studied.

Fresh guavas are also potassium-rich fruits and their consumption has benefited patients with hypertension [33]. As the drying process of dried guavas promotes loss of vitamins and minerals [22], their potential hypotensive role has yet to be elucidated.

24.2.5 Commercial products from dried guavas

Dried guava slices are available in the markets in Southeast Asia [5, 14]. Osorio *et al.* [16] developed two types of guava powders by hot-air drying and freeze drying. Both powders were rich in pectin and had a good retention of guava aroma. These two guava powders can be used as value-added food products and as alternatives to fresh guavas. Dry guava by-products can be used in poultry feed and also as value-added products [19]. Krasaekoopt and Suthanwong [34] developed probiotic food by fortifying probiotics (*Lactobacillus casei*, 10^8–10^9 log cfu/g) in partially dried guava pieces. Ready to use dehydrated products such as slices and leather was developed by Sagar and Kumar [35]. Osmotic drying considerably increased the sugar content and reduced the acidity without causing any significant changes in color, texture, and original flavor of the slices. Guava leather was also prepared from guava pulp [35].

Tea prepared from guavas can be consumed as a safe drink as well as a chemopreventing agent [26]. Cookies enriched with guava fibers are developed with a high rate of acceptability in flavor relation [20]. A clarified guava juice powder was obtained using different drying methods such as freeze drying, spray drying, and tunnel drying. The freeze-dried product had a superior quality; however, the spray-dried product was more stable and may be economical for industrial exploitation [36].

24.3 Papayas

Papayas (*Carica papaya* L.) belong to the family of *Caricaceae*. They are an important fruit crop grown in the tropical and subtropical regions of the world. Their production is next to mangos, bananas, citrus fruits, and pineapples at a global level [37]. The fruits are very nutritious (rich in vitamins and carotenoids) and have good taste, color, and aroma. They are usually eaten fresh, or are processed into juice, jam, jellies, fruit salad, ice cream, canned, or frozen forms [38–40].

24.3.1 Nutritional composition of dried papayas

Effect of osmotic pre-treatment on vitamin C, texture, and color of osmotically dehydrated papayas was studied by Lemus-Mondaca *et al.* [41]. Treatment with a 40–50% sucrose solution enhanced retention of vitamin C with improved color and texture quality. The

Table 24.5 Nutritional composition of fresh, sliced, and dried papaya fruits (values in per 100 grams edible portion)

Samples	Units	Juice	Fresh slices	Dried slices
Energy	kJ	836	867	163
Fat	g	0.51	0.53	0.10
Fiber	g	1.8	1.6	2.6
Vitamin A (RAE)	μg	85.6	380	8.8

Source: Adapted with permission from Gouado *et al.* [43].
Data expressed as means ($n = 3$) on a fresh weight basis.
RAE, retinol activity equivalents.

nutritional quality of dried cereal flakes containing ripe papayas is as follows: moisture (1.08%), acidity (0.34), total sugars (60%), reducing sugars (56%), ascorbic acid (241 mg/100 g), and carotenoids (7.3 mg/100 g) [42]. Nutritional composition of fresh, sliced, and dried papayas are presented in Table 24.5 [43]. Dried papaya slices had higher fiber content than fresh slices.

Vitamin C content has been reported to be higher in freeze-dried (88%), vacuum-dried (86%), carbon dioxide-dried (82%), and nitrogen-dried (80%) than fresh papaya fruits [23]. The nutritional qualities of freeze-dried papayas have been found to be lower than fresh ones (Table 24.6) [10]. Kamaruzzman *et al.* [44] investigated the nutritional composition of dried papaya skin and found that protein (25.2%) and sugars (65%) were dominant components. Aspartic acid and glycine were the major amino acids found in dried papaya skin (Table 24.7). Proximate composition of dried papaya skin was also evaluated by Fouzder *et al.* [45]. Effect of drying on vitamin A, vitamin C, total sugars, and moisture content of dried papaya powders was studied by Mugula *et al.* [39]. A significant loss in vitamin A (97%), vitamin C (98%), total sugars (86.8%), and moisture content (84.2%) was noticed in dried papaya powders when compared to those of fresh papayas (Table 24.8).

24.3.2 Phytochemicals in dried papayas

Dried papaya in cereal flakes are rich in ascorbic acid (241 mg/100 g) and low in carotenoids (7.3 mg/100 g) [42]. Carotenoid composition of fresh, sliced, and dried papayas is given

Table 24.6 Nutritional characteristics of fresh and freeze-dried papaya pulps

	Nutritional composition		Mineral composition	
Samples	Moisture (g/100 g)	Vitamin C (mg/100 g)	Phosphorus (mg/100 g)	Calcium (mg/100 g)
Fresh papayas	89.8	87.2	9.3	13.8
Freeze-dried papayas	0.8	66.1	6.6	13.9

Source: Adapted from Marques *et al.* [10].
Data expressed as means ($n = 3$) on a fresh weight basis.

Table 24.7 Nutritional characteristics of dried papaya skin

Nutrient	(g/100 g)
Proximate composition	
Moisture	91.0
Crude protein	25.2
Crude fiber	6.7
Lipids	2.1
Ash	1.0
Nitrogen free extract	65.0
Amino acids	
Alanine	0.62
Arginine	0.48
Aspartic acid	2.60
Glutamic acid	1.50
Glycine	3.00
Isoleucine[a]	0.41
Leucine[a]	0.77
Lysine[a]	0.79
Phenylalanine[a]	0.39
Serine	0.60
Threonine[a]	0.36
Tryptophan[a]	0.02
Tyrosine	0.29
Valine[a]	0.52

Source: Adapted with permission from Kamaruzzaman *et al.* [44].
Data expressed as means ($n = 3$) on a dry weight basis.
[a]Indispensable amino acids.

in Table 24.9. Among them, lycopene is the predominant one followed by β-carotene, cryptoxanthin, α-carotene, and zeaxanthin [43]. Sian and Ishak [46] studied carotenoid concentration of dried papayas at various blanching temperatures and found that they decreased with increase in blanching temperature. The carotenoid concentration ranged from 99.4 to 85.8 μg/g dry weight. In addition, anthocyanin content was also reduced from 9.2 to 6.7 μg/g, with increase in blanching temperature.

Table 24.8 Effect of drying on nutritional characteristics of fresh and dried papaya powders

Samples	Units	Fresh papaya	Sun dried	Oven dried
Moisture	g/100 g	93.8	15.24 (83.8)	14.85 (84.2)
Total sugar	g/100 g	171.7	22.7 (86.8)	22.9 (86.9)
Vitamin A (RAE)	mg/100 g	13.39	0.41 (97)	0.93 (92)
Vitamin C	mg/100 g	812.9	20.3 (97.5)	5.1 (98.1)

Source: Adapted with permission from Mugula *et al.* [39].
Numbers in parentheses represent percentage nutrient loss compared to fresh sample.
Data expressed as means ($n = 3$) on a dry weight basis.
RAE, retinol activity equivalents.

Table 24.9 Carotenoid compositions of fresh, sliced, and dried papaya fruits

Samples	Units	Juice	Fresh slices	Dried slices
α-Carotene	μg/100 g	368	83	nd
β-Carotene	μg/100 g	3427	668	93.7
Lycopene	μg/100 g	5081	581	188
Cryptoxanthin	μg/100 g	1885	635	24
Zeaxanthin	μg/100 g	83	99	tr

Source: Adapted with permission from Gouado *et al.* [43].
Data expressed as means ($n = 3$) on a fresh weight basis.
nd, not detected; tr, trace.

24.3.3 Antioxidant activity of dried papayas

Antioxidant activity (*in vitro* and *in vivo*) of dried papaya juices was also evaluated by Mehdipour *et al.* [47]. Dried papaya juices exhibited excellent 1,1-diphenyl-2-picrylhydrazyl (DPPH) scavenging activity in a dose-dependent manner. The highest inhibition was 80% at a concentration of 17.6 mg/mL. In addition, using animal models, lipid peroxidation was reduced by 35, 39, and 40% at dosages of 100, 200, and 400 mg/kg per day, respectively. Furthermore, total antioxidant power in the blood was also significantly improved (11–24%) as compared to the control and vitamin E (18%) groups [47].

24.3.4 Health benefits of dried papayas

Similar to the guava fruits, papayas have also been widely used in traditional medicine to improve digestion problems, as a wound-healing agent and as a folk remedy for contraception and abortion. In a survey conducted by Adebisi and Bello [48], they found that dried papayas have been the primary herbs used as a traditional male contraception method in South-West Nigeria.

Aqueous and ethanolic extracts of dried papaya fruits have some hepatoprotective capability in animal models [49]. These extracts inhibited harmful effects of carbon tetrachloride that may induce hepatotoxicity [49]. This protection effect could be due to the decreasing levels of serum bilirubin and several liver-related enzymes such as serum alanine aminotransferase, aspartate aminotransferase, and alkaline phosphatase. Interestingly, consumption of dried papaya extracts showed remarkable reduction in necrosis and degeneration of liver cells, when compared with the control in histopathological studies on rat liver [49, 50].

Indeed, combined mixtures of dried papayas and dried kales at a ratio of 40:60% produced a positive prebiotic and intestinal anti-inflammatory effect in a rat colitis model [51]. The prebiotic effect was achieved by increasing the colony count of *Lactobacillus* and *Bifidobacterium*. On the other hand, intestinal anti-inflammatory activity of the same mixture played a significant role for necrosis reduction in colitis of rats [51].

In addition to the above, regular consumption of papayas and their products (in the form of either juices or dried fruits) would improve vitamin A status of human beings. Gouado *et al.* [43] demonstrated that frequent consumption of fresh, juice, and dried papayas increased serum retinol levels in the Cameroon population. However, as compared to dried papayas, fresh fruits and juice forms were found to be more beneficial and this could be due to the bioavailability of carotenoids which was better in the fresh forms than in the dried products.

The anti-ulcer activity of alcoholic extracts of dried papayas has been investigated in an animal model [52]. In this study, improvement in ulcer was assessed by measuring the volume of gastric secretion, free acidity, total acidity, and ulcer index. This study showed that dried papaya extracts had anti-ulcer effect as indicated by decreasing the volume of gastric acid secretion, free acidity, total acidity, and ulcer index with respect to the control group [52].

24.3.5 Commercial products from dried papayas

Cereal flakes were developed from ripe papayas [42]. They had a good shelf life of 60 days (at refrigerated storage) with good sensory qualities. Dried pickles were developed by Trongpanich *et al.* [53]. Fernandes *et al.* [54] used ultrasonic pre-treatment to remove sugars for production of dehydrated papayas. Krasaekoopt and Suthanwong [34] developed probiotic food by fortifying probiotics (*Lactobacillus casei*, 10^8–10^9 log cfu/g) in partially dried papaya pieces. Sulfur dioxide and glycerin-treated dried papayas were developed by Buenz and Fuatai [13] and these have the potential to be profitable value-added agricultural products. Restructured fruit was made from concentrated papaya pulps, with the addition of other ingredients such as sugar, sodium alginate, pectin, and glycerol. They were cut in a solid cylindrical form, dehydrated, and covered with icing sugar [55]. In addition, papaya cubes [56] and papaya powders [39] have been developed. The powder can be used as a beverage drink with acceptable sensory qualities. Dry papaya skins can also be used as a dietary ingredient for broiler chickens [45].

24.4 Conclusions

Dried fruits help in reduction of fruit wastage and improve their shelf life. They are handy when needed and are packed with nutrients, specifically fiber and vitamin C, which could play a role to overcome nutrient deficiency. Since many of the dried fruits appear wrinkled with loss of valuable nutrients during the drying process, further studies are warranted to determine the most novel food processing method in minimizing nutrient loss while at the same time retaining their phytochemicals. Health claims of dried guavas and papayas have been extensively documented; however, proper studies either in animals or humans using these dried fruits in comparison with their fresh counterparts for the prevention and the management of chronic diseases are needed.

Acknowledgments

The authors would like to acknowledge the financial support under the Science Fund-MOSTI, No 5450555; Research University Grants Scheme (RUGS) No 9192600 and 02-01-09-0703RU under Research Management Center, UPM.

References

1. Khoo, H.E., Prasad, K.N., Kong, K.W., Jiang, Y. & Amin, I. (2011) Carotenoids and their isomers. Colorful pigments in fruits and vegetables. *Molecules*, **16**, 1710–1738.

2. Prasad, N.K., Aradhya, S.M. & Shivamurthy, G.R. (2008) *Ipomoea aquatica*—an underutilized green leafy vegetable—review. *International Journal of Botany*, **4**, 93–99.
3. WHO (2002) *Global Strategy on Diet, Physical Activity and Health*; *World Health Assembly, Resolution 55.23*. Published online at: http://www.who.int/dietphysicalactivity/strategy/eb11344/strategy_english_web.pdf, last accessed November 11, 2011.
4. Steffen, L.M., Jacobs, J.D.R., Stevens, J., Shahar, E., Carithers, T. & Folsom, A.R. (2003) Associations of whole-grain, refined-grain, and fruit and vegetable consumption with risks of all-cause mortality and incident coronary artery disease and ischemic stroke: the atherosclerosis risk in communities (ARIC) study. *American Journal of Clinical Nutrition*, **78**, 383–390.
5. Strawn, L.K., Schneider, K.R. & Danyluk, M. (2011) Microbial safety of tropical fruits. *Critical Reviews in Food Science and Nutrition*, **51**, 132–145.
6. FAO (2011) *Global Food Losses and Food Waste: Extent, Causes, and Prevention*. Published online at: http://www.fao.org/fileadmin/user_upload/ags/publications/GFL_web.pdf, last accessed November 11, 2011.
7. Golan, R.B. (2001) Subtropical and tropical fruits. *Postharvest Diseases of Fruits and Vegetables*, 268–293.
8. Karim, M.A. & Hawlader, M.N.A. (2005) Mathematical modeling and experimental investigation of tropical fruits drying. *International Journal of Heat and Mass Transfer*, **48**, 4914–4925.
9. FAO (2003) *Medium-Term Prospects for Agricultural Commodities—Projections to the Year 2010*. Published online at: http://www.fao.org/docrep/006/y5143e/y5143e00.htm, last accessed August 8, 2011.
10. Marques, L.G., Silveira, A.M. & Freire, J.T. (2006) Freeze-drying characteristics of tropical fruits. *Drying Technology*, **24**, 457–463.
11. Marques, L.G., Prado, M.M. & Freire, J.T. (2009) Rehydration characteristics of freeze-dried tropical fruits. *LWT—Food Science and Technology*, **42**, 1232–1237.
12. Ganjyal, G.M., Hanna, H.M. & Devadattam, D.S.K. (2003) Processing of sapota (*Sapodilla*): drying. *Journal of Food Science*, **68**, 517–520.
13. Buenz, E.J. & Fuatai, L. (2007) A potential dried fruit market in independent Samoa. *Journal of Sustainable Agriculture*, **31**, 17–51.
14. Fernandes, F.A.N., Rodrigues, S., Law, C.H. & Majumdar, S. (2011) Drying of exotic tropical fruits: a comprehensive review. *Food and Bioprocess Technology*, **4**, 163–185.
15. Kong, K.K., Rajab, N.F., Prasad, K.N., Ismai, A., Markom, M. & Tan, C.P. (2010) Lycopene-rich fractions derived from pink guava by-product and their potential activity towards hydrogen peroxide-induced cellular and DNA damage. *Food Chemistry*, **123**, 1142–1148.
16. Osorio, C., Carriazo, J.G. & Barbosa, H. (2011) Thermal and structural study of guava (*Psidium guajava* L.) powders obtained by two dehydration methods. *Quimica Nova*, **34**, 636–640.
17. Munhoz, C.L., Sanjinez-Argandoña, E.J. & Soares-Júnior, M.S. (2010) Extraction of pectin from dehydrated guava. *Ciencia e Tecnologia de Alimentos*, **30**, 1.
18. Tsai, P.J., Sun, Y.F. & Hsiao, S.H. (2010) Strengthening the texture of dried guava slice by infiltration of phenolic compounds. *Food Research International*, **43**, 825–830.
19. El-Deek, A.A., Hamdy, S.F., Attia, Y.A. & El-Sahat, A.M. (2009) Guava by product meal processed in various ways and fed in differing amounts as a component in laying hen diets. *International Journal of Poultry Science*, **8**, 866–874.
20. Matias, M.D.F.O., Oliveira, D., Gertrudes, E.L. & Dos Anjos Magalhães, M.M. (2005) Use of fibres obtained from the cashew (*Anacardium occidentale* L.) and guava (*Psidium guajava*) fruits for enrichment of food products. *Brazilian Archives of Biology and Technology*, **48**, 143–150.
21. Sanjinez-Argandona, E.J., Cunha, R.L., Menegalli, F.C. & Hubinger, M.D. (2005) Evaluation of total carotenoids and ascorbic acid in osmotic pretreated guavas during convective drying, *Italian Journal of Food Science*, **17**, 305–314.
22. Queiroz, V.A.V., Berbert, P.A., De Molina, M.A.B., De Gravina, G.A., Queiroz, L.R. & Da Silva, J.A. (2008) Nutritional quality of osmo-convective dried guavas. *Ciencia e Tecnologia de Alimentos*, **28**, 329–340.
23. Hawleder, M.N.A., Perera, C.O., Tian, M. & Yeo, K.Y. (2006) Drying of guava and papaya: impact of different drying methods. *Drying Technology*, **24**, 77–87.
24. Chua, K.J., Chou, S.K., Ho, J.C., Mujumdar, A.S. & Hawlader, M.N.A. (2000) Cyclic air temperature drying of guava pieces: effects on moisture and ascorbic acid contents. *Food and Bioproducts Processing: Transactions of the Institution of Chemical Engineers, Part C*, **78**, 72–78.

25. Queiroz, V.A.V., Berbert, P.A., De Molina, M.A.B., Gravina, G.D.A. & Deliza, R. (2007) Osmotic dehydration and convective drying of guava. *Pesquisa Agropecuaria Brasileira*, **42**, 1479–1486.
26. Lee, B.H., Hsieh, M.C., Chiang, T.C. & Wu, S.C. (2007) The antioxidative and antimutagenic properties of guava tea treated by different drying processes. *Taiwanese Journal of Agricultural Chemistry and Food Science*, **45**, 247–256.
27. Gutierrez, R.M.P., Mitchell, S. & Solis, R.V. (2008) *Psidium guajava*: a review of its traditional uses, phytochemistry and pharmacology. *Journal of Ethnopharmacology*, **117**, 1–27.
28. Chen, H.Y. & Yen, G.C. (2007) Antioxidant activity and free radical scavenging capacity of extracts from guava (*Psidium guajava* L.) leaves. *Food Chemistry*, **101**, 686–694.
29. Sarma, D.S., Mallick, A.R. & Ghosh, A.K. (2010) Free radicals and their role in different clinical conditions: an overview. *International Journal of Pharma Sciences and Research*, **1**, 185–192.
30. Norshazila, S., Syed Zahir, I., Mustapha, S.K., Aisyah, M.R. & Kamarul, R.K. (2010) Antioxidant levels and activities of selected seeds of Malaysian tropical fruits. *Malaysian Journal of Nutrition*, **16**, 149–159.
31. Rahmat, A., Bakar, M.F.A. & Hambali, Z. (2006) The effects of guava (*Psidium guajava*) consumption on total antioxidant and lipid profile in normal male youth. *African Journal of Food Agriculture Nutrition and Development*, **6**, 1–12.
32. Norazmir, M.N. & Ayub, M.Y. (2010) Beneficial lipid-lowering effects of pink guava puree in high fat diet induced-obese rats. *Malaysian Journal of Nutrition*, **16**, 171–185.
33. Singh, R.B., Rastogi, S.S., Singh, R., Ghosh, S., Gupta, S. & Niaz, M.A. (1993) Can guava fruits decrease blood pressure and blood lipids? *Journal of Human Hypertension*, **7**, 33–38.
34. Krasaekoopt, W. & Suthanwong, B. (2008) Vacuum impregnation of probiotics in fruit pieces and their survival during refrigerated storage. *Kasetsart Journal—Natural Science*, **42**, 723–731.
35. Sagar, V.R. & Kumar, P.S. (2007) Processing of guava in the form of dehydrated slices and leather. *Acta Horticulturae*, **735**, 579–589.
36. Chopda, C.A. & Barrett, D.M. (2001) Optimization of guava juice and powder production. *Journal of Food Processing and Preservation*, **25**, 411–430.
37. Rajarathnam, S. (2010) Perspectives of processing papaya (*Carica papaya*) fruit: national and international strategies. *Acta Horticulturae*, **851**, 547–554.
38. Furtado, J., Siles, X. & Campos, H. (2004) Carotenoid concentration in vegetables and fruits common to Costa Rican diet. *International Journal of Food Science and Nutrition*, **55**, 101–113.
39. Mugula, J.K., Kessy, F.L. & Lyimo, M.H. (1994) Production and storage stability of non-alcoholic pawpaw beverage powder. *Plant Foods for Human Nutrition*, **46**, 167–173.
40. Wall, M.M. (2006) Ascorbic acid, vitamin A and mineral composition of banana (*Musa* sp.) and papaya (*Carica papaya*) cultivars grown in Hawaii. *Journal of Food Composition and Analysis*, **19**, 434–445.
41. Lemus-Mondaca, R., Miranda, M., Andres Grau, A., Briones, V., Villalobos, R. & Vega-Galvez, A. (2009) Effect of osmotic pretreatment on hot air drying kinetics and quality of Chilean papaya (*Carica pubescens*). *Drying Technology*, **27**, 1105–1115.
42. Rai, S. & Chauhan, A.S. (2008) Quality attributes of drum-dried papaya-cereal flakes developed from ripe papaya (*Carica papaya* L.). *Electronic Journal of Environmental, Agricultural and Food Chemistry*, **7**, 2914–2931.
43. Gouado, I., Schweigert, F.J., Ejoh, R.A., Tchouanguep, M.F. & Camp, J.V. (2007) Systemic levels of carotenoids from mangoes and papaya consumed in three forms (juice, fresh and dry slice). *European Journal of Clinical Nutrition*, **61**, 1180–1188.
44. Kamaruzzaman, M., Chowdhury, S.D., Podder, C.K. & Pramanik. M.A.H. (2005) Dried papaya skin as a dietary ingredient for broiler chickens. *British Poultry Science*, **46**, 390–393.
45. Fouzder, S.K., Chowdhury, S.D., Howlider, M.A.R. & Podder, C.K. (1999) Use of dried papaya skin in the diet of growing pullets. *British Poultry Science*, **40**, 88–90.
46. Sian, N.K. & Ishak, S. (1991) Carotenoid and anthocyanin contents of papayas and pineapple: influence of blanching and predrying treatments. *Food Chemistry*, **39**, 175–185.
47. Mehdipour, S., Yasa, N., Dehghan, G., Khorasani, R., Mohammadirad, A., Rahimi, R. & Abdollahi, M. (2006) Antioxidant potentials of Iranian *Carica papaya* juice *in vitro* and *in vivo* are comparable to α-tocopherol. *Phytotherapy Research*, **20**, 591–594.
48. Adebisi, I.M. & Bello, S.O. (2011) An ethnobotanical survey of herbal male contraceptives used in South-West Nigeria. *African Journal of Pharmacy and Pharmacology*, **5**, 289–291.
49. Sadeque, Z. & Begum, Z.A. (2010) Protective effect of dried fruits of *Carica papaya* on hepatotoxicity in rat. *Bangladesh Journal of Pharmacology*, **5**, 48–50.

50. Rajkapoor, B., Jayakar, B., Kavimani, S. & Murugesh, N. (2002) Effect of dried fruits of Carica papaya LINN on hepatotoxicity. *Biological and Pharmaceutical Bulletin*, **25**, 1645–1646.
51. Lima de Albuquerque, C., Comalada, M., Camuesco, D., Rodriguez-Cabezas, M.E., Luiz-Ferreira, A., Nieto, A., de Souza Brito, A.R.M., Zarzuelo, A. & Gálvez, J. (2010) Effect of kale and papaya supplementation in colitis induced by trinitrobenzenesulfonic acid in the rat. *e-SPEN, the European e-Journal of Clinical Nutrition and Metabolism*, **5**, e111–e116.
52. Rajkapoor, B., Jayakar, B., Anandan, R. & Murugesh, N. (2003) Antiulcer effect of dried fruits of *Carica papaya* L. in rats. *Indian Journal of Pharmaceutical Sciences*, **65**, 638–639.
53. Trongpanich, K., Stonsaovapak, S., Puminat, W. & Yunchalad, M. (2008) Quality determination of Thai dried pickled fruits and vegetables. *Acta Horticulturae*, **787**, 363–371.
54. Fernandes, F.A.N., Oliveira, F.I.P. & Rodrigues, S. (2008) Use of ultrasound for dehydration of papayas. *Food and Bioprocess Technology*, **1**, 339–345.
55. Grizotto, R.K., Berbari, S.A.G., De Moura, S.C.S.R. & Claus, M.L. (2006) Shelf life studies of restructured and dried fruit made from concentrated papaya pulp. *Ciencia e Tecnologia de Alimentos*, **26**, 709–714.
56. Kaleemullah, S., Kailappan, R. & Varadharaju, N. (2002) Studies on osmotic-air drying characteristics of papaya cubes. *Journal of Food Science and Technology*, **39**, 82–84.

25 Dried mangoes: phytochemicals, antioxidant properties, and health benefits

Fouad Abdulrahman Hassan, Sadeq Hasan Al-Sheraji, and Amin Ismail

25.1 Introduction

Mangoes are one of the most popular tropical fruits in the world. They ranked fifth in total world production among the major fruit crops with world's annual production of 33.5 million metric tonnes (MT) in 2009 [1] and have been listed at the top of the superfruit list because of their health-promoting properties [2]. India is the world's largest mango producer followed by China, Thailand, Indonesia, Pakistan, Mexico, and Brazil. Most mangoes are consumed uncooked as a dessert fruit, the rest of them being processed into various products, such as nectar, juice powder, dried mangoes, canned mango slices in syrup, chutneys, and pickles, among others [3].

The majority of the mango species are indigenous and underutilized and have the potential to be further exploited and utilized commercially. Among the edible species are *Mangifera indica*, *M. caesia*, *M. foetida*, *M. odorata*, and *M. pajang*. Most of the fruit commonly known as mango belongs to the species *M. indica*. Other species such as *M. pajang*, an underutilized fruit, are found in Malaysia, Brunei, and Indonesia with size three times larger than commercial mangoes (*M. indica*). The physical characteristics of mangoes are almost similar. Several studies have reported that *Mangifera* species are rich in dietary fiber and phytochemicals in all their parts [4–8] and play an important role as antioxidants [9, 10].

It has been widely accepted that increased consumption of fruits, such as mangoes, could reduce the risk of cardiovascular diseases (CVD) and certain cancers, thus preserving the quality of life [11, 12]. In general, most of the fresh fruits contain more than 80% water that limits their shelf life and hence are more susceptible to deterioration. As such, preservation technique such as drying is most suitable to remove the water content of the fruits. The appearance of such dehydrated fruits has continually grown in national and international markets due to improved quality characteristics such as retention of the natural health-promoting compounds, prevention of the growth of microorganisms, and reduction of storage and transportation costs. This chapter reviews nutritional characteristics, antioxidant capacity, and health-promoting properties of dried mangoes (pulp and peels).

Dried Fruits: Phytochemicals and Health Effects, First Edition. Edited by Cesarettin Alasalvar and Fereidoon Shahidi.

25.2 Compositional and nutritional characteristics of dried mangoes

25.2.1 Compositional characteristics

Table 25.1 shows the compositional characteristics of fresh and dried mango varieties. The average moisture content of *M. indica* powder has been reported to be 1% [13] and 4% [14]. *M. pajang* fibrous powder prepared from fibrous pulp contains less than 5% moisture [7]. For every 100 g edible portion *M. foetida* flesh contains 78 g moisture [15]. *M. pajang* peel powder has a relatively low moisture content of 4% [8]. The differences observed in the moisture content may be due to different drying methods used, variety/cultivar, and treatment processing for preparation of the powder.

Ash content of *M. indica* L. fiber (2.8%) [5] was higher than *M. pajang* fiber pulp of 0.84% [7]. Meanwhile the ash content of *M. pajang* juice powder was 3.3% [16], which is three times higher than *M. indica* juice powder [13]. This indicates that the intake of dried mango products will provide a good source of minerals.

Fresh mangoes contain about 2 g protein/100 g edible portion. Protein content of *M. pajang* juice powder (3.8%) [16] was higher than *M. indica* juice powder [13]. Furthermore, protein contents in *M. pajang* fiber and peel powder were 3.7% [7] and 4.6% [8], respectively. Edible portion of *M. foetida* flesh contained 0.8% protein [15].

Mangoes are one of the tropical fruits that contain relatively very low amounts of fat and are generally low in calories. Fat contents of *M. pajang* pulp, *M. pajang* juice powder, and *M. pajang* fibrous pulp are 1.98, 1.75, and 0.79% of dry sample, respectively. The fat content of *M. pajang* peel powder (2.9%) [8] was similar to those reported previously for *M. indica* peels [4, 5, 9].

The carbohydrate of *M. indica* L. (32.6% of dry sample) [5] was higher than 0.18% reported for *M. foetida* flesh [15], 4.1% for *M. pajang* pulp [7], 7.3% for *M. pajang* peel powder [8], and 29% for Indian mango peels [9]. Among fruits, *M. indica* had the highest amount of carbohydrate as compared to grape skins (3%) [20], citrus peel fiber (27%), and guava fiber (28%) [21].

Table 25.1 Chemical composition (%) of fresh and dried mango varieties

Proximate composition	*M. indica* (fresh) [17, 18]	*M. indica* (powder) [13]	*M. pajang* (fresh) [16]	*M. pajang* (powder) [16]	*M. indica* (edible portion) [19]	*M. pajang* (peel powder) [8]	*M. pajang* (fibrous) [7]
Moisture	79.1	0.8	86.8	10.0	88.2	3.9	4.65
Protein	0.98	1.3	1.13	3.78	0.7	4.6	3.37
Fat	0.32	0.1	1.98	1.75	0.1	2.9	0.79
Ash	0.5	0.7	0.43	3.3	nd	2.7	0.84
Carbohydrate	15.6	95.8	21.0	76.1	11.6	7.3	4.02
SDF	1.0	nd	0.42	0.68	nd	33.4	9.05
IDF	1.0	nd	4.84	0.12	nd	38.8	78.5
Energy (kcal/100 g)	nd	389	429	335	nd	nd	nd

SDF, soluble dietary fiber; IDF, insoluble dietary fiber; nd, not detected.

Table 25.2 Dietary fiber content of mango varieties and their valuable findings

Variety	Part	Valuable findings	Dietary fiber	Reference
M. indica	Peel	Rich source of dietary fiber	70.1	[4]
M. indica	Peel with pulp	Antiradical efficiency	28.1	[5]
M. indica	Peel	Improved antioxidant properties	51.2	[6]
M. indica	Whole fruit	Equal ratio of SDF/IDF	7.9	[22]
M. indica	Whole fruit	Rich source of soluble fiber	10.0	[18]
M. indica	Unripe, ripe peel	Highest DF found in ripe peel	45–78	[23]
M. pajang	Powdered juice	Rich in SDF	0.8	[16]
M. indica	Powdered juice	High content of fiber	1.4	[13]
M. pajang	Fibrous pulp	Rich source of insoluble fiber	88.0	[7]
M. pajang	Peel	Balance ratio of SDF/IDF		[8]
M. indica	Pulp	High content of crude fiber	3.7	[24]
M. indica	Peel	High pectin (SDF)	11–21	[25]
M. indica	Green, ripe fruit	Green rich in fiber	10.7	[26]

SDF, soluble dietary fiber; IDF, insoluble dietary fiber; DF, dietary fiber.

Ripe and unripe mangoes represent good sources of dietary fiber for consumers and the food industry. Table 25.2 lists the dietary fiber content in different parts of mangoes with their valuable finding. Dietary fiber fractions, neutral sugars (monosaccharides), uronic acids, and Klason lignin of *M. pajang* fiber pulp and powdered peel are shown in Table 25.3. The presence of uronic acids together with monosaccharides reported in mangoes indicated the presence of hemicelluloses (arabinoxylans, glucuronoxylans, and xyloglucans) and pectic substances associated with the cell wall matrix in *M. pajang* fibrous and peel powders.

Table 25.3 Monosaccharide compositions of SDF and IDF in dried products prepared from mangoes of *M. pajang*

	***M. pajang* fibrous pulp (% dry weight)**		***M. pajang* peel (% dry weight)**	
Monosaccharide	**SDF**	**IDF**	**SDF**	**IDF**
Erythrose	0.14 ± 0.01	nd	nd	nd
Glucose	0.39 ± 0.01	4.46 ± 0.11	2.49 ± 0.23	1.15 ± 0.15
Galactose	0.05 ± 0.01	1.20 ± 0.06	0.80 ± 0.06	0.17 ± 0.01
Rhamnose	0.16 ± 0.01	1.65 ± 0.04	0.44 ± 0.02	0.20 ± 0.02
Arabinose	0.72 ± 0.02	18.47 ± 0.19	4.89 ± 0.28	3.05 ± 0.13
Mannose	1.51 ± 0.03	3.15 ± 0.12	12.49 ± 0.56	4.87 ± 0.22
Xylose	0.04 ± 0.01	0.99 ± 0.02	0.40 ± 0.14	0.09 ± 0.01
Fructose	nd	nd	0.15 ± 0.01	0.10 ± 0.01
Fucose	0.01 ± 0.01	0.26 ± 0.02	nd	nd
Neutral sugars	3.02 ± 0.11	30.18 ± 0.56	21.66 ± 0.32	9.63 ± 0.53
Uronic acids	5.83 ± 0.13	15.51 ± 0.17	11.75 ± 0.23	7.60 ± 0.11
Klason lignin	nd	33.11 ± 0.72	nd	21.51 ± 0.47
Total NSP	8.85 ± 0.24	78.80 ± 0.96	33.41 ± 0.32	17.26 ± 0.52

Source: Adapted from Al-sheraji *et al.* [7] and Hassan *et al.* [8].
Data expressed as means ($n = 3$) on a dry weight basis.
SDF, soluble dietary fiber; IDF, insoluble dietary fiber; NSP, non-starch polysaccharide (neutral sugars + uronic acid); nd, not detected.

Table 25.4 Mineral and vitamin contents of *M. indica* (values per 100 g edible portion)

Nutrient	Unit	Fresh mangoes [27]	Dried mangoes [13, 23]
Minerals			
Calcium	mg	11	48.4
Copper	mg	0.11	0.3
Iron	mg	0.16	2.2
Magnesium	mg	10	16.7
Manganese	mg	0.06	nd
Phosphorus	mg	14	nd
Potassium	mg	168	486
Selenium	μg	0.6	nd
Sodium	mg	1	732
Zinc	mg	0.09	0.12
Vitamins			
Vitamin C	mg	36.4	31.5
Vitamin E (ATE)	mg	0.90	33.7

ATE, α-tocopherol equivalents; nd, not detected.

In addition, these dried mangoes have considerable amounts of mannose, arabinose, and galactose, which may be suggestive of the presence of arabinomannose, galactomannas, or other pectic monosaccharides.

25.2.2 Nutritional characteristics

The mineral and vitamin compositions of fresh and dried mangoes are listed in Table 25.4. Mangoes serve as a good source of potassium, phosphorus, calcium, magnesium, sodium, copper, iron, zinc, manganese, and selenium. Dried mangoes are rich in sodium, potassium, calcium, magnesium, sodium, copper, iron, and zinc [13].

Ascorbic acid content of *M. pajang* juice powder (132 mg/100 g) was higher than that of *M. pajang* pulp (46 mg/100 g) [16], *M. foetida* (47 mg/100 g) [15], *M. indica* pulp (18–65 mg/100 g), and *M. indica* pulp powder (63 mg/100 g) [14, 28]. Ascorbic acid content in *M. pajang* pulp and juice powder was similar to that previously reported for *M. indica* pulp [29]. During the drying process of fruits, several factors can affect the degradation of ascorbic acid such as pH, temperature, light, oxygen, metal catalysts, presence of enzymes, and drying methods [30]. Conventional drying method affects the quantity of ascorbic acid more than that of the freeze drying method [31].

25.3 Phytochemicals and antioxidant activity of dried mangoes

25.3.1 Antioxidant activity

Antioxidant components that include polyphenols, carotenoids, and ascorbic acid possess bioactivities. The antioxidant activity of *M. pajang* peel powder as determined by the 1,1-diphenyl-2-picrylhydrazyl (DPPH) and ferric-reducing antioxidant power (FRAP) assays

Table 25.5 Antioxidant compounds of fresh and freeze-dried mangoes

	Phenolic content (mg of GAE/100 g)	Ascorbic acid (mg/100 g)	β-Carotene (μg/100 g)	Reference
Fresh	99.7 ± 8.7	8.36 ± 2.33	660 ± 61	[23]
Freeze-dried	76.6 ± 8.1	8.34 ± 1.74	487 ± 29	[35]

GAE, gallic acid equivalents.

exhibited a strong potency corresponding to that of butylated hydroxytoluene (BHT) and ascorbic acid. The antioxidant power of *M. pajang* peel powder was 1248 μg/mL compared to ascorbic acid (1318 μg/mL) [32], and the DPPH scavenging activity determined as IC_{50} was 44 μg/mL [8]. Gorinstein *et al.* [33] found that commercial mango (*M. indica*) had higher antioxidant properties compared to other fruits. Fresh *M. foetida* fruit exhibited the highest antioxidant activity based on the FRAP and trolox equivalent antioxidant capacity (TEAC) assays followed by its fiber and powder [34]. Freeze-dried mangoes have low antioxidant activity as compared to the fresh pulp [35]. In contrast, Soong and Barlow [36] reported that freeze-dried mango flesh (*M. indica*) had a higher antioxidant activity as compared to that of its fresh counterpart. This could be due to differences in applying antioxidant assays, maturity of the fruit, and sample preparation. Mohamad Shofian *et al.* [35] have reported a comparison of antioxidant compounds (phenolic content, ascorbic acid, and β-carotene) between fresh and freeze-dried mangoes as shown in Table 25.5. They found no significant change ($P > 0.05$) in ascorbic acid content of the fresh and dried mangoes.

Mangoes exhibited strong antioxidant activity due to the presence of associated total phenolics. Total phenol content (16.1 mg of gallic acid equivalents (GAE)/g) in *M. indica* L. [5] was lower than that reported for *M. pajang* fibrous (30 mg of GAE/g). *M. indica* had a lower amount of total phenolics than those in Mexican lime (10.55 and 19.9 mg of GAE/g) [37] and guava (58.7 mg of GAE/g) [38], but higher than that of apple (3 mg of GAE/g) [4]. Total phenol content was different among the Ubá, Palmer, Haden, and Tommy Atkins cultivars grown in Brazil, being higher in Ubá mango pulp at 220 mg of GAE/g compared with the Palmer, Haden, and Tommy Atkins cultivars (130, 60, and 50 mg of GAE/g, respectively) [28]. The mango cultivars contained total phenol content that may contribute to increased antioxidant intake in the human diet, since the intake of polyphenolic compounds in the diet is estimated to range between 0.15 and 1.0 g/day [39, 40]. Total phenol content of *M. pajang* juice powder (0.19 mg of GAE/g) was lower than that of *M. pajang* pulp (0.26 mg of GAE/g) [16]. *M. pajang* pulp had a low phenolic content (48–87%) as compared to *M. indica* reported by Ribeiro *et al.* [28]. As reported earlier, the total phenol content for *M. pajang* was higher in peels and kernel fractions [10].

The total extractable polyphenol (TEP) content in the *M. pajang* peel powder was 98 mg of GAE/g [8], which is higher than that of Haden mango peel fiber (70 mg of GAE/g) [4], Mexican lime peels (19 mg of GAE/g) [37], guava peels (59 mg of GAE/g) [38], grape skins (37.6–52.2 mg of GAE/g) [20], and unripe *M. pajang* fruit peels (24 mg of GAE/g) [10], but similar to the TEP content of Indian mango peels (96.2 mg of GAE/g) [9]. The high level of TEP, which is distinct to *M. pajang* peel, had potential antioxidant activity. Freeze-dried mangoes had 23% lower TEP content compared to the fresh ones [35].

25.3.2 Phenolic acids

Plant foods are rich in phenolic acids. Hassan *et al.* [32] reported that phenolic acids, such as gallic acid and methyl gallate, have antimicrobial activity, controlling dental caries and periodontal disease. Other studies have reported that ellagic acid exhibits antimutagenic, antiviral, antitumor, and antioxidant properties, along with the ability to whiten the skin [32]. Among those identified in mangoes are gallic acid, benzoic acid, 3,4-dihydroxybenzoic acid, gallic acid methyl ester, gallic acid propyl ester, and benzoic acid propyl ester [41]. In mango pulp, gallic acid, *m*-digallic acid, gallotannin, mangiferin, and an unknown hydrolyzable tannin were detected [42]. During all stages of mango ripening, total phenolic acids were higher in the peels than in the pulp [43]. El-sissi *et al.* [44] identified some major components of phenolic acids (hydrolyzable tannins) in mango parts (pulp, peels, seeds, leaf, and stem bark extracts) which included gallic acid, methyl gallate, digallic acid, ellagic acid, β-glucogallin and α-gallotannin at a low concentration.

Mangoes are a rich source of various phenolic acids [45]. The major polyphenols identified in mango (*M. indica*) pulp are mangiferin, gallic acids (*m*-digallic and *m*-trigallic acids), ellagic acid, and β-glucogallin derivative [42]. Gallic acid has been identified as the major polyphenol present in mango fruit pulp [46]. *M. pajang* peels had numerous phenolic acids such as gallic acid, ellagic acid, protocatechuic acid, 4-hydroxybenzoic acid, chlorogenic acid, vanillic acid, *p*-coumaric acid, ferulic acid, methyl gallate, and ethyl gallate [8]. Gallic acid was the major polyphenol in mango fruits [8, 47], with the highest amount in ripe mangoes as compared with other fruits [33]. The main phenolic acids found in mangoes are shown in Figure 25.1.

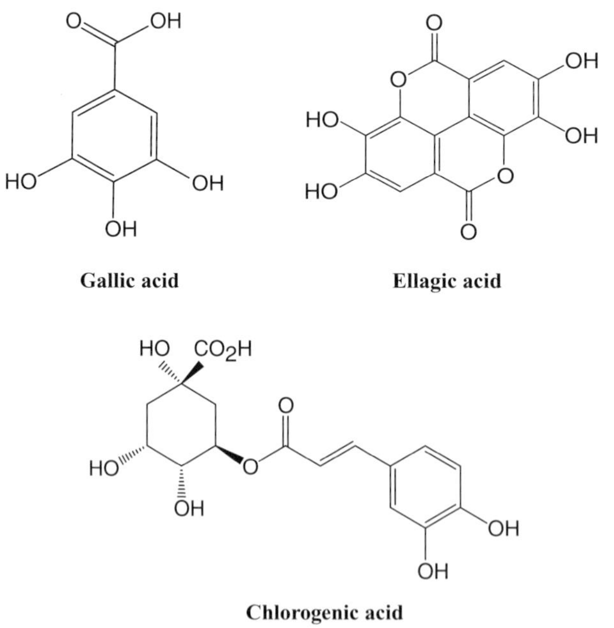

Figure 25.1 Chemical structures of major phenolic acids detected in mangoes.

25.3.3 Flavonoids and xanthones

Flavonoids can be divided into several classes according to the degree of oxidation of the oxygen heterocycle: flavones, flavanones, flavonols, flavanols, isoflavones, anthocyanins, and proanthocyanidins. The flavonoids identified in mangoes are reported by Schieber *et al.* [42], who demonstrated the presence of quercetin and related glycosides in mango pulps, with the predominant flavonol glycoside, namely quercetin 3-galactoside, at 22.1 mg/kg, followed by quercetin 3-glucoside (16.0 mg/kg) and quercetin 3-arabinoside (5.0 mg/kg). The amount of quercetin aglycone was 3.5 mg/kg. Other flavonol glycosides, such as kaempferol, were also present in trace amounts.

Mangoes are a rich source of flavonoids and xanthones such as mangiferin, catechins, quercetin, kaempferol, and rhamnetin [45], flavonol, and xanthone [48]. Quercetin and isoquercetin are reported in *M. indica* pulp [42]. While quercetin, kaempferol, rhamnetin, mangiferin, mangiferin gallate, isomangiferin, and isomangiferin gallate can be found in *M. indica* peels [46], in *M. pajang* peels, flavonoids such as (+)-catechin, rutin, morin, daidzein, kaempferol, and xanthone mangiferin are present [8].

A study on mangoes by Berardini *et al.* [48] demonstrated that peels of the cultivar Tommy Atkins contained a large number of flavonol *O*- and xanthone *C*-glycosides including isomangiferin, mangiferin gallate, isomangiferin gallate, quercetin 3-*O*-diglycoside, quercetin3-*O*-galactoside, quercetin 3-*O*-glucoside, quercetin 3-*O*-xyloside, quercetin 3-*O*-arabinopyranoside, quercetin 3-*O*-arabinofuranoside, quercetin 3-*O*-rhamnoside, kaempferol 3-*O*-glucoside, rhamnetin 3-*O*-galactoside, rhamnetin 3-*O*-glucoside, quercetin (aglycone), mangiferin, isomangiferin, mangiferin gallate, and isomangiferin gallate . Ribeiro *et al.* [49] identified some flavonol and xanthone compounds, including mangiferin, isomangiferin, mangiferin gallate, kaempferol, quercetin, and their related glycosides. Structures of some flavonoids present in mangoes are shown in Figure 25.2.

Catechins are also reported in some varieties of mangoes such as (+)-catechin, (–)-epicatechin, (–)-epigallocatechin, (–)-epicatechin gallate, and (+)-gallocatechin. The *M. indica* extract was rich mainly in catechin and epicatechin [41, 46, 50].

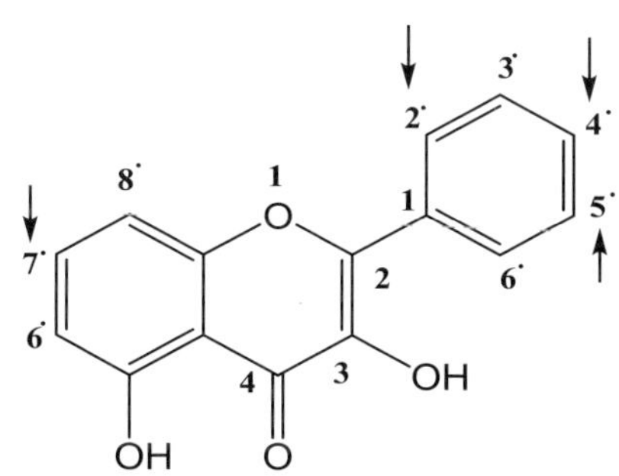

2 = OH, 4 = OH, 7 = OH: morin
4 = OH, 5 = OH, 7 = OH: rotin
4 = OH, 7 = OH: kaempferol

Figure 25.2 Chemical structures of flavonoids found in dried mangoes.

The most prominent flavonoids present were anthocyanins which are universal plant colorants responsible for the red, purple, and blue color in many foods with over 600 structures identified [51]. In mango peels, anthocyanins have been estimated at 203–565 mg/100 g dry matter depending on the variety and stage of maturity [52].

25.3.4 Carotenoids

Carotenoids are classified into two main groups:

(1) Carotene which are hydrocarbon molecules comprising only carbon and hydrogen atoms; this group includes β-carotene and lycopene [53].
(2) Xanthophylls which are oxygenated derivatives of the carotenes and include lutein, zeaxanthin, isozeaxanthin, capsanthin, capsorubin, cryptoxanthin, astaxanthin, 3-epilutein, and canthaxanthin [53].

Mangoes are rich in carotenoids which are responsible for the yellow to orange color of ripe mango pulp and provide a high provitamin A and antioxidant activity. As shown in Table 25.6, β-carotene is generally the predominant carotenoid in mango pulps, comprising 48–84% of the total carotenoids [57]. Among the wide variety of carotenoids in mangoes, *all-trans*-violaxanthin is the predominant carotenoid followed by all-*trans*-β-carotene [58]. The Haden variety had a total carotenoid significantly lower than that of the Tommy Atkins and Palmer varieties [28]. β-Carotene content of Haden mangoes is similar to the one described in studies with the same variety cultivated in Brazil (494 μg/100 g) [57]. β-Carotene content of the evaluated mango pulps was higher than those described for other mangoes (49.8 μg/100 g) [59]. Values obtained for total carotenoids for varieties of Tommy Atkins and Palmer were lower than those reported previously with the same mango cultivars obtained under experimental field conditions in Brazil [60].

Considering the vitamin A value expressed in retinol activity equivalents (RAE) [61], mangoes serve as a very good source of provitamin A. The RAE values for Haden, Tommy

Table 25.6 Carotenoids found in mangoes and their byproducts

Variety	Type of carotene	Part	Content (mg/100 g)	Reference
M. indica	Total carotenoids	Edible portion	0.13	[54]
M. indica	Total provitamin	Edible portion	0.63	[54]
M. indica	Total carotenoids	Powdered peel	309	[6]
M. indica	Total carotenoids	Powdered peel	334	[9]
M. indica	Total carotenoids	Ripe powdered peel	395	[55]
M. indica	Total carotenoids	Unripe powdered peel	140	[55]
M. foetida	Total carotenoids	Edible portion	0.26	[15]
M. pajang	β-Carotene	Powdered pulp	42.21	[16]
M. pajang	β-Carotene	Powdered juice	35.60	[16]
M. pajang	α- and β-carotenes	Powdered pulp	8 and 20	[56]
M. pajang	α- and β-carotenes	Powdered peel	4 and 13	[56]
M. indica	β-Carotene	Pulp	0.66–2.29	[28]
M. indica	Total carotenoids	Pulp	1.91–2.63	[28]
M. indica	β-Carotene	Pulp	0.49	[57]

Atkins, Palmer, and Ubá cultivars were 74, 51, 55, and 185 μg/100 g, respectively. Mangoes of *M. foetida* contained 255 μg carotenes/100 g. β-Carotene content of *M. pajang* pulp, juice, and peel powder [16] was relatively high when compared to that reported for *M. indica* by El-sissi *et al.* [44]. A previous report showed that β-carotene was relatively unstable and often degraded during processing and storage [14]. α- and β-carotenes in *M. pajang* pulp were 8 and 20 mg/100 g, and 4 and 13 mg/100 g in its peels, respectively (Table 25.6). In adult men, the daily intake of one serving of pulp equivalent to 150 g of Haden, Tommy Atkins, Palmer, and Ubá varieties supplies 12, 6, 6, and 21%, respectively, of the recommended dietary allowance (RDA) of vitamin A [61].

25.4 Health benefits of dried mangoes

The dried mango fruits are generally stored for periods between 1 and 18 months [62, 63]. Fruits and vegetables consumption is associated with a reduced risk of heart disease and certain cancers [64, 65]. Mangoes have been reported to be rich in dietary fiber [4, 7–9] with associated antioxidants such as carotenoids and phenolic compounds [32, 42, 56]. The dietary fibers of mangoes are of interest to the food industry because of their health-promoting properties. It has been shown that high consumption of dietary fiber is associated with reduced incidence of disorders and diseases that are common in developed nations such as chronic bowel disorders, obesity, diabetes, CVD, and cancer [64]. Several studies reported that dietary fiber and naturally occurring antioxidants (obtained from fruits) are dietary factors involved in CVD risk reduction [66, 67]. Polyphenols are bioactive compounds that have protective effect against diseases such as coronary heart disease (CHD), cancer, and neurodegenerative disorders, mostly through their antioxidant properties [68]. The presence of mangiferin in mangoes has a wide range of pharmacological effects including hypolipidemic, antidiabetic, anti-HIV, antitumor, immunomodulatory, and antioxidant activities [32]. The chemical structure of mangiferin is presented in Figure 25.3.

25.4.1 Anticancer

Dietary fiber of mangoes has been associated with a number of health benefits. Soluble and insoluble dietary fibers are reported to have positive effects in reducing the incidence and severity of gastrointestinal disease and cancer [69]. Data on 25 plants including mangoes indicated strong antimutagenic properties in the plants [70]. The presence of polyphenols,

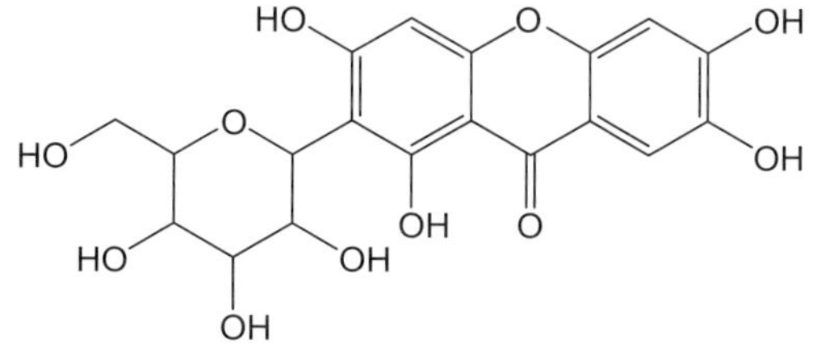

Figure 25.3 Chemical structure of xanthone (mangiferin) detected in dried mango pulp, peels, and seeds.

carotenoids, and antimutagens in the mangoes suggests their potential anticancer activity. Whole mango juice and its extract demonstrate anticancer activity by inhibiting the growth cycle in an immortal cancer cell line. The 50% methanolic extract of mangoes was the only fraction that significantly arrested cells in the G0/G1 phase of the cell cycle [65].

25.4.2 Cardiovascular protective

Ascorbic acid presented in mangoes provides protection against oxidative stress-related diseases such as CVD and in respiratory infection [71]. Among different food varieties, β-carotene in mangoes provides the highest vitamin A activity [28], which contributes to the protection against free radical-induced diseases. A study found that β-carotene inhibits the progression of atherosclerosis and cancer [72]. Flavonoids from *M. indica* successfully reduced lipid levels in serum and tissues of induced hyperlipidemia models. In addition, the plasma lecithin cholesterol acyltransferase (LCAT) level was elevated [73]. The activities of free radical-scavenging enzymes were significantly elevated and products of lipid oxidation significantly decreased in flavonoid-treated hypercholesterolemic rats [74].

It has been reported that *M. indica* L. (Vimang) increased extracellular superoxide dismutase activity and serum total antioxidant status. It also decreased serum thiobarbituric acid-reactive substances (TBARS). Components of the mango extract could be utilized by blood and endothelial cells [75].

Mangiferin extracted from *M. indica* decreased plasma total cholesterol triacylglycerols (TAG) and low-density lipoprotein (LDL) cholesterol and elevated high-density lipoprotein (HDL) cholesterol [76]. It has been found that mangiferin and epigallocatechin gallate (EGCG) isolated from *M. indica* protect erythrocytes from reactive oxygen species (ROS) production, contributing to the decrease in TBARS production [77]. The protective effect might be related to the strong radical-scavenging ability of mangiferin and EGCG. In addition, *M. indica* elevated bile acids in fecal excretion that reduced total cholesterol and LDL due to its mangiferin [78].

25.5 Conclusions

Dried mangoes have valuable nutritional, antioxidant, and health-promoting properties comparable to their fresh counterparts. The dried products prepared from mangoes are of interest to the food industry as a food supplement, including prospective applications as functional food ingredients in cookies, crackers, confectioneries, and low-calorie products. Besides fresh mangoes, the dried forms could also serve as a rich source of nutrients and antioxidants. These health-promoting compounds present in dried mangoes could help managing and reducing the risk factors of degenerative diseases.

References

1. FAO (2011) *Food and Agricultural Commodities Production*. Published online at: http://faostat.fao.org./site/339/default.aspx, last accessed November 27, 2011.
2. Gross, P. (2010) *Superfruits*, 1st edn. CRC Press, McGraw-Hill Comp., New York, NY.
3. Sahni, C.K. & Khurdiya, D.S. (1989) Effect of ripening and storage on the quality of mango nectar. *Indian Food Packer*, **43**, 5–11.

4. Larrauri, J.A., Rupbrez, P., Bravo, L. & Saura-Calixto, F. (1996) High dietary fibre peels: associated powders from orange and lime polyphenols and antioxidant capacity. *Food Research International*, **29**, 757–762.
5. Vergara-Valencia, N., Granados-Pérez, E., Agama-Acevedo, E., Tovar, J., Ruales, J. & Bello-Pérez, L.A. (2007) Fibre concentrate from mango fruit: characterization, associated antioxidant capacity and application as a bakery product ingredient. *Lebensmittel-Wissenschaft und-Technologie*, **40**, 722–729.
6. Ajila, C.M., Leelavathi, K. & Rao, U.J.S.P. (2008) Improvement of dietary fibre content and antioxidant properties in soft dough biscuits with the incorporation of mango peel powder. *Journal of Cereal Science*, **48**, 319–326.
7. Al-sheraji, S.A, Ismail, A, Manap, M.Y., Mustafa, S., Yusof, R.M. & Hassan, F.A. (2011) Functional properties and characterization of dietary fiber from *Mangifera pajang* Kort. fruit pulp. *Journal of Agricultural and Food Chemistry*, **59**, 3980–3985.
8. Hassan, F.A., Ismail, A., Abdul Hamid, A., Azlan, A. & Al-sheraji, S.H. (2011) Characterisation of fibre-rich powder and antioxidant capacity of *Mangifera pajang* K. fruit peels. *Food Chemistry*, **126**, 283–288.
9. Ajila, C.M., Aalami, M., Leelavathi, K. & Prasada Rao, U.J.S. (2010) Mango peel powder: a potential source of antioxidant and dietary fiber in macaroni preparations. *Innovative Food Science and Emerging Technologies*, **11**, 219–224.
10. Abu Bakar, M.F., Mohamed, M., Rahmat, A. & Fry, J. (2009) Phytochemicals and antioxidant activity of different parts of bambangan (*Mangifera pajang*) and tarap (*Artocarpus odoratissimus*). *Food Chemistry*, **113**, 479–483.
11. Halvorsen, B.L., Carlsen, M.H., Phillips, K.M., Bøhn, S.K., Holte, K., Jacobs, D.R. & Blomhoff, R. (2006) Content of redox-active compounds (i.e. antioxidants) in foods consumed in the United States. *American Journal of Clinical Nutrition*, **84**, 95–135.
12. Vasco, C., Ruales, J. & Kamal-Eldin, A. (2008) Total phenolic compounds and antioxidant capacities of major fruits from Ecuador. *Food Chemistry*, **111**, 816–823.
13. Prasad, N.N., Siddalingaswamy, M., Parameswariah, P.M., Radhakrishna, K., Rao, R.V., Viswanathan, K.R. & Santhanam, K. (2000) Proximate and mineral composition of some processed traditional and popular Indian dishes. *Food Chemistry*, **68**, 87–94.
14. Hymavathi, T.V. & Khader, V. (2005) Carotene, ascorbic acid and sugar content of vacuum dehydrated ripe mango powders stored in flexible packaging material. *Journal of Food Composition and Analysis*, **18**, 181–192.
15. Tee, E.S., Mohd Ismail, N., Mohd Nasir, A. & Khatijah, I. (1997) *Nutrient Composition of Malaysian Foods*. Institute Medical for Research, KL, Malaysia.
16. Ibrahim, M., Prasad, K.N., Ismail, A., Azlan, A. & Abd Hamid, A. (2010) Physiochemical composition and antioxidant activities of underutilized *Mangifera pajang* fruit. *African Journal of Biotechnology*, **9**, 4392–4397.
17. Mamiro, P., Fweja, L, Chove, B., Kinabo, J., George, V. & Mtebe, K. (2007) Physical and chemical characteristics of off vine ripened mangoes (*Mangifera indica* L.) fruit (Dodo). *African Journal of Biotechnology*, **7**, 65–72.
18. Ramulu, P. & Rao, P.U. (2003) Total, insoluble and soluble dietary fibre contents of Indian fruits. *Journal of Food Composition and Analysis*, **16**, 677–685.
19. Paul, D.K. & Shaha, R.K. (2004) Nutrients, vitamins and minerals in common citrus fruits in the northern region of Bangladesh. *Pakistan Journal of Biological Sciences*, **7**, 238–242.
20. Bravo, L. & Saura-Calixto, F. (1998) Characterization of dietary fiber and the *in vitro* indigestible fraction of grape pomace. *American Journal of Enology and Viticulture*, **49**, 135–141.
21. Chau, C. & Huang, Y. (2003) Comparison of the chemical composition and physicochemical properties of different fibres prepared from de peel of *Citrus sinensis* L. cv. Liucheng. *Journal of Agricultural and Food Chemistry*, **51**, 2615–2618.
22. Chang, S., Lee, M., Lin, C. & Chen, M. (1998) Dietary fiber content and composition of fruits in Taiwan. *Asian Pacific Journal of Clinical Nutrition*, **7**, 206–210.
23. Ajila, C.M., Bhat, S.G. & Prasada Rao, U.J.S. (2007) Valuable components of raw and ripe peels from two Indian mango varieties. *Food Chemistry*, **102**, 1006–1011.
24. Peter, M.F., Leonard, C., Bernard, K., Joyce, G. & Victor Kaswija, M. (2007) Physical and chemical characteristics of off vine ripened mango (*Mangifera indica* L.) fruit (Dodo). *African Journal of Biotechnology*, **6**, 2477–2483.

25. Sirisakulwat, S., Nagel, A., Sruamsiri, P., Carle, R. & Neidhart, S. (2008) Yield and quality of pectins extractable from the peels of Thai mango cultivars depending on fruit ripeness. *Journal of Agricultural and Food Chemistry*, **56**, 10727–10738.
26. Mahattanatawee, K., Manthey, J.A., Luzio, G., Talcott, S.T., Goodner, K. & Baldwin, E.A. (2006) Total antioxidant activity and fiber content of select Florida-grown tropical fruits. *Journal of Agriculture and Food Chemistry*, **54**, 7355–7363.
27. USDA (2011) *National Nutrient Database for Standard Reference, Release* 24. Published online at: http://www.nal.usda.gov/fnic/foodcomp/search/, last accessed December 1, 2011.
28. Ribeiro, S.M.R., De Queiroz, J.H., Lopes, M.E., De Queiroz, C.F.M. & Santana, H.M.P. (2007) Antioxidant in mango (*Mangifera indica* L.) pulp. *Plant Foods for Human Nutrition*, **62**, 13–17.
29. FAO/WHO/UNU (1981) *Energy and Protein Requirements*. Technical Reports Series 724, Geneva, Switzerland.
30. Santos, P.H.S. & Silva, M.A. (2008) Retention of vitamin C in drying processes of fruits and vegetables—a review. *Drying Technology*, **26**, 1421–1437.
31. Marques, L.G., Silveria, A.M. & Freire, J.T. (2006) Freeze-drying characteristics of tropical fruits. *Drying Technology*, **24**, 457–463.
32. Hassan, F.A., Ismail, A., Abdulhamid, A. & Azlan, A. (2011) Identification and quantification of phenolic compounds in bambangan (*Mangifera pajang* Kort.) peels and their free radical scavenging activity. *Journal of Agricultural and Food Chemistry*, **59**, 9102–9111.
33. Gorinstein, S., Zemser, M., Haruenkit, R., Chuthakorn, R., Grauer, F., Martin-Belloso, O. & Trakhtenberg, S. (1999) Comparative content of total polyphenols and dietary fiber in tropical fruits and persimmon. *Journal of Nutritional Biochemistry*, **10**, 367–371.
34. Tyug, T.S., Johar, M.H. & Ismail, A. (2010) Antioxidant properties of fresh, powder, and fiber products of mango (*Mangifera foetida*) fruit. *International Journal of Food Properties*, **13**, 682–691.
35. Mohamad Shofian, N., Abdul Hamid, A., Osman, A., Saari, N., Anwar, F., Pak Dek, M.S. & Hairuddin, M.R. (2011) Effect of freeze-drying on the antioxidant compounds and antioxidant activity of selected tropical fruits. *International Journal of Molecular Sciences*, **12**, 4678–4692.
36. Soong, Y.Y. & Barlow, P.J. (2004) Antioxidant activity and phenolic content of selected fruit seeds. *Food Chemistry*, **83**, 411–417.
37. Ubando-Rivera, J., Navarro-Ocaña, A. & Valdivia-López, M.A. (2005) Mexican lime peel: comparative study on contents of dietary fibre and associated antioxidant activity. *Food Chemistry*, **89**, 57–61.
38. Jiménez-Escrig, A., Rincón, M., Pulido, R. & Saura-Calixto, F. (2001) Guava fruit (*Psidium guajava* L.) as a new source of antioxidant dietary fiber. *Journal of Agricultural and Food Chemistry*, **49**, 5489–5493.
39. Stahl, W., Van Den Berg, H., Arthur, J., Bast, A., Dainty, J., Faulks, R.M., Gärtner, C., Haenen, G., Hollman, P., Holst, B., Kelly, F.J., Polidori, M.C., Rice-Evans, C., Southon, S., van Vliet, T., Viña-Ribes, J., Williamson, G. & Astley, S.B. (2002) Bioavailability and metabolism. *Molecular Aspects of Medicine*, **23**, 39–100.
40. Scalbert, A. & Williamson, G. (2000) Dietary intake and bioavailability of polyphenols. *Journal of Nutrition*, **130**, 2073S–2085S.
41. Rastraelli, L., Selles, A.J.N., Castro, H.T.V., Aguero-Aguero, J., Gonzalez-Gonzalez, J., Naddeo, F. & Simone, F.D. (2002) Isolation and quantitative analysis of phenolic antioxidants, free sugars, and polyols from mango (*Mangifera indica* L.) stem bark aqueous decoction used in Cuba as a nutritional supplement. *Journal of Agricultural and Food Chemistry*, **50**, 762–766.
42. Schieber, A., Ullrich, W. & Carle, R. (2000) Characterization of polyphenols in mango puree concentrate by HPLC with diode array and mass spectrometric detection. *Innovative Food Science and Emerging Technologies*, **1**, 161–166.
43. Lakshminarayana, S., Subhadra, N.V. & Subramanyam, H. (1970) Some aspects of developmental physiology of mango fruit. *Journal of Horticultural Science*, **45**, 133–142.
44. El-sissi, H., Ishak, S., Wahid, M.S.A.E. & El-Ansari, MA. (1971) The gallotannins of *Rhus coriaria* and *Mangifera indica*. *Planta Medica*, **19**, 342–351.
45. Berardini, N., Carle, R. & Schieber, A. (2004) Characterization of gallotannins and benzophenone derivatives from mango (*Mangifera indica* L. cv. "Tommy Atkins") peels, pulp and kernels by high-performance liquid chromatography/electrospray ionization mass spectrometry. *Rapid Communication in Mass Spectrometry*, **18**, 2208–2216.
46. Masibo, M. & He, Q. (2008) Major mango polyphenols and their potential significance to human health. *Comprehensive Reviews in Food Science and Food Safety*, **7**, 309–319.

47. Kim, Y., Brecht, J.K. & Talcott, S.T. (2007) Antioxidant phytochemical and fruit quality changes in mango (*Mangifera indica* L.) following hot water immersion and controlled atmosphere storage. *Food Chemistry*, **105**, 1327–1334.
48. Berardini, N., Knödler, M., Schieber, A. & Carle, R. (2005) Utilization of mango peels as a source of pectin and polyphenolics. *Innovative Food Science and Emerging Technologies*, **6**, 443–453.
49. Ribeiro, S.M.R., Barbosa, L.C.A., Queiroz, J.H., Knodler, M. & Schieber, A. (2008) Phenolic compounds and antioxidant capacity of Brazilian mango (*Mangifera indica* L.) varieties. *Food Chemistry*, **110**, 620–626.
50. Scartezzini, P. & Speroni, E. (2000) Review on some plants of Indian traditional medicine with antioxidant activity. *Journal of Ethnopharmacology*, **71**, 23–43.
51. Konczak, I. & Zhang, W. (2004) Anthocyanins more than nature's colours. *Journal of Biomedical Biotechnology*, **5**, 239–240.
52. Berardini, N., Schieber, A., Klaiber, I., Beifuss, U., Carle, R. & Conrad, J. (2005) 7-*O*-Methylcyanidi 3-*O*-β-D-galactopyranoside, a novel anthocyanin from mango (*Mangifera indica* L.) cv. 'Tommy Atkins' peels. *Chemical Science*, **60**, 801–804.
53. Olivera, R.M., Monterrey, N.L., Quiroga, J.T. & Garcia, G. (2008) *Absorption and Bioavailability of Carotenoids, Formulation and Applications*. United States Patent. P.N. US, **7**,435, 846, B2.
54. Gouado, I., Schweigert, F.J., Ejoh, R.A., Tchouanguep, M.F. & Camp, J.V. (2007) Systemic levels of carotenoids from mangoes and papaya consumed in three forms (juice, fresh and dry slice). *European Journal of Clinical Nutrition*, **61**, 1180–1188.
55. Davis, B.H. (1976) Carotenoids. In: *Chemistry and Biochemistry of Plant Pigments* (ed. T.W., Goodwin). London Press, London, UK, pp. 38–166.
56. Khoo, H.-E., Prasad, K.N., Ismail, A. & Mohd-Esa, N. (2010) Carotenoids from *Mangifera pajang* and their antioxidant capacity. *Molecules*, **15**, 6699–6712.
57. Godoy, T.H. & Rodriguez-Amaya, D.B. (1989) Carotenoid composition of commercial mangoes from Brazil. *Lebensmittel-Wissenschaft und-Technologie*, **22**, 100–103.
58. Mercadante, A.Z. & Rodriguez-Amaya, D.B. (1998) Effects of ripening, cultivar differences, and processing on the carotenoid composition of mango. *Journal of Agricultural and Food Chemistry*, **46**, 128–130.
59. Burns, J., Fraser, P.D. & Bramley, P.M. (2003) Identification and quantification of carotenoids, tocopherols and chlorophylls in commonly consumed fruits and vegetables. *Phytochemistry*, **62**, 939–947.
60. Carvalho, C.R.L., Rosseto, C.J., Mantovani, D.M.B., Morgano, M.A., Castro, J.V. & Botoletto, N. (2004) Evaluation of mango cultivars selected by "Instituto Agronomico de Campinas" compared to others of commercial importance. *Revista Brasileira de Fruticultura*, **26**, 264–271.
61. Institute of Medicine (2000) *Dietary Reference Intakes for Vitamin C, Vitamin E, Selenium, and Carotenoids*. National Academy Press, Washington, DC.
62. Kadzere, I., Hove, L., Gatsi, T., Masarirambi, M.T., Tapfumaneyi, L., Maforimbo, E., Magumise, I., Sadi, J. & Makaya, P.R. (2001) *Post-Harvest Fruit Handling Practices and Traditional Processing of Indigenous Fruits in Zimbabwe Annual Report*. ICRAF, Harare, Zimbabwe.
63. Saka, J.D.K., Rapp, I., Ndolo, V., Mhango, J. & Akinnifesi, F.K. (2007) A comparative study of the physicochemical and organoleptic characteristics of *Uapaca kirkiana*, *Strychnos cocculoides*, *Adansonia digitata* and *Mangifera indica* products. *International Journal of Food Science & Technology*, **42**, 836–841.
64. Johnson, I.T. (2004) New approaches to the role of diet in the prevention of cancers of the alimentary tract. *Mutation Research*, **551**, 9–28.
65. Percival, S.S., Talcott, S.T., Chin, S.T., Mallak, A.C., Lounds-Singleton, A. & Pettit-Moore, J. (2006) Neoplastic transformation of BALB/3T3 cells and cell cycle of HL-60 cells are inhibited by mango (*Mangifera indica* L.) juice and mango juice extracts. *Journal of Nutrition*, **136**, 1300–1304.
66. Arts, I.C. & Hollman, P.C. (2005) Polyphenols and disease risk in epidemiologic studies. *American Journal of Clinical Nutrition*, **78**, 559S–569S.
67. He, Z. & Xia, W. (2007) Analysis of phenolic compounds in Chinese olive (*Canarium album* L.) fruit by RPHPLC-DAD-ESI-MS. *Food Chemistry*, **105**, 1307–1311.
68. Wan, Y., Vinson, J.A., Etherton, T.D., Proch, J., Lazarus, S.A. & Kris-Etherton, P.M. (2001) Effects of cocoa powder and dark chocolate on LDL oxidative susceptibility and prostaglandin concentrations in humans. *American Journal of Clinical Nutrition*, **74**, 596–602.

69. Rosamond, W.D. (2002) Dietary fiber and prevention of cardiovascular disease. *Journal of the American College of Cardiology*, **39**, 57–59.
70. Botting, K.J., Young, M.M., Pearson, A.E., Harris, P.J. & Ferguson, L.R. (1999) Antimutagens in food plants eaten by Polynesians: micronutrients, phytochemicals and protection against bacterial mutagenicity of the heterocyclic amine 2-amino-3- methylimidazo[4,5-f] quinoline. *Food and Chemical Toxicology*, **37**, 95–103.
71. Khaw, K. & Woodhouse, P. (1995) Interrelation of vitamin C, infection, heamostatic factors and cardiovascular disease. *British Medical Journal*, **310**, 1559–1562.
72. Krinsky, N.I. & Johnson, E.J. (2005) Carotenoid actions and their relation to health and disease. *Molecular Aspects of Medicine*, **26**, 459–516.
73. Anila, L. & Vijayalakshmi, N.R. (2002) Flavonoids from *Emblica officinalis* and *Mangifera indica* effectiveness for dyslipidemia. *Journal of Ethnopharmacology*, **79**, 81–87.
74. Anila, L. & Vijayalakshmi, N.R. (2003) Antioxidant action of flavonoids from *Mangifera indica* and *Emblica officinalis* in hypercholesterolemic rats. *Food Chemistry*, **83**, 569–574.
75. Pardo-Andreu, G.L., Philip, S.J., Riaño, A., Sánchez, C., Carmen Viada, C., Núñez-Sellés, A.J. & Delgado, R. (2006) *Mangifera indica* L. (Vimang) protection against serum oxidative stress in elderly humans. *Archives of Medical Research*, **37**, 158–164.
76. Muruganandan, S., Srinivasan, K., Gupta, S., Gupta, P.K. & Lal, J. (2005) Effect of mangiferin on hyperglycemia and atherogenicity in streptozotocin diabetic rats. *Journal of Ethnopharmacology*, **97**, 497–501.
77. Rodríguez, J., Di Pierro, D., Gioia, M., Monaco, S., Delgado, R., Coletta, M. & Marini, S. (2006) Effects of a natural extract from *Mangifera indica* L. and its active compound, mangiferin, on energy state and lipid peroxidation of red blood cells. *Biochimica et Biophysica Acta*, **1760**, 1333–1342.
78. Akila, M. & Devaraj, H. (2008) Synergistic effect of tincture of *Crataegus* and *Mangifera indica* L. extract on hyperlipidemic and antioxidant status in atherogenic rats. *Vascular Pharmacology*, **49**, 173–177.

26 Phytochemicals and health applications of dried passion and pineapple fruits

Jian Sun, Li Li, Xiangrong You, Changbao Li, Zhichun Li, and Fen Liao

26.1 Introduction

Passion flower, a vine species in the *Passifloraceae* family, is native to Paraguay, Brazil, and northern Argentina, which is now cultivated in most tropical and subtropical areas of the world as well as in a few temperate zones [1]. Eight countries (United States, Australia, Papua New Guinea, Fiji, South Africa, Kenya, Colombia, and Sri Lanka) account for 80–90% of the world production of passion fruits [2]. Passion fruit is round or ovaloid and apricot-sized (4–6 cm in diameter) at maturity, with a soft to firm, juicy, and aromatic interior filled with numerous seeds [1, 3, 4]. There are two recognized varieties, purple (*Passiflora edulis* Sims.) and yellow (*Passiflora edulis* Sims. f. *flavicarpa* Degener). Passion fruits are usually used for the processing of pulp and juice, and sometimes they are produced into dried fruit powder and dried fruit bar [5–12].

Pineapples (*Ananas comosus* (L.) Merr.) are another tropical and subtropical plant widely cultivated in many places including Hawaii, Philippines, Thailand, Malaysia, Mexico, South Africa, and China [13]. This plant is indigenous to South America and is by far the most economically important one in the *Bromeliaceae* family. Its juicy and aromatic fruit is oval to cylindrical-shaped, developing from many small fruits fused together. Pineapples are well known for their nutritive and health-promoting properties [14]. The fruit can be consumed fresh, but most of the production is used in processing. Canned slices, juices, concentrates, salads, and jams are common commercial products of pineapples in many countries [15, 16]. Some dried fruit products such as dried fruit pieces (chunks or dices) are also widely available in the market [17, 18].

This chapter reviews compositional and nutritional characteristics, phytochemicals, health benefits, commercial products, and industrial applications of the dried passion and pineapple fruits.

Dried Fruits: Phytochemicals and Health Effects, First Edition. Edited by Cesarettin Alasalvar and Fereidoon Shahidi.

26.2 Compositional and nutritional characteristics of dried passion and pineapple fruits

Although dried passion fruits are not so frequently found in the market as are dried pineapples, they are good sources of nutrients. The basic nutritional compositions have been detected by several research groups (Table 26.1). Elmadfa *et al.* [19] determined the contents of proteins, fats, total sugars, fiber, vitamin C, some minerals (Fe, K, and Na), and phosphates in lyophilized passion fruits from Spain. Romero-Rodriguez *et al.* [4] further comprehensively analyzed nutritional compositions of lyophilized passion fruits from northwest of Spain. They reported that this lyophilized fruit contained proteins, fats, sugars (sucrose, glucose, and fructose), fiber, organic acids, vitamin C, and some minerals (Ca, Cu, Fe, K, Mg, Mn, Na, and Zn), and phosphates. From their investigation, passion fruits possessed relatively high content of dietary fiber with known health benefits. In the seeds of passion fruits, dietary fiber was especially high [20]. Twenty minerals in passion fruits (obtained from northeast China) after heated drying were also determined by Wang *et al.* [21] using inductively coupled plasma-emission spectroscopy (ICP-ES). Amongst them, Na (100 μg/g), K (10,810 μg/g), Ca (1410 μg/g), Mg (3160 μg/g), Zn (37.65 μg/g), Fe (93.96 μg/g), and Ge (82.12 μg/g) were abundant, while five common toxic elements, Cd, Hg, As, Cr, and Pb, were not detected.

Table 26.1 The nutritional composition of lyophilized passion fruits

		References	
Composition	**Unit**	**[19]**	**[4][a]**
Proteins	%	2.80	3.0
Fats	%	0.40	0.12
Total sugars	%	13.4	–
Glucose	%	–	2.1
Fructose	%	–	2.1
Sucrose	%	–	2.9
Fiber	%	1.50	12.8
Ash	%	–	0.5
Organic acids			
Citric acid	%	–	3.0
Malic acid	%	–	0.3
Vitamin C	mg/100 g	20.0	23.3
Phosphates	mg/100 g	54.0	63.8
Minerals			
Na	mg/100 g	28.0	8.0
K	mg/100 g	350	208
Ca	mg/100 g	–	6.8
Mg	mg/100 g	–	27.9
Fe	mg/100 g	1.1	0.6
Cu	mg/100 g	–	0.2
Zn	mg/100 g	–	0.5
Mn	mg/100 g	–	0.2

[a]Results are mean values of replicated determinations of six samples.

In the processing of pineapple fruits, osmotic dehydration (OD) is a common method for producing dried fruit products, because other drying techniques such as sun drying, heated air drying, or mechanical drying often influence sensory attributes (e.g., color, flavor, and texture) and even compromise the nutritional value of these products [22–24]. OD is widely used for the partial removal of water from plant tissues by immersion in a hypertonic (osmotic) solution. The driving force for the diffusion of water from the tissue into the solution is provided by the higher osmotic pressure of the hypertonic solution. The diffusion of water is accompanied by simultaneous counter diffusion of solute from the osmotic solution into the tissue [22, 25–27]. A most commonly used hypertonic solution is sucrose syrup during OD. Zheng [28] reported that fresh pineapple fruits contained approximately 13% sugars, 0.6% proteins, 0.6% organic acids, 0.3% fiber, 0.02% Ca, 0.9% Fe, and 0.01% P. The contents of vitamin A (60 IU), vitamin B_2 (120 mg/100 g), vitamin C (63 mg/100 g), Ca, Fe, and P are relatively high in passion fruits [28]. Additionally, pineapple fruits are also rich in a commercially important enzyme, bromelain, which possesses potent anticancer effects [29, 30]. After drying by OD, some nutritional substances can be lost from pineapple fruit. Peiró-Mena *et al.* [31] studied some compositional changes after 15 successive OD cycles and found that citric acid, majority of minerals (Ca, Mg, K, P, Na), and galacturonic acid (AGU) (related to pectin) in pineapple fruits decreased after each cycle (Table 26.2). Pineapple fruit bromelain is a cysteine endopeptidase which is unstable and prone to loss of activity after drying treatment [29].

26.3 Phytochemicals in dried passion and pineapple fruits

There is an absence of report on phytochemicals in dried passion and pineapple fruits. However, some important chemical constituents in the two fresh fruits, such as carbohydrates, carotenoids, phenolic compounds, and other phytochemicals, have already been isolated, identified, and quantified. These chemical constituents may exist in the dried form as well.

Carbohydrates constitute a major group of nutrients in passion fruits. Pruthi and Lal [32] reported that three sugars (glucose, fructose, and sucrose) together composed approximately 86.3% of the total carbohydrate makeup, the rest being starch. Cillie and Joubert [33] established that the passion fruit starch, belonging to the waxy type, was rather unusual in that it contained only 1–2% amylose and as such was almost a pure amylopectin in which the side chains had an average length of 17 glucose residues. Pectins mostly exist in passion fruit rinds and their chemical constituents mainly include D-galacturonic acid, L-arabinose, and D-galactose, among others [34]. Passion fruits have an orange-yellow color of juice which is due mostly to a complex mixture of carotenoid pigments [35]. Pruthi and Lal [36] reported variation range of the major groups of carotenoid pigments in purple passion fruits: free xanthophylls at 10.3–21.5%, xanthophyll esters at 11.1–34.6%, and epiphasic nonsaponifiables (mostly carotenes) at 45.7–76.3% of the total carotenoid pigments. No material difference was noticed in the carotenoid makeup of yellow and purple passion fruits except that total carotenoids and xanthophyll esters are usually higher in the former [37]. Mercadante *et al.* [38] further identified 13 carotenoids (phytoene, phytofluene, ζ-carotene, neurosporene, β-carotene, lycopene, prolycopene, monoepoxy-β-carotene, β-cryptoxanthin, β-citraurin, antheraxanthin, violaxanthin, and neoxanthin) from yellow passion fruits. Their chemical

Table 26.2 Organic acids and mineral content of fresh pineapples (FP) and osmodehydrated pineapples (ODP)

Sample	Citric acid	Galacturonic acid	Ca	Mg	K	P	Na
$FP_1{}^a$	1016 ± 144	746 ± 5	26 ± 2	17 ± 1	167 ± 5	8.1 ± 0.4	3.4 ± 0.3
$ODP_1{}^a$	698 ± 28	542 ± 35	5 ± 1	11 ± 1	111 ± 14	5.4 ± 0.3	1.6 ± 0.2
$ODP_1{}^b$	790 ± 32	614 ± 40	6 ± 1	13 ± 1	126 ± 16	6.1 ± 0.4	1.8 ± 0.2
$Loss_1{}^c$	30 ± 13	27 ± 5	81 ± 6	32 ± 10	33 ± 8	33 ± 5	51 ± 8
$FP_3{}^a$	994 ± 50	857 ± 53	28 ± 2	19.5 ± 0.4	168 ± 7	6.8 ± 0.1	3.2 ± 0.5
$ODP_3{}^a$	674 ± 94	618 ± 66	18 ± 4	14.8 ± 0.2	71 ± 7	5.1 ± 0.3	2.4 ± 0.2
$ODP_3{}^b$	741 ± 103	699 ± 75	20 ± 5	16.3 ± 0.2	78 ± 8	5.6 ± 0.4	2.7 ± 0.2
$Loss_3{}^c$	32 ± 9	26 ± 3	36 ± 15	24 ± 1	58 ± 3	24 ± 5	23 ± 16
$FP_5{}^a$	1342 ± 59	751 ± 19	35 ± 3	20 ± 1	198 ± 5	9.8 ± 0.3	3.7 ± 0.5
$ODP_5{}^a$	884 ± 38	655 ± 45	6 ± 1	11.7 ± 0.3	93 ± 13	6.2 ± 0.2	1.9 ± 0.1
$ODP_5{}^b$	988 ± 42	741 ± 51	7 ± 1	13.4 ± 0.4	107 ± 15	7.1 ± 0.3	2.2 ± 0.1
$Loss_5{}^c$	34 ± 3	14 ± 7	82 ± 3	42 ± 3	53 ± 7	37 ± 3	48 ± 8
$FP_7{}^a$	1043 ± 148	804 ± 39	22.4 ± 0.2	18.5 ± 0.4	184 ± 11	6.1 ± 0.2	2.8 ± 0.4
$ODP_7{}^a$	684 ± 108	679 ± 49	9 ± 2	15 ± 1	142 ± 1	5.4 ± 0.4	1.8 ± 0.2
$ODP_7{}^b$	770 ± 118	768 ± 56	10 ± 2	17 ± 1	156 ± 2	5.9 ± 0.4	2.0 ± 0.2
$Loss_7{}^c$	34 ± 8	13 ± 10	61 ± 8	18 ± 3	23 ± 4	12 ± 9	36 ± 5
$FP_9{}^a$	942 ± 33	888 ± 62	29 ± 3	18 ± 1	195 ± 2	9.9 ± 0.3	3.0 ± 0.4
$ODP_9{}^a$	638 ± 39	662 ± 50	12 ± 1	12.8 ± 0.3	98 ± 9	5.9 ± 0.4	1.9 ± 0.1
$ODP_9{}^b$	729 ± 45	749 ± 57	13 ± 1	14.5 ± 0.3	112 ± 10	6.7 ± 0.4	2.2 ± 0.2
$Loss_9{}^c$	32 ± 2	26 ± 3	60 ± 9	29 ± 3	47 ± 3	40 ± 3	35 ± 13
$FP_{12}{}^a$	970 ± 27	932 ± 35	23 ± 1	17.3 ± 0.4	161 ± 3	7 ± 1	3.2 ± 0.2
$ODP_{12}{}^a$	660 ± 64	722 ± 37	16 ± 1	13 ± 2	92 ± 6	5.7 ± 0.3	1.9 ± 0.2
$ODP_{12}{}^b$	750 ± 73	817 ± 42	18 ± 1	14 ± 2	105 ± 7	6.4 ± 0.4	2.1 ± 0.2
$Loss_{12}{}^c$	32 ± 9	23 ± 1	28 ± 4	28 ± 10	43 ± 3	15 ± 6	41 ± 5
$FP_{15}{}^a$	994 ± 133	630 ± 50	25 ± 1	17 ± 1	151 ± 4	7.4 ± 0.3	3 ± 1
$ODP_{15}{}^a$	559 ± 44	571 ± 65	13 ± 3	15 ± 1	61 ± 8	4.3 ± 0.2	1.4 ± 0.1
$ODP_{15}{}^b$	634 ± 50	646 ± 73	15 ± 3	17 ± 1	69 ± 9	4.9 ± 0.3	1.5 ± 0.1
$Loss_{15}{}^c$	44 ± 4	10 ± 4	46 ± 12	11 ± 3	60 ± 4	42 ± 6	57 ± 11
Mean FP^a	1043 ± 158	780 ± 112	27 ± 4	18 ± 1	174 ± 17	8 ± 1	3.2 ± 0.2
Mean ODP^a	685 ± 112	615 ± 83	11 ± 5	13 ± 2	96 ± 25	5 ± 1	1.9 ± 0.3
Mean ODP^b	772 ± 112	696 ± 93	12 ± 5	16 ± 2	108 ± 27	6 ± 1	2.1 ± 0.3
Mean $loss^c$	34 ± 8	21 ± 9	59 ± 21	26 ± 10	44 ± 13	30 ± 12	41 ± 13

Source: Adapted with permission from Peiró-Mena *et al.* [31].
Data are expressed as means ± standard deviation ($n = 3$).
[a] mg component/100 g FP.
[b] mg component/100 g ODP.
[c] mg component loss/100 mg initial component.

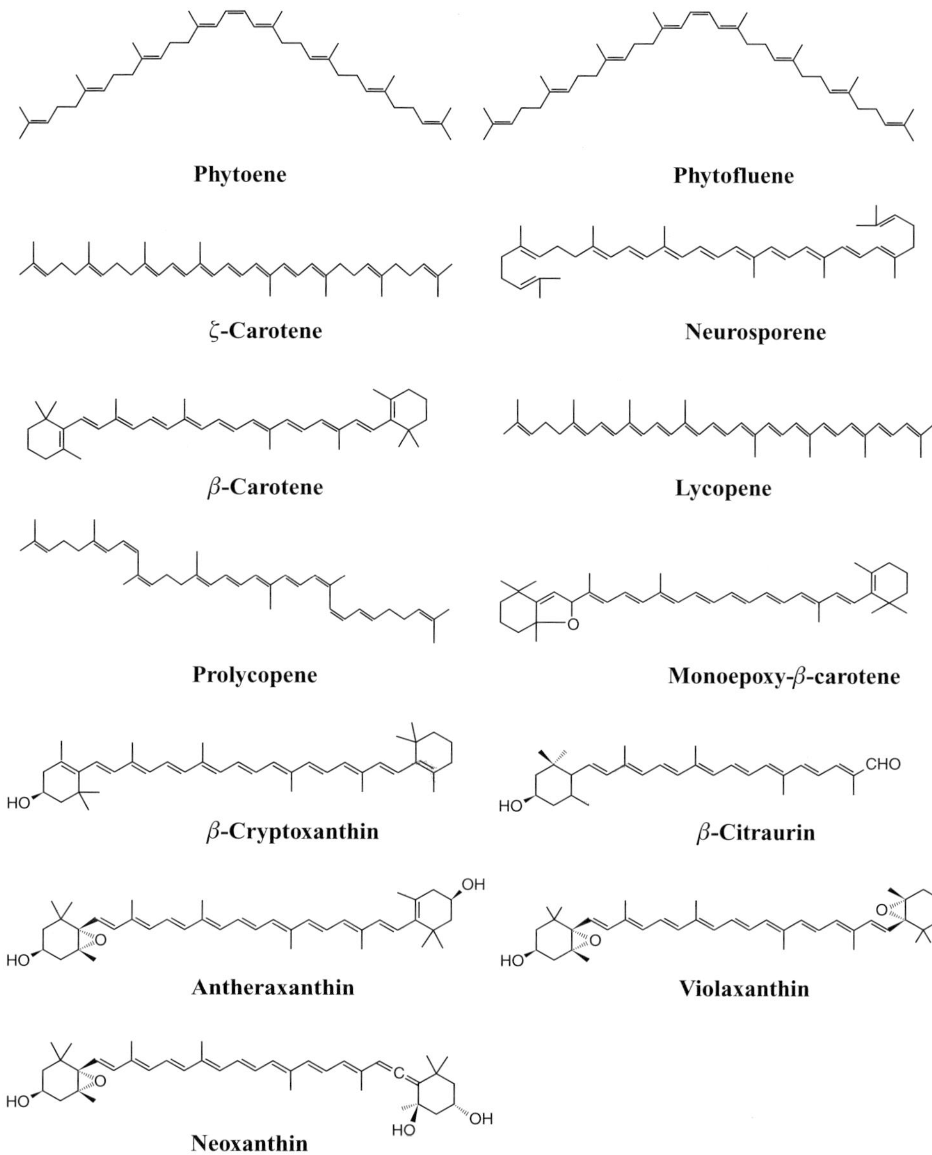

Figure 26.1 Chemical structures of carotenoids from yellow passion fruits.

structures are shown in Figure 26.1. A previous report [39] has described the presence of flavonoids as one of the major constituents in passion fruits, mainly *C*-glycosylflavones. The flavonoids schaftoside, isoschaftoside, isoorientin, orientin, isovitexin, luteolin-6-*C*-chinovoside, and luteolin-6-*C*-fucoside (Figure 26.2) have been identified from passion fruits. Polyphenols (Figure 26.3) mainly exist in passion fruit rinds (e.g., anthocyanins in purple passion fruit rinds: cyanidin-3-glucoside, quercetin-3-glucoside, pelargonidin-3,5-diglucoside, cyanidin-3–6″-malonyl glucoside, cyanidin-3-*O*-β-glucopyranoside, and

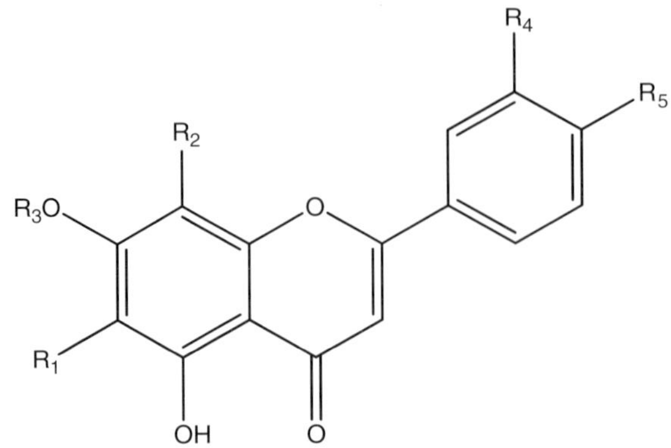

	R_1	R_2	R_3	R_4	R_5
Schaftoside	glc	ara	H	H	OH
Isoschaftoside	ara	glc	H	H	OH
Isoorientin	glc	H	H	OH	OH
Orientin	glc	H	H	OH	OH
Isovitexin	glc	H	H	H	OH
Luteolin-6-*C*-chinovoside	chinovose	H	H	OH	OH
Luteolin-6-*C*-fucoside	fucose	H	H	OH	OH

glc = β-D-glucopyranosyl; ara = α-L-arabinopyranosyl

Figure 26.2 Flavonoids identified from passion fruits

cyanidin-3-*O*-β-galactopyranoside) and seeds (e.g., piceatannol and scirpusin B) [40, 41]. Fruit pulps only contain a few polyphenols, of which tannic acid has been found to be the major constituent [34]. In addition to the above-mentioned chemical constituents, there are other important phytochemicals in passion fruits. Two ionones I and II (Figure 26.4) have been isolated from passion fruits [41]. Additionally, Winterhalter [42] has identified a wide range of terpenoids (Figure 26.5) in purple passion fruits, including linalool, 4-hydroxy-β-ionol, 4-oxo-β-ionol, 4-hydroxy-7,8-dihydro-β-ionol, 4-oxo-7,8-dihydro-β-ionol, 3-oxo-α-ionol, isomeric 3-oxoretro-α-ionols, 3-oxo-7,8-dihydro-α-ionol, 3-hydroxy-1,1,6-trimethyl-1,2,3,4-tetrahydronaphthalene, vomifoliol, dehydrovomifoliol, and some monoterpenoids, among others.

Pineapples contain unusual fiber components with specific physiological properties. Vidal-Valverde *et al.* [43] reported contents of in dietary fiber, cellulose, hemicellulose, lignin, and pectic substances of pineapples, describing that they are very rich in hemicelluloses (41.8%). Polysaccharides isolated from pineapple flesh were chemically analyzed by Smith and Harris [44]. Glucuronoarabinoxylans were the major noncellulosic polysaccharides in pineapples. Xyloglucans were also present, together with small amounts of

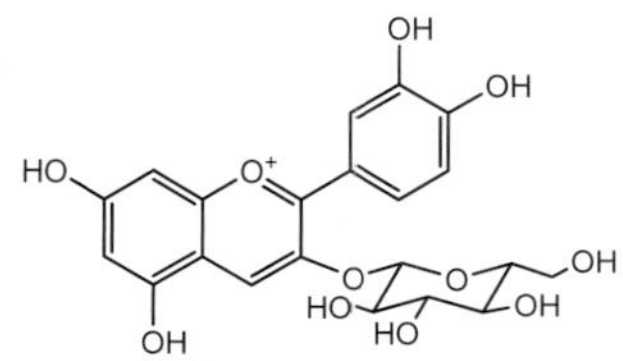

Cyanidin-3-glucoside

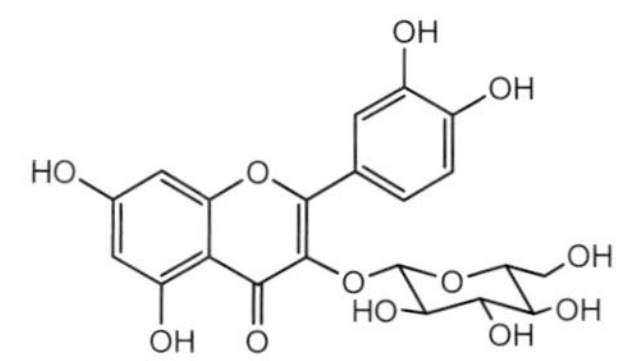

Quercetin-3-glucoside

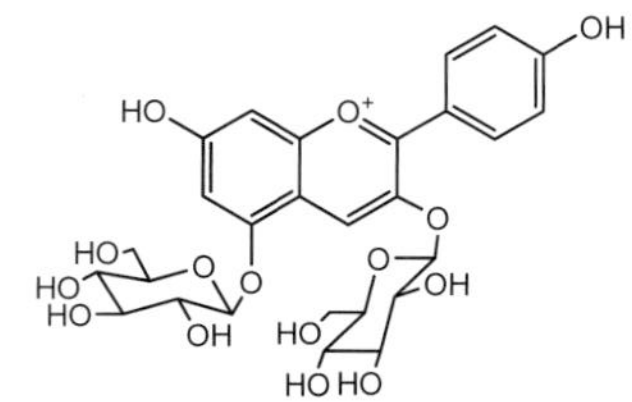

Pelargonidin-3,5-diglucoside

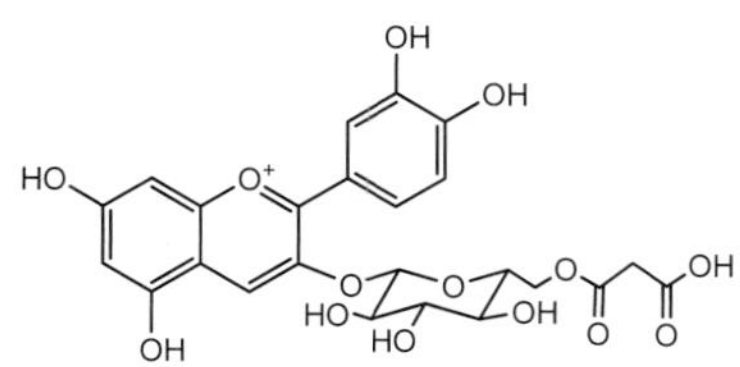

Cyanidin-3-6''-malonyl glucoside

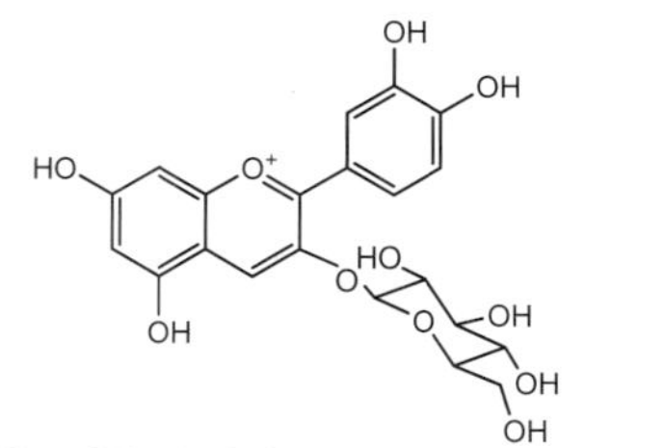

Cyanidin-3-*O*-β-glucopyranoside

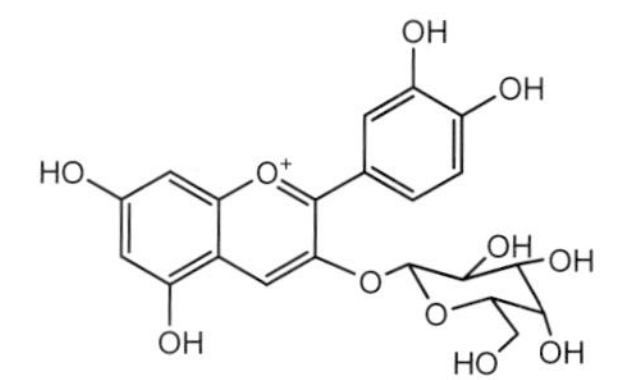

Cyanidin-3-*O*-β-galactopyranoside

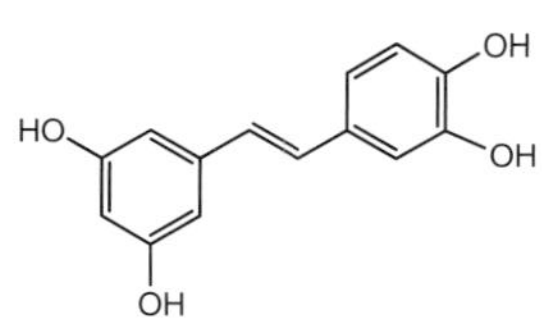

Piceatannol

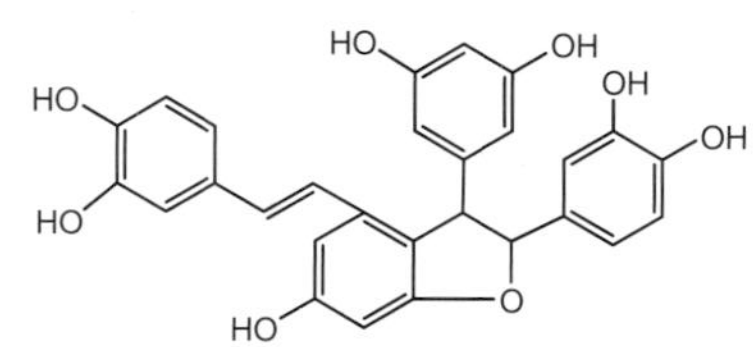

Scirpusin B

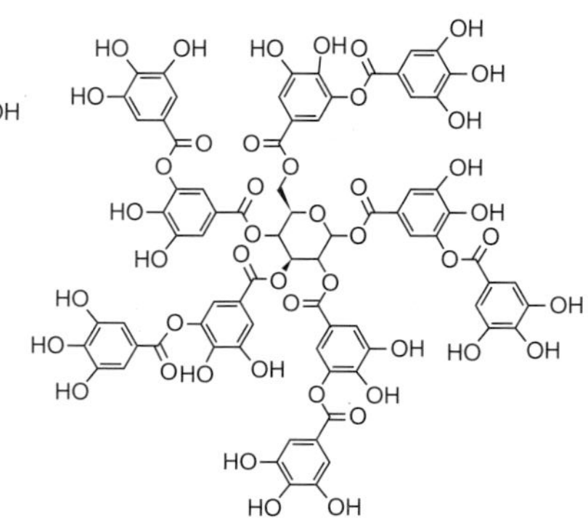

Tannic acid

Figure 26.3 Chemical structures of polyphenols from passion fruits.

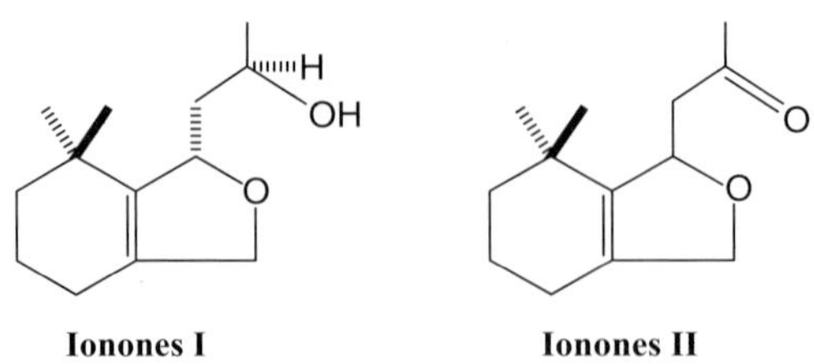

Figure 26.4 Chemical structures of ionones I and II from passion fruits.

pectic polysaccharides and glucomannans (or galactoglucomannans). They further indicated that ferulic acid was ester-linked to glucuronoarabinoxylans, and the most abundant feruloyl oligosaccharide released was *O*-[5-*O*-(*E*-feruloyl)-α-L-arabinofuranosyl](1→3)-*O*-β-D-xylopyranosyl-(1→4)-D-xylopyranse (FAXX) (Figure 26.6) [45]. A "neutral" polysaccharide and an "acidic" one were also isolated by Bartolomé and Rupérez [46] from pineapples. The "neutral" fraction was composed of xylose, arabinose, glucose, galactose, and minor quantities of mannose. Yapo [47] reported that pectic polysaccharides in pineapple flesh consisted of homogalacturonans as well as type I and II arabinogalactan side chain-containing rhamnogalacturonan I. From pineapples, Zheng *et al.* [13] isolated and identified many phenolic compounds (Figure 26.7), such as (*S*)-2-amino-5-((*R*)-1-carboxy-2-((*E*)-3-(4-hydroxy-3-methoxyphenyl)allylthio)ethylamino)-5-oxopentanoic acid (compound **1**), and (*S*)-2-amino-5-((*R*)-1-(carboxymethylamino)-3-((*E*)-3-(4-hydroxyphenyl)allylthio)-1-oxopropan-2-ylamino)-5-oxopentanoic acid (compound **2**), *N*-[*N*-L-γ-glutamyl-*S*-[3-(4-hydroxy-3-methoxyphenyl)-2-propenyl]-L-cysteinyl]-glycine (compound **3**), *N*-L-γ-glutamyl-*S*-sinapyl-L-cysteine (compound **4**), *S*-sinapylglutathione (compound **5**), *S*-sinapyl-L-cysteine (compound **6**), sinapoylglucose (compound **7**), 3-(4β-D-glucopyranosyloxy-3,

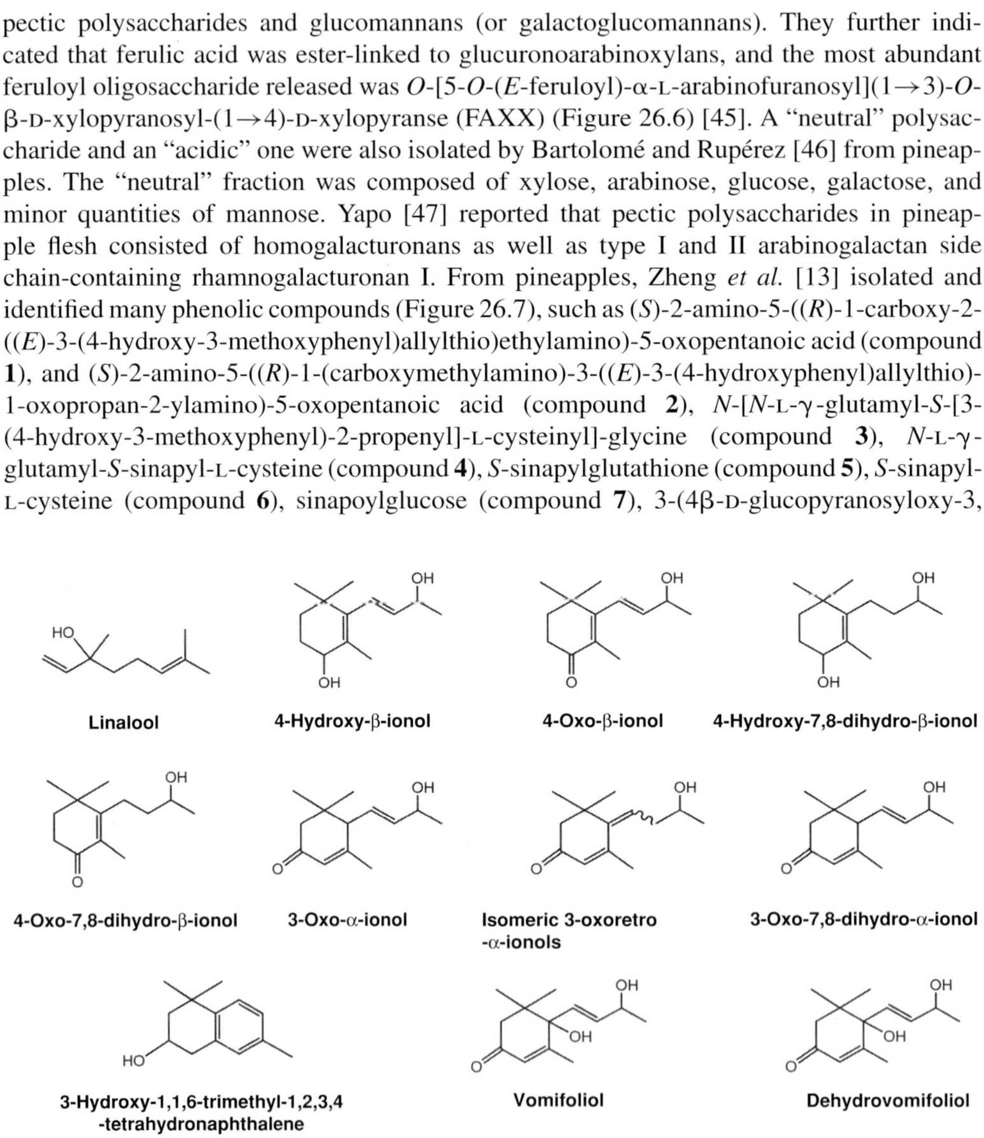

Figure 26.5 Chemical structures of terpenoids from purple passion fruits.

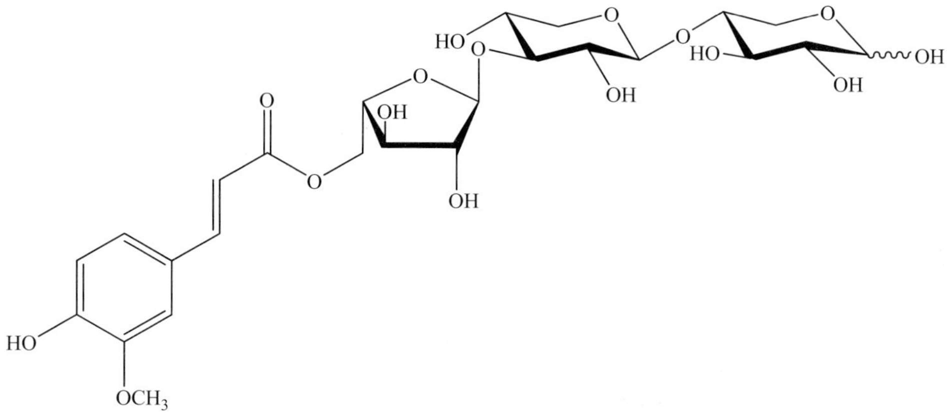

Figure 26.6 Chemical structures of FAXX from pineapples.

5-dimethoxy)-phenyl-2*E*-propenol (compound **8**), erigeside C (compound **9**), *p*-coumaryl alcohol β-D-glucopyranoside (compound **10**), caffeoyl glucose (compound **11**), 1-*O*-feruloyl-β-D-glucose (compound **12**), sinapyl aldehyde (compound **13**), triandrin (compound **14**), citrusin D (compound **15**), icariside H1 (compound **16**), ananasate (compound **17**) and so on, as well as one 1*H*-pyrrole-2-carboxylic acid derivative, 6-(3-(1*H*-pyrrole-2-carbonyloxy)-2-hydroxypropoxy)-3,4,5-trihydroxy-tetrahydro-2*H*-pyran-2-carboxylic acid (compound **18**). In addition, pure bromelain from pineapples is needed for the preparation of health-care products. Li and Lee [48] isolated and purified α- and β-D-mannopyranosidases from crude pineapple bromelain.

26.4 Health benefits of dried passion and pineapple fruits

Dried passion and pineapple fruits are good natural sources for human health. Epidemiological and medical anthropological investigations have suggested that these two fruits have positive effects in the prevention or treatment of many diseases.

Passion fruits are known for many ethno-pharmacological properties. Dhawan *et al.* [41] summarized the research status of the genus *Passiflora* and described that passion fruits were regarded as digestive stimulant and were used as remedy for gastric carcinoma in Madeire, while this fruit was eaten for relief from constipation in Nagaland (India). Bezerra *et al.* [49] reported that the pulps of yellow passion fruits were commonly used as sedative, pain killer, and anti-inflammatory agent and also for the treatment of skin wounds and the healing of colonic anastomoses. Puricellia *et al.* [50] found that the decoction of purple passion fruits inhibited the activity of gelatinase matrix metalloproteinase-2 (MMP-2) and matrix metalloproteinase-9 (MMP-9), two metalloproteases involved in tumor invasion, metastasis, and angiogenesis. Additionally, a series of investigations have revealed that oral administration of purple passion fruit rind extracts (a mixture of bioflavonoids, phenolic acids, and anthocyanins) can attenuate blood pressure in hypertensive patients [51], reduce wheeze and cough and improve shortness of breath in adults with asthma [52], and reduce pain and

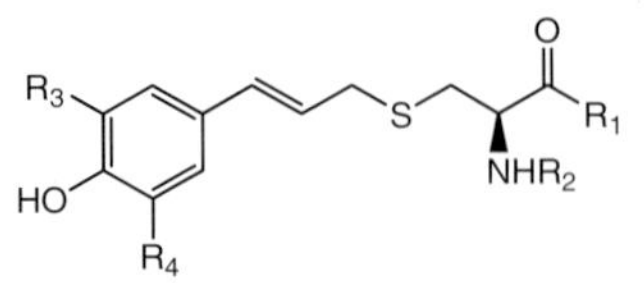

Compounds	R_1	R_2	R_3	R_4
1	OH	—C(=O)—CH$_2$—CH$_2$—CH(NH$_2$)—COOH	OMe	H
2	—NH—CH$_2$—COOH	—C(=O)—CH$_2$—CH$_2$—CH(NH$_2$)—COOH	H	H
3	—NH—CH$_2$—COOH	—C(=O)—CH$_2$—CH$_2$—CH(NH$_2$)—COOH	OMe	H
4	OH	—C(=O)—CH$_2$—CH$_2$—CH(NH$_2$)—COOH	OMe	OMe
5	—NH—CH$_2$—COOH	—C(=O)—CH$_2$—CH$_2$—CH(NH$_2$)—COOH	OMe	OMe
6	OH	H	OMe	OMe

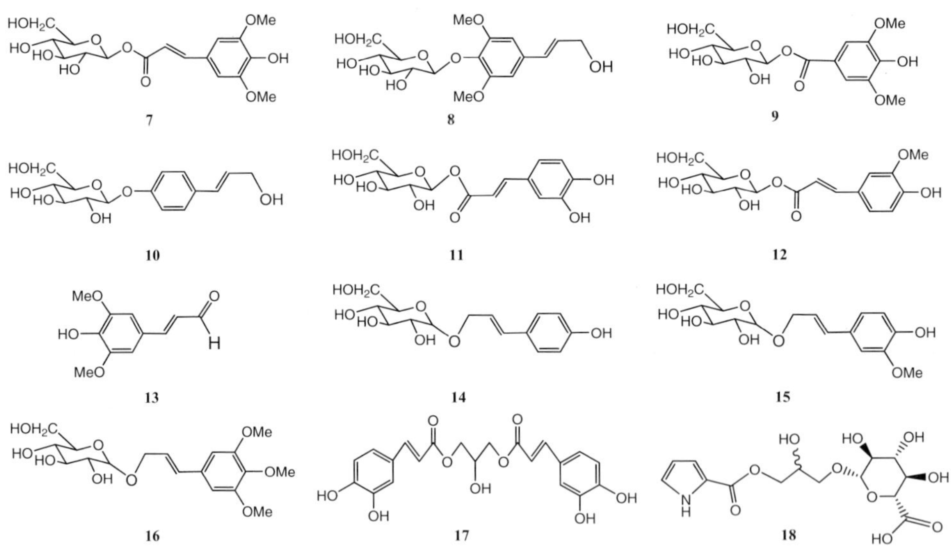

Figure 26.7 Chemical structures of phenolic compounds from pineapples.

stiffness and improve physical function in adult patients with knee osteoarthritis [53]. On the other hand, Rebello *et al.* [54] reviewed the health benefits of yellow passion fruit rind extracts (containing pectin, tryptophan, grax acid, and amino acids) and reported that the studies in human beings suggested an effect of pectin on gastric emptying, protein metabolism in the digestive tract, reduction of cholesterol, glucose blood levels, and intestinal glucose absorption. Wen *et al.* [55] found that passion fruit rind extracts exhibited strong antioxidant capacity, which could effectively scavenge 1,1-diphenyl-2-picryl hydrazyl (DPPH) radical and hydroxyl radical. Besides pulps and rinds, the seeds in dried passion fruits also exhibit a good bioactivity. An antifungal protein, passiflin, was isolated from seeds by Lam and Ng [56], which impeded mycelial growth in *Rhizoctonia solani* and potently inhibited proliferation of Michigan Cancer Foundation-7 (MCF-7) breast cancer cells. Pelegrini *et al.* [57] also purified an antifungal peptide, *Passiflora edulis* antifungal peptide-1 (Pe-AFP1), from seed. *In vitro* assays indicated that Pe-AFP1 was able of inhibiting the development of the filamentous fungi *Trichoderma harzianum*, *Fusarium oxysporum*, and *Aspergillus fumigatus*.

In dried pineapples, polysaccharides, phenolic substances, and bromelain are important bioactive constituents. Guo [58] found that pineapple polysaccharides were good natural antioxidants which had the ability to scavenge hydroxyl radical and superoxide anion. Furthermore, he studied multiplication effects of pineapple polysaccharides on active *Bacillus bifidus* and found that these polysaccharides could adjust intestinal flora and improve intestinal environment. Dried pineapples contain dietary fiber which possesses some medical functions such as intestinal disease prevention, coronary heart disease (CHD) prevention, blood sugar level regulation, and cholesterol reduction, among others [59]. Binding of bile acids and increasing their fecal excretion has been hypothesized as a possible mechanism by which dietary fiber lowers cholesterol. Kahlon and Smith [60] indicated that pineapples presented a high bile acid binding and therefore possessed health-promoting potential. Phenolic substances in pineapples are known to promote human health, mainly due to their antioxidant activity. Mhatre *et al.* [61] revealed that pineapple core/pulp extracts exhibited strong antioxidant activity as evidenced by DPPH radical-scavenging, oxygen radical absorbance capacity (ORAC), and lipid peroxidation assays. Hossain and Rahman [62] also reported that flavonoids and other phenolics from pineapples exhibited high radical-scavenging activity. Table 26.3 shows total phenolic content and antioxidant activities of passion fruits and pineapples. Bromelain from pineapples is an especially important bioactive constituent which

Table 26.3 Total phenolic content and antioxidant activities of passion fruits and pineapples

Samples	Total phenolic	DPPH radical (%)	References
Passion fruits[a]			
Rind	28.11 ± 0.22	91.62 ± 1.60	[55]
Pulp	41.20 ± 4.20	89.51 ± 5.00	[63, 64]
Pineapples[b]			
Rind	0.95 ± 0.01	87.60 ± 6.90	[65, 66]
Pulp	0.42 ± 0.00	93.70 ± 0.58	[65, 67]

DPPH, 1,1-diphenyl-2-picryl hydrazyl.
[a]Total phenolics are expressed as milligrams of gallic acid equivalents (GAE) per gram.
[b]Total phenolics are expressed as milligrams of tannic acid equivalents (TAE) per gram.

has diverse biological properties. Fruit bromelain can be used in various medical applications including prevention of diarrhea, as digestive aids, as antithrombotic, treatment of edema and osteoarthritis, and promotion of absorption of antibiotic drugs, and additionally it also regulates and activates various immune cells and their cytokine production [68, 69]. Existing evidence indicates that bromelain serves as a promising candidate for the development of future oral enzyme therapies for oncology patients [70].

26.5 Commercial products and industrial applications of dried passion and pineapple fruits

Some commercial products of the dried passion fruits, such as freeze-dried fruits, dehydrated fruits, dried fruit powders, or dried fruit bars, have been found in markets in the United States, Brazil, India, China, Thailand, Colombia, and South Africa, among others. These products can be used by the food industry to produce functional foods and beverages, snack foods, sport drinks, as well as additives or flavorings for fruit jellies, sauces, candies, desserts, breakfast cereals, and yogurts, among others. The common commercial products of the dried pineapples in international markets include dried fruit pieces (chunks or dices), dried fruit rings, dried fruit powders, honey-dipped dried fruit slices, freeze-dried fruits, sweetened pineapple titbits, and chocolate-covered pineapples, among others. The major applications of these dried pineapple products are for healthy snacks, additives in breakfast cereals, trail mix, beverages, pancakes, and breads, as well as ingredients in recipes. Sometimes, they are added to leafy salads or for baking. Otherwise, there is currently an increasing interest on dried passion and pineapple fruits in pharmaceutical industry because of the health benefits arising from their phytochemicals.

26.6 Conclusions

The consumption of dried tropical fruits is on the rise due to not only their good taste and aroma, but also the health benefits from their bioactive components. The investigations on nutritional compositions and phytochemicals have indicated that dried passion and pineapple fruits contain certain nutritional ingredients and bioactive constituents which are of great benefits to human health. Many compounds have been isolated and identified from these two fruits. Epidemiological and pharmacological evaluations have confirmed their positive effects on the prevention or treatment of some diseases. Several commercial products of both dried fruits have been sold in markets or used in food industry as additives and flavorings. Both dried fruits may also be potentially used by the pharmaceutical industry and beyond.

Acknowledgments

The authors appreciate the financial supports from the National Natural Science Foundation of China (Grant No. 31000927, 31160407), Guangxi Scientific Research and Technological Development Projects (Grant No. 10100009–2), Guangxi Natural Science Foundation (Grant No. 2010GXNSFC013010, 2011GXNSFB018049), and Fundamental Research Funds from Gangxi Academy of Agricultural Sciences (2011YZ24).

References

1. Sale, P.R. (1988) *Passion Fruit Culture*. Ministry of Agriculture and Fisheries, Wellington, New Zealand.
2. Okoth, M.W., Kaahwa, A.R. & Imungi, J.K. (2000) The effect of homogenisation, stabiliser and amylase on cloudiness of passion fruit juice. *Food Control*, **11**, 305–311.
3. Genders, R. (1988) *Plantas Silvestres Comestibles*. Editorial Blume, Barcelona, Spain.
4. Romero-Rodriguez, M.A., Vazquez-Oderiz, M.L., Lopez-Hernandez, J. & Simal-Lozano, J. (1994) Composition of babaco, feijoa, passionfruit and tamarillo produced in Galicia (North-west Spain). *Food Chemistry*, **49**, 23–27.
5. Mouquet, C., Aymard, C., Guilbert, S., Cuvelier, G. & Launay, B. (1997) Influence of initial pH on gelation kinetics of texturized passion fruit pulp. *Lebensmittel-Wissenschaft und-Technologie*, **30**, 129–134.
6. Calle, E.V., Ruales, J., Dornier, M., Sandeaux, J., Sandeaux, R. & Pourcelly, G. (2002) Deacidification of the clarified passion fruit juice (P. *edulis* f. *flavicarpa*). *Desalination*, **149**, 357–361.
7. Vera, E., Dornier, M., Ruales, J., Vaillant, F. & Reynes, M. (2003) Comparison between different ion exchange resins for the deacidification of passion fruit juice. *Journal of Food Engineering*, **57**, 199–207.
8. Vera, E., Ruales, J., Dornier, M., Sandeaux, J., Persin, F., Pourcelly, G., Vaillant, F. & Reynes, M. (2003) Comparison of different methods for deacidification of clarified passion fruit juice. *Journal of Food Engineering*, **59**, 361–367.
9. Sandi, D., Chaves, J.B.P., De Souza, A.C.G., Parreiras, J.F.M., Da Silva, M.T.C. & Constant, P.B.L. (2004) Hunter color dimensions, sugar content and volatile compounds in pasteurized yellow passion fruit juice (*Passiflora edulis* var. flavicarpa) during storage. *Brazilian Archives of Biology and Technology*, **47**, 233–245.
10. Laboissière, L.H.E.S., Deliza, R., Barros-Marcellini, A.M., Rosenthal, A., Camargo, L.M.A.Q. & Junqueira, R.G. (2007) Effects of high hydrostatic pressure (HHP) on sensory characteristics of yellow passion fruit juice. *Innovative Food Science and Emerging Technologies*, **8**, 469–477.
11. Cal-Vidal, J. & Falcone, M. (1985) Processing conditions affecting the hygroscopic behavior of freeze-dried passion fruit juice. *Journal of Food Science*, **50**, 1238–1241.
12. Reqinatto, F.H., De-Paris, F., Petry, R.D., Quevedo, J., Ortega, G.G., Gosmann, G. & Schenkel, E.P. (2006) Evaluation of anxiolytic activity of spray dried powders of two South Brazilian *Passiflora* species. *Phytotherapy Research*, **20**, 348–351.
13. Zheng, Z.-P., Ma, J., Cheng, K.-W., Chao, J., Zhu, Q., Chang, R.C.-C., Zhao, M., Lin, Z.-X. & Wang, M. (2010) Sulfur-containing constituents and one 1H-pyrrole-2-carboxylic acid derivative from pineapple [*Ananas comosus* (L.) Merr.] fruit. *Phytochemistry*, **71**, 2046–2051.
14. Mortan, J.F. (1987) Pineapple. In: *Fruits of Warm Climates* (ed. J.F. Mortan). Florida Flair Books, Miami, FL, pp.18–28.
15. Elkins, E.R., Lyon, R., Huang, C.J. & Matthys, A. (1997) Characterization of commercially produced pineapple juice concentrate. *Journal of Food Composition and Analysis*, **10**, 285–298.
16. Nigam, J.N. (1999) Continuous ethanol production from pineapple cannery waste. *Journal of Biotechnology*, **72**, 197–202.
17. Lombard, G.E., Oliveira, J.C., Fito, P. & Andrés, A. (2008) Osmotic dehydration of pineapple as a pre-treatment for further drying. *Journal of Food Engineering*, **85**, 277–284.
18. Saxena, S., Mishra, B.B., Chander, R. & Sharma, A. (2009) Shelf stable intermediate moisture pineapple (*Ananas comosus*) slices using hurdle technology. *LWT-Food Science and Technology*, **42**, 1681–1687.
19. Elmadfa, I., Aign, W., Muskat, E., Fritzsch, D. & Cremer, H.D. (1989) *La Gran Guia de la Eomposicion de los Alimentos*. Oassis, Barcelona, Spain.
20. Chau, C.F. & Huang, Y.L. (2004) Characterization of passion fruit seed fibres—a potential fibre source. *Food Chemistry*, **85**, 189–194.
21. Wang, Y., Liu, Z., Kang, W., Wu, Y., Zhao, D. & Zhang, J. (2009) Analysis of mineral elements in *Passiflora* fruits. *Food Science*, **30**, 328–330.
22. Rastogi, N.K. & Raghavarao, K.S.M.S. (2004) Mass transfer during osmotic dehydration of pineapple: considering Fickian diffusion in cubical configuration. *LWT-Food Science and Technology*, **37**, 43–47.
23. Rahman, M.S. & Lamb, J. (1991) Air drying behaviour of fresh and osmotically dehydrated pineapple. *Journal of Food Process Engineering*, **14**, 163–171.
24. Silveira, E.T.F., Rahman, M.S. & Buckle, K.A. (1996) Osmotic dehydration of pineapple: kinetics and product quality. *Food Research International*, **29**, 227–233.

25. Dixon, G.M. & Jen, J.J. (1977) Changes of sugar and acid in osmovac dried apple slices. *Journal of Food Science*, **42**, 1126–1131.
26. Lerici, C.L., Pinnavaia, G., Dalla Rosa, M. & Bartolucci, L. (1985) Osmotic dehydration of fruit: influence of osmotic agents on drying behaviour and product quality. *Journal of Food Science*, **50**, 1217–1219.
27. Giangiacomo, R., Torreggiani, D. & Abbo, E. (1987) Osmotic dehydration of fruit. Part I: sugar exchange between fruit and extracting syrup. *Journal of Food Processing and Preservation*, **11**, 183–195.
28. Zheng, Z.Q. (1986) Nutritional compositions and applications of pineapple. *Fujian Science & Technology of Tropical Crops*, **4**, 25 (in Chinese).
29. Rowan, A.D. & Buttle, D.J. (1994) Pineapple cysteine endopeptidases. *Methods in Enzymology*, **244**, 555–568.
30. Bhui, K., Tyagi, S., Prakash, B. & Shukla, Y. (2010) Pineapple bromelain induces autophagy, facilitating apoptotic response in mammary carcinoma cells. *Biofactors*, **36**, 474–482.
31. Peiró-Mena, R., Camacho, M.M. & Martínez-Navarrete, N. (2007) Compositional and physicochemical changes associated to successive osmodehydration cycles of pineapple (*Ananas comosus*). *Journal of Food Engineering*, **79**, 842–849.
32. Pruthi, J.S. & Lal, G. (1960) Carbohydrates in passion fruit juice. *Indian Journal of Horticulture*, **17**, 133.
33. Cillie, C.G. & Joubert, F.J. (1950) Occurrence of an amylopectin in fruit of the grenadilla (*Passiflora edulis*). *Journal of the Science of Food and Agriculture*, **1**, 355.
34. Pruthi, J.S. (1963) Physiology, chemistry, and technology of passion fruit. *Advances in Food Research*, **12**, 203–282.
35. Pruthi, J.S. & Lal, G. (1958) Carotenoids in passion fruit juice (*Passiflora edulis*). *Food Research*, **23**, 505.
36. Pruthi, J.S. & Lal, G. (1958) Fortification of passion fruit squash with L-ascorbic acid. *Indian Food Packer*, **12**, 9.
37. Pruthi, J.S. (1959) Physico-chemical composition of passion fruit (*P. edulis*). III. Effect of degree of maturity. *Indian Journal of Horticulture*, **16**, 110.
38. Mercadante, A.Z., Britton, G. & Rodriguez-Amaya, D.B. (1998) Carotenoids from yellow passion fruit (*Passiflora edulis*). *Journal of Agricultural and Food Chemistry*, **46**, 4102–4106.
39. Zeraik, M.L. & Yariwake, J.H. (2010) Quantification of isoorientin and total flavonoids in *Passiflora edulis* fruit pulp by HPLC-UV/DAD. *Microchemical Journal*, **96**, 86–91.
40. Sano, S., Sugiyama, K., Ito, T., Katano, Y. & Ishihata, A. (2011) Identification of the strong vasorelaxing substance scirpusin B, a dimer of piceatannol, from passion fruit (*Passiflora edulis*) seeds. *Journal of Agricultural and Food Chemistry*, **59**, 6209–6213.
41. Dhawan, K., Dhawan, S. & Sharma, A. (2004) *Passiflora*: a review update. *Journal of Ethnopharmacology*, **94**, 1–23.
42. Winterhalter, P. (1990) Bound terpenoids in the juice of the purple passion fruit (*P. edulis Sims*). *Journal of Agriculture and Food Chemistry*, **38**, 452–455.
43. Vidal-Valverde, C., Herranz, J., Blanco, I. & Rojas-Hidalgo, E. (1982) Dietary fiber in Spanish fruits. *Journal of Food Science*, **47**, 1840–1845.
44. Smith, B.G. & Harris, P.J. (1995) Polysaccharide composition of unlignified cell walls of pineapple [*Ananas comosus* (L.) Merr.] fruit. *Plant Physiology*, **107**, 1399–1409.
45. Smith, B.G. & Harris, P.J. (2001) Ferulic acid is esterified to glucuronoarabinoxylans in pineapple cell walls. *Phytochemistry*, **56**, 513–519.
46. Bartolomé, A.P. & Rupérez, P. (1995) Polysaccharides from the cell walls of pineapple fruit. *Journal of Agriculture and Food Chemistry*, **43**, 608–612.
47. Yapo, B.M. (2009) Pineapple and banana pectins comprise fewer homogalacturonan building blocks with a smaller degree of polymerization as compared with yellow passion fruit and lemon pectins: implication for gelling properties. *Biomacromolecules*, **10**, 717–721.
48. Li, Y.-T. & Lee, Y.C. (1972) Pineapple α- and β-D-mannopyranosidases and their action on core glycopeptides. *Journal of Biological Chemistry*, **247**, 3677–3683.
49. Bezerra, J.A., Campos, A.C., Vaconcelos, P.R., Nicareta, J.R., Ribeiro, E.R., Sebastião, A.P., Urdiales, A.I., Moreira, M. & Borges, A.M. (2006) Extract of *Passiflora edulis* in the healing of colonic anastomosis in rats: a tensiometric and morphologic study. *Acta Cirúrgica Brasileira*, **21** (Suppl. 3), 16–25.

50. Puricellia, L., Dell'Aica, I., Sartor, L., Garbisa, S. & Caniato, R. (2003) Preliminary evaluation of inhibition of matrix-metalloprotease MMP-2 and MMP-9 by *Passiflora edulis* and *P. foetida* aqueous extracts. *Fitoterapia*, **74**, 302–304.
51. Zibadi, S., Farid, R., Moriguchi, S., Foo, L.Y., Tehrani, P.M., Ulreich, J.B. & Watson, R.R. (2007) Oral administration of purple passion fruit peel extract attenuates blood pressure in female spontaneously hypertensive rats and humans. *Nutrition Research*, **27**, 408–416.
52. Watson, R.R., Zibadi, S., Rafatpanah, H., Jabbari, F., Ghasemi, R., Ghafari, J., Afrasiabi, H., Foo, L.Y. & Faridhosseini, R. (2008) Oral administration of the purple passion fruit peel extract reduces wheeze and cough and improves shortness of breath in adults with asthma. *Nutrition Research*, **28**, 166–171.
53. Farid, R., Rezaieyazdi, Z., Mirfeizi, Z., Hatef, M.R., Mirheidari, M., Mansouri, H., Esmaelli, H., Bentley, G., Lu, Y., Foo, Y. & Watson, R.R. (2010) Oral intake of purple passion fruit peel extract reduces pain and stiffness and improves physical function in adult patients with knee osteoarthritis. *Nutrition Research*, **30**, 601–606.
54. Rebello, B.M., Moreno, S.R.F. & Godinho, C.R. (2008) Effects of *Passiflora edulis flavicarpa* on the radiolabeling of blood constituents, morphology of red blood cells and on the biodistribution of sodium pertechnetate in rats. *Applied Radiation and Isotopes*, **66**, 1788–1792.
55. Wen, L., Mao, H., Zhang, Y. & Li, Y. (2008) Study on compositions and antioxidant activity of *Passiflora edulis* rind. *Food Science*, **29**, 54–58 (in Chinese).
56. Lam, S.K. & Ng, T.B. (2009) Passiflin, a novel dimeric antifungal protein from seeds of the passion fruit. *Phytomedicine*, **16**, 172–180.
57. Pelegrini, P.B., Noronha, E.F., Muniz, M.A., Vasconcelos, I.M., Chiarello, M.D., Oliveira, J.T. & Franco, O.L. (2006) An antifungal peptide from passion fruit (*Passiflora edulis*) seeds with similarities to 2S albumin proteins. *Biochimica et Biophysica Acta*, **1764**, 1141–1146.
58. Guo, Q. (2010) *Study on the Extraction of Pineapple Polysaccharide and Its Biological Activity*. MSc Thesis, Fujian Agriculture and Forestry University, Foochow, Fujian, China.
59. Shi, J. (2010) *Study on Dietary Fiber and Other Functional Components in Pineapple Fruit*. MSc Thesis, Hainan University, Haikou, Hainan, China.
60. Kahlon, T.S. & Smith, G.E. (2007) *In vitro* binding of bile acids by bananas, peaches, pineapple, grapes, pears, apricots and nectarines. *Food Chemistry*, **101**, 1046–1051.
61. Mhatre, M., Tilak-Jain, J., De, S. & Devasagayam, T.P.A. (2009) Evaluation of the antioxidant activity of non-transformed and transformed pineapple: a comparative study. *Food and Chemical Toxicology*, **47**, 2696–2702.
62. Hossain, M.A. & Rahman, S.M.M. (2011) Total phenolics, flavonoids and antioxidant activity of tropical fruit pineapple. *Food Research International*, **44**, 672–676.
63. de Oliveira, A.C., Valentim, L.B., Silva, C.A., Bechara, E.J.H., de Barros, M.P., Mano, C.M. & Goulart, M.O.F. (2009) Total phenolic content and free radical scavenging activities of methanolic extract powders of tropical fruit residues. *Food Chemistry*, **115**, 469–475.
64. Zeraik, M.L., Serteyn, D., Deby-Dupont, G., Wauters, J.-N., Tits, M., Yariwake, J.H., Angenot, L. & Franck, T. (2011) Evaluation of the antioxidant activity of passion fruit (*Passiflora edulis* and *Passiflora alata*) extracts on stimulated neutrophils and myeloperoxidase activity assays. *Food Chemistry*, **128**, 259–265.
65. Wang, Z., Li, Q., Yang, C., Yao, W., He, N., Wang, Y. & Sun, D. (2007) Polyphenol contents in eight fruits and their antioxidant activities. *Natural Product Research and Development*, **19**, 1040–1043 (in Chinese).
66. Wen, L., Wang, W. & Mai, Q. (2008) Study on microwave extraction and its antioxidant activities of antioxidant components from pineapple peel. *Food Science*, **29**, 210–214 (in Chinese).
67. Alothman, M., Bhat, R. & Karim, A.A. (2009) Antioxidant capacity and phenolic content of selected tropical fruits from Malaysia, extracted with different solvents. *Food Chemistry*, **115**, 785–788.
68. Maurer, H.R. (2001) Bromelain: biochemistry pharmacology and medical use. *Cellular & Molecular Life Science*, **58**, 1234–1245.
69. Brien, S., Lewith, G., Walker, A., Hicks, S.M. & Middleton, D. (2004) Bromelain as a treatment for osteoarthritis: a review of clinical studies. *Evidence-Based Complementary and Alternative Medicine*, **1**, 251–257.
70. Chobotova, K., Vernallis, A.B. & Majid, F.A.A. (2010) Bromelain's activity and potential as an anti-cancer agent: current evidence and perspectives. *Cancer Letters*, **290**, 148–156.

Index

Dried Fruits: Phytochemicals and Health Effects, First Edition. Edited by Cesarettin Alasalvar and Fereidoon Shahidi.

Introduction

1) The legitimacy gap

In 430 BCE, amid the early years of the Peloponnesian War, the Athenian statesman Pericles stood before his people to honour the fallen. He put a pivotal question to his fellow Athenians: What makes Athens worth fighting for? Its distinctive constitutional system, Pericles answered—a system that not only governed the lives of its own people but also stood as a 'model'[1] to others. 'Our constitution is called a democracy', he proclaimed, 'because power is in the hands not of a minority but of the whole people'.[2] Although democracy in Athens died a century later, Athenian democracy became a paradigm that has inspired political theorists and constitution designers through the ages.[3] Today, however, the Athenian system has largely lost its lustre.[4] As has become all too clear, it did not in fact put power in the hands of the whole people, as Pericles claimed. Quite the opposite, it excluded women, slaves, foreigners, as well as Athenian males born out of wedlock or under the age of eighteen. Of a total population of approx. 200,000–250,000 people, only around 30,000 adult males were in fact enfranchised.[5] From a modern perspective, a system in which only 12–15 per cent of the people have a say in the decisions governing them hardly seems legitimate.[6] Certainly, it is more akin to an oligarchy than a democracy.[7]

[1] Athens was not the only democratic city-state, however. See John Ma, *Polis: A New History of the Ancient Greek City-State from the Early Iron Age to the End of Antiquity* (Princeton UP 2024) ch 9.

[2] Thucydides, *History of the Peloponnesian War* (Rex Warner tr, Penguin 1954) bk 2, [37], 145.

[3] Dino Piovan and Giovanni Giorgini, 'Introduction' in Dino Piovan and Giovanni Giorgini (eds), *Brill's Companion to the Reception of Athenian Democracy: From the Late Middle Ages to the Contemporary Era* (Brill 2020) 1.

[4] With the exception of direct democracy scholarship. See eg Matt Qvortrup, *Direct Democracy: A Comparative Study of the Theory and Practice of Government by the People* (Manchester UP 2013) 12.

[5] Mogens Herman Hansen, 'The Nature of Athenian Democracy' in Dino Piovan and Giovanni Giorgini (eds), *Brill's Companion to the Reception of Athenian Democracy: From the Late Middle Ages to the Contemporary Era* (Brill 2020) 30–31 (the data are from the age of Demosthenes).

[6] We understand legitimacy in a normative sense, referring to a standard of justification that a constitutional system needs to meet, rather than a descriptive sense that draws on people's beliefs. See generally Fabienne Peter, 'Political Legitimacy' in Edward N Zalta and Uri Nodelman (eds), *The Stanford Encyclopedia of Philosophy* (Winter 2023, Metaphysics Research Lab, Stanford University 2023) <https://plato.stanford.edu/archives/win2023/entries/legitimacy/> accessed 26 October 2024.

[7] See Ellen Meiksins Wood, *Citizens to Lords: A Social History of Western Political Thought from Antiquity to the Late Middle Ages* (Verso 2008) 38–39; Christian Meier, *The Greek Discovery of Politics*

Animals and the Constitution. John Olusegun Adenitire and Raffael Fasel, Oxford University Press.
 DOI: 10.1093/9780198910534.001.0001

Contemporary constitutions are decidedly more inclusive. Voter protections can often do with improvement, it is true. But there is no democratic system in the world today that does not at least claim to enfranchise women. Some even give political rights to foreigners and, while too many humans still find themselves in modern forms of slavery, the formal status of chattel slaves has been abolished.[8] And yet, it would be misguided to think that democratic constitutions have solved the legitimacy problems that marred Greek and later democracies. As this book shows, current constitutional systems suffer from a legitimacy gap that is similar, if not worse, than that which plagued the Attican system. Athens' constitution governed the lives of up to 88 per cent of the population without giving them a say in the decisions affecting them. Modern-day constitutions have put power in significantly more hands, but they still exclude an overwhelming number of those whom they govern. This becomes clear once we expand our constitutional horizon, as this book proposes we should, from humans to all sentient animals.

Consider the United Kingdom. As we explain in Chapter 2, even on a conservative estimate a country of the United Kingdom's size counts at least 2.4 billion sentient non-human animals (henceforth called 'animals'[9]). Compared to its human population of 67 million, animals thus make up 98 per cent of those living under the UK constitution. Of course, most if not all of these animals may lack the capacity to understand and abide by the rules of human constitutions and legal systems in ways that rational adult humans can.[10] Animals are therefore not *subject* to those legal rules in the sense of being the normative addressees who can be expected to comply with the rules and who can be punished for failing to do so.[11] But, importantly, that does not mean they are merely *regulated* by the rules in the way inanimate objects such as cars, websites, or buildings are. Instead, animals are *governed* by the rules.

(David McLintock tr, Harvard UP 1990) 154. For a wider critique, see Michael G Hanchard, *The Spectre of Race: How Discrimination Haunts Western Democracy* (Princeton UP 2018).

[8] See Jean Allain, 'Slavery and Its Obligations Erga Omnes' (2019) 36 Australian Year Book of International Law 83, 87–95. On contemporary forms of slavery, see Siddharth Kara, *Modern Slavery: A Global Perspective* (Columbia UP 2017).

[9] Although there is a danger of reinforcing the idea that humans are separate from animals, for the sake of simplicity we will be using the term 'animals' to refer to 'non-human animals', except in contexts where the difference between human and non-human animals needs highlighting to avoid confusion.

[10] Of course, this is not to say that they cannot follow the norms of their own communities. See Kristin Andrews, Simon Fitzpatrick, and Evan Westra, 'Human and Nonhuman Norms: A Dimensional Framework' (2024) 379 Philosophical Transactions of the Royal Society B: Biological Sciences 1.

[11] This has not stopped legal systems from treating animals as something akin to legal subjects in the past: Katie Sykes, 'Human Drama, Animal Trials: What the Medieval Animal Trials Can Teach Us About Justice for Animals' (2011) 17 Animal Law Review 273.

By 'governed' we mean that the legal and political frameworks and rules in place in a constitution have effective control over the lives, bodies, or habitats of beings who have intrinsic moral value. As we will show, a key reason why many animals have intrinsic moral value is because they are sentient.[12] Being 'sentient' means having the capacity to have valenced experiences, that is, experiences which feel good or bad.[13] This capacity includes the ability to feel pain and joy, but it can also encompass a much richer set of emotions, including empathy,[14] jealousy,[15] disgust,[16] grief,[17] indignation,[18] and even love.[19] However, animals' sentience has not stopped constitutional systems from governing their lives, often in intrusive ways. To give only a few examples, constitutions determine the lives of animals in comprehensive ways, as does, for example, Article 381(1) of the Constitution of Bolivia (2009), according to which '[n]ative animal and vegetable species are natural assets. The State shall establish the measures necessary for their conservation, exploitation and development'.[20] In a similar vein, some constitutions, such as that of the Bahamas (1973), state that animals are property. Some govern animals in more specific ways, such as Article 111 of the Constitution of the Republic of Congo (2001), which determines that animals can be used for husbandry. And sometimes even constitutional rules meant to protect animals—such as Article 83(6) of the Ecuadorian Constitution which imposes a duty on citizens to respect the rights of nature and to use natural resources responsibly—have been interpreted to justify their exploitation, such as for the purposes of pest control, food production, clothing, transportation, and even leisure.[21]

[12] See Chapter 2 Subsection 1.A.

[13] Jonathan Birch, *The Edge of Sentience: Risk and Precaution in Humans, Other Animals, and AI* (OUP 2024) 26.

[14] Frans de Waal, 'Evidence Implies That Animals Feel Empathy' (*Scientific American*, 1 September 2015) <https://www.scientificamerican.com/article/evidence-implies-that-animals-feel-empathy/> accessed 26 October 2024.

[15] Christine E Webb and others, 'Jealous Behavior in Chimpanzees Elicited by Social Intruders' (2020) 1 Affective Science 199.

[16] Trevor I Case and Richard J Stevenson, 'Evaluating the Presence of Disgust in Animals' (2024) 14 Animals 264.

[17] Lucy Rutherford and Lindsay E Murray, 'Personality and Behavioral Changes in Asian Elephants (Elephas Maximus) Following the Death of Herd Members' (2021) 16 Integrative Zoology 170.

[18] Sarah F Brosnan and Frans de Waal, 'Monkeys Reject Unequal Pay' (2003) 425 Nature 297.

[19] Marc Bekoff, 'Animal Emotions: Exploring Passionate Natures' (2000) 50 BioScience 861.

[20] 'Bolivia (Plurinational State of) 2009' <https://www.constituteproject.org/constitution/Bolivia_2009> accessed 26 October 2024. Unless otherwise indicated, all references to constitutional provisions in this book are drawn from the constituteproject.org website.

[21] See 'Estrellita', Constitutional Court, no. 253-20-JH/22, 27 January 2022, [106]—[110] <https://animal.law.harvard.edu/wp-content/uploads/Sentencia_253-20-JH-20_Derechos_Naturaleza_y_Animales.pdf> accessed 26 October 2024.

As beings who are governed by constitutions, we argue, animals should be recognized as having 'constitutional worth', that is, as having interests that are of constitutional importance. Constitutions derive their legitimacy at least partly from how well they cater for those who have constitutional worth. Yet, as we will demonstrate, contemporary constitutions fail to cater for sentient animals.[22] To be sure, around two-thirds of all states have anti-cruelty laws on the books which prohibit the imposition of certain types of cruel treatment on some animals. Around one-third even have animal welfare laws, which set more 'positive' requirements on how animals ought to be kept or cared for. And a few states even have constitutional provisions mandating the protection of animals.[23] But no constitution to date gives due concern to animals. As we define it, giving 'due concern' means showing respect for the intrinsic value of a being, which requires protecting their basic interests, such as their interest in life, bodily integrity, or liberty. As we will show, even the most demanding constitutions fall markedly short of this standard.

The reason why they fall short is as plain as it is passed over: current constitutions are, for the largest part, focused on the interests of human beings. In other words, they are *anthropocentric*. As such, they privilege human interests over the interests of non-humans, treating non-humans largely as a means for human ends. This anthropocentrism manifests itself, for instance, in the fact that animals lack legal personhood and fundamental rights and that they are excluded from democratic decision-making processes. In fact, as we will see, constitutions are not just centred around human beings, but they are centred around adult human beings who are considered rational political and legal agents. In both constitutional thinking and practice, these humans are treated as 'paradigmatic'[24] examples of their species, with the effect of marginalizing 'non-paradigmatic' humans such as infants, persons with severe cognitive disabilities, or persons who are senile, who are seen to lack the same rational and agential qualities. While different to non-human animals in obvious ways, these non-paradigmatic human beings sometimes share the former's fate of being governed by constitutional systems without being given due concern.

[22] See Chapter 1 Section 3.

[23] See Raffael N Fasel and Sean C Butler, *Animal Rights Law* (Hart Publishing 2023) 18–22, 25–27.

[24] For the use of this term, see eg Paola Cavalieri, *The Animal Question: Why Non-Human Animals Deserve Human Rights* (Catherine Woollard tr, OUP 2001) 76; Anna Grear, 'Deconstructing Anthropos: A Critical Legal Reflection on "Anthropocentric" Law and Anthropocene "Humanity"' (2015) 26 Law and Critique 225, 230. We will adopt this use, employing 'paradigmatic' and 'non-paradigmatic' in a descriptive sense to refer to how these human beings are generally conceived in classic constitutionalism. We do not believe that any groups of humans should be considered paradigmatic.

Current constitutions therefore face a legitimacy gap. They govern sentient animals without respecting their intrinsic value and without protecting their basic interests. What is more, they often do so in ways that marginalize non-paradigmatic human beings. Considering that, in a country like the United Kingdom, animals and non-paradigmatic humans make up around 98 per cent of all governed beings, this gap is in fact a gaping chasm. And this chasm calls into question the legitimacy of modern constitutions, much like we have come to doubt the legitimacy of the Athenian constitution.

2) Constitutionalisms of the future

The aim of this book is to fill this legitimacy gap. It does so by developing a more inclusive account of the constitution that gives due concern to all sentient animals, including paradigmatic and non-paradigmatic humans. We call this *sentience-based constitutionalism*.

The book joins an increasingly loud chorus of voices calling for a change to the dominant anthropocentric model of constitutionalism. The dominant model has been challenged, among others, by feminist thinkers,[25] critical race theorists,[26] and postcolonial scholars.[27] Constitutionalism's anthropocentrism has recently also come under attack by environmental thinkers who are proposing 'environmental constitutionalism' or 'green constitutionalism' to counter the threat that the current model poses to the planet and its ecosystems.[28] Our proposal for sentience-based constitutionalism follows these existing critiques in trying to develop what Louis J Kotzé has called 'constitutionalism of the future': a more legitimate model of constitutionalism that overcomes the deficiencies of the dominant anthropocentric model.[29]

Although inspired by these alternative visions for constitutionalism, ours contrasts and complements existing proposals by focusing on the need for constitutions to give sentient animals their due. Here, our theory forms part of a broader trend often referred to as the 'animal turn': a shift in thinking that has occurred in many disciplines towards 'viewing animals as "selves" whose

[25] See eg Catherine A MacKinnon, *Toward a Feminist Theory of the State* (Harvard UP 1989).

[26] See eg Glenn E Bracey, 'Toward a Critical Race Theory of State' (2015) 41 Critical Sociology 553.

[27] See eg Duncan Ivison, 'Decolonizing the Rule of Law: Mabo's Case and Postcolonial Constitutionalism' (1997) 17 Oxford Journal of Legal Studies 253.

[28] James R May and Erin Daly, *Global Environmental Constitutionalism* (CUP 2014); Lael Weis, 'Theorising a Green Constitutionalism', *Public Law Conference, Dublin 2022* (on file with authors).

[29] Louis J Kotzé, *Global Environmental Constitutionalism in the Anthropocene* (Hart Publishing 2016) 134.

lived experiences are morally, socially, politically, and even legally significant', as Jessica Eisen describes it.[30] Take political philosophy, for instance, where recent work by Alasdair Cochrane, Sue Donaldson, Robert Garner, Eva Meijer, Will Kymlicka, and others has greatly raised the profile of animals, giving rise to a fast-growing literature on the status of animals in political communities.[31] Moral philosophy, too, has witnessed an animal turn. A majority of moral philosophers today hold that animals have moral rights and are deserving of protection.[32] Scholarship in law and legal theory has been no exception either. This is illustrated by the fast-growing community of legal scholars who have turned their attention to analysing the legal rules governing animals,[33] as well as the many theorists exploring animals' legal status from legal-philosophical perspectives.[34]

Recently, the animal turn has also arrived in constitutional law and constitutional theory literature. There is an increasingly robust body of scholarship critiquing domestic constitutions' protections of animals, a body that often highlights the need for significant reforms.[35] In addition, there are valuable cross-country studies that compare how different constitutions approach animals and what protections they (fail to) provide them.[36] Finally, a number of

[30] Jessica Eisen, 'Animals in the Constitutional State' (2017) 15 International Journal of Constitutional Law 909, 912.

[31] Sue Donaldson and Will Kymlicka, *Zoopolis: A Political Theory of Animal Rights* (OUP 2011); Robert Garner and Siobhan O'Sullivan (eds), *The Political Turn in Animal Ethics* (Rowman & Littlefield 2016); Alasdair Cochrane, *Sentientist Politics: A Theory of Global Inter-Species Justice* (OUP 2018); Eva Meijer, *When Animals Speak: Toward an Interspecies Democracy* (New York UP 2019); Alasdair Cochrane, *Should Animals Have Political Rights?* (Polity Press 2020).

[32] See Richard Healey and Angie Pepper, 'Interspecies Justice: Agency, Self-Determination, and Assent' (2021) 178 Philosophical Studies 1223–24.

[33] See eg Anne Peters, Saskia Stucki, and Livia Boscardin, 'The Animal Turn: What Is It and Why Now?' (*Verfassungsblog*, 14 April 2014) <https://verfassungsblog.de/the-animal-turn-what-is-it-and-why-now/> accessed 26 October 2024.

[34] See eg Matthew H Kramer, 'Do Animals and Dead People Have Legal Rights?' (2001) 14 Canadian Journal of Law & Jurisprudence 29; Visa AJ Kurki, *A Theory of Legal Personhood* (OUP 2019).

[35] See eg Kate M Nattrass, '"und die Tiere": Constitutional Protection for Germany's Animals' (2004) 10 Animal Law 283; Margot Michel and Evelyne Schneider Kayasseh, 'The Legal Situation of Animals in Switzerland: Two Steps Forward, One Step Back—Many Steps to Go' (2011) 7 Journal of Animal Law 1; Tagore Trajano de Almeida Silva, 'The Constitutional Defense of Animals in Brazil' in Deborah Cao and Steven White (eds), *Animal Law and Welfare: International Perspectives* (Springer 2016); Mathilde Cohen, 'Of Milk and the Constitution' (2017) 40 Harvard Journal of Law and Gender 115; Elien Verniers, 'Animal Constitutionalism: Paving the Way for Animal Inclusion in the Belgian Constitution' (2022) 10 Global Journal of Animal Law 1.

[36] See eg Jessica Eisen and Kristen Stilt, 'Protection and Status of Animal', *Max Planck Encyclopedia of Comparative Constitutional Law* (2016) <https://oxcon.ouplaw.com/display/10.1093/law-mpeccol/law-mpeccol-e71> accessed 26 October 2024; Mathilde Cohen, 'The Comparative Constitutional Law of Cows and Milk: India and the United States' (2017) 7 Indian Journal of Constitutional Law 1; Maneesha Deckha, 'Constitutional Protections for Animals: A Comparative Animal-Centred and Postcolonial Reading' in Kelly Struthers Montford and Chloë Taylor (eds), *Colonialism and Animality: Anti-Colonial Perspectives in Critical Animal Studies* (Routledge 2020); Cochrane, *Should Animals Have Political Rights?* (n 31) 37–51.

scholars have begun to ask how constitutions could be reimagined to better protect animals. David Bilchitz has taken some of the earliest steps in this direction in his explorations of how the South African constitution—in light of its commitment to 'transformative constitutionalism'[37]—could be interpreted to protect animals' dignity, legal personhood, and rights.[38] Taking domestic constitutional developments as her starting point, Eisen's trailblazing scholarship has made the case for a more general rethinking of constitutionalism:

> Constitutional animal protection requires more of constitutional theory than a simple widening of the circle of dignity-bearers or correction of democratic process deficiencies. Instead, constitutional animal protection requires an account of whether and how constitutionalism might embrace the interests of subjects who will not and cannot speak in the language of law with their own voices.[39]

Eisen herself proposes a vulnerability-based approach to constitutionalism, an idea that is echoed in the work of Maneesha Deckha, who has critiqued anthropocentric legal systems and argued that specific constitutional principles such as the rule of law may be employed to serve animals' interests.[40]

Attempts to make constitutions more inclusive of animals' interests have not been limited to the academic literature, either. In recent decades, animal rights advocates in numerous countries have pushed for better protection for animals. Activist organizations, such as Direct Action Everywhere, Animal Rising, and Animal Rebellion, have been involved in rescues of farmed animals or have staged acts of civil disobedience such as blockading streets to raise public awareness and bring an end to animal exploitation.[41] Other groups have

[37] See Michaela Hailbronner, 'Transformative Constitutionalism: Not Only in the Global South' (2017) 65 American Journal of Comparative Law 527.

[38] David Bilchitz, 'Moving Beyond Arbitrariness: The Legal Personhood and Dignity of Non-Human Animals' (2009) 23 South African Journal on Human Rights 38; David Bilchitz, 'Does Transformative Constitutionalism Require the Recognition of Animal Rights?' in Stu Woolman and David Bilchitz (eds), *Is This Seat Taken? Conversations at the Bar, the Bench and the Academy About the South African Constitution* (Pretoria University Law Press 2012).

[39] Eisen (n 30) 910.

[40] See Maneesha Deckha, 'Initiating a Non-Anthropocentric Jurisprudence: The Rule of Law and Animal Vulnerability Under a Property Paradigm' (2013) 50 Alberta Law Review 783; Maneesha Deckha, *Animals as Legal Beings: Contesting Anthropocentric Legal Orders* (University of Toronto Press 2021); Maneesha Deckha, 'Supplanting Anthropocentric Legalities: Can the Rule of Law Tolerate Intensive Animal Agriculture?' in Matilda Arvidsson and Emily Jones (eds), *International Law and Posthuman Theory* (Routledge 2024). See also Janneke Vink, *The Open Society and Its Animals* (Palgrave Macmillan 2020) (who rethinks liberal democratic institutions with a view to better protecting animals' interests).

[41] Direct Action Everywhere, 'Until Every Animal Is Free' <https://www.directactioneverywhere.com/> accessed 26 October 2024; Animal Rising, 'Transformative Campaigns for All Life' <https://

aimed for reform by leveraging legal tactics. A particularly prominent such strategy has been the pursuit of *habeas corpus* rights for animals. This strategy, which was first attempted in Brazil in 1972 and has more recently made the headlines with the Nonhuman Rights Project's (NhRP) litigation efforts on behalf of chimpanzees and elephants in the United States, seeks to have captive animals recognized as legal persons and moved to sanctuaries.[42] A *habeas corpus* petition was also the trigger for a groundbreaking decision in 2022 by the Ecuadorian Constitutional Court which held that individual wild animals are beneficiaries of fundamental rights under the Ecuadorian Constitution's rights of nature provision.[43]

In parallel to such litigation efforts, animal rights defenders have sought to amend constitutions. In 2017, the Swiss NGO Sentience Politics, inspired by proposals of the Great Ape Project,[44] launched a popular initiative in the Canton of Basel-Stadt with the aim of adding the right to life and the right to bodily and mental integrity for all non-human primates to the Constitution's bill of rights. A quarter of citizens approved the initiative at the ballot box.[45] In 2019, the Finnish Animal Rights Lawyers Society proposed a new Chapter 2A for the Finnish Constitution. Chapter 2A contains an extensive catalogue of constitutional rights, including, among others, wild animals' rights to life and to live freely in their natural habitats and the rights to life, shelter, food, drink, medical care, and freedom from pain for animals who are dependent on humans.[46]

Despite these developments and the valuable scholarly contributions mentioned above, the discourse on animal constitutionalism still lacks a study that scrutinizes anthropocentric constitutionalism, explains the need for an animal-focused framework, and surveys its implications in detail. This book takes up these challenges and advances the conversation by offering the first extensive treatment of constitutionalism for all sentient beings. It deepens our understanding of the flaws of current constitutional models by revealing

www.animalrising.org> accessed 26 October 2024; Animal Rebellion, 'Creating A Plant-Based Future' <https://animalrebellion.org/> accessed 26 October 2024.

42 See Nonhuman Rights Project, 'Litigation' <https://www.nonhumanrights.org/litigation/> accessed 26 October 2024. For an overview of this litigation, see Fasel and Butler (n 23) 124–34.

43 See Estrellita (n 21).

44 Paola Cavalieri and Peter Singer (eds), 'A Declaration on Great Apes', *The Great Ape Project: Equality Beyond Humanity* (St Martin's Griffin 1993).

45 Sentience Politics, 'Fundamental Rights for Primates: First Direct Democratic Vote on Animal Rights Worldwide' (*Initiative Grundrechte für Primaten*) <https://sentience.ch/en/project/initiative-basic-rights-for-primates/> accessed 31 October 2024.

46 Finnish Animal Rights Lawyers Society, 'Proposal: The Constitution of Finland' (2020) <https://www.elaintenvuoro.fi/english/> accessed 26 October 2024.

the problematic ways in which they exclude animals and marginalize non-paradigmatic humans, and it makes an original case for why constitutions should promote the interests of all sentient beings they govern. In addition, it illustrates how four central principles of constitutionalism—fundamental rights, proportionality, rule of law, and democracy—can be rethought to cater for the interests of all sentient beings. These explorations can serve as a practical roadmap for constitutional lawyers, judges, and policymakers to reinterpret and redesign constitutions to overcome at least some of the injustices that mar contemporary constitutions.

As our analysis will make clear, even seemingly animal-friendly constitutions suffer from significant, even extreme legitimacy deficits. In other words, constitutions have a long way to go to become sentience-based constitutions. The problem of illegitimate constitutions is not new, however, and it is certainly no reason for despondency.[47] Indeed, as we will explain in Chapter 2, legitimacy is a matter of degree: the more a constitution can be changed to give due concern to all governed sentient beings, the more legitimate it will become. The constitution of ancient Athens was only minimally legitimate because of the many beings it excluded. Modern-day constitutions—some of them, anyway—are more legitimate because they give due concern to significantly more human beings. And future constitutions may be more legitimate still by giving due concern to governed sentient animals, too. As Chapters 3–6 will illustrate, there are numerous ways in which constitutions can fill the current legitimacy gap. These include extending fundamental rights to animals, making the proportionality principle more amenable to their interests, accommodating them under the rule of law, and recognizing their equal right to democratic participation.

3) A realistic utopia for sentient beings

Some clarifications regarding our focus and methodology will be helpful at this point.

First, the book's *method*. This is a work of constitutional theory, but it adopts a grounded approach that draws on constitutional texts and practices to inform its normative arguments. It also crosses disciplinary boundaries into political philosophy and political theory when dealing with the exclusion of animals

[47] See eg Karl Marx, *A Contribution to the Critique of Political Economy* (NI Stone tr, Charles H Kerr 1904); Karl Marx, *The German Ideology* (Prometheus Books 1998) 'Theses on Feuerbach', 571.

and non-paradigmatic humans from the polity. Our approach is rooted in the liberal tradition of these disciplines, not least because the dominant model of constitutionalism is a liberal one, as we will explain below. However, we will also draw on critical traditions to question and rethink the classic liberal approach to constitutionalism.

We are both based in the United Kingdom and will occasionally focus on the United Kingdom to illustrate our points, as we have done above. However, nothing in this book should be interpreted to limit the applicability of sentience-based constitutionalism to any one jurisdiction. To drive this point home, the book will draw on examples from constitutions around the world to scrutinize the *status quo* of animals and to unpack how the principles of constitutionalism can be rethought for animals. In keeping with the inclusivity of sentience-based constitutionalism, we endeavour to adopt an inclusive approach in our constitutional comparisons by drawing on a variety of oft-neglected constitutions, rather than just the 'usual suspects',[48] from both the civil and common law world.

Second, an explanation of our *focus* on constitutionalism and sentient beings. Starting with the former, it is important to note that, unlike some of the accounts we mentioned in the previous section, our theory does not challenge constitutionalism with a view to overcome it. We start from the premise that, although it needs to be revised, constitutionalism is a useful approach in principle. In other words, we take constitutionalism to be both our starting and end point. As Eisen has observed, every theory has a 'bedrock'—things it takes to be inevitable—and 'linchpins'—things it thinks can be transformed.[49] One of the bedrocks of this book is constitutionalism, which we understand as the idea that public powers ought to be limited, including by a constitution meeting certain requirements, such as the curtailment of arbitrary power.[50] The anthropocentrism of current constitutionalism, on the other hand, is its linchpin. The former we take as a given, the latter we believe must change.

Constitutionalism is our bedrock for two reasons. First, because it is the bedrock of many others, too. As Bruce Ackerman has argued, the twentieth century has seen the rise of 'world constitutionalism'.[51] And although the beginning of

[48] Ran Hirschl, *Comparative Matters: The Renaissance of Comparative Constitutional Law* (OUP 2014) 192.

[49] Jessica Eisen, 'Of Linchpins and Bedrock: Hope, Despair, and Pragmatism in Animal Law' (2022) 72 University of Toronto Law Journal 468.

[50] For more detail, see Chapter 2 Subsection 1.B. Even though animals may possess their own social norms (see Andrews, Fitzpatrick, and Westra (n 10)), they most likely lack constitutionalism in the sense described here, as indeed have many human societies historically. But cf also Stephen Cornell, '"Wolves Have a Constitution": Continuities in Indigenous Self-Government' (2015) 6 International Indigenous Policy Journal 1.

[51] Bruce Ackerman, 'The Rise of World Constitutionalism' (1997) 83 Virginia Law Review 771.

the twenty-first century has witnessed mounting challenges to constitutionalism,[52] even its latest detractors acknowledge that constitutionalism is 'rapidly becoming the world's most influential contemporary philosophy of government'.[53] Given its prevalence in both practice and theory, reforming constitutionalism has better prospects than abandoning it in favour of some other as-of-yet untested idea. Our book aims to develop a theory that, in John Rawls' words, is 'realistically utopian'.[54] That is, our goal is not to present an account of constitutionalism that is perfectly just but stands no chance of being put into practice. Nor do we wish to fall into the opposite trap of defending an account so conservative that it sacrifices necessary change on the altar of practicability. Caught between the devil of utopianism and the deep blue sea of realism, we chart a course that pursues ideal lands while remaining anchored in firm ground. This approach, we hope, grants sentience-based constitutionalism a viable path towards adoption, even if that moment lies beyond the immediate horizon.

Second, we do not think it would be desirable to abandon constitutionalism, even if it was feasible to do so. This is because constitutionalism is a normatively attractive idea. Among its central virtues are its ability to challenge arbitrary exercises of power and to help protect fundamental rights.[55] Of course, constitutionalism is not without its flaws, neither on paper nor on the ground.[56] One of the criticisms levelled at constitutionalism, for instance, is that it is based on liberal concepts such as rights, legal personhood, and the rule of law. As Anna Grear has pointed out, these concepts have tended to 'prioriti[ze] a highly particular, historically dominant, identifiable beneficiary: the paradigmatic European property-owning male citizen'.[57] However, in our view, it would be a category mistake to infer from this gloomy history that the future of these concepts is equally bleak. As this book will show, it is possible to reclaim constitutionalism and to harness concepts such as the

[52] See eg Andrew Arato and Jean L Cohen, *Populism and Civil Society: The Challenge to Constitutional Democracy* (OUP 2021).

[53] Martin Loughlin, *Against Constitutionalism* (Harvard UP 2022) x.

[54] John Rawls, *Justice as Fairness: A Restatement* (Harvard UP 2001) 4. For this, and many other things, we take inspiration from Cochrane: Cochrane, *Sentientist Politics* (n 31) 11.

[55] See eg Charles Howard McIlwain, *Constitutionalism: Ancient and Modern* (Liberty Fund 1947) 21; Dieter Grimm, *Constitutionalism: Past, Present, and Future* (OUP 2016) 364; Maria Tzanakopoulou, *Reclaiming Constitutionalism: Democracy, Power and the State* (Hart Publishing 2018) 200. For a nuanced empirical analysis of constitutional rights, see Adam Chilton and Mila Versteeg, *How Constitutional Rights Matter* (OUP 2020).

[56] See eg Loughlin (n 53); Rosalind Dixon and David Landau, *Abusive Constitutional Borrowing: Legal Globalization and the Subversion of Liberal Democracy* (OUP 2021).

[57] Anna Grear, 'Legal Imaginaries and the Anthropocene: "Of" and "For"' (2020) 31 Law and Critique 351, 353.

rule of law and democracy to allow all governed sentient animals—and not just humans—to savour constitutionalism's fruits.

But why focus on *sentient* animals rather than all animals? Indeed, why not cast constitutionalism's net even more widely to include plants, ecosystems, and other natural entities? As we pointed out above, we view the theory of sentience-based constitutionalism as complementary to other proposals for future-oriented constitutionalisms, including models of environmental constitutionalism, which aim to extend constitutional protections to a larger set of natural entities. As we will see in Chapter 2, we take sentience merely to be a sufficient condition, not a necessary condition, for possessing inherent moral value. As such, we are open to the idea that a wider range of natural entities might be considered governed by constitutions and thus potentially have constitutional worth.

At the same time, there are good reasons for focusing on sentient animals. There is widespread recognition today that many animals are sentient and have complex needs that ought to be protected by the law. As noted above, sentient beings possess the remarkable ability to have valenced experiences. They have at least a minimal degree of awareness of the things that are happening around them and the capacity to cognitively process them.[58] Simply put, they can *feel things*. As Thomas Nagel has famously described it, for an organism that has such experiences, 'there is something it is like to *be* that organism'.[59] Sentience thus distinguishes its holders from insentient beings such as bacteria or jellyfish, and inanimate entities such as tables and rocks, which all lack these capacities.[60] One of the advantages of focusing on sentience is that it can be measured scientifically.[61] Today, there is compelling scientific evidence that at least all vertebrate animals—a category that includes humans and other mammals, as well as amphibians, reptiles, and fish—are sentient. It is also widely accepted that some invertebrates such as cephalopods and decapod crustaceans possess sentience.[62]

As we will see, sentient animals' special place in nature is also recognized in law.[63] Many states base their animal protection laws on a recognition

[58] Cochrane, *Sentientist Politics* (n 31) 15; Donald M Broom, *Sentience and Animal Welfare* (CABI 2014) xiii.

[59] Thomas Nagel, 'What Is It Like to Be a Bat?' (1974) 83 The Philosophical Review 435, 436 (emphasis in original).

[60] On jellyfish, see Birch (n 13) 245.

[61] Broom (n 58) 15.

[62] Helen Proctor, 'Animal Sentience: Where Are We and Where Are We Heading?' (2012) 2 Animals 628, 632–35; Broom (n 58) 122.

[63] Chapter 2. See generally Jane Kotzmann and MB Rodriguez Ferrere, *The Legal Recognition of Animal Sentience: Principles, Approaches and Applications* (Hart Publishing 2024).

of animals' ability to have valenced experiences. As Charlotte Blattner has pointed out, there is 'an emergent universal consensus among state actors that sentience serves as a gateway to according animals protection under the law'.[64] To give just one example, New Zealand's Animal Welfare Act (1999) states that it is an Act 'to reform the law relating to the welfare of animals and the prevention of their ill-treatment; and, in particular, (i) to recognize that animals are sentient'.[65] While it can be the case that certain laws also protect other characteristics of animals, such as their vulnerability,[66] the legal and political consensus on the importance of sentience appears to be the most universal. As such, sentience is uniquely suited for a theory like ours, which aspires to be utopian *and* realistic.

What about plants? Are they sentient?[67] In recent times, the view has become more prominent that they, too, may possess some kind of sentience. Alessandro Pelizzon and Monica Gagliano, for example, have argued that 'plants exhibit sophisticated abilities to actively perceive, assess, learn, remember, solve problems, make decisions and communicate with each other by acquiring information from their environment, all indicators of human and non-human sentience'.[68] Like-minded authors have proposed that we should see plants as 'persons',[69] or as having 'subjectivity'.[70] However, these claims have largely remained contested. Specifically, the view that plants have sentience in a comparable sense to that of humans and animals is rejected by a majority of plant biologists.[71]

What about robots and artificial intelligences (AIs)? Some scholars have already considered the relationship between AIs and constitutionalism, albeit with a focus on how constitutionalism can be used to mitigate perceived risks posed by AIs, rather than to protect AIs themselves.[72] And, as with plants, there

[64] Charlotte E Blattner, 'The Recognition of Animal Sentience by the Law' (2019) 9 Journal of Animal Ethics 121, 125.

[65] Long title of the Animal Welfare Act 1999 (NZ).

[66] See Martha Fineman Albertson and Anna Grear (eds), *Vulnerability: Reflections on a New Ethical Foundation for Law and Politics* (Routledge 2013); Eisen (n 30) 940–50.

[67] For a discussion of this and other questions about the boundaries of sentience, see Birch (n 13).

[68] Alessandro Pelizzon and Monica Gagliano, 'The Sentience of Plants: Animal Rights and Rights of Nature Intersecting?' (2015) 11 Australian Animal Protection Law Journal 5, 11.

[69] Matthew Hall, *Plants as Persons: A Philosophical Botany* (SUNY Press 2011).

[70] Michael Marder, *Plant-Thinking: A Philosophy of Vegetal Life* (Columbia UP 2013) 8.

[71] For an overview, see Jon Mallatt and others, 'Debunking a Myth: Plant Consciousness' (2021) 258 Protoplasma 459.

[72] See eg Carlo Casonato, 'AI and Constitutionalism: The Challenges Ahead' in Bertrand Braunschweig and Malik Ghallab (eds), *Reflections on Artificial Intelligence for Humanity* (Springer 2021); Hans-W Micklitz and others (eds), *Constitutional Challenges in the Algorithmic Society* (CUP 2021); Toni M Massaro and Helen Norton, 'Siri-Ously? Free Speech Rights and Artificial Intelligence' (2016) 110 Northwestern University Law Review 1169.

is increasing interest in whether AIs are or could potentially become sentient at some point. Undoubtedly, AIs are becoming better and better at pretending to be sentient. For example, Shelly, a robot tortoise designed to teach children not to abuse robots, hides in its plastic shell when beaten.[73] However, the consensus among many experts is that AIs are not currently sentient,[74] and some doubt that they ever will or can be.[75]

Certain authors in science and technology studies and actor-network theory go even further by proposing to include inanimate objects in our politics, too.[76] Bruno Latour, for example, has advanced the idea of a 'parliament of things'.[77] According to him, not only humans and other sentient animals are agents to be included in this 'parliament', but a much wider range of entities too, including intestinal bacteria, mushrooms, and soils.[78] While thought-provoking, we will not discuss these proposals any further in this book because they are not centred on sentient beings. Writing in the liberal tradition, we adopt a less sweeping understanding of who is a relevant agent—or, in our terminology, who is governed—in a constitution.

For these reasons, we will not discuss non-sentient animals, plants, AIs, or inanimate objects, although there is nothing in principle that would prevent our account from applying to other beings should they be shown to be sentient at some point in the future. Going forward, we will centre on sentient animals. However, as will have become obvious from the discussion above, our book also devotes space to discussing non-paradigmatic humans, even though our main focus will be on animals. As we will see, the same theories and practices that exclude animals from constitutional protection are often the very ones that also marginalize non-paradigmatic humans. As such, it is at times helpful to consider the two together.

[73] Evan Ackerman, 'Robotic Tortoise Helps Kids to Learn That Robot Abuse Is a Bad Thing' *IEEE Spectrum* (14 March 2018) <https://spectrum.ieee.org/shelly-robotic-tortoise-helps-kids-learn-that-robot-abuse-is-a-bad-thing> accessed 26 October 2024.

[74] See John Basl and Joseph Bowen, 'AI as a Moral Right-Holder' in Markus D Dubber, Frank Pasquale, and Sunit Das (eds), *The Oxford Handbook of Ethics of AI* (OUP 2020) 297–300; Martin Gibert and Dominic Martin, 'In Search of the Moral Status of AI: Why Sentience Is a Strong Argument' (2022) 37 AI & Society 319.

[75] See eg Yoshija Walter and Lukas Zbinden, 'The Problem with AI Consciousness: A Neurogenetic Case Against Synthetic Sentience' (arXiv, 7 December 2022) <http://arxiv.org/abs/2301.05397> accessed 26 October 2024.

[76] See eg Gunther Teubner, 'Rights of Non-Humans? Electronic Agents and Animals as New Actors in Politics and Law' (2006) 33 Journal of Law and Society 497; Andreas Fischer-Lescano, 'Nature as a Legal Person: Proxy Constellations in Law' (2020) 32 Law & Literature 237; Donna J Haraway and Cary Wolfe, *Manifestly Haraway* (University of Minnesota Press 2016) 'A Cyborg Manifesto: Science, Technology, and Socialist-Feminism in the Late Twentieth Century', 5.

[77] Bruno Latour, *We Have Never Been Modern* (Catherine Porter tr, Harvard UP 1993) 142.

[78] Bruno Latour, *Down to Earth: Politics in the New Climatic Regime* (Polity 2018) 87, 124–25.

4) How our analysis unfolds

The book has two parts. Part I examines *what* status animals currently occupy in constitutional theory and practice, revealing that the focus is primarily on human beings and often centred around paradigmatic humans. It then shows *why* this anthropocentrism is problematic and why we should adopt sentience-based constitutionalism instead. Part II builds on the foundations laid in Part I and explores *how* current constitutions can be transformed into sentience-based constitutions. To do so, it proposes ways of rethinking four cardinal principles of constitutionalism: fundamental rights, proportionality, rule of law, and democracy.

Part I opens with Chapter 1, which addresses the *status quo* of constitutional theory and practice. As we will show, legal systems are generally anthropocentric in at least four ways: they only recognize human beings as having (full) intrinsic value; they operate with a 'use paradigm' that justifies the exploitation of animals; they treat animals as ownable things; and they operate with human-centric concepts and/or interpret concepts that are potentially neutral in an anthropocentric way. Turning to constitutions specifically, we analyse how anthropocentrism manifests itself in influential theories of constitutionalism as well as how it operates in the interpretation and application of key constitutional principles such as fundamental rights, proportionality, rule of law, and democracy. The chapter ends by arguing that the constitution-level protections that certain animals have received in some countries fall well short of overcoming anthropocentrism.

Chapter 2 addresses *why* it is necessary to replace anthropocentric constitutionalism with sentience-based constitutionalism. Anthropocentric constitutionalism's main problem is that its dominant legitimating principle—the (hypothetical) consent of the governed—excludes governed sentient animals and marginalizes non-paradigmatic human beings. As such, anthropocentric constitutionalism struggles to justify exercises of public power governing animals and non-paradigmatic humans. This is the legitimacy gap we identified above. To fill this gap, we make the case for sentience-based constitutionalism, which is based on a more inclusive legitimating principle: due concern for the interests of all those who are governed (the All Governed Interests Principle). This principle, we argue, has no difficulty justifying exercises of public power over sentient animals and non-paradigmatic humans. Finally, we defend sentience-based constitutionalism against two objections that could be raised against it.

The chapters in Part II imagine *how* a sentience-based constitution would look. Chapter 3 begins with fundamental rights. It asks: which fundamental rights would be recognized for animals in a sentience-based constitution, and what shape would rights catalogues have in a constitution devoted to the protection of all sentient beings, human and non-human alike? We answer these questions by arguing that, in many cases, the creation of new rights for animals is not necessary. This is because existing fundamental rights practice has features—such as its progressive expansionism, the absence of a formal hierarchy between rights-holders, the over-inclusiveness of existing bills of rights, and the generalized way in which these rights are formulated—which make it possible for existing catalogues to be extended to animals, thus obviating the need for an animal bill of rights. We exemplify how such an extension of fundamental constitutional rights can work by focusing on the right to life, the prohibition of torture and inhuman or cruel treatment, and the right to freedom.

Once animals are recognized as holders of fundamental rights, then the question arises of how potential conflicts between their rights and humans' fundamental rights should be resolved. Chapter 4 focuses on the proportionality test, which is used nearly universally by courts to deal with conflicts of rights. We argue that this test can also be applied to resolve conflicts in a sentience-based constitution. In fact, as we show, harnessing the proportionality test for the rights of all governed animals helps identify areas for improvement in existing conceptualizations of the test, some of which are problematically centred on paradigmatic humans. We also maintain that the test comes with a minimum commitment to what we call the Non-Instrumentalization Principle, which requires that all beings with intrinsic value must not be treated as mere means to others' ends. The chapter shows what difference this principle makes to how the proportionality test is commonly understood and provides a step-by-step analysis of how the test can resolve conflicts of rights that might arise under a sentience-based constitution.

Chapter 5 sheds new light on the rule of law. We start from the finding that several of the leading rule of law theories are premised on the basic idea that the law should be capable of guiding the actions of legal subjects—a basic idea that is problematically centred on paradigmatic human beings. We challenge this commonplace assumption and argue that the rule of law's basic idea should instead be understood as the restraint of arbitrary power to protect those who can be victims of such power. This basic idea is more inclusive, covering not only paradigmatic humans but also animals and non-paradigmatic humans, for the straightforward reason that all of them can be victims of arbitrary power. The chapter then spells out how this inclusive approach transforms formal, procedural, and substantive accounts of the rule of law. While protections will vary depending on the type of account that is employed, the protection of animals will no longer be a contingent matter under any account.

Chapter 6 asks what the proper place is for animals in democracy. We take issue with animals' current exclusion from the *demos* and propose that, in a sentience-based democracy, animals would have an equal right to democratic participation. This does *not* mean that animals should be granted formal political rights such as the right to vote. Rather, it means that democratic processes ought to be restructured in ways that afford due concern to the interests of all sentient beings whom these processes govern. The chapter demonstrates how this can be done, highlighting four key ingredients. First, human representatives are required to channel animals' interests into formal democratic processes. Second, human representatives can only effectively exercise this role if animals participate in their representation through processes that take seriously their informal political agency. Third, a combination of both legal and political mechanisms is necessary to protect animals' right to democratic participation. Finally, the chapter returns to Athens in at least one respect by moving beyond the literature's current focus on indirect democratic institutions and by arguing that, to give animals their full due, these institutions should be complemented by direct democratic mechanisms such as 'sentient beings' initiatives'.

The book concludes by discussing the broader implications of the proposed shift to sentience-based constitutionalism, how its ideas might be used, and what directions of research it opens.

PART I

1

Anthropocentric constitutionalism

Introduction

We live in constitutional systems that are deeply anthropocentric. That is, constitutional thinking and practices are centred on human beings and privilege their interests over the interests of non-human beings. Our constitutions speak of 'we the people', grant fundamental rights to people, and enable people to use the natural world as a resource for human purposes. Although this human-centeredness is one of the—if not *the*—defining mark of current constitutionalism, it is rarely remarked upon in mainstream discourse. Anthropocentrism is right under our human nose.

The aim of this chapter is to unearth the anthropocentric presuppositions on which the current model of constitutionalism is based. We will do so by exploring the assumptions behind current constitutional thinking as well as demonstrating how these assumptions manifest themselves in constitutional practices. As will emerge from this overview of the *status quo* of anthropocentric constitutionalism, constitutional orders are primarily focused on the interests of human beings and exclude those of animals. As we will show, this exclusion of animals often occurs in ways that marginalize those human beings who are not considered to be paradigmatic examples of their species, usually because they are believed to lack the rationality of 'normal' adult legal agents.

There are already prominent critiques of law's focus on rationality and on other 'paradigmatic' features that serve to marginalize certain human beings.[1] As mentioned in the Introduction, law's anthropocentrism has been subject to critical scrutiny too. For example, it has been interrogated by writers in the environmental tradition.[2] Animal law and critical animal studies scholars,

[1] For an overview of feminist critiques, see eg Rosemary Hunter, 'Contesting the Dominant Paradigm: Feminist Critiques of Liberal Legalism' in Vanessa E Munro and Margaret Davies (eds), *The Ashgate Research Companion to Feminist Legal Theory* (Routledge 2018). For critical race theories, see Richard Delgado, Jean Stefancic, and Angela Harris, *Critical Race Theory: An Introduction* (3rd edn, New York UP 2017) 3. On postcolonial theories: Alpana Roy, 'Postcolonial Theory and Law: A Critical Introduction' (2008) 29 Adelaide Law Review 315. And on disability legal studies: Anna Lawson, 'Disability Law as an Academic Discipline: Towards Cohesion and Mainstreaming?' (2020) 47 Journal of Law and Society 558.

[2] See eg Cormac Cullinan, *Wild Law: A Manifesto for Earth Justice* (Green Books 2011); Louis J Kotzé and Duncan French, 'The Anthropocentric Ontology of International Environmental Law and

Animals and the Constitution. John Olusegun Adenitire and Raffael Fasel, Oxford University Press.
 DOI: 10.1093/9780198910534.003.0001

too, have challenged law's human focus.[3] Our chapter will build on these and other accounts to focus on how anthropocentrism operates in contemporary constitutions and constitutional thinking. As we will demonstrate, anthropocentrism can be found, among other things, in how four central principles of constitutionalism—fundamental rights, proportionality, the rule of law, and democracy—are interpreted and applied; and it can be found at the heart of both classic and more recent theories of constitutionalism.

Even though this chapter will demonstrate that current constitutional thinking and practices serve to exclude or marginalize animals and, sometimes, non-paradigmatic humans, it will not criticize these exclusions or marginalizations. Instead, its purpose and method are expository: it will unearth and present the anthropocentrism that is prevalent in today's legal and, in particular, constitutional systems. When reading this chapter, a sceptical reader might therefore wonder whether there is anything wrong with the manifestations of anthropocentrism that we are describing, given that law and constitutions are, at the end of the day, human creations. This reader will find their answer in Chapter 2, which offers a critique of anthropocentric constitutionalism and makes the case for sentience-based constitutionalism.

The chapter is structured as follows. Because constitutions are set in broader legal systems, Section 1 discusses anthropocentrism in law generally by defining the term 'anthropocentrism' and outlining the four most important ways in which current legal systems are anthropocentric. Section 2 then focuses specifically on constitutional anthropocentrism and reveals how the four above-mentioned constitutional principles are interpreted and applied anthropocentrically. We also discuss the theories of constitutionalism which have influenced constitutional practices and we shed light on their anthropocentrism. Finally, in Section 3, we consider an emerging trend towards recognizing animals' interests in constitutions. Pointing to the shortcomings of existing constitutional protections of animals, we argue that this trend is far from supplanting anthropocentric constitutionalism.

the Sustainable Development Goals: Towards an Ecocentric Rule of Law in the Anthropocene' (2018) 7 Global Journal of Comparative Law 5.

[3] See eg Maneesha Deckha, 'Critical Animal Studies and Animal Law' (2012) 18 Animal Law 207; Maneesha Deckha, *Animals as Legal Beings: Contesting Anthropocentric Legal Orders* (University of Toronto Press 2021); Jessica Eisen, 'Animals in the Constitutional State' (2017) 15 International Journal of Constitutional Law 909; Challie Facemire and others, 'Law Through the Eyes of Animals' in Michael J Glover and Les Mitchell (eds), *Animals as Experiencing Entities: Theories and Historical Narratives* (Palgrave Macmillan 2024).

1) Legal anthropocentrism

In this section, we provide a brief overview of the ways in which law in general is anthropocentric. This will set the context for our discussion of anthropocentrism in constitutional law and constitutional thinking in the next section. To reveal the anthropocentrism of current legal systems, it is helpful first to explain what we mean by 'anthropocentrism' (Subsection A). We will then shed light on the four most important ways in which anthropocentrism manifests itself in law (Subsection B).

A) Anthropocentrism and law

The Cambridge mathematician and philosopher Frank Ramsey once quipped: 'My picture of the world is drawn in perspective, and not like a model to scale. The foreground is occupied by human beings and the stars are as small as threepenny bits.'[4] Although made in a different context, Ramsey's statement encapsulates the central characteristics of anthropocentrism: an outlook on the world that focuses on the human being and eclipses anything that is not human. What Ramsey's quote perhaps only implies is the strong evaluative component that generally marks anthropocentrism. As Peter Burdon describes it, anthropocentrism is 'a mode of thinking that separates human beings from nature and regards the latter as existing for the use and exploitation of the former'.[5] Ben Mylius, who distinguishes between different kinds of anthropocentrism, refers to this evaluative element as 'normative anthropocentrism'.[6] According to Mylius, normative anthropocentrism is a paradigm that either passively privileges the human species or actively makes prescriptions based on assertions about 'the superiority of *Homo sapiens*, its capacities, the primacy of its values'.[7] Importantly, as Mylius argues, there are also non-normative forms of anthropocentrism such as 'perceptual anthropocentrism'[8] (the fact that humans have no choice but to perceive things through human eyes, think things through human minds, etc) and 'descriptive anthropocentrism'[9] (a

[4] Frank Plumpton Ramsey, *The Foundations of Mathematics and Other Logical Essays* (Routledge 1931) 291.

[5] Peter D Burdon, 'Ecological Law in the Anthropocene' (2020) 11 Transnational Legal Theory 33, 38.

[6] Ben Mylius, 'Three Types of Anthropocentrism' (2018) 15 Environmental Philosophy 159, 159.

[7] ibid 184–85 (emphasis in original).

[8] ibid 167.

[9] ibid 169.

way of viewing the world that centres around the human, without necessarily privileging it normatively).

The kinds of anthropocentrism that are important in law—and therefore to our analysis here—are descriptive and normative anthropocentrism. Law operates with descriptive paradigms that revolve around humanity. And these paradigms, in turn, underpin law's normative anthropocentrism when it communicates and enforces prescriptions that promote the human species and its interests. 'Legal texts and institutions', as Irus Braverman puts it, 'tend toward and prescribe certain human–animal ideologies.'[10] These ideologies are often viewed to entail a kind of human chauvinism, which 'at its core . . . involves the planetary-scale subordination of nonhuman organisms that denies they have value in their own right'.[11] Historically, the primary reason for the subordination of animals has been the belief that, particularly in dominant Western traditions, only humans are considered to have intrinsic value—which itself was explained by reference to allegedly unique human properties such as rationality, culture, language, or the possession of a soul. Because animals were believed to lack these properties, they were—and often still are—not considered to have intrinsic value.[12] In Ramsey's words, they are 'as small as threepenny bits'.[13]

It is important to note that while anthropocentrism in law privileges the human species as a whole, it also operates with a certain built-in hierarchy.[14] For it tends to privilege the viewpoints and interests of some human beings over others. The reason for this internal hierarchy is directly connected to the above-mentioned justifications adduced for human superiority. Specifically the view that human beings are rational autonomous agents[15]—able to forge their own paths in life and to exercise their rights and take responsibility for their actions—has led to a marginalization of human beings who, for whatever reason, lack this capacity (or are believed to lack it).[16] Anna Grear has made this point forcefully when critiquing the 'hierarchies operative within the "anthropocentrism" of law'.[17] Law, Grear argues, is based on the 'construct

[10] Irus Braverman, 'Law's Underdog: A Call for More-than-Human Legalities' (2018) 14 Annual Review of Law and Social Science 127, 140.

[11] Helen Kopnina and others, 'Anthropocentrism: More than Just a Misunderstood Problem' (2018) 31 Journal of Agricultural and Environmental Ethics 109, 115.

[12] For a brief overview of the intellectual history of anthropocentrism, see Alexander Gillespie, *International Environmental Law, Policy, and Ethics* (2nd edn, OUP 2014) 4–10.

[13] Ramsey (n 4) 291.

[14] On hierarchies more generally, see Claire Jean Kim, *Dangerous Crossings: Race, Species, and Nature in a Multicultural Age* (CUP 2015) 15–21.

[15] Gillespie (n 12) 13.

[16] See eg Peter Halewood, 'Law's Bodies: Disembodiment and the Structure of Liberal Property Rights' (1996) 81 Iowa Law Review 1331, 1343.

[17] Anna Grear, 'Deconstructing Anthropos: A Critical Legal Reflection on "Anthropocentric" Law and Anthropocene "Humanity"' (2015) 26 Law and Critique 225, 227.

of a paradigmatic "rational human subject"'.[18] According to Grear, law's focus on those who are viewed as paradigmatic human subjects not only prioritizes those humans over animals (creating 'inter-species hierarchies'[19]) but also prioritizes the interests of paradigmatic humans over non-paradigmatic humans' such as children, persons with cognitive disabilities, or persons who are senile (leading to 'intra-species hierarchies'[20]). Hence, some humans are seen as 'others'[21] who, despite belonging to the human species, are not considered its prime examples. As Maneesha Deckha summarizes the point, '[l]aw is an anthropocentric terrain. Not only is law the product of human actors, it entrenches the interests of humans over virtually all others and centres the reasonable human person as a main legal subject'.[22]

B) Manifestations of legal anthropocentrism

How exactly does this normative anthropocentrism show itself in law? We focus here on the inter-species hierarchies between humans and animals, and identify the four most significant manifestations.[23] As we will see, these manifestations in part also demonstrate the intra-species hierarchies within anthropocentrism, but they establish a particularly sharp contrast between humans and the non-human world.

First, anthropocentric legal systems generally *only recognize human beings as having (full) intrinsic value*. Environmental advocates in particular have pressed the point that laws are human-centric in that they tend to ignore that nature can have intrinsic value too.[24] As Joshua Bruckerhoff has argued, '[m]ost environmental laws are anthropocentric: they focus primarily on preventing and remedying only those environmental problems that directly affect humans'.[25] There are some exceptions to this. For instance, some legal systems

[18] ibid 230.

[19] ibid.

[20] ibid.

[21] ibid 231. On 'Othering' more generally, see Poul Rohleder, 'Othering' in Thomas Teo (ed), *Encyclopedia of Critical Psychology* (Springer 2014).

[22] Maneesha Deckha, 'Initiating a Non-Anthropocentric Jurisprudence: The Rule of Law and Animal Vulnerability Under a Property Paradigm' (2013) 50 Alberta Law Review 783, 784.

[23] For a similar but partially divergent list, see Peter D Burdon, 'The Great Jurisprudence' in Peter D Burdon (ed), *Exploring Wild Law: The Philosophy of Earth Jurisprudence* (Wakefield Press 2011) 60–63 (who also emphasizes the anthropocentric character of mainstream legal theories).

[24] ibid 63.

[25] Joshua J Bruckerhoff, 'Giving Nature Constitutional Protection: Less Anthropocentric Interpretation of Environmental Rights' (2008) 86 Texas Law Review 615, 618.

have recognized rivers and mountains as having intrinsic value.[26] In even more legal systems, animals have for some time now benefited from anti-cruelty laws that are enacted to protect them from suffering, arguably in recognition of their intrinsic value.[27] Numerous countries even have more wide-ranging animal welfare laws that go beyond the mere protection of animals against cruelty and impose positive obligations on human to cater for the well-being of animals.[28]

However, not only do more than a fifth of all states still lack animal protection laws altogether,[29] but it is also widely recognized among experts that the laws in countries that do possess such laws still have serious shortcomings. These shortcomings include, but are not limited to, the fact that most animal protection laws exclude large categories of sentient beings from their scope of protection (eg farmed animals or invertebrates like cephalopods);[30] allow for suffering to be imposed on many animals that appears incompatible with their intrinsic value;[31] and are often underenforced.[32] This contrasts with the strong rights protections that humans enjoy for their basic needs and interests, which are generally much more stringently enforced than animal welfare protections.

Second, and connected, legal anthropocentrism *operates with a 'use paradigm'*,[33] that is, a paradigm that regards animals as resources that can be exploited and killed for human purposes. Examples include the 'production' of animals for food, their use in scientific research, sports, and entertainment. To be sure, this paradigm generally requires that animals be treated 'humanely' when they are used. Whether or not animals are indeed treated humanely is often assessed in virtue of an 'unnecessary suffering' test, which requires a balancing of human interests against animals' interests. However, not only would it be unthinkable to subject humans to similar treatments as long as they are considered 'humane'. But, as Gary Francione has pointed out, the unnecessary

[26] Peter D Burdon and Claire Williams, 'Rights of Nature: A Constructive Analysis' in Douglas Fisher (ed), *Research Handbook on Fundamental Concepts of Environmental Law* (Edward Elgar 2016). For an overview of rights of nature laws, see 'Rights of Nature Law Library' (*Center for Democratic and Environmental Rights*) <https://www.centerforenvironmentalrights.org/rights-of-nature-law-library> accessed 31 October 2024.

[27] Charlotte E Blattner, 'The Recognition of Animal Sentience by the Law' (2019) 9 Journal of Animal Ethics 121, 125–26.

[28] Raffael N Fasel and Sean C Butler, *Animal Rights Law* (Hart Publishing 2023) 14.

[29] ibid.

[30] Blattner (n 27) 124–27.

[31] ibid 127–29.

[32] For an overview of these problems, see Fasel and Butler (n 28) 23–25.

[33] Charlotte E Blattner, *Protecting Animals Within and Across Borders: Extraterritorial Jurisdiction and the Challenges of Globalization* (OUP 2019) 339. For an illustration of the related 'welfare paradigm', see Richard L Cupp, 'Animals as More Than "Mere Things", But Still Property: A Call for Continuing Evolution of the Animal Welfare Paradigm' (2016) 19 Pepperdine University School of Law, Legal Studies Research Paper Series 1, 1027.

suffering test and the use paradigm ('welfarism', in his terms) more generally fall far short of actually protecting animals' basic interests. As he argues,

> Legal welfarism requires that we 'balance' the interests of humans and animals in order to decide what constitutes 'humane' treatment and 'unnecessary' suffering. The problem is that the framework of legal welfarism contains numerous normative considerations that render any attempt to 'balance' as an empty exercise for the most part—at least as far as animal interests are concerned. The result of legal welfarism is that in many instances, a relatively trivial human interest is balanced against an animal's most fundamental interests in not experiencing pain or death, and the human interest nevertheless prevails.[34]

Closely connected to the use paradigm is legal anthropocentrism's third feature: it *treats animals as ownable things*.[35] It is this feature which explains why animals can legally be used in the ways described above. Every legal system allows and makes it possible for humans to have ownership over animals. As a result, animals can be purchased, sold, rented out, destroyed, and so on—much like inanimate pieces of property such as books or tables. This status of animals as ownable objects or things has a long pedigree in Western law. The law of ancient Rome, for example, famously already classified animals as things rather than persons.[36] The legal status of animals as things starkly contrasts with that of human beings. Ownership over human beings, though once common especially over non-paradigmatic human beings, has at least formally been outlawed in all legal systems.[37]

To be sure, in recent years, a number of legal systems have explicitly recognized that animals are more than simple property.[38] However, these legal systems have not reclassified animals as legal persons. The consensus view is that these recognitions—often called 'dereifications'[39]—have been largely

[34] Gary L Francione, 'Animals, Property and Legal Welfarism: Unnecessary Suffering and the Humane Treatment of Animals' (1994) 46 Rutgers Law Review 721, 723.

[35] See generally Wendy A Adams, 'Human Subjects and Animal Objects: Animals as Other in Law' (2009) 3 Journal of Animal Law and Ethics 29.

[36] See Steven M Wise, 'The Legal Thinghood of Nonhuman Animals' (1996) 23 Boston College Environmental Affairs Law Review 471, 493.

[37] See generally Jean Allain, *The Law and Slavery: Prohibiting Human Exploitation* (Brill Nijhoff 2015).

[38] Fasel and Butler (n 28) 12–13.

[39] Tomasz Pietrzykowski, 'The Idea of Non-Personal Subjects of Law' in Visa AJ Kurki and Tomasz Pietrzykowski (eds), *Legal Personhood: Animals, Artificial Intelligence and the Unborn* (Springer 2017) 53.

symbolic. This is because the legal systems that have formally changed animals' legal status still apply the property regime to them.[40]

Finally, anthropocentric legal systems *operate with human-centric concepts and/or interpret even potentially neutral concepts in an anthropocentric way*. Let us first consider some key concepts that are explicitly human-centric. Current legal systems often employ standards that use openly anthropocentric metrics. One prominent example is that of the 'reasonable man' or 'reasonable person'. This concept has been used among other things to determine negligence in tort law, to identify standards of care in contract law, and to determine culpability in criminal law. As critics have pointed out, the 'reasonable man' has served to legally cement existing inequalities. Mayo Moran, for example, has shown that 'reasonableness' is often interpreted as 'ordinariness', with the effect that ordinary discriminations and exclusions are sanctioned by the law.[41] Such discriminations have been not only directed against animals but also served to marginalize non-paradigmatic humans who fall short of being counted as reasonable men or persons.[42]

Now consider concepts and terms that are neutral on their face but are still interpreted in anthropocentric ways. One example is that of the 'legal person'. As Ngaire Naffine has argued, there are possible interpretations of this concept which understand it as an artificial construct that law can apply to any entity it wants (Naffine calls this the 'Legalist' position).[43] However, as she emphasizes, legal personhood is consistently tied up with concrete conceptions of the *person*. One prominent conception is that of the 'Rationalists' whose 'paradigmatic legal person is the rational actor, what once was called the "man of reason" ';[44] another is the 'Religionist' position that connects legal personhood to a religious view of the sacredness of all human life.[45] According to Naffine, conceptions such as the Rationalists' and Religionists' have become fundamental to Western law, creating patterns of thinking that prioritize (paradigmatic) humans.[46]

Even mundane terms such as 'everyone' or 'victim' are interpreted in anthropocentric ways, despite their susceptibility to also being applied to

[40] Janneke Vink, *The Open Society and Its Animals* (Palgrave Macmillan 2020) 86–88.

[41] Mayo Moran, *Rethinking the Reasonable Person: An Egalitarian Reconstruction of the Objective Standard* (OUP 2003) 14.

[42] ibid.

[43] Ngaire Naffine, *Law's Meaning of Life: Philosophy, Religion, Darwin and the Legal Person* (Hart Publishing 2009) 21–22.

[44] ibid 22–23.

[45] ibid 23.

[46] ibid 166. See also Anna Grear, 'Law's Entities: Complexity, Plasticity and Justice, Jurisprudence' (2013) 4 Jurisprudence 76, 90, 95.

animals.[47] For example, pursuant to Article 6(1) of the European Convention on Human Rights (ECHR), '[i]n the determination of his civil rights . . . *everyone* is entitled to a fair and public hearing', while Article 34 states that '[t]he Court may receive applications from any person ... claiming to be the *victim* of a violation by one of the High Contracting Parties of the rights set forth in the Convention or the Protocols thereto'.[48] However, when faced with a law suit that claimed that the chimpanzee Hiasl had not been granted a right to a fair trial in Austria and risked being sold to animal researchers,[49] the European Court of Human Rights declined to view Hiasl as being part of 'everyone' or a 'victim' of a violation of Article 6 ECHR and thus dismissed the case.[50]

2) Constitutional anthropocentrism

Having provided an overview of legal anthropocentrism and its most important manifestations, we can now turn to the question of how anthropocentrism operates on the constitutional level. As is true for legal systems more generally, not all constitutions are equally anthropocentric. Rather, to speak with Bruckerhoff, '[c]urrent constitutional provisions reflect a spectrum of anthropocentrism'.[51] However, as we show in this section, there is no doubt that despite some variations in the level of protection afforded to animals, constitutions and constitutionalism more broadly remain staunchly anthropocentric. In fact, due to the constitution's place at the top of the legal hierarchy, it is constitutional anthropocentrism that makes possible the various forms of legal anthropocentrism we have described in the previous section.

As with legal anthropocentrism, there are numerous ways in which the anthropocentrism of constitutions and constitutionalism manifests itself. In this section, we first focus on four central constitutional principles and show how, in law and practice, they are being interpreted and applied in anthropocentric ways (Subsection A). We then turn to constitutional theory and shed light on

[47] See eg David Bilchitz, 'Does Transformative Constitutionalism Require the Recognition of Animal Rights?' in Stu Woolman and David Bilchitz (eds), *Is This Seat Taken? Conversations at the Bar, the Bench and the Academy About the South African Constitution* (Pretoria University Law Press 2012) 133–88.

[48] Emphasis added.

[49] Alexia Staker, 'Should Chimpanzees Have Standing? The Case for Pursuing Legal Personhood for Non-Human Animals' (2017) 6 Transnational Environmental Law 485, 14–15; on the background of the case, see Martin Balluch and Eberhart Theuer, 'Trial on Personhood for Chimp "Hiasl"' (2007) 7 ALTEX 335.

[50] *Balluch v Austria*, no 26180/08, 15 January 2010; *Stibbe v Austria*, no 26188/08, 15 January 2010.

[51] Bruckerhoff (n 25) 630.

the anthropocentrism that underlies classic and more recent theories of constitutionalism (Subsection B).

A) Manifestations of anthropocentrism in constitutional law and practice

To assess the anthropocentrism of current constitutional law and practice, it is helpful to focus on the human-biased interpretation and application of four key constitutional principles: fundamental rights, proportionality, rule of law, and democracy. We will consider these principles—which also form the template of Part II of our book—in turn.

First, with the exception of corporations and institutions such as churches, constitutions confer *fundamental rights* only on human beings.[52] This anthropocentric focus of rights is so obvious that many constitutions do not explicitly state whom their rights catalogues are intended to protect. For instance, as David Bilchitz has pointed out in the context of the South African Constitution (1996), there is 'a clear omission in the application clause of the Bill of Rights, in that it does not define who, in general, are the beneficiaries of the rights contained therein'.[53]

One of the characteristic features of an anthropocentric distribution of rights is the explicit recognition of human dignity that numerous constitutions contain as their guiding principle.[54] For example, pursuant to Article 1 of the Constitution of Peru (1993), '[t]he defence of the human person and respect for his dignity are the supreme purpose of the society and the State'. Numerous constitutions specifically state that human dignity is the principle on which fundamental constitutional rights are based. Article 30 of the Polish Constitution (1997) serves as an illustration: 'The inherent and inalienable dignity of the person shall constitute a source of freedoms and rights of persons and citizens. It shall be inviolable'. In addition to being a source of fundamental rights, human dignity can also serve as an aid in interpreting rights provisions.[55]

[52] Jessica Eisen and Kristen Stilt, 'Protection and Status of Animal', *Max Planck Encyclopedia of Comparative Constitutional Law* (2016) <https://oxcon.ouplaw.com/display/10.1093/law-mpeccol/law-mpeccol-e71> accessed 26 October 2024 [2].

[53] Bilchitz (n 47) 183.

[54] See generally Jacob Weinrib, *Dimensions of Dignity: The Theory and Practice of Modern Constitutional Law* (CUP 2016) 1–3.

[55] Dieter Grimm, 'Dignity in a Legal Context: Dignity as an Absolute Right' in Christopher McCrudden (ed), *Understanding Human Dignity* (OUP 2014) 385.

By contrast, no constitutional provision currently confers fundamental rights on animals. That is, no bill of rights explicitly states that animals have fundamental constitutional rights such as the right to life, bodily and mental integrity, freedom, or any other fundamental right.[56] Constitutional rights practice mirrors this exclusion of animals from fundamental rights, with only a few limited exceptions.[57] The first exception can be found in Argentina, where a local court issued a writ of *habeas corpus* on behalf of the chimpanzee Cecilia—which it considered to be a 'constitutional mechanism to protect the freedom of movement'.[58] The other exception is Ecuador, where the Constitutional Court recognized that wild animals are holders of certain fundamental rights under Chapter 7 of its Constitution, which protects rights of nature.[59]

Second, the requirement that state action be *proportionate* only fully applies to humans. 'Proportionality' comes in roughly two forms in liberal constitutions. First, it is a general constitutional principle applying to any state action, requiring it to be proportionate to the goals pursued. Second, it is a principle that applies specifically to state actions and laws that limit constitutional rights.[60] In this second, more specific context, the proportionality test, which has been adopted in constitutional adjudication around the world,[61] is used to determine whether the limitation of a given constitutional right is justified. To pass the proportionality test, state action must generally be aimed at achieving a legitimate goal, be suitable to achieving the goal, be necessary for achieving the goal, and not impose a disproportionate burden on constitutional rights.[62] This last stage—determining whether a burden is disproportionate—requires the balancing of individuals' clashing constitutional rights, or an individual's constitutional rights and the public interest, as the case may be.[63]

[56] Eisen and Stilt (n 52) [2]. See also World Animal Protection, 'Animal Protection Index' (2020) <https://api.worldanimalprotection.org/indicators> accessed 31 October 2024 (which does not list constitutional rights protection as one of its indicators).

[57] Despite some judgments in India and Pakistan suggesting that animals have constitutional rights, our view is that these are not genuine instances of rights. For a full argument, see John Olusegun Adenitire, 'Are There Any Animal Rights?' in John Olusegun Adenitire and Raffael Fasel (eds), *Fundamental Rights for Non-Humans: Foundations, Faults, and Futures* (Hart Publishing 2026).

[58] Tercer Juzgado de Garantías, Mendoza, 'Chimpanzee "Cecilia", P-72.254/15' (Ana María Hernández tr, 3 November 2016) 17 <https://www.nonhumanrights.org/content/uploads/2016/12/Chimpanzee-Cecilia_translation-FINAL-for-website.pdf> accessed 31 October 2024.

[59] 'Estrellita', Constitutional Court, no. 253-20-JH/22, 27 January 2022 <https://animal.law.harvard.edu/wp-content/uploads/Sentencia_253-20-JH-20_Derechos_Naturaleza_y_Animales.pdf> accessed 31 October 2024. For a discussion of this and other cases, see Fasel and Butler (n 28) ch 6.

[60] Aharon Barak, *Proportionality: Constitutional Rights and Their Limitations* (Doron Kalir tr, CUP 2012) 2.

[61] See Chapter 4.

[62] Kai Möller, 'Proportionality: Challenging the Critics' (2012) 10 International Journal of Constitutional Law 709, 711.

[63] ibid 719. For a critical discussion, see Chapter 4.

In both of its forms, the proportionality principle is currently applied in ways that privilege human interests. Take the principle's first and more general form: proportionality outside of constitutional rights adjudication. If applied to animals, this principle would require that any state action or law governing animals must be proportionate to the goals it pursues. However, this is rarely the case. For example, as we saw above, many animal protection laws enshrine a requirement to not impose 'unnecessary suffering' on animals. Although not identical with the proportionality principle, what is 'unnecessary' must at least in part be determined by considering whether the suffering imposed is proportionate to its goals.[64] In the case of farmed animals, for instance, the primary goal is to raise and kill them to produce food for humans. In principle, only actions that are proportionate to this goal would be allowed. However, practices such as the debeaking of chickens, the docking of pigs' tails, and the cramming together of most industrially produced animals are widespread even though they cause suffering without being 'necessary' in any meaningful sense of the term.[65]

In its second form—proportionality in constitutional rights adjudication—the principle is also applied in anthropocentric ways. The primary problem is that, because animals do not have constitutional rights, they lack the kinds of rights that could be balanced against countervailing human rights—an issue we will return to in the next section.[66] However, even in the rare instances when courts are able to give weight to animals' interests when assessing whether the limitation of a human's countervailing constitutional right is justified, the animals' interests are generally outweighed. As Saskia Stucki observes,

> not all legally recognised interests count equally when balanced against each other, and rights-empowered interests typically take precedence over or are accorded more weight than unqualified competing interests. At present, the structural makeup of the balancing process governing human–animal conflicts is predisposed towards a prioritisation of human over animal interests.[67]

[64] See eg UK's Animal Welfare Act 2006 art 4(3)(d).

[65] See RSPCA, 'What Is Beak Trimming and Why Is It Carried Out?' (2 March 2020) <https://kb.rspca.org.au/knowledge-base/what-is-beak-trimming-and-why-is-it-carried-out/> accessed 31 October 2024; RSPCA, 'What Are the Animal Welfare Issues with Piglet Husbandry Procedures?' (3 August 2020) <https://kb.rspca.org.au/knowledge-base/what-are-the-animal-welfare-issues-with-piglet-husbandry-procedures/> accessed 31 October 2024.

[66] See Subsection 3.B.

[67] Saskia Stucki, 'Towards a Theory of Legal Animal Rights: Simple and Fundamental Rights' (2020) 40 Oxford Journal of Legal Studies 533, 554.

Consider the following case from Switzerland. In 2014, researchers applied for a permit to carry out basic research (ie research that is not applied and has no certain practical benefit) on four rhesus macaques that entailed procedures reaching the highest degree of severity on the applicable scale. The research involved surgically fixing a mount on the skulls of the primates to track their brain activity whilst being confined to a so-called primate chair with their heads fixated for several hours over a period of up to two years, at which point the animals are euthanized.[68] After the cantonal commission on animal experimentation had granted the permit for the experiments, a minority of the commission's members appealed the decision, as well as its subsequent approval by the cantonal executive branch, to the Cantonal Administrative Court of Zurich.[69] The Court had to apply the proportionality test and answer the question: Is a limitation of the researchers' constitutionally guaranteed right to scientific freedom (Article 20 of the Swiss Constitution (1999)) justified? The Court recognized that animal protection is considered a public interest in Switzerland. However, even though many commentators regarded the researchers' interests in carrying out this basic research as being less weighty than the severe infringement of the animals' vital interests in their lives and bodily integrity, the Court dismissed the appeal, deciding that the infringement of the animals' interests was indeed proportionate.[70]

Third, the *rule of law* only extends its protection to human beings. The rule of law—roughly, the idea that law imposes limits on public powers and individuals—has been widely accepted in constitutions around the world. At least 116 constitutions make explicit reference to it. Some link the rule of law explicitly to humanity. For instance, the Constitution of Barbados (1966) states in its Preamble:

> Now, therefore, the people of Barbados . . .
>
> b) affirm their belief that men and institutions remain free only when freedom is founded upon respect for moral and spiritual values and the rule of law;

Other constitutions are less explicit about the connection between rule of law and humanity, but draw it, nevertheless. One example is the US Constitution (1789), which is firmly rooted in the idea of 'a government of laws, not of

[68] Verwaltungsgericht des Kantons Zürich, no VB.2016.00048, 5 April 2017, [I].
[69] ibid [II]–[III].
[70] ibid [10].

men'.[71] Even though the US Constitution does not explicitly link the rule of law to human beings, that connection becomes obvious when we consider the historical reasons behind building the Constitution on the principles of the rule of law. As James McClellan puts it, the aim was to create a system in which 'no [human] person, however powerful or talented, can be allowed to act as if he were superior to the law of the land. Public decisions must be made upon the basis of law, and the laws must be general rules that everybody obeys'.[72]

This anthropocentric focus of the rule of law is also evinced in rule of law practice. For example, the leading index on the practical application of the rule of law—the World Justice Project Rule of Law Index—does not mention the term 'animal' once in its extensive reports. Instead, it focuses entirely on human beings and human metrics. For instance, it defines the rule of law as 'a durable system of laws, institutions, norms, and community commitment that delivers' the four principles of 'accountability', 'just laws', 'open government', and 'accessible and impartial dispute resolution'.[73] Regarding 'just laws' in particular, the report's anthropocentric focus becomes obvious by concentrating on laws that 'protect fundamental rights, including the security of persons and contract, property, and human rights'.[74]

Finally, constitutions only permit *democratic participation* of human beings. Animals are excluded from 'the people' that has the right to participate in democratic rule-making. According to *The Economist*'s 2023 Democracy Index, more than 45 per cent of the world's population lives in democratic regimes.[75] Hence, although democratic participation is sometimes flawed due to corruption or election irregularities in some of these countries, humans are at least given the possibility to contribute to the selecting of representatives and/or to decision-making directly. By contrast, no democracy in the world includes animals in the democratic process.

Now, the fact that democracies focus on human beings may be no surprise given that the very term 'democracy' refers to rule (*kratos*) by the people (*dēmos*) and has its origins in popular self-rule in Greek city-states.[76] The US Supreme Court, when interpreting the concept of political equality in the US Declaration of Independence and the Fifteenth, Seventeenth, and Nineteenth

[71] James McClellan, *Liberty, Order, and Justice: An Introduction to the Constitutional Principles of American Government* (3rd edn, Liberty Fund 2000) 387.

[72] ibid 349–50 (emphasis added).

[73] World Justice Project, 'Rule of Law Index' (2020) 10 <https://worldjusticeproject.org/sites/default/files/documents/WJP-ROLI-2020-Online_0.pdf> accessed 31 October 2024.

[74] ibid.

[75] The Economist Intelligence Unit, 'Democracy Index 2023: Age of Conflict' (2024) 4.

[76] For a useful overview, see Robert A Dahl, 'Democracy', *Encyclopedia Britannica* (2021) <https://www.britannica.com/topic/democracy> accessed 31 October 2024.

Amendments of the US Constitution, encapsulated this human-centrism in its famous 'one person, one vote' formula.[77] Of course, who exactly forms part of the 'demos' or is considered a 'person' with rights to political participation has been at the heart of many struggles of enfranchisement, such as those of Black people and women who were (and to some degree still are) excluded from democratic participation. However, what has never been in question, at least since the dawn of democracy, is that *at least some* and *only* human beings have a say in political matters.

This holds true both for democracies that allow direct popular participation in law-making as well as for constitutional systems based entirely on representative democracy. The Constitution of Albania (1998), for example, provides for direct democratic participation of the people in constitutional amendments through referendums and initiatives (Article 150(1)). But neither in this provision nor in any other provision or constitution are animals considered to be part of the electorate or the people who are able to participate in direct ways in the popular decision-making process.

The same holds for more indirect forms of democratic participation such as through the election of representatives. Even constitutions that apportion seats in parliament based on the idea of representing different groups in society do so solely with a focus on human beings. The Constitution of Ethiopia (1994), for example, tries to do justice to the multi-ethnic composition of the Ethiopian state in its apportionment of seats in both houses of parliament: the House of Peoples' Representatives and the House of the Federation. Focusing on the former, Article 54 states that:

> 1. Members of the House of Peoples' Representatives shall be elected by the People for a term of five years on the basis of universal suffrage . . .
> 2. . . .
> 3. Members of the House, on the basis of population and special representation of minority Nationalities and Peoples, shall not exceed 550; of these, minority Nationalities and Peoples shall have at least 20 seats. Particulars shall be determined by law.
> 4. Members of the House are representatives of the Ethiopian People as a whole.[78]

[77] US Supreme Court, *Gray v Sanders* 372 US 368 (1963), [5d]; *Reynolds v Sims* 377 US 533 (1964).

[78] For a similar quota-based provision for the House of the Federation, see Ethiopia's Constitution art 61.

This provision is representative of other constitutional systems that provide for the representation of human interests in parliament, sometimes even by reserving quota for specific human groups, but that do not create any representatives for animals.

B) Anthropocentrism in constitutional theory

The previous subsection has explored the human-centric interpretation and application of four main constitutional principles: fundamental rights, proportionality, rule of law, and democracy. In so doing, we have shown how the prevalent anthropocentrism of current constitutional law and practices excludes or marginalizes the interests of animals.

That current constitutional principles are formed the way they are is not a coincidence. Rather, they are the product of specific theories of constitutionalism that have influenced these principles and their practical application. As Nick Barber has pointed out, it is our 'model of constitutionalism' that 'shapes our understandings of the principles of constitutionalism'.[79] It is therefore important to shed light on the models of constitutionalism that underpin anthropocentric constitutional principles and that can thus help us better understand anthropocentric constitutionalism more broadly.

The first thing to note is that not only constitutional theory in general, but also the theory of constitutionalism more specifically, is centred on the human being. In fact, as Jessica Eisen has pointed out, theories of constitutionalism often operate with a focus on paradigmatic human beings: 'constitutional theory widely presupposes a particular kind of necessarily human agent as the reference point for legitimate constitutionalism'.[80] We propose to parse this point, first, by focusing on how the theory of classic constitutionalism is anthropocentric. We then consider two recent alternative theories of liberal constitutionalism—Barber's positive constitutionalism and Catherine Dupré's dignity-based constitutionalism—which, as we will see, are anthropocentric too.

Let us begin with *classic constitutionalism*, which emerged in the late eighteenth century and remains the dominant model of constitutionalism. According to Martin Loughlin, this conception 'has almost become

[79] Nick W Barber, *The Principles of Constitutionalism* (OUP 2018) 4.
[80] Eisen (n 3) 925.

synonymous with the modern concept of the constitution itself'.[81] As Loughlin defines it, classic

> [c]onstitutionalism is a theory of limited government and is concerned mainly with the norms which modern constitutions should contain. These norms not only impose limits on the exercise of public power but also on the procedures through which such power should be exercised. Its key principles are independence of the judiciary, separation of governmental powers, respect for individual rights, and the promotion of the judiciary's role as guardians of constitutional norms.[82]

As this definition illustrates, classic constitutionalism is a liberal theory that emphasizes the need to protect individuals' rights by limiting public powers.[83] But classic constitutionalism is not aimed at protecting the rights of just *any* individuals. Rather, it focuses on rational, free, and equal individuals—that is, on what we have been calling paradigmatic human beings.

The reason for this focus on paradigmatic human beings is that classic constitutionalism emerged as a response to constitutional systems that were based on the traditional notion of monarchs' divine right to rule, which did not require the consent of the ruled.[84] As Dieter Grimm emphasizes, and as we will see in more detail in Chapter 2, this traditional approach was rejected in the American and French Revolutions and replaced with a constitutional approach that put the idea of popular sovereignty—the idea that all constitutional authority should emanate from the will of the people—front and centre.[85] Despite what the idea of popular sovereignty might suggest, however, classic constitutional theory rarely focuses on the people as a whole and instead singles out specific members of the human species. This has to do with the device that classic constitutionalists employ to explain constitutional systems that are based on the consent of the ruled: the social contract. In a nutshell, the social contract imagines a hypothetical situation—the state of nature—in which all individuals are rational, free, and equal. As such, they are able to discuss and decide what it would take for them to consent to leaving the state of nature and submitting themselves to a constitutional system.

[81] Martin Loughlin, 'What Is Constitutionalisation?' in Petra Dobner and Martin Loughlin (eds), *The Twilight of Constitutionalism?* (OUP 2010) 48.

[82] ibid 55.

[83] ibid 59.

[84] ibid 48.

[85] Dieter Grimm, 'The Achievement of Constitutionalism and Its Prospects in a Changed World' in Petra Dobner and Martin Loughlin (eds), *The Twilight of Constitutionalism?* (OUP 2010) 5–7.

The focus on paradigmatic humans in classic constitutionalism manifests itself in two ways. First, although the device of the social contract itself is a hypothetical one, it only works if one presupposes rational, free, and equal agents—paradigmatic human beings—who are capable of negotiating and agreeing on a constitution that would warrant leaving the state of nature.[86] Second, to the extent that a 'contract' is actually made in the form of a written constitution,[87] only paradigmatic human beings participate in the making (and changing) of these constitutions. That is, only human beings who are rational agents take part in constitutional assemblies or constitutional amendments.

To be sure, some have proposed more inclusive accounts of constitutionalism, which do not restrict their focus to paradigmatic human beings. For instance, Grimm identifies the following as one of the seven features of constitutionalism: 'The constitutional constraints must act to the benefit of all persons subject to rule, and not only privileged groups'.[88] However, the fact remains that not all human beings can actually participate in the creation of rules or the ruling itself. As a result, they depend on paradigmatic humans' willingness to shape the rules to everyone's benefit. Often, the consequence of this focus on paradigmatic humans has been that it is primarily the interests and needs of paradigmatic human beings that have determined the constitution.

The next chapter will probe the history and theoretical justifications of classic constitutionalism in more detail in order to challenge it. For our purposes here, it is sufficient to note that, whatever its merits, classic constitutionalism excludes all animals and, by focusing on rational agents, also marginalizes non-paradigmatic human beings.

Despite its prominence, classic constitutionalism is not the only theory of constitutionalism. In recent years, other liberal constitutional theories have been put forward that take a wider approach than classic constitutionalism. We will focus on two prominent ones. The first is *positive constitutionalism*. Barber, who defends this model of constitutionalism, does not reject classic constitutionalism, but thinks it needs to be complemented. According to Barber, classic constitutionalism—'negative constitutionalism',[89] as he calls it—is too narrow because it only focuses on limiting public powers in order to protect individual rights and liberties from external interference. This focus, Barber argues, misses something important: 'Constitutionalism also requires

[86] ibid 7.

[87] Loughlin (n 81) 47.

[88] Dieter Grimm, *Constitutionalism: Past, Present, and Future* (OUP 2016) 22.

[89] See also Walter F Murphy, *Constitutional Democracy: Creating and Maintaining a Just Political Order* (Johns Hopkins University Press 2007) 10.

the creation of an effective and competent set of state institutions; it has a positive dimension.'[90] The reason why state powers should be not only limited but also effective, according to Barber, is because they ought to 'help bring about the common good'.[91] Drawing on Aristotle, Barber argues that the purpose of the state, and its constitution, is 'to advance the well-being of its members'.[92] Hence, on Barber's model of positive constitutionalism, the constitutional state is not reduced to posing and preventing threats to individuals' rights, but it is also regarded as the primary vehicle to 'work for the good of its people'.[93]

Due to its focus on well-being and the common good, positive constitutionalism is not centred on paradigmatic human beings in the same way that classic constitutionalism is. Even more promising, the concepts of 'well-being' and 'common good' have the potential of being interpreted non-anthropocentrically, much like the concept of 'legal person' that we have encountered above. However, when Barber talks of 'our well-being' and the 'well-being of [the constitutional state's] members',[94] he only seems to have human beings in mind, as animals never enter the picture. Hence, also Barber's positive constitutionalism is anthropocentric.[95] For while his theory could potentially be interpreted to apply to animals, there seems to be no evidence that he himself is adopting this view.[96]

The anthropocentric focus in more 'positive' understandings of constitutionalism and constitutional theory is not exceptional. As Burdon has pointed out, 'mainstream theories [of law] . . . are concerned ultimately with human beings and human relationships', and that 'when law is constructed as purposive and reference is made to the "common good", as is common in natural law theory, the term is expressly limited to human good'.[97]

Another recent alternative model to classic constitutionalism is *dignity-based constitutionalism*. That human dignity is at the heart of many constitutions is well established.[98] However, Catherine Dupré has made the case that dignity is also central to constitutionalism itself, especially to European constitutionalism. As she has argued, in Europe,

[90] Barber (n 79) 1.
[91] ibid.
[92] ibid 5.
[93] ibid 19.
[94] ibid 10.
[95] See Burdon (n 23) 61 (critiquing the prevalent anthropocentric interpretations of the 'common good').
[96] We have found no evidence in Barber's writings that he thinks positive constitutionalism applies to animals.
[97] See Burdon 'The Great Jurisprudence' (n 23) 61.
[98] See Aharon Barak, 'Human Dignity: The Constitutional Value and the Constitutional Right' in Christopher McCrudden (ed), *Understanding Human Dignity* (OUP 2014) ch 4.

> human dignity is not only the foundation of human rights and democracy in each constitutional order considered separately, it is what brings all the constitutional (at national and supranational levels) orders together, giving European constitutionalism its unique identity and dynamic.[99]

Hence, according to Dupré, dignity is the ground of constitutional commitments in the European Union (EU) Member States and their constitutions, as well as in regional bodies such as the EU itself and the ECHR. At the heart of all these constitutional systems, Dupré holds, is a foundational commitment to the inviolability of human dignity.[100] As such, dignity-based constitutionalism explicitly claims to take an inclusivist approach. That is, dignity-based constitutions claim to focus on humanity as a whole rather than solely paradigmatic human beings. As Dupré emphasizes, 'human dignity protects humanity, understood as the humanity that we equally share as human beings'.[101]

However, despite its goal to cover all human beings, Dupré's dignity-based constitutionalism is not free from a certain bias towards paradigmatic human beings. This becomes clear in her account of the history of dignity-based constitutionalism. As she points out, 'over the centuries, human dignity has captured three broad ideals that have sustained its construction: freedom, autonomy and equality'.[102] Dupré specifically draws on the theories of Giovanni Pico della Mirandola and Immanuel Kant, who explained the uniqueness and dignity of man by pointing to the ability of self-determination and rationality, respectively.[103] Dupré is aware of the potentially marginalizing character of a focus on the capacity to reason, but she insists that 'it has remained crucial in shaping the concept of dignity'.[104] Nonetheless, she believes that the focus on paradigmatic capacities such as reason or autonomy do not lead to an exclusion of humans who lack them: 'in the dignity paradigm human beings deprived of autonomy, because they are, for instance, too young or too old, severely disabled or in a persistent vegetative state, are not treated as an exception when it comes to determining the scope of their human rights and to protecting them'.[105]

[99] Catherine Dupré, *The Age of Dignity: Human Rights and Constitutionalism in Europe* (Hart Publishing 2015) 8.
[100] ibid 4–5.
[101] ibid 22.
[102] ibid 31.
[103] ibid 30–36.
[104] ibid 34.
[105] ibid 22.

In the next chapter, we will return to the question of whether Dupré's account manages to include non-paradigmatic human beings without marginalizing them. What we can note here is that, even if dignity-based constitutionalism did not marginalize any human beings (which we believe it does), it certainly excludes all animals. In fact, as Dupré emphasizes, the exclusion of animals is no mere side effect of dignity-based constitutionalism but rather a core component. The reason for this is that dignity-based constitutionalism explains humanity's special constitutional status by distinguishing it from the status of animals. 'The legal system of human rights protection in Europe (and more generally in the West)', Dupré notes, 'rests on the assumption that, as human beings, we are born with the unique quality of dignity that distinguishes us from other beings (primarily animals), justifying and explaining the special protection of our rights.'[106] Drawing on Kant, Dupré argues that, on this model of constitutionalism, 'dignity is used to define humanity . . . by distinction from other beings which only have a "relative worth", namely, animals or things'.[107] In a sense, then, the very purpose of dignity-based constitutionalism is to make sure that humans cannot be 'reduce[d] . . . to the condition of a thing or an animal'.[108]

As this shows, dignity-based constitutionalism shares with the theories of classic and positive constitutionalism a focus on human beings and human interests. In fact, given its explicit reliance on the distinction between the human and the animal,[109] it is even more overtly committed to anthropocentrism than other theories of constitutionalism.[110]

3) Emerging constitutional concern for animals: A paradigm shift?

The previous sections have revealed the often-unnoticed anthropocentric assumptions behind law generally and constitutions in particular and have explored the main ways in which these assumptions manifest themselves in constitutional law and theory. In this last section, we address a relatively recent

[106] ibid 28.
[107] ibid 35.
[108] ibid 50. See more generally on this argument Raffael N Fasel, 'The Old "New" Dignitarianism' (2019) 25 Res Publica 531.
[109] A distinction that is often viewed as being imbued with problematic racial, gender, ability, and other connotations: see eg Kim (n 14) and Aph Ko and Syl Ko, *Aphro-Ism: Essays on Pop Culture, Feminism, and Black Veganism from Two Sisters* (Lantern Books 2020).
[110] See Dupré (n 99) 177.

trend in constitutional law and practice towards enshrining concern for animals in constitutions.[111] The purpose of this section is to do justice to the diversity of constitutional approaches to animals and thus to present current constitutionalism in its best light. However, by discussing these latest developments we also aim to show that, while they reflect a unique trend that supports our argument, they are far from toppling constitutionalism's anthropocentrism.

We begin by giving an overview of the recent trend to protect animals' intrinsic value through the constitutional recognition of their interests (Subsection A). We then shed light on the limits inherent in these recognitions and argue that they have yet to supplant constitutionalism's anthropocentric focus (Subsection B).

A) Constitutional recognition and protection for animals

As mentioned above, numerous legal systems have statutory protections for animals. Of the 193 member states of the United Nations, nearly four-fifths have some form of animal protection law.[112] However, as we also saw, the current legal protections of animals have serious shortcomings. That is why, in recent decades, there has been a shift in some countries towards recognizing animals' interests on a constitutional level. As Anne Peters has put it, there is a trend towards 'constitutionalisation' as 'a tiny but growing number of states have elevated concern for animals to the status of constitutional provisions'.[113] In animal law scholarship, this trend has coincided with an increased interest of scholars in the shortcomings of welfare laws, specifically the fact that they are lower in the legal hierarchy than constitutional rights and principles.[114]

An oft-mentioned example in this respect is Germany, which amended Article 20a of the Basic Law in 2002. Article 20a now states: '[m]indful also of its responsibility towards future generations, the state shall protect the natural foundations of life *and animals* by legislation and, in accordance with law and justice, by executive and judicial action, all within the framework of

[111] There are also indigenous legal systems that are on the whole non-anthropocentric but which are outside the focus of our discussion. See eg Maneesha Deckha, 'Unsettling Anthropocentric Legal Systems: Reconciliation, Indigenous Laws, and Animal Personhood' (2020) 41 Journal of Intercultural Studies 77.

[112] Fasel and Butler (n 28) 14.

[113] Anne Peters, 'Introduction to Symposium on Global Animal Law (Part I): Animals Matter in International Law and International Law Matters for Animals' (2017) 111 American Journal of International Law 252, 252.

[114] See eg Vink (n 40) 211.

the constitutional order.'[115] Prior to this constitutional amendment, Germany had general animal welfare laws in place since 1933.[116] However, these laws were statutes enacted by parliament and therefore did not possess constitutional rank. The problem this caused was that the welfare laws had particular difficulty meeting their goal of protecting animals when that protection conflicted with constitutionally protected rights of human beings, such as the rights to artistic and scientific freedom, freedom of profession, or freedom of religion.[117] Kate Nattrass provides the following example of a circuit court decision dealing with a conflict between statutory animal welfare protections and the constitutionally protected artistic freedom:

> In a case brought to the circuit court of Kassel, an artist was charged with animal cruelty on the basis of . . . the *Tierschutzgesetz* [the German animal protection act]. The artist had bound a bird in a glue-like substance made of egg yolk and ground sausage in order to represent how humans are often mishandled in modern society. The bird, unable to fly, was forced to hobble about the exhibit in obvious distress. While the Kassel court agreed that the bird had suffered and experienced pain, it ruled that the constitutionally protected freedom of artistic expression took precedence over the animal protection law.[118]

The problem with the legal situation the court was confronted with—a legal situation that is still prevalent in most countries—was that constitutional doctrine did not allow the court to give weight to animals' interests when they conflict with constitutionally protected human interests. As Bilchitz points out,

> [e]ven if ultimately the human interests in such an instance would outweigh those of the animals, the problem lies in the fact that without some form of constitutional protection, there was no balancing at all to be done in legal terms. The difficult moral evaluation in question could not be articulated in constitutional terms.[119]

[115] 'Basic Law for the Federal Republic of Germany' <https://www.btg-bestellservice.de/pdf/80201000.pdf> accessed 31 October 2024 (emphasis added).
[116] Kate M Nattrass, '"und die Tiere": Constitutional Protection for Germany's Animals' (2004) 10 Animal Law 283, 286.
[117] ibid 292.
[118] ibid 293–94. For a similar example, see Constitutional Tribunal of Poland, Judgment, 10 December 2014, no K 52/13.
[119] Bilchitz (n 47) 191.

An increasing number of states have taken steps in recent times to include animals in their constitutions. Such inclusions have taken roughly three forms: provisions that regulate which governmental authority is competent in animal protection matters (competence norms); provisions that make animal protection a constitutional objective (state objectives); and provisions that are aimed more directly at protecting animals and their interests (protection norms).[120] Only state objectives and protection norms are relevant for our purposes here because mere competence norms do not necessarily signal a commitment to consider and protect animals as beings with intrinsic value.[121]

Apart from Germany's Article 20a Basic Law, only a few other constitutions have been amended to include state objectives to protect animals. These are the constitutions of Austria, Brazil, Egypt, Luxembourg, Slovenia, and India.[122] The Constitution of Egypt (2014), for example, provides in Article 45 that:

> The state also commits to the protection and development of green space in urban areas; the protection of plants, livestock and fisheries; the protection of endangered species; and the prevention of cruelty to animals. All the foregoing takes place as regulated by law.

The constitutional provisions these states have enacted are characterized by their commitment to recognize animal protection as a state goal but generally without imposing any immediately binding obligations on the state, let alone granting fundamental rights to animals.

Examples of constitutions that contain norms that protect animals in more direct ways are even more limited in number. Arguably the most prominent example is Article 120(2) of the Swiss Constitution (1999), which holds that:

> The Confederation shall legislate on the use of reproductive and genetic material from animals, plants and other organisms. In doing so, it shall take account of the dignity of living beings as well as the safety of human beings, animals and the environment, and shall protect the genetic diversity of animal and plant species.

The protection of animals' dignity is today recognized as a general constitutional principle that applies throughout the Swiss legal order, rather than

[120] Fasel and Butler (n 28) 25–29.
[121] For a list of constitutions with only competence norms, see Eisen and Stilt (n 52) [9].
[122] ibid [11]–[65].

simply in the context of reproductive and genetic materials.[123] Although it is a contested matter in Swiss doctrine whether animal dignity is comparable to human dignity, there is widespread agreement that Article 120(2) recognizes the inherent value of animals and aims to protect them for their own sake.[124]

B) The limits of current constitutional recognitions and protections for animals

At first glance, the above-mentioned constitutional recognitions and protections of animals seem to put in question our suggestion in the previous sections that current constitutionalism is anthropocentric. After all, how could an anthropocentric constitution recognize or even directly protect animals as having inherent value and interests that ought to be protected regardless of their instrumental use for humans? Upon closer examination, however, it becomes clear that while the trend towards including animals in constitutions is certainly noteworthy, it has not fundamentally altered the anthropocentric paradigm.[125]

To see this, we need to understand that even the most advanced constitutional protections have some serious shortcomings. Arguably the most important shortcoming is that, as we saw above, no constitution explicitly grants constitutional rights to animals, and only two countries have extended limited constitutional rights to some specific animals through judicial rulings.[126] This means that the fundamental interests that many animals have in life, bodily and mental integrity, liberty, or family, are not protected constitutionally.

Someone might object that affording animals fundamental constitutional rights is only one possible way in which constitutions can shed their anthropocentrism. Another way might be to protect their inherent value and interests through the state objectives and protection norms mentioned above. Take Article 20a of the German Basic Law. Could it not be argued that its purpose

[123] Gieri Bolliger, 'Legal Protection of Animal Dignity in Switzerland: Status Quo and Future Perspectives' (2016) 22 Animal Law 311, 328.

[124] ibid 326.

[125] For a similar verdict, see Maneesha Deckha, 'Constitutional Protections for Animals: A Comparative Animal-Centred and Postcolonial Reading' in Kelly Struthers Montford and Chloë Taylor (eds), *Colonialism and Animality: Anti-Colonial Perspectives in Critical Animal Studies* (Routledge 2020).

[126] See Gillespie (n 12). While in *Estrellita* the Constitutional Court appears to recognize constitutional rights for all animals (at [80]–[83]), we do not believe that it in fact conferred genuine rights on domesticated animals. This is because, as the Court recognizes, their apparent rights can be easily overridden for human purposes. See *Estrellita* (n 59) [108–10]. For a fuller argument, see Adenitire (n 57) pt III.

is for animals' interests to be considered as part of the 'public interest' when they are balanced against human interests? There are two problems with this objection that illustrate that even animal-friendly constitutions are still firmly anthropocentric: there is a lack of a level playing field in the balancing test, and there is a lack of equal opportunities to trigger the balancing test.[127]

First, while some constitutions allow for animals' interests to factor into the balancing test, this does not mean that they put these interests on an equal footing with the interests of humans against which they are measured in cases of conflict. In fact, the interests of animals are habitually given less weight than comparable human interests. As a result, animals' interests often lose out in the balancing test, even if they conflict with relatively trivial human interests. Perhaps the most obvious example in this respect is the industrial-scale production and killing of animals such as chickens, turkeys, cows, pigs, horses, and others for the purposes of food. Though rarely challenged in a legal forum, human rights such as the freedom of profession and the right to food have been invoked as potential justifications for these practices.[128] However, at least in countries where sufficient alternative food sources are available to humans, it is difficult to see how their interests in raising and killing animals for culinary purposes could weigh more heavily than those animals' vital interests in being alive, not to suffer, and to live in a healthy environment.[129]

As has been emphasized in the literature, it is not a coincidence that the current constitutional weighing of interests is biased towards human interests. Rather, it is a product of the more general anthropocentrism in law and legal thinking that leads judges and other legal officials to apply the 'ostensibly apolitical "balancing" of interests'[130] in an anthropocentric way.[131]

However, even if a fairer balancing process was possible within the constraints of the current constitutional protections for animals, animals face a second structural problem: they cannot demand that a balancing test be conducted to determine the legality of how they are treated. In contrast to what one might expect, the reason for this is not that they cannot themselves file a complaint against violations of their interests, as this does not prevent human

[127] For additional problems, see eg Jessica Eisen, 'Constitutional Animal Protection: Written Provisions and Unwritten Principles' in Anne Peters, Kristen Stilt, and Saskia Stucki (eds), *Oxford Handbook of Global Animal Law* (OUP 2025) (forthcoming) (arguing that there are unwritten constitutional principles of animal exploitation).

[128] See Nattrass (n 116) 293; Becky L Witt, 'Towards a Human Right to Food: Implications for Urban Growing in Baltimore City, Maryland' (2016) 43 Fordham Urban Law Journal 405, 415.

[129] See also Vink (n 40) 231.

[130] Eisen (n 3) 947.

[131] See also Vink (n 40) 194.

beings who lack legal capacity from taking legal action with the aid of legal guardians. The problem, rather, is that even if animals had such guardians, which they currently lack,[132] these guardians would not have standing required to take legal action on the animals' behalf.[133]

Consider the US case *Tilikum v Sea World*, which was brought in 2012 by the People for the Ethical Treatment of Animals (PETA).[134] PETA argued that Tilikum, Katina, Corky, Kasatka, and Ulises—five orcas kept in tanks and used for shows in the Sea World park in Orlando, Florida—are protected from slavery and involuntary servitude under Section 1 of the Thirteenth Amendment of the US Constitution, and that this protection is violated by their confinement at Sea World.[135] The US District Court for the District of Southern California decided that it did not have jurisdiction to entertain the case because the plaintiffs—the five orcas—lacked standing according to Article III of the US Constitution.[136] According to this provision, which determines standing in US federal courts, standing is only available if it is 'likely, as opposed to merely speculative, that the injury will be redressed by a favourable decision'.[137] Focusing on this requirement, the Court reasoned that because the Thirteenth Amendment only applied to humans and not to animals, the orcas' alleged injury would not be redressed by a favourable decision, which is why they did not have standing in the first place.[138]

The *Tilikum* judgment brings to the fore a fundamental dilemma that the current constitutional protection of animals faces. The availability of standing is often directly linked to the applicability of substantive laws such as the constitutional protection against slavery. This creates a catch-22 situation for animals. Most substantive laws were not created to protect them and are therefore not considered applicable to them, with the consequence that animals do not have standing to enforce these laws; and because animals

[132] On some limited exceptions, see Eva Bernet Kempers, 'Transition Rather Than Revolution: The Gradual Road Towards Animal Legal Personhood Through the Legislature' (2022) 11 Transnational Environmental Law 581. On the former 'Tieranwalt' (animal attorney) in the Canton of Zurich, see Vanessa Gerritsen, 'Animal Welfare in Switzerland—Constitutional Aim, Social Commitment, and a Major Challenge' (2013) 1 Global Journal of Animal Law 1, 13–14. On the Royal Society for the Prevention of Cruelty to Animals (RSPCA) in the United Kingdom, see Jerry L Anderson, 'The Origins and Efficacy of Private Enforcement of Animal Cruelty Law in Britain' (2012) 17 Drake Journal of Agricultural Law 263.

[133] On *habeas corpus* as an exception to these standing difficulties, see Steven M Wise, 'Nonhuman Rights to Personhood' (2013) 30 Pace Environmental Law Review 1278.

[134] *Tilikum et al. v Sea World Parks & Entertainment Inc.* 842 F Supp 2d 1259 (SD Cal 2012).

[135] ibid 2–3.

[136] See Cass R Sunstein, 'Standing for Animals (with Notes on Animal Rights) A Tribute to Kenneth L. Karst' (2000) 47 UCLA Law Review 1333.

[137] *Tilikum et al.* (n 134) 3.

[138] ibid 7.

are considered to lack standing, they cannot request that these laws be made applicable to them.

In fact, even in the limited cases where constitutional provisions were created to protect animals, animals are not provided with the standing required to enforce these provisions. [139] To see this, consider again the Swiss case of the rhesus macaques we encountered in Subsection 2.A. The contested decision by the Cantonal Administrative Court of Zurich, which held that the experiments were proportionate and therefore legal, was not, in fact the first decision on this type of experiment. A few years before this decision, the same researchers had filed forms to conduct an almost identical set of experiments on rhesus macaques. However, this set of experiments was declared unconstitutional by the Swiss Federal Supreme Court in 2009. In a landmark decision, the Court exceptionally found that the constitutional dignity protection of the macaques (and their statutory protection against suffering) outweighed the researchers' interests in conducting the experiments whose expected utility was uncertain.[140]

So why were the researchers able to go ahead with conducting virtually identical experiments only a few years later? The reason is that animals do not have legal standing to enforce their constitutionally protected interests. The first case reached the Supreme Court only because the researchers had made an appeal against a refused permit. In the second case, they had no need to appeal because their permit was granted. And that permit could not be appealed any further because the animals themselves could not bring action to have their constitutional dignity protected, and the humans who had previously been acting in the interests of the animals (members of the cantonal animal experimentation commission) did not have standing to bring a federal appeal.

If animals were on an equal footing in the balancing test, then it would seem reasonable to expect that they would have the same opportunities to trigger that test. However, as the Swiss case illustrates, there is currently no such equality. Instead, balancing tests are only ever carried out when human beings believe that their interests have been violated by the state. For these reasons, despite what is a remarkable trend towards including animals, constitutional systems and constitutionalism more generally are still firmly focused on human beings and privilege their interests over the interests of animals.

[139] But see (n 140) on standing.

[140] Swiss Federal Supreme Court, Judgment no 135 II 405, 7 October 2009, [4.3.4].

Conclusion

This chapter has answered the question of *what* the *status quo* is of current constitutional law and theory. As we have argued, that status is firmly anthropocentric in that it prioritizes human interests at the exclusion of animals' interests, and in ways that often marginalize the interests of non-paradigmatic humans. To demonstrate this, we have first highlighted four central ways in which legal systems generally can be considered anthropocentric. Turning to constitutional anthropocentrism specifically, we then exposed how four key principles of constitutionalism—fundamental rights, proportionality, rule of law, and democracy—are interpreted and applied in ways that favour (paradigmatic) human beings. As we have argued, this exclusion of animals is also mirrored in both classic and more modern theories of liberal constitutionalism.

Some might be hopeful that the recent trend towards enshrining animal protection in constitutions, which has occurred in some countries, reverses or at least counteracts the anthropocentric *status quo*. This chapter, however, tempers that optimism. Direct protections are few and far between, and even when there are provisions that could afford substantial protections to animals, they are interpreted in ways that are biased towards human interest. As Deckha has summarized the problem:

> because almost all of these initiatives have been interpreted through an anthropocentric lens they have had limited substantive impact; where the constitutional provisions have informed litigation or legislative reform the resulting changes, if any, have occurred at a symbolic level. The underlying welfarist cultural milieu influencing the interpretation of these constitutional texts dilutes their potential to favour animals.[141]

For these reasons, this trend, though laudable, falls well short of displacing anthropocentric constitutionalism.

[141] Deckha, 'Constitutional Protections for Animals' (n 125) 220.

2

Sentience-based constitutionalism

Introduction

The previous chapter has demonstrated that current constitutional law and practices as well as theories of constitutionalism are anthropocentric. That is, they revolve around and privilege the interests of human beings, often favouring what are viewed as paradigmatic rational human beings.[1] Having established *what* the *status quo* of constitutionalism is, we turn in this chapter to *why* anthropocentric constitutionalism is problematic and why it should be replaced with a more capacious understanding of constitutionalism, which we term 'sentience-based constitutionalism'. According to sentience-based constitutionalism, those exercising public powers ought to give due concern to the interests of all sentient beings they govern. We call this the All Governed Interests Principle. As we define it, 'due concern' entails respect for the intrinsic value of a being and protection of their basic interests.[2] By 'governed' we mean that the legal and political frameworks and rules of a constitution have effective control over the lives, bodies, or habitats of beings who possess intrinsic moral value.

As we saw in Chapter 1, there are countless ways in which anthropocentric constitutions fail to give due concern to the sentient animals they govern and can thus be said to be lacking in legitimacy. For instance, they do not afford fundamental rights to animals, they do not give animals' interests enough weight when they conflict with human interests, they fail to harness the rule of law for animals, and they do not give a say to animals in democratic decision-making. Part II of this book will show how sentience-based constitutionalism gives due concern to animals and non-paradigmatic humans in all these

[1] As a reminder, we use the term 'paradigmatic' in a descriptive sense, referring to how these human beings are generally conceived in classic constitutionalism. We do not believe that any groups of humans should be considered paradigmatic.

[2] What 'due concern' means concretely for fundamental rights, proportionality, rule of law, and democracy will be explored in Part II of the book. This form of respect is what Stephen Darwall calls 'recognition respect'. See Stephen L Darwall, 'Two Kinds of Respect' (1977) 88 Ethics 36.

Animals and the Constitution. John Olusegun Adenitire and Raffael Fasel, Oxford University Press.
 DOI: 10.1093/9780198910534.003.0002

domains of the constitution, thereby addressing the legitimacy deficits of the anthropocentric model.

The present chapter begins one step earlier by exposing a problem that lies at the foundations of the dominant anthropocentric model and that is a leading cause for its defects in the above-mentioned respects. This model, which has its roots in the American and French Revolutions, has justified constitutional governance mainly by reference to the (hypothetical) consent of the governed.[3] As we will show, this makes constitutional legitimacy dependent on paradigmatic human beings. And this, in turn, excludes animals and marginalizes non-paradigmatic humans—a category that, as we saw in Chapter 1, includes infants, young children, persons with severe mental disabilities, persons in a coma, as well as persons who are senile. This creates a legitimacy gap because these beings are governed by constitutions yet have no say in what makes them legitimate. Sentience-based constitutionalism, far from being at odds with constitutionalism, has a more suitable principle of constitutional legitimacy at its heart: the All Governed Interests Principle.[4] As we will argue, this principle helps fill current constitutions' legitimacy gap by basing their legitimacy on due concern for *all* sentient beings they govern.

The chapter is structured as follows. In Section 1, we identify the shortcomings of anthropocentric constitutionalism. As we point out, one of anthropocentric constitutionalism's structural problems is that its dominant legitimating principle—(hypothetical) consent by the governed—struggles to justify exercises of public power governing the interests of non-paradigmatic human beings and sentient animals. To remedy this problem, Section 2 proposes a more inclusive constitutionalism—sentience-based constitutionalism—which is based on a more capacious legitimating principle: due concern for the interests of all the governed. Finally, in Section 3, we defend sentience-based constitutionalism against two potential objections: one arguing that constitutional legitimacy depends on participation in constitution-making processes; the other challenging sentience-based constitutionalism on the grounds that it is itself too anthropocentric. The conclusion recapitulates the chapter's main insights.

[3] As we discuss in Section 3, there are other accounts of what makes constitutions legitimate. However, as we will see, at least some of these accounts suffer from similar problems regarding the exclusion of animals and the marginalization of non-paradigmatic humans.

[4] To use an alternative framing, one could say that we are arguing for constitutions to follow a different, sentience-based 'constitutional identity'. On this concept, see Gary Jeffrey Jacobsohn, *Constitutional Identity* (Harvard UP 2010). We are grateful to Lulu Weis for suggesting this to us.

1) The problems of anthropocentric constitutionalism

As we saw in Chapter 1, anthropocentric constitutionalism largely excludes the interests of animals from constitutional consideration, and it often marginalizes the interests of non-paradigmatic human beings. However, we have not yet explained why exactly anthropocentric constitutionalism's focus on (paradigmatic) human beings is so problematic. A sceptic might think that this focus is simply a natural feature of constitutionalism because constitutionalism is a human idea and practice. Hence, the sceptic might retort that the anthropocentric model of constitutionalism is not in need of change.

In this section, we respond to this, first, by demonstrating that anthropocentric accounts are problematic because they fail to give due concern to the intrinsic moral value of a large number of beings (Subsection A). We then show that this is not merely a peripheral problem for constitutionalism, but one that affects its very foundations. This is because it exposes a substantial deficiency in anthropocentric constitutionalism's dominant legitimating principle: (hypothetical) consent by the governed (Subsection B).

A) The moral relevance of sentient animals and non-paradigmatic humans

The first step in understanding why anthropocentric constitutionalism is problematic is to see that animals and human beings who do not conform to the rationalist paradigm have morally relevant interests that ought *prima facie* to be considered by humans, especially those exercising public powers.

Let us begin with *animals*, where the case that they have intrinsic value may seem less straightforward than for non-paradigmatic humans. As we saw in Chapter 1, it is widely accepted that humans have intrinsic value or worth. In other words, humans are viewed as ends in themselves, meaning that they matter morally regardless of their potential instrumental value to others. However, even though humans are biologically animals,[5] the view that other animals may also have intrinsic value was not widely accepted for a long time, at least in the Western world. This was due in no small part to the prevailing

[5] Ian Tattersall, 'Homo Sapiens', *Encyclopedia Britannica* (2020) <https://www.britannica.com/topic/Homo-sapiens> accessed 31 October 2024.

belief that humans are separate from animals, who were believed to lack souls or minds or other features thought to make humans special.[6]

The idea that humans are separate from animals has been challenged at least since the seventeenth century.[7] However, it is only in recent decades that we have witnessed a significant change in thinking.[8] An important contributing factor to this shift is that the study of animals' welfare and behaviour started to be placed on a more rigorous scientific footing.[9] As we saw in the Introduction, it is widely recognized today that many animals are sentient. This means that they have the capacity to have valenced experiences, that is, 'experiences that can feel bad or feel good to the subject'.[10] But while pain and pleasure are important, sentience should not be reduced to such experiences.[11] As Martha Nussbaum has cautioned, such reductionism obscures the fact that 'animals are complex creatures striving for their characteristic ends, with the aid of perception/imagination/thought and of desires and emotions of many kinds'.[12] As we pointed out, animals can feel a variety of different things, including empathy, jealousy, disgust, grief, indignation, and love.[13] Sentience thus distinguishes its holders from insentient beings such as jellyfish and bacteria, and inanimate entities such as tables and rocks, which all lack these capacities. Today, there is compelling scientific evidence that at least all vertebrate animals—a category that includes humans and other mammals, as well as amphibians, reptiles, and fish—are sentient. It is also widely accepted that some invertebrates such as cephalopods and decapod crustaceans possess sentience.[14]

What does this mean for how we ought to treat these beings? According to a widely adopted position, sentience confers moral status on its holders.[15] More specifically, sentient beings are considered to have *intrinsic* value. As beings

[6] See generally James Rachels, *Created from Animals: The Moral Implications of Darwinism* (OUP 1999) 173–223.

[7] Justin EH Smith, *Nature, Human Nature, & Human Difference: Race in Early Modern Philosophy* (Princeton UP 2015) 7–8, 18.

[8] David DeGrazia, 'Animal Ethics Around the Turn of the Twenty-First Century' (1999) 11 Journal of Agricultural and Environmental Ethics 111.

[9] Donald M Broom, *Sentience and Animal Welfare* (CABI 2014) 22–27.

[10] Jonathan Birch, *The Edge of Sentience: Risk and Precaution in Humans, Other Animals, and AI* (OUP 2024) 26 (emphasis removed).

[11] cf Peter Singer, *Animal Liberation: A New Ethics for Our Treatment of Animals* (HarperCollins 1975) 8.

[12] Martha C Nussbaum, *Justice for Animals: Our Collective Responsibility* (Simon & Schuster 2022) 125.

[13] See Introduction, Section 1.

[14] Helen Proctor, 'Animal Sentience: Where Are We and Where Are We Heading?' (2012) 2 Animals 628, 632–35; Broom (n 9) 122.

[15] Colin Allen and Mark Bekoff, 'Animal Minds, Cognitive Ethology, and Ethics' in Klaus Petrus and Markus Wild (eds), *Animal Minds & Animal Ethics: Connecting Two Separate Fields* (Transcript 2013) 37. For a classic defence, see Singer (n 11).

with intrinsic value, they have a direct interest in faring well rather than ill. Their interests are 'direct' because they themselves experience negative sensations like pain if these interests are not met, and positive sensations when they are.[16] As Alasdair Cochrane emphasizes, sentient beings 'are concerned with how their own lives are going—because they have a stake in their own fates—they possess a moral worth *of their own* which cannot be reduced to their usefulness for others'.[17]

Cochrane goes so far as to argue that sentience is a necessary and sufficient condition for having any interests.[18] This view has been challenged by authors who believe that most, if not all, forms of life can be said to possess interests. For example, plants could be described as having an interest in being watered. Bacteria could have an interest in finding a suitable host organism. And ecosystems might be said to have an interest in having keystone species protected.[19] A related challenge has come from critical animal studies scholars, some of whom prefer the concept of vulnerability to that of sentience. Roughly, vulnerability can be defined as the susceptibility of corporeal beings standing in relationships with other such beings of having their interests frustrated, such as by being harmed. As Maneesha Deckha points out, 'it is possible to be vulnerable through one's body without being sentient. There are living beings that can die (a form of vulnerability) without going through what we normally characterize as sentient experience: plants and embryos . . . are ready examples'.[20] We remain agnostic here as to whether sentience is a necessary condition for a being to have interests. There is widespread consensus, including among proponents of plant interests and of the vulnerability approach, that sentience is at least a *sufficient* condition.[21] This is enough to support our claim that sentient animals themselves have morally relevant interests.

Of course, the fact that animals have such interests would make little difference if it did not change our normative landscape. In other words, animals' interests would be inconsequential for our purposes if they did not affect what morally should and should not be done. The step from animals' interests to

[16] Alasdair Cochrane, *Sentientist Politics: A Theory of Global Inter-Species Justice* (OUP 2018) 15.

[17] ibid (emphasis in original).

[18] Alasdair Cochrane, 'From Human Rights to Sentient Rights' (2013) 16 Critical Review of International Social and Political Philosophy 655, 657.

[19] For an overview, see Ronald Sandler, 'Intrinsic Value, Ecology, and Conservation' (*Nature Education Knowledge*, 2021) <https://www.nature.com/scitable/knowledge/library/intrinsic-value-ecology-and-conservation-25815400/> accessed 31 October 2024.

[20] Maneesha Deckha, *Animals as Legal Beings: Contesting Anthropocentric Legal Orders* (University of Toronto Press 2021) 145.

[21] See eg Gary L Francione, 'Taking Sentience Seriously' in Klaus Petrus and Markus Wild (eds), *Animal Minds & Animal Ethics: Connecting Two Separate Fields* (Transcript 2013) 250.

moral obligations requires philosophical argument that has already been made by other authors.[22] We agree with the widely accepted premise here, which is also consistent with common sense understandings of morality, that because all sentient beings can feel, we have at least *prima facie* duties to not violate their interests. As Lori Gruen points out, this is because '[w]hen we deny sentient beings their basic interests we cause them to suffer emotionally, physically, or both, and we are causing them harm'.[23] In fact, our moral obligations are not exhausted in simply refraining from directly causing harm to sentient beings. Like human beings, Gruen stresses, non-human sentient beings can only thrive if they have 'adequate nutrition and hydration, relative health and bodily integrity, shelter from the elements, a non-toxic living environment, freedom of movement, social engagement (for social beings), and freedom of expression in its various forms'.[24] While some believe that our positive responsibilities to help other sentient beings thrive may be more limited than our direct obligations not to harm them,[25] sentience establishes at least a *prima facie* demand on others to help all sentient beings where possible.

Recall from the Introduction that this recognition of sentience's normativity is not limited to moral philosophy. It is also widespread in law, with many states basing their animal protection laws on a recognition of animals' sentience.[26] We already encountered an example in the long title of New Zealand's Animal Welfare Act (1999), which states that it is an Act to 'to reform the law relating to the welfare of animals and the prevention of their ill-treatment; and, in particular, (i) to recognize that animals are sentient'. As in New Zealand, most sentience recognitions can be found on the statutory level. However, there are a few countries in which sentience is even recognized on the constitutional level. For instance, Article 41 of the Constitution of Luxembourg (1868) recognizes since 2023 that animals are sentient beings, as does since 2024 Article 7bis of the Belgian Constitution (1831).[27] In fact, not only states but also international organizations have recognized sentience as a basis for moral and legal respect.

[22] See eg Singer (n 11) 1–22.

[23] Lori Gruen, *Ethics and Animals: An Introduction* (CUP 2011) 33.

[24] ibid 32.

[25] For a summary of this view, see Sue Donaldson and Will Kymlicka, *Zoopolis: A Political Theory of Animal Rights* (OUP 2011) 5–6 (who themselves focus also on positive obligations).

[26] Charlotte E Blattner, 'The Recognition of Animal Sentience by the Law' (2019) 9 Journal of Animal Ethics 121, 125. See also Jane Kotzmann and MB Rodriguez Ferrere (eds), *The Legal Recognition of Animal Sentience: Principles, Approaches and Applications* (Hart Publishing 2024).

[27] Ministère d'État, 'Version consolidée applicable au 01/07/2023: Constitution du Grand-Duché de Luxembourg' (2023) <https://legilux.public.lu/filestore/eli/etat/leg/constitution/1868/10/17/n1/consolide/20230701/fr/pdf/eli-etat-leg-constitution-1868-10-17-n1-consolide-20230701-fr-pdf.pdf> accessed 31 October 2024; 'De belgische grondwet' (2024) <https://www.senate.be/doc/const_nl.html> accessed 31 October 2024.

For instance, the Treaty on the Functioning of the European Union provides in Article 16 that

> [i]n formulating and implementing the Union's agriculture, fisheries, transport, internal market, research and technological development and space policies, the Union and the Member States shall, since animals are sentient beings, pay full regard to the welfare requirements of animals.[28]

In a similar vein, the Council of Europe's Convention for the Protection of Vertebrate Animals used for Experimental and Other Scientific Purposes (1986) recognizes in its preamble 'that man has a moral obligation to respect all animals and to have due consideration for their capacity for suffering'.[29]

As this shows, it is widely recognized that we have moral obligations to sentient animals, and many legal systems have created rules that are aimed at protecting these animals.[30] However, it is worth keeping in mind that, despite these existing moral and legal protections, sentient animals are still far from receiving due concern for their interests. As we saw in Chapter 1, existing animal welfare laws and even the most progressive constitutional provisions fall well short of protecting animals' basic interests.

Let us now turn to *non-paradigmatic human beings*, specifically to humans who are not considered to be paradigmatic rational agents.[31] As suggested above, the view that all human beings have intrinsic moral value and deserve to have their basic interests protected by the law is much more widely recognized than the view that all sentient beings matter in this way. Since the end of the Second World War, all humans have lived under the banner of human dignity and equality, even if this promise has remained unfulfilled for many in practice. The Universal Declaration of Human Rights, adopted by the UN General Assembly in 1948, makes this clear in its first preambular paragraph, which states that 'recognition of the inherent dignity and of the equal and inalienable rights of all members of the human family is the foundation of freedom, justice and peace in the world'.[32]

[28] Treaty on the Functioning of the European Union (2009) OJ C 326.

[29] European Convention for the Protection of Vertebrate Animals used for Experimental and Other Scientific Purposes (1986) ETS No 123.

[30] Blattner (n 26) 131.

[31] Of course, even humans who are considered not to have paradigmatic rationality can have certain forms of agency, see eg Geraldine Boyle, 'Recognising the Agency of People with Dementia' (2014) 29 Disability & Society 1130.

[32] 'Universal Declaration of Human Rights' (1948) <http://www.un.org/en/universal-declaration-human-rights/> accessed 31 October 2024.

Moral justifications abound for why all human beings—including non-paradigmatic ones—matter.[33] In fact, the mere attempt to investigate these justifications may appear insulting to some.[34] However, we believe that debunking some of the existing notions about the purportedly protective effects of anthropocentrism will help lay the groundwork for a more inclusive foundation that can better protect the interests of non-paradigmatic human beings and sentient animals. We will focus here on outlining the challenges faced by what we take to be the main justifications that exist within anthropocentric thinking for taking the interests of non-paradigmatic humans seriously.[35]

Arguably, the foremost anthropocentric justification for the moral status of non-paradigmatic humans is, quite simply, the fact that they are human beings.[36] Whether expressed in terms of their inherent 'dignity', 'equality', or similar notions, the principal idea is that these human beings matter morally because they belong to the species *Homo sapiens.* To use a phrase common in human rights discourse, non-paradigmatic humans have moral standing, rights, and dignity 'simply in virtue of being human'.[37] The central problem with this justification is that it does not explain *why* membership in the human species is morally relevant. Surely, the mere fact of being human cannot reasonably be invoked as a ground for moral status. This is because, as Peter Singer has emphasized, other general categories—such as being white or being male—would not be considered legitimate grounds for moral status either.[38] According to Singer, membership in the human species does not seem any more morally relevant than being a member of a certain race or sex. This is why using it to exclude members of other species is to engage in 'speciesism', akin to how excluding beings based on race or sex would be considered racism and sexism, respectively.[39]

To avoid the charge of speciesism, some advocates of anthropocentrism have tried to justify the moral status of humanity in a foundationalist way. That is to say that they have tried to identify features that only human beings possess and that make the human species special in a morally relevant way. Features such

[33] For an overview, see David Wasserman and others, 'Cognitive Disability and Moral Status' in Edward N Zalta (ed), *The Stanford Encyclopedia of Philosophy* (Fall 2017) <https://plato.stanford.edu/archives/fall2017/entries/cognitive-disability/> accessed 31 October 2024.

[34] This is the case for some foundationalists, a view we discuss below. See text from around n 39.

[35] For more detail, see Raffael N Fasel, '"Simply in Virtue of Being Human"? A Critical Appraisal of a Human Rights Commonplace' (2018) 9 Jurisprudence 1.

[36] See Agnieszka Jaworska and Julie Tannenbaum, 'The Grounds of Moral Status' in Edward N Zalta (ed), *The Stanford Encyclopedia of Philosophy* (Spring 2018) <https://plato.stanford.edu/archives/spr2018/entries/grounds-moral-status/> accessed 31 October 2024.

[37] See generally Fasel (n 35).

[38] Singer (n 11) 9.

[39] ibid 6.

as rationality, autonomy, language, culture, or empathy have been put forward as possible candidates.[40] However, this approach of grounding moral status in 'showpieces of man',[41] as Friedrich Nietzsche called them, faces two problems from an anthropocentric point of view. First, the approach risks being *overinclusive*. If there are non-human beings that possess the relevant features, then not *only* human beings would have moral status but also non-humans. For example, if one takes language as the ground for moral status, then animals such as whales whom many scientists believe to have languages would have to be included too. In fact, some animals could possess the relevant features to an even greater degree than humans. Compare for instance an adult humpback whale in the prime of her linguistic abilities with a human being who, due to a congenital disease, lacks the ability to communicate.[42] Second, grounding human moral status by reference to ostensibly special human features can risk defining the relevant features so narrowly that not *all* human beings possess them. This especially threatens to exclude non-paradigmatic humans. For example, to exclude smart non-human animals from moral status, 'rationality' could be defined as practical reason in the Kantian sense, that is, as the capacity to act according to principles.[43] But this would also exclude those non-paradigmatic humans who are incapable of acting based on principles. In other words, the approach is also in danger of being *underinclusive*.

As a way of rescuing the argument, it has been suggested that even if non-paradigmatic humans do not possess the relevant feature or features that make humans special and deserving of moral status, they at least possess the *potential* of developing that feature or, alternatively, that *normal* (ie paradigmatic) members of the human species possess it. Rebuttals of these arguments have become legion in the literature on non-human rights.[44] In a nutshell, the *potentiality* approach is problematic because not all non-paradigmatic human beings may actually possess the potential of developing these features.[45] In everyday contexts, too, one's mere potential to have or do something does not confer on one the rewards that come with one's actual achievement or possession of that thing either. Just because I have the potential of passing the driving

[40] See eg Jeremy Waldron, *One Another's Equals: The Basis of Human Equality* (Belknap Press of Harvard UP 2017) (on rationality).
[41] Friedrich Nietzsche, *On the Genealogy of Morals* (Douglas Smith tr, OUP 1996) Second Essay, II, 44.
[42] For an accessible introduction to whale communication see Tom Mustill, *How to Speak Whale: A Voyage into the Future of Animal Communication* (Grand Central Publishing 2022).
[43] Immanuel Kant, *Critique of Practical Reason* (Mary J Gregor tr, CUP 1997).
[44] See eg Cochrane (n 16) 20–21.
[45] Raffael N Fasel, 'More Equal Than Others: Animals in the Age of the Human Rights Aristocracy' (University of Cambridge 2019) 147.

test, for example, does not mean that I should be allowed to drive before I have actually passed the test. The *normality* approach is deficient among other reasons because it fails to make clear why the status of human beings should be determined in relation to what is normal for the human species rather than what is normal for other groups to which these beings also belong, such as the great apes or the kingdom of animals.[46]

A very different anthropocentric strategy to secure the status of non-paradigmatic humans consists in entirely abandoning the foundationalist search for features that ground their moral status and to instead simply assume or decide that they do possess that status.[47] This assumption or decision is guided by the view that humanity is unique and therefore all humans ought to have a special status. We cannot discuss the virtues and vices of this anti-foundationalist approach in detail here.[48] For our purposes, it will be sufficient to flag two issues.

First, this approach struggles to give a compelling answer to those who do not share the assumption or decision, or are at least not willing to accept it unquestioningly. Second, and more problematically, the approach inherently tends to marginalize beings who lack the 'showpieces' that serve to motivate our assumptions or decisions that humans should have a special status. The reason for this is that the accentuation of features that are thought to make humanity special, such as rationality or autonomy, necessarily creates a standard which some humans fail to meet. While different in other regards, the anti-foundationalist approach has this problem in common with the foundationalist approach. As we saw, this approach marginalizes those who lack the special features that are supposed to ground humanity's moral status. If, as seems to be the case on both the foundationalist and anti-foundationalist approaches, what ultimately grounds or guides our assumptions about moral status is the normal and actual possession of a feature, then any being that does not possess this feature must necessarily be regarded as *falling short*. This is illustrated, for example, in the striking language that defenders of the special moral status of humans use when talking about persons with severe cognitive disabilities. For instance, according to Jeremy Waldron, these human beings are 'misfortunes',[49] 'tragedies',[50] or 'heartbreaking cases'.[51] Both the foundationalist and

[46] ibid 147–48.

[47] See eg Anne Phillips, *The Politics of the Human* (CUP 2015).

[48] For more detail, see Raffael N Fasel, *More Equal Than Others: Humans and the Rights of Other Animals* (OUP 2024) 69–71.

[49] Waldron (n 40) 246.

[50] ibid 225.

[51] ibid 216. For a critique, see Raffael N Fasel, 'One Another's Equals: The Basis of Human Equality by Jeremy Waldron' (2018) 77 Cambridge Law Journal 230.

anti-foundationalist approaches, then, only manage to lift humanity at the cost of excluding or marginalizing non-paradigmatic human beings who pale by comparison to members of their species who are seen as actually rational and therefore 'normal'. Note the bitter irony here. The declared aim of many who adopt these approaches is to protect vulnerable non-paradigmatic humans.[52] However, because of the inherent flaws of their position, they risk demoting rather than promoting the interests of these human beings. As Will Kymlicka has put it in his critique of anthropocentric human rights theory, 'adults are the "centre" of human rights, and the rights of everyone else are left hanging by a thread'.[53]

In our view, a non-anthropocentric criterion such as sentience serves as a more suitable justification for the moral status of non-paradigmatic human beings.[54] On the sentience-based approach, non-paradigmatic beings are intrinsically valuable and have interests that ought to be given due concern because of their sentience. On this approach, whether or not these human beings are rational or have language or culture is irrelevant to their moral status. Thus, non-paradigmatic beings are included and counted morally without the risk of being marginalized as 'tragic' or 'unfortunate' cases because they fail to live up to standards of paradigmatic human beings. Furthermore, as indicated above, our sentience-based approach only takes sentience to be a sufficient condition for moral status. It therefore does not stand in the way of other approaches, such as those that draw on vulnerability and relationality to explain how we ought to treat non-paradigmatic humans.[55]

Let us now consider a possible objection that could be raised at this stage. Even if we accept that anthropocentric approaches struggle to accommodate sentient animals and marginalize non-paradigmatic human beings, this need not concern constitutionalists—or so the objection goes—because these beings are only a problem at the margins of society. In other words, even if they matter, they do not matter enough to be worth the time and effort of constitutional thinkers. A brief look at the statistics reveals the flaw in this objection. Focusing on the United Kingdom as an example, we can see that 20.6 per cent

[52] eg Waldron (n 40) 30, 223.

[53] Will Kymlicka, 'Human Rights Without Human Supremacism' (2018) 48 Canadian Journal of Philosophy 763, 778.

[54] See also below Subsection 3.B.

[55] See eg Angharad E Beckett, *Citizenship and Vulnerability: Disability and Issues of Social and Political Engagement* (Palgrave Macmillan 2006) (for a vulnerability-based approach to citizenship for persons with disabilities); Rosie Harding, *Duties to Care: Dementia, Relationality and Law* (CUP 2017) (for a relational approach to law and people with dementia).

(13.9 million) of the population of England and Wales was underage in 2022.[56] In addition, an estimated 3.5 per cent (2 million) of the population fall under the protection of the Mental Capacity Act (2005) because of cognitive restrictions that prevent them from making the kinds of decisions that paradigmatic legal subjects can make.[57] Another 5 per cent (3.4 million) of the population was aged 80 or older.[58] Although it is of course true that not all of the human beings in these groups are devoid of rational agency, it is possible that a significant number would not be classified as paradigmatic rational human beings.

In addition to these human beings, who make up around a third of the UK population, there is a vast number of sentient non-human beings in the country. In 2022, more than 2.5 million mice, fishes, rats, birds, cats, dogs, horses, primates, and other species were used for scientific experiments.[59] In 2024, the United Kingdom also counted 36 million companion animals—a category that includes cats, dogs, rabbits, birds, guinea pigs, snakes, horses, and rats, among others.[60] Though numerous, animals used in experiments and companion animals are dwarfed by farmed animals. In 2023, livestock was estimated at 4.7 million pigs, 9.2 million cows, bulls, and calves, 21.2 million sheep and lambs, and 178 million chickens, ducks, geese, turkeys, and other forms of poultry.[61] In the case of pigs, chickens, and other species of poultry, even these impressive figures do not represent the actual number of animals that lived in the country during that year. This is because these animals are slaughtered before the age of one, meaning that over the course of a year a much greater number of them are reared and killed.[62] The total number of chickens

[56] Office for National Statistics, 'Estimates of the Population for the UK, England, Wales, Scotland and Northern Ireland' (26 March 2024) <https://www.ons.gov.uk/peoplepopulationandcommunity/populationandmigration/populationestimates/bulletins/annualmidyearpopulationestimates/mid2022> accessed 28 August 2024.

[57] The latest estimate can be found in: HM Government, 'The National Mental Capacity Forum: Chair's Annual Report 2017' (2017) 38 <https://assets.publishing.service.gov.uk/government/uploads/system/uploads/attachment_data/file/708149/nmcf-annual-report-2017.pdf> accessed 31 October 2024.

[58] Office for National Statistics (n 56).

[59] Office for National Statistics, 'Statistics of Scientific Procedures on Living Animals, Great Britain: 2022' (9 November 2023) <https://www.gov.uk/government/statistics/statistics-of-scientific-procedures-on-living-animals-great-britain-2022/statistics-of-scientific-procedures-on-living-animals-great-britain-2022#key-results> accessed 29 August 2024. The numbers are for Great Britain.

[60] UK Pet Food, 'UK Pet Population' (21 March 2024) <https://www.ukpetfood.org/information-centre/statistics/uk-pet-population.html> accessed 29 August 2024.

[61] Department for Environment, Food and Rural Affairs, 'Livestock Populations in the United Kingdom at 1 December' (28 March 2024) <https://www.gov.uk/government/statistics/livestock-populations-in-the-united-kingdom/livestock-populations-in-the-united-kingdom-at-1-december--2> accessed 29 August 2024.

[62] See on chickens European Food Safety Authority, 'Scientific Opinion on the Influence of Genetic Parameters on the Welfare and the Resistance to Stress of Commercial Broilers' (2010) 8 EFSA Journal 1, 38.

slaughtered over the course of 2023, for example, was 1.118 billion.[63] In addition, it is estimated that 876 million fishes were caught or farmed.[64] Finally, there are great numbers of wild land animals. These animals are not generally recorded in official statistics, but a back-of-the-envelope calculation suggests that wild land vertebrates in the United Kingdom range somewhere between 180 million and 180 billion animals.[65]

Even on a conservative estimate, then, roughly 2.4 billion non-human vertebrates live in the United Kingdom in a given year and are therefore at least in principle governed by its constitution. If we add the more than 15 million humans who may be viewed as lacking paradigmatic rational agency, we arrive at a total of around 2.415 billion beings. By comparison, the approximately 50 million paradigmatic human beings make up only 2 per cent of the total number of potentially sentient beings living under the UK constitution. At least in numerical terms, then, animals and non-paradigmatic humans are not the exception. They are the rule and thus deserving of our attention.

B) The constitutional relevance of sentient animals and non-paradigmatic humans

The previous section has shown that sentient animals and non-paradigmatic humans have intrinsic value simply by virtue of their sentience and are therefore owed due concern. We also saw that these beings far outnumber paradigmatic human beings and are thus not a marginal phenomenon that can be ignored. What we have not yet shown, however, is that these beings matter *for the purposes of constitutionalism*, and not just morally. After all, a constitutionalist might well agree that sentience is a sufficient criterion for intrinsic moral value. But that does not mean that they need to believe that any changes are required to how they understand constitutionalism. In this section, we show that

[63] Department for Environment, Food and Rural Affairs, 'Latest Poultry and Poultry Meat Statistics' (22 August 2024) <https://www.gov.uk/government/statistics/poultry-and-poultry-meat-statistics> accessed 29 August 2024.

[64] See 'Annual UK Slaughter Statistics' (*Animal Clock UK*) <http://animalclock.org/uk> accessed 31 October 2024; Alison Mood, 'Worse Things Happen at Sea: The Welfare of Wild-Caught Fish' (fishcount.org.uk 2010) 72 <http://www.fishcount.org.uk/published/standard/fishcountfullrptSR.pdf> accessed 31 October 2024; fishcount.org.uk, 'Numbers of Farmed Fish Slaughtered Each Year' <http://fishcount.org.uk/fish-count-estimates-2/numbers-of-farmed-fish-slaughtered-each-year> accessed 31 October 2024.

[65] See Brian Tomasik, 'How Many Wild Animals Are There?' (2019) <https://reducing-suffering.org/how-many-wild-animals-are-there/> accessed 31 October 2024. Our ballpark figure for the United Kingdom is based on Tomasik's global estimate and the United Kingdom's share of the world's land mass (0.184 per cent).

the requirement to give due concern to all sentient beings is not only of some constitutional relevance but touches constitutionalism's core. This is because it reveals a legitimacy gap created by the shortcomings of the dominant legitimating principle of constitutions: consent by the governed.[66]

Before we embark on this demonstration, it is important to note that there is more than one view of what makes a constitution—and constitutionalism more generally—legitimate. According to one popular approach, constitutions receive their legitimacy procedurally by dint of their creation through a constituent power (we will return to this view in Section 3). On another view, defended by Ronald Dworkin, constitutions gain their legitimacy by protecting individuals' rights.[67] In what follows, we will focus on what we regard as the dominant account of legitimacy that has underpinned constitutionalism since its foundation, especially in its classic form: the principle that a constitution is legitimate if it is based on the (hypothetical) consent of those who are governed by it.[68]

To understand this principle, it is necessary to consider in more detail classic constitutionalism's history and *raison d'être*, which we could only sketch in Chapter 1. There are different accounts of the origins of classic constitutionalism,[69] but a particularly influential view is that presented by Dieter Grimm, who locates the beginnings of constitutionalism in the second half of the eighteenth century. Specifically, according to Grimm, constitutionalism emerged with the American and French Revolutions.[70] While the concept of

[66] See also Jessica Eisen, 'Animals in the Constitutional State' (2017) 15 International Journal of Constitutional Law 909, 911; Janneke Vink, *The Open Society and Its Animals* (Palgrave Macmillan 2020) 2.

[67] Ronald Dworkin, *Freedom's Law: The Moral Reading of the American Constitution* (OUP 1996).

[68] An anonymous reviewer suggested that our approach to constitutional legitimacy operates with a mixed input and output notion of constitutional legitimacy, perhaps similar to that adopted in Fritz Scharpf, *Governing in Europe: Effective and Democratic?* (OUP 1999), 6–21. We are indeed critical of pure input-based theories of constitutional legitimacy. As we say in Subsection 3.A, if constitutional legitimacy depended solely on participation in input constitutional processes (eg constituent assemblies, referenda, amendments), very few constitutions would be legitimate because these input processes are not undertaken often and some of them, eg constituent assemblies, require the participation of very few constitutional subjects. Accordingly, the All Governed Interests Principle, which is our approach to determining constitutional legitimacy (set out in Section 2.B), says that a constitution is more legitimate the more it respects the intrinsic value of sentient beings and the more it protects their basic interests. However, that is not to say that input does not matter at all. As we say in Section 2.B and explore in more detail in Chapter 6, constitutional subjects have a right to political participation, including participation in processes of legal and constitutional change. So, to the extent that people cannot participate eg in exercises of constituent power then that too can reduce the legitimacy of a constitution.

[69] See eg Horst Dippel, 'Modern Constitutionalism: An Introduction to a History in Need of Writing' (2005) 73 Tijdschrift voor rechtsgeschiedenis 153.

[70] Dieter Grimm, 'The Achievement of Constitutionalism and Its Prospects in a Changed World' in Petra Dobner and Martin Loughlin (eds), *The Twilight of Constitutionalism?* (OUP 2010) 3.

the constitution existed well before the eighteenth century, Grimm argues that it was only then that it started to be used to justify constitutional governance:

> Different from the older legal bonds of political power, the new constitutions did not modify a pre-existing right to rule: they preceded the ruler's right to rule. They created this right, determined the procedure in which individuals were called into office, and laid down the conditions under which they were entitled to exercise the power given to them.[71]

As Charles Howard McIlwain put this thought, with the Revolutions the constitution became 'antecedent to the government'.[72] This shift towards normative constitutions that do not just describe public powers but provide a normative basis for them correlated with the search for a justification of public power. Because the right to rule could no longer be presupposed or explained as God-given, it became necessary to find another source for the constitution's legitimacy. That source was the people.[73] The simple idea was that those who are ruled should get to determine who rules them and how. In the words of Thomas Jefferson in the Declaration of Independence (1776), 'Governments are instituted among Men, deriving their just powers from the consent of the governed'.[74] According to Martin Loughlin, this modern idea of the constitution 'reject[ed] traditional orderings based on status and hierarchy [and] expressed the conviction that government, being an office established for the benefit of the people, must be based on their consent'.[75] The 'legitimating principle'[76] of this new order—to use Grimm's term—was '[p]opular sovereignty'.[77] As Ilan Wurman put this idea in the context of the US Constitution: '"We the People" enacted the Constitution in a public act of ratification, and because the Constitution is thus clothed with the consent of the governed, we must continue to adhere to it today'.[78]

On this account, it would at first glance appear to be the people in its entirety that confer on the constitution its legitimacy by giving their actual consent to

[71] ibid 7.
[72] Charles Howard McIlwain, *Constitutionalism: Ancient and Modern* (Liberty Fund 1947) 10, 15 (internal quotation marks omitted). The term is Paine's: Thomas Paine, *Rights of Man, Common Sense, and Other Political Writings* (OUP 2008), 122.
[73] Grimm (n 70) 7.
[74] 'Declaration of Independence: A Transcription' (*National Archives*, 1 November 2015) <https://www.archives.gov/founding-docs/declaration-transcript> accessed 27 October 2024.
[75] Martin Loughlin, 'What Is Constitutionalisation?' in Petra Dobner and Martin Loughlin (eds), *The Twilight of Constitutionalism?* (OUP 2010) 48.
[76] Grimm (n 70) 8. We are using the terms 'legitimating principle' and 'justification' synonymously.
[77] ibid.
[78] Ilan Wurman, *A Debt Against the Living: An Introduction to Originalism* (CUP 2017) 59.

being governed by it.[79] However, there is an obvious problem with making constitutional legitimacy dependent on the actual consent of the governed: many beings are not able to consent to constitutional arrangements governing them, and even those who are do not always get the chance to consent to a specific arrangement. Non-paradigmatic humans and sentient animals belong to the first group. Because of the cognitive constraints that some of them face, they may lack the ability to understand the constitutional framework that governs their lives, and to meaningfully consent or refuse to give consent to it. As a result of this inability, the legitimacy of a constitution based on the legitimating principle of consent by the governed does not take into account how well or poorly it treats non-paradigmatic humans.

The dominant legitimating principle furthermore struggles to accommodate the second group of beings: paradigmatic humans who, although able to give valid consent to a constitution, simply never get the opportunity to do so, at least not in a formal way. This is not an exceptional scenario: many paradigmatic humans do not have the chance to consent to their constitution but are nonetheless governed by it. Take the US Constitution. This Constitution only received the formal consent of a small number of delegates in state conventions.[80] As was already recognized by some at the time, the consent these delegates gave to ratify the Constitution between 1787 and 1790 could hardly be viewed as binding on future generations of US Americans who could not consent to the Constitution themselves, at least not in the same way as the delegates could.[81]

To circumvent this problem, classic constitutionalists resort to the argument that it is not the *actual* consent that serves to legitimate a constitution but people's *hypothetical* consent. In other words, legitimate constitutional rule is determined by asking under which conditions people *would* give consent, if they had the chance.[82] Eighteenth-century thinkers invoked the now well-known device of the state of nature to accomplish this constitutional thought experiment. As we saw in Chapter 1, the state of nature is a hypothetical, not a real situation: most people are not actually asked to give their consent to the constitution. Rather, it is asked under which conditions they would consent were they given the opportunity to do so.

[79] Dieter Grimm, *Constitutionalism: Past, Present, and Future* (OUP 2016) 7.

[80] John R Vile, *The Writing and Ratification of the U.S. Constitution: Practical Virtue in Action* (Rowman & Littlefield Publishers 2012) 143–46.

[81] See, famously, Letter from Thomas Jefferson to James Madison, 6 September 1789' <http://founders.archives.gov/documents/Madison/01-12-02-0248> accessed 31 October 2024.

[82] Grimm (n 70) 7.

The move from actual to hypothetical consent may be able to overcome the issue that not all paradigmatic humans actually get the chance to consent to the reigning constitutional arrangement. However, it does not overcome the problem of excluding non-paradigmatic humans and animals. Not everyone is allowed into the hypothetical state of nature as it is commonly understood. Rather, only paradigmatic human beings are granted access.[83] This is because, by its very conception, the state of nature presupposes a particular kind of participant: persons who are free, equal, and rational. Only paradigmatic humans can freely decide under which conditions it would make sense for them to leave the state of nature, and only they can consent to submitting themselves to the constitutional governance of a new system in the civil state. As Walter Murphy has put it, '[c]onsent can function as a legitimator only to the extent that the people agree as reasoning, morally autonomous, and responsible human beings'.[84] Grimm also underscores the focus on rationality in state of nature arguments:

> Leaving the state of nature became a dictate of reason. Given the equal freedom of all individuals, the step from the state of nature to government called for a general agreement. Legitimacy could be acquired only by a government based on the consent of the governed. It was also up to the governed to determine the conditions under which political power could be exercised.[85]

The takeaway for us here is that, since the American and French Revolutions, a constitution's legitimacy has been seen as hinging on whether it provides the kind of public power that paradigmatic human beings would consent to in the state of nature.[86] As Grimm points out, this classic understanding of constitutionalism continues to be the dominant 'model for the organisation and legitimation of public power'—more than two centuries after its emergence.[87]

[83] See Tom Huffman, 'Animals, Mental Defectives, and the Social Contract' (1993) 9 Between the Species 20. Some social contract theories explicitly consider animals to live in the state of nature: see eg Jean-Jacques Rousseau, *A Discourse on Inequality* (Maurice Cranston tr, Penguin 1984) 243–44. However, even in such theories, animals are not regarded as belonging to the relevant (human) community of rational beings who can consent to leaving the state of nature. Hence, their interests do not factor into the legitimacy of a constitutional system either.

[84] Walter F Murphy, 'Consent and Constitutional Change: Essays in Honour of Brian Walsh' in James O'Reilly (ed), *Human Rights and Constitutional Law* (Round Hall Press 1992) 142.

[85] Grimm (n 70) 7.

[86] See also Mattias Kumm, 'Constituent Power, Cosmopolitan Constitutionalism, and Post-Positivist Law' (2016) 14 International Journal of Constitutional Law 697, 706.

[87] Grimm (n 70) 3.

Many have hailed this model of constitutionalism as a success story.[88] We agree that the shift it achieved towards a normative conception of the constitution, justified by virtue of popular sovereignty, is laudable. So too is the curtailing of arbitrary exercises of power and the protection of fundamental human rights to which it has contributed. However, classic constitutionalism's success story has also clouded our view of the problem that lies at its heart: it explains the legitimacy of a constitution mainly by virtue of the consent of the governed, but it cannot apply this to all those who are governed by that constitution. By interpreting consent of the governed as hypothetical consent of paradigmatic human beings, it makes constitutional legitimacy dependent only on what paradigmatic human beings might want. This creates a legitimacy gap: the constitution applies not only to paradigmatic humans but also to sentient animals and non-paradigmatic humans. It therefore marginalizes them by failing to directly consider their interests as relevant for constitutional legitimacy.

Of course, one could argue that the hypothetical consent of paradigmatic humans should be dependent on the constitution respecting the interests of non-paradigmatic humans or animals. That is, one could assume that paradigmatic humans would not consent to constitutions that disregard the interests of other sentient beings. This way, constitutional legitimacy would at least depend indirectly on the interests of non-paradigmatic humans and sentient animals. However, this would still marginalize these beings as the protection of their basic interests would depend on a contingent state of affairs.[89] In fact, it is highly questionable whether paradigmatic humans would consent to constitutional governance only if the constitution respected the interests of these beings. After all, as we note below, paradigmatic humans have enabled anthropocentric constitutions that, in many ways, disregard the basic interests of animals and sometimes also non-paradigmatic humans.

Because sentient animals and non-paradigmatic humans make up around 98 per cent of those governed by a constitution like the United Kingdom's, the exclusion and marginalization of these beings should worry anyone interested in legitimate *constitutional* rule. This is particularly concerning considering that constitutional systems generally govern the lives of animals and non-paradigmatic humans in much more sweeping ways than they govern the lives of paradigmatic humans. Animals, for instance, are subject to constitutional

[88] See eg Bruce Ackerman, 'The Rise of World Constitutionalism' (1997) 83 Virginia Law Review 771–72.
[89] See Vink (n 66) 70–72.

systems that grant them no fundamental rights and consider them to be objects over which humans can have property rights. For example, constitutional systems permit and regulate the raising and killing of animals for food, they allow animals to be used in medical testing, and they permit the hunting of wild animals. They also accord no relevance to the suffering and death of animals that result indirectly from human activity, such as through loss of habitat or road traffic. As Irus Braverman puts it, animals are, without a doubt, 'objects of governance'.[90]

Non-paradigmatic human beings also frequently figure as objects of intrusive governance. The reality is that non-paradigmatic human beings have at times been, and sometimes still are, subjected to treatments that violate their basic interests in life, bodily and mental integrity, and family life, among others.[91] According to a report by Human Rights Watch, people with disabilities continue to 'face significant barriers to realizing their human rights, including discrimination in education, employment, housing and transport; denial of the right to vote; and being stripped of the right to make decisions about their own lives, including their reproductive choices.'[92] Hence, both the sheer quantity of sentient animals and non-paradigmatic humans as well as the poor quality of their constitutional treatment reveal that the legitimacy gap from which anthropocentric constitutionalism suffers is no mere crack—it is a canyon.

To be sure, as we showed in Chapter 1, classic constitutionalism is not the only model of constitutionalism, nor is it the only path to constitutional legitimacy, as mentioned above. Other models include Nick Barber's positive constitutionalism, according to which the purpose of constitutionalism is the protection of (human) well-being generally,[93] and Catherine Dupré's dignity-based constitutionalism, which aims to give all human beings a special constitutional place.[94] However, even if these alternative models succeeded in not marginalizing non-paradigmatic humans—which, as we saw, is questionable at least in the case of dignity-based constitutionalism—they still exclude all animals when it comes to determining what makes a constitution legitimate.[95]

[90] Irus Braverman, 'Law's Underdog: A Call for More-than-Human Legalities' (2018) 14 Annual Review of Law and Social Science 127, 131.

[91] See also Donaldson and Kymlicka (n 25) 113.

[92] Human Rights Watch, 'One Billion Forgotten: Protecting the Human Rights of Persons with Disabilities' (2014) 3 <https://www.hrw.org/sites/default/files/related_material/2014%20disabilities_program_low.pdf> accessed 31 October 2024.

[93] Nick W Barber, *The Principles of Constitutionalism* (OUP 2018).

[94] Catherine Dupré, *The Age of Dignity: Human Rights and Constitutionalism in Europe* (Hart Publishing 2015).

[95] See Chapter 1 Section 2.B.

To recapitulate, then, the dominant model of constitutionalism has at its core a legitimating principle—(hypothetical) consent of the governed—which fails to justify constitutions for the vast majority of those they govern. Non-paradigmatic humans and sentient animals all have intrinsic moral value, but their interests do not count in determining what their governance ought to look like. Instead, only free, equal, and rational human beings matter when it comes to determining what is constitutionally legitimate.

2) Towards sentience-based constitutionalism

For the reasons pointed out in the previous section, the legitimating principle of consent of the governed should be replaced with a more capacious principle. As we argue in this section, sentience-based constitutionalism offers such a more capacious principle. According to sentience-based constitutionalism, a constitution's legitimacy depends not on the (hypothetical) consent of paradigmatic human beings but on whether the constitution gives due concern to the interests of all governed sentient beings.[96] This principle—which we call the All Governed Interests Principle—fills the legitimacy gap that characterizes anthropocentric constitutionalism. Constitutional systems that satisfy this new standard of legitimacy recognize that all sentient beings have what we call 'constitutional worth'. That is, they recognize that sentient animals possess interests that are of constitutional importance and require respect for their intrinsic value as well as protection of their basic interests through constitutional norms.[97]

This idea is not as far-fetched as it may seem. In 2022, the Animal Welfare (Sentience) Act entered into force in the United Kingdom to set up an Animal Sentience Committee. The purpose of this Committee is to report on new government policies with a view to scrutinizing whether 'the government has all due regard to the ways in which the policy might have an adverse effect on the welfare of animals as sentient beings'.[98] The Committee can also make legally non-binding recommendations to the government about how to achieve this goal. Implicitly, the Act recognizes that sentient animals have interests of moral significance, which government policies ought to give due regard to be

96 See also Vink (n 66) 7.

97 See also Eisen (n 66) 910.

98 UK Animal Welfare (Sentience) Act 2022, Section 2(4). For analysis, see Rachel Dunn and Joshua Jowitt, 'The Animal Sentience Committee: Evolution or Revolution in the Legal Recognition of Animal Sentience in the UK?' in Jane Kotzmann and MB Rodriguez Ferrere (eds) *The Legal Recognition of Animal Sentience: Principles, Approaches and Applications* (Hart Publishing 2024).

legitimate.[99] To be sure, the Animal Welfare (Sentience) Act differs from our proposal in that it does not require that animals and their basic interests are given due concern in the constitution. It applies neither to all state authorities nor to non-paradigmatic humans. Nevertheless, it shows that the idea of making governmental legitimacy dependent on respect for the interests of sentient animals has already found its way into law in one country.

To make the case for sentience-based constitutionalism, we first show why constitutional legitimacy should be determined by focusing on interests rather than consent (Subsection A). We then discuss the implications of this proposal by developing the All Governed Interests Principle (Subsection B).

A) Legitimating constitutions: From consent to interests

As we showed in the previous section, proponents of anthropocentric constitutionalism generally adopt the view that constitutions are justified insofar as they are based on the hypothetical consent of those whom they govern. This legitimating principle is preferable to the requirement of actual consent. As we saw, the requirement of actual consent excludes not only non-paradigmatic human beings and sentient animals who cannot give consent in the relevant ways but also paradigmatic humans who lack opportunities to consent formally. However, the concept of hypothetical consent is not devoid of problems either. Its primary problem is that it is based on the presumption that those participating in the state of nature are free, equal, and rational human beings. This has served to exclude sentient animals and has marginalized non-paradigmatic humans who are not as free, equal, and rational as paradigmatic humans.

Given these problems, it is best to look for a more capacious legitimating principle; a principle that manages to accommodate more than just paradigmatic humans. When looking for such an alternative principle, however, it is important that we do not throw out the baby with the bathwater. Specifically, consent and popular sovereignty are crucial and hard-earned ideas. We should therefore be careful not to abandon them simply because they struggle to justify constitutions to those lacking the ability or opportunity to consent. Instead, our goal should be to find a principle that can make room both for those who can consent as well as for those who cannot.

[99] See eg Tim Farron (18 January 2022), Animal Welfare (Sentience) Bill [Lords] [Hansard].

We believe that a more inclusive principle emerges once we put consent under the microscope. Although consent is frequently discussed as a free-standing concept, scrutiny reveals that it is not as independent as it may seem. Particularly, consent is neither the only nor the most fundamental source of constitutional legitimacy. Rather, the requirement to seek consent from who can consent presupposes a more fundamental source of legitimacy. That source is the *interests* of those from whom consent is sought. The idea is this: we do not ask for people's consent for no reason. Instead, we believe there is something special about people that makes it the case that a constitution governing such people must, to be legitimate, obtain their (hypothetical) approval. And that 'something special' is that they possess morally relevant interests in the first place, such as in virtue of being sentient and autonomous. Put differently, both actual and hypothetical consent only matter if the being required to give consent matters. And whether a being matters in the first place depends on whether they matter morally, that is, whether they are considered to have morally relevant interests. As we explain below, this does not mean that everything can be boiled down to the interests that people have in giving consent, thus making consent superfluous. Suppose I consent to having my aching tooth extracted because I have an interest in relieving my pain. If the dentist were to extract the tooth without my consent, she would not violate my interest in pain relief, as that outcome would still be achieved. However, she would disregard my autonomy and thus the fact that, as a sentient being, I have morally relevant interests.

This shift from consent to interest raises important questions, such as how to identify interests and determine whose interests count. Before we address these questions in the next section, it is worth highlighting that basing constitutional legitimacy on the protection of interests of beings with intrinsic moral value has two main advantages when compared to the (hypothetical) consent principle. First, it is more inclusive and makes constitutional legitimacy dependent not just on paradigmatic humans but also on non-paradigmatic humans and sentient animals. They, too, are governed by the constitution and have morally relevant interests in being governed well. By incorporating these otherwise excluded or marginalized beings, the move from consent to interests fills the legitimacy gap that bedevils anthropocentric constitutionalism.

Importantly, however, focusing on interests achieves this feat without jettisoning consent. This is because—and this brings us to the second point—basing constitutional legitimacy on the protection of interests ensures that consent is protected in beings who have the ability to consent. For beings who possess that ability, being given the chance to consent (or refuse to consent) to the constitutional structures that govern them is in their interest. Carving out such a space for consent is vital in an account of constitutional legitimacy. For those beings who can consent, consent is an inherently valuable expression of their autonomy.[100] It is also instrumentally useful because it helps us identify what a person wants or does not want. Although people can make bad decisions, they generally do not consent to things that are not in their interest. And even when they do consent to things that are not in their interest, respecting those decisions is key to respecting their autonomy. For these reasons, not losing sight of consent is imperative as it serves as a safeguard against autocratic constitutional systems, where rulers claim to be acting in their subjects' interest, but often lack their actual and hypothetical consent.

As we will see below, the difficulty of applying consent to sentient animals and some non-paradigmatic animals can make it more challenging to determine what is in their interest. But that does not mean that they lack morally relevant interests in being governed well. Basing constitutional legitimacy on interests accommodates these beings because it does not view consent as the only or most fundamental source of legitimacy. Rather, respecting consent is only *one* way of giving due concern to the interest of sentient animals. There are other ways that this can be accomplished, which are not mediated through consent and are appropriate for beings that lack the ability to consent.

This shift from focusing on consent to focusing on interests raises numerous questions: how are a being's interests to be determined? Whose interests count when we are dealing with constitutional legitimacy? What does it mean to be governed by a constitution? And which interests count? It is to these questions that we now turn.

B) The All Governed Interests Principle

The aim of this sub-section is to defend the All Governed Interests Principle by answering the questions raised above. According to this principle, all sentient beings who are governed by a constitution should be recognized as having constitutional worth. As explained above,[101] beings with constitutional worth have interests that are of importance to the constitution. This means that constitutional governance of these beings is only justified if their intrinsic value is

[100] Jessica W Berg and others, *Informed Consent: Legal Theory and Clinical Practice* (2nd edn, OUP 2001) 22.

[101] See text above around n 96.

respected and their basic interests protected. If this is not the case, and to the extent that it is not, the constitution lacks legitimacy. Let us consider this principle in more detail by asking how we can determine a being's interests, whose interests count, what it means to be governed by a constitution, and which interests need to be protected.

Starting with the first, and perhaps most fundamental question: How can one determine what is in a being's interests? As we define it, an interest is something that promotes the well-being of its holder when it is satisfied.[102] Well-being should be understood in a broad sense, covering not only the absence of negative interferences with one's bodily integrity and physical health but also the provision of positive opportunities to flourish. This idea is captured by Nussbaum's Capabilities Approach. According to her, all sentient animal species have an interest in flourishing in accordance with what is normal for their species.[103] Despite species-based differences, there are some capabilities that all animals (including humans) have in common, including bodily liberty, play, and others.[104] Animals flourish when these capabilities can be expressed, and they struggle to flourish when the capacities are frustrated.

In the human case, we tend to have a good understanding of what falls within our interests. This is because we have direct access to our personal well-being and because we have an effective way of determining the interests of other people: by asking them. If I ask you what you want and you tell me that you would like to eat something, then it is reasonable to assume that eating is in your interest. If your dentist asks you whether you consent to having your teeth pulled, then your consenting to the procedure tells us that undergoing this procedure is in your interest. Of course, people sometimes want things that are not actually in their interest. Perhaps you want to eat an unhealthy burger full of trans fats. Or perhaps you have been badly informed and do not actually need to have your teeth extracted. To put it in more technical terms, not everything that accords with people's preference interests aligns with their welfare interests.[105] But in many circumstances, verbal communication and consent are good indicators of what is in someone's interests.

Things get more difficult once we turn to non-paradigmatic humans and animals with whom it may be more difficult or impossible to communicate verbally. Verbal communication is an efficient means of conveying information about interests, and it is more difficult for us to identify the interests of

[102] Joseph Raz, *The Morality of Freedom* (OUP 1988) 180.

[103] Nussbaum (n 12) 95–96.

[104] ibid 88–89.

[105] On this distinction, see Tom Regan, *The Case for Animal Rights* (University of California Press 1983) 86–87.

beings who lack that ability. However, that does not mean that it is impossible to determine what their interests are. Quite the contrary, there are many non-verbal ways in which non-paradigmatic humans and animals can and do express their interests.

Starting with non-paradigmatic humans, alternative forms of communication that allow these humans to communicate their interests are available. For example, sign language has long served as a powerful way for humans to communicate non-verbally.[106] Some non-paradigmatic humans find technological aids to be a useful way of expressing their needs.[107] For non-paradigmatic humans who lack the ability to communicate using sign language or technological tools, other forms of expressing their interests, including through changes to their behaviour, are available. For example, a study conducted on children undergoing needle procedures before surgery has shown that children express their fear and pain non-verbally in numerous different ways. Among their nonverbal cues of communication are screaming, shaking, foot kicking, eye-closing, frowning, grimacing, and freezing.[108] Finally, in addition to non-paradigmatic humans telling us what they want through verbal, sign, or other forms of communication, we have another effective way of assessing their interests: physiology. Aside from age or disability-related factors, their physiology is, of course, the same as that of paradigmatic humans. As such, it allows us to draw inferences by comparing how their body responds to stimuli with how paradigmatic humans respond to the same stimuli. For example, the skin of non-paradigmatic humans will respond in the same way to burning as that of paradigmatic humans, and it is therefore reasonable to conclude that they have the same interest in not being burnt.

Similar means are available for determining the interests of non-human animals. Scientific curiosity in the physiology and behaviour of animals goes back centuries, but since the 1960s a distinct scientific discipline called animal welfare studies has emerged which studies animal well-being.[109] Although, like other disciplines, animal welfare studies is ever evolving, it has identified robust methods of determining what interests animals have. For one thing, like humans, animals have an interest in physical health and bodily integrity. The state of their physical well-being can be measured, for example, by considering changes in an animal's physiology as well as behaviour. For example, a starving

[106] H-Dirksen L Bauman, *Open Your Eyes: Deaf Studies Talking* (University of Minnesota Press 2007) 5.

[107] Meryl Alper, *Giving Voice: Mobile Communication, Disability, and Inequality* (MIT Press 2017).

[108] Ida Kleye and others, 'Children's Communication of Emotional Cues and Concerns During a Preoperative Needle Procedure' (2022) 105 Patient Education and Counselling 1518.

[109] Donald M Broom, 'A History of Animal Welfare Science' (2011) 59 Acta biotheoretica 121–25.

animal shows signs of physiological change, such as emaciation. The animal will also change their behaviour, such as by spending more time looking for food or attempting to consume things they do not normally eat.[110]

Like non-paradigmatic humans, however, animals have an interest in not only being physically intact but also flourishing in positive ways. As Mariam Stamp Dawkins puts it, to determine an animal's well-being, we need to consider their physical health *as well as* 'what the animal wants'.[111] Despite the fact that most animals are unable to communicate their interests verbally,[112] animals have other ways of telling us what they want. Many of us will have everyday life experience of communicating non-verbally with animals. For example, a cat sitting in front of his empty food bowl does not need to use words to communicate effectively that they have an interest in eating. The same goes for a dog waiting restlessly in front of a patio door: here, too, no words are needed for us to understand that she wants to go to the garden. Some animals have even learned to communicate with humans using sign language.[113]

There are also scientific methods for determining what animals want, beyond everyday communication. One effective method is so-called choice tests. Choice tests aim to determine the interests of an animal by giving them different options to choose from, such as different types of food. Some choice tests can also involve an operant element, meaning that animals need to work for what they want, which can serve as a measure of how great their interest in something is. Take the operant choice test that was conducted on American mink who are farmed for their fur. The mink were placed in cages that provided access to different compartments containing features such as tunnels, toys, an alternative nesting site, and one containing a small pool for swimming. The doors to enter these compartments were made successively heavier to gauge how badly the mink wanted to access what was in the relevant compartment. The study found that the mink showed strongest interest in the compartment with the small pool, as they were willing to push open a door weighing up to 1.25 kg (about as much as they can push).[114] Of course, just because an animal

[110] Peter Fordyce, 'Suffering in Non-Human Animals: Perspectives from Animal Welfare Science and Animal Welfare Law' (2017) 5 Global Journal of Animal Law 12, 23.

[111] Marian Stamp Dawkins, *Animal Welfare: The Science and Its Words* (OUP 2021) 8.

[112] For an exception, see eg the parrot Alex who mastered complex concepts and used speech to communicate with humans: Irene Maxine Pepperberg, *The Alex Studies: Cognitive and Communicative Abilities of Grey Parrots* (Harvard UP 2000).

[113] For a famous example, see eg the gorilla Koko, who learned to communicate using American Sign Language: John D Bonvillian and Francine GP Patterson, 'Sign Language Acquisition and the Development of Meaning in a Lowland Gorilla' in Charlotte Mandell and Allyssa McCabe (eds), *The Problem of Meaning: Behavioral and Cognitive Perspectives* (Elsevier 1997).

[114] Georgia J Mason, Jonathan Cooper, and Catherine Clarebrough, 'Frustrations of Fur-Farmed Mink' (2001) 410 Nature 35.

chooses a particular option does not mean that that option fully promotes the animal's well-being. It may simply be the least bad option. For example, even with access to a small pool, which most fur farmed mink lack, they are still far from having their basic interests met.[115]

Nonetheless, choice tests are useful because they can improve our understanding of the interests animals have. One particularly valuable feature of choice tests is that animals themselves are given a say. They allow animals to communicate what they want directly to us. As such, if they are well-designed, they can help animals express their own agency, which is inherently valuable.[116] As Richard Healey and Angie Pepper argue, for instance, 'in virtue of possessing . . . agential capacities, animals have a significant interest in determining what happens in a variety of domains, and so in being able to exert control over the contents of their lived experience.'[117] Furthermore, as with asking for paradigmatic humans' consent and listening to non-paradigmatic humans' expressions of agency, giving a say to animals is instrumentally valuable. This is because these mechanisms reduce the risk of paternalistic or anthropomorphic determinations of their interests, which tend to mistake what humans want for what the animals themselves need.

Emphasis on animals' agency has become widespread in recent years, with seminal work by Sue Donaldson and Will Kymlicka urging us to be attentive to the 'various forms of agency'[118] that animals exhibit. As they note, if animals are provided with sufficient opportunities, they can choose where they want to live, which (if any) humans they want to enter social relationships with, and which members of their own species they want to engage with, among many other things.[119] Some believe animals' agency goes as far as giving them the ability to express consent to certain things such as to work,[120] while others argue that animals at least have the ability to express assent and dissent, that is, to affirm or resist a certain state of affairs through their behaviour in domains where the animal has sufficient understanding, such as joining a human companion to go for a run.[121] For our current purposes, it is sufficient to note

[115] For an overview of the welfare problems of fur farmed mink, see European Commission, Scientific Committee on Animal Health and Animal Welfare, 'The Welfare of Animals Kept for Fur Production' (2001) 178–79.

[116] For what this means in the context of democracy, see Chapter 6.

[117] Richard Healey and Angie Pepper, 'Interspecies Justice: Agency, Self-Determination, and Assent' (2021) 178 Philosophical Studies 1223, 1227.

[118] Donaldson and Kymlicka (n 25) 65.

[119] ibid 65–66.

[120] See eg Charlotte E Blattner, 'Animal Labour: Toward a Prohibition of Forced Labour and a Right to Freely Choose One's Work' in Charlotte E Blattner, Kendra Coulter, and Will Kymlicka (eds), *Animal Labour: A New Frontier of Interspecies Justice?* (OUP 2019).

[121] See eg Healey and Pepper (n 117) 1233–36. We return to this debate about animal consent and assent in Chapters 3 and 6.

that listening to animals' own choices is an important way to determine their interests.

Of course, even if we listen carefully, questions may remain about what animals truly want. Not every species has the same interests, and interests can differ on an individual level as well. Moreover, there are some inherent limitations to what a species like ours can know about the interests of another species.[122] But this is not a reason to give up on *trying* to identify animals' interests. Rather, it is a reason to stay updated on evolving scientific insights as well as to learn to cultivate our ability to listen to what animals themselves can tell us. Importantly, even though there are some questions about animals' natures that may remain unanswered for the foreseeable future, and potentially forever, we do have certainty on key facts relating to their basic interests. Today, it cannot be doubted that sentient animals have basic interests in physical and mental integrity as well as in flourishing more positively, such as through social contacts, play, and physical movement.

Let us now turn to the second question: whose interests count when it comes to determining constitutional legitimacy? In our discussion so far, we have intentionally left that question open and simply referred to the interests of beings with intrinsic moral value. To defend the All Governed Interests Principle, we need to consider this issue in more detail. Are the interests of any sentient being relevant for determining the legitimacy of a constitution, or only the interests of some and not others?

On a *sentience*-based account of constitutionalism, it would seem at first glance that *all* sentient beings are potential candidates for constitutional worth. Should we therefore support an All Sentient Interests Principle, as one might call it? The obvious issue with such a principle is that it would place constitutional legitimacy on an unrealistically broad footing. For instance, it would make the Senegalese Constitution's legitimacy dependent on whether or not it respects the interests of Patron, the Jack Russell terrier employed by the Ukraine military to sniff out bombs. Not only does Patron have no obvious connection to the Senegalese Constitution, but, even if he did, it is not clear what the Senegalese constitutional system could do to give due respect to his interests, let alone all other sentient beings around the world.

A narrower principle is available that would seem more suitable because it does not base constitutional legitimacy on all sentient animals' interests.

[122] For a famous exploration of these limitations, see Thomas Nagel, 'What Is It Like to Be a Bat?' (1974) 83 The Philosophical Review 435. For an accessible account, see Frans de Waal, *Are We Smart Enough to Know How Smart Animals Are?* (WW Norton & Company 2016).

According to the All Affected Interests Principle, not just any sentient being but only sentient beings whose interests are *affected* by a constitutional system should be given due concern by it.[123] This principle is popular among political philosophers, some of whom defend versions of it to argue that animals' interests should be considered politically if they are affected by a political decision. After all, as was pointed out above, political decisions often affect animals, generally in ways that are more invasive than for human beings.[124] Should we follow this line of thinking and adopt the view that all *affected* interests matter for determining constitutional legitimacy?

While it is preferable to the All Sentient Interests Principle, there are good reasons that speak against adopting the All Affected Interests Principle for our purposes. As critics have pointed out in the context of political decision-making, one problem with the All Affected Interests Principle is that—depending on how it is interpreted—it, too, is impractically broad.[125] As Ben Saunders puts it, the principle raises the issue that one would 'need to enfranchise almost everyone in almost every decision'.[126] The same problem would arise if we made constitutional legitimacy dependent on any sentient being whose interests are affected by a constitutional system. This is because, in a globalized world, a constitutional system can impact beings not only in neighbouring countries but also around the earth.

Consider the following example. Article 112(2) of Iraq's Constitution provides that the federal government 'shall undertake the management of oil and gas extracted from present fields, provided that it distributes its revenues in a fair manner in proportion to the population distribution in all parts of the country'. Given the negative impact of fossil fuels on the climate, would this mean that the legitimacy of Iraq's Constitution depends on its respect for, say, polar bears in the Arctic whose habitats are deteriorating because of climate change? Should these bears (through their representatives) be enabled to, for example, file a claim for judicial review in the Federal Supreme Court of Iraq to get the government to divest from fossil fuels, or at least give the bears a share of the revenues?

One problem of the All Affected Interests Principle, then, is that it is not practicable enough: it would be very difficult, if not impossible, to identify all

[123] See generally Robert E Goodin, 'Enfranchising All Affected Interests, and Its Alternatives' (2007) 35 Philosophy & Public Affairs 40.

[124] See Robert Garner, 'Animals, Politics and Democracy' in Robert Garner and Siobhan O'Sullivan (eds), *The Political Turn in Animal Ethics* (Rowman & Littlefield 2016) 115.

[125] On other problems, see David Miller, 'Democracy's Domain' (2009) 37 Philosophy & Public Affairs 201, 214–18.

[126] Ben Saunders, 'Defining the Demos' (2011) 11 Politics, Philosophy & Economics 280, 281.

the beings whose interests are affected by a constitutional system, what causal role that system has played in affecting the interests of these beings (compared to other constitutional systems), and how all these beings' interests should be taken into account to determine the legitimacy of the relevant constitution.[127] Hence, while the All Affected Interests Principle may appeal to those with more idealist predilections, our realistic utopian approach counsels against it.

In order to avoid this problem, some democracy theorists have opted for a narrower principle known as the All Subjected Interests Principle. According to this principle, all who are *subjected* in relevant ways to decisions ought to have their interests considered.[128] This approach has also found its way into animal law scholarship. Janneke Vink, for example, proposes on practical grounds that only the interests of sentient beings who reside on the *territory* of a state are subjected in the relevant way and must be protected by that state.[129] Does Vink's residency-based approach fare better? While narrower and therefore more practicable than the untrimmed All Affected Interests Principle, her approach, which we might call the All Resident Interests Principle risks not casting its net wide enough. According to Vink's principle, the legitimacy of a constitution *x* does not depend on the interests of an animal *a* who does not currently reside on *x*'s territory. This is a problem in the event that *x*'s authorities nevertheless manage to exercise control over *a*, or over the territory in which *a* lives. For example, an animal living in an occupied territory does not reside on the formal territory of the occupying state, but it is still subject to the control of that state. Vink's principle would not apply to such an animal, even though the animal is effectively governed by the occupying state.

Robert Garner has proposed another version of the All Subjected Interests Principle which includes all beings *coerced* by a decision. We might call it the All Coerced Interests Principle. As he suggests, 'animals . . . qualify because they are coerced into serving our needs, whether it is for our desire to eat their flesh or our desire to use them as laboratory tools'.[130] Garner draws on the work of David Miller who defines coercion as follows:

> first, . . . to be coerced is to be made to pursue some reasonably specific course of action that follows the intentions of the coercer; second, . . . this is not a

[127] See generally Robert A Dahl, *After the Revolution? Authority in a Good Society* (Yale UP 1970) 66–67.

[128] See eg David Owen, 'Constituting the Polity, Constituting the Demos' (2012) 5 Ethics & Global Politics 129.

[129] Vink (n 66) 47.

[130] Garner (n 124) 115–16.

> course of action that the agent, left to her own devices, would pursue; third, . . . the coercer applies 'sufficient pressure', meaning that it is no longer reasonable for the agent to do other than what the coercer demands.[131]

This is a promising approach that can be applied to the question of who counts as 'governed' for the purposes of constitutionalism's legitimating principle. However, at least in Miller's formulation, the All Coerced Interests Principle seems unduly focused on autonomous and rational agents able to give or refuse consent to be coerced. As Miller emphasizes: '[t]he key point is that coercion undermines autonomy'.[132] As such, it risks excluding sentient animals and marginalizing non-paradigmatic humans.

The All Governed Interests Principle is superior to all the above-discussed alternatives. It is more realistic than the All Sentient Interests Principle and the All Affected Interests Principle. At the same time, it is more attractive than the two alternative variants of the All Subjected Interests Principle: it is superior to the All Resident Interests Principle because it does not ignore the interests of those who are outside the formal territory of a constitution but are still effectively governed by it; and it has the edge over the All Coerced Interests Principle because it does not exclude or marginalize those who are not paradigmatic rational and autonomous agents.

This brings us to the third question: what does it mean to be governed by a constitutional system? Firstly, we need to distinguish being 'governed' from being 'subjected' on the one hand and being 'regulated' on the other. As we define it, being *subjected* to a legal rule means being the normative addressee of the rule who is expected to comply with it and who can be punished for failing to do so. To be subject to a legal rule in this sense requires the capacity to understand and abide by human legal rules. At present, only rational adult humans seem to have that ability. To be sure, there are historical examples of animals being subjected to legal rules and tried for their violation. This happened, for instance, in 1394 in France when a pig was found guilty and hanged for having murdered a child.[133] However, the fact that pigs and other animals are, to the best of our knowledge, unable to understand and abide by human legal rules such as the prohibition of murder shows that it makes little sense to consider them subjected to legal rules. But that does not mean that they are merely *regulated* by legal rules. We understand being 'regulated' as being shaped or

131 Miller (n 125) 219.
132 ibid.
133 EP Evans, *The Criminal Prosecution and Capital Punishment of Animals* (William Heinemann 1906) 141.

otherwise impacted by legal rules while lacking intrinsic value, such as on the grounds of sentience. Things regulated in this sense include all kinds of inanimate objects such as bicycles, windows, or clocks as well as, potentially, non-sentient animals and plants.[134] Sentient animals are neither subjected to legal rules nor simply regulated by them. They are governed by them.

Secondly, as we suggested, governing a sentient being means that those exercising constitutionally sanctioned public powers have effective control over it. The European Court of Human Rights (ECtHR) and the International Court of Justice (ICJ) have developed the concept of 'effective control' in their jurisprudence as a way of determining the extraterritorial applicability of law. Inspired by this concept, we define 'effective control' as the exercise of public power over a certain territory or being.[135] In the case of animals, being governed thus means that the organs of a constitutional system (eg the branches of government, agencies, officials) can effectively control either the habitat or environment of the being, or the being itself. This can occur for example through direct control over an animal's life, body, or health.

This definition covers sentient animals who are most often used for human purposes, such as all animals used in the production of food, for research, or for entertainment. Even when private individuals or companies use animals for these purposes, the organs of the constitutional system generally maintain effective control over the animals. This control manifests itself, for instance, through legislation, licensing, inspections, and the ability of officials to intervene directly and seize or kill animals, such as when they are considered a threat to public health.

Of course, there will be borderline cases in which it is less clear that authorities have effective control over a sentient being. This is especially the case in larger countries like Russia, Australia, or Canada, where some sentient wild animals may not always be under the effective control of the state's organs despite living on the state's territory. Yet, as numerous so-called culling campaigns of non-native species have shown, even wild animals in remote locations can often fall under the effective control of those states (or hunters commissioned by them).[136] In fact, Nussbaum goes as far as to claim that, today, 'there is no

[134] 'Potentially' because this depends on whether one considers non-sentient animals and plants to have intrinsic value: see Section 1.A.

[135] See ICJ, Case Concerning Military and Paramilitary Activities in and Against Nicaragua (Merits), Judgment, 27 June 1986; *Al-Skeini and Others v United Kingdom* App no 55721/07 (ECtHR, 7 July 2011); *Bankovic and Others v Belgium and Others* App no 52207/99 (ECtHR, 12 December 2001).

[136] See eg Livia Albeck-Ripka and Matthew Abbott, 'Majestic Icon or Invasive Pest? A War Over Australia's Wild Horses' *The New York Times* (28 June 2020) <https://www.nytimes.com/2020/06/28/world/australia/brumbies-horses-culling.html> accessed 31 October 2024.

such thing as "the wild", no space, that is, that is not controlled by humans.'[137] One need not go as far as Nussbaum, however, to acknowledge that humans have effective control over at least some wild animals.

Lastly, there is the question of which interests a constitution needs to give due concern to in order to be legitimate. Humans and animals have interests in a plethora of different things. For example, cats tend to enjoy having their belly rubbed. Does that mean that the constitutions that govern cats are only legitimate if they protect their interest in having their bellies rubbed? Or take the example of humans who like a good foot massage. If they cannot get a foot massage, does that mean the legitimacy of constitutions that govern them is diminished?

To answer this question, it is again helpful to draw on Nussbaum's Capabilities Approach. According to Nussbaum, not every capability of a human or animal needs to be protected but only what she identifies as the ten basic capabilities.[138] Moreover, as she argues, society does not have a duty to ensure that every individual can maximally flourish. Rather, it must only ensure that a minimum threshold is guaranteed that allows animals to lead a dignified life, even if they cannot maximize all their capabilities. Building on this approach, we propose to distinguish between a being's basic and non-basic interests. Basic interests are those which a being needs to have fulfilled to have a minimally flourishing life—a life of dignity in Nussbaum's terminology. For most sentient animals, such basic interests include protection of life, bodily and mental integrity, freedom, and social affiliations. By contrast, non-basic interests are the kind of interests that allow an animal to flourish more fully if met but are not necessary for a dignified life. These interests include belly rubs for cats and foot massages for humans. On the All Governed Interests Principle that we propose, constitutional legitimacy only depends on the safeguarding of basic interests, not of non-basic interests.

The focus on basic interests also helps to explain why current constitutional provisions governing animals fall short. As we saw in Chapter 1, some states have put in place constitutional rules relating to animals as part of a trend towards constitutionalization in animal law. However, as we showed, these existing provisions fail to give animals due concern. For while they may promote the interests of animals, they are insufficient to secure their basic interests. As such, they do not recognize animals as having constitutional worth.

[137] Nussbaum (n 12) 224.

[138] These capabilities are: life; bodily health; bodily integrity; senses, imagination, and thought; emotions; practical reason; affiliation; other species; play; and control over one's environment. See ibid 88–89.

To recapitulate, we have argued that to fill the legitimacy gap created by the dominant principle of constitutional legitimacy—the principle of (hypothetical) consent—one should replace it with the All Governed Interests Principle. According to this more capacious principle, a constitution's legitimacy depends on whether it recognizes all sentient animals it governs as having constitutional worth. This recognition is achieved by giving due concern to the basic interests (including the interests to give consent) of all these beings. To the extent that it fails to do so, it lacks legitimacy.

The purpose of this section was to lay the foundation for sentience-based constitutionalism. In Part II of the book, we will explore the implications of this theory in more detail and show how constitutions can become more fully legitimate by including animals in fundamental rights, proportionality, rule of law, and democracy.

3) Two objections to sentience-based constitutionalism

Given the unorthodox nature of our theory of sentience-based constitutionalism, objections could come from a range of directions. In this final section, we discuss two central objections that someone might raise. We begin by considering an objection that could come from anthropocentric constitutionalists. Non-human animals have never made a constitution themselves or cooperated with humans in constitution-making, so why should humans have to accommodate animals' interests? We call this the 'Only Humans Make Constitutions' objection (Subsection A). We then turn to a potential objection coming from the opposite camp—those critical of anthropocentric constitutionalism. By focusing on sentience and embracing liberal notions such as constitutionalism, are we not building our theory on problematic anthropocentric concepts? We call this the 'Your Account is Also Anthropocentric' objection (Subsection B).

A) The 'Only Humans Make Constitutions' Objection

The first objection takes us back to one of the alternative approaches to constitutional legitimacy we have encountered above: that constitutions obtain their legitimacy in virtue of their creation by a constituent power.[139] According to this approach, the sentience-based model of constitutionalism can be criticized

[139] See text above around n 67.

on the grounds that it makes the legitimacy of constitutions dependent on whether they give due concern to the interests of sentient non-humans even though these beings do not form part of the constituent power.[140] The reason for this is that animals are not seen as belonging to 'the people' because they cannot participate in constitution-making assemblies, constitutional referenda, or any other processes that create and shape a constitution. To incorporate their interests into a constitution would therefore contradict the idea that constitutions derive from popular sovereignty, that is, that they are the product of cooperative self-authorship by a people. As Loughlin and Neil Walker argue, the modern idea that constitutions are authored by a people ties constitutions to a 'collective political identity'.[141] Constitutional authority is only possible, on their account, if it reflects that identity. Animals, it could be argued, lack the ability to engage in this process of self-authorship and are therefore not part of the collective political identity. Hence, while the humans who do participate in constitution-making can and perhaps are morally obliged to protect animals, the legitimacy of their constitution does not depend on it.

Although intuitively appealing, there are two reasons why this objection fails. First, it conflates two important questions: 'Who makes constitutions?' and 'Who lives under constitutions?'. Put another way, it does not distinguish between those who govern and those who are governed. While most, if not all, of those who are involved in formulating constitutional rules are also governed by them (to the extent that they are still alive), others are also governed by these rules without having been involved in their making. As Nussbaum has pointed out, this conflation is common in contract-based approaches in political philosophy:

> For a contractarian . . . the question 'Who makes the laws and principles?' is treated as having, necessarily, the same answer as the question 'For whom are the laws and principles made?' That conflation is dictated by the theory's account of the purposes of social cooperation.[142]

We agree with her that these two questions should not be run together.[143] Yet, as we have shown in Chapter 1, anthropocentric constitutional systems

[140] See John Rawls, *Political Liberalism* (Columbia UP 2011) 20.

[141] Martin Loughlin and Neil Walker, 'Introduction' in Martin Loughlin and Neil Walker (eds), *The Paradox of Constitutionalism: Constituent Power and Constitutional Form* (OUP 2008) 2.

[142] Martha C Nussbaum, 'Beyond "Compassion and Humanity": Justice for Nonhuman Animals' in Cass R Sunstein and Martha Nussbaum (eds), *Animal Rights: Current Debates and New Directions* (OUP 2005) 306.

[143] ibid 302.

currently tend to conflate them. As we surmised in Section 1.B above, this conflation may be at the root of why these systems often govern animals in more intrusive ways than paradigmatic humans. By excluding the interests of all animals from constitutionalism's legitimating principle, the objection allows for the governance of sentient beings in ways that violate their interests and that are incompatible with their moral status.

But what if this conflation of governors and governed was justified—whatever the moral consequences—precisely because animals do not share in the constituent power? First and foremost, this retort is problematic because it ignores the many ways in which animals already cooperate in political communities. Donaldson and Kymlicka, for example, employ a wide range of studies and anecdotes to illustrate that dogs are not only able 'to cooperate and please their human companions, but they are also quite prepared to assert their own preferences, and to (re)negotiate terms of cooperation—in short, to exercise a form of citizenship involving both rights and responsibilities'.[144]

However, even if we assume that animals are incapable of cooperating politically, or at least not in ways relevant to the formulation of constitutional rules, the objection would still fail. This is because it leads to a repugnant conclusion: non-human animals are not the only ones who cannot participate in constitution-making; some human beings cannot either. The obvious cases, of course, are non-paradigmatic human beings, some of whom lack the ability to participate in the formulation of constitutions in constitutional assemblies or in votes on constitutional amendments. According to the proponents of the objection, these humans would have to be excluded from the legitimating principle because they cannot take part in constitutional self-authorship and therefore do not form part of the constituent power. What rears its head here is the problem of under-inclusiveness, which we discussed in Section 1.A. By employing a narrow standard—cooperation in constitution-making—the objection can exclude animals from the legitimating principle but only at the cost of excluding some human beings too.

In fact, if one takes actual participation in constitution-making or governance more generally to be the defining criterion, then one also risks excluding many paradigmatic humans. The reason for this is that, at least in constitutional systems without regular referendums or other modes of direct popular participation, many paradigmatic humans do not actively participate in constitution-making and constitution-changing processes. Most paradigmatic humans also

[144] Donaldson and Kymlicka (n 25) 120. See also Chapter 6.

never become an actual part of the government, the legislature, the judiciary, or wield any other public powers. Only a handful of paradigmatic humans co-operate in constitutional self-authorship and are thus governors and governed, rather than just governed. That is why the concept of 'the people' is generally regarded as a fiction.[145] It relies on the assumption that, even though most humans do not participate in the exercise of constituent power, the constitutional system still somehow reflects their collective political identity and serves their interests. The objection has a hard time accommodating this fiction due to its emphasis on self-authorship. However, even if it did not, it is unclear why the fiction could not also be applied to animals. For once self-authorship is understood to include humans who do not and even those who cannot cooperate in constitution-making, then there is no reasonable basis for stopping at the boundary of *Homo sapiens*.

B) The 'Your Account is Also Anthropocentric' Objection

The second potential objection comes not from anthropocentric constitutionalists but their adversaries. The problem with our proposal for sentience-based constitutionalism, the proponents of this objection hold, is not that it tries to include animals. Rather, it is that it does *not do enough to overcome anthropocentrism*. Specifically, the objection could build on critiques prominent in parts of the critical animal studies literature as well as in feminist, critical race, disability, and postcolonial theory. It could do so to challenge, first, our focus on sentience and, second, our attempt to develop a more inclusive account of constitutionalism.[146]

First, regarding *sentience*, it could be objected that this is not as unproblematic and non-anthropocentric a criterion for moral status and constitutional worth as we have suggested. Not only is the line between sentient and non-sentient beings not as clear as it might seem but using sentience does not solve the problem of dominating non-humans. It simply draws the line in a different place: from human/animal to sentient/non-sentient.[147] As such, our theory could be associated with other problematic binaries such as white/black,

[145] See eg Randy E Barnett, 'Constitutional Legitimacy' (2003) 103 Columbia Law Review 111, 115. This is not to say that the idea of constitutional self-authorship by a people plays no role in determining constitutional legitimacy. However, it is not the only legitimating factor. See on this n 68 above.

[146] For an overview, see Deckha (n 20) 10–15.

[147] Taimie L Bryant, 'Animals Unmodified: Defining Animals/Defining Human Obligations to Animals' [2006] The University of Chicago Legal Forum 137, 165.

man/woman, rational/irrational, civilized/primitive, and able/disabled.[148] As Susanne Kappeler has argued, an approach that simply extends consideration 'downward along the ladder'[149] does not challenge the problematic hierarchy that enables oppression in the first place. For this reason, critics like her might view our approach as being 'no different'[150] from the anthropocentric or paradigmatic human-centric approaches we have criticized above.

Furthermore, even if our approach were different, it could be argued that distinguishing between sentient and non-sentient beings misunderstands the relationship of dependency that exists between the two. For example, Taimie Bryant has opposed the sentient/non-sentient dichotomy on the grounds that it 'excludes the means by which animals survive in the world'.[151] In other words, it fails to protect the non-sentient beings, plants, and even bacteria with whom all sentient beings are intertwined.[152] Finally, in addition to maintaining a problematic binary, proponents of the objection might argue that using sentience as a legitimacy criterion is anthropocentric. The reason for this is that, although sentience is presented as an objective, scientific criterion for considering animals' interests, it is in fact a political criterion that is determined by human values.[153] Humans value sentience because they themselves are sentient and decide who qualifies as being sentient. As Kappeler contends, such a 'self-appointed assembly of human experts'[154] will naturally be inclined to test sentience using human-centric yardsticks and therefore likely find sentience in animals that are more like human beings.[155]

This multipronged objection rightly highlights the imperfections inherent in many categorizations that the law employs. Legal categories and concepts, such as personhood, breach of contract, or intent tend to be rough-hewn. They apply easily to a broad range of cases but also have a grey area—or 'penumbra of uncertainty'[156]—where it is less clear if they apply. Perhaps even more problematically, legal categorizations have been created by humans and therefore tend to be informed by the inevitably limited and often biased viewpoints that

[148] See Susanne Kappeler, 'Speciesism, Racism, Nationalism . . . or the Power of Scientific Subjectivity' in Carol J Adams and Josephine Donovan (eds), *Animals and Women: Feminist Theoretical Explorations* (Duke UP 1995) 323–24.

[149] ibid 330–31 (internal quotation marks omitted).

[150] ibid 330.

[151] Bryant (n 147) 166.

[152] Taimie L Bryant, 'Sacrificing the Sacrifice of Animals: Legal Personhood for Animals, the Status of Animals as Property, and the Presumed Primacy of Humans' (2008) 39 Rutgers Law Journal 247, 168.

[153] Kappeler (n 148) 332.

[154] ibid 334.

[155] ibid 332–33.

[156] HLA Hart, *The Concept of Law* (3rd edn, OUP 2012) 12.

humans have. However, these limitations do not necessarily undermine our focus on sentience. This is for four main reasons.

First, as we have pointed out in Section 1.A, our approach only views sentience as a *sufficient* condition for moral status, not a necessary one. Hence, our theory is open to other approaches that try to explain on different grounds, such as relationality or vulnerability, why animals (or other beings) matter morally.[157] It may also be compatible with other alternative models of constitutionalism such as environmental constitutionalism, which aims to protect non-animal organisms and whole ecosystems of which animals form a part. To be sure, that compatibility has its limits. Environmental frameworks that, for example, would sacrifice the fundamental rights of sentient animals to safeguard an ecosystem would fail to pay due regard to the constitutional worth of these animals and would therefore not be compatible with sentience-based constitutionalism.[158] Accordingly, while we are open to frameworks that confer moral status based on criteria other than sentience, such frameworks still ought to take seriously the constitutional worth of individual sentient beings.

Second, it is no secret that the line between sentient and non-sentient beings is not clear cut and is often being redrawn.[159] Most recently, for example, scientists have started discussing whether insects are sentient.[160] That there is fuzziness at the margins of sentience, however, does not change the fact that it is a scientifically and legally widely accepted concept. It is also a considerably more inclusive criterion for moral status than membership in the human species or the possession of certain paradigmatic human features. As Blattner points out, despite its shortcomings, '[s]entience is by far the most objective and reasonable criterion to determine the moral and legal status of someone'.[161]

Third, while it is true that sentience is a concept that humans have developed and test from a human perspective (eg by looking for a central nervous system[162]), this does not mean that sentience as a phenomenon is an arbitrary human construct. As we discussed in Section 1.A, sentience describes the capacity to have valenced experiences, which tracks a number of measurable properties that cannot be found only in humans but also in many animals—a

[157] See text above around n 20.

[158] The possibility of this conflict is well outlined in Mark Sagoff, 'Animal Liberation and Environmental Ethics: Bad Marriage, Quick Divorce' 22 Osgoode Hall Law Journal 304.

[159] Broom (n 9) 5.

[160] See Brian Tomasik, 'The Importance of Insect Suffering' (25 April 2016) <https://reducing-suffering.org/the-importance-of-insect-suffering/> accessed 31 October 2024.

[161] Blattner (n 26) 131.

[162] Lynne U Sneddon, 'Evolution of Nociception in Vertebrates: Comparative Analysis of Lower Vertebrates' (2004) 46 Brain Research Reviews 123.

fact for which there are evolutionary reasons.[163] Humans cannot avoid what Ben Mylius has called 'perceptual anthropocentrism', that is, they necessarily see the world through human eyes, thinking things through a human mind. However, that does not entail that humans can only describe and value what is human.[164] As Blattner rightly argues, the human biases that can plague the use of the concept of sentience should not be taken as a reason to abandon the concept but rather as a ground for trying to prevent these biases as far as possible, and to make sure that the scientific tests for sentience and the laws that are informed by these tests are reviewed regularly.[165] The fact that other human-created categorizations such as race or sex have been used on pseudo-scientific grounds to justify morally objectionable policies does not entail that sentience cannot be an overall useful criterion for protecting the interests of humans and animals.[166] This is supported by the fact that even authors who adopt approaches that are sceptical of binaries arrive at the conclusion that sentience is a useful metric for determining whether a being should be protected.[167] For example, even Deckha who, as we saw in Subsection 1.A, believes that sentience and vulnerability are not conceptually synonymous, argues that privileging sentience makes sense 'in many instances because of sentience's correlation with vulnerability and suffering'.[168]

Finally, even if one granted that it would be better, in an ideal world, to abandon the sentient/non-sentient categorization, it is not clear that this should lead us to abandon that categorization in the here and now. As the previous sections have made clear, our account focuses not only on abstract theories of justice, where there might be more room for radical proposals to rethink relationships between humans and animals. Rather, it focuses on constitutional systems which are inherently bound up with concrete legal and political realities. Our account, as we said, is *realistically* utopian. And, to us, it is far from clear whether it is possible at all for a constitutional system to operate without categories. But assume for a moment that it was. Proposals to do away entirely with such categories or, for that matter, with humans acting as 'self-appointed' judges would still not succeed in practice. As Kappeler herself acknowledges, radical critiques along the lines proposed by the objection 'would

[163] Donald M Broom, 'Evolution of Pain' in Lord Soulsby and David Morton (eds), *Pain: Its Nature and Management in Man and Animals* (Royal Society of Medicine Press 2001).

[164] Ben Mylius, 'Three Types of Anthropocentrism' (2018) 15 Environmental Philosophy 159, 159.

[165] Blattner (n 26) 126–30.

[166] cf Kappeler (n 148) 337.

[167] See eg Joan Dunayer, 'The Rights of Sentient Beings: Moving Beyond Old and New Speciesism' in Raymond Corbey and Lanjouw Annette (eds), *The Politics of Species: Reshaping our Relationships with Other Animals* (CUP 2013) 38.

[168] Deckha (n 20) 152.

imply the necessity of abolishing power and violence' altogether.[169] Some believe that steps in that direction are already available to us.[170] Perhaps this is true. But even if it is, radical change will not happen overnight. If it is possible at all, then until it is those who aspire to such change will require less arbitrary standards that can help protect the interests of those who are currently excluded or marginalized.[171] And sentience—however anthropocentric one might regard it—provides such a less arbitrary standard. As Eva Meijer has put it, '[c]hange will need to start with humans, and existing concepts and institutions can help us to think and act, provided that we also address the anthropocentrism inherent in them'.[172]

This brings us to the objection's second target: *constitutionalism*. Pointing to the history of constitutionalism outlined in Section 1.B, it could be objected that constitutionalism is a deeply liberal concept and, as such, modelled on a particular view of the paradigmatic human being as rational, independent, adult, and so on. Because non-paradigmatic humans and animals do not fit this description, the argument goes, constitutionalism is antithetical to their interests. In fact, as authors such as Anna Grear contend, not only is constitutionalism inherently liberal and anthropocentric but (Western) law as a whole is. 'Central to law's imaginary, in the liberal legal tradition,' she holds, 'is the potent founding fiction of the liberal social contract, folded into the foundations of which is a highly particular subject—a specific kind of person—the "man of property/classic contractor." '[173] In her view, this means that all those who are not paradigmatic constitutional contractors 'do not—and possibly never *can*—figure as fully recognisable instances of the dominant subject ... of law'.[174] Hence, the objection concludes that we should abandon rather than reimagine constitutionalism.

The problem with this objection is that it overlooks both the flexibility of concepts like constitutionalism and the drawbacks of external critiques. First, while it is certainly important not to lose sight of the liberal and rationality-based genealogy of constitutionalism, it is a category error to conclude from this genealogy that constitutionalism is tainted in its essence and incapable of change.[175] That more inclusive accounts of constitutionalism which focus not

[169] Kappeler (n 148) 335–36.

[170] See eg Matthew Calarco, *Beyond the Anthropological Difference* (CUP 2020) 28–31.

[171] See also David Nibert, *Animal Rights/Human Rights: Entanglements of Oppression and Liberation* (Rowman & Littlefield 2002) 184–88.

[172] Eva Meijer, *When Animals Speak: Toward an Interspecies Democracy* (New York UP 2019) 144.

[173] Anna Grear, 'Legal Imaginaries and the Anthropocene: "Of" and "For" ' (2020) 31 Law and Critique 351–52.

[174] Anna Grear, 'Deconstructing Anthropos: A Critical Legal Reflection on "Anthropocentric" Law and Anthropocene "Humanity" ' (2015) 26 Law and Critique 225, 236 (emphasis in original).

[175] See also Vink (n 66) 276.

only on rational men's negative liberties are indeed possible is illustrated by the fact that such accounts have already garnered significant attention. As we saw above and in Chapter 1, positive constitutionalism and dignity-based constitutionalism already go beyond the protection of negative liberties. And environmental constitutionalism goes further still and rethinks constitutionalism in ways that protect the environment for its own sake. In fact, as Grear herself acknowledges, the artificial nature of law confers on its concepts a certain 'innate plasticity'.[176] In other words, legal concepts can extend beyond their narrowly human and rationality-centric origins. This has been the case, for instance, with the concept of legal personhood, which has been extended beyond the human species to entities as diverse as corporations, idols, rivers, and mountains.[177]

Second, external critiques of widely accepted concepts have a notoriously low chance of success. Not only do we not know whether more attractive alternatives to the liberal legal order and constitutionalism are actually available and practicable.[178] But we are also faced with the problem of ongoing (and perhaps even greater) injustice if—as is possible—a complete change of systems fails. Hence, rather than 'trashing'[179] constitutionalism, we believe it is preferable to rethink it *from within*, by drawing on and reimagining its own core principles. Our case for an internal reform of constitutionalism is also in line with positions adopted in parts of critical scholarship. For example, as Kimberlé Williams Crenshaw has argued in the context of racial oppression, 'the challenge ... is to create a counter-hegemony by manoeuvring within and expanding the dominant ideology to embrace the potential for change'.[180] This process is not as radical as that of abandoning constitutionalism, but it is all the more promising for it.

Conclusion

This chapter has explained why the current anthropocentric model of constitutionalism is problematic and why we should instead adopt a sentience-based

[176] Grear (n 175) 246.

[177] See Ngaire Naffine, *Law's Meaning of Life: Philosophy, Religion, Darwin and the Legal Person* (Hart Publishing 2009) 166.

[178] Kimberlé Williams Crenshaw, 'Race, Reform, and Retrenchment: Transformation and Legitimation in Antidiscrimination Law' (1988) 101 Harvard Law Review 1331, 1366.

[179] Mark G Kelman, 'Trashing' (1984) 36 Stanford Law Review 293.

[180] Williams Crenshaw (n 178) 1386.

model. Anthropocentric accounts of constitutionalism justify constitutions based on (hypothetical) consent of paradigmatic human beings. This excludes all sentient animals and marginalizes non-paradigmatic human beings, which creates a legitimacy gap. This is because these beings have intrinsic value and interests that can be and often are directly affected by the constitution governing them yet have no impact on the legitimacy of that constitution. As Eisen has put this point,

> prevailing accounts of constitutionalism . . . marginalise or erase the experiences of constitutional subjects who are not able to engage with law in its own languages. The case of animals might be cast as an extreme example of a broader category of constitutional subjects who have been excluded from law's meanings.[181]

To remedy this situation, we made the case for sentience-based constitutionalism. We have argued that constitutions should not be understood as being justified by their respect for hypothetical consent of paradigmatic humans. Rather, they are justified by their giving of due concern to the interests of all the sentient beings whom they govern (the All Governed Interests Principle). *Governing* a sentient being means having effective control over their life, body, or habitat. And giving *due concern* to a sentient being means respecting their intrinsic value by protecting their basic interests. This is necessary to recognize the constitutional worth of sentient beings, that is, as beings who have basic interests that the constitution should safeguard. As we have argued, basing constitutionalism on the All Governed Interests Principle is preferable to alternative principles because it makes constitutional legitimacy dependent on not only the interests of paradigmatic humans but also those of governed animals and non-paradigmatic humans. Moreover, it manages to do so without jettisoning the importance of consent, which, as we argued, is ultimately grounded in paradigmatic humans' interests. For these reasons, the All Governed Interests Principle can fill the legitimacy gap created by anthropocentric constitutionalism.

This chapter has thus explained why we ought to abandon our anthropocentric understanding and practices of constitutionalism and instead adopt sentience-based constitutionalism. However, we have not yet shown how exactly sentience-based constitutionalism would reform existing constitutional principles and practices and how, concretely, giving due concern to all

[181] Eisen (n 66) 925.

governed beings remedies existing legitimacy deficits. In other words, we have only made the case for the 'why' but not the 'how'. The aim of Part II of this book is to build on the foundations laid in Part I and to spell out how we should rethink constitutionalism to respect the constitutional worth of all governed sentient beings.

PART II

3
Fundamental rights

Introduction

In 2022, the Constitutional Court of Ecuador recognized that wild animals enjoy an extensive list of constitutional rights, including the right to life and the right to freedom, derived from the constitutional recognition of the rights of nature.[1] Two years earlier, in 2020, the Swiss Federal Supreme Court gave the green light for a citizens' initiative from the Canton of Basel-Stadt which aimed to include the right to life and the right to physical and mental integrity for non-human primates in the Constitution of the Canton.[2] Similarly, the Finnish Animal Rights Lawyers Society has put forward a proposal on fundamental rights for animals to be included in the Constitution of Finland.[3] In contrast to the Ecuadorian Constitutional Court's decision, the Finnish proposal would grant more rights to domesticated animals than to wild animals. These developments follow from earlier failed attempts to provide constitutional rights for animals. For example, in 2014, the Giordano Bruno Foundation petitioned an amendment to Article 20a of the German Basic Law to include the following: 'The right of great apes to personal freedom, to life and bodily integrity is protected'. The petition was rejected by the Petitions Committee of the German Bundestag on the basis that it did not have any chance of success.[4]

Inspired by these efforts, we answer the following question in this chapter: What fundamental rights would be recognized for animals in a sentience-based constitution? We argue that already existing constitutional

[1] The writing of this chapter was substantively concluded prior to the publication of Saskia Stucki, *One Rights: Human and Animal Rights in the Anthropocene* (Springer 2023). While Stucki's book does not focus on constitutionalism, her arguments are complementary to those that we offer in this chapter. Therefore, we invite readers who are interested in the themes of this chapter to consult her excellent book; Constitutional Court, no 253-20-JH/22 (*Estrellita*) (Ecuador Constitutional Court).

[2] Constitutional Court of the Canton Basel-Stadt, no VG.2018.1 (primate rights initiative), 15 January 2019. The citizens' initiative was unsuccessful as only around 25 per cent of the electorate voted in its favour on 13 February 2022.

[3] Finnish Animal Rights Lawyers Society, 'Proposal: The Constitution of Finland' (2020) <https://www.elaintenvuoro.fi/english/> accessed 26 October 2024.

[4] This attempt is chronicled by the Foundation at <https://www.giordano-bruno-stiftung.de/meldung/verfassungsziel-speziesismus> accessed 1 November 2024.

Animals and the Constitution. John Olusegun Adenitire and Raffael Fasel, Oxford University Press.
 DOI: 10.1093/9780198910534.003.0003

catalogues of rights[5] should be extended to include animals. This is because familiar catalogues of constitutional rights can protect the overlapping interests and capacities that humans and animals share.[6]

There are several corollaries to the view we defend in this chapter. First, in contrast to the Ecuadorian, Swiss, or German efforts referenced above, constitutional rights should be available for all sentient animals rather than only for specific animal species, for example, great apes or wild animals. Second, in contrast to the Ecuadorian and Finnish approaches, no general differentiation should be made based on whether animals are domesticated or living in the wild. However, as we discuss in Section 2.B, we grant that certain rights will be applicable to certain species and not others. Finally, unlike the proposals discussed above, protecting the fundamental rights of animals does not, in principle, require new constitutional clauses or chapters, including rights of nature. As we explain in Section 2.A, existing constitutional catalogues of rights can be interpreted to cover animals, just as they are currently interpreted to cover certain non-humans such as corporations and churches. However, we are aware that some fundamental rights catalogues are very short (eg the Constitution of the Australian Commonwealth contains only a handful of express rights). Consequently, we acknowledge that there may be instances in which new rights would need to be added to the text of a catalogue of rights to adequately protect both human and non-human animals.

This chapter is structured as follows. Section 1 explains our methodology. Section 2 then argues that features of existing fundamental rights practice—particularly rights' progressive expansionism, the absence of a hierarchy between rights-holders, as well as the overinclusiveness and generality of fundamental rights catalogues—entail that these catalogues can be extended to cover animals. These features of fundamental rights practice are analysed in turn. Section 3 then exemplifies how the extension of constitutional rights would work. We consider a small but illustrative selection of fundamental

[5] This chapter uses the term 'fundamental rights' for basic rights protected at either constitutional or international level. It uses 'constitutional rights' for those basic rights protected at constitutional level. It uses 'human rights' for those basic rights of humans protected at either constitutional or international level. In some instances, such as in the United Kingdom, the protection of human rights is undertaken through statutory means having constitutional significance (eg the UK Human Rights Act 1998). For the purposes of this chapter, these constitutional statutes count as catalogues of constitutional rights. The concept to be captured by the term 'basic rights' is the familiar institution of catalogues of rights protected in constitutional documents, some of which are modelled after international documents such as the International Covenant on Civil and Political Rights or the International Covenant on Economic, Social and Cultural Rights.

[6] This view has been endorsed by Paola Cavalieri, *The Animal Question: Why Non-Human Animals Deserve Human Rights* (Catherine Woollard tr, OUP 2001); Alasdair Cochrane, 'From Human Rights to Sentient Rights' (2013) 16 Critical Review of International Social and Political Philosophy 655.

rights familiar from constitutional rights practice—the right to life, the prohibition of torture and inhuman or cruel treatment, and the right to freedom—and show how and to what extent they can be extended to animals.

1) Methodology: Between theory and practice

In terms of the method employed in this chapter, we will be drawing on the current practice of fundamental rights both in constitutional doctrine and in international human rights law, especially the practice of the European Convention on Human rights (ECHR). Despite the fact that the main focus of this book is on constitutional law and theory, we think it is useful to occasionally refer to international human rights practice for several reasons.

First, both international human rights law and constitutional rights law are concerned with the same subject matter (ie guaranteeing fundamental rights). To be sure, the institutional contexts and doctrinal tools employed by each are remarkably different.[7] However, given that this chapter's main focus is on the content of the catalogues of rights rather than the impact of contexts and doctrinal tools, we feel confident that drawing from each is appropriate.

Second, the ECHR, which is the main international human rights law instrument we draw on, has a significant impact on the constitutional rights doctrine of Member States of the Council of Europe and beyond. In Member States, ECHR rights form part of domestic law (be it directly, as in so-called monist countries, or indirectly such as in the countries like the United Kingdom that have incorporated them through legislation).[8] Either way, international instruments like the ECHR have considerable influence on the way domestic constitutional rights are interpreted in Europe.[9]

Most importantly, reliance on the *practice* of fundamental rights also helps us avoid some of the stalemates of academic discourse on animal rights, which is predominantly focused on the moral rights of animals. This focus on moral as opposed to legal rights sometimes entails that certain conditions to enjoying

[7] eg in the ECHR context, the doctrine of the margin of appreciation, which requires deference to differing member states' approaches to a Convention right, often determines whether a particular right has been infringed. The level of deference narrows if there is enough consensus within member states' practice regarding a particular right. There is no immediate counterpart to this doctrine in constitutional law. The concept of the margin of appreciation started developing in *Handyside v United Kingdom* [1976] ECHR 5.

[8] The United Kingdom incorporated many of the ECHR rights into domestic law via the Human Rights Act 1998.

[9] For applications in a UK, German, and French context, see Eirik Bjorge, *Domestic Application of the ECHR: Courts as Faithful Trustees* (OUP 2015).

rights are set in moral theory which legal practice would not even entertain. This is the case, for example, with respect to the question of whether animals have a right to life. Some have argued that animals do not have the capacity to conceptualize themselves across time (ie that they live only in the present moment) or that they do not have the capacity to conceptualize the wish not to die. This is said to entail that animals cannot be harmed at all by a painless death or, all things being equal, that a painless death harms them much less than it does beings, such as able-bodied human adults, who have such mental capacities.[10] Reference to fundamental rights practice helps us bypass these disputes. No specific conceptual capacities are required for a human to benefit from the right to life under Article 2 ECHR. Indeed, Article 2 provides a measure of protection even in extreme cases, such as that of Charlie Gard, where a human infant had no signs of normal brain activities such as responsiveness, interaction, or crying. In that case, the United Kingdom was required to satisfy several procedural guarantees before life-sustaining treatment could be withdrawn in a manner compatible with Article 2.[11] This entails that the right to life applies even in cases, such as in Gard's, where the rights-holder had no ability to conceptualize anything given that he did not have a functioning brain. One could therefore argue that it applies even more fully to animals that have normal brain activities, even on the assumption that they do not have the capacity to conceptualize themselves across time or to conceptualize the wish not to die.

However, this reliance on the practice of fundamental rights also has some disadvantages. For example, it might be criticized for being unduly anthropocentric. It is of course true that seeking to understand what fundamental rights other animals have from our current understanding of the practice of fundamental rights for humans is anthropocentric. However, this criticism ignores that humans are animals too, meaning that humans may share several interests and capacities with related animals. By exploring the practice of fundamental rights, we seek to identify these shared interests and capacities and to extend the protection of constitutional rights to animals where appropriate. Still, we must attend to the differences between different species, in a similar way to how we should attend to the differences between different human groups (who differ by gender, sexual orientation, beliefs, etc) and make provisions for their

[10] Belshaw advances a version of the argument that it is not wrong to painlessly kill an animal in Christopher Belshaw, 'Death, Pain, and Animal Life', *The Ethics of Killing Animals* (OUP 2015). Singer advances a version of the argument that it is less wrong to painlessly kill certain animals than a normal human adult under certain circumstances in Peter Singer, *Practical Ethics* (2nd edn, CUP 1993) 132–34. Singer's view has been criticized in Gary Francione, *Introduction to Animal Rights: Your Child or the Dog?* (Temple UP 2000) 135–48.

[11] *Gard v United Kingdom (admissibility decision)* (2017) 65 EHRR 81 (ECtHR) [80].

protection (eg through measures prohibiting discrimination based on these protected traits). In Section 3, when attempting to show how certain constitutional rights can be extended to animals, we therefore pay attention to the differences in capacities and interests that may justify withholding or extending certain rights for animals. It is this combination of appreciating interspecies commonality, while factoring in inter and intra-species differences, which rescues our project from undue anthropocentrism.[12]

Another objection could be that our reliance on the practice of fundamental rights places excessive emphasis on legal practice at the expense of moral theory. The charge correctly highlights that a prioritization of practice over theory, while helpful in certain domains, does not guarantee correctness. Indeed, if taken to the extreme, disproportionate reliance on practice would doom the whole project of constitutional rights for animals. This is because, with the limited exceptions mentioned in Chapter 1,[13] the vast majority of constitutional practice does not recognize that animals have fundamental rights. Hence, if legal practice was unduly prioritized, it would follow that animals should not be recognized as possessing fundamental rights.

The correct approach is therefore not to uncritically prioritize legal practice of fundamental rights over its theory. Rather, we should take the practice as a starting point, investigate its underlying values and commitments, and question to what extent they are compelling. This is clearly a theoretical endeavour. Once, and to the extent, we find the values and commitments of fundamental rights practice to be morally compelling, we should then seek to extend those values and commitments to animals where they are applicable. This process is therefore not based on an undue prioritization of the practice of fundamental rights. Rather, it is one which excavates the underlying moral values of human rights practice and seeks to put those values to good use in the service of interspecies justice.[14]

2) Principles for extending constitutional rights to animals

The idea that constitutional catalogues of rights can be extended to protect animals might sound implausible to some. This may be because most people

[12] Our project may therefore be said to follow the approach suggested by Matthew Calarco, *Thinking Through Animals: Identity, Difference, Indistinction* (Stanford UP 2015).

[13] See Chapter 1 Section 2.A.

[14] This method is similar to Dworkin's interpretative methodology. See Ronald M Dworkin, *Law's Empire* (Hart Publishing 1998).

understand fundamental rights to be exclusively reserved for humans—a belief that could be rooted in the view that fundamental rights are based, at least in part, on *human* dignity. The South African Constitution, for example, explicitly bases its Bill of Rights on this value. Article 7.1 states that '[t]his Bill of Rights . . . enshrines the rights of all people in our country and affirms the democratic values of human dignity, equality and freedom'. In a similar vein, pursuant to Article 40 of the Constitution of Cuba '[h]uman dignity is the supreme value that underpins the recognition and exercise of the rights and duties enshrined in the Constitution, treaties, and laws'. If, encouraged by such clauses, we equate fundamental rights with human rights, the idea of animals possessing fundamental rights would indeed appear to be a conceptual impossibility. In fact, how could *non-humans* possess *human* rights? Similarly, the notion that a catalogue of fundamental rights founded on human dignity could be used, as we argue in this chapter, to protect non-human animals would seem implausible.

However, this claim sounds implausible only if it is considered in the abstract. Once we pay attention to the way fundamental rights work in practice, in both a constitutional and international law context, some of this implausibility quickly dissipates. As we show in this section, fundamental rights practice has long recognized that fundamental rights are not limited to human beings. Rather, it has embraced the view that certain non-human entities, such as churches and corporations, possess fundamental rights. For this reason, it is not implausible to think that other non-humans, such as animals, can possess fundamental rights (Subsection A). Once the implausibility of extending fundamental rights to animals has been eased, we can see that other features of fundamental rights practice—their non-hierarchical nature (Subsection B), their over-inclusiveness (Subsection C), and their generality (Subsection D)—strengthen the case for the view that constitutional catalogues of rights can be further extended to protect animals. We discuss these principles in turn.

A) Progressive expansionism

Why are existing catalogues of constitutional rights sufficient to protect humans and animals? One compelling reason is that fundamental rights have long been characterized by the phenomenon of what we call 'progressive expansionism' of rights-holders. Take, for example, the right to vote in the United States. First, it was a privilege of property-owning white males. It then became available to white males who did not own property, before being extended, at least formally, to non-white males through the Fifteenth Amendment. Male Native Americans were then allowed to vote before the right was also extended to women by the Nineteenth Amendment. The Constitution eventually lowered the voting eligibility age from twenty-one to eighteen.[15] Some states, though not all, have lifted restrictions on voting eligibility of certain criminals.[16] This progressive expansionism of the right to vote exemplifies the general expansionism of fundamental rights-holders.

However, it need not, and it has not, stopped at humans. Indeed, fundamental rights have also been recognized for non-humans in legal practice. For example, the US Supreme Court has recognized that corporations enjoy several constitutional rights, including due process,[17] free speech,[18] and equal protection.[19] Also, the South African Bill of Rights, despite its explicit grounding in the value of human dignity, declares in Article 8.4 that 'A juristic person is entitled to the rights in the Bill of Rights to the extent required by the nature of the rights and the nature of that juristic person'. To be sure, non-human entities are not recognized everywhere as fundamental rights-holders. For example, Article 1.2 of the American Convention on Human Rights (American Convention) explicitly restricts the benefits of its rights to human beings.[20] However, while the practice may not be uniform, it does show that fundamental rights are not necessarily a prerogative of human beings. It is therefore not conceptually implausible to argue that animals can also enjoy fundamental rights.

[15] These histories are recounted in 'Voting in the U.S.A.' (*Encyclopedia Britannica*) <https://www.britannica.com/story/voting-in-the-usa> accessed 1 November 2024; 'Voting Rights Act: Major Dates in History' (*American Civil Liberties Union*) <https://www.aclu.org/voting-rights-act-major-dates-history> accessed 1 November 2024.

[16] A useful guide is provided in National Conference of State Legislatures, 'Felon Voting Rights' <https://www.ncsl.org/research/elections-and-campaigns/felon-voting-rights.aspx> accessed 1 November 2024.

[17] *Minneapolis & St Louis Ry Co v Beckwith* 129 US 26, 28 (1889) (recognizing that corporations receive protection for the 'enjoyment of property', including the ability to challenge 'legislation injuriously affecting it').

[18] Citizens United, 130 S Ct at 903 ('[T]he First Amendment does not allow political speech restrictions based on a speaker's corporate identity').

[19] See *Metro Life Ins Co v Ward* 470 US 869, 881 n 9 (1985) (citing the 'well established' Equal Protection Clause rights of corporations); *Santa Clara Cnty v S Pac RR Co* 118 US 394, 396 (1886) (noting that the Court agreed that the Equal Protection Clause applied to corporations at issue).

[20] Organization of American States, American Convention on Human Rights 'Pact of San Jose, Costa Rica', <https://www.oas.org/dil/treaties_b-32_american_convention_on_human_rights.pdf> accessed 30 October 2024. This provision says that '[f]or the purposes of this Convention, "person" means every human being'. This provision, however, has not stopped corporations from indirectly benefiting from the Convention's protection through claims made by corporations' shareholders (who must be humans). However, the rights of the shareholder must be clearly distinguished from the rights of the corporation. See *Case of Perozo et al (Venezuela)* (2009), Preliminary Objections, Merits, Reparations and Costs, Inter-Am Ct HR (Ser C) No 195, at [399]–[403].

It could be objected that fundamental rights practice sometimes holds that non-humans such as churches or corporations have rights because to suggest otherwise would frustrate the fundamental rights of certain humans. In *Ashers Baking*, for example, the UK Supreme Court held that imposing liability on a baking company when it refused to provide a cake with the logo 'support gay marriage' would frustrate the right to freedom of religion of its two human owners. Lady Hale, writing for the Court, explicitly said '[i]n holding that the company is not liable, this court is not holding that the company has rights under article 9 [freedom of religion]; rather, it is upholding the rights of the McArthurs under that article'.[21] Similarly, in *Kustannus*, the European Commission on Human Rights (a body that preceded the European Court of Human Rights (ECtHR)), stated the following:

> The Commission has repeatedly held that a church body or an association with religious and philosophical objects is capable of possessing and exercising the right to freedom of religion, since an application by such a body is in reality lodged on behalf of its members.[22]

We acknowledge that the rationale adopted by legal practice for recognizing fundamental rights of corporations and churches is sometimes human-centric. However, this does not negate the fact that the relevant rights-holder is not always the human behind the institution (as happened in the UK *Ashers* case). Rather, as happens in South Africa, in the United States, and under the ECHR, the relevant rights-holder is the church body or association itself rather than its constituent or controlling human members.

But is the fact that corporations and churches only enjoy fundamental rights in order to protect the interests and capacities of humans not an obstacle to extending fundamental rights to animals? After all, it is not clear that extending fundamental rights to animals is also needed to further protect humans.[23] We propose that the need to extend fundamental rights to animals need not accrue any further protections to humans. This is for two reasons. First, as argued in Chapter 2, sentient humans and animals should be understood as possessing

[21] *Lee (Respondent) v Ashers Baking Company Ltd and others (Appellants) (Northern Ireland)* [2018] UKSC 49 [57].

[22] *Kustannus OY Vapaa Ajattelija AB v Finland* (1996) 22 EHRR CD 69.

[23] Note, however, that the Islamabad High Court has explicitly employed this rationale. It argued that non-human animals should enjoy certain rights because this would safeguard certain rights of humans, especially humans' right to life. See *Islamabad Wildlife Management Board v Metropolitan Corporation Islamabad* [2020] WP No 11552019 F.

constitutional worth so that constitutionalism is devoted to safeguarding their own well-being.

Second, the need to extend fundamental rights to animals follows if we accept that fundamental rights protect certain interests and capacities, such as life, physical and mental integrity, and community, which are shared by humans and animals. We elaborate on this in Section 3 below when we look at the specific interests and capacities protected by several fundamental rights and demonstrate how they can be understood to be shared by humans and animals. We can grant that fundamental rights for corporations and churches are merely a means to protecting the interests and capacities of humans. However, we can show that animals have interests and capacities that are worth protecting for their own sake. Indeed, if the law can grant fundamental rights to corporations and churches to protect the interests and capacities of humans, it can, *a fortiori*, also grant fundamental rights to animals given that their interests and capacities are worth protecting for their own sake.

B) Non-hierarchical nature

Another compelling reason why existing catalogues of fundamental rights can be extended to animals is that the practice of fundamental rights aims to be non-hierarchical. Leave aside corporations and churches which hold fundamental rights and focus on humans as rights-holders.[24] According to fundamental rights practice, no one group of human rights-holders is considered to be more important or more deserving of fundamental rights. While humans are often divided into different groups (eg on the basis of their gender, beliefs, etc), constitutional rights catalogues are meant to apply universally across human groups, irrespective of any such categorizations. Of course, there may be specific constitutional provisions that create special protections for certain categories of humans. The Indian Constitution, for example, establishes in Article 51A(e) that every Indian citizen has a duty 'to renounce practices derogatory to the dignity of women'. Furthermore, in Article 243D, it establishes quotas for women at certain levels of government. However, despite having

[24] We set them aside merely for analytical simplicity. The non-hierarchical notion applies also to them. This is the case because the fundamental rights of corporations and churches can and do prevail over conflicting rights of humans. See eg *Ashers Baking* (n 21). So even though corporations and churches may be said to enjoy rights only in order to benefit their human members, it is still the case that their rights are not inferior to those of humans.

such gender-specific provisions, the Indian Constitution clearly does not have separate catalogues of rights divided by gender or any other status.

A constitution devoted to the well-being of humans and animals should embrace the non-hierarchical nature of fundamental rights practice along two dimensions. The first is that constitutional rights should be applicable to both humans and animals. This follows from the fact that, as argued in Chapter 2, not only humans but also sentient animals should be recognized as having constitutional worth with equally valid claims to rights based on their sentience. So, just like several constitutions recognize that everyone should enjoy constitutional rights without distinction based on colour, gender, and so on,[25] a catalogue of constitutional rights in a sentience-based constitution should also ensure that everyone enjoys its guaranteed rights without distinction based on species-membership. If this view is accepted, then there is no reason why, as the Finnish Animal Rights Lawyers Society's proposal suggests, a separate catalogue of constitutional rights for animals would be required. As we discuss further in Subsections C and D and exemplify in Section 3, existing catalogues of fundamental rights can protect, when properly interpreted, the fundamental interests and capacities of animals.

This is not to deny that, just like the Indian constitutional provisions for women, there may sometimes be appropriate circumstances for species-specific provisions (eg to remedy historic disadvantages that animals have endured or to protect exceptional species-specific rights). However, a single catalogue of constitutional rights for non-human and human animals would reinforce the view that the constitution is equally devoted to the well-being of all sentient animals, whether they are human or not.

The second way in which the non-hierarchical nature of fundamental rights practice should be developed is that, as much as possible, there should be no unnecessary divisions between different groups of humans and non-humans in a sentience-based catalogue of rights. Approaches that advocate for constitutional rights only for certain species of animals potentially fall short of this. Consider, for example, the work of the Nonhuman Rights Project (NhRP). The NhRP has petitioned for *habeas corpus* protection for certain species of animals, such as primates and elephants, who have cognitive abilities that are more developed than other animals'. While the NhRP would likely suggest that

[25] See eg Article 28(2) of the Constitution of Cyprus which provides that 'Every person shall enjoy all the rights and liberties provided for in this Constitution without any direct or indirect discrimination against any person on the ground of his community, race, religion, language, sex, political or other convictions, national or social descent, birth, colour, wealth, social class, or on any ground whatsoever, unless there is express provision to the contrary in this Constitution'.

this is a necessary division, some commentators have argued that this strategy might make it more difficult for other animals to receive fundamental rights as they do not have the same cognitive capacities.[26]

The second dimension of the non-hierarchical nature of fundamental rights takes sentience as a sufficient condition and assigns fundamental rights to all those who possess this capacity. This is in line with the practice of fundamental rights which, as the *Charlie Gard* case shows, does not assign rights based on the complexity of cognitive capacities of putative rights-holders. However, we recognize that current and ongoing attempts to provide for constitutional rights for only certain animals (such as primates or elephants) are chiefly motivated by pragmatic considerations of progressive expansion of constitutional legitimacy for all animals. In a sentience-based constitution, this progressive expansion of rights-holders would have already been achieved, and the second dimension of the non-hierarchical desideratum would be fully realized.

C) Over-inclusiveness

Animals within and across species differ in many of their interests and capacities. Dolphins have a vital interest in a habitable ocean, while eagles have an interest in non-toxic air. These differences may appear to threaten the project of extending fundamental rights to animals. If fundamental rights protect specific interests and capacities, and if these differ within and across species, how can we hope to use the same catalogue of rights for all sentient animals, human and non-human alike?

Although these differences make the present task more demanding, the overinclusiveness of existing catalogues of fundamental rights goes a long way to mitigating concerns in this regard. By overinclusiveness, we mean that not all rights contained in a catalogue of rights are applicable to all right-bearers. To illustrate this, consider the case of human babies and human adults with severe cognitive disabilities who do not have the capacity to vote and will not be able to exercise the right to vote as long as they lack that capacity. Similarly, men will not be able to enjoy the special constitutional protection of women's rights (eg those in the Indian Constitution). The fact that catalogues of rights

[26] Martha C Nussbaum, *Justice for Animals: Our Collective Responsibility* (Simon & Schuster 2023) 37–56; Joan Dunayer, 'The Rights of Sentient Beings' in Raymond Corbey and Annette Lanjouw (eds), *The Politics of Species: Reshaping Our Relationships with Other Animals* (CUP 2013). While Dunayer does not specifically address the NhRP, she addresses similar arguments, such as that offered by Peter Singer, that would confer the right to life only to species with higher cognitive abilities.

are overinclusive in this sense allows them to be employed for rights-holders with differing interests and capacities. This approach to overinclusiveness should also be adopted in a catalogue of rights in a sentience-based constitution. On this approach, what matters is that animals' fundamental interests and capacities are adequately protected, and not if all animals have the relevant interests or capacities to benefit from the protection of all the rights in the catalogue of rights.

However, even if we embrace overinclusiveness in the project of extending fundamental rights to animals, a challenge remains. Humans often do not know enough about other animals to be sure that a catalogue of rights for all sentient animals will sufficiently protect the fundamental interests and capacities of all these animals. For one, new animal species are discovered every year. Some of them might have previously unknown interests or capacities. Also, even in relation to the species that are known to humans, history reveals that we do not always have a perfect understanding of the capacities of other animal species. Indeed, a few remain sceptical to this day about whether animals can feel pain,[27] or have a form of language.[28] However, these are increasingly fringe positions. Today, it is uncontroversial from a natural science perspective that many animals can feel pain.[29] It is also clear that animals communicate with each other and that certain primates can learn and use sign language.[30] Still, in formulating a catalogue of rights, it is important to remain open to the view that such a catalogue is revisable. The more humans learn about the interests and capacities of other animals, the more complex and, hopefully, complete the catalogue of rights will become.

D) Generality

The worry highlighted at the end of the last section can to some extent be mitigated by another feature of the practice of fundamental rights. Our incomplete knowledge of other animals is mitigated by the fact that fundamental rights are

[27] A sceptical account of pain in non-human animals is provided by Peter Harrison, 'Do Animals Feel Pain?' (1991) 66 Philosophy 25.

[28] George Kateb, *Human Dignity* (Harvard UP 2010).

[29] We put aside exceptional genetic conditions, such as congenital analgesia, which prevent some individual sentient animals, including humans, from feeling pain. For an account and possible treatment of this condition in humans and mice, see Michael S Minett and others, 'Endogenous Opioids Contribute to Insensitivity to Pain in Humans and Mice Lacking Sodium Channel Nav1.7' (2015) 6 Nature Communications 8967.

[30] R Allen Gardner, Beatrix T Gardner, and Thomas E Van Cantfort (ed), *Teaching Sign Language to Chimpanzees* (SUNY Press 1989).

often formulated in very general terms which can be interpreted to give rise to a variety of specific rights, depending on the time and context. What does this mean and how will this help to address the worry about our knowledge gaps when it comes to other animals?

In relation to the first query, we aim to determine whether existing catalogues of fundamental rights in their most abstract and general formulations can be said to apply to all sentient animals. So, we will be asking whether animals can benefit from generally formulated rights such as the right to life or the right to health rather than more specific and derived entitlements that these general rights provide.

Take, for example, the right to access a clean toilet which the Bombay High Court declared in 2015 to be a constitutional right, especially for women. The Court derived this particular right from a conjoined reading of the constitutional right to life under Article 21 and from the right to health under Article 47 of the Indian Constitution. The High Court found that the inability of women to access free, safe, and clean toilets undermined their constitutional right to health and to live with dignity. It then gave specific instructions to local authorities to build and maintain free and safe toilets every few kilometres.[31] The right to access a free, safe, and clean toilet may seem too specific a right to be protected by constitutional law. But the existence of such a right need not surprise us if we see constitutional rights as general principles which need to be applied to specific circumstances. This process of concretization certainly requires an interpretative body, such as a constitutional court, to tailor the more general constitutional principle (eg right to health and dignity) to the occasion (eg access to toilets).

How does the generality of fundamental rights help minimize the worry about our sometimes insufficient understanding of other animals' capacities and interests? While we can be generally confident that animals, by virtue of their sentience, have an interest in maintaining their lives and their health, we cannot be sure that none of them have the relevant privacy interests which warrant the right to sanitation. To be clear, the doubt is not whether, similarly to humans, other animals have the need to dispose of their bodily waste. Rather, when we think of sanitation in the human context, we do not usually just consider it as a place where human bodily waste can be disposed of. Rather, we think of it as a place where this can be done in relative privacy.[32]

[31] *Milun Saryajani v Pune Municipal Corporation and Others* [2015] AIR 361 (Bombay High Court).

[32] Inga Winkler, 'The Human Right to Sanitation' (2016) 37 University of Pennsylvania Journal of International Law 76. The first line of the abstract of this paper accordingly says: 'Sanitation is a very personal and private matter, inextricably linked to human dignity'.

While we know that some domesticated animals, such as cows, do not usually have such privacy interests, we cannot generalize with confidence across the entire animal kingdom. And, in fact, we do not need to, given that the right to sanitation is a right derived from more abstract rights which we can be more confident in applying to animals. Because the chief question is whether animals can benefit from those more abstract rights, we need not worry too much if we do not know whether all other animals are like humans or are closer to cows in their privacy interests. And if there comes a time when we know that some animals are closer to humans than to cows in this regard, then we can be confident that the more abstract right to health can be suitably interpreted to afford a more specific right to sanitation for those animals that have the relevant interest.

3) Fundamental rights for animals

Having addressed the general features of the practice of fundamental rights that permit their extension to animals, we now consider how this extension would look like. In this section, we consider a limited selection of fundamental rights to exemplify how familiar rights can be extended to animals. This part considers the right to life, the prohibition of torture, and the right to freedom. This exercise should be understood only as an illustration of how existing rights can be extended to animals. We leave to another occasion a fuller analysis of how a more exhaustive compendium of constitutional rights can be so extended.

A) The right to life

The right to life is the right which has most routinely been denied to animals. In fact, it is estimated that between 1 and 3 trillion animals are killed by humans for food each year.[33] We have shown in Section 1 that moral theorists

[33] Matt Zampa, 'How Many Animals Are Killed for Food Every Day?' (Sentient Media, 16 September 2018) <https://sentientmedia.org/how-many-animals-are-killed-for-food-every-day/> accessed 1 November 2024. Sentient Media, an animal rights organization, reports that 'More than 200 million animals are killed for food around the world every day—just on land. Including wild-caught and farmed fishes, we get a total closer to 3 billion animals killed daily. That comes out to 72 billion land animals and over 1.2 trillion aquatic animals killed for food around the world every year.'

continue to debate whether animals have the relevant interests and capacities to enjoy the right to life. Some, such as Peter Singer, argue that many animals may not be able to conceptualize themselves across time (ie they live only in the present) or that they do not have the capacity to conceptualize the wish not to die. These thinkers then contend that the lack of these cognitive abilities entails that the animals suffer no harm when they are painlessly killed. However, we argued that if we are to take the continuity between human rights and animal rights seriously, such cognitive abilities are irrelevant to the enjoyment of the right to life because human rights law does not require the possession of such abilities (recall *Charlie Gard*).

So, on what basis should humans and other animals enjoy the right to life? We argue, based on our discussion in Chapter 2 of sentience as a sufficient basis for moral status, that the right to life should be enjoyed on the ground that the continued existence of a particular individual is of inherent value because that individual is sentient. From this view, it follows that the right to life should not depend on any particular cognitive capacity (eg the ability to conceptualize oneself across time) but is merely a reflection of inherent value. This, we argue, is the best moral justification for why legal practice does not require that individuals should possess certain types of cognitive capacities in order to deserve the right to life.

To be sure, this moral justification of the right to life is not perfectly reflected in constitutional practice. In relation to human foetuses, for example, the practice is sometimes overinclusive. Our current scientific understanding is that foetuses are not sentient until around the third trimester.[34] On our justification of the right to life, it should follow that human foetuses' right to life is protected from around the third trimester onwards. However, there are several constitutional texts, such as the constitutions of Chile and the Dominican Republic, that protect the foetus' right to life from conception.[35]

We should not be too concerned with the fact that the moral justification of the right to life is not perfectly reflected in the practice of that right. As already discussed in the methodology section of this chapter, our analysis of the practice of fundamental rights chiefly serves to identify the underlying moral values of those rights so that they can be extended to animals. While sentience

[34] The evidence is summarized in Royal College of Obstetricians and Gynaecologists, 'RCOG Fetal Awareness Evidence Review' (December 2022) <https://www.rcog.org.uk/media/gdtnncdk/rcog-fetal-awareness-evidence-review-dec-2022.pdf> accessed 1 November 2024.

[35] See Chile's Constitution art 19.1 and Dominican Constitution art 37.

should be sufficient to endow animals with the right to life, we remain open to the view that other justifications may be adduced to grant that right to non-sentient human foetuses. Several theological and philosophical arguments have been made in this regard, but we express no views on these.[36]

If the right to life is to be extended to animals, we need to explore the scope of this right. In line with fundamental rights practice, the right to life has a negative and positive aspect. Let us first consider the former. The negative aspect requires the state to refrain from generally taking the life of a constitutional subject. Understandably, there may be particular instances in which a life is taken by the state without breach of this right. The ECHR, for example, sets out three such circumstances under Article 2(2):

> Deprivation of life shall not be regarded as inflicted in contravention of this Article when it results from the use of force which is no more than absolutely necessary:
> (a) in defence of any person from unlawful violence;
> (b) in order to effect a lawful arrest or to prevent the escape of a person lawfully detained;
> (c) in action lawfully taken for the purpose of quelling a riot or insurrection.

In a similar vein, the American Convention and the African Charter on Human and Peoples' Rights (African Charter) both emphasize that no one is to be arbitrarily deprived of the right to life.[37]

The key idea behind these provisions is that the state has a *prima facie* duty not to kill and can only take life in exceptional circumstances. What these circumstances are will depend on the values to which a jurisdiction is committed. However, in a sentience-based constitution, in line with Article 2(2) ECHR, the main circumstance in which the state may kill animals is in proportionate action aimed at defending human and other animals from violence. The idea of proportionality, which we explore in more detail in the next chapter, is crucial in this regard. Consider, for example, semi-wild bears who visit urban

[36] For an overview of some of the arguments for and against the right to life of human foetuses, see Kate Greasley and Christopher Kaczor, *Abortion Rights: For and Against* (CUP 2017). For the relationship between foetuses' rights and animal rights, see Sherry Colb and Michael Dorf, *Beating Hearts: Abortion and Animal Rights* (Columbia UP 2016).

[37] American Convention art 4(1); African Charter art 4: African Union, African Charter on Human and Peoples' Rights, <https://au.int/en/treaties/african-charter-human-and-peoples-rights> accessed 30 October 2024.

centres to eat food from dumpsters.[38] Some such bears have attacked humans and other animals. The default current practice is to kill the bears even when they do not pose any immediate danger to the life of others.[39] However, more proportionate means will sometimes be available. For example, to the extent that these animals continue to pose a threat to others, they may permissibly be contained and relocated to areas which are further away from urban centres.

Consider also instances where non-humans pose a threat to the lives of humans by spreading infectious diseases. Denmark, for example, killed 17 million minks as a preventative measure to avoid the spread of a variation of Covid-19 detected in some minks kept in fur farms.[40] This measure was disproportionate for two reasons. First, it was not targeted to the mink population that had been infected with the new strain of Covid-19 or even restricted to mink farms that were close by. Second, non-lethal measures, such as quarantines and vaccinations,[41] could have been administered or at least explored. It seems, then, that such disproportionate action clearly violates the right to life of minks even if carried out with the intention of protecting human life.

The mink example leads to a more challenging set of reflections. If the state is bound to respect the right to life of animals, it follows that it must not kill, or allow the killing of, sentient animals for the purpose of food, clothing, entertainment, or research. These purposes are not compatible with the practice of fundamental human rights, as Article 2(2) ECHR suggests. Should they be compatible with the practice of fundamental rights in a sentience-based constitution? No. Remember that, as argued in Chapter 2, a sentience-based constitution is devoted to the well-being of all governed sentient animals, human and non-human alike. While the right to life may be violated for the non-arbitrary reason of protecting others from violence, we argue that killing animals for the purpose of food, clothing, entertainment, or research will not count as such a reason in most, if not all, cases.

The killing of animals for food in economically well-off societies, for example, is generally unnecessary given the wide availability of plant-based foods

[38] John Roach, 'Black Bears Adapting to City Living, Study Says' (*National Geographic*, 26 November 2003) <https://www.nationalgeographic.com/animals/2003/11/black-bears-adapt-cities-animals/> accessed 1 November 2024.

[39] This was the approach taken in the account provided in Sarah Gibbens, 'Why a Bear Attacked a Teen in His Sleep' (*National Geographic*, 10 July 2017) <https://www.nationalgeographic.com/news/2017/07/black-bear-attack-campers-head-colorado-spd/> accessed 1 November 2024. See particularly the section titled 'one of many'.

[40] 'Coronavirus: Denmark Shaken by Cull of Millions of Mink' *BBC News* (11 November 2020) <https://www.bbc.com/news/world-europe-54890229> accessed 1 November 2024.

[41] Maria Vasilyeva, 'Russia Trials COVID-19 Vaccine for Domestic Animals, Including Mink' *Reuters* (18 December 2020) <https://www.reuters.com/article/health-coronavirus-russia-animal-vaccine-idUSKBN28O1HU> accessed 1 November 2024.

and cheap supplements which can satisfy humans' dietary needs.[42] Equally, the killing of animals for clothing is unnecessary given the worldwide adoption of synthetic and plant-based clothing (eg polyester and cotton). Similarly, the killing of animals for entertainment, such as recreational hunting or bullfights, is completely unnecessary given the many alternatives for entertainment at humans' disposal (eg videogames, including ones that simulate hunting).

Finally, could it at least be argued that there is justification for lethal medical research on animals if this would bring some important benefit to human life? Even though this may be permissible under some utilitarian theories, it would not be acceptable in a rights-based account like ours.[43] What is more, fundamental rights practice rejects such utilitarian reasoning given that it would much more readily justify deadly human experimentation. In fact, experimentation on humans would produce much more reliable medical insights for humans than experimentation on animals ever could.[44] Yet, as we show in Subsection B below, constitutional rights practice consistently prohibits such experimentation. In a sentience-based constitution, we should extend this prohibition to sentient animals as well.

Let us now consider the positive duties entailed by the right to life.[45] The state has an obligation to protect the right to life of all the animals under its control. This entails that the state has a duty to prohibit, usually through the criminal law, the taking of the lives of animals. It also entails a duty to enforce and investigate instances of breaches of such prohibitions. Current practices of humans killing animals for the purpose of food, entertainment, clothing, or research would therefore be prohibited and prosecuted.

The view that the state has a positive duty to protect the right to life of all sentient animals has significant implications. Of course, it is currently politically, economically, and socially unthinkable for an anthropocentric state to

[42] Marco Springmann and others, 'Health and Nutritional Aspects of Sustainable Diet Strategies and Their Association with Environmental Impacts: A Global Modelling Analysis with Country-Level Detail' (2018) 2 The Lancet Planetary Health e451. The study compared a healthy vegan diet to other dietary patterns and found that it was the most effective in providing most nutrients while reducing premature mortality (reduction of 19 per cent [95 per cent CI 18–20] for the flexitarian diet to 22 per cent [18–24] for the vegan diet).

[43] Peter Singer, *Animal Liberation* (2nd edn, Pimlico 1995) 82–83. According to Singer, such experiments on animals would be acceptable in theory as long as experimenters were also willing to carry out the procedures on humans with similar mental capacities and social relations as their non-human counterparts. Given that experimenters and the general public would almost never want to carry out these procedures on humans, in practice the experiments on non-human animals would not be justifiable.

[44] Aysha Akhtar, 'The Flaws and Human Harms of Animal Experimentation' (2015) 24 Cambridge Quarterly of Healthcare Ethics 407.

[45] For recognition of this positive duty in an ECHR context, see *Valentin Câmpeanu v Romania* (2014) 37 BHRC 423.

abolish established practices of killings of animals. Remember, however, that our discussion around the right to life is premised on a sentience-based constitution that gives equal consideration to the well-being of both humans and animals. On this premise, the recognition of the right to life for all sentient animals would also entail, at least theoretically, the more demanding responsibility on the state to prevent killings between animals. Indeed, the right to life of a chicken is equally violated by a human farmer as by a fox killing the bird for food. However, the state can more easily control the actions of human farmers than the actions of wild animals given that humans, unlike wild animals, are capable of understanding and complying with legal prohibitions of such killings.

Yet, our point here is not only about feasibility. It may also be undesirable for the state to protect certain animals from other animals. The lives of carnivorous animals, for example, may be jeopardized by human policies that hinder them from having access to their usual preys. So, there may be real conflicts of the right to life on top of pragmatic considerations if the state intervened to stop killings between animals.

These issues of conflicts of rights are of immense difficulty and cannot be fully explored here.[46] What we can say, however, is that, in line with human rights law,[47] the positive aspect of the right to life does not impose a duty on the state to prevent all instances of interpersonal killings. Rather, it is only incumbent on the state to *take appropriate steps*, including thorough investigations, to safeguard the lives of those within its jurisdiction in the context of any activity, public or not, in which the right to life may be at stake. This is an especially pressing duty in those contexts, such as prisons and hospitals, where vulnerable persons are institutionalized by the state.[48] Because regulation of interpersonal killings between animals will be ineffective, given that animals cannot normally follow legal rules, it would be unreasonable for the state to regulate such killings. Nonetheless, duties can be imposed on humans who have a special relationship with certain animals (eg those who have companion animals) to safeguard those animals from attacks from others.

Furthermore, given that in many instances the right to life of an animal depends on their ability to access prey, the state would not be allowed to prioritize

[46] eg McMahan has argued that there are reasons to desire the extinction of carnivorous wild animals in Jeff McMahan, 'The Meat Eaters' *The New York Times* (19 September 2010). Korsgaard has opposed this view in Christine M Korsgaard, *Fellow Creatures: Our Obligations to the Other Animals* (OUP 2018) ch 11.

[47] See for further information n 44 above.

[48] *Valentin Câmpeanu v Romania* (n 45) [130].

the life of one over the other. Such situations are generally therefore best left unregulated by the state. However, in contexts where the state has direct control over the living conditions of animals (eg in state-run conservation areas), it would be incumbent on it to make provisions so that the right to life of each species is respected. This may entail erecting barriers to protect prey animals from their predators and providing adequate nutrition to the predators (eg through a combination of plant-based meals, faux-meats, and dietary supplements).[49]

As should be apparent from the above discussion, the extension of the right to life in a constitution for all sentient animals would profoundly transform current practices in anthropocentric systems. As mentioned, entire human industries would need to be abolished or significantly reformed with severe economic, social, and political repercussions. While this prospect may appear daunting, we should remember that it is inherent in the trajectory of fundamental rights. Recognition of hitherto denied fundamental rights in the human context has always had large-scale consequences. For example, in the United States, the abolition of non-penal slavery by the Thirteenth Amendment was the product of the American Civil War and brought about a revolutionary change of social, political, and economic practices. Extending constitutional rights to animals may change the fabric of society in similar ways.

B) Prohibition of torture and inhuman or degrading treatment

Several constitutional catalogues of rights prohibit interferences with the physical integrity of their subjects. Accordingly, Article 19(1) of the Chilean Constitution, for instance, states that it guarantees the right to 'the physical and mental integrity of the person'. However, this is a less common version of the more widespread prohibition of torture and inhuman or degrading treatment or punishment.[50] Article 34(1)(a) of the Nigerian Constitution, for example, provides that 'no person shall be subject to torture or to inhuman or degrading treatment'. Article 28G of the Indonesian Constitution similarly declares that '[e]very person shall have the right to be free from torture or inhumane and

[49] We consider these issues again in the next chapter on the proportionality test, especially Subsection 3.A.

[50] According to the database of Constituteproject.org, only twenty-two constitutions worldwide use the term 'physical integrity'. On the other hand, 160 constitutions worldwide use the more standard prohibition of torture, cruel, and unusual treatment. Some constitutional provisions, such as Article 25(5) of the Constitution of Timor-Leste, use both.

degrading treatment'. As we show in the next paragraphs, the prohibition of torture and inhuman or degrading treatment is essentially a prohibition of interferences with the right to physical and mental integrity (torture causing the highest level of pain and suffering). However, given the relative dominance of the prohibition of torture and inhuman or degrading treatment over the right to physical integrity, we shall examine whether the terms of the former can be extended to animals. So, we will ask whether animals can really be 'tortured', and whether they can be subjected to 'inhuman' and 'degrading' treatment. As we will argue, once the rationale and scope of the prohibition of torture and inhuman or degrading treatment are duly considered, it can and should be extended to animals.

Human rights practice provides the following rationale for the prohibition of torture: it undermines the core of the physical and mental integrity of a sentient individual and causes the highest level of severe pain and suffering. The ECtHR has held that

> the special stigma of 'torture' [should] attach only to deliberate inhuman treatment causing very serious and cruel suffering . . . In addition to the severity of the treatment, there is a purposive element, as recognised in the United Nations Convention against Torture and Other Cruel, Inhuman or Degrading Treatment or Punishment . . . which defines torture in terms of the intentional infliction of severe pain or suffering with the aim, *inter alia*, of obtaining information, inflicting punishment or intimidating.[51]

It should be noted that, perhaps counterintuitively, human rights practice does not consider that torture is committed only when it is done to obtain information. Indeed, as the quotation above indicates, all that is required is the intentional infliction of severe pain or suffering with the aim to punish, intimidate, or obtain information. Animals can be, and often are, punished. Think, for example, of negative and positive punishment methods in canine training.[52] Likewise, it is uncontroversial to state that animals, like humans, can be intimidated. For example, studies show that males within baboon communities wildly use sexual intimidation towards females.[53] Accordingly, there

[51] *El-Masri v Macedonia* (2013) 57 EHRR 25 [197].

[52] EF Hiby, NJ Rooney, and JWS Bradshaw, 'Dog Training Methods: Their Use, Effectiveness and Interaction with Behaviour and Welfare' (2004) 13 Animal Welfare 63. The authors note that, historically, pet dogs have been trained using mainly negative punishment techniques.

[53] Alice Baniel, Guy Cowlishaw, and Elise Huchard, 'Male Violence and Sexual Intimidation in a Wild Primate Society' (2017) 27 Current Biology 2163.

is good reason to think that the prohibition of torture could be extended to protect animals from the infliction of pain or suffering designed to punish or intimidate.

What about the prohibition of 'inhuman or degrading treatment or punishment'? Can it also be extended to animals? Some aspects of the practice of fundamental rights provide reasons to doubt this. The ECtHR, for example, has provided a human-dignity centred rationale for this right. It has said that inhuman or degrading treatment or punishment involves

> [s]ubjecting a person to ill-treatment that attains . . . a minimum level of severity [which] usually involves actual bodily injury or intense physical or mental suffering. However, even in the absence of those characteristics, where treatment humiliates or debases an individual, showing a lack of respect for or diminishing his or her human dignity, or arouses feelings of fear, anguish or inferiority capable of breaking an individual's moral and physical resistance, it may be characterised as degrading and also fall within the prohibition set forth in Article 3. It may well suffice that the victim is humiliated in his own eyes, even if not in the eyes of others.[54]

The explicit mention of 'human dignity' coupled with the idea of a victim feeling humiliated in his own eyes may give rise to the objection that the prohibition of inhuman or degrading treatment cannot apply to animals. This is because, according to this objection, animals do not possess human dignity or the necessary cognitive abilities to have self-respect or to feel humiliated.[55] The ECtHR has accepted that a single slap administered by the police to a human during questioning is sufficient to find a violation of the prohibition of degrading treatment. This was because, *inter alia*, the slap would arouse feelings of humiliation in the victim.[56]

We would need some solid ethological evidence on self-respect and humiliation in animals to counter this objection. Until then, we need not insist that this self-respect-centred aspect of the prohibition should be extended to animals. As we argued in Section 2.C, fundamental rights may be overinclusive, entailing that not all aspects of a right need to be applicable to all rights-bearers.

[54] *Nicolae Virgiliu Tanase v Romania* [2019] ECHR 491 (ECtHR) [118].

[55] S Camenzind, '7. Three the Hard Way: Humiliation of Animals in the Swiss Animal Welfare Act', *Sustainable Governance and Management of Food Systems* (Wageningen Academic Publishers 2019). The author rejects the idea that non-human animals can be humiliated because they do not possess the necessary cognitive abilities for self-respect.

[56] *Bouyid v Belgium* (2016) 62 EHRR 32 [105].

However, this does not mean that other aspects of the prohibition of inhuman or degrading treatment cannot be extended to animals. As the ECtHR quote above shows, the presence of subjectively humiliating, disrespectful, or undignified treatments is sufficient but not necessary to determining whether inhuman or degrading treatment or punishment has been inflicted. The presence of sufficiently severe actual bodily injury or intense physical or mental suffering alone is enough.

This becomes clearer when we look at the prohibition of inhuman or degrading treatment or punishment as regards to very young children. Like animals, young children may not possess the necessary cognitive abilities to have self-respect or to be humiliated; yet they enjoy this right. The UN Committee on the Rights of the Child, in its General Comment on Article 37 UNCRC (which prohibits torture, inhuman and degrading treatment or punishment), has categorized all forms of corporal punishment of children, irrespective of age, as a form of inhuman and degrading treatment, and consequently prohibited them. This suggests that the prohibition of inhuman and degrading treatment, even if animals lack the ability to have self-respect or to be humiliated, may be extended to animals in a sentience-based constitution.

The conclusion that the prohibition of inhuman and degrading treatment may be extended to animals might still be resisted on the basis that the prohibition is grounded, as the ECtHR's quotation above suggests, in human dignity and as such cannot apply to animals. However, dignity is a concept that is not only applicable to humans. In fact, some constitutional texts and doctrines already recognize that animals possess dignity. Article 120(2) of the Swiss Constitution, for example, recognizes that animals and other living beings possess dignity. The idea of animal dignity is defined in the Swiss Animal Welfare Act as 'the inherent worth of the animal that must be respected when dealing with it'.[57] According to the Act, animal dignity is violated not only when unjustified suffering is caused to animals but also when they are exposed to anxiety or humiliation, or when there is major interference with their appearance, or when they are excessively instrumentalized.[58] Activities that disregard their dignity are prohibited, unless explicitly authorized.[59] The Act accepts animal husbandry, trade, and research as some of the instances where some degree of suffering of animals may be justified.[60]

[57] Animal Welfare Act 2005 art 3.
[58] ibid.
[59] ibid art 4.
[60] See ibid arts 6, 13, and 17.

Similarly, in *Nagaraja*, the Indian Supreme Court accepted that animals possess dignity. It has held that the constitutional right to life for humans under Article 21 also includes a right to dignified existence for animals. It said:

> So far as animals are concerned, in our view, life means something more than mere survival or existence or instrumental value for human-beings, but to lead a life with some intrinsic worth, honour and dignity . . . Right [sic] to dignity and fair treatment is, therefore, not confined to human beings alone, but to animals as well . . . Animals have also [*sic*] a right against the human beings [*sic*] not to be tortured and against infliction of unnecessary pain or suffering.[61]

Similarly to the Swiss provisions, the Indian Supreme Court accepted that the right to dignified existence of animals can be limited in certain instances that accrue important benefits to humans, such as for food and medical research,[62] but not in others, such as the traditional bull cart race on which the case centred.[63]

Existing constitutional rights doctrine therefore provides some precedent for the view that invocations of dignity can protect animals from interference with their right to physical and mental integrity. Admittedly, this precedent is limited because it recognizes that the inherent worth of animals can be outweighed, for example, to provide humans with food and for research purposes. However, as we consider in the next few paragraphs, in a sentience-based constitution, the prohibition of torture and inhuman or degrading treatment would not be outweighed in these circumstances.

What would having a right to physical and mental integrity entail for animals? We propose that, in line with human rights practice, this right would have both a positive and negative aspect.[64] Under the latter, the state is prohibited from inflicting torture and inhuman or degrading treatment on a sentient being. States undermine the physical and mental integrity of non-human subjects through, for example, state-approved medical research on animals. In the United Kingdom, for instance, the state grants licences to

[61] *Animal Welfare Board of India v A Nagaraja* (2014) 7 SCC 547 [62]. This statement has been narrowly reinterpreted in *Animal Welfare Board of India et al v Union of India et al* [2023] Writ Petition (Civil) No 23 of 2016 (Supreme Court of India) [24]. The Supreme Court now considers the earlier statements that animals enjoy constitutional rights merely as 'a judicial suggestion'.

[62] *Animal Welfare Board of India v A Nagaraja* (n 61) [60].

[63] ibid [61]. Note, however, that this kind of sport was subsequently found to be permissible by the Indian Supreme Court in *Animal Welfare Board of India et al v Union of India et al* (n 61).

[64] For the recognition of this positive duty in an ECHR context, see *Assenov v Bulgaria* (1998) 28 EHRR 652 [102].

laboratories to carry out clinical research on animals under the Animals (Scientific Procedures) Act 1986. Hence, states often authorize painful procedures, which usually culminate in death, to be carried out on animals for research and educational purposes. These procedures should be understood as instances of torture or inhuman or degrading treatment (depending on their level of severity, torture being the highest). In fact, several constitutions and international instruments specify that medical procedures and experiments may be categorized as instances of torture or inhuman or degrading treatment if they are carried out without the consent of the subject. Accordingly, Article 8 of the Constitution of Madagascar states that

> No one may be submitted to torture or to cruel, inhuman or degrading penalties or treatments. In particular, it is prohibited to submit a person without their free consent to a medical or scientific experiment.

Similar language is adopted in the constitutions of, among others, Rwanda,[65] Serbia,[66] Fiji,[67] South Africa,[68] Somalia,[69] Russia,[70] and Turkey.[71] Given that animals cannot normally consent to these procedures and given that, in any case, painful scientific experimentation on animals is usually not carried out for the benefit of the animal being experimented upon (in fact, they are usually killed at the end of such experiments), a sentience-based constitution could not condone painful scientific experimentation.

The right to physical and mental integrity also has a positive aspect. This requires the state to prohibit (including by criminalizing) the commission of acts by private actors that would constitute torture and inhuman or degrading treatment. It also has a duty to put in place procedures to prevent, investigate, and prosecute such acts. This obligation would not be particularly problematic as most legal systems already have legislation in place to outlaw cruel practices towards animals. As mentioned in Chapter 1, only a few jurisdictions completely lack animal protection laws. However, in many jurisdictions, a wide array of animals—especially aquatic animals and birds—are exempted from this protection. These exemptions are often granted to enable animal farming,

[65] Rwanda's Constitution art 15.
[66] Serbia's Constitution art 25.
[67] Fiji's Constitution art 11.
[68] South Africa's Constitution art 12.
[69] Somalia's Constitution art 15(3).
[70] Russian Federation's Constitution art 21.
[71] Turkey's 1982 Constitution art 17. See also Convention on the Rights of Persons with Disabilities ('UNCRPD') art 15.

which we have already argued would contravene the right to life. The prohibition of torture and inhuman or degrading treatment therefore complements that right insofar as the exploitation of animals leads to their death and causes physical and psychological harm, which would amount to torture and inhuman or degrading treatment.

As discussed in relation to the right to life, the state must protect animals from torture and inhuman or degrading treatment committed by it or by human persons. However, it will likely be impracticable for it to protect animals, especially those living in the wild, from instances of interpersonal violence. We therefore propose that a duty to protect animals from violence caused by other animals should arise principally only where there is a special relationship (eg care of companion animals) or where the state has direct control over the living conditions of animals (eg in state-run conservation areas). Additionally, it would seem practicable to us that state bodies should have the duty to care for animals who have been left injured by their predators and whom state bodies encounter during the performance of their normal duties.

C) The right to freedom

Catalogues of fundamental rights usually provide for two liberty-based rights. The first is a general guarantee of freedom or liberty and the second is freedom of movement within the jurisdiction. For example, Article 12(1) of the South African Constitution provides that 'Everyone has the right to freedom' and Article 21(1) states that 'Everyone has the right to freedom of movement'. This distinction between liberty-based rights is also found in several other constitutions, such as those of India,[72] Russia,[73] and Canada.[74] Can these liberty-based rights, in either form, be extended to animals under a constitution for all sentient animals? We suggest that they generally can. We will consider these separate rights under a broader right to freedom. In a constitution for all sentient animals, the more restrictive the interference with individual freedom, the harder it will be for the state to justify such interference.

[72] Personal liberty is protected under Article 21, while freedom of movement is protected under Article 19(1)(d).

[73] Personal liberty is protected under Article 22(1), while freedom of movement is protected under Article 27.

[74] Personal liberty is protected under Article 7 of the Canadian Charter, while freedom of movement is protected under Article 6.

Let us first consider the consequences of recognizing this right for animals before analysing on what basis it can be extended and what limitations may be justifiably imposed on it. Animals are usually kept in captivity to facilitate their exploitation and killing for agriculture, clothing, entertainment, and research. For example, intensive farms in the United Kingdom typically have about fifteen chickens per square metre, or an area about the size of an A4 sheet of paper for each bird.[75] We saw in the previous sections that these industries routinely violate animals' right to life and the prohibition of torture and inhuman or cruel treatment. We argue in this section that these industries also violate the right to freedom of animals. This additional violation comes on top of the already existing reasons why these industries would have to be abolished in a constitution for all governed sentient animals.

Why are animals entitled to a fundamental right to freedom? Some philosophers have denied that animals have an interest in freedom. Alasdair Cochrane, for example, has argued that liberty is only intrinsically valuable for autonomous beings, that is, those who have the capacity to devise and revise a conception of the good life and who therefore need the freedom to pursue it.[76] If this is true, then, to the extent that an animal is not autonomous in this sense, the right to freedom is not valuable for them. To be sure, Cochrane admits that liberty may be *instrumentally* valuable for animals to the extent that certain forms of confinement, such as those in factory farms, are necessarily harmful and eventually lead to death. A right to freedom in these circumstances would therefore be instrumental in avoiding breaches of their right to life and the prohibition of torture and inhuman or cruel treatment. However, on his account, confinement that does not cause them suffering and death, such as in certain laboratories where non-invasive and non-lethal experiments are conducted, is permissible.[77]

The controversy over whether animals have an interest in liberty has received some attention in courts. Advocates of animals across the globe have filed *habeas corpus* petitions in order to free captive animals, usually kept in zoos or research laboratories. So far, only few of these claims have succeeded.[78]

[75] 'The Rise of the "Megafarm": How British Meat Is Made' (*The Bureau of Investigative Journalism*) <https://www.thebureauinvestigates.com/stories/2017-07-17/megafarms-uk-intensive-farming-meat> accessed 1 November 2024.

[76] Alasdair Cochrane, *Animal Rights without Liberation: Applied Ethics and Human Obligations* (Columbia UP 2012) 26–29. See also Alasdair Cochrane, *Sentientist Politics: A Theory of Global Inter-Species Justice* (OUP 2018) 40.

[77] Cochrane, *Animal Rights without Liberation* (n 76) 72–76.

[78] The Nonhuman Rights Project has launched several *habeas corpus* cases on behalf of elephants and chimpanzees, see <https://www.nonhumanrights.org/litigation/> accessed 1 November 2024. All those cases have failed so far. The leading case is *People ex rel Nonhuman Rights Project, Inc v Lavery* (2014) 124 AD 3d (NY: Appellate Div, 3rd Dept). A successful *habeas corpus* case was decided in favour

However, if Cochrane is correct, some of these claims, to the extent they invoke a right to liberty, are badly formulated in the first place because they seek to vindicate an interest which animals (most of them anyway) do not have.[79]

Consistently with the method of this chapter, we can resolve this dispute by reference to human rights practice, which shows us that non-autonomous agents, such as humans with severe cognitive disabilities, have a right to liberty. The wording of Article 5(1) ECHR, for example, provides that 'everyone' is entitled to its protection. This includes individuals who the Convention calls 'persons of unsound mind'. Under Article 5(1)(e), the freedom of an individual belonging to this category can be restricted, among other conditions, only if 'the person needs therapy, medication or other clinical treatment to cure or alleviate his condition, but also where the person needs control and supervision to prevent him from, for example, causing harm to himself or other persons'.[80] From this follows that even humans with severe cognitive disabilities who are not considered autonomous are nevertheless entitled to freedom, and any restrictions ought to be justified by cogent reasons. So, in line with human rights practice, we can say that the right to liberty should not be a prerogative of autonomous agents alone. Indeed, for the purposes of human rights practice, it is irrelevant whether a rights-holder is autonomous to enjoy the right to freedom. Consequently, in a sentience-based constitution, given the continuity of the rights of humans and non-humans, non-autonomous humans and animals alike should enjoy the right to freedom.

This approach of human rights practice can be justified by reference to the importance of the well-being of sentient beings. Non-autonomous humans and animals have desires and wishes (eg what to eat, where to go, and with whom to associate), the satisfaction of which contributes to their well-being. Without the necessary freedom, they would not be able to satisfy these desires and wishes. The contribution that freedom makes to their well-being rather than to their autonomy is why they should enjoy a fundamental right to freedom. To be sure, these desires may be bad for them or for others. Furthermore, there might be alternative ways in which those same desires and wishes could be

of the chimpanzee Cecilia in Argentina. See *Judicial Power of Mendoza, Third Court of Guarantees, Re Cecilia*, P-72 254/15, 3 November 2016.

[79] We speak of only some because many of these *habeas corpus* cases have been undertaken on behalf of great apes and some cetaceans. See the previous note. Cochrane is open to the view that great apes and cetaceans might be autonomous beings given their high level of cognitive abilities. See Cochrane, *Sentientist Politics* (n 76) 117. However, some of these *habeas corpus* cases, eg that involving *Estrellita* (a Woolly Monkey) or involving elephants, would, on Cochrane's account, be inappropriate because he does not recognize that these animals are autonomous beings.

[80] *Ilnseher v Germany* [2018] ECHR 991 (ECtHR) 133.

satisfied. Nevertheless, this at most suggests that the right to freedom should be circumscribed rather than, as Cochrane tells us, that these non-autonomous beings do not have an interest in freedom at all.

What could potentially justify limitations to the right to freedom in a sentience-based constitution? It is here that we can draw on some aspects of Cochrane's theory, which also provides an example of the overinclusiveness of rights in a sentience-based constitution. Autonomous beings have a general interest in not being treated paternalistically because their autonomy entails the ability to make informed choices and bearing the personal responsibility for these. The state disrespects this capacity whenever it tries to dictate how they should live and when it restricts their freedom accordingly. In a constitution for all sentient animals, the liberty of autonomous beings (whether humans or not) cannot be restricted for purely paternalistic considerations, such as harm to oneself or on the basis that certain ways to live are better than others.

The flipside of this is that non-autonomous beings can be treated paternalistically, and their freedom restricted if doing so promotes their well-being. Human children, for example, can be compelled to go to school even though they may prefer to spend the entirety of their days playing instead. Some severely cognitively disabled individuals could be provided care in mental health institutions to ensure their well-being. In a similar way, animals' freedom may be restricted to protect them from harming themselves or others. All of these restrictions on the right to freedom should of course be compatible with other fundamental rights. It goes without saying that non-autonomous beings' freedom cannot be permissibly restricted in order to put them in situations that would violate their right to life or the prohibition of torture and inhuman or degrading treatment (eg in factory farms).

That said, the restriction of the freedom of both autonomous and non-autonomous beings in order to protect the right of others is ordinarily justifiable. Such restriction of freedom ought to still be compatible with the rights entailed by the rule of law (which will be explored in depth in Chapter 5). It is worth noting that restrictions on the freedom of certain animals will be the default way of protecting humans and animals under human care who would be harmed by other animals. Recall our discussion on the right to life of semi-wild bears attacking humans when searching for food or minks suffering from diseases which can spread to humans. In these scenarios, we argued that restricting their freedom is more proportionate than killing them.

While the right to freedom appears to be skewed in favour of agential humans (because it protects their more commonly present interest in autonomy), there is another aspect of this right that instead favours animals. Constitutions do not recognize a general right for non-citizens to move freely across borders without restrictions,[81] although some jurisdictions, like France, recognize a constitutional right to asylum from persecution for non-citizens.[82] By contrast, some animals seem to have a fundamental interest in moving freely across borders in situations not comparable to asylum. Consider, for example, Atlantic salmon who will spawn only in the freshwaters where they were born and travel hundreds of kilometres from the ocean to their original homes.[83] Or consider the thousands of species of birds that travel from one continent to another to secure their food-source as seasons change.[84] One cannot simply argue that these animals should be treated like human refugees as their interest in moving freely across borders is not dictated by the extraordinary need to flee persecution. It is instead based on a routine and natural need to procreate and obtain access to their usual sources of nutrition. Accordingly, we propose that the right to freedom for these animals should be more expansively interpreted to entail a right to move freely across borders.

This conclusion has two consequences. The first is that constitutional states that take seriously the well-being of all sentient animals should take measures to ensure that barriers erected to limit human crossing of borders (eg border walls) are designed in such ways as not to frustrate the freedom of movement of animals who have a fundamental interest in moving across borders. In some instances, this may entail that such barriers should not be erected at all, especially given the damage they cause to entire animal species and ecosystems.[85] Second, the recognition of a more expansive right to freedom for certain animals indicates that the overinclusiveness of fundamental rights practice is not a feature that exclusively grants more rights to humans. That overinclusiveness can run the other way too. Certain animals will have more fundamental rights given that they have different interests and capacities from humans which warrant more extensive protection.

[81] Freedom of movement in Article 9 of the Indian Constitution is reserved for citizens; Article 27(2) of the Russian Constitution reserves the right to freely return to Russia only to citizens; Article 6(1) of the Canadian Charter reserves the right to enter, remain in, and leave Canada only to citizens.

[82] Preamble to the Constitution of 27 October 1946. See also Mexican Constitution art 11.

[83] US Fish and Wildlife Service, 'Fish Migration—Atlantic Salmon' <https://www.fws.gov/sites/default/files/documents/Fish-Need-to-Move-Atlantic-salmon.pdf> accessed 1 November 2024.

[84] Royal Society for the Protection of Birds, 'Migration' <https://www.rspb.org.uk/birds-and-wildlife/bird-migration> accessed 1 November 2024.

[85] Brian Owens, 'Trump's Border-Wall Pledge Threatens Delicate Desert Ecosystems' (2016) 536 Nature News 260.

Conclusion

We have argued in this chapter that existing constitutional catalogues of human rights can be extended to cover animals in a sentience-based constitution. This is because familiar catalogues of fundamental rights can protect overlapping interests and capacities that humans and animals share. We have shown that the practice of fundamental rights currently grants rights to certain non-humans, such as churches and corporations. In a constitution that recognizes animals as equal constitutional subjects, fundamental rights should also be extended to animals. In fact, this chapter has shown that the progressive expansionism, the non-hierarchical nature, the overinclusiveness, and the generality of the practice of fundamental rights for humans, strengthen and make plausible this chapter's argument. Finally, to show how constitutional rights can be extended to animals, we considered some familiar constitutional rights in Section 3 and analysed how humans and animals alike can benefit from them.

4

Proportionality

Introduction

In the summer of 2021, the US army and its allies, including the United Kingdom, left Afghanistan after years of conflict. People who had collaborated with the Western troops desperately sought to evacuate the country which was being taken over by the Taliban. Thousands failed to board the last few flights leaving Afghanistan. International news outlets showed harrowing scenes of humans futilely clinging to departing planes—and falling to their deaths. Amid all the chaos, the UK Prime Minister allegedly authorized Paul Farthing, founder of the animal charity Nowzad, to leave Kabul with around 150 rescued dogs and cats on a private plane. Once this made the news, many quickly condemned the use of the plane space for animals, arguing that it could have been reserved to humans.[1] As a senior adviser to the British troops claimed:

> People have died in Afghanistan, because of their connections to the UK, but the prime minister was just allowing animals to get out of there. It is like a joke. I see now that animals have a higher value than us.[2]

While the way these events were portrayed in the media remains disputed,[3] in this chapter we use this version of events as one of several examples to ask how a sentience-based constitution would resolve instances of conflicts of rights between humans and animals, and between different animals, using the principle of proportionality.

To be sure, this principle has different uses in law and is not only a principle of constitutional law. It is employed, for example, in criminal law (eg in relation

[1] 'Boris Johnson Authorised Afghan Animal Evacuation, Leaked Email Suggests' *BBC News* (26 January 2022) <https://www.bbc.com/news/uk-politics-60148635> accessed 4 November 2024.

[2] Emma Graham-Harrison and Amelia Gentleman, '"It's a Joke": Left-behind Afghans Despair at Dog Rescue Revelations' *The Guardian* (26 January 2022) <https://www.theguardian.com/world/2022/jan/26/its-a-joke-left-behind-afghans-despair-at-dog-rescue-revelations> accessed 4 November 2024.

[3] Nowzad itself has challenged the prevalent account of the events at 'Our History' <https://www.nowzad.com/our-work/our-history> accessed 4 November 2024.

Animals and the Constitution. John Olusegun Adenitire and Raffael Fasel, Oxford University Press.
 DOI: 10.1093/9780198910534.003.0004

to self-defence),[4] in public international law (eg in relation to *jus ad bellum*),[5] and in natural law reasoning (eg in the context of the principle of double effect).[6] Even within constitutional law, different jurisdictions use it for different purposes. Sometimes it serves as a constitutional ground of administrative decisions. For example, section 19(1) of the Constitution of the Cayman Islands and the same section in the Constitution of the Turks and Caicos Islands state that decisions of public officials must be 'lawful, rational, *proportionate* and procedurally fair'.[7] In this chapter, we confine ourselves to theorizing the proportionality test as a constitutional principle used to resolve conflicts of non-absolute rights.[8]

We assume that many instances of conflicts such as the one described at the outset will be unavoidable. For example, there is an inevitable conflict between the right to life of carnivorous animals and the right to life of prey animals or between the right to scientific freedom of humans and the right to bodily integrity of animals used in medical experiments. How would a sentience-based constitution resolve these and other pressing and inevitable conflicts? As argued in Chapter 3, we take the view that a sentience-based constitution would recognize fundamental rights for all sentient animals, both human and non-human. It is therefore incumbent on us to say how these conflicts would be resolved. While we have already suggested in Chapter 3 how some conflicts would be resolved,[9] in this chapter, we provide a more detailed account of the process of resolution.

In a nutshell, we argue that conflicts of rights in a sentience-based constitution can be resolved using the proportionality test.[10] As Matthias Klatt and Moritz Meister have explained,

[4] George Letsas, 'Proportionality as Fittingness: The Moral Dimension of Proportionality' (2018) 71 Current Legal Problems 53, 76–80.

[5] Thomas Hurka, 'Proportionality in the Morality of War' (2005) 33 Philosophy & Public Affairs 34.

[6] Martin Luterán, 'The Lost Meaning of Proportionality' in Bradley W Miller, Grant Huscroft, and Grégoire Webber (eds), *Proportionality and the Rule of Law: Rights, Justification, Reasoning* (CUP 2014).

[7] Emphasis added. For a commentary, see Bilika H Simamba, 'Proportionality as a Constitutional Ground of Judicial Review with Special Reference to Human Rights' (2016) 16 Oxford University Commonwealth Law Journal 125.

[8] In Section 3.B, we reject the Dworkinian view that all constitutional rights are absolute rights. In this chapter, we do not analyse which rights should be considered absolute and which should be considered qualified. Our aim, instead, is to argue that the proportionality test can be a useful tool for resolving clashes between qualified rights. However, as we argue in Section 3.B, we oppose the view that there are only absolute rights.

[9] See eg Chapter 3 Section 3.

[10] For avoidance of confusion, we refer to the last stage of the proportionality principle as the balancing stage rather than, as it is sometimes called in the literature, proportionality *stricto sensu*.

> [i]n various jurisdictions worldwide, and across a broad range of areas of the law, there is a firm consensus that the proportionality test plays an indispensable role in constitutional rights reasoning. It is often assigned the central task of reconciling conflicting rights, interests, and values.[11]

We agree with them that the proportionality test 'stands out well as a suitable candidate for construing a global grammar of constitutionality' given its ubiquitous use in constitutional doctrine.[12] As we propose, the proportionality test can fruitfully be used not only in anthropocentric constitutions but also to adjudicate instances of conflicts of rights in a sentience-based constitution.

As we show in this chapter, there are several theories of the proportionality test. We therefore need to specify which theory we think best suits a sentience-based constitution. We will defend a theory of the test which is committed to what we call the Non-Instrumentalization Principle. This principle arises from our view that using the proportionality test as a tool to adjudicate conflicts of rights presupposes a moral commitment to the rights of humans and animals alike. Such a commitment carries with it the Kantian notion that holders of rights are inherently valuable and cannot be used as a mere means for others' purposes. Consequently, the Non-Instrumentalization Principle bars utilitarian trade-offs between a constitutional subject and the good of the greatest number. This theory contrasts with some scholarly views which claim that the proportionality test is neutral between different ethical theories, such as different versions of liberalism.[13] It also contrasts with morally committed theories of the proportionality test such as George Letsas' position that the test is about respecting constitutional subjects' ethical independence or Kai Möller's view that proportionality is about managing autonomy interests.[14] As we will explain fully below, these theories of proportionality unhelpfully exclude animals from being counted as subjects of rights and are therefore unsuitable for a sentience-based constitution.

[11] Matthias Klatt and Moritz Meister, *The Constitutional Structure of Proportionality* (OUP 2012) 1.

[12] ibid 3.

[13] ibid 56. The authors derive this view from Robert Alexy: see Robert Alexy, *A Theory of Constitutional Rights* (Julian Rivers tr, OUP 2009); Robert Alexy, 'Thirteen Replies' in Matthias Klatt (ed), *Institutionalized Reason: The Jurisprudence of Robert Alexy* (OUP 2012) 344.

[14] George Letsas, 'Rescuing Proportionality' in Rowan Cruft, S Matthew Liao, and Massimo Renzo (eds), *Philosophical Foundations of Human Rights* (OUP 2015); Letsas, 'Proportionality as Fittingness' (n 4); Kai Möller, *The Global Model of Constitutional Rights* (OUP 2015) 134–206. For reason of space, we defer engagement with Jacob Weinrib's moral reading of proportionality, which he bases in human dignity. See Jacob Weinrib, *Dimensions of Dignity: The Theory and Practice of Modern Constitutional Law* (CUP 2016).

At times, the proportionality test has been subject to scathing criticism. Among other things,[15] it has been alleged that the test's use undermines a serious commitment to rights;[16] that its appearance of neutrality and objectivity is an excuse for unconstrained judicial discretion;[17] and that the language of balancing is misleading at best because conflicts of rights often involve choices between incommensurable values that cannot be weighed against each other.[18] As we will show, adopting the Non-Instrumentalization Principle, combined with other values foundational to sentience-based constitutionalism, can help provide answers to some of these objections.

The chapter proceeds as follows. In Section 1, we start by outlining the constituent parts of the proportionality test as commonly used around the globe. We then show how some authors have argued that it is a formal principle that needs to be complemented by external principles of political morality. We argue that this view is mistaken given that the use of proportionality in rights adjudication itself presupposes as a minimum a commitment to rights. We unpack the consequences of this commitment to show that it should be adopted together with the Non-Instrumentalization Principle. Section 2 defends this principle and demonstrates how it influences the four stages of the proportionality test. We explain how this principle differs from others proposed in the literature, before shedding light on how several conflicts of rights would be resolved in a sentience-based constitution. In Section 3, we tackle some of the objections that scholars have levelled against the use of balancing in the proportionality test. Finally, we summarize our findings in Section 4.

1) The proportionality test and its internal morality

In this section, we outline what we term the ecumenical understanding of the proportionality test (Subsection A). This understanding describes the proportionality test as used by constitutional judges around the world. It is

[15] A useful catalogue of criticisms is provided in Matthias Jestaedt, 'The Doctrine of Balancing—Its Strengths and Weaknesses' in Matthias Klatt (ed), *Institutionalized Reason: The Jurisprudence of Robert Alexy* (OUP 2012).

[16] Grégoire Webber, 'On the Loss of Rights' in Bradley W Miller, Grant Huscroft, and Grégoire Webber (eds), *Proportionality and the Rule of Law: Rights, Justification, Reasoning* (CUP 2014); Stavros Tsakyrakis, 'Proportionality: An Assault on Human Rights?' (2009) 7 International Journal of Constitutional Law 468, 489–90.

[17] Grégoire CN Webber, *The Negotiable Constitution: On the Limitation of Rights* (CUP 2009); Bradley W Miller, 'Majoritarianism and Pathologies of Judicial Review' in Grégoire Webber and others (eds), *Legislated Rights: Securing Human Rights Through Legislation* (CUP 2018).

[18] Timothy Endicott, 'Proportionality and Incommensurability', *Proportionality and the Rule of Law* (CUP 2014); Tsakyrakis talks about comparing 'apples to oranges' in Tsakyrakis (n 16) 491.

'ecumenical' (ie inclusive of different churches or schools of thought) in two senses. First, this understanding describes the judicial practice of the proportionality test in a wide variety of jurisdictions. Second, scholars belonging to different theoretical schools of thought agree that this understanding captures the judicial practice of the proportionality test. However, as we show, scholars offer competing theoretical accounts to explain or justify this judicial practice. This raises the question of which theoretical conception of the proportionality test is best suited to a sentience-based constitution. In this section and the next, we will show that the proportionality test is not morally neutral and needs to be read in conjunction with the Non-Instrumentalization Principle. We will also explore, and reject, alternative morally committed views of the test (Subsection B).

A) The ecumenical view of the proportionality test

As suggested above, our focus in this chapter is on how the proportionality test is used to resolve conflicts between non-absolute rights. In this domain, the test is commonly, but not always, characterized as entailing an analysis of the following questions when rights conflict:

(1) Does the legislation (or other government action) establishing the right's limitation pursue a legitimate objective of sufficient importance to warrant limiting a right? ['legitimate goal']
(2) Are the means in service of the objective rationally connected (suitable) to the objective? ['suitability']
(3) Are the means in service of the objective necessary, that is, minimally impairing of the limited right, taking into account alternative means of achieving the same objective? ['necessity']
(4) Do the beneficial effects of the limitation on the right outweigh the deleterious effects of the limitation; in short, is there a fair balance between the public interest and the private right? ['balancing'][19]

We take this characterization of the four stages of the proportionality test from the work of Grant Huscroft, Bradley Miller, and Grégoire Webber, who, despite being ardent critics of the test, provide a good account of how it is

[19] Grant Huscroft, Bradley W Miller, and Grégoire Webber, 'Introduction', *Proportionality and the Rule of Law* (CUP 2014) 2.

generally understood around the world. As they note, judicial formulations of the test differ and, in some instances, some of the components are omitted altogether.[20] Nevertheless, their above-cited version is generally representative of the global practice of the test.[21]

Let us apply these stages of the proportionality test to see how they structure the conflict of rights in the Nowzad case. Under the legitimate goal stage, we would ask whether the UK Government's alleged decision to approve the rescue of the animals and leave some humans behind was in pursuit of a legitimate goal. Under a sentience-based constitution, all sentient animals have fundamental rights. Accordingly, the state may be acting permissibly by aiding the evacuation of the animals if done with the goal of protecting their right to life. Under the second stage, the suitability stage, we would ask whether evacuating the animals was a suitable means of protecting their right to life. Under the third stage, the necessity stage, we would ask whether the evacuation of the animals minimally impaired the humans' rights, especially their right to life, considering available alternatives to the evacuation. Finally, under the last stage, the balancing stage, we would assess whether the relevant interests of the animals and humans were properly weighed so that the evacuation struck, all things considered, a fair balance between these interests.

Applying this structured test is less challenging than asking an alternative and more abstract question of whether it is was *right* for the UK Government to approve the evacuation of the animals at the cost of leaving some humans behind.[22] While the outcome of the proportionality test effectively provides an answer to this more general question, the test allows us to address it in more bite-sized portions. Importantly, if any of the questions (say the suitability question) are answered negatively, one need not proceed to consider the next questions. This makes the proportionality test particularly suitable for judicial analysis. Judges in constitutional courts are not in a moral philosophy seminar and need a more structured way of answering such difficult moral questions.[23] The proportionality test is an institutionally expedient tool because it makes

[20] They cite *de Freitas v Permanent Secretary of Ministry of Agriculture, Fisheries, Lands and Housing* (1999) 1 AC 6 (Judicial Committee, Privy Council) which omits the last question.

[21] See also Tom Hickman, 'Proportionality: Comparative Law Lessons' (2007) 12 Judicial Review 31; Aharon Barak, *Proportionality: Constitutional Rights and Their Limitations* (CUP 2012); Vicki C Jackson, 'Constitutional Law in an Age of Proportionality' (2015) 124 Yale Law Journal 2680.

[22] Kai Möller suggests that this is essentially the question that proportionality helps to answer, see Möller (n 14) 179.

[23] A particularly illuminating contribution on this issue is Jeremy Waldron, 'Judges as Moral Reasoners' (2009) 7 International Journal of Constitutional Law 2. We take the view that, although proportionality in constitutional rights adjudication is a legal test, it deals with moral issues, including the issue of which rights ought to be prioritized. This view is not limited to natural lawyers: see eg Scott Shapiro, *Legality* (Harvard UP 2013) 173.

engaging with these questions more manageable. Specifically, it promotes economy of reasoning insofar as it allows judges to skip subsequent stages of the test when prior stages are not satisfied.

We have set out the ecumenical understanding of the proportionality test and shown how it would structure engagement with the Nowzad scenario. In Section 2, we will try to apply the test to the stipulated facts in the scenario.[24] Before we do so, however, it is important to note that different scholars have developed different accounts of the proportionality test. While most scholars agree that it has the four stages we have set out, there is disagreement about how best to conceptualize the different elements. The answers to several fundamental questions depend on which theoretical conception is chosen: Whose rights are subject to the proportionality test? Can animals' interests be counted within that test? And what interests can outweigh the rights of animals in the proportionality test? We therefore need to expound our theory of proportionality and show how it provides answers to these and other pressing questions. In what follows, we will differentiate our account from the influential morally neutral account of the proportionality test championed by Robert Alexy. We will also differentiate our account from the morally committed accounts by Letsas and Möller. We engage in this discussion to identify what theoretical conception of proportionality best suits a sentience-based constitution.

B) The internal morality of the proportionality test

Alexy has proposed that the balancing stage of the proportionality test has a formal structure which needs to be supplemented with moral arguments drawn from a more general theory of legal argumentation. As he puts it, '[p]roportionality analysis is, as the weight formula shows, a formal structure that essentially depends on premises provided from outside'.[25] Accordingly, the weight formula—Alexy's mathematical formula for the balancing stage—requires moral reasoning which is not inherent to the proportionality test. Importantly, Alexy himself does not provide a theory of political morality which is to be used in proportionality reasoning.[26] Klatt and Meister propose

[24] See text around n 71.

[25] Alexy, 'Thirteen Replies' (n 13) 344. The weight formula (in its complete version) is represented mathematically in Alexy, 'The Weight Formula' (n 25) 172 as follows: $W_{i,j} = \frac{I_i W_i}{I_j W_j}$.

[26] Lorenzo Zucca, *Constitutional Dilemmas: Conflicts of Fundamental Legal Rights in Europe and the USA* (OUP 2008) 20.

that this is a strength, not a weakness of Alexy's framework. According to them, the weight formula provides for legal certainty because it identifies the issues which need assessment. Importantly, they submit that the weight formula can make a plausible claim to neutrality as it is compatible with different moral frameworks.[27]

However, the claim that the proportionality test is a morally neutral principle is, in our view, mistaken. The proportionality test, as a tool for adjudicating conflicts of fundamental rights, cannot be neutral about different moral frameworks. Specifically, there is one framework it cannot be neutral about: that of rights. This is because, since the purpose of the proportionality test is to adjudicate conflicts of rights, it presupposes a commitment to rights. This is no trivial commitment. Some theories, such as certain versions of utilitarianism,[28] postcolonial,[29] and postmodernist[30] theories, are not committed to rights at all. Furthermore, several moral theories deny the desirability of animal rights.[31] The proportionality test's necessary commitment to rights heavily qualifies the claim that the test is compatible with different moral frameworks: it cannot be neutral about the desirability of rights and, in the context of sentience-based constitutionalism, about the view, defended in the previous chapter, that animals should have fundamental rights. Let us illustrate this by outlining the contrast between utilitarianism, which is incompatible with the test as we understand it, and our Kantian rights-based ethical framework (ie the Non-Instrumentalization Principle).

The core commitment of utilitarianism is the principle of utility.[32] According to the version of this principle made famous by Jeremy Bentham, the morally required act is that which produces the greatest amount of good for the greatest number. The 'good' here is understood in hedonistic terms, meaning

[27] Klatt and Meister (n 13) 56.

[28] Philip Schofield, Catherine Pease-Watkin, and Cyprian Blamires (eds), 'Nonsense upon Stilts, or Pandora's Box Opened', *The Collected Works of Jeremy Bentham: Rights, Representation, and Reform: Nonsense upon Stilts and Other Writings on the French Revolution* (OUP 2002).

[29] Makau Mutua, *Human Rights: A Political and Cultural Critique* (University of Pennsylvania Press 2013).

[30] Zühtü Arslan, 'Taking Rights Less Seriously: Postmodernism and Human Rights' (1999) 5 Res Publica 195.

[31] Some contemporary accounts that deny animal rights are Peter Harrison, 'Do Animals Feel Pain?' (1991) 66 Philosophy 25; Peter Carruthers, *The Animals Issue: Moral Theory in Practice* (CUP 1992); Roger Scruton, *Animal Rights and Wrongs* (Continuum 2006). For an overview of earlier accounts that deny animal rights, including those by Immanuel Kant, René Descartes, and Thomas Aquinas, see Scott D Wilson, 'Animals and Ethics' (*Internet Encyclopedia of Philosophy*) <https://iep.utm.edu/animals-and-ethics/> accessed 4 November 2024.

[32] Julia Driver, 'The History of Utilitarianism' in Edward N Zalta and Uri Nodelman (eds), *The Stanford Encyclopedia of Philosophy* (Winter 2022, Metaphysics Research Lab, Stanford University 2022) <https://plato.stanford.edu/archives/win2022/entries/utilitarianism-history/> accessed 4 November 2024.

that an action is morally mandatory if it promotes the greatest pleasure for the greatest number and/or if it minimizes the most pain for the greatest number. Under this view, the welfare of individuals can be sacrificed if it accords with the principle of utility.[33] Famously, for Bentham, the idea that there are moral rights protecting individuals from being sacrificed for the good of the greatest number is 'nonsense upon stilts'.[34] Accordingly, he also took a sceptical stance towards bills of fundamental rights such as the French Declaration of the Rights of Man and the Citizen.[35] Furthermore, even though he considered animals as entitled to moral consideration,[36] his utilitarianism excluded the possibility that animals could have rights.[37]

Now contrast Bentham's view with the Kantian rights-based view that individuals cannot be used as mere means for others' purposes, including to promote the public good.[38] Immanuel Kant, in one of his formulations of the categorical imperative, instructs us to '[a]ct so that you use humanity, as much in your own person as in the person of every other, always at the same time as end and never merely as means'.[39] As we have made clear in earlier chapters, we challenge the notion that humans or individuals endowed with 'humanity' (which for Kant required rational capacities) are the only beings who ought never to be treated merely as a means. Sentient non-human animals and non-paradigmatic humans who are not endowed with rationality should never be treated in this way either.[40] This expanded Kantian view holds that individuals (whether human or non-human) can never be sacrificed for the pursuit of the good of the greatest number. In other words, this view bars the application of the principle of utility. It does so by holding that individuals have inviolable rights which cannot be surrendered for the good of the greatest number.[41]

This contrast between utilitarianism and our Kantian view explains why the proportionality test cannot be neutral between different moral frameworks. That is, because the use of proportionality arises in the context of adjudicating

[33] See Jeremy Bentham, *An Introduction to the Principles of Morals and Legislation* (Mews Gate 1789).

[34] Schofield, Pease-Watkin, and Blamires (n 28).

[35] For a commentary, including the view that Bentham did not oppose rights in all instances, see Philip Schofield, 'Jeremy Bentham's "Nonsense upon Stilts"' (2003) 15 Utilitas 1.

[36] Bentham (n 33) 283 note b.

[37] Some have argued, based on unpublished work by Bentham, that he believed that 'any form of animal use that increases human welfare is permissible, as long as it is not wanton'. See Johannes Kniess, 'Bentham on Animal Welfare' (2019) 27 British Journal for the History of Philosophy 556, 561.

[38] Immanuel Kant, *Groundwork for the Metaphysics of Morals* (Allen W Wood ed, Yale UP 2018).

[39] This is the so-called formula of humanity, ibid 42.

[40] For a Kantian approach to animal rights, see Christine M Korsgaard, *Fellow Creatures: Our Obligations to the Other Animals* (OUP 2018).

[41] For an application of this principle to animal rights, see Tom Regan, *The Case for Animal Rights* (University of California Press 2004) 312–15.

between conflicting rights, proportionality presupposes a commitment to rights which is incompatible with utilitarianism.[42] Furthermore, in the context of sentience-based constitutionalism, the use of proportionality requires a commitment to the idea that animals can also hold fundamental rights which can conflict with the rights of humans. Consequently, contrary to what some authors have suggested,[43] it is implausible to view any of the stages of the proportionality test as requiring utilitarian reasoning. On the contrary, because proportionality entails adjudication between conflicting rights, the resolution of such conflicts bars utilitarian reasoning, given the incompatibility between utilitarianism (of the type advocated by Bentham) and rights.

To summarize, we have showed that proportionality in constitutional rights adjudication is not morally neutral. Using the proportionality test presupposes a commitment to rights and, in the context of sentience-based constitutionalism, presupposes a commitment to the view that animals can hold fundamental rights which can conflict with the rights of humans. The rights-based view we have been advocating for is a Kantian one, where sentient animals, human and non-human alike, cannot be treated merely as a means. Their interests cannot be sacrificed for the greatest good of the greatest number. The internal morality of proportionality in constitutional rights adjudication is that rights-holders, human and animals alike, cannot be used as a mere means for others' purposes. In other words, proportionality should be read in light of the Non-Instrumentalization Principle.

2) The Non-Instrumentalization Principle and the stages of the proportionality test

As we saw in the previous section, the Non-Instrumentalization Principle embodies the Kantian idea that rights-holders must not be treated as mere means to others' ends. However, as we explore in this section, a sentience-based constitution pairs the Non-Instrumentalization Principle with other moral

[42] As should be apparent, we are here setting a contrast between Kantian deontology and Benthamite utilitarianism. Utilitarianism (like deontology) is a wide church. Some versions of it, such as that championed by John Stuart Mill, are compatible with a rights-based morality. Mill, for example, is a utilitarian whose framework explicitly incorporates rights. See John Stuart Mill, *On Liberty, Utilitarianism and Other Essays* (Mark Philp and Frederick Rosen eds, 2nd edn, OUP 2015) 166–77. Because of the broadness of utilitarianism, any references to it in this chapter are to utilitarianism as championed by Bentham.

[43] Tsakyrakis (n 16) 472. Klatt and Meister read Tsakyrakis as claiming that proportionality requires a commitment to utilitarianism in order to be workable. We are not so confident that this is Tsakyrakis' argument given that he is generally sceptical about proportionality. Klatt and Meister (n 13) 64.

commitments, including due concern and respect for constitutional subjects.[44] Understanding these commitments will help resolve some of the pressing cases of rights conflicts which may arise in a sentience-based constitution. In this section, we investigate how the Non-Instrumentalization Principle shapes the stages of the proportionality test: the legitimate goal stage (Subsection A), the suitability stage (Subsection B), the necessity stage (Subsection C), and the balancing stage (Subsection D). In this analysis, we will also engage with alternative moral readings of proportionality proposed by other scholars. Our aim is to show that the Non-Instrumentalization Principle helpfully guides each stage of the proportionality test so that constitutional subjects, human and animal alike, cannot be used merely as a means for the greater good of the greatest number.

A) The legitimate goal stage

The position that rights-holders must not be sacrificed for the greatest amount of good for the greatest number shapes the legitimate goal stage. In our view, this disallows infringing constitutional rights for the purposes of promoting the common welfare if we understand the common welfare as an aggregate of individual people's interests. Permitting this would allow rights-holders to be sacrificed for the purposes of the good of the greatest number. Hence, for the legitimate goal stage not to boil down to utilitarianism, the only legitimate goal which may justify infringing a constitutional right is another constitutional right.

To be sure, legitimate goals are often fashioned in terms of generalized public interests, not constitutional rights. In the Indian Constitution, for example, it is possible to limit freedom of speech and expression 'in the interests of the sovereignty and integrity of India, the security of the State, friendly relations with foreign States, public order, decency or morality, or in relation to contempt of court, defamation or incitement to an offence'.[45] Understood broadly, these purposes, however laudable, do not seem to take the notion of rights seriously enough. In fact, as Webber has argued, allowing rights to be infringed for the pursuit of these public interests would result in a 'loss of rights'.[46] Webber is

[44] We distinguish between 'due concern and respect' and the idea of 'equal concern and respect'. We adopt the former in Section 2.D but reject the latter in Section 3.B. We make this distinction to distance ourselves from the Dworkinian view, championed by Letsas, that there are only absolute rights. We consider this view in Section 3.B below.

[45] Constitution of India, Art 19(2).

[46] Webber (n 16).

not exaggerating his case. In fact, some scholars, chiefly among them Möller, have argued that 'rights do not seem to enjoy any special or elevated status over public interests, but rather operate on the same plane as policy considerations'.[47] However, if this reading of Möller is correct, then it is not clear how the proportionality test could prevent individual rights from being sacrificed, in utilitarian fashion, for the general welfare.[48]

To maintain the importance of rights, the only legitimate purpose which may justify their limitation under the legitimate goal stage is the protection of others' rights. One could object that this stance flies in the face of widespread legal practice. After all, constitutional courts around the globe allow a wide range of public interests to justify the limitation of rights. Taking the Indian example, protecting public order seems to be a sufficiently important interest used to justify interfering with freedom of expression without making rights subject to utilitarian trade-offs. However, our view is compatible with this practice. This is because, when constitutional documents and judges talk about the public interest, this can generally be interpreted as being, directly or indirectly, about other people's constitutional rights.

Take, for example, people's interest in public order which the Indian Constitution allows as a limitation to freedom of expression. Public order here can be understood as an abstract way of talking about a combination of the different rights of a number of people, such as their right to the peaceful enjoyment of their property or their right to bodily integrity. Accordingly, freedom of expression may conflict with these rights in cases where the expression consists of acts of vandalism against private property and risks endangering other people.

To be sure, not all public interests may be reconstructed as aggregated rights. For example, it is unclear how 'friendly relations with foreign States' could be boiled down to constitutional rights. Perhaps, what is indirectly involved here is a right to bodily integrity of citizens which would be violated if a state was to go to war with another. However, if this was the focus, then 'friendly relations with foreign States' would be better reconstructed under the heading of 'public order'. Under our approach, then, it seems that certain limitation clauses of current anthropocentric constitutions would not be legitimate under a sentience-based constitution unless they can be reconstructed, either directly or indirectly, as involving constitutional rights. Still, because such

[47] Möller (n 14) 15. See also David Beatty, *The Ultimate Rule of Law* (OUP 2005) 171.

[48] Below we consider further Möller's theory. In particular, we explore the view that he is in fact committed to the position that the public interest is also a matter of constitutional rights (or autonomy interests, as he would put it).

reconstruction is often possible, our approach is compatible with widespread constitutional practice.

Our argument that only a constitutional right can permit the infringement of another constitutional right needs to be further qualified. In a sentience-based constitution, the recognition of certain constitutional rights which would totally extinguish the rights of sentient animals would be illegitimate. Certain anthropocentric constitutions, such as Ecuador's, have recognized that sentient animals have the constitutional rights to life and to physical integrity, which they derive from the constitutional recognition of the rights of nature.[49] However, although the Ecuadorian Constitutional Court found that wild animals have these individual rights, it also held that domesticated animals could be legitimately exploited by humans for food, leisure, and other purposes.[50] It held these exploitative practices to be legitimate based on Article 74 of the Ecuadorian Constitution which recognizes that 'people, communities, nations, and peoples have the right to benefit from the environment and natural wealth'.

Accordingly, even though the Court justified limiting animal rights based on another constitutional right, its interpretation of Article 74 effectively eviscerated domesticated animals' right to life and to physical integrity. Such an interpretation of Article 74 would not be possible under a sentience-based constitution, especially given that it would otherwise allow using sentient animals as mere means for the purposes of others. The Non-Instrumentalization Principle, therefore, needs to be considered in both a formal and a substantive way. In its formal dimension, the principle requires that only a constitutional right can justify infringing another constitutional right so that individuals are not sacrificed for the common welfare. In its substantive dimension, it prohibits granting constitutional right status to legal instruments that would authorize the instrumentalization of sentient beings.

While we have explained our view that only constitutional rights can justify infringing another constitutional right (the formal dimension of the Non-Instrumentalization Principle), there might still be an objection to this view. The objection says that we do not take the public interest seriously enough. Call this the 'Public Interest Matters' objection. To understand this objection, we need to consider Möller's morally laden theory of proportionality. Möller is

[49] 'Estrellita', Constitutional Court, no. 253-20-JH/22, 27 January 2022, [90–99] <https://animal.law.harvard.edu/wp-content/uploads/Sentencia_253-20-JH-20_Derechos_Naturaleza_y_Animales.pdf> accessed 26 October 2024.

[50] ibid [106–10]. The Court found that wild animals have different sets of rights which narrow the scope of activities through which humans can legitimately exploit wild animals. ibid [111–21].

sometimes believed to give up on the importance of rights and to readily sacrifice them for the public interest.[51] But, in fact, the more accurate picture is that Möller does not downgrade constitutional rights so that they inhabit the plane of policy considerations. Rather, he elevates policy considerations so that they become matters of constitutional rights.

Indeed, he has defended the view that both constitutional rights and the public interest derive from a super-right to personal autonomy, which grants individuals whatever they need to be able to freely direct their lives.[52] In his view, the role of proportionality is to specify 'the spheres of autonomy of [constitutional subjects] in a way which respects their status as equals'.[53] Building on Möller's theory, the Public Interest Matters objection can be phrased in the following manner: the view that only constitutional rights can justify infringing another constitutional right does not sufficiently acknowledge that both constitutional rights and the public interest are about advancing personal autonomy. The public interest should be capable of justifying infringing constitutional rights because both are manifestations of a super-right to personal autonomy.

What should we make of the Public Interest Matters objection? We should reject it because the idea of a super-right to personal autonomy is under-inclusive of sentient beings, human and non-human alike, who can have constitutional rights even if they do not have autonomy interests. As we discussed in the previous chapter, constitutional rights practice has long recognized a range of rights which have nothing to do with the personal autonomy of the rights-holder. For example, in contemporary human rights practice, whether at constitutional or international level, individuals with severe cognitive disabilities who might lack autonomy interests because they do not have the cognitive capacities to make choices for themselves still possess fundamental rights.[54] In a similar vein, some constitutions recognize that corporations have fundamental rights despite themselves lacking the capacity to have autonomy interests in any straightforward sense. For instance, as we saw in Chapter 3, the US Supreme Court protects corporations' right to due process,[55] free speech,[56] and equal protection.[57]

[51] Webber (n 16).

[52] Möller (n 14) 58, 84. See also Kai Möller, 'Proportionality and Rights Inflation' in Bradley W Miller, Grant Huscroft, and Grégoire Webber (eds), *Proportionality and the Rule of Law: Rights, Justification, Reasoning* (CUP 2014) 156.

[53] Möller (n 14) 112.

[54] See the discussion in Chapter 3 Section 1.

[55] *Minneapolis & St Louis Ry Co v Beckwith* 129 US 26, 28 (1889).

[56] *Citizens United v Federal Election Commission* 558 US 310, 347 (2010).

[57] See *Metro Life Ins Co v Ward* 470 US 869, 881 n.9 (1985); *Santa Clara Cnty v S Pac RR Co* 118 US 394, 396 (1886).

Furthermore, and importantly, this autonomy infused version of the Public Interest Matters objection is not morally attractive. This is because its implication is that only paradigmatic humans, because they have the capacity for personal autonomy, possess rights and interests. As we have explained at various junctures now, this view is unduly anthropocentric and excludes, without a compelling reason, sentient animals who possess not only interests but also constitutional rights.[58] We can therefore reject the Public Interest Matters objection and, with it, Möller's moral reading of the proportionality test as a way of specifying the spheres of autonomy of constitutional subjects. Accordingly, we can insist that the Non-Instrumentalization Principle correctly requires that, under the legitimate goal stage, only constitutional rights can justify infringing another constitutional right.

B) The suitability stage

The Non-Instrumentalization Principle also clarifies what is at stake in the suitability stage of the test. Here, the government is asked to prove that the means taken to interfere with a constitutional right are suitable for the achievement of an identified legitimate goal. Clearly, if an interference with a constitutional right does not pursue the protection of another right, then an individual is being sacrificed for no compelling reason.

Take, for example, the killing in the 1950s of 40,000 wild elephants in what was then Rhodesia. The killing occurred based on the mistaken belief that the herd movements and grazing by these animals significantly contributed to the desertification of their habitats, a view that was propagated by Allan Savory, a land degradation expert.[59] We can assume that desertification would have caused a decrease in food sources for elephants and other species, resulting in the loss of animal life. On this assumption, the killing of 40,000 elephants would have been suitable to protect the right to life of various species, including other elephants. However, as Savory later publicly admitted,[60] the killing did

[58] See especially the more robust rights that wild animals have in comparison to domesticated animals in Ecuador. Constitutional Court, no 253-20-JH/22 (*Estrellita*) (n 49) [111–21].

[59] Colin Sullivan, 'Can Livestock Grazing Stop Desertification?' *Scientific American* (5 March 2013) https://www.scientificamerican.com/article/can-livestock-grazing-stop-desertification/ accessed 4 November 2024. Some have argued that the desertification was instead caused by fires in the dry season. See RM Lawton and Mary Gough, 'Elephants or Fire—Which to Blame?' (1970) 10 Oryx 244.

[60] Allan Savory, 'How to Fight Desertification and Reverse Climate Change' (*Ted Talk*, February 2013) <https://www.ted.com/talks/allan_savory_how_to_fight_desertification_and_reverse_climate_change> accessed 4 November 2024.

not prevent desertification. In fact, it accelerated it. If this is right, then the 40,000 elephants would have been needlessly killed. Consequently, the means chosen—killing the elephants—was not suitable to achieving the legitimate objective, that is, protecting the right to life of different species. It therefore impermissibly instrumentalized the 40,000 elephants without achieving a legitimate goal.[61]

C) The necessity stage

As we saw, at the necessity stage, a court is required to determine whether the means chosen to pursue the legitimate goal are minimally impairing of the right that is being interfered with. If this question is answered in the negative, then the rights of a constitutional subject have been needlessly violated. As we argued in Chapter 2, all constitutional subjects have constitutional worth, and this worth is respected by governments among other things when they respect the rights of these subjects. Governments fail to do so when they interfere with these rights more than it is necessary.

Take the Indian Supreme Court case *Animal Welfare Board of India v A Nagaraja* as an example. In that case, the Court stated that Article 21 of the Indian Constitution, which protects life and personal liberty, grants animals 'a right to life and security, subject to the law of the land'.[62] In the same sentence, however, the Court also said that this right includes the deprivation of the life of animal species out of human necessity.[63] According to the Court, human necessity includes 'dehorning of cattle, castration, branding, and nose-roping of any animal; killing of stray dogs; killing of any animal for any legally sanctioned reason; killing of animals for food; and subjecting animals to scientific experiments'.[64] In a sentience-based constitution, few of these 'human necessities' would be permissible as goals for limiting animal rights, given that they are not matters of constitutional rights. For the sake of argument, however, let us assume that killing animals for food or for medical experiments is in

[61] It should go without saying that, even if the killing of the elephants had been suitable for the aim of protecting the right to life of other animals, this does not mean that the killing would have been permissible overall under the proportionality test. Also, as we discuss below in Section 3.A, the state should take any reasonable steps available to it to avoid collisions between rights.

[62] *Animal Welfare Board of India v A Nagaraja* (2014) 7 SCC 547. This statement has been narrowly reinterpreted in *Animal Welfare Board of India et al v Union of India et al* [2023] Writ Petition (Civil) No 23 of 2016 (Supreme Court of India) [24]. The Supreme Court now considers the earlier statements that animals enjoy constitutional rights merely as 'a judicial suggestion'.

[63] *Animal Welfare Board of India v A Nagaraja* (n 62) [62].

[64] ibid 32–33.

pursuit of the human right to life, given the need for nutrition and safe drugs. In a sentience-based constitution, killing animals for such purposes would still be impermissible in almost all circumstances because of the availability of less intrusive alternatives, such as plant-based foods for nutrition, and computer models and volunteer studies for drug testing. Accordingly, killing animals cannot be considered minimally impairing since there are less intrusive ways of achieving the same legitimate end of safeguarding the human right to life via adequate nutrition and testing.

A sceptic might object that, while it is true that killing animals for food and drug testing is not strictly necessary to preserve human life, the alternatives we have identified are not as effective at achieving these purposes. This objection fails for two reasons. First, as we have explained in the previous chapter,[65] it is now well established that humans can thrive without consuming animal foods. To be sure, in certain countries that experience famine or are in economic disarray, it may be impossible to avoid animal products for sustenance. We consider this issue in more detail below.[66] Nevertheless, in many circumstances, humans can thrive without relying on animal foods.[67] Similarly, while experimentation on animals has arguably produced some benefits in terms of drug safety, we have shown in the previous chapter that a more reliable method for drug testing would be human testing.[68] The sceptic who is concerned solely with the most effective means of drug testing would therefore have to prioritize human testing over animal testing.

D) The balancing stage

The second reason why the sceptic's objection fails is because, even if killing animals for food and drug testing was more effective than its alternatives, doing so would nevertheless be impermissible in most circumstances, at the balancing stage. This is the final stage of the proportionality test, where the question is asked whether the government action strikes a fair balance between the conflicting rights. No such fair balance exists if one of the rights was

[65] Chapter 3 Subsection 3.A.

[66] See Subsection 3.A of this chapter.

[67] Marco Springmann and others, 'Health and Nutritional Aspects of Sustainable Diet Strategies and Their Association with Environmental Impacts: A Global Modelling Analysis with Country-Level Detail' (2018) 2 The Lancet Planetary Health e451.

[68] Aysha Akhtar, 'The Flaws and Human Harms of Animal Experimentation' (2015) 24 Cambridge Quarterly of Healthcare Ethics 407. Discussed in Section 3.A of Chapter 4. See also Richard J Miller, *The Rise and Fall of Animal Experimentation* (OUP 2023).

completely extinguished for a gain that is not strictly required for the other right. Hence, even if killing animals for food and drug testing was more effective than alternatives such as plant-based foods for nutrition and computer models and volunteer studies for drug testing, animals' right to life would be completely extinguished for a mere improvement in humans' right to life. In our view, such prioritizations of the right to life of humans over the right to life of animals would be impermissible. This is because, in a sentience-based constitution, an automatic prioritization of human interests is incompatible with the idea that every sentient being has equal constitutional worth and rights.

Recall that some, such as Alexy, argue that balancing is merely a formal principle in need of being supplemented by external moral principles. In our view, balancing requires the state to consider whose rights to prioritize by reference to the constitutive moral values of the constitutional system. We have already considered some of these values, including the view—inherent to the proportionality test—that sentient beings cannot be merely instrumentalized and the related view that all sentient beings have equal constitutional worth. Recall from our discussion in Chapter 2 that constitutional worth entails not only a commitment to the rights of governed sentient beings, but it demands due concern for them more generally. In the context of the proportionality test, we take this to mean that all sentient beings need to be respected for who they are, based on their specific interests and capacities.[69] This idea of due concern and respect can help us make decisions about whose rights to prioritize when we need to balance them.[70]

Consider the scenario with which we began this chapter. Was it right for the UK Government to have (allegedly) prioritized the evacuation of animals over that of humans from Kabul?[71] Were animal lives illegitimately put before human lives? In a sentience-based constitution, it is impermissible for state actors to value one sentient being more than another. Accordingly, the UK Government would have acted impermissibly if, as alleged, they had approved the flight. This is not because human life should be automatically prioritized over animal life. Rather, it is because it is arguable that humans had stronger interests in being rescued from the Taliban in the given circumstances. After

[69] This would be a form of recognition respect in the terminology of Stephen L Darwall, 'Two Kinds of Respect' (1977) 88 Ethics 36.

[70] In Subsection 3.A below, we consider Letsas' reading of the idea of 'equal concern and respect' which we differentiate from our idea of 'due concern and respect'.

[71] As stated in the Introduction, this version of events is disputed. The charity maintains that the animals were stored in the cargo hold of an empty aircraft and that the UK government had refused to fill the flight's empty seats.

all, it is not clear that, compared to the lives of humans who collaborated with the old regime, the lives of the animals were any more at risk under the Taliban regime than the previous regime. To be sure, dogs are sometimes considered impure under certain interpretations of Islam.[72] But, as investigation by *The Economist* concerning stray animals in Afghanistan attests, it is doubtful whether the lives of stray or domesticated animals were more at risk simply because they now lived under the Taliban. In fact, *The Economist* reports that the Taliban have encouraged animal charities, including Nowzad, to continue their work of rescuing stray animals in Kabul.[73] By contrast, humans who collaborated with the US and British armies and who are still in Afghanistan continue to fear for their lives and to seek asylum abroad.[74]

Furthermore, the rescued animals are unlikely to be aware of the political circumstances from which they were taken. They are unlikely to ever know about the illiberalism of the Taliban and the different political system that is now in place.[75] To be sure, they will be aware of the more immediate circumstances in which they find themselves. Perhaps being relocated to Britain has allowed them to be adopted by a well-resourced family rather than to continue to be housed in the more limited facilities of Nowzad in Kabul. In such circumstances, they will naturally be aware of the new people and places with which they interact. Nevertheless, had the animals remained in Afghanistan, they would likely not have sufferered from psychological anguish due to the fact of a change in the Afghan regime. This is different for many of the humans who did not manage to escape. Their awareness of the change in regime, in particular the fear that they will be harmed or killed by the Taliban, is part of what is causing their anguish.[76]

[72] Alan Mikhail, 'Dogs in Ancient Islamic Culture' (*Oxford University Press Blog*, 13 July 2017) <https://blog.oup.com/2017/07/dogs-ancient-islamic-culture/> accessed 4 November 2024.

[73] 'The Taliban Go Big on Animal Welfare' (*The Economist*, 18 May 2023) <https://www.economist.com/asia/2023/05/18/the-taliban-go-big-on-animal-welfare> accessed 4 November 2024.

[74] Holly Bancroft, 'Abandoned by UK: Afghans with Taliban Targets on Their Backs Denied Sanctuary' *The Independent* (4 June 2023) <https://www.independent.co.uk/news/uk/home-news/afghans-british-army-arap-rejected-taliban-b2348067.html> accessed 4 November 2024.

[75] As we explore in Chapter 6, we believe that due concern for all governed beings entails living in a constitutional democracy with an array of rights of political participation which apply to animals, even though they may not be aware of them. We acknowledge that living in an illiberal theocracy under the Taliban would violate these rights. The arguments we make in this section need to be read in light of those in Chapter 6.

[76] Of course, there are humans, such as infants, who are not aware of such a change of circumstances either. Nevertheless, they are in a different situation to the Nowzad animals because the Taliban regime is targeting them and their families due to their families' collaboration with the UK and US armies. As far as we are aware, no animals were specifically targeted by the Taliban because of their military involvement.

It seems to us then that this case called for the prioritization of the humans over the dogs and cats. This is not because the humans deserved automatic prioritization over the animals—something that would have been incompatible with animals' equal constitutional worth. Rather, the constitutional value of respect for constitutional subjects demands giving due concern to the weight of their respective interests and the relevance of their capacities when determining whose rights should be prioritized in the circumstances.

We acknowledge that others may reach a different conclusion than us on how the Nowzad case should have been resolved. Contrary to what some scholars claim,[77] applying the proportionality test is not a mechanistic and value-free exercise. Reasonable people may place greater weight on some considerations over others. They might say, for example, that the Nowzad plane was the only plane devoted to rescuing animals. If animals matter morally, then it is right that at least one plane should be used to rescue them. Our aim here is not to convince everyone that the scenario should be decided the way we propose. Rather, our aim is to show that our theory of the proportionality test can be helpful in structuring the moral enquiry involved in conflicts of rights. Our theory can also reveal the values we should appeal to (eg due concern and respect) in reaching any conclusion.

3) Countering objections against balancing in the proportionality test

What if a sceptic not only objected to how we resolved the Nowzad scenario but took aim more fundamentally at the usefulness of the proportionality test as such? In this section, we seek to counter objections levelled against the test. All the objections considered here are aimed against the balancing stage of the test. Indeed, this stage is the one that finds itself in the critics' crosshairs most often, perhaps because it involves a more open-ended enquiry. We focus on two objections: that balancing requires comparing incommensurable values (Subsection A), and that balancing abandons the idea of rights as trumps (Subsection B). As we shall see, the Non-Instrumentalization Principle will help us to counter both these objections.

[77] See the discussion in Section 3.A below.

A) Balancing and value-incommensurability

Stavros Tsakyrakis has argued that balancing is often portrayed as 'objective, neutral, and totally extraneous to any moral reasoning'.[78] However, he argues that this is impossible given that balancing

> risks neglecting the complexity of moral evaluation and, especially, the complexity of rights. More specifically, it tends to overlook . . . the fact that our moral universe includes ideas not amenable to quantification, with the result that these ideas are not given due consideration in our reasoning.[79]

Yet, this way of portraying balancing is off the mark. Not only does balancing require moral reasoning, as the Nowzad scenario just discussed shows, but it also requires reasoning based on the moral values that are foundational to the constitutional system. In the context of sentience-based constitutionalism, these values include the Non-Instrumentalization Principle as well as due concern and respect for the constitutional subject. This is a nuanced exercise that is neither mechanistic nor straightforwardly numerical.

Once we accept that the balancing stage involves appeal to moral values, another objection might arise concerning the fact that balancing requires weighing incommensurable values. According to Timothy Endicott, this is for two reasons. First, he says that balancing requires weighing incommensurable values because the interests protected by rights cannot be independently weighed.[80] As he puts it, there is no 'scale of measurement for the gravity' of the interference with one right.[81] However, it is unclear whether this is true. As Endicott himself admits, it is possible for interests 'to be critically important or somewhat serious or minor or trivial'.[82] For example, in the Nowzad scenario, it seems clear that the interests in life of the humans who were left behind were critically important because, unlike the animals, they were at risk of being killed. To be sure, any kind of numerical quantification of the gravity of an interference will be merely a tool to convey the same idea of importance. Accordingly, 'scale of measurement' is only metaphorical speech. But if the crux of this version of the incommensurability objection is that the metaphor of balancing is misleading because it suggests exact numerical measurements,

[78] Tsakyrakis (n 16) 474.
[79] ibid 475.
[80] Endicott (n 18) 316.
[81] ibid.
[82] ibid.

then we can easily use another metaphor, or none. The metaphor of balancing is only in place to suggest that we should prioritize the more important interests of a constitutional subject over the less important ones of another.

The second part of Endicott's critique represents a more serious challenge. According to Endicott, the interests at stake in rights conflicts are radically different so that one cannot rationally chose between one and the other. However, this need not be so. The nature of the interests at stake may be identical as the Nowzad scenario shows. What was at stake there was the interest in life of humans over the interest in life of cats and dogs. Given that the right to life was at issue for both, we can rationally choose whose right to life to prioritize based on the considerations we have outlined above. For example, as we have said, the humans who collaborated with the Western troops were in imminent mortal danger, whereas no similar threat to life existed for the animals.

But even if we assumed that the interests at stake were different, there would still be rational ways to choose whom to prioritize. Say, for example, that the choice was between giving the space on the plane to humans who would otherwise be killed by the Taliban and giving it to the animals who would have received a chance to be rehomed in a facility that is more comfortable than the Nowzad Kabul facilities. Here, the conflict would be between an interest in life for the humans and an interest in better housing for the animals. In such circumstances, one would have to prioritize the former, given that life is a prerequisite for enjoying other interests. In contrast to what Endicott suggests, this would be a rational choice, and perhaps even an obligatory one.

More generally, the reason why incommensurability is not a fatal objection to balancing, at least in our theory, is because we accept that only rights can justify interferences with other rights. In contrast to Endicott,[83] we do not accept that generalized public interests can justify interferences with rights. Given that we adopt an interest theory of rights, it is possible in many cases to determine whose rights need to be prioritized. This is because an interest theory of rights is based on the view that interests are components of a subject's well-being and, consequently, it is often possible to determine whose well-being is more seriously encroached upon in cases of conflicts of rights. Given the idea of equal constitutional worth in a sentience-based constitution, which demands due concern and respect for constitutional subjects, priority ought to be given to the subject whose well-being is more seriously encroached upon. Accordingly, choice in these circumstances is not arbitrary but instead dictated by constitutional values.

[83] ibid 313–14.

However, there will likely remain situations where there appears to be no rational choice that can be made. This is especially the case in scenarios of mortal conflicts. Consider the conflict between the rights to life of prey animals and obligate carnivores. Or take the conflict between the right to life of humans who cannot choose to forgo meat because of a famine and the right to life of the animals who are eaten by these humans. In such scenarios, no prioritization is possible because the facts dictate that one life will be extinguished if the other is not. As we saw, the requirement to show due concern and respect prohibits giving automatic prioritization to one constitutional subject over another. This leads to a stalemate where neither right can be prioritized without extinguishing the other.

To be sure, if the facts can be changed, such as if steps can be undertaken to eliminate the famine for humans or to provide faux meats to obligate carnivores, then this must be done to prevent the conflict from arising in the first place. Nevertheless, there will be situations where such steps are either impossible or extremely difficult. None of this, however, means that balancing is irrational because of the problem of incommensurability. Rather, it means that sometimes there are no rational solutions to moral conflicts. Governments, in these scenarios, could not be held responsible for failing to protect the rights of their constitutional subjects.

B) Balancing and rights as trumps

Another objection to balancing is that it betrays the idea of rights as trumps. This conception of rights, made famous by Ronald Dworkin,[84] has been applied to the proportionality test by Letsas.[85] To better understand this objection, we need to understand Letsas' own moral reading of the test. Letsas bases his theory on a more general understanding of the nature of rights. Unlike us, he rejects interest theories of rights in favour of what he calls a 'reason-blocking theory of right'. In this theory,

> the most fundamental moral right individuals have against the government, from which more specific requirements follow, is the right to be treated with equal respect and concern, in Ronald Dworkin's famous formulation . . . rights

[84] Ronald Dworkin, *Taking Rights Seriously* (Harvard UP 1978) xi.

[85] George Letsas, *A Theory of Interpretation of the European Convention on Human Rights* (OUP 2007) ch 5; Letsas, 'Rescuing Proportionality' (n 14); Letsas, 'Proportionality as Fittingness' (n 4).

> tell us which facts do not constitute valid reasons for depriving someone of a liberty or an opportunity or for imposing a risk on him.[86]

While Letsas' commitment to equal concern and respect may appear similar to our idea of due concern and respect, it is in fact set in a different context and has a different meaning. As we will see shortly, his idea of equal concern and respect actually excludes animals and non-paradigmatic humans from being considered subject of rights. In the context of constitutional rights, Letsas says that the proportionality test requires that government officials comply with the specific reasons that bear on the officials' legitimacy, which is 'to show equal concern for the fate of every person over whom [they claim] dominion . . . [and to] respect fully the responsibility and right of each person to decide for himself how to make something valuable of his life'.[87] On Letsas' theory, the demands of the proportionality test are not, then, about deciding how to prioritize conflicting rights. They are more broadly about ascertaining whether a government has respected the most fundamental right to equal concern and respect. In particular, government agents are barred from infringing people's constitutional rights on the basis of a judgment that

> our plan of life is impoverished or immoral; that we are despised by the majority; that our views shock and offend others; that we should be forced to lead a particular kind of life; and that our life is worth less than the life of others.[88]

On this view, constitutional rights are absolute rather than subject to balancing. Or, to put it in Dworkin's words, rights are trumps.[89] Letsas explains it as follows:

> Rights thus understood are absolute: it can never become justified for the government to restrict my liberty for the reasons just mentioned. When these reasons are present, balancing is morally inappropriate. When these reasons are absent, we should not say that we have a right which is not absolute and whose limitation is justified. Rather, we should say that we had no right in the first place.[90]

[86] Letsas, 'Rescuing Proportionality' (n 14) 329.
[87] Ronald Dworkin, *Justice for Hedgehogs* (Harvard UP 2013) 2.
[88] Letsas, *A Theory of Interpretation of the European Convention on Human Rights* (n 85) 117.
[89] Dworkin (n 84) xi.
[90] Letsas, *A Theory of Interpretation of the European Convention on Human Rights* (n 85) 117.

What are we to make of Letsas' moral reading of proportionality? Is accepting, as we do, the balancing stage a betrayal of the very idea of rights? Our view is that we ought not to accept the theory of rights as trumps for three reasons. First, rejecting the balancing stage altogether contradicts how proportionality is widely used in constitutional adjudication. As we mentioned above, the proportionality test is used by judges globally to balance conflicting non-absolute rights. On the view of rights as trumps, all these judges would be mistaken about what they are doing.[91] But that would go against what we called the ecumenical view of proportionality, which gives us a strong presumptive reason to reject this view.

The idea of rights as trumps also faces a second issue: the blocking theory of rights can easily be incorporated into the legitimate goal stage of the ecumenical understanding of proportionality.[92] As we have interpreted that stage, governments can interfere with a constitutional right only on the basis of protecting another constitutional right. Accordingly, the kinds of unacceptable reasons ruled out by the reason-blocking theory would be filtered out at this stage. Consider, for example, the unacceptable reason that a particular way of life is despised by the majority (say same-sex partnerships in a sexually conservative regime). There is no moral right to despise minority groups.[93] Accordingly, if a government sought to infringe a constitutional right (eg the right to non-discrimination) on the basis that a group is despised by the majority, there would be no need to go beyond the legitimate goal stage to declare such governmental action impermissible because it is based on an illegitimate goal. Therefore, there is no need to deny the work done in the balancing stage when the main thrust of the idea of rights as trumps can be incorporated into the legitimate goal stage.

The final and most important reason for rejecting the idea of rights as trumps is that it is paired up with an overly restrictive conception of the nature of the constitutional subject: it is focused on paradigmatic humans and leaves out non-paradigmatic humans and animals. Following Dworkin, Letsas thinks that the only general right that constitutional subjects have is the right against the government disrespecting their responsibility to decide for themselves how to make something valuable out of their life.[94] If one has a responsibility to

[91] Letsas, 'Rescuing Proportionality' (n 14) 326. He says that 'judges and constitutional lawyers are not moral philosophers and we should be charitable towards what they say'.

[92] This is a view also held by Möller. See Möller (n 14) 183–93.

[93] For a defence of this view in the context of conscientious objection to same-sex marriage, see John Adenitire, *A General Right to Conscientious Exemption: Beyond Religious Privilege* (CUP 2020) ch 9.

[94] Letsas, *A Theory of Interpretation of the European Convention on Human Rights* (n 85) 117.

decide one's own life plan, one also needs to be capable of making and undertaking such plans by, for example, rationally choosing between different alternative plans and then pursuing them by acting on them. Strictly speaking, then, only paradigmatic humans can have rights on Letsas' account because only they have the required capacities which government ought to respect. As discussed in previous chapters, this view is morally unattractive and incompatible with sentience-based constitutionalism. Hence, it should be rejected.

To avoid the anthropocentrism of the view of rights as trumps, it is preferable to interpret the demand for due concern and respect as entailing that governments ought to pay due concern to the interests of all sentient beings and that governments ought to respect sentient beings for the kind of beings they are. Some of them will be rational and fully autonomous beings who can take moral and legal responsibility for their own lives. Others cannot be held responsible for all their actions, but they nevertheless have rights which governments must safeguard. This more inclusive interpretation of due concern and respect avoids the shortcomings of the view of rights as trumps.

In conclusion, we have not found that the idea of rights as trumps offers any cogent reason to abandon the use of the balancing stage. On the contrary, this view unnecessarily dismisses the ecumenical view of the proportionality test because its main concern—prohibiting the government from acting out of a negative moral judgment about constitutional subjects' way of life—can be incorporated within the legitimate goal stage. Furthermore, because this insistence about the ethical independence of constitutional subjects betrays an anthropocentric bias, we ought to reject it as particularly unsuitable for a sentience-based constitution. Under a sentience-based constitution, both humans and animals can have rights that can come into conflict. Balancing will therefore often be necessary. But balancing the rights of humans and animals can be conducted in a morally attractive way when done in accordance with the Non-Instrumentalization Principle and with due concern and respect for all constitutional subjects.

Conclusion

In this chapter, we have argued that the proportionality test should be used to adjudicate rights conflicts in a sentience-based constitution. To reach that conclusion, we argued that the various stages of the proportionality test need to be read in conformity with the Non-Instrumentalization Principle. According to this principle, both human and non-human holders of rights must not be used as mere means for others' purposes. The proportionality test, therefore, forbids utilitarian trade-offs that allow sacrificing sentient beings for the greater good. Hence, the proportionality test is not neutral about different moral frameworks as it is based on a Kantian view of rights.

We then illustrated the implications of this moral reading of the proportionality test for a constitution that is fit for all sentient beings. First, it entails that only other rights can permissibly be grounds for restricting a constitutional right. The public interest can permissibly be used as a ground to restrict a constitutional right only if that public interest is an indirect way of recognizing constitutional rights. Furthermore, we saw that the Non-Instrumentalization Principle prohibits granting constitutional right status to legal instruments that would authorize the instrumentalization of sentient beings. Our account upholds the normative significance of rights and, importantly, avoids sacrificing rights-holders for the good of the greatest number.

As we showed, both the suitability and necessity stages of the proportionality test need to be read in conformity with the Non-Instrumentalization Principle. This means that measures that restrict a fundamental right need to be suitable and necessary for promoting another constitutional right (whether directly or indirectly) so that no constitutional subject is unduly instrumentalized. We also showed that our moral reading of the proportionality test requires that the balancing stage be read in light of constitutional values, especially the idea of due concern and respect for all sentient beings. This prohibits that the rights of a class of sentient beings (eg humans) are automatically prioritized over the rights of other beings (eg other sentient animals). Rather, a context-sensitive analysis of the relevant interests and capacities needs to be undertaken to determine whose well-being is more heavily impacted by a specific governmental measure.

Finally, we showed how our analysis can help respond to certain objections about the usefulness of balancing. That stage of the proportionality test is not mechanistic and value-free. On the contrary, as we have argued, it requires deep engagement with the values to which a sentience-based constitution is committed, including the Non-Instrumentalization Principle and the idea of due concern and respect for the constitutional subject. While not all conflicts of rights can be resolved through proportionality reasoning—especially mortal conflicts—many conflicts can be rationally resolved through balancing. We also rejected the Dworkinian theory that balancing is inappropriate because rights are trumps. While rights do inhabit an elevated plane of normative importance, they are often not absolute. We found that this theory is also linked to an unattractive view of paradigmatic humans as the only kind of rights-holders. Accordingly, we ought to reject it in favour of the view that animals and non-paradigmatic humans are also legitimate rights-holders who should be treated with due concern and respect, and who should therefore not be treated as mere means for the benefit of others.

5
Rule of law

Introduction

In the case of *Jackson*,[1] English hunters challenged the validity of the Hunting Act 2004, which outlawed the traditional English sport of hunting wild mammals—typically foxes—with dogs on horseback. They claimed that the procedure used to enact the statute—involving only the House of Commons and not the House of Lords—rendered the Act invalid. The United Kingdom's highest court rejected the unconstitutionality claim and reasserted the principle of Parliamentary Sovereignty: the UK Parliament, as long as it respects established legislative procedures, can make and unmake any law whatsoever and courts cannot strike down those laws.[2]

However, worried about the extreme implications of Parliamentary Sovereignty, such as the UK Parliament willingly breaching fundamental rights, some of the judges sought to soften its reach. They suggested, in *obiter dicta*, that the rule of law imposes limits on the law-making power of the UK Parliament. Lord Hope, for example, said that the rule of law 'protects the individual from arbitrary government [and] is the ultimate controlling factor on which our constitution is based'.[3] Interestingly, however, none of the judges considered whether the 'individuals' who need protection from arbitrary government could include animals. In *Jackson*, there were at least three types of individuals hiding in plain sight: the horses whom hunters rode to chase their prey; the hounds used as an instrument of distress; and, perhaps most important of all, the foxes who were trying to flee the hunters and their hounds. Hence, even though *Jackson* was a case about hunting animals and the rule of law, none of the judges considered that animals might also require protection through the rule of law.

As we show in this chapter, this case is representative of what happens in constitutional theory more broadly. In all prevailing theories of the rule of

[1] *Regina (Jackson and others) v Attorney General* [2005] UKHL 56 (UK House of Lords).
[2] Albert V Dicey, *Introduction to the Study of the Law of the Constitution* (Roger E Michener ed, Liberty Fund 1982).
[3] *Regina (Jackson and others) v Attorney General* (n 1) [107].

Animals and the Constitution. John Olusegun Adenitire and Raffael Fasel, Oxford University Press.
 DOI: 10.1093/9780198910534.003.0005

law, the individual protected from arbitrariness by the rule of law is always the human individual. Specifically, we demonstrate that influential contemporary legal theorists have built their rule of law theories on the idea of a legal subject who is capable of understanding and following legal norms and who can be held responsible for breaches in compliance. Typically, on these theories, legal subjects are paradigmatic human beings. An exclusive focus on such paradigmatic humans may be understandable if one subscribes to the view that the basic idea of the rule of law[4] is that the law should be capable of guiding the actions of legal subjects. The law can only guide those who can understand and comply with legal norms, that is, rational beings. As we put it in Chapter 2, these beings are *subjected* to legal rules, rather than simply being governed or regulated by them.

This chapter will argue against this view of the rule of law. Building on the case for sentience-based constitutionalism developed in Part I of this book, we propose that judges and scholars should view the basic idea of the rule of law, in line with an important tradition in scholarship,[5] as restraining arbitrary power to protect those who can be victims of such power. The reason for this proposal is that paradigmatic human beings are not the only potential victims of arbitrary power. Non-paradigmatic human beings such as infants, young children, individuals with cognitive disabilities, and individuals who are senile can also be victims of arbitrary power. What is more, not only humans but sentient animals can become victims too. If the basic idea of the rule of law is to restrain arbitrary power, then any potential victims of such power, whether they are human or animal, ought to be protected by the rule of law. This chapter develops an inclusive account of the rule of law which can protect all governed human and sentient animals from arbitrary power.

To make our case, we show how three influential theories of the rule of law have been built on the idea of paradigmatic human beings. For this purpose, we follow the common classification of theories of the rule of law into formal, procedural, and substantive accounts.[6] Our focus will be on the accounts of

[4] This chapter uses the term 'basic idea of the rule of law' in the same way as Ronald Dworkin and Jeremy Waldron to indicate the most abstract formulation of a concept, which, once properly analysed by any particular theorist, results in different conceptions. 'Basic idea' is therefore meant to capture the uncontroversial formulation of a particular concept which different theorists should be able to agree on. However, as this chapter suggests, the basic idea of the rule of law adopted by many theorists is in fact highly contestable because it presupposes a certain kind of legal subject. On Dworkin's use of this method, see Ronald M Dworkin, *Law's Empire* (Hart Publishing 1998) 93. On Waldron's use, see Jeremy Waldron, 'Is the Rule of Law an Essentially Contested Concept (in Florida)?' (2002) 21 Law and Philosophy 137, pt VI.

[5] See below nn 33–37.

[6] See Brian Z Tamanaha, *On the Rule of Law: History, Politics, Theory* (CUP 2004). Jeremy Waldron, 'The Rule of Law and the Importance of Procedure' in James Fleming (ed), *Getting to the Rule of Law*

Lon Fuller, Jeremy Waldron, and TRS Allan as champions of each of these traditions. Of course, theirs are not the only influential rule of law theories of these types, and we could have chosen others to make our argument.[7] However, we believe that their accounts are illustrative of the wider pattern of constructing rule of law theories that are exclusively focused on paradigmatic human beings.

We show that a straightforward reading of the texts of these theorists reveals that their accounts do not include animals and non-paradigmatic humans. Sometimes, it is possible to charitably extract something valuable for animals and non-paradigmatic humans from these rule of law accounts. However, we argue that any value for animals and non-paradigmatic humans will be contingent on the existence of specific legal rules which protect them. Such legal rules may not exist and there is nothing in these accounts that calls for their existence. Consider, for example, animal protection laws which prohibit causing unnecessary suffering to animals. While these laws are widespread around the world, we saw in Chapter 1 that there are jurisdictions where these laws do not exist at all,[8] or where the laws apply only to an insignificant number of animals. In places like these, rule of law theories such as Fuller's, Waldron's, and Allan's which focus on paradigmatic humans will not be beneficial to animals. In countries where they do apply, animals may indirectly benefit from their theories, but only to the extent that the theorists insist that all laws should be properly enforced.

Once we move to inclusive accounts that explicitly recognize that the rule of law should protect both humans and animals from arbitrary power, animals will always be afforded some degree of protection by the rule of law. The extent of the protection will vary depending on the nature of the inclusive account. It will be weakest on an inclusive formal account, stronger on an inclusive procedural account, and the strongest on an inclusive substantive account that requires the recognition of a complete catalogue of fundamental rights for animals. On any inclusive account, however, the protection will never simply depend on the contingent existence of any specific laws.

In this chapter, we remain agnostic as to whether an inclusive theory of the rule of law should be formal, procedural, or substantive. To be sure, we have argued in Chapter 2 that a legitimate constitutional order gives due concern

(New York UP 2011). For criticism, see eg John Gardner, 'The Supposed Formality of the Rule of Law' in *Law as a Leap of Faith* (OUP 2012).

[7] See eg Joseph Raz, 'The Rule of Law and Its Virtue' in *The Authority of Law* (OUP 1979); Joseph Raz, 'The Law's Own Virtue' (2019) 39 Oxford Journal of Legal Studies 1; Tom Bingham, *The Rule of Law* (Penguin 2011).

[8] See also Raffael N Fasel and Sean C Butler, *Animal Rights Law* (Hart Publishing 2023) chs 1 and 7.

to the basic interests of sentient animals. In Chapter 3, we also argued that familiar catalogues of fundamental rights for humans should be interpreted in a sentience-based constitution to also protect animals. However, these arguments do not mean that one necessarily has to embrace a substantive account of the rule of law that legitimizes the constitutional order, and which recognizes fundamental rights for all governed sentient beings. This is because those who prefer formal or procedural rule of law accounts can justify sentience-based constitutional legitimacy and constitutional rights for animals independently from the rule of law.

The chapter proceeds as follows. In Section 1, we set out three anthropocentric theories of the rule of law—formal, procedural, and substantive—and show why and how these accounts exclude non-paradigmatic humans and animals from the rule of law's scope of protection. In Section 2, we build on these critiques and set forth our own inclusive formal, procedural, and substantive conceptions of the rule of law. However, we do not jettison totally the theories analysed in Section 1. Rather, we show how they should be rethought to be fully inclusive of all sentient animals, humans and animals alike.

1) Anthropocentric theories of the rule of law

This section critically examines three anthropocentric theories illustrating the three different rule of law approaches: Fuller's formal theory (Subsection A), Waldron's procedural theory (Subsection B), and Allan's substantive conception (Subsection C). In relation to each theory, we identify what the theorist takes to be the basic idea, requirements, and moral value of the rule of law.

A) Fuller's formal theory

Fuller's theory of the rule of law is one of the best known within the Anglophone world.[9] He famously held that the *basic idea* of the rule of law is 'to subject *human* conduct to the governance of rules'.[10] As he explicitly stated,

[9] It should be noted that Fuller refers to his own account as a procedural account. Lon L Fuller, *The Morality of Law* (Yale UP 1977) 96–97. However, by contrast to Waldron's account, his does not specifically focus on legal procedures which regulate the workings of courts. It is perhaps for this reason that his account is usually designated as a formal one in the literature.

[10] ibid 106 (emphasis added).

> legal morality . . . cannot be neutral in its view of *man* himself. To embark on the enterprise of subjecting *human conduct* to the governance of rules involves of necessity a commitment to the view that *man is, or can become, a responsible agent, capable of understanding and following rules, and answerable for his defaults.*[11]

As this quote makes clear, Fuller's basic idea of the rule of law does not just presuppose any kind of human subject. Rather, it is based on the idea of a human who is a rational agent.

This reveals that Fuller's account is doubly exclusionary. First, it is anthropocentric because it assumes that the rule of law is to be theorized from the perspective of a human person. However, as we have argued in Chapter 2, there is no good reason why the principles of constitutionalism, including the rule of law, should not pertain to animals. Indeed, as we show in Section 2 below, there is a compelling theory of the rule of law which is applicable to human and non-human animals alike. The second problem with Fuller's account is that it is exclusionary even within its anthropocentricism. This is because there are human beings who are not rational agents. Due to their age or disability, for example, some human beings cannot understand or follow legal norms and cannot therefore be held responsible for breaching legal norms.

Fuller's exclusion of animals and non-paradigmatic humans also permeates his conception of the requirements and moral value of the rule of law. Let us start with Fuller's *requirements* of the rule of law. In Chapter II of *The Morality of Law*, he argues that, given that the rule of law is to be understood as the enterprise of subjecting human conduct to the governance of rules, there are eight ways in which a system can fail to live up to its demands. To identify and illustrate these eight potential failures, Fuller famously employs the fable of an imaginary king, Rex, who wishes to set up a new legal system but fails in his attempts. This well-known fable will not be rehearsed here. Its lesson can be summarized as follows: the rule of law demands that laws be (1) general, (2) publicly promulgated, (3) clear, (4) non-contradictory, (5) possible to comply with, (6) relatively constant over time, (7) non-retroactive, and (8) that there be congruence between official action and declared norm.[12]

[11] ibid 162–63 (emphasis added).

[12] ibid 39. See also Kristen Rundle, 'Fuller's Internal Morality of Law' (2016) 11 Philosophy Compass 499, 500.

Fuller's requirements of the rule of law fall short of a theory that incorporates animals and non-paradigmatic humans. For example, it is not clear how Fuller's second demand (ie that laws should be publicly promulgated) serves their well-being. Because animals and some humans cannot comprehend legal norms, whether a particular law is publicly promulgated or is instead kept secret makes no difference *to them*. To be sure, it makes a considerable difference to those humans who can comprehend legal norms and can plan their lives on this basis. And if such laws happen to protect the interests of animals and non-paradigmatic humans, then public knowledge of these laws is of course also essential to protecting their interests. However, as explained in the Introduction to this chapter, any benefits that this account might yield for animals and non-paradigmatic humans are contingent on the existence of specific laws protecting their interests. Such laws may in fact not exist and their existence is not required by Fuller's account. By contrast, Fuller's account always affords some degree of due concern for paradigmatic humans. Public promulgation of the law, for example, serves their interest in planning their lives independently of whether or not such laws confer any more specific benefits. This illustrates how Fuller's theory is formulated to always provide some measure of protection for paradigmatic humans. Animals and non-paradigmatic humans, on the other hand, can only benefit from Fuller's account if the law already protects their interests. By comparison, a sentience-based approach, as we will show in Section 2.A below, will always provide some degree of protection from arbitrary power to humans and sentient non-humans.

Fuller's narrow focus is mirrored in what he takes to be the *moral value* of the rule of law. That value, he argues, is the exercise of political power in a way that demonstrates respect for *human agency and responsibility*, which he understands to be the basis of human dignity. Fuller's requirements of the rule of law set an ethos of political governance which limits the forms through which political power may be justifiably exercised. Political power that ignores the rule of law often takes the form of norms that are secret, retrospective, unclear, and so on. According to Fuller, such exercises of political power undermine the status of the legal subject as a responsible and autonomous agent who can be guided by norms and who is answerable for breaching them. That status, in Fuller's theory, is so highly valued that it sets the benchmark of legitimacy of the legal order. Only if political power is exercised in compliance with the rule of law's demands—thereby respecting human agency and responsibility—can those who hold political power legitimately demand allegiance to the law. In Fuller's words, those subject to law only have a duty of 'fidelity to law'[13] if political power sufficiently complies with the rule of law.[14]

It is not clear, however, why legal legitimacy should depend only on law's respect for the agency and responsibility of paradigmatic humans. We do not believe that there is a compelling theory of legal legitimacy that permits the mistreatment of non-paradigmatic humans. The Nazi regime, for example, persecuted both fully able-bodied and disabled individuals, including people with severe cognitive disabilities.[15] On our account, the Nazi regime was less legitimate (if it had any legitimacy at all) because it persecuted people with severe cognitive disabilities. Fuller's account is not able to accommodate this relatively uncontroversial position because of his argument that legal legitimacy depends only on respect for individuals capable of understanding and being guided by law.

It seems, then, that Fuller's exclusive focus on paradigmatic humans undermines his account of the moral value of the rule of law by making the conditions of legal legitimacy unduly narrow. This narrowness is the result of his problematic premise that the basic idea of the rule of law is to enable the governance of human conduct. As explained above, animals and non-paradigmatic humans will only contingently be protected under a conception built on this basic idea.

B) Waldron's procedural theory

Let us now turn to Waldron's procedural theory of the rule of law. As we show in the following, the same exclusionary presupposition that affects Fuller's formal theory also troubles Waldron's procedural theory.[16]

To begin, it is important to note that Waldron does not repudiate Fuller's formal version of the rule of law and his argument that the *basic idea* of the rule of law is the guidance of conduct.[17] Rather, Waldron argues that formal accounts need additional emphasis on the principles of natural justice, that is, principles requiring fair trials run by independent judges who are responsive to parties' legal arguments. According to Waldron, there is a close conceptual link between the idea of a legal system and a fully functional judicial system which

[13] Lon L Fuller, 'Positivism and Fidelity to Law—A Reply to Professor Hart' (1958) 71 Harvard Law Review 630, 631.

[14] Fuller (n 9) 39–41.

[15] The Wiener Holocaust Library, 'Disabled People—The Holocaust Explained' <https://www.theholocaustexplained.org/life-in-nazi-occupied-europe/oppression/disabled/> accessed 4 November 2024.

[16] Waldron (n 6). On whether Fuller is also a procedural theorist, see n 9 above.

[17] Waldron (n 6) 4.

respects the principles of natural justice. Most people, he believes, would not consider a system without judicial hearings and impartial proceedings to be a legal system at all.[18]

Based on what we have said so far, Waldron's theory could be seen as having some relevance for animals and non-paradigmatic humans. Animals, for example, can benefit from this account if judicial institutions impartially administer laws, such as animal protection laws, that aim to protect them. However, as noted, this benefit is contingent on the existence of such laws. Animal protection laws may not exist or may only apply to a narrow class of animals. Furthermore, there is nothing in Waldron's account that calls for their existence.

Crucially, however, it is not clear on Waldron's theory whether animals and non-paradigmatic humans can be part of the legal process and so whether his procedural account is even applicable to them. To understand this, we must analyse his position on the *requirements* of the rule of law. According to Waldron,

> no one should have any penalty, stigma, or serious loss imposed upon him- or herself by government except as the upshot of procedures that involve:
>
> A. A hearing by an impartial tribunal that is required to act on the basis of evidence and argument presented formally before it in relation to legal norms that govern the imposition of penalty, stigma, loss, and so forth;
> B. A legally trained judicial officer whose independence of other agencies of government is ensured;
> C. A right to representation by counsel and to the time and opportunity required to prepare a case;
> D. A right to be present at all critical stages of the proceeding;
> E. A right to confront witnesses against the detainee;
> F. A right to an assurance that the evidence presented by the government has been gathered in a properly supervised way;
> G. A right to present evidence on one's own behalf;
> H. A right to make legal argument about the bearing of the evidence and about the bearing of the various legal norms relevant to the case;
> I. A right to hear reasons from the tribunal when it reaches its decision that are responsive to the evidence and arguments presented before it; and
> J. Some right of appeal to a higher tribunal of a similar character.[19]

[18] ibid 13.
[19] ibid 6.

Are animals and non-paradigmatic humans counted among those who, according to Waldron, 'should [not] have any penalty, stigma, or serious loss imposed upon by government except as the upshot of procedures' that meet the above-mentioned requirements? It seems not. Consider requirements E-I. These are premised on the idea of an agent who can judge the propriety of the process, who has a view as to how their case ought to be argued, and who has an interest in understanding the reasons for the outcome. Animals and some humans do not straightforwardly have these capacities. To be sure, legal representatives could be appointed on their behalf to represent their interests. In the United Kingdom, for example, children in family courts can be independently represented by the Children and Family Court Advisory and Support Service.[20] Waldron's account, focused as it is on paradigmatic humans who can independently engage in the legal process, does not explicitly consider this issue. A sentience-based theory, as we show in Section 2, explains how animals can be empowered to participate in the legal process.

Waldron's anthropocentricism is most evident in how he conceptualizes the *moral value* of the rule of law. He explicitly follows Fuller when connecting the rule of law with respect for human dignity. As Waldron holds: 'I think what Fuller said about the connection between his formal principles and dignity can be said even more about the connection between procedure and dignity.'[21] According to Waldron, respect for human dignity is shown by institutions that take seriously the evidence and legal arguments provided by the parties and attend to those views by issuing a judgment that responds to the evidence and arguments. As he summarizes this point, 'law is a mode of governing people that treats them with respect, as though they had a view or perspective of their own to present on the application of the norm to their conduct and situation.'[22]

Crucially, while Waldron does not tell us how his account applies to non-paradigmatic humans, it seems sufficiently clear that they are excluded from it. As he claims, '[a]pplying a norm to a *human individual* is not like deciding what to do about a *rabid animal* or a *dilapidated house*. It involves paying attention to a point of view and respecting the personality of the entity one is dealing with.'[23] This exclusion of animals is made even more explicit in

[20] The powers of this statutory body are set out in Chapter II of the Criminal Justice and Court Services Act 2000.

[21] Waldron (n 6) 15.

[22] ibid 16. See also Jeremy Waldron, 'How Law Protects Dignity' (2012) 71 Cambridge Law Journal 200.

[23] Waldron (n 22) 210 (emphasis added).

Waldron's latest work, where he explicitly argues against recognizing the applicability of dignity, rights, and equality to animals, while maintaining it for non-paradigmatic humans. Although on different topics, this work reinforces the view that his rule of law account does not apply to animals either.[24] It suggests that Waldron's procedural rule of law theory cannot serve the interests of animals because they are not capable of applying legal norms to their conduct. Waldron may be right that most animals are not capable of self-application of legal norms. However, unlike dilapidated houses, they still have a point of view (at least in the sense that they have preferences) and have personalities.[25] As we put it in Chapter 2, animals are governed and not simply regulated by the law. So why should the legal process not pay regard to their point of view and respect their personalities? Again, on Waldron's account, this cannot even figure as a problem because he seems to assume that the legal process should treat animals no different than dilapidated houses. In Section 2, we will propose modifications to Waldron's account to avoid these problematic assumptions.

C) Allan's substantive theory

Let us now turn to Allan's substantive account of the rule of law. Allan's premise is his Dworkinian interpretivism which sees law as a means to resolve disputes in accordance with standards of justice and fairness.[26] On this view, the *basic idea* of the rule of law is that 'the law should embody a coherent and consistent scheme of justice, treating all alike according to the standards of that scheme'.[27] In essence, the rule of law is supposed to ensure adjudication according to pre-existing moral rights. Accordingly, the *requirements* of the rule of law include giving effect to those moral rights in addition to the formal and procedural demands already encountered above. As Allan points out:

> The law must preserve a sphere of independent decision and action, securing as far as possible an equal freedom for all. Such familiar constitutional

[24] Jeremy Waldron, *One Another's Equals: The Basis of Human Equality* (Harvard UP 2017). For a critique, see Raffael N Fasel, 'One Another's Equals: The Basis of Human Equality by Jeremy Waldron' (2018) 77 Cambridge Law Journal 230.

[25] See eg Oliver P John, Richard W Robins, and Lawrence A Pervin, *Handbook of Personality, Third Edition: Theory and Research* (Guilford Press 2008); Samuel D Gosling and Oliver P John, 'Personality Dimensions in Nonhuman Animals: A Cross-Species Review' (1999) 8 Current Directions in Psychological Science 69.

[26] TRS Allan, 'The Rule of Law' in David Dyzenhaus and Malcolm Thorburn (eds), *Philosophical Foundations of Constitutional Law* (OUP 2016) 201. His account draws chiefly on Dworkin (n 4).

[27] Allan (n 26) 203.

> freedoms as those of conscience, speech, association, and personal liberty supplement the protections for person and property in the ordinary civil law.[28]

It is important to note that, on Allan's account, the moral rights that the law must give effect to are not only familiar constitutional rights. In light of his insistence that the rule of law requires determination of moral entitlements from the state and private actors, these moral rights also include those protected by the ordinary criminal law and private law (eg the prohibition of homicide and the protection of private property). Furthermore, his account has profound implications for a general theory of adjudication and for several principles of constitutionalism such as the separation of powers, legislative supremacy, and judicial review.[29]

Even though Allan's theory is more comprehensive than Fuller's, it nonetheless inherits Fuller's undue focus on paradigmatic human beings. This becomes clear when Allan clarifies the *moral value* of the rule of law. He does not reject the Fullerian view that a formal account of the rule of law serves the value of freedom and autonomy. Allan accordingly concedes that a formal account of the rule of law serves the republican idea of freedom as independence.[30] However, he insists that his substantive conception too serves that value. As he states,

> [t]here can be no restriction of the central freedoms of thought, speech, conscience, and association that would curtail a person's ability to form and pursue her own ethical ideals . . . It is an affront to people's dignity to deny them liberty to conduct their own lives in accordance with their convictions, assuming that they do not impede an equal liberty enjoyed by others.[31]

It follows that the moral value of Allan's substantive rule of law is not different from that of Fuller's formal conception: the preservation of human autonomy (now understood as freedom as independence) and human dignity. That value, however, necessarily presupposes an agent who has the necessary cognitive abilities to form and pursue a life in accordance with their ethical convictions. This excludes animals and non-paradigmatic humans.

28 ibid 204.

29 Allan sets these out in ibid 211–18. For a fuller account, see TRS Allan, *The Sovereignty of Law: Freedom, Constitution, and Common Law* (OUP 2015).

30 Allan (n 26) 205. Allan is here influenced by Simmonds' view in Nigel Simmonds, *Law as a Moral Idea* (OUP 2008).

31 Allan (n 26) 2019.

However, if, according to Allan, the rule of law and justice are intrinsically linked, then there is no compelling reason why the rule of law should not include the interests of all sentient animals. Because these beings have inherent moral value and morally relevant interests, as we have argued in Chapter 2, a substantive conception of the rule of law needs to cater to their rights too, and protect them from arbitrary power. Once Allan's anthropocentric and rationality-centric presuppositions are exposed, it becomes clear that his theory of the rule of law also excludes animals and non-paradigmatic humans. Therefore, those who wish to advocate for a substantive theory of the rule of law that does not have this exclusionary focus need to look elsewhere.

2) The rule of law for sentient animals

As the previous section has demonstrated, current formal, procedural, and substantive rule of law conceptions have a problematically narrow focus on rational human beings and therefore exclude animals from their scope and marginalize non-paradigmatic humans. Thus, to accommodate the interests of these beings, we need a more inclusive theory of the rule of law.

In the following sections, we will primarily focus on animals. We therefore do not focus on developing inclusive accounts of the rule of law that explicitly includes non-paradigmatic human beings. This is because other scholars and practitioners have already carried out important work on including non-paradigmatic humans in the legal and political community.[32] However, many of the upshots of our discussion relating to animals and the rule of law will, *mutatis mutandis*, also apply to non-paradigmatic human beings.

As we will show, our inclusive conception of the rule of law does not require discarding altogether the theories which we have shown in the previous section to be chiefly focused on paradigmatic humans. We need only discard their narrow focus. This section does so by developing three more capacious accounts of the rule of law that track the formal (Subsection A), procedural (Subsection B), and substantive conceptions (Subsection C) of the theorists investigated in Section 1.

[32] See eg the various contributions in Barbara Arneil and Nancy J Hirschmann (eds), *Disability and Political Theory* (CUP 2016); Camilla Lundberg and Eva Simonsen, 'Disability in Court: Intersectionality and Rule of Law' (2017) 17 Scandinavian Journal of Disability Research 7; Jonathan Todres and Shani M King (eds), *The Oxford Handbook of Children's Rights Law* (OUP 2020).

A) A formal theory of the rule of law for animals

We have suggested in the Introduction that there is a more inclusive *basic idea* of the rule of law: the law should curtail arbitrary power. We saw some reference to this more inclusive basic idea in *Jackson* when Lord Hope said that the rule of law 'protects the individual from arbitrary government [and] is the ultimate controlling factor on which our constitution is based'.[33] In fact, this basic idea is well recognized within constitutional practice and legal theory more generally. The Supreme Court of India, for example, has found that '[t]he absence of arbitrary power is the first postulate of rule of law upon which our whole constitutional edifice is based'.[34] In a similar vein, the influential constitutional scholar AV Dicey's first formulation of the rule of law is that 'the rule of law is contrasted with every system of government based on the exercise by persons in authority of wide, arbitrary, or discretionary powers of constraint'.[35] Even Joseph Raz, once a leading adherent of the view that the basic idea of the rule of law is the guidance of conduct,[36] has more recently progressed to seeing the main aim of the rule of law as being the avoidance of arbitrary government.[37]

We propose that this more inclusive basic idea of the rule of law should be preferred over the rival one of guidance of conduct. As we have shown in Section 1, this rival idea excludes or marginalizes animals and non-paradigmatic humans insofar as they cannot be guided by legal norms. By contrast, the more inclusive basic idea is open to protecting anyone who can be a victim of arbitrary power. This means that not only paradigmatic humans but also animals and non-paradigmatic humans can be included in rule of law theories, including formal conceptions of the rule of law.

Let us unpack what we mean when we advocate that the basic idea of the rule of law is to curtail arbitrary power. By 'arbitrary power' we mean power that is unconstrained by any standard external to the power-holder and, importantly, which is not subject to accountability external to the power-holder. In other words, the individual wielding arbitrary power does not have to justify to anyone else that the power has been exercised according to the applicable standards.

33 *Regina (Jackson and others) v Attorney General* (n 1) [107].
34 *Som Raj v State of Haryana* (1990) 1 SCR 535 [5].
35 Dicey (n 2) 109.
36 Raz, 'The Rule of Law and Its Virtue' (n 7) 212.
37 Raz, 'The Law's Own Virtue' (n 7) 5.

Some theorists of the rule of law usually require that only state officials be subject to the rule of law. As a result, they worry only about arbitrary state or governmental power. The main reason for this narrow focus is that they seek to uphold individual autonomy and fear excessive legal regulation of affairs between private individuals.[38] We do not believe that the rule of law's focus should only be on governmental power. This is because anyone in a constitutional system, whether a private or public actor, can potentially exercise arbitrary power. Furthermore, some private actors, such as multinational corporations and wealthy individuals, may have comparable or even more power than constitutional actors over ordinary citizens and may exercise it just as arbitrarily.[39]

While we recognize that our view entails a potential limitation of individual autonomy in return for a gain in control over arbitrary power, we do not see this as an undue sacrifice. Indeed, we believe that the autonomy-focused rationale for restricting the rule of law's scope to governmental actors is premised on the undue privileging of paradigmatic humans, which we have been criticizing in this chapter. After all, it is paradigmatic humans who have the autonomy that is vital for understanding and following legal requirements. Consequently, restricting the scope of the rule of law to governmental actors only out of respect for individual autonomy perpetuates the exclusion of animals and the marginalization of non-paradigmatic humans, which this chapter has argued against.

What is more, there is evidence that a rule of law account which extends to interactions between private actors would not necessarily cause excessive legal regulation. Indeed, legal practice already provides ample illustrations of straightforward instances when the law constrains the power private actors exercise over each other. For example, ordinary criminal law, tort law, contract law, and private law more generally fulfil the rule of law's basic idea of constraining arbitrary power between private actors. Our aim is not to provide a blanket defence of any particular legal doctrine which limits the power of private actors. Without a doubt, there are existing legal doctrines which should be repealed in favour of individual autonomy.[40] Still, legal practice shows that

[38] Gardner explicitly defended this in Gardner (n 6). See also Raz, 'The Rule of Law and Its Virtue' (n 7) 212.

[39] This suggestion is in line with the view expressed in Robin West, 'The Limits of Process' in James Fleming (ed), *Getting to the Rule of Law* (New York UP 2011). It is also in line with Allan's view already analysed in Section 1. Several scholars now support the applicability of the rule of law in private law settings. See the chapters by Postema, Lucy, Allan, Dyzenhaus, Fox-Decent, and Zipursky in *Private Law and the Rule of Law* (OUP 2014).

[40] An example in this regard could be the criminalization of assisted dying for individuals who are unable to commit suicide without assistance because of a physical disability. See eg John Adenitire, 'A Conscience-Based Human Right to be "Doctor Death"' (2016) 4 Public Law 613.

there are many instances in which the law, while limiting individual autonomy, appropriately governs the exercise of power between private actors.

This conclusion helps us understand how the basic idea of the rule of law will always provide a degree of protection to animals, even on a purely formal account. Recall that, on Fuller's account, the basic idea of the rule of law is to enable the guidance of conduct so as to serve human autonomy and human dignity. Animals could only benefit from this account by requiring enforcement of legal norms that serve their interests (eg animal protection laws), if they happen to exist. However, there are some jurisdictions where there are no such norms, and many others where they only apply to a small minority of animals. By contrast, the account we put forward here provides a consistent degree of protection to animals by requiring that every exercise of power over them be non-arbitrary.

Indeed, animals (and non-paradigmatic humans) can be victims of arbitrary power of both public and private actors. As we saw in Chapter 2, animals in particular may be harmed, captured, put in captivity, killed for food, and used for entertainment and research. And all this may be done arbitrarily. For example, in those jurisdictions with animal protection provisions that only apply to some animals, such as the United States and China, there are no legal standards applying to the treatment of animals who are not covered by animal protection laws. Such unconstrained and non-accountable power is inimical to the rule of law under the basic idea we endorse.

Reorientating the basic idea of the rule of law as curtailing arbitrary public and private power therefore opens the possibility that formal accounts of the rule of law will always provide a degree of protection to animals. But it is worth noting that because formal accounts do not require the protection of fundamental rights, it is consistent with such accounts that public and private power may be constrained in ways that still allow for animals' exploitation and killing. What *is* inconsistent with the basic idea of an inclusive formal account, however, is that power exercised over animals remains totally uncontrolled by law, as happens to be the case in jurisdictions where animal protection provisions do not exist at all or, where they do, only apply to limited categories of animals. While inclusive formal accounts of the rule of law are not committed to the full protection of the interests of animals, they are at least committed to the view that the law must always regulate how they are treated, however poorly or maliciously.

Having established the basic idea of a more inclusive formal theory of the rule of law, we can now identify the *requirements* of such a theory. We propose the following criteria for the formal rule of law and explain them in turn below:

1. The law should monopolize the use of force.
2. The law should regulate interactions between private and public actors in those areas where they can cause significant harm to each other.
3. Legal norms should generally possess the following qualities. They should be:
   (a) general;
   (b) public;
   (c) clear;
   (d) prospective;
   (e) non-contradictory;
   (f) stable; and
   (g) possible to comply with.
4. Enforcement officers should faithfully enforce the law as interpreted by judicial bodies.
5. Judicial bodies should be independent and impartial and should interpret the law according to their conscience without fear or favour.
6. There should be a well-functioning legal profession.
7. The exercise of discretion by public bodies (in particular of bodies charged with investigating and prosecuting breaches of the law) should be set within clear parameters and be reviewable by independent judicial bodies.
8. The law should control instances of its suspension (eg during situations of public emergencies).

The idea behind this list is that a constitutional system should be ruled by law and not by personal whim. Importantly, the ability of public and private actors to cause each other significant harm should be regulated by law. Interpersonal use of force should be monopolized by the law because of its propensity to cause harm (requirement (1)). To be sure, this should not be understood as a requirement that the law should protect individuals from any harm that other actors (both public and private) may commit. That would move us towards a more substantive understanding of the rule of law. All that is required under this formal conception is that the use of force and other ways in which individuals can cause harm to each other should be subjected to legal regulation. On this account, the law may permit individuals to harm each other (eg by allowing the exploitation and killing of animals). However, such harm cannot be committed arbitrarily (ie outside of the law) and individuals can be held accountable if they exceed the terms of the law.

We leave requirement (2) about the legal regulation of interpersonal harm somewhat abstract to acknowledge that there are instances where individuals harm each other, yet legal intervention may be inappropriate. A falling out with a friend can cause considerable psychological harm. However, it seems uncontroversial to say that the law would overreach if it regulated who should be friends. Nevertheless, as the discussion above on the need for the rule of law to apply between private actors has shown, ordinary criminal law, tort law, contract law, and private law more generally suggest that it is already recognized that there are appropriate instances where legal regulation is warranted even between private individuals.

Requirement (3) essentially reproduces Fuller's list of the internal morality of law, bar his last requirement, which is reproduced—with added clarity—in the separate requirement (4). The rationale behind incorporating Fuller's requirements is in line with the basic idea of curtailing arbitrary power. If the law is to curtail power by setting standards external to public and private actors, then those standards should be public, understandable, and possible to comply with. Of course, not all actors are able to align their actions with legal standards. Animals and non-paradigmatic humans fall in this category. Here, it is important to recall that the account we put forward constrains arbitrary power of paradigmatic humans to benefit *all* humans—regardless of whether they are paradigmatic—and sentient animals. As long as arbitrary power is constrained against all, it does not pose a problem for our account that some of its requirements are specifically focused on and directed towards paradigmatic humans.

Requirements (4) to (7) aim to enable legal accountability. They assume that a functioning legal system has a doctrine of separation of powers and that judicial review of executive powers is available. Even though the requirements do not mention judicial review of legislation, the account assumes that such review will ordinarily be available, given that this is a commonplace in the vast majority of constitutional systems.[41] After all, lawmakers can exercise power arbitrarily just as—if not more than—ordinary individuals. The requirements therefore place emphasis on the judicial branch to ensure legal accountability, especially over other branches of government. Yet, this should not be understood as a call for judicial supremacy. In a constitutional order dedicated to the

[41] Tom Ginsburg and Mila Versteeg, 'Why Do Countries Adopt Constitutional Review?' (2014) 30 The Journal of Law, Economics, and Organization 587, 587–88. Ginsburg and Versteeg show that around 83 per cent of world constitutions make some form of judicial review of legislation available. They base this on the Constitute.net dataset available at <https://www.constituteproject.org/constitutions?lang=en&key=uncon&status=in_force&status=is_draft> accessed 4 November 2024.

rule of law, no one branch of government is supreme.[42] Rather, the law itself is the highest authority, and judges exercise their interpretive powers within the limits set by law and legal tradition. Accordingly, each branch of government should be understood to be collaborating to bring about the supremacy of law in a constitutional system in which arbitrary public and private power is appropriately curtailed.

The last requirement—requirement (8)—according to which the law should control instances of its suspension, builds on the notion that government by law is of great moral value (on which we will say more shortly). Is this a requirement that could make governmental action unnecessarily complicated in emergency situations? There may indeed be some situations where inexpediency caused by adherence to law is too high of a price to pay. Public emergencies such as war or natural disasters may necessitate expedient reaction to protect the public good. In these cases, acting outside the fetters of the rule of law might therefore be said to directly serve the common good.

However, given the danger that power is particularly likely to be abused in such situations, the law itself should regulate when, how long, and to what extent it can be suspended.[43] More would need to be said about exactly how this would work and what it would entail. Due to the confines of this chapter, these questions cannot be addressed here. What we can say, at least, is that the idea of a permanent state of emergency which requires the broadest suspension of the law would be incompatible with the rule of law.[44]

Finally, what is the *moral value* of this formal account of the rule of law? Given that the account incorporates all of Fuller's requirements, it also incorporates all of the moral values which that account serves. Accordingly, it serves human autonomy, human dignity, and a republican conception of freedom.[45] However, the account goes beyond Fuller's requirements in a way that also makes it valuable to animals. In particular, requirements (1) and (2)—according to which the law should monopolize the use of force and regulate instances of interpersonal harm—entail that there will always be a degree of legal protection for animals: they can no longer be arbitrarily harmed by private or public actors. The law must be clear when it permits such harm, and it is

[42] Philip A Joseph, 'Parliament, the Courts, and the Collaborative Enterprise' (2004) 15 King's Law Journal 321.

[43] See also David Dyzenhaus, *The Constitution of Law: Legality in a Time of Emergency* (CUP 2006) 196–220.

[44] Alan Greene, *Permanent States of Emergency and the Rule of Law: Constitutions in an Age of Crisis* (Hart Publishing 2018).

[45] Nigel E Simmonds, 'The Nature of Law: Three Problems with One Solution' (2011) 12 German Law Journal 601.

possible that the law may only allow certain kinds of harm or prohibit harm altogether. Whenever protections are in place, the rule of law also dictates that they be properly enforced—something which many legal systems with animal protection laws currently fail to live up to.[46]

This account of the moral value of the rule of law for animals may leave a bitter taste because it is a moral value with limits. The account frees animals from arbitrary power, but it puts them into the vagaries of whether legal power will protect their fundamental interests. Indeed, under this account, mass exploitation and mass killings of animals are perfectly acceptable so long as they are explicitly legally authorized. While this is true, it affects formal rule of law accounts more generally. For instance, as Raz points out,

> [a] non-democratic legal system, based on the denial of human rights, on extensive poverty, on racial segregation, sexual inequalities, and religious persecution may, in principle, conform to the requirements of the rule of law better than any of the legal systems of the more enlightened Western democracies.[47]

Likewise, even on an inclusive formal account of the rule of law that incorporates animals, a legal system that lives up to the requirements of this account may continue to systematically exploit and kill animals. Such exploitation and killing would need to be explicitly legally authorized. But animals' fate would not be substantively improved.

B) A procedural theory of the rule of law for animals

Recall from the discussion in Section 1.B that procedural theories of the rule of law share the same basic idea as formal theories, while emphasizing the importance of independent courts and judicial procedures. Consequently, on our inclusive procedural account, the *basic idea* of the rule of law is the same as on our inclusive formal account: the curtailment of arbitrary power. We need to emphasize, however, the necessity for external accountability of power through a system of independent courts and fair legal procedures. This is already to some degree implicit in requirements (5) to (7) set out above in our inclusive formal account. However, we make them explicit here by wholly

[46] See eg Leslie Rudloff, 'Failure to Launch: The Lack of Implementation and Enforcement of the Animal Welfare Act' (2017) 67 Syracuse Law Review 18.

[47] Joseph Raz, *The Authority of Law: Essays on Law and Morality* (OUP 2009) 211.

incorporating the *requirements* of Waldron's procedural rule of law account, while making some modifications and adding two more requirements which remedy his anthropocentricism:

> No one, including animals, should have any legal penalty, stigma, or serious loss imposed upon them by a public or private actors except as the upshot of procedures that involve:
>
> 9. A hearing by an impartial tribunal that is required to act on the basis of evidence and argument presented formally before it in relation to legal norms that govern the imposition of penalty, stigma, loss, and so forth.
> 10. A legally trained judicial officer whose independence of other agencies of government is ensured.
> 11. A right to legal representation and to the time and opportunity required to prepare a case.
> 12. A right to be present at all critical stages of the proceeding.
> 13. A right to confront witnesses against the detainee.
> 14. A right to an assurance that the evidence presented by the government has been gathered in a properly supervised way.
> 15. A right to present evidence on one's own behalf.
> 16. A right to make legal argument about the bearing of the evidence and about the bearing of the various legal norms relevant to the case.
> 17. A right to hear reasons from the tribunal when it reaches its decision that are responsive to the evidence and arguments presented before it.
> 18. Some right of appeal to a higher tribunal of a similar character.

To this (modified) list, we add the following two requirements:

19. Animals enjoy legal standing.
20. The legal rights of animals can be enforced by any legal representative that acts in their best interest.

The above account avoids Waldron's anthropocentricism by explicitly incorporating animals within a procedural theory of the rule of law. As already argued, animals are potential victims of arbitrary power too, which is why the rule of law should require their non-contingent protection. Making room for animals in such a more inclusive theory ensures that no harm may be done to them without the legal accountability mechanisms that humans already enjoy.

While it may at first glance seem odd to consider cows, pigs, chickens, and other animals as being entitled to participation in a fair judicial process to defend their legal entitlements, this is not unfamiliar terrain in some legal traditions. For instance, there have been several *habeas corpus* cases where animal advocates have sought to free animals from captivity by having courts recognize their legal personhood.[48] In the Colombian case of the Andean bear Chucho, for example, several courts heard arguments for and against granting a writ of *habeas corpus* for Chucho and moving him from a zoo to a natural reserve. The Constitutional Court even heard from non-Colombian lawyers who argued in favour of granting the writ to the bear.[49] Even though, so far, only one *habeas corpus* case for an animal has been successful,[50] there is precedent for the view that fair legal procedures should be applied before animals can be harmed.

We have added requirements (19) and (20) to Waldron's list as a response to the rationale underlying some animal protection cases involving standing issues. In many jurisdictions, animals do not have standing to enforce laws that protect them.[51] Furthermore, humans seeking to enforce animal protection laws often have to show that they have some particular legal interests in enforcing the relevant statute. The fact that an animal has been injured in breach of a statute is often insufficient to grant standing to any general member of the public.

For example, in the *Animal Lovers* case from the United States, an animal protection association sought to prevent the killing of a colony of wild goats living on a military enclave that was not publicly accessible. The organization sought to argue that the killing of the goats was in contravention of federal law. The possible breach of the law was alone not sufficient to give standing to the organization. Consequently, to have standing to enforce the statute, the plaintiff resorted to arguing that its members would suffer 'aesthetic injury' from the killing of the goats. That is, the death of the goats would result in 'direct sensory impact' on the members' physical environment. However, even this ingenious argument to enforce the law failed. This is because the members of the organization could not satisfy this test given that they had no right to access the

[48] See eg the US case *People ex rel Nonhuman Rights Project, Inc v Lavery* (2014) 124 AD 3d (NY: Appellate Div, 3rd Dept). See generally Fasel and Butler (n 8) ch 7.

[49] Colombian Constitutional Court, no T-6.480.577 (Chucho), SU-016/20, 23 January 2020.

[50] Judicial Power of Mendoza, Third Court of Guarantees, Re Cecilia, P-72 254/15, 3 November 2016. Even though, strictly speaking, the *habeas corpus* petition in *Estrellita* was unsuccessful, the Constitutional Court in Ecuador recognized animals as rights-holders and emphasized that animals have procedural rights to secure the substantive rights granted by the Constitution. See 'Estrellita', Constitutional Court, no. 253-20-JH/22, 27 January 2022, [181] <https://animal.law.harvard.edu/wp-content/uploads/Sentencia_253-20-JH-20_Derechos_Naturaleza_y_Animales.pdf> accessed 26 October 2024.

[51] For an overview in the US context, see Cass R Sunstein, 'Can Animals Sue?' in Cass R Sunstein and Martha Nussbaum (eds), *Animal Rights: Current Debates and New Directions* (OUP 2005).

military enclave where the goats would be killed. Accordingly, they were denied standing to enforce the statute. This was despite the organization having been set up specifically to protect the targeted colony of goats.[52]

As this illustrates, current standing requirements, due to their anthropocentrism, entail that too limited a class of individuals can enforce laws that protect animals. The consequence is that breaches of the law will often be left unaddressed. If animals enjoyed legal standing, they could bring legal suits themselves with the aid of a legal representative.[53] This would result in a rule of law gain, given that it would substantially increase the possibility of redressing breaches of the law.

Who should be entitled to act as a legal representative of animals in such cases? The answer is likely to vary depending on the constitutional system and legal subject matter. Consider, for example, criminal cases involving cruelty to animals. While, as a general rule, it is public prosecutors who have the authority to initiate legal proceedings in criminal matters,[54] some jurisdictions have specific statutory bodies which investigate and prosecute animal cruelty cases. In South Africa, for instance, the National Society for the Prevention of Cruelty to Animals (NSPCA) is statutorily recognized in the Societies for the Prevention of Cruelty to Animals Act (SPCA Act).[55] The South African Constitutional Court has affirmed, despite opposition from the National Prosecuting Authority, that the NSPCA is empowered to privately prosecute instances of animal cruelty.[56]

In other countries, non-statutory bodies can enforce and prosecute criminal cases involving animal abuse. The Royal Society for the Prevention of Cruelty to Animals (the oldest animal protection organization in the world), for example, is a charity in England and Wales which has historically investigated and prosecuted instances of animal cruelty criminal laws. It has been able to prosecute such cases because English law recognizes a right of private prosecution of criminal statutes.[57] However, there are many jurisdictions where no similar right exists.[58]

In France, Article 2-13 of the Code of Penal Procedure makes it possible for registered animal protection organizations to become a civil party to criminal

[52] *Animal Lovers Volunteer Association v Weinberger* (1985) 765 F 2d 937 (Court of Appeals, 9th Circuit).

[53] See eg Cass R Sunstein, 'Can Animals Sue?' in Cass R Sunstein and Martha Nussbaum (eds), *Animal Rights: Current Debates and New Directions* (OUP 2005).

[54] Jamil Ddamulira Mujuzi, 'The Right to Institute a Private Prosecution: A Comparative Analysis' (2015) 4 International Human Rights Law Review 222.

[55] Section 6 of the SPCA Act 1993.

[56] *National Society for the Prevention of Cruelty to Animals v Minister of Justice and Constitutional Development and Another* (2016) 2016 ZACC 46.

[57] Section 6(1) of the Prosecution of Offences Act (POA) 1985.

[58] Mujuzi (n 54).

cases against individuals accused of animal abuse. This means that French law recognizes that these organizations can recover compensation which animal victims would have been able to recover had they been human.

This brief overview of existing mechanisms to legally represent animals in litigation shows that there are several options for appointing legal representatives for animals in a constitutional system that lives up to the requirements of an inclusive procedural rule of law. Legal representatives could be statutorily designated bodies, charities making use of the power of private prosecution, or animal protection organizations taking advantage of a statutory power to claim damages on behalf of victimized animals. No doubt other options exist too.[59]

While we do not advocate for any particular mechanism, we argue that procedures to designate legal representatives should not frustrate, but rather facilitate, the possibility of legal accountability for breaches of laws protecting animals. Our discussion above of anthropocentric standing requirements has revealed that narrow standing can lead to a lack of legal accountability for breaches of animal protection laws. We have consequently argued that animals themselves should enjoy legal standing. Procedures to appoint legal representatives, independently of their specifics, should therefore make real the view that the legal process takes seriously the enforcement of the rights and duties created by law to protect animals.

What is the *moral value* of this inclusive procedural account? Because it incorporates all of Waldron's requirements, it serves the same anthropocentric value identified by Waldron: respect for human dignity. However, the non-anthropocentric components of the inclusive account extend its moral value to animals and equally serve their dignity, in the sense that animals are now members of the legal community who have a right to address and be addressed by the law through their representatives. This is especially so if requirements (19) and (20) are respected. Standing for animals and the availability of legal representatives to defend their interests signifies that animals' legal interests are to be taken seriously by the legal process. Like humans, they can have their day in court to uphold whatever protection the law has granted them. This way, we can now extend Waldron's idea of dignity to animals as, under this inclusive conception, the law has to respect that animals too are entitled (through their legal representatives) to enforce the legal norms applicable to them.

In extending not only the formal but also procedural benefits of the rule of law to animals, the inclusive procedural account provides more protections for

[59] See eg Kimberly K Smith, *Governing Animals: Animal Welfare and the Liberal State* (OUP 2012) 118–23.

them than the inclusive formal account. However, it is worth emphasizing that the inclusive procedural theory does not require the protection of any substantive rights for these beings. Hence, a legal system that explicitly permits the mass exploitation and mass killing of animals is still consistent with it. Still, at least such permission would have to be granted in a way that respects the procedural rights of animals and, in particular, our inclusive conception would demand that their representatives are able to test the legality of such permission in an independent and fair court.

C) A substantive theory of the rule of law for animals

An inclusive substantive theory of the rule of law requires further specifying the *basic idea* of the rule of law which we have been defending in this chapter. That basic idea remains the curtailment of arbitrary power. However, in addition to requiring accountability to standards set by law, a substantive theory demands accountability to standards which themselves are non-arbitrary. This means that standards set by a constitutional system to curtail arbitrary power have to be themselves accountable to the non-arbitrary standards of justice. In other words, the basic idea of a substantive theory of the rule of law is to curtail arbitrariness in all possible ways: in relation to power exercised by public and private actors *and* in relation to legal standards themselves. Here, the chain of accountability goes all the way up. Individuals are accountable to legal standards, and legal standards are accountable to justice.

Because the basic idea of this substantive conception is not divorced from the other accounts, its *requirements* can incorporate all the requirements of the formal and procedural accounts defended above. However, we also add the following requirements:

21. The law should regulate, in a way consistent with the demands of justice, interactions between private actors in those areas where they can cause significant harm to each other;
22. The law should regulate the ways public actors may exercise their powers over private actors by ensuring they respect private actors' fundamental rights.

These requirements are framed in more general terms than those of the formal and procedural accounts. Their essence can be reduced to the necessity that the law should comply with the requirements of justice, which is not to reduce the rule of law to the rule of justice. There are aspects of justice that do not apply to the rule of law and aspects of the rule of law that are not directly dictated by justice. For example, who should act as one's best man at a wedding is responsive to the demands of justice. It would be unfair to choose a distant relative over a close friend who has provided constant help and support and who wishes to act as best man. However, out of respect for personal autonomy, it would be inappropriate for the law to regulate this aspect of an individual's life. Conversely, justice can constrain the legitimacy of the road traffic law (eg people should not be prohibited from driving based on their skin colour), but it does not dictate many other aspects which the law is also expected to regulate (eg which side of the road people should drive on). As this illustrates, requiring the law to comply with justice does not entail that justice and the rule of law become the same ideal.

Requirement (21) of our inclusive substantive conception builds on the notion already developed above—that the rule of law demands the legal recognition of the ordinary criminal law, tort law, contract law, and private law more generally, in order to constrain the arbitrary power of private actors over each other. However, the recognition of such rules is not by itself sufficient on a substantive account. The rules must also be compatible with justice. Put simply, not just any rules will do. The rules must afford due concern to the intrinsic value of sentient beings by protecting their basic interests. More work would be needed to provide detailed guidance on the rule of law requirements in these areas of law, but that goes beyond the scope of this chapter. However, given this chapter's focus, no account would be complete if it did not include animals within its scope. In particular, no account would be compatible with a substantive conception of the rule of law if it disregarded the intrinsic moral value of animals by allowing them to be exploited and killed by private actors for human uses such as food, entertainment, or medical research. This is because such uses would be incompatible with the moral rights which animals have due to their sentience.

We have already defended in Chapter 3 the view that in a sentience-based constitution there would be a continuity between fundamental rights for humans and the fundamental rights that animals should have in law. In particular, we argued that the same catalogue of fundamental rights could be used to protect the basic interests of humans and animals. We gave some examples in that chapter of how some well-established rights—the right to life, the right to physical and moral integrity, and the right to freedom—could be extended to non-human animals. This entails that on our substantive account of the rule of law, the law cannot permit animals to be exploited or killed in contravention of these rights. Consequently, private law that constrains arbitrary power between private actors must ensure that these fundamental rights are not violated.

What does this mean practically? It means that the law, through the ordinary criminal and civil law, cannot allow the killing of animals for food because this would interfere with animals' fundamental rights to life, physical and mental integrity, and—due to the confinements of animals in animal farming—freedom. This is especially the case because such killing is often unnecessary since plant-based foods and cheap supplements which can meet humans' dietary needs are widely available, certainly in economically affluent countries.[60] Similarly, the killing of animals for entertainment, such as recreational hunting or bullfights, not only violates animals' fundamental right to life and physical integrity. It is also unnecessary given the many alternatives for entertainment at humans' disposal (eg theatre, sports, videogames). Finally, while some utilitarian moral theories permit lethal medical research on animals if this brings some important benefit to other sentient life,[61] our account finds this to be inconsistent with animals' fundamental rights to life, physical and mental integrity, and with their right to freedom. It follows that an inclusive substantive theory of the rule of law requires that practices of exploitation and killing of animals be prohibited by the law which regulates interpersonal harm (ie criminal law, tort law, etc).[62]

Requirement (22) builds on requirement (21). An inclusive substantive conception of the rule of law does not only require that the basic interests of humans and animals be protected from the arbitrary power of private actors. It also requires the protection of animals' interests through the institution of fundamental rights which bind the state and its organs. We have already said that familiar catalogues of rights can and should be meaningfully extended to animals to ensure the protection of their basic interests and capacities. But does this entail that a substantive account of the rule of law calls for the recognition of all possible fundamental rights? Or is it that a substantive theory of the rule of law requires recognition of only certain rights?

Take, for example, the right to use any language, guaranteed under Article 18 of the Swiss Constitution. Does a substantive rule of law call for the recognition

[60] See eg Marco Springmann and others, 'Health and Nutritional Aspects of Sustainable Diet Strategies and Their Association with Environmental Impacts: A Global Modelling Analysis with Country-Level Detail' (2018) 2 The Lancet Planetary Health e451.

[61] For a useful overview, see Christine M Korsgaard, *Fellow Creatures: Our Obligations to the Other Animals* (OUP 2018) ch 12.

[62] We discuss these issues more extensively in Chapters 3 and 4.

of this right? On Allan's substantive theory, for example, the right to use any language would arguably not be a direct requirement of the rule of law, though it could be an indirect requirement. According to Allan's theory, the rule of law requires the recognition of those rights that ensure that a citizen can form a conception of the good life and live that conception independently of the will of another. This is why Allan emphasizes those rights, such as free speech, conscience, religion, association, and assembly, which enable citizens to form and live out—alone or in groups—their conception of the good life.[63] A right to speak any language, on Allan's account, would arguably only be required if it was a way to specify the more general rights of, for example, speech or assembly.

Would an inclusive substantive conception also focus on those constitutional rights that enable individuals to form and live, alone or in groups, their conception of the good life? While the inclusive account is compatible with such a focus, it cannot exclusively have this focus. This is because our account is inclusive of those beings, such as animals and non-paradigmatic humans, who do not have the capacity to form and pursue conceptions of the good life. Animals, of course, have desires and preferences and act to satisfy them; they have agency and are able to assent or dissent to certain things.[64] However, most of them do not normally have the relevant capacities to devise and revise a sufficiently complex conception of the good life. It is for this reason that our inclusive substantive account also puts emphasis on fundamental rights (such as life, physical and mental integrity, and freedom) that do not require their holders to have the capacity of forming and pursuing sufficiently complex conceptions of the good life.

Finally, we need to reflect on the *moral value* served by an inclusive substantive account of the rule of law. Given that the account incorporates formal and procedural requirements, it serves the moral values that those theories serve. Accordingly, it promotes the values of human autonomy, human dignity, and a republican conception of freedom. Furthermore, considering that it is an inclusive account, our theory non-contingently protects animals and humans from arbitrary public and private power and upholds their dignity in the legal process. Importantly, its substantive nature also means that our theory demands that the law can no longer authorize the harming of animals and humans in ways which undermine their fundamental rights. This means that the

[63] Allan (n 26) 204 and 219.

[64] Richard Healey and Angie Pepper, 'Interspecies Justice: Agency, Self-Determination, and Assent' (2021) 178 Philosophical Studies 1223, 1233–41.

law ought to be compatible with (among others) the right to life, the right to physical and mental integrity, and the right to freedom. Additionally, for those actors who can formulate and pursue a conception of the good life, the law must protect those rights that enable them to do so, such as the rights to free speech, conscience, religion, association, and assembly.

The outcome of this substantive theory of rule of law is a situation where the law gives due concern to the interests of all sentient beings governed by the constitution. When a constitutional system lives up to the demands of the substantive rule of law, all sentient animals and humans have what we called in Chapter 2 'constitutional worth'. That is, the constitution recognizes that they have interests of constitutional importance, demands respect for their intrinsic worth, and ultimately protects their basic needs through fundamental rights. In Chapter 2, we also argued that the protection of constitutional worth of all sentient animals and humans is a condition of constitutional legitimacy. It follows that a constitutional system that complies with an inclusive substantive account of the rule of law fills the current legitimacy gap which current anthropocentric constitutions suffer from. Hence, the ultimate moral value of the inclusive substantive rule of law is to legitimize the whole constitutional order: it leads us to sentience-based constitutionalism.

Conclusion

In this chapter, we have argued for more inclusive accounts of the rule of law. Prevailing theories of the rule of law assume that it ought to chiefly serve the well-being of paradigmatic humans. As we have shown, animals and non-paradigmatic humans do not feature in these theories and, where they do, only contingently benefit from these accounts. Once this undue focus on paradigmatic humans is swept aside, existing theories of the rule of law can be remodelled to become more inclusive. We proposed that the basic idea of the rule of law—whether on formal, procedural, or substantive conceptions—should be the curtailment of arbitrary power which private or public actors can exercise. This opens the possibility that animals and non-paradigmatic humans are protected by the rule of law because they are possible victims of arbitrary power.

As we have shown in Section 2, the scope of the protection offered by the rule of law to animals will vary depending on which account one endorses. It will be thinnest on a formal account, stronger on a procedural account, and thickest on a substantive account. As stated in the introduction to this chapter, we remain agnostic as to whether an inclusive theory of the rule of law should be formal, procedural, or substantive. While we have shown that embracing an inclusive substantive account of the rule of law leads to embracing sentience-based constitutionalism, we realize that some may find that our substantive conception unduly transforms the rule of law into the rule of good—or legitimate—law.[65] Such sceptical readers may elect to commit only to our inclusive formal or procedural accounts while rejecting our substantive account of the rule of law.

Ultimately, however, given our argument that constitutional orders must be legitimate for all governed sentient beings, such readers will have to embrace the arguments put forward in Chapter 2. Furthermore, given our argument that constitutional orders should extend fundamental rights to animals, readers who are sceptical of our inclusive substantive account will have to accept the arguments put forward in Chapter 3. While they may reject the substantive rule of law path to sentience-based constitutionalism explored here, they cannot escape the final goal of embracing a constitutional order that pays due concern to all sentient beings.

[65] John Tasioulas, 'The Rule of Law' in John Tasioulas (ed), *The Cambridge Companion to the Philosophy of Law* (CUP 2020) 117–22.

6
Democracy

Introduction

In discussions regarding the rights of animals, sceptics often invoke the prospect that animals might get a right to vote as a *reductio ad absurdum* against the entire animal rights idea.[1] Even animal rights advocates have distanced themselves from the notion that animals could have political rights. Peter Singer, for instance, contends that '[s]ince a pig can't vote, it is meaningless to talk of its right to vote',[2] and Cass Sunstein observes that the argument for legal animal rights '[o]f course . . . does not mean that those animals can vote or run for office'.[3] However, this exclusion of animals from democratic rights and representation has recently come under reconsideration with the 'political turn'[4] in animal rights scholarship. In a nutshell, the political turn has put centre stage the role politics plays for animals and the role animals play in politics. For example, according to Sue Donaldson, one of the leading voices of the political turn, animals' interests should not only be taken into account by political decision makers but animals should *themselves* be viewed as political participants who co-author such decisions.[5] In a similar vein, Ioan-Radu Motoarcă has recently argued that 'governments should recognize a political right to vote for at least certain categories of animals'.[6] Proposals such as these have raised the question of what the proper place for animals is in a democracy.

In this chapter, we argue that, in a sentience-based democracy, non-human animals have an equal right to democratic participation as humans. This does not mean that animals need to be treated the same as humans. Specifically, it does not give animals formal political rights such as a right to vote or to run

[1] See eg Simon Jenkins, 'Votes for Dogs Appeals, but Giving Animals Rights Is Moral Chaos' *The Guardian* (18 March 2010) <https://www.theguardian.com/commentisfree/2010/mar/18/animal-rights-moral-chaos> accessed 4 November 2024.

[2] Peter Singer, 'All Animals Are Equal' in Tom Regan and Peter Singer (eds), *Animal Rights and Human Obligations* (2nd edn, Prentice-Hall 1976) 150.

[3] Cass R Sunstein, 'The Rights of Animals' (2003) 70 The University of Chicago Law Review 387, 399.

[4] Tony Milligan, 'The Political Turn in Animal Rights' (2015) 1 Politics and Animals 6.

[5] Sue Donaldson, 'Animal Agora: Animal Citizens and the Democratic Challenge' (2020) 46 Social Theory and Practice 709.

[6] Ioan-Radu Motoarcă, 'Animal Voting Rights' (2024) 84 Analysis 56, 56.

Animals and the Constitution. John Olusegun Adenitire and Raffael Fasel, Oxford University Press.
 DOI: 10.1093/9780198910534.003.0006

for office. This is because, to the best of our knowledge, animals lack the capacity to understand formal political processes such as elections and the powers and responsibilities that come with holding political office, and they therefore lack the ability to exercise the corresponding rights in a meaningful way. Importantly, however, this does not mean that animals' interests matter less in a democracy than humans'. Recognizing animals' right to democratic participation means structuring democratic processes in ways that afford due concern to the interests of all sentient beings whom these processes govern. As will be recalled from Chapter 2, giving 'due concern' to animals entails respecting their intrinsic value and protecting their basic interests. This chapter will spell out what giving 'due concern' to animals means in the context of constitutional democracy. As we will argue, it entails at least four things: first, human representatives are required so that animals' interests can be channelled into formal decision-making processes; second, these representatives can only effectively exercise their role if animals participate in their representation; third, a combination of legal and political mechanisms is necessary to give animals their due democratic concern; and, finally, direct democratic processes such as 'sentient beings' initiatives' are required to complement indirect democratic processes. These four elements ensure that, while animals themselves cannot meaningfully participate in formal democratic processes, they can still participate democratically through the aid of human representatives attuned to animals' informal political agency.

The chapter is structured as follows. In Section 1, we make the case that constitutional systems should extend an equal right to democratic participation to all sentient beings whom they govern. In Section 2, we argue that effectively protecting animals' right to democratic participation necessitates the adoption of an integrated approach that harnesses both legal and political mechanisms. Section 3 builds on the proposal of an integrated approach and argues that it is necessary to go beyond the literature's current focus on indirect democratic institutions. Instead, we propose direct democratic mechanisms, specifically 'sentient beings' initiatives', to complement indirect democratic approaches. Finally, the conclusion highlights areas where more research is required.

1) An equal right to democratic participation

In Plato's *Republic*, Socrates tells the story of a democratic city in which freedom is so widespread that women are as free as men, slaves as free as masters, and domesticated animals as free as humans. In this city, 'horses and mules are accustomed to go about in total freedom, always haughtily barging into those who get in their way, unless they step aside, and everything else is full of freedom like this'.[7] The veracity of this allegorical account is, of course, doubtful. But what is beyond doubt is that there is no democracy today that includes non-human animals in law- and policy-making processes. We will begin this section by emphasizing the anthropocentric *status quo* of current democracies (Subsection A), before making the case for animals' equal right to democratic participation (Subsection B).

A) Democracy's anthropocentric *status quo*

As already outlined in Chapter 1, contemporary democracies are firmly anthropocentric. This can best be seen in legal provisions setting out citizens' political rights, which are often located in the constitutions of democratic states. Consider two examples. The Constitution of the United Republic of Tanzania, under the heading 'Claim of universal suffrage', states in Article 5.1 that '[e]very citizen of the United Republic who has attained the age of eighteen years is entitled to vote in any election held in Tanzania'. And according to Article 23(2) of the Constitution of Bhutan,

> A person shall have the right to vote by direct adult suffrage through secret ballot at an election if the person is:
> A Bhutanese citizen as evidenced by a Citizenship Card;
> Not less than eighteen years of age;
> Registered in the civil registry of that constituency for not less than one year, prior to the date of the election; and
> Not otherwise disqualified from voting under any law in force in Bhutan.

None of these provisions contemplate animals as holders of political rights or as being deserving of democratic inclusion in some other fashion. This illustrates what is true in constitutions around the world: animals lack institutionalized political representation of their interests.[8] Hence, the claim to 'universal

[7] Plato, *Republic*, vol 1 (Chris Emlyn-Jones and William Preddy trs, Harvard UP 2013) Book VIII, 563c. Plato himself, of course, was sceptical of democracy: see eg Malcolm Schofield, *Plato: Political Philosophy* (OUP 2006) ch 3.

[8] On the relatively recent phenomenon of political parties centred on animal advocacy, see Paul Lucardie, 'Animalism: A Nascent Ideology? Exploring the Ideas of Animal Advocacy Parties' (2020)

suffrage' of many of these constitutions needs to be taken with a fistful of salt. Not only is participation in formal democratic processes limited to adult human beings who meet additional conditions such as not being disqualified from voting. But further, constitutions do not envisage forms of democratic participation for animals.[9] In other words, animals are not part of the 'demos' in current democracies. That is, they do not form part of 'the people' who participate in the making of the rules that govern them.

This anthropocentric *status quo* of democracies is untenable because it violates animals' equal right to democratic participation. But what exactly is this right and how can it be justified? To understand the argument we are making, it will be helpful to first recall our case for sentience-based constitutionalism in Chapter 2. As we proposed there, the All Governed Interests Principle is the most compelling basis of constitutional legitimacy. This is because it can fill the legitimacy gap created by human-focused principles of constitutional legitimacy that do not accommodate the interests of all governed sentient beings. The All Governed Interests Principle remedies this by requiring that those exercising public powers ought to give due concern to the interests of all sentient beings they govern. As mentioned above, 'due concern' involves respect for the intrinsic value of sentient beings and the protection of their basic interests, whereas 'governing' a being means having effective control over the lives, bodies, or habitats of such beings, not merely affecting their interests in a looser way.

It is helpful to separate two different questions here: *why* animals' interests should be included in democracies, and *what* such inclusion would look like. We will answer these questions in the next subsection. In Sections 2 and 3, we will then address *how* exactly the inclusion of animals can be operationalized.

B) The case for animals' right to democratic participation

Let us begin with the 'why' question. Our argument for extending the right to democratic participation to animals builds on the more general case for

25 Journal of Political Ideologies 212. Such parties can play an important role but—being dependent on sufficient votes in elections—they cannot guarantee the inclusion of animal rights in the democratic process.

[9] On related debates in the context of political participation of children and people with disabilities, see Harry Pearse, 'Children, Voting, and the Meaning of Universal Suffrage' (2023) 22 Political Studies Review 821; Linda Barclay, 'Cognitive Impairment and the Right to Vote: A Strategic Approach' (2013) 30 Journal of Applied Philosophy 146.

sentience-based constitutionalism, which we developed in Chapter 2. The short answer is this: sentient animals should be included in democratic decisions because they are governed by democratic processes. In democracies, those exercising public powers hold their powers on democratic grounds. That is, depending on the system in place, they may either be directly elected by the people (as is the case with members of parliament and presidents), be appointed by or be accountable to directly elected representatives (members of the executive and regulatory agencies), or serve a role that has wider democratic legitimation (judges). The idea behind the democratic basis of these powerholders is as straightforward as it is known: the people should rule themselves.[10] Such self-rule is said to have both intrinsic and instrumental value. It is intrinsically valuable because it treats people as equals who all get an equal say in decisions affecting them.[11] It is instrumentally valuable because it can help prevent tyranny and have other desirable consequences such as protecting personal interests and increasing prosperity.[12]

As we emphasized in Chapter 2, human beings are not the only ones who are governed by human laws and policies. Countless sentient non-human animals are also governed. For example, when a democratic parliament enacts a law permitting the export of live animals for slaughter,[13] then it governs the lives of these animals. And when an executive agency decides that animals need to be killed for health and safety reasons, then that too governs the lives of animals.[14] Because animals' basic interests are directly at stake in such laws, policies, and decisions, they should play a role in the formation of these laws, policies, and decisions, and they should be able to reap some of the instrumental rewards arising from democratic processes. In fact, because animals are just as governed by democratic processes as human beings, they have an equal right to democratic participation as human beings. In this sense, governed sentient animals should be seen as part of the 'demos', that is, they are part of the people who have a right to rule themselves.

To be sure, as we explain below when addressing the 'what' question, an *equal right* to democratic participation, which animals possess, should not be conflated with a right to be treated *equally* to humans, which animals do

[10] Robert A Dahl, *On Democracy* (2nd edn, Yale UP 2015) 90 (using the term 'polyarchy').

[11] Thomas Christiano, *The Constitution of Equality: Democratic Authority and Its Limits* (OUP 2008).

[12] See Dahl (n 10) 45.

[13] See eg Jordan M Sosnowski, 'Detailed Discussion of Australian Live Export Laws' (*Animal Legal & Historical Center* 2013) <https://www.animallaw.info/article/detailed-discussion-australian-live-export-laws> accessed 4 November 2024.

[14] See eg Alys Davies, 'Danish Mink Cull: PM Frederiksen and Officials Heavily Criticised' *BBC News* (30 June 2022) <https://www.bbc.com/news/world-europe-62001162> accessed 4 November 2024.

not possess. Animals need to be given due concern in accordance with their capacities and interests, properly understood in collaboration with animals, but they need not and ought not be treated identically to humans in matters of political participation. Before we turn to this issue, however, it is worth noting that ours is not a radically new idea. While not all scholars agree with us that animals have a right to democratic *participation*—a proposal we will justify below—there is scholarly precedent for the view that animals' interests matter democratically. For example, Alasdair Cochrane has advocated for what he calls 'sentientist democracy'. Such a democracy, he proposes, would protect the 'equal worth and rights of sentient creatures'[15] through democratic mechanisms such as animal representatives, an idea we return to in Section 2. More recently, Cochrane has proposed that 'animals must have the right to democratic representation',[16] providing for non-contingent ways of representing animals in political processes. In a similar vein, Janneke Vink has argued that sentient animals have a 'consideration right'. This, she explains, is a 'political right to have one's interests considered by the political community's rulers'.[17]

Yet, while there is increasing agreement in scholarship that animals possess a right to have their interests considered democratically, there is disagreement about *what* such a right would entail. Specifically, a divide has emerged around the issue of whether animals should be viewed as political patients who are dependent on human beings to represent their interests in politics, or political agents who are able to participate in politics themselves.[18] Those who regard animals as lacking political agency—a view we refer to as the Political Patients Approach—tend to understand agency as a cognitively demanding concept.[19] For instance, as Angie Pepper proposes, political agency requires '(i) the capacity to intend to affect institutions; (ii) the capacity to engage in complex forms of shared intentionality; and (iii) the capacity to imagine alternative political futures'.[20] This approach is challenged by scholars who propose

[15] Alasdair Cochrane, *Sentientist Politics: A Theory of Global Inter-Species Justice* (OUP 2018) 36.

[16] Alasdair Cochrane, *Should Animals Have Political Rights?* (Polity Press 2020) 93.

[17] Janneke Vink, *The Open Society and Its Animals* (Palgrave Macmillan 2020) 32.

[18] For the view that animals are political patients, see eg Cochrane (n 16); Vink (n 17); Bernd Ladwig, *Politische Philosophie der Tierrechte* (Suhrkamp 2020); Angie Pepper, 'Political Agency in Humans and Other Animals' (2021) 20 Contemporary Political Theory 296. For the view that animals are political agents, see eg Sue Donaldson and Will Kymlicka, *Zoopolis: A Political Theory of Animal Rights* (OUP 2011); Eva Meijer, *When Animals Speak: Toward an Interspecies Democracy* (New York UP 2019); Charlotte E Blattner, Sue Donaldson, and Ryan Wilcox, 'Animal Agency in Community: A Political Multispecies Ethnography of VINE Sanctuary' (2020) 6 Politics and Animals 1.

[19] See also Vink (n 17) 24 (using similar terminology).

[20] Pepper (n 18) 298.

that we should see animals as political agents.[21] We can call this the Political Agents Approach.[22] Donaldson, for instance, argues that we should understand agency in a relational sense as this better captures how both humans and non-humans rely on their social environment and its 'uptake'[23] to develop and exercise agency. According to her, this broader view of agency allows us to see animals as exercising political agency in the private and public spaces in which they live, and it also encourages us to co-create with animals 'new spaces of citizenship'[24] that allow genuine inclusion of animals in the political community.

While this debate captures important differences in outlook and methodology, its importance for constitutional theory should not be overstated. This is because there is some unspoken common ground between authors on both sides: they all acknowledge that formal political processes, such as law-making in parliament or the creation of policies in ministries, continue to be the domain of human beings, and so, any inclusion of animals in these processes will have to rely on human representatives. This is obvious in the work of proponents of the Political Patients Approach who argue that animals lack the capacities to intentionally affect institutions and engage in complex processes such as elections to pursue their own political futures. However, perhaps surprisingly, it is also an assumption that underpins the Political Agents Approach. For example, as Donaldson clarifies, political participation for animals is primarily possible in 'wider non-electoral dimensions',[25] which precisely go beyond the 'vote-centric conceptions of democracy'[26] that exclude animals because they cannot vote in the ordinary sense.

Although the point should not be taken too far, there is thus an important respect in which authors of the two approaches are talking past one another: proponents of the Political Patients Approach tend to focus on *formal* political processes like elections and voting, whereas advocates of the Political Agents Approach tend to focus on *informal* processes such as how animals and humans can co-create norms in private and public spaces. The takeaway for us is this: when it comes to rethinking constitutional democracy, we do not have

[21] For an informative exchange on this, see Sue Donaldson, Janneke Vink, and Jean-Paul Gagnon, 'Realizing Interspecies Democracy: The Preconditions for an Egalitarian, Multispecies, World' (2021) 8 Democratic Theory 71.

[22] Scholars adopting this approach question the very distinction between agents and patients: see Will Kymlicka and Sue Donaldson, 'Rethinking Membership and Participation in an Inclusive Democracy: Cognitive Disability, Children, Animals' in Barbara Arneil and Nancy J Hirschmann (eds), *Disability and Political Theory* (CUP 2016).

[23] Sharon R Krause, 'Beyond Non-Domination: Agency, Inequality and the Meaning of Freedom' (2013) 39 Philosophy & Social Criticism 187, 197.

[24] Donaldson (n 5) 713.

[25] ibid 711. See also ibid 730.

[26] Blattner, Donaldson, and Wilcox (n 18) 711.

to reject the fact that it is humans, and more specifically adult human beings with the relevant capacities, who will remain the primary subjects engaged in formal political processes. Indeed, far from having to reject it, it is a point that we need to embrace. For animals' interests to have any chance of being effectively channelled into democratic politics, there need to be humans serving as their representatives. This is the first important point to note.

But there is also a second one: human representatives cannot accomplish this task on their own—they require the participation of animals. To understand this, it is important to recognize that just because human representatives are required to channel animals' interests into formal politics does not mean that animals lack political agency altogether, and that their representatives can therefore do without animals' participation.

Above, we suggested that the debate between the two approaches is not so crucial for our purposes. However, this is not to say that there are no important differences between the Political Agents Approach and Political Patients Approach. Specifically, according to adherents of the Political Patients Approach, animals possess a right to have their interests considered, but they do not have a right to *participate* in democratic processes themselves. For instance, according to Pepper, 'only political agents have rights to democratic political participation, including the right to vote, and to freedoms of speech and association'.[27] Cochrane argues in a similar vein that, while animals have a right to democratic representation, they do not have 'rights of participation'.[28] Cochrane holds that this is because animals—or most animals anyway—lack the capacity to engage in, and therefore lack an interest in, collective self-government.[29] In other words, they have no interest in being part of the demos. Nonetheless, he believes that it is 'possible to represent animals effectively without assigning them rights to participation' as long as we 'ensure that the representatives of animals are attentive to the agency, voice and complex interests of animals'.[30]

We agree with proponents of the Political Patients Approach that only humans can meaningfully hold formal political rights such as the right to vote. However, it is a mistake to infer from this, as proponents of this approach have, that animals cannot participate politically *at all*, and that it is therefore sufficient to have their interests incorporated through human representation in which animals do not get to participate.[31] Vink, for example, claims that the

[27] Pepper (n 18) 296.
[28] Cochrane (n 16) 104.
[29] ibid 104–05.
[30] ibid 108.
[31] On the importance of effective participation, see generally Dahl (n 10) 37–39.

agency that animals express in private and public spaces, such as stray dogs choosing to ride on the subway in some cities, is 'irrelevant'[32] when it comes to politics, and that animals are therefore 'merely undergoers of political action'[33] by humans.[34] However, views such as this rely on an overly narrow conception of politics and political participation which excludes all forms of activities that are not centred around elections and voting. Such a narrow conception is unappealing for at least three reasons: it is based on an unattractive account of democracy, it is epistemically wanting, and it is unnecessary.

First, the narrow conception is based on an unattractive account of democracy. A conception of politics and, more specifically democracy, which focuses on elections and voting is premised on a problematic majoritarian conception of democracy, which reduces democracy to collective decision-making by majorities.[35] As we will show in Section 2, a consequence of this majoritarian understanding is that judicial review of legislation by courts can be seen as illegitimate. For example, on a majoritarian understanding, a court that invalidates a law that unjustifiably discriminates against a certain group of people or that is plainly irrational would be acting undemocratically because it frustrates majority will. However, this majoritarian understanding is in tension with the widespread practice of courts around the world which exercise some form of judicial review over legislation. The best example of this is what Mark Tushnet has called 'strong-form review',[36] that is, judicial review in which the judiciary has the last word over whether or not a law enacted by parliament is constitutional. However, also forms of 'weak-form review'[37]—which allows other branches to respond to judicial review decisions—are in tension with a purely majoritarian account of democracy on which there can be no limits to majority will.[38] The Political Patients Approach, centred as it is on elections and voting, does not regard judicial review of either kind as 'democratic' insofar as they do not deal with elections and voting. Yet, there is an attractive sense in which

[32] Vink (n 17) 30.

[33] ibid 24 (emphasis removed).

[34] A more nuanced formulation is adopted by Lucy Parry, who notes that animals 'are unable to participate *directly*' (our emphasis): Lucy J Parry, 'Deliberative Democracy and Animals: Not So Strange Bedfellows' in Robert Garner and Siobhan O'Sullivan (eds), *The Political Turn in Animal Ethics* (Rowman & Littlefield 2016) 146. If one understands 'direct participation' in the sense of 'participation in formal democratic processes', then that formulation strikes us as correct.

[35] Ronald Dworkin, *Freedom's Law: The Moral Reading of the American Constitution* (OUP 1996) 15–16.

[36] Mark Tushnet, *Weak Courts, Strong Rights: Judicial Review and Social Welfare Rights in Comparative Constitutional Law* (Princeton UP 2008) 33.

[37] ibid 23.

[38] Ronald Dworkin refers to this as the constitutional conception of democracy: Dworkin (n 35) 17.

courts act democratically when they review laws and, by inference, participate in politics.[39]

Similarly, there is an attractive sense in which animals' actions can be interpreted as political in the broader sense. For instance, Donaldson proposes the creation of 'microboards' consisting of humans whom an animal trusts and who would be responsible for identifying that animal's interests in the workplace or at home, and whose 'knowledge and power . . . could then be fed upstream into more collective decision making (eg union policies or family laws)'.[40] In this way, even a seemingly 'unpolitical' private act like a dog's choosing to stay with the parent who did not purchase them in case of a divorce can be read as a political act that can translate, through microboards and more formal democratic processes, into a law that requires judges to take into account the best interest of animals in cases of divorce.

Second, the narrow conception is wanting from an epistemic point of view.[41] This is because it is doubtful whether human representatives can 'represent animals effectively without assigning them rights to participation', as Cochrane and other proponents of the Political Patients Approach believe. Their position seems compelling at first glance: as long as human representatives are sufficiently attentive to the interests of the animals they represent—that is, as long as they become good listeners and seek out interactions with their animal constituents—then why would it be necessary for animals to participate politically themselves?[42] However, as Donaldson has pointed out, there is an issue with this position: most animals cannot successfully communicate their interests without being given the chance to explore a range of different options. This is because they cannot express their interests in an abstract way but only through embodied forms of communication. In other words, they need to be able to explore different situations so as to be able to choose one over the other, thus effectively expressing their interests. As Donaldson puts the point, '[u]nlike many humans, DAs [domesticated animals] cannot articulate political imaginaries, realistic utopias, and visions of justice. They need to actually

[39] Such a broader reading of democracy and political participation is also more in tune with the recent shift away from narrow principal-agent-based accounts of political representation: Svenja Ahlhaus, 'Political Representation of Animals' in Anne Peters, Kristen Stilt, and Saskia Stucki (eds), *Oxford Handbook of Global Animal Law* (OUP 2025) 2 (forthcoming; on file with the authors). In particular, it is also more in line with the push to wider understandings of political participation in the context of children's rights and rights of people with disabilities: see eg Harri Raisio, Katja Valkama, and Elina Peltola, 'Disability and Deliberative Democracy: Towards Involving the Whole Human Spectrum in Public Deliberation' (2014) 16 Scandinavian Journal of Disability Research 77.

[40] Donaldson (n 5) 723.

[41] Donaldson refers to this as a matter of 'epistemic injustice': ibid 717–18. See also Miranda Fricker, *Epistemic Injustice: Power and the Ethics of Knowing* (OUP 2007).

[42] Cochrane (n 16) 109.

experience alternative possibilities.'[43] This has a critical implication: because animals need to be able to participate in different possibilities to truly express their interests, a non-participatory approach cannot fully capture those interests. And, as a result, it is wanting epistemically. Human representatives can only effectively represent animals politically if animals participate in their representation.

Finally, the conception is also unnecessary. The central worry that seems to undergird the Political Patients Approach is that political rights and political concepts lose their meaning once we say that animals have a right to participate democratically or politically.[44] However, as should have become clear by now, that worry is unfounded. Even if we recognize animals' equal right to democratic participation, that does not mean that animals get the same right to democratic participation as humans. As we showed, even advocates of the Political Agents Approach concede that, in formal democratic processes, human representatives are still required, and these representatives possess a different set of rights and responsibilities than animals who lack their capacities. Put simply, just because animals are enabled to participate democratically in some settings does not mean that they receive the right to vote or to run for office, let alone that humans will lose any of their rights. Hence, it is not necessary to reject the idea that animals have political agency and a right to participate democratically to maintain the meaning and value of forms of democratic participation that are characteristic of adult human beings.

To recapitulate, this section has first shown that constitutions around the world still adopt an anthropocentric approach to democracy. We then made the case for granting all governed sentient animals an equal right to democratic participation. This does not mean that animals and humans should be treated identically when it comes to political rights. Only certain human beings have the capacity to engage in formal democratic processes, and only they should therefore possess the right to vote and to run for office. For the same reason, it is inevitable that human representatives are required to feed animals' interests into formal democratic processes. Yet, that does not prevent us from viewing animals as political agents with a right to democratic participation. Indeed, centring animals' participation is critical to 'move from thinking about other animals to thinking and working with them'.[45]

[43] Donaldson (n 5) 724.
[44] See eg Pepper (n 18) 313.
[45] Meijer (n 18) 138.

Having responded to the question *why* animals should be involved in democratic processes and *what* their right to participate would look like, Sections 2 and 3 will turn to outlining *how* this right should be put into practice.

2) A legal *and* political approach

One of the most extensive treatments of how animals should be included in the democratic process can be found in Vink's *The Open Society and Its Animals.* Vink argues that it is through legal rather than through political institutions that animals' interests should be protected. This section starts by challenging Vink's position by pointing out the shortcomings of a purely legal approach and highlighting the advantages of the political approach (Subsection A). Pulling these threads together, we then make the case for an integrated approach that combines legal and political mechanisms (Subsection B).

A) The shortcomings of a purely legal approach

As we saw in Section 1, Vink proposes that sentient animals should have a 'consideration right', that is, a right to have their interests considered by those in charge of the community. To determine how this right can best be protected, Vink explores and contrasts two avenues: the political and the legal.[46] The political avenue focuses on protecting animals' interests through legislative and executive institutions. The legal avenue, by contrast, involves the creation of legal protections for animals, and the enforcement of these protections through the courts. Vink mentions various political options, such as introducing an ombudsperson for animals, reserving seats in parliament for animal representatives, and creating a committee on animal welfare. However, she finds that all these options face insurmountable challenges. Among these challenges are the difficulty that representatives cannot easily identify what the interests of their animal constituents are, the risk that representatives abuse their powers, and the challenge of identifying the right amount of power for representatives to hold in the first place. Because of these challenges, Vink concludes that the political avenue is unsuitable for animals. Instead, she argues that legal institutionalization of animals' consideration right can do the job. Specifically, she proposes two legal pathways: the creation of constitutional state objectives that

[46] Vink (n 17) 125.

impose duties on the community to protect animals and the introduction of fundamental legal rights for animals.[47]

Vink's analysis of the challenges of the political avenue is valuable. However, her argument that animals can be given due concern in democracies through the legal avenue alone is not compelling. To see what we mean, and to understand why we argue for an integrated approach that utilizes legal *and* political mechanisms, it will be helpful to shed light on how Vink underestimates the weaknesses of the legal avenue and how she exaggerates the shortcomings of the political avenue. Vink underestimates the deficiencies of the legal avenue in at least four respects.

First, Vink argues that judges can approach state objectives or fundamental legal rights for animals in an 'objective' way, thus overcoming what she views as the 'anthropocentric biases'[48] that cloud the judgment of political representatives. While Vink is certainly right that judges operate within constraints such as legal hierarchies and are tasked to apply rules and principles using established methods of interpretation, her claim that 'judges can, and professionally are required to, prevent irrational biases' from influencing their judgments seems overly optimistic. As empirical studies have revealed, judges are not free from bias either.[49] And much like other public officials, they too can abuse their powers.[50]

Second, not all laws and policies that are created touch on state objectives or fundamental legal rights, especially rights of the 'negative'[51] sort that Vink envisages. Take fundamental rights, which are the most powerful instrument in Vink's legal toolset. Vink states that negative rights require a 'back-off'[52] attitude from government. Consider the right to life and the right not to be tortured which Vink discusses.[53] Public officials must not kill sentient animals or subject them to torture, but, according to her, the officials have no positive duty towards animals, such as the duty to create oversight bodies that monitor places where animals are at risk of being killed or tortured. We think that Vink construes these rights too narrowly because, as we argued in Chapter 3 Section 3, the right to life and the right not to be tortured have both negative and positive aspects. However, on Vink's view that these rights simply require a 'back-off'

[47] ibid 206.

[48] ibid 319.

[49] See eg Linda G Mills, *A Penchant for Prejudice: Unraveling Bias in Judicial Decision Making* (University of Michigan Press 1999).

[50] See eg Rosalind Dixon and David Landau, *Abusive Constitutional Borrowing: Legal Globalization and the Subversion of Liberal Democracy* (OUP 2021) ch 5.

[51] Vink (n 17) 269.

[52] ibid.

[53] ibid 283.

attitude from officials, they would not be sufficiently protective. This is because there would be too many laws and policies that would fall outside the scope of these rights. For example, a law could fall outside the scope if it had nothing to do with torture or a threat to life. Suppose a law was passed that legalized certain fertilizers for crops that are known to cause sterility in the wild animals that feed on them. Courts could not bring animals' negative rights to life and not to be tortured to bear on such a law because it does not relate to the killing or torturing of animals. Furthermore, even a law that related to the killing or torturing of animals could fall outside the scope of these rights if the state does not violate its duty to 'back-off'. Imagine a company that provides donkey rides for children was to beat and shoot all the new donkeys that turn out not to be docile enough for the job. Since it is not the state killing the donkeys, but a private company, the courts would again be powerless and could not enforce the donkey's negative rights to life and not to be tortured.

Third, even for laws and policies that *do* fall within the scope of fundamental legal rights or state objectives, there is the problem that animals' interests are not factored into the making of the laws and policies. Cochrane has drawn the helpful distinction in this context between 'framing' and 'constraining' policy.[54] Fundamental legal rights and other duties imposed on public officials *constrain* what laws and policies can be enacted. For instance, if animals have a constitutional right to life, then parliament cannot permissibly create a law that violates that right. However, the legal avenue Vink prioritizes does not guarantee that animals' interests will be considered in the *framing* of laws. This is problematic for, as Cochrane points out, 'it is perfectly clear from the experience of women, minority, and other disenfranchised groups that simply relying on policymakers to constrain their actions is insufficient for equal consideration'.[55]

Finally, even if we ignored the issue of law and policy framing, there remains the further problem that not all laws and policies are judicially reviewable. Even if a given law or policy fell within the scope of animals' right to life or freedom from torture, there is no guarantee that the courts can actually enforce these rights. This is because courts in some jurisdictions lack the relevant powers that would allow them to invalidate unconstitutional acts of parliament in a 'strong-form review' sense, or at least exercise judicial review of such laws in a 'weak-form' sense.[56] Vink suggests we can be 'confident'[57] that alternative

[54] Cochrane (n 15) 42.
[55] ibid.
[56] See Tom Ginsburg and Mila Versteeg, 'Why Do Countries Adopt Constitutional Review?' (2014) 30 The Journal of Law, Economics, and Organization 587, 587–88.
[57] Vink (n 17) 307.

ways of protecting fundamental rights can fill the gap. In the Netherlands, she argues, the lack of formal review powers is remedied by the application of international human rights documents by national courts as well as the fact that the Council of State, an advisory body, comes in when legislation is drafted and emphasizes the importance of protecting constitutional rights.[58] The problem is that, even if this provided sufficiently non-contingent protection of constitutional rights in the Netherlands,[59] there are countries for which this is not true, or where there is at the very least a live debate about whether rights are sufficiently entrenched, or whether the courts should be given more review powers.[60]

Vink also exaggerates the shortcomings of the political avenue. First, it is important to note that the political avenue has benefits that the legal avenue cannot provide. As indicated above, one key benefit is that it allows for animals' interests to be factored into the process of framing laws and policies. This is beneficial from an outcome-based perspective in that incorporating animals' interests into the shaping of laws and policies can impact what these laws and policies will look like. But in addition, the political avenue can also improve the process in which laws and policies are made. As Robert Goodin has argued, the incorporation of animal interests in the political process can have beneficial effects on the political deliberations themselves. Not only will the presence of animal representatives in institutions such as parliament require that other representatives justify their positions, but it can also lead to a process of 'anticipatory internalisation'[61] whereby others will suppress self-interested reasons knowing they have to justify themselves. As a result, some may internalize critical positions in anticipation of these very criticisms.

Second, the shortcomings that Vink identifies are not as problematic as she makes them out to be. Of course, there are difficulties that beset political options for incorporating animals' interests into the democratic process. It cannot be denied, for example, that representatives of humans will have an easier time communicating with and establishing the interests of their constituents than animal representatives of theirs.[62] What is more, there are certainly risks that

[58] ibid.

[59] See eg Gerhard van der Schyff, 'The Prohibition on Constitutional Review by the Judiciary in the Netherlands in Critical Perspective: The Case and Roadmap for Reform' (2020) 21 German Law Journal 884.

[60] For a canonical work in the UK context, see TRS Allan, *The Sovereignty of Law: Freedom, Constitution, and Common Law* (OUP 2013).

[61] Robert E Goodin, 'Enfranchising the Earth, and Its Alternatives' (1996) 44 Political Studies 835, 847.

[62] This is relative, of course. As Christopher Stone's has remarked: 'I am sure I can judge with more certainty and meaningfulness whether and when my lawn wants (needs) water, than the Attorney General can judge whether and when the United States wants (needs) to take an appeal from an adverse

unchecked representatives could abuse their powers, and that we do not get the balance of powers between animal and human representatives right.

However, Vink overlooks a crucial point: the different options of incorporating animal interests politically (animal representatives in parliament, animal ombudspersons, etc) are not mutually exclusive, but work best in concert. Indeed, when combined with one another, their respective risks may well be cancelled out or at least mitigated. For example, while unchecked animal representatives in parliament may be a danger, this changes once other forms of checks are brought into play, such as formal checks by parliamentary committees or informal checks through ombudspersons. As Vink herself acknowledges in the context of the Netherlands, political bodies such as the Council of State can exert political pressure and provide different forms of accountability. Svenja Ahlhaus makes this point forcefully:

> Even though it is possible to discuss the merits of elected, appointed, and self-appointed representatives separately, I think that it is most promising to see them as complementary. Just like human representation is not limited to a single representative relationship, animal concerns need to be treated in different forms of institutionalised and informal representation.[63]

B) An integrated legal-political approach

Ahlhaus is right to argue for what she calls the 'dynamic interplay between differently positioned representatives'.[64] But in our view she does not go far enough. It is not only necessary that there is an interplay between different ways of representing animals' interests politically, but we also need an interplay between political *and legal* institutions.[65] We call this the integrated approach to sentience-based democracy. The idea here is that animals' equal right to democratic participation can only be effectively accommodated if both political and legal avenues are pursued.

A sceptic might ask whether it is not a stretch to include legal institutions in an account of democracy. We believe not, and to see this, we can return

judgment by a lower court': Christopher D Stone, 'Should Trees Have Standing—Toward Legal Rights for Natural Objects' (1972) 45 Southern California Law Review 450, 471.

[63] Ahlhaus (n 39) 14. She uses the term 'surrogate accountability' to describe the idea that there are political, not only legal ways, of holding representatives to account: ibid 10.

[64] ibid 14.

[65] See also Cochrane (n 15) 37–38.

to our discussion of the problems of the Political Patients Approach's overly narrow understanding of democracy. As we saw in Section 1, proponents of that approach have difficulty explaining why a court exercising judicial review over a problematic law, and potentially striking it down, may be acting democratically. The reason is that the Political Patients Approach presupposes an unattractive majoritarian conception of democracy, which only views as relevant to democracy what goes on in elections and parliament. However, there is a broader and more attractive sense of democracy: Dworkin's constitutional conception of democracy. This conception requires 'that collective decisions be made by political institutions whose structure, composition, and practices treat all members of the community, as individuals, with equal concern and respect'.[66] In other words, democracy still involves collective political decisions, but it subjects these decisions to constraints which can be enforced by courts. While Dworkin was, in our view, wrong to focus only on human beings, he was right to suggest that a successful democracy requires the interplay of not only political bodies, such as parliament and the executive, but also legal bodies, such as courts. In fact, we can cast our democratic net even more widely and include more informal processes, too. Donaldson puts this thought well when noting that '[t]he legitimacy and quality of democratic decision-making depends on dense horizontal and vertical webs of participation, deliberation, representation and accountability that operate alongside and in-between moments of periodic elections'.[67]

How would such an integrated approach work? In other words, how exactly can we ensure that democracies can give due concern to all governed sentient animals? We do not have the space here to explore all the different possibilities in detail, and we will therefore limit ourselves to outlining what we believe to be some of the most effective ways of including animals' interests—both legally and politically. As we will emphasize in the book's Conclusion, more research is needed not only to add substance to the proposals we mention but also to explore new options.

We can keep the discussion of legal options for including animals' interests in democracy short. This is not because these options are less important than political options, but because we have already discussed them in more detail elsewhere in the book. The two legal options we propose are: fundamental political rights for animals and legal standing for animals.

[66] Dworkin (n 35) 17. In Chapter 4 Section 3.B, we distanced ourselves from Dworkin's 'equal concern and respect' because it entails the existence of only absolute rights. We instead used the concept of 'due concern and respect'.

[67] Donaldson (n 5) 711.

First, sentience-based democracies require the recognition of entrenched *fundamental political rights* for sentient animals, and they require a mechanism of enforcing these entrenched rights, such as constitutional judicial review. As we saw above, fundamental rights that are judicially enforced can serve as an important constraint on law-making and policy-making processes. In Chapter 3, we argued that existing bills of rights that protect human interests can be reimagined to also protect the interests and capacities of animals. The same argument we make there for interpreting existing rights like the right to life so that it extends to animals can also be applied to existing political rights. These rights, we propose, should be interpreted in a way that protects a right to democratic participation not only for humans but for all governed sentient animals.

Second, legal mechanisms need to be put in place to allow the enforcement of animals' political rights when they are violated. A tried and tested way of achieving this in the human context is the granting of *legal standing* to humans whose rights have been violated. Similar proposals have been made for animals to have legal standing exercised through human guardians. In Section 2.B of Chapter 5, we discussed how such standing could work and who would be suitable guardians.

Let us now turn to political pathways of including animals' interests. We propose three such mechanisms: a reserved number of seats in legislative bodies for sentient beings' representatives, sentient beings' committees inside and outside of parliament, and sentient beings' initiatives. We will sketch the first two mechanisms here and will introduce the last mechanism in Section 3.

First, legislative bodies in democratic states should have a *reserved number of seats* for humans who are tasked with representing the interests of sentient beings. We call them 'sentient beings' representatives'. Sentient beings' representatives differ from members of existing parties that promote animal rights in that these parties provide no guarantee that animals' interests will be factored in.[68] This is because these parties may or may not get elected. Also, even if they do get elected, it is entirely up to the members of these parties to decide if and how they want to protect animals. By contrast, sentient beings' representatives would ensure that law-making bodies always contain a minimum number of representatives whose responsibility it is to bring in animals' voices.[69] Sentient beings' representatives should exist in all legislative bodies: national parliaments and, where they exist, subnational parliaments, and even local councils

[68] See above n 8.
[69] See Ahlhaus (n 39) 11.

that have law-making powers. The reason for this is that sentient animals are potentially governed by laws and policies that are created at all levels of the state. It is therefore necessary to ensure that they are given due concern in all these processes.[70]

Who gets to elect sentient beings' representatives? This is not easy to answer. As Vink has highlighted, it is difficult to identify criteria for determining suitable representatives, and there is the above-mentioned danger that unchecked representatives misuse their powers. The main reason these problems exist is because animals cannot themselves elect and—just as importantly—unelect representatives.[71] However, just because these are difficult issues does not mean that they cannot be resolved. In our view, Cochrane has developed the most promising answer. According to him, sentient animals should have 'dedicated animal representatives'.[72] These are human representatives whose job it is to 'ensure that animal interests are heard, are considered and weighed appropriately in the framing and formulation of public policies'.[73] The dedicated animal representatives would be elected by 'deliberative selection assemblies'[74] rather than by the citizenry as a whole. These assemblies would consist of a subset of voters who are chosen by lottery from the citizenry and whose role it is to 'hear and take evidence from the various candidates, and from relevant experts, and to deliberate with each other before casting a vote as to who should be selected'.[75] Importantly for our purposes, animal participation could come in at several points here. For instance, the above-mentioned microboards could be part of the relevant experts that advise the deliberative selection assemblies and that would draw on the participation of animals themselves. Also, members of microboards would be prime candidates to run for office as sentient beings' representatives.

Cochrane's model is promising because it can resolve the above-mentioned difficulties. It accommodates the worry that there is no check on animal representatives since animals cannot themselves influence their representatives at the ballot box. In Cochrane's model, the deliberative selection assemblies serve as such a check: they get to decide whom to elect, whom not to elect, and—if given the power—even whom to remove from office. What is more, the lottery

[70] We leave open here the question of whether ministerial departments should also contain (appointed) sentient beings' representatives. Considering the significant policy-making powers that executive bodies can have, this would seem desirable. However, because the executive is, at least in many states, accountable to the legislature, we focus here on the latter.

[71] Vink (n 17) 149.

[72] Cochrane (n 16) 101.

[73] ibid.

[74] ibid 103.

[75] ibid.

system solves the problem of having to identify suitable criteria for selecting representatives; and the information provided to members of the selection assemblies by candidates and experts ensures that they can represent animals' interests in an educated way. Finally, the lottery system also preserves political equality among citizens because it gives everyone an equal chance to be drawn by lot to serve in an assembly.[76]

Lastly, how many reserved seats should there be? There would seem to be an obvious and radical answer to this question: in Chapter 2, we argued that in a constitutional system such as that of the United Kingdom, around 98 per cent of governed sentient beings are non-human. Should we therefore reserve 98 per cent of the seats in parliament for sentient beings' representatives? Not necessarily. Such a direct correlation between representatives and constituents is not required, and perhaps not desirable, because sentient beings' representatives would not be pure delegates of those who elected them. A delegation-based model, on which representatives effectively become their constituents' mouthpieces, is difficult to implement in the case of non-human animals who cannot communicate their political preferences in the ways many humans can.[77]

This means that the number need not be 98 per cent, but nor should it be too low. As Ahlhaus notes, 'the degree of effectivity of animal representation will obviously vary with their number',[78] so a number needs to be identified to ensure a sufficient degree of effectiveness. While the exact figure may differ from system to system, we propose that, as a pragmatic and realistic first step, at least 30 per cent of legislative seats should be occupied by sentient beings' representatives. This number is low enough to allow human beings to retain a majority of seats, thus making it more likely for sentient beings' representatives to be created in the first place. At the same time, 30 per cent is not too low either. As the experience of parties with a substantial but below-majority number of seats in parliaments has shown, this share of seats would allow sentient beings' representatives to exercise significant influence such as by playing 'kingmaker' in coalition negotiations; exerting policy concessions for animals from coalition partners; blocking problematic legislation, especially legislation that requires a supermajority; shaping policies through exercising influence in committees; and influencing public opinion through presence in the media.[79] Nonetheless,

[76] Ahlhaus (n 39) 12.
[77] Cochrane (n 15) 46.
[78] Ahlhaus (n 39) 19.
[79] On the kingmaker role of small parties, see Amir Abedi and Alan Siaroff, 'The Kingmaker Is Dead, Long Live the Kingmaker: Examining the Degree of Influence of Small Parties in the Coalition-Formation Process in Germany' (2011) 20 German Politics 243. On policy concessions, see Heike

30 per cent is not a hard-and-fast rule, and it is far from set in stone. If a system moves to a higher percentage of sentient beings' representatives, then this will help boost its legitimacy.

Second, there is a need for *extra- and intra-parliamentary committees* to channel animals' interests into the law-making and policy-making process. Firstly, an extra-parliamentary committee should be created to provide an external check on the activities of parliament and the executive. As Ahlhaus proposes, such an independent committee could consist of appointed experts whose role is to produce reports and to advise lawmakers on scientific and legal matters relating to sentient beings.[80] Although giving this committee the power to veto legislation or to propose binding amendments would be undesirable from a democratic point of view, it could still exert significant political pressure on law- and policymakers and provide them with helpful additional expert knowledge on relevant issues.

In addition, it is important to put in place mechanisms internal to parliament that ensure that sentient animals' interests are factored into the law-making process beyond the resort to sentient beings' representatives. For this purpose, we propose the creation of an intra-parliamentary sentient beings' committee. Parliaments around the world already rely on the work of a diverse range of committees which are staffed not with external experts but with legislators who either have or can develop specialized expertise in the committee's area. The role of such committees is to provide specialized scrutiny on proposed legislation and governmental policy, as well as to examine and discuss other political matters. In the United Kingdom, for example, the Joint Committee on Human Rights (JCHR) consists of members appointed from both the House of Commons and House of Lords who examine governmental bills on their conformity with human rights. The JCHR issues reports that have influenced parliamentary debates and, in some cases, helped defeat problematic bills.[81] In a similar way, the intra-parliamentary sentient beings' committee we propose

Klüver, Hanna Bäck, and Svenja Krauss, 'Coalition Agreements and Ministerial Portfolios' in Heike Klüver, Hanna Bäck, and Svenja Krauss (eds), *Coalition Agreements as Control Devices: Coalition Governance in Western and Eastern Europe* (OUP 2023). On the blocking of legislation, see Ian Shapiro, 'On Political Parties' 2 <https://jackson.yale.edu/wp-content/uploads/2022/08/Shapiro-2022.pdf> accessed 4 November 2024. On the influence in committees, see Joseph Crupi, 'Including Minority Parties in Policymaking: A Legislative Requirement to Address Member Interests' (2023) 122 Michigan Law Review 12, 16.

[80] Ahlhaus (n 39) 13.

[81] UK Parliament, 'Human Rights (Joint Committee)' (21 March 2024) <https://committees.parliament.uk/committee/93/human-rights-joint-committee> accessed 4 November 2024. See Michael C Tolley, 'Parliamentary Scrutiny of Rights in the United Kingdom: Assessing the Work of the Joint Committee on Human Rights' (2009) 44 Australian Journal of Political Science 41, 53.

can be tasked with reviewing bills and policies that govern sentient animals, and to produce reports with their findings.[82] To ensure that animals' interests are sufficiently factored into such a committee, a rule could be put in place requiring that a minimum percentage of committee seats must be filled with sentient beings' representatives.

To sum up, this section has argued that approaches like Vink's, which focus solely on the law to include animals' interests in democracy, are insufficient. Instead, we proposed an integrated approach that combines legal and political mechanisms. On the legal side, we proposed that animals should receive fundamental rights protection and the standing to enforce these rights. On the political side, we argued that sentient beings' representatives should hold reserved seats in legislatures and that intra- and extra-parliamentary sentient beings' committees should be created. The third political mechanism we propose are 'sentient beings' initiatives' to which we turn now.

3) Sentient beings' initiatives: Beyond indirect democracy

Although the literature on the democratic inclusion of animals is still relatively new, it is notable that existing proposals focus exclusively on ways to make representative democracy more attuned to animals. As such, there is no discussion of how direct democratic instruments could be utilized to give animals their due concern. This is surprising considering that many countries possess direct democratic instruments,[83] and that these instruments can be a potent way of including animals' interests that complements indirect democratic mechanisms. In several jurisdictions, such direct democratic instruments have been used to call for better protection of animals' interests, such as for example in a recent popular initiative in Switzerland demanding the abolition of factory farming.[84] However, the question of how sentient beings themselves could participate in such processes has so far been ignored.

[82] For a similar institution, see the Animal Sentience Committee, introduced by the UK's Animal Welfare (Sentience) Act 2022. We briefly discuss this in Section 2 of Chapter 2. For an analysis, see Rachel Dunn and Joshua Jowitt, 'The Animal Sentience Committee: Evolution or Revolution in the Legal Recognition of Animal Sentience in the UK?' in Jane Kotzmann and MB Rodriguez Ferrere (eds), *The Legal Recognition of Animal Sentience: Principles, Approaches, and Applications* (Hart Publishing 2024).

[83] The Navigator to Direct Democracy, 'Legal Designs in the World—Navigator' <https://www.direct-democracy-navigator.org/countries> accessed 4 November 2024.

[84] The Federal Council, 'Factory Farming Initiative' (2022) <https://www.admin.ch/gov/en/start/dokumentation/abstimmungen/20220925/iniziativa-sull-allevamento-intensivo.html> accessed 4 November 2024.

In this section, we begin by arguing that our integrated legal-political approach extends to both indirect *and direct* democratic mechanisms, and we explore the main functions of popular initiatives—a powerful type of direct democratic instrument (Subsection A). We then propose the creation of 'sentient beings' initiatives' to leverage these functions for animals (Subsection B).

A) Animals and direct democracy

Goodin once observed that '[w]e cannot literally enfranchise nature and let it tend its interests for itself politically. A direct democracy of the birds and bees and boulders is simply not on the cards'.[85] On the face of it, this view is compelling, and it may explain why scholars exploring the political representation of animals have so far exclusively focused on indirect democratic mechanisms. However, upon closer inspection, there is no good reason for shying away from direct democratic mechanisms when it comes to animals. Quite the contrary, direct democratic instruments are useful supplementary tools in an integrated approach to sentience-based democracy.

It is first helpful to see why Goodin is not entirely right when he suggests that a 'direct democracy for the birds and bees' is not 'on the cards'. He is right, of course, that we cannot expect birds and bees themselves to vote on direct democratic proposals. However, the same is true for representative democracy: birds and bees will not themselves sit in—or rather hover over—the reserved seats in parliament or in the intra- or extra-parliamentary committees. Rather, human representatives will be sitting there on the animals' behalf. In the context of indirect democratic processes, this limitation does not stop Goodin from proposing that animals' interests can be incorporated, and it should not stop us from incorporating them in direct democratic processes either.

In fact, direct democratic processes are not only 'on the cards', but they can also be a potent means of including animals' interests in democracies—a means that can complement the indirect democratic mechanisms discussed in Section 2. One of the key direct democratic instruments on which we focus in this section are popular initiatives, which can come in two forms: rejective and proactive.[86] Rejective initiatives are put forward in response to a proposed

[85] Goodin (n 61) 841.

[86] Other direct democratic mechanisms include agenda initiatives, executive-initiated referendums, and law-initiated referendums. For a useful overview, see Defining the Limits of Direct Democracy (LIDD), 'The LIDD Project' <https://lidd-project.org/research/> accessed 4 November 2024.

law or constitutional amendment and allow citizens to trigger a popular vote aimed at preventing the relevant law or amendment from passing. Conversely, a proactive initiative allows citizens to propose a new law or constitutional amendment and to trigger a binding vote on their proposal.[87] Both initiatives can have legally binding results, if adopted. As one of us has explained elsewhere, popular initiatives have several advantages.[88] We focus on four advantages here: they can force a vote on a specific issue (Agenda-Setting Function); they can introduce the citizenry to new ideas (Catalyst Function); they can bypass deadlocks in parliament or the judiciary (Bypass Function); and they are more receptive to animal participation than indirect democratic processes (Participation Function).

First, popular initiatives have an *Agenda-Setting Function*. As mentioned above, to give animals' interests due concern, it is necessary to include them when laws are framed, not only when they are constrained. Popular initiatives can play an important role in this because they allow animals' interest to determine what gets on the agenda of citizens and legislators, thus enhancing their influence on policy-framing beyond the legislature.[89] Proactive popular initiatives illustrate this particularly well because they allow citizens to directly intervene in the law-making process and to propose new laws or constitutional amendments. But rejective initiatives can give voters control over the agenda, too, by forcing a vote over a proposed bill or amendment. Moreover, initiatives of both types can be a nudge for parliament and the government to propose new laws or make changes to existing proposals, in the hopes that voters will adopt their version of the law instead of the initiators'.[90]

Second, the *Catalyst Function*.[91] Popular initiatives have the advantage of introducing innovative proposals to society. Importantly, popular initiatives can do this even if they are ultimately unsuccessful at the ballot box. Popular initiatives can thus serve as catalysts for novel ideas. This can lay the foundation for future political processes in which the proposed laws, or some version of them, may eventually become reality.[92] This is particularly interesting in the

[87] We follow Daniel Moeckli's terminology and definitions: Defining the Limits of Direct Democracy (LIDD), 'LIDD Data Dashboard' <http://lidd-project.org/data/> accessed 4 November 2024 (click on Documentation, then Items & Definitions).

[88] Raffael N Fasel, 'Direct Democracy and Animal Rights' in Anne Peters, Kristen Stilt, and Saskia Stucki (eds), *Oxford Handbook of Global Animal Law* (OUP 2025) 6–13 (forthcoming).

[89] On the importance of agenda control, see Dahl (n 10) 38.

[90] Wolf Linder, *Swiss Democracy: Possible Solutions to Conflict in Multicultural Societies* (3rd edn, Palgrave Macmillan 2010) 108.

[91] See Wolf Linder, *Schweizerische Demokratie: Institutionen, Prozesse, Perspektiven* (Paul Haupt 1999) 260–61.

[92] See ibid 261, 263.

context of sentience-based democracy. The idea of conferring a right to democratic participation to animals, for example, is still new and is bound to face resistance at the ballot box. Yet, that is not a reason not to put it forward in an initiative. After all, such an initiative in itself will help introduce the public to an innovative idea, which can then set a political process in motion that may, ultimately, lead to the enactment of the proposal in one form or another.[93]

Third, the *Bypass Function*. Popular initiatives are appealing because they can force a vote on an issue without having to go through parliament. This ability to bypass the legislature is particularly attractive in the case of bills that may find support with a majority of voters but that are deadlocked in parliament because the ruling party is opposed to them. What is more, popular initiatives cannot only serve to circumvent the legislature, but they can also bypass courts. This can be helpful in situations where courts are opposed to the idea of extending rights to animals[94] but where a majority of citizens might be in favour of it.

Finally, the *Participation Function*. As Robert Dahl has pointed out, effective participation is one of the hallmarks of democracy,[95] and popular initiatives help make participation more effective in at least two ways, one general and one specific to animals. First, the general point: popular initiatives give voters the opportunity to participate democratically beyond casting a vote in elections every four to five years. Second, popular initiatives lend themselves to the participation of animals. This is because popular initiatives of both the rejective and proactive sort are about concrete—and often discrete—subject-matters. They concern specific topics voters feel passionately about, such as the legalization of marijuana, the raising of the minimum wage, or access to abortion.[96]

This contrasts with elections where voters are faced with candidates standing for an often-diverse set of abstract values and manifesto commitments. Because of the concreteness and the narrow character of popular initiatives, it is easier for sentient beings' representatives and other humans working

[93] See also Fasel (n 88) 8–9.

[94] For an overview of sceptical court decisions from the United States, see eg Raffael N Fasel and Sean C Butler, *Animal Rights Law* (Hart Publishing 2023) 125–34.

[95] Dahl (n 10) 37.

[96] On marijuana, see eg 'Maryland Question 4, Marijuana Legalization Amendment (2022)' (*Ballotpedia*) <https://ballotpedia.org/Maryland_Question_4,_Marijuana_Legalization_Amendment_(2022)> accessed 1 November 2024. On minimum wage increase, see eg 'Nebraska Initiative 433, Minimum Wage Increase Initiative (2022)' (*Ballotpedia*) <https://ballotpedia.org/Nebraska_Initiative_433,_Minimum_Wage_Increase_Initiative_(2022)> accessed 1 November 2024. On abortion, see eg 'California Proposition 1, Right to Reproductive Freedom Amendment (2022)' (*Ballotpedia*) <https://ballotpedia.org/California_Proposition_1,_Right_to_Reproductive_Freedom_Amendment_(2022)> accessed 1 November 2024.

with animals to identify what animals' views on these initiatives are, than it is to establish where these animals stand with respect to political candidates who represent different sets of abstract political values. This is certainly true for popular initiatives that directly concern animals, such as recent initiatives on whether farmed animals should be provided with more space;[97] whether non-human primates should be granted rights to life and bodily and mental integrity;[98] or whether factory farming should be abolished.[99] However, to a lesser degree, it also holds true for proposals that only tangentially concern animals, say, initiatives to improve gun control. Because in initiatives the issues that are voted on are narrower, sentient beings' representatives and others can more readily identify what animals would want. As a result, it becomes easier for animals to participate in the political process through their representatives.

B) Sentient beings' initiatives

This brings us to our proposal for two new direct democratic instruments: rejective sentient beings' initiatives and proactive sentient beings' initiatives. The aim of these instruments is to harness the above-mentioned functions of popular initiatives to help give due concern to animals' interests.

Let us begin with *rejective sentient beings' initiatives.* Recall from the previous section that rejective popular initiatives can be launched in response to a proposed legislative reform or constitutional amendment. The initiative must be brought within a certain timeframe and is aimed at preventing the given reform or amendment from passing. Rejective sentient beings' initiatives would serve the same purpose as rejective popular initiatives, but they would only be available in the case of laws and amendments that would govern animals. For instance, if parliament wanted to pass a constitutional amendment removing some fundamental rights from animals, or if it passed a bill to make it more difficult for animals' representatives to bring legal cases, then a rejective sentient beings' initiative could be brought to trigger a popular vote on the issue. However, if the proposed law was concerned with funding for human education, then the animals' representatives could not bring a rejective initiative. Of

97 'California Proposition 12, Farm Animal Confinement Initiative (2018)' (*Ballotpedia*) <https://ballotpedia.org/California_Proposition_12,_Farm_Animal_Confinement_Initiative_(2018)> accessed 1 November 2024.

98 'Fundamental Rights for Primates' (*Initiative Grundrechte für Primaten*) <https://www.primaten-initiative.ch/en/> accessed 1 November 2024.

99 The Federal Council (n 84).

course, even where they could bring an initiative, there would be no guarantee that the vote would turn out to be in favour of animals, but it could still be a way of setting the agenda, of raising public awareness on an issue important for sentient animals, and of allowing these animals to participate politically.

Who should be allowed to bring a rejective initiative for sentient beings? Here, we encounter the same issue that we saw in the context of sentient beings' representatives in parliament. If all citizens were allowed to bring a rejective sentient beings' initiative, then there is the risk that some may abuse this tool to prevent a bill or amendment from passing that they are opposed to independently of whether it is harmful to animals. What is more, there is the problem that not all citizens will be equally or adequately informed about the needs and rights of animals. However, as in the case of sentient beings' representatives, there is a solution to this problem: the initiatives could be brought by members of the deliberative selection assemblies. Recall that members of these assemblies are drawn by lot, which reduces the risk that they are an interest group bent on abusing the initiative. What is more, members of these assemblies possess specialized knowledge that makes them more qualified at deciding when an initiative may be required.

Of course, much depends on how large the relevant assemblies are and how many signatures an initiative requires to lead to a popular vote. If the bar is too low, then there is a risk that too many initiatives are brought, thus potentially overloading the system. If the bar is too high, then the initiatives may fail to meet their purpose and won't be brought in the first place. It seems reasonable that, to trigger a popular vote, one should require not only a supermajority of members of the deliberative selection assembly but also a certain number of signatures from other citizens. Take the example of Switzerland where currently 50,000 signatures by citizens (out of a total of 5.5 million citizens[100]) are required to trigger a rejective popular initiative.[101] Suppose Switzerland had a deliberative selection assembly made up of 500 members. To trigger a popular vote on a proposed law, one could require a three-fifths supermajority (300 members) in the assembly as well as 10,000 signatures by citizens.

However, rejective sentient beings' initiatives could only respond to existing reforms concerning animals and not put forward proposed laws that would serve animals. This would be a particular problem if sentient beings' representatives would not succeed at winning majorities for their proposals

[100] Bundesamt für Statistik, 'Stimmbeteiligung' (2023) <https://www.bfs.admin.ch/bfs/de/home/statistiken/politik/abstimmungen/stimmbeteiligung.html> accessed 1 November 2024.

[101] Swiss Constitution (1999) art 141.

in parliament, or if the legislature is deadlocked for other reasons. *Proactive sentient beings' initiatives* should therefore also be available in sentience-based democracies. Like their rejective counterparts, proactive sentient beings' initiatives are modelled after popular initiatives, specifically proactive popular initiatives. As such, those eligible to bring them do not have to wait to respond to a bill or constitutional amendment, but they can proactively bring about a binding vote by the citizenry on a proposal they put forward. As in the case of rejective sentient beings' initiatives, proactive sentient beings' initiatives could only be brought to propose laws and constitutional amendments that directly relate to animals. Here, too, it would be members of deliberative selection assemblies who are eligible to bring proactive sentient beings' initiatives. The involvement of deliberative selection assemblies helps ensure that those launching such initiatives have the relevant expertise, and it reduces the risk that they abuse such initiatives for purposes having nothing to do with animals.

As in the case of rejective sentient beings' initiatives, we propose that the threshold for bringing proactive sentient beings' initiatives should be lower than the threshold for proactive popular initiatives. However, it should be higher than for rejective sentient beings' initiatives. This is because proactive initiatives allow the enactment of a law or constitutional amendment in the fast lane, as it were: there would be no need to go through the usual steps of executive drafting and parliamentary readings, which are common for laws created through the legislature. Take again the Swiss case. For a proactive popular initiative to bring about a vote by the citizenry, 100,000 signatures need to be collected, so double the number that is required for rejective initiatives.[102] If we apply a similar logic to proactive sentient beings' initiatives, we could require that, for such an initiative to trigger a popular vote, it would require endorsement by four-fifths of the member of the deliberative selection assembly (so at least 400 members) as well as the signatures of 20,000 other citizens.

To recapitulate, this section has started to fill the gap in scholarship on the role that direct democratic instruments can play for animals. Focusing on popular initiatives, we started out by arguing that these direct democratic mechanisms have four key functions that can be beneficial to sentient beings. We then proposed the introduction of rejective and proactive 'sentient beings' initiatives' and have outlined how these initiatives could be operationalized.

[102] Swiss Constitution (1999) arts 138 and 139.

Conclusion

This chapter started with the observation that even in animal rights discourse the idea that animals could receive political rights is often dismissed as infeasible at best and absurd at worst. As we have argued, this response is both correct and incorrect. It is correct insofar as, even in a sentience-based constitution, animals would not receive formal political rights themselves. This is because, to the best of our knowledge, animals do not have the capacity to understand formal political processes, such as elections and the powers and responsibilities that office-holding entails. It is in this sense that the Political Patients Approach—which says that animals lack political agency—is right. However, as we have demonstrated, this does not mean that the interests of animals matter any less in a democracy than human interests. Quite the contrary, animals, like humans, have an equal right to democratic participation. As such, it is incorrect to suggest that animals should not have any political rights. This is where we follow the Political Agents Approach, according to which animals can be seen as political agents, at least in informal political contexts.

To give sentient animals due concern in democracies, we must therefore recognize and respect their equal right to democratic participation. But what does that mean concretely? To flesh this out, we have explored four requirements. First, we argued that human representatives are still needed so that animals' interests can be channelled into formal decision-making processes. Second, we demonstrated that these representatives can only effectively exercise their role if animals participate in their representation. Third, we established that a combination of both legal and political mechanisms is necessary to give animals their due democratic concern. Finally, we argued that direct democratic processes are required to complement indirect democratic processes. Meeting these requirements ensures that animals can participate democratically with the help of human representatives attuned to their informal political agency, even if the animals themselves lack the ability to participate in formal democratic processes.

Conclusion

1) Sentience-based constitutionalism: Filling the legitimacy gap

This book began with Pericles galvanizing his fellow Athenians to continue their fight against Sparta. The Athenian constitution is worth fighting for, he declared, 'because power is in the hands not of a minority but of the whole people'.[1] As we have shown, Pericles was not the only one misguided about the legitimacy of his own constitution. This book has demonstrated that there is a wide legitimacy gap affecting all constitutions existing today. Just like the Attican constitution excluded from its scope a significant proportion of the human population—women, slaves, foreigners, and others—current constitutions do not pay due concern to the vast majority of beings they govern. In particular, we have shown in Chapter 1 that, due to their anthropocentrism, existing constitutions do not pay due concern to animals. That is, no constitution respects the intrinsic value of animals, which would require protecting their basic interests, such as their interests in life, bodily integrity, or liberty.

Animals, we argued in Chapter 2, are intrinsically valuable because they are sentient. They have the capacity to have valenced experiences, that is, experiences which feel good or bad. As we saw, this capacity includes the ability to feel pain and pleasure, but it can also include a much richer set of emotions, such as empathy, jealousy, disgust, grief, indignation, and even love.[2] Because they are sentient, animals are *governed* rather than merely *regulated* by constitutions. As we defined it, entities that have no intrinsic moral value, such as tables and boulders, are merely regulated by constitutions. By contrast, inherently valuable beings, such as humans and other sentient animals, are governed by a constitution when it has effective control over their lives, bodies, or habitats.

Yet, constitutions often treat animals like tables and boulders. In fact, constitutions deem animals as property as a matter of practice, as a matter of formal

[1] Thucydides, *History of the Peloponnesian War* (Rex Warner tr, Penguin 1954) Pericles' Funeral Oration, Book II, [37], 145.

[2] See Section 1 of the book's Introduction.

classification, or both. For example, Article 381(1) of the Constitution of Bolivia (2009) states that '[n]ative animal and vegetable species are natural assets'. Similarly, Article 27(2) of the Constitution of the Bahamas recognizes that 'property [can] consist of an animal'. Even the most animal-friendly constitutions, such as that of Ecuador, which has been interpreted to recognize animals as 'sentient beings' and 'subjects of rights',[3] still permits their systematic exploitation, such as for the purposes of pest control, food production, clothing, transportation, and even leisure.[4]

At various points in this book we examined how constitutional regimes such as these are problematic because they fail to adequately cater to sentient animals.[5] This affects their legitimacy because legitimacy is at least partly based on how well constitutions cater to those who have constitutional worth. As we proposed, as beings who are governed by constitutions, animals should be recognized as having 'constitutional worth', that is, as having interests that are of constitutional importance. In failing to cater to sentient animals, who form the vast majority of those governed, constitutions face a deep legitimacy gap.

To overcome that gap, we advocated moving towards sentience-based constitutionalism—a model of constitutionalism that gives due concern to all sentient beings that constitutions govern. We characterized sentience-based constitutionalism as a realistic utopia.[6] This means that, while we propose to radically rethink existing constitutional principles to make them inclusive of sentient animals, our analysis remained grounded in existing practices and pragmatic considerations. We focused on four existing principles of constitutionalism—fundamental rights, proportionality, rule of law, and democracy—and illustrated how these can be reimagined to cater to the interests of all sentient beings. We hope these explorations will serve as a compass that can help constitutional lawyers, judges, policymakers, activists, and scholars chart constitutions of the future.

We started our exploration with fundamental rights. In Chapter 3, we argued that, in many cases, the creation of new fundamental rights specifically for animals will not be necessary. This is because existing fundamental rights practice has features—such as its progressive expansionism, the absence of a formal hierarchy between right-holders, overinclusiveness, and generality—that

[3] 'Estrellita', Constitutional Court, no. 253-20-JH/22, 27 January 2022, [77]—[89] <https://animal.law.harvard.edu/wp-content/uploads/Sentencia_253-20-JH-20_Derechos_Naturaleza_y_Animales.pdf> accessed 26 October 2024.

[4] See ibid [106]–[110].

[5] See eg Chapter 1 Section 3.

[6] See the Introduction to the book, Section 3.

make it possible for existing catalogues of rights to be extended to animals, thus obviating the need for an animal bill of rights. We exemplified how such an extension of fundamental rights can work by focusing on the right to life, the prohibition of torture and inhumane or cruel treatment, and the right to freedom.

We then addressed the proportionality test, which is used nearly universally by courts to deal with conflicts of rights. We argued that this test can also be applied to resolve conflicts in a sentience-based constitution. In fact, in Chapter 4, we showed that harnessing the proportionality test for the rights of all governed animals helps us improve our existing theories of the test, some of which are problematically centred on paradigmatic humans. We defended the view that the test comes with a commitment to what we call the Non-Instrumentalization Principle, which requires that all beings with intrinsic value must not be treated as mere means to others' ends. The chapter provided several examples of how the Non-Instrumentalization Principle helps resolve conflicts between the rights of animals and the rights of humans that might arise under a sentience-based constitution.

We subsequently shifted our focus to advocate for an inclusive conception of the rule of law in Chapter 5. We started from the finding that several of the leading rule of law theories are premised on the law being capable of guiding the actions of legal subjects. This basic idea, we argued, is problematically centred on paradigmatic human beings. Instead, we offered an alternative idea: the rule of law should be understood as restraining arbitrary power to protect those who can be victims of such power. This basic idea is more inclusive, covering not only paradigmatic humans but also animals and non-paradigmatic humans, for the simple reason that all of them can be victims of arbitrary power. The chapter then spelt out how this inclusive approach transforms formal, procedural, and substantive accounts of the rule of law. While protections will vary depending on the type of account, the protection of animals will no longer be a contingent matter under any account.

We finished our consideration of constitutional principles by focusing on democracy in Chapter 6. We took issue with animals' current exclusion from the *demos* and proposed that, in a sentience-based democracy, animals have as equal of a right to democratic participation as humans. This does not mean that animals should be granted formal political rights such as a right to vote. Rather, it means that democratic processes ought to be restructured in ways that afford due concern to the interests of all sentient beings whom these processes govern. We argued that this entails at least four things: first, human representatives are necessary to channel animals' interests into formal democratic processes; second, human representatives can only effectively exercise this role if animals participate in their representation through processes that take seriously their informal political agency; third, a combination of both legal and political mechanisms is necessary to protect animals' right to democratic participation; finally, these institutions should be complemented by direct democratic mechanisms such as 'sentient beings' initiatives'.

2) Future directions

The road to sentience-based constitutionalism may be long and arduous, but this book has provided a map that will make reaching that destination easier by charting the central features of this new constitutional landscape. To be sure, there are still many details, and perhaps some more hills, to be mapped. This is because, among other things, we could not provide an exhaustive account of all the existing principles of constitutionalism and show how they should be rethought to recognize sentient animals as having constitutional worth. However, we were able to identify the problem—anthropocentrism in constitutional thought and practice—and lay the foundation for resolving it by advocating for a reconceptualization of four cardinal principles of constitutionalism: fundamental rights, proportionality, rule of law, and democracy.

There are other principles of constitutionalism that are yet to be addressed. While there is no authoritative list of constitutional principles, important contenders include sovereignty,[7] equality,[8] secularism,[9] dignity,[10] fraternity,[11] and liberty.[12] It seems to us that the road to sentience-based constitutionalism requires considering how all of these principles, and perhaps more, can be transformed to promote the interests of all sentient beings. In what follows, we offer

[7] Martin Loughlin, 'Why Sovereignty?' in Richard Rawlings, Peter Leyland, and Alison Young (eds), *Sovereignty and the Law: Domestic, European and International Perspectives* (OUP 2013).

[8] Sandra Fredman, 'Substantive Equality Revisited' (2016) 14 International Journal of Constitutional Law 712; Wojciech Sadurski, *Equality and Legitimacy* (OUP 2008).

[9] András Sajó, 'Preliminaries to a Concept of Constitutional Secularism' (2008) 6 International Journal of Constitutional Law 605; Lorenzo Zucca, *A Secular Europe: Law and Religion in the European Constitutional Landscape* (OUP 2012); Andrew Copson, *Secularism: A Very Short Introduction* (OUP 2019).

[10] Aharon Barak, *Human Dignity: The Constitutional Value and the Constitutional Right* (CUP 2015); Jacob Weinrib, *Dimensions of Dignity: The Theory and Practice of Modern Constitutional Law* (CUP 2016); Catherine Dupré, *The Age of Dignity: Human Rights and Constitutionalism in Europe* (Hart Publishing 2015).

[11] Farrah Ahmed, 'Fraternity' (2 January 2024) <https://papers.ssrn.com/abstract=4871125> accessed 4 October 2024.

[12] Friedrich A Hayek, *The Constitution of Liberty* (The University of Chicago Press 1978); TRS Allan, *The Sovereignty of Law: Freedom, Constitution, and Common Law* (OUP 2015).

some brief remarks on three of these principles—sovereignty, equality, and secularism. Our aim is to ask some of the relevant questions and to sketch tentative answers which may help guide future research.

A) Sovereignty

Let us start with sovereignty. In classic liberal thought, it is the people who are considered as the source of the state's sovereignty.[13] This contrasts with the premodern notion of the divine right of kings.[14] As we explored in Chapter 6, anthropocentric constitutions exclude animals from 'the people'. This means that, currently, state sovereignty is not legitimated by reference to animals. However, as we argued, animals should be included in 'the people' by granting them equal rights to political participation. This suggests that the idea of sovereignty in modern constitutionalism is not best understood as 'popular sovereignty' because it excludes animals from being counted as part of the people. Sovereignty should, rather, be reimagined more inclusively as what we might call 'sentience-based sovereignty'.

But should all animals who are governed by a constitution be included within sentience-based sovereignty? One possible answer, which seems to follow naturally from the All Governed Interests Principle, is that they should be. After all, all animals who are governed by a constitution matter for the legitimacy of a constitution, and they can contribute to that legitimacy, among other things, by being part of the sentience-based sovereign. However, some scholars have argued that wild animals should be considered independent sovereign groups because they are self-sufficient and, as such, entitled to govern themselves independently from humans.[15] While this view has its appeal,[16] it raises the question of whether it is compatible with the idea of a sentience-based sovereignty.

[13] One of the earliest formulations of this idea can be found in the writings of Roger Williams who said, in his *The Bloudy Tenent of Persecution for Cause of Conscience* (1644), 'the sovereign, original, and foundation of civil power lies in the people'. See James Calvin Davis (ed), *On Religious Liberty: Selections from the Works of Roger Williams: 96* (Harvard UP 2008) 132.

[14] Martin Loughlin, 'What Is Constitutionalisation?' in Petra Dobner and Martin Loughlin (eds), *The Twilight of Constitutionalism?* (OUP 2010) 48.

[15] Sue Donaldson and Will Kymlicka, *Zoopolis: A Political Theory of Animal Rights* (OUP 2013) 175–76. For a critique of this view, see Alasdair Cochrane, *Sentientist Politics: A Theory of Global Inter-Species Justice* (OUP 2018) 81.

[16] For an early view of one of us building on this insight, see John Adenitire, 'Review of Sentientist Politics' (2019) 10 Jurisprudence 588, pt III.

One potential avenue that could be explored further in this context is that of viewing wild animal sovereignty as a federated sovereignty that forms part of a larger sentience-based sovereignty. After all, while all governed animals contribute to sentience-based sovereignty, they may be regarded as societies that belong to a federated state to the extent that they are self-sufficient communities. This, of course, raises questions as to what areas of competence such animal communities should enjoy. Arguably, it would depend on the characteristics of such groups. One could imagine, for example, that a pod of orcas is more self-sufficient[17] than a murder of crows, considering that the latter have become more dependent on human communities.[18] The orcas could therefore be understood as being primarily in charge of their own health, nutrition, living environment, and so on. By contrast, to the extent that the crows rely on human communities, the state may need to take direct measures in relation to their well-being by, for example, mandating bird-friendly buildings to avoid collisions. Of course, all groups of animals would need to have their fundamental rights, including the right to political participation, guaranteed by the state. This would ensure that all governed groups are truly considered as being within a unified constitutional order.

This sketch of sentience-based sovereignty raises several questions. Is the concept of federalism, arguably itself a distinct constitutional principle, genuinely applicable to communities of animals? If so, does a differentiated set of competences, which vary according to groups of animal species, truly respect the capacities and interests of all animals? To what extent (if any) should humans be allowed to intervene in the affairs of self-sufficient animal communities? If it seems likely that sovereign animal communities might not protect the fundamental rights of their own members, what then?

B) Equality

What about equality? There are at least two plausible aspects to equality as a constitutional principle: legal equality and equality as a right. Our book has already provided some answers in relation to the former in Chapter 5. There, we argued that a substantive inclusive conception of the rule of law requires,

[17] Natasha Daly, 'Orcas Don't Do Well in Captivity. Here's Why' (*National Geographic*, 25 March 2019) <https://www.nationalgeographic.com/animals/article/orcas-captivity-welfare> accessed 5 October 2024.

[18] Tove Danovich, 'Crows Are the New Pigeons' (*The Atlantic*, 15 March 2024) <https://www.theatlantic.com/science/archive/2024/03/crows-urban-city-habitat/677766/> accessed 5 October 2024.

first, that animals are equally protected, just like humans, from arbitrary public and private power; second, that they should be equally and fully included within the legal process; and, third, that they should be equally protected by a catalogue of fundamental rights which they share with humans (as argued in Chapter 3).

Consider equality now as a constitutional right which prohibits unjustified differential treatment by the state or by private parties. Current anthropocentric constitutions prohibit discrimination based on a number of traits, such as race, gender, sex, pregnancy, and marital status.[19] But no constitution or statute currently prohibits discrimination on the basis of species membership. Yet, if we investigate the rationales for why states prohibit discrimination on the basis of commonly found traits, it is not clear why species membership should be excluded. Sometimes, these traits are said to be protected because they are immutable or mutable only at a very high cost.[20] Clearly, species membership is immutable, so the immutability rationale would fully apply to protect animals from discrimination. Others have suggested that the rationale for protected traits is that they pick out groups that are socially disadvantaged in abiding, pervasive, and substantial ways.[21] Clearly, given the widespread exploitation of animals in human societies for food, experimentation, and entertainment, animals are severely socially disadvantaged and should be protected from discrimination. Or, perhaps, discrimination law picks out traits that are socially salient?[22] If so, being an animal is salient as this is very likely to indicate exploitation in a large number of social interactions.

On first look, then, equality appears to be a constitutional right that should fully apply to animals to protect them from unjustified discrimination. Of course, one would need to specify what constitutes unjustified discrimination against animals. Arguably, animals could still legitimately be denied certain rights, such as the right to vote or to run for public office (as we proposed in Chapter 6), while these rights remain available to humans. Similarly, it seems that private individuals should be able to discriminate against animals in

[19] Article 9(3) of the Constitution of South Africa, for example, prohibits the state and individuals from unfairly discriminating against anyone on any of the following grounds: 'race, gender, sex, pregnancy, marital status, ethnic or social origin, color, sexual orientation, age, disability, religion, conscience, belief, culture, language, and birth.'

[20] *Corbiere v Canada (Minister of Indian and Northern Affairs)* (1999) 2 SCR 203 [13]. The Supreme Court of Canada said that the constitutional provision of discrimination under section 15 of the Canadian Charter 'targets the denial of equal treatment on grounds that are actually immutable, like race, or constructively immutable, like religion.'

[21] Tarunabh Khaitan, *A Theory of Discrimination Law* (OUP 2015) 35–36.

[22] Kasper Lippert-Rasmussen, *Born Free and Equal?: A Philosophical Inquiry into the Nature of Discrimination* (OUP 2013) 30–36. He says that 'A group is socially salient if perceived membership of it is important to the structure of social interactions across a wide range of social contexts.' ibid 30.

employment, housing, and other sectors. For example, it would be absurd for discrimination law to prohibit hiring doctors or teachers who are not human.

Therefore, arguing that the right to equality should fully apply to animals raises more questions than it answers. What justifies differential treatment of animals when such treatment would be wrongful towards humans (eg discrimination in employment)? If, as seems probable, discrimination against animals can be justified in a considerable range of circumstances, in what ways are animals then really protected by the right to equality? Would it not be easier to say that a sentience-based constitution guarantees the fundamental rights of animals but does not see them as entitled to a right to equality?

C) Secularism

Secularism may be understood as a constitutional principle which demands a neutral relationship between state power and 'religion'. The Canadian Supreme Court once held that:

> The pursuit of secular values means respecting the right to hold and manifest different religious beliefs. A secular state respects religious differences, it does not seek to extinguish them. Through this form of neutrality, the state affirms and recognizes the religious freedom of individuals and their communities.[23]

Much of the value and content of secularism stands and falls with the disputed meaning of 'religion'.[24] If by 'religion' we mean some form of theism, then it is not clear that secularism should be valued in a sentience-based constitution. This is because the most popular theistic traditions are taken by a large portion of their participants to justify animal exploitation. To be sure, there are doctrines within these traditions which could be amplified to require a greater degree of animal protection.[25] However, few theistic traditions actively oppose the exploitation of animals such as for food or experimentation. According to the Pew Research Center, the majority of the world's population will be

[23] *Loyola High School v Quebec (Attorney General)* (2015) 1 SCR 613 [43]–[44].

[24] Some have argued that legal definitions of 'religion' are arbitrary and that it would be best for the law to do without the legal category of 'religion'. See Winnifred Fallers Sullivan, 'Why Distinguish Religion, Legally Speaking?' (2014) 51 San Diego Law Review 1121, 1132.

[25] This is the central claim of Joyce D'Silva, *Animal Welfare in World Religion: Teaching and Practice* (Routledge 2023).

either Christian (31%) or Muslim (30%) by 2050.[26] Most interpretations of Christianity and Islam are theistic, and most interpretations do not oppose animal exploitation for food or experimentation.[27] If secularism demands respect for these forms of theism which are often compatible with animal exploitation, then sentience-based constitutionalism would seem to require rejecting the concept of secularism.

Perhaps we can avoid this conclusion if we reconceptualize secularism. Secularism should not be understood, as the Canadian Supreme Court held, as entailing respect for the right to practice theistic principles. Rather, it means, more restrictively, that the state should not be based on specifically theistic principles. Instead, it should be based on principles which do not unfairly discriminate against any version of the good life, whether theistic, atheistic, or non-theistic.[28] These principles, to be sure, will not be neutral towards versions of the good life that seek to exploit animals. As we have argued throughout this book, in the context of sentience-based constitutionalism, the state ought to be committed to upholding the constitutional worth of animals by, among other things, protecting them from arbitrary power of public and private entities and by guaranteeing their fundamental rights.

An objection to this may be that sentience-based constitutionalism is inherently illiberal—oppressive even—because it cannot tolerate ideological diversity regarding the moral status of animals. It requires everyone to subscribe to a disputed moral view: that animals have equal constitutional worth just like humans. No doubt this will conflict with the views of many belief systems, including many theistic traditions.

Is sentience-based constitutionalism illiberal? We do not think so. Liberalism has always had to define the scope of the liberal subject. Contemporary forms of liberalism focus on 'free and equal citizens' who have the capacity to have a sense of justice and a conception of the good.[29] In short, the liberal subject for whom the basic structure of society is designed is the paradigmatic human. As we saw in Chapter 2, there are good reasons for rejecting such a restrictive conception. Animals and non-paradigmatic humans should also be considered liberal subjects who should be protected by the rule of law and fundamental

[26] Reem Nadeem, 'Key Findings from the Global Religious Futures Project' (*Pew Research Center*, 21 December 2022) <https://www.pewresearch.org/religion/2022/12/21/key-findings-from-the-global-religious-futures-project/> accessed 5 October 2024.

[27] D'Silva (n 25) chs 2–3.

[28] Effectively, this means that secularism is nothing else than a version of liberalism. The version of liberalism we have in mind is liberal pluralism as set out in John Adenitire, 'Conscientious Exemptions: Is Liberal Neutrality Possible?' in Cecile Laborde, Micah Schwartzman, and Nelson Tebbe (eds), *Discrimination By/Against Religion* (OUP 2025).

[29] John Rawls, *A Theory of Justice* (Harvard UP 1999).

rights, including political rights.[30] Consequently, sentience-based constitutionalism is not illiberal. Quite the contrary, it better protects all liberal subjects.

Is sentience-based constitutionalism oppressive? Not necessarily. The requirement to give due concern to animals can, in theory, be applied in oppressive ways. For example, public officials would act oppressively if they curtailed the freedom of speech of those who are opposed to sentience-based constitutionalism. However, this potential for oppression is not unique to sentience-based constitutionalism, as it also affects other models of benign constitutionalism. The key challenge will be to ensure that constitutions protect animals while guaranteeing freedoms, in terms of free speech, conscience, and association, for those opposed to sentience-based constitutionalism. In other words, sentience-based constitutionalism needs a theory of how to accommodate those who are opposed to its foundational values.

This tentative account raises several questions, of course. Can sentience-based constitutionalism be sufficiently accommodating of difference? What specifically do the rights to free speech, conscience, and association afford to those who oppose the values of sentience-based constitutionalism? For example, is hate speech directed at animals allowed?[31] In what ways can those who support sentience-based constitutionalism convince others of the truth of their values? Does sentience-based constitutionalism support a right to civil disobedience for those opposed to its values?

3) Towards sentience-based constitutionalism

This is the subtitle of our book, and we chose it deliberately. The account we have offered in this book of sentience-based constitutionalism makes no claim to being the last word on the topic. Rather, as we explained in this Conclusion, our book only intends to lay the groundwork for sentience-based constitutionalism and to explore its most prominent features. As the previous section acknowledged, many questions remain to be answered in relation to other principles of constitutionalism. Other things, too, will be required for a fuller account. For example, even once scholarship has developed a comprehensive

[30] For a recent critique of Rawls' position on animals, see Guy Baldwin, 'Rawls and Animal Moral Personality' (2023) 13 Animals 1238. For earlier critiques, see Robert Garner, *A Theory of Justice for Animals: Animal Rights in a Nonideal World* (OUP 2013).

[31] Josh Milburn and Alasdair Cochrane, 'Should We Protect Animals from Hate Speech?' (2021) 41 Oxford Journal of Legal Studies 1149.

account of all existing principles of constitutionalism for sentience-based constitutionalism, an account is still necessary of how constitutional actors, such as constitution makers, judges, litigants in constitutional courts, activists, and academics can take practical steps to achieving sentience-based constitutionalism. This task is beyond our own expertise. However, it is required for sentience-based constitutionalism to be realized.

We have noted in the Introduction of this book how several actors have already taken steps which will help fill the legitimacy gap that we have identified in current constitutional practices. The Nonhuman Rights Project, for example, is undertaking litigation efforts on behalf of chimpanzees and elephants in the United States so that they are recognized as legal persons.[32] Similarly, the Finnish Animal Rights Lawyers Society has been lobbying for a new chapter of the Finnish Constitution which contains an extensive catalogue of constitutional rights for animals. We cannot predict whether such litigation efforts and proposals for constitutional amendment will be successful and, if they are, whether they will spark similar efforts in other places or remain isolated successes.[33] We invite work by others with experience in public policy to explore what strategies are most promising when it comes to realizing sentience-based constitutionalism.

Finally, and relatedly, the role that emotions and culture play in this context will need to be investigated. Constitutions and constitutional doctrine are merely written on paper. They cannot translate into real-life practices unless they are sustained by a public culture of adherence to constitutional norms.[34] Such a culture is not only institutional, that is, a matter for public officials. Rather, the general public, too, needs to sustain that culture. It needs to have confidence in the legitimacy of the constitutional order and be able and willing to defend it from internal and external threats. A fuller account of sentience-based constitutionalism will therefore need to study the role that emotions and culture can play to realize its goals. Alasdair Cochrane has started pivotal work on this in the context of interspecies solidarity. He has argued for what he calls 'sentientist solidarity', which is 'a citizenry with feelings of shared affiliation with sentient non-human animals, as well as a commitment to the institutions

[32] See Nonhuman Rights Project, 'Litigation' <https://www.nonhumanrights.org/litigation/> accessed 4 November 2024. For an overview of this litigation, see Raffael N Fasel and Sean C Butler, *Animal Rights Law* (Hart Publishing 2023) 124–34.

[33] Finnish Animal Rights Law Society, 'Proposal: The Constitution of Finland' (2020) <https://www.elaintenvuoro.fi/english/> accessed 4 November 2024.

[34] For analysis in a US historical context, see Jason Mazzone, 'The Creation of a Constitutional Culture' (2013) 40 Tulsa Law Review 671. See also on the concept of the material constitution, Marco Goldoni and Michael A Wilkinson, 'The Material Constitution' (2018) 81 Modern Law Review 567.

designed for their protection'.[35] Together with Mara-Daria Cojocaru, he has inspired others to think more deeply about this idea in a recent volume.[36] We hope others will follow in their footsteps to develop accounts of emotions, such as love, compassion, and trust, that can sustain a culture necessary for sentience-based constitutionalism.

Our own goal in this book has been to encourage others to take sentience-based constitutionalism seriously, and to join the conversation for a more legitimate constitutional order in which all sentient beings are accorded constitutional worth. This is only the beginning of an exciting journey. While many questions have yet to be answered, its destination—sentience-based constitutionalism—is a constitutional order truly worth fighting for.

[35] Cochrane (n 15) 120.

[36] Alasdair Cochrane and Mara-Daria Cojocaru (eds), *Solidarity with Animals: Promises, Pitfalls, and Potential* (OUP 2024).

Index

For the benefit of digital users, indexed terms that span two pages (e.g., 52–53) may, on occasion, appear on only one of those pages.

The manufacturer's authorised representative in the EU for product safety is Oxford University Press España S.A. of el Parque Empresarial San Fernando de Henares, Avenida de Castilla, 2 – 28830 Madrid (www.oup.es/en or product.safety@oup.com). OUP España S.A. also acts as importer into Spain of products made by the manufacturer.

www.ingramcontent.com/pod-product-compliance
Lightning Source LLC
LaVergne TN
LVHW011756200926
848484LV00003B/718

* 9 7 8 0 1 9 8 9 1 0 5 0 3 *